VOLUME SIXTY-THREE

DEVELOPMENTS IN SEDIMENTOLOGY

Deep-Sea Sediments

VOLUME SIXTY-THREE

Developments in Sedimentology

Deep-Sea Sediments

EDITORS

HEIKO HÜNEKE
Institut für Geographie und Geologie
Universität Greifswald
Greifswald

THIERRY MULDER
Université Bordeaux
Talence Cedex
France

ELSEVIER

AMSTERDAM • BOSTON • HEIDELBERG • LONDON
NEW YORK • OXFORD • PARIS • SAN DIEGO
SAN FRANCISCO • SINGAPORE • SYDNEY • TOKYO

Elsevier
Radarweg 29, PO Box 211, 1000 AE Amsterdam, The Netherlands
Linacre House, Jordan Hill, Oxford OX2 8DP, UK

Library of Congress Cataloging-in-Publication Data
A catalog record for this book is available from the Library of Congress

British Library Cataloguing-in-Publication Data
A catalogue record for this book is available from the British Library

For information on all Academic Press publications
visit our website at elsevierdirect.com

ISBN: 978-0-444-53000-4
ISSN: 0070-4571

Printed and bound by CPI Group (UK) Ltd, Croydon, CR0 4YY

Transferred to Digital Print 2011

Finally, we are particularly grateful to our families and friends whose enduring support and forbearance has sustained us over the years that "the book" has been in preparation

to my parents, my wife Dagny and my children, Ragnar and Lukas (H. H.)

to my parents, my wife Claire and my children, Lucy, Clothilde, Romaric and Lorraine-Marie (T. M.)

Full colour versions of the figures in this book can be found at
http://www.elsevierdirect.com/companion.jsp?ISBN=9780444530004

Contents

CONTRIBUTORS

Torsten Bickert
Zentrum für Marine Umweltwissenschaften, Universität Bremen, Germany

Steven N. Carey
Graduate School of Oceanography, University of Rhode Island, Narragansett, Rhode Island, USA

Jean-Claude Faugères
Université de Bordeaux, UMR CNRS 5805 EPOC, Talence Cedex, France

Rüdiger Henrich
Department of Sedimentology and Paleoceanography, Faculty of Geosciences, University of Bremen, klagenfurter Straße, Bremen, and Fachbereich Geowissenschaften, Universität Bremen, Germany

Reinhard Hesse
Earth and Planetary Sciences, McGill University, Montreal, Quebec, Canada

Heiko Hüneke
Institut für Geographie und Geologie, Universität Greifswald, Jahn-Strasse 17a, D–17487 Greifswald, Germany

Patrice Imbert
Total, CSTJF, Avenue Larribau, 64000 Pau, France

Thierry Mulder
Université de Bordeaux, UMR CNRS 5805 EPOC, Avenue des facultés, 33185 Talence Cedex, France

Ulrike Schacht
Australian School of Petroleum, The University of Adelaide, Adelaide, SA, Australia

Jean-Luc Schneider
Université Bordeaux 1, Observatoire Aquitain des Sciences de l'Univers, CNRS-UMR EPOC, Talence Cedex, France

A. Stadnitskaia
Department of Marine Organic Biogeochemistry, Royal Netherlands Institute for Sea Research (Royal NIOZ), Texel, The Netherlands

A. Uchman
Institute of Geological Sciences, Jagiellonian University, Kraków, Oleandry 2a, Poland

A.J. Van Loon
Geological Institute, Adam Mickiewicz University, Maków Polnych 16, 61–606 Poznan, Poland

Helmut Weissert
Department of Earth Sciences, ETH-Z, Zürich, Switzerland

A. Wetzel
Geologisch-Paläontologisches Institut der Universität, Basel, Switzerland

A.J. Wheeler
School of Biological, Earth & Environmental Sciences and Environmental Research Institute, University College Cork, Cork, Ireland

PREFACE

There are many reasons for the fast-growing understanding of deep-marine sedimentary processes during the past few decades. Research has benefited greatly from a number of newly developed, highly sophisticated exploration techniques and comprehensive data sets, thanks to the immense industrial interest in deep-sea sediments. Multidisciplinary research, in addition, has shed increasingly more light on the complex biogeochemical processes driving and controlling productivity and, thus, an important part of the deep-sea sedimentation. Moreover, deep-sea sediments have been recognized as archives of information about the changing boundary conditions in the oceans' histories and in the evolution of life. They also became of particular interest as keys for unravelling present-day climate changes, which challenge modern society.

This book grew out of our desire to keep up with this rapidly expanding area of knowledge and to integrate the main process-based aspects of siliciclastic, biogenic and volcaniclastic deposits of both modern and ancient deep-marine sedimentation into one single, unified and comprehensive text. The volume is structured to follow the various sedimentary depositional processes in the deep sea, from sediment gravity flows and contour currents to pelagic settling and hemipelagic advection, periplatform settling, planktic and benthic bioproductivity, and volcanic activity. In addition, the relationships between depositional environment and endobenthic organisms, as well as early-diagenetic processes at and within the deep-sea floor are dealt with. The book, finally, includes an introduction to the climatic interpretation of the various proxies that reveal global changes during the Mesozoic greenhouse and Neogene icehouse conditions, and it addresses the specific interest of the hydrocarbons industry in deep-water sediments. While each chapter is self-contained, they are interrelated, thus reflecting the complexity of the subject, spanning flow transformation of sedimentary density flows and currents, bentho-pelagic coupling, changes in sea-water chemistry, major innovations in organism evolution, and changes in external controls on sedimentation and productivity.

The book is an attempt to bring together the knowledge both of scientists working in the present-day deep oceans and geologists studying ancient deposits of deep-marine environments now exposed on land. The main advantage of the actualistic point of view is, of course, that the processes driving the production, the supply and the deposition of sedimentary particles accumulating in the deep sea can be qualified and quantified more or less precisely. The fossil record, on the other hand, if successfully deciphered holds valuable clues about the changing boundary conditions, controlling the sedimentation in deep-marine environments. This is important in particular where evolutionary processes are involved in the formation of deep-sea sediments.

We have endeavoured to produce a well-balanced book without important omissions. We attempt to summarize the current factual knowledge in the field of deep-sea sedimentation and the application of this knowledge to a variety of scientific and applied problems. We invited authors from both academia and industry to contribute to this book, thus striving for different viewpoints on the various aspects of deep-marine sedimentation. Considering the rather broad topic of the book, however, we cannot exclude that some gaps may be found. We hope there are not too many.

This book will be of interest to undergraduates taking specialist courses or simply orientating themselves with respect to the largest depositional setting on Earth: the deep sea. Postgraduates and professional geologists concerned with deep-sea research will find it useful for understanding specific aspects of deep-sea sedimentology, or as an introduction to regional considerations. Oceanographers, geochemists, biologists, palaeontologists, geophysicists, palaeoclimatologists and structural geologists will also find the book useful as a reference for understanding the sedimentological aspects of the deep sea.

First of all, we thank our authors, who not only kept up a very high standard of contribution, but also stuck (fairly closely) to the guidelines imposed by us. This also concerns our reviewers, chosen from various countries, who deserve considerable praise for their efforts in providing quick and fair critical comments on the contributions.

The editors gratefully express their thanks also to Tom van Loon, the series editor who encouraged the publication of a volume on this rather broad topic and gave us longstanding valuable support during the preparation. It was a pleasure to work with you. We also thank the staff of Elsevier for their help in organizing this book, in particular Anita Koch, development editor, Derek Coleman, senior developmental editor, Mageswaran BabuSivakumar, project manager, and Karishma Rathore, rights administrator. Furthermore, we would like to thank Heike Sengpiehl and Dagmar Lau from the Geological Department at the University of Greifswald, who did a large part of the high-quality figure drawing for many chapters.

Heiko Hüneke and Thierry Mulder

CHAPTER 1

Progress in Deep-Sea Sedimentology

Thierry Mulder,* Heiko Hüneke,† and A.J. Van Loon‡

Contents

1. Introduction

In this book, all marine domains extending seaward of the shelf break are considered as deep-sea. This domain represents 63.6 % of the Earth's surface (the ocean in its entirety covers 361×10^6 km^2 or 70.8% of the Earth's surface, including continental shelves). From a stricter geological point of view, the oceanic domain would begin at the boundary between the high-density (3.25 on average), usually thin (5 km in average) oceanic crust and the thick (30 km on average) low-density (2.7 in average) continental crust. A transitional crust may exist in between.

The study of deep-sea sediments benefited greatly from recent improvements in technologies. These improvements have been driven by academic needs (most of the sea floor remains unexplored in detail and most of the topography of abyssal plains has not been mapped with accurate tools) and

* Université de Bordeaux, UMR CNRS 5805 EPOC, Avenue des Facultés, 33185 Talence Cedex, France
† Institut für Geographie und Geologie, Universität Greifswald, Jahn-Strasse 17a, D–17487 Greifswald
‡ Geological Institute, Adam Mickiewicz University, Maków Polnych 16, 61–606 Poznan, Poland

Developments in Sedimentology, Volume 63
ISSN 0070-4571, DOI: 10.1016/S0070-4571(11)63001-X

by economic needs, such as the demand for mineral deposits (metal-bearing nodules, exploration of ultra-deep offshore oil). These newly-developed technologies benefited from both *in situ* data collection and data interpretation in laboratory.

In terms of data collection, this includes:

- Sea-floor morphology (multibeam bathymetry),
- subsurface investigation (seismic tools),
- high-resolution echosounders,
- 3-D tools,
- sampling gear (interface corer).

In terms of data interpretation in the laboratory, this includes:

- core scanners for measurement of geotechnical and physical properties, X-ray, geochemistry,
- the development of biological tracers and biomarkers for palaeoenvironmental reconstruction,
- the improvement and development of stratigraphic tools and dating methods based on radiogenic and non-radiogenic elements (especially for the Quaternary), the development of micro-lithostratigraphy (IRD, tephra recognition) and magnetostratigraphy.

1.1. Scope of the book

The chapters of this book have the following objectives:

- to explain the formation and supply of sedimentary particles by continental erosion (river load, ice or wind transport), coastal erosion, current-induced winnowing, through volcanic and authigenic processes, and by means of biogenic productivity;
- to describe the way the sediments are transported from the source area (continental edge, slope, surface water) to the accumulation zone in the deep-sea;
- to present the early geochemical transformations affecting the particles in the water column or the sediments as soon as they are produced and accumulate on the sea floor;
- to show how sediments are preserved on the sea floor despite erosion and dissolution;
- to present the characteristic features and main changes in worldwide ocean sedimentation with focus on "modern" oceans that have been formed since the disintegration of Pangaea (Mesozoic-Cenozoic);
- to discuss major changes in biogenic productivity, sea-water chemistry, and external controls of deep-sea sedimentary processes, depending on long-term trends in ocean history;

– to present the academic (e.g., palaeoclimatic studies), societal (e.g., natural hazards) and industrial interests (e.g., the exploration for mineral resources) in the study of deep-sea sediments.

2. What are Deep-Sea Sediments?

The sea-water environment can be subdivided into shallow (epicontinental) seas and deep seas. The morphology of modern oceans and marginal seas is based on the water depth and on changes in the slope gradient (Fig. 1.1). Using a classical cross-section through a passive continental margin, the shallowest environment is the continental shelf (or platform), which extends in the continental domain from the shoreline to the shelf break. It represents 26×10^6 km^2 (7.2% of the marine area). In this area, the sea-floor gradient is $< 0.5°$. In offshore direction, the water depth extends down to 100–110 m such as on the north-western African margin (Seibold and Hinz, 1974) or 200 m on most of the continental margins, including the northern European and the North-American Atlantic margins. Its extent can be from several hundreds of kilometres (1500 km for the Siberian shelf, > 600 km for the southern Argentina–Patagonian Shelf) to a few kilometres (off Nice in the Mediterranean). Active continental margins, such as the South-American Pacific margin, are usually only a few kilometres wide.

The continental shelf is exposed to numerous oceanographic processes that are absent in deep seas. Most of them are related to atmospheric processes. They include swell and storm waves that generate oscillatory motions in the water column (producing specific sedimentary structures such as hummocky cross-stratifications), tides, shallow contour currents, as well as shelf and coastal currents, including littoral drift. The continental shelf is separated from the continental slope by the shelf break, which is defined by a change in the slope gradient. The slope steepens from a gradient $< 1°$ on the shelf to 3–5° in average along the slope, to sometimes more than 20° in areas where canyons are incised the slope and the shelf. Further downslope, it passes into the continental rise at a water depth of about 2500 m. The continental slope corresponds approximately to the bathyal zone (200–3000 m). On the rise, the slope gradient decreases to 1–2° and the relief becomes smoother. Because of this change, the continental rise is the preferential area for final deposition of terrigenous sediment that bypassed the shelf and slope area. Together, the continental shelf, slope and rise form the continental margin. The margin can be passive and tectonically quiescent (North Atlantic margin) or active and tectonically dynamic (circum-Pacific margins).

At about 5000 m water depth, the rise passes into the abyssal plain. Abyssal plains represent the largest oceanic domains with a mean water

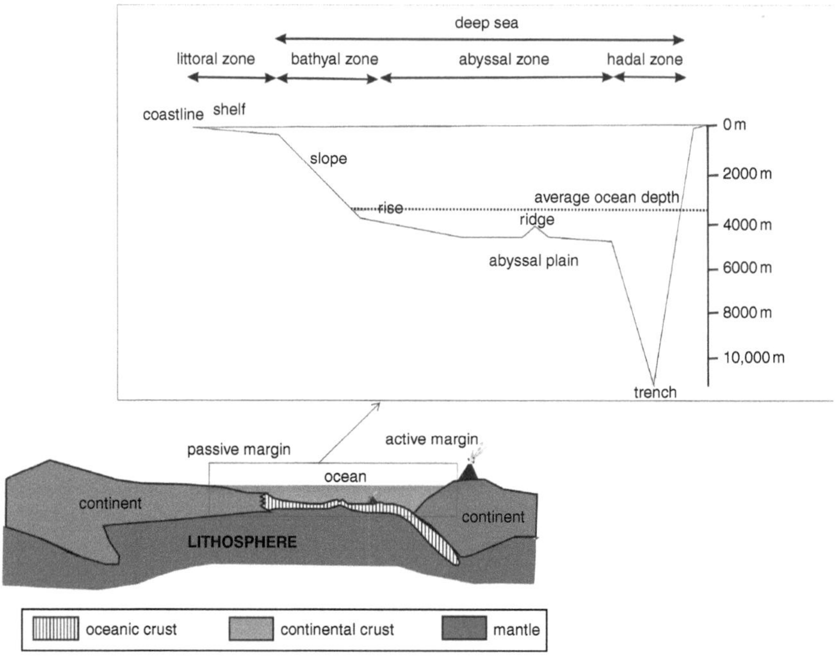

Figure 1.1 Cross-section through an ocean, showing the various deep-sea environments and domains. Lithosphere includes upper part of upper mantle plus oceanic or continental crust. (A multi-colour version of this figure is on the included CD-ROM.)

depth of ~ 3800 m. Abyssal plains are "flat" at a large scale. A closer look reveals, however, that their "flatness" is disrupted by tectonic and volcanic features: transform faults at different scales and strike-like faults with hanging walls of several hundreds of metres or even several kilometres in height and related local sedimentary basins. There are, in addition, hot-spot-related volcanic mounds and islands, volcano alignments forming the oceanic ridges, channels and thick and extensive accumulations of sediments forming drifts, and levees, gypsum diapirs; there are also dissolution structures. The continental rise and abyssal plains constitute the abyssal domain (3000–6000 m). Only 2% of the total ocean surface is deeper than 6000 m (hadal domain). In subduction areas, the presence of a subduction trench generates the deepest oceanic environments, down to 11,020 m (Mariana Trench). There, the presence of an accretionary prism can generate important relief-forming processes, such as mud diapirs and volcanoes (which may be related to the upward motion of deep fluids) and pockmarks, which are due to liquefaction related to fluid escape.

The sediments in the deep-sea consist of (1) clastic particles derived from eroded rocks and sediments outcropping either on the emerged continents or previously deposited in a marine environment, (2) particles formed by volcanic eruptions, (3) particles formed by living organisms, including organic matter, skeletal hard parts of calcareous, opaline or phosphatic composition, and faecal particles, and (4) particles formed by chemical precipitation of the elements contained in the salty sea water (average concentration of dissolved salts in sea water is 35.5 g/l). Most of chemical processes include microbiotic reactions and are thus grouped under the term "biochemical processes".

The term "sedimentation" describes the process of accumulating sediments in the form of layers or beds and includes all events that take place during particle formation (by weathering, erosion or biogenic production), through transport to final deposition of the sedimentary particles. It also includes all the consolidation processes (such as dewatering) occurring either during the deposition or shortly after, as well as the associated biochemical and chemical changes occurring in the sediment just after deposition and favouring particle bonding (cementation) through a variety of processes summarized under the term "diagenesis" that finally transforms the (soft) sediment into an (indurated) rock. Sedimentation also includes biological processes that rework sediments early after deposition (bioturbation) and that favour early diagenesis through improvement of fluid circulation. Despite wind and atmospheric transport, which are responsible for a small part of oceanic sediment-particle transport (wind-driven dust, volcanic ashes), water should be considered as the main agent of particle transport to and within the deep-sea.

3. Tools Used for Deep-Sea Sediment Investigations

3.1. Geophysics

The deep-sea can be investigated by both indirect and direct measurements from a boat or a vessel. During indirect measurements, a signal (usually acoustic) is emitted towards the sea-floor. It can be reflected at the sea-water/sea-floor interface or it penetrates into the sediment before it is reflected at a bedding plane or any other disconformity. Whatever the path is, parts of the signal come back to the boat and are recorded to be subsequently processed and studied (Fig. 1.2).

During direct measurements, a submersible or an ROV (Remotely Operated Vehicle) is sent along the sea floor, and a sampling device or any probe penetrates into the sea floor.

In all cases, the quality and reproducibility of measurements along the sea floor have been drastically improved during the last decade because of the enhanced positioning with the development of the GPS (Global Positioning System).

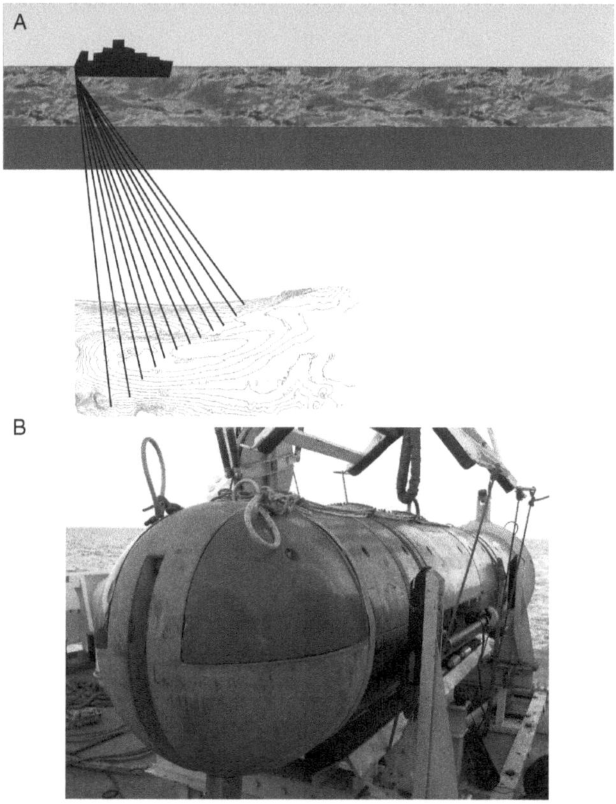

Figure 1.2 Geophysical research of the deep-sea floor. (A) Principle of multibeam bathymetry survey (modified from Ifremer's internet website) (http://www.ifremer.fr/anglais/). (B) The French side-scan sonar SAR (Système Acoustic Remorqué = Acoustic Towed System) operated by Genavir. Picture by T. Mulder. (A multi-colour version of this figure is on the included CD-ROM.)

3.1.1. Tools measuring bathymetry

Multibeam echosounders allow measuring the bathymetry (direct distance between the acoustic source and the sea floor) on a strip parallel to the boat track with a width of typically 120°–150°, in order to provide high-precision (0.5 m resolution) bathymetric maps (Figs. 1.2A and 1.3). Because of the high density of data collected within a survey, this tool is well-suited to provide 3-D views of the sea-floor topography. The insonified stripe has a width that corresponds to approx. 5–7 times the water depth. Most of the multibeam bathymetry gears are permanently embedded on the boat hull and a few are trailed behind the boat. Most of them can be operated at high speed (10 knots ≈ 18.5 km per hour).

At the same time, the sounder provides a backscatter of the sea floor that can be related to its sedimentary characteristics (e.g., grain size, porosity, water content) (Table 1.1). The systems operate at frequencies varying from 12 to 500 kHz.

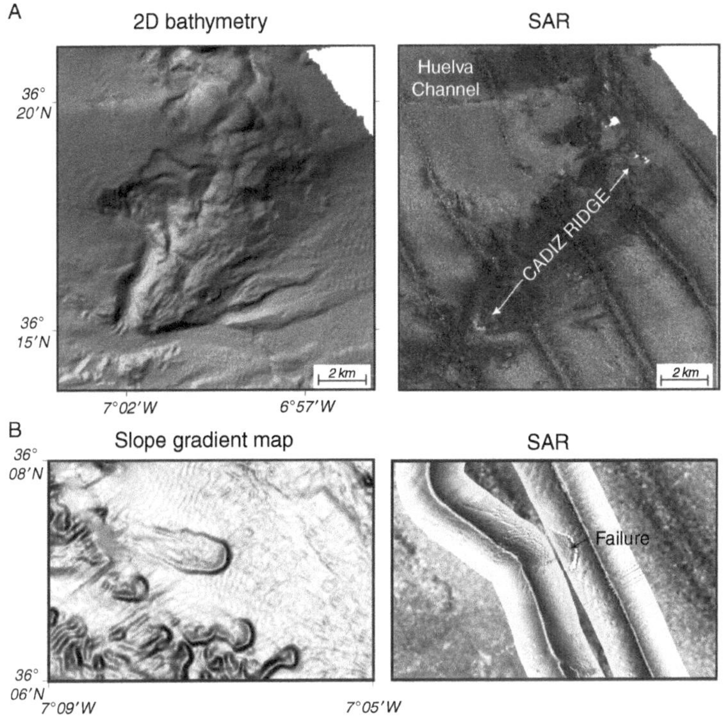

Figure 1.3 Example of Simrad EM 300 bathymetry and corresponding SAR image in the Gulf of Cadiz. (A) Cadiz Channel. (B) Slump along the giant contouritic levee. (A multi-colour version of this figure is on the included CD-ROM.)

3.1.2. Side-scan sonar

A side-scan sonar (Figs. 1.2B and 1.3) is a deep-towed acoustic system that are used mainly to map the morphology and composition of the sea floor. This equipment is essential to identify small (metre-range) sedimentary features. They either record the returned signal from an acoustic beam transmitted by the tool, or the backscatter from the sea floor. The backscatter signal is a function of the topography and particularly of the sea-floor slope, which influences the angle of incidence and the nature of the sea floor. The main types of a side-scan sonar devices used for sedimentological investigation operate at frequencies from 65 to 500 kHz and are listed in Table 1.2.

3.1.3. Seismic tools

Artificial seismics are based on the measurement of the travel time of acoustic waves generated by a non-natural source. We will restrict ourselves here to seismic reflection, which is the method most used in sedimentary

Table 1.1 Main features of multibeam echosounders (from Masson, 2003).

Echosounder	Frequency (Hz)	Maximum swath width (km)	Resolution (m)	Water depth (m)
Low frequency	12–24	20	7 (cross-track), 60–200 (along track)	> 2500
Middle frequency	300	4–5		1000–2500
High frequency	100–1000	1	0.2–0.4% of water depth	5–800

Table 1.2 Main features of a side-scan sonar (from Masson, 2003).

Side-scan sonar	Frequency (Hz)	Swath (km)	Resolution (m)	Towing speed (knots)
Low frequency	6–12	up to 45	few 10's (cross-track), 10's–100's (along track)	10
Middle frequency	30	2–6	1–2 (cross track), 10–40 (along track)	<2–3
High frequency	10–500	0.1–1.5	1 (cross- and along track)	<2

studies. The wave source can be high-pressure air or water (air/water guns) or an electric pulse (sparker). It generates a signal with a well-characterised amplitude, frequency and duration. Acoustic pulses are sent towards the sea floor at constant time intervals; they and reflected by natural interfaces (reflectors) separating beds of different acoustic impedance (sound velocity times density). The reflected waves are collected by a receiver, which is constituted of a series of hydrophones forming a streamer (Fig. 1.4). The time taken for the sound to travel and return from a given reflector (two-way travel time, TWTT) makes it possible to calculate the reflector depth below the sea floor. Seismic reflection thus allows the recognition of the 2-D structure, as well as the geometry of geological layers below the sea floor. Seismic refraction makes it possible to calculate the longitudinal velocity of the generated waves. Seismic reflection is essentially used for the study of sedimentary drapes on the ocean floor. It comprises mono-trace seismics with low-penetration and high-resolution characteristics, as well as multi-trace seismics with high-penetration and low–resolution characteristics. The type of the seismic device is chosen on the basis of the nature of the target. With increasing frequency, the resolution increases and the penetration below the sea floor decreases (Table 1.3; Fig. 1.5). Very-high-resolution systems only penetrate the uppermost layers of sediments. These

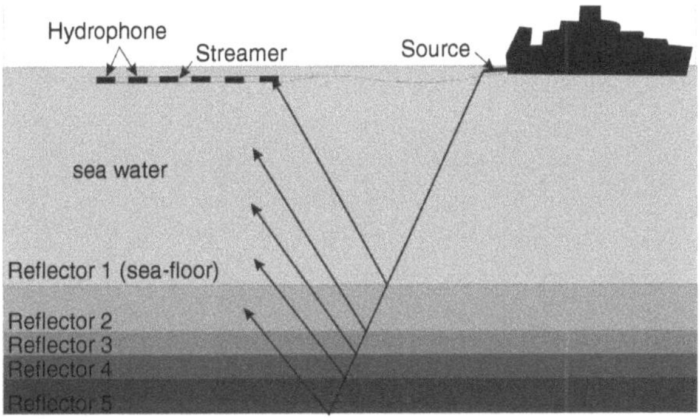

Figure 1.4 General principle of seismic reflection (modified from Ifremer's internet website) (http://www.ifremer.fr/anglais/).

Table 1.3 Types of seismic profilers (from Masson, 2003).

Type of seismics	Frequency (Hz)	Penetration (km)	Resolution (m)	Objective
Natural seismicity	0.1–5			Earth structure
Low frequency	5–80	30	>100	Basin analysis
High resolution (HR)	50–600	1	10	Sedimentary system
Very high resolution (VHR)	300–5000	<0.05	1	Sedimentary body; recent sedimentation
Very high resolution (VHR)	5000–10000	<0.03	<1	Sedimentary body; Recent sedimentation

systems are also called 'sediment sounder'. In contrast, low-frequency systems can penetrate the entire oceanic crust. For very high-frequency systems, the penetration is less in coarse or chaotic sediments than in fine-grained and well-stratified sediments.

More sophisticated seismic devices such as multiple coverage and 3-D seismics are used occasionally.

Multiple-coverage seismics includes an array of parallel hydrophone streamers towed behind the ship. 3-D seismics is mainly used by industrial companies, despite some academics recently developed such tools. It allows a 3-D visualisation of the investigated sedimentary succession in the form of a cube which can be split-up into 'time slices.

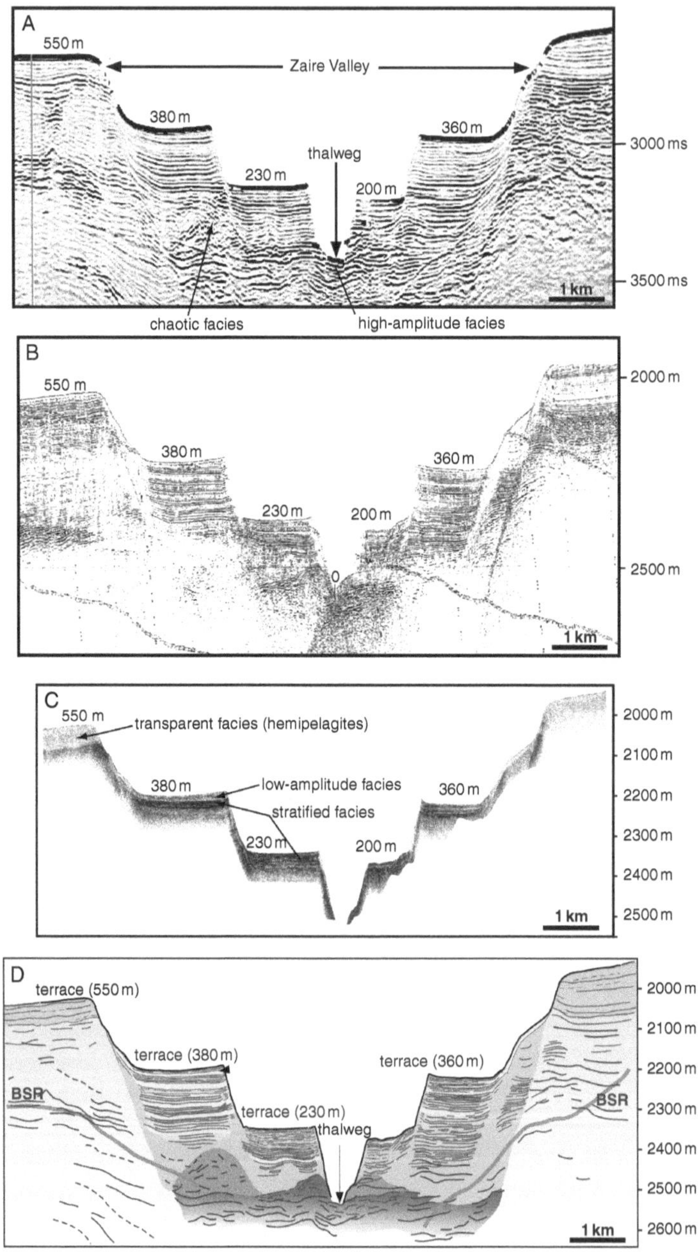

Figure 1.5 Comparison between different types of seismic tools on the Zaire channel (from Babonneau, 2002). Data have been collected by Ifremer and Total during the Zaiango Project. Published with permission of Ifremer, Total and N. Babonneau. (A) Industrial seismics. (B) Multichannel seismics. (C) Boomer seismics. (D) Very-high-resolution echo-sounder. (A multi-colour version of this figure is on the included CD-ROM.)

The French side-scan sonar SAR is operated with a near-bottom seismic streamer called 'PASISAR' (SIsmic PASsenger of SAR; Savoye et al., 1995). This allows reducing noise during the return travel of the waves. Several institutions develop presently very-high-resolution seismic gears with both a source and a receiver near the bottom. The system has been recently improved using both source and reception along the sea floor (SISYF tool: Marsset et al., 2010) (Fig. 1.6A)

3.2. Geotechnic tools

Geotechnical tools can be used for *in situ* measurement of physical parameters below the sea floor and to evaluate the mechanical behaviour of sediments, for example by estimating the non-drained shear resistance.

Figure 1.6 Sediment-sampling devices. (A) SISYF tool (Copyright Ifremer, A. Massol). (B) PENFELD penetrometer (picture courtesy of D. Leynaud). (A multi-colour version of this figure is on the included CD-ROM.)

The penetrometre PENFELD operated by Genavir (Fig. 1.6B), for example, can measure profiles resting on the sea floor at depth up to 6,000 m. Its rod can penetrate 30 m below the sea floor, using an instrumented tip. It measures gamma ray, lateral and tip stress, induced pressure and dip values. These tools are particularly well-suited to identify interbedded sand and clay layers, quick clay layers, or to evaluate their liquefaction potential. Piezometers can be positioned on the sea floor to identify water fluxes.

3.3. Sediment sampling

3.3.1. Wells

Well drilling in submarine sedimentary deposits is essentially done by the oil industry and by the International Ocean Drilling Program (IODP) consortium. Wells can be drilled with a length of several hundreds of metres to a few kilometres and well-logged geophysical data (e.g., gamma-ray, resistivity, neutron, sonic) can thus be obtained. Whereas most of the industrial data are confidential, IODP data collected by the JOIDES Resolution drilling ship can be easily used for scientific research.

3.3.2. Cores

3.3.2.1. Cylindrical corer A classical cylindrical-corer device includes a nose to facilitate sediment penetration, a core catcher to prevent sediment escape during the lift-up of the corer, and a tube with a length varying between 1 and 70 m for giant piston corers. Gravity corers are mainly used to obtain short cores. The corer penetrates under the influence of its own weight using a free-fall method (Fig. 1.7).

Piston corers are the most commonly used devices for sediment sampling at any ocean depth. The used systems are based on the device invented by Kullenberg (1947), with substantial modifications. The main difference with a simple gravity corer is the presence of a piston that remains attached to the wire during the free fall. The depression facilitates penetration of sediment in the core by reducing friction. The vacuum suction created by the non-moving piston inside the tube and the downward motion of the tube penetrating the sediment allows the sediment to enter the tube. The system facilitates the recovery of a long core but a significant problems are the core deformation along the sides (stretching), liquefaction and flow of porous water-saturated sediment (sands), and elongation of the sediment pile in the upper, more deformable part of the core.

In sandy and consolidated sediments of the continental shelf, a vibrocorer can be used to recover cores with a length of generally less than 5 m.

3.3.2.2. Interface corer Interface corers usually integrate a frame that lies on the sediment. One or several tubes can then penetrate the sediment simultaneously, usually by gravity. They penetrate less than 1 metre of

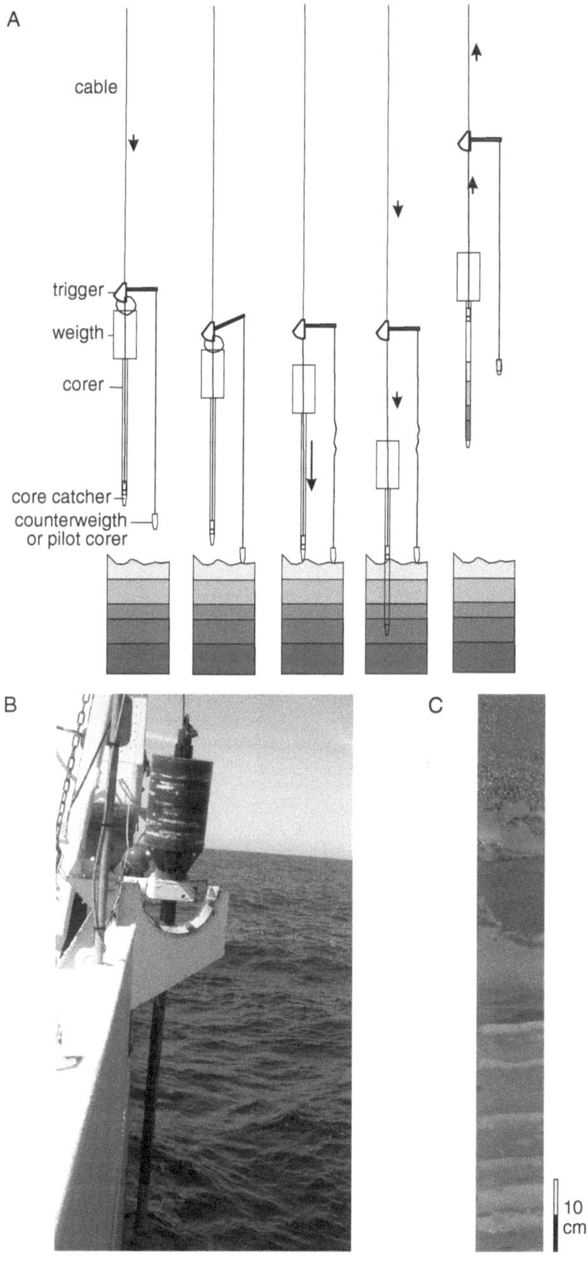

Figure 1.7 Core sampling. (A) Principle of core collection with a gravity corer (modified from Ifremer website). (B) The corer. (C) Example of a sediment core; turbidites recovered from a terrace within the Timiris Canyon off Mauritania (courtesy of R. Henrich). (A multi-colour version of this figure is on the included CD-ROM.)

sediment and thus recover only the very top of the sedimentary succession, including sea water and the water/sediment interface.

3.3.2.3. Box corer A box corer and grad are used for sampling a volume of surface sediment ranging from a few cubic centimetres to about a cubic meter. They are used for fast sampling of the surface deposits, including the water/sediment interface, in coarse-grained material where an interface corer has difficulty to penetrate. They are also extensively used to obtain quickly samples that can serve as a reference for the interpretation of acoustic imagery, in order to map the lithology of the surface sediments.

3.3.3. Dredging Dredging is the most archaic system to collect sediment or rock samples from the sea floor. It is used to grab consolidated sediments or rock fragments. Establishing the sample location is less accurate than for coring.

3.4. Submersible systems

Submersible systems (see Masson, 2003) include submersibles with a crew. Examples are the French Nautile, the Russian MIR I and II and the Japanese Shinkai (all with a maximum depth of 6,000 m), and the American Alvin (maximum depth 4,500 m). Manned submersibles allow direct observation of the sea floor by one or two scientists. They are usually equipped with a coring system, an articulated arm to grab samples and a video camera.

The manned submersibles are now challenged by remotely operated vehicles that have no people onboard but that are equipped with sampling tools, cameras, micro-gravimetry and micro-bathymetry survey gear. Some systems include only a video recorder (e.g., the French Scampi).

3.5. Current meters and particle traps

Deep-sea current meters are systems anchored on the sea floor. A current meter can record parameters such as current velocity, current direction, current density, and gear inclination (under the influence of currents or flows). Several current meters can be associated in a vertical line to measure current velocities at several altitudes above the sea floor. They can be connected to turbidity meters and sediment traps.

3.6. Laboratory analyses

Laboratory techniques for sedimentological studies of deep-sea sediments include as a first step a non-destructive analysis using core scanners (St-Onge et al., 2007). This allows a measurement of the sediment properties over the entire length of a core sample, including colour reflectance and other physical properties (gamma densimetry, magnetic susceptibility, P-wave velocity), geochemistry of the main chemical elements by X-ray fluorescence (XRF), and 3-D analysis of sedimentary structures (X-ray scan). The study of indurated sediments or rocks is more complicated and requires first of all the preparation of thin sections.

Three sets of investigations can subsequently be carried out: (1) analysis of the principal sediment composition, (2) analysis of palaeoenvironmental proxies, and (3) stratigraphy.

3.6.1. Sediment composition

Sediment analysis is primarily based on the recognition of the main sedimentary components, including the identification of heavy minerals and clay minerals for provenance studies (Weltje and Von Eynatten, 2004). Textural and structural analyses are based on standard routines and techniques used in sedimentology. Colour reflectance is usually performed using a spectrocolorimeter (hand-held or scanned) just after core opening, so as to avoid rapid changes in colour that occur quickly afterwards because of oxidation and dewatering. It provides information on the mineralogy, in particular on the presence of beds with a high content of iron oxides, as well as information on early diagenesis. The calculation of element ratios in cores, using XRF spectrometry methods, can help to distinguish phases of increased terrigenous input to the ocean: elements such as Al, Ti, and Fe are typical of the siliciclastic fraction, whereas elements such as Ca and Sr are typical of the biological fraction (St-Onge et al., 2007).

Other measurements, for instance of the magnetic susceptibility anisotropy, can provide information on current directions. Grain-size analysis (sortable silt of McCave et al., 1995, and of McCave, 2007) provides information on the hydraulic energy of the depositional environment (and on submarine current velocities).

3.6.2. Palaeoenvironmental proxies

Several sediment properties deliver information that helps to reconstruct specific conditions of the depositional environment. Interpretation of the deep-sea palaeoenvironment requires first of all recognition of the fauna and flora, i.e., macro- and microfossils, in particular calcareous benthic and planktic foraminifers, nannoflora, dinoflagellates, ostracods and siliceous microfossils such as diatoms and radiolarians. Analysis of the variations in the ratios of stable isotopes ($\delta^{13}C$, $\delta^{18}O$) from the various inorganic and organic sedimentary particles and the analysis of the organic matter itself can help, for example, to reconstruct the productivity and oceanic circulation patterns and to track anomalies in the carbon cycle (anoxic events). The ratio of stable-oxygen isotopes ($\delta^{18}O$) is important for the reconstruction of palaeotemperatures and palaeoclimates. Reconstructions of the oceanic circulation patterns can rely on isotope signatures (Nd/Pb) as well as on mineralogical tools (clay: Fagel, 2007).

A series of palaeoceanographic proxy variables have been studied and successfully calibrated over the last decades, i.e. parameters which can be employed to approximate unobservable environmental variables such as export production and surface-water temperature (see Wefer et al., 1999).

3.6.3. Stratigraphy

The stratigraphical analysis of deep-sea sediments is essentially based on biostratigraphy, lithostratigraphy, physical stratigraphy and radiostratigraphy. Biostratigraphy includes the recognition of index fossils and fossil associations, based on macro- and micro-fauna, as well as on flora. The lithostratigraphy is based on the recognition of marker beds and their characteristics, including high organic-matter contents (black shales and sapropelitic limestones), tephra (usually associated with increased magnetic susceptibility), clay-rich beds, fossil-rich beds, and detritus-rich beds (Quaternary Heinrich layers). The physical stratigraphy is essentially based on palaeomagnetism at different time scales, including the recognition of reversal periods, changes in palaeointensities and secular variations of the magnetic field (Stoner and St-Onge, 2007).

Stable isotopes ($\delta^{18}O$ and $\delta^{13}C$) can be used as stratigraphic tools in ancient successions as well as palaeoceanographic tools for modern sediments. The following radiogenic isotopes are now largely used; they cover almost the full range of Earth's history:

^{137}Cs (for the time-span during which atmospheric nuclear tests were carried out, and for recognition of sediments that accumulated shortly after the Chernobyl accident)
$^{210}Pb_{exc}$ (half-life of 22.2 years)
^{14}C (half-life of 5730 $\pm$ 40 years)
^{235}U – ^{207}Pb (half-life of 0.7×10^9 years)
^{40}K – ^{40}Ar (half-life of 1.3 $\times10^9$ years)
^{238}U – ^{206}Pb (half-life of 4.5 $\times10^9$ years)
^{232}Th – ^{208}Pb (half-life of 14 $\times10^9$ years)
^{87}Rb – ^{87}Sr (half-life of 49 $\times10^9$ years).

4. Structure of the Book

Following the above-mentioned scope of this book, it covers, in twelve chapters, all aspects of particle supply or production, deep-sea sediment transport from source to sink, deposition, and early-diagenetic transformations (Fig. 1.8). It discusses the relationships between sedimentation and marine life, and it reviews the past biogenic sedimentation depending on key functional groups of marine organisms that evolved during

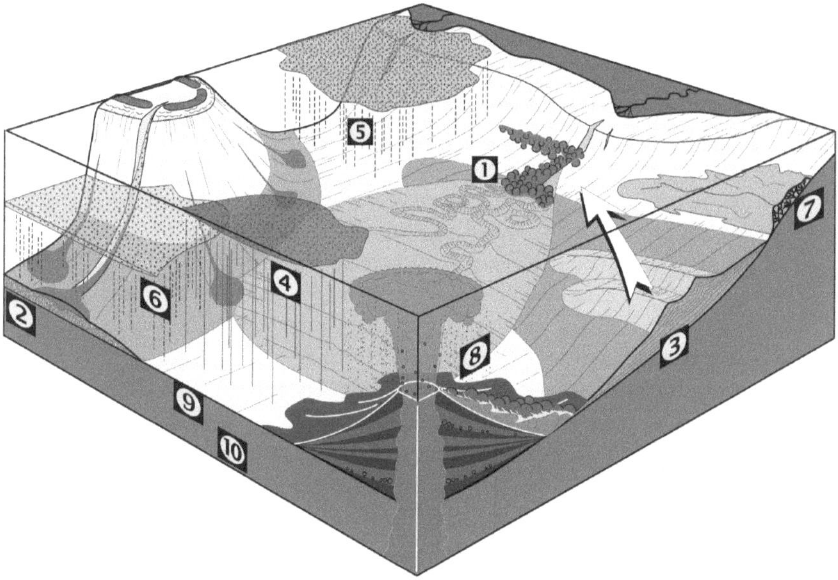

Figure 1.8 Sedimentary processes in the deep-sea (see text for further explanation): 1, gravity-flow processes (deep-sea fan); 2, rock fall and gravity-flow processes (slope-apron); 3, bottom current (contourite drift); 4, pelagic sedimentation; 5, hemipelagic advection; 6, periplatform sedimentation; 7, benthic carbonate production (cool-water coral reefs); 8, volcaniclastic processes (seamounts); 9, bioturbation; and 10, early diagenesis. (A multi-colour version of this figure is on the included CD-ROM.)

Earth's history. Special emphasize is given to climatic aspects of marine archives and to industrial interests. The reader who is interested most in a historical introduction to research on deep-sea sediments should begin with Chapter 11 before continuing with the next two chapters.

The two chapters following this introductory chapter devoted mainly to the most abundant sediments in deep seas: siliciclastic deposits.

The chapter on gravity processes and deposits (Chapter 2) deals with the various modes of gravity-driven sediment transport that occurs in more or less close contact to the sea floor. It reviews the various deep-sea deposits, i.e. the typical sediments on the lower slope and rise down to abyssal plains (Mulder, 2011, this volume). A feasible classification is outlined that sheds light on, in particular, the various types of sediment gravity flows and their deposits. Section 2.1 describes the downslope gravity processes, from the shelf break to the abyssal plain. It presents the main characteristics of the initiation and transport of sediment gravity-flows and their progressive transformation along a uniform continental slope or their rapid transformation due to sea-floor roughness. Special attention is given to hyperpycnal flows initiated at river mouths, since rivers represent the main way of sediment input in the marine realm. Section 2.2 presents the main types of deposits resulting from submarine downslope gravity flows and their diagnostic criteria. It details the difficulties met when relating the sedimentary deposits to flow processes. Section 2.3 describes the sedimentary systems and structures shaping continental margins, from the shelf break to the abyssal plains, in particular deep-sea turbidite systems (deep-sea fans). Results provided by the numerous cruises dedicated to their study during the last twenty years make detailed descriptions possible of their overall morphology and geometry but also of the shape of their architectural elements: canyons, channel/levees and lobes. The most recent result on the stratigraphy of these deep-sea depositional systems, their periods of activity and their evolution with time according to external forcing parameters is provided. It shows the new concepts that are currently applied to these systems since the early developments of sequence stratigraphy.

The chapter on contour currents and contourite drifts (Chapter 3) discusses the impact of bottom currents on deep-sea sedimentation and describes the genesis of contourites, which are the sediments deposited by, or significantly affected by, bottom currents (Faugères and Mulder, 2011, this volume). It presents the main characteristics of geostrophic currents flowing essentially along-slope following the thermohaline circulation. It shows the interaction between contour currents and the nepheloid layer and how they imprint the sedimentary record through erosion and dissolution processes, as well as the deposition of the resulting, mainly fine-grained sediments called "contourites". The importance of these processes is considerable because of the global thermohaline belt that exists in the modern oceans; the sea floor is always and everywhere affected by water currents that can rework sedimentary deposits. The authors of the chapter review the contourite deposits in both ancient and modern environments as well as their stratigraphic successions, which may consist of large sediment accumulations termed "contourite drifts". The interaction between contour currents (along-slope currents) and downslope gravity flows is also reviewed. The chapter outlines, finally, the palaeoceanographic significance of contourites. The interest in these deposits continuously increases since they represent valuable markers of palaeocirculation events, and since they occur frequently in relation with palaeoclimatic and eustatic changes.

The chapter on pelagic sedimentation (Chapter 4) reviews all processes of sediment formation and particle settling in the oceanic water column as it can be studied in the modern oceans, and it gives an overview of the pelagic sedimentation in the earth history (Hüneke and Henrich, 2011, this volume). It reviews all modern biological groups that contribute to the pelagic sediment factory, comprising phytoplankton (calcareous coccolithophorids and siliceous diatoms) and zooplankton (calcareous planktic foraminifers and pteropods and siliceous radiolarians and dinoflagellates). The authors track the way of the particle types of the two main mineralogies (calcareous and siliceous tests) from production in the marine realm to their final deposition on the sea floor. A synthesis of the parameters controlling the dissolution of carbonate and opal tests and consequently their preservation potential in the sedimentary record is also provided. An important point of this chapter concerns the production, fluxes and preservation of organic matter (biological pump), which usually represent a small percentage in sediments but are an important proxy, marker and a factor of high economical value. The last section of this chapter presents the history of ancient pelagic sediment factories since the beginning of life on Earth (Archaean). It shows how marine plankton evolution impacted the Earth productivity since life appeared on Earth. It further shows how the biosphere development influenced the whole earth system through changes in ocean chemistry and how these changes, in turn, controlled the evolution of life. Their impact on the oxygenation of the seawater and atmosphere is important. The carbonate and opal factories are reviewed, and the chapter ends with a discussion of the factors controlling this biological pump and of the resulting pelagic sediments: they appear to be related to large geodynamic events and climatic fluctuations at various frequencies.

The chapter on hemipelagic advection and periplatform sedimentation (Chapter 5) gives an overview of the processes that cause the widespread dilution of the open-ocean pelagic sedimentation by fine-grained terrigenous detritus supplied from clastic shelves and by neritic particles derived from shallow-water carbonate platforms (Henrich and Hüneke, 2011, this volume). Since this siliciclastic and carbonate detritus is commonly introduced into the upper part to the oceanic water column, these particles rain down through the water column in concert with pelagic biogenic material and reach the deep-sea floor by settling. The contribution of fluvial supply, terrigenous dust, and glaciomarine material to hemipelagic sediments is reviewed, and much attention is paid to methods to quantify these shelf sources. The distinction between bank-derived and open-ocean carbonate materials, in addition, is dealt with. The chapter gives special attention to compositional variations in the shallow-water-derived materials since these mainly reflect changes in the source/provenance areas or in the bypass area on the shelf. For the accumulation of both types of sediments, hemipelagic and periplatform oozes, the impact of sea-level fluctuations and climatic changes is very high, and this is reflected by variations in the composition, the accumulation patterns, and the sedimentation rates.

The chapter on benthic deep-sea carbonates (Chapter 6) explains the formation and preservation of biogenic and authigenic carbonates within the darkness of the deep sea, depending on biogenic productivity (in surface water), the variations in supply of siliciclastic detritus and of dissolution parameters (Wheeler and Stadnitskaia, 2011, this volume). The authors distinguish between carbonate sediments of which the carbonate is predominantly in the form of biological remains, and authigenic carbonate accumulations that form by the chemical precipitation of carbonates from seawater under certain conditions. The contribution focuses on the recent findings in the field of cold-water coral reefs, the authigenic carbonate formation related to cold seeps of methane (mud volcanoes and pockmarks) and the new ideas about mound initiation and development. Sessile calcareous reefal organisms may accumulate significant deposits of carbonate remains composed not only of their own remains but also of the calcareous remains of other organisms living in the reef habitat. Deep-sea cold seeps fuel the formation of authigenic carbonates through microbial consumption of escaping methane. The state-of-the-art knowledge on the chemistry involved in the transformation of methane into carbonates through microbial activity is presented.

The chapter on volcaniclastic processes (Chapter 7) shows how volcanic materials, which can be produced both under submarine and subaerial conditions, are locally important in shaping the sea floor (Carey and Schneider, 2011, this volume). The chapter emphasizes the particularities of volcaniclastic transport and deposition, and the possible convergence with classical sedimentary gravity processes. The important submarine environments in which volcaniclastic processes occur are reviewed: mid-ocean ridges, seamounts and large oceanic islands, as well as subduction zones. Particular attention is given to the description and evolution of volcaniclastic aprons, using the Hawaiian islands as an example. The unexpected economic importance of volcaniclastic deposits for oil prospection are detailed, and the obvious impact of volcaniclastic processes on large marine natural hazards is also dealt with.

The chapter on deep-sea ichnology (Chapter 8) outlines the relationship between depositional environment and endobenthic organisms (Uchman and Wetzel, 2011, this volume). It focuses on bioturbation, i.e. the process of sediment reworking by living organisms on and within the sea floor that disturb the primary sedimentary structures and that occurs before, during and after depositions in which traces of bioturbation can be found. The authors present the most recent ideas about the classification of trace fossils, ichnofacies, ichnoassemblages and ichnofabrics, their evolution through the Earth history and their up-to-date interpretation. They explain the concept of ichnofacies (and ichnosubfacies), which distinguishes between the various trace-fossil associations characterizing specific bathymetric ranges, and they deal with critical aspects such as the impact of taphonomy, lithological variability and other ecological factors on the diversity and the composition of trace-fossil associations. The chapter shows how ichnofabrics can be of valuable help for the interpretation of deep-marine environments (oxic/anoxic, low- or high-energy currents). It also discusses examples of ichnofabrics from high-energy deep-sea environments with contour or turbidity currents.

The chapter on early diagenesis (Chapter 9) discusses the various processes by which changes in deep-sea sediments are brought about after their deposition, i.e. as soon as they have become buried under younger layers of sediment at relatively low temperatures (Hesse and Schacht, 2011, this volume). The general strategy of the authors is to gain first insight into the diagenesis of the various classes of deep-sea sediments by looking at typical pore-water profiles, and to depict the mineral reactions that are associated with changes in the pore-water chemistry. The chapter highlights the importance of water circulation (advection and diffusion) through the sediments' pores and the reactivity of the different sediment constituents (calcareous, siliceous, volcanogenic, organic, etc.) for diagenetic processes. It then reviews diagenetic processes and mineral formations in brown abyssal clay, biogenic siliceous and carbonate sediments, hemipelagic, and anoxic sediments. Some sections are devoted to the highly reactive volcaniclastic environments and to gas-hydrate bearing sediments. Clues are presented for the study of the early diagenesis in active-margin settings with important fluid-flow activity, and of diagenetic effects of hydrothermal and igneous activity on sediment-covered mid-ocean ridges.

The chapter on the industrial interest in deep-sea sedimentological research (Chapter 10) focuses on the specificity of the oil and gas industry (Imbert, 2011, this volume). The quest for hydrocarbons in particular has led to the development of specific investigation techniques that in turn provide a great dataset for understanding deep-water geology at all scales. The author of the chapter presents the various types of deep-sea sediments forming source rocks, seals and traps, but mainly concentrates on the question how some of the rocks will form excellent reservoirs and why other types will not. The chapter also presents the specific tools used by the oil industry to obtain information from sedimentary deposits buried deeply below the sea floor. Special sections are devoted to the two main techniques used in oil exploration: 3-D seismics and wells. The chapter outlines how architectural elements are identified and how their reservoir properties are defined in such a way that the appropriate decision can be taken. Particular attention is given to gas hydrates and to environmental hazards related to the exploration and exploitation of deep-sea reservoirs on inclined sea floors (slope stability).

The two last chapters deal with the significance of deep-sea sediments as archives for ocean and climate history. Both highlight the study of climate processes under boundary conditions that were different from today; this is a particularly complex geoscientific challenge. Chapter 11 focuses on Mesozoic climates, dominated by greenhouse conditions, and shows how the marine pelagic and hemipelagic sediments record perturbation in the carbon cycle and in the global oceanic circulation pattern (Weissert, 2011, this volume). Mesozoic black shales are of particular interest. They serve as proxies for peculiar conditions in palaeoceanography and they record globally occurring "Oceanic Anoxic Events", possibly triggered by changes in atmospheric CO_2 concentrations, as recorded in the carbon-isotope signature of marine carbonates. The chapter gives, in addition, an introduction to the history of deep-sea research (on land and under the sea), which has been made possible by international research projects (DSDP, ODP and IODP) and which benefited greatly from the commissioning of the famous research vessels *HMS Challenger, Glomar Challenger,* and *JOIDES Resolution.*

Chapter 12 presents the record of Cenozoic climates in deep-sea sediments; they were increasingly dominated by icehouse conditions (Bickert and Henrich, 2011, this volume). It discusses how these climate developments were primarily controlled by the geodynamics on Earth, setting up a changing geometry of ocean basins and the redistribution of land masses relative to climatic zones, which included the opening and closing of ocean gateways and ultimately controlled the thermohaline circulation. The authors show that climates are a complex interplay of global processes such as mountain building (orogenesis), atmospheric circulation, oceanic circulation, precipitation, weathering of the continents, and the activity of the biosphere. These processes finally depend on the interrelations between the internal and external geosphere, the hydrosphere, the atmosphere and the biosphere. The chapter highlights how these processes and interrelations explain the climatic history of the Cenozoic: the Palaeocene-Eocene thermal maximum, the Eocene cooling and the growth of the Antarctic ice shield, the middle Eocene climate transition and the emplacement of the climate during the Neogene that finally evolved towards the ice age of the Pleistocene.

The book contains a CD-ROM with files of all the figures that are included in the individual chapters. In many cases, the reader will find here additional colour versions of figures that display the content in a much more instructive and attractive way.

REFERENCES

Babonneau, N., 2002. Mode de fonctionnement d'un chenal turbiditique méandriforme: cas du système turbiditique actuel du Zaïre. Unpubl. Ph.D. Univ.1, Bordeaux, p. 308.

Bickert, T., Henrich, R., 2011. Climate records of deep-sea sediments: towards the Cenozoic ice house. In: Hüneke, H., Mulder, T. (Eds.), Deep-Sea Sediments. Developments in Sedimentology, Elsevier, Amsterdam, vol. 63, pp. 793–823.

Carey, S.N., Schneider, J.-L., 2011. Volcaniclastic processes and deposits in the deep-sea. In: Hüneke, H., Mulder, T. (Eds.), Deep-Sea Sediments. Developments in Sedimentology, Elsevier, Amsterdam, vol. 63, pp. 457–515.

Fagel, N., 2007. Clay minerals, deep circulation and climate. In: Hillaire-Marcel, C., de Vernal, A. (Eds.), Proxies in Late Cenozoic Paleoceanography, Developments in Marine Geology, Elsevier, Amsterdam, pp. 139–184.

Faugères, J.-C., Mulder, T., 2011. Contour currents and contourite drifts. In: Hüneke, H., Mulder, T. (Eds.), Deep-Sea Sediments. Developments in Sedimentology, Elsevier, Amsterdam, vol. 63, pp. 149–214.

Henrich, R., Hüneke, H., 2011. Hemipelagic advection and periplatform sedimentation. In: Hüneke, H., Mulder, T. (Eds.), Deep-Sea Sediments. Developments in Sedimentology, Elsevier, Amsterdam, vol. 63, pp. 353–396.

Hesse, R., Schacht, U., 2011. Early diagenesis of deep-sea sediments. In: Hüneke, H., Mulder, T. (Eds.), Deep-Sea Sediments. Developments in Sedimentology, Elsevier, Amsterdam, vol. 63, pp. 557–714.

Hüneke, H., Henrich, R., 2011. Pelagic sedimentation in modern and ancient oceans. In: Hüneke, H., Mulder, T. (Eds.), Deep-Sea Sediments. Developments in Sedimentology, Elsevier, Amsterdam, vol. 63, pp. 215–351.

Imbert, P., 2011. Industrial application of deep-sea sediments. In: Hüneke, H., Mulder, T. (Eds.), Deep-Sea Sediments. Developments in Sedimentology, Elsevier, Amsterdam, vol. 63, pp. 715–764.

Kullenberg, B., 1947. The piston core sampler. In: Svenska Hydrografisk.-Biologiska Kommissionens Skrifter s.3, 1–46.

Marsset, T., Marsset, B., Ker, S., Thomas, Y., Le Gall, Y., 2010. High and very high resolution deep-towed seismic system: Performance and examples from deep water geohazard studies. Deep Sea Res. I: Oceanographic Res. Papers 57, 628–637.

Masson, D.G., 2003. Summary of geophysical techniques. In: Mienert, J., Weaver, P.P.E. (Eds.), European Margin Sediment Dynamics: Side-scan Sonar and Seismic Images. Springer, Berlin, pp. 9–16.

McCave, I.N., 2007. Deep-sea sediment deposits and properties controlled by currents. In: Hillaire-Marcel, C., de Vernal, A. (Eds.), Proxies in Late Cenozoic Paleoceanography. Developments in Marine Geology, Elsevier, Amsterdam, pp. 19–62.

McCave, I.N., Manighetti, B., Robinson, S.G., 1995. Sortable silt and fine sediment size/composition slicing: Parameters for palaeocurrent speed and palaeoceanography. Paleoceanography 10, 593–610.

Mulder, T., 2011. Gravity processes on continental slope, rise and abyssal plains. In: Hüneke, H., Mulder, T. (Eds.), Deep-Sea Sediments. Developments in Sedimentology, Elsevier, Amsterdam, vol. 63, pp. 25–148.

Sortable silt and fine sediment size/composition slicing: Parameters for palaeocurrent speed and palaeoceanography.

Savoye, B., Leon, P., de Roeck, Y.H., Marsset, B., Lopes, L., Herveou, J., 1995. PASISAR: a new tool for near-bottom very high-resolution profiling in deep water. First Break 13, 253–258.

Seibold, E., Hinz, K., 1974. Continental slope construction and destruction, West Africa. In: Burk, C.A., Drake, C.L. (Eds.), The Geology of Continental Margins. Springer Verlag, New York, pp. 179–196.

St-Onge, G., Mulder, T., Francus, P., Long, B., 2007. Continuous physical properties of cored marine sediments. In: Hillaire-Marcel, C., de Vernal, A. (Eds.), Proxies in Late Cenozoic Paleoceanography. Developments in Marine Geology. Elsevier, Amsterdam, pp. 63–98.

Stoner, J.S., St-Onge, G., 2007. Magnetic stratigraphy in paleoceanography: Reversals, excursions, paleointensity, and secular variation. In: Hillaire-Marcel, C., de Vernal, A. (Eds.), Proxies in Late Cenozoic Paleoceanography. Developments in Marine Geology, Elsevier, Amsterdam, pp. 99–138.

Uchman, A., Wetzel, A., 2011. Deep-sea ichnology: The relationships between depositional environment and endobenthic organisms. In: Hüneke, H., Mulder, T. (Eds.), Deep-Sea Sediments. Developments in Sedimentology, Elsevier, Amsterdam, vol. 63, pp. 517–556.

Wefer, G., Berger, W.H., Bijma, J., Fischer, G., 1999. Clues to ocean history: a brief overview of proxies. In: Fischer, G., Wefer, G. (Eds.), Use of Proxies in Paleoceanography: Examples from the South Atlantic. Springer, Berlin, pp. 1–68.

Weltje, G.J., Von Eynatten, H., 2004. Quantitative provenance analysis of sediments: review and outlook. Sed. Geol. 171, 1–11.
Weissert, H., 2011. Mesozoic pelagic sediments – archives for ocean and climate history during green-house conditions. In: Hüneke, H., Mulder, T. (Eds.), Deep-Sea Sediments. Developments in Sedimentology, Elsevier, Amsterdam, vol. 63, pp. 765–792.
Wheeler, A.J., Stadnitskaia, A., 2011. Benthic deep-sea carbonates: reefs and seeps. In: Hüneke, H., Mulder, T. (Eds.), Deep-Sea Sediments. Developments in Sedimentology, Elsevier, Amsterdam, vol. 63, pp. 397–455.

CHAPTER 2

Gravity Processes and Deposits on Continental Slope, Rise and Abyssal Plains

Thierry Mulder

Contents

Université de Bordeaux, UMR CNRS 5805 EPOC, Avenue des Facultés, 33185 Talence Cedex, France
E-mail address: thierry.mulder@u-bordeaux1.fr

Developments in Sedimentology, Volume 63
ISSN 0070-4571, DOI: 10.1016/S0070-4571(11)63002-1

1. Gravity Processes on Continental Slope, Rise and Abyssal Plains

1.1. Introduction

Gravity processes in deep-sea environments have never been directly observed. They can be detected by their impact on human infrastructures such as cable failures (Heezen et al., 1964, in Congo-Zaire; Gennesseaux et al., 1980, offshore the Var River), *in situ* measurements (current meters, turbidity meters, particle traps), and they are infrequently recorded on seismic lines. They are mainly interpreted from the nature of their deposits. Natural analogues on land are also used to understand the behaviour of a submarine process. For example, pyroclastic surges or snow avalanches can represent an acceptable analogue for turbidity currents. However, the problem of the supporting fluid (air or water) makes a strong difference between subaerial and submarine processes. In water, the apparent density ($\Delta\rho$) of a flow is

$$\Delta\rho = \rho_{\mathrm{f}} - \rho_{\mathrm{w}},$$

where ρ_{f} is the density of the flow and ρ_{w} is the density of the water.

For any calculation, the reduced density (g') is used:

$$g' = g\Delta\rho/\rho_{\mathrm{f}},$$

where g is the gravitational acceleration (9.81 m s^{-1}).

In addition, in a volcanic setting, the heat and the orientation of the initial blast add to the differences between a volcanic and a sedimentary process. To have a better knowledge of hydrodynamic processes, scale experiments and numerical modelling are also used. However, the scaling of the parameters, the space for the experiment and the duration of the processes differ drastically between laboratory experiments and natural processes. To make a comparison between laboratory experiments and natural phenomena, non-dimensional numbers are used, such as the Froude number, which is defined as the ratio between inertial and gravity forces:

$$Fr = \frac{U}{\sqrt{g'H}}.$$

The Froude number defines the state of the flow: critical if $Fr = 1$, subcritical if $Fr < 1$ and supercritical if $Fr > 1$, where U is the mean flow body velocity and H is the thickness of the flow body.

A hydraulic jump forms when Froude number initially > 1 becomes < 1.

The Reynolds number is the ratio between inertial and viscous forces:

$$Re = \frac{UH}{\nu},$$

where U is the kinematic viscosity of the flow.

The Reynolds number defines the flow regime: laminar if $Re < 500$, transitional if $500 < Re < 2000$ (Lowe and Guy, 2000) and turbulent if $Re > 2000$.

The Richardson number (Ri) describes the stability of a flow interface

$$Ri = 1/Fr^2$$

and is used for quantification of the water entrainment within a flow.

1.2. Classification of submarine gravity processes

Submarine gravity processes are classified according to the mechanical behaviour of the process, the particle-support mechanism, the concentration or the longitudinal change in their deposits.

Classification based on the mechanical behaviour (rheology) of the processes was developed by Dott (1963), Mulder and Cochonat (1996) and Shanmugam (2000) (Fig. 2.1). This type of classification is particularly well suited for flow processes. The most crucial point is to assess the viscosity of the flow (Locat and Demers, 1988). Several types of viscoplastic behaviour can be defined and used for flow modelling, such as Bingham (Johnson, 1970) and Coulomb-viscous (Hampton, 1972; Schwab et al., 1996).

Middleton and Hampton (1973), Nardin et al. (1979), Lowe (1979, 1982), de Vries Klein (1982) and Stow et al. (1996) used the particle-support mechanism to classify gravity processes. This is the most commonly used classification in literature for the present-day deep-sea environment. The classification distinguishes four types of particle-support mechanism: matrix strength, grain-to-grain interactions, fluid support and turbulence.

Classification using flow concentration includes characteristics coming from the observation of flow deposits (Mulder and Alexander, 2001a).

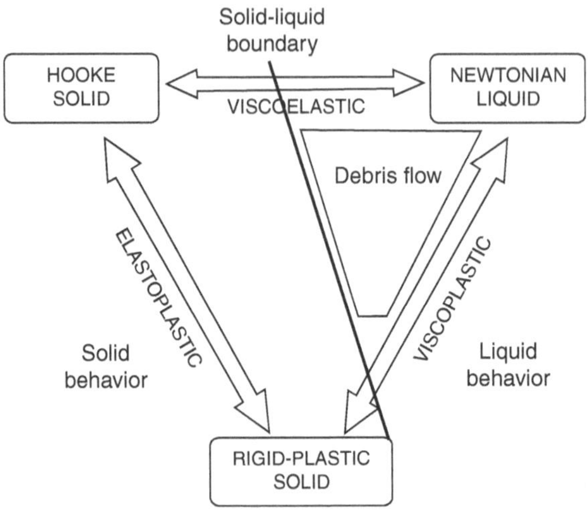

Figure 2.1 Solid/liquid rheology (from Hugot, 2000).

It principally takes into account the flow transformation with space and time and its progressive dilution. Because of fluid entrainment, flows transform from hyperconcentrated to concentrated flow and finally to turbidity current.

The classification based on deposits (Mutti and Ricci Lucchi, 1975; Pickering et al., 1989) is extensively used in research on ancient environments and in the oil industry. It is based on the sedimentary facies and their evolution along the pathway of the flow.

1.3. Main types of flow processes

The main types of flow processes are rock or consolidated material avalanching, creeping and failures, slides and slumps, flows (cohesive and non-cohesive), water-dominated flow and turbulent flow.

1.3.1. Rock avalanching

The avalanche of discrete blocks of large size (metres to hundreds of metres) occurs only in places where consolidated sediment or rocks outcrop, forming steep slopes or undersea cliffs. The most frequent sedimentary environment where rock avalanching occurs is the volcaniclastic environment (see Carey and Schneider, 2011, this volume, Chapter 7) despite examples of largest sedimentary slope failures (called mass-transport complexes (MTCs)) belong to this category. Recent examples of these large failures affecting deep-sea turbidite systems can be seen along North America margin (Tripsanas et al., 2008), in the Espirito Santo Basin, offshore

Brazil (Alves and Cartwright, 2009) or in the Nile deep-sea turbidite system (Rouillard, 2010).

1.3.2. Creeping and failures

Creeping is a long-term deformation of sediments submitted to a constant load on a gentle slope. The velocity of deformation is low (Mulder and Cochonat, 1996). Creeping can eventually lead to failure (Nardin et al., 1979). The process thus becomes a slide or a slump.

Sediment failures can be divided into slides and slumps. Both of them correspond to the motion of large volumes of sediment or rock along a failure surface. Sediment failures show typical morphological characteristics in addition to the presence of a failure surface: the head of the failure shows extension structures (tension cracks, Fig. 2.2A and D, normal faults). The toe of the failure shows compressive structures such as folds (that can be mistaken for sediment waves), inverse faults and overthrusts.

The failed surface can reach hundreds or thousands of square meters in size (Stow et al., 1996). Horizontal displacement does not exceed a few kilometres. As the failed sediment moves downslope, internal deformation increases and the original internal structures such as stratification are progressively deformed (de Vries Klein, 1982).

1.3.3. Slides and slumps

Rotational slides are called slumps (Fig. 2.2A, B and C). They have a curved surface of failure, being concave upward. This leads to the back tilt to the slipping mass. Slumps are usually deep rooted with a D/L (where D is the maximum depth of the slip surface, and L is the total length of the slump) ratio between 0.15 and 0.33 (Skempton and Hutchinson, 1969).

Translational slides, simply called "slides" have a D/L ratio generally <0.15. According to the size of the slide, the process can be named block or slab sliding (or gliding). For this kind of failure, the sliding surface is predetermined and corresponds usually to a layer with a low shear resistance such as a permeable sand layer or quick clays, for example, a structural heterogeneity or a change in the nature of sediment (e.g. sand/clay alternation). High-porosity sand layers can be evidenced using geotechnical tools. Quick-clay layers can be detected using a submarine piezometer to indicate permeability anomalies and submarine sources. A quick-clay layer is responsible for the failure surface of the Nice airport slump in 1979 (Habib, 1994; Piper and Savoye, 1993; Piper et al., 1992). In general, the smaller the depth to the failure surface, the greater the translational element will be (Skempton and Hutchinson, 1969). Slide spreading can generate internal vertical tension faults that cut the original slide into several slabs or blocks.

Slides and slumps are usually not isolated structures (Fig. 2.2D and E). They form complex structures with multiple phases of failure. The most common are the multiple retrogressive failures that form because of an

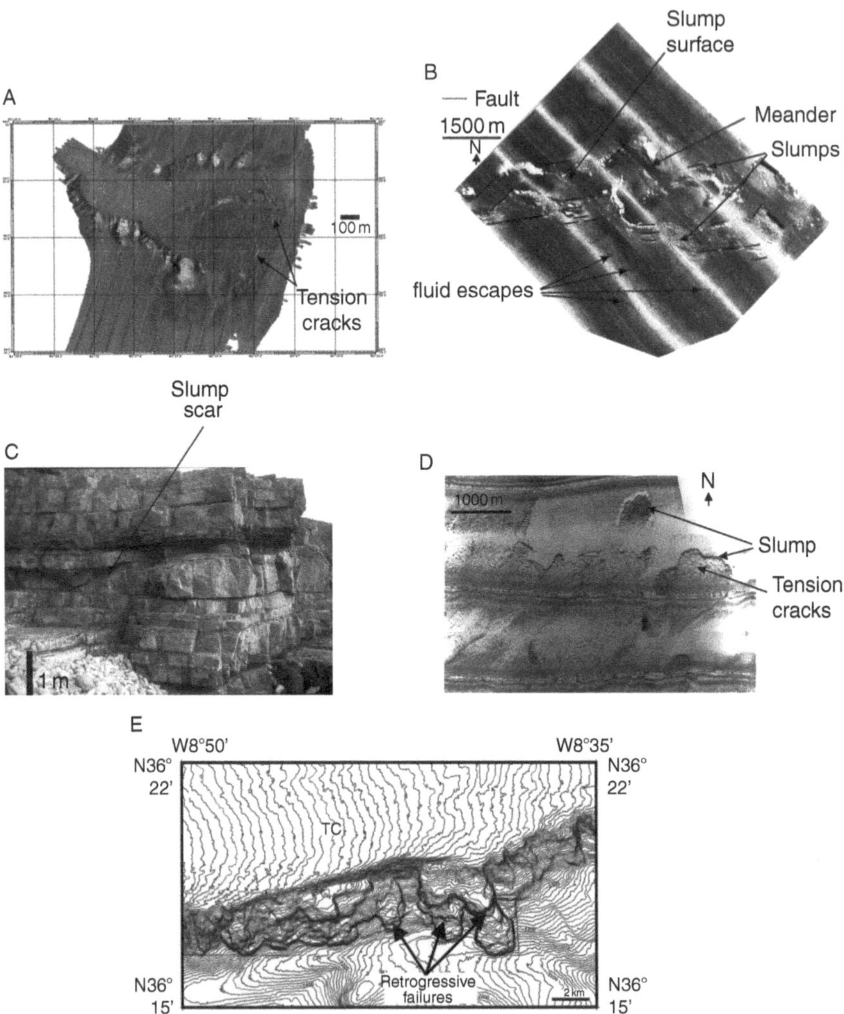

Figure 2.2 Examples of deep-sea sediment failures. (A) Retrogressive (backward) slumps at the head of the Capbreton Canyon (Gaudin et al., 2006a,b; Reproduced with kind permission of Springer Science and Business Media). (B) Slumps on the wall of the deep channel of the Orinoco (Faugères et al., 1997). (C) Slump in Silurian deposits (Aberystwyth Grits Formation at New Quay/Aberarth, West Wales). (D) Slumps in the source area of the Grand Banks turbidity current (photograph courtesy P. Cochonat). (E) Retrogressive slumps on the wall of a channel (Gulf of Cadiz). (A multi-colour version of this figure is on the included CD-ROM.)

upslope propagation of the failure (Fig. 2.2E). This is related to the formation of a steep escarpment at the failure head with tensional faults remoulding the sediment.

The other, less-frequent case of multiple-stage failures are the successive slumps or slides (domino-like slumps of Mulder and Cochonat, 1996). These are characterized by a downslope propagation of the failure. Each failed mass overloads the sediment located at its toe, generating a new failure. This kind of failure was inferred in the area of the Titanic wreck in the North Atlantic (Savoye et al., 1990).

1.3.4. Flows (general)

Flows can be subdivided into cohesive and non-cohesive (frictional) flows (Mulder and Alexander, 2001a). Cohesive flows have a matrix (cohesive) strength that results from cohesion between fine particles (clay and fine silt). Frictional flows are made of discrete particles. The space between grains can generate a high porosity filled with sea water in subaqueous environments.

Iverson (1997) suggested that it might be possible to define the division between viscous flows and frictional flows under subaerial conditions by using non-dimensional numbers that describe five parameters of momentum transport that can simultaneously affect a steady shear flow: inertial grain collision, grain-contact friction, viscous shear, inertial (turbulent) fluid-velocity fluctuations and solid–fluid interactions. The numbers describe the following processes and characteristics:

- the tendency for interstitial pressure developed between moving grains to buffer grain interactions;
- the ratio of inertial grain stress to viscous shear stress;
- the ratio of solid inertia to fluid inertia;
- the ratio of inertial shear stress associated with grain collisions to quasi-static shear stress associated with the weight and friction of the granular mass;
- the shear stress borne by sustained grain contacts to viscous shear stress.

If the inertial grain stress is more than 200 times larger than the viscous shear stress, collisional stresses dominate over viscous stresses (so the flow becomes a frictional flow). If the ratio is 200–450, the flow becomes a more dilute flow (concentrated flow or turbidity current). In these flows, the Reynolds numbers is always below 100, suggesting laminar conditions.

The behaviour of frictional flows thus is related directly to the relative proportion of grains and water. If porosity is low, grain-to-grain interaction dominates (grain flow of Nardin et al., 1979). If porosity is high, water forces dominate (liquefied and fluidized flows of Nardin et al., 1979).

1.3.5. Cohesive flows

Cohesive flows (Fig. 2.3) can be subdivided using the proportion of silt and clay they contain, forming mudflows and siltflows. Flows with less than 25% clay might be called silty mudflows and those with over 40% clay grade may be classified as clay-rich mudflows.

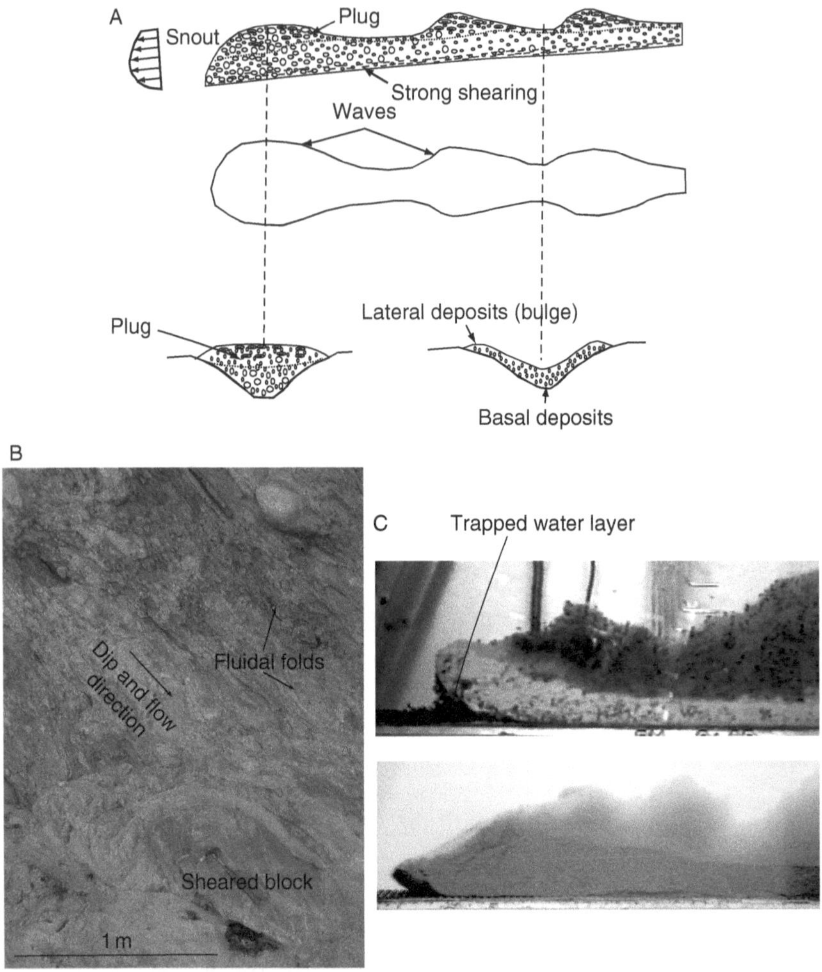

Figure 2.3 Characteristics of debris flows. (A) Longitudinal morphology of a partially channelized debris flow (modified from Johnson, 1984). Longitudinal structure mimics sediment waves. Flow head (snout) mimics a depositional lobe. Lateral deposits (bulge) mimic sedimentary levee; decametre to hectometre scale. (B) Synsedimentary deformation in a cohesive debris flow due to laminar flow regime: sheared (sigmoid) deformation and fluidal (sheath) folds (Cretaceous flysch of the Basque Country, Baie de Loya, France). (C) Hydroplaning forming at the base of a flow during a laboratory experiment (Mohrig et al., 1998). Flow head is decimetre thick. Picture courtesy D. Mohrig. (A multi-colour version of this figure is on the included CD-ROM.)

Cohesive flows usually show a relatively thick head (snout; Fig. 2.3A). The motion of cohesive flows is due to the low permeability of the matrix. The fine matrix imparts to the flow a pseudoplastic rheology and hence reduces the rate at which it can dilute. The density of the flow and its capacity are maintained (Stow et al., 1996). A water layer is trapped under the nose and the first part of the flow body. This reduces the flow resistance at the flow/seafloor interface, favours the flow motion and reduces the erosional power of the flow. This phenomenon is called "hydroplaning" by Mohrig et al. (1998; Fig. 2.3C).

Because of the existence of a cohesive matrix and the viscous strength, debris flows may transport boulder-size clasts of soft sediment or rock that float close to the upper surface of the flow, and very large rafts or olistoliths (Fig. 2.3; Johnson, 1970, 1984; Leigh and Hartley, 1992; Rodine and Johnson, 1976; van Weering et al., 1998). Hampton (1972) showed that the size of the carried blocks was inversely related to the density of the matrix. In most cases, the core of the flow moves as a rigid plug that forms when the internal shear stress overcomes the matrix strength (freezing; Fig. 2.3A). Shear zones form on both flow lateral edges.

1.3.6. Non-cohesive (granular) flows

In granular flows, particle support is due to dispersive pressures resulting from grain-to-grain interactions (Middleton and Hampton, 1973). This process occurs uniquely in sand or coarse silts and necessitates steep slopes ($>18°$) to be maintained. Flow energy is important and seafloor erosion due to these flows is intense. Grain-to-grain interaction can become too high because of increasing concentration due to seafloor erosion and flow collapse due to spreading. The velocity decreases primarily due to a decrease in slope.

Erosion structures are frequent at the base of these flows. Reverse grading can also be observed due to the strong velocity gradient between seafloor and the flow body in addition to dispersive pressure acting preferentially on larger particles.

1.3.7. Water-dominated flows

Water-dominated flows are the results of the destruction of the sediment fabric due to an increase in the interstitial pressure (de Vries Klein, 1982; Middleton and Hampton, 1973). Particles are supported by the upward force of the interstitial fluid (hindered settling).

Liquefied flows are defined by Nardin et al. (1979) as cohesionless flows supported by the upward displacement of fluid in a loosely packed structure. This kind of flow is usually generated by a forcing event that suddenly increases pore pressure (surge, waves, earthquake or upward motion of fluid). It produces sediment with a high porosity (e.g. quick sands). The increase in pore pressure is caused by destruction of the sediment fabric that no longer forms a rigid framework (Middleton and Hampton, 1973). Fluidization describes the transformation of a fine-grained material towards a liquid state under the effect of gas. It is frequent in subaerial volcanic environments (e.g. nuée ardente). Nichols (1995) calls "liquefication" the transformation of a solid into a liquid. This includes fluidization when the transformation results from pore fluid movement, liquefaction when it is caused by the grain agitation during cyclic shear stress and shear liquefication, which results from grain movement during the application of a shear stress across a sandbody. In this book, we follow Middleton and Hampton (1973) who used the terms fluidization and liquefaction as synonymous.

The transport capacity of the flow is maintained as long as the pore pressure is larger than the hydrostatic pressure. The dissipation of the pore pressure occurs rapidly during the motion. It depends on the permeability (essentially due to the grain-size distribution) of the carried sediment.

1.3.8. Turbulent flow

Turbulent flows are usually termed "turbidity currents". The latter term is, however, commonly misused. It designates a flow that is neither turbulent nor (sometimes) a current.

The etymology of "turbidity current" means a turbid flow (i.e. a flow with suspended particles) but widely accepted definitions of turbidity currents *sensu lato* (Middleton and Hampton, 1973) state that they are sediment gravity flows in which the sediment is supported mainly by the upward component of fluid turbulence. "Mainly" means that within the same flow, there may be other particle-support processes acting near the bed. Bagnold (1962) adds a limit of 9% sediment concentration by volume for full turbulent support of sediment. This is used as the threshold value to define turbidity currents *sensu stricto*.

The term "current" refers to the continuous movement of a fluid body in a given direction with a uniform velocity (Bates and Jackson, 1980). Using the terminology of Lüthi (1980), Laval (1988), Laval et al. (1988) and Ravenne and Beghin (1983), turbulent flows can be subdivided using flow uniformity into instantaneous (turbulent) surges (range of hours to days), longer duration surge-like flows and quasi-steady flows or turbidity currents (range of weeks to months; Fig. 2.4).

Consequently, a turbidity current is defined as a current where fluid turbulence is the main particle transport mechanism, although other mechanisms may also operate to varying degrees. Common usage would also include surge-like flows in the term "turbidity current". Because turbulent flows are important features with great impact on deep-sea sedimentation, their relevant characteristics will be dealt with in Section 1.4 in detail.

1.4. Characteristics of turbulent flows

1.4.1. Morphology of turbulent flows

The longitudinal morphology of turbulent flows has been described using observations of atmospheric currents, pyroclastic surges and scaled experiments (Fig. 2.5A). Turbulent flows consist of a head, a neck (not always

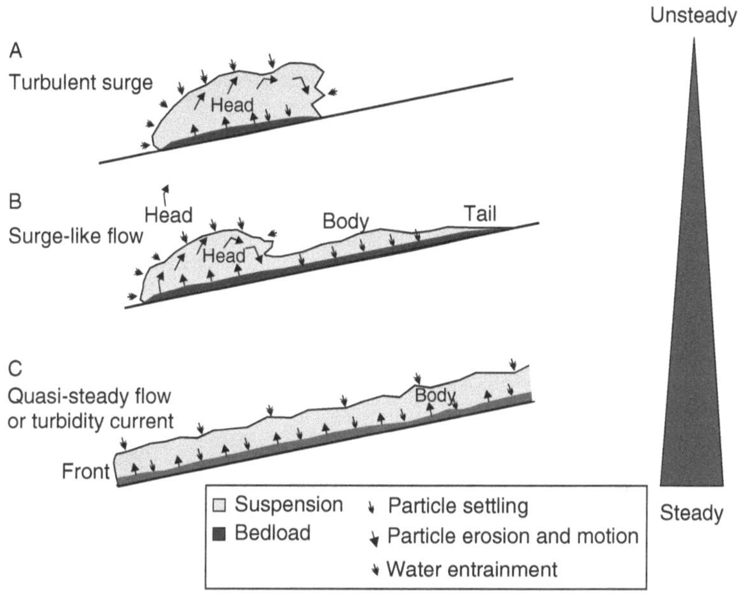

Figure 2.4 Turbulent-flow terminology (modified from Laval, 1988; Laval et al., 1988; Lüthi, 1980). (A) Turbulent surge (modified from Pickering et al., 1989). (B) Surge-like flow. (C) Quasi-steady flow or turbidity current.

observed), a body and a tail (Fig. 2.5A). The head of a turbidity current has a particular bulge-shaped form due to its flow dynamics. It is lobate with divergences from the main flow directions. Inside the head, the fluid has a circular motion towards the front and the top of the head. Coarser grains tend to concentrate in this part of the current. The head is mainly erosional. The body is located behind the head. It can be separated from the head by a transitional part (neck). The body is mainly depositional. The tail is the diluted back part of the flow that quickly thins.

The vertical morphology of turbidity currents show that they are stratified. Both flow density and competence decrease upward. This is related to the vertical decrease in flow velocity with velocity maximum that occurs in the lower part of the flow (Fig. 2.30B). This feature is particularly important to explain flow spilling and stripping on obstacles such as sedimentary levees (see Section 3.3.5) and for meander formation (see Section 3.3.8).

1.4.2. Velocity of turbulent flows

The velocity of turbulent flows can be quantified using numerical modelling, and compared to average velocities calculated on time records of submarine cable failures. The first historical example is the Grand Banks

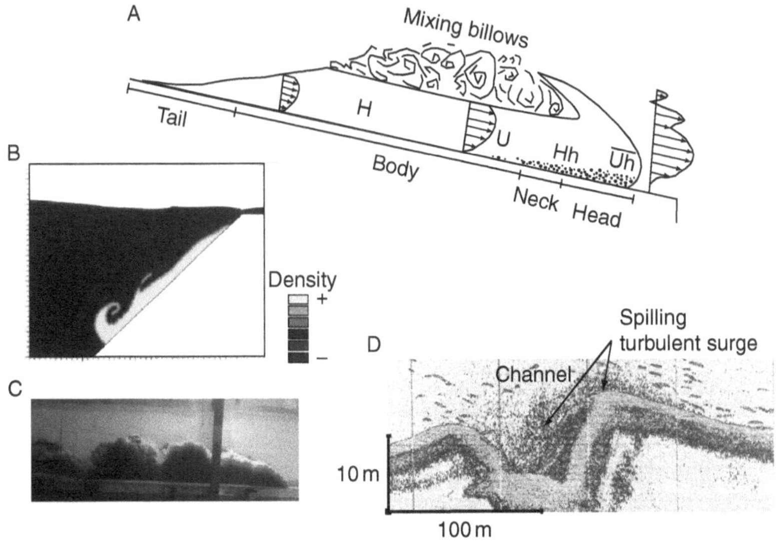

Figure 2.5 Examples of turbulent-flow morphology. (A) Morphology of a turbulent surge (modified from Pickering et al., 1989). (B) Numerical simulation of a turbulent surge (from Assier-Rzadkiewicz, 1997). (C) Basin experiment of a turbulent surge. Flow is 10 to 20 cm thick. (D) Very high-resolution seismic profile showing a turbidity current spilling over a channel wall (Hay, 1987). Copyright 1987 American Geophysical Union. Reproduced with permission from American Geophysical Union. (A multicolour version of this figure is on the included CD-ROM.)

turbidity current that occurred in 1929 (Heezen and Ewing, 1952; Kuenen, 1952). A magnitude 7.2 earthquake close to the slope break on the upper St. Pierre slope in the neighbourhood of the Laurentian channel, east of Newfoundland, generated several shallow slides (Fig. 2.2D) that transformed into a turbidity current that carried sediment down to the Sohm abyssal plain. The most distal deposits were located at about 1000 km from the source. The current reached a maximum velocity of about 19 m s^{-1} (67 km h^{-1}). The second example occurred in October 1979 seaward of the Nice Airport. The initial slide was triggered on the steep slopes off the Nice Airport during works for the extension of the airport, after a period of intensive rain that generated a flood of the Var River (Mulder et al., 1998a). No earthquake was recorded in the area during this period. The velocity of the flow was of about 20 m s^{-1} (72 km h^{-1}) on the continental slope; the flow stabilized at about 5 m s^{-1} (18 km h^{-1}) on the continental rise. The current covered a distance of about 230 km to supply the lobes of the Var turbidite system that lies off the western margin of Corsica.

The numerical modelling of the velocity of turbulent flows can be assessed using a simple 1D approach. Most frequently used is the Chézy equation that was originally derived to calculate the mean velocity of a fluvial current. Daly (1936) applied it to a submarine flow using the density contrast ($\Delta\rho$) between the flow and the ambient water. The initial Daly equation is not applicable if the density difference between the flow and the ambient fluid is small, which is the case in dilute turbulent flows. To solve this problem, he included the friction coefficient along the bed (C_f), and Middleton (1966b) included a friction coefficient (α) that is related to both Reynolds and Froude numbers. C_f varies from 0.035 to 0.005 (Bowen et al., 1984; Komar, 1977). This gives the classic formulation of the Chézy equation (Komar, 1977):

$$U^2 = \frac{cg' H \sin s}{(1+\alpha)C_f}$$

where s is the slope of the seafloor and c is the volume concentration of sediment.

The Chézy equation is simple but applies only to uniform and stationary 1D flows. The natural surges are highly non-uniform and non-stationary. In addition, the Chézy equation cannot be used when the slope = 0. However, as most of the parameters in the equation are very difficult to measure or at least quantify (e.g. flow concentration or density, flow thickness), this equation can be used as a first approach to provide a rough quantitative estimate of the flow velocity. The consistency of this estimate can thus be verified by comparing the velocity values with the size of the bedforms related to the flow or to the maximum size of carried particles. For example, Normark et al. (1980) suggested that the wavelength of sediment waves is related by the following equation for subcritical flows:

$$H = \frac{L}{2\pi Fr^2}.$$

Using the Froude number definition, Middleton (1966a) experimentally verified that the velocity of the flow head (U_f) was

$$U_h = C_h\sqrt{g' H_h},$$

where H_h is the mean thickness of the flow head and C_h is a friction coefficient. The value of C_h is estimated to be 0.71 (Keulegan, 1938) to 0.74 (Benjamin, 1968).

Bowen et al. (1984) and Klaucke et al. (1997) estimated the velocity of the flow using the grain-size distribution inside the flow. The flow velocity is then

$$U^2 = \frac{U_*^2}{C_f}.$$

Van Tassel (1981) and Reynolds (1987) suggest that $U_\star = w_s$ is a good approximate; w_s is the settling velocity of particles calculated using Stokes' law ($w_s = g'D^2\Delta\rho/18\nu$, where D is the particle diameter). Bowen et al. (1984) suggested that $U_\star = kw_s$, where k is a coefficient depending on the transport, bedload, suspended load and washload (see Fig. 2.11)

Another method to calculate the flow velocity is the energetic approach used by Bagnold (1954, 1962). Using the conservation of energy law, Bagnold (1954) defined a criterion called "autosuspension", which predicts whether a current will be able to self-sustain its motion. This criterion can be simply written as

$$U \sin s > w_s,$$

where w_s is the fall velocity of a particle that can be estimated using Stokes' law.

Autosuspension is criticized by several authors as an infrequent phenomenon in natural flows. It might occur only in thick, dilute turbulent flows transporting fine particles on steep slopes (Middleton, 1966c; Pantin, 1979, Pantin and Franklin, 2009). In fact, only a small part of the energy (2% according to Pickering et al., 1989) is used for sediment support. The rest is dissipated to counterbalance friction at the upper and lower interfaces of the flow and to produce turbulence. Parker (1982) demonstrated that ignition that occurs only in supercritical flows (Pantin and Franklin, 2009) needs to be added to autosuspension to self-sustain turbulent surges. Ignition predicts that a flow will collapse and become depositional below a density threshold that depends on slope, grain size, flow thickness and friction. Above this threshold and if the grain size is smaller than fine sand, the flow expands, and increases both its density and velocity.

More recent models for the calculation of flow velocity include non-stationary models (Beghin, 1979), stationary non-uniform models based on the theory of boundary layers and integration methods (Hinze, 1960; Plapp and Mitchell, 1960), mixing models based on Navier–Stokes equations (Assier-Rzadkiewicz, 1997; Naaim, 1995; Fig. 2.5B) and cellular automata models (D'Ambrosio et al., 2003).

1.4.3. Erosion by turbulent flows

Giant scours (Hughes-Clarke et al., 1990) and erosional surfaces at the base of turbidite beds suggest that erosion by turbulent flows is important. However, it is not clear whether the observed erosion is an effect of the turbulent flow itself or is due to the concentrated flow forming the base of bipartite flows. In addition, historical events show that the volume of a turbidite deposited by one event can be several times larger than the volume involved in the initial failure. For example, the initial slide in the case of the Nice event was 0.08 km^3 and the turbidite related to this event by Piper and Savoye (1993) represents approximately 2 km^3 (D. Piper, personal communication). For the 1929 Grand Banks event, initial failures represent approximately 20 km^3 and the related turbidite in the Sohm abyssal plain has a volume of 200 km^3 (Nisbet and Piper, 1998; Piper and Aksu, 1987). Seafloor erosion could be substantially increased by conduit flushing (Piper and Normark, 2009).

A simple way to estimate erosion is to relate it to the shear resistance of the sediment. For homogeneous sediment, the undrained shear resistance increases linearly with depth below sea floor (i.e. increases with consolidation). This allows to model erosion following the intuitive concept that the more the flow has eroded at a given location, the less it can further erode.

1.5. Initiation of gravity flows

Processes for flow initiation always include a gravity component. These processes can be initiated by ocean current of three types: (1) oceanic processes acting, for example, in canyon heads, (2) transformation of a submarine slide in a laminar and then turbulent flow and (3) direct flux from continental rivers to submarine environment (concentrated sediment-suspension injection of Normark and Piper, 1991; Piper and Normark, 2009). The latter case is an important issue because rivers represent the main entry point for terrigeneous material provided by erosion on the continents towards sedimentation areas, lakes or marine sedimentary basins. The difference between slide transformation and direct injection is fundamental in terms of dynamics, as the first has an interstitial fluid with the same density as the ambient fluid in the basin, whereas the second has an internal fluid with a density lower than the ambient fluid.

Most of the processes at the origin of gravity flows decrease the shear resistance (strength) of the sediment by increasing the pore pressure:

- dynamic processes on the continental shelf: tides, storm waves, swells;
- earthquake shaking;
- ice loading;
- fluid escape within the seabed; the fluid can be water, thermogenic methane, biogenic methane or methane coming from the phase change of gas hydrates (Locat and Lee, 2000; See Wheeler and Stadnitskaia (2011, this volume, Chapter 6.4.2.1), Hesse and Schacht (2011, this volume, Chapter 9.7.1) and Imbert (2011, this volume, Chapter 10.4.2.3.3)).

1.5.1. Gravity flows related to oceanic processes

The tide effect on slide initiation is unclear. Bjerrum (1971) reports subaqueous slope failures in Norwegian fjords that occurred during low-tide conditions or during exceptional low tides. It is difficult to state if the overpressure due to emersion of a part of the sediment or the reduction of the hydrostatic pressure could act in addition to overloading.

Prior et al. (1989) showed accelerometer and pore-water pressure sensor records of storm-wave reactivation of ancient slide deposits on the Huanghe Delta. Reactivation involves the loss of bearing capacity and shows gentle and gradual motion, in contrast to what is observed for initial failures. The behaviour could be interpreted either as a kind of liquefaction process or as thixotropic behaviour. Mulder et al. (2001c) described a turbidite deposited in the Capbreton Canyon just after the violent storm that hit the north European Atlantic coast in late December 1999. Although there is a temporal correspondence between the turbidite deposition and the storm event, the exact triggering cause (or combination) of the turbidity current is uncertain: slump in the canyon head (Fig. 2.2A) due to excess pore pressure related to swell and wave stress, or oceanic processes (Piper and Normark, 2009) including either acceleration of the southward coastal drift or dissipation of the water bulge accelerating the nepheloid layer (see Section 3).

Implication of oceanic processes on the continental shelf such as littoral drift in canyon supply is also suggested for Monterey Fan (Klaucke et al., 2004) and Ascension Canyon during low sea-level periods (Gardner et al., 1996) and more generally in canyon off California (e.g. La Jolla; Piper and Normark, 2009) with their head crossing the surf zone. Storm suspension or resuspension is suspected in Logan and Dawson canyons along the Scotian Slope (SE Canadian Margin; Mosher et al., 2004) or in the Eel River (N. California). Fluid-mud suspension forms (Puig et al., 2003) by the combination of storm-wave action and high fluvial discharge and is captured by the Eel Canyon where ignition occurs (Lamb and Parsons, 2005).

One of the mechanisms explaining transfer of sediments from shelf to deep sea under the action of oceanic processes is the plunging (cascading) of dense water. This happens, for example, in Mediterranean during winter eastern storms (Gaudin et al., 2006a; Palanques et al., 2006). The enhanced velocity along the seafloor (up to 80 cm^{-1}) allows resuspension of sand deposited during sea-level lowstand and transgressive periods on the outer shelf (Berné et al., 2001).

1.5.2. Failure-related gravity flows

Failure-related gravity flows are related to either external sedimentary cycles (including continent denudation and sediment accumulation along continental margins and depending on climate changes) or global plate tectonics through earthquakes or volcanic shaking.

Several dynamic processes can add stress leading to failure. Some of them add to the shear stress:

- overloading in areas with high sedimentation rates, or loading by a non-buoyant ice cap;
- oversteepening in rapidly prograding areas.

Overloading and oversteepening are usually associated processes. Both have been observed by Bornhold et al. (1994) in British Columbia (Canada) fjords. The sediment flows were recorded by current meters and correlated with flooding conditions at the river mouth, but the sediment concentrations were too low to be explained by a continuum between the river flow and the submarine flow. The only explanation was failure in rapidly prograding bars at the river mouth during periods of high sedimentation rates (floods). Similar interpretation including measurement of sediments fluxes using sediment traps is provided in Itirbilung Fjord by Syvitski and Hein (1991).

Earthquake shaking is the process that is suggested when no other process could explain the triggering of a submarine failure and flow. This is particularly true for ancient gravity processes for which the clues for the initiation are partially missing. For recent gravity processes, the temporal relation between an earthquake and a submarine process can be easily established. For example, submarine cable failures occurred a few hours after the earthquake in the case of the 1929 Grand Banks event. Other historical examples of earthquake-induced failures have been described from the Algerian Atlantic coast (Orléansville-El Asnam: Bourcart and Glangeaud, 1958), in the Marmara Sea (Beck et al., 2003), Algeria (Giresse et al., 2004) and Sumatra (Singh, 2005). Clear evidence of gravity motion triggered by an earthquake is the presence of simultaneous independent failures over a large (regional) area. It seems that the alternation of silty and clay layers is a parameter favouring the development of liquefaction (Dan et al., 2007).

Earthquakes are one of the most efficient processes for liquefying sediment by increasing pore-water/gas pressure. This leads to an upward motion of water and gas, which generates sediment collapse, leaving a circular depression called a "pockmark" (see Sections 5.4.2 and 9.4.2.3).

Low-magnitude earthquakes can also be generated by upward motion of magma in magmatic chambers of volcanoes, especially hot spot volcanoes, and in volcanic arcs. These earthquakes affect unstable freshly settled material on steep slopes and can lead to giant tsunamogenic landslides such as in Hawaii (see Carey and Schneider, 2011, this volume Chapter 7). In addition, earthquakes produce vertical and horizontal ground accelerations that reduce the shear resistance of the sediment.

Fluid escape has been recently recognized as a frequent phenomenon on the sea floor. It can create either liquefaction structures or simply vents of fluid. Pockmarks (see Sections 5.4.2 and 9.4.2.3) are related to fluid escape. They were recognized for the first time by King and Mac Lean (1970) around Nova Scotia (eastern Canada). Pockmarks form depressions with a sub-circular to elliptical shape and are bordered by steep escarpments; they have a flat bottom, and preferentially form in a thin sedimentary cover (Rise et al., 1999). They can be isolated but they can also form clusters or alignments. They may be elongated if they are associated with bottom currents (Hovland, 1983; Josenhans et al., 1978). Their shape and size depend on the characteristics of the bottom sediments, but they are encountered in sediments with grain sizes ranging from clay to sand (Hovland and Judd, 1988). Their diameter varies from a few metres to several hundreds of metres. Pockmarks are small in fine cohesive sediments and have an irregular shape in coarse sediments (Uchupi et al., 1996); rarely they reach a diameter of 1000 m, such as in the Skagerrak (North Sea: Bøe et al., 1998) and in north Gabon (Le Moigne, 1999). Their depth is generally less than 10 m and depends on the thickness of the sedimentary cover and the sediment texture (Josenhans et al., 1978). They are usually associated with sedimentary or tectonic structures (faults, diapirs, synclines and anticlines). On seismic lines, pockmarks are associated with acoustic anomalies (turbidity zones or acoustic masks) suggesting the presence of gas (Baraza and Ercilla, 1996).

Several theories exist for the genesis of pockmarks:

- fresh-water seepage (King and Mac Lean, 1970);
- gas escape (Baraza and Ercilla, 1996; Hovland, 1981; Nelson et al., 1979), essentially thermogenic or biogenic methane;
- dissipation of excess pore pressure during sediment compaction (Harrington, 1985);
- state transformation of gas hydrates (Paull et al., 1995) (see Wheeler and Stadnitskaia (2011, this volume, Chapter 6.4.2), Hesse and Schacht (2011, this volume, Chapter 9.7.1) and Imbert (2011, this volume, Chapter 10.4.2.3));
- escape of ice-clasts due to fresh-water freezing at high latitudes (Paull et al., 1999).

The formation of a pockmark; see Imbert (2011, this volume, Chapter 10.4.2.3.4) might be initiated by the formation of a positive morphology due to overpressure below the sea floor due to gas accumulation. The gas escapes when the excess pore pressure exceeds the shear resistance of the sediment. Fine particles are expulsed and settle in the neighbourhood of the structures. The finest particles can be transported in a preferential certain direction if a bottom current is locally active. The pockmark will thus have an elongated shape. The elliptic shape of the pockmark can be intensified by the erosion of the edge of the pockmark by bottom currents.

Mud volcanoes (see Wheeler and Stadnitskaia (2011, this volume, Chapter 6.4.2.1) and Imbert (2011, this volume, Chapter 10.4.2.3.5)) are also fluid-escape structures. In contrast to pockmarks, they form a positive topography. Their morphology can be elongated, elliptical or conical with a flat top. They can be simple, but most of them are compound and formed from coalescent volcanoes. The size of a volcano can range from several tens of metres to several hundreds of kilometres. The material escaping from the vents consists essentially of fluid mud with a mud content of approximately 99%, supporting rounded to angular rock clasts with a size varying between several millimetres and several tens of centimetres. The mud is mixed with fluids (water, gas, oil).

This mixture is called "mud breccia" (Dimitrov, 2002). This mud breccia is expulsed from a main chimney called a "feeder channel". Its external extremity is the main crater, which usually has a caldera shape.

More frequent is methane escape from the sea bottom. This can be due to not only a steady seep of biogenic methane but also more intense methane-mud eruptions at mud volcanoes. Eruption of mud volcanoes leads to mud flows restricted to the flanks of the volcano.

One of the best examples of a causal relationship between fluid escape and a major slide is the Storegga slide that is exposed on the continental margin of Norway (Mienert et al., 2005). This slide is the largest in the world with a total slipped surface of 112,000 km^2 and a volume of about 5580 km^3 (Bugge, 1983). The edge wall is nearly 300 km long (Bryn et al., 2005). Slide debris form a 450-km-long deposit with a maximum spreading distance of nearly 800 km. The slide shows three main failure events, dated to 7300 ^{14}C years BP (8100 cal. years BP; Haflidason et al., 2001). Acoustic images of the area located just north of the northern slide escarpment show circular patches with a diameter of 200–400 m, which are interpreted as pockmarks or mud volcanoes (Mienert et al., 1998). These fluid-escape structures are associated with a well-defined bottom simulating reflector (BSR) identified on seismic lines from the NE flanks of the current slide scar (Bugge, 1983). These characteristics suggest that the three-stage slide is related to fluid escape due to gas-hydrate destabilization and to earthquake activity (Bugge, 1983). In this case, the slide collapse could have released 5 GT of methane in the atmosphere (Judd et al., 1997). Such a volume can have an important impact on the greenhouse effect for a few years.

1.5.3. Submarine flows as a continuum of a river flow

Rivers are the main transport agent of sediment from the continent to the ocean, bringing 89% of the total sediment load to the global ocean (25×10^9 tonnes per year). Most of this load is supplied as suspended load (64%) or dissolved load (18%). The proportion of bedload is small (7%). Rivers are therefore much larger sediment suppliers than glaciers, sea ice and icebergs, which bring jointly only 2×10^9 tonnes per year (7% of the total sediment load to ocean), whereas the wind supplies 0.7×10^9 tonnes per year (2.5%) and coastal erosion is responsible for 0.4×10^9 tonnes per year (1.5%) (Syvitski, 2003).

At a river mouth, flow evolution depends on the density difference between the flow (ρ_f) and the surrounding water (ρ_w) (Fig. 2.6). This allows definition of three flow types (Bates, 1953): overflow (hypopycnal flow) if $\rho_f < \rho_w$, homopycnal flow if $\rho_f = \rho_w$, and underflow (hyperpycnal flow) if $\rho_f > \rho_w$. Mulder and Alexander (2001a) add the term mesopycnal flow (intraflow or intrusive flow) if ρ_f is between the density of two layers (ρ_{w1} and ρ_{w2}). This kind of flow frequently occurs in areas with hypersaline depressions, such as the eastern Mediterranean (Rimoldi et al., 1996), or

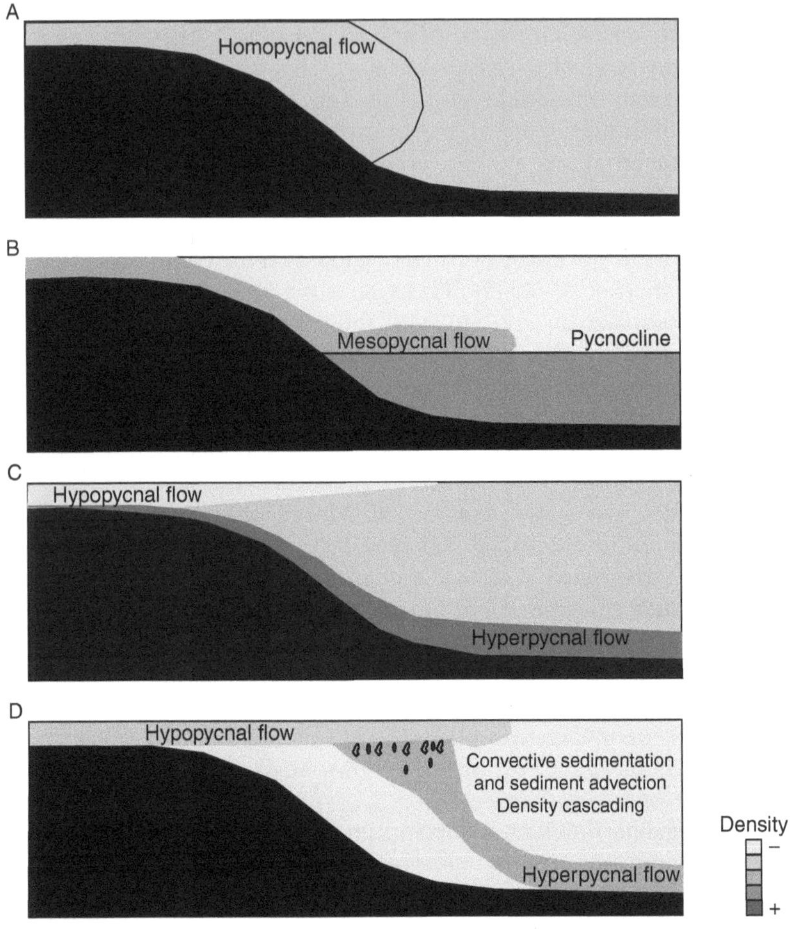

Figure 2.6 Types of density flows (modified from Bates, 1953; Mulder and Alexander, 2001a). ρ_f = density of flow; ρ_w = density of ambient fluid (ρ_{w1}, ρ_{w2}: densities of water in stratified body). (A) Homopycnal flow: $\rho_f = \rho_w$. (B) Mesopycnal flow: $\rho_{w1} < \rho_f < \rho_{w2}$. (C) Hypopycnal flow: $\rho_f < \rho_w$ and hyperpycnal flow $\rho_f > \rho_w$, formed by direct plunging. (D) Hyperpycnal flow formed by density cascading generated by both double diffusion and settling convection. Reproduced with permission from John Wiley and Sons.

with well-stratified water masses. The flow travels above a pycnocline. This stratification and the existence of several superposed pycnoclines generating several superposed mesopycnal flow is particularly important in the density cascading process.

Using a simple definition, any flow moving over the basin floor is a hyperpycnal flow. However, we suggest restricting the definition of a hyperpycnal flow to the original definition of Bates (1953), that is, to a flow that is the direct continuation of a river flow. In this sense, we can distinguish two types of hyperpycnal flows: (1) hyperpycnal flows *sensu stricto* (Mulder and Syvitski, 1995), which are synonymous with suspended-load-dominated hyperpycnal turbidity currents (we will use the term "hyperpycnal turbidity currents *sensu stricto*" for these flows; using this definition, hyperpycnal is used to mean "above a density threshold" and not simply "high density"), and (2) bedload-dominated hyperpycnal flows, which form at stream mouths and cannot be considered as turbidity currents (Mutti et al., 1996). These flood-related flows are frequent in tectonically active basins where steep slopes are present (Mutti et al., 1996, 2000). These flows behave as hyperconcentrated flows. They can transform into a classical continuum of hyperconcentrated and concentrated flows, and finally into a classical turbidity current.

Hyperpycnal flows *sensu stricto* can be termed quasi-steady flows since the flow is fed by prolonged river flow with a duration of hours to months (Mulder and Syvitski, 1995, 1996) so that the deposit volume mostly represents body conditions, whereas the flow front is unimportant with respect to sediment deposition. This suggests that hyperpycnal turbidity currents are the only true particulate gravity flow that should be termed "currents" that form along the sea floor.

Quasi-steady hyperpycnal turbidity currents were first reported in lakes (Forel, 1885, 1892), where they develop frequently. In fresh-water basins, very little suspended sediment is needed in the fluvial effluent to produce excess density. When rivers discharge into marine basins, depending on the temperature and salinity at the river mouth, 36–44 kg m^{-3} of suspended sediment is required to produce a hyperpycnal plume (Mulder and Syvitski, 1995, 1996; Table 2.1). In contrast, turbidity currents generated within the marine environment, for example, by a sediment failure, have saline interstitial water and sediment concentrations as low as 1–2 kg m^{-3} and are sufficient to maintain a current on a slope (e.g. as reported in the Var Canyon, Gennesseaux et al., 1971).

Table 2.1 Average temperature, salinity (from Kennish 1989) and density of sea water for different climates, and the corresponding critical particle concentration (Cc) to overcome the difference between fresh and salt water assuming a particle density of 2650 kg m^{-3}

	Temperature (°C)	Salinity (‰)	Density (10^{-3} kg m^{-3})	Cc (kg m^{-3})
(1)	27	34.75	1.02257	36.25
(2)	24	35.75	1.02424	38.93
(3)	13	35.25	1.02661	42.74
(4)	1	33.75	1.02708	43.49

(1): equatorial (latitude < 10°); (2): tropical and subtropical (latitude 10–30°); (3): temperate (latitude 30–50°); (4): subpolar (latitude > 50°). Modified from Mulder and Syvitski (1995).

1.6. Hyperpycnal flows

1.6.1. The formation of hyperpycnal flows

The formation of a hyperpycnal flow can occur in different river settings and due to various types of external forcing. Hyperpycnal flows can be either dominated by turbulence and transport by suspension and thus correspond to hyperpycnal turbidity currents of Mulder and Syvitski (1995) or non-turbulent and dominated by bedload transport and thus correspond to inertia flows of Bates (1953) or hyperpycnal flows of Mutti et al. (1996).

Bedload-dominated hyperpycnal flows are mainly generated by catastrophic (outsized) events such as flash floods under hot arid climates, rapid ice melting in periglacial streams or sudden dam break. The examples are essentially related to natural or artificial dam break (Mulder et al., 2003, 2009a), the entering of a mass flow in a subaquatic basin or flash floods. Suspended-load-dominated hyperpycnal flows are generated by slow erosion of natural dams, particular geologic conditions and climates with periods of intense rainfall or sustained precipitation.

(1) Hot arid climates

Streams located in arid hot climates have an intermittent flow regime. The stream bed might stay dry during months or years. Water supply is sporadic, short and intense. Wadis (oueds) in North Africa are active after heavy rains. A similar behaviour is shown by Californian and west Mexican streams after cyclones or hurricanes (Warrick and Milliman, 2003).

(2) Periglacial streams

Streams under cold climates experience two ways of generating hyperpycnal flows: (a) ice melting due to the alternation of warm/cold conditions or to catastrophic ice melting, and (b) alternation of warm/cold conditions at seasonal frequencies (winter/summer variations) or at millennial scale (orbital forcing). These seasonal changes could be the origin of laminated beds that frequently occur at high latitudes. Such beds have been described by Hesse et al. (1996) and Hesse and Khodabakhsh (1998) from the NAMOC (North Atlantic Mid-Ocean Channel). In periglacial streams, the dominance of bedload transport is demonstrated by the formation of large sandurs (e.g. Skeidarársandur in Iceland) and anastomosed river systems (Blum and Törnqvist, 2000; Gomez et al., 2000) such as the Yukon or Copper rivers in Alaska or the Waimakariri River in New Zealand. Anastomosed channel networks covered by coarse sand and gravels such as in the Var Canyon (Parize et al., 1989; Savoye and Piper, 1991) suggest the importance of bedload transport in parts of deep submarine environment.

(3) Catastrophic ice melting

This occurs, among other conditions, when a glacier covers active volcanoes. Melting of a large volume of ice can result from a subglacial volcanic eruption, which can give rise to the formation of a subglacial lake. If the ice wall—or the substratum—confining the lake breaks, millions of cubic metres of fresh water mixed with volcanic and glacial deposits flow to the ocean. This phenomenon is frequent in Iceland where it is named "jökulhlaup". Jökulhlaups are short-lived and violent phenomena lasting only a few hours to a few days. Jökulhlaups are also frequent in Alaska (Baker, 1995). Such a jökulhlaup formed in November 1996 because of the eruption of the Grimsvötn volcano below the Vatnajökull glacier (Einarsson et al., 1997; Grönvold and Jóhannesson, 1984; Gudmunsson et al., 1997). Peak discharge reached 50,000 $m^3\ s^{-1}$ where the flow crossed the Skeidararsandur and reached the ocean after travelling <70 km. A total water volume of 3 km^3 including clay to boulders and ice blocks were transported to the ocean during 2 days. During these jökulhlaups, particle concentration is very high. The short duration of the phenomenon (range of hours to days) explains the very high instantaneous discharges and flow velocities capable of transporting large blocks and destroying roads and bridges. When entering the sea, the flow plunges quickly.

At geological scale, catastrophic drainage of proglacial lakes (moraine-dammed-lakes) can occur at the end of glacial phases. Brunner et al. (1999) and Zuffa et al. (2000) described deposits related to hyperpycnal outburst of glacial Lake Missoula (Utah, USA) during late Pleistocene (17–12 ka). The proglacial lake had a volume of 2100 km^3 and was drained 40–80 times by the breaking of the moraine front during the melting of the North American ice sheet. The pluvial Lake Bonneville formed 32,000 years ago and probably emptied and filled several times according to wet/dry climatic cycles. Its estimated final drainage 14,500 years ago produced a flood with an estimated peak discharge of $425 \times 10^3\ m^3\ s^{-1}$ (Malde, 1968; Malde and Powers, 1962) that used the present Snake River system in USA. Similarly, the outburst flood of the Agassiz-Ojibway subglacial lake, 8470 years ago (Barber et al., 1999; Lajeunesse and St-Onge, 2008) resulting also from the melting of the Laurentide Ice sheet produced a flux of fresh water in the Atlantic Ocean sufficient to perturb the thermohaline circulation in the Atlantic Ocean and produce the cold climatic event at 8.2 ky (Broecker et al., 1989).

(4) Erosion of natural dams

This way of triggering hyperpycnal flows occurs in areas where several natural processes are combined on the continent, in particular, a high frequency of slides in fine-grained formations and a long period of severe floods. The long period of floods can be due to either ice melting or a seasonal dry period. A good example is provided by the AD 1663 flood in the Saguenay Fjord, a tributary of the St Lawrence River (eastern Canada). In AD 1663, a high-magnitude earthquake shook the neighbouring area of the Saguenay Fjord and generated approximately 3 km^3 of landslides and submarine slides (Schafer and Smith, 1987). This area is prone to landslides as the top of the sedimentary pile is formed by the Lahave clay formation. This clay was deposited in a shallow gulf that filled the isostatic depression that was present in Quebec just after the melting of the inland ice of the last glaciation. One of the terrestrial slides, with a volume of 0.2 km^3, dammed the Saguenay River. During the following spring, ice melting generated a large flood reaching an estimated 9000 $m^3\ s^{-1}$ peak discharge (Syvitski and Schafer, 1996) and breaching the dam. A hyperpycnite of 2–16 m thick, representing a volume of 0.31 km^3 resulted from the flood-generated hyperpycnal flow and high particle concentrations generated by erosion of the natural dam. Six hyperpycnites were deposited in the fjord during the last 7200 years by a similar process (St-Onge et al., 2004).

A similar history explains the increase in frequency of hyperpycnal flows after large earthquakes at the mouth of Taiwanese rivers (Dadson et al., 2005). In this case, slides of the river banks are generated by the earthquake but the material that slid down is easily removed during the typhoon season.

(5) Particular geological conditions

Some areas in the world are covered by extensive soft and easily erodible deposits, such as the wind-transported loess in China that supplies fine particles for the most-loaded rivers of the world: the Daling, Haile and Huanghe Rivers. The loess is intensively eroded during the monsoon rains, generating unusual suspended-particle concentrations at the river mouths. These rivers generated several-month-long hyperpycnal flows, which have been recorded by Wright et al. (1986, 1988, 1990). In a similar way, easily erodible black shales in the Alps may account for hyperpycnal-flow formation at the Var River mouth.

Volcanic ashes also constitute an excellent fine-grained material to generate a fast increase in suspended-sediment concentration when eroded by intense rains. Lahars are water-rich hyperconcentrated flows with poorly sorted volcanic material. These phenomena are frequent in Indonesia, from where the term originates. Lahars follow the hydrographic network on the continent. When arriving at sea, they can transform into hyperpycnal flows.

(6) Climate with high precipitation and the floods of small- and medium-sized rivers

The most common cause of hyperpycnal-flow formation at a river mouth is a flood. The rating curve (the curve relating the water discharge to the suspended-sediment discharge or concentration) is the best tool for predicting suspended load at a river mouth. Because rating curves are exponential laws, almost all the suspended load is supplied at a river mouth during floods. Consequently, measurements of mean values or measurements at constant frequency severely underestimate the supply of suspended load at a river mouth. The permanent monitoring of suspended load of rivers in the world, and sampling during peak flood conditions or other extreme events, is still rare.

Using the rating curve and a database of 147 rivers in the world, Mulder and Syvitski (1995) showed that 55% of the world rivers can generate hyperpycnal floods with an average frequency of <100 years, and that 71% can form a hyperpycnal flow with a frequency of <1000 years (Table 2.2).

Rivers that can generate hyperpycnal flows are small to medium size with an average annual discharge <380–460 $m^3\ s^{-1}$. The ability to produce a hyperpycnal flow increases with high relief (Milliman and Syvitski, 1992).

Table 2.2 Examples of large rivers that cannot produce hyperpycnal flows

River	Q_{av} ($m^3\ s^{-1}$)	Cs_{av} (kg m^{-3})	Cc (kg m^{-3})	Cs_{flood} (kg m^{-3})
Orinoco (Venezuela)	34,500	0.14	36.2	<1
Mississippi (USA)	15,500	0.8	42.7	11
Amazon (Brazil)	17,500	2.0	36.2	14
Paraná (Argentina)	13,600	0.18	38.9	3
Columbia (USA)	7960	0.06	42.7	2
Mekong (Vietnam)	14,800	0.3	38.9	4
Danube (Romenia)	6420	0.3	42.7	25
Yukon (USA)	6120	0.3	42.7	26
Zambezi (Mozambique)	17,600	0.09	38.9	<1
MacKenzie (Canada)	9750	0.14	43.6	4
Amur (Russia)	10,600	0.16	42.7	4
Zaire (= Congo) (Zaïre)	41,200	0.03	36.2	<1
Pechora (Russia)	3370	0.06	43.6	13
Niger (Nigeria)	6140	0.21	36.2	17
Volga (Russia)	17,200	0.03	42.7	0.4
Ob (Russia)	10,300	0.05	43.6	1
Lena (Russia)	16,200	0.02	43.6	0.3
Yenisey (Russia)	18,000	0.02	43.6	0.2
S. Dvina (Russia)	3660	0.04	43.6	8
Kolyma (Russia)	2840	0.07	43.6	25
Sao Francisco (Brazil)	3040	0.06	36.2	21
St. Lawrence (Canada)	14,900	0.01	42.7	0.1

Cs_{av}, average annual suspended-particle concentration values; Q_{av}, average annual discharge values; Cc, concentration threshold to generate hyperpycnal flows (Mulder and Syvitski, 1995); Cs_{flood}, maximum flood concentration in suspended particle. Rivers that easily form hyperpycnal flows by re-concentration of a hyperpycnal plume are underlined.

In contrast, "giant" rivers, that is, rivers with an average annual discharge >380–460 $m^3 s^{-1}$ such as the Nile, the Mississippi and all the Siberian rivers have a maximum flood particle concentration far below the concentration threshold that would generate direct hyperpycnal flows (i.e. without taking into account re-concentration processes). This is explained by two main reasons: (a) the particle load is diluted by their considerable volume of water thus reducing drastically the suspended-particle concentration, and (b) giant rivers currently trap much of their sediment load within their flood plains and subaerial deltas, and only small amounts reach the deep sea (Table 2.3).

In addition, several processes can intensify the formation of hyperpycnal flows at river mouths: (a) the dilution of sea water by fresh water during long-duration floods (range of weeks to months) that can decrease the concentration threshold to initiate hyperpycnal flows, and (b) erosion of mouth bars that can substantially increase the fine sand content within a flow.

Recent data from Taiwanese rivers show that hyperpycnal flows are more frequent after typhoons (Milliman and Kao, 2005; Milliman et al., 2007). Typhoons generate landslides along river banks. Erosion of these slides increases sediment concentration in rivers that in turn generates hyperpycnal flows according to a process similar to (4).

(7) Sediment reconcentration

A last but significant process resulting in submarine flows, and particularly suspended-load-dominated hyperpycnal flows, is sediment reconcentration. Surge-like turbulent flows may be generated by collapse of a suspension cloud due to dynamic processes on the continental shelf. Myrow et al. (2002) and Lamb et al. (2008) described Pennsylvanian deposits in Colorado (USA) deposited by storm-influenced hyperpycnal flow. In that case, the meteorological process (storm) is at the origin of both storm waves and flooding rivers. Prior et al. (1989)

Table 2.3 Dispersal of suspended-sediment load at the mouth of giant world rivers (from Meade, 1996)

River	Sediment load (10^{12} kg/year)	Delta (%)	Coast and continental shelf (%)	Deep sea (%)
Amazon (Brazil)	1	20	80	0
Ganges and Brahmaputra (India)	1.1	55	36	9
Huanghe (China)	1.1	82	18	0

observed a particle/water suspension on the shelf due to intense storm activity. Postma (1969) and Wilson and Roberts (1995) described the reconcentration of a nepheloid layer by a process called "density cascading". Oceanic processes (Piper and Normark, 2009) occurring in canyon heads are in this category, such as the storm-related turbidite in the Canyon of Capbreton in 1999 (Mulder et al., 2001c). The 1–2 m bulge forming the SE end of the Bay of Biscay during periods of exceptionally low atmospheric pressure associated with eastward winds could dissipate by entering the canyon, the Canyon of Capbreton, accelerating the nepheloid layer over the sea bottom.

The importance of reconcentration is demonstrated by two processes that can affect hypopycnal flows: double diffusion sedimentation and convective sedimentation (Fig. 2.7). Both have been described from laboratory experiments simulating the discharge of a fresh-water particle-laden flow above a cooler and denser brine. Another process that affects fine surface sediment at a river mouth is flocculation. This process is important for plume dynamics but plays a minor part in reconcentration processes.

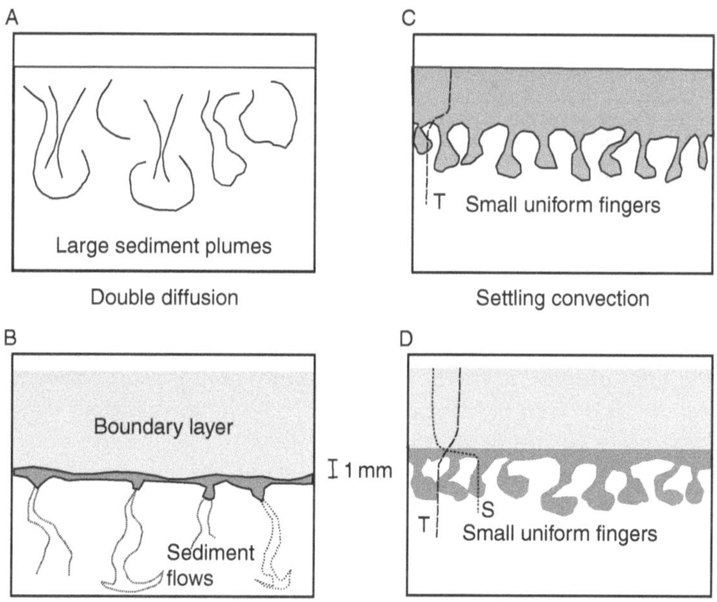

Figure 2.7 Comparison of reconcentration processes observed during laboratory experiments (from Parsons et al., 2001). Darker grey tone indicates increasing density. (A, B) Double diffusion. (C, D) Settling convection. T: temperature; S: salinity. Reproduced with permission from John Wiley and Sons.

Double diffusion (Fig. 2.7A and B) is equivalent to salt fingering of the thermohaline circulation that occurs when a warm brine becomes situated above a colder and denser water mass (Hoyal et al., 1999). The millimetre-large fingers are due to the difference of diffusivity between heat and salt. In the sedimentation process, the fine particles replace the salt.

Settling convection (Fig. 2.7C and D) has been described by Chikita (1991) and Parsons et al. (2001). They simulated the discharge of a fresh-water particle-laden flow above a cooler and denser brine, thus forming hypopycnal flows that quickly developed convective instabilities that took the form of sediment-laden fingers descending from the base of the horizontal interface between the two flows. The convection is due to the gradient in particle concentration that forms gravity instabilities (Hoyal et al., 1999). Settling convection and double diffusion differ by the size of the convective fingers and the space between each finger. Maxworthy (1999) has demonstrated that settling convection is not significant for flows with a high particle concentration. Convection is important for flow concentration of 1 kg m^{-3} and sediment-laden fingers appear for concentrations of >5 kg m^{-3}. These experiments suggest that the critical concentration for hyperpycnal-flow formation could be as low as 5 kg m^{-3}. If this is true, 84% of the world rivers can generate hyperpycnal flows in the marine environment with a frequency of more than once every 100 years.

1.6.2. Plunging and persistence of a hyperpycnal flow

Kassem and Imran (2001) successfully predict the development of the entire process from the free-surface-flow condition at the upstream end to the formation of the turbidity current using a numerical model. The plunging acts in four parts:

(1) The sediment-laden water flows in with a dominant dynamic force; the ambient water is pushed forward and the interface between the flow and the ambient fluid becomes pronounced.
(2) When the pressure at the bottom becomes significant, it accelerates the flow; the flow moves faster at the bottom than at the top.
(3) As the pressure continue to grow, the flow plunges to the bottom and begins to move as an underflow; at this stage, the velocity at the surface of the ambient water is still significant enough to move the plunging point forward.
(4) When equilibrium is reached, the velocity at the top disappears, and a stable plunge point forms. The hyperpycnal turbidity current moves forward with its classical bulge-shaped head and an elongated body.

In hyperpycnal flows, initial internal fluid is fresh water. This necessitates the maintenance of negative buoyancy by suspended sediment for plunging.

Using numerical modelling, Skene et al. (1997) and Mulder et al. (1998b) showed that a hyperpycnal flow is maintained along the seafloor because (1) entrainment of sea water into the flow progressively increases the density of the water phase while dilution of the suspended-particle concentration decreases the internal friction and (2) erosion of the seafloor increases flow density (driving force). Along the travel path of the AD 1663 Saguenay event, the density of the modelled interstitial fluid reached 1020 kg m^{-3} at 5 km from the river mouth and 1027 kg m^{-3} (i.e. the density of the ambient ocean water) at 22 km, where deposition began.

Absence of flow entrainment or reduced entrainment rate simultaneously to particle deposition should increase flow buoyancy. In that case, the forward motion of the flow decreases and the flow might become an ascendant plume loosing contact with the seafloor (flow lift-off or flow lofting; Sparks et al., 1993). This phenomenon occurs also in case of contour currents (See Faugères and Mulder, 2011, this volume, Chapter 3).

1.7. Differences between quasi-steady flows and surge-like flows

Several differences exist between quasi-steady flows generated by the continuation of a subaerial stream and surge-like flows initiated by a sediment failure (Table 2.4). The initial presence of fresh water in hyperpycnal flows strongly reduces the density difference between flow and ambient water. As demonstrated in laboratory hyperpycnal flows (Alexander and Mulder, 2002), hyperpycnal flows are much slower than turbulent surges moving on the same slopes.

Turbulent surges have a strong concentration gradient (Kneller and Buckee, 2000; Stacey and Bowen, 1987). This suggests that the base of the flow can behave differently from the top. Hyperpycnal flows have a more gradual vertical gradient in particle concentration (Alexander and Mulder, 2002; Mulder and Alexander, 2001a). Turbulent surges are typical unsteady flows that accelerate shortly after the triggering and then decelerate rapidly. They therefore require more complex models (Pratson et al., 2001). The duration of a surge-like flow can be a few tens of hours in a rare large event, taking into account the period from the initial slide to final deposition in the most distal part (Hughes-Clarke et al., 1990).

Hyperpycnal flows are quasi-steady flows (currents). That means their velocity increases and decreases slowly with time in relation to their flood origin. The process appears to be steady for a prolonged period, particularly if the flood hydrograph rises and falls gradually. Consequently, a quasi-steady turbidity current can persist for several days or weeks, depending on the flood duration at the river mouth (Mulder et al., 1998b; Skene et al., 1997). Quasi-steadiness allows more simple numerical models to be applied to hyperpycnal flows.

Table 2.4 Behaviour of hyperpycnal and slide-induced surges

	Quasi-steady flow	Surge
Minimum threshold of particle concentration for triggering	Yes	No
Initial concentration	5–200 kg m^{-3}	<1–1500 kg m^{-3}
Flow velocity	<2 m s^{-1}	<4 m s^{-1} to >10 m s^{-1} on steep slopes
Flow morphology	Front and body (type c in Fig. 2.4)	Well-defined head + body + tail (type (A) or (B) in Fig. 2.4)
Flow structure	Quite homogeneous; bedload transport at the base of flow	Strong vertical gradient. Concentrated flow at the base and turbulent flow at the top
Flow behaviour	Quasi-steady	Unsteady
Duration	Minutes to weeks	Minutes to hours
Deposits	Hyperpycnites	Turbidites (Bouma sequence)
Bedload transport (single event)	Over long distance	Over short distance

Synthesis of *in situ* observations (Gennesseaux et al., 1971), laboratory experiments (Alexander and Morris, 1994; Alexander and Mulder, 2002; Garcia, 1994; Laval et al., 1988; Lüthi, 1980; Middleton, 1967; Mulder and Alexander, 2001a; Ravenne and Beghin, 1983) and numerical modelling (Mulder et al., 1998b; Skene et al., 1997).

The steadiness of the flow has a fundamental impact on the development of bedload motion and sand transport. Turbulent unsteady surge cannot develop bedload transport over long distances. Sand can be eroded and put into suspension but quickly settles due to the short duration of the flow. For the same reason, the bedload transport is short. Conversely, long-duration quasi-steady flow can develop longer duration sand transport. Once the sand is eroded, the long flow body can allow bedload transport along considerable distance. Suspended transport of coarse particles by hyperpycnal flows is, however, not frequent because of their low competence. This suggests that sand transport along the distance of a whole turbidite system by a single event is not realistic.

As a conclusion (see also Table 2.4), hyperpycnal flows are slow and turbulent flows with their density largely remaining low along their travel path. They fit the description of the "low-density turbidity currents" of Nardin et al. (1979), Lowe (1982) and Mulder and Cochonat (1996). In contrast, slide-induced surge turbidity currents represent the transformation of fast-moving flows through ignition (Parker, 1982). The high velocity is due to the initial driving force and possibly also to the reduction of basal friction through hydroplaning (Mohrig et al., 1998). In this series of transformations, flow concentration and density both constantly decrease due to water entrainment. Some of the members of this type of flow are regarded as the "high-density turbidity currents" of Nardin et al. (1979), Lowe (1982) and Mulder and Cochonat (1996). This highlights the meaning of the term "hyperpycnal". "Hyperpycnal" means "above a density threshold" and not "high density".

A particular case concerns the coalescent surge flows generated by retrogressive slides or slumps (Fig. 2.2E). They form long turbulent surges (true turbidity currents). If they are maintained over sufficiently long periods (i.e. if sediment supply through sediment failure is maintained for a sufficiently long time), they can form long-duration flows (range of weeks to months) sharing the same characteristic as quasi-steady flows.

1.8. Flow transformation

Flow transformations are still poorly understood although they constitute the basis of the genetic classification of deposits and have been suspected for a long time (Heezen and Ewing, 1952). In addition, these transformations seem not to be progressive, as they intensify during change in flow behaviour due to, for example, a change in seafloor topography (presence of an obstacle, or slope reduction generating a hydraulic jump; Weirich, 1988).

Five types of flow transformation from a laminar to a turbulent regime have been defined by Fisher (1983) (Fig. 2.8).

(1) Body transformation from a laminar to a turbulent state without significant change in flow volume (no entrainment of the surrounding fluid or loss of the interstitial fluid; Fig. 2.8A); this process seems to be rare: it could be involved in flow transformation occurring when a hydraulic jump forms, but hydraulic jumps are usually associated with an increase in flow volume (Komar, 1971, 1977).
(2) Transition from laminar to turbulent with increase in flow volume (Fig. 2.8B); this case appears at slope breaks and generates a hydraulic jump; the increase in turbulence after the hydraulic jump maintains the particles in suspension within the flow (Garcia and Parker, 1989). The difference with surface transformation (type 4 below) is the shortness of transformation inducing both erosion (close to the slope break) and then quick deposition (slope-break deposits).
(3) Gravitational transformation of a flow leading to the formation of a basal laminar part due to the increase in flow concentration in a flow that was initially fully turbulent (Fig. 2.8C); the upper part remains

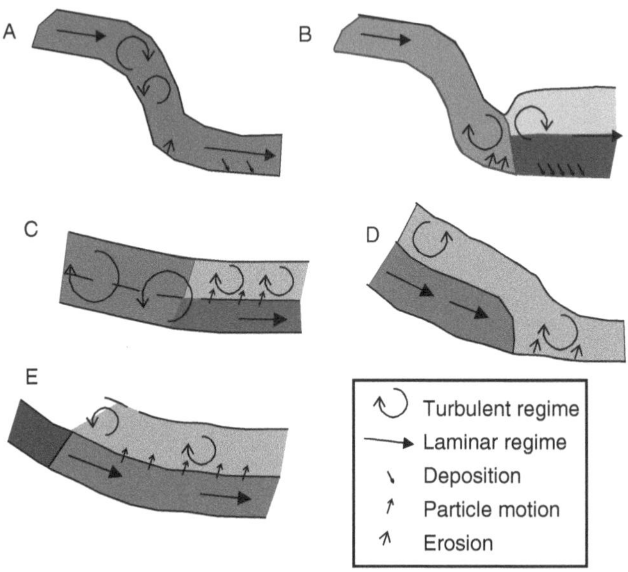

Figure 2.8 Hydrodynamic transformation of a flow (modified from Fisher, 1983 with the authorization of the Geological Society of America). (A) Transition laminar/turbulent without change in volume of interstitial flow (body transformation). (B) Transition laminar/turbulent with increase in volume of interstitial flow (hydraulic jump). (C) Gravitational transformation: formation of a high-concentration (hyperconcentrated or concentrated) laminar basal flow and a top concentrated flow by grain-size segregation under the action of gravity. (D) Surface transformation: progressive formation of a turbulent upper part in a flow by dilution and flow entrainment. (E) Elutriation: formation of a turbulent, low-concentration upper part in a flow by upward motion of fine particles and interstitial fluid.

diluted and turbulent. This process is involved in the formation of the Lowe sequence (Lowe, 1982; Postma et al., 1988).

(4) Surface transformation by entrainment of the surrounding fluid and progressive mixing (Fig. 2.8D) as shown by laboratory experiments (Mohrig et al., 1999); the flow becomes progressively more dilute and its concentration decreases until turbulence can develop.
(5) Transformation by elutriation and progressive fluidization (Fig. 2.8E); Elutriation is a process that removes fine particles in a sediment suspension by combination of sustained washing and suspension fallout (Bates and Jackson, 1980). Fine particles move toward the top of the flow and are extracted from the dense basal part of the flow to supply the dilute turbulent cloud forming above it.

Water entrainment and fine particle elutriation are both involved in the transformation of a dense laminar flow to a dilute turbulent flow (turbidity current). Consequently, fine particles move upward in the suspension and coarse particles are left behind. The conditions for flow transformation are described in the ignition concept of Parker (1982).

Scale experiment results show that a change in slope significantly perturbs the flow dynamics whatever the scale is. At a regional scale, a decrease in slope generates a hydraulic jump in the flow (Alexander and Morris, 1994; Garcia, 1994; Garcia and Parker, 1989; Komar, 1971; Mulder and Alexander, 2001b; Ravenne and Beghin, 1983). The thickness of the flow increases and entrainment of surrounding water intensifies. Flow velocity decreases and coarser particles settle rapidly just down the place where the slope decreases. The resulting deposits are called "slope-break deposits" (Mulder and Alexander, 2001b). Their thickness and extent both depend on the velocity of the flow before the slope break and on the size of the particles carried. The corresponding depression between the slope break and the slope-break deposits could be at the origin of large depressions called "plunge pools". These are observed along continental margins, at the toe of the continental slope (Farre and Ryan, 1985).

Motion of the flow and complex seafloor topography suggest that any flow may progressively change its character along the transport path, with transformation primarily due to decrease in sediment concentration through progressive entrainment of surrounding fluid and/or sediment deposition (Fig. 2.9). The parameters that control the flow transformation are flow-internal parameters such as the rate of fluid entrainment depending on flow velocity or flow thickness, and consequently flow transformation or external parameters such as slope gradient, lateral confinement or bed roughness.

Flow transformation from a homogeneous sediment failure generates rapidly stratified flows with a strong vertical velocity gradient. During most of its pathway, a gravity flow is bipartite with a basal concentrated to hyperconcentrated laminar flow and a top turbulent flow (Sanders, 1965). A fining-upward vertical grain-size distribution progressively forms in the most diluted turbulent part of the flow while the basal part remains ungraded or coarsening-upward.

Flows with high and low sediment concentrations may thus co-exist in one transport event because of downflow transformations, flow stratification or shear-layer development of the mixing interface with the overlying water (mixing cloud formation). Deposits of an individual flow event at one site may therefore form from a succession of different flow types, and this introduces considerable complexity into the classification of flow events and flow types from the deposits.

At a local scale, a change in slope and obstacles also modify the flow behaviour (Normark, 1985). Scaled-experiments (Kneller and McCaffrey, 1999; McCaffrey and Kneller, 2004) show that flow with small Froude number tends to divide when encountering an obstacle. If flow height is large when compared to obstacle height, flow stripping occurs and the upper dilute flow part spills over the obstacle. If not, the flow is partly

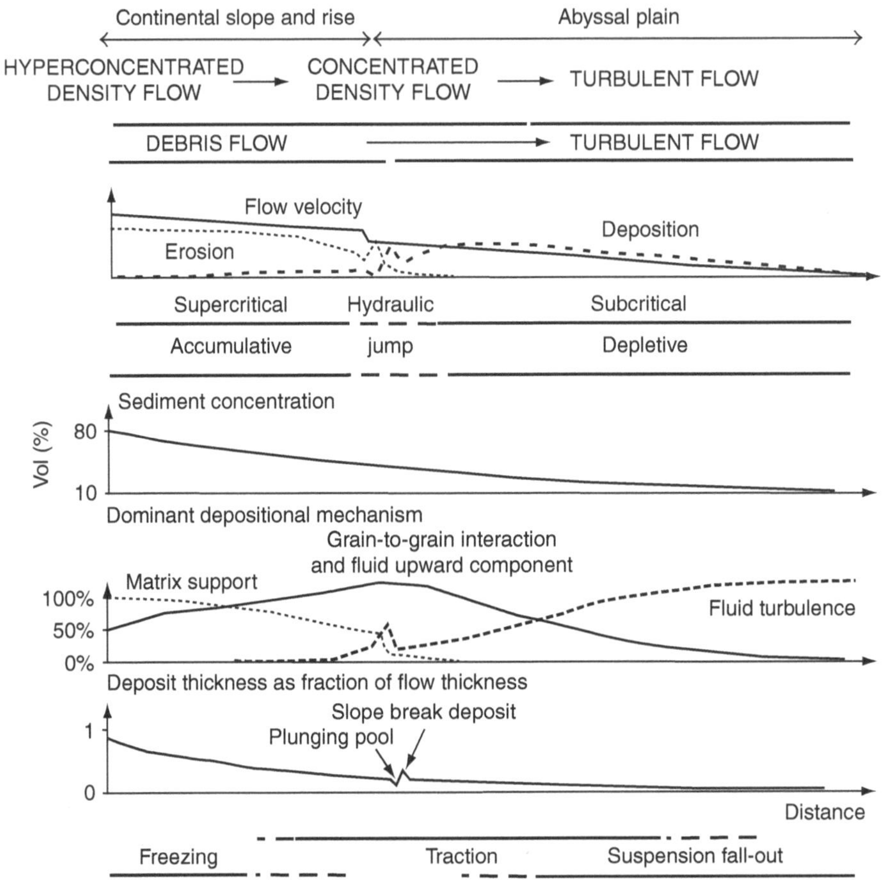

Figure 2.9 Flow transformations (modified from Mulder and Alexander, 2001a). Transformation of sediment flows on a typical continental slope. Slump-generated hyperconcentrated density flows or debris flows are transformed into concentrated density flows and then into surge-like turbulent flows. The sediment concentration decreases progressively with distance, although the shape of the curve presented is speculative (it may be steeper, e.g. at the transition from hyperconcentrated to concentrated density flow). The graph also shows the major particle-support mechanisms operating and demonstrates part of the basics to distinguish between flow types. A hydraulic jump can form at the base of the continental rise, where the slope gradient diminishes abruptly. It increases erosion and generates slope-break deposits. This increases fluid turbulence. Reproduced with permission from John Wiley and Sons.

reflected against the obstacle. For flow with large Froude numbers, either there is total flow reflection (obstacle height ≫ flow height) or overflow is total (flow height ≫ obstacle height). The flow stratification is particularly important for controlling the interaction between the flow and the obstacle (Kneller and Buckee, 2000). Pantin and Leeder (1987), Alexander and

Morris (1994) and Kneller and Buckee (2000) showed that large obstacles such as sedimentary levees or diapirs can disturb flow dynamics by generation and migration of hydraulic jumps, reflection of solitary waves and intensification of mixing vortices. These hydrodynamic changes have an impact on sediment deposition, for example, by increasing erosion on top of topographic highs, generating a better grain-size sorting or by the deposition of a thick sediment ridge in front of the obstacle due to the formation of a standing billow. Pochat (2003) showed that small obstacles such as synsedimentary fault scarps could also induce changes in flow dynamics generating sediment sorting at the top of the obstacle and scouring due to the formation of a hydraulic jump behind the obstacle. At a regional scale, ocean floor features have been interpreted to result from flow transformation to explain several phenomena such as those associated with the 1929 Grand Banks event (Piper et al., 1985, 1999a), Quaternary features on the Scotian Slope (Piper et al., 1992) and Madeira abyssal plain (Rothwell et al., 1992), or along the runout of the Nice submarine event in 1979 (Piper and Savoye, 1993). Debris flows are more likely to transform into hyperconcentrated density flows or concentrated density flows if they contain high proportions of particles larger than medium silt (Hampton, 1972, 1975).

2. Gravity-Fall and Gravity-Flow Deposits

2.1. Rock-fall and slope-failure deposits

Rock-fall deposits included in debris avalanches are poorly sorted deposits made of clasts of various size (millimetres to hectometres) with little matrix. They are frequent in volcaniclastic environments along oceanic islands (see Carey and Schneider, 2011, this volume, Chapter 7). The blocky structure leads to hyperbolic facies on seismic profiles (e.g. Hawaii, Moore et al., 1989, 1994, 1995; La Fournaise, Réunion Island, Ollier et al, 1998; El Hierro Canary Island, Masson, 1996; Urgeles et al., 2003).

Slide and slump deposits are recognized by the shape of the failure surface. As transport is short, the deposits show little internal deformation (pre-existing bedding is preserved). Plastic deformation can occur at the base of the failed deposit (Mulder and Cochonat, 1996; Nardin et al., 1979).

2.2. Gravity-flow deposits

Gravity-flow deposits have been studied in the field (e.g. Bouma, 1962; Mutti, 1979, 1985, 1992) and using high-resolution seismics (Damuth and Hayes, 1977) or acoustic imagery. Several classifications have attempted to order deposits from gravity flows. They are either descriptive (Bouma, 1962; Guibaudo, 1992; Pickering et al., 1989; Piper, 1978; Stow and

Shanmugam, 1980) or genetic (Mutti, 1985, 1992; Mutti and Ricci Lucchi, 1975; Nardin et al., 1979; Walker, 1978).

Particle deposition by sediment gravity flow can occur by freezing, traction and suspension fallout. Freezing or "en masse" deposition occurs when the flow resistance exceeds the driving shear force in a cohesive flow; the flow then stops abruptly (cohesive freezing of Middleton and Hampton, 1973; Lowe, 1979, 1982). Freezing frequently occurs in cohesive and hyperconcentrated flows.

Traction is a synonym of bedload transport (Bates and Jackson, 1980). It is responsible for all the sedimentary structures that form by particle motion at the bottom boundary layer. In a sufficiently dilute flow, suspension fallout allows all particles to reach the bottom boundary layer and then tractional structures can form. Traction is frequent in concentrated flows.

Suspension deposition represents the fallout of particles. It is at the origin of most of the vertical normal grading (fining-up). It is typical of dilute turbulent flows.

2.3. Freezing of cohesive and of frictional flows

Freezing reduces the sorting out of the transported clasts (Nardin et al., 1979). The base of the flow can show crude imbrication of clasts (long axis of clasts parallel to, and dipping in, the flow direction; Lindsay, 1968), or inverse grading due to shearing along the seafloor, dispersive pressure or kinetic sieving (smaller particles fall between larger ones). The clast fabric is cyclic along the flow body and develops preferentially when the flow velocity decreases (Lindsay, 1968). Its preservation depends on the time when a part of a flow freezes. The flow matrix can also show fabric and oriented particles.

Debris-flow deposits (debrites) have generally a poor grading and fabric. They form massive beds with some blocks supported by the plug and transported at the top of the flow. The result is a typical, hummocky surface on the seafloor. On seismic profiles, debrites have usually a convex-up shape with low amplitude to transparent facies (Embley, 1976). The presence of blocks generates hyperbolic reflectors on seismic profiles.

In cohesive flows, the internal shearing related to the laminar regime induces some synsedimentary deformation structures in the deposits such as sigmoidal deformations of plastic blocks (sheath folds; Fig. 2.3B). Internal shearing can also produce crude imbrication (Nardin et al., 1979) at the base of the flow, or inverse grading. When channelled, debris flows can generate undulating deposits that mimic crude sediment waves (Fig. 2.3A). When the flow spreads, rapid freezing generates lobate deposits (Fig. 2.3A). Freezing on lateral thin edges of the flow generates thick lateral bulges that look like crude levees but usually are made of coarse particles. These levees contribute to channelize the debris flow.

The process is similar for frictional flows. When the energy dissipated by grain-to-grain interactions is larger than the kinetic energy of the flow, the flow stops (frictional freezing). A crude inverse grading can appear at the base of a frictional-flow deposit and a crude fabric of blocks (*a*-axis parallel to the flow motion) can develop. On high-resolution single-channel seismic profiles, hyperconcentrated-flow deposits generate hyperbolic facies if blocks are present. Concentrated sandy flow deposits (Ta Bouma interval) generate discontinuous reflectors with high acoustic impedance.

For flows with a high water content (liquefied flows), deposition occurs as successive freezing from the base to the top of the flow (freezing upward of Middleton and Hampton, 1973). This upward motion is responsible for the fluid-escape structures such as dish structures, pipe structures (Fig. 2.10E) and the vertical orientation of clasts (Fig. 2.10D). Dish structures (Fig. 2.10B and C) preferentially form when the water is regularly distributed in the flow body. Pipes preferentially form in low-permeability flows, for example, when a wedge of water is pinched below the flow (hydroplaning). In this case, when the flow stops, the water tends to escape vertically in areas of less resistance, forming low-diameter (pipes) or large-diameter (pillars) conduits. Convolute lamination also indicates synsedimentary dewatering (Fig. 2.10A). This lamination is probably due to poor initial packing since they are common in environment where hydroplaning does not occur such as tidal channels (D. Piper, personal communication).

2.4. Traction-freezing deposition: the Lowe sequence

The Lowe sequence (Lowe, 1982; Figs. 2.11 and 2.12A) represents deposition by a bipartite current. The basal part is a hyperconcentrated flow with a laminar regime and the upper part is a more dilute flow in which turbulence progressively develops. It corresponds to the coarse-grained turbidite depositional sequence of Nardin et al. (1979). Bedload along the sea floor reworks coarse material, pebbles or gravels and forms gravel waves with oriented clasts (unit R1 in Fig. 2.11).

The basal flow has a strong vertical velocity gradient. Because of friction along the seabed, the velocity at the base of the hyperconcentrated flow is close to zero. It increases progressively upward but flow lines remain parallel. Consequently, the competence of the flow increases upward. An inverse grading (coarsening-up) forms within the moving basal flow. In the more dilute upper flow, flow lines are not parallel because of turbulence. In this upper flow, suspension fallout generates a particle flux from the upper to the basal flow. Particle load in the basal flow increases its concentration. If a concentration threshold is reached, the flow freezes. The internal inverse grading is thus preserved (traction carpet of unit R2 in Fig. 2.11). Particles at

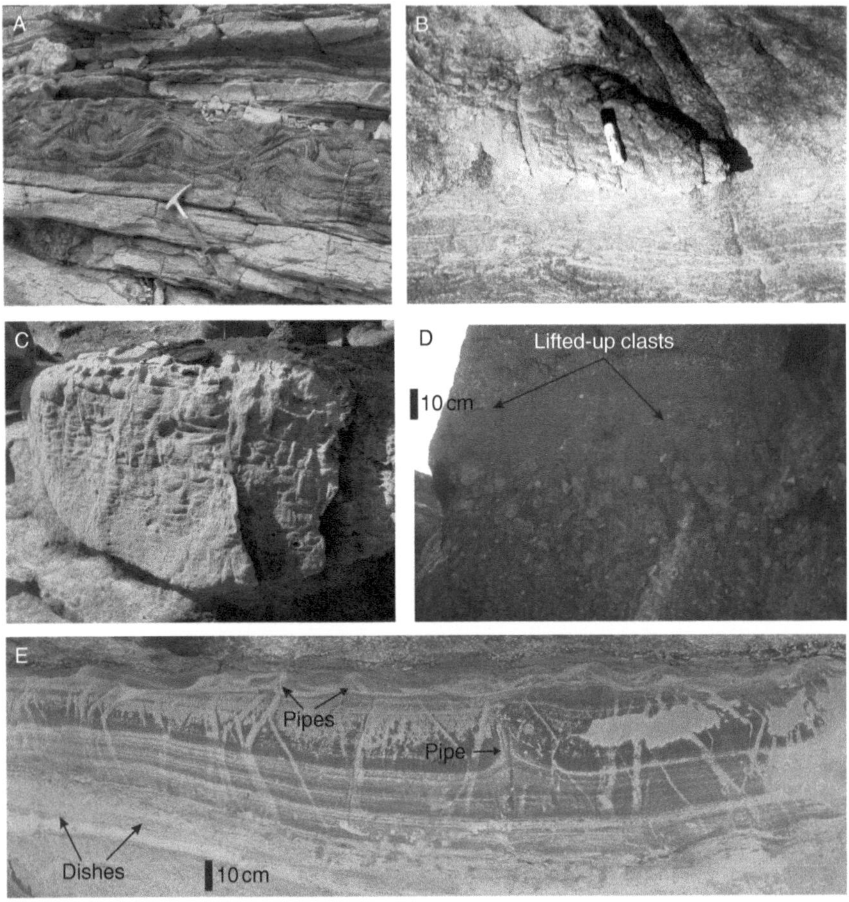

Figure 2.10 Examples of fluid-escape structures. (A) Convolute structures (Cretaceous Basque Flysch, Guéthary, Basque Country, SW France). (B, C) Dish structures (Cabo Higuer, Tertiary Flysch, Basque Country, northern Spain). (D) Vertically oriented pebbles underlying the top of the flow in a hyperconcentrated flow (Late Ordovician, Cwmdauddwr, Rayader, Powys, Wales). (E) Pipes and dishes (Cabo Higuer, Tertiary Flysch, Basque Country, northern Spain). Note the small sand volcanoes formed around the pipe mouth. (A multi-colour version of this figure is on the included CD-ROM.)

the top of the basal flow settle by suspension fallout forming the unit R3 with a crude normal grading. During the first part of the flow motion, coarse particle are abundant and initial beds (units R1–R2 in Fig. 2.11) are massive or crudely graded (R3) beds. Because coarse particles quickly settle, coarse particles become less abundant in the flow. Particle transport mechanisms that formed R1 and R2 still act but they form lenses rather than massive beds (units S1 and S2 in Fig. 2.11).

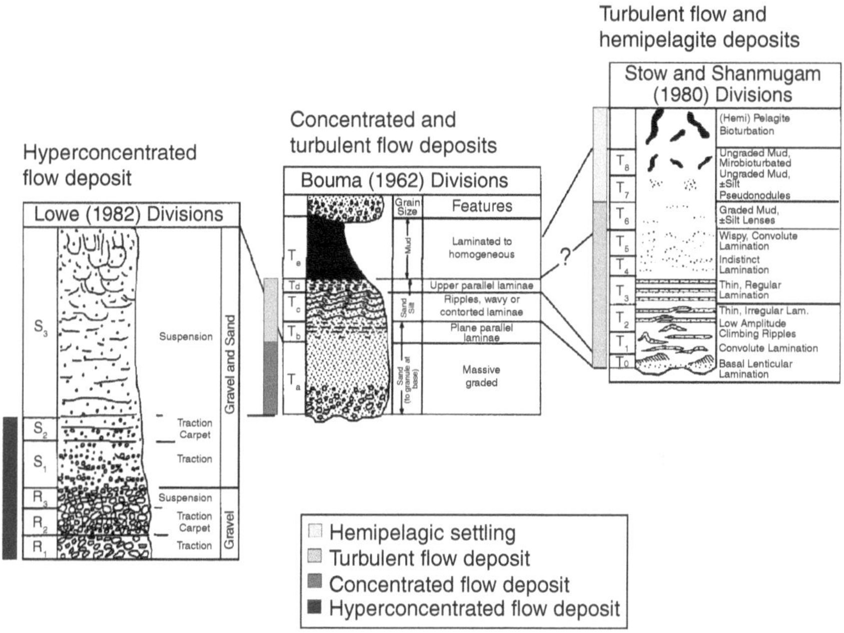

Figure 2.11 Continuity of deposits from a hyperconcentrated flow (base of Lowe sequence: Lowe, 1982), a concentrated flow (top of Lowe sequence and Ta interval of Bouma sequence: Bouma, 1962) and a turbulent flow (turbidite) (Tb-e interval of Bouma sequence, and Stow and Shanmugam sequence: Stow and Shanmugam, 1980) (modified from Shanmugam, 2000 and reproduced with permission from Elsevier). Equivalence with terminology of Mulder and Alexander (2001a).

2.5. Traction–suspension deposition: the Bouma sequence

Kuenen and Migliorini (1950) firstly related graded bed sequences to turbidity currents. They interpreted the grading as related to the fallout of particles in a single waning turbidity current (i.e. with velocity decreasing from head to tail). This relation is also used by Kuenen (1967) and Shanmugam (2002).

The first model of gravity-flow deposition is the Bouma sequence (Bouma, 1962; Figs. 2.11 and 2.12B and E), corresponding to classical turbidite deposition. It is a typical model of a traction/suspension fallout process. Several facies with typical sedimentary structures indicate a bottom-to-top decrease in the energy level. The sequence is normally graded, indicating suspension fallout.

The sedimentary structures in the Bouma sequence are interpreted by analogy with structures formed during laboratory experiments to simulate the bedforms formed in river beds under decreasing energy conditions.

At high energy, flows are erosive. The erosional surface at the base of the sequence with erosional features was caused by the passage of the energetic head of the current or by the passage of several currents (including reflected flows) as indicated by several groove directions in some turbidite deposits (Mulder et al., 2002). Erosion can be due to direct digging by the turbulent flow (scours, gouges and flute marks and casts). It can also be due to the drag (groove and brush marks and casts) or the impact (prod and bounce marks and casts) or rolling of a clast along the basin floor (see synthesis in

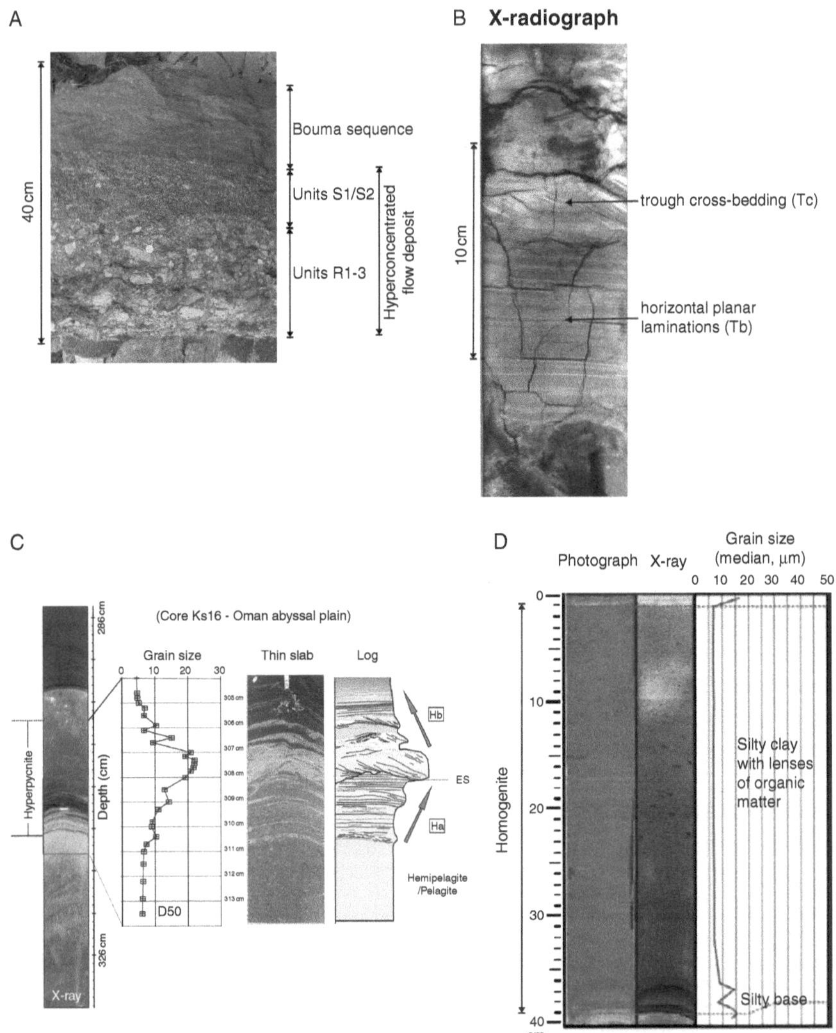

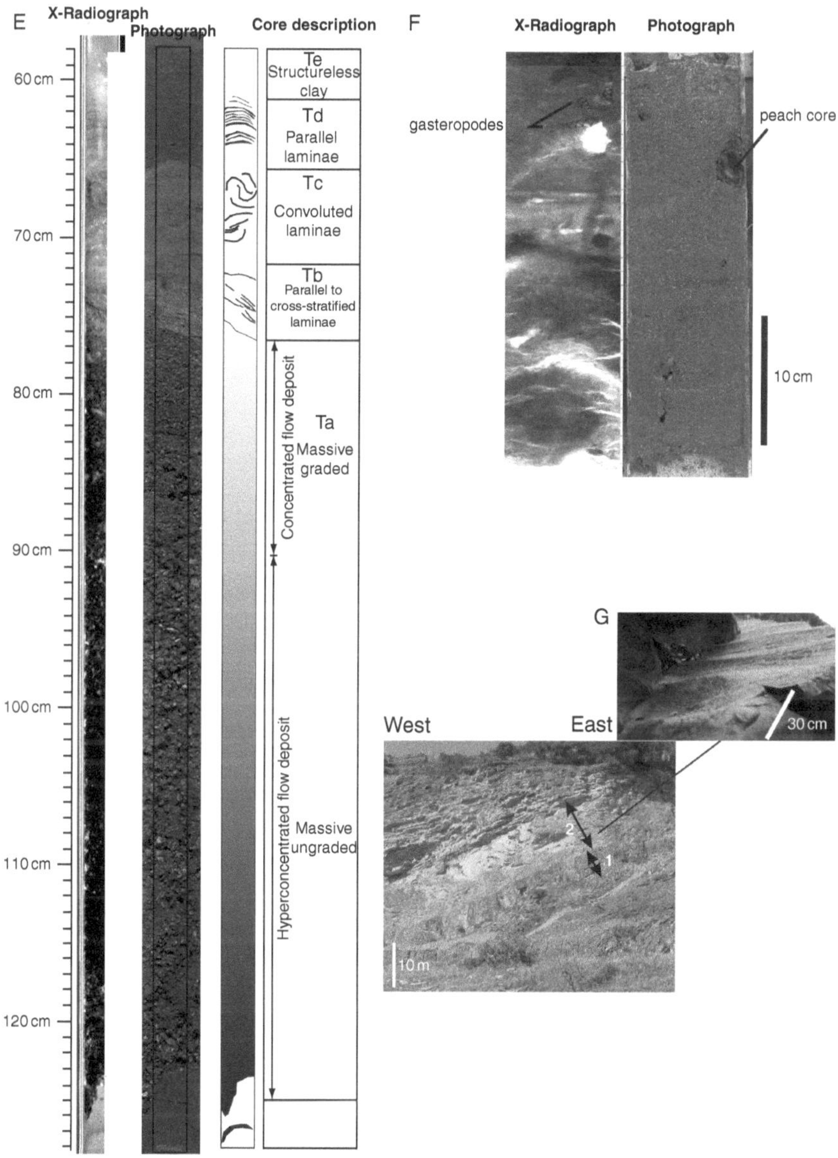

Figure 2.12 (Continued)

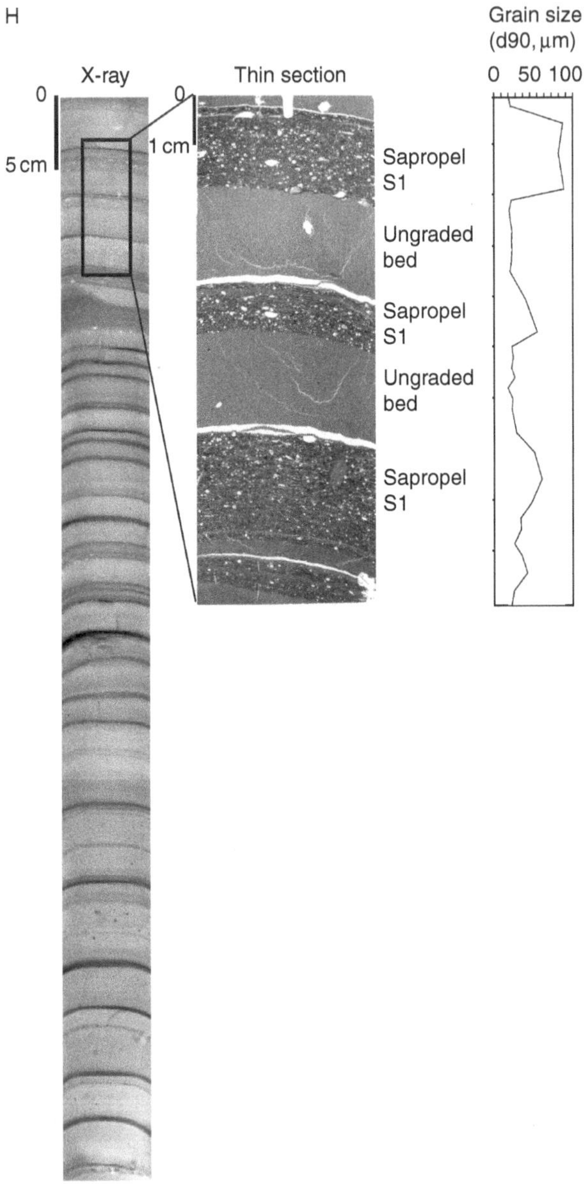

Figure 2.12 Examples of gravity-flow deposits. (A) Outcrop example of a hyperconcentrated-flow deposit (Lowe sequence). Cretaceous Basque Flysch, Baie de Loya, Basque Country, SW France. (B) X-ray image of a classical turbidite bed (Bouma sequence) in the Mahakam deep-sea turbidite system (Faugères et al., 1989). (C) Vertical grain-size trend and thin-section photograph of a hyperpycnite in the Oman Basin (Bourget, 2009). ES: erosion surface. (D) Core photograph, X-ray image and

Lanteaume et al., 1967). Erosion can also be marked in deposits by the presence of rip-up clasts in the sequence or clasts collapsed from undercut banks.

The concentrated character of the basal part of the current is the reason of the poorly graded Ta interval of the turbidite (Fig. 2.12A). The concentration is high and turbulence cannot develop (Fig. 2.11). The crude grading suggests that suspension fallout has no time to occur because of the coarse size of the particles. The Ta interval can be interpreted as a concentrated flow deposit if the grading is crude (Fig. 2.12A), or as a hyperconcentrated flow deposit if no grading is visible. This interval frequently shows floating mud clasts derived from the erosion of a muddy substratum or canyon/channels sidewalls, or reverse grading due to either the presence of mud clasts or a positive vertical velocity gradient at the seafloor. Both suggest a laminar regime in the basal flow. In the last case, the basal part of the deposit corresponds to the S2 unit of the Lowe sequence.

The other intervals of the Bouma sequence suggest the passage of the flow body with a turbulent regime and a progressively decreasing energy (Fig. 2.12B).

The first part of the body is energy-rich, depositing planar lamination (Tb interval), ascribed to deposition under upper-flow-regime plane-bed conditions. The middle part of the body has less energy, forming sediment ripples with cross-lamination (Tc interval). The boundary between the Tb and Tc intervals marks the transition from upper- to lower-flow regime. The distal part of the body has a low energy forming planar laminations (Td), ascribed to deposition under lower-flow-regime plane-bed conditions. The body shows a transition from sedimentation dominated by traction (middle and medium parts) to sedimentation dominated by suspension fallout (distal part). The tail of the flow is dominated by particle fallout (Te). Silty laminations are sparse and crude.

The suspension fallout-dominated intervals (Td and Te) have been detailed by Stow and Shanmugam (1980). They represent the fine-grained parts of the turbidite sequence. They show an intensification of bioturbation

vertical grain-size trend of a turbidite-homogenite bed in the Danube deep-sea turbidite system (Jermannaud, 2004). (E) X-ray Image, core photograph and description of a turbidite with well-developed Ta unit (concentrated flow deposit) in the Armorican turbidite system (Bay of Biscay) (Zaragosi, 2001; Zaragosi et al., 2001). (F) X-ray image and core photograph of a bedload-dominated hyperpycnal-flow deposit (corresponding to a hyperconcentrated to concentrated flow deposit). Silty sand deposit from the Malpasset Dam break in the Bay of Fréjus (Mediterranean Sea) (Mulder et al., 2009a). (G) Outcrop picture of the North Pyrenean Megaturbidite showing the well-graded basal part (1) and the slightly graded upper part (2; homogenite; Mulder et al., 2009b). Detail showing the antidune structures. (H) Ungraded mud beds interpreted as the deposition of reconcentrated hypo- and mesopycnal flows during a period of high stratification of the Eastern Mediterranean (Ducassou et al., 2008). The mud beds are intercalated in a single sapropel (S1; 9–6.8 ka). (A multi-colour version of this figure is on the included CD-ROM.)

towards the top, including distinct burrows and perturbation of lamination. Te can be subdivided into Tet (turbiditic Te, = T6 and T7 of Stow and Shanmugam, 1980 and Stow and Piper, 1984) and Teh (hemipelagic Te, = T8 of Stow and Shanmugam, 1980). T6 and T7 have been interpreted as resulting from the interaction of (hemi)pelagic sedimentation with the fine particle fallout from the turbulent cloud. This corresponds to the thick, fine-grained, bioturbated hemiturbidites of Stow and Wetzel (1990) or unifites (thick ungraded mud beds) of Stanley (1981). These thin units can also be deposited in restricted ponded (or contained) basins (i.e. small confined basins with a width of tens of kilometres) where they form acoustically layered deposits corresponding to the reflection and/or oscillation of the distal part of "trapped" turbidity currents. In ponded deposits of the Cloridorme succession (Canada), Pickering and Hiscott (1985) described particular Tc intervals containing cross-stratification indicating opposite flow direction and repeated reversal of the flow. They interpret this feature as the result of the repeated reflection of the ponded flow, at 180°. Haughton (1994) also related particular deposits observed in the Sorbas Basin (Spain) to ponded turbidites. The deposits show two characteristics. (1) The proximal deposits show basal fining-up turbidite deposits with frequent internal grain-size breaks suggesting erosion of repeated reflected flows. (2) Deposits become thicker distally suggesting the intensification of ponding in distal fine-grained turbidity currents. (3) The usual normally graded beds at the base of the turbidite are capped by a several-metre-thick ungraded mudstone that shows a convergence of facies with homogenites (Fig. 2.12D) or seiche deposits (see Section 2.7). The distinction between Tet and Teh (hemi)pelagic intervals is easy for very recent sediments (oxidized level) but is not simple for ancient sediments (presence of pelagic fauna). As the occurrence of sedimentary features in each facies depends on the grain-size distribution within the flow, the lack of certain intervals in the Bouma sequence is frequent. Flow transformation explains the lack of intervals in the Bouma sequence at the top or at the base of the sequence (top- or base-missing sequences, respectively).

The Bouma sequence represents the typical deposit of a turbidity current, *sensu* Middleton and Hampton (1973), that is, a turbidite. On seismic profiles, coarse-grained turbidites (Ta-b) appear as discontinuous high-reflectivity beds. Fine-grained turbidites appear as continuous or onlapping layered facies with high acoustic impedance.

2.5.1. Longitudinal differentiation

Flow transformation with distance is the basis of the genetic classification of deposits. Shanmugam (2000) has given the first synthesis of the horizontal evolution of a gravity flow. A dynamic classification that can be related to the evolution of the grain support mechanisms was published by Mulder and Alexander (2001a).

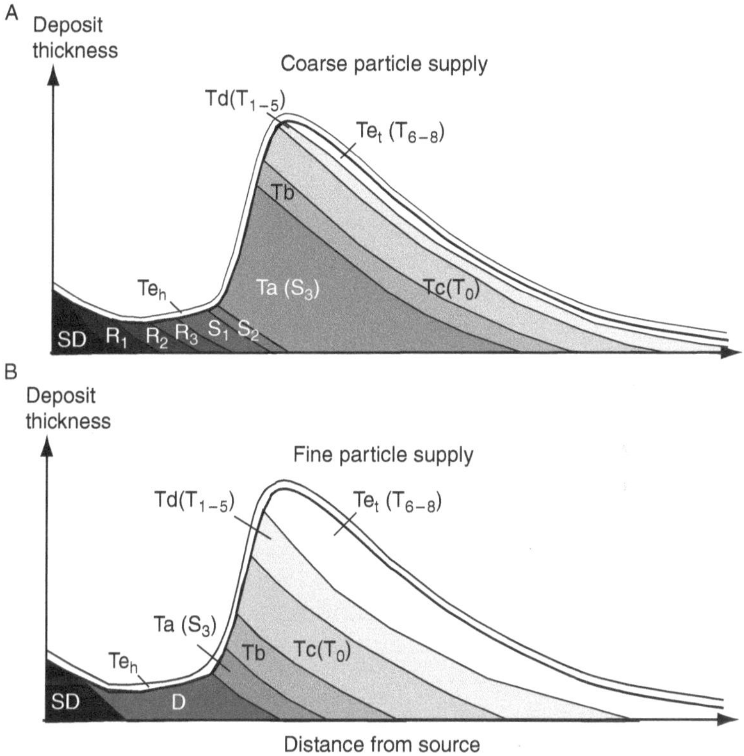

Figure 2.13 Longitudinal evolution of gravity-flow deposits. (A) Depositional setting dominated by coarse-particle supply. (B) Depositional setting dominated by fine particle supply. SD: slump deposits. Facies R, S and T are explained in Fig. 2.11.

In sand-rich source environments (Fig. 2.13A), hyperconcentrated flows first deposit massive and coarse beds (Lowe sequences) with few sedimentary structures except fluid-escape structures. Progressive flow dilution generates the motion of individual particles along the seafloor (R1 facies of Lowe, 1982). Grain-to-grain interaction becomes the dominant particle-support mechanism. As dilution proceeds, turbulence develops and suspension fallout generates normally graded deposits. Massive hyperconcentrated deposits, Lowe, Bouma and fine-grained turbidite sequences form a continuum.

In mud-dominated environments (Fig. 2.13B), debrites are deposits resulting from the transformation of mass-failures. They represent usually proximal deposits but in some case, flow deposits have been observed 1500 km away from their potential source (Talling et al., 2007). As the coarse fraction (pebbles to sand) is reduced, the elutriation rapidly generates a dilute clay- and silt-rich turbidity current. After the freezing of the debris flow, no hyperconcentrated or concentrated flow forms. Consequently, the

Lowe sequence and the base of the Bouma sequence can be lacking and may be replaced by a bypass or erosion zone. The top intervals of the Bouma sequence are well developed.

2.6. Particular flood-generated turbidites: Hyperpycnal deposits (hyperpycnites)

The hyperpycnal sequence (hyperpycnite) has been defined by Mulder et al. (2001a, 2002). It corresponds to suspended-load-dominated hyperpycnal-flow deposits. The complete sequence (Fig. 2.12C) is explained by the changes of the discharge in time. During increasing discharge at the river mouth, the flow competence increases. The hyperpycnal flow will develop a coarsening-upward basal unit, Ha. During decreasing discharge at the river mouth, the competence of the hyperpycnal flow decreases and the flow will deposit a fining-upward top unit, Hb. The complete hyperpycnite is formed by these two stacked units. The Ha/Hb transition is characterized by a maximum grain size and marks approximately the peak of the flood, that is, the period of maximum energy (discharge) at the river mouth, except when the discharge follows an irregular pattern, for example, with a plateau in discharge of particle concentration. Mutti et al. (2002) suggest that preservation of the Ha unit deposited by the waxing flow is rare in ancient deposits. St-Onge et al. (2004), however, show well-preserved Ha units in the Saguenay sequences, but it shows only horizontal parallel lamination. The most common structure found in the Hb unit is climbing-ripple lamination (Migeon et al., 2001; Mulder et al., 2001a, 2002; Mutti et al., 2002). The presence of both sedimentary structures and a clear sorting suggests that the flows are low concentrated and that traction and particle settling act simultaneously.

Pulses of the hyperpycnal flow due to variations in the discharge of the river could be the origin of intrasequence erosion at the base of coarse-to-fine laminae doublets. Duringer et al. (1991) produced such intrasequence erosional contacts during laboratory experiments. These authors show facies in ancient deposits that could be interpreted as hyperpycnites, with alternating arenite/siltite beds and lenses, laminated fine-grained sequences and numerous intrasequence erosional contacts. Similar beds have been recognized in the Var and Zaire turbidite systems and in ancient systems in the Apennines (Mavilla, 2000; Mutti et al., 1996, 2000, 2002, 2003) and in the Annot Sandstones (Joseph and Lomas, 2004).

One of the major unsolved questions is the limit at which the impact of the discharge hydrograph at the river mouth (superposed coarsening- and fining-up units) is recorded in sedimentation. Before the limit, the hyperpycnal flow forms a classical hyperpycnite; after it, it should form a classical (flood-induced) turbidite. This limit should depend on the flood characteristic (duration, intensity), the grain-size distribution of transported particles and the morphology of the seafloor (slope). When arriving at the shelf-slope transition, the flow becomes accumulative and may transform into a true turbidity current, generating classical turbidites (Piper and Normark, 2009). However, Nakajima (2006) described classical hyperpycnites in the Toyama deep-sea fan, at more than 700 km from the river mouth.

Bedload-dominated hyperpycnal-flow deposits are drastically different from suspended-load-dominated hyperpycnal-flow deposits. In the Astoria deep-sea turbidite system, Brunner et al. (1999) and Zuffa et al. (2000) described submarine deposits related to hyperpycnal outburst of glacial Lake Missoula (Utah). They are constituted either by several-metre-thick massive sand beds with no grading or by massive sand beds with a crude grading occurring at the top. These beds correspond respectively to hyperconcentrated and concentrated flow deposits (Mulder and Alexander, 2001a). The Malpasset Dam break deposits in the Fréjus Bay (Mediterranean Sea, SE France) show similar facies with a few-decimetre-thick organic-rich, fining-up sandy-silt layer with no or crude grading developing at the top (Fig. 2.12F; Mulder et al., 2009a).

2.7. Homogenites

Homogenites are homogeneous mud bed with a thickness of several centimetres to several metres (Fig. 2.13D). No bioturbation is present. The base of the bed shows an erosive to sharp contact with a normally graded base composed of sand to sandy silt (Cita et al., 1984). On seismic profiles, they appear as an acoustically transparent facies (Chapron, 1999). On field outcrops, the massive beds show rare sedimentary structures including HCS (hummocky cross-stratification)-like structures interpreted as antidune depositin (Mulder et al., 2009b; Prave and Duke, 1990). They have been interpreted as the result of erosion and liquefaction of a superficial sedimentary bed after the passage of a tsunami wave (Cita et al., 1996). Megabeds deposited consecutively to the Santorini eruption in Cretan Basin (South Aegean Sea; Anastasakis, 2006) lead to the interpretation of similar beds as the distal part of megaturbidites. A similar interpretation with the addition of ponding in the restricted Flysch Basque Basin is provided by Mulder et al. (2009b) for the North Pyrenean Megaturbidite (SW France; Fig. 2.12G). In a restricted basin, the sediment is put into suspension by oscillation of the whole water column by the ponding of turbidity currents (internal seiche). Some of the hemiturbidites of Stow and Wetzel (1990) or unifites (Stanley, 1981) could be some homogenites. The main difference between homogenites and fine-grained turbidites is usually the thickness of the deposits, in particular, the fine ungraded upper part. Mc Cave and Jones (1988) suggested that deposition could occur by freezing in these thick dense mud suspensions by suppression of the turbulence.

2.8. Deposits related to flow reconcentration

Ducassou (2006) and Ducassou et al. (2008) described fine-grained homogeneous beds containing structureless, slightly graded clastic mud (grain size < 50 μm), ligneous fragments, amorphous organic matter patches, without bioturbation and almost azoic. These beds are interpreted as convective sedimentation deposits related to successive reconcentration processes from a hypopycnal (surface) plume (Parsons et al., 2001) occurring during periods of increased stratification of the eastern Mediterranean Basin (Fig. 2.12H). This more intense stratification is related to more intense Nile floods during periods of increased monsoon and related to sapropel formation. The formation of a fresh-water layer in surface and the resulting increase in water stratification in the Eastern Mediterranean prevent direct hyperpycnal-flow formation. The grading of the beds suggests that particle fallout is the most important process acting in the formation of these beds suggesting that these beds are fine-grained flood-related turbidites.

Gaudin et al. (2006a) described massive sand beds in the Bourcart canyon (Western Mediterranean) interpreted as the result of deposition by plunging of dense water massive during winter storms. These deposits called cascadites seem to be restricted to periods of sea-level highstand. They show two grain-size modes and share diagnostic facies with shallow-water contourites of Viana et al. (1998).

2.9. Facies convergences of deep-sea deposits

Strong similarity can exist between facies associations in deep-sea deposits. This hinders their interpretation in terms of processes significantly. In addition, bed truncations can add to the resemblance between depositional sequences.

For example, the vertical evolution of grain size in a hyperpycnite (coarsening and fining upward) can be mistaken with the same grading in contourite beds as defined by Gonthier et al. (1984; see Section 3.4.1.2), particularly if bioturbation is intense. In both cases, the facies associations reflect the acceleration and subsequent slowing down of a flow at a given location. Only the duration of deposition is different.

Similar to Bouma sequences, hyperpycnite sequences can be truncated. However, Bouma sequences are truncated because of the longitudinal evolution of the flow, or by erosion by more recent flows. They may also result from temporal variations, for example, if simultaneous earthquake-triggered slumps convert at different times in different tributary valleys. Hyperpycnal sequences are truncated because of temporal variations of velocity in the flow. The discharge and velocity reached during the flood peaks can be high enough to prevent deposition or to initiate erosion. Deposition occurs only when the velocity drops during decreasing discharge. Thus, the contact between Ha and Hb is either sharp or erosional.

This erosion generates an intrasequence erosional contact. During peak conditions of a severe flood, the Ha unit can be completely eroded. In this case, there is a sequence convergence between a base-truncated hyperpycnite and a classical Bouma-like turbidite sequence deposited by a turbulent surge. The Var example shows that a medium-size river can generate alternatively hyperpycnal turbidity current and classical turbidity current (slide-initiated) depending on the flood magnitude (Mulder et al., 2001b).

Another case of facies convergence is between homogenites, top facies of base-missing Bouma sequences and hemipelagites. In this case, only a detailed analysis of the grain-size distribution and the nature of the grains can help to interpret the depositional process.

The main diagnostic features for differentiating between hyperpycnites, contourites and classical turbidites are summarized in Table 2.5.

2.10. Genetic classification of gravity-flow deposits

Mutti and Ricci Lucchi (1975), Mutti (1979, 1992) and Mutti et al. (1999) built a classification based on detailed field observations, facies analyses and stratigraphic correlations between facies of gravity-flow deposits (Fig. 2.14). This genetic classification is based on facies evolution interpreted as the consequence of the main hydrodynamic transformation that occurs in a gravity flow along transport. The classification is based on a flow with its velocity decreasing with both time and space (waning depletive flow of Kneller and Branney, 1995). This model relies on the transport-efficiency concept (Mutti, 1979; Pickering et al., 1989).

2.11. Flow competence, capacity and efficiency

Flow competence represents the maximum particle size that a flow can carry. It depends on the slope gradient and on the particle-support mechanisms.

Flow capacity represents the total amount of sediment a flow can carry.

Flow efficiency (Mutti, 1979; Pickering et al., 1989) represents the ability of a flow to carry sediment according to its clay content. High-efficiency flows can transport sediment over long distances and form well-sorted deposits as elutriation acts during transport. Low-efficiency flows transport particles over short distance and generate poorly sorted deposits.

2.12. The theoretical prediction of deposits: the velocity matrix

Kneller (1995) and Kneller and Branney (1995) have demonstrated that the vertical and horizontal evolution of a flow depends on the variations in velocity with time and distance, respectively (Fig. 2.15).

Table 2.5 Criteria for the recognition of contourites, turbidites (surge deposits), homogenites and hyperpycnites (modified from Mulder et al., 2002)

Bed type	Turbidite sequence (Bouma-like)	Homogenite	Hyperpycnal turbidite sequence (hyperpycnite)	Contourite sequence
Flow type	Turbulent surge	Oscillatory flow	Turbidity current	Contour current
Flow behaviour	Unsteady; mainly waning	Mainly steady	Mainly steady, waxing then waning	Almost completely steady; waxing then waning
Dominant flow regime	Turbulent	Turbulent	Turbulent	Turbulent
Source	Failure	Tsunami or slump (more rarely)	Flood	Thermohaline current
Flow duration and time for deposition	Minutes to days	Hours to weeks	Hours to weeks	Episodic within thousands of years
Base contact	Erosive to sharp	Erosive to sharp	Gradational	Gradational
Top contact	Gradational	Sharp to gradational	Gradational	Gradational
Intrabed contact	Infrequent between facies	None	Erosive to sharp	None
Grading	Clear, normal	Only at base, none at top	Clear, inverse then normal	Crude, inverse then normal
Bioturbation	Absent to intense	Absent	Absent to intense	Thorough and intense
Ichnofacies	Few	None	Few	Many
Structures	Well-developed parallel and cross-bedding lamination, convolute lamination	None	Well-developed parallel and cross-bedding lamination; climbing lamination frequent	Crude and sparse parallel and cross-bedding lamination; frequent mottles and lenses
Fauna/flora	Allochthonous, mainly marine	Parautochthonous	Allochthonous, mainly continental; frequent plant and wood fragments	Mainly autochthonous

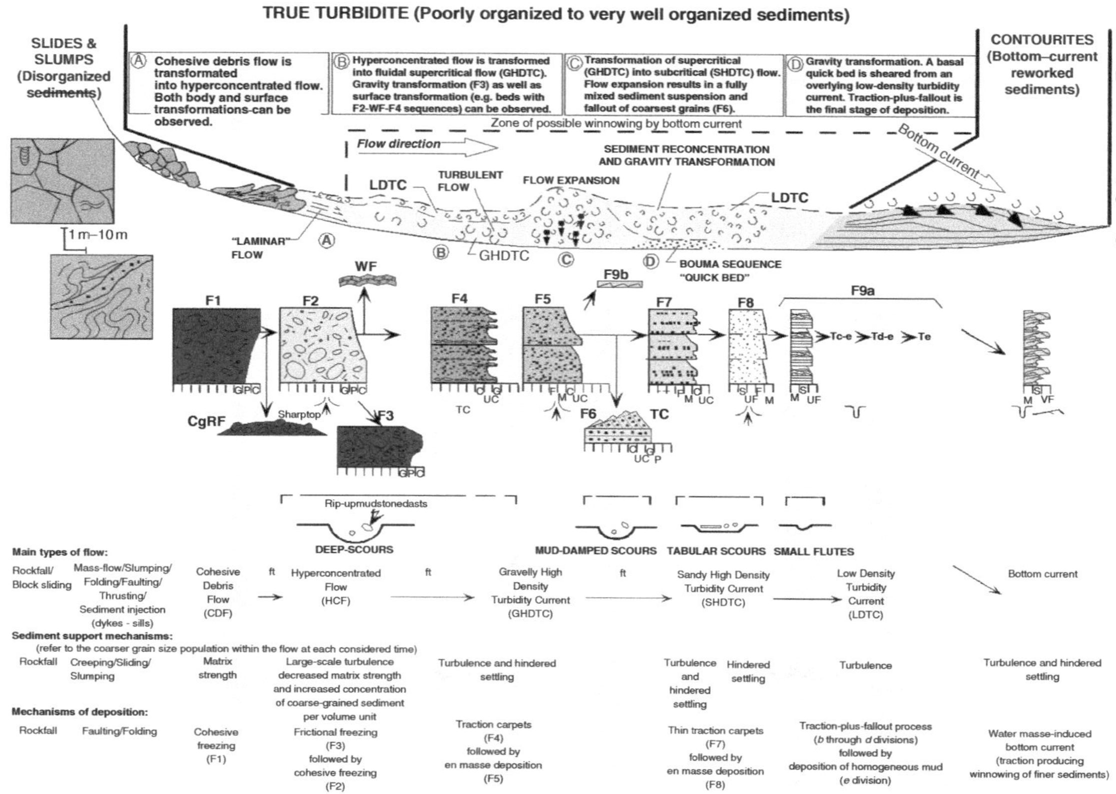

Figure 2.14 Classification of facies evolution (from Gérard et al., 2000; Mutti, 1992; Reproduced with permission from E. Mutti). (A multi-colour version of this figure is on the included CD-ROM.)

The total acceleration is the product of the variation of the velocity with time ($\delta u/\delta t$) at a given location and its variation with distance (Fig. 2.15A). If velocity decreases or increases with time ($u\ \delta u/\delta x$), the flow is waning or waxing, respectively (Fig. 2.15B). If it is constant with time, the flow is steady. If velocity decreases or increases with distance, the flow is depletive or accumulative, respectively (Fig. 2.15C). If it is constant with time, the flow is uniform. Typically, a flow is depletive on a concave-up slope, or when it laterally expands at a channel mouth or after a channel avulsion (Fig. 2.15D). In that case, the flow tends to collapse (thickness abruptly decreases) as it spreads, and to reconcentrate. Grain-to-grain interaction becomes the most dominant particle-support process (concentrated flow) and the flow quickly freezes. Conversely, a flow is accumulative on a convex-up slope or when the flow section is restricted due to channelling (Fig. 2.15E).

In the classical case of a turbulent surge moving on a concave-up slope, for example, a channel bottom of a continental slope, the flow will be waning and accumulative. The deposit will be fining-up. With distance, the deposits become finer and thinner than in the proximal part. This corresponds to the vertical and longitudinal evolution of a classical Bouma sequence (Figs. 2.11 and 2.12B and E).

The matrix allows predicting the vertical grain size trend of hyperpycnites (Mulder et al., 1998b). Assuming a concave-upward slope (depletive flow conditions), the flow that originates from a river flood will be first waxing (during increasing discharge at the river mouth) and then waning (during decreasing discharge at the river mouth). The corresponding deposits will be first coarsening-upward (facies Ha), then fining-upward (facies

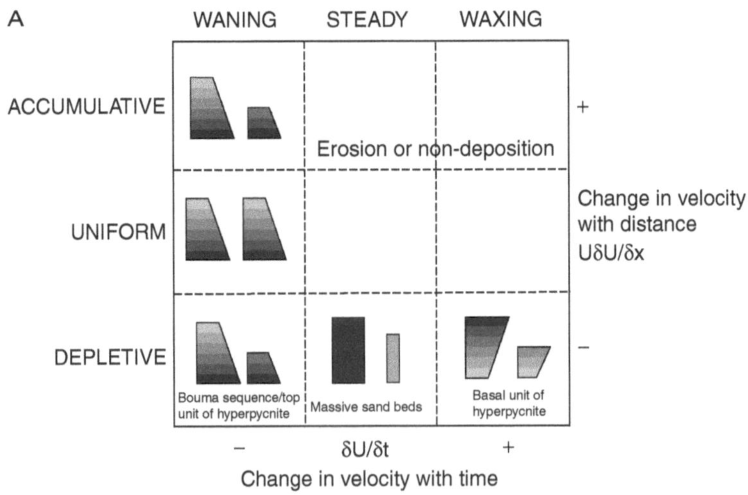

Figure 2.15 (Continued)

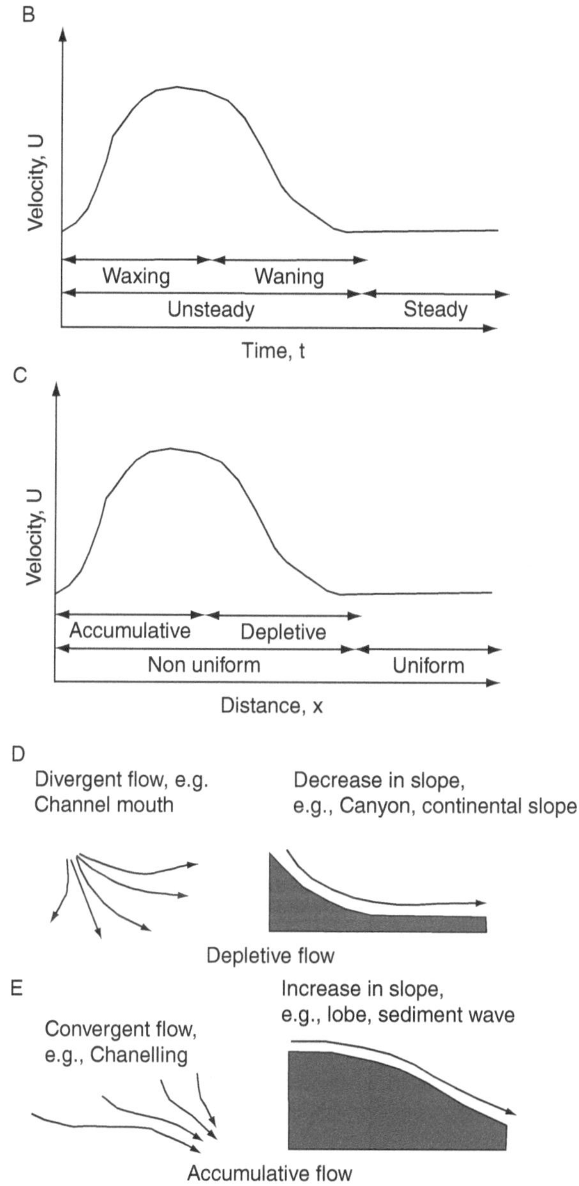

Figure 2.15 The velocity matrix (from Kneller, 1995; Kneller and Branney, 1995). (A) Changes in velocity with both time and distance. (B–E) Examples: the flow is waxing when its velocity increases with time, for example, during increasing discharge in a flooding river (B). The flow is waning when its velocity decreases with time, for example, during decreasing discharge in a flooding river (B). The flow is steady when its velocity is constant in time. The flow is depletive when its velocity decreases with distance (C), for example, when the flow spreads or when it moves on a concave-up slope (D). The flow is uniform when its velocity is constant with distance. The flow is accumulative when its velocity increases with distance (C), for example, when the flow is confined or channelized, or on a convex-up slope (E). Reproduced with permission from Geological Society, London.

Hb: Fig. 2.12C). As the flow is depletive, deposits are finer and thinner in the distal part than in the proximal part.

According to the interpretation of Kneller and Branney (1995), the only way to deposit massive sand beds is by steady depletive flows.

3. Deep-Sea Turbidite Systems

3.1. Historical background of model types

Shepard et al. (1966) and Shepard (1967) made early observations of depositional dynamics in submarine canyons and channels on the Californian coast. The first general model of a recent deep-sea system was published by Normark (1970), and worked out further by Normark (1978). This model synthesizes the study of Pleistocene Californian fans (San Lucas and Astoria). It was based on a morphological analysis of the system, identifying three parts:

- the upper fan, including a canyon and a 1–5 km-large valley with a sinuous talweg;
- the middle fan corresponding to a depositional lobe (called suprafan) forming a topographic bulge beginning at the mouth of the valley, which is covered with sinuous channels less than 1 km wide;
- the lower fan, without channels and covered only by fine-grained deposits.

The second model for fan morphology was defined by Mutti and Ricci Lucchi (1972) using the data from Cretaceous and Tertiary turbidite systems in the Apennines and Spanish Pyrenees. This model is at the origin of the genetic model of gravity-deposits evolution developed by Mutti (1985, 1992).

Walker (1978), Mutti (1979) and Shanmugam and Moiola (1991) proposed an improved version of the Normark (1970) model, which includes the facies concepts of Mutti and Ricci Lucchi (1975). This first predictive model for deep-sea depositional systems integrates the progradation of the whole system. The progradational sequence begins with thin-bedded turbidites (Mutti, 1977), followed by lower-fan deposits and middle- and upper-fan deposits. It ends with the filling of the main feeder channel by coarse-grained hyperconcentrated-flow deposits.

The Walker model is limited in use because of its hybrid definition and because the suprafan definition of the Normark (1978) model is not consistent with the most recent data. In addition, the scale of the fans studied by Normark and Walker for their models is not suitable for the interpretation of more recent data collected on giant muddy fans located on passive margins.

To reduce the gap between studies on ancient and recent deep-sea fans, Barnes and Normark (1985) showed a set of comparative diagnostic criteria, including the shape and size of the fan, the type of basin, the nature of the source, and the age.

The maps show that turbidite systems vary largely in size and shape (Fig. 2.16). Their length varies from 100 to 3000 km, depending on the volume of sediment supplied (Wetzel, 1993). The largest deep-sea fans are the Bengal (3×10^6 km^2), the Indus (10^6 km^2), and the Amazon, Mississippi and Zaire fans (0.3×10^6 km^2). A few have a "fan" shape, justifying the use of the term "turbidite system". Key parameters selected for classification are as follows: (1) the source feature including the dominant grain size supplied and its shape (point or line source, single or multiple, stable or migrating) and (2) the geodynamic context of the basin (active or passive margin). The parameters have a clear impact on the shape of the turbidite system. Radial systems (La Jolla, Redondo and Navy) have a dominant sand supply. In contrast, elongate fans such as those of the Bengal, Indus, Amazon, Zaire and Mississippi have a dominant clay supply.

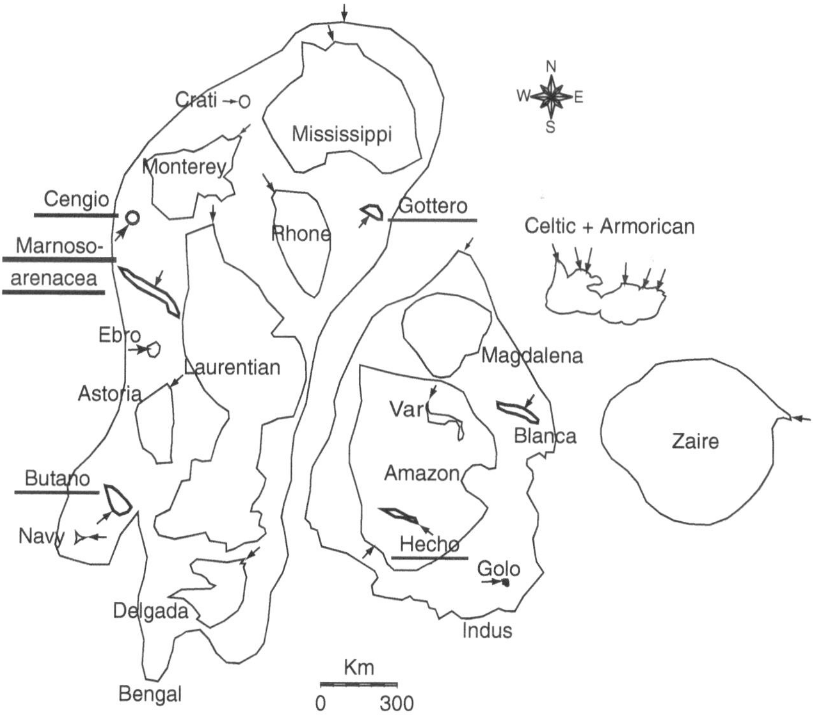

Figure 2.16 Compared areal extent of ancient (underlined) and recent deep-sea turbidite systems (modified from Babonneau, 2002; Barnes and Normark, 1985; Shanmugam and Moiola, 1988; Zaragosi, 2001). Modified with permission from Elsevier.

The most recent classifications try to make a synthesis between sedimentary processes, depositional environment and sedimentary facies. They include control factors such as the sedimentary source, the tectonic context and the eustatic changes. Characteristics of the source and tectonic context lead to more or less elaborated classifications of systems. Eustasy adds a stratigraphic framework for the system development. It represents a step towards a sequence stratigraphic model for deep-sea turbidite systems. All these classifications emphasize the importance of grain size.

The most recent classifications of deep-sea turbidite systems are those of Shanmugam and Moiola (1988; Fig. 2.17) and Reading and Richard (1994; Fig. 2.18).

Shanmugam et al. (1988) and Shanmugam and Moiola (1988) published a classification essentially based on the geodynamic context (Fig. 2.17). Using the classification of Bally and Snelson (1980), they propose two extreme types of turbidite systems: (1) passive-margin fans (Fig. 2.17B) including both the immature passive-margin fans (North Sea) showing small sandy systems with well-developed lobes and the mature passive-margin fans (Atlantic type) of Bally and Snelson (1980) showing large muddy systems with small lobes and (2) active-margin fans (Fig. 2.17A) including both active-margin fans (Pacific type) showing small sandy systems with large lobes and the mixed settings of Bally and Snelson (1980) including, for example, the Bengal and Indus systems. As the two models are from composite data, they result from the interaction of several parameters and lead to inconsistency: for example, the Magdalena fan is located on an active margin and shows the characteristics of a passive-margin turbidite system.

Reading and Richards (1994) provided a classification based essentially on the type of the source and the dominant lithology (Fig. 2.18). Their matrix-shaped classification includes results provided by both recent and ancient systems. The grain size is on the *X*-axis. It is divided in four classes: mud-rich, sand-rich, mixed (mud/sand-rich) and gravel-rich. The *Y*-axis houses the shape of the source: point (submarine-fan point source), linear (slope apron linear source), or multiple (multiple source ramp). Consequently, twelve categories are obtained. This classification is probably the most complete environmental classification. However, it has several drawbacks. The terminology is considered confusing by Bouma (2000), as most ancient systems belong to the mixed mud/sand-rich systems. This point of criticism can partially be explained by the difference in the global tectonic context between ancient (Mesozoic and Cenozoic fans) and recently active fans. Ancient fans developed in a context of either early drifting (with the formation of young and restricted oceans) or collision (with rare occurrences of large passive margins). In such active tectonic settings, the marine sedimentary basins were filled under either extensional or compressive conditions, but they were always confined. In contrast, most recent fans

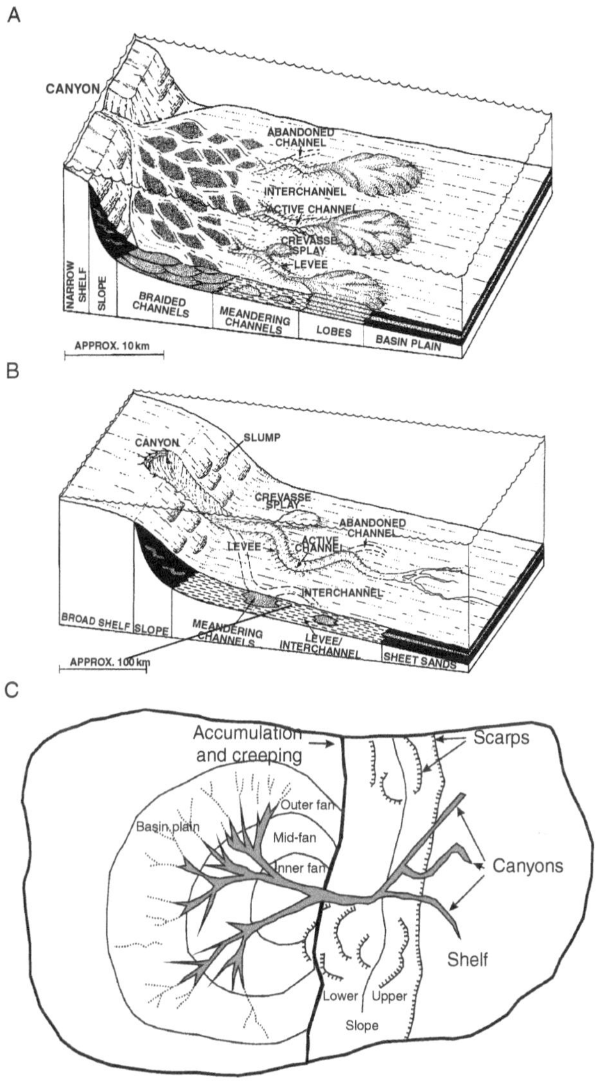

Figure 2.17 Fan morphology models/deep-sea gravity-flow systems. (A, B) Shanmugam model (Shanmugam et al., 1988; Reproduced with permission from AAPG). (A) Active-margin fan type. (B) Passive-margin fan type. (C) Mutti model (Mutti, 1985, 1992; Reproduced with permission from E. Mutti).

develop along large passive margins without confinement. Bouma (2000) proposed therefore to differentiate between coarse-grained sand-rich fans, fine-grained sand-rich fans, and fine-grained mud-rich fans.

Mutti (1979) followed a different logic. Using the efficiency concept, he differentiated between low-efficiency and high-efficiency fan systems

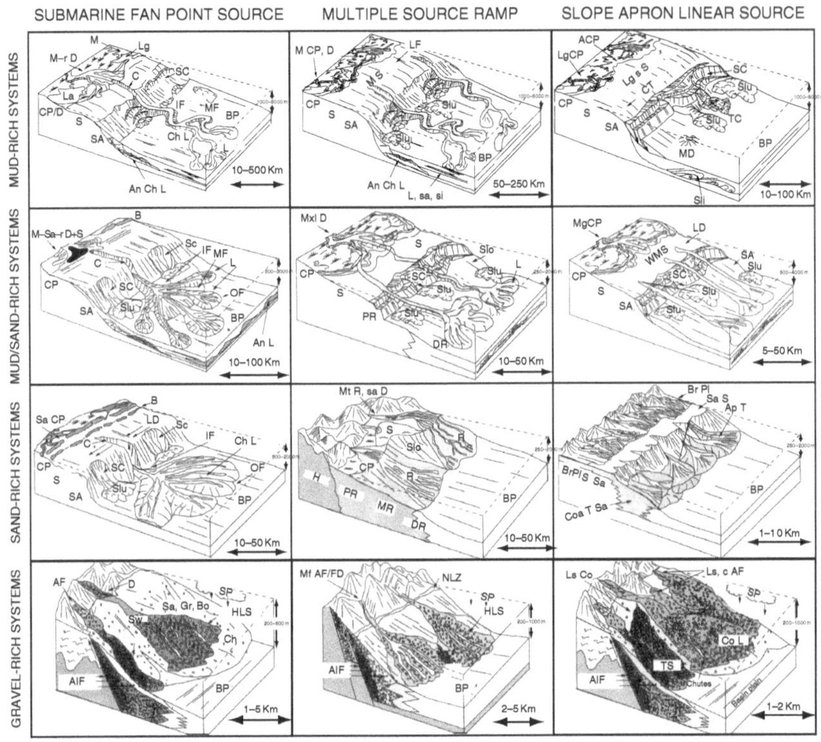

Figure 2.18 Reading and Richard's model of deep-sea gravity-flow systems (from Reading and Richard, 1994; Reproduced with permission from AAPG). (Ls), (C) AF: (line source) (coalescing) alluvial fan; AIF: avalanching inertia flow; Ap T: apron turbidites; B: barrier; Bo: boulders; BP: basin plain; Br Pl: braided plain; C: canyon; Ch: chutes; (An) Ch L: (ancient) channel–levee; (A) (Lg) (mg) (M) (Sa) CP: (arid), (low gradient), (medium gradient), (muddy), (sandy) coastal plain; (Ls) Co: (line source) cone; Coa T Sa: coalescing turbidite sands; (Mxl) (M/Sa-r) D: (mixed-load) (mud/sand-rich) delta; DR: distal ramp; (Mf) FD: mountain-fed fan delta; Gr: gravels; H: hinterland; HLS: hummocks, lobes and splay; IF: inner fan; (An) (Ch) (Co) L: (ancient) (channelized) (coalescing) lobe; La: lake; LD: longshore drift; LF: lateral feeding; Lg: lagoon; M: marsh; MD: mud diapir; MF: mid fan; MR: medial ramp; m-r: mud-rich; m-s-r: mud/sand-rich; NLZ: narrow littoral zone; OF: outer fan; PR: proximal ramp; R: rams; Mt R: multiple rivers; (lg), (M), (S) (Sa) S: (low gradient), (muddy) (starved) (sandy) shelf; Sa: sand; SA: slope apron; Sc: slide or slump scar; Sh: sheets; Si; silt; Sli: slide; Slo: slope; Slu: slump; SP: surface plume; Sw: swales; TC: turbidity current; TS: talus slope; WMS: wide mud-sand shelf.

(Fig. 2.19). Low-efficiency fan systems are those where gravity flows are not capable of transporting sand over long distances. They are sand-rich systems (Fig. 2.19B). In contrast, high-efficiency fan systems are those where mud-rich gravity flows are capable of transporting sand over long distances. They are mud-rich systems (Fig. 2.19A). Modern mud-rich systems (Fig. 2.19C)

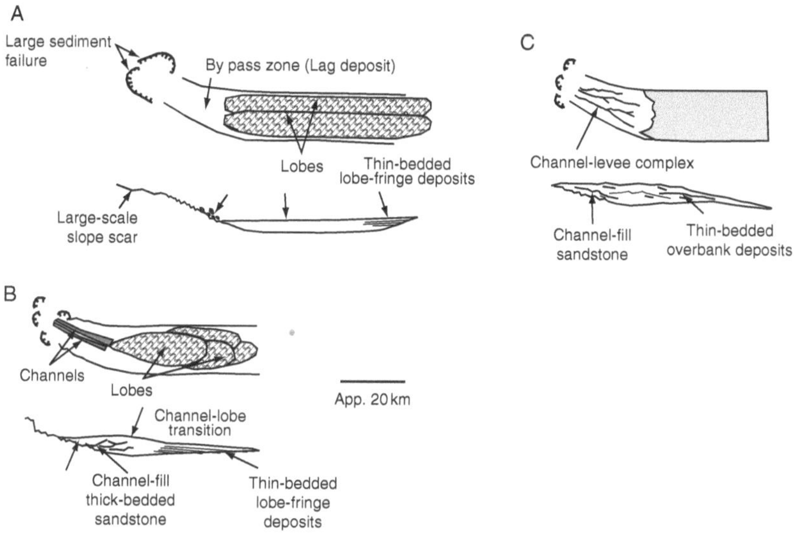

Figure 2.19 Model of gravity-flow system evolution of Mutti (1985; Reproduced with permission from E. Mutti). (A) Type 1 systems with high transport capacity. (B) Type 2 systems with low transport capacity. (C) Type 3 systems corresponding to large modern mud-rich fans.

are considered as a separate group because they have no equivalent in ancient deposits. The *Flysch à helminthoides* in the Alps could, however, be a good analogue.

Analysis of sedimentation rates in deep-sea turbidite systems does not make sense for short durations because they include both vertical (hemipelagic) fallout and horizontal (mass transport) deposition. Both types of sedimentation should be separated. Over long timespans, they provide an estimate of the sediment accumulation rate and of the system construction. For the Amazon, this rate varies between 1 and 50 m ka^{-1} for intervals of activity (glacials) and between 0.05 and 0.1 m ka^{-1} for intervals of inactivity (interglacials) (Mikkelsen and Maslin, 1997).

3.2. The architectural-elements concept

Because of the difficulty in conceiving a unique fan model, the most recent approach uses an analytical method that was previously applied to river systems (Miall, 1985). This method is based on the organization of the turbidite system into sedimentary bodies that form the individual elements of the system. These bodies are called "architectural elements". Once these architectural elements are identified, their nature and architecture allows the interpretation of the sedimentary processes that formed them.

Four main features of a typical architectural element have been defined by Miall (1985):

(1) the nature of its base and top surface (erosive or gradual, planar or curved, regular or irregular);
(2) its external geometry (finger-like lenses, sheet, wedge, concave-up erosional base, U-filling);
(3) its 3D geometry (thickness, lateral and longitudinal extent);
(4) its internal geometry: longitudinal and lateral facies associations (sequence analysis), sedimentary structures.

Using this framework, Mutti and Normark (1987) defined five architectural elements: channels, lobes, channel/lobe transition zones (CLTZ), overbank deposits, and major erosional features (see Imbert, 2011, this volume, section 4.2.1). They also established a spatial and temporal hierarchy with five nested orders:

first order: turbidite complex;
second order: turbidite system;
third order: turbidite stage;
fourth order: turbidite sub-stage;
fifth order: turbidite bed.

This hierarchy allows a link between observations at outcrop scale (ancient systems) and observations of recent systems (using seismic or acoustic imagery) at a geophysics scale.

Pickering et al. (1995) improved the method. Architectural elements are interpreted by him using a hierarchy of the surfaces, identifying lithological or seismic facies, and reconstructing the 3D geometry. The classification of Pickering et al. (1995) proposes a framework for the interpretation of architectural elements according to the description of these parameters based on well-defined characteristics. This framework can be applied at different scales.

This method was intensively used and improved by the oil industry, as it can be directly applied to the interpretation of geophysical data, as it is based on the morphology of elements. It can be used for the modelling of reservoir properties (Weimer et al., 2000). For this application, Prather et al. (2000) defined a hierarchy based on seismic, well log and outcrop studies. Wireline logs, in particular, the "gamma-ray" log recording natural radioactivity, allow calibration of seismic facies and estimates of their sand/clay ratio.

The main problem of this method is that each local or regional study defines a hierarchy and a terminology of the architectural elements. Architectural elements can therefore be difficult to compare from a study to another. Stow and Mayall (2000) tried to inventory all architectural elements (Fig. 2.20). In addition, the 3D geometry of architectural elements is

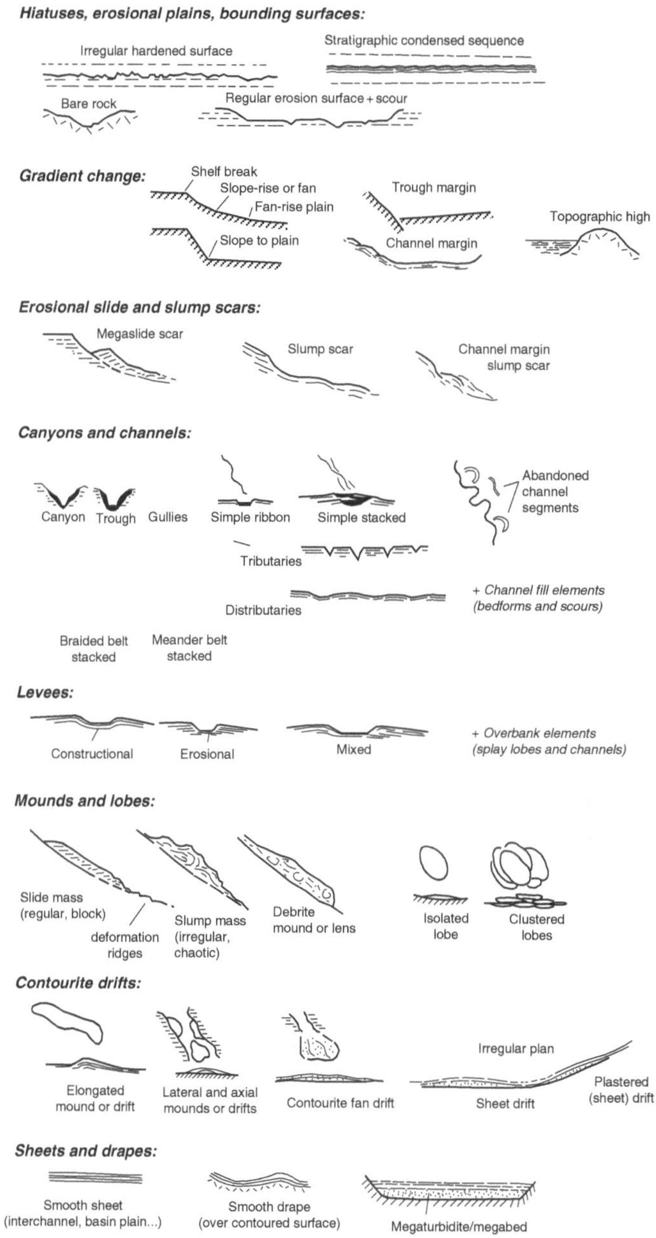

Figure 2.20 Inventory of architectural elements in deep-sea systems (Stow and Mayall, 2000). Reproduced with permission from Elsevier.

sometimes difficult to define and depends on the conditions of outcrops or on the resolution of geophysical data (Rigollet, 2001). To summarize, it can be said that the architectural-element method is a powerful descriptive and interpretative tool, but that it does not allow prediction as a conceptual theory.

3.3. Main architectural elements of deep-sea turbidite systems

The best description of deep-sea turbidite systems is provided by studies of recent systems, in particular, the Amazon (Flood and Damuth, 1987; Flood et al., 1991). These studies are consistent in distinguishing four parts in a turbidite system: (1) a canyon, (2) an upper fan consisting of an erosional channel and thick levees, (3) a middle fan with a depositional channel and thin levees and (4) a lower fan with distal lobes.

3.3.1. Submarine valleys, canyons and gullies

Submarine valleys and, particularly, canyons on continental slopes are the main conduits for sediment transport from continent to ocean (Fig. 2.21). For example, the Amazon lobes are located at 1100 km from the canyon head (Jegou, 2008; see Imbert (2011, this volume, Chapter 10.4.2.1.1)).

Canyons represent the most impressive structures that shape the present morphology of passive continental margins. They are deep (hundreds of metres, frequently more than a kilometre) and narrow (a few kilometres) submarine valleys with steep walls, mainly formed by erosion of the continental slope and outer shelf (Shepard, 1981). Along the northern Spanish shelf, the structurally controlled south wall of the Capbreton Canyon reaches more than 2000 m at 133 km from the coastline (Cirac et al., 2001; Gaudin et al., 2006b; Shepard and Dill, 1966) (Fig. 2.22). The flanks show numerous furrows, gullies, slump scars and tributaries.

Recent results of Gaudin (2006) and Gaudin et al. (2006b) show that, although erosion dominates in canyons, the canyon shape is the result of alternating stages of filling and incision. A 39-m-long core collected on a terrace at 425 m water depth in the Capbreton Canyon (Bay of Biscay) shows a ^{14}C age of 2390 years BP at 22.8 m below the seafloor, suggesting a high sediment accumulation rate during the latest Holocene. In the Bourcart Canyon (Gulf of Lyons), and in the Monterey fan (Gardner et al., 1996; Klaucke et al., 2004), phases of incision and filling alternate according to sea-level fluctuations and to a direct connection or disconnection of the canyon head to the river mouth.

Transverse cross sections in sinuous or meandering submarine channels are identical to channel cross sections in sinuous or meandering subaerial rivers (Kuenen, 1950), as documented for the La Jolla fan valley and for the

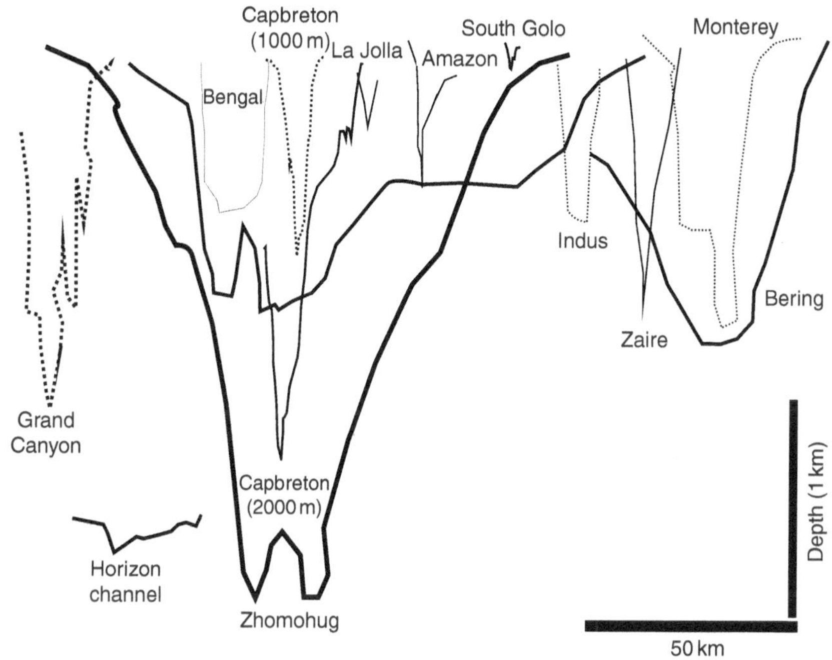

Figure 2.21 Canyon morphology and detailed morphology of meandering channels and terraces. Am: abandoned meander. Compared cross sections of submarine canyons with one continental canyon (Grand Canyon). Vertical exaggeration = 48.5. Modified from Normark and Carlson (2003).

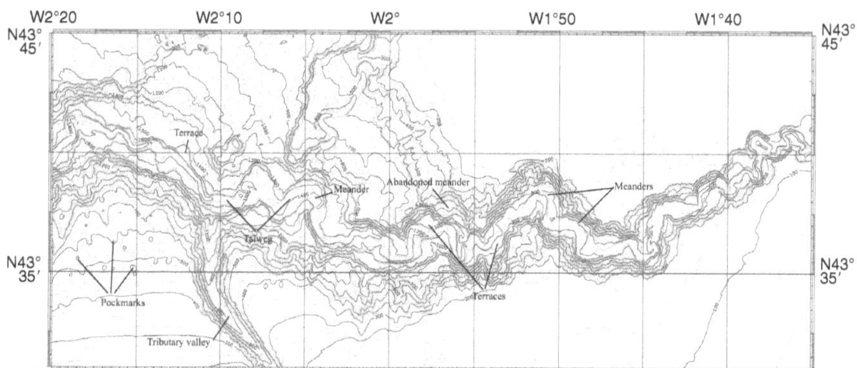

Figure 2.22 Upper part of Capbreton Canyon (Gaudin et al., 2006b). Isobaths are in metres. Reproduced with kind permission of Springer Science and Business Media.

Zaire deep-sea channel (Rigaut, 1997). Many submarine canyons clearly show an abrupt erosive flank and a smoother depositional flank, where terraces form. The erosive flank is deeper than the depositional flank. This morphology makes the canyon asymmetrical. The talweg is not located in the middle of the canyon but is situated at the side of the erosive flank. In contrast, straight submarine channels show either a symmetric V-shaped (e.g. the Hudson: Shepard and Dill, 1966) or a U-shaped valley (e.g. the Var: Piper and Savoye, 1993).

Canyons frequently occur at the mouth of rivers, being the continuation of a subaerial stream; this holds for the Zaire, Amazon and Orinoco in the Atlantic, for the Indus in the Indian Ocean, and for the Var, Rhône and Ebro in the Mediterranean. Other canyons are not connected with any river mouth, such as Dume and La Jolla canyons off southern California.

The canyon head is sometimes located very close to the continent. The head of the Capbreton Canyon in the Bay of Biscay starts at only 250 m from the coastline (Fig. 2.2A and Fig. 2.22). The head of the Zaire Canyon starts even within the river estuary. The path of the canyon can be straight to sinuous. Sinuosity is defined as the ratio between the length of the talweg axis and the down-valley distance (Leopold and Wolman, 1957). A valley is straight if sinuosity = 1, sinuous if sinuosity is between 1 and 1.5 and meandering if sinuosity >1.5.

3.3.2. Canyon formation

The origin of canyon formation and persistence is still an unsolved question. Four processes produce canyons (Mulder et al., 2004): they may form (1) after a subaerial phase of erosion by a river system, but they can also develop solely from (2) backward (retrogressive) submarine erosion, (3) forward erosion by continuous steady flows and (4) bypassing on rapidly prograding margins.

(1) Subaerial initiation of present-day submarine canyons has been demonstrated for many canyons in the Mediterranean Sea. There, canyons began to form by subaerial river erosion during the drastic fall of the sea level that occurred during the Messinian salinity crisis (Clauzon, 1978).

(2) Canyons are usually connected to a river, and rivers are the main source of sediment supply from continent to ocean. The frequency of cable failures in the Zaire Canyon suggests frequent energetic gravity flows (Droz et al., 1996). At a canyon head, the rates of sediment supply and accumulation are large. This suggests that canyons are both formed and maintained by retrogressive failures. Submarine retrogressive erosion is the simplest explanation for canyon formation (Guillocheau et al., 1983). The canyon head thus moves progressively landwards. The numerous slump scars observed at some canyon heads suggest that this process affects most canyons (Fig. 2.2A). This hypothesis is consistent

with the alignment of pockmarks in the landward extension of such valley heads. This has been observed in tributary valleys of the Capbreton Canyon (Cirac et al., 2001; Fig. 2.22) and the Zaire Canyon (Le Moigne, 1999; see Imbert (2011, this volume, Chapter 10.4.2.3.4)). These pockmarks suggest deep disorganization of the sediment due to the upward movement of pore water. It is not clear yet if pockmarks are the cause or the result of this disorganization. Effectively, these pockmarks could either result from the sediment disorganization in the direction of retrogression or indicate weakness zones that direct retrogression. The latter would be consistent with the fact that canyons such as Capbreton follow regional tectonic directions (Cirac et al., 2001).

(3) Canyons can also result from intense and frequent or continuous erosion by erosive sediment-laden flows (Pratson and Haxby, 1996), such as hyperpycnal flows (Mulder et al., 2004). The discovery of hyperpycnites on a terrace located in the Var Canyon (Mulder et al., 2001b) is consistent with this hypothesis. *In situ* measurements of a sporadic or continuous activity of particle-laden flows in submarine canyons and channels (Gennesseaux et al., 1971; Inman, 1970; Shepard and Dill, 1966; Shepard et al., 1977, 1979) also support this hypothesis. The presence of a shelf delta (Hueneme; Normark et al., 2009) favours a frequent supply by floods. Mitchell (2005) evidenced active erosion and channel formation solely by hyperpycnal turbidity currents activity in a dump site of dredge spoil of the Puyallup River (Washington State, USA).

(4) Canyons can be "constructed structures". In fact, they may be areas of low sedimentation rates or "bypass" areas in a regionally overall prograding margin. They then represent the narrow areas where margins do not prograde as fast as along the rest of the margin (Pratson et al., 1994). This hypothesis seems to be valid only for margins with a high progradation rate.

Most of the ancient canyons are filled. The stop of canyon activity is related to the disconnection of the canyon head with the sediment source, either because of tectonics or by a change in the relative sea level.

3.3.3. Gullies

Gullies are small-size submarine valleys, usually straight, with a width between 100 and 250 m and a depth between 5 and 50 m. They are bordered by steep flanks with numerous failure scars, and they show a morphology similar to subaerial badlands. They usually form a connected network, incising the continental slope and shelf. They seem to originate either from retrogressive erosion or from density cascading as shown by the Eel River example (Palanques et al., 2008).

3.3.4. Shelf-break deltas

Shelf-break deltas are located at the transition between continental shelf and slope. Although they are preferentially formed during sea-level fall and low-stand, particularly during forced regressions (regressions induced by eustatic fall; Porebski and Steel, 2003; Posamentier et al., 1992), they also can persist during highstand such as the present Mississippi (Coleman et al., 1998) or the Nile (Rouillard, 2010) deltas. Their can spread over 90 km (present Mississippi). The thickness of deposits can reach 140 m. Their internal structure usually shows large prograding sigmoid clinoforms thinning in basin direction. A typical prograding sequence begins with turbidites and mass-flow deposits, then mouth bar, delta-front deposits; it ends with river deposits (Sydow and Roberts, 1994). Gravity processes include various types of slump and mudflow (rounder, elongate, bottlenecked, coalescing; Coleman et al., 1983) that affect the rapidly prograding delta front and river mouth bar (oversteepening and overloading). Because of the direct link with the river system, turbidite formation includes hyperpycnal processes. A major difference with deeper mass-flow deposits is the influence of storm waves, swell and potentially tide in the shallowest part. These processes can transform down the continental slope and supply a deep-sea turbidite system.

3.3.5. Channel and levees

Deep-sea channel morphology (see Imbert (2011, this volume, Chapter 10.4.2.1.2)) is controlled by erosion and deposition by turbidity currents (Fig. 2.23). Commonly, channel erosion tends to decrease from upstream to downstream, while deposition increases. Two end-members can be observed: incising (erosive) channels (Zaire, Fig. 2.24B: Babonneau, 2002; Bengal: Hübsher et al., 1997) and aggrading (constructive) channels (Amazon, Fig. 2.24A: Flood et al., 1991). Aggradation means that sediment piles up vertically. Erosional (incised) channels are deep, with small or no lateral levees, because flow spilling is insufficiently frequent. The channel bottom deepens by eroding the sediment covering the channel floor. In contrast, in constructive channels, the channel floor aggrades simultaneously to levee growth. The channel morphology depends on the nature of the transported sediment. Channels in environments with coarse-particle supply are wide, straight and shallow. In areas where bypass dominates, gravel deposits in the proximal part can form gravel waves, such as in the Var Valley. The levees cannot be very high, as the shear resistance of the granular material is low. In contrast, channels supplied with fine-grained sediments are narrow, deep, and sinuous or meandering, with well-developed levees, as the walls of the levees are maintained by particle cohesion. If coarse-sediment supply is sufficient, the channel bottom is filled with sand and with coarse silt units that form layered to discontinuous high-amplitude seismic reflectors called HARs (high-amplitude reflectors; Flood et al.,

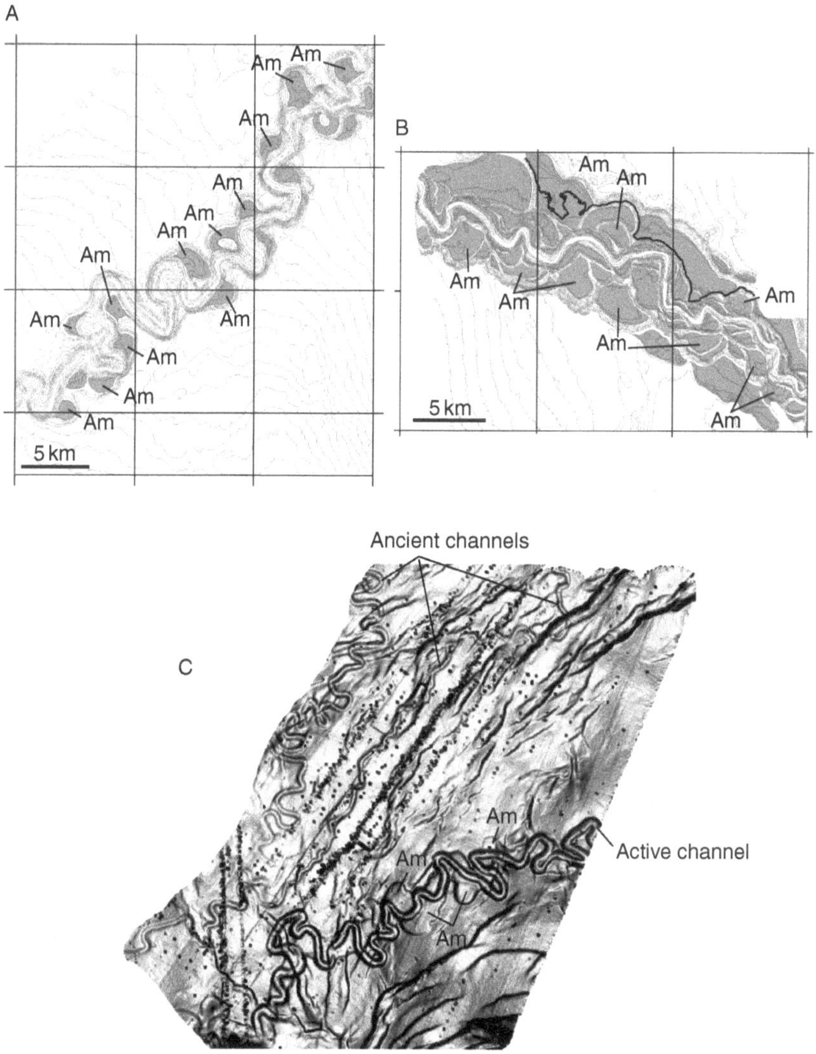

Figure 2.23 Deep-sea fans. (A, B) Zaire deep-sea fan. (A) EM12 bathymetric map; (B) EM300 bathymetric map (Babonneau, 2002; Babonneau et al., 2004). Terraces are in grey. Most of them are abandoned meanders acting as nested levees. Reproduced with permission from Geological Society, London. (C) Indus deep-sea fan (Garlan, 2004).

1991; Kastens and Shor, 1985; Weimer, 1990). Incising channels have no HARs. In channels, the distinction of deposits from individual events is made difficult because of amalgamation. Sequences with poor vertical grain-size changes are stacked without lithological changes between each individual sequence. In the Zaire channel, massive sands have been cored

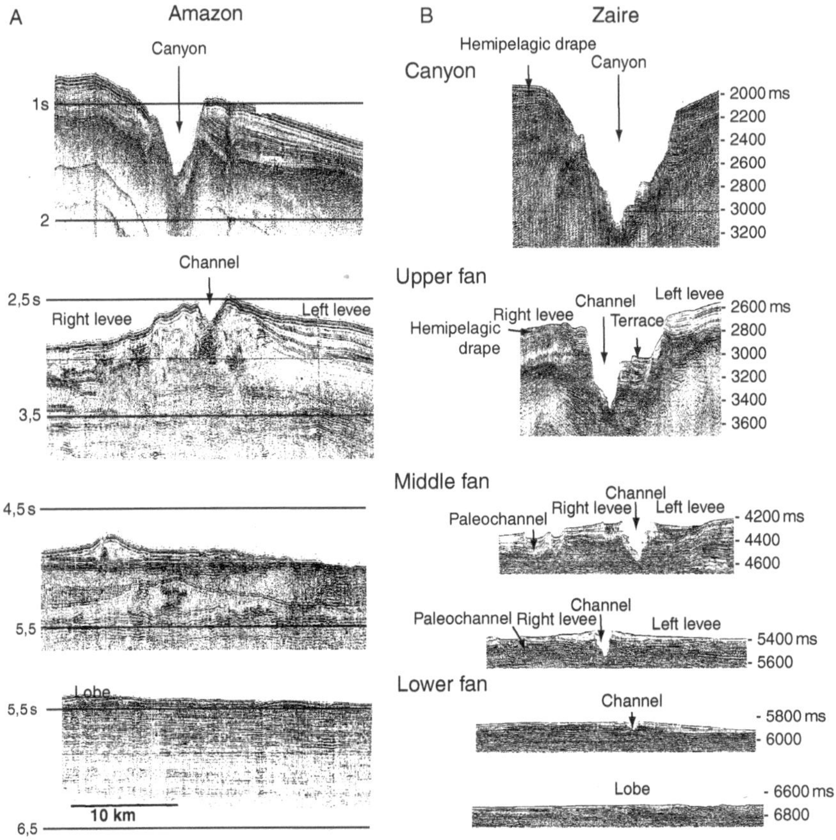

Figure 2.24 Longitudinal evolution of the seismic structure of modern deep-sea fans. (A) Amazon (Flood et al., 1991). Reproduced with permission from John Wiley and Sons. (B) Zaire (Babonneau, 2002; Babonneau et al., 2004). Reproduced with permission from Geological Society, London.

over a few metres (Babonneau, 2002). In ancient lithified outcrops, this suggests that the same bed can contain several sequences, each resulting from the deposition during an individual event (amalgamated beds).

In contrast to canyons, which are almost permanent structures on continental margins, channels are temporary features. They can migrate suddenly, through a phenomenon called "avulsion" (Fig. 2.25). Avulsion occurs because of failure of a levee or a levee spillover after initial breaching. Avulsion can occur because of a zone of weakness in the channel levee, for example, due to the presence of a fault (Ferry, 2004). On the Amazon fan, avulsion occurs preferentially in areas with the steepest slope (Beaubouef and Friedmann, 2000; Pirmez and Flood, 1995; Pirmez et al., 2000). Once avulsion has taken place, the next turbidity current follows the new path and

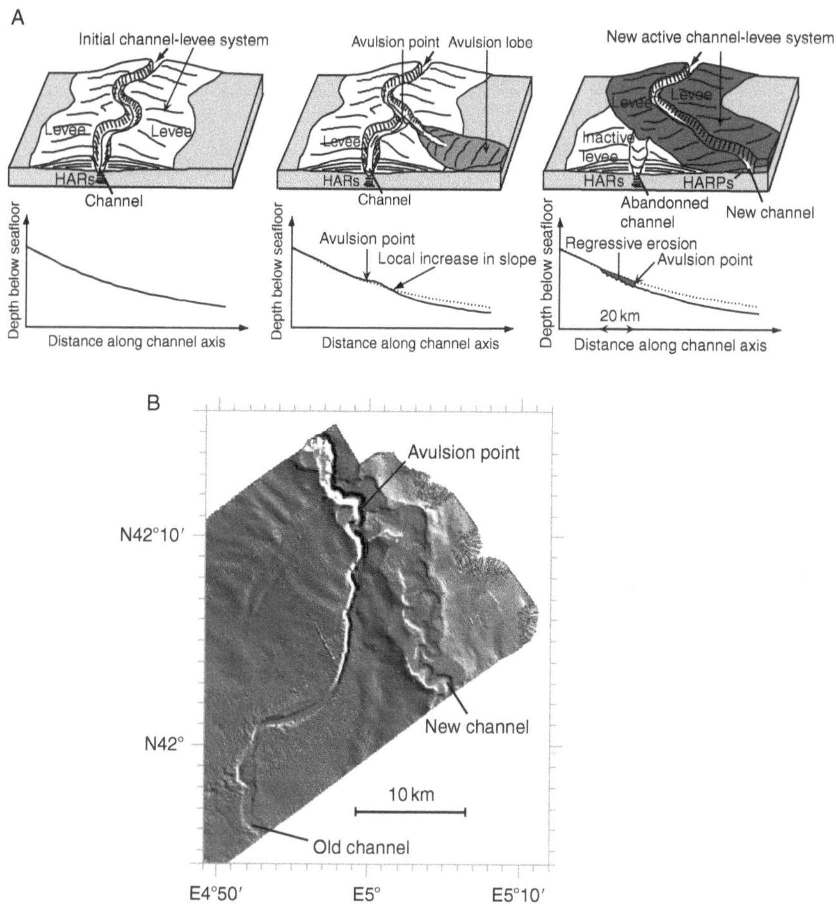

Figure 2.25 Channel avulsion. (A) Channel avulsion and evolution of the downslope morphology of a channel during and after avulsion (from Babonneau, 2002). Bold arrow indicates direction of the channelized flow. Reproduced with permission from N. Babonneau. (B) Last avulsion point of the Rhône deep-sea fan (that led to the formation of the Rhone neofan) (from Bonnel et al., 2005). Reproduced with permission from Elsevier.

moves downwards into the topographic low bordering the levee. As the current is no longer channelized, it tends to spread. It collapses and reconcentrates, and finally transforms into a concentrated flow that eventually quickly freezes and deposits tabular sandy packages called "sheet sands", which form well-stratified, HARs called HARPs (high-amplitude reflection packets). These flat, lobate structures are called "avulsion lobes" (Flood et al., 1991). They form the base of the next levee deposits and have a morphology and geometry very close to those of terminal lobes. ODP cores on the Amazon fan showed that HARPs are made of poorly sorted sand

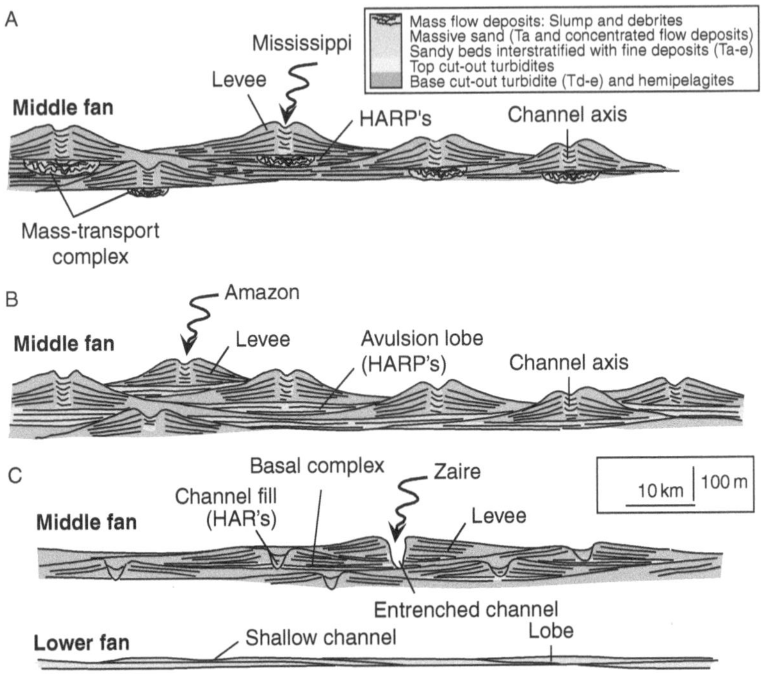

Figure 2.26 Channel–levee structure and facies in deep-sea fans with high sand load (Mississippi, A; Amazon, B) or with low sand load (Zaire, C). High-amplitude reflectors (HARs) are present at the base of levees. High-amplitude reflection packets (HARPs) represent channel fill (from Babonneau, 2002). Reproduced with permission from N. Babonneau. (A multi-colour version of this figure is on the included CD-ROM.)

(Fig. 2.26B) (Pirmez et al., 1997). In addition to the Amazon, HARPs were recognized on the Danube fan (Popescu et al., 2001), the Indus (Kenyon et al., 1995), the Congo (Droz et al., 1996) and the Mississippi (Weimer, 1990). In systems where the coarse-particle load is small (Zaire), HARPs are not present. They are replaced by a thin sole of sandy-silt deposits (Fig. 2.26C). In the Mississippi, and to a lesser degree in the Amazon (Flood et al., 1991), HARPs can be partially replaced by mass-flow deposits (Fig. 2.26A) called Mass Transport Complex (MTC) or "basin-floor fan" (Weimer, 1990; see Imbert (2011, this volume, Chapter 10.4.2.1.3)). Avulsion leads to the formation of an abrupt change in local slope at the location of the avulsion (knick point lip: Gardner, 1983). Consequently, the equilibrium profile of the channel will tend towards a smoother shape by retrogressive erosion that can propagate over tens of kilometres.

The avulsion rate tends to decrease downstream. Avulsion never occurs in canyons or at a timescale spanning the whole evolution of a margin. It happens infrequently in the upper channel–levee complex (in the range of tens to hundreds of thousands of years), becomes frequent in the lower channel–levee complex (in the range of thousands to tens of thousands of years) and becomes very frequent in lobe channels. This constitutes a problem for the analysis of the longitudinal sequence: an erosional surface in a canyon does not represent the same ordering (and time length) as an erosional surface downstream (Ferry, 2004). Several incision surfaces in the lower part of the system are nested in a major surface (with a regional significance) upstream.

Levees are topographic highs beginning approximately at a canyon's mouth and bordering the side of the channel. The levee elevation (height from the channel to the top of the levee) decreases from a few hundreds of metres to several tens of metres from upstream to downstream. Considering that both channel and levees aggrade, the channel–levee complex can represent a total sediment thickness of 1000–1500 m of sediment (Damuth and Flood, 1985). The lateral extent of levees can reach 50 km on each side of the channel.

Because of the Coriolis force, levees at mid-latitude are usually asymmetrical, with the right-hand levee more developed than the left one (looking downflow, in the northern hemisphere; e.g. Cap-Ferret: Cremer, 1983; St Lawrence: Piper et al., 1984; Var: Savoye et al., 1993; NAMOC: Klaucke et al., 1998; Hueneme: Piper et al., 1999b). The opposite is true for the southern hemisphere (Komar, 1969). The impact of the Coriolis force on current flow is inversely proportional to flow velocity (Cremer, 1983).

On seismic profiles, levees are easily recognizable by their topography and their characteristic "bird-wing" or "moustache" shape. The internal well-stratified reflectors remain continuous and parallel over long distances and pinch out (downlap) on the external levee sides.

Levees are built by the spillover of the upper part of turbidity currents. The spilling can occur either because the flow thickness is greater than the levee height (overbank or overspill: Hiscott et al., 1997; Fig. 2.27A) or due

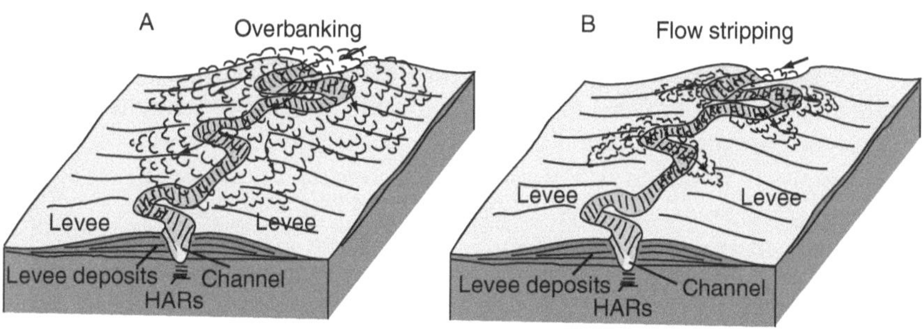

Figure 2.27 (A) Overbanking or overspilling (Hiscott et al., 1997). (B) Flow stripping (Piper and Normark, 1983). Bold arrow indicates direction of the channelized flow (from Babonneau, 2002). Reproduced with permission from N. Babonneau.

to the centrifugal force when the channel changes abruptly its direction, for example, in a bend or a meander (flow stripping: Piper and Normark, 1983; Fig. 2.27B). Flow stripping can induce separation into bipartite flows (see Sections 1.4 and 1.8). The dense concentrated bottom flow proceeds as a channelized flow, while the dilute turbulent upper part spills and spreads over the levee. In contrast to channels, where erosion and amalgamation are important, sedimentary levees record the gravity processes and the evolution of the turbidite system over long timespans. When the frequency of spilling turbidity currents is low, turbidite spill is separated from the next one by a hemipelagic drape that allows the build-up of a good stratigraphic framework. Spillover occurs preferentially in channel bends or meanders. In areas where spillover is intensive, secondary channels can form (Masson et al., 1995). In areas where the levee is low as well as in meanders, outer levee erosion during spillover can form lobate deposits called "crevasse-splay deposits" by analogy with river systems (Masson et al., 1995). At this stage, avulsion can proceed, forming a crevasse-splay lobe and finally an avulsion lobe. At this location, the channel width varies from a few metres to a few hundreds of metres, and channel depth does not exceed ten metres. As turbidity currents have an internal normal grain-size distribution, only the finest particles of the turbidity current spill over the levee at a given point. This induces several characteristics with respect to the lateral and vertical evolution of levees.

As levees grow vertically and provided that the type (volume, intensity) and frequency of turbidity currents remain constant, spillover by turbidity current becomes more and more difficult. Both the volume and the grain size of the sediment that spills over decrease. Consequently, levees are made of thinning and fining-upward successions. If the type, frequency or characteristics of turbidity currents change, the levee records both allocyclic and autocyclic signals, and the thinning/thickening, fining/coarsening trend is more difficult to decipher, such as for the Amazon (Hiscott et al., 1997; Piper and Deptuck, 1997).

As a turbidity current moves down a channel, the successive spillover progressively diminishes the proportion of fine particles in the turbidity current. The turbidity current therefore gradually becomes coarser and tends to reconcentrate and to thin. Longitudinally, levees become thinner and coarser because both the spilling-over of the turbidity current becomes increasingly difficult, and the ever more dominant presence of coarse sediment diminishes the shear resistance in the levee deposits.

Because of the proportional increase of coarse material in the channelized turbidity current due to the progressive spillover of the upper part of the current, the longitudinal evolution of channel and levees is as follows: in the proximal part, levees are thick, fine-grained and the channel is narrow and sinuous. Characteristic deposits are base-missing silty/clay Bouma intervals (Tde: Hesse and Chough, 1980; Piper and Deptuck, 1997).

In the distal part, the channel tends to become straight and large, and the levees become thin and coarse-grained (Tbe Bouma intervals). Spilling over levees acts as a sorting machine. This sorting is particularly efficient when the height of the levees is small, such as in the case of the Amazon turbidite system. When the channel is deeply incised and the height of the levees from the channel bottom upwards is high, spillover is difficult and sorting is poor (Table 2.6). Babonneau (2002) showed that there is huge difference in sand trapping inside the channelized part of turbidite systems. In the Amazon, channel fill by sand is very important and thick, while the sandy channel fill in the Zaire system is very thin and sand occurs only in the part of the channel connected to the lobes.

The internal side (channel side) of the levee is steep and can be affected by slope failures (slumps) towards the channel (Fig. 2.2B). Seafloor observation by ROV and SAR in the Zaire channel showed numerous fan-shaped block clusters accumulating at the base of the channel flank and generated by the erosion of the flanks by the dense part of channelized flows. The failed material is thus eroded by following flows. Changes in flow dynamics by incorporation of failed sediment have been successfully modelled by Salles (2006) and Salles et al. (2007) using cellular automata. The induced increase in flow velocity can explain the presence of internal erosion surfaces in the Bouma sequences. When the incision exceeds 100 m, incisions seem to be preponderant over lateral erosion (Babonneau, 2002). The external flank of the levee also show slumps but has usually gentle slopes and is frequently covered with sediment waves (Migeon, 2000; Migeon et al., 2001; Skene, 1998). A specific characteristic of these sedimentary bodies is their progradation upstream and upslope. In contrast to what happens in contourite environments, turbidity-current-generated

Table 2.6 Main differences between a system with good efficiency for sand sorting and a system with low efficiency (from Babonneau, 2002; Bonnel, 2005)

Type of system	High efficiency: Amazon	Low efficiency: Zaire
Sediment supply	Large	Medium
Canyon head	Close to the shelf break	On the shoreline
Type of channel	Aggradational (with HARs)	Incising
Depth of the channel	Low	High
Flow spilling on levees	Important	Small
Lobes	Sandy	Muddy (and debris flows)
Lobe migration	Progradation, retrogradation and avulsion	Progradation
System activity	Lowstand	Lowstand and highstand

sediment waves always have a crest direction perpendicular to the flow direction. Turbiditic sediment waves are, in general, asymmetrical, with an upstream (progradational) flank that is steeper and shorter than the downstream flank (antidune morphology). These sediment waves can be initiated by oscillatory internal waves forming within flows passing an obstacle such as a levee (Allen, 1982; Flood, 1988; Kneller and Buckee, 2000). Their growth is explained by the preferential deposition by the flows on the upstream flank. The downstream flank is subjected to erosion and less deposition (Migeon, 2000; Migeon et al., 2001). This is due to the change of the flow velocity along the topographic relief formed by the wave (Fig. 2.28). On the downstream flank of the wave, the flow accelerates (waxing flow). According to the nature of the seafloor and the flow velocity, either erosion or some deposition can take place. After passing the interwave low, the flow moves along the steep upstream flank of the wave. It decelerates rapidly (waning flow), and an important volume of sediment is then deposited. Behind the wave crest, the flow accelerates again along the downstream flank.

Base-missing turbidites in sediment waves show typically laminated beds with an alternation of silty and clay laminae. The number of silt and clay couplets is only 6 or 7 in the proximal part of the levee, but can reach 50 in the distal part (Fig. 2.29; Migeon, 2000; Migeon et al., 2001). The silty laminae are fining and thinning upwards, whereas the clay laminae are thickening upwards. On a grain-size/thickness diagram, the successions show a typical Christmas-tree shape (Migeon, 2000; Migeon et al., 2001;

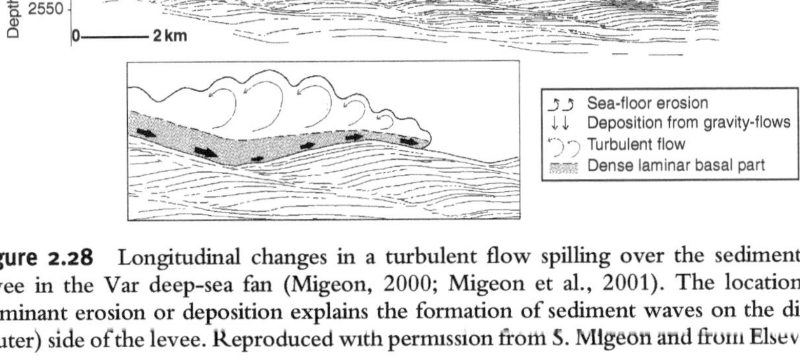

Figure 2.28 Longitudinal changes in a turbulent flow spilling over the sedimentary levee in the Var deep-sea fan (Migeon, 2000; Migeon et al., 2001). The location of dominant erosion or deposition explains the formation of sediment waves on the distal (outer) side of the levee. Reproduced with permission from S. Migeon and from Elsevier.

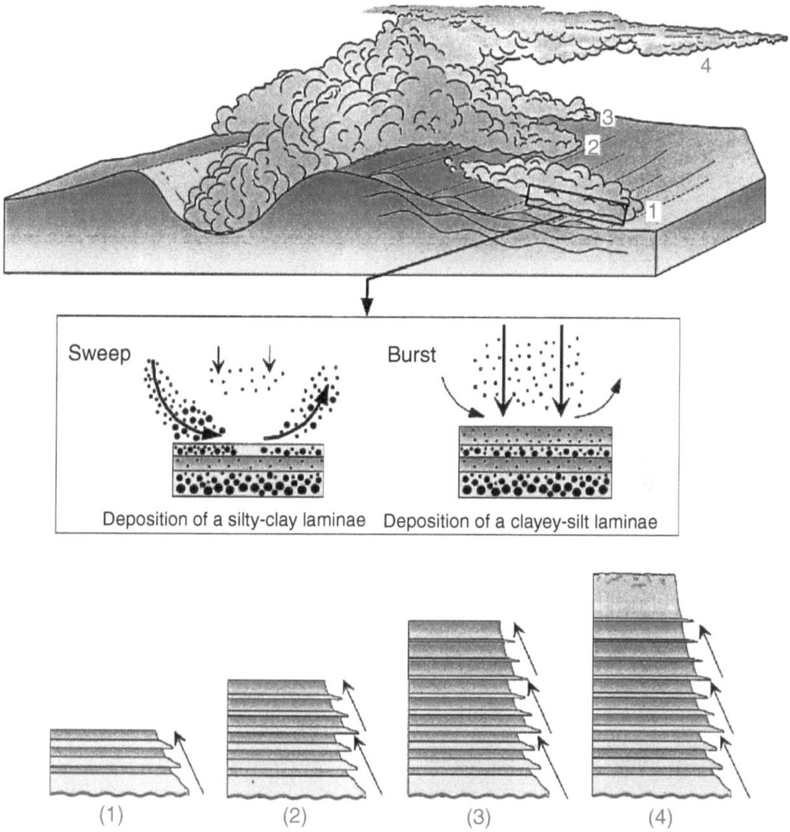

Figure 2.29 Spilling over a levee of a channelized turbidity current. Transversal scale is few kilometres to few tens of kilometres. Deposition by several pulses (1, 2 and 3) and eventually the final suspension fallout (4) produce a fining-upward bed constituted by several sub-beds. Burst and sweep processes produce the alternation of silt- and clay-dominated laminae (Migeon, 2000; Shanmugam et al., 1993). Reproduced with permission from S. Migeon and from AAPG. (A multi-colour version of this figure is on the included CD-ROM.)

Piper and Deptuck, 1997). These laminated successions are explained by the fast variations in velocity and the related shear stress in the boundary layer at the bottom of the flow, forming alternative bursts and sweeps (Hesse and Chough, 1980). During a burst, the particle is subject to a resulting downward directed force. Both silt and clay particles settle. During the following burst, only the finer fractions of the particles that have just been deposited are put into suspension so that a coarse silt lamina forms. Suspension fallout of the fine particles forms the clay lamina. As the spilling current wanes, its energy, capacity and competence decrease and the coarse lamina becomes thinner and finer as suspension fallout predominates.

3.3.6. Confined (inner) levees

Small talweg channels commonly construct inner levees within an ancestral larger channel–levee system. The seismic facies inside flat-topped terraces that are the morphological expression of inner levees is layered, suggesting particle fallout. Analysis of sedimentary facies from cores (Gaudin, 2006) yields top-missing Bouma intervals and a fast sedimentation rate (1 m ka^{-1}). This suggests that the levee grows vertically by spilling of the top of the turbidity currents. On seismic profiles (Deptuck et al., 2003; Pichevin et al., 2003), the terraces follow the lateral migration of the talweg. The seismic facies inside this kind of terrace is chaotic at the base and layered at the top, suggesting that these terraces are initially formed due to slump failures and persist as confined sedimentary levees. Turbidity currents moving into the canyon spill over the slump deposit. The finest particles transported in the top part of the turbidity current settle and form the layered facies.

3.3.7. Terraces in submarine canyons and channels

Sinuous submarine canyons and valleys show frequently terraces, that is, flat structures several metres to several tens of metres above the talweg (Figs. 2.22, 2.23 and 2.30). Terraces can be superimposed on one another, suggesting that canyons can undergo several phases of incision (Cirac et al., 2001). Deptuck et al. (2007) suggest that the evolution of submarine channels occurs by "plug and cut migrations", that is, alternating phases of channel fill and incision. The following four processes are described for submarine terrace formation.

(1) Failures of the canyon side walls (Berné et al., 1999; Fig. 2.30A and D).
(2) Meander cut-off and abandonment forming horseshoe terraces. The path of the abandoned and filled talweg is usually clearly visible (Tranier, 2002; Fig. 2.31). Meander cut-off has been detailed by Babonneau (2002) and results of flow stripping. Before total abandonment, the lower part of the flow still moves in the ancient meander, while the upper part moves straight fully and continues to erode the new path. Deptuck et al. (2007) suggest to use the term "pseudo-meander cut-off" because the fill of the meander occurs prior to cut-off, that is, the opposite as in fluvial systems.
(3) Confined levees (Babonneau et al., 2004; Pichevin et al., 2003) due to the lateral shift of the channel (Fig. 2.30B–D and F), forming flat terraces with or without a topographic high on its internal side. Whether the origin of the terrace is slump, meander cut or channel shift, all terraces finally act as confined levees. For Deptuck et al. (2007), inner levees with vertical aggradation would form only in case of instantaneous channel migration.
(4) Incision due to avulsion upstream or to meander abandonment (Figs. 2.30E–G).

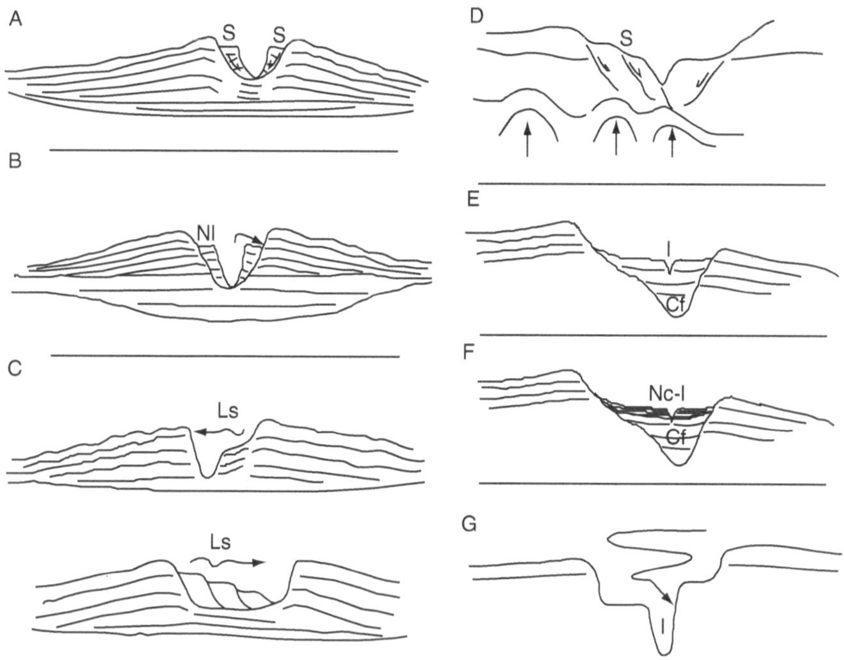

Figure 2.30 Models for terrace interpretation in deep-sea turbidite channels. (A) Slumps (S) on canyon/channel flanks (e.g. Indus Channel: Kenyon et al., 1995). (B) Formation of nested levees (Nl) during sea-level rise (e.g. Bengal Channel: Hübsher et al., 2007a,b; Hueneme Channel: Piper et al., 1999b). (C) Filling of the canyon/channel because of lateral shifting (Ls) of the canyon/channel or of decrease in the volume of the channelized flows (e.g. Amazon Channel: Damuth et al., 1988; Toyama Channel: Nakajima et al., 1998; Indus Channel: Mc Hargue and Webb, 1986). (D) Deformation and slumping (S) due to gravitational tectonics (e.g. Kaoping Canyon: Liu et al., 1993; Zaire Canyon: Cramez and Jackson, 2000). (E) Incision (I) of a talweg in the canyon/channel fill (Cf) because of an avulsion that occurred downstream (e.g. Rhône Canyon: O'Connel et al., 1995). (F) Formation of a nested channel–levee (Nc-l) system in the canyon/channel fill (Cf) (e.g. Rhône Canyon: Torres et al., 1997). (G) Meander abandonment and talweg incision (I) (e.g. Basin Arequipa Canyon: Hagen et al., 1994; Indus Canyon: Von Rad and Tahir, 1997). From Babonneau et al. (2004). Reproduced with permission from Geological Society, London.

Abreu et al. (2003) published a seismic profile that shows the internal structure of a levee with foresets dipping towards the talweg axis. This suggests a genesis of the terrace similar to the genesis of subaerial river terraces. In that case, the terrace would consist of a series of prograding accretionary bars called "lateral accretionary packages" (LAPs: Abreu et al., 2003). However, this way of formation of submarine terraces seems very rare. Deptuck et al. (2007) suggest that channel bank accretion (forming inner levees analogous to submarine point bars) could occur during phases of abrupt (not instantaneous) channel migration.

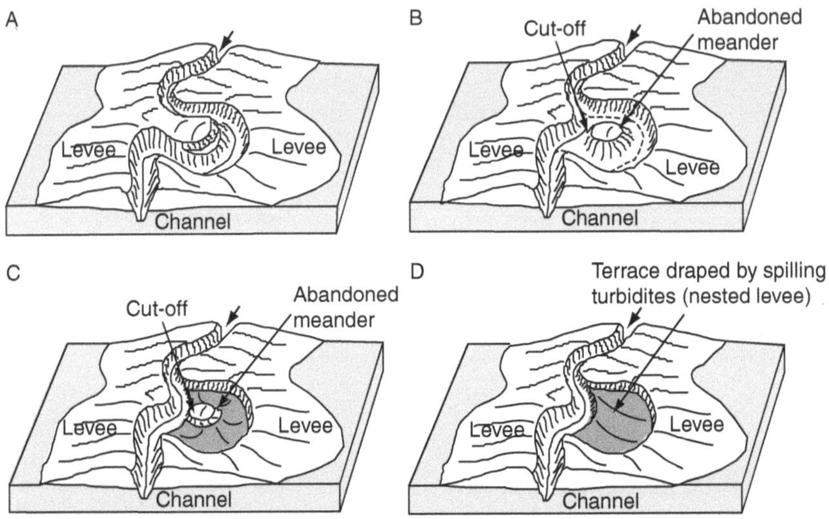

Figure 2.31 Scenario for meander abandonment (Babonneau, 2002; Babonneau et al., 2004). (A) Lateral shift of a talweg. A meander forms. (B) Meander cut-off. (C) Beginning of the filling by spilling of turbidity current in the abandoned meander. (D) Formation of a terrace that acts as a nested levee. Bold arrow indicates direction of the channelized flow. Reproduced with permission from Geological Society, London.

3.3.8. Meandering of submarine channels

Clark et al. (1992) and Clark and Pickering (1996) showed morphological convergence between submarine and subaerial channels. The variation of channel sinuosity with increasing channel slope has been observed in both rivers and scaled experiments. A similar correlation exists between submarine-channel sinuosity and channel slope. In addition, the general distribution of width/depth ratios (aspect ratio) in submarine channels varies between 10:1 and 100:1 in both modern and ancient settings. For river examples provided by Leopold et al. (1964), this ratio varies between 5:1 for small rivers and 50:1 for large rivers such as the Potomac. The only important morphological difference seems to be that submarine channels are substantially larger than subaerial channels (Rigaut, 1997).

However, several lines of evidence suggest that the morphological resemblance of subaerial and submarine channels is not the result of an identical genesis:

(1) fluvial terraces are essentially point bars, but point bars are very rare in submarine environments;
(2) meander cut-off is frequent in subaerial streams: it occurs at least twice as often as in submarine channels (Klaucke and Hesse, 1996);
(3) the wavelength of meanders is larger in the submarine environment (several kilometres) than in river systems (a few hundreds of metres);

(4) the grain size of transported particles is larger in rivers than in submarine environments, and river systems are dominated by bedload transport, whereas submarine channels are dominated by suspended-load transport; in addition, Peakall et al. (2000) showed that the evolution of submarine channels is highly different from the evolution of subaerial channels;
(5) submarine channels migrate laterally but do not prograde downslope (Fig. 2.32A), whereas river meanders migrate both transversally and longitudinally with respect to the channel; Posamentier and Walker (2006) showed that longitudinal migration also exist in submarine channels;
(6) aggradation is very important for submarine meanders and insignificant for river meanders as suggested by Deptuck et al. (2007) on the Niger slope. Numerical modelling results of Inram et al. (1999) suggest that erosion plays a minor role in the evolution of submarine channels. Meander migration seems to be due to either flow restriction related to an increase in the sinuosity, or the aggradation of the channel. However, Deptuck et al. (2007) suggest that narrow channels with important lateral migration and low vertical aggradation show a similar behaviour as fluvial channels, including the ability to develop point bars.

Using examples of high-latitude channels, Peakall (2007a,b) suggested that channel shape varies according to three factors: (1) the high-frequency variation in flow types, (2) the nature of sediment supply controlling channel-bank stability (high-latitude channels receive mainly coarse-grained sediments, and equatorial systems are supplied mainly with fine-grained material) and (3) the Coriolis force, which is inversely correlated to the sinuosity.

Corney et al. (2006) and Keevil et al. (2007) suggested that the differences between subaerial (river) and submarine meanders and their related deposits are partially related to the difference in the velocity profiles (Fig. 2.32B). A vertical profile within a subaerial flow shows that the velocity increases upward in a linear, logarithmic or exponential manner. Consequently, the helical motion in a meander of a fluvial system induces a surface flow towards the outer flank and a basal flow towards the inner flank. In contrast, the vertical decrease in flow velocity in a submarine channel (see Section 1.4.2) induces a helical motion in a meander with a surface flow towards the inner flank and a basal flow towards the outer flank. In addition, the higher buoyancy of submarine flows and the variability of discharge make them thicker and slower than river flows. Consequently, suspension is more important in submarine flows than in most rivers and more sediment is deposited from suspension within submarine channels, generating confined, nested and stepped fine-grained levees with a flat internal layering. This confirms the result that channel morphology would be correlated to flow type and nature of

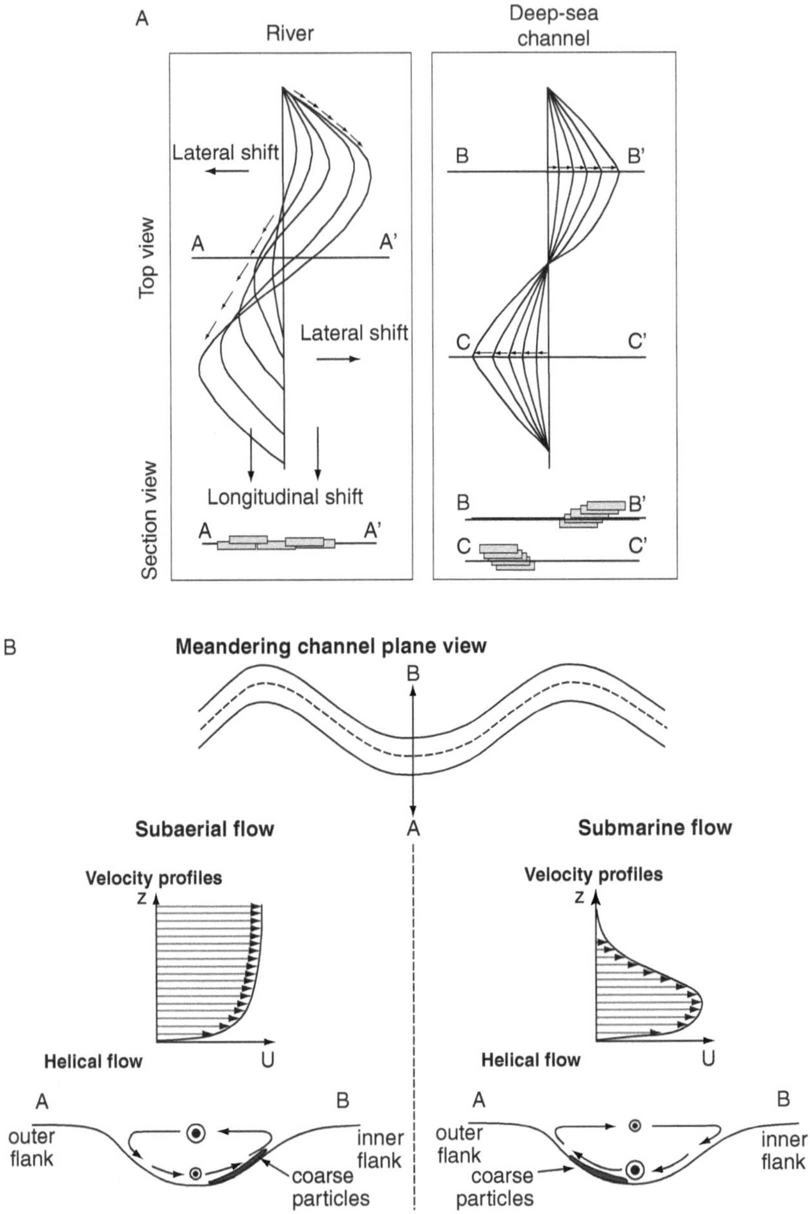

Figure 2.32 Comparison of meander evolution for subaerial rivers and deep-sea channels. (A) A river channel is initially straight and meanders and enlarges both by lateral shift (swing) and longitudinal shift (sweep). Deep-sea channels only migrate laterally (from Peakall et al., 2000). Reproduced with permission from SEPM. (B) Comparison of vertical velocity profiles, helical flow motion and deposition of particles in a channel meander related to a subaerial river flow (left) and a submarine flow (right) (Corney et al., 2006).

transported particles. The recent evolution of the Celtic Fan (see Faugères and Mulder (2011, this volume, Chapter 3.4)) confirms this hypothesis.

The present Zaire system shows a morphology that perfectly mimics river systems on land, with a large valley limited by meander bends (meander belts) and a talweg incising the valley. The formation of such a morphology is illustrated in Fig. 2.33 (Babonneau, 2002; Babonneau et al., 2004):

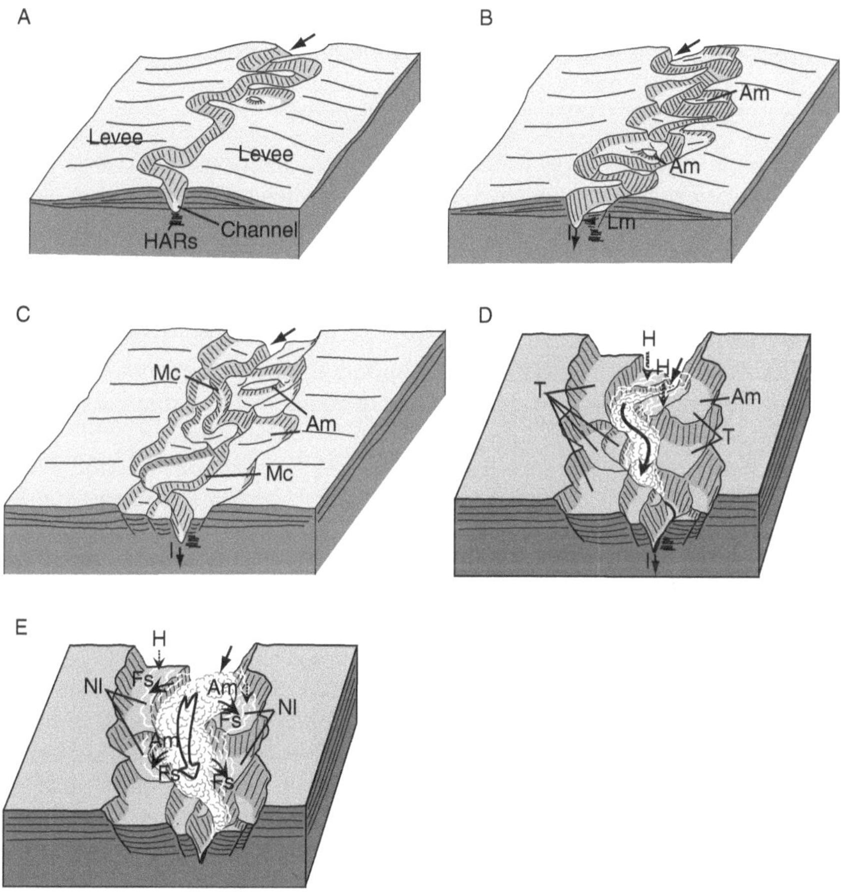

Figure 2.33 Recent evolution of the Zaire Valley (from Babonneau, 2002; Babonneau et al., 2004). Reproduced with permission from Geological Society, London. (A) Sinuous channel/levee system with few meander cut-offs (Mc). (B) Simultaneous channel-floor incision (I) and lateral migration (Lm) of meanders. Meander cut-off generates abandoned meanders (Am) and terraces (T). (C) Continuation of both incision and lateral migration of the channel. (D, E) Present time: entrenchment of the talweg inside the deep valley. (D) Low-volume channelized flows. (E) Large-volume, spilling flows (Fs and arrows) forming a nested levee (Nl). Bold arrow indicates direction of the channelized flow. Dashed vertical arrow indicates hemipelagic (H) sedimentation.

(1) an initial sinuous channel–levee system settles, and meanders are locally abandoned by cut-off (Fig. 2.33A);
(2) the talweg incises and migrates laterally, so that meanders become abandoned, then filled up, thus resulting in flat terraces, the upper parts of which are confined levees (Fig. 2.33B);
(3) incision and migration of the talweg proceed to form a meander belt; accumulation of fine-grained turbidites is reduced on the terraces and replaced by hemipelagic settling (Fig. 2.33C);
(4) the present situation: low-volume flows remain channelized and cannot spill over the terraces; hemipelagites drape the terraces (Fig. 2.33D); large-volume flows can spill over the terraces depositing fine-grained turbidites (Fig. 2.33E); nested levees aggrade.

Babonneau (2002) suggests that local flow accelerations due to meanders would be more important than the slope gradient in maintaining flow energy and velocity.

3.3.9. Mass-transport complexes (MTCs)

Most of MTCs (see Imbert (2011, this volume, Chapter 10.4.2.1.3)) or mass-transport deposits (MTDs) correspond to debris avalanches with a volume varying from a few cubic metres to thousands of cubic kilometres. The largest MTCs have volumes of hundreds to thousands of cubic kilometres: Storegga (5500 km^3: Bugge et al., 1987); Hinlopen (1350 km^3: Vanneste et al., 2006); Traenadjupet (760 km^3: Laberg and Vorren, 2000); Nile (18 MTDs for the last Pliocene-Quaternary succession with a volume in the range 1–1345 km^3: Rouillard, 2010). In the Mississippi turbidite system, where they have been defined, they represent 10–20% of the stratigraphic sequences (Weimer, 1989, 1990). In the Nile turbidite system, the thicknesses of the MTC deposits vary between 55 and 425 m, suggesting that the base of the transported material involves consolidated sediments or rocks. In the Amazon turbidite system, two major MTDs have been identified with a probable simultaneous triggering (Western and Eastern MTDs: Piper et al., 1997). The volume of the western MTD is close to 2000 km^3, and its thickness varies between 100 and 200 m (Damuth and Embley, 1981; Piper et al., 1997). The volume of the eastern MTD is close to 1500 km^3 (Piper et al., 1997). On the Rhône system, the most recent avulsion that led to the construction of the last lobe (neofan) formed over two MTDs called the "Eastern" and the "Western MTDs" (Bonnel et al., 2005; Droz and Bellaiche, 1985) extending over 150 km (length) and 50 km (width) with maximum thickness of 100 m and a volume of 260 km^3 (Gaullier et al., 1998).

MTDs are usually associated with spoon-shaped failure scars (slump scars) and limited by frontal and side scarps; they lie over a basal scour (Frey-Martinez et al., 2005; Gee et al., 2005; 2006). The seismic facies of MTDs is usually chaotic and/or discontinuous with low amplitude (Collot et al., 2001; Garziglia et al., 2008). Most of MTDs move over a short distance (a few metres to a few kilometres) with exceptions for the largest ones, which can move over hundreds of kilometres.

The position of MTDs in the stratigraphic architecture is still unclear. In the Mississippi, the seismic units that may contain MTC at their base are related to 100 ka orbital cycles. On the Nile, MTCs of the last 200 ka are predominantly related to sea-level fall, whereas some are related to sea-level rise (Rouillard, 2010). They also seem to be predominantly related to pluvial phases (wet climatic phases over the Nile headwaters and high river discharges). In the Amazon system, the MTDs are dated between 45 and 41 ka for the oldest and between 37 and 35 ka for the youngest (highstand and early fall, respectively: Maslin et al., 2005). The Rhône Western MTD is dated between 19.9 and 21.1 ka cal. year BP (glacial maximum and lowstand: Dennielou et al., 2009).

The analysis of microfauna in debrite clasts and matrix (Ducassou et al., 2010) suggest that the source of Nile MTDs is on the upper continental slope (200–1000 m water depth). The source of Amazon MTDs corresponds to scars between 700 and 1500 m water depth (Damuth and Embley, 1981; Piper et al., 1997).

The initiation of MTDs is still under debate and depends on their position in stratigraphic cycles. If they are more frequent and larger during sea-level fall and lowstand than during sea-level rise (Rouillard, 2010), they could be related to a combination of processes including oversteepening, overloading (intensified by high sedimentation and progradation rates during pluvial phases) and earthquakes. The role of upward moving fluids can also be very important, particularly the global destabilization of gas hydrates. Fluid-escape structures, including gas chimneys, mud volcanoes (Sharp and Samuel, 2004) and free gas, were evidenced close to the source of the MTDs in the Nile system (Bayon et al., 2009; Dupré et al., 2007; Huguen et al., 2009) and in the source area of Storegga (Mienert et al., 2005). Whatever their origin is, MTDs completely reshape the morphology of deep-sea turbidite systems. They are very good stratigraphic markers. Their surface morphology controls the development of the following channel–levee complexes by controlling the topography (Rouillard, 2010).

3.3.10. Lobes

A lobe represents the distal part of a channel–levee system corresponding to the initial definitions of suprafan lobe and lower fan (Normark, 1970, 1978).

The term "lobe" (Normark, 1970) represents several concepts and describes several types of sedimentary structures and sedimentary bodies. In their synthesis of architectural elements, Stow and Mayall (2000) keep only two items: lobes and clustered lobes. The suprafan lobe (Normark, 1970) is a coarse-grained sedimentary body forming in the channel axis of the middle fan and is now regarded as an obsolete term. The depositional lobe of Mutti and Ricci Lucchi (1972) has been defined for ancient systems but is used for both ancient and recent environments. It represents the whole system of non-channelized sedimentary bodies with a very large lateral extension developing at the mouth of channels and made of fine and medium sand (sheet sands). The fan lobe of Bouma (1985) describes the whole deep-sea turbidite system of the Mississippi. This concept introduces confusion between the whole turbidite system and the architectural element.

Mutti and Normark (1987, 1991) initially defined depositional lobes as non-channelized sand bodies with a large spatial extent.

The existence of channelling in lobes (channelized lobes) was demonstrated by Nelson et al. (1985) and by Normark and Gutmacher (1985) (leveed-valley lobes). The term "ponded lobe" has been used by Nelson et al. (1985) for the Ebro turbidite system to represent flat apron deposits at the toe of a slope failure. Mounded lobes (Galloway, 1998) are sedimentary bodies made of amalgamated beds of coarse material. Sheet lobes (Galloway, 1998) essentially consist of fine-grained deposits. They have a larger lateral extent than mounded lobes but are flatter. All the terms cited above should not be used for the following reasons: (1) they are related to outdated concepts (suprafan lobe), (2) they concern a whole system (fan lobe) or (3) they concern only a part of the lobe features (sheet lobe, mounded lobe) which is inconsistent with the concept of architectural elements.

A lobe initially forms when a levee is breached (crevasse-splay lobe). Once the levee fails, the flow path is deviated from the existing channel and water flows into the inter-channel space (avulsion lobes of Flood et al., 1991) generating concentrated flow deposits forming HARPs (see Section 3.3.5).

Recent works show that lobes are sedimentary bodies that form high-reflectivity patches on acoustic-reflectivity maps (Fig. 2.34). The lobes have the following features.

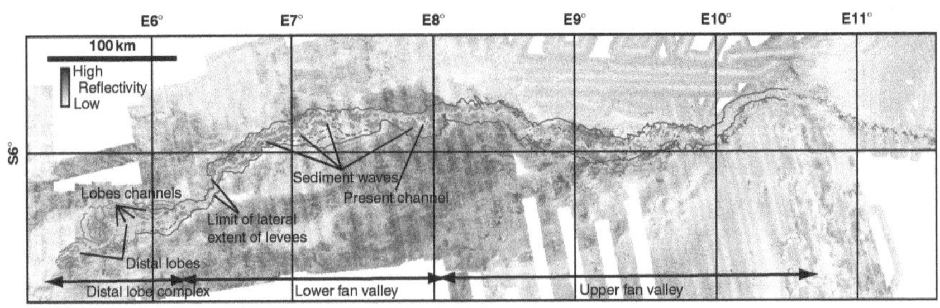

Figure 2.34 Acoustic imagery (EM12) of the present Zaire deep-sea fan, including the mapping and imagery of main sedimentary bodies (Babonneau, 2002; Babonneau et al., 2004). Reproduced with permission from Geological Society, London. (A multi-colour version of this figure is on the included CD-ROM.)

(1) They form at the mouth of a submarine channel or valley after a transitional zone characterized by erosional (scours and furrows) and bypass processes. They are associated with megaripple or dune-type bedforms (Mutti's F6 facies) generating a chaotic seismic facies (Gervais, 2002; Gervais et al., 2004), and with channels filled with a layered high-amplitude facies. This zone is called the "CLTZ": Bouma et al., 1985; Jegou, 2008; Wynn et al., 2002; Fig. 2.35). The CTLZ exists only for turbidite systems with a high sand/clay ratio (Navy Channel: Normark et al., 1979; Laurentian Fan: Normark et al., 1983; Agadir Channel, Valentia Fan: Morris et al., 1998; Wynn et al., 2002). These lobes thus correspond to the detached lobes of Mutti (1992). For mud-dominated turbidite systems such as those of the Zaire (Savoye et al., 2000) or the Mississippi (Twitchell et al., 1991), the CLTZ can be missing. These lobes thus correspond to the attached lobes of Mutti (1992). The CLTZ corresponds to the bypass zone of Mutti and Normark (1987). It shows successively, from proximal to distal, furrows oriented parallel to the flow direction, large

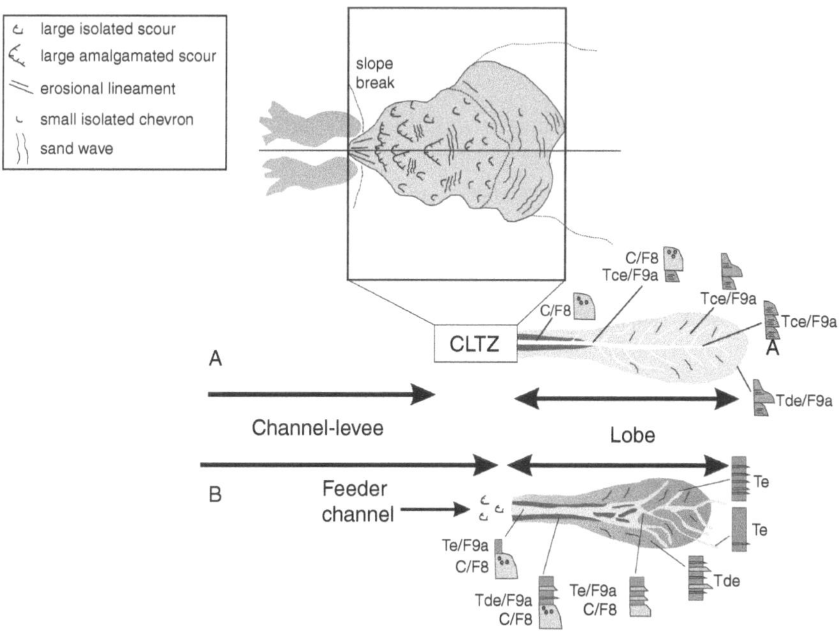

Figure 2.35 Subdivisions used to describe lobe and channel–lobe transition zone morphology and depositional facies in lobes. See Fig. 2.14 for facies description. (A) sandy lobe; (B): muddy lobe. CLTZ: channel-lobe transition zone. (C) concentrated flow deposits (from Wynn et al., 2002; Bonnel, 2005). Reproduced with permission from AAPG and C. Bonnel. (A multi-colour version of this figure is on the included CD-ROM.)

amalgamated scours, large isolated scours and small isolated chevrons, spoon-shaped scours and, finally, sand waves oriented perpendicular to the flow direction. The CLTZ was recognized essentially in ancient deposits (Mutti, 1985) and is more difficult to find in modern environments (Bonnel, 2005). In particular, scouring can appear outside the CLTZ. The CLTZ would appear preferentially when flows are confined to the channel mouth (Zaire type) rather than when the flows spread (Var type) (Bonnel, 2005).

(2) They have an ovoid shape with a lateral extent of a few tens of metres to a few tens of kilometres.

(3) They have a convex-up (positive) topography and a relief that is generally less that 25 m above the surrounding sea floor. The lobes' slopes are very smooth. This topography is more marked for lobes in small turbidite systems (Navy and San Lucas turbidite systems: Normark, 1970), corresponding to mounded lobes (Galloway, 1998). In contrast, sheeted lobes (Galloway, 1998) would correspond to lobes with finer material. The elevation remains always small in comparison to their spatial extent: examples of lobe thicknesses are 10–40 m (Mississippi), 5–10 m (Zaire), 12 m (Rhône), 60 m (South Golo), 20–30 m (Var) and 10–25 m (Amazon: Jegou, 2008).

(4) They are mainly depositional areas (depositional lobes of Mutti and Ricci Lucchi, 1972) with a large lateral extent.

(5) Recent progress in acoustic imagery and multibeam bathymetry shows that channels are present over almost all the lobes, but channel size and organization depend on the sediment nature of the lobe (Bonnel, 2005). The lobes can be subdivided into a proximal, channelized part that forms the continuation of channel–levee systems (Fig. 2.36A), a non-channelized lobe and a much reduced distal part (lobe fringe) without channels and with a smooth surface (Fig. 2.36B). In the proximal part of the Monterey System (Gardner et al., 1996), the channel depth does not exceed 10 m (for a width of about 100 m), and decreases progressively downstream. In the Var and Zaire fans, the main channel can be several kilometres wide and 20 m deep in the upper part (Bonnel, 2005). Secondary channels are shallower (a few metres in depth), less large (a few hundreds of metres at maximum). Crude levees can appear close to the end of the channel. Channels are usually formed by a main branch in the axis of the channel of the channel–levee complex. Secondary channels are connected to this main branch (Bonnel, 2005). Most of the channels have a very short life span. Some of them could be formed by the front part of a flow and are filled by the tail of the same flow (Bonnel, 2005). During the phase of channel formation, the channel erodes and extends forward from its main branch. During channel regression, the channel erodes backwards (headwards) and also fills backwards (Bonnel, 2005). The presence of

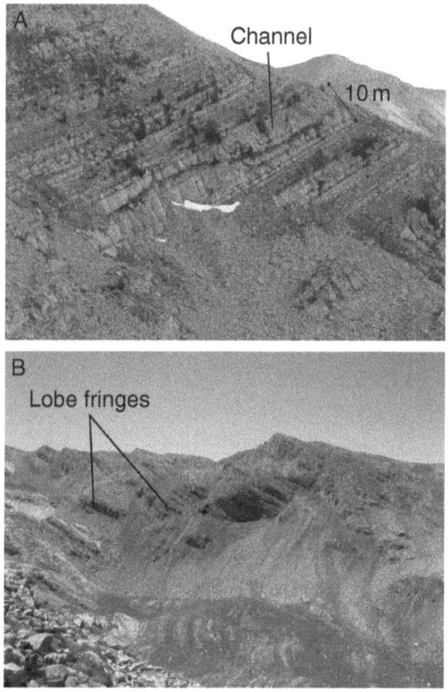

Figure 2.36 Channelized lobes: Trois Evéchés, Annot Sandstone, SE France. Cliffes at lobe fringe location are decametre-high. Lobe fringes: Lac du Lauzanier, Annot Sandstone, SE France. In both cases, a massive sand bar represents one lobe element. (A multi-colour version of this figure is on the included CD-ROM.)

channels suggests that sheet sands do not represent a sand volume as large as initially suspected by Mutti and Ricci Lucchi (1972). The non-channelized lobe occupies a small surface made of a sandy sheet resulting from the rapid deposition of flows reconcentrating at the mouth of lobe channels. In the Zaire system, lobes end with tongue-morphology bodies interpreted by Jegou (2008) as lobe precursors. The lobe fringe is not a part of the turbidite system, as it is an area covered by fine-grained deposits made of the interbedding of rare thin fine-grained turbidites and hemipelagites with the silt/clay ratio decreasing downflow as the turbidite/hemipelagite ratio decreases.

(6) Lobes are mainly composed of sand if the channel is not incised, because sorting of coarse material along the channel is efficient. White areas representing low-reflectivity areas on the acoustic-reflectivity map on the Var (Bonnel, 2005; Unterseh, 1999) or on the Celtic Turbidite System (Auffret et al., 2000) represent high-porosity sand deposits. The Amazon lobe contains 50–80% of sand (Bonnel, 2005; Jegou, 2008), mainly constituted of massive fine to very fine sand. If the channel is deep and incised, the clay content can be important (as in the Zaire:

Babonneau, 2002; Bonnel, 2005; Table 2.6). In this case, sand is restricted to the small lobe channels (fine to medium sand). The Zaire lobe body is mainly dominated by silty-mud turbidites. The high sand content in Amazon-type lobes is due to the progressive elutriation and spillover of the clay fraction over levees. However, it is not realistic to assume that sand travels directly from the source (river mouth or shelf edge) in one single flow event, at least in the case of mud-, silt- and mud/silt-dominated systems that extend longitudinally over hundreds of kilometres. Even if long-duration quasi-steady flows dominate, sand moves as bedload over a limited distance that depends on the nature of the flow (surge-like or quasi-steady). Any sand grain needs probably numerous passages of successive flows to reach its most distal settling.

(7) Confinement has a strong influence on lobe shape. The lobe shape is roughly rounded to ovoid if the spatial confinement is low. The length of the Zaire lobes is 20–40 km for a width of 10–40 km (Jegou, 2008). In contrast, lobes are elongated if lobe confinement is high. The length of the Amazon lobes is 21–83 km for a width of 6.5–25 km. These lobes preferentially settled in topographic lows located between old channel–levees forming topographic highs (Jegou et al., 2008). The Monterey lobe has a length of 130 km for a width of 45 km. The lobes of the Rhône Neofan are confined along the Pyreneo-Languedocian Ridge and the older channel–levee complexes of the Rhône turbidite system. Their lengths vary between 25 and 125 km and their widths vary between 5 and 35 km (Jegou, 2008).

(8) On the Zaire lobe and the Rhône Neofan, lobes are separated from the channel mouth by a scoured area (Bonnel, 2005). This suggests that the lobe would form first by a single event (lobe precursor of Jegou, 2008). Then, retrogressive erosion in the scouring zone would allow the connection of the lobe channel with the feeder channel.

Along slope aprons, lobes can form after a crude channel resulting from the pathway of mass-failure deposits (Reading and Richards, 1994). They thus correspond to ponded lobes of Nelson et al. (1985).

Lobes are built by the activity of concentrated flows that form at the end of the channel–levees. Just before the CLTZ, the slope is smooth, but it becomes steeper at the CLTZ. Flows that arrive there, converge and accelerate (accumulative flows). They are energetic and erosive, and form erosive channels. Flows remain erosive in the proximal part of the lobe. However, because of the convex-up shape of the lobe, the flows begin to spill over, inducing the migration of the lobe. In the distal part of the lobe (fringe), flows are divergent. They decelerate and their competence decreases (depletive flows). Flows are therefore essentially depositional in this part.

Because of the particle sorting in the channel–levee complex, lobes were early recognized as constituted by coarse-grained deposits (Normark, 1970).

In the Mississippi (Nelson et al., 1992) and Monterey fans (Gardner et al., 1996), two sedimentary facies dominate: silty clay beds alternating with silty beds generating high-backscatter bodies (Gardner et al., 1996), and metre-thick sand beds corresponding to low-backscatter areas.

Lobe deposits have been described by Bonnel (2005) as a continuum with the following two end-members (Fig. 2.35).

(1) Large mud-dominated systems (Zaire type). In the upper and middle parts of the channel axis, deposits are made of massive sand and debrites with mud clasts derived from eroded channel sides by the flow before entering the lobe. Sequences with inverse grading generated by hyper-concentrated laminar flow are frequent. The rest of the lobe is composed of fine sediments deposited by turbulent flows spilling over the channel sides; the spilling cloud forms terraces on the channel side in the proximal part of the lobe and fills the abandoned channels. Although the sand fraction dominates in the deposits, the mud fraction can occasionally be important. In these systems, the lobe passes through different stages: crevasse-splay lobe, avulsion lobe and terminal lobe.

(2) Sand-dominated systems (Var type). In the upper part of the channel axis, concentrated flows deposit amalgamated beds of massive sand with mud clasts derived from the erosion by the flow of the channel before entering the lobe. In the distal part of the channel and in secondary channels, fining-up sequences exist above massive sand layers, indicating a progressive dilution of the flow and development of turbulence downstream. However, even in small "sandy" systems such as the Navy Fan (Piper and Normark, 1983) and the Hueneme Fan (Piper et al., 1999b) massive sand beds always alternate with clayey levels. In small sandy system or in system supplied by a carbonate shelf, no canyon or channel–levee complex exists and the lobe extends directly at the toe of the source area (ramp-flat lobe).

Gervais (2002), Gervais et al. (2004) and Deptuck et al. (2008) showed that the confined lobes of the eastern Corsica margin do not consist of massive sheet sands (Fig. 2.37) but have a complex nested internal architecture. The lobes usually lie on an erosion surface with numerous scours corresponding to the progradation over the CLTZ. The hierarchy has been built on the same ordering as the architectural-element nested hierarchy (see Section 3.2).

What is usually called a "lobe" corresponds to an architectural element with a length and width of the order of 10–20 km and a thickness of several metres (Fig. 2.37; Prelat et al., 2009). It is composed of lobe elements corresponding to lobes units and sub-units of Gervais (2002) and Gervais et al. (2006). Lobe elements have a length and width of kilometre scale and a thickness of 1–2 m. The lobe elements are made of lobe beds, each corresponding to individual depositional beds (scale of depositional process).

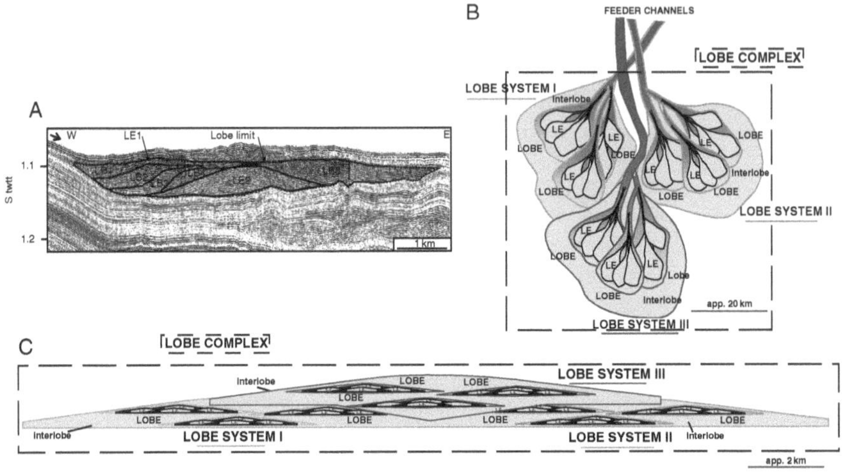

Figure 2.37 Terminology used for lobe geometry. Lobes correspond to nested depositional bodies from lobe complex (basin scale) to lobe bed (depositional process). Modified from Prelat et al. (2009), and Mulder and Etienne (2010). (A) Structure of the South Golo lobe, E. Corsica (Gervais, 2002; Gervais et al., 2005). The lobe is composed of three elements (LE1, LE2 and LE3). LE1 is composed of four lobe beds (LB1, LB2, LB3 and LB4). Reproduced with permission from Elsevier. (B) Plane view of lobe hierarchy in a lobe complex. LE: lobe elements. (C) Transverse view of a lobe complex. (A multi-colour version of this figure is on the included CD-ROM.)

Lobes related to a joint feeder channel form a lobe system with length and width of several tens of kilometres and a thickness of tens of metres. The entire lobe complex consists of the stack of lobe systems (Fig. 2.37). They are perennial, representing a several million year scale.

Stow and Mayall (2000) keep only two items: lobes and clustered lobes, corresponding to the lobe system. The internal structures within lobes could perturb the lateral and vertical connectivity and the oil-reservoir potential of lobes.

In the proximal part of the lobe, lobe elements prograde by lateral and longitudinal migration. In the distal part, the elements mainly retrograde. A similar nested architecture is described for the Zaire by Babonneau et al. (2002) and Droz et al. (2003). Lobe elements migrate by topographic compensation (Mutti and Sonnino, 1981).

Both because lobe channels are very shallow and crude levees are made of coarse material, avulsion of lobe channels is frequent. Lobe elements shift rapidly by avulsion and fill nearby topographic lows, particularly in sand-dominated turbidite systems (Amazon: Jegou, 2008). The channel network is more perennial in mud-dominated turbidite systems (Zaire; Fig. 2.35B).

It appears that the lobe activity is very important for a satisfactory explanation of the extension of recent turbidite systems along passive margins. To extend over hundreds of kilometres, a system must prograde intensively. This progradation acts particularly in the very distal part of the system, that is, in the lobes. The following two end-members for progradation can be distinguished.

(1) In a system without topographic confinement (Zaire), the original lobe deposit fills a topographic depression located at the feeder channel mouth. When the depression is filled, avulsion of the main feeder channel takes place, and the channel progrades rapidly along the margin of the original lobe (lobe evitment or avoidance of Turakiewicz, 2004). The timespan between two lobe avulsions is long, because the space to fill is large. Because modern lobes are not covered by more recent gravity deposits, they can be observed easily on the seafloor (Fig. 2.34).
(2) In a confined system (Var, South Golo in E. Corsica), lobes still prograde by topographic compensation, but avulsion and progradation are very fast. Lobes pile up, forming a lobe carpet above which the channel–levee complex progrades. The channel/levee deposits quickly bury the lobes and only the most recent lobe is visible (Bonnel, 2005).

The lobe progradation rate varies with time and space. Jegou, (2008) showed that a lobe initially forms by the deposition of a single event occurring after avulsion and forming a lobe precursor. Just after avulsion, lobes prograde and aggrade rapidly because they are located in the most proximal area of their lifespan and are confined between ancient levee deposits. This rapid progradation/aggradation leads to a flattening of the channel that may later be incised again (Bonnel, 2005). More distally, the progradation rate decreases and distal lobes show an alternation of progradation and retrogradation phases (Jegou, 2008), depending on the supply of sediment. For example, the Rhône Neofan shows the four oldest lobe elements with a clearly prograding trend, and the seven youngest lobe elements with a clear retrograding trend (Jegou, 2008). This progradation of lobe produces coarsening and thickening-upward sequences (Mutti, 1977).

3.4. Sequence stratigraphy in deep-sea turbidite systems and its controlling factors

The stratigraphy of deep-sea turbidite systems is organized in nested sequences of different orders. Their succession in superimposed sedimentary units is the result of cyclic control under the influence of forcing factors.

In a basin, both facies distribution over time and bed architecture are influenced by basin physiography, sediment supply and relative sea-level changes. The relative sea level, also called "total accommodation", is measured from a reference point located on the basin substratum before the sedimentation begins (Posamentier and Vail, 1988). It represents the volume between the substratum of the basin (that can move downward because of subsidence, or upward because of uplift) and the absolute sea level (eustasy). It is also the water depth (bathymetry or "unfilled accommodation" or "volume available for sedimentation") plus the thickness of the accumulated sediment ("effective accommodation").

Whether the accommodation space becomes filled depends on the sedimentation rate and the sediment supply. These parameters are highly variable, as they depend on tectonic movements on the continent, activity of oceanic currents, river load, climate (rainfall, temperature), morphology of the drainage basin (size, maximum elevation), nature of the bedrock in the drainage basin and extent of the vegetation cover (Milliman and Syvitski, 1992; Schumm, 1963). The balance between the sedimentation rate (*S*) and the relative sea level or total accommodation (*A*) determines the stratigraphic architecture and the spatial distribution of the deposits (Jervey, 1988; Posamentier and Allen, 1999). According to Homewood et al. (2000),

if $A/S > 1$, the architecture is retrograding;
if $A/S = 1$, the architecture is only aggrading (vertical piling);
if $1 > A/S > 0$, there is creation of available space, and the architecture is prograding and aggrading;
if $A/S = 0$, the available space is constant and the architecture is only prograding;
if $A/S < 0$, there is a rapid downward shift as a response to available space destruction.

The important boundary between a progradation/retrogradation sequence is the maximum flooding surface (MFS) corresponding to the deepest sedimentary facies in the basin (condensed interval). It is particularly easy to recognize in outcrops, sediment cores and well logs (deepest depositional environment). In this framework, a rise of the absolute sea level can lead either to retrogradation or to progradation, depending on whether the accommodation is larger or smaller than the sediment load, respectively. Similarly, a decrease in accommodation combined with termination of the sediment supply leads to a deepening of the sedimentary depositional environment.

The A/S concept of Homewood et al. (2000) is suitable for shallow environments in which subtle changes in relative sea level are well recorded in sedimentary facies. However, in the deep-sea environment (depth of several kilometres), relative sea-level changes (a few hundreds of metres maximally) have a small impact, and the progradation/retrogradation trends are mainly related to sediment load. Consequently, an analysis based on the equilibrium-profile concept such as developed by Ferry (2004) and Ferry et al. (2005) is more appropriated.

The original model developed by Exxon based on seismic data (Vail et al., 1987) had the objective to build a stratigraphic framework at a global scale by defining time lines that could be used to correlate basin infillings worldwide. This model defined the base concepts of sequence stratigraphy but suffered of too many simplifications, such as a presumed constant subsidence and sedimentation rate, thus overemphasizing the impact of eustasy on sequence control and leading to a dogmatic interpretation by the early users of the model.

Mutti (1985) proposed a model (Fig. 2.38E) that describes the stratigraphic evolution of deep-sea turbidite systems under the control of sediment supply. He demonstrated that a cycle origin higher than of the third order is tectonic. Using the example of the Hecho Group (Eocene, N Spain), he identified three types of turbidite systems corresponding to successive stages of evolution in response to relative sea-level fluctuations: (1) type I represents systems with high transport capacity (Fig. 2.19A); large failures located on the shelf generate channelized flows that supply thick sand-rich detached lobes in the distal part of the system; these lobes are separated from the source by a zone of bypass or erosion; (2) type II is made of amalgamated channels and sandy lobes that correspond to systems of low transport capacity (Fig. 2.19B) and sand-rich systems on active margins (Mutti, 1985); this system has much in common with the model of Normark (1978); type III corresponds to large modern mud-rich fans with well-developed channel and levees, but without lobes (Fig. 2.19C). Sand is concentrated in the proximal par of channels.

Mitchum et al. (1977) defined a sequence as a succession of genetically linked strata bounded at both base and top by an unconformity. Vail and Mitchum (1977) related this sequence to a regression/transgression cycle. The sequence boundaries correspond to the maximum fall of the sea level and the maximum slow down of the sea-level rise. Intrasequence unconformities can form during relative sea-level rise: transgression and flooding surfaces (Posamentier and Allen, 1999). These unconformities are easier to recognize on seismic profiles than the MFS.

A complete sequence is divided into genetic units called "system tracts" (ST) (Emery and Myers, 1996): highstand ST, lowstand ST, transgression interval and shelf-margin wedge. In this model, which is based on the hypothesis that subsidence and sediment load are small compared to eustasy, and correspond to a sedimentary cycle in the range of a million years, deep-sea turbidite systems develop during lowstand STs. The development of the system is divided in the following three periods, which have been identified by Bouma et al. (1989), and Posamentier and Vail (1988) for the Mississippi turbidite system (Fig. 2.38A–D): the basin-floor fan (sea-level lowstand), the slope fan and the lowstand prograding wedge (early sea-level rise).

(1) A deep-sea turbidite system with a siliciclastic supply grows in the basin during the beginning and the maximum fall of sea level (Fig. 2.38A). During the beginning of sea-level fall, the turbidite system forms

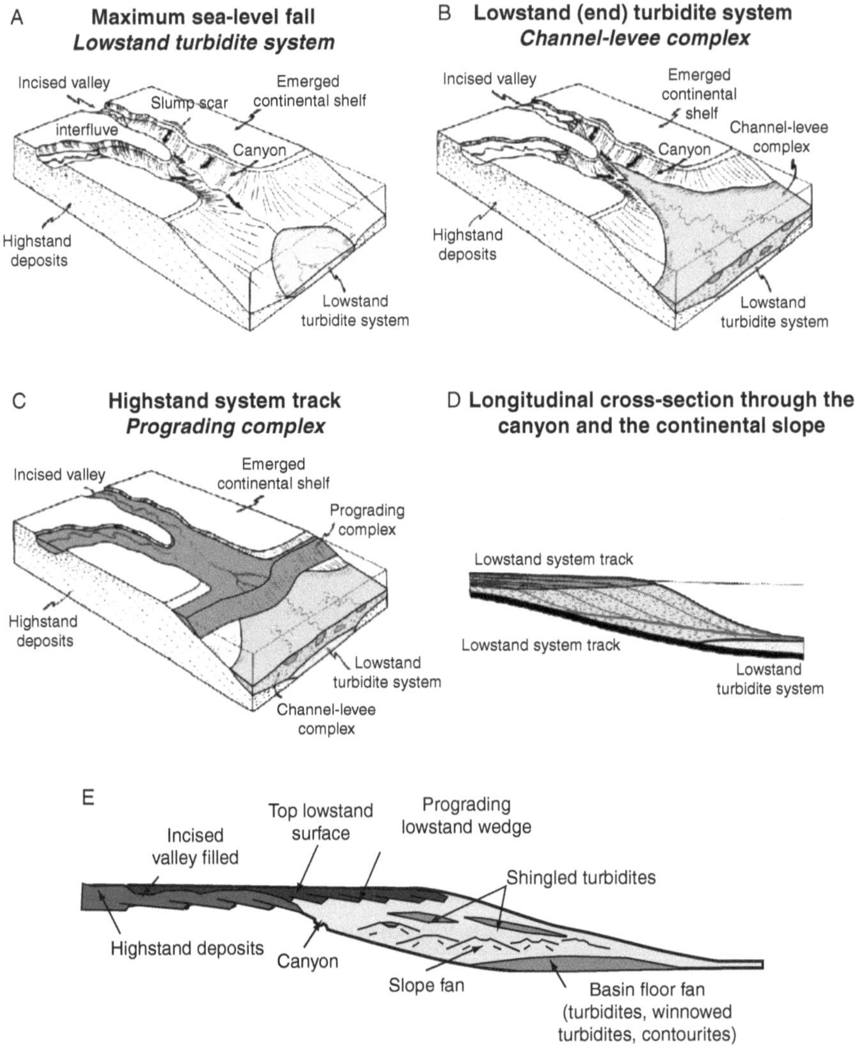

Figure 2.38 Comparison of conceptual sequence stratigraphy models. (A-D) Posamentier and Vail (1988) model for sequence stratigraphy. (A) Fall of sea level: maximum turbidite-system development. (B) Sea-level lowstand: muddy channel-levee complex development. (C) Sea-level rise: formation of prograding complexes. (D) Cross-section to compare with E. (E) Mutti model for sequence stratigraphy (Mutti, 1985). Reproduced with kind permission of Springer Science and Business Media. (A multi-colour version of this figure is on the included CD-ROM.)

simultaneously with the erosion of the shelf due to incision of canyons and river valleys and slope failures. Erosion increases mainly because riverbeds tend to reach their equilibrium profile. Experimental models

(Schumm, 1993) confirm that incision occurs when the sea level drops below the shelf edge. Deposits that will form the base of the new system are formed over this major erosion surface. In the case of the Mississippi, they are constituted by MTCs in the upper fan and less concentrated mass-flow deposits in the middle fan. Whatever their stratigraphic position, MTCs completely reshape the system topography (Rouillard, 2010). Jegou (2008) also notes that the morphology of the lobe elements constituting the Rhône Neofan is defined by the morphology of the top of the underlying MTC (Western MTD). In the middle fan of the Rhône turbidite system, mass-flow deposits are formed by spreading turbidite deposits forming HARPs (Droz and Bellaiche, 1985); they result from the spreading of concentrated flows on early lobes. If sediment supply is channelized through a canyon, sedimentation occurs mainly in the deep basin, forming large lobate sand deposits (basin-floor fan of Posamentier and Vail, 1988). During the latter phase of sea-level fall, the erosion rate decreases on the continent and deep-sea sedimentation is dominated by fine-grained gravity deposits on the continental slope (slope fan of Posamentier and Vail, 1988).

(2) When the sea level stabilizes, before the lowstand, a mud-dominated turbidite system grows with well-developed channel and levees (Fig. 2.38B).

(3) When the sea level rises, the lowstand wedge (or complex) begins to prograde and fills incised valleys and canyons (Fig. 2.38C). Destabilization of the front of the prograding wedge generates shingled turbidites in the most distal part. The deep-sea system becomes disconnected from the continental system. When the sea level stabilizes at the maximum of the transgression, only hemipelagites accumulate, forming a condensed interval (MFS). Normark and Piper (2009) noted that the sea level defines which canyons remain active or have an increased activity, in addition to the control of sediment supply.

During the end of a sea-level rise and highstand, deep-sea systems are not active. Hemipelagic deposits drape the system and form a condensed horizon.

Brami et al. (2000) proposed the following alternative, more detailed model for the Orinoco turbidite system.

(1) Emplacement of MTCs and debrites during the sea-level fall in response to increased erosion of the continents.

(2) If the continental shelf is emerged, rivers connect to the drainage system at the shelf break and small sand-rich turbidite systems form. For the Nile system, Rouillard (2010) shows that, during sea-level fall, the rate of gravity flow on the slope intensifies progressively and simultaneously with the rapid progradation of the deltaic system on the median and

outer continental shelf. The delta progradation can increase during humid phases with increased sediment input at the Nile river mouth. The rate of gravity processes on the slope is also increased.

(3) Incision of deep canyons by fluvial systems. Small sandy channels form and supply lobe complexes to the continental slope.

(4) Development of a large mud-dominated channel–levee complex at the end of the sea-level fall. The morphology of the system is constrained by the surface morphology of MTDs on the Nile system (Rouillard, 2010).

(5) Sedimentation stops at the beginning of the sea-level rise. Sediments aggrade on the continental shelf and in riverbeds. On the Nile system (Rouillard, 2010), the shelf-break delta retrogrades and transgressive sand can be deposited. The delta can persist on the outer shelf if the climate is humid because sediment supply compensates for the effect of sea-level rise. This delta can supply sediment for mud-rich mass-flow processes on the slope. The deep-sea turbidite system can still be fed, and lobe and levee complexes can continue to aggrade along with a meandering channel. The stop of the combined activity of the shelf delta and deep-sea turbidite system occurs only at the end of the pluvial interval (Ducassou, 2006; Rouillard, 2010).

These developments show that for high-frequency cycles (frequency$<$400 ka or order $>$4), sediment supply and climatic variations are important for the stratigraphic model in the deep sea. An example is also provided by the Amazon Turbidite System. The activity of the last system and lobe construction were emphasized at the end of the Younger Dryas (11.3 ka cal. BP) because of a discharge event generated by the first melting of the Andean ice sheet (Maslin et al., 2000).

Most recent developments in sequence stratigraphy in carbonate-dominated environments (Razin, 2008) show that, for some turbidite systems settled on a carbonate ramp, the activity of the systems occurs when the biogenic productivity is maximum, that is, when the carbonate shelf is flooded (highstand).

3.4.1. Sedimentary cycles

Sequences are defined whatever the timescale is. Consequently, the sedimentary succession is made of nested sequences formed during timespans of different length (Fig. 2.39). These timespans can be roughly grouped in three categories (Table 2.7):

(1) timespans $> 5 \times 10^5$ years are controlled by global tectonics; they form cycles of orders 1–3;

(2) timespans in the range of 2×10^4–5×10^5 years are controlled by astronomic cycles; they form cycles of orders 4–6; they correspond to the channel story and channel phases of Navarre et al. (2002).

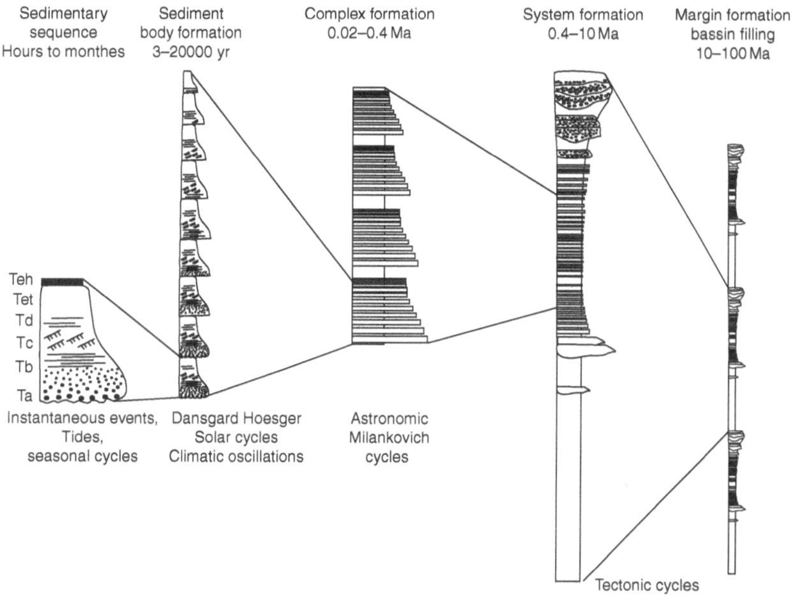

Figure 2.39 Concept of nested sequences and inferred forcing parameters.

(3) timespans in the range of a few years to 2×10^4 years are controlled by infra-Milankovitch processes with an impact on climate and autocyclic processes; they form cycles of orders >6. These infra-Milankovitch processes include (a) abrupt climatic changes, (b) Dansgaard Oeschger/Bond stadials and interstadials (~1500 years; Bond et al., 1997, 2001), solar cycles ($T = 2300$ years (Hallstatt cycle; Vasiliev and Dergachev, 2002), 210 years (Suess cycle; Braun et al., 2005), 88 years (Gleissberg cycle), 22 years (Hale cycle) and 11 years; Lantos, 1994), ENSO (El Nino Southern Oscillation) and NAO (North Atlantic Oscillation) (2–7 years). Autocyclic processes include, for example, seasonal cycles and tidal cycles. They correspond to the facies association of Navarre et al. (2002).

In terms of natural-resources exploitation, orders 2–3 correspond to basin-scale exploration, whereas orders 3–4 correspond to the prospect phase and orders >5 are at the scale of a reservoir model (Navarre et al., 2002).

Seismic stratigraphy studies cycles of orders 1–3. Field work and core studies indicate the impact of Milankovitch cycles on foredeep basins such as the Annot Sandstone (du Fornel et al., 2004).

Seismic stratigraphy has been applied mainly to ancient rocks. For Pleistocene and Holocene fans (Amazon, Mississippi, Indus and Astoria: Shanmugam et al., 1985; Rhône: Droz and Bellaiche, 1985; Nile: Bellaiche and Mear, 1995; Hudson and Navy: Nelson and Kulm, 1973),

Table 2.7 Controls and frequency of sedimentary cycles

Order	1	2	3	4	5	6	>6
Timespan (years)	100×10^6	$3–50 \times 10^6$	$0.5–3 \times 10^6$	95×10^3–410×10^3	41×10^3	19×10^3 and 23×10^3	$<10^3$
Spatial extent	Regional	Regional	Local	Global	Global (except mid-latitudes)	Global	
Origin	Continental fragmentation, crustal extension	Increase in accretion and subsidence rates	Regional tectonics: folding, faulting, magmatism, diapirism	Eccentricity	Obliquity	Precession	Infra-Milankovitch
Cause	Tectonics			Glacio-eustacy (astronomical parameters)			Autocyclic and regional climate
Impact	Basin shape and depth	Basin shape and depth	Basin shape and depth, sediment supply	Change in volume of sea water/change in sediment supply	Change in volume of sea water/change in sediment supply	Change in volume of sea water/change in sediment supply	Change in sediment supply
Research tool	Seismic studies			Field and core studies			

the activity is related to eustatic changes. These fans have been inactive during the two last highstands (present–11,000 BP and 120,000–127,000 BP). The sedimentation rate was low and only pelagic sediments accumulated. In contrast, the sedimentation rate was high during lowstands (e.g. 11,000–120,000 BP), essentially due to turbidite deposition. However, several deep-sea turbidite systems show an important activity during highstands, such as the Bengal (Weber et al., 1997), the Congo (Heezen et al., 1964), the Var (Migeon, 2000) and the Celtic Fan (Zaragosi et al., 2003). For the two first-mentioned fans, a substantial change in sediment supply took place between high and low sea-level stands. For the Amazon, the last channel–levee complex and the eight most recent systems have been active between 19,700 year cal. BP and 10,389 year cal. BP (Jegou, 2008; Jegou et al., 2008). The two last lobes have been active during 3000 years, that is, an average value of 1500 years. This is consistent with the activity of the last Zaire lobe (1000 years; Jegou, 2008). The Rhône Neofan was active during 2300–2500 years. These values give an estimate of the duration of a channel/levee, lobe activity. However, this duration seems highly variable, depending on the sediment supply.

The Amazon turbidite system became inactive at 11 ka cal. BP, when the littoral drift constituted by the North Brazil Coastal Current became active after the flooding of the continental shelf. Then, the drifts totally pirated the sediment load of the Amazon River and the deep-sea turbidite system switched off.

The recent lobe progradation during the Holocene in the Zaire turbidite system (Babonneau, 2002; Bonnel, 2005; Jegou, 2008) is consistent with the frequent cable failures (Droz et al., 1996). The most proximal last lobe formed between 10,000 ^{14}C BP and 4000 ^{14}C BP, the intermediate lobe formed at $\sim$4000 ^{14}C BP and the most distal lobe was still active during the past 1000 years (Jegou, 2008). The Rhône Neofan was active only during 2500 years (20.8–21 ka cal. year BP to 18.5 ka cal. BP), corresponding to the Last Glacial Maximum and sea-level lowstand and early rise. During 18.5 and 15 ka cal. year BP, sporadic turbidity currents have been triggered by dense-water cascading, reworking pro-deltaic sands (Dennielou et al., 2009). However, the disconnection between the Rhône deltaic complex on the internal continental shelf and the Petit Rhône canyon head does not allow a constant turbidite activity (Berné et al., 2007; Jouet, 2007).

Sediment for the Celtic fan was supplied during the last lowstand directly by the palaeo-Manche River (Lericolais, 1997), a river with an estimated discharge close to the present discharge of the McKenzie (Toucanne, 2008). The activity of the Manche Palaeoriver is correlated with the paroxysm of ice melting of the European ice sheet and a high rate of turbidite deposition in the Whittard channel and levee (between 18.3 and 17 ka; Toucanne, 2008). Identical high turbidite activity (one turbidite deposited every 1–10 years on the Whittard levee; Toucanne, 2008) is also observed for

MIS 12, 6 and 2. During these intervals, a large volume of coarse and poorly sorted material was deposited on the deep-sea system by large-scale gravity processes. However, because of their small residence time on the shelf, the turbidites deposited during this interval have a low sand/clay ratio. During lowstands and glacial maximum (20–18.3 ka), the turbidite activity was low (one turbidite deposited every 50–300 years on the Whittard levee: Toucanne, 2008) because the ice prevented the Manche Palaeoriver to settle. The turbidites accumulating in the Armorican deep-sea turbidite system were initiated from failures along the continental shelf and had a high sand/clay ratio. During highstand, the sedimentation rate is very low in deep-sea systems: one turbidite deposited every 500–2000 years on the Whittard levee, Armorican deep-sea turbidite system (Toucanne, 2008). The turbidity currents are generated by the transformation of failures occurring on the sand banks located on the shelf of the Channel, which are the remnants of the Channel River delta (Lericolais, 1997; Reynaud, 1996). However, because the material has been reworked during their long residence time on the shelf by hydrodynamic processes, the turbidites have a very high sand/clay ratio.

Normark and Piper (2009) noted that half of the major submarine canyons of the California continental borderland remained active during the Holocene highstand (e.g. Hueneme, Redondo, La Jolla, Newport) by pirating sand transported by littoral hydrodynamic cells. High sedimentation rates are even encountered in the Mugu and Dume fans in which sediment from the Santa Clara delta is transported southwards by the littoral drift (Normark and Piper, 2009).

Piper and Savoye (1993) suggested that, during the Late Pleistocene lowstands, gravity processes occurring on the Var deep-sea fan were essentially large, thick frictional and debris flows. During the Holocene, the gravity processes were only small turbulent surges and hyperpycnal flows (Mulder et al., 1998a,b). The activity during highstand at the Var River mouth can also be related to the absence of a continental shelf at the river mouth and to the direct connection of the river with the steep canyon.

The present review of factors controlling the deposition in deep-sea turbidite systems shows that identification of forcing parameters is complex (Table 2.8) because flow initiation and system evolution result from global factors interacting with regional and local factors, acting at different time-scales. These processes include the geodynamic context (basin shape and bathymetry, drainage-basin shape and elevation, margin shape and continental shelf width, subsidence, isostasy), structural context and neotectonics (earthquake frequency and magnitude, seepage of deep and shallow fluids, etc.), eustatic changes (storage location on the continent margin, existence of shelf processes, bathymetry, etc.) and climatic changes (erosion, dissolution process, vegetal cover on the continent, flood rate and magnitude, etc.).

Table 2.8 Controls on the morphology of deep-sea turbidite systems

Margin type	Active	Passive
Tectonic influence	Very high	Low
Eustatic influence	Variable	Very high
Shelf extent	Small	Important sediment storage
Basin area	Small (confined)	Large (not confined)
Extent of deep-sea turbidite system	Tens to hundreds of km^2	Thousands to millions of km^2
Transport distance	Tens of km	Thousands of km
Volume of particle load	Important (presence of nearshore reliefs)	Variable
Slope	Steep (5–10°)	Gentle (<3°)
Sand/clay ratio	Considerable (coarse sediment)	Low (fine-grained sediment)
Morphology	Rounded or according confinement	Elongate
Example	Navy, East Corse, Ligurian and Provence margin, Makran	Amazon, Mississippi, Zaire, Nile
Drainage basin	small (10^2–10^3 km^2)	Large (10^4–10^7 km^2)
Grain size	Coarse-grained	Fine-grained
Flow concentration	Considerable (hyperconcentrated, concentrated flows	Low (turbidity currents)

Some of these processes are cyclic (eustatic changes, climate); some are not (earthquakes). However, the clue to decipher the parameters controlling deep-sea gravitational processes triggering and evolution of deep-sea turbidite systems is time. Identification of the control parameters at a given timescale will require improvement of the dating methods and the development of a combination of methods to obtain a good stratigraphy (good resolution over a large timescale) in deep-sea turbidite systems; such methods should include biostratigraphy, chemiostratigraphy (stable and radiogenic isotopes) and lithostratigraphy (Heinrich layers, sapropels, tephras and other marker beds).

REFERENCES

Abreu, V., Sullivan, M., Pirmez, C., Mohrig, D., 2003. Lateral accretion package (LAPs): an important reservoir element in deep water sinuous channels. Mar. Pet. Geol. 20, 631–648.

Alexander, J., Morris, S., 1994. Observations on experimental, non-channelized, high-concentration turbidity currents and variations in deposits around obstacles. J. Sediment. Res. A 64, 899–909.

Alexander, J., Mulder, T., 2002. Experimental quasi-steady density currents. Mar. Geol. 186, 195–210.
Allen, J.R.L., 1982. Sedimentary structures: their characters and physical basis, in Developments in Sedimentology, v I and II. Elsevier, Amsterdam.
Alves, T., Cartwright, J.A., 2009. Volume balance of a submarine landslide in the Espírito Santo Basin, offshore Brazil: quantifying seafloor erosion, sediment accumulation and depletion. Earth Planet. Sci. Lett. 288 (3–4), 572–580.
Anastasakis, G., 2006. The anatomy and provenance of thick volcaniclastic flows in Cretan Basin, South Aegean Sea. Mar. Geol. 240, 113–135.
Assier-Rzadkiewicz, S., 1997. Etude numérique et expérimentale d'un glissement de sédiments le long d'une pente sous-marine et des vagues générées. Thèse de doctorat, Univ. Montpellier II.
Auffret, G., Zaragosi, S., Voisset, M., Droz, L., Loubrieu, B., Pelleau, P., et al., 2000. Premières observations de la morphologie et des processus sédimentaires récents de l'éventail Celtique. Oceanol. Acta 23 (1), 109–116.
Babonneau, N., 2002. Mode de fonctionnement d'un chenal turbiditique méandriforme: cas du système turbiditique actuel du Zaïre. Unpubl. Ph.D., Univ. Bordeaux 1, 308pp.
Babonneau, N., Savoye, B., Cremer, M., Klein, B., 2002. Morphology and architecture of the present channel system of Zaire Deep-Sea Fan. Mar. Pet. Geol. 19, 445–467.
Babonneau, N., Savoye, B., Cremer, M., Bez, M., 2004. Multiple terraces within the deep incised Zaire Valley (ZaiAngo Project): Are they confined levees. In: Lomas, S., Joseph, P. (Eds.), Confined Turbidite Systems, Geological Society, London, Special Publication, 222, 91–114.
Bagnold, R.A., 1954. Experiments on a gravity-free dispersion of large solid spheres in a Newtonian fluid under shear. Proc. R. Soc. Lond. A 225, 49–63.
Bagnold, R.A., 1962. Auto-suspension of transported sediment: turbidity currents. Proc. R. Soc. Lond. A 265, 315–319.
Baker, V., 1995. Surprise endings to catastrophism and controversy on the Columbia. GSA Today 15 (9), 169–173.
Bally, A.W., Snelson, S., 1980. Realms of subsidence. In: Miall, A.D. (Ed.), Fact and principles of word petroleum occurrence, Canadian Society of Petroleum Geologists Memoir 6, 9–94.
Baraza, J., Ercilla, G., 1996. Gas-charged sediments and large pockmark-like features on the Gulf of Cadiz slope (SW Spain). Mar. Pet. Geol. 13 (2), 253–261.
Barber, D.C., Dyke, A.S., Hillaire-Marcel, C., Jennings, A.E., Andrews, J.T., Kerwin, M. W., et al., 1999. Forcing of the cold event of 8, 200 years ago by catastrophic drainage of Laurentide lakes. Nature 400, 344–348.
Barnes, P.M., Normark, W.R., 1985. Diagnostic parameters for comparing modern submarine fans and ancient turbidite systems. In: Bouma, A., Normark, W., Barnes, N. (Eds.), Submarine Fans and Related Turbidite Systems. Springer-Verlag, New York, pp. 13–14.
Bates, C.C., 1953. Rational theory of delta formation. Bull. Am. Assoc. Pet. Geol. 37 (9), 2119–2162.
Bates, R.L., Jackson, J.A., 1980. Glossary of Geology, second ed. American Geologicate Institute, Falls Church, Virginia 751p.
Bayon, G., Loncke, L., Dupré, S., Caprais, J.-C., Ducassou, E., Duperron, S., et al., 2009. Multi-disciplinary investigation of fluid seepage on an unstable margin: the case of the Central Nile deep sea fan. Mar. Geol. 261 (1–4), 92–104.
Beaubouef, R.T., Friedmann, S.J., 2000. High resolution Seismic/Sequence Stratigraphic Framework for the Evolution of Pleistocene Intra Slope Basins, Western Gulf of Mexico: Depositional Models and Reservoir Analogs. In: Weimer, P., Slatt, R.M., Coleman, J., Rosen, N.C., Nelson, H., Bouma, A.H., Styzen, M.J., Lawrence, D.T. (Eds.), Deep-Water

Reservoirs of the World. GCSSEPM Foundation 20th Annual Research Conference, Dec. 3-6, Houston, Gulf Coast Section-SEPM Special Publication, pp. 40–60.
Beck, C., Schneider, J.-L., Cremer, M., MercierdeLépinay, B., Cagatay, N., Labeyrie, L., et al., 2003. Late Pleistocene major sedimentary reworking event (homogeneite) in Marmara Sea Central basin: preliminary results of giant piston-coring and high-resolution seismic reflection. In: EGS-AGU joint meeting 6–11 April, 2003, Nice, France.
Beghin, P., 1979. Etude des bouffées bidimensionnelles de densité en écoulement sur pente avec application aux avalanches de neige poudreuse. Thèse de doctorat, Université Scientifique et Médicale de Grenoble et Institut National Polytechnique de Grenoble, 120pp.
Bellaiche, G., Mear, Y., 1995. Morphostructure, growth pattern and tectonic control of the Rhone and Nile deep-sea fans: a comparison. Am. Assoc. Pet. Geol. Bull. 79 (2), 259–284.
Benjamin, T.B., 1968. Gravity currents and related phenomena. J. Fluid Mech. 31, 209–248.
Berné, S., Loubrieu, B., THE CALMAR Shipboard Party, 1999. Canyons et processus sédimentaires récents sur la marge occidentale du Golfe du Lion. Premiers résultats de la campagne Calmar. C. R. Acad. Sci. Paris 328, 471–477.
Berné, S., Carré, D., Loubrieu, B., Mazé, J.-P., Normand, A., 2001. Carte morpho-bathymétrique du Golfe du Lion. Ifremer-Région Languedoc-Roussillon, Brest.
Berné, S., Jouet, G., Bassetti, M.A., Dennielou, B., Taviani, M., 2007. Late glacial to pre-boreal sea-level rise recorded by the Rhône deltaic system (NW Mediterranean). Mar. Geol. 245 (1–4), 65–88.
Bjerrum, L., 1971. Subaqueous slope failures in Norwegian fjords. Norge Geotech. Bull. no. 88 pp. 1–8.
Blum, M.D., Törnqvist, T.E., 2000. Fluvial responses to climate and sea-level change: a review and look forward. Sedimentology 47 (Suppl. 1), 2–48.
Bøe, R., Rise, L., Ottesen, D., 1998. Elongated depressions on the southern slope of the Norwegian Trench (Skagerrak): morphology and evolution. Mar. Geol. 146 (1–4), 191–203.
Bond, G., et al., 1997. A pervasive millennial-scale cycle in North Atlantic Holocene and glacial climates. Science 278 (5341), 1257–1266.
Bond, G., et al., 2001. Persistent solar influence on North Atlantic climate during the Holocene. Science 294 (5549), 2130–2136.
Bonnel, C., 2005. Mise en place des lobes distaux dans les systèmes turbiditiques actuels: analyse compare des systèmes du Zaïre, Var et Rhône. Unpubl. Ph.D. thesis, Univ. Bordeaux 1, 293pp.
Bonnel, C., Dennielou, B., Droz, L., Mulder, T., Berné, S., 2005. Architecture and depositional pattern of the Rhone neofan and recent gravity activity in the Gulf of Lions (Western Mediterranean). Mar. Pet. Geol. 22, 827–843.
Bornhold, B.D., Ren, P., Prior, D.B., 1994. High-frequency turbidity currents in British Columbia fjords. Geo-Mar. Lett. 14, 238–243.
Bouma, A.H., 1962. Sedimentology of some Flysch deposits. A Graphic Approach to Facies Interpretation. Elsevier, Amsterdam 168pp.
Bouma, A.H., 1985. Introduction to submarine fans and related turbidite systems. In: Bouma, A., Normark, W., Barnes, N. (Eds.), Submarine Fans and Related Turbidite Systems. Springer-Verlag, New York, pp. 3–6.
Bouma, A.H., 2000. Coarse-grained and fine-grained turbidite systems as end member models: applicability and dangers. Mar. Pet. Geol. 17, 137–143.
Bouma, A., Normark, W., Barnes, N. (Eds.), 1985. Submarine Fans and Related Turbidite Systems. Springer-Verlag, New York 351pp.
Bouma, A.H., Coleman, J.M., Stelting, C.E., Kohl, B., 1989. Influence of relative sea level changes on the construction of the Mississippi fan. Geo-Mar. Lett. 9, 161–170.

Bourcart, J., Glangeaud, L., 1958. Perturbations sous-marines et courant de turbidité résultant du tremblement de terre d'Oléansville. Bull. d'information du Comité Central d'Océanographie et d'Etude des Côtes 10 (10).
Bourget, J., 2009. Les systèmes du golfe d'Oman et de la marge est-africaine: architecture, évolution des apports au quaternaire terminal et impact de la distribution sédimentaire sur les propriétés géoacoustiques des fonds. Unpubl. Ph.D. thesis, Univ. Bordeaux 1 no. 3915, 376pp.
Bowen, A.J., Normark, W.R., Piper, D.J.W., 1984. Modelling of turbidity currents on Navy Submarine Fan, California Continental Borderland. Sedimentology 31, 169–185.
Brami, T.R., Pirmez, C., Archie, C., Heeralal, S., Holman, K.L., 2000. Late Pleistocene deep-water stratigraphy and depositional processes, offshore Trinidad and Tobago. In: 20th Annual GCSEPM Foundation Bob F. Perkins Research Conference "Deep-Water reservoirs in the World", Houston GCSEPM Special Volume.
Braun, H., Christl, M., Rahmstorf, S., Ganopolski, A., Mangini, A., Kubatzki, C., et al., 2005. Possible solar origin of the 1, 470-year glacial climate cycle demonstrated in a coupled model. Nature 438 (7065), 208–211.
Broecker, W.S., Kennett, J.P., Flower, B.P., Teller, J.T., Trumbore, S., Bonami, G., et al., 1989. Routing of meltwater from the Laurentide Ice Sheet during the younger Dryas cold episode. Nature 341, 318–321.
Brunner, C.A., Normark, W.R., Zuffa, G.G., Serra, F., 1999. Deep-sea sedimentary record of the late Wisconsin cataclysmic floods from the Columbia River. Geology 27 (5), 463–466.
Bryn, P., Berg, K., Forsberg, C.F., Solheim, A., Kvalstad, T.J., 2005. Explaining the Storegga slide. Mar. Pet. Geol. 22 (1–2), 11–19.
Bugge, T., 1983. Submarine slides on the Norwegian continental margin, with special emphasis on the Storegga area. IKU Publ. no. 110, 152pp.
Bugge, T., Befring, S., Beldersson, R.H., Eidvin, T., Jansen, E., Kenyon, N.H., et al., 1987. A giant three-stage submarine slide off Norway. Geo-Mar. Lett. 7, 191–198.
Carey, S.N., Schneider, J.-L., 2011. Volcaniclastic processes and deposits in the deep-sea. In: Hüneke, H., Mulder, T. (Eds.), Deep-Sea Sediments. Developments in Sedimentology, Elsevier, Amsterdam, vol. 63, pp. 457–515.
Chapron, E., 1999. Contrôles climatique et sismo-tectonique de la sédimentation lacustre dans l'Avant-Pays Alpin (Lac du Bourget, Léman) durant le Quaternaire récent. Géologie Alpine. Mémoire Hors-Série no. 30, 265pp.
Chikita, K., 1991. Dynamic processes of sedimentation by river induced turbidity currents. II. Application of a two-dimensional, advective diffusion model. Jpn. Geomorphol. Union Trans 13-1, 1–18.
Cirac, P., Bourillet, J.-F., Griboulard, R., Normand, A., Mulder, T., The ITSAS Shipboard Scientific Party, 2001. Le canyon de Capbreton: nouvelles approches morphostructurales et morphosédimentaires Premiers résultats de la campagne Itsas. C. R. Acad. Sci. Paris 332, 447–455.
Cita, M.B., Beghi, C., Carmelenghi, A., Kastens, K.A., McKoy, F.W., Nosetto, A., et al., 1984. Turbidites and megaturbidites from the Herodotus Abyssal Plain (Eastern Mediterranean) unrelated to seismic events. Mar. Geol. 55, 79–101.
Cita, M.B., Carmelenghi, A., Rimoldi, B., 1996. Deep-sea tsunami deposits in the eastern Mediterranean: new evidence and depositional model. Sediment. Geol. 104, 155–173.
Clark, J.D., Pickering, K.T., 1996. Submarine Channels: Processes and Architecture. Vallis Press, London 231pp.
Clark, J.D., Kenyon, N.H., Pickering, K.T., 1992. Quantitative analysis of the geometry of submarine channels: implications for the classification of submarine fans. Geology 20, 633–636.

Clauzon, G., 1978. The Messinian Var Canyon (Provence, southern France). Paleogeographic implications. Mar. Geol. 27 (3/4), 231–246.

Coleman, J.M., Prior, D.B., Lindsay, J.F., 1983. Deltaic influences on shelf edge instability processes. In: Stanley, D.J., Moore, G.T. (Eds.), The Shelf break, Critical Interface on Continental Margins, 121–137.

Coleman, J.M., Roberts, H.H., Stone, G.W., 1998. Mississippi river delta: an overview. J. Coast. Res. 14 (3), 698–716.

Collot, J.-Y., Lewis, K., Lamarche, G., Lallemand, S., 2001. The giant Ruatoria debris avalanche on the northern Hikurangi margin, New Zeland: result of oblique seamount subduction. J. Geophys. Res. 106 (9), 19271–19297.

Corney, R.K.T., Peakall, J., Parsons, D.R., Elliot, L., Amos, K.J., Best, J.L., et al., 2006. The orientation of helical flow in curved channels. Sedimentology 53, 249–257.

Cramez, C., Jackson, M.P.A., 2000. Superposed deformation straddling the continental-oceanic transition in deep-water Angola. Mar. Pet. Geol. 17, 1095–1109.

Cremer, M., 1983. Approches sédimentologiques et géophysiques des accumulations turbiditiques. L'éventail profond du Cap-Ferret (Golfe de Gascogne), La série des grès d'Annot (Alpes de Hautes Provence). Unpubl. thesis, Univ. Bordeaux 1, 344pp.

Dadson, S., Hovius, N., Pegg, S., Dade, W.B., Horng, M.J., Chen, H., 2005. Hyperpycnal river flows from an active mountain belt. J. Geophys. Res. 110, F04016, doi: 10.1029/2004JF000244.

Daly, R.A., 1936. Origin of submarine canyons. Am. J. Sci. Ser. 5 31, 401–420.

D'Ambrosio, D., Di Gregorio, S., Iovine, G., 2003. Simulating debris flows through a hexagonal Cellular Automata model: Sciddica S3-hex. Nat. Hazards Earth Syst. Sci. 3, 545–559.

Damuth, J.E., Embley, R.W., 1981. Mass-transport processes on the Amazon Cone: Amer. Assoc. Petrol. Geol. Bull. 65, 629–643.

Damuth, J.E., Flood, R.D., 1985. Amazon Fan, Atlantic Ocean. In: Bouma, A., Normark, W., Barnes, N. (Eds.), Submarine Fans and Related Turbidite Systems. Springer-Verlag, New York, pp. 97–106.

Damuth, J.E., Hayes, D.E., 1977. Echo character of the East Brazilian continental margin and its relationship to sedimentary processes. Mar. Geol. 24 (2), 73–95.

Damuth, J.E., Flood, R.D., Kowsmann, R.O., Belderson, R.H., Gorini, M.A., 1988. Anatomy and growth pattern of amazon deep-sea fan as revealed by long-range side-scan sonar (GLORIA) and high-resolution seismic studies. Am. Assoc. Pet. Geol. Bull. 72 (8), 885–911.

Dan, G., Sultan, N., Savoye, B., Déverchère, J., Yelles, K., 2007. Quantifying the role of sandy-silty sediments in generating slope failures during seismic shaking: example from the Algerian margin. Int. J. Earth Sci. Eurodom, special issue.

De Vries Klein, G., 1982. Sandstone Depositional Models for Exploration for Fossil Fuels. Geological Sciences Series 209pp.

Dennielou, B., Jallet, L., Sultan, N., Jouet, G., Berné, S., Voisset, M., et al., 2009. Post-glacial persistence of turbiditic activity within the Rhône deep-sea turbidite system (Gulf of Lions, Western Mediterranean): linking the outer shelf and the basin sedimentary record. Mar. Geol. 257, 65–86.

Deptuck, M.E., Steffens, G.S., Barton, M., Pirmez, C., 2003. Architecture and evolution of upper fan channel-belts on the Niger Delta slope and in the Arabian Sea. Mar. Pet. Geol. 20 (6–8), 649–676.

Deptuck, M.E., Sylvester, Z., Pirmez, C., O'Byrne, C., 2007. Migration–aggradation history and 3-D seismic geomorphology of submarine channels in the Pleistocene Benin-major Canyon, western Niger Delta slope. Mar. Pet. Geol. 24, 406–433.

Deptuck, M.E., Piper, D.J.W., Savoye, B., Gervais, A., 2008. Dimensions and architecture of late Pleistocene submarine lobes off the northern margin of East Corsica. Sedimentology 926, 1–32.

Dimitrov, L.I., 2002. Mud volcanoes—the most important pathway for degassing deeply buried sediments. Earth Sci. Rev. 59 (1–4), 49–76.
Dott, R.H., 1963. Dynamics of subaqueous gravity depositional processes. Am. Assoc. Pet. Geol. Bull. 47, 104–128.
Droz, L., Bellaiche, G., 1985. Rhone Deep-Sea Fan: morphostructure and Growth Pattern. Am. Assoc. Pet. Geol. Bull. 69 (1), 460–479.
Droz, L., Rigaut, F., Cochonat, P., Tofani, R., 1996. Morphology and recent evolution of the Zaire turbidite system (Gulf of Guinea). Geol. Soc. Am. Bull. 108, 253–269.
Droz, L., Marsset, T., Savoye, B., Spy-Anderson, F.L., 2003. Architecture of an active mud-rich turbidite system: the Zaire Fan (Congo-Angola margin, southeast Atlantic): results from ZaiAngo 1 and 2 cruises. Am. Assoc. Pet. Geol. Bull. 87 (7), 1145–1168.
Du Fornel, E., Joseph, P., Desaubliaux, G., Eschard, R., Guillocheau, F., Lerat, O., et al., 2004. The southern Grès d'Annot outcrops (French Alps): an attempt at regional correlation. In: Joseph, P., Lomas, S. (Eds.), Deep-Water Sedimentation in the Alpine Foreland Basin of the SE France: New Perspectives on the Grès d'Annot and Related Systems, Geological Society, Special Publication, 221, 137–160.
Ducassou, E., 2006. Evolution du système turbiditique profond du Nil au cours du Quaternaire récent. Unpubl. Ph.D. thesis, Université Bordeaux 1, no. 3263, 336pp.
Ducassou, E., Mulder, T., Migeon, S., Gonthier, E., Murat, A., Capotondi, L., et al., 2008. Nile floods recorded in deep Mediterranean sediments. Quatern. Res. 70, 382–391.
Ducassou, E., Migeon, S., Capotondi, L., Mascle, J., 2010. Les coulées de débris dans un système turbiditique profond. Environnement de dépôt, origine et mise en place. 23ème Réunion des Sciences de la Terre, Bordeaux, October 25-29th, Abstract Book, p. 103.
Dupré, S., Woodside, J., Foucher, J.-P., de Lange, G., Mascle, J., Boetius, A., et al., 2007. Seafloor geological studies above active gas chimneys off Egypt (Central Nile Deep Sea Fan). Deep Sea Res. I 54 (7), 1146–1172.
Duringer, P., Paicheler, J.-C., Schneider, J.-L., 1991. Un courant d'eau peut-il générer des turbidites Résultats d'expérimentations analogiques. Mar. Geol. 99, 231–246.
Einarsson, P., Brandsdóttir, B., Tumi Gudmunsson, M., Björnsson, H., Grönvold, K., Sigmundsson, F., 1997. Center of the Iceland hotspot experiences volcanic unrest. EOS Transactions Am. Geophysical Union 78 (35), 369–375.
Embley, R.W., 1976. New evidence for occurrence of debris flow deposits in the deep sea. Geology 4, 371–374.
Emery, D., Myers, K.J., 1996. Sequence Stratigraphy. Blackwell Science ed, London 297pp.
Farre, J.A., Ryan, W.B., 1985. Three-dimensional view of erosional scars on the U.S. Mid-Atlantic continental margin. Am. Assoc. Pet. Geol. Bull. 69, 923–932.
Faugères, J.-C., Gayet, J., Gonthier, E., 1989. Microphysiographie des dépôts quaternaires dans le détroit de Makassar (océan Indien). Opposition entre une marge stable (Bornéo, Kalimantan) et une marge active (Célèbes, Sulawesi). Bull. Soc. Géol. Fr. 8 (V-4), 807–818.
Faugères, J.-C., Gonthier, E., Bobier, C., Griboulard, R., 1997. Tectonic control on sedimentary processes in the southern termination of the Barbados prism. Mar. Geol. 140, 117–140.
Faugères, J.-C., Mulder, T., 2011. Contour currents and contourite drifts. In: Hüneke, H., Mulder, T. (Eds.), Deep-Sea Sediments. Developments in Sedimentology, Elsevier, Amsterdam, vol. 63, pp. 149–214.
Ferry, J.-N., 2004. Développement et géométrie des corps sableux de pente, de glacis et du bassin profond, sur le prisme sédimentaire progradant (Oligocène Pliocène) de la marge ouest africaine entre le Sud-Gabon et l'Angola. Unpubl. Ph.D. thesis, Univ. Bordeaux 1, 482pp.
Ferry, J.-N., Mulder, T., Parize, O., Raillard, S., 2005. Equilibrium profile in deep-water turbidite systems; effects of local physiographic changes on deposits geometries and

nature of sedimentary process. In: Hodgson, D.M., Flint, S.S. (Eds.), Submarine Slope Systems. Processes and Products, Geol. Soc., London, Special Publication, 244, 181–193.
Fisher, R.V., 1983. Flow transformations in sediment gravity flows. Geology 11, 273–274.
Flood, R.D., 1988. A lee wave model for deep-sea mudwave activity. Deep-Sea Res. 35, 973–983.
Flood, R.D., Damuth, J.E., 1987. Quantitative characteristics of sinuous distributary channels on the Amazon deep-sea fan. Geol. Soc. Am. Bull. 98, 728–738.
Flood, R.D., Manley, P.L., Kowsmann, R.O., Appi, C.J., Pirmez, C., 1991. Seismic facies and late Quaternary growth of Amazon submarine fan. In: Weimer, P., Link, M.H. (Eds.), Seismic Facies and Sedimentary Processes of Modern and Ancient Submarine Fans and turbidite systems. Frontiers in sedimentary geology. Springer-Verlag, New York Inc., 415–433.
Forel, F.A., 1885. Les ravins sous-lacustres des fleuves glaciaires. C. R. Acad. Sci. Paris 101 (16), 725–728.
Forel, F.A., 1892. Le Léman: Monographie Limnologique 1. F. Rouge, Lausanne 543p.
Frey-Martinez, J., Cartwright, J., Hall, B., 2005. 3D seismic interpretation of slump complexes: examples from the continental margin of Israel. Basin Res. 17, 83–108.
Galloway, W.E., 1998. Siliciclastic slope and base-of-slope depositional systems: component facies, stratigraphic architecture, and classification. Am. Assoc. Pet. Geol. Bull. 82, 569–595.
Garcia, M.H., 1994. Depositional turbidity currents laden with poorly sorted sediment. J. Hydraul. Eng. 120, 1240–1263.
Garcia, M.H., Parker, G., 1989. Experiments on hydraulic jumps in turbidity currents near a canyon-fan transition. Science 245, 393–396.
Gardner, T.W., 1983. Experimental study of knickpoint and longitudinal profile evolution in cohesive, homogeneous material. Geol. Soc. Am. Bull. 94, 664–672.
Gardner, J.V., Bohannon, R.G., Field, M.E., Masson, D.G., 1996. The morphology, processes, and evolution of Monterey Fan: a revisit. In: Gardner, J.V., Field, M.E., Twichell, D.G. (Eds.), Geology of the United States' Seafloor. The View from GLORIA. Cambridge University Press, Cambridge, UK, pp. 193–220.
Garlan, T., 2004. Apports de la modélisation dans l'étude de la sédimentation marine récente. Mémoire d'Habilitation à Diriger les Recherches, Université des Sciences et Techniques de Lille 158pp.
Garziglia, S., Migeon, S., Ducassou, E., Loncke, L., Mascle, J., 2008. Mass-transport deposits on the Rosetta province (NW Nile deep-sea turbidite system, Egyptian margin): characteristics, distribution, and potential causal processes. Mar. Geol. 250 (3–4), 180–198.
Gaudin, M., 2006. Processus et enregistrements sédimentaires dans les canyons sous-marins Bourcart et de Capbreton Durant le dernier cycle clymatique. Unpubl. Ph.D. thesis, Univ. Bordeaux 1, no. 3122, 296pp.
Gaudin, M., Berné, S., Jouanneau, J.-M., Palanques, A., Puig, P., Mulder, T., et al., 2006a. Massive sand beds attributed to deposition by dense water cascades in the Bourcart canyon head, Gulf of Lions (northwestern Mediterranean Sea). Mar. Geol. 234, 111–128.
Gaudin, M., Mulder, T., Cirac, P., Berné, S., Imbert, P., 2006b. Past and present sedimentary activity in the Capbreton Canyon, southern Bay of Biscay. In: Mulder, T. (Ed.), Geo-Marine Letters, vol. 26(6), 331–345 Special Issue on "Deep-Sea Turbidite Systems on French Margins".
Gaullier, V., Antonini, E., Benkhelil, J., Got, H., 1998. Recent gravity-driven sedimentary bodies in the North-Balearic basin geometry and quantification. C. R. Acad. Sci. IIA, Earth Planet. Sci. 327 (10), 677–684.
Gee, M.J.R., Gawthorpe, R.L., Friedmann, J.S., 2005. Giant striations at the base of a submarine landslide. Mar. Geol. 214, 287–294.
Gee, M.J.R., Gawthorpe, R.L., Friedmann, J.S., 2006. Triggering and evolution of a giant submarine landslide, offshore Angola, revealed by 3D seismic stratigraphy and geomorphology. J. Sediment. Res. 76, 9–19.

Gennesseaux, M., Guibout, M., Lacombe, H., 1971. Enregistrement de courants de turbidité dans la vallée sous-marine du Var (Alpes-Maritimes). C. R. Acad. Sci. Paris D 273, 2456–2459.

Gennesseaux, M., Mauffret, A., Pautot, G., 1980. Les glissements sous-marins de la pente continentale niçoise et la rupture de câbles en mer Ligure (Méditerranée Occidentale). C. R. Acad. Sci. Paris D 290, 959–962.

Gérard, J., Dupont, G., Moron, J.M., Masse, P.L., Pourtoy, D., Bromley, R.G., 2000. From a refined chronostratigraphic framework to seismic interpretation: integration of quantitative palynology and micropaleontology, ichnology, sedimentology and seismic from the Lower Congo Tertiary basin (Angola). In: AAPG Annual Meeting and Exhibition. New Orleans, Louisiana Poster session.

Gervais, A., 2002. Analyse multi-échelles de la morphologie, de la géométrie et de l'architecture d'un système turbiditique sableux profond (système du Golo, marge est-Corse, mer Méditerranée). Unpubl. Ph.D. thesis, Univ. Bordeaux 1, 288pp.

Gervais, A., Savoye, B., Mulder, T., Piper, D.J.W., Cremer, M., Pichevin, L., 2004. Present morphology and depositional architecture of a sandy submarine system: The Golo turbidite system (Eastern margin of Corsica). In: Lomas, S., Joseph, P. (Eds.), Confined Turbidite Systems, Geological Society, London, Special Publication, 222, 59–89.

Gervais, A., Mulder, T., Savoye, B., 2006. Sandy modern turbidite lobes: a new insight from high resolution seismic data. Mar. Pet. Geol. 23, 485–502.

Giresse, P., Pauc, H., Deverchère, J., Savoye, B., The MARADJA Scientific Party, 2004. Gravity-induced transport on the Algerian margin. In: EGU 1st meeting Nice, France, 25–30, April, 2004.

Gomez, B., Smith, L.C., Magilligan, F.J., Mertes, L.A.K., Smith, N.D., 2000. Glacier outburst floods and outwash plain development: Skeidarársandur, Iceland. Terra Nova 12 (3), 126–131.

Gonthier, E., Faugères, J.-C., Stow, D.A.V., 1984. Contourite facies of the Faro drift, Gulf of Cadiz. In: Stow, D.A.V., Piper, D.J.W. (Eds.), Fine-Grained Sediments: Deep Water Processes and Facies, Geol. Soc. London, Spec. Publ. 15, 245–256.

Grönvold, K., Jóhannesson, H., 1984. Eruption in Grímsvötn 1983, course of events and chemical studies of the tephra. Jökul 34, 1–11.

Gudmunsson, M.T., Sigmundsson, F., Björnsson, H., 1997. Ice-volcano interaction of the 1996 Gjálp subglacial eruption, Vatnajökull, Iceland. Nature 389, 954–957.

Guibaudo, G., 1992. Subaqueous sediment gravity flow deposits: practical criteria for their field description and classification. Sedimentology 39, 423–454.

Guillocheau, F., Pautot, G., Auzende, J.-M., 1983. Les canyons du Var et du Paillon (marge des Alpes méridionales-Méditerranée occidentale): une origine quaternaire par glissement. C. R. Acad. Sci. Paris 296, 91–96.

Habib, P., 1994. Aspects géotechniques de l'accident du nouveau port de Nice. Rev. Fr. Géotech. 65, 3–15.

Haflidason, H., Sejrup, H.P., Bryn, P., Lien, R., 2001. The Storegga Slide; chronology and flow mechanism. In: EUG XI April 8–12, Journal of Conference Abstracts, Strasbourg p. 740.

Hagen, R.A., Bergersen, D.D., Moberly, R., Coulbourn, W.T., 1994. Morphology of a large meandering submarine canyon system on the Peru-Chile forearc. Mar. Geol. 119, 7–38.

Hampton, M.A., 1972. The role of subaqueous debris flow in generating turbidity currents. J. Sediment. Petrol. 42, 775–793.

Hampton, M.A., 1975. Competence of fine-grained debris flows. J. Sediment. Petrol. 45, 834–844.

Harrington, P.K., 1985. Formation of pockmarks by pore-water escape. Geo-Mar. Lett. 5, 193–197.

Haughton, P.D.W., 1994. Deposits of deflected and ponded turbidity currents, Sorbas basin, Southeast Spain. J. Sediment. Res. A 64, 233–246.
Hay, A.E., 1987. Turbidity currents and submarine channel formation in Rupert Inlet, British Columbia 2. The roles of continuous and surge-type flow. J. Geophys. Res. 92, 2883–2900.
Heezen, B.C., Ewing, M., 1952. Turbidity currents and submarine slumps, and the 1929 Grand Banks earthquake. Am. J. Sci. 250, 849–873.
Heezen, B.C., Menzies, R.J., Schneider, E.D., Ewing, W.M., Granelli, N.C.L., 1964. Congo submarine canyon. Am. Assoc. Pet. Geol. Bull. 48 (7), 1126–1149.
Hesse, R., Chough, S.K., 1980. The Northwest Atlantic Mid-Ocean Channel of the Labrador Sea: II. Deposition of parallel laminated levee-mud from the viscous sublayer of low-density turbidity currents. Sedimentology 27, 697–711.
Hesse, R., Khodabakhsh, S., 1998. Depositional facies of late Pleistocene Heinrich events in the Labrador Sea. Geology 26 (2), 103–106.
Hesse, R., Klauke, I., Ryan, W.B.F., Edwards, M.B., Piper, D.J.W., NAMOC Study Group, 1996. Imaging Laurentide ice sheet drainage into the deep sea: impact on sediments and bottom water. Geol. Soc. Am. Today 3–9.
Hesse, R., Schacht, U., 2011. Early diagenesis of deep-sea sediments. In: Hüneke, H., Mulder, T. (Eds.), Deep-Sea Sediments. Developments in Sedimentology, Elsevier, Amsterdam, vol. 63, pp. 557–714.
Hinze, J.O., 1960. On the hydrodynamics of turbidity currents. Geol. Mijnbouw 39, 18–25.
Hiscott, R.N., Hall, F.R., Pirmez, C., 1997. Turbidity current overspill from the Amazon Channel: texture of the silt/sand load, paleoflow from anysotropy of magnetic susceptibility, and implications for flow processes. In: Flood, R.D., Piper, D.J.W., Klaus, A., Peterson, L.C. (Eds.), Proc. ODP, Sci. Results, 155, 53–78, College Station, TX.
Homewood, P., Mauriaud, P., Lafont, F., 2000. Vade-Mecum de stratigraphie séquentielle, Elf EP éditions Mémoire, 25 81pp.
Hovland, M., 1981. Characteristics of pockmarks in the Norvegian Trench. Mar. Geol. 39, 103–117.
Hovland, M., 1983. Elongated depressions associated with pockmarks in the western slope of Norwegian Trench. Mar. Geol. 51, 35–36.
Hovland, M., Judd, A.G., 1988. Seabed Pockmarks and Seepages. Graham and Trotman, London, 293pp.
Hoyal, D.C., Bursik, M.I., Atkinson, J.F., 1999. Settling-driven convection: a mechanism of sedimentation from stratified fluids. J. Geophys. Res. 104, 7953–7966.
Hübsher, C., Spiess, V., Breitzke, M., Weber, M.E., 1997. The youngest channel-levee system of the Bengal Fan: results from digital sediment echosounder data. Mar. Geol. 141, 125–145.
Hughes-Clarke, J.E., Shor, A.N., Piper, D.J.W., Mayer, L.A., 1990. Large-scale current-induced erosion and deposition in the path of the 1929 Grand Banks turbidity current. Sedimentology 37, 613–629.
Hugot, A., 2000. Modélisation des écoulements gravitaires catastrophiques par une approche objet dynamique: erosion—transport—dépôt. Unpubl. Ph.D. thesis, Univ. Paris 6, 386pp.
Huguen, C., Foucher, J.-P., Mascle, J., Ondreas, H., Thouement, M., Gontharet, S., et al., 2009. Menes caldera, a highly active site of brine seepage in the Eastern Mediterranean sea:"in situ" observations from the NAUTINIL expedition (2003). Mar. Geol. 261 (1–4), 138–152.
Imbert, P., 2011. Industrial application of deep-sea sediments. In: Hüneke, H., Mulder, T. (Eds.), Deep-Sea Sediments. Developments in Sedimentology, Elsevier, Amsterdam, vol. 63, pp. 715–764.
Inman, D.L., 1970. Strong currents in submarine canyons. Abstr. Trans. Am. Geophys. Union 51, 319.

Inram, J., Parker, G., Pirmez, C., 1999. A nonlinear model of flow in meandering submarine and subaerial channels. J. Fluid Mech. 400, 295–331.
Iverson, R.M., 1997. Physics of debris flows. Rev. Geophys. 35, 245–296.
Jegou, I., 2008. Etude de la transition chenal/levee/lobe dans les systèmes turbiditiques récents. Application à l'éventail turbiditique de l'Amazone et au néofan du Petit-Rhône. Unpubl. Ph.D. thesis, Univ. Bretagne Occidentale 1, 351pp.
Jegou, I., Savoye, B., Primez, C., Droz, L., 2008. Channel-mouth lobe complex of the recent Amazon Fan: the missing piece. Mar. Geol. 252, 62–77.
Jermannaud, P., 2004. Etude sédimentologique et sismique des dépôts quaternaires de l'éventail turbiditique du Danube. Unpubl. Master thesis, Univ. Bordeaux 1, 31pp.
Jervey, M.T., 1988. Quantitative geological modelling of siliciclastic rock sequences and their seismic expression. In: Wilgus, C., Hastings, B., Posamentier, H., van Wagoner, J., Ross, C., Kendall, G. (Eds.), Sea-Level Changes: An Integrated Approach, SEPM Spec. Publ., 42, 48–69.
Johnson, A.M., 1970. Physical Processes in Geology. Freeman, Cooper and Co., San Francisco, 577pp.
Johnson, A.M., 1984. Debris flow. In: Brunden, D., Prior, D.B. (Eds.), Slope Instability. Wiley, Toronto, pp. 257–362.
Josenhans, H.W., King, L.H., Fader, G.B., 1978. A side-scan sonar mosaic of pockmarks on the Scotian Shelf. Can. J. Earth Sci. 15, 831–840.
Joseph, P., Lomas, S., 2004. Deep-water sedimentation in the alpine foreland basin of the SE France: new perspectives on the Grès d'Annot and related systems—an introduction. In: Joseph, P., Lomas, S. (Eds.), Deep-Water Sedimentation in the Alpine Foreland Basin of the SE France: New Perspectives on the Grès d'Annot and Related Systems, Geological Society, Special Publication, 221, 1–16.
Jouet, G., 2007. Enregistrements stratigraphiques des cycles climatiques et glacio-eustatiques du Quaternaire terminal. Modélisation de la marge continentale du golde du Lion. Unpubl. Ph.D. thesis, Univ. Bretagne Occidentale, vol. I, 443pp.
Judd, A., Davies, G., Wilson, J., Homes, R., Baron, G., Bryden, I., 1997. Contributions to atmospheric methane by natural seepages on the UK continental shelf. Mar. Geol. 137 (1–2), 165–189.
Kassem, A., Imran, J., 2001. Simulation of turbid underflow generated by the plunging of a river. Geology 29 (7), 655–658.
Kastens, K.A., Shor, A.N., 1985. Depositional processes of a meandering channel on Mississippi Fan. Am. Assoc. Pet. Geol. Bull. 69 (2), 190–202.
Keevil, G.M., Peakall, J., Best, J.L., Amos, K.J., 2007. The influence of scale, slope and channel geometry on the flow dynamics of submarine channels. Mar. Petrol. Geol. 24, 487–503.
Kennish, M.J., 1989. Practical Handbook of Marine Science. CRC Press Inc., Boca Raton, Florida, 710pp.
Kenyon, N.H., Amir, A., Cramp, A., 1995. Geometry of the younger sediment bodies of the Indus Fan. In: Pickering, K.T., Hiscott, R.N., Kenyon, N.H., Lucchi, Ricci, Smith, R. D.A. (Eds.), Atlas of Deep-Water Environments: Architectural Style in Turbidite Systems. Chapman et Hall, London, pp. 89–93.
Keulegan, G.H., 1938. Laws of turbulent flow in open channels. J. Res. Natl Bur. Stand. Research Paper 1151. Washington, District of Columbia. United States Department of Commerce.
King, L.H., Mac Lean, B., 1970. Pockmarks on the Scotian shelf. Geol. Soc. Am. Bull. 81, 3141–3148.
Klaucke, I., Hesse, R., 1996. Fluvial features in the deep-sea: new insight from the glacigenic submarine drainage system of the Northwest Atlantic Mid-Ocean Channel on the Labrador Sea. Sediment. Geol. 106, 223–234.

Klaucke, I., Hesse, R., Ryan, W.B.F., 1997. Flow parameters of turbidity currents in a low-sinuosity giant deep-sea channel. Sedimentology 44, 1093–1101.

Klaucke, I., Hesse, R., Ryan, W.B.F., 1998. Seismic stratigraphy of the Northwest Atlantic Mid-Ocean Channel: growth pattern of a mid-ocean channel-levee complex. Mar. Pet. Geol. 15, 575–585.

Klaucke, I., Masson, D.G., Kenyon, N.H., Gardner, J.V., 2004. Sedimentary processes of the lower Monterey Fan channel and channel-mouth lobe. Mar. Geol. 206 (1–4), 181–198.

Kneller, B.C., 1995. Beyond the turbidite paradigm: physical models for deposition of turbidites and their implications for reservoir prediction. In: Hartley, A.J., Prosser, D.J. (Eds.), Characterization of Deep Marine Clastic Systems, Geol. Soc. London, Spec. Publ. 94, 31–49.

Kneller, B.C., Branney, M.J., 1995. Sustained high-density turbidity currents and the deposition of thick massive beds. Sedimentology 42, 607–616.

Kneller, B.C., Buckee, C., 2000. The structure and fluid mechanics of turbidity currents: a review of some recent studies and their geological implications. Sedimentology 57 (Suppl. 1), 62–94.

Kneller, B.C., McCaffrey, W.D., 1999. Depositional effects of flow non-uniformity and stratification within turbidity currents approaching a bounding slope: deflection, reflection and facies variation. J. Sediment. Res. A Sediment. Petrol. Process. 69, 980–991.

Komar, P.D., 1969. The channelized flow of turbidity currents with application to Monterey deep sea fan channel. J. Geophys. Res. B 74, 4544–4558.

Komar, P.D., 1971. Hydraulic jumps in turbidity currents. Geol. Soc. Am. Bull. 82, 1477–1488.

Komar, P.D., 1977. Computer simulation of turbidity current flow and the study of deep-sea channels and fan sedimentation. In: Goldberg, E.D., McCave, I.N., O'Brien, J.J., Steele, J.H. (Eds.), The Sea, v.6, Marine Modelling. John Wiley, New York, pp. 603–621.

Kuenen, P.H., 1950. Turbidity currents of high density. In: Int. Geol. Congr., 18th, London, 1948, 44–52, rept. 8.

Kuenen, P.H., 1952. Estimated size of the Grand Banks turbidity current. Am. J. Sci. 250, 849–873.

Kuenen, P.H., 1967. Emplacement of flysch-type sand beds. Sedimentology 9, 203–243.

Kuenen, P.H., Migliorini, C.I., 1950. Turbidity currents as a cause of graded bedding. J. Geol. 58, 91–127.

Laberg, J.S., Vorren, T.O., 2000. The Trenadjupet Slide, offshore Norway—morphology, evacuation and triggering mechanisms. Mar. Geol. 171 (1–4), 95–114.

Lajeunesse, P., St-Onge, G., 2008. The subglacial origin of the Lake Agassiz-Ojibway final outburst flood. Nat. Geosci. 1, 184–188.

Lamb, M.P., Parsons, J.D., 2005. High-density suspensions formed under waves. J. Sediment. Res. 75, 386–397.

Lamb, M.P., Myrow, P.M., Lukens, C., Houck, K., Strauss, J., 2008. Deposits from wave-influenced turbidity currents: Pennsylvanian Minturn Formation, Colorado, USA. J. Sediment. Res. 78, 480–498.

Lanteaume, M., Beaudoin, B., Campredon, R., 1967. Figures sédimentaires du flysch 'Grès d'Annot' du synclinal de Peïra-Cava, Editions du CNRS. Paris, 97pp.

Lantos, P., 1994. Le Soleil. Que Sais-Je, first ed. Presses Universitaires de France, 108Bd Saint-Germain, Paris, 128pp.

Laval, A., 1988. Modélisation d'écoulements de type bouffée de densité. Application à l'interprétation des dépôts turbiditiques. Unpubl. Ph.D. thesis, Univ. Bordeaux I, 262pp.

Laval, A., Cremer, M., Beghin, P., Ravenne, C., 1988. Density surges: two-dimensional experiments. Sedimentology 35, 73–84.

Le Moigne, M., 1999. Compréhension des mécanismes de formation des pockmarks sur la pente du Golfe de Guinée. Unpubl. Master thesis, Univ. Sci. Tech Lille 1.

Leigh, N.S., Hartley, A., 1992. Mega-debris flow deposits from the Oligo-Miocene Pindos foreland basin, western mainland Greece: implications for transport mechanisms in ancient deep marine basins. Sedimentology 39, 1003–1012.

Leopold, L.B., Wolman, M.G., 1957. River channel pattern: braided, meander, and straight. In: US Geological Survey Professional Paper, 282-B, 39–84.

Leopold, L.B., Wolman, M.G., Miller, J.P., 1964. Fluvial Processes in Geomorphology. Dover Publications, Inc., New York, 522pp.

Lericolais, G., 1997. Evolution plio-quaternaire du fleuve Manche. Stratigraphie et géomorphologie d'une plate-forme continentale en régime périglaciaire. Unpubl. Ph.D. thesis, Univ. Sci. Tech Lille 1, 265pp.

Lindsay, J.F., 1968. The development of clast fabric in mudflows. J. Sediment. Res. 38 (4), 1242–1253.

Liu, C.S., Lundberg, N., Reed, D.L., Lu, H.Y., 1993. Morphological and seismic characteristics of the Kaoping submarine canyon. Mar. Geol. 111, 93–108.

Locat, J., Demers, D., 1988. Viscosity, yield stress, remolded strength, and liquidity index relationships for sensitive clays. Can. Geotech. J. 25, 799–806.

Locat, J., Lee, H.J., 2000. Submarine landslides: advances and challenges. In: Proceedings of the 8th International symposium on Landslides, Cardiff.

Lowe, D.R., 1979. Sediment gravity flows: their classification and some problems of application to natural flows and deposits. In: Doyle, N.J., Pilkey, O.H. (Eds.), Geology of Continental Slopes, SEPM Special Publication 27, 75–82, Tulsa.

Lowe, D.R., 1982. Sediment gravity flows: II. Depositional models with special reference to the deposits of high-density turbidity currents. J. Sediment. Petrol. 52, 279–297.

Lowe, D.R., Guy, M., 2000. Slurry-flow deposits in the Britannia Formation (Lower Cretaceous), North Sea: a new perspective on the turbidity current and debris flow problem. Sedimentology 47, 31–70.

Lüthi, S., 1980. Some new aspects of two-dimensional turbidity currents. Sedimentology 28, 97–105.

Malde, H.E., 1968. The catastrophic late Pleistocene Bonneville flood in the Snake River Plain, Idaho. Geol. Survey Prof. Paper 596U. S. Government printing Office, Washington 69pp.

Malde, H.E., Powers, H.A., 1962. Upper Cenozoic stratigraphy of the western Snake River plain, Idaho. Geol. Soc. Am. Bull. 73 (10), 1197–1219.

Maslin, M., Durham, E., Barns, S.J., Plazmn, E., Grootes, P., Greig, S.E.J., et al., 2000. Paleoreconstruction of the Amazon River freswater and sediment discharge using sediments recovered at site 942 on the Amazon Fan. J. Quat. Sci. 15 (4), 419–434.

Maslin, M., Vilela, C., Mikkelsen, N., Grootes, P., 2005. Causes of catastrophic sediment failures of the Amazon Fan. Quatern. Sci. Rev. 24, 2180–2193.

Masson, D.G., 1996. Catastrophic collapse of the volcanic island of Hierro 15 ka ago and the history of landslides in Canary Islands. Geology 24 (3), 231–234.

Masson, D.G., Kenyon, N.H., Gardner, J.V., Field, M.E., 1995. Monterey fan: channel and overbank morphology. In: Pickering, K.T., Hiscott, R.N., Kenyon, N.H., Ricci, Luchi, Smith, R.D.A. (Eds.), Atlas of Deep-Water Environments: Architectural Style in Turbidite Systems. Chapman et Hall, London, pp. 74–79.

Mavilla, N., 2000. Stratigraphie et analyse de faciès de la succession d'âge oligo-miocène du bassin tertiaire piémontais (Italie nord occidentale). Unpubl. Ph.D. thesis, Univ. Bordeaux 1, 225pp.

Maxworthy, T., 1999. The dynamics of sedimenting surface gravity currents. J. Fluid Mech. 392, 27–44.

Mc Cave, I.N., Jones, K.P.N., 1988. Deposition of ungraded muds from high-density non-turbulent turbidity currents. Nature 333, 250–252.
Mc Hargue, T., Webb, J.E., 1986. Internal geometry, seismic facies and petroleum potential of canyons and inner fan channels of the Indus submarine fan. Am. Assoc. Pet. Geol. Bull. 70, 161–180.
McCaffrey, W.D., Kneller, B.C., 2004. Scale effects of non-uniformity on deposition from turbidity currents with reference to the Gres d'Annot of SE France. In: Lomas, P.J., Lomas, S.A. (Eds.), International Workshop on Confined Turbidite Systems, 301–310 AAPG Bookstore, PO Box 979, Tulsa, OK 74101-0979, USA.
Meade, R.H., 1996. River-sediment inputs to major deltas. In: Milliman, J.D., Haq, B.U. (Eds.), Sea-Level Rise and Coastal Subsidence. Kluwer Academic Publishers, Dordrecht, pp. 63–85.
Miall, A.D., 1985. Architectural-element analysis: a new method of facies analysis applied to fluvial deposits. Earth Sci. Rev. 22, 261–308.
Middleton, G.V., 1966a. Experiments on density and turbidity currents: I. Motion of the head. Can. J. Earth Sci. 3, 523–546.
Middleton, G.V., 1966b. Experiments on density and turbidity currents: II. Uniform flow of density currents. Can. J. Earth Sci. 3, 627–637.
Middleton, G.V., 1966c. Small scales models of turbidity currents and the criterion for auto-suspension. J. Sediment. Petrol. 36, 202–208.
Middleton, G.V., 1967. Experiments on density and turbidity currents: III. Deposition of sediment. Can. J Earth Sci. 4, 475–505.
Middleton, G.V., Hampton, M.A., 1973. Sediment gravity flows: mechanics of flow and deposition. In: Middleton, G.V., Bouma, A.H. (Eds.), Turbidity and Deep Water Sedimentation. SEPM, Pacific Section, Short Course Lecture Notes, Anaheim, pp. 1–38.
Mienert, J., Posewang, J., Bauman, M., 1998. Geophysical signature of gas hydrates along the north-eastern Atlantic Margins: possible hydrate-bound margin instabilities and possible transfer of methane from oceanosphere to atmosphere. In: Henriet, J.-P., Mienert, J. (Eds.), Gas hydrates: Relevance to World Margin Stability and Climate Change, Geol. Soc. London, Spec. Publ. 137, 275–292.
Mienert, J., Buenz, S., Guidard, S., Vanneste, M., Berndt, C., 2005. Ocean bottom seismometer investigations in the Ormen Lange area offshore mid-Norway provide evidence for shallow gas layers in subsurface sediments. Mar. Pet. Geol. 22 (1–2), 11–19.
Migeon, S., 2000. Dunes Géantes et levées Sédimentaires en domaine marin profond: approches morphologique, sismique et sedimentologique. Thèse, Université Bordeaux I, 288pp.
Migeon, S., Savoye, B., Zanella, E., Mulder, T., Faugères, J.-C., Weber, O., 2001. Detailed seismic-reflection and sedimentary study of turbidite sediment waves on the Var sedimentary Ridge (SE France): significance for sediment transport and deposition and for the mechanisms of sediment-waves construction. Mar. Pet. Geol. 18, 179–208.
Mikkelsen, N., Maslin, M., Giraudeau, J., Showers, W., 1997. Biostratigraphy and sedimentation rates of the Amazon Fan. In: Flood, R.D., Piper, D.J.W., Klaus, A., Petersen, L.C. (Eds.), Proceedings of the Ocean Drilling Program, Scientific Results, 155, College Station TX, 577–594.
Milliman, J.D., Kao, S.-J, 2005. Hyperpycnal discharge of fluvial sediment to the ocean: impact of Super-Typhoon Herb (1996) on Taiwanese rivers. J. Geol. 113, 503–516.
Milliman, J.D., Syvitski, J.P.M., 1992. Geomorphic/tectonic control of sediment discharge to the ocean: the importance of small mountainous rivers. J. Geol. 100, 525–544.
Milliman, J.D., Lin, S.W., Kao, S.J., Liu, J.P., Liu, C.S., Chiu, J.K., et al., 2007. Short-term changes in seafloor character due to flood-derived hyperpycnal discharge: Typhoon Mindulle, Taiwan. Geology 35, 779–782.

Mitchell, N., 2005. Channelled erosion through a marine dump site of dredge spoils at the mouth of the Puyallup River, Washington State, USA. Mar. Geol. 220 (1–4), 131–151.
Mitchum, R.M., Vail, P.R., Thompson, S., 1977. Seismic stratigraphy and global changes of sea level, part 2: the depositional sequence as a basic unit for stratigraphic analysis. In: Payton, C.E. (Ed.), Seismic Stratigraphy—Application to Hydrocarbon Exploration, American Association of Petroleum Geologists Memoir 26, 53–62.
Mohrig, D., Whipple, K.X., Hondzo, M., Ellis, C., Parker, G., 1998. Hydroplaning of subaqueous debris flows. Geol. Soc. Am. Bull. 110 (3), 387–394.
Mohrig, D., Elverhoi, A., Parker, G., 1999. Experiments on the relative mobility of muddy subaqueous and subaerial debris flows, and their capacity to remobilize antecedent deposits. Mar. Geol. 154, 117–129.
Moore, J.G., Clague, D.A., Holcomb, R.T., Lipman, P.W., Normark, W.R., Torregan, M. E., 1989. Prodigious submarine landslides on the Hawaiian Ridge. J. Geophys. Res. 94, 17,465–17,484.
Moore, J.G., Normark, W.R., Holcomb, R.T., 1994. Giant Hawaiian landslides. Annu. Rev. Earth Planet. Sci. 22, 119–144.
Moore, J.G., Bryan, W.B., Beeson, M.H., Normark, W.R., 1995. Giants blocks in the south of Kona landslide, Hawaii. Geology 23, 125–128.
Morris, W.R., Kenyon, N.H., Limonov, A., Alexander, J., 1998. Downstream changes of large-scale bedforms in turbidites around the Valencia channel mouth, north-west Mediterranean: implications for paleoflow reconstruction. Sedimentology 45 (2), 365–377.
Mosher, D.C., Piper, D.J.W., Campbell, D.C., Jenner, K., 2004. Near surface geology and sediment-failure geohazards of the central Scotian Slope. Am. Assoc. Pet. Geol. Bull. 88, 703–723.
Mulder, T., Alexander, J., 2001a. The physical character of sedimentary density currents and their deposits. Sedimentology 48, 269–299.
Mulder, T., Alexander, J., 2001b. Abrupt change in slope causes variation in the deposit thickness of concentrated particle-driven density currents. Mar. Geol. 175, 221–235.
Mulder, T., Cochonat, P., 1996. Classification of offshore mass movements. J. Sediment. Res. 66, 43–57.
Mulder, T., Etienne, S., 2010. Lobes in deep-sea turbidite systems State of the art. In: Mulder, T. (Ed.), Sedimentary Geology 211, 75–80 Special Issue on Lobes in Deep-Sea Turbidite Systems.
Mulder, T., Syvitski, J.P.M., 1995. Turbidity currents generated at river mouths during exceptional discharges to the world oceans. J. Geol. 103, 285–299.
Mulder, T., Syvitski, J.P.M., 1996. Climatic and morphologic relationships of rivers: implications of sea level fluctuations on river loads. J. Geol. 104, 509–523.
Mulder, T., Savoye, B., Piper, D.J.W., Syvitski, J.P.M., 1998a. The Var submarine sedimentary system: understanding Holocene sediment delivery processes and their importance to the geological record. In: Stoker, M., Evans, M., Cramp, A. (Eds.), Geological Processes on Continental Margins: Sedimentation, Mass-Wasting and Stability, Geological Society, London, Special Publication, 129, 145–166.
Mulder, T., Syvitski, J.P.M., Skene, K.I., 1998b. Modelling of erosion and deposition by turbidity currents generated at river mouths. J. Sediment. Res. 68, 124–137.
Mulder, T., Migeon, S., Savoye, B., Faugères, J.-C., 2001a. Inversely-graded turbidite sequences in the deep Mediterranean. A record of deposits by flood-generated turbidity currents. Geo-Mar. Lett. 21 (2), 86–93.
Mulder, T., Migeon, S., Savoye, B., Jouanneau, J.-M., 2001b. Twentieth century floods recorded in the deep Mediterranean sediments. Geology G (11), 1011–1014.

Mulder, T., Weber, O., Anschutz, P., Jorissen, F.J., Jouanneau, J.-M., 2001c. A few months-old storm-generated turbidite deposited in the Capbreton Canyon (Bay of Biscay, S-W France). Geo-Mar. Lett. 21 (3), 149–156.

Mulder, T., Migeon, S., Savoye, B., Faugères, J.-C., 2002. Inversely-graded turbidite sequences in the deep Mediterranean. A record of deposits by flood-generated turbidity currents Reply. Geo-Mar. Lett. 22 (2), 112–120.

Mulder, T., Syvitski, J.P.M., Migeon, S., Faugères, J.-C., Savoye, B., 2003. Hyperpycnal turbidity currents: initiation, behaviour and related deposits. A review. In: Mutti, E., Steffens, G.S., Pirmez, C., Orlando, M., Roberts, D. (Eds.), Marine and Petroleum Geology, vol. 20/6–8, 861–882 Special issue: turbidites: models and problems.

Mulder, T., Cirac, P., Gaudin, M., Bourillet, J.-F., Tranier, J., Normand, A., et al., 2004. Understanding continent-ocean sediment transfer. EOS American Geophysical Union Transaction 257, 261–262.

Mulder, T., Zaragosi, S., Jouanneau, J.-M., Bellaiche, G., Guérinaud, S., Querneau, J., 2009a. Deposits related to the failure of Malpasset Dam in 1959. An analogous for hyperpycnal deposits from jökulhlaups. Mar. Geol. 260, 81–89.

Mulder, T., Zaragosi, S., Razin, P., Grelaud, K., Lanfumey, V., Bavoil, F., 2009b. A new conceptual model for the deposition process of homogenite: application to a cretaceous megaturbidite of the western Pyrenees (Basque region, SW France). Sediment. Geol. 222, 263–273.

Mutti, E., 1977. Distinctive thin-bedded turbidites facies and related depositional environments in the Eocene Hecho Group (South-central Pyrenees, Spain). Sedimentology 24, 107–131.

Mutti, E., 1979. Turbidites et cônes sous-marins profonds. In: Homewood, P. (Ed.), Sédimentation Détritique (Fluviatile, Littorale et Marine), Short Course: Fribourg, Institut de Geologie de 1'Universite de Fribourg, 353–419.

Mutti, E., 1985. Turbidite systems and their relations to depositional sequences. In: Zuffa, G. (Ed.), Provenance of Arenites, D. Reidel Publishing Co, Dordrecht, Holland, 65–93.

Mutti, E., 1992. Turbidite Sandstones. Agip, Instituto di Geologia, Università di Parma, San Donato Milanese 275pp.

Mutti, E., Ricci Lucchi, F., 1972. Le torbiditi dell'Appennino settentrionale: introductione all'analisi di facies. Mem. Soc. Geol. It. 11, 161–199.

Mutti, E., Normark, W.R., 1987. Comparing examples of modern and ancient turbidite systems: problems and concepts. In: Legget, J.K., Zuffa, G.G. (Eds.), Marine Clastic Sedimentology. Graham and Trotman, London, pp. 1–38.

Mutti, E., Normark, W.R., 1991. An integrated approach to the study of turbidite systems. In: Weimer, P., Link, M.H. (Eds.), Seismic Facies and Sedimentary Processes of Submarine Fans and Turbidite Systems. Springer-Verlag, New-York, pp. 75–106.

Mutti, E., Ricci Lucchi, F., 1975. Turbidite facies and facies association. In: Mutti, E., Parea, G.C., Ricci Lucchi, F., Sagri, M., Zanzucchi, G., Ghibaudo, G., Iaccarino, S. (Eds.), Example of Turbidite Facies Associations from Selected Formation of Northern Apennines, Congres I.A.S, Nice21–36.

Mutti, E., Sonnino, M., 1981. Compensation cycles: a diagnostic feature of turbidite sandstone lobes. In: Int. Ass. Sediment., 2nd Regional European Meeting, Bologna, Italy, 120–123.

Mutti, E., Davoli, G., Tinterri, R., Zavala, C., 1996. The importance of ancient fluvio-deltaic systems dominated by catastrophic flooding in tectonically active basins. Sci. Géol. Mém. 48, 233–291.

Mutti, E., Tinterri, R., Remacha, E., Mavilla, N., Angella, S., Fava, L., 1999. An introduction to the Analysis of ancient Turbidite Basins from an Outcrop Perspective. American Association Petroleum Geologists Continuing Education Course Note Series 39, 61pp.

Mutti, E., Tinterri, R., di Biase, D., Fava, L., Mavilla, N., Angella, S., et al., 2000. Delta-front facies associations on ancient flood-dominated fluvio-deltaic systems. Rev. Soc. Geol. Espana 13 (2), 165–190.

Mutti, E., Ricci Lucchi, F., Roveri, M., 2002. Revisiting turbidites of the Marnoso-arenacea Formation and their basin-margin equivalents: problems with classic models. In: Excursion Guidebook of the Turbidite Workshop, Parma, Italy, 21–22 May, 2002.

Mutti, E., Tinterri, R., Benevelli, G., di Biase, D., Cavanna, G., 2003. Deltaic, mixed and turbidite sedimentation of ancient foreland basins. In: Mutti, E., Steffens, G.S., Pirmez, C., Orlando, M., Roberts, D. (Eds.), Turbidites: Models and Problems, Marine and Petroleum Geology, Special issue, 20(6–8), 733–755.

Myrow, P.M., Fisher, W., Goodge, J.W., 2002. Wave-modified turbidites: combined-flow shoreline and shelf deposits, Cambrian, Antarctica. J. Sediment. Res. 72, 641–656.

Naaim, M., 1995. Modélisation numérique des avalanches aérosols. Houille Blanche 6 (5), 56–62.

Nakajima, T., 2006. Hyperpycnites deposited 700 km away from river mouths in the central Japan Sea. J. Sediment. Res. 76 (1), 60–73.

Nakajima, T., Satoh, M., Okamura, Y., 1998. Channel-levee complexes, terminal deep-sea fan and sediment wave fields, associated with the Toyama Deep-Sea Channel system in the Japan Sea. Mar. Geol. 149, 25–41.

Nardin, T.R., Hein, F.J., Gorsline, D.S., Edwards, B.D., 1979. A review of mass movement processes, sediment and acoustic characteristics, and contrasts in slope and base-of-slope systems versus canyon-fan-basin floor systems. In: Doyle, L.J., Pilkey, O.H. (Eds.), Geology of Continental Slopes, SEPM, Spec. Publ. 27, 61–73.

Navarre, J.-C., Claude, D., Librelle, F., Safa, P., Villon, G., Keskes, N., 2002. Deepwater turbidite system analysis. West Africa: sedimentary model and implications for reservoir model construction. Lead. Edge 21, 1132–1139.

Nelson, C.H., Kulm, V.D., 1973. Submarine fans and channels: turbidites and deep water sedimentation. Soc. Econ. Paleontol. Miner. 39–78 Short course.

Nelson, C.H., Thor, D.R., Sanstrom, M.W., 1979. Modern biogenic gas-generated craters (sea-floor "pockmarks") on the Bering Shelf, Alaska. Geol. Soc. Am. Bull. 90, 1144–1152.

Nelson, C.H., Maldonado, A., Coumes, F., Got, A., Monaco, A., 1985. Ebro Fan, Mediterranean. In: Bouma, A.H., Normark, W., Barnes, N. (Eds.), Submarine Fans and Related Turbidite Systems. Springer-Verlag, New York, pp. 121–127.

Nelson, C.H., Twitchell, D.C., Schwab, W.C., Lee, H.J., Kenyon, N.H., 1992. Upper Pleistocen turbidite sand beds and chaotic silt beds in the channelized distal, outer-fan lobes of the Mississippi fan. Geology 20, 693–696.

Nichols, R.J., 1995. The liquification and remobilization of sandy sediments. In: Hartley, A. J., Prosser, D.J. (Eds.), Characterization of Deep Marine Clastic Systems, Geol. Soc. Lond., Spec. Publ., 94, 63–76.

Nisbet, E.G., Piper, D.J.W., 1998. Giant submarine landslides: slumps, megaturbidites and their impact. Nature 392, 329–330.

Normark, W.R., 1970. Growth patterns of Deep-Sea Fans. Am. Assoc. Pet. Geol. Bull. 54 (11), 2170–2195.

Normark, W.R., 1978. Fan valleys, channels, and depositional lobes on modern submarine fans: characters for recognition of sandy turbidite environments. Am. Assoc. Pet. Geol. Bull. 62 (6), 912–931.

Normark, W.R., 1985. Local morphologic controls and effects of basin geometry on flow processes in deep marine basins. In: Zuffa, G. (Ed.), Provenance of Arenites, D. Reidel Publishing Co, Dordrecht, Holland, pp. 65–93.

Normark, W.R., Carlson, P.R., 2003. Giant submarine canyons: is size any clue to their importance in the rock record. In: Chan, M.A., Archer, A.W. (Eds.), Extreme Depositional Environments: Mega End Members in Geologic Time, Geological Society of America, Special Paper, 1–15.

Normark, W.R., Gutmacher, C.E., 1984. Delgada Fan, preliminary interpretation of channel development. Geo-Marine Letters 3 (2–4), 79–83.
Normark, W.R., Piper, D.J.W., 1991. Initiation processes and flow evolution of turbidity currents: implications for the depositional record. SEPM Special Publication 46 pp. 207–229.
Normark, W.R., Piper, D.J.W., Hess, G.R., 1979. Distributary channels, sand lobes, and mesotopography of navy submarine fan, California Borderland, with application to ancient fan sediments. Sedimentology 26, 749–774.
Normark, W.R., Hess, G.R., Stow, D.A.V., Bowen, A.J., 1980. Sediment waves on the Monterey fan levee: a preliminary physical interpretation. Mar. Geol. 37, 1–18.
Normark, W.R., Piper, D.J.W., Stow, D.A.V., 1983. Quaternary development of channels, levees, and lobes on middle Laurentian Fan. Am. Assoc. Pet. Geol. Bull. 67 (9), 1400–1409.
Normark, W.R., Piper, D.J.W., Romans, B.W., Covault, J.A., Dartnell, P., Sliter, R.W., 2009. Submarine canyons and fan systems of the California Borderland. The Geol. Soc. America, Spec. Paper 454, pp. 141–168.
O'Connel, S., McHugh, C., Ryan, W.B.F., 1995. Unique fan morphology in an entrenched Talweg channel on the Rhone Fan. In: Pickering, K.T., Hiscott, R.N., Kenyon, N.H., Lucchi, Ricci, Smith, R.D.A. (Eds.), Atlas of Deep-Water Environments; Architectural Style in Turbidite Systems. Chapman et Hall, London, pp. 80–83.
Ollier, G., Cochonat, P., Lénat, J.-F., Labazuy, P., 1998. Deep-sea volcaniclastic sedimentary systems: an example from La Fournaise volcano, Réunion Island, Indian Ocean. Sedimentology 45, 293–330.
Palanques, A., Durrieu de Madron, X., Puig, P., Fabres, J., Guillen, J., Calafat, A., et al., 2006. Suspended sediment fluxes and transport processes in the Gulf of Lions submarine canyons. The role of storms and dense water cascading. Mar. Geol. 234, 43–61.
Palanques, A., Guillén, J., Puig, P., Durrieu de Madron, X., 2008. Storm-driven shelf-to-canyon suspended sediment transport at the southwestern Gulf of Lions. Cont. Shelf Res. 28 (15), 1947–1956.
Pantin, H.M., 1979. Interaction between velocity and effective density in turbidity flow: phase plane analysis, with criteria for autosuspension. Mar. Geol. 31, 59–99.
Pantin, H.M., Franklin, M.C., 2009. Predicting autosuspension in steady turbidity flows: ignition points and their relation to Richardson numbers. J. Sediment. Res. 79, 862–871.
Pantin, H.M., Leeder, M.R., 1987. Reverse flow in turbidity current. The role of internal solitons. Sedimentology 34, 1143–1156.
Parize, O., Savoye, B., Sahabi, M., Cochonat, P., 1989. Observation d'un réseau d'origine sous-marine de chenaux en tresses sur le fond de la tête du canyon du Var (Nice, France). In: Origine et Implications, 2ème Congrès Français de Sédimentologie, Livre des résumés, Publications ASF 10, Paris227–228.
Parker, G., 1982. Conditions for the ignition of catastrophically erosive turbidity currents. Mar. Geol. 46, 307–327.
Parsons, J.D., Bush, J., Syvitski, J.P.M., 2001. Hyperpycnal flow formation with small sediment concentrations. Sedimentology 48, 465–478.
Paull, C.K., Ussler, W.L., Borowski, W.S., Speiss, F.N., 1995. Methane-rich plumes on the Carolina continental rise: associations with gas hydrates. Geology 23, 89–92.
Paull, C.K., Ussler, W.I.I.I., Borowski, W.S., 1999. Freshwater ice rafting: an additional mechanism for the formation of some high-latitude submarine pockmarks. Geo-Mar. Lett. 19, 164–168.
Peakall, J., Amos, K.J., Keevil, G.M., Bradbury, P.W., Gupta, S., 2007a. Flow Processes and sedimentation in submarine channel bends. Mar. Petrol. Geol. 24, 470–486.
Peakall, J., Ashworth, P.J., Best, J.L., 2007b. Meander-bend evolution, alluvial architecture, and the role of cohesion in sinuous river channels. A flume study. J. Sediment. Res. 77, 197–212.

Peakall, J., McCaffrey, B., Kneller, B.C., 2000. A Process Model for the Evolution, Morphology, and Architecture of Sinuous Submarine Channels. J. Sediment. Res. 70 (3), 434–448.
Pichevin, L., Mulder, T., Savoye, B., Cremer, M., Piper, D.J.W., Gervais, A., 2003. The Golo submarine system (east-Corsica margin): turbidite system morphology and processes of terrace formation from high-resolution seismic Boomer profiles. Geo-Mar. Lett. 23 (2), 117–124.
Pickering, K.T., Hiscott, R.N., 1985. Contained (reflected) turbidity currents from the Middle Ordovician Cloridorme Formation, Quebec, Canada: an alternative to the antidune hypothesis. Sedimentology 32, 373–394.
Pickering, K.T., Hiscott, R.N., Hein, F.J., 1989. Deep Marine Environments: Clastic Sedimentation and Tectonics. Unwin Hyman, London 416pp.
Pickering, K.T., Clark, J.D., Smith, R.D.A., Hiscott, R.N., Ricci Lucchi, F., Kenyon, N. H., 1995. Architectural element analysis of turbidite systems, and selected topical problems for sand-prone deep-water systems. In: Pickering, K.T., Hiscott, R.N., Kenyon, N.H., Lucchi, Ricci, Smith, R.D.A. (Eds.), Atlas of Deep-Water Environments; Architectural Style in Turbidite Systems. Chapman et Hall, London, pp. 1–10.
Piper, D.J.W., 1978. Turbidites, muds and silts on deep-sea fans and abyssal plains. In: Stanley, D.J., Kelling, G. (Eds.), Sedimentation in Submarine Canyons, Fans and Trenches. Dowden, Hutchinson and Ross, Stroudsburg, PA, pp. 163–176.
Piper, D.J.W., Aksu, A.E., 1987. The source and origin of the 1929 Grand Banks turbidity current inferred from sediment budgets. Geo-Mar. Lett. 7, 177–182.
Piper, D.J.W., Deptuck, M., 1997. Fine-grained turbidites of the Amazon Fan: facies characterization and interpretation. In: Flood, R.D., Piper, D.J.W., Klaus, A., Peterson, L.C. (Eds.), Proc. ODP, Sci. Results, 155, 79–108 College Station, TX.
Piper, D.J.W., Normark, W.R., 1983. Turbidite depositional patterns and flow characteristics. Navy Submarine fan, California Borderland. Sedimentology 30, 681–694.
Piper, D.J.W., Normark, W.R., 2009. Processes that initiate turbidity currents and their influence on turbidites: a marine geology perspective. J. Sediment. Res. 79 (5), 347–362.
Piper, D.J.W., Savoye, B., 1993. Processes of late Quaternary turbidity current flow and deposition on the Var deep-sea fan, north-west Mediterranean Sea. Sedimentology 40, 557–583.
Piper, D.J.W., Stow, D.A.V., Normark, W.R., 1984. Laurentian Fan: Sohm Abyssal Plain. Geo-Mar. Lett. 3, 141–146.
Piper, D.J.W., Farre, J.A., Shor, A.N., 1985. Late Quaternary slumps and debris flows on the Scotian Slope. Geol. Soc. Am. Bull. 96, 1508–1517.
Piper, D.J.W., Cochonat, P., Ollier, G., Le Dreezen, E., Morrison, M., Baltzer, A., 1992. Evolution progressive d'un glissement rotationnel en un courant de turbidité: cas du séisme de 1929 des Grand Bancs (Terre Neuve). C. R. Acad. Sci. Paris II 314, 1057–1064.
Piper, D.J.W., Pirmez, C., Manley, P.L., Long, D., Flood, R.D., Normark, W.R., et al., 1997. Mass-transport deposits of the Amazon Fan. In: Flood, R.D., Piper, D.J.W., Klaus, A., Peterson, L.C. (Eds.), Proceedings of the Ocean Drilling Program, Scientific Results. College Station TX, 109–146.
Piper, D.J.W., Cochonat, P., Morrison, M., 1999a. The sequence of events around the epicentre of the 1929 Grand Banks earthquake: initiation of debris flows and turbidity current inferred from sidescan sonar. Sedimentology 46, 79–97.
Piper, D.J.W., Hiscott, R.N., Normark, W.R., 1999b. Outcrop scale acoustic facies analysis and latest Quaternary development of Hueneme and Dune submarine fans, offshore California. Sedimentology 46, 47–78.
Pirmez, C., Flood, R.D., 1995. Morphology and structure of Amazon Channel. Proceedings of the Ocean Drilling Program, Initial Reports 155, 23–45.

Pirmez, C., Hiscott, R.N., Kronen Jr., J.K., 1997. Sandy turbidite successions at the base of channel-levee systems of the Amazon Fan revealed by FMS logs and Cores: unraveling the facies architecture of large submarine fans. In: Flood, R.D., Piper, D.J.W., Klaus, A., Peterson, L.C. (Eds.), Proc. ODP, Sci. Results, 155, 7–33 College Station, TX.

Pirmez, C., Beauboueff, R.T., Friedmann, S.J., Mohrig, D.C., 2000. Equilibrium Profile and Baselevel in Submarine Channels: Examples from Late Pleistocene Systems and Implications for the architecture of Deep-Water Reservoir. In: Weimer, P., Slatt, R.M., Coleman, J., Rosen, N.C., Nelson, H., Bouma, A.H., Styzen, M.J., Lawrence, D.T. (Eds.), Deep-Water Reservoirs of the World, GCSSEPM, Foundation 20th Annual Research Conference, Dec. 3-6, Houston, Gulf Coast Section-SEPM Special Publication, pp. 782–805.

Plapp, J.E., Mitchell, J.P., 1960. A hydrodynamic theory of turbidity currents. J. Geophys. Res. B. 65 (3), 983–992.

Pochat, S., 2003. Escarpement de faille syssédimentaire. Perturbation deséoulements gravitaires sous-marins et détermination de la cinétique des failles. Ph.D. Thesis, Univ. Rennes 1, Mémoires Géosciences Rennes no. 105, 283pp.

Popescu, I., Lericolais, G., Panin, N., Wong, H.K., Droz, L., 2001. Late quaternary channel avulsions on the Danube deep-sea fan, Black Sea. Mar. Pet. Geol. 179, 25–37.

Porebski, S.J., Steel, R.J., 2003. Shelf-margin deltas: their stratigraphic significance and relation to deepwater sands. Earth Sci. Rev. 62 (3–4), 283–326.

Posamentier, H.W., Allen, G.P., 1999. Siliciclastic Sequence Stratigraphy—Concepts and Applications. SEPM Concepts in Sedimentology and Paleontology no. 7 210pp.

Posamentier, H.W., Vail, P.R., 1988. Eustatic controls on clastic deposition II—sequence and systems tract models. In: Wilgus, C., Hastings, B., Posamentier, H., Van Wagoner, J., Ross, C., Kendall, G. (Eds.), Sea-Level Changes: An Integrated Approach, SEPM, Special Publication 42, 125–154.

Posamentier, H.W., Walker, R.G., 2006. Facies Models Revisited. SEPM, Special Publication, 84 Tulsa, Oklahoma, USA, 531pp.

Posamentier, H.W., Erskine, R.D., Mitchum, R.M., 1991. Models for submarine-fan deposition within a sequence-stratigraphic framework. In: Weimer, P. (Ed.), Seismic Facies and Sedimentary Processes of Submarine Fans and Turbidite Systems. Springer-Verlag, New York, pp. 127–136.

Posamentier, H., Allen, G.P., James, D.P., Tesson, M., 1992. Forced regressions in a sequence stratigraphic framework: concepts, examples and exploration significance. Am. Assoc. Pet. Geol. Bull. 76, 1687–1709.

Postma, H., 1969. Suspended matter in the marine environment. Morning Review, lectures of the Second International Oceanographic Congress, Moscow, 1966 213–219.

Postma, G., Nemec, W., Zachariasse, W.J., 1988. Large floating clasts in turbidites: a mechanism for their emplacement. Sediment. Geol. 58, 47–61.

Prather, B.E., Keller, F.B., Chapin, M.A., 2000. Hierarchy of Deep-Water Architectural Elements with Reference to Seismic Resolution: Implications for Reservoir Prediction and Modeling. In: Weimer, P., Slatt, R.M., Coleman, J., Rosen, N.C., Nelson, H., Bouma, A.H., Styzen, M.J., Lawrence, D.T. (Eds.), Deep-Water Reservoirs of the World, GCSSEPM Foundation 20th Annual Research Conference, Dec. 3-6, Houston, Gulf Coast Section-SEPM Special Publication, pp. 817–835.

Pratson, L.F., Haxby, W.F., 1996. What is the slope of the U.S. continental slope. Geology 24, 3–6.

Pratson, L.F., Ryan, W.B.F., Mountain, G.S., Twichell, D.C., 1994. Submarine canyon initiation by downslope-eroding sediment flows: evidence in late Cenozoic strata on the New Jersey continental slope. Geol. Soc. Am. Bull. 106, 395–412.

Pratson, L., Imran, J., Hutton, E., Parker, G., Syvitski, J.P.M., 2001. BANG1D: a one-dimensional, Lagrangian model of turbidity current mechanics. Comput. Geosci. 27 (6), 701–716.

Prave, A.R., Duke, W.L., 1990. Small-scale hummocky cross-stratification in turbidites: a form of antidune stratification. Sedimentology 37 (3), 531–539.

Prelat, A., Hodgson, D.M., Flint, S.S., 2009. Evolution, architecture and hierarchy of distributary deep-water deposits: a high-resolution outcrop investigation from the Permian Karoo Basin, South Africa. Sedimentology 56, 2132–2154.

Prior, D.B., Suhayda, J.N., Lu, N.-Z., Bornhold, B.D., Keller, G.H., Wiseman, W.J., et al., 1989. Storm wave reactivation of a submarine landslide. Nature 341, 47–50.

Puig, P., Ogston, A.S., Mullenbach, B.L., Nittrouer, C.A., Sternberg, R.W., 2003. Shelf-to-canyon sediment-transport processes on the Eel continental margin (northern California). Mar. Geol. 193, 129–149.

Ravenne, C., Beghin, P., 1983. Apport des expériences en canal à l'interprétation sédimentologique des dépôts de cônes détriques sous-marines. Rev. Inst. Fr. Pétrol. 38, 279–297.

Razin, P., 2008. Les systèmes de plate-forme-bassin. Architecture, dynamique et facteurs de contrôle. Vol. 1 Synthèse des Travaux. Mémoire d'Habilitation à Diriger les Recherches, Univ. Bordeaux 3 411pp.

Reading, H.G., Richard, M.T., 1994. The classification of deep-water siliciclastic depositional systems by grain size and feeder systems. Am. Assoc. Pet. Geol. Bull. 78, 792–822.

Reynaud, J.-Y., 1996. Architecture et évolution d'un banc sableux de mer celtique méridionale. Unpubl. Ph.D. thesis, Univ. Sci. Tech. Lille 1, 185pp.

Reynolds, S., 1987. A recent turbidity current event, Hueneme Fan, California: reconstruction of flow properties. Sedimentology 34, 129–137.

Rigaut, F., 1997. Analyse et évolution récente d'un système turbiditique méandriforme: l'éventail profond du Zaïre. Unpubl. Ph.D. thesis, Univ. Bretagne Occidentale., 209pp.

Rigollet, C., 2001. Valorisation sédimentologique de l'information sismique: application au comblement des bassins profonds de la Marge Atlantique Nord Européenne du Crétacé à l'actuel. Thèse, Ecole des Mines de Paris, 507pp.

Rimoldi, B., Alexander, J., Morris, S.A., 1996. Experimental turbidity currents entering density-stratified water: analogues for turbidites in Mediterranean hypersaline basins. Sedimentology 43, 527–540.

Rise, L., Saettem, J., Thorsnes, T., Ottesen, D., Bø, R., 1999. Sea-bed pockmarks related to fluid migration from Mesozoic bedrock strata in the Skagerrak offshore Norway. Mar. Pet. Geol. 16 (7), 619–631.

Rodine, J.D., Johnson, A.M., 1976. The ability of debris, heavily freighted with coarse clastic materials, to flow on gentle slopes. Sedimentology 23, 213–234.

Rothwell, R.G., Pearce, T.J., Weaver, P.P.E., 1992. Late quaternary evolution of the Madeira Abyssal Plain, Canary Basin, NE Atlantic. Basin Res. 4, 103–131.

Rouillard, P., 2010. Modèle architectural et lithologique du système de Rosetta (Delta du Nil, Méditerranée Orientale): implications pour un analogue actuel de réservoir pétrolier. Unpubl. Ph.D. thesis, Université de Nice Sophia-Antipolis, 375pp.

Salles, T., 2006. Modélisation numérique du remplissage sédimentaire des canyons et chenaux sous-marins par approche génétique. Unpubl. Ph.D. thesis, Univ. Bordeaux 1, no. 3254, 215pp.

Salles, T., Mulder, T., Gaudin, M., Cacas, M.C., Lopez, S., Cirac, P., 2007. Simulating the 1999 turbidity current occurred in Capbreton canyon through a Cellular Automata model. Geomorphology doi:10.1016/ 2007.09.005.

Sanders, J.E., 1965. Primary sedimentary structures formed by turbidity currents and related resedimentation mechanisms. In: Middleton, G.V. (Ed.), Primary Sedimentary Structures and their Hydrodynamic Interpretation, SEPM Special Publication 12, 192–219.

Savoye, B., Piper, D.J.W., 1991. The Messinian event on the margin of the Mediterranean Sea in the Nice area, southern France. Mar. Geol. 97, 279–304.
Savoye, B., Cochonat, P., Piper, D.J.W., 1990. Seismic evidence for a complex slide near the wreck of Titanic: model of an instability corridor for non-channelled gravity events. Mar. Geol. 91, 281–298.
Savoye, B., Piper, D.J.W., Droz, L., 1993. Plio-pléistocène evolution of the Var deep-sea fan on the French Riviera. Mar. Pet. Geol. 10, 550–571.
Savoye, B., Cochonat, P., Apprioual, R., Bain, O., Baltzer, A., Bellec, V., et al., 2000. Structure and recent evolution of the Zaire deep-sea fan: preliminary results of the Zaïango 1 & 2 cruises (Angola-Congo margin). C. R. Acad. Sci. IIA 331 (3), 211–220.
Schafer, C.T., Smith, J.N., 1987. Hypothesis for a submarine landslide and cohesionless sediment flows resulting from a 17th century earthquake-triggered landslide in Quebec, Canada. Geo-Mar. Lett. 7, 31–37.
Schumm, S.A., 1963. Disparity between modern rates of denudation and orogeny. USGS Prof. Paper 454-H 13pp.
Schumm, S.A., 1993. River response to base level change: implication for sequence stratigraphy. J. Geol. 101, 279–294.
Schwab, W.C., Lee, H.J., Twichell, D.C., Locat, J., Nelson, C.H., McArthur, W.G., et al., 1996. Sediment mass-flow processes on a depositional lobe, outer Mississippi fan. J. Sediment. Res. 66, 916–927.
Shanmugam, G., 2000. 50 years of the turbidite paradigm, (1950s-1990s): deep-water processes and facies models—a critical perspective. Mar. Pet. Geol. 17, 285–342.
Shanmugam, G., 2002. Ten turbidite myths. Earth Sci. Rev. 58, 311–341.
Shanmugam, G., Moiola, R.J., 1988. Submarine fans: characteristics, models, classification, and reservoir potential. Earth Sci. Rev. 24, 383–428.
Shanmugam, G., Moiola, R.J., 1991. Types of submarine fan lobes: models and implications. Am. Assoc. Pet. Geol. Bull. 75 (1), 156–179.
Shanmugam, G., Moiola, R.J., Damuth, J.E., 1985. Eustatic control of submarine fan development. In: Bouma, A., Normark, W., Barnes, N. (Eds.), Submarine Fans and Related Turbidite Systems. Springer-Verlag, New York, pp. 23–28.
Shanmugam, G., Moiola, R.J., McPherson, J.G., O'Connel, S., 1988. Comparison of modern Mississippi fan with selected ancient fans. Gulf Coast Assoc. Geol. Soc. Trans. 38, 157–165.
Shanmugam, G., Spalding, T.D., Rofheart, D.H., 1993. Process sedimentology and reservoir quality of deep-marine bottom current reworked sands (sandy contourites): an example from the Gulf of Mexico. Am. Assoc. Pet. Geol. Bull. 77, 1241–1259.
Sharp, A., Samuel, A., 2004. An example study using conventional 3D seismic data to delineate shallow gas drilling hazards from West Delta Deep Marine concession, offshore Nile Delta, Egypt. Pet. Geosci. 10, 121–129.
Shepard, F.P., 1967. Submarine Geology, second ed. Harper and Row, New York 557pp.
Shepard, F.P., 1981. Submarine canyons: multiple causes and long-time persistence. Am. Assoc. Pet. Geol. Bull. 65, 1062–1077.
Shepard, F.P., Dill, R.F., 1966. Submarine Canyons and other sea-valleys. Rand McNally, Chicago, IL 381pp.
Shepard, F.P., Dill, R.F., von Rad, U., 1966. Physiography and sedimentary processes of La Jolla submarine fan and fan valley, California. Am. Assoc. Pet. Geol. Bull. 53, 390–420.
Shepard, F.P., McLoughlin, P.A., Marshall, N.F., Sullivan, G.G., 1977. Current-meter recordings of low-speed turbidity currents. Geology 5, 297–301.
Shepard, F.P., Marshal, N.F., McLoughlin, P.A., Sullivan, G.G., 1979. Currents in submarine canyons and other sea valleys. American Association of Petroleum Geologists, Studies in Geology, No. 8, 173pp.
Singh, S.C., 2005. Sumatra earthquake research indicates why rupture propagated northward. EOS Trans. Am. Geophys. Union 86 (48), 497–502.

Skempton, A.W., Hutchinson, J.N., 1969. Stability of natural slopes and embankment foundations. State-of-the-art report. In: Proc. 7th Int. Conf. SMFE, Mexico City, vol. 2291–335.

Skene, K.I., 1998. Architecture of submarine channel levees. Unpubl. Ph.D. thesis, Dalhousie University, Halifax, 413pp.

Skene, K.I., Mulder, T., Syvitski, J.P.M., 1997. INFLO1: a model predicting the behaviour of turbidity currents generated at river mouths. Comput. Geosci. 23 (9), 975–991.

Sparks, R.S.J., Bonnecaze, R.T., Huppert, H.E., Lister, J.R., Hallworth, M.A., Mader, H., et al., 1993. Sediment-laden gravity currents with reversing buoyancy. Earth Planet. Sci. Lett. 114, 243–257.

St-Onge, G., Mulder, T., Piper, D.J.W., Hillaire-Marcel, C., Stoner, J.S., 2004. Earthquake and flood-induced turbidites in the Saguenay Fjord (Québec): a Holocene paleosismicity record. Quat. Sci. Rev. 23, 283–294.

Stacey, M.W., Bowen, A.J., 1987. The vertical structure of density of turbidity currents: theory and observations. J. Geophys. Res. 93, 3528–3542.

Stanley, D.J., 1981. Unifites: structureless muds of gravity-flow origin in Medoterranean Basins. Geo-Mar. Lett. 1, 77–83.

Stow, D.A.V., Mayall, M., 2000. Deep-water sedimentary systems: new models for the 21st century. Mar. Pet. Geol. 17, 125–135.

Stow, D.A.V., Piper, D.J.W., 1984. Deep-water fine-grained sediments: facies model. In: Stow, D.A.V., Piper, D.J.W. (Eds.), Fine-Grained Sediments: Deep-Water Processes and Facies, Geological Society of London, Special Publication, 15, 611–645.

Stow, D.A.V., Shanmugam, G., 1980. Sequence of structures in fine-grained turbidites: comparaison of recent deep-sea and ancient flysch sediments. Sediment. Geol. 25, 23–42.

Stow, D.A.V., Wetzel, A., 1990. Hemiturbidite: a new type of deep-water sediment. In: Cochran, J.R., Stow, D.A.V. et al., Proc. ODP, Sci. Results, 116, 25–34.

Stow, D.A.V., Reading, H.G., Collison, J.D., 1996. Deep seas. In: Reading, H.G. (Ed.), Sedimentary Environments: Processes, Facies and Stratigraphy. Blackwell Science, Oxford, pp. 395–453.

Sydow, J., Roberts, H.H., 1994. Stratigraphic framework of a Late Pleistocene shelf-edge delta, northeast Gulf of Mexico. Am. Assoc. Pet. Geol. Bull. 78, 1276–1312.

Syvitski, J.P.M., 2003. Sediment fluxes and rates of sedimentation. In: Middleton, G.V. (Ed.), Encyclopedia of Sediments and Sedimentary Rocks. Kluwer Academic Publishers, Dordrecht, Netherlands, pp. 600–606.

Syvitski, J.P.M., Hein, F.J., 1991. Sedimentology of an arctic basin: Itirbilung Fjord, Baffin Island, Northwest Territories. Geol. Survey of Can., Paper, 91 67pp.

Syvitski, J.P.M., Schafer, C.T., 1996. Evidence for an earthquake-triggered basin collapse in Saguenay Fjord, Canada. Sediment. Geol. 104, 127–153.

Talling, P.J., Wynn, R.B., Masson, D.G., Frenz, M., Cronin, B.T., Schiebel, R., et al., 2007. Onset of submarine debris flow deposition far from original giant landslide. Nature 450, 541–544.

Torres, J., Droz, L., Savoye, B., Terentieva, E., Cochonat, P., Kenyon, N.H., et al., 1997. Deep-sea avulsion and morpho-sedimentary evolution of the Rhone Fan Valley and Neofan during the Late Quaternary (north-western Mediterranean Sea). Sedimentology 44, 457–477.

Toucanne, S., 2008. Reconstruction des transferts sédimentaires en provenance du système glaciaire de Mer d'Irlande et du paléofleuve Manche au cours des derniers cycles climatiques. Unpubl. Ph.D. thesis, Univ. Bordeaux 1, no. 3699, 340pp.

Tranier, J., 2002. Etude des processus sédimentaires dans la partie supérieure du canyon de Capbreton. Origine des terrasses. Unpubl. Master thesis, Univ.Bordeaux 1, 31pp.

Tripsanas, E.K., Piper, D.J.W., Jenner, K., Bryant, W.R., 2008. Submarine mass-transport facies: new perspectives on flow processes from cores on the eastern North American margin. Sedimentology 55, 97–136.
Turakiewicz, G., 2004. Mécanismes forçants dans les éventails turbiditiques des marges matures: exemple de l'éventail quaternaire du Congo. Unpubl. Ph.D. thesis, Univ. Montpellier II, 367pp.
Twitchell, D.C., Kenyon, N.H., Parsons, L.M., Mc Gregor, B.A., 1991. Depositional patterns of the Mississippi Fan surface: evidence from GLORIA II and high-resolution seismic profiles. In: Link, M.L. (Ed.), Seismic Facies and Sedimentary Processes of Submarine Fans and Turbidite Systems. Springer-Verlag, New York, pp. 349–363.
Uchupi, E., Swift, S.A., Ross, D.A., 1996. Gas venting and late quaternary sedimentation in the Persian (Arabian) Gulf. Mar. Geol. 129 (3–4), 237–269.
Unterseh, S., 1999. Cartographie et caractérisation du fond marin par sondeur multifaisceaux. Unpubl. Ph.D. thesis, Univ. nancy, 234pp.
Urgeles, R., Canals, M., Masson, D.G., Gee, M.J.R., 2003. El Hierro: shaping of an oceanic island by mass wasting. In: Mienert, J., Weaver, P.P.E. (Eds.), European Margin Sediment Dynamics, Side-Scan Sonar and Seismic Images. Springer-Verlag, New York, pp. 301–305.
Vail, P.R., Mitchum, R.M., 1977. Seismic stratigraphy and global changes of sea level, part 1: overview. In: Payton, C.E. (Ed.), Seismic Stratigraphy—Application to Hydrocarbon Exploration, American Association of Petroleum Geologists Memoir 26, 51–52.
Vail, P.R., Colin, J.-P., Jan du Chene, R.J., Kuchly, J., Mediavilla, F., Trifilieff, V., 1987. La stratigraphie séquentielle et son application aux corrélations chronostratigraphiques dans le Jurassique du bassin de Paris. Bull. Soc. Geol. Fr. 7, 1301–1321 t. III.
van Tassel, J., 1981. Silver abyssal plain carbonate turbidite: flow characteristics. J. Geol. 89, 317–333.
van Weering, T.C.E., Nielsen, T., Kenyon, N.H., Akentieva, K., Kuijpers, A.H., 1998. Large submarine slides on the NE Faeroe continental margin. In: Stoker, M.S., Evans, D., Cramp, A. (Eds.), Geological Processes on Continental Margins: Sedimentation, Mass-Wasting and Stability, Geol. Soc. London, Spec. Publ. 129, 5–17.
Vanneste, M., Mienert, J.R., Bãinz, S., 2006. The Hinlopen slide: a giant, submarine slope failure on the northern Svalbard margin, Arctic Ocean. Earth Planet. Sci. Lett. 245 (1–2), 373–388.
Vasiliev, S.S., Dergachev, V.A., 2002. The 2400-year cycle in atmospheric radiocarbon concentration: bispectrum of 14 C data over the last 8000 years. ANGEO 20 (1), 115–120.
Viana, A.R., Faugères, J.-C., Stow, D.A.V., 1998. Bottom-current-controlled sand deposits—a review of modern shallow-to deep-water environments. Sediment. Geol. 115 (1–4), 53–80.
Von Rad, U., Tahir, M., 1997. Late Quaternary sedimentation on the outer Indus shelf and slope (Pakistan): evidence from high-resolution seismic data and coring. Mar. Geol. 138, 193–236.
Walker, R.G., 1978. Deep-water sandstone facies and ancient submarine fans: models for exploration for stratigraphic traps. Am. Assoc. Pet. Geol. Bull. 62 (6), 932–966.
Warrick, J.A., Milliman, J.D., 2003. Hyperpycnal sediment discharge from semiarid southern California rivers: implications for coastal sediment budgets. Geology 31 (9), 781–784.
Weber, M.E., Wiedicke, M.H., Kudrass, H.R., Hübscher, C., Erlenkeuser, H., 1997. Active growth of the Bengal Fan during sea-level rise and highstand. Geology 25, 315–318.
Weimer, P., 1989. Sequence stratigraphy of the Mississippi Fan (Plio–Pleistocene), Gulf of Mexico. Geo-Mar. Lett. 9, 185–272.
Weimer, P., 1990. Sequence stratigraphy, facies geometries, and depositional history of the Mississippi Fan, Gulf of Mexico. Am. Assoc. Pet. Geol. Bull. 74 (4), 425–453.

Weimer, P., Slatt, R.M., Coleman, J., Rosen, N.C., Nelson, H., Bouma, A.H., et al., 2000. Deep-Water Reservoirs of the World. In: Gulf Coast Society of Economic Paleontologists and Mineralogists Foundation, 20th Annual Conference, Houston.

Weirich, F.H., 1988. Field evidence for hydraulic jumps in subaqueous sediment gravity flows. Nature 332, 626–629.

Wetzel, A., 1993. The transfer of river load to deep-sea fans: a quantitative approach. Am. Assoc. Pet. Geol. Bull. 77, 1679–1692.

Wheeler, A.J., Stadnitskaia, A., 2011. Benthic deep-sea carbonates: reefs and seeps. In: Hüneke, H., Mulder, T. (Eds.), Deep-Sea Sediments. Developments in Sedimentology, Elsevier, Amsterdam, vol. 63, pp. 397–455.

Wilson, P.A., Roberts, H.H., 1995. Density cascading: off-shelf transport, evidence and implications, Bahama Banks. J. Sediment. Res. A65, 45–56.

Wright, L.D., Yang, Z.-S., Bornhold, B.D., Keller, G.H., Prior, D.B., Wisenam Jr., W.J., 1986. Hyperpycnal flows and flow fronts over the Huanghe (Yellow River) delta front. Geo-Mar. Lett. 6, 97–105.

Wright, L.D., Wisenam Jr., W.J., Bornhold, B.D., Prior, D.B., Suhayda, J.N., Keller, G.H., et al., 1988. Marine dispersal and deposition of Yellow River silts by gravity-driven underflows. Nature 332, 629–632.

Wright, L.D., Wiseman, W.J., Yang, Z.-S., Bornhold, B.D., Keller, G.H., Prior, D.B., et al., 1990. Processes of marine dispersal and deposition of suspended silts off the modern mouth of the Huanghe (Yellow River). Cont. Shelf Res. 10, 1–40.

Wynn, R.B., Kenyon, N.H., Masson, D.G., Stow, D.A.V., Weaver, P.P.E., 2002. Characterization and recognition of deep-water channel-lobe transition zones. Am. Assoc. Pet. Geol. Bull 86 (8), 1441–1446.

Zaragosi, S., 2001. Les systèmes turbiditiques profonds de la marge celtique–armoricaine (Golfe de Gascogne): physiographie et évolution au cours des derniers 30 000 ans. Unpubl. Ph.D. thesis, Univ. Bordeaux 1, no. 2264, 291pp.

Zaragosi, S., Le Suave, R., Bourillet, J.-F., Auffret, G.A., Faugères, J.-C., Pujol, C., et al., 2001. The deep-sea Armorican depositional system (Bay of Biscay), a multiple source, ramp model. Geo-Mar. Lett. 20 (4), 219–232.

Zaragosi, S., Auffret, G.A., Voisset, M., Garlan, T., 2003. Morphology and depositional processes of Celtic Fan, Bay of Biscay. In: Mienert, J., Weaver, P.P.E. (Eds.), European Margin Sediment Dynamics, Side-Scan Sonar and Seismic Images. Springer-Verlag, New York, pp. 239–243.

Zuffa, G.G., Normark, W.R., Serra, F., Brunner, C.A., 2000. Turbidite megabeds in an oceanic rift valley recording jökulhlaups of Late Pleistocene Glacial Lakes of the Western United States. J. Geol. 108, 253–274.

CHAPTER 3

Contour Currents and Contourite Drifts

Jean-Claude Faugères*,[1] *and* Thierry Mulder*

Contents

* Université de Bordeaux, CNRS 5805 EPOC, Talence Cedex, France
[1] Corresponding author.
E-mail address: afaugeres@numericable.fr

Developments in Sedimentology, Volume 63
ISSN 0070-4571, DOI: 10.1016/S0070-4571(11)63003-3

1. Introduction

The geostrophic circulation is the main feature of the present-day world oceans. It includes two components: the surface and the deep geostrophic circulation. These form at different water depths and by different physical processes.

The oceanic surface circulation results from the winds and is known for centuries. It generates currents that reach a few tens of cm s^{-1} to a few m s^{-1} in velocity. In contrast, the deep circulation has been understood only since the middle of the twentieth century and the development of physical and chemical oceanography. It operates in the deepest ocean basins and is called *thermohaline circulation* because it is driven by density differences between water masses due to variations in water temperature and salinity. The thermohaline circulation forms a large network of slowly moving currents.

The surface and deep oceanic circulation play a major role in deep-sea erosion, transport, distribution, and deposition of sedimentary particles, whatever their origin (continental or oceanic) is. The thermohaline circulation is the most important process that controls deposition by bottom currents, whereas the surface, wind-derived circulation rather controls pelagic and hemipelagic deposition, and are also important for iceberg transport. However, the surface currents may also be involved in processes that affect the sea-floor; this includes currents that are driven by oceanographic forces such as tides and waves.

The occurrence of bottom currents was first pointed out during the 1960s with the pioneer works of Heezen et al. (1966) and Heezen and Hollister (1971). By comparing sea-floor photographs and cored deposits from the continental rise in the western Atlantic, they understood the role of bottom currents in the transport of deep-sea sedimentary particles and in the shaping of the continental rise, and they proposed the original concepts of contour currents and their deposits: contourites. At that time, these concepts suffered criticism from numerous physical oceanographers, particularly regarding the current velocity and the currents' ability to transport sand and silt. A controversy arose on the relative importance of persistent contour currents versus sporadic turbidity currents regarding sediment deposition along continental margins (Hollister and Heezen, 1972).

The late 1970s were mainly focused on the recognition of contourites (Stow and Lowell, 1979) and on the description of general patterns of large contourite accumulations ('Outer Ridge' or 'Drift'). Most of the research was conducted in the Atlantic Ocean, but local studies in the Indian and Pacific Oceans focused on passages or gaps that funnelled deep currents, and on basin floors with manganiferous deposits.

In the 1980s, numerous investigations were performed on contourite deposits all over the world's oceans. Since that time, these benefit greatly from the International Deep-Sea Drilling Programme. Significant advances occurred with respect to the contourite facies (Gonthier et al., 1984; Stow, 1982), bedforms, the growth and distribution of drifts (McCave and Tucholke, 1986), the sedimentary processes of contour currents (HEBBLE project; Nowell and Hollister, 1985) and the decoding of global palaeoceanographic changes (Tucholke and Mountain, 1986).

During the following years, the interest in contourite drifts increased again, as the interests of the oil industry in exploration activities shifted towards deeper water (Shanmugan et al., 1993). Numerous works show clearly that a large part of the sediments along modern margins consist of intercalating gravity deposits, pelagites and contourites, and that turbiditic and contouritic activities frequently interact during depositional processes. Consequently, significant research from the last 15 years dealt in detail with margins where both processes are active. The results of these studies allowed to define diagnostic features for distinguishing between both types of deposits (facies, bedforms, morphology and geometry of the sedimentary units), and they helped to understand the interactions between the depositional processes of contourites and turbidites (a.o. Carter and McCave, 1994, 2002; Faugères et al., 1999; Howe, 1996; Locker and Laine, 1992; McMaster et al., 1989; Rebesco et al., 1996, 2002; Shanmugan et al., 1993; Stoker et al., 1998; Stow et al., 2002a; Tucholke and Mountain, 1986). In addition, palaeoceanographic and palaeoclimatic research projects focused on drift deposits. Relationships between current velocity and grain size of the contourites were used to decipher the hydrodynamic variations in the sedimentary record (Robinson and McCave, 1994). Numerous detailed sedimentological, chemical and micro-palaeontological studies of contourites and pelagites were performed in the framework of international programmes (IODP, IMAGES), and resulted in distinctly increased knowledge of the palaeoceanographic global changes.

Today, contourite identification in modern and ancient series still remains a difficult challenge. To succeed in decoding such units, it is necessary to gather a complete set of data regarding lithofacies, bedform morphology and geometry at various scales (Faugères and Stow, 2008; Hüneke and Stow, 2008; Nielsen et al., 2008; Stow and Faugères, 2008; Stow et al., 1998, 2002b; Wynn and Masson, 2008). Whatever the recent progress may be, more detailed studies are still required for a better recognition of the interactive roles played by contour and turbidity currents. This will be useful for the improvement of interpretations of ancient sedimentary series for oil prospecting and for deciphering the palaeoclimatic indications in contourites.

Originally, contourites were defined as the sediment deposited by thermohaline bottom currents flowing parallel to bathymetric contours

(Heezen et al., 1966). These authors considered contourites as deep-sea deposits, as they were studied only on the continental rise. During 1980s, the term lost its original meaning; it is now used for all bottom-current derived sediments deposited in environments ranging from shallow to deep-seas. Some contourites have also been described from lakes (Flood and Johnson, 1984). Consequently, the origin of the current is not the key parameter that defines contourites (Lowel and Stow, 1981), and a wide range of bottom currents might be involved in contourite deposition: currents generated by internal waves, tides, wind-derived surface currents and up and downwelling. Faugères and Stow (1993) have proposed to restrict the term 'contourite' to the original concept of Hollister and Heezen (1972), for both modern and ancient environments. The term 'contourite' should, thus, be used exclusively for sediments in relatively deep water (deeper than ~500 m), deposited or significantly reworked by stable geostrophic currents. This definition is now widely accepted. However, 'stable geostrophic currents' imply either thermohaline geostrophic currents or wind-driven surface currents. Part of the thermohaline currents and most of the wind-driven currents may act on a shallow sea-floor and be responsible for contourite deposition in shallow-water environments such as the continental shelf edge and the uppermost part of the slope (Carter, 2007; Fulthorpe and Carter, 1991, Viana and Faugères, 1998; Viana et al., 2002). Such contouritic deposits have been called 'shallow-water contourites' (Viana et al., 1998).

Large deep-sea accumulations of sediments consisting primarily of contourites that were originally named sediment drifts or ridges, are now called 'contourite drifts' (or 'shallow contourite drifts' if located in shallow water) to prevent any confusion with other oceanic drifts.

In this chapter, we first provide an overview of the global geostrophic circulation, and the characteristics of contour currents. Then, we discuss the sedimentary processes controlled by contour currents, the facies variability of contourite deposits, and the associated bedforms. A following section outlines the features of the sedimentary bodies built by contour currents: contourite drifts and other combined contourite/turbidite systems.

2. Oceanic Geostrophic Circulation and Contour Currents

2.1. Surface geostrophic circulation

The direction of large wind-driven currents developed at the ocean surface is controlled by the atmospheric cells, the Coriolis force due to the Earth's spin and the ocean morphology. The water molecules are transported into a direction and with a velocity mainly controlled by the wind energy and the Coriolis force. On the northern hemisphere, the Coriolis force deflects any flow, including ocean surface currents, to the right of the wind direction; this is to the left on the southern hemisphere. The velocity decreases with water depth due to increasing friction, which deflects increasingly water layers with depth. Water thus takes a spiral motion with depth, called the 'Ekman spiral'. The maximum depth for wind influence is highly variable, but can reach down to thousands of metres. These currents follow cyclonic or anticyclonic gyres (Fig. 3.1). The mechanisms of water transport are complex and beyond the scope of this chapter. The surface water is not usually driven only by the wind. The pathway followed by the surface water is never straight. Remote sensing and *in situ* investigations demonstrated that a major surface current like the Gulf Stream or the Agulhas Current (south of South Africa) have 'spaghetti' and gyre-like shapes (Fig. 3.2).

Some of the currents forming parts of the gyres of the global surface circulation are almost permanent and represent the motion of an enormous flux of water with a high velocity (from tens of cm s^{-1} to more than 1 m s^{-1}). Consequently, they have a strong impact on deep-sea environments. That is certainly so for the Gulf Stream, which has a water flux of 120 Sv [1 Sv (Sverdrup) = 10^6 m^3 s^{-1}]. This represents more than 100 times the combined discharge of all rivers on earth. Comparable situations exist for the Brazil Current, the Kuroshio Current off Japan and the Antarctic Circumpolar Current driven by the western wind. However, these currents do not follow isobaths over a long distance, and they cross all the oceanic environments, from the continental shelf to the deep-sea. The deepest part of such currents usually sweeps the deep-sea bottom along

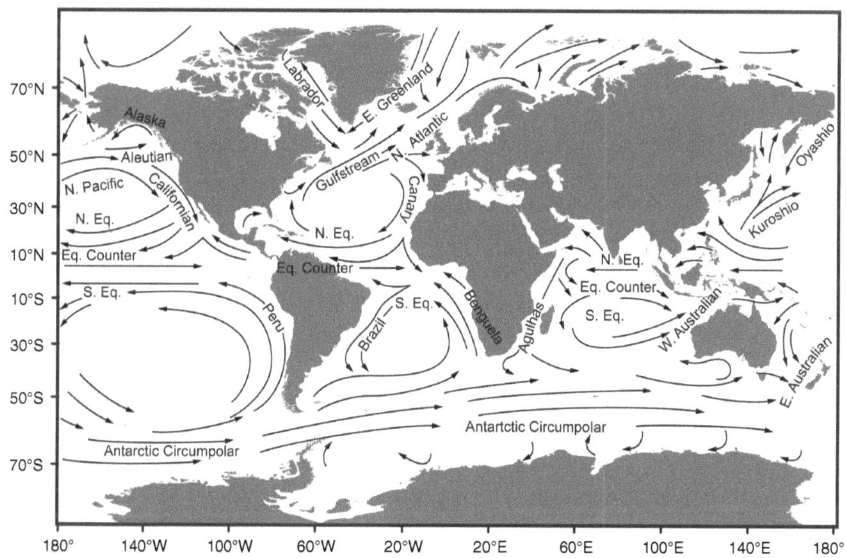

Figure 3.1 Major surface currents of the oceans.

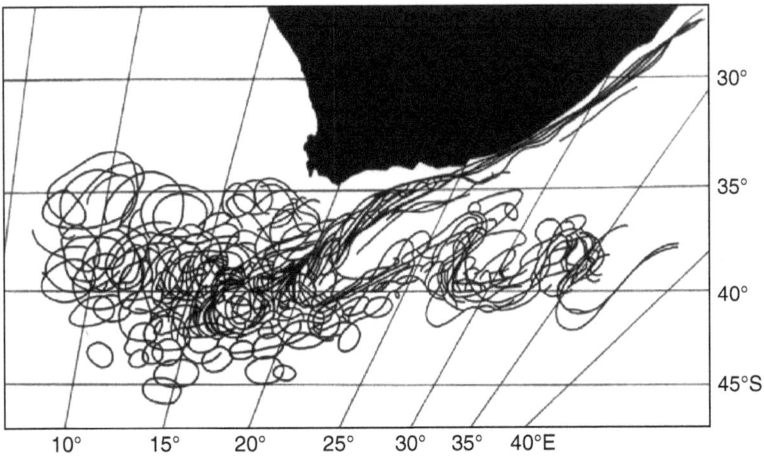

Figure 3.2 Edges of the Agulhas Current for the December 1984 to December 1985 period (modified from Lutjeharms, 1996).

the continental slope. There, they form bottom currents with erosion, transport and depositional processes that are similar to those generated by the thermohaline circulation. In addition, when they cross an underflowing (deeper) contour current, the resulting interaction may be responsible for contour-current shift or abyssal storms. This process is discussed in more detail in the following sections.

2.2. Geostrophic thermohaline circulation

The sinking of dense surface water at high latitudes forms the modern deep ocean water (Fig. 3.3). The high density of this water is due to cooling of the polar water and its increased salinity as a result of sea-ice formation. This process produces distinct water masses, with subtle differences in density and chemical signature. The dense-water masses are stratified according to their density, which depends on temperature and salinity variations. A given water mass can be traced over considerable distance. For example, the Mediterranean Outflow Water (MOW) that forms in the Alboran Sea (western Mediterranean) can be found in the Rockall area (Scotland). In general, a water mass is found at a water depth that corresponds to its density equilibrium in the water column. The kinetic energy acquired during the sinking of the water allows the flow to move slowly over thousands of kilometres from the source area. The stratification is a stable element of the global ocean, but the amplitude of the density contrast, the mixing at the boundary between two water masses, the interactions with surface currents and the effects of the sea-floor topography result in a complex deep-circulation pattern (Fig. 3.4).

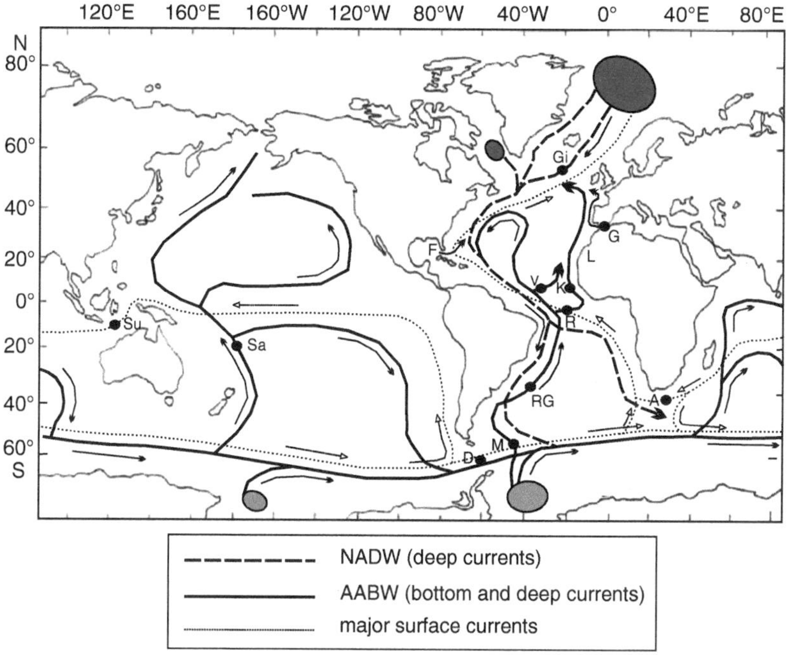

Figure 3.3 Schematic global thermohaline circulation and major currents. A, Agulhas Gateway; D, Drake Passage; F, Florida Strait; G, Gibraltar Strait; Gi, Gibbs fracture zone; K, Kane Gap; M, Falklands Channel; RG, Rio Grande Rise and Vema deep channel; R, Romanche Gap; S, Samoan Gateway; Su, Sumba Strait; V, Vema Fracture Zone.

In the modern oceans, the polar regions are the two major sources for the thermohaline circulation (Fig. 3.3). The coldest waters with the highest density form in the southernmost polar ocean around Antarctica. They represent the largest source of bottom currents for the global ocean. Major sources are particularly localized in the Weddell Sea and the Ross Sea. The former mostly supplies water to the Atlantic Ocean. The latter feeds the Indian and South Pacific Oceans. The water mass formed in the Arctic Ocean has a slightly lower density. It forms the North Atlantic Deep Water (NADW) that flows southward above the northward flowing Antarctic Bottom Water (AABW), and finally participates to the deep circulation in the Indian and Pacific Oceans (Fig. 3.4). The origin of the NADW is fairly complex as it is composed of cold saline water that has sunk in the Norwegian and Labrador Seas, some AABW and a component of warm saline water derived from the Mediterranean Sea.

Some minor sources are also involved in the thermohaline circulation. The Mediterranean Sea is an area where intense evaporation is responsible for an increase of the surface-water salinity and for the formation of dense,

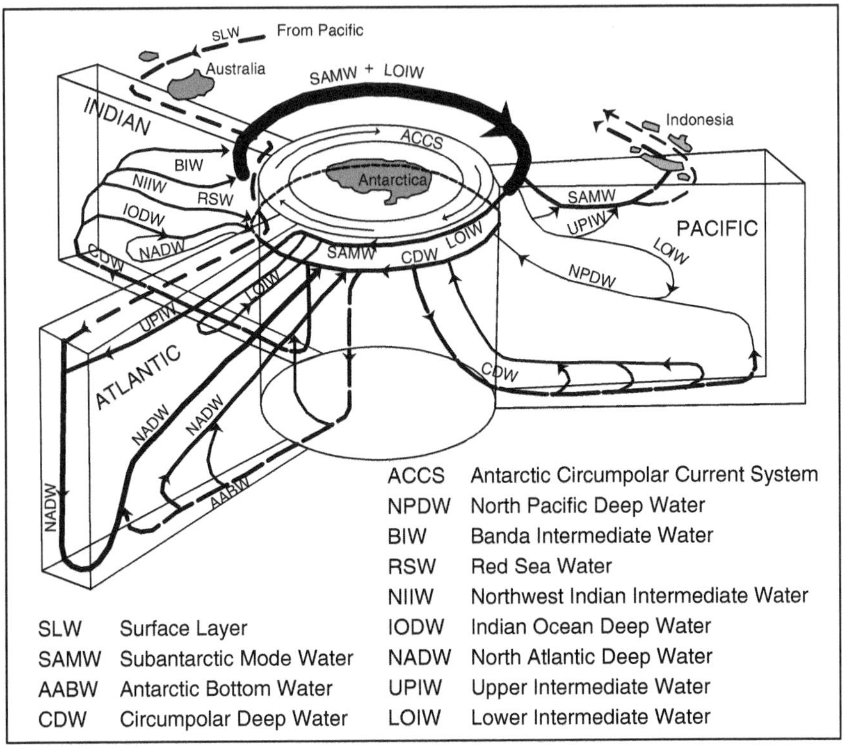

Figure 3.4 The thermohaline circulation. The Atlantic, Pacific and Indian Oceans are shown essentially as continent-enclosed arms radiating from the central Southern Ocean (modified from Schmitz, 1996).

warm saline water. This water flows across the Gibraltar Strait into the Atlantic Ocean, where it forms the MOW. The Caribbean Sea represents a similar but less active source of warm, saline water. This water is injected into the Northwest Atlantic Ocean circulation across the Florida Strait (Fig. 3.3).

The Atlantic Ocean has an N–S elongated shape with two major and two minor sources. For these reasons, it is an ocean that is highly suited to study the thermohaline circulation and its impact on the deep-sea sedimentation. The dense waters flowing from the south and north polar sources form stacked water masses flowing slowly (1–2 cm s^{-1}) towards the north and the south, respectively (Figs. 3.3 and 3.4).

This pattern of the thermohaline circulation is fairly recent. It progressively settled during the Cenozoic, following the development of ice sheets in the polar areas, and the associated global climatic cooling. It is not realistic to assume a similar pattern for older parts of the geological history like the Cretaceous, when warm and saline bottom waters formed on shallow shelf seas drove deep circulation. Such global changes in the thermohaline circulation also result from different positions of the continental plates.

2.3. Thermohaline bottom currents: Contour currents

The thermohaline bottom currents can be compared to submarine rivers that flow at different levels within the water column with a width that varies from a few to tens of kilometres, and a flux that more often largely surpasses that of the biggest rivers in the world (NADW: 18 Sv; Intermediate Antarctic Water, IAAW: 30 Sv; AABW: 1 Sv).

2.3.1. Definition and geographic distribution of contour currents

The thermohaline bottom currents have a complex distribution in the world ocean due to the existence of several controlling factors: the position of the dense-water sources, the Coriolis force, the topography of the sea-floor that controls the current paths in the deep basins and the shearing along other water masses.

The thermohaline bottom currents do not all follow the bathymetric contours. For instance, there are downslope currents that sink from the source areas towards the deep ocean. As soon as such a current reaches its equilibrium depth, the Coriolis force plasters it against the western continental margin of the ocean basins. Therefore, it flows parallel to the bathymetric contours of the margin and becomes a true contour current (Heezen et al., 1966). They are restricted and intensified along the margin, forming the Western Boundary Undercurrents (WBUCs), like along the northwest Atlantic rise. Finally, the water flows along-slope over a long distance, and part of the current water spreads slowly eastwards. That is more particularly the case in the equatorial regions, where the Coriolis force is hardly noticeable. In the equatorial Atlantic Ocean, the AABW is not constrained along the margin and flows eastward across the Mid-Atlantic Ridge, feeding the north-eastern Atlantic basins. On its way, the current loses energy and mixes with shallower waters; its density decreases and then it moves up to the surface, where it becomes part of the wind-driven circulation. This process results in a gentle, uniform upward diffusion, except in some coastal areas or larger oceanic regions such as the Eastern Equator Ocean, where enhanced upwellings are observed. That is also the case for the NADW, in the South Atlantic, where this current flows north to south. Its depth progressively decreases because its density decreases. On its way, it meets the dense AABW that sinks northward. Both the NADW and parts of the AABW are constrained to move upwards at a latitude of about 60°S (Antarctic Divergence).

A generalized pattern of deep currents that agree with the models of Stommel (1957) and Stommel and Aarons (1960) is presented in Fig. 3.3. Despite the difficulty of such a reconstruction, fairly detailed maps have been proposed for the Atlantic Ocean (Faugères et al., 1993; McCave and Tucholke, 1986), showing the great complexity of the current pathways (Fig. 3.5). All these documents highlight the role of the topography in the current distribution and show the pathway followed by the contour currents that form the WBUC along the South and North American margins.

In the Atlantic Ocean, volcanoes and seamounts that are aligned perpendicular to the trend of the oceanic margins (like the Malvinas Rise, Rio Grande Rise, Sierra Leone Rise, Walvis Ridge, in the South Atlantic) form topographic barriers for the currents. The presence of such topographic highs has two consequences:

(1) they make difficult the communication between the deep sub-basins, and a part of the flow can be trapped on the basin floor (this happens in the Argentine Basin, where a major part of the AABW flux is trapped and forms a complex system of gyres);
(2) most of the time, narrow, deep passages allow the communication between two adjacent basins. The current there becomes intensively restricted and intensified (this happens in the Vema Channel that cuts into the Rio Grande Rise between the Argentine and Brazilian Basins).

Such deep passages are also called 'channels' (Vema channel), 'gaps' (Kane Gap), 'seaways' or 'gateways'. Shallower passages are called 'strait' (Gibraltar Strait). These passages are frequent in the mid-oceanic ridge, where they correspond to the Fracture-Zone gaps (F.Z.). In the case of the Atlantic Ocean (Figs. 3.3 and 3.5), they allow the deep circulation to flow from the eastern to the western basins (the NADW through the Gibbs F.Z.) or from the western to the eastern basins (AABW through the Romanche F.Z.). The most important passages in the world are the gateways between the various oceans (the Drake Passage that connects the Pacific and Atlantic Oceans, and the Tasmanian and Indonesian seaways between the Indian and Pacific Oceans). They represent the location of the largest transfer of oceanic waters and show the major effects of the thermohaline circulation on the sea-floor.

2.3.2. Current velocity

The bottom-current velocity is the parameter that controls the impact of the flow on the sea-floor. The current velocity was first estimated semi-quantitatively from current bedforms observed on bottom photographs (Heezen and Hollister, 1971). Today, numerous long-duration direct current-metre measurements, recorded in various deep-oceanic domains, show a great variability in current velocity and direction.

Although much of the deep-sea-floor is swept by low-velocity currents (order of a few centimetres per second), higher current velocities are locally measured: 10–20 cm s^{-1} along the WBUC with peaks up to 70 cm s^{-1}

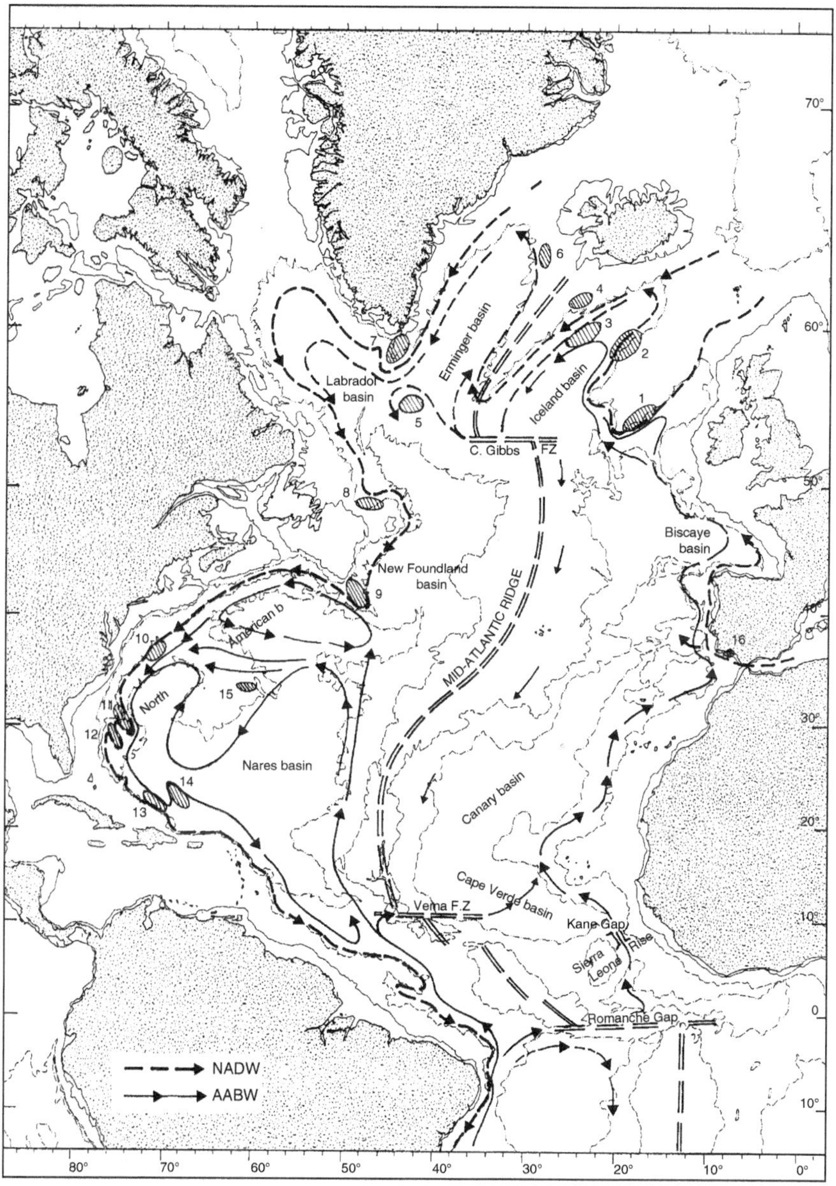

Figure 3.5 North Atlantic Ocean sub-basins, thermohaline deep circulation and locations of major drifts (from Faugères et al., 1993). NADW, North Atlantic Deep Water; AABW, Antarctic Bottom Water; 1, Feni; 2, Hatton; 3, Gardar; 4, Bjorn; 5, Gloria; 6, Snorri; 7, Eirik; 8, Sackville Spur; 9, Newfoundland; 10, Hatteras; 11, Blake; 12, Bahamas; 13, Caicos; 14, Greater Antilles; 15, Northern Bermuda; 16, Faro.

(Nowell and Hollister, 1985). Higher velocities are measured for down-welling currents and currents confined by sea-floor topography.

The velocity is variable from one current to another. For a given current, it is highly variable in both space and time. The average velocity decreases as the current flows away from the source area. For example, the MOW velocity decreases from up to 2.5 m s^{-1} west of the Gibraltar Strait to 10 cm s^{-1} at the western end of the Gulf of Cadiz. Along the current pathway, the irregularities of the sea-floor topography modify the velocity. Deep passages are locations where large velocity increases are observed. In places where the current is plastered against sea-floor topography by the Coriolis force, variations in the dip of the slopes are responsible for velocity changes: the higher the dip, the higher the velocity (Fig. 3.6). In reverse, currents spread and slow down when reaching flat abyssal plains or oceanic basins. In addition, the velocity can also change along a vertical section as a water mass can be formed by the superposition of several sub-currents with different densities. That is the case for the MOW in the Gulf of Cadiz, which is built by three superimposed currents (Ambar and Howe, 1979; O'Neil-Baring and Price, 1999).

The thermohaline bottom-current activity may also vary with time. At short time-scales (days to years), variations related to tidal or seasonal periodicity, including current reversals, have been recorded (DeMadron

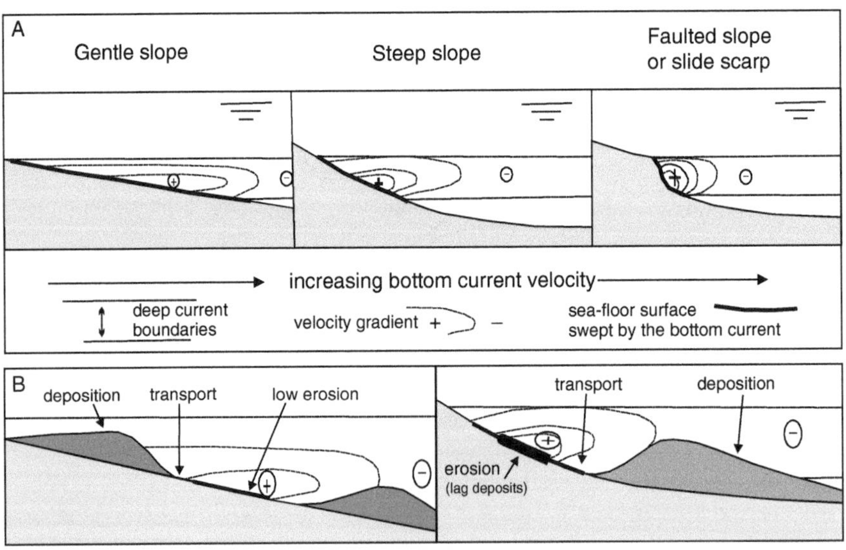

Figure 3.6 Relationship between the dip of the sea-floor and contour-current velocity (A) and sedimentary processes (B). (A) The steeper the slope, the higher the current velocity as an effect of the Coriolis force (here, current flowing towards the back of the page in the southern hemisphere). (B) Distribution of the sedimentary processes.

and Weatherly, 1994; Hogg et al., 1996). Other variations are related to internal waves (Hosegood and Van Haren, 2004) or to 'abyssal storms' (Nowell and Hollister, 1985) during which large-scale water eddies peel off and move at a right angle off the main flow direction, with velocities higher than during normal conditions (see Section 3.3). At a geological time-scale, and more particularly since the Eocene, the thermohaline bottom circulation experienced large variations, which are recorded in deep-sea deposits (Kennett, 1982). They are characterized by periods of slow and stable circulations that favour depositional processes and sediment accumulation in the deep ocean, alternating with short episodes of intense erosion and with the formation of widespread sedimentary hiatuses and erosional surfaces. These episodes are caused by 'hydrologic events' (intense formation of dense water) that are related to maximum ice-sheet extensions in the polar regions. They are related to global climatic changes that result from global tectonic events and re-ordering of the ocean structure and circulation (Bickert and Henrich, 2011, this volume, Chapter 12). Such variations have no regular periodicity. Since the end of the Neogene, the thermohaline bottom-current variations have a higher frequency but a lower amplitude (erosion processes are dominant only at local scales). They seem to follow the rapid and regular 'glacial' and 'interglacial' cyclicity, and thus to be controlled mainly by astronomical parameters. However, the exact relationship between current variations and global climate changes is still under debate.

3. Sedimentary Processes Related to Contour Currents

The sedimentary processes related to contour currents are also called 'along-slope processes' in order to distinguish them from the turbiditic downslope processes. Contour currents are efficient agents of sediment transport and deposition, and play a major role in shaping the deep-sea morphology. There is, however, not yet a complete understanding of the mechanisms that are responsible for the erosion and deposition by these currents. Consequently, we will highlight in this section only some of the problems that are being debated.

The main forces involved in the sedimentary processes of contour currents are the fluid thermohaline force, the wind force that transfers surface potential energy downwards through the water column during major storms and the Coriolis force. Chemical forces like those responsible for dissolution and cementation processes are also involved. In addition, the sedimentary processes also depend on the volume of the available sediments, that is, the amount of particles that reaches the deep-sea and forms a 'nepheloid layer'.

3.1. The nepheloid layer and sediment supply

It has been observed for long that high-velocity contour currents like the WBUC are frequently associated with a deep and thick (500 m to more than 1500 m) layer of turbid bottom water, called the 'nepheloid layer' (Biscaye and Eitreim, 1977; Ewing and Thorndike, 1965; Jerlov, 1953). This deep-sea nepheloid layer is well developed along the Western Atlantic margins (Nova Scotia and Argentine margins; Fig. 3.7) but also along the Antarctic and northern Pacific margins. It shows a downward increasing concentration (10 to more than 300 μg l^{-1}; Fig. 3.8) of suspended particles (about 12 μm in size). These particles remain in suspension because of the turbulent eddy diffusion for a period (residence time) of about 1 to several years.

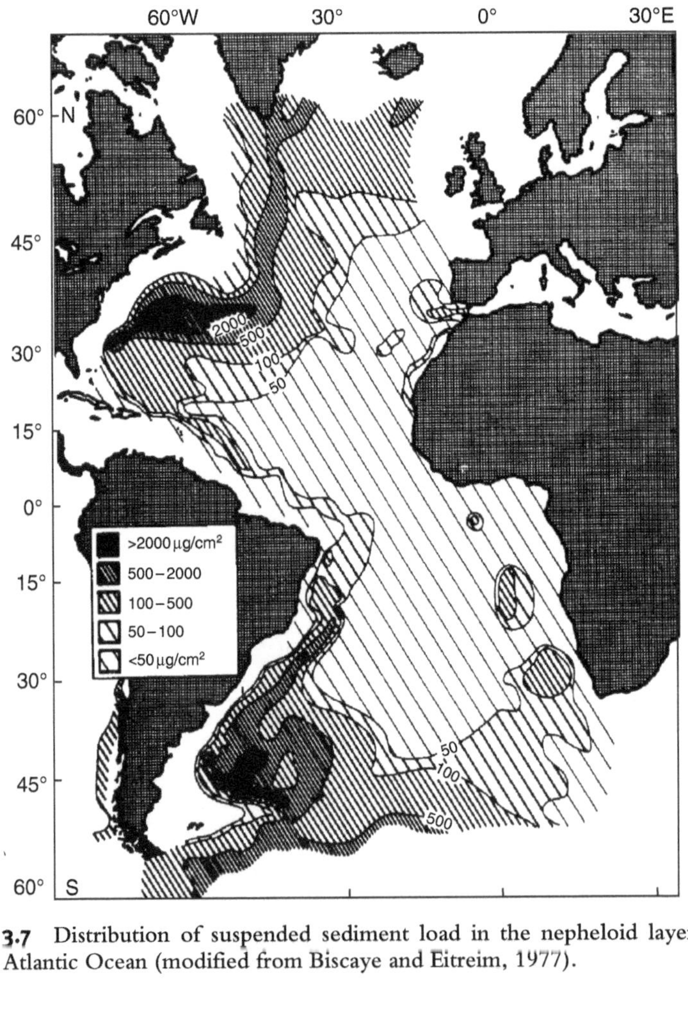

Figure 3.7 Distribution of suspended sediment load in the nepheloid layer of the abyssal Atlantic Ocean (modified from Biscaye and Eitreim, 1977).

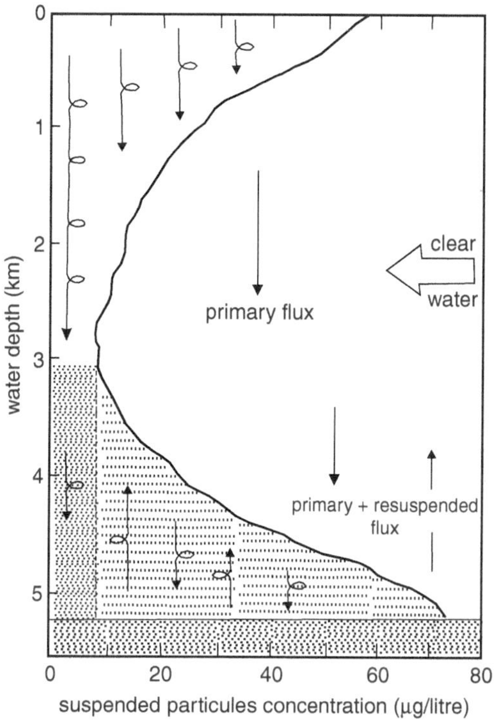

Figure 3.8 Typical nephelometer profile from an area with a well-developed nepheloid layer (redrawn and modified from Biscaye and Eitreim, 1977).

The structure and dynamic nature of the nepheloid layer have not yet been well understood. Rapid inversions of the vertical distribution of densities are frequently observed; they generate concentration discontinuities (McCave, 1986). Therefore, the nepheloid layer appears to be constituted of several superimposed beds, each of them showing a gentle upward density decrease. The discontinuities occur at certain depths (Fig. 3.9), forming isopycnal surfaces (pycnoclines) that trap the suspended sediment particles (the 'separated mixed layer model' of McCave (1986)). The fine-grained particles can be transported downslope by low-density turbidity currents (tens of metres thick on average). When a sea-floor current meets a pycnocline, the most diluted part of the flow detaches and moves laterally (lateral advection) along the surface (lutite flows: McCave, 1972). Reconcentration processes (Parsons et al., 2001) can generate density cascading. Locally, active internal waves may also be involved in the formation of dense layers of suspended particles. Such processes create a sharp layer with a higher particle concentration within the nepheloid layer. If several superimposed isopycnal surfaces are present, they may be responsible for the presence of several density peaks.

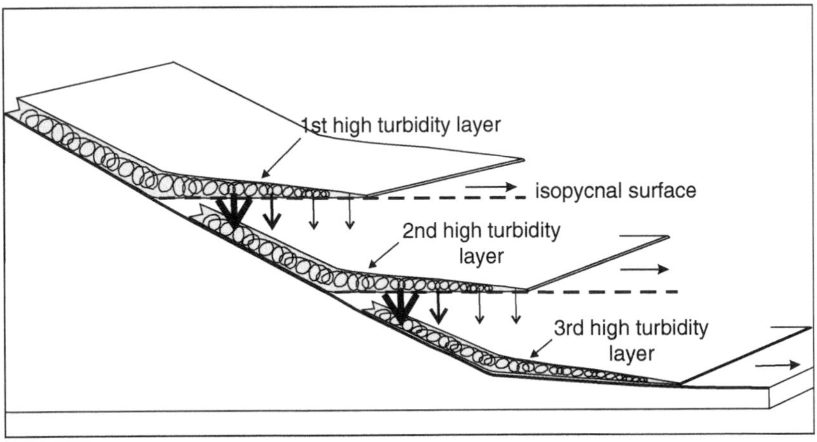

Figure 3.9 Schematic particle cascading and turbidity variations in the nepheloid layer about 900-m thick (redrawn and modified from McCave, 1986, 1972).

The settling of the suspended particles down the nepheloid layer is not well understood. It depends on the particle size. The largest silt particles (10–20 μm) show rapid settling, whereas the sinking of large clay particles (2–4 μm) is slower. The finest particles (0.5–1 μm) have a much lower accumulation rate. Particles with low settling velocities can be transported over long distances before reaching the sea-floor. For example, a 100-μm particle will sink in 5 days from the surface to a 4000-m deep abyssal plain, whereas a 10-μm and a 1-μm particle will reach the sea-floor in 1.5 and 15.0 years, respectively. Within a constant 0.1 m s^{-1} current, a 100-, 10- and 1-μm particle will theoretically move horizontally over 4,400 and 40,000 km, respectively, from the place at the ocean's water where it began to sink until it reaches the sea-floor. However, aggregation of fine particles (clay and fine silt) is a process that strongly changes the particle composition in the nepheloid layer, and that increases the accumulation rate. Alternation of slow settling of fine-grained particles and more rapid deposition of coarser-grained aggregates and particles are possibly the processes responsible for the formation of lamination in silty to muddy contourites.

The particles feeding the nepheloid layer have several sources. Turbidity currents are the most efficient agents. Although most of the coarse material is directly deposited in turbidite systems, the fine-grained particles remain in suspension in the upper part of turbidity currents. They can thus become redistributed by contour currents and transported inside the nepheloid layer. Another important process is resuspension of bottom sediments by erosion induced by contour currents. This process directly injects particles into the nepheloid layer. Suspended particles can also be advected from the continental slope along pycnoclines and be transported into the nepheloid layer by cascading (Fig. 3.9). Other particles come from settling of biogenic particles and pellets. A last process is the resuspension of particles by burrowing activity of benthic organisms. Deep-sea bottoms swept by bottom currents are oxygen- and nutrient-rich environments and show a flourishing fauna. Consequently, this process is probably not negligible. These multiple sources suggest that the volume of particles available for contour-current transport and deposition varies largely with climatic and sea-level fluctuations.

3.2. Processes at the deep-sea-water–sediment interfaces: The benthic boundary layer

The sedimentary processes controlled by thermohaline bottom currents mainly occur at the sea-water–sediment interface that acts as a critical transition zone. Because the sedimentary surface is variable in terms of compaction, cohesion, roughness, slope and particle grain size (soft spongy mud, gravely/sandy sediments, consolidated sea-floor, rock outcrops), the mechanisms of the sedimentary processes have not yet been clearly established.

At the very base of the nepheloid layer, there is a few metre-thick water layer in which the physical and chemical behaviour remains constant. This layer is named the 'BBL' (Fig. 3.10). In this layer, the flow is turbulent because of the irregularities in the sea-floor topography (including micro-irregularities). Because of the turbulent mixing, the values of the flow temperature, salinity and turbidity remain constant ('mixed layer'). Above the BBL, the temperature increases upward and the turbidity and salinity both decreases. In contrast, the velocity of the current increases from the

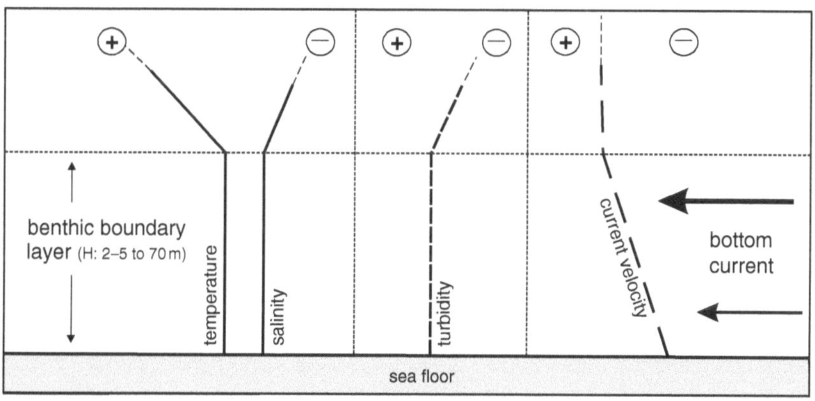

Figure 3.10 The benthic boundary layer and variations of physical and chemical parameters.

base to the top of the BBL and remains almost constant above it. Because the velocity increases as a log function of the distance from the sea-floor, this layer is also called the 'logarithmic layer'. Consequently, the abrupt change in temperature, salinity and velocity marks the upper limit of the BBL. Any change in the flow intensity affects the water turbulence and the characteristics of the benthic layer. Thus, a drastic increase in current velocity and in turbulence during benthic storms (see Fig. 3.12) will cause a strong thickening of the BBL (up to tens of metres). Such conditions simultaneously generate a high shear stress and increase the erosion of the sea-floor.

The critical contour-current shear velocity required to erode and transport sediment particles from and over a flat sea-floor determines the type of sediment transport (bedload or suspended load). Each sediment grain deposited at the sediment–water interface is subjected to a drag force and a lift force due to the current motion. When the fluid shear stress increases, these combined forces can exceed the grain weight due to gravity. The critical shear-stress threshold for grain movement is reached at a certain moment, and the grain then begins to move. Several experimental studies focused on the relationship between the current velocity that is just sufficient to induce motion for different grain sizes and types (Hjulstrom, 1939; Miller et al., 1977; Postma, 1967). The relationship is simple for coarse silt, sand and coarser particles, because it depends only on the particle size. However, it is poorly established for fine-grained particles (mud and fine silt), mainly because of the action of cohesive forces and the formation of grain aggregates (McCave, 1984, 2008). McCave (1984) proposed a diagram giving grain size as a function of critical shear velocity (Fig. 3.11). It shows transport/deposition limits for fine sediment and transport/erosion limits for sandy and coarser particles.

The experimental shear velocity (U^*) has been obtained as a function of the flow velocity (U) and the deposit's grain size (cohesion and roughness). It appears that there is no direct relationship between U^* and U. U^* also depends on the fluid viscosity:

$$\frac{U}{U^*} = \frac{5.75\,\log Y}{\left(0.108\frac{V}{U^*} + 0.033\mathrm{ks}\right)}$$

where Y is the height of the measurement point with respect to the sea-floor and V is the kinematic viscosity; the term ks represents a constant.

Hence, the erosion process does not only depend on flow velocity and particle grain size. As soon as the process is initiated, it increases exponentially with $U^\star$. For deep-sea muddy sediments that are mostly deposited by contour currents, the role of the high sediment cohesion is difficult to assess. Using results from scaled-tank experiments, velocities of 15–35 cm s^{-1} and of 1–1.5 m s^{-1} have been proposed to erode a carbonate ooze and a cohesive mud, respectively. Erosion velocities for unconsolidated silts and

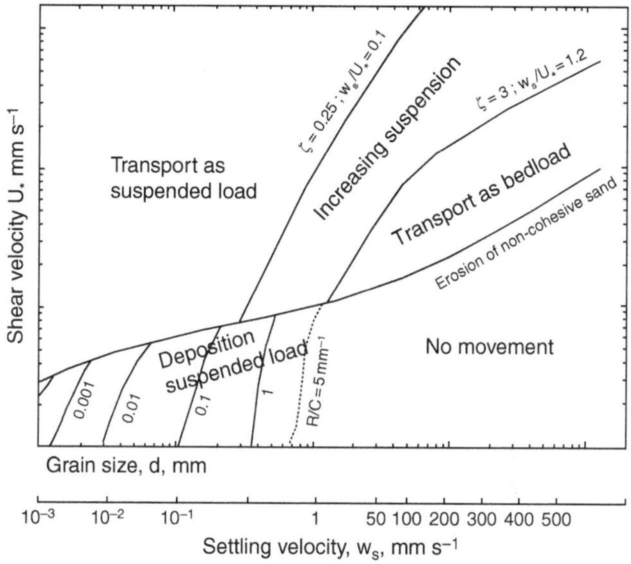

Figure 3.11 Erosion, transport and deposition of deep-sea sediments (modified from McCave, 1984). Reproduced with permission from Geological Society, London.

for unconsolidated sands are about 10–20 cm s^{-1} and about 20–40 cm s^{-1}, respectively. However, such velocities are exceptions in deep-sea environments.

3.3. Benthic storms: The results of the HEBBLE programme

In 1978, the High Energy Benthic Boundary Layer Experiment (HEBBLE) was initiated. The goal of this 10-year multidisciplinary programme was to increase understanding of the physical and biological response of deep-sea sediments to high-energy flows, using *in situ* observations, experiments and modelling, and to predict the sea-floor response to high-stress events (Hollister and McCave, 1984; Nowell and Hollister, 1985). The study area was located at a water depth of about 4800 m along the lower Scotian Rise (NE USA margin), downstream the Laurentian Fan. This area is swept by the active WBUC (velocity up to 70 cm s^{-1}). It is supplied with abundant terrigenous sedimentary particles and has a dense nepheloid layer with particle concentrations up to 120 μg l^{-1}. During the experiment, parameters including temperature, salinity, turbidity and current directions have been measured continuously and simultaneously with observations of the sea-floor (photographs) to assess variations in bedforms and biologic activity. Surface-sediment box cores have been collected regularly to establish the changes in sediment structure and texture, and to estimate the local erosion or sedimentation rates.

The major result of the programme was the determination of the occurrence of rapid variations in kinetic energy conditions along the sea-floor. Episodes of high-energy processes occurred and were characterized by high current velocities and dominant erosion processes. They lasted a few days to weeks (8–10 events per year). These episodes are called 'deep-sea storms' or 'abyssal storms' or 'benthic storms'. They alternate with long intervals (weeks to months) of moderate-energy conditions, low-current velocities and predominantly depositional processes.

During benthic storms, the mean current velocity is over 15 cm s^{-1}, with peaks over 40 cm s^{-1}. The current direction varies largely and often with current reversals. It suggests a gyre-like water rotation with the gyres moving along the rise. The thickness of the BBL increases from a few metres to several tens of metres (up to 70 m), and the turbidity becomes very high (300–10,000 μg/l^{-1} instead of 80–100 μg/l^{-1}). Winnowed sands and numerous dynamic figures form scattered patches on the sea-floor. A surface layer of a few millimetres to a few centimetres thick can be eroded. These episodes are thus characterized by significant sea-floor erosion and resuspension. Consequently, a large volume of resuspended particles is injected into the nepheloid layer, and available to be transported over long distances by the WBUC, before deposition takes place again.

During the low-energy intervals, the current velocity is much lower (5–10 cm s^{-1}). The turbidity within the nepheloid layer decreases drastically, because of the rapid settling of the suspended particles. Accumulation rates of up to 1.4 cm per month occur. Much organic debris is present on the sea-floor, becoming covered by a layer of soft mud that is intensively bioturbated. The only visible current bedforms are centimetre-long, elongated mud mounds. Deposition and bioturbation are thus the predominant processes.

The surface deposits collected in box cores show two superimposed beds, dated both as Holocene. The upper bed is composed of a soft silty/clayey mud bed of a few (0.5–12.5 cm) centimetres thick, which has a low-shear resistance. The lower bed is more than 30-cm thick. It consists of a compact silty/clayey mud with a high content of large foraminifers. The shear resistance is 10 times higher in this bed than in the upper one. These observations suggest that the upper bed consists of reworked deposits that have suffered several cycles of erosion and deposition. The lower bed corresponds to older deposits that have not been affected by the modern erosion and sedimentation processes. During a longer time span, such as the Holocene, a net sedimentation rate of 5.5 cm per 1000 years is observed on this part of the continental rise. The balance of the alternating depositional and erosional processes is positive and some material progressively aggraded at the top of the lower bed. Such values are consistent with the sedimentation rate estimated for larger geological intervals (2–10 cm per 1000 years) for the giant contourite drifts in the North Atlantic Ocean.

The HEBBLE results clearly demonstrated that the processes that are responsible for benthic storms are due to interplay of the thermohaline bottom currents and the wind-driven surface currents. Atmospheric high-energy conditions increase the surface-current energy, propagate downwards and induce high-energy conditions on the deep-sea-floor, thus generating a benthic storm (Fig. 3.12). Areas of high-kinetic energy at the ocean surface are linked to the major surface currents driven by major winds (such as the trade winds) above the equatorial Atlantic Ocean, the Antarctic circumpolar wind (or West wind drift) and the Kuroshio off Japan in the Pacific Ocean. The active atmospheric system and the oceanic surface circulation produce large energetic and rotating eddies that persist for weeks to months, and that are frequent under local low atmospheric pressure. Such conditions are observed in a few areas such as along the eastern USA and Argentine margins, around South Africa, in the circum-Antarctic regions and off Japan (Fig. 3.13). These areas coincide with the areas where dense nepheloid layers, active deep circulation and significant contourite sedimentation occur. They also correlate with areas of important benthic eddies derived from variations in current velocity (Schmitz, 1984, 1988), and with the global distribution of benthic storms, as established from bedform photographs (Hollister, 1993).

Benthic storms may occur at various places along these areas. Consequently, a particle can be resuspended several times before being definitively buried. Each time, it is mixed with particles injected into the nepheloid layer from new sources. Thus, the sediment particles may have experienced

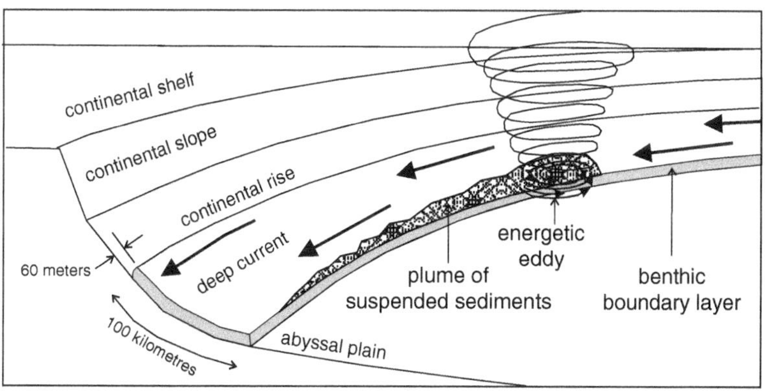

Figure 3.12 Schematic benthic storm on the lower continental rise of the USA. Atlantic margin: a mesoscale eddy interacts with the southward-flowing abyssal current. Such abyssal eddies (in the form of an ellipse of approx. 30 km long and 5 km wide) are correlated with the presence at the sea surface of mesoscale eddies shed by the Gulf Stream (modified from Hollister et al., 1984).

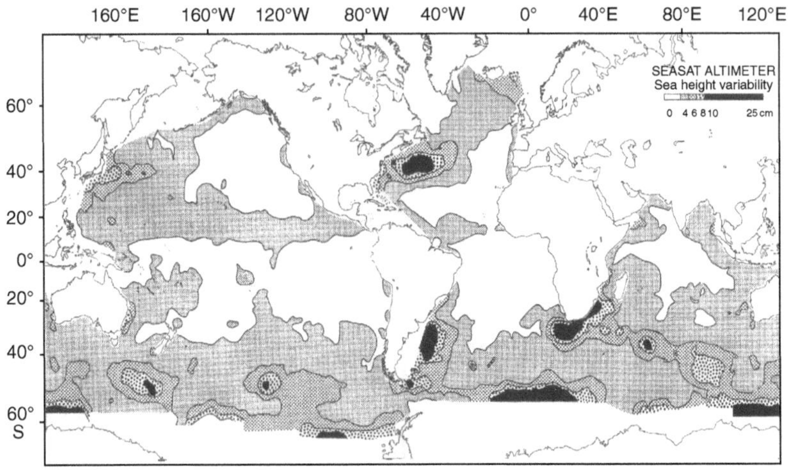

Figure 3.13 Variability in the height of the sea surface. Note that a high variability (up to 4 cm) is observed in only a few areas: along the eastern USA and Argentine margins, around South Africa, in the circum-Antarctic regions and off Japan (modified from Cheney et al., 1983).

a long transport with successive phases of deposition, and they may be derived from multiple sources before final burial.

A last HEBBLE result is that rapid and large variations in current direction and velocity occur, as shown by current measures and bedforms. Contour currents are not really stable as thought previously. Consequently, sedimentologists must be cautious with palaeocurrent reconstructions on the basis of bedforms that have been observed in local outcrops only.

3.4. Erosion and dissolution processes at the scale of oceanic basins

3.4.1. Erosional surfaces and sedimentary discontinuities

Erosion of the deep-sea deposits by contour currents is mainly recorded by erosion surfaces or discontinuities marked by a hiatus associated with overconsolidated sediments (hardground), and sometimes an unconformity. Such surfaces may be recorded at the scale of oceanic basins and oceans (Kennett, 1982). They result from exceptional high-energy events. They may also have a smaller extent and be controlled by a sea-floor topography that locally favours an increase in current velocity.

Some sedimentary facies and bedforms are typical of current-induced erosion of the deep-sea bottom, like coarse-grained lag deposits that are derived from previously deposited sediments by current winnowing and erosion of the fine material; and such as shale clasts in a mud breccia, furrows, elongated patches of sands, elongated ripples, grooves and craig-and-tail features.

Major erosional discontinuities (Fig. 3.14) extend over the whole Atlantic and the West Pacific Ocean margins. They are well documented from seismic profiles and DSDP (IODP) wells crossing the main contourite drifts (e.g. Duarte and Viana, 2007; Faugères et al., 1999; Gambôa et al., 1983; Johnson, 1972; Kennett, 1982; Mézerais et al., 1993; Stoker, 1998; Tucholke and Mountain, 1986, among others). They are associated with important sedimentary gaps representing hiatuses of several (up to tens) millions of years, but it is difficult to find out whether the missing sediments result from erosion or non-deposition. Erosion associated with hiatuses has been estimated to have removed several hundreds of metres (Johnson, 1972). The most important discontinuities are found:

- at the Palaeocene/Eocene boundary;
- at the Eocene/Oligocene boundary, when the Antarctic ice sheet had a large extent, initiating a huge increase of the AABW;
- at about the Middle-Late Miocene boundary (about 14 Ma), when the Iceland Ridge sank, making possible the overflow of a large volume of dense Arctic water to the Atlantic Ocean, and intensifying the NADW circulation (simultaneously, most of the Antarctic ice sheet was formed, increasing the AABW production);

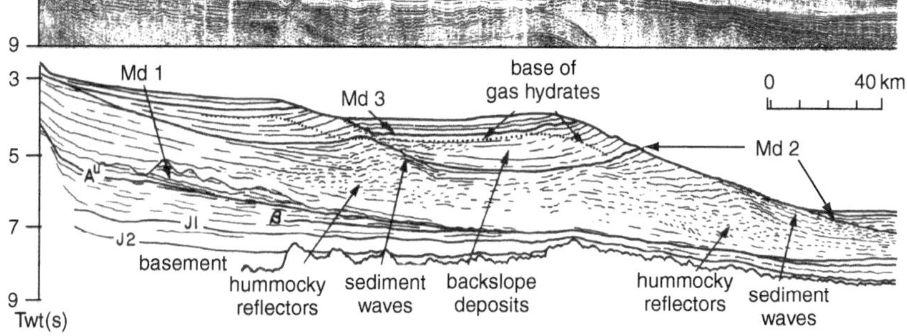

Figure 3.14 Seismic line crossing the Blake Outer Ridge (SE USA margin) and showing three major discontinuities (Md1, Md2, Md3) that are related to global hydrological events (modified after Mountain and Tucholke, 1985).

– during the late Pliocene, at about 2.5 Ma, when the Central American Seaway ('Panama Strait') closed, generating the modern Gulf Stream current, the formation of the Arctic ice-sheet and then the initiation of the modern NADW circulation.

Therefore, the surfaces representing hiatuses are caused by major hydrological changes in the deep circulation pattern related to plate-tectonic readjustments, inducing major changes in the configuration of ocean basins. However, many questions concerning their formation remain unsolved. The end of an erosion interval can be determined unless the overlying material has been eroded by a younger erosion event. The initiation time is even more difficult to determine, because the thickness of the removed deposits is usually unknown. Whatever the genesis and timing of the surfaces are, their existence means that no particles were deposited on most of the continental rises and in oceanic basins over long time spans. These intervals might be associated with the formation of a particularly thick and concentrated nepheloid layer.

Erosion and bypass surfaces of shorter duration and extent are associated with local sea-floor highs that cause current narrowing and velocity increase, as previously discussed.

In deep channels, truncations of seismic reflectors form evidence of intense erosion of the channel flanks. In addition, a surface pavement of manganiferous material with patches of coarse sand and gravel reflects the dominance of transport processes along the channel floor, with deposition only at the downstream issue of the channel, where the velocity suddenly drops.

When the Coriolis force plasters a contour current along a steep slope, the velocity is relatively high along the slope and decreases seaward. Consequently, on the upslope side of the flow, erosion and transport are dominant and a moat (marginal) channel forms. At the other side, the flow velocity is low, and deposition progressively increases. Thus, a contourite drift develops. If these processes are active for a long time, the channel will deepen simultaneously with the thickening of the drift. Consequently, the current becomes more and more confined. Its velocity increases and erosion processes on the sea-floor are intensified (Fig. 3.6). Tectonic processes like faulting may also facilitate the marginal erosion and channel formation. The resulting sedimentary systems are called 'moat drift' systems or 'contourite channel–levee' systems or 'separated drifts' (drift terminology of McCave and Tucholke, 1986, see section 5-1).

An isolated topographic high on the sea-floor, such as an abyssal hill, may be located on the pathway of a contour current. In this case, local erosion affects the sea-floor upstream and on both sides of the high. Erosion is thus associated with moat formation and contourite accumulation downstream of the high, in the 'hydrodynamic shadow' (Davies and Laughton,

1972). Larger areas may experience continuous erosion or non-deposition at a geological time-scale when active currents are trapped along the floor of confined basins. This occurred in the Cap Basin where the AABW circulation has prevented deposition all along the rise of the margins of the basin since the Miocene (Tucholke and Embley, 1984).

3.4.2. Dissolution and discontinuities

Some of the major discontinuities in the sedimentary successions of the modern oceans are not the result of active contour currents. They may be caused by dissolution processes (chemical erosion, corrosion), due to a specific chemistry of the deep water with low terrigenous load.

Below the Carbonate Compensation Depth (CCD), the water masses have a strong dissolution capacity with respect to the biogenic carbonate particles that sink from the surface. These carbonates are largely dissolved in the water column and/or on the sea-floor, except where high sedimentation rates result in fast burial (Hüneke and Henrich, 2011, this volume, Chapter 4). When the sediment supply is dominated by pelagic shells, and when the currents are active enough to slow down the particle vertical settling, dissolution is the prevailing process. That prevents deposition and thus, a gap in the sedimentary succession results. Because the sediment–water interface may not witness accumulation for a long time, such processes favour diagenetic alteration of the sea-floor that will protect the underlying deposits from further erosion.

In marine successions, such discontinuities record major chemical changes in the deep-water masses, and they are related with major hydrological events. For example, the rise of the CCD related to an increased productivity is responsible for the formation of dissolution-derived discontinuities in pelagic successions that were previously situated above the CCD. Such processes are particularly well recorded from the Cenozoic in the Pacific Ocean (Haq, 1991; Kennett, 1982). They are generally consistent with major sea-level fluctuations but do not fit with the major Cenozoic hydrological events. They are also known from the Quaternary, for instance, from the glacial stages in the South Atlantic Basin (Bickert and Wefer, 1996), where they are due to an increased formation of the cold and CO_2-rich AABW. Such high-frequency CCD variations are responsible for small hiatuses and rapid cyclic variations of the carbonate content in deep Quaternary deposits, for instance, in the Pacific Ocean.

3.4.3. Indurated sediment surfaces and condensed surfaces: Authigenesis

Other discontinuities may result from the interaction of various forces: bottom-current activity, chemical forces, sediment supply and eustatic sea-level fluctuations. Bottom-current activity favours transport processes together with a limited sediment supply. This generates longer exposure of the sea-floor and thus a relatively high exchange of dissolved chemical species at the sediment–water interface. Such a condition leads to early diagenesis and authigenic precipitation of minerals such as carbonates, phosphates, ferro-manganese oxides on the sea-floor surface or in the underlying surface sediments. That results in a consolidated sea-floor, ranging from a slight cementation to a hard crust. Such consolidated layers are frequently associated with major discontinuities and local erosional surfaces (e.g. Hüneke, 2006).

Authigenic processes occur frequently on the abyssal floor of the deepest oceanic basins that receive low terrigenous input due to the long distance from the continental margins. These basins are commonly swept by more or less active bottom currents and located below the CCD (Hallbach et al., 1988; Lonsdale, 1976). Fields of ferro-manganese micro- and macro-nodules and metallic crusts develop largely in the areas of red clay deposition. However, not all the fields of manganese nodules are associated with bottom-current activity (Kennett, 1982), because they need a specific (e.g. volcanic) chemical environment to be formed.

Condensed surfaces (or intervals, or successions) defined in the concept of seismic stratigraphy (Vail et al., 1977) show authigenesis (formation of carbonates, phosphate, glauconite, iron, manganese minerals, Hesse and Schacht, 2011, this volume, Chapter 9). These areas are present on the seaward shelf border and upper marginal slopes, during stages of increasing accommodation. Condensed intervals are retrograding and diachronous along the upper part of the margin, which makes a difference with discontinuities due to the increased activity of currents that are synchronous over large areas.

3.5. Transportation and depositional processes

Particles transported by contour currents are mainly either supplied by downslope processes or winnowed from older deposits. They are mainly terrigenous (clays and fine silts) or biogenic (mollusc shells, foraminifers and coccoliths, diatoms, radiolarians). These fine-grained particles are suspended in the nepheloid layer and transported over thousands of kilometres along the continental margins. Material originally supplied from the Norwegian Sea can be transported down to the Blake–Bahamas drift system, after a 15-year transport of about 6500 km (McCave, 1986). Similarly, chlorite- and diatom-rich sediments formed at high latitudes in the southwestern Atlantic are transported over more than 3000 km northward along the Argentine and Brazilian margins by the AABW (Massé et al., 1996). The HEBBLE results show that this transport is the result of alternating long intervals of motion and deposition, and short phases of erosion and particle resuspension. Consequently, the contourites forming the drifts have multiple sources.

HEBBLE also showed that coarse particles can be reworked and transported by the currents during benthic storms. However, it seems that these particles are, as a rule, rapidly deposited without significant horizontal transport. Finally, transport of coarse-grained particles occurs only where 'stable' high-velocity currents exist. These currents occur in specific morphological and/or hydrological environments such as the outer shelf and upper slope, deep moats and channel systems, areas of sinking dense water or local deep-sea areas of active bottom currents (Viana et al., 1998). There, bedload transport of sand is associated with winnowing and forms sandy sheet-like or patchy deposits associated with sandy ribbons, transversal and barchanoid ripples. If the current energy is high enough, the sea-floor is swept and winnowed repeatedly. Sandy particles are removed and a gravely lag deposit remains, associated with a ferro-manganese pavement.

Contour-current deposition occurs as soon as the current velocity decreases. The current velocity controls the flow competency and capacity and thus the horizontal and vertical variations in the contourite facies. The flow velocity thus controls the global drift morphology and geometry and the associated bedforms.

3.6. Parameters controlling contourite deposition

The oceanic circulation is the most important control factor regarding contourite deposition (Figure 3.15). Secondary factors include sea-level changes and the nature and volume of sediment supply. These factors depend on the astronomic cycles and high-frequency (<20,000 years) climatic changes (Milankovitch, 1941; Mulder, 2011, this volume, Chapter 2.3.4.1), and the morphological and geological background (size and shape of the sedimentary basin, tectonic environment). These factors will not be discussed in detail, but the impact of the eustatic sea-level changes on the distribution of contourite deposits will be briefly discussed underneath in order to compare it with its impact on turbidites.

Sea-level high-stands are associated with a low terrigenous load, reduced oxygenation of sea water, rising of the CCD, an increased sedimentation rate on the shelves and upper continental margins, and reduced deposition in the deep-sea with major mud-drape facies and condensed successions. Turbidite deposition is limited (Mulder, 2011, this volume, Chapter 2.1.3.2). In contrast, sea-level low-stands are associated with a high terrigenous load, widespread erosional discontinuities on the shelves and slopes, and active development of large turbiditic systems on the continental rise and abyssal plains.

There are no uniquivocal data that permit to make such a direct link between the sea level and the rates of drift accumulation or destruction, and certainly not at the scale approaching that of glacial–interglacial sea-level fluctuations. Significant accumulation on contourite drifts is mostly favoured by a moderate intensity of bottom current, and needs not too

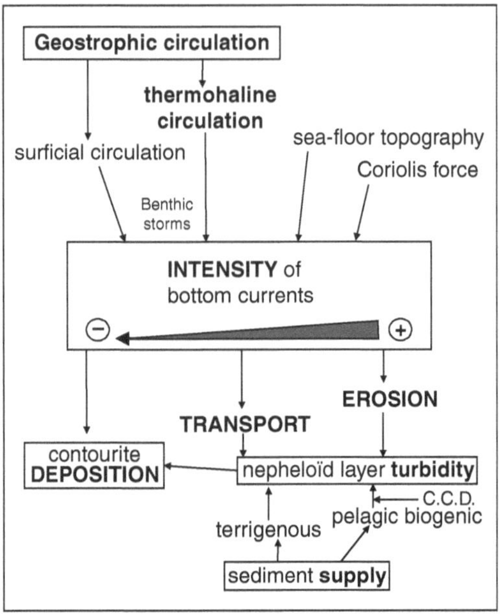

Figure 3.15 Factors controlling the sedimentation by contour currents (from Faugères et al., 1993).

high rates of sediment supply via turbidity currents or other mass flows (otherwise that would mask the contourite sedimentation). As there is still much uncertainty today in relating particular climatic conditions (and hence sea level) to particular intensities of the thermohaline circulation, and as different water masses may behave differently during the same climatic episode, we might therefore conclude that, at the scale of glacial–interglacial cycles or longer during the Neogene, there are more or less random variations in drift growth related to current intensity and sea-level changes.

4. Contourite Facies and Bedforms

4.1. Contourite facies in recent deposits

The question of which diagnostic lithofacies characterize contourites has been addressed first by Hollister and Heezen (1972) and Stow and Lowell (1979), and has only partially been answered up till now. Following the large number of drilling wells and gravity cores collected in contourite drifts from all the world oceans, significant progress has been made in the characterization of modern contourites, and reliable facies models have been proposed. However, because contourites can resemble other deep-sea deposits, their identification remains difficult by the naked eye, and therefore requires additional analyses, including grain-size analysis, X-ray radiographs (Fig. 3.16; Migeon et al., 1999) and image analysis of thin sections of indurated sediments (Fig. 3.17). That is particularly true for the fine-grained contourites that may be misinterpreted as fine-grained turbidites or

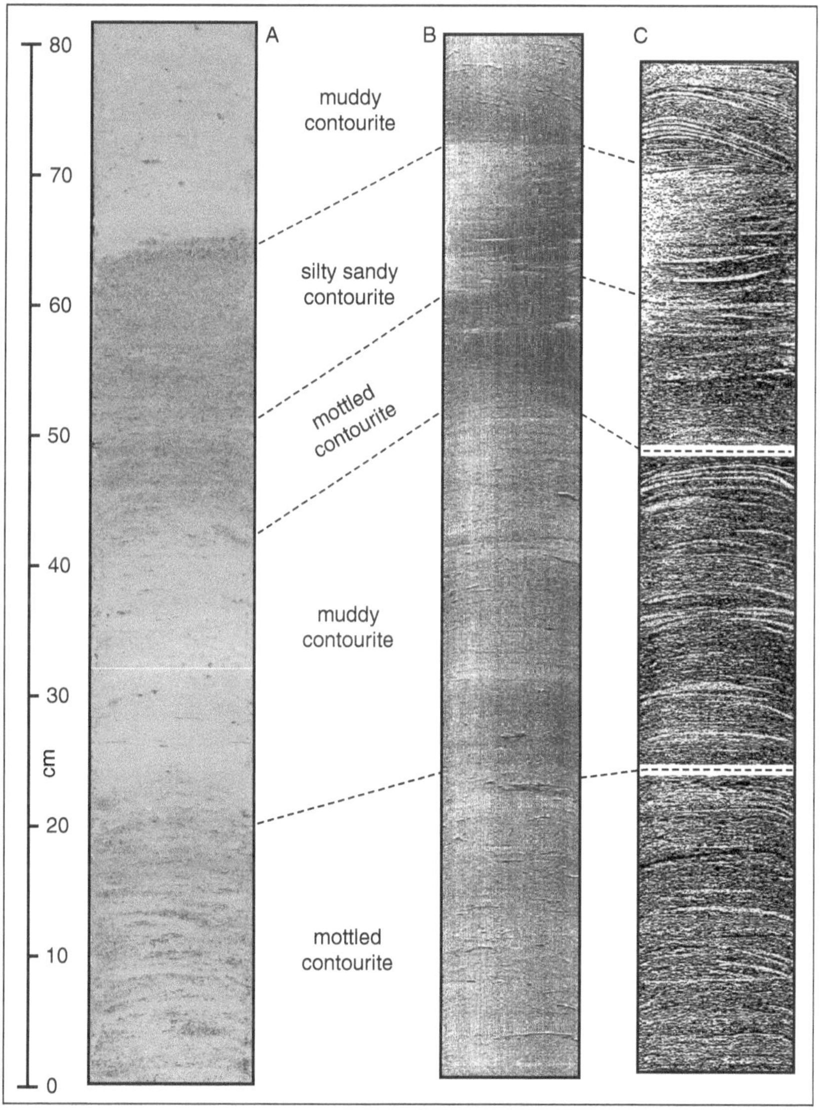

Figure 3.16 Contourite facies (Faro drift, Gulf of Cadiz). (A) Core photograph; (B) X-radiograph; (C) processed X-ray image (modified from Migeon, unpublished data). (A multi-colour version of this figure is on the included CD-ROM.)

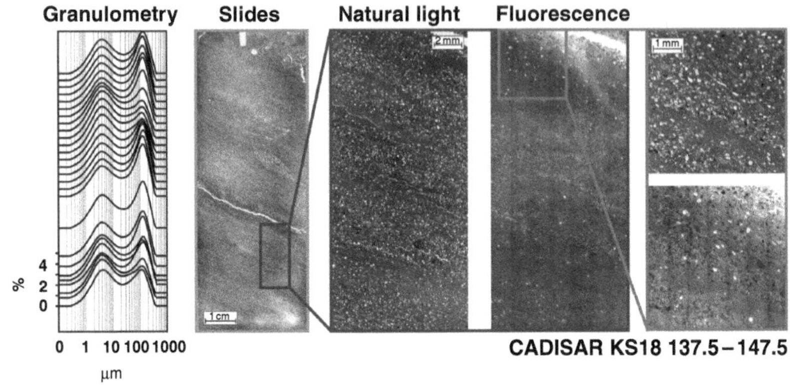

Figure 3.17 Silty/sandy contourite (Gulf of Cadiz.): high-resolution analyses using thin sections of indurated sediments and microscope-image analysis (fully automated Leica DM6000 B Digital Microscope). Note the bimodal grain-size distribution with a fine-sand mode (100 µm) that results from current sorting and the presence of laminated structures. The surprising texture of the sand-rich laminations that still contain abundant clayey matrix (4-µm mode) shows the concomitant deposits of fine-grained and coarse-grained particles. (A multi-colour version of this figure is on the included CD-ROM.)

hemipelagites (see Table 2.5 in Mulder, 2011, this volume, Chapter 2.9). In addition, the reworking of turbidite facies by bottom currents results in lithofacies similarities.

4.1.1. Contourite facies

According to the flow characteristics, the sediment nature, texture and origin, a large variety of contourite facies have been recognized in the modern oceans (Stow and Faugères, 2008; Stow et al., 1996, 2002b). Depositional processes depend on the bottom-current velocity: slow currents allow vertical settling of the suspended particles from the nepheloid layer; high-velocity currents have more bedload transport and deposition; high-velocity currents generate large-scale winnowing and erosion, finally resulting in a lag deposit. Chemical processes (dissolution and authigenesis) can accompany the physical processes.

The contourite grain size varies from sand (more rarely gravel) to mud. The nature of the sediment can be terrigenous siliciclastic, volcaniclastic or biogenic (siliceous or calcareous), and is commonly mixed. Whatever the grain size and nature are, the various contourite facies present three common diagnostic features (Faugères et al., 1984; Gonthier et al., 1984).

(1) Abundant bioturbation with several types of burrows because the benthic fauna depends on the current intensity. This bioturbation may help in distinguishing muddy contourites from hemipelagites

with homogenous burrows and from fine-grained turbidites where bioturbation is usually restricted to the uppermost part of the bed (Wetzel et al., 2008).

(2) Bad preservation of the dynamic sedimentary structures, because of the bioturbation. These structures are well preserved in fine-grained turbidites.

(3) Irregular vertical variation in grain size. Superposition of well-developed coarsening-upward and fining-upward units is the main difference with fining-upward turbidites, but could be mistaken for hyperpycnites (Mulder et al., 2003).

The most frequent contourite facies, summarized in Table 3.1 and Fig. 3.16, are the following:

The *muddy contourite facies* is mainly homogeneous with rare or absent bedding. Mottling and distinct burrows scattered throughout the facies are common. Laminations are not commonly visible for the naked eye, because primary structures are often reworked or destroyed by bioturbation. The most frequent structures are irregular concentrations of winnowed silty

Table 3.1 Main types of contourite facies

		Contourite component origin			
Contourite facies : main types	Contou -rites	**Clastic contourites (silici-, volcani-)**	**Biogenic contourites (calcareous or siliceous)**	**Mixed clastic-biogenic contourites**	**Chemogenic**
	Sandy	i.e. quarztitic sandy contourite	i.e. foraminifera sandy contourite	i.e. terrigenous-calcareous sandy contourites	
	Muddy	i.e. clayey muddy contourites	i.e. radiolarian and (or) diatom silty contourite	i.e. mixed terrigenous-calcareous muddy contourites (the most frequent)	i.e manganiferous muddy contourite
	Mottled silty-muddy	i.e. mottled clastic contourites	?	i.e. mottled terrigenous-calcareous contourites	
	Shale -clast muddy	i.e. micro-brecciated clastic muddy contourite		i.e. micro-brecciated clastic-biogenic muddy contourite	
	Lag-deposit	Any composition i.e. gravel- and sand-rich contourites			
	Reworked Turbidites	Any composition			

sand. Muddy contourites are made of silty clay with a siliciclastic terrigenous or a mixed (including calcareous biogenic) nature. They are the most frequent contourite facies in the modern oceans, where they form the major part of the giant contourite drifts in the North Atlantic Ocean and are described as hemipelagic carbonaceous silty clay. However, some layers in these Neogene drift deposits (like in Hatteras drift or Blake–Bahama Outer Ridge) are formed almost completely of biogenic muddy contourites that resemble pelagites.

The *sandy contourite facies* shows an irregular alternation of strongly bioturbated sandy and silty/muddy beds. Horizontal and cross-lamination are rarely visible for the naked eye. Irregular erosional contacts and concentrations of coarse sand (lag deposits) may be preserved. High-resolution analyses of the grain size usually show a bimodal distribution with a predominant silt/fine-grained sand fraction (Fig. 3.17). The sorting is bad because of the intense burrowing. The sediments are either siliciclastic or bioclastic and have commonly a local source. The sandy contourites have a particular distribution in the oceans (Viana et al., 1998). They are relatively rare in the deep-sea, because they are restricted to environments with active bottom currents (see Section 3.5). They form centimetre-thick to decimetre-thick layers interbedded with muddy contourites, pelagites and sometimes, turbidites. At these locations, they are mainly composed of bioclastic material. They occur more frequently along the upper continental margins, where along-slope or downwelling currents are active, and where they can form large sheets lying along the continental slope, banks, terraces or channel flanks. These sheets are formed by decimetre-thick to metre-thick mixed (siliciclastic/bioclastic) sand layers associated with muddy/silty contourites. Shallow-water sandy contourites are well developed on the outer continental shelf and upper slope, where they constitute thick sheets (average thickness of sand layers is up to 20 m) covered by fields of sand dunes and ribbons. The nature of these deposits is mainly siliciclastic, and they are sometimes interbedded with inner shelf deposits or slope hemipelagites. These shallow sandy contourites constitute the most significant potential oil reservoir in ancient contourite successions.

The *mottled silty/muddy contourite facies* show a truly irregular arrangement of mud, silt or sandy silt in pockets, lenses and streaks, and less commonly, a rapid alternation of thin irregular layers of these three sediment types. They constitute an important part of the deep-sea drifts. Due to the mottling generated by bioturbation, it is difficult to distinguish the primary structures. Crude, more or less continuous, irregular or wavy lamination may be visible. This facies forms centimetre-thick to metre-thick beds with alternating coarsening-upward and fining-upward units. They can be interbedded with both muddy and sandy contourites. The top and bottom contacts of the mottled intervals are erosional to gradational.

The *shale-clast* or *shale-chip contourite facies* occur as centimetre-thick to decimetre-thick layers made of muddy clasts with a diameter ranging from a few millimetres to a few centimetres. These layers are interbedded with homogeneous contouritic muds of the same composition as the clasts. The clasts are interpreted as resulting from the bottom-current erosion of consolidated mud deposits. Such facies form in areas with a low sedimentation rate controlled by active bottom currents that favour rapid consolidation of the sediment. Such contourites have been recognized in a few deep locations, including the south Brazilian basin (Faugères et al., 2002a; Mézerais et al., 1993).

The *contourite lag facies* results from winnowing and reworking processes by powerful bottom currents. It contains a large range of grain sizes and compositions. They form irregular centimetre-thick to metre-thick beds of poorly sorted sediments, mainly coarse sands, gravels and pebbles. Terrigenous and biogenic components are common and frequently coated by iron–manganese precipitates. They are most often deposited on the bottom of shallow straits, narrow moats and deep channels swept by high-velocity bottom currents.

The *manganiferous muddy contourite facies* commonly show ferro-manganiferous-rich horizons. This metal enrichment occurs either as very fine particles, micro-nodules and nodules dispersed in silty/clayey mud (forming red clay) or as encrusted, centimetre-thick to decimetre-thick laminated horizons. The induration of these horizons depends on the metal concentration and transforms soft layers into hard crusts. Such facies are controlled by bottom currents generating very low sedimentation rates (bypass) and favour chemical transformations along the sea-floor. They have been frequently observed in both siliciclastic and biogenic deposits in large abyssal drifts and sheets swept by low velocity currents. They are particularly frequent in settings located below the CCD.

The *bottom-current-reworked turbidite facies* results from the *in situ* reworking of sandy turbidites (Mutti et al., 1992; Shanmugan et al., 1993; Stanley, 1993). The reworking affects the top of the turbidite bed over a thickness depending on the current intensity. The reworking facies can be capped by a sandy contourite facies. A depositional model, such as defined for ancient turbidites, seems rarely applicable to modern environments (Massé et al., 1998) and is still under debate (see Section 6).

4.1.2. Contourite sequence and record of current velocity and paleoclimate

The vertical succession of contourite facies is characterized by frequent and irregular changes in facies type. Alternations of coarsening- and fining-upward intervals are frequent. The coarsening-upward succession is formed by (1) a muddy contourite, (2) mottled silt and mud with increasing grain size and (3) a silty/sandy contourite. The fining-upward succession is composed of the same facies, but in the reverse order. Such centimetre-thick to decimetre-thick successions may be truncated. In that case, some facies are missing and sharp, erosional or gradational contacts are present between the facies. A standard contourite sequence corresponding to this succession has been defined on the basis of contourite deposits in the Gulf of Cadiz (Faugères et al., 1984; Gonthier et al., 1984; Fig. 3.18). The sequence thickness varies from a few decimetres to about 3 m and represents a time interval from about 2000 to 10,000 years. This is a major difference with turbidites and hyperpycnites, which result from more or less instantaneous deposition. The vertical facies succession and grain-size variations in a complete contourite sequence are interpreted as the consequences of a steadily increasing and then decreasing current velocity. In the sequence of Fig. 3.18, the velocity varies from 5 to 10 cm s^{-1} at the base and top of the sequence to 18–25 cm s^{-1} in the coarsest (middle) part. In the case of relatively fast currents or currents with rapid changes in velocity, thinner and truncated sequences with erosional surfaces are formed.

Using the relationship between contourite deposition and current patterns, contourites can be used as a palaeoclimatic tool (Ledbetter, 1984, 1986). By using simple grain-size parameters like the mean grain size of the

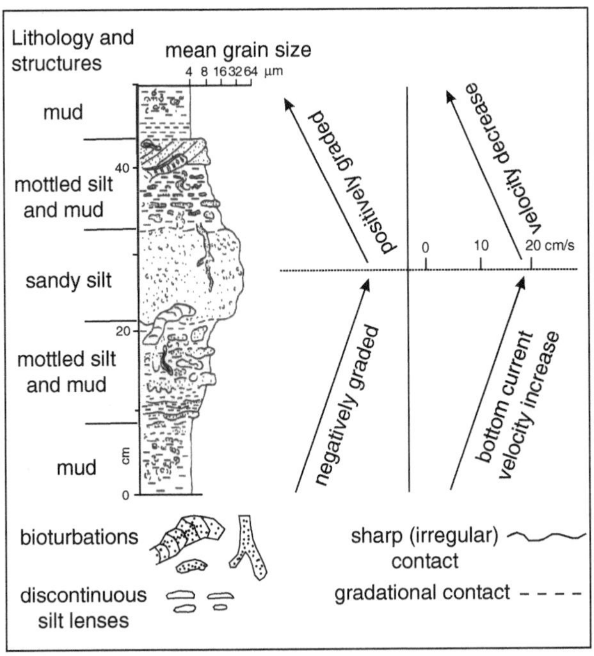

Figure 3.18 Contourite facies: the ideal contourite sequence formed by a coarsening-upward interval followed by a fining-upward interval (after Faugères et al., 1984).

contourite silt fraction (McCave et al., 1995), the cyclic variations in palaeocurrent velocity related to astronomical cycles become apparent (Fig. 3.19). It is not always possible, however, to use grain-size parameters as indicators of palaeocurrent variations, because other factors such as the terrigenous sediment supply and the bioproductivity may also affect the contourite grain size. In this case, a multivariable analysis is necessary to interpret the origin of the sediment (components, source, duration of transport, rate of deposition, distribution: Faugères et al., 1994).

Somoza et al. (1997) and Llave et al. (2006) showed that contourite deposits in the Gulf of Cadiz present a sequence architecture with periodicities of 22–23 and 100–110 ka, consistent with Milankovitch forcing. Llave et al. (2005) showed that high-resolution seismic facies alternate with the

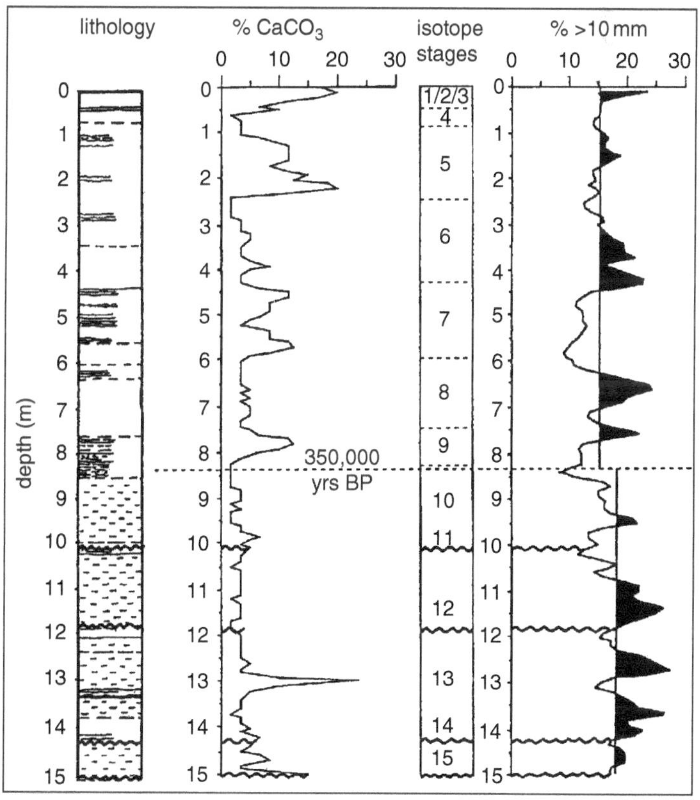

Figure 3.19 Grain-size variations in a deep core collected in the south Brazilian Basin (modified from Massé et al., 1994). The shaded areas in the curve of the percentage variations of particles coarser than 10 μm indicate phases of inferred increased AABW flow. Note that the average flow velocities (vertical lines on the grain-size curve indicating the average grain-size values) were higher before 350 ka BP than afterwards.

frequencies of Dansgaard–Oeschger (1–1.5 ka) and Bond (10–15 ka) cycles. Using sediment cores and grain size, magnetic susceptibility, foraminifera and stable isotopes analysis, Voelker et al. (2006) and Toucanne et al. (2007) showed that the lower core of the MOW was intensified during cold intervals (e.g. Younger Dryas), and particularly during the formation of Heinrich layers in the Atlantic and during Dansgaard–Oeschger stadials. During these intervals, cold MOW formation is intensified in the Alboran Sea (Eastern Mediterranean) and the MOW velocity increases in the Gulf of Cadiz. During warm intervals (e.g. Bölling–Allerod, Dansgaard–Oeschger interstadial and the early Holocene), MOW formation was reduced and MOW velocity decreased in the Gulf of Cadiz; only fine-grained contourites were then deposited.

4.2. Bedforms

The impact of deep bottom currents on sediments was first deduced from abyssal sea-floor photographs showing various sedimentary bedforms (a.o. Heezen and Hollister, 1971; Hollister and McCave, 1984; McCave and Tucholke, 1986). The shape, scale and genesis of these bedforms depend on the current velocity. Whatever the process involved is, the bedform wavelength ranges from centimetres to kilometres.

4.2.1. Erosion, bedforms and current velocity

The absence of clear current-induced bedforms corresponds to low-intensity currents and is marked by the presence of abundant burrows, tracks and mounds on a muddy, soft sea-floor. The size of the sedimentary structures increases with the flow velocity. Small-scale (centimetre long) bedforms generally result from slow to moderate current activity. They develop either as depositional features, perpendicular (ripples) or parallel (longitudinal ripples, small lineations) to the flow direction, or as gentle erosional features (grooves, scour-and-tail features, small furrows). Large-size structures formed by high-velocity currents are perpendicular to the flow direction, such as large sand ripples and sediment waves, or parallel to the flow direction such as large erosional furrows. However, the bedform formation also depends on the volume of sediment available and the sediment types (muddy to sandy). Giant mud waves can be formed by relatively low-velocity currents. High-velocity currents generate erosional surfaces. These surfaces can be relatively flat and regularly sculpted by small-scale erosional and depositional lineations, or irregular and marked by a gravely to pebbly lag deposit and abundant large-scale erosional features (moats, scours, furrows and lineaments). Charts relating small-scale bedform associations and current velocity have been proposed to visually estimate the velocity of the contour currents (Fig. 3.20). Bedform distribution at the

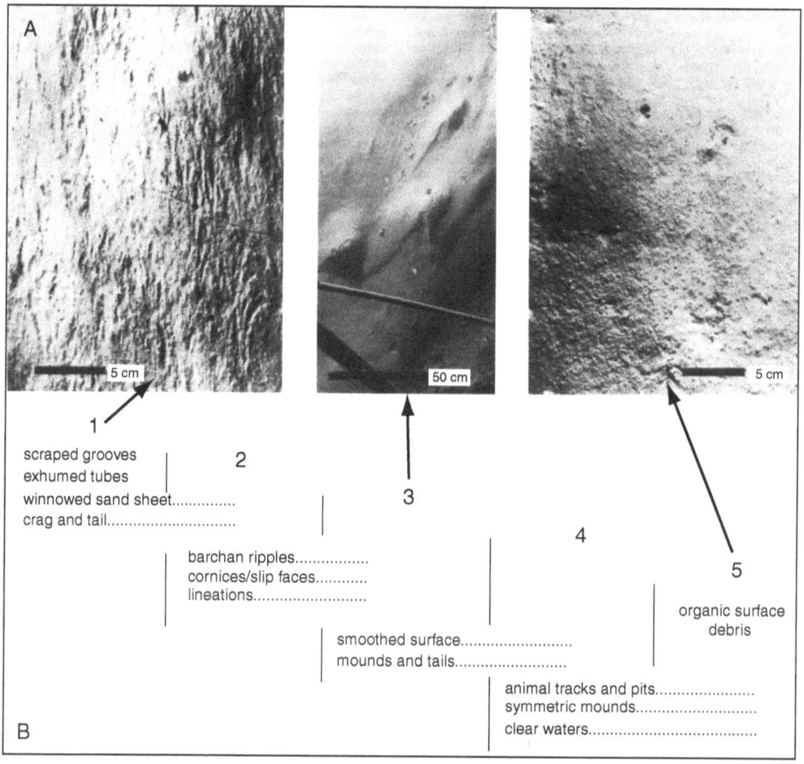

Figure 3.20 Small-scale bedforms (A) and charts (B) of bedform associations varying with current velocity, and formerly proposed to visually assess the velocity of the contour currents from photographs (modified from Hollister and McCave, 1984). (A multi-colour version of this figure is on the included CD-ROM.)

scale of a contourite drift reflects the evolution of the current intensity along it (Stow et al., 1996).

4.2.2. Erosional bedforms: Furrows

In modern oceans, the most important concentration of bedforms and the largest bedforms, including extensive discontinuities and moats, are observed along the WBUC axis, in channels linking oceanic basins and in ocean gateways. One of the most spectacular erosional bedforms is formed by large furrows at the surface of giant drifts (Embley et al., 1980). These furrows are erosional troughs up to several kilometres long, a few metres wide and 1–20-m deep. They are regularly spaced with intervals of 10–100 m and extend parallel to the flow direction. They seem to develop particularly in fine-grained cohesive sediments. Although the velocity to form such elongated bedforms is unknown, it seems that such furrows are relicts from current velocities that once were higher than those at present. They may have been caused by narrow high-energy threads of water separated by wide bands of lower flow energy. As they have been observed in areas of overall deposition, they would be maintained because the furrow topography funnels and increases the velocity of any existing current.

4.2.3. Depositional bedforms: Sediment waves

Depositional bedforms tend to occur at the sides of the current pathway, and they are present mainly at the surface of giant contourite drifts (along the western continental margins of the oceans or downstream deep channels). One of the most striking and frequent large-scale depositional bedforms in deep-sea environments are the sediment waves. However, their origin is difficult to interpret because they can result from different processes without significant distinctive patterns (size, morphology and geometry; Wynn and Stow, 2002). The three genetic interpretations of sediment waves are (1) deposition by a contour current, (2) deposition by a turbidity current and (3) soft-sediment deformation resulting in a 'wave-like' bedform (Faugères et al., 1999). Sometimes, interaction of two or three processes can be involved in the building of a single wave, making its genesis still more difficult to interpret (Faugères et al., 2002b).

4.2.3.1. Contouritic sediment waves Contouritic sediment waves are frequent in deep-seas because they can form on a relatively flat sea-floor, whatever the water depth is. They are giant depositional bedforms with a wavelength ranging from 0.5 to 10 km and a height ranging from 10 to 150 m. They form widespread undulated fields at the top of most of the giant-mounded contourite drifts and flat contourite sheets. The waves can also cover terraces located along deep channel flanks or continental slopes. They are mostly muddy. Contouritic sediment waves may be parallel, perpendicular or oblique to the current direction. On a relatively flat sea-floor, they are more or less perpendicular to the current direction (transversal waves), with a symmetrical or asymmetrical cross-section. Propagation of asymmetrical waves can be either downstream or upstream. If the sea-floor is moderately steep, the wave is often elongated, with the longer axis extending parallel to both the current direction and the slope contour. The wave migrates upslope in the same way as an antidune. The variability in sediment-wave orientation suggests that different depositional mechanisms are involved in the wave construction. These mechanisms are controlled by parameters such as current velocity, sea-floor topography, Coriolis force and the amount and nature of transported sediment. Their formation is still not well understood because only a few long-duration *in situ* observations and measurements have been performed.

The internal geometry of a sediment wave results from the combination of aggradation and progradation (Fig. 3.21).

The *stable (or standing) sediment waves* have a symmetrical cross-section and result only from aggradation. Aggradation may be either uniform (the sediment layers have a constant thickness throughout the wave) or non-uniform, with thicker layers on the top than on the flanks, or just the reverse. In the former case, the aggradation is a passive process and corresponds generally to the deposition of pelagic or hemipelagic sediments that drape pre-existing sediment waves. In the latter case, a bottom-flow process is responsible for a non-uniform deposition of the sediments.

The *migrating sediment waves* have an asymmetrical cross-section with a long, gently dipping flank on the upcurrent side, and a short, steep flank on the downcurrent side. They result from interaction between progradation and aggradation. According to the relative importance of progradation versus aggradation, various distributions and geometries are present in these waves between two end-members (Fig. 3.21-1):

(1) slightly migrating waves show more or less sigmoid-shaped layers that are continuous all over the wave section but with thickness variations; they are thin on the long flank, the thickness increases on the short flank and decreases again in the wave trough;
(2) strongly migrating waves show erosion and/or bypass both on the long flank and in the wave trough, and deposition of thick downlapping oblique deposits on the short flank because of the dominant progradation.

4.2.3.2. Other types of sediment waves Because of the large variability in wave geometry (Fig. 3.22), it remains difficult to distinguish contouritic sediment waves from sediment waves formed by other processes, unless the morphological and hydrological context bears clear evidence of the controlling process.

Turbiditic sediment waves (Fig. 3.22A; e.g. Migeon et al., 2000, 2001; Normark et al., 1980) are less frequent, as they are generally restricted to the backside of turbiditic levees associated with a major channel in deep-sea fans (Mulder, 2011, this volume, Chapter 2.3.3.5).

'Wave-like' bedforms that resemble true depositional sediment waves are due to synsedimentary deformation of deposits during gravity mass-transport events (Fig. 3.22B and C; Correggiari et al., 2001; Lee et al., 2002; Mulder, 2011, this volume, Chapter 2.3.4). They are frequently present on steep to gentle slopes from the upper-shelf slopes to the continental margin and on the flank of volcanic seamounts (Carey and Schneider, 2011, this volume, Chapter 7). They result from constant strain such as compaction and creep, or from downslope mass-transport processes such as sliding and slumping. The recognition of these wave-like bedforms is even more complex (Fig. 3.22D)

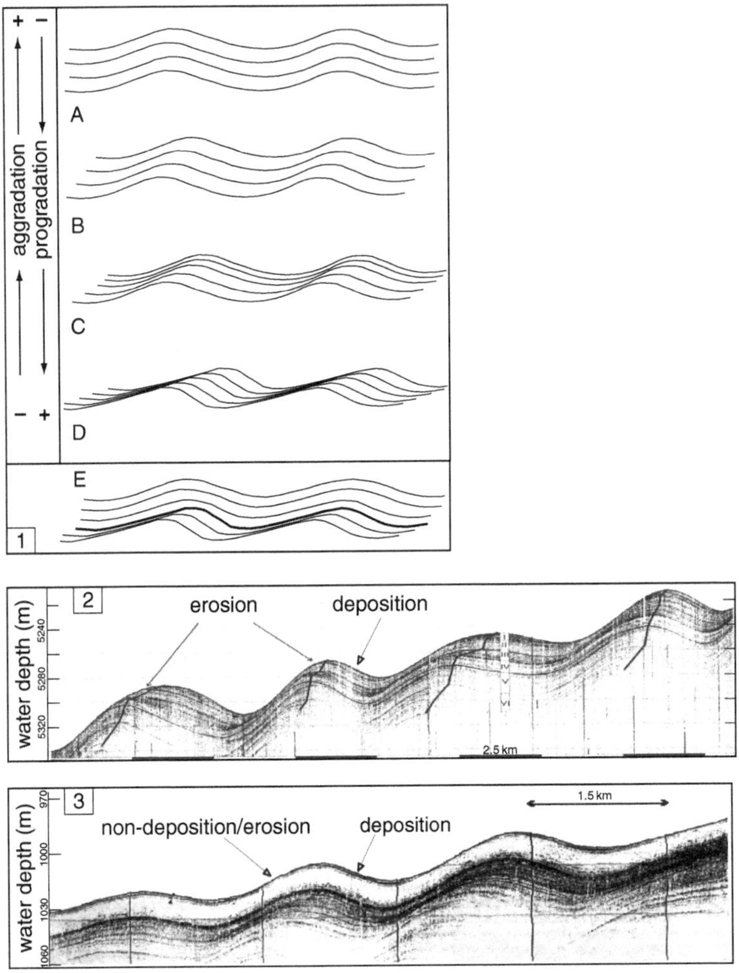

Figure 3.21 Contourite sediment waves. 1A–1D: Different wave geometries (A: standing waves; B–D: waves with various migrating directions) according to the dominant process (aggradation versus progradation). 1E: Vertical change of the wave geometry due to a change in the sedimentary processes (a basal layer of rapidly migrating waves built by active currents is overlain by apparently standing waves that correspond either to a drastic decrease in bottom-current activity or to the draping (pelagic–hemipelagic) cover of the undulated surface inherited from the basal layer. 2: High-resolution seismic profile across sediment waves in the Argentine Basin (from Von Lom-Keil et al., 2002). Note the sediment layers that show alternatively slow and rapid wave-crest migration (thick line). 3: High-resolution seismic profile across sediment waves in the northern Rockall Trough (from Howe, 1996; Howe et al., 1994). Note the transparent layer with slow wave migration overlying rapidly migrating waves.

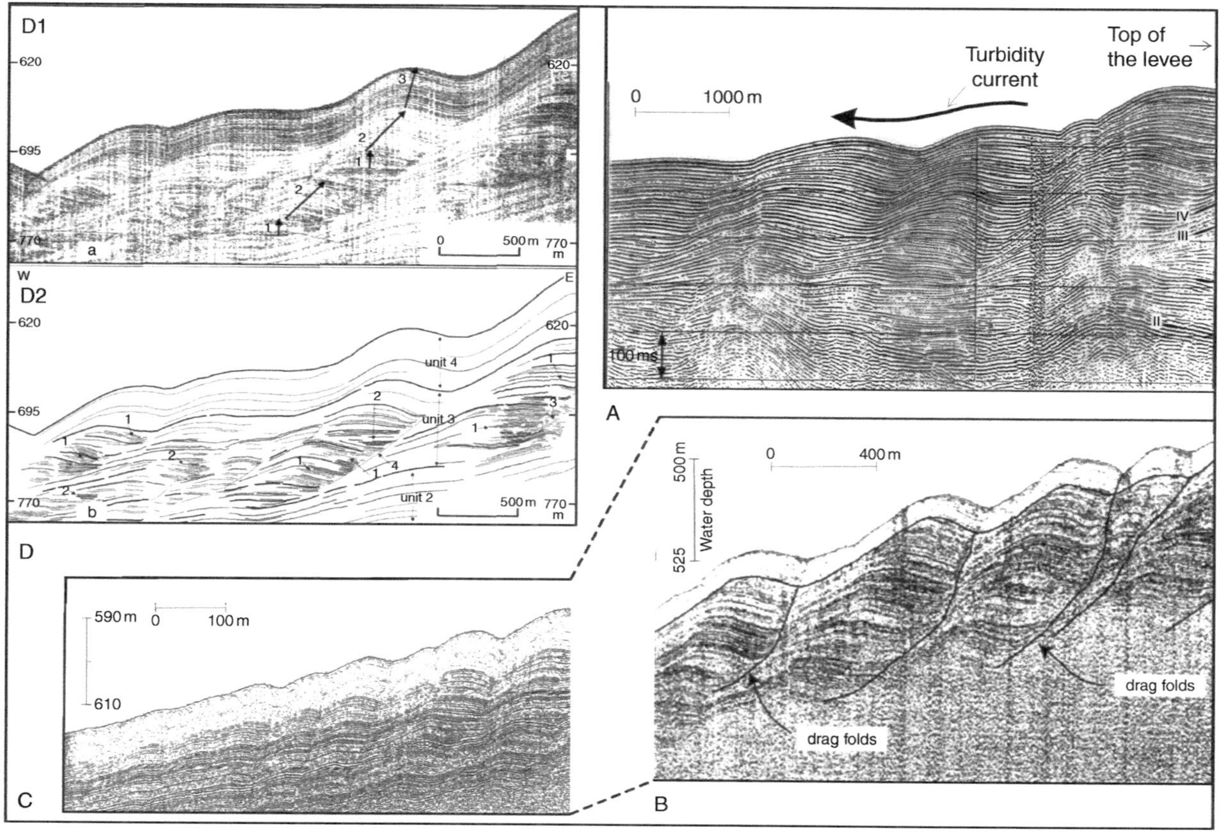

Figure 3.22 Different types of sediment waves according to the process of formation. A, Turbiditic sediment waves on the Var levee (air gun seismic line, from Migeon et al., 2000); note the upslope migration of the wave opposite to the turbidity current trend. B and C, portions of two high resolution seismic lines crossing the Humboldt Slide and showing sediment wave-like structures due to gravity deformation of the

when they result from the interaction between deformational and depositional processes (Faugères et al., 2002b; Lee et al., 2002).

4.2.4. Bedforms and echofacies

High-resolution low-penetrating seismic gears (3.5 kHz and chirp echo profilers) that penetrate the upper few metres to tens of metres of sediments have been widely used since the pioneer works of Damuth (Damuth, 1975, 1980; Damuth and Hayes, 1977). They can be used to document the sea-floor morphology, nature of sediment, detailed geometry of the sub-recent deposits and inferred sedimentary processes. An abundant literature relates echofacies' types or associations of echofacies to sedimentary environments (e.g. channel, levee, rocky sea-floor and unstable slopes) and transport and deposition processes (contourites, turbidites, pelagic settling).

In contouritic systems (Fig. 3.23), the following echo types are commonly recognized. The equivalent echo types defined by Damuth (1975) are provided between brackets:

(1) sharp distinct echo or more or less prolonged echo without any sub-bottom reflectors (type IA to IIA) related to an erosive environment (channel) filled by a more or less indurated coarse-grained lag deposit;
(2) distinct echo with multiple parallel sub-bottom reflectors or medium prolonged echo with discontinuous sub-bottom reflectors (type IB to IIB) formed in muddy or more sandy/silty depositional environments, respectively;
(3) medium- and small-scale hyperbolic echo with more or less irregular hyperbole elevation (type IIIC) formed in environments with large erosional bedforms (furrows) and/or irregular depositional bedforms;
(4) hyperbolic echo with hyperboles tangent to the sea-floor (type IIID), associated with flat erosional surfaces and fairly active and regular bottom currents;
(5) regular to irregular wavy echo (standing to steadily migrating sediment waves) with sub-bottom reflectors (type IIIB) formed by fields of sediment waves.

deposits (from Gardner et al., 1999); note the transition from folded and back-rotated slide blocks (B: drag folds) to gently folded then almost undeformed sequences at the end of the slide D. Multi-process-generated sediment waves on the Landes Plateau (sparker seismic line, (from Faugères et al., 2002a); in D1, note the vertical evolution of an individual wave from symmetric (1) to asymmetric migration (2), and the draping deposit at the top (3); in D2, details of the deposit geometry showing evidences of both depositional processes. (1): toplap–downlap reflector termination and sigmoidal reflections; (2): ponded horizontal deposits; (3) deformation gravity processes; (4): discontinuity that limits two adjacent waves. (A multi-colour version of this figure is on the included CD-ROM.)

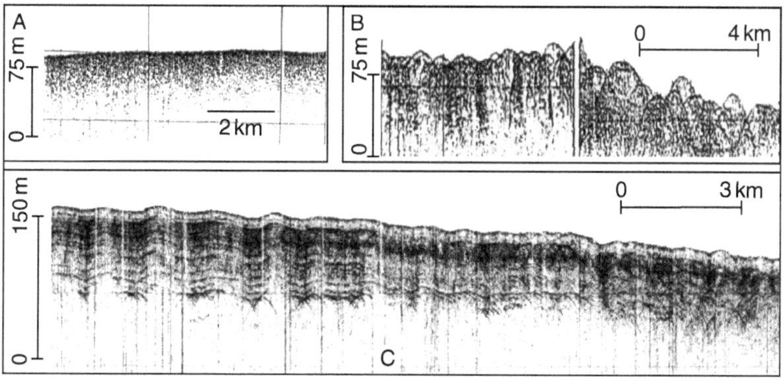

Figure 3.23 Examples of echofacies (3.5-kHz seismic profiles) in the South Brazilian Basin (from Faugères et al., 2002c; Mézerais et al., 1993). (A) Prolonged echo without any sub-bottom reflectors (type IIA) from an erosive environment (channel) with coarse-grained lag deposit and/or manganiferous more or less indurated floor. (B) Medium- and small-scale hyperbolic echoes with more or less irregular hyperbole elevation (type IIIC), in environments with large erosional bedforms (furrows) and irregular depositional bedforms. (C) Hyperbolic echoes with hyperboles tangent to the sea-floor associated with fairly active and regular bottom currents. (A multi-colour version of this figure is on the included CD-ROM.)

Distinguishing contourite systems from other systems using only echofacies is however, difficult. For example, all the echoes described earlier have also been recognized in turbidite systems, where they have also been defined initially. Only a few echofacies are characteristic of deposits related to a single sedimentary process. These include draping, stratified low-amplitude echo (type IB) related to pelagic–hemipelagic deposits, well-stratified, horizontal and parallel, high-amplitude echoes (types IB–IIB) related to confined turbidites and more or less prolonged echoes with discontinuous sub-bottom reflectors formed in turbidite channels filled with coarse-grained sediments. The echofacies that seems the most characteristic of contouritic deposits shows well-stratified layers and regular, intense, overlapping hyperbolae with vertices tangent to the sea-floor or sub-bottom reflectors (Fig. 3.23C).

Consequently, reliable identification of contourite deposits cannot be based only on an echofacies pattern but needs a set of data as large as possible, including data about sediment nature and distribution, bedforms and geometry at various scales.

5. Contourite Drifts

Contour-current activity builds giant accumulations of sediment known as 'sediment ridges' or 'drift' and named today 'contourite drifts' (Faugères and Stow, 2008; Faugères et al., 1993, 1999; McCave and

Tucholke, 1986). Giant contourite drifts are found in all oceans from the continental slope/shelf edge (shallow contourite drifts) to abyssal plains. They are as frequent as deep-sea turbidite systems (Mulder, 2010) along most of modern continental margins. They are particularly abundant and well-developed along the passive margins of the North, West and South Atlantic, Antarctica, the southwest Indian Ocean and north-eastern Australia. They form sporadically along active margins.

The sizes of contourite drifts are similar to those of deep-sea turbidite systems. They range from small patch drifts (about 100 km^2) to giant elongated drifts (> 100,000 km^2). In the Argentine Basin, the area covered by contourite deposits is even larger, close of 1,000,000 km^2. Drifts commonly have an elongated mounded shape (length and width reaching 500 and 100 km, respectively), and may form positive relief up to 1500–2000-m high).

Contourite drifts have been identified for the first time only by their typical mounded morphology with their elongation parallel to the continental margins. However, the recognition of contourite drifts, either in the sea-floor topography or on seismic profiles, is not easy, because their forms can strongly vary from the classical mounded–elongated shape and because of the difficulties in distinguishing between contourites and turbidites. This difficulty is emphasized in areas where the two types of deposits interplay.

5.1. Drift morphology and large-scale deposit geometry

The formation and geometry of contourite drifts are controlled by four major factors:

(1) morphological and bathymetric context;
(2) current velocity and variability;
(3) amount and type of available sediment;
(4) persistence of an active bottom current.

Consequently, four major types of drift morphology have been recognized in the modern oceans (Faugères et al., 1999; McCave and Tucholke, 1986; Stow et al., 1996); they are presented in the following subsections (Fig. 3.24). These contourite-drift types have also been identified in sedimentary successions buried below the present-day sea-floor by turbiditic or pelagic deposits.

5.1.1. Contourite sheeted drifts

Contourite sheets form a layer of more or less constant thickness (up to a few hundreds of metres) covering large areas, especially in abyssal basins (Fig. 3.24). The internal seismic facies has typically low-amplitude, discontinuous reflectors or is, sometimes, more or less transparent. They are fully or partially covered with fields of sediment waves. These accumulations are present in different hydrological and morphological contexts.

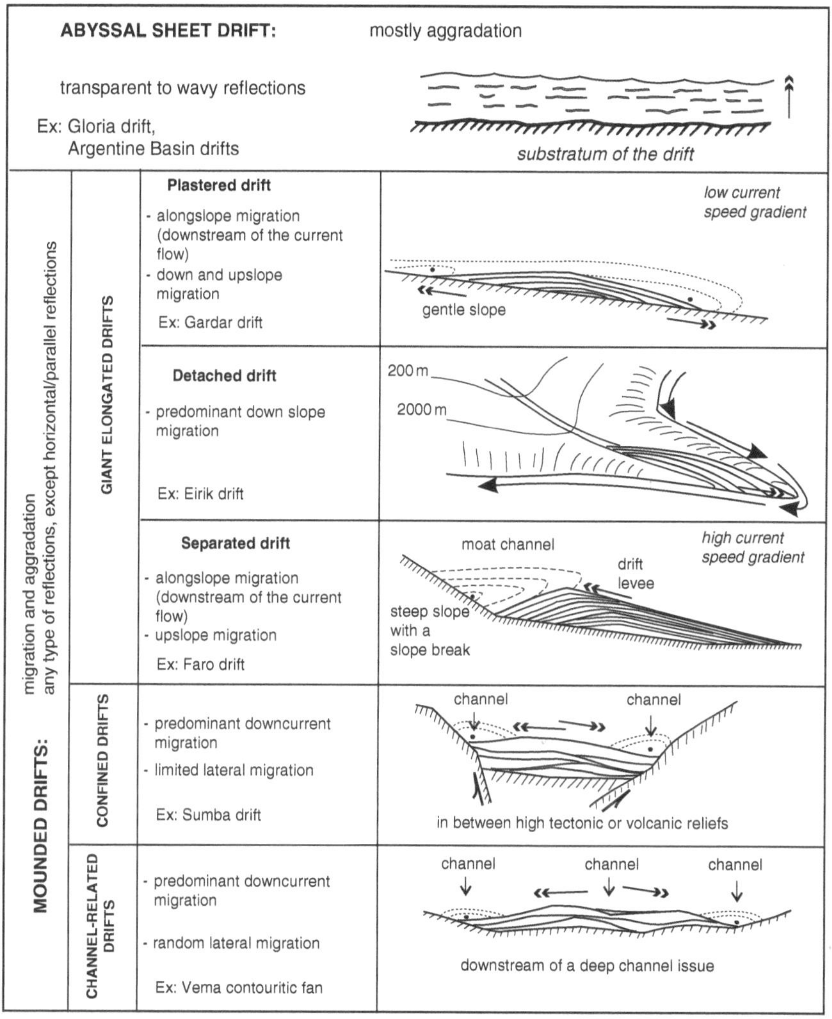

Figure 3.24 Different types of contourite drifts (from Faugères et al., 1999), showing the general drift geometry and the direction of migration–aggradation (double black arrow).

Abyssal contourite-sheeted drifts are the most impressive. They cover extensive areas on the floors of abyssal plains, such as the south Brazilian Basin (Damuth, 1975; Damuth and Hayes, 1977; Mézerais et al., 1993) and the Mozambique Basin (Ben-Avraham et al., 1994; Kolla et al., 1980), or they cover deep basins with a complex gyre-type circulation responsible for the deposition of giant elongate-bifurcated drifts such as the Irminger (Gloria drift: Egloff and Johnson, 1975, 1978) and Argentinian Basins (Ewing drift,

Zapiola drift: Flood and Shor, 1988; Shipboard Scientific Party, 1976; Fig. 3.25).

Slope contourite-sheeted drifts occur either near the base of continental slopes where outflow and downwelling bottom currents exist (like in the Gulf of Cadiz; Faugères et al., 1985a; Kenyon and Belderson, 1973; Nelson et al., 1993), or plastered against the slope at various water depths (e.g. Hebrides margin: Howe et al., 1994; Stoker, 1998; Stoker et al., 1998; Chatham Rises: a.o. Wood and Davy, 1994). They cover a smaller surface area than abyssal sheets and can even have a patchy shape with a wide low-mounded geometry. The depositional units that form the sheet have a fairly regular thickness over the whole area swept by the currents. They are mainly aggrading without significant lateral migration.

5.1.2. Elongate mounded drifts

This type of contourite drift is distinctly mounded and elongated (Figs. 3.24 and 3.26) with a variable length that ranges from a few tens of kilometres to over 1000 km, a length–width ratio varying from 2 to 10, and thicknesses reaching several hundreds of metres. They may occur anywhere from the outer shelf/upper slope down to abyssal plains. The elongated drifts formed in channels or in confined basins are considered separately.

Elongate drifts are clearly prograding. However, the directions of their elongation and progradation can vary according to the shape of the continental margins or basins, and depend on the interaction between the morphology (i.e. slope gradient or shape, sea-floor roughness), the current intensity and the Coriolis force.

The plastered drifts are located on gentle regular slopes swept by low-velocity currents, with sediment deposition on one side or both sides of the flow; upslope or upslope–downslope lateral migration takes place as well as a downcurrent migration (Fig. 3.26A).

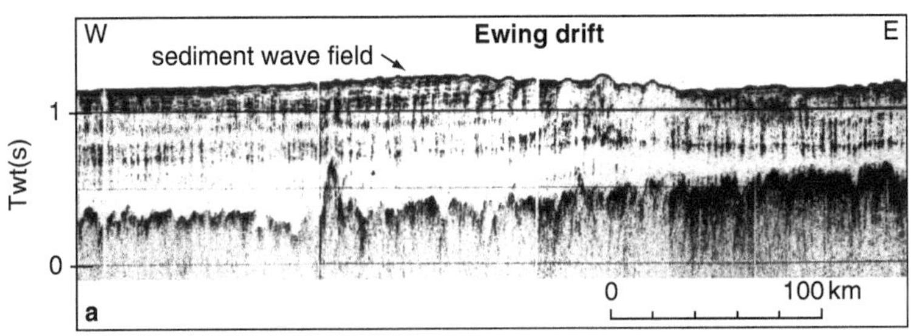

Figure 3.25 Abyssal contourite sheet drift linked to bottom currents trapped on the bottom of a basin: single-channel airgun seismic line crossing the North Argentine Basin (Ewing drift) (modified from Shipboard Scientific Party, 1976).

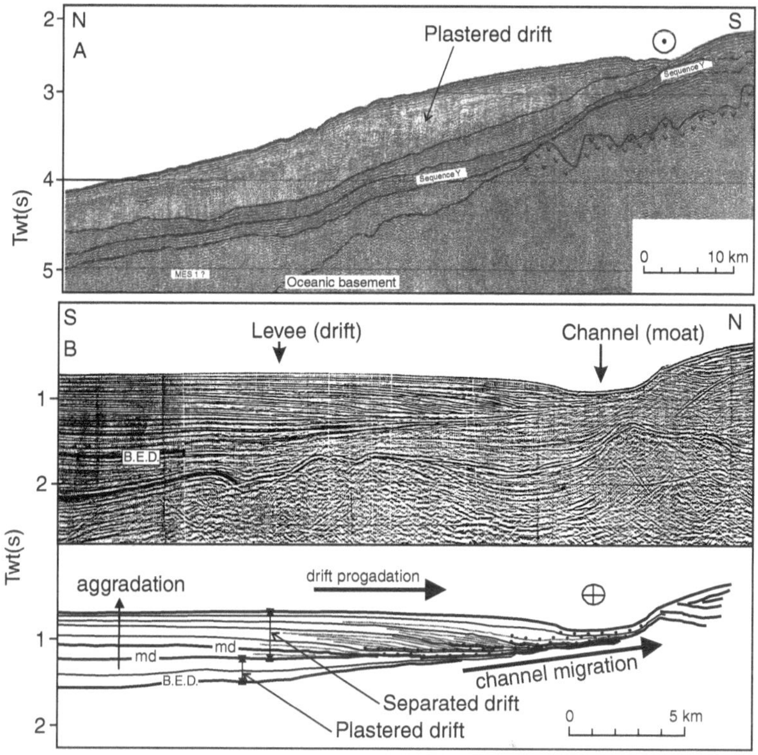

Figure 3.26 Seismic profiles crossing elongate-mounded drifts. (A) Airgun profile crossing the north-eastern Chatham rise (eastern New Zealand margin) and showing a typical geometry of a plastered drift (from Wood and Davy, 1994). Note the upslope migration of the reflectors without any channel between the drift deposit and the continental slope. (B) Multichannel seismic profile crossing the Faro drift (Gulf of Cadiz), showing a separated drift (modified from Faugères et al., 1999). Note that the initial plastered drift evolves as a separated drift with a channel and levee, and is characterized by an upslope progradation (B.E.D: basal erosional discontinuity of the drift; md: major discontinuity).

The *separated drifts* are located at the base of steep regular slopes where the contour-current section is restricted because of the Coriolis force. Erosion is predominant in the channel (moat) along the right side of the current (northern hemisphere) and deposition occurs along the left side of the current, where the velocity decreases. Such a drift shows upslope lateral migration marked by oblique or sigmoid reflectors on seismic profiles, as illustrated by the Faro drift (Fig. 3.26B), in the Gulf of Cadiz (Faugères et al., 1985b; Faugères et al., 1999); downcurrent drift migration is also present.

For plastered and separated drifts, elongation is generally parallel or almost parallel to both the basin's margin and the contour-current direction because of the downcurrent drift progradation (McCave and Tucholke, 1986).

The *detached drifts* form where the margin bends (Eirik drift: Arthur et al., 1989; Fig. 3.27D) or where surface and bottom currents interact (Cape Hatteras and Blake–Bahamas drifts: McCave and Tucholke, 1986; Fig. 3.27E). They typically prograde basinwards, more or less perpendicularly to the margin (Fig. 3.24). Contourite drifts perpendicular to the margin may also be deposited in a particular morphological and hydrological context. When downslope turbidity currents occur across a margin swept by strong along-slope bottom currents, elongate levees that are mainly constituted of contourites may be deposited (Antarctic margins: Rebesco et al., 1996; Weber et al., 1994; Figs. 3.27F and 3.30). Downwelling bottom currents may be responsible for the erosion of downslope channels bordered by levee-like elongate contourite drifts such as those of the Chatham Rise off NE New Zealand (Barnes, 1992, 1994) and the Gulf of Cadiz (Faugères et al., 1985b; Hanquiez et al., 2010; Fig. 3.27G). However, this unusual morphology seems rare compared to the more frequent mixed turbidite/contourite systems (see Section 5.3).

5.1.3. Channel-related drifts

This drift type is related to deep channels or gateways (Fig. 3.24) where the bottom currents are constrained and flow velocities are substantially increased (e.g. Vema Channel, Kane Gap, Samoan Passage, Almirante

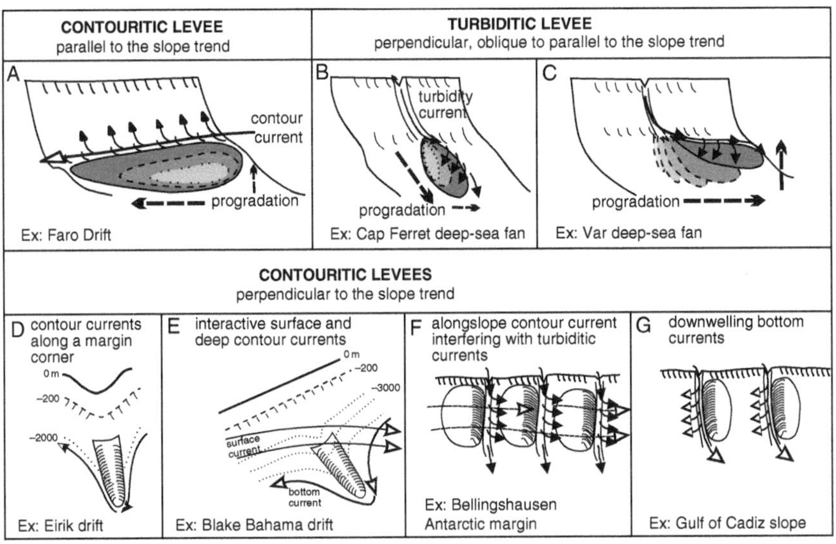

Figure 3.27 Summary of the relationship between the directions of contouritic and turbiditic levees and the direction of the margin along which the levees are developed (from Faugères et al., 1999). (A) Usual trend for contouritic drift. (B) Usual trend for turbiditic levee. (C) Variations in direction during the growth of turbiditic levees. (D–G) Scenarios for contouritic levees perpendicular to the slope direction.

Passage, Sand Dune Valley: Johnson et al., 1983; Lonsdale, 1981; Lonsdale and Malfait, 1974; Meinert, 1986). In addition to significant erosion and scouring along the gateway floor, irregular and discontinuous sediment bodies are deposited along the floor and side of the channel forming gently mounded axial and lateral patch drifts or moat-related drifts (Howe et al., 1994; McCave and Carter, 1997), and at the channel mouth, forming a contourite fan (Maldonado et al., 2003; Mézerais et al., 1993).

Patch drifts are typically small (a few tens of km^2, 10–150 m thick) contourite accumulations preserved along a channel floor. They are either irregular in shape or elongated in the flow direction. They can be reflector-free or show a chaotic seismic facies similar to that of debris-flow deposits. Contourite fans such as the Vema fan (Mézerais et al., 1993) are much larger, cone-shaped deposits, more than 100-km long and 300-m thick (Fig. 3.28). They resemble small- and medium-sized turbidite fans, and can even contain distinct channel–overbank units. Contourite fans are composed of aggrading, flat, irregular, lens-shaped units of small extent and bounded by major erosional surfaces. Minor lateral deposit migration can occur within the fan, while the topmost unit shows downflow progradation.

5.1.4. Confined drifts

There are relatively few examples of contourite drifts confined in small basins or troughs. They typically form in tectonically active areas, such as the Louisville Drift in the deep part of the eastern New Zealand margin

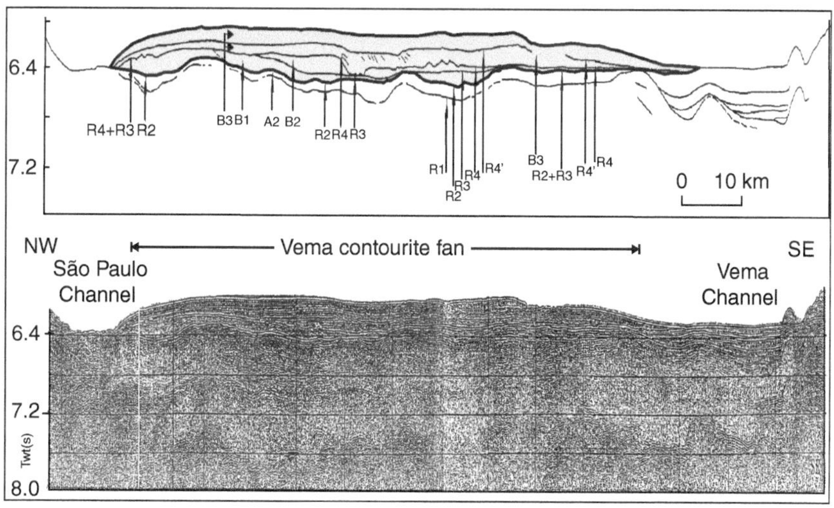

Figure 3.28 Mounded, channel-related drift. Watergun seismic profile crossing the Vema contourite fan (South Brazilian Basin, Mézerais et al., 1993; modified from Faugères et al., 1999). Note the major widespread discontinuities (R1–R4′) and the drift's lenticular depositional units (B1–B3); R2 = basal erosional surface of the drift.

(Carter and McCave, 1994), the Sumba Drift in the Sumba forearc basin of the Indonesian arc system (Reed et al., 1987), the Meiji Drift in the Aleutian Trench (Scholl et al., 1977) and an unnamed drift in the Falkland Trough (Cunningham and Barker, 1996; Cunningham et al., 2002). The gross seismic character appears similar to elongate-mounded drifts (Faugères et al., 1999). However, a major different feature is the presence of distinct moats along both sides of the drift (Fig. 3.24).

5.2. Seismic features of contourite drifts: Contourite versus turbidite

The four described drift types and geometries form end-members of a continuous spectrum of morphologies and geometries. An evolution from slope-sheet to plastered and then separated drift may sometimes be observed within one single contouritic system. Although these types may, in some cases, be distinguished on seismic records, they are not exclusive to contourite systems, adding to the difficulty of contourite-drift recognition on seismic lines.

The principal features to distinguish contourite drifts on seismic profiles were defined by Faugères et al. (1999), Rebesco and Stow (2001), Stow et al. (2002b) and Maldonado et al. (2005). Contourite drifts are characterized by widespread discontinuities that can be followed across the whole drift and that are marked by a continuous high-amplitude reflector. These discontinuities typically result from erosion phases due to increased bottom-current intensity, but may also be caused by sharp changes in grain size or nature. *Depositional* (seismic) units in contourite drifts are generally lens-shaped, with a convex-up geometry that is not parallel to the discontinuity formed by the previous erosion event. The stacking of depositional units (*progradation/aggradation*) reveals the general migration of the drift. This direction is different for different drift types. Downcurrent progradation or oblique progradation is the most common. *Lateral migration*, with a downlapping (toplapping) sigmoid reflector, can sometimes be used to distinguish moat/contourite drifts from turbidite channel–levee complexes (Faugères et al., 1999). The recognition of seismic facies and echofacies may also help.

The geometry of elongate-mounded drifts can particularly be confused with turbidite levees (Fig. 3.27): (1) both show channels or moats with an elongate mound or drift developed preferentially along a channel flank; (2) drifts can be elongated downslope like turbiditic levees; (3) turbidite channel–levee systems, which are usually elongated downslope, can sometimes be elongated along-slope under the influence of the Coriolis force or because of structural control (Fig. 3.27C); (4) mixed levee-drift systems can form under the combined action of contour and turbidity currents. Two criteria can be used as diagnostic features: (1) some contourite mounds are clearly isolated from downslope supply, and (2) contourite drifts commonly lie on a more or less flat, major erosional surface corresponding to an isochronous time-surface (Figs. 3.14, 3.26 and 3.28). Such a surface does not occur at the base of a turbiditic levee.

Confusion is particularly easy between a turbidite 'channel–levee' system and a separated drift. However, it is sometimes possible to distinguish them by considering their progradation direction in relation to the overall trend of the margin and the flow direction (Faugères et al., 1999). Distinction among slope-sheeted contourite patches, small-plastered drift and turbiditic lobes is possible if the morphological context is taken into account and if the seismic facies is analysed. Turbiditic fan lobes typically have continuous, parallel, high-amplitude reflectors and a smooth surface. In contrast, contourite drifts usually show low-amplitude reflectors and wavy bedforms.

5.3. Turbidite/contourite mixed systems

The interaction of down-slope and along-slope processes, at various scales of time and space, is a permanent feature along most of the modern ocean margins and, more particularly, along the western ocean margins and the Antarctic continental margin. Contourite facies, global drift morphology and deposit geometry can, therefore, be drastically modified by the interaction of these sedimentary processes.

Process interaction has been clearly demonstrated for limited time-spans and at a local scale by interbedded turbidite and contourite layers, and bottom-current reworking of turbidites (Massé et al., 1998; Shanmugan et al., 1993; Stanley, 1993; Stow et al., 1998). At a geological time-scale and at a large scale in space, this interaction is evidenced by lateral and vertical transitions between contouritic and turbiditic bodies in the sedimentary record (e.g. Carter and McCave, 1994, 2002; Faugères et al., 2002c; Mulder et al., 2008; Stoker, 1998; Tucholke and Mountain, 1986, among others).

5.3.1. Morphology and overall system geometry

Various scenarios for interacting downslope and along-slope processes have been described (see Faugères et al., 1999; Stow et al., 2002b) to explain the large variety of mixed system morphologies (Fig. 3.29).

Contour currents crossing a deep-sea turbidite system may induce a lateral shift of the turbiditic levee. The importance of the deflection depends on the velocity of the contour currents and can potentially lead to levee erosion. The shifted levees are composed of turbiditic deposits, contour-current-reworked turbidites and contourites. The contourites are made of material taken from turbidity currents and transported by contour currents. They form the 'fan drift' of Carter and McCave (1994). In a similar hydrodynamic context, downcurrent progradation of a mounded-elongate drift may induce a

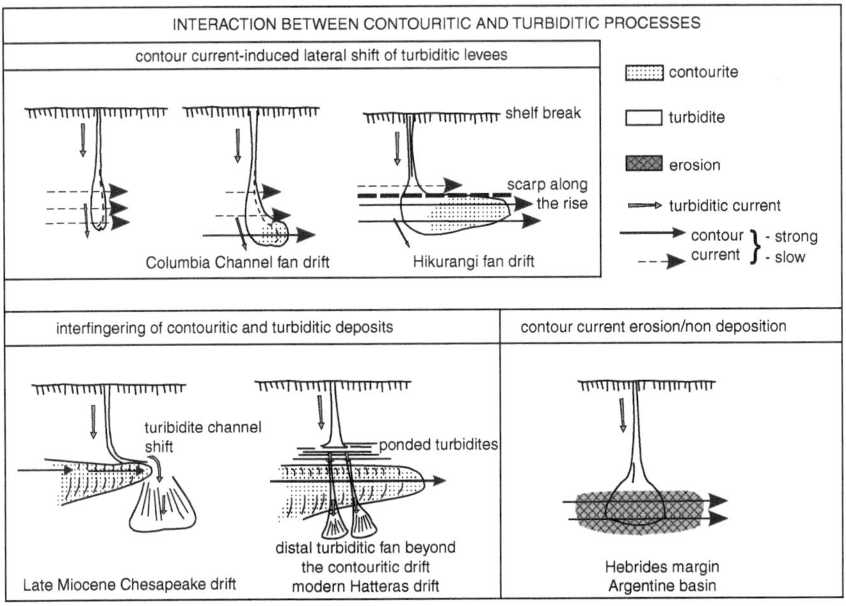

Figure 3.29 Scenarios for interacting downslope and along-slope processes (from Faugères et al., 1999).

deflection of turbiditic channels, levees and lobes. The maximum complexity in deposit geometry occurs if a drift crosses the whole deep-sea turbidite system. In this case, ponded turbidites settle on the back side of the drift. Erosive turbidite channels cross the drift and feed the troughs of the contouritic sediment waves by overspilling or overbank flow (Mulder, 2011, this volume, Chapter 2.3.3.5), and finally built a distal turbidite system beyond the drift (Tucholke and Laine, 1983). Continental slopes with both downslope canyons or channels and active contour currents show downslope mounded–elongated sediment bodies composed mainly of contourites, but containing also a few turbidites (Rebesco et al., 1996, 1997, 2002; Fig. 3.30).

5.3.2. Deposit geometry

Over a long time, the two interactive processes may be active simultaneously or alternatively. In the first case, the two types of sediment bodies are strongly imbricated with large areas of sediment interfingering (Massé et al., 1998; Shanmugan et al., 1993). In the second case, the alternation of processes can result from variations in climate, sea level and bottom circulation coupled with basin morphology and margin topography (Faugères and Stow, 1993). This has been particularly true since the late Eocene onset of the phase with intense thermohaline circulation, and with the abrupt

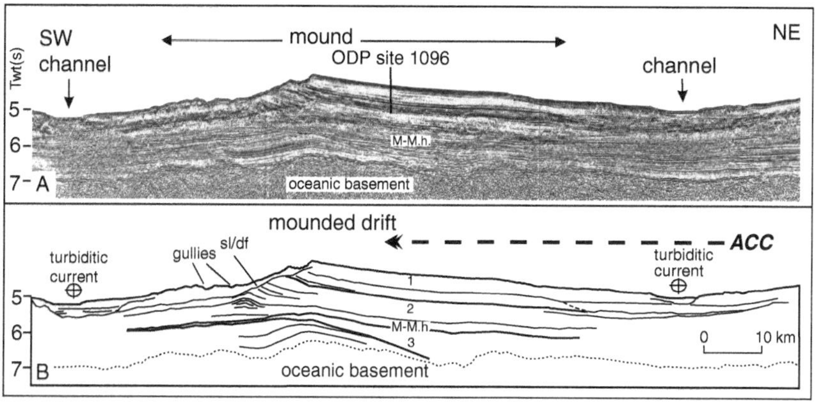

Figure 3.30 Profile showing turbidite/contourite mixed systems. (A) Airgun seismic line over the upper continental rise of the western margin of the Antarctica peninsula (after Rebesco et al., 1996). The mounded drift is elongated downslope between two feeding turbiditic channels. It shows an asymmetric section with a gentle northern flank and a steep southern flank. (B) profile interpretation. The mound slightly thickens downstream of the N–S bottom currents (ACC: Antarctic Circumpolar Current), and the deposits mainly consist of a mixed muddy contourite/fine-grained turbidite succession near the crest of the mound (ODP Site 1096) and a largely turbiditic succession, at the seaward termination of the same drift, in the vicinity of the NE turbiditic channel (ODP Site 1095). The southern flank of the mound has been intensely eroded, probably by gravity mass-flow processes (1: drift-maintenance stage; 2: drift-growth stage; 3: pre-drift stage; M-M.h.: Middle Miocene hiatus, sl/df:slump/debris flow).

alternation of depositional patterns reflecting glacial–interglacial cycles during the past 2 Ma. Two examples illustrate the nature and complexity of these interactions recorded in seismic profiles:

(1) The eastern margin of the USA, off Cape Hatteras (Locker and Laine, 1992; McMaster et al., 1989), is dissected by numerous turbiditic canyons and channels that supply the continental rise with sediment. The along-slope WBUC sweeps the lower rise. Both turbidity and contour currents are active simultaneously, but with variable intensities. The result is a succession of intervals during which the processes dominate alternately; the deposits from these successive phases are separated by major discontinuities related to global hydrological events. On seismic profiles, the result is a quite complex imbrication of deposits that has been called a 'companion drift-fan'.
(2) Multichannel seismic reflection profiles from the continental rise west of the Antarctic Peninsula reveal the presence of eight large sediment mounds, elongated perpendicular to the margin and separated by turbidity-current channels (Fig. 3.30; Rebesco et al., 1996, 2002). The asymmetry of these mounds suggests that their construction occurs by

entrainment (pirating) of the suspended load of downchannel turbidity currents by the south-westerly bottom currents and its deposition downcurrent. Simple Coriolis deflection of turbidity-current material would result in the opposing asymmetry.

6. Ancient Contourites

The application of contourite facies models and drift models derived from studies of modern systems to land-based studies is not convincing (Stow et al., 1998). In the fossil contourites reported in the literature, there are few examples that are reliable and fit the diagnostic criteria developed for modern contourites (Fig. 3.31).

The lack of progress in the recognition of ancient contourites is due in part to diagenesis, as this process increases the possible confusion with other deep-water sedimentary facies like fine-grained turbidites and hemipelagic shale deposits, and partly to outcrop discontinuities and tectonic deformation of ancient rock units (Hüneke and Stow, 2008). This makes the reconstruction of the deposit geometry and the reconstruction of the palaeoenvironmental background (palaeodepth, morphology and hydrology) often speculative. The most convincing examples of relatively deep contourite successions are often located in peri-Tethys palaeogeography (Duan et al., 1993; Kindler et al., 1995; Stow and Faugères, 1993, 1998; Stow et al., 2002a; Vilard, 1991). The best candidates for ancient rocks containing contourites are probably in Neogene successions deposited when climatic instability and the development of polar ice caps led to enhanced geostrophic bottom-water circulation.

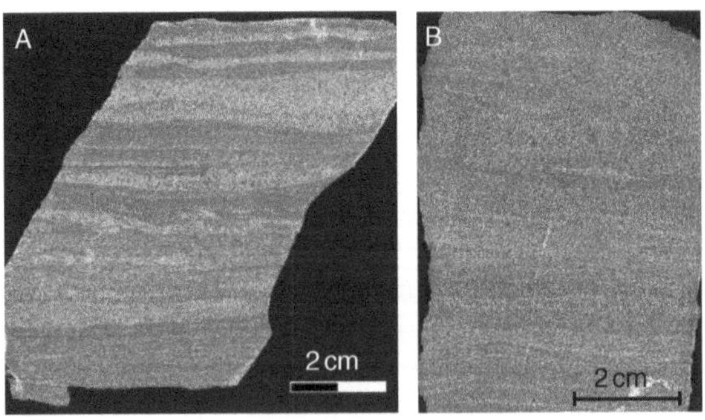

Figure 3.31 Ancient contourite (calcarenite from the Palaeogene Ultrahelvetic Prealps) (Kindler et al., 1995). Note the laminated structure (alternation of coarse- and fine-grained lamination), erosional contacts, traction features and bioturbation. (A, B) Coarse- and finer-grained contourites, respectively (original photos from P. Kindler). (A multi-colour version of this figure is on the included CD-ROM.)

Other examples come from older rocks and are deposited in shallow water, like the Ordovician contourites (Stow et al., 1998), or in oceanic passages, like the Devonian contourites (Hüneke, 2006, 2007).

6.1. Bottom-current-reworked turbidites

The reworking of turbidites by contour currents is not well known from modern environments (Massé et al., 1998). The attempt to address the problem solely on the basis of studies of ancient turbidites (Mutti et al., 1992; Shanmugan et al., 1993; Stanley, 1993) can lead to some misinterpretation. Most of the characteristics of bottom-current reworking provided by the just-mentioned authors are typical of fine-grained turbidites in both modern and ancient series.

Two characteristics are less common in turbidites and could be used as diagnostic features: (1) reverse grading in lamina or beds, and (2) the presence of bi-directional cross-laminations. However, reverse grading is frequent in coarse-grained turbidites (Lowe, 1982; Mulder, 2011, this volume, Chapter 2.2.4) and in fine-grained turbidites and hyperpycnites (Mulder et al., 2003; Mulder, 2011, this volume, Chapter 2.2.6). Bi-directional laminations have often been interpreted as indicators of bottom-current reworking. However, HCS-like structures associated with bipolar structures have been interpreted as features formed by turbidity currents under specific conditions (Mulder et al., 2009). In addition, bi-directional cross-laminations have also been observed in reflected ponded turbidites like at Peïra Cava in the Annot Sandstone (Amy, 2000).

The following three additional features can indicate bottom-current reworking:

(1) The sequence of fine-grained turbidites (Stow et al., 1996) differs considerably from the classical Bouma sequence (Bouma, 1962). The absence of these structures or the complete bioturbation of the muddy Te unit could indicate interaction of a contour current.
(2) Sharp contacts at tops of silty or sandy intervals are frequent in ancient turbidites. However, the presence of these features cannot be taken as a definitive proof of reworking, because turbidity currents are able to erode the beds that they have deposited just before. In addition, turbidites can result from several alternations of erosion and deposition (Salles et al., 2008). However, sharp or erosive contacts within turbidite sequences, coupled with different characteristics above and below the contact, such as an anomalous grain size or bioturbation, may be an indication of reworking by a contour current.
(3) The abundance of truncated fine-grained turbidites, or the abundance of clean, thin cross-laminated silts without a muddy top-division, can

suggest that the finest turbidity current load may have been picked up from a bottom current.

In any case, intensification of a bottom current is required to produce significant reworking of turbidites. Particular environments are more susceptible than others to current reworking; examples are moats, channels, gateways and any steep part of ancient continental margins.

6.2. Shallow-water ancient contourites

Shallow-water deposits similar to modern shallow-water drifts may have built large sediment accumulations on ancient outer shelves and on the upper parts of continental slopes. However, the recognition of the ancient analogues of such modern drifts deposited under the action of wind-driven surface currents or thermohaline bottom currents is not easy, because numerous bottom currents can be involved in shallow-water bedload deposition in addition to the currents related to geostrophic circulation patterns: meteorological currents (including surface waves, storm rip currents, local wind-driven currents), tidal currents, internal waves, clear-water (=without significant amounts of sedimentary particles) canyon currents and currents related to upwelling systems. The recognition problem is due to the fact that sedimentary structures do not vary significantly with the type of current. A major confusion could be between contourites with various ripple types and ancient tempestites with more or less well-preserved HCS structures (due to diagenesis or metamorphosis). A good example is provided by some Ordovician sandy/silty siliciclastic rocks. Despite similar facies characteristics in the various areas where these rocks crop out, they have been interpreted as contourites, turbidites and tempestites, according to their geographic location (e.g. Guillocheau, 1983; Stow et al., 1998, 2002b). As the palaeogeography of the Ordovician oceans is badly constrained, it is difficult to conclude whether these various interpretations reflect the existence of various deposits or that they are misinterpretations caused by the lack of reliable criteria.

It is obvious that the recognition of contourites in the field is not an easy task. Any attempt to identify these deposits should be supported by a set of criteria as large as possible. The following three-stage approach for the recognition of ancient contourites has been suggested (Stow et al., 1998):

(1) At a small scale (field, borehole or core), detailed analyses of the sedimentary facies and sequences, as well as of temporal variations, must be carried out to evaluate the possible exclusive activity of bottom currents or the joint occurrence of several processes responsible for deposition. In addition, data must be studied to verify the consistency between deposits and hydrodynamics of the bottom current that is supposed to have controlled the deposition, or to make likely

the importance of additional terrigenous inputs or biogenic productivity.

(2) At an intermediate scale (sedimentary successions), the regional directions of palaeocurrents, facies distribution, sediment textures, mineralogical (clays) and geochemical tracers must be studied to reveal possible evidence of bottom-current activity. Evidence can be found in the form of the presence of isochronous unconformities, condensed surfaces, drift geometry, elongation of sedimentary bodies and direction of progradation parallel to the margin.

(3) At a large scale (sedimentary basin), conclusions of (1) and (2) must be compared and integrated in an overall reconstruction of the basin geometry and palaeoceanographic conditions to check whether a permanent bottom-current system existed in the study area at the time of deposition.

7. Conclusions

Bottom currents and contourites represent important components of the deep ocean basins and their margins. The processes of sediment deposition and drift construction are, however, still partially unknown. Physical oceanographers have worked independently of geologists on the nature and variability of bottom currents. Interaction between the two disciplines is necessary to successfully study and understand the processes at the sediment–sea-water interface.

The study of mixed turbidite/contourite systems and shallow-water contourite systems should be more particularly the target of future research, because these systems are not well known in terms of depositional processes or sedimentary facies and geometry. This may be why industrial interest in contourites as hydrocarbon-bearing reservoirs is not as high as might be expected.

Recent results show that contourite deposits are a good sedimentological tool to decipher global changes in palaeocirculation and palaeoclimate. Improvements in the characterization of these deposits and a better understanding of transport/deposition processes will be necessary and can be obtained through integrated studies of sedimentology, geochemistry, micro-palaeontology and physical and biological oceanography.

REFERENCES

Ambar, I., Howe, M.R., 1979. Observations of the Mediterranean outflow—II. The deep circulation in the vicinity of the Gulf of Cadiz. Deep Sea Res. A 26, 555–568.

Amy, L.A., 2000. Architectural analysis of a sand-rich confined turbidite basin: the Grès de Peïra Cava, south-east France. Ph.D. thesis, University of Leeds, UK.
Arthur, M.A., Srivastava, S.P., Kaminski, M., Jarrard, R., Osler, J., 1989. Seismic stratigraphy and history of deep circulation and sediment drift development in Baffin Bay and the Labrador Sea. In: Shrisvastava, S.P., Arthur, M., Clement, B. et al., Initial Reports of the Ocean Drilling Program, Scientific Results, ODP, vol. 105B. 891–922.
Barnes, P.M., 1992. Mid-bathyal current scours and sediment drifts adjacent to the Hikurangi deep-sea turbidite channel, eastern New Zealand: evidence from echo character mapping. Mar. Geol. 106, 169–187.
Barnes, P.M., 1994. Pliocene–Pleistocene depositional units on the continental slope off central New Zealand; control by slope currents and global climate cycles. Mar. Geol. 117, 155–175.
Ben-Avraham, Z., Niemi, T.M., Hartnady, C.J.H., 1994. Mid-Tertiary changes in deep ocean circulation patterns in the Natal Valley and Transkei basin, Southwest Indian Ocean. Earth Planet. Sci. Lett. 121, 639–646.
Bickert, T., Henrich, R., 2011. Climate records of deep-sea sediments: towards the Ice House. In: Hüneke, H., Mulder, T. (Eds.), Deep-Sea Sediments. Developments in Sedimentology, Elsevier, Amsterdam, vol. 63, pp. 793–823.
Bickert, T., Wefer, G., 1996. Late quaternary deep water circulation in the south Atlantic: reconstruction from carbonate dissolution and benthic stable isotopes. In: Wefer, G., Berger, W.H., Siedler, G., Webb, D.J. (Eds.), The South Atlantic: Present and Past Circulation. Springer-Verlag, Berlin, Heidelberg, pp. 599–620.
Biscaye, P.E., Eitreim, S.L., 1977. Suspended particulate loads and transport in the nepheloid layer of the abyssal Atlantic Ocean. Mar. Geol. 23, 155–172.
Bouma, A.H., 1962. Sedimentology of Some Flysch Deposits. A Graphic Approach to Facies Interpretation. Elsevier, Amsterdam, vol. 63, 168pp.
Carey, S.N., Schneider, J.-L., 2011. Volcaniclastic processes and deposits in the deep sea. In: Hüneke, H., Mulder, T. (Eds.), Deep-Sea Sediments. Developments in Sedimentology, Elsevier, Amsterdam, vol. 63, pp. 457–515.
Carter, R.M., 2007. The role of intermediate-depth currents in continental shelf-slope accretion: Canterbury Drifts, Southwest Pacific Ocean. In: Viana, A.R., Rebesco, M. (Eds.), Economic and Palaeoceanographic Significance of Contourite Deposits, Geol. Soc. London, Sp. Pub. 276, 129–154.
Carter, L., McCave, I.N., 1994. Development of sediment drifts approaching an active plate margin under the SW Pacific deep western boundary undercurrent. Paleoceanography 9 (6), 1061–1085.
Carter, L., McCave, I.N., 2002. Eastern New Zealand drifts, Miocene-Recent. In: Stow, D. A.V., Pudsey, C., Howe, J.A., Faugères, J.-C., Viana, A.R. (Eds.), Deep-Water Contourite Systems: Modern Drifts and Ancient Series, Seismic and Sedimentary Characteristics, Geol. Soc. London Mem. 22, 385–407.
Cheney, R.E., Marsh, J.G., Beckley, B.D., 1983. Global mesoscale variability from collinear tracks of SEASAT altimeter data. J. Geophys. Res. 88, 4343–4354.
Correggiari, A., Trincardi, F., Langone, L., Roveri, M., 2001. Styles of failure in late Holocene highstand prodelta wedges on the Adriatic shelf. J. Sediment. Res. 71 (2), 218–236.
Cunningham, A.P., Barker, P.F., 1996. Evidence for westward-flowing Weddell Sea deep water in the Falkland Trough, western South Atlantic. Deep Sea Res. Part I Oceanogr. Res. Pap. 43 (5), 643–654.
Cunningham, A.P., Howe, J.A., Barker, P.F., 2002. Contourite sedimentation in the Falkland Trough, western South Atlantic. In: Stow, D.A.V., Pudsey, C., Howe, J.A., Faugères, J.-C., Viana, A.R. (Eds.), Deep-Water Contourite Systems: Modern Drifts and Ancient Series, Seismic and Sedimentary Characteristics, Geol. Soc. London Mem. 22, 337–352.

Damuth, J.E., 1975. Echo character of the western equatorial Atlantic floor and its relationship to the dispersal and distribution of terrigenous sediments. Mar. Geol. 18, 17–45.
Damuth, J.E., 1980. Use of high-frequency (3.5–12 kHz) echograms in the study of near-bottom sedimentation processes in the deep-sea: a review. Mar. Geol. 38, 51–75.
Damuth, J.E., Hayes, D.E., 1977. Echo character of the east Brazilian continental margin and its relationship to sedimentary processes. Mar. Geol. 24, 73–95.
Davies, T.A., Laughton, A.S., 1972. Sedimentary processes in the North Atlantic. In: Laughton, A.S., Berggren, W.A. et al., Initial Reports of the Deep Sea Drilling Project, DSDP 12. U.S. Govt. Printing Office, Washington, pp. 905–934.
DeMadron, X.D., Weatherly, G., 1994. Circulation, transport and boundary layers of the deep currents in the Brazil basin. J. Mar. Res. 52, 583–638.
Duan, T., Gao, Z., Zeng, Y., Stow, D., 1993. A fossil carbonate contourite drift on the lower Ordovician paleocontinental margin of the middle Yangtze Terrane, Jiuxi, northern Hunan, southern China. Sediment. Geol. 82 (1–4), 271–284.
Duarte, C.S.L., Viana, A.R., 2007. Santos Drift System: stratigraphic organization and implications for late Cenozoic palaeocirculation in the Santos Basin. In: Viana, A.R., Rebesco, M. (Eds.), Economic and Palaeoceanographic Significance of Contourite Deposits, Geol. Soc. of London, Sp. Pub. 276, 171–198.
Egloff, J., Johnson, G.L., 1975. Morphology and structure of the southern Labrador Sea. Can. J. Earth Sci. 12, 2111–2133.
Egloff, J., Johnson, G.L., 1978. Erosional and depositional structures of the south-west Iceland insular margin. In: Watkins, J.S., Montadert, L., Dikerson, J.W. (Eds.), Geological and Geophysical Investigations of Continental Margins, Amer. Assoc. Petrol. Geologists Mem. 29, 43–63.
Embley, R.W., Hoose, P.J., Lonsdale, P., Mayer, L., Tucholke, B.E., 1980. Furrowed mud waves on the Bermuda rise. Geol. Soc. Am. Bull. 91, 731–740 , Part 1.
Ewing, M., Thorndike, E.M., 1965. Suspended matter in deep ocean water. Science 147, 1291–1294.
Faugères, J.-C., Stow, D.A.V., 2008. Contourite drifts: nature, evolution and controls. In: Rebesco, M., Camerlenghi, A. (Eds.), Contourites. Developments in Sedimentology, vol. 60. Elsevier, Amsterdam, pp. 257–288.
Faugères, J.-C., Stow, D.A.V., 1993. Bottom-current controlled sedimentation: a synthesis of the contourite problem. Sediment. Geol. 82 (1–4), 287–297.
Faugères, J.-C., Gonthier, E., Stow, D.A.V., 1984. Contourite drift molded by deep Mediterranean outflow. Geology 12, 296–300.
Faugères, J.-C., Frappa, M., Gonthier, E., Grousset, F., 1985a. Impact de la veine d'eau méditerranéenne sur la sédimentation de la marge sud et ouest ibérique au Quaternaire récent. Bull. Inst. Géol. Bassin d'Aquitaine, Talence, France 37, 259–287.
Faugères, J.-C., Cremer, M., Monteiro, H., Gaspar, L., 1985b. Essai de reconstitution des processus d'édification de la ride sédimentaire de Faro (marge sud-portugaise). Bull. Inst. Géol. Bassin d'Aquitaine, Talence, France 37, 229–258.
Faugères, J.-C., Mézerais, M.L., Stow, D.A.V., 1993. Contourite drift types and their distribution in the North and South Atlantic Ocean basins. Sediment. Geol. 82 (1–4), 189–203.
Faugères, J.-C., Gonthier, E., Monteiro, H., Vergnaud-Grazzini, C., 1994. Sediment records of deep currents: an example, the Mediterranean Outflow in the Quaternary. Comm. Inst. Geol. e Mineiro, Lisbonne 80, 71–88.
Faugères, J.-C., Stow, D.A.V., Imbert, P., Viana, A., 1999. Seismic features diagnostic of contourite drifts. Mar. Geol. 162, 1–38.
Faugères, J.-C., Gonthier, E., Mulder, T., Kenyon, N., Cirac, P., Berné, S., Lesuavé, R., 2002a. Multi-process generated sediment waves on the Landes Plateau (Bay of Biscay, North Atlantic). Mar. Geol. 182, 279–302.

Faugères, J.-C., Zaragosi, S., Mezerais, M.L., Massé, L., 2002b. The Vema contourite fan in the south Brazilian basin. In: Stow, D.A.V., Pudsey, C., Howe, J.A., Faugères, J.-C., Viana, A.R. (Eds.), Deep-Water Contourite Systems: Modern Drifts and Ancient Series, Seismic and Sedimentary Characteristics, Geol. Soc. London Mem. 22, 223–238.

Faugères, J.-C., Franca Lima, A., Massé, L., Zaragosi, S., 2002c. The Columbia channel–levee system: a fan drift in the southern brazil basin. In: Stow, D.A.V., Pudsey, C., Howe, J.A., Faugères, J.-C., Viana, A.R. (Eds.), Deep-Water Contourites: Modern Drifts and Ancient Series, Seismic and Sedimentary Characteristics, Geol. Soc. London Mem. 22, 223–238.

Flood, R.D., Johnson, T.C., 1984. Side scan targets in Lake Superior, evidence for bedforms and sediment transport. Sedimentology 31, 311–333.

Flood, R.D., Shor, A.N., 1988. Mud waves in the Argentine basin and their relationship to regional bottom circulation patterns. Deep Sea Res. Part I Oceanogr. Res. Pap. 35, 943–972.

Fulthorpe, C.S., Carter, R.M., 1991. Continental shelf progradation by sediment drift accretion. Geol. Soc. Am. Bull. 103 (2), 300–309.

Gambôa, L.A., Buffler, R.T., Barker, P.F., 1983. Seismic stratigraphy and geologic history of the rio Grande Gap and Southern Brazil basin. In: Barker, P.F., Carlson, R.L., Johnson, D.A. (Eds.), Initial Reports, DSDP 72. US Goverment Printing Office, Washington, DC, pp. 481–498.

Gonthier, E., Faugères, J.-C., Stow, D.A.V., 1984. Contourite facies of the Faro drift. Geol. Soc. London, Sp. Pub. 15 pp. 275–292.

Guillocheau, F., 1983. Les dépôts de tempêtes: le modèle de l'Ordovicien moyen ouest-armoricain. Thèse 3ème cycle, Université de Bretagne Occidentale, Brest, France.

Hallbach, P., Friedrich, G., Stackelberg, U.V. (Eds.), 1988. Manganese nodule belt of the Pacific Ocean. Springer Verlag, Stuttgart 254pp.

Hanquiez, V., Mulder, T., Toucanne, S., Lecroart, P., Bonnel, C., Marchès, E.,, Gonthier, E.,, 2010. The sandy channel-lobe depositional systems in the Gulf of Cadiz: gravity processes forced by contour current processes. Sed. Geol. 229, 110–123.

Haq, B.U., 1991. Sequence stratigraphy, sea-level change, and significance for the deep-sea. Int. Assoc. Sediment., Sp. Pub. 12 pp. 3–39.

Heezen, B.C., Hollister, C.D., 1971. The Face of the Deep. Oxford University Press, London 659pp.

Heezen, B.C., Hollister, C.D., Ruddiman, W.F., 1966. Shaping of the continental rise by deep geostrophic bottom currents. Science 152, 502–508.

Hesse, R., Schacht, U., 2011. Early diagenesis of deep-sea sediments. In: Hüneke, H., Mulder, T. (Eds.), Deep-Sea Sediments. Developments in Sedimentology, Elsevier, Amsterdam, vol. 63, pp. 557–714.

Hjulstrom, F., 1939. Transportation of detritus by moving water. In: Trask, P.D. (Ed.), Recent Marine Sediments: A Symposium, Soc. Econ. Paleont. Mineral., SEPM, Tulsa, Sp. Pub. 4, 5–31.

Hogg, N.G., Owens, W.B., Siedler, G., Zenk, W., 1996. Circulation in the deep Brazil Basin. In: Wefer, G., Berger, W.H., Siedler, G., Webb, D.J. (Eds.), The South Atlantic: Present and Past Circulation. Springer-Verlag, Berlin, Heidelberg, pp. 13–44.

Hollister, C.D., 1993. The concept of deep-sea contourites. Sediment. Geol. 82, 5–11.

Hollister, C.D., Heezen, B.C., 1972. Geological effects of ocean bottom currents: western north Atlantic. In: Gordon, A.L. (Ed.), Studies in Physical Oceanography, vol. 2. Gordon and Breach, NY, pp. 37–66.

Hollister, C.D., McCave, I.N., 1984. Sedimentation under deep-sea storms. Nature 309 (5965), 220–225.

Hollister, C.D., Nowell, A., Jumars, P., 1984. Courants et tempêtes dans les abysses. Pour Sci. 79, 20–33

Hosegood, P., van Haren, H., 2004. Near-bed solibores over the continental slope in the Faroe-Shetland Channel. Deep Sea Res. Part II Top. Stud. Oceanogr. II-51, 2943–2971.

Howe, J., 1996. Turbidite and contourite sediment waves in the northern Rockall Trough, north Atlantic Ocean. Sedimentology 43, 210–234.

Howe, J., Stoker, M.S., Stow, D.A.V., 1994. Late Cenozoic sediment drift complex, northeast Rockall Trough. Paleoceanography 9 (6), 989–999.

Hüneke, H., 2006. Erosion and deposition from bottom currents during the Givetian and Frasnian: response to intensified oceanic circulation between Gondwana and Laurussia. Palaeogeogr. Palaeoclimatol. Palaeoecol. 234, 146–167.

Hüneke, H., 2007. Pelagic carbonate ooze reworked by bottom currents during Devonian approach of the continents Gondwana and Laurussia. In: Viana, A.R., Rebesco, M. (Eds.), Economic and Palaeoceanographic Significance of Contourite Deposits, Geol. Soc. London, Sp. Pub. 276, 299–328.

Hüneke, H., Henrich, R., 2011. Pelagic sedimentation in modern and ancient oceans. In: Hüneke, H., Mulder, T. (Eds.), Deep-Sea Sediments. Developments in Sedimentology, Elsevier, Amsterdam, vol. 63, pp. 215–351.

Hüneke, H., Stow, D.A.V., 2008. Identification of ancient contourites: problems and palaeoceanographic significance. In: Rebesco, M., Camerlenghi, A. (Eds.), Contourites. Developments in Sedimentology, vol. 60. Elsevier, Amsterdam, pp. 323–344.

Jerlov, N.G., 1953. Particle distribution in the ocean. Rep. Swed. Deep Sea Expedition 3, 73–97.

Johnson, D.A., 1972. Ocean-floor erosion in the Equatorial Pacific. Geol. Soc. Am. Bull. 83, 3121–3144.

Johnson, D.A., Ledbetter, M.T., Damuth, J.E., 1983. Neogene sedimentation and erosion in the Amirante Passage, Western Indian Ocean. Deep Sea Res. A 30 (2A), 195–219.

Kennett, J.P., 1982. Marine Geology. Englewood Cliffs, Prentice Hall 813 pp.

Kenyon, N.H., Belderson, R.H., 1973. Bedforms of the Mediterranean undercurrent observed with side scan sonar. Sediment. Geol. 9 (2), 77–100.

Kindler, P., Ujetz, B., Charollais, J., Wernly, R., 1995. Submarine resedimentation of Cretaceous deposits during the Paleogene: the "Formation gréso-glauconieuse" from the Ultrahelvetic Prealps (Haute-Savoie, France). Bull. Soc. Geol. France 166 (5), 507–515.

Kolla, V., Eittrem, S., Sullivan, L., Kostecki, J.A., Burckle, L.H., 1980. Current-controlled abyssal microtopography and sedimentation in Mozambique basin, Southwest Indian Ocean. Mar. Geol. 34, 171–206.

Ledbetter, M.T., 1984. Bottom-current speed in the Vema Channel recorded by particle size of sediment fine-fraction. Mar. Geol. 58, 137–149.

Ledbetter, M.T., 1986. Bottom-current pathways in the Argentine basin revealed by mean silt particle size. Nature 321 (6068), 423–425.

Lee, H.J., Syvitski, J.P.M., Parker, G., Orange, D., Locat, J., Hutton, E.W.H., et al., 2002. Distinguishing sediment waves from slope failure deposits: field examples, including the "Humboldt slide" and modelling results. Mar. Geol. 192 (1–3), 79–104.

Llave, E., Hernandez-Molina, F.J., Somoza, L., Diaz del Rio, V., Stow, D.A.V., Maestro, A., 2005. Seismic stacking patterns of the Faro-Albufeira contourite system (Gulf of Cadiz): a quaternary record of paleogeographic and tectonic influences. Mar. Geophys. Res. 22, 487–508.

Llave, E., Schönfeld, J., Hernandez-Molina, F.J., Mulder, T., Somoza, L., Diaz del Rio, V., et al., 2006. High-resolution stratigraphy of the Mediterranean outflow contourite system in the Gulf of Cadiz during the Late Pleistocene. The impact of Heinrich events. Mar. Geol. 227 (3–4), 241–262.

Locker, S.D., Laine, P., 1992. Paleogene-Neogene depositional history of the middle U.S. Atlantic continental rise: mixed turbidite and contourite depositional systems. Mar. Geol. 103, 137–164.

Lonsdale, P., 1976. Abyssal circulation of the Southeastern Pacific and some geological implications. J. Geophys. Res. 81 (6), 1163–1176.

Lonsdale, P., 1981. Drifts and ponds of reworked pelagic sediment in part of the southwest Pacific. Mar. Geol. 43, 153–193.

Lonsdale, P., Malfait, B., 1974. Abyssal dunes of foraminiferal sand in Carnegie Ridge. Geol. Soc. Am. Bull. 83, 289–316.

Lowe, D.R., 1982. Sediment gravity flows: II. Depositional models with special reference to the deposits of high-density turbidity currents. J. Sediment. Petrol. 52, 279–297.

Lowel, J.P.B., Stow, D., 1981. Identification of ancient sandy contourites. Geology 9, 347–349.

Lutjeharms, J.R.E., 1996. The exchange of water between the South Indian and South Atlantic oceans. In: Wefer, G., Berger, W.H., Siedler, G., Webb, D.J. (Eds.), The South Atlantic: Present and Past Circulation. Springer-Verlag, Berlin, Heidelberg, pp. 125–162.

Maldonado, A., Barnolas, A., Bohoyo, F., Galindo-Zaldivar, J., Hernandez-Molina, J., Lobo, F., Rodriguez-Fernandez, J., Somoza, L., Vasquez, J.-T., 2003. Contourite deposits in the central Scotia Sea: the importance of the Antarctic Circumpolar Current and the Weddell Gyre flows. Palaeogeogr. Palaeoclimatol. Palaeoecol. 198, 187–221.

Maldonado, A., Barnolas, A., Bohoyo, F., Escutia, C., Galindo-Zaldivar, J., Hernandez-Molina, J., Jabaloy, A., Lobo, F.J., Nelson, C.H., Rodriguez-Fernandez, J., Somoza, L., Vasquez, J.-T., 2005. Miocene to recent contourite drift development in the northern Weddell Sea (Antarctica). Glob. Planet. Change 45, 99–129.

Massé, L., Faugères, J.-C., Bernat, M., Pujos, A., Mézerais, M.-L., 1994. A 6,00, 000 yr record of Antactic Bottom Water activity inferred from sediment textures and structures in a sediment core from the Southern Brazil Basin. Paleoceanography 9 (6), 1017–1026.

Massé, L., Faugères, J.-C., Pujol, C., Pujos, A., Labeyrie, L.D., Bernat, M., 1996. Sediment flux distribution in the Southern Brazil Basin during the late Quaternary: the role of deep-sea currents. Sedimentology 43, 115–132.

Massé, L., Hrovatin, V., Faugères, J.-C., 1998. The interplay between turbidite and contouritic processes in the Southern Brazil Basin. Sediment. Geol. 115 (1–4), 111–132.

McCave, I.N., 1984. Erosion, transport and deposition of fine-grained marine sediments. Geol. Soc. London, Sp. Pub. 15 pp. 35–69.

McCave, I.N., 1986. Local and global aspects of the bottom nepheloid layers in the world ocean. J. Sea Res. (Netherlands) 20 (2–3), 167–181.

McCave, I.N., 2008. Size sorting during transport and deposition of fine sediments: sortable silt and flow speed. In: Rebesco, M., Camerlenghi, A. (Eds.), Contourites. Developments in Sedimentology, vol. 60. Elsevier, Amsterdam, pp. 121–142.

McCave, I.N., Manighetti, B., Beveridge, N.A.S., 1995. Circulation in the glacial north Atlantic inferred from grain-size measurements. Nature 374, 149–152.

McCave, I.N., 1972. Transport and escape of fine-grained sediment from shelf areas. In: Swift, D.J.P., Duane, D.B., Pilkey, O.H. (Eds.), Shelf Sediment Transport: Process and Pattern. Hutchinson and Ross, Stroudsburg, PA, pp. 225–248.

McCave, I.N., Tucholke, B.E., 1986. Deep current-controlled sedimentation in the western North Atlantic. In: Vogt, P.R., Tucholke, B.E. (Eds.), The Geology of North America. The Western North Atlantic Region, Decade of North America Geology, vol. M. Geological Society of America, Boulder, CO, pp. 451–468.

McCave, I.N., Carter, L., 1997. Recent sedimentation beneath the Deep Western Boundary Current off northern New Zealand. Deep Sea Res. I 44 (7), 1203–1237.

McMaster, R.L., Locker, S.D., Laine, E.P., 1989. The early Neogene continental rise off the eastern United States. Mar. Geol. 87, 137–163.

Meinert, J., 1986. Akustostratigraphie im äquatorialen Ostatlantik: zur Entwicklung der Tiefenwasserzirkulation der letzten 3, 5 Millonen Jahre. Meteor Forsch.-Ergebnisse, C 40, 19–86.
Mézerais, M.L., Faugères, J.-C., Figueiredo, A., Massé, L., 1993. Contour current accumulation off Vema Channel mouth, southern Brazil basin: pattern of a "contourite fan" Sediment. Geol. 82 (1–4), 173–188.
Migeon, S., Weber, O., Faugères, J.-C., Saint-Paul, J., 1999. SCOPIX: a new X-ray imaging system for core analysis. Geo Mar. Lett. 18, 251–255.
Migeon, S., Savoye, B., Faugères, J.-C., 2000. Quaternary development of migrating sediment waves in the Var deep-sea fan: distribution, growth pattern and implication in levee evolution. Sediment. Geol. 133, 265–293.
Migeon, S., Savoye, B., Zanella, E., Mulder, T., Faugères, J.-C., Weber, O., 2001. Detailed seismic-reflection and sedimentary study of turbidite sediment waves on the Var Sedimentary Ridge (SE France): significance for sediment transport and deposition and for the mechanisms of sediment-wave construction. Mar. Petrol. Geol. 18, 179–208.
Milankovitch, M.M., 1941. Kahnon der Erdbestrablung und seine Anwendung auf das Eiszeitenproblem. Ed. Sp. Acad Royale Serbe, Belgrade 633 pp.
Miller, M.A., McCave, I.N., Komar, P.D., 1977. Threshold of sediment motion under unidirectional currents. Sedimentology 24, 507–528.
Mountain, G.S., Tucholke, B.E., 1985. Mesozoic and Cenozoic geology of the U.S. Atlantic continental slope and rise. In: Poag, C.W. (Ed.), Geologic Evolution of the United States Atlantic Margin. New York, Van Nostrand Reinhold Company, pp. 293–491.
Mulder, T., 2011. Sedimentary processes and their deposits. In: Hüneke, H., Mulder, T. (Eds.), Deep-Sea Sediments. Developments in Sedimentology, Elsevier, Amsterdam, vol. 63, pp. 25–148.
Mulder, T., Syvitski, J.P.M., Migeon, S., Faugères, J.-C., Savoye, B., 2003. Marine hyperpycnal flows: initiation, behavior and related deposits. A review. Mar. Petrol. Geol. 20, 861–882.
Mulder, T., Faugères, J.-C., Gonthier, E., 2008. Mixed turbidite–contourite systems. In: Rebesco, M., Camerlenghi, A. (Eds.), Contourites. Developments in Sedimentology, vol. 60. Elsevier, Amsterdam, pp. 435–456.
Mulder, T., Razin, P., Faugères, J.-C., 2009. Hummocky cross-stratification-like structures in deep-sea turbidites: upper Cretaceous Basque basins (Western Pyrenees, France). Sedimentology 56 (4), 997–1015.
Mutti, E., Davoli, G., Segadelli, S., Cavalli, C., Carminatti, M., Stechi, S., et al., 1992. Turbidite Sandstone. AGIP/Instituto di Geologia, Universita di Parma, Parma, 236pp.
Nelson, C.H., Baraza, J., Maldonado, A., 1993. Mediterranean undercurrent "contourites" in the eastern Gulf of Cadiz. Sediment. Geol. 82 (1–4), 103–132.
Nielsen, T., Knutz, P.C., Kuijpers, A., 2008. Seismic expression of contourite depositional systems. In: Rebesco, M., Camerlenghi, A. (Eds.), Contourites. Developments in Sedimentology, vol. 60. Elsevier, Amsterdam, pp. 301–321.
Normark, W.R., Hess, G.R., Stow, D.A.V., Bowen, A.J., 1980. Sediment waves on the Monterey fan levee: a preliminary physical interpretation. Mar. Geol. 37, 1–18.
Nowell, A.R.M., Hollister, C.D. (Eds.), 1985. Deep Ocean Sediment Transport, In: Preliminary Results of the High Energy Benthic Boundary Layer Experiment, Mar. Geol. 66, 409pp.
O'Neil-Baring, P., Price, J.F., 1999. A review of the physical oceanography of the Mediterranean outflow. Mar. Geol. 155, 63–82.
Parsons, J.D., Bush, J., Syvitski, J.P.M., 2001. Hyperpycnal flow formation with small sediment concentrations. Sedimentology 48, 465–478.
Postma, H., 1967. Sediment transport and sedimentation in the estuarine environments. In: Lauff, G.H. (Ed.), Estuaries, vol. 83. Am. Assoc. Adv. Sci, Washington, pp. 158–179.

Seismic Expression of Contourites and Related Deposits 2001. In:. Rebesco, M., Stow, D. A.V. (Eds.), 2001. In: Mar. Geophys. Res., 22 (5–6), 303–521 (special issue).

Rebesco, M., Larter, R.D., Camerlenghi, A., Barker, P.F., 1996. Giant sediment drifts on the continental rise west of the Antarctic Peninsula. Geo Mar. Lett. 16, 65–75.

Rebesco, M., Larter, R.D., Barker, P.F., Camerlenghi, A., Vanneste, L.E., 1997. The history of sedimentation on the continental rise west of the Antarctic Peninsula. In: Cooper, A.K., Barker, P.F. (Eds.), Geology and Seismic Stratigraphy of the Antarctic Margin, Part 2, American Geophysical Union, Antarctic Research Series, 71, 29–49.

Rebesco, M., Pudsey, G.J., Canals, M., Camerlenghi, A., Barker, P.F., Estrada, F., et al., 2002. Sediment drifts and deep-sea channel systems, Antarctica Peninsula Pacific margin. In: Stow, D.A.V., Pudsey, C., Howe, J.A., Faugères, J.-C., Viana, A.R. (Eds.), Deep-Water Contourites: Modern Drifts and Ancient Series, Seismic and Sedimentary Characteristics, Geol. Soc. London Mem. 22, 353–372.

Reed, D.L., Meyer, A.W., Silver, E.A., Prasetyo, H., 1987. Contourite sedimentation in an intraoceanic forearc system: eastern Sunda Arc, Indonesia. Mar. Geol. 76 (3–4), 223–242.

Robinson, S.G., McCave, I.N., 1994. Orbital forcing of bottom currents enhanced sedimentation on Feni drift, NE Atlantic, during the mid-Pleistocene. Paleoceanography 9 (6), 943–972.

Salles, T., Mulder, T., Gaudin, M., Cacas, M.C., Lopez, S., Cirac, P., 2008. Simulating the 1999 turbidity current occurred in Capbreton canyon through a Cellular Automata model. Geomorphology 97 (3–4), 516–537.

Schmitz, J.W., 1984. Abyssal eddy kinetic energy in the North Atlantic. J. Mar. Res. 42, 509–536.

Schmitz, J.W., 1988. Exploration of the eddy field in the mid-latitude North Pacific. J. Phys. Oceanogr. 18 (3), 459–468.

Schmitz, W.J., 1996. On the World Ocean Circulation. The Pacific and Indian Oceans. A Global Update. Technical Report, WHOI-96-08, vol. II. .

Scholl, D.W., Hein, J.R., Marlow, M., Buffington, E.C., 1977. Meiji sediment tongue: North Pacific evidence for limited movement between the Pacific and north American plates. Geol. Soc. Am. Bull. 88, 1567–1576.

Shanmugan, G., Spalding, T.D., Rofheart, D.H., 1993. Process, sedimentology and reservoir quality of deep-marine bottom-current reworked sands (sandy contourites): an example from the Gulf of Mexico. AAPG Bull. 77 (7), 1241–1259.

Shipboard Scientific Party, 1976. Site 331. In: Barker, P.F., Dalziel, I.W.D. et al., 1976. Init. Repts of the DSDP, 36. U.S. Government, Printing Office, Washington, pp. 259–265.

Somoza, L., Hernandez-Molina, F.J., De Andres, J.R., Rey, J., 1997. Continental shelf architecture and sea-level cycles: late Quaternary high-resolution stratigraphy of the Gulf of Cadiz, Spain. Geo. Mar. Lett. 17, 133–139.

Stanley, D.J., 1993. Model for turbidite-to-contourite continuum and multiple process transport in deep marine settings: examples in the rock record. Sediment. Geol. 82, 241–256.

Stoker, M.S., 1998. Sediment-drift development on the Rockall continental margin, off NW Britain. In: Stoker, M.S., Evans, D., Cramp, A. (Eds.), Geological Processes on Continental Margins: Sedimentation, Mass-Wasting and Stability, Geol. Soc. of London, Sp. Pub. 129, 229–254.

Stoker, M.S., Akhurst, C., Howe, J.A., Stow, D.A.V., 1998. Sediment drifts and contourites on the continental margin, off northwest Britain. Sediment. Geol. 115 (1–4), 33–52.

Stommel, H., 1957. A survey of ocean currents theory. Deep Sea Res. 4, 149–184.

Stommel, H., Aarons, A.B., 1960. On the abyssal circulation of the world ocean—II. An idealized model of the circulation pattern and amplitude in oceanic basins. Deep Sea Res. 6, 217–233.

Stow, D.A.V., 1982. Bottom currents and contourites in the North Atlantic. Bull. Inst. Geol. Bassin d'Aquitaine, Talence, France 31, 151–166.
Stow, D.A.V., Faugères, J.-C. (Eds.), 1993. Contourites and Bottom Currents, In: Sediment. Geol., 82, 310pp.
Stow, D.A.V., Faugères, J.-C., 1998. Contourites, Turbidites, and Interaction Processes, Sediment. Geol., 115 (1–4), 384 pp.
Stow, D.A.V., Faugères, J.-C., 2008. Contourite facies and the facies model. In: Rebesco, M., Camerlenghi, A. (Eds.), Contourites. Developments in Sedimentology, vol. 60. Elsevier, Amsterdam, pp. 223–256.
Stow, D.A.V., Lowell, J.P.B., 1979. Contourites: their recognition in modern and ancient sediments. Earth Sci. Rev. 14, 251–291.
Stow, D.A.V., Reading, H.G., Collinson, J., 1996. Deep seas. In: Reading, H.G. (Ed.), Sedimentary Environments, third ed. 395–453.
Stow, D.A.V., Faugères, J.-C., Viana, A., Gonthier, E., 1998. Fossil contourites, a critical review. Sediment. Geol. 115 (1–4), 3–32.
Stow, D.A.V., Howe, J.A., Pudsey, C.J., Faugères, J.-C., Viana, A.R. (Eds.), 2002a. Deep-Water Contourites: Modern Drifts and Ancient Series, Seismic and Sedimentary Characteristics, In: Geol. Soc. London, Mem. 22, 464pp.
Stow, D.A.V., Faugères, J.-C., Howe, J.A., Pudsey, C.J., Viana, A.R., 2002b. Bottom currents, contourites and deep-sea sediment drifts: current state-of-art. In: Stow, D.A.V., Pudsey, C., Howe, J.A., Faugères, J.-C., Viana, A.R. (Eds.), Deep-Water Contourites: Modern Drifts and Ancient Series, Seismic and Sedimentary Characteristics, Geol. Soc. London, Mem. 22, 7–20.
Toucanne, S., Mulder, T., Schönfeld, J., Hanquiez, V., Gonthier, E., Duprat, J., et al., 2007. Contourites of the Gulf of Cadiz: a high-resolution record of the paleocirculation of the Mediterranean outflow water during the last 50, 000 years. Palaeogeogr. Palaeoclimatol. Palaeoecol. 246, 354–366.
Tucholke, B.E., Mountain, G.S., 1986. Tertiary paleoceanography of the western North Atlantic Ocean. In: Vogt, P.R., Tucholke, B.E. (Eds.), The Geology of North America. The Western North Atlantic Region, vol. M. Geol. Soc. Am, pp. 631–650 (Chapter 38).
Tucholke, B.E., Embley, R.W., 1984. Cenozoic regional erosion of the abyssal sea floor off South Africa. In: Schlee, J.S. (Ed.), Interregional Unconformities and Hydrocarbon Accumulation, AAPG Memoir 36, 145–164.
Tucholke, B.E., Laine, E.P., 1983. Neogene and Quaternary development of the lower continental rise off the central US coast. In: Watkins, J.S., Drake, C.L. (Eds.), Studies in Continental Margin Geology, AAPG Memoir 34, 295–305.
Vail, P.R., Mitchum, R.M., Todd, R.G., 1977. Seismic stratigraphy and global changes of sea level. In: Payton, C.E. (Ed.), Seismic Stratigraphy—Applications to Hydrocarbon Exploration, AAPG Memoir 26, 49–212.
Viana, A.R., Faugères, J.-C., 1998. Upper slope sand deposits: the example of Campos Basin, a latest Pleistocene/Holocene record of the interaction between along and across slope currents. In: Stoker, M.S., Evans, D., Cramp, A. (Eds.), Geological Processes on Continental Margins: Sedimentation, Mass-Wasting and Stability, Geol. Soc. London, Sp. Pub. 129, 287–316.
Viana, A.R., Faugères, J.-C., Stow, D.A.V., 1998. Bottom current controlled sand deposits—a review from modern shallow- to deep-water environments. Sediment. Geol. 115 (1/4), 53–80.
Viana, A.R., de Almeida Jr., W., de Almeida, C.W., 2002. Upper slope sands: late Quaternary shallow-water sandy contourites of Campos Basin, SW Atlantic margin. In: Stow, D.A.V., Pudsey, C., Howe, J.A., Faugères, J.-C., Viana, A.R. (Eds.), Deep-Water Contourites: Modern Drifts and Ancient Series, Seismic and Sedimentary Characteristics, Geol. Soc. London Mem. 22, 261–270.

Vilard, F., 1991. Evolution paléogéographique du domaine delphino-helvétique (entre Chartreuse et Morcles) au Crétacé supérieur (Turonien-Maastrichtien): biostratigraphie, sédimentologie et dynamique sédimentaire sur une rampe carbonatée10, Publ. Dépt. Géol. et Paléont. Univ. Genève 173pp.

Voelker, A.H.L., Lebreiro, S., Schönfeld, J., Cacho, I., Exlenkenser, H., Abrantes, F., 2006. Mediterranean outflow strengthens northern hemisphere cooling. Salt source for the glacial Atlantic. Earth Planet. Sci. Lett. 245, 39–55.

von Lom-Keil, H., Spies, V., Hopfauf, V., 2002. Fine-grained sediment waves on the western flank of the Zappiola drift, Argentine Basin: evidence for variations in Late Quaternary bottom flow activity. Mar. Geol. 192 (1–3), 239–258.

Weber, M.E., Bonani, G., Futterer, K.D., 1994. Sedimentation processes within channel–ridge systems, southeastern Weddell Sea, Antarctica. Paleoceanography 9 (6), 1027–1048.

Wetzel, A., Werner, F., Stow, D.A.V., 2008. Bioturbation and biogenic sedimentary structures in contourites. In: Rebesco, M., Camerlenghi, A. (Eds.), Contourites. Developments in Sedimentology, vol. 60. Elsevier, Amsterdam, pp. 183–202.

Wood, R., Davy, B., 1994. The Hikurangi Plateau. Mar. Geol. 118, 153–173.

Wynn, R.B., Masson, D.G., 2008. Sediment waves and bedforms. In: Rebesco, M., Camerlenghi, A. (Eds.), Contourites. Developments in Sedimentology, vol. 60. Elsevier, Amsterdam, pp. 289–300.

Wynn, R.B., Stow, D.A.V. (Eds.), 2002. Recognition and Interpretation of Deep-Water Sediment Waves: Implications for Paleoceanography, Hydrocarbon Exploration and Flow Process Interpretation, In: Mar. Geol., 192 (1–3), 333pp.

CHAPTER 4

Pelagic Sedimentation in Modern and Ancient Oceans

Heiko Hüneke*[,1] *and* Rüdiger Henrich[†]

Contents

1. Oceanic Provinces and Sediment Factories: An Overview

The main controls on surface circulation of the ocean are direct wind shear and the Coriolis force. Hence, the main oceanic surface currents are strongly related to the planetary wind system. In the tropics, the omnipresent trade winds thus generate the North and South Equatorial Currents which carry warm waters westwards and towards the poles into higher

* Institute of Geography and Geology, University of Greifswald, Greifswald, Germany
† Department of Sedimentology and Palaeoceanography, Faculty of Geosciences, University of Bremen, Bremen, Germany
1 Corresponding author.
E-mail address: hueneke@uni-greifswald.de

Developments in Sedimentology, Volume 63
ISSN 0070-4571, DOI: 10.1016/S0070-4571(11)63004-5

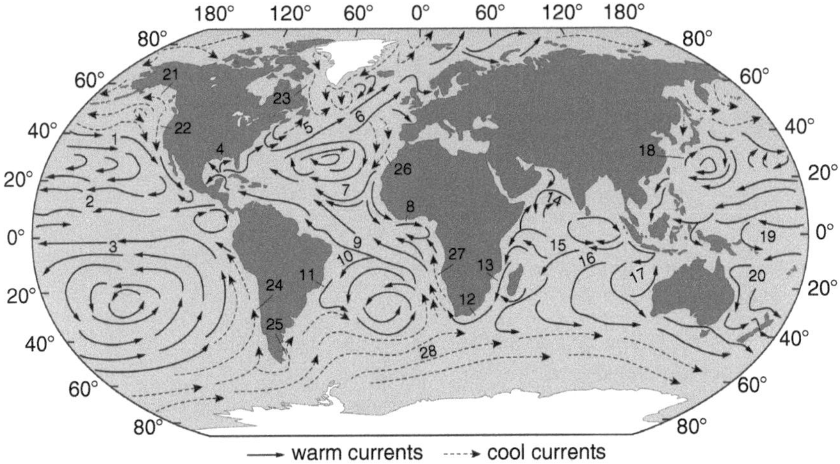

Figure 4.1 Global surface-water circulation (see text for further explanation). 1, North Pacific Current; 2, Equatorial Countercurrent; 3, South Equatorial Current; 4, Florida Current; 5, Gulf Stream; 6, North Atlantic Current; 7, North Equatorial Current; 8, Guinea Current; 9, Equatorial Countercurrent; 10, South Equatorial Current; 11, Brazil Current; 12, Agulhas Current; 13, Mozambique Current; 14, North Equatorial Current; 15, Equatorial Countercurrent; 16, South Equatorial Current; 17, West Australian Current; 18, Kuroshio Current; 19, Equatorial Countercurrent; 20, Australian Current; 21, Alaska Current; 22, California Current; 23, Labrador Current; 24, Humboldt Current; 25, Falkland Current; 26, Canary Current; 27, Benguela Current; 28, West-wind drift.

latitudes on the western margins of the oceans. Here, entering the west wind belt, the surface currents are directed to the eastern sides of the ocean basins, where a return flow of cooled water towards the equator is initiated. This large-scale motion creates the so-called subtropical gyres in each ocean basin (Fig. 4.1). The wind force around these anticyclonic highs causes Ekman transport of surface waters towards the centres of the gyres, which creates a positive topography of up to 1.4 m, with an apex located in the western sector of the gyre. A striking feature is the intensification of surface flow at the western margin of the gyres, which is manifested as the large, strong warm western boundary currents such as the Gulf Stream and the South Brazil Current in the Atlantic, the Pacific Kuro Shio Current and the Agulhas Current in the Indian Ocean. Spectacular eddy motions, meanders and the formation of deep-reaching warm and cool water rings inducing the so-called deep-sea storms (Hollister and McCave, 1984) are prominent elements for sediment dynamics and nutrient mixing within these current systems (McGillicuddy et al., 1998). In addition, along the inner tropical zone of convergent trade winds north and south of the equator, Ekman transport towards the poles by the westward flowing equatorial currents causes a steady equatorial upwelling of about 1 m per day.

On account of the distinctly different physico-chemical properties of the surface waters, contrasting neritic and pelagic environmental regimes can be distinguished in each region of the huge subtropical gyres, for example, (1) in the tropical to subtropical belt warm water carbonate factories including reefs and lagoons on the western shelves versus cool water, kelp-dominated carbonate factories on the eastern shelves of the ocean basins (Henrich et al., 1995), (2) an extensive belt of coastal upwelling induced by the trade winds along the eastern margin of the gyres contrasting with oligotrophic, low-productivity regimes with a deep thermocline in the western sector of the equatorial belt, (3) elevated biogenic pelagic production in the mesotrophic to eutrophic surface waters of the equatorial upwelling zone and (4) strongly oligotrophic low-productivity regimes in the central parts of the gyres, in many studies characterised as the central oceanic deserts. Pole-wards of the huge subtropical gyres, convergence systems occur on both hemispheres, for example, the subpolar gyres in the Norwegian Sea and Weddell Sea. At the Antarctic Polar Front (APF), marked by the border between cold Antarctic surface waters and much warmer sub-Antarctic waters, parts of the cold Antarctic surface waters descent, forming Antarctic Intermediate Water (AAIW). This descent is accompanied by upwelling of nutrient-rich warm waters inducing massive plankton productivity, for example, in the circum-Antarctic belt of siliceous oozes dominated by diatoms, and in the coastal upwelling systems of the Pacific and Atlantic. Waters upwelled in the North Pacific have a higher Si:N ratio than waters of the North Atlantic (Codispoti, 1983; Kamykowski and Zentara, 1989). Such a difference has been attributed to Si recycling, which tends to enrich the deep waters in Si with respect to other nutrients, and leads to basin-wide differences in Si:C-flux ratios measured in sediment traps (Ragueneau et al., 2000).

The main features of global surface water circulation are nicely portrayed by satellite images displaying the chlorophyll concentration of surface waters (Fig. 4.2).

Intermediate- and deep-water circulation in the ocean is mainly driven by density gradients arising from differences in temperature and salinity. This thermohaline circulation (Fig. 4.3), driven by transport of heat, salt and fresh water at the ocean surface, is a significant mechanism of the global oceanic salinity conveyor belt (Broecker and Peng, 1982), which carries about 50 times the volume of water supplied by the average discharge of the worlds rivers. Because of their great influence on the global atmospheric CO_2 concentration, the dynamics of deep-water circulation has become a central theme of the recent climate discussion. The major aspect is the huge volume of CO_2 stored in the ocean in deep and intermediate waters, as stressed already in 1982 by Broecker. Even small volumetric changes in the deep-water CO_2 reservoir and dynamics have the capacity to influence global climate considerably. GEOSECS data in the modern ocean clearly

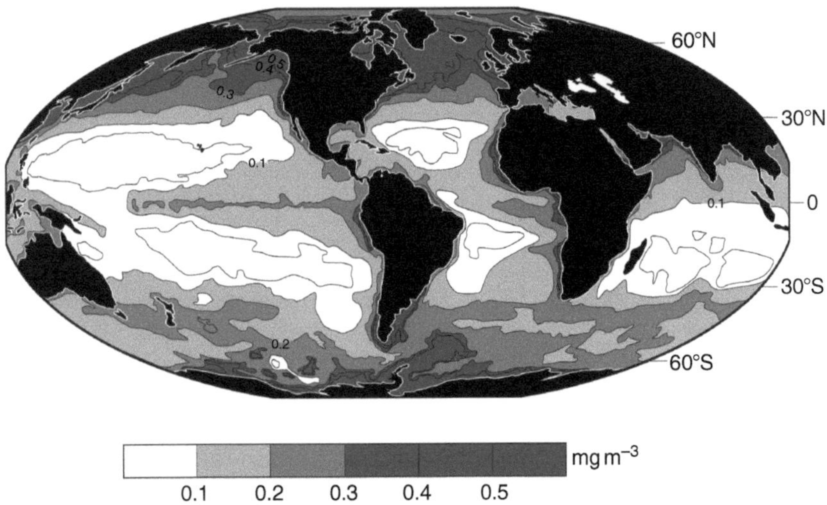

Figure 4.2 Satellite image of global chlorophyll concentration in surface-ocean water (redrawn and modified from Smetacek, 1991). (A multi-colour version of this figure is on the included CD-ROM.)

indicate that the highest CO_2 contents are found in deep- and intermediate-water masses from southern source regions, which are the Antarctic Bottom Water (AABW) and the AAIW. In contrast, the Nordic intermediate and deep waters, for example, Upper and Lower North Atlantic Deep Water (UNADW, LNADW), are young, oxygen-rich and CO_2-poor. In the modern ocean, about 10–13 Sv (Dickson and Brown, 1994; Worthington, 1976) of dense and oxygen-rich North Atlantic Deep Water (NADW) originates from this area. It is compensated by an equally large northward flow at thermocline depths (Schmitz and McCartney, 1993). The northern source region is the most sensitive and vulnerable system as compared to AABW. Studies by Dickson and Brown (1994) reveal at least four different source processes contributing to the export of NADW from the northernmost Atlantic: (1) deep convection in the Greenland Sea, (2) brine formation on the Arctic shelves during sea ice formation, (3) intermediate water formation by mid-gyre convection in the Iceland and Labrador Seas and (4) admixture of Mediterranean outflow water and long-travelled and re-circulated AABW into NADW.

The South Atlantic around Antarctica plays a complementary key role in the global thermohaline circulation; the NADW is fed there into the Antarctic Circumpolar Current (ACC), and the second major deep-water source is located there. Dense Weddell Sea Deep Water (WSDW), which is derived from super-cooled surface waters which became even more saline during sea ice formation, and slightly less dense Circumpolar Deep Water

Figure 4.3 Deep-water circulation of the Atlantic (compiled by Gröger, 2002; etopo1 data after Amante and Eakins, 2009). Water-mass structure and pathways of main deep-water masses in the western and eastern Atlantic sub-basins displayed in schematic cross-sections. AABW, Antarctic Bottom Water; NADW, North Atlantic Deep Water; UCDW, Upper Circumpolar Deep Water; AAIW, Antarctic Intermediate Water. (A multi-colour version of this figure is on the included CD-ROM.)

(CDW), which is derived from upwelled currents re-circulated around Antarctica (Foldvik and Gammelrød, 1988; Rhein et al., 1996), occur here. The density characteristics in the southern South Atlantic north of the ACC are such that, at about 45°S, the NADW divides the CDW into an upper and a lower branch. Consequently, the relatively warm and saline NADW occupies the depth interval between 1500 and 4000 m, while lower CDW (LCDW = AABW) is encountered below 4000 m. Further to the North, in the deep western South Atlantic, the main path flow of deep water is dominated by the interaction between NADW flowing to the South at a higher sub-bottom level, and AABW flowing to the North as a bottom layer. For the eastern basins of the South Atlantic, the inflow of AABW is restricted by the Mid-Atlantic Ridge (MAR) in the West and the Walvis Ridge in the South to only small quantities passing the sills through the Romanche Fracture Zone and the Walvis Passage. Therefore, the deepest parts of the eastern basins are filled almost exclusively by NADW, causing a pronounced asymmetry of deep-water distribution in the South Atlantic under modern conditions (Fig. 4.3).

A major consequence of the global deep-water circulation is the pronounced basin-to-basin fractionation between the Atlantic "carbonate-rich ocean" with its typical anti-estuarine circulation and the Pacific "silicate-rich ocean" with an estuarine circulation mode. As Berger (1970) pointed out, ocean basins having an estuarine circulation, such as the North Pacific, enrich silica and dissolve high amounts of CO_2, a situation which favours preservation of opal and dissolution of $CaCO_3$.

As a result of the global surface- and deep-water circulation, nutrients needed for phytoplankton growth, for example, nitrate, phosphate and silicate, are abundant in the surface waters of the subarctic Pacific, equatorial Pacific and Southern Oceans. In addition, there is compelling evidence that phytoplankton growth is limited by iron availability in these regions. A lack of iron prevents the complete biological utilisation of the ambient nitrate and influences phytoplankton species composition in these open ocean "high-nitrate, low-chlorophyll" (HNLC) regimes (Hutchins and Bruland, 1998; Takeda, 1998).

Because opal recycles more slowly than organic matter, silicic acid tends to accumulate along the path of deep-water flow through the oceans (Broecker and Peng, 1982). Consequently, the nutrient source waters in the Pacific have, on average, higher ratios of silicic acid to nitrate and phosphate than in the Atlantic (see Codispoti, 1983; Kamykowski and Zentara, 1989).

2. Modern Pelagic Factories: An Overview

Pelagic sediments are produced in open ocean surface waters inhabited by planktic and nektic organisms. In converting inorganic substances into organic matter via oxygenic photosynthesis, the free-floating phytoplanktic organisms serve as the basis of the marine food chain and control the biological productivity of the entire marine community. Thus, the production of organic matter and mineralised skeletal elements is intimately tied to the euphotic environment with light, nutrients, temperature and population dynamics as main primary controlling factors. In the modern oceans, primary producers (phytoplankton) having mineralised tests are the coccolithophorids (calcareous) and diatoms (siliceous). The major consumers (zooplankton) are planktic foraminifers and pteropods (calcareous) and radiolarians (siliceous). Dinoflagellates belong to different trophic levels (phyto- and zooplankton) and are represented in the fossil record only through their cysts.

Coccolithophores are important primary producers which inhabit the sunlit layer of the world's oceans. These unicellular algae secrete minute calcite plates, called coccoliths (see Wheeler and Stadnitskaia, 2011, this volume, Fig. 6.2). Because of their small size, coccolithophores are members of the nannoplankton group (2–20 μm). Coccolithophores affect the climate system via the biological and carbonate pumps and through the emission of dimethyl sulphide (Fig. 4.4) (Keller et al., 1989; Westbroek et al., 1994). They respond sensitively to changes within the photic zone, like nutrient availability, temperature, salinity and water column stability (Brand, 1994; Winter et al., 1994). Owing to rapid bottom-ward transport of coccoliths via faecal pellets and marine "snow" (Honjo, 1976), the preserved imprint of the ocean floor coccolith assemblages can be related to present-day conditions within the upper water column (McIntyre and Bé, 1967; Roth, 1994). The limiting factors for phytoplankton growth are light availability and iron (Boyd et al., 2002).

Planktic foraminifers are unicellular organisms that secrete low-Mg calcite tests in a variety of shapes up to 1 mm in size (see Wheeler and Stadnitskaia, 2011, this volume, Fig. 6.2). Depending on reference, there are some 30 living species which can be grouped into five major zoogeographic provinces (Bé and Hutson, 1977; Hemleben et al., 1989) (Fig. 4.5). From plankton net investigations, Bé and Tolderlund (1971) found that each province is represented by at least one principal species, which actually is not necessarily limited to that area (Fig. 4.5). Additionally, most living species can be attributed to distinct water depths (Kemle-von Mücke and Oberhänsli, 1999; Oberhänsli et al., 1992; Ravelo et al., 1990). Due to access of light and food, the majority of the spinose species are surface dwellers, or at least prefer to live in the upper part of the euphotic zone (e.g. *Globigerinoides* spp.; Fig. 4.5). Non-spinose species preferentially live at subsurface depths below 50 m (e.g. some *Globorotalia* ssp.; Fig. 4.5) and a few species inhibit depths below 100 and 200 m (e.g. *Globorotalia theyeri*; Fig. 4.5). However, some species prefer thermocline conditions (e.g. *Neogloboquadrina dutertrei*) and consequently follow the seasonally changing thermocline depth; other species (e.g. *Glogerina bulloides*, Fig. 4.5) also correlate

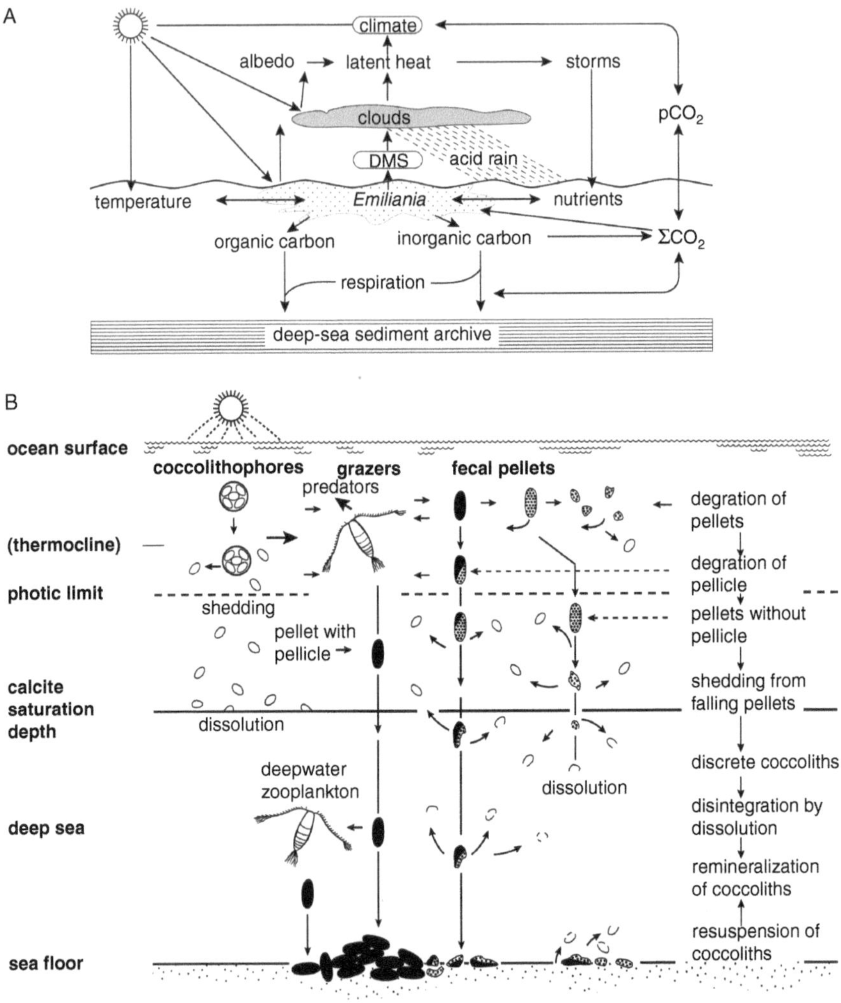

Figure 4.4 Coccoliths. (A) Coccoliths and climate (redrawn after Westbroek et al. (1993)). (B) Schematic cartoon displaying production and flux of coccoliths through the water column (slightly redrawn and modified after Honjo, 1976). Sinking-rate estimates for coccoliths are 160 m per day in a pellet and 0.15 m per day as discrete coccoliths.

with high nutrient supply or low temperature (*Neogloboquadrina pachyderma*). In conclusion, planktic foraminifers are excellent palaeoceanographical tools to reconstruct physical–chemical properties and the vertical structure of the surface water masses. Overviews supplying evaluations of environmental preferences of species and of the various proxies that are derived from planktic foraminifers are given by Mix and Morey (1996), Wefer et al. (1996), Kemle-von Mücke and Oberhänsli (1999) and Mulitza et al. (2003).

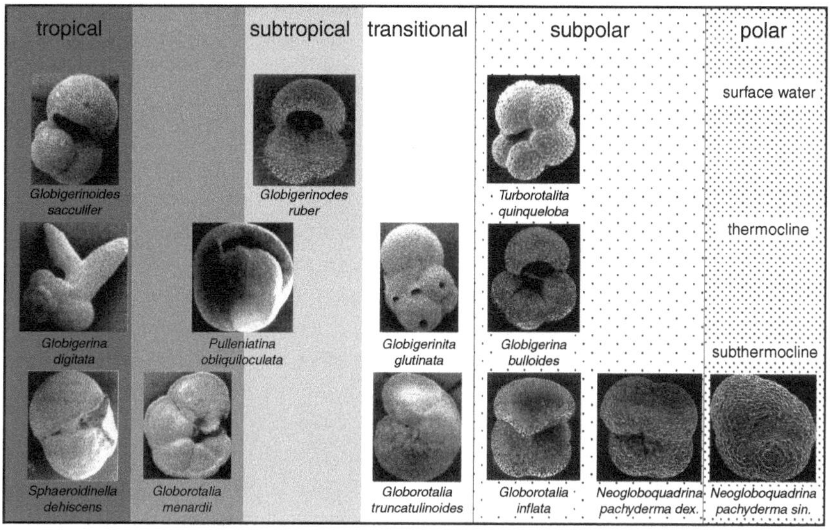

Figure 4.5 Planktic foraminifer provinces and depth habitats (redrawn and modified from Dittert, 1998).

Pteropods are marine gastropods adapted to pelagic life (see Wheeler and Stadnitskaia, 2011, this volume, Fig. 6.2). The fin-shaped wings enable these animals to swim actively in the uppermost 500 m of the water column, migrating diurnally from great depths (during day time) to shallow positions (at night) (Herman, 1978). Pteropods are widely distributed and abundant in all oceans, although most species seem to prefer the tropical and subtropical regions. Euthecosomatous pteropods feed on phyto- and zooplankton (Bé and Gilmer, 1977; Herman, 1978). An upper food size limit of approximately 200 μm was determined by Gilmer (1974) for the large pteropods *Creseis acicula* (shell length and diameter up to 33 and 1.5 mm, respectively; Van der Spoel, 1972) and *Cavolinia uncinata* (shell length 5.5–7.5 mm, width 4.0–6.6 mm; Van der Spoel, 1972). Moreover, it has been observed that a close relationship exists between pteropod abundance, seasonal phytoplankton blooms and nutrient levels (Bé and Gilmer, 1977). In the Bermuda and Barbados regions, for instance, it seems that pteropods are most abundant between March and June, whereas the January to February period is the least productive (Almogi-Labin, 1982).

Most diatoms are unicellular, although some form chains or simple colonies. They are encased within a unique frustule made of opaline silica, composed of two valves that fit together like a pillbox. New valves are constructed during cell division, followed by sequential deposition of the girdle bands (Martin-Jézéquel et al., 2000). Once completed, the mature valve is extruded to the exterior of the cell wall and coated with polysaccharides, protecting it from dissolution (Pickett-Heaps et al., 1990). As elements of the phytoplankton, diatoms are confined to the photic zone. However, they show a wide range of habit adaptation occurring as planktic or benthic, attached or mobile. The reproduction of diatoms can be very rapid, with populations capable of doubling in about 1 day. In times of low nutrients, poor sunlight or other stresses, diatoms may form metabolically inactive spores called resting spores. Diatoms are dominant in the equatorial and coastal upwelling regions and the fertile high-latitude oceans, and wherever else the supply of silicic acid, from which they make their opal frustules, is abundant. They dominate the global production of opal in modern oceans (Falkowski, 2002; Lisitzin, 1972; Nelson et al., 1995) and play a key role in the downward organic matter export from the surface of the ocean (Buesseler, 1998).

Diatoms are classified as either centric or pinnate forms. The pinnate forms tend to be more heavily silicified and hence fossilise more easily. Diatom frustules vary in size between 2 μm and 2 mm, but most forms range between 10 and 100 μm. Biology, taxonomy and biostratigraphic use are summarised by Round et al. (1996). An overview of the environmental applications of diatoms can be found in Stoermer and Smol (2001).

Diatom communities are strongly related to specific water masses (Schröder-Ritzrau et al., 2001). During summer, for example, they characterise the various environments in the northern North Atlantic and allow distinguishing Atlantic, Arctic and polar water masses as well as small-scale hydrographical features such as oceanographic fronts and ice margins. Because of their small size, small mass and hydrodynamic properties, diatoms are readily transported by oceanic bottom currents (see Chapter 3) and can be used as tracers of the currents, particularly those from high- to low-latitude regions (Burckle and Stanton, 1975). Many species of high-latitude diatoms either live attached to the bottom side of sea ice or are ice-related forms, found most abundantly in leads around the sea ice edge (see Horner et al., 1988, for terminology).

Within deep-sea sediments beneath the Antarctic Circumpolar Current, diatom frustules are preserved more readily than other calcareous plankton; they have therefore been used to reconstruct past variations in sea surface temperatures and sea ice extent (e.g. Gersonde et al., 2005). Recently proposed biogeographical schemes of major diatom taxa from the Southern Ocean sediments distinguish between sea ice, open ocean and tropical- and subtropical-related species (e.g. Armand et al., 2005; Crosta et al., 2005; Romero et al., 2005). The sedimentary assemblages preferentially preserved in surface sediments are different from the surface water communities, due to the preferential dissolution and lateral advection of thinly silicified diatoms (Fig. 4.6). Heavily silicified diatoms are generally enriched.

Radiolarians are a diverse group of planktic protozoans that are most abundant in the upper few hundred metres of the open oceans, but they

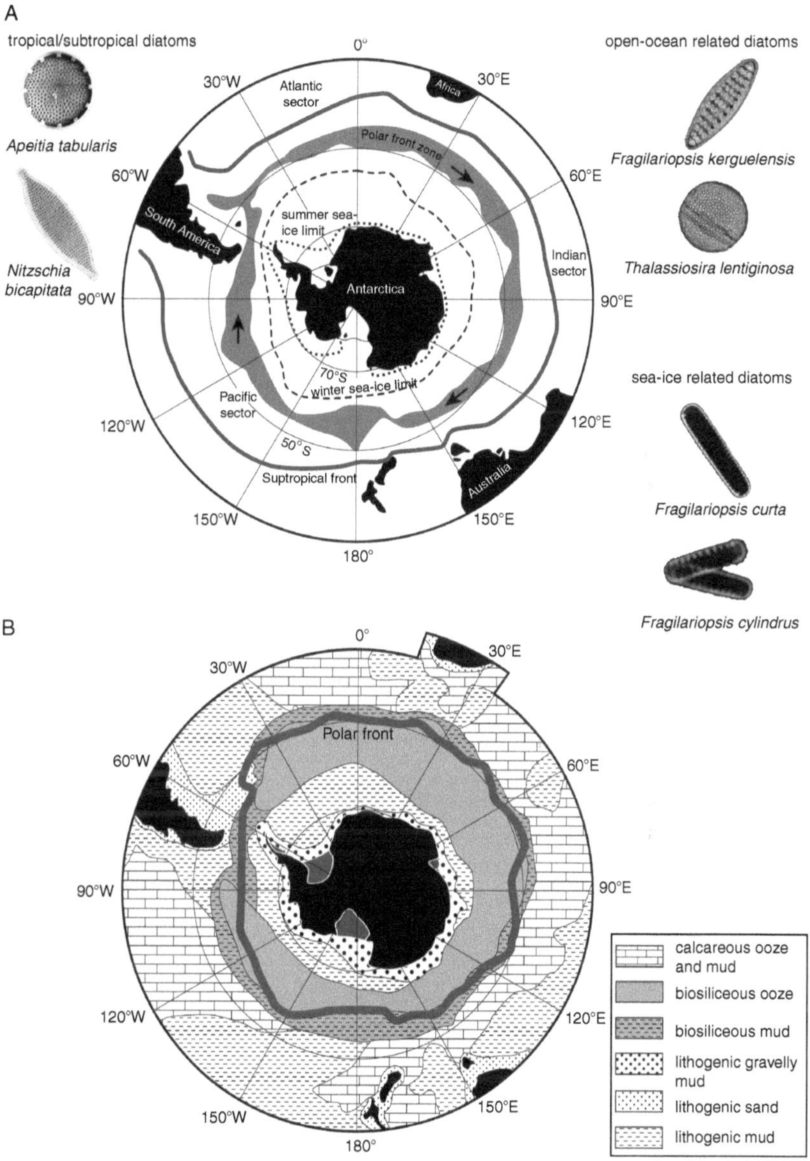

Figure 4.6 Diatom biogeography from deep-sea sediments of the Southern Ocean (redrawn and modified from Burckle et al., 1982; Diekmann, 2007). (A) Oceanographic frontal system of the Antarctic Circumpolar Current and the seasonal sea-ice limits. The majority of sea-ice-related species reveal a distribution limited and decreasing in abundance northwards by the maximum averaged winter sea-ice extent. Open-ocean related diatoms occur with maximum relative abundances from the maximum winter sea-ice edge to the Polar Front Zone. The abundance of the tropical–subtropical diatom species abruptly increase at sea-surface temperatures

have been reported from all depths, including deep trenches of the Pacific. Most radiolarians are unicellular solitary forms, of which the maximum dimension varies from 30 μm to 2 mm. Some groups of radiolarians may form macroscopic colonies consisting of hundreds of cells interconnected by cytoplasm strands and enclosed with a gelatinous envelope (De Wever et al., 2001). The food of radiolarians consists mainly of diatoms, coccolithophores, dinoflagellates and several zooplankton groups. They may also consume bacteria and organic detritus.

There are two major groups of radiolarians (De Wever et al., 2001). The first group consists of Polycystina, which have a central capsule that is either completely perforated (spumellarians), or mainly perforated towards one of its extremities (nassellarians). As a consequence, spumellarians are mainly spherical forms, whereas nassellarians are ring- or cap-shaped. Both types construct robust-spined skeletons composed of pure amorphous silica. The skeleton, being embedded in the cytoplasm, is not in contact with the sea water and is consequently not subjected to dissolution during the life of the cell. The second group consists of Phaeodarians, which have a central capsule that is perforated by only three pores. Their skeletons are made of agglutinated particles (silica and organic matter) and usually disintegrate after death.

Since radiolarians are heterotrophic, they are widely distributed in the water column, with different species often inhabiting different depth horizons (e.g. Abelmann and Gowing, 1996, 1997; Abelmann and Nimmergut, 2005; Itaki, 2003; Kling and Boltovskoy, 1995). Most of the deep-living species are adapted to specific water masses and water-mass structures; the morphologies of radiolarian test are clearly adapted to maintain buoyancy at desired water column levels. The highest standing stock generally occurs in the vicinity of the thermocline (Gowing, 1986; Takahashi, 1983, 1991). Because spumellarians and nassellarians frequently live in symbiosis with algae (including dinoflagellates), which are contained in cytoplasmic vacuoles, they prefer to spend at least daylight hours within the photic zone (50–200 m). These surface species can be grouped into assemblages that have geographical boundaries similar to those of the surface water masses.

Radiolarians are the second major producers of opal. They occur from the Arctic to the Antarctic, being most abundant and diverse in the equatorial zone (Boltovskoy, 1998). The most diverse radiolarian assemblages occur in the Pacific. The APF is a distinct boundary at which both the quantitative distribution and the qualitative composition of radiolarians

above 11 °C, which appear northwards of the subtropical front. (B) Facies of modern deep-sea sediments in the Southern Ocean. The spatial distribution of biosiliceous oozes and muds constitutes the Circumpolar Opal Belt. Diatom images are kindly provided by Xavier Crosta, Oscar Romero, and Leanne Armand.

change sharply (see Fig. 4.6). In tropical areas, radiolarians generally outnumber diatoms in deep-sea sediments, while diatoms are the predominant opal skeletons elsewhere in the ocean (Lisitzin, 1972).

Surface and subsurface distributions of radiolarian species in modern oceans are affected by a special set of ecological conditions, with biogeographical provinces characteristically mirroring the surface and subsurface water masses they live in. The present-day distribution of radiolarians in deep-sea sediments depicts the general patterns of the major water masses, current systems and production zones (e.g. Lisitzin, 1972, 1985; Molina-Cruz et al., 1999). Based on these observations, the study of fossil assemblages permits to trace the position of water masses and current systems through earth history.

Silicoflagellates are unicellular planktic organisms that live both phototrophic and heterotrophic. Their internal opal skeletons are composed of a network of bars, resembling those of radiolarians. Usually contributing less than 5% of the siliceous components in marine oozes, silicoflagellates are less important with respect to modern biogenic silica production (Tréguer et al., 1995). They are nevertheless widely distributed throughout the world's oceans, and certain species are very useful palaeoceanographic proxies. Like diatoms, silicoflagellates preferentially populate fertile areas of the oceans with moderate-to-high nutrient levels (McCartney, 1993).

Most other organisms that constitute a substantial part of the pelagic production are very small (0.2–2 μm). The tiny size of this picoplankton (and the lack of ballasting tests), however, limits its capability to sink and to be efficient in exporting particulate matter from the photic zone (Sarmiento and Gruber, 2006). Recent estimates suggest that picoplankton may account for more than half of all photosynthetic activity in warm waters of low productivity, whereas the larger plankton groups portrayed above can dominate the biomass in highly productive polar and coastal ecosystems. Their biomass is, however, usually not available to larger consumers in the food chain because larger zooplankton is unable to separate these small organisms from the sea water. Instead, microflagellates (protozoans) effectively graze autotrophic picoplankton as quickly as they are produced and even smaller heterotrophic bacteria promptly use the products of their photosynthetic activity. This microbial loop accounts for the highly efficient recycling of the organic matter produced in the euphotic zone under steady conditions (Falkowski et al., 2003; Sarmiento and Gruber, 2006).

Under natural conditions, the phytoplankton communities develop characteristic ecological successions (Young, 1994, Fig. 4.7). These are strongly related to gradients in the nutrient levels in surface waters, as may be displayed by (1) a temporal succession of a bloom with an early, a later and a post-bloom situation, (2) the lateral transition from mesotrophic conditions in fringing areas of upwelling to eutrophic environments in the upwelling core, and, finally, to oligotrophic open ocean conditions and (3)

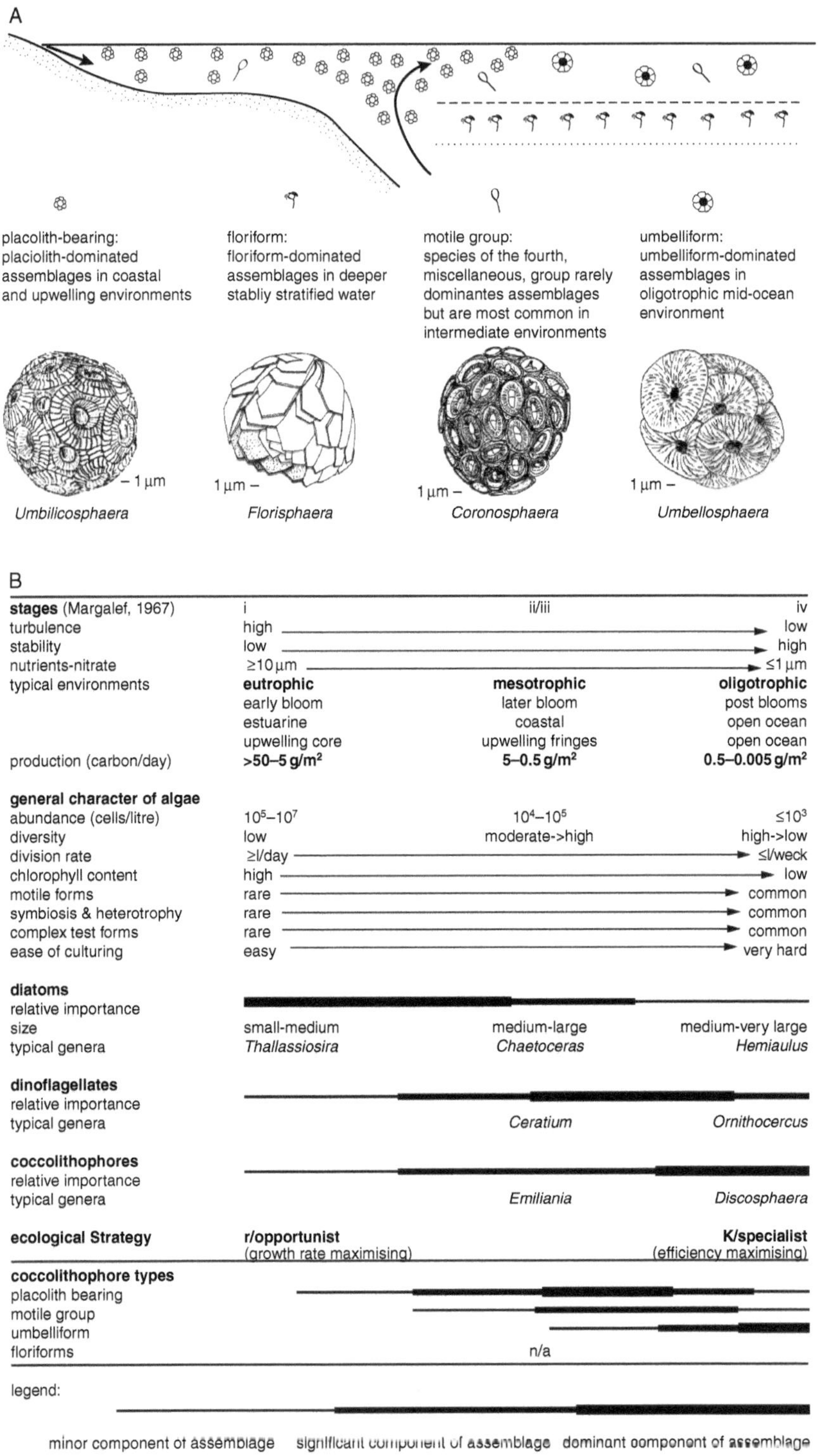
A
placolith-bearing:
placiolith-dominated
assemblages in coastal
and upwelling environments
floriform:
floriform-dominated
assemblages in deeper
stabliy stratified water
motile group:
species of the fourth,
miscellaneous, group rarely
dominantes assemblages
but are most common in
intermediate environments
umbelliform:
umbelliform-dominated
assemblages in
oligotrophic mid-ocean
environment
– 1 µm
1 µm –
1 µm –
1 µm –
Umbilicosphaera
Florisphaera
Coronosphaera
Umbellosphaera
B
stages (Margalef, 1967) i ii/iii iv
turbulence high low
stability low high
nutrients-nitrate ≥10 µm ≤1 µm
typical environments eutrophic mesotrophic oligotrophic
early bloom later bloom post blooms
estuarine coastal open ocean
upwelling core upwelling fringes open ocean
production (carbon/day) >50–5 g/m² 5–0.5 g/m² 0.5–0.005 g/m²
general character of algae
abundance (cells/litre) 10⁵–10⁷ 10⁴–10⁵ ≤10³
diversity low moderate->high high->low
division rate ≥l/day ≤l/weck
chlorophyll content high low
motile forms rare common
symbiosis & heterotrophy rare common
complex test forms rare common
ease of culturing easy very hard
diatoms
relative importance
size small-medium medium-large medium-very large
typical genera Thallassiosira Chaetoceras Hemiaulus
dinoflagellates
relative importance
typical genera Ceratium Ornithocercus
coccolithophores
relative importance
typical genera Emiliania Discosphaera
ecological Strategy r/opportunist (growth rate maximising) K/specialist (efficiency maximising)
coccolithophore types
placolith bearing
motile group
umbelliform
floriforms n/a
legend:
minor component of assemblage
significant component of assemblage
dominant component of assemblage

an estuarine versus a coastal versus an open ocean configuration. Under coastal and upwelling eutrophic environmental conditions, diatoms and placolith-bearing coccolithophorids dominate, whereas a mesotrophic configuration commonly reveals high amounts of dinoflagellates and the group of miscellaneous coccoliths. Finally, in open ocean oligotrophic settings, umbelliform coccolithophorids and distinct species of dinoflagellates are most common (Young, 1994) (Fig. 4.7).

The biogenic particles produced in surface waters, together with fine-grained terrigenous materials supplied from various sources on land, settle down to greater depth or they may be dispersed laterally at various levels in the water column and at the sea floor by currents. Depending on size, density and shape, settling particles sink at rates ranging from <1 to >1000 m per day (Fowler and Knauer, 1986). Only the largest particles, with a diameter of >100 μm (such as some diatoms and foraminifers), sink at rates that are sufficiently rapid to contribute significantly to the deep-sea sedimentation (see Sarmiento and Gruber, 2006; Turner, 2002). The longer a particle settles through the water column, and the longer it stays at the sea floor, the more extensive the remineralisation of organic matter and the dissolution of opaline and calcareous tests are. Smaller particles would not sink sufficiently fast to really arrive at the sea floor, due to their low equivalent particle radius, as can be calculated from Stokes law. Tests made of opal, calcite or aragonite are major ballasting constituents of sinking biogenic particles. Being more dense than both seawater and typical organic matter, these minerals normally provide a large part of the density difference needed to allow these particles to sink (Honjo, 1996).

Sediment-trap experiments reveal, in addition, that the vertical transport is strongly accelerated by the formation of (1) faecal pellets and (2) aggregates, for example, marine "snow" or by (3) rapidly growing phytoplankton blooms that descent to the sea floor without entering the water column consumer food web (see reviews by Turner, 2002; Turner and Ferrante, 1979). Perhaps the most drastic evidence for particle flux accelerated by biological activity was the finding that radioactive fallout from the Chernobyl disaster rapidly passed the oceanic water column (e.g. Bacon, 1987). Fowler et al. (1987) recorded peaks of the ^{141}Ce and ^{144}Ce radionuclides signature from Chernobyl in zooplankton faecal pellets in time-series sediment traps at 200-m water depth in the Mediterranean off Corsica

Figure 4.7 Plankton communities (redrawn and modified after Young, 1994). (A) Oligotrophic versus eutrophic plankton communities. The ecological distribution of coccolithophore types is shown in a transect from continental shelf to mid-ocean at low latitudes. (B) Summary of phytoplankton assemblages developed during ecological successions.

from 8 to 15 May, within a few days after delivery of these radionuclides to the ocean surface by a period of rainfall on 4–5 May.

All zooplankton and nekton produce faecal pellets, which range in size from a few μm, for example, the mini pellets of protozoans, to several mm-sized pellets produced by large crustaceans, gelatinous zooplankton and fish (Fowler and Knauer, 1986). Faecal material can occur as relatively loose, amorphous aggregates, which readily disintegrate, or as dense, tightly packed pellets that maintain their integrity over long distances. Many pellets are already reprocessed in the water column by microbial decomposition or coprophagy (see Turner, 2002). Faecal pellets sink rapidly, ranging from tens to more than two thousand of metres per day (Table 4.1).

Macroscopic aggregates (>500 μm) consisting of organic matter, living organisms and inorganic detritus are named "marine snow" (Alldredge and Silver, 1988; Suzuki and Kato, 1953). The structure of such aggregates varies along a continuum from fragile, porous, loose associations of smaller particles and organisms to highly cohesive, robust, gelatinous structures produced by zooplankton. There are three general genetic modes of marine

Table 4.1 Sinking rates of zooplankton faecal pellets, marine snow and phytodetritus (from Turner, 2002; references therein)

Particle types	Sinking rate (m per day)	Source
Faecal pellets of		
Copepods	5–220	Smayda (1971), Turner (1977), Honjo and Roman (1978), Paffenhöfer and Knowles (1979), Small et al. (1979), Bienfang (1980), Yoon et al. (2001)
Euphausiids	16–862	Fowler and Small (1972), Youngbluth et al. (1989), Yoon et al. (2001)
Doliolids	41–504	Bruland and Silver (1981), Deibel (1990)
Appendicularians	25–166	Gorsky et al. (1984)
Chaetognaths	27–1313	Dilling and Alldredge (1993)
Pteropods	120–1800	Silver and Bruland (1981), Yoon et al. (2001)
Heteropods	120–646	Yoon et al. (2001)
Salps	43–2700	Madin (1982), Yoon et al. (2001)
Marine "snow"	16–368	Alldredge (1979), Shanks and Trent (1980), Silver and Alldredge (1981), Taguchi (1982b), Gorsky et al. (1984), Asper (1987), Alldredge and Gotschalk (1988)
Phytodetritus	100–200	Billett et al. (1983), Lampitt (1985), Honjo et al. (2000), Nelson et al. (2002)

snow (Alldredge and Silver, 1988; Alldredge et al., 1993; Kiørboe and Hansen, 1993; Passow, 2000): (1) aggregates of living marine organisms may form gelatinous sheaths which decay to marine snow, (2) marine snow can result from the biologically enhanced physical aggregation of smaller component particles and (3) organic aggregates known as transparent exopolymer particles (TEP) may be particularly important in the flocculation and mass sinking of diatom blooms (Passow et al., 2001; Schuster and Herndl, 1995). As reviewed by Alldredge and Silver (1988), these aggregates have significance both as transport vehicles and as unique, partially isolated microenvironments. Organisms associated with diver- or submersible-collected marine snow (from various sites and depths) include a range of photosynthetic picoplankton, diatoms, diatom resting spores, dinoflagellates, coccolithophorids, amoeboids, ciliates, other various invertebrate zooplankton and even metazoans (see summary in Turner, 2002).

The sedimentary flux of sinking phytoplankton, faecal pellets and marine snow from the photic zone is an important component of the biological pump that not only transports and recycles materials to greater depths, but also may help to remove greenhouse gases from the atmosphere. Microbes, zooplankton and other filtre-feeding animals consume, however, most organic components of the sinking particles within the first 1000 m of their journey (the twilight zone). In this way, marine snow in particular constitutes the foundation of deep-sea mesopelagic and benthic ecosystems. As sunlight cannot reach them, deep-sea organisms rely heavily on marine snow as an energy source. The small percentage of material not consumed in the upper water column is incorporated into the biogenic ooze blanketing the deep-sea floor, where it is further decomposed through biological activity (see Wheeler and Stadnitskaia, 2011, this volume). The availability of marine snow changes with seasonal fluctuations in photosynthetic activity and ocean currents. Thus, marine snow is heavier in spring, and the reproductive cycles of some deep-sea animals are synchronised to take advantage of this. This relationship is termed "bentho-pelagic coupling" (Graf, 1989; Ritzrau et al., 2001a; see Wheeler and Stadnitskaia, 2011, this volume, section 2).

2.1. Production and sedimentary regimes: Processes and proxies

In modern oceans, the overall pattern of pelagic sedimentation corresponds closely to the productivity in the surface water (Asper et al., 1992; Deuser et al., 1990; Honjo and Manganini, 1993; Lampitt and Antia, 1997). The link between productivity in the sunlit zone and the accumulation of pelagic sediments at deep-sea floor is the working of the biological pump, which pulls nutrients and carbon out of surface waters to form organic matter and to pass it on to greater depths (see Armstrong and Jahnke, 2001;

Lochte et al., 2003). The strength of this pump depends on the leakage of organic matter out of the pelagic food web, by exporting it into the thermocline and deep ocean. The resulting downward flux of organic matter is accompanied by inorganic skeletal parts (mainly carbonate and opal tests). Many empirical relationships have been established to relate the accumulation of organic matter with primary production, organic-carbon rain rate to the sea floor, sedimentation rate, organic carbon degradation rate and bottom-water oxygen concentration, as summarised below.

Sediment distribution in modern ocean basins is, however, not a simple mirror of productivity in surface waters, since many other controls restrain the fate of organic and inorganic particles. Most importantly, the majority of it is remineralised during transport of organic matter into the deep sea, and only a minor part is buried after delivery of the remainder to the ocean floor. In addition, dissolution of carbonate and opal tests determines sediment accumulation. While biogenic carbonate preservation is largely controlled by the CO_3^{2-} concentration in the deep ocean, the preservation of biogenic opal is mediated by rapidly commencing dissolution already in the upper ocean due to severe undersaturation of silicic acid with respect to opal (see Van Cappellen et al., 2002). The resulting preservation patterns are summarised in Section 2.2.

Sarmiento and Gruber (2006) have developed a decision-tree chart that illustrates the interplay of the most important factors determining production and export of organic matter, recycling efficiency of the ecosystems and the influence of the latter on the surface nutrient concentration (Fig. 4.8). The availability of the inorganic nutrients phosphate and nitrogen [PO_4^{3-}, NO_3^-, NH_4^+] and micronutrients such as iron [Fe^{2+}, Fe^{3+}] largely controls and limits the rate of photosynthetic fixation of carbon (e.g. Martin and Fitzwater, 1988; Martin et al., 1993; Moore et al., 2002). Biogenic opal-forming organisms depend, in addition, on the availability of silicic acid [$Si(OH)_4$]. The input of nutrients comes primarily from the upwelling of nutrient-rich thermocline or deeper water in Ekman divergent regions along the shelf margin or near the equator, and by deep-reaching winter convection and mixing. Eddy activity has also been identified to supply nutrients to the euphotic zone (McGillicuddy et al., 1998, 2003; Oschlies, 2002). Additional sources of nutrients are riverine water, aeolian dust, volcanic ash and meltwater from glacial ice and sea ice.

Where the supply of nutrients is at an optimum, light (and to some extent temperature) greatly influence the amount of biomass produced through photosynthesis (primary production) (Finkel et al., 2004). Polar regions temporarily covered by sea ice are an extreme example of light- and temperature-controlled production in surface water (e.g. Diekmann, 2007; Hebbeln and Wefer, 1991). In very turbid environments, in which the depth of the euphotic zone can be as shallow as a few metres, the

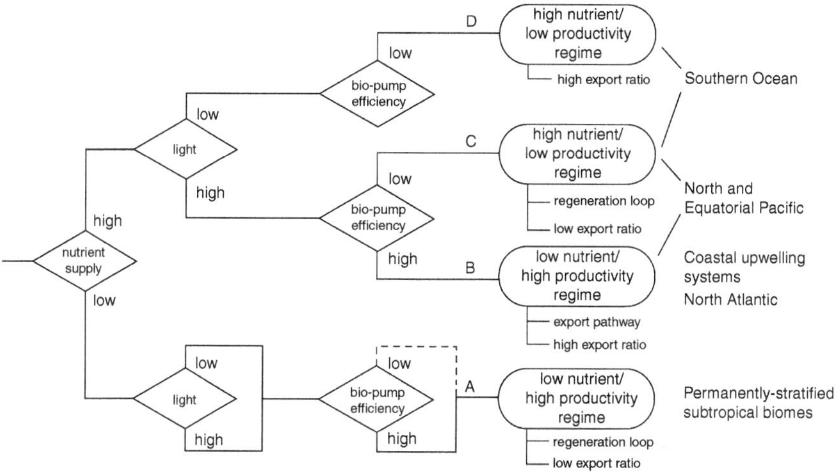

Figure 4.8 Schematic diagram illustrating the primary processes that determine production and export of organic matter from ocean surface waters (redrawn and slightly modified from Sarmiento and Gruber, 2006). The regeneration loop is made up primarily of the microbial loop consisting of heterotrophic bacteria and picoplankton, and is very efficient at recycling organic matter. The export pathway usually exists as an add-on to the regeneration loop and consists of large phytoplankton and zooplankton that jointly are very inefficient in recycling organic matter.

phytoplanktic organisms will spend part of their time in darkness, even during daytime, without synthesising organic matter.

In addition, intrinsic biological processes also control pelagic production and sedimentation, in particular, the configuration of the food web. Falkowski et al. (2003) distinguish between balanced and perturbed states of pelagic production. A picoplankton-based system is practically always at equilibrium and results in a quasi-constant biomass by recycling the organic matter highly efficiently, whereas the larger plankton preferentially thrives when the food web has been perturbed. Perturbations may be caused by seasonal changes in irradiance, wind-mixing events and seasonal or pulsed inputs of nutrients. In oceanic surface water, the balanced microbial loop is always operational, while during non-steady-state conditions, the perturbed state with larger plankton is added on. As a result, the larger organisms are highly efficient in exporting organic and inorganic particulate matter by generating large and dense particles that can sink rapidly (Sarmiento and Gruber, 2006).

Perturbed states of pelagic production now rely heavily on the metabolic processes of modern phytoplankton. A major phylogenetic achievement of diatoms is that they can assimilate nutrients rapidly under physically highly dynamic conditions and store the nutrients in vacuoles for later cell growth. This divests competing groups of phytoplankton of essential nutrients while allowing diatoms to grow rapidly, forming blooms (e.g. Dunne et al., 1999).

Unlike diatoms, coccolithophorids do not store nutrients very efficiently, but they can bloom when nutrients are provided at low rates. Therefore, coccolithophorids thrive at low abundance in tropical and subtropical seas, and at higher concentrations at high latitudes in midsummer, following diatom blooms. Caused by the consumption of nutrients, blooms of diatoms in spring at high latitudes can pass into a phase of rapid coccolithophore growth later in the year (Samtleben and Bickert, 1990; Smetacek, 1991).

In addition, zooplankton contributes to pelagic production. The biomass of zooplankton is on average 10% of that of the phytoplankton. Most abundant are tiny crustaceans called copepods (1–2 mm long), accounting for more the 50% of the zooplankton in modern oceans. The majority of smaller copepods feed directly on the phytoplankton. In the southern high-latitude oceans, another important primary consumer is a pelagic crustacean known as krill (1–2 cm long), which grazes on the omnipresent diatoms in the Antarctic biogenic opal belt. In turn, krill are eaten in tremendous amounts by seabirds, squids, fishes and whales. Rather small primary consumers but important sediment producers are planktic foraminifers and radiolarians.

A puzzling feature concerning the control of production and export in modern oceans is the existence of vast regions that have unexpectedly low biological pump efficiencies in spite of a high nutrient supply and adequate light availability (see Fig. 4.8). Some of these areas, which have long been identified as HNLC regions by biological oceanographers, are (1) the Southern Ocean, including the marginal sea ice zone, the subpolar gyre and the seasonally stratified tropics, (2) the Equatorial Pacific and (3) the seasonally stratified North Pacific (Sarmiento and Gruber, 2006).

The ecosystems of the HNLC regions may be limited by the availability of iron, by grazing herbivores or by both (e.g. Morel et al., 1991). The micronutrient iron functions as a cofactor in enzymes required to utilise nitrate and nitrite. A major source of iron for the open oceans is aeolian dust. In regions that are not supplied by dust-carrying winds (e.g. the surface waters of the Southern Ocean), the lack of iron can severely limit the phytoplankton growth and leaves a surplus of other nutrients. Several iron fertilisation experiments in the northwest Pacific, the equatorial Pacific and the Southern Ocean all confirm that iron addition leads to a significant increase in phytoplankton biomass, with diatoms generally showing the greatest response (e.g. Gervais et al., 2002; Tsuda et al., 2003).

2.1.1. Organic matter

2.1.1.1. Production rates (primary production) Global marine primary production, which is in the range of 45–57 × 10^{15} g of carbon per year, contributes approximately one-half of the total net primary productivity on earth (Field et al., 1998). In contrast to terrestrial ecosystems, where carbon can be "stored" in living organic matter (e.g. forests), the carbon fixed by marine phytoplankton is rapidly consumed by grazers and may be exported from the surface to the ocean interior without delay.

Primary production varies with the seasons, over geological periods of time and within or between latitudes (see Fig. 4.9) (Antoine et al., 1996; Behrenfeld and Falkowski, 1997; Field et al., 1998; Longhurst et al., 1995).

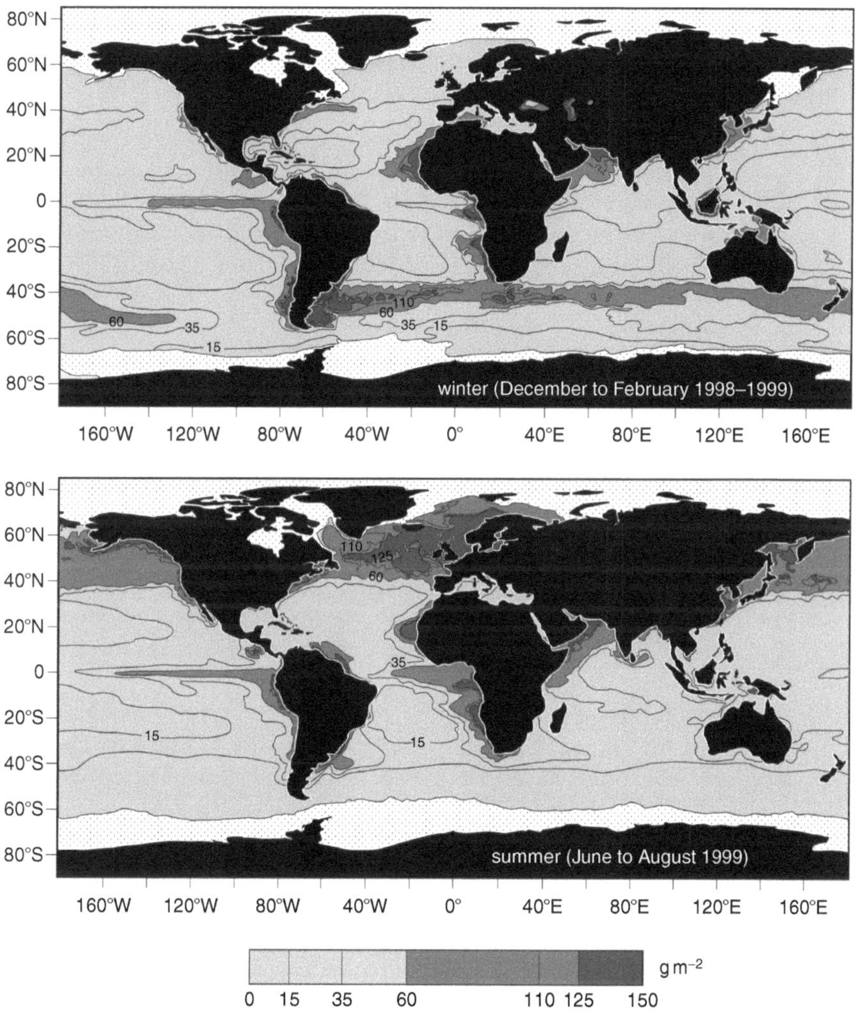

Figure 4.9 Seasonal primary production of carbon in surface water of modern oceans based on chlorophyll concentration (redrawn and simplified from SeaWiFS data: http://marine.rutgers.edu/opp/). Oceanic chlorophyll fields are derived from standard SeaWiFS ocean-colour algorithms collected during 1998–1999. These data were used to generate net primary-production fields using the algorithm described in Behrenfeld and Falkowski (1997). The dotted signature outlines missing data field. (A multi-colour version of this figure is on the included CD-ROM.)

Typical values for the open ocean range from 40 g of carbon per square metre and year in the centres of the subtropical gyres to 400 g of carbon per square metre and year in coastal ocean regions with strong mixing processes, showing a range of a factor of 10 (cf. Berger et al., 1989).

Chlorophyll maps from satellite ocean-colour sensors and maps of primary productivity derived from satellite data show patterns of biological activity that closely match the pattern of ocean surface nutrients (see Schlitzer, 2004; Schlitzer et al., 2003).

The modern configuration of global marine organic carbon production displays the following elements as portrayed by Garrison (2007).

(1) *Along the equator*: The open ocean along the equator is an area with high annual primary production (~250–300 g of carbon per square metre and year). Slim cold fingers of high productivity pointing westwards from South America and Africa are a result of wind-induced upwelling due to Ekman transport on either side of the equator. Despite perennial stable warm conditions in surface waters, seasonal variations can be high, shifting between 5 and 50 g per square metre and month. El Niño events may temporarily cause strong inter-annual variations in the Pacific.

(2) *Within the subtropical gyres*: The large subtropical gyres receive abundant sunlight and the surface waters are characterised by a high pCO_2. Because of a perennially stable deep thermocline, however, an overall deficiency in nutrients results in extraordinary low values of primary production. The tropical gyres of the ocean basins far away from land are therefore often considered as oceanic deserts. Here, productivity rarely exceeds 30 g of carbon per square metre and year and seasonal fluctuation is low. Production and consumption are almost permanently in a balanced state.

(3) *In the northern temperate and subpolar zones*: The northern temperate and subpolar zones experience the highest productivity. Thanks to the reliable light and moderate nutrient supply, the annual production in these zones is the highest of any open ocean area (carbon production amounts to ~200–400 g m^{-2} a^{-1}). Winter storms cause instability of the thermocline and give way to seasonal mixing of nutrients into the euphotic zone, leading to widespread blooms due to an increasing irradiation by the sun starting in spring. Under optimum conditions, seasonal productivity can exceed 150 g of carbon per square metre during summer (June–August) and may drop below 60 g of carbon per square metre during the winter season (December–February). Over a whole year, the biogenic production changes between balanced and perturbed states.

(4) *In the polar regions*: At very high latitudes during the winter darkness, productivity is severely limited, in particular in areas of extensive sea ice coverage. In contrast, minimum sea ice extension and the presence of

upwelled and meltwater-derived nutrients can lead to spectacular plankton blooms lasting for a few weeks during polar summer. These short-lived summer peaks of organic carbon production ($\sim$50 g m^{-2} per month in June) contrast with a long unproductive wintertime ($<$5 g m^{-2} per month). The average productivity at very high northern latitudes tends to be lower than at high southern latitudes, since the Arctic Ocean is bounded by landmasses that limit water circulation. Hence, nutrients supplied by melting sea ice are quickly used up. The Southern Ocean, on the contrary, is steadily supplied with nutrients from upwelling waters that replace the sinking AABW. Biogenic production is here in perturbed state only during short summer blooms.

(5) *On and off the continental shelves*: Because of coastal upwelling, riverine and aeolian supply, nutrient levels can be very high near the continents. Under such conditions, plankton is most abundant, and annual primary productivity is highest (carbon production amounts to $>$400 g m^{-2} a^{-1}). Biogenic production is nearly always in a perturbed state.

The recent configuration of global marine primary production can be used as a reference point for interpreting basic productivity patterns of ancient ocean basins, having in mind that all the above places and mechanisms of nutrient supply, which are largely governed by ocean circulation, have certainly been different in the past, and that ancient pelagic ecosystems followed different rules (see Chapter 3). In particular, our knowledge of pre-Mesozoic primary producers is incomplete, and the understanding of food webs or even production modes is only rudimentary.

2.1.1.2. *Organic matter export from surface waters: Controls and flux-rate estimates*

2.1.1.2.1. Controls on organic matter export The question "what controls the carbon export flux" is still an open and active area of research (Sarmiento and Gruber, 2006). Undoubtedly, the food web structure has a prominent role in determining the extent to which organic carbon and associated nutrients are recycled in or exported from surface water (Michaels and Silver, 1988; Peinert et al., 1989). The type of phytoplankton plays a crucial role in utilising the available nutrients, as discussed above. The amount and temporal variation of export production can be described as depending on the following two modes of the production states (Buesseler, 1998; Falkowski et al., 2003; Goldman, 1993):

(1) *In a balanced state of the system*, the organic matter produced by photoautotrophs in the euphotic zone is consumed in the same region by heterotrophs. Picophytoplankton, the dominant producer in this state, contributes little to export. Small, rapidly growing zooplanktic organisms extensively graze on the picophytoplankton, thus preventing the picophytoplankton from blooming. Sarmiento and Gruber (2006) use

the term "regeneration loop" (Fig. 4.8) for such a system stage, which is characterised by the fact that little organic matter leaves the system.

(2) *In a perturbed state of the system*, however, the mechanisms controlling production and consumption of organic matter are decoupled, resulting in a rapidly growing biomass and an increase in export flux. These sites of high export are often dominated by the larger phytoplankton, particularly diatoms. Large zooplanktic organisms grazing on diatoms have low growth rates, particularly at low temperatures, making it difficult for them to respond quickly to diatom blooms. Therefore, diatom blooms often result in large export events due to aggregation and rapid sinking. This export pathway is an add-on to the regeneration loop, which is always operating (Sarmiento and Gruber, 2006). It requires abundant supply of macronutrients as well as iron.

The diatom-dominated export in present-day oceans can result in a rapid loss of silicic acid in the surface layer during the productive season. As a consequence, the (limited) availability of silicic acid may control the contribution of diatoms to the total production and hence, the export fluxes of biogenic matter out of the photic layer. This mechanism is termed "silicate pump" (Dugdale et al., 1995) and its importance has been demonstrated for the silica-depleted Pleistocene as well as Holocene oceans (e.g. Dugdale and Wilkerson, 1998; Pollock, 1997). Its impact arose with the increasing utilisation of silicic acid from seawater by diatoms during its ecological rise in the Palaeogene (see Chapter 3).

2.1.1.2.2. Flux-rate estimates of organic matter export Schlitzer (2004) summarised the different experimental approaches that have been used to estimate the export flux of particulate carbon and nutrients from the euphotic zone: (1) the direct measurement of the vertical particle fluxes with moored or drifting sediment traps (e.g. Honjo et al., 1995; Wefer et al., 1982), (2) the estimation of primary or export production from surface water chlorophyll concentrations through satellites (e.g. Antoine et al., 1996; Arrigo et al., 1998; Behrenfeld and Falkowski, 1997; Laws et al., 2000; Longhurst et al., 1995), (3) the determination of oxygen-utilisation rates in tracer-dated thermocline, intermediate- and deep-water masses (e.g. Feely et al., 2004; Jenkins, 1987) and (4) the inverse modelling of biogeochemical processes from observed distributions of oxygen, nutrients and carbon distributions in the oceans (Schlitzer, 2002; Usbeck et al., 2003).

The export production might reach 20–40% of the recent estimates of the global marine primary production, although it is largely constrained to areas where the production and consumption of organic matter are decoupled. Schlitzer (2002) estimated the global marine export flux of biogenic carbon out of the euphotic zone to be $\sim$10 Gt a^{-1}. This value is in good agreement with other model results (Antoine et al., 1996;

Yamanaka and Tajika, 1996) and is very similar to the value of ~11 Gt a^{-1} by Laws et al. (2000), which is based on monthly mean total production maps estimated from SeaWIFS mission data.

The utilisation of dissolved nutrients, carbon and oxygen distributions fields by Schlitzer (2002, 2004) and Schlitzer et al. (2003) yield large-scale and long-term carbon flux estimates (Fig. 4.10). A large particulate carbon export predominantly occurs (1) in coastal areas, mainly close to upwelling regions (e.g. off Namibia, Chile–Peru and western North America) (2) a broad belt centred at about 50°SL, roughly coinciding with the course of the Polar Front and the Antarctic Circumpolar Current and (3) two narrow belts close to the equatorial upwelling in the western Pacific. Low export fluxes are found (1) in the centres of the subtropical gyres at about 30°SL and 30°NL, where surface waters are nutrient-depleted and (2) in polar regions (Weddell Sea, Ross Sea and Arctic Ocean), where permanent or seasonal ice coverage and presumably the unavailability of micronutrients such as iron (De Baar et al., 1995) limit biological productivity.

These patterns of export production, derived by inverse modelling, are in good agreement with ocean surface chlorophyll as detected by satellite sensors and with primary productivity estimates based on the satellite data (Fig. 4.11). An exception is the Southern Ocean, where the inverse model fluxes are systematically higher than the satellite-based values by factors between 2 and 5 (e.g. Arrigo et al., 1998; Behrenfeld and Falkowski, 1997; Falkowski et al., 1998, 2003; Longhurst et al., 1995).

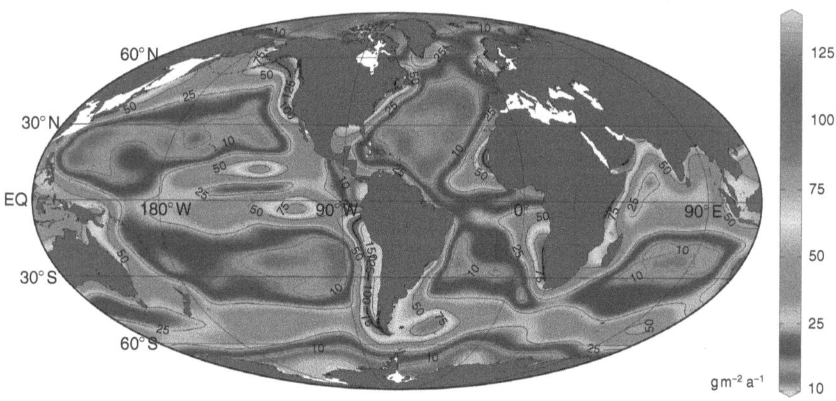

Figure 4.10 Annual average export production of particulate organic carbon (POC) in the world oceans determined by an inverse model from water-column oxygen, nutrient and carbon distributions (modified from Schlitzer, 2000). (A multi-colour version of this figure is on the included CD-ROM.)

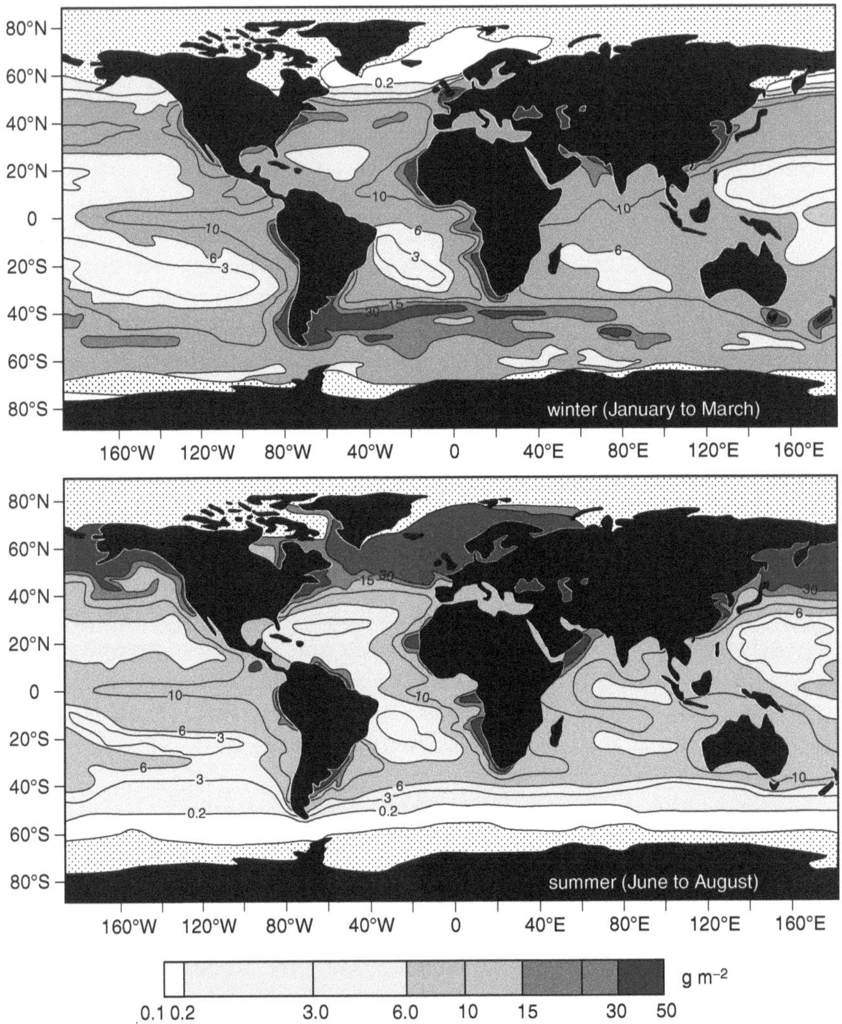

Figure 4.11 Seasonal variation of export production of particulate organic carbon (POC) in modern oceans (redrawn and simplified from Falkowski et al., 2003), as calculated from the estimates of temperature and net photosynthesis by Laws et al. (2000). Net photosynthesis was estimated on a monthly basis using data collected by the SeaWiFS satellite from 1998 to 2001. The dotted signature outlines missing data fields. (A multi-colour version of this figure is on the included CD-ROM.)

The large temporal variability seen on earth is better revealed by export flux estimates based on satellite productivity maps and sediment-trap time series. The model by Laws et al. (2000) is suitable to highlight a seasonal time frame of weeks and months (Fig. 4.11; Falkowski et al., 2003). The patterns of export production in modern oceans vary temporarily by more than a factor of 50. The high organic carbon flux related with the Polar Front and the Antarctic Circumpolar Current occurs seasonally during the austral summer. Furthermore, the model displays a high seasonal export production in nearly all northern temperate and subpolar regions (45–65°N) during the northern hemisphere summer.

In general, the regional and temporal variation in export flux of organic matter reflects the pattern of primary production within the oceans' surface waters. The variability of the export flux can even amplify the regional differences in primary production by a factor of seven. The export ratio (ef-ratio), which scales the export production (or new production) to total primary production, can be described as a function of mainly two variables, primary production per unit volume and temperature (Laws et al., 2000). According to this model, export ratios are clearly lower for oligotrophic ($ef = 0.15$) and mesotrophic regions ($ef = 0.18$) than for eutrophic regions ($ef = 0.36$). On average, export ratios are low in tropical and subtropical latitudes and highest in relatively shallow northern temperate and polar seas, which show distinct seasonality. Polar areas with seasonal shelf or sea ice, such as the Antarctic Weddell Seas and the Arctic Greenland Sea, can export more than 50% of the organic matter produced during the short summer period ($ef > 0.5$).

For shorter time spans, when the food web in surface water can be temporarily decoupled as discussed above, sediment-trap studies provide more reliable export flux estimates. In particular time series, studies are suited for investigations of subtle habitat changes, irregularly spaced stochastic events and complex interdependent ecological phenomena that affect the biogeochemical cycles in the world's ocean (Fig. 4.12). One of the first findings of moored sediment-trap studies was the discovery of strong seasonality in deep-ocean fluxes, indicating a close linkage between processes in the ocean interior and at the sea surface via a rapidly sinking rain of material from above (Deuser and Ross, 1980). Seasonality in particle fluxes to the deep sea has been observed to be a nearly global characteristic of both eutrophic and oligotrophic regions (e.g. Honjo et al., 1995; Van der Loeff and Berger, 1991; Wefer and Fischer, 1993). Reliable export flux estimates can only be obtained, however, if appropriate flux calibrations using radionuclides are applicable, because sediment-trap data seem to underestimate fluxes systematically (Usbeck et al., 2003).

2.1.2. Carbonate

The fact that the ocean contains approximately 60 times more CO_2 than the atmosphere (Berger, 1985; Broecker and Peng, 1987) underlines its important role in global CO_2 budgeting. Biological productivity, as driven by photosynthesis, is one of the primary mechanisms responsible for partitioning carbon within the ocean (Berger et al., 1987). The efficiency of this

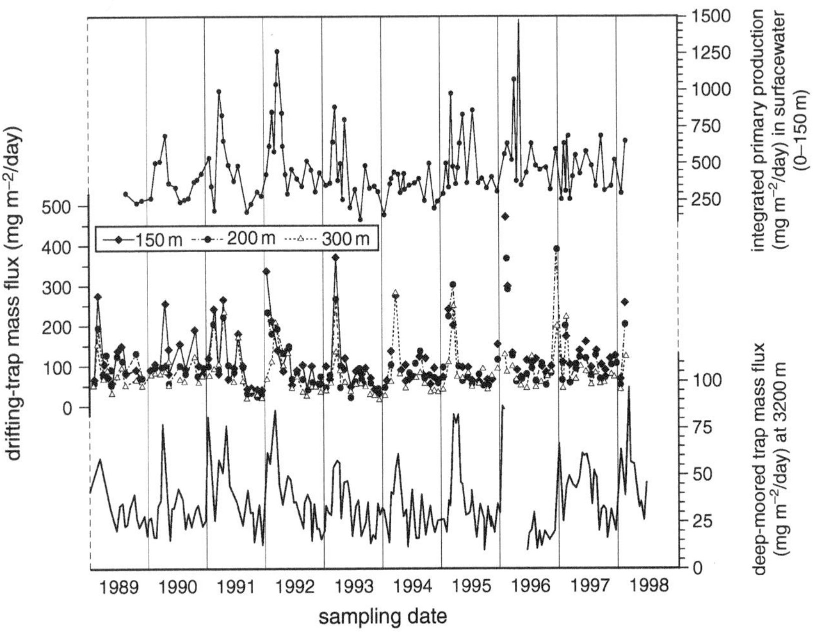

Figure 4.12 Comparison of upper-ocean primary productivity and 3D-integrated mass fluxes measured in drifting sediment traps at the Bermuda-Atlantic Time Series site with a biweekly integrated mass flux at 3200 m depth measured in deep-moored sediment traps in the Sargasso Sea near Bermuda by the Oceanic Flux Programme (redrawn and slightly modified from Conte et al., 2001). Note the high coherence and lack of any significant time lag between the upper-ocean and deep mass-flux records. Relative flux variations measured by drifting traps at 150–300 m depths are strongly correlated with relative flux variations in the deep-moored trap. The peaks in the drifting-trap mass flux correspond to most of the peaks observed in the deep-trap mass-flux record. The coherence, however, between the upper-ocean primary production and the mass flux at 3200 m is weak. The magnitudes of primary production and of deep fluxes in a given year are generally consistent (e.g. 1992 vs. 1993), but beyond this, there is no correlation between production and deep flux. In fact, the peak in deep flux appears to precede the peak in primary productivity in some years. This weak correlation illustrates that the question "what controls carbon export flux?" is still an open area of research.

biological pump is affected by the amount of carbon buried in the sediment as particulate organic carbon or in the form of carbonate. Generally, organic carbon is efficiently demineralised in the water column and on the sea floor, so that only about 0.3% by weight of the production in the photic zone is preserved in the sediment (Berger et al., 1989). In contrast, biogenic carbonate is much better preserved in the sediments than the organic carbon accompanying it. The present-day production of $CaCO_3$ in the world's ocean is calculated to be at about 5×10^9 tons per year. There is general consensus that, in the present-day cycle, more than half of the global carbonate budget accumulates on the shelves, even though their surface area represents only 7.5% of the global ocean (Milliman, 1993; Wollast, 1994). This pattern is clearly related to interglacial sea level highstands. However, the database used for budgeting modern open ocean carbonate production rates is sparse, and estimates differ significantly between various models (Milliman and Droxler, 1996). Geochemists have calculated a mean global pelagic carbonate production of 21–24 g m^{-2} per year by modelling the alkalinity anomalies and mean residence times of the various water masses (Morse and MacKenzie, 1990). In contrast, the global average fluxes of pelagic carbonate ranging from only 8 g m^{-2} per year (Milliman, 1993) to 10–12 g m^{-2} per year (Milliman and Droxler, 1996) were derived from long-term sediment-trap measurements. Regionally, the carbonate production ranges from about 2.5 g m^{-2} per day in the central ocean gyres to up to about 30 g m^{-2} per day in highly productive areas (Milliman, 1993). Generally, calcite–carbonate producers such as coccolithophorids and foraminifers are most important for the long-term storage of carbon in open ocean sediments, whereas the contribution of aragonite by pteropods to the total $CaCO_3$ production has been estimated at about 10–12% on average (Berner and Honjo, 1981; Fabry and Deuser, 1991, 1992). However, approximately 90% of the aragonite flux is already remineralised in the upper 2200 m of the water column (Betzer et al., 1984).

Since, in general, CO_2 is released during calcification to the surface waters and thus to the atmosphere, the pelagic carbonate production has to be considered as a major source of CO_2. However, coccolithophorids also consume CO_2 during their photosynthetic activity. Furthermore, the CO_2 that is released into the cell during formation of the calcite platelets of the coccoliths may be used to stimulate photosynthesis. As a result, at least small-sized coccolithophorid species may act neutral with respect to the CO_2 source–sink phenomenon.

In addition, another problem exists regarding the formation of calcareous plankton tests. Higher concentrations of CO_2 in the atmosphere cause surface seawater to become more acidic, lowering the calcium carbonate saturation through the consequent decrease in carbonate ion concentration (Broecker and Peng, 1982). This, in turn, may hamper marine calcification, for example, by lowering the weight of planktic foraminifer shells (Barker and Elderfield, 2002). A high sensitivity to changes in seawater carbonate ion concentration has been demonstrated for coccolithophorids (Riebesell et al., 2000), and culture experiments using the planktic foraminifer *Orbulina universa* have been shown to produce thicker shell walls at higher seawater carbonate ion concentrations (Bijma et al., 1999). As a result, changes in the atmospheric CO_2 provide a negative feedback to the influence of marine calcification and thus are essential for the practical application of palaeoceanographic proxies (Barker and Elderfield, 2002).

Hence, it appears to be obvious that climate modellers need separate quantitative carbonate flux rates for the different calcareous plankton groups. Until recently, such data had been compiled only in selected case studies for individual groups, such as for coccolithophorids (e.g. Broerse et al., 2000; Sprengel et al., 2002; Young and Ziveri, 2000; Ziveri and Thunnell, 2000), calcareous dinocysts (Broerse et al., 2000) and pteropods (Betzer et al., 1984; Kalberer et al., 1993). In the following, we summarise new quantitative data compiled for South Atlantic surface sediments using innovative sedimentological and micropalaeontological approaches which have a high application potential for other regions of the world's oceans and in the fossil record (e.g. at least in the time span since the Early Miocene).

Recently, a first integrated case study has been compiled for South Atlantic surface sediments by Baumann et al. (2003), providing quantitative data for all calcareous plankton groups. These authors presented new carbonate calculations for the content of coccoliths, calcareous dinocysts, planktic foraminifers and pteropods. In general, the carbonate input of the various groups of organisms is highly variable although commonly dominated by planktic foraminifers and coccolithophorids. The contribution of coccolith carbonate is highest in the oligotrophic gyres of the South Atlantic, whereas planktic foraminifers attain the highest values in regions of elevated fertility, like the mesotrophic equatorial divergence zone (Fig. 4.13). In contrast, calcareous dinocysts attain only values of maximally some 5% of carbonate, and thus are insignificant for the carbonate budget. Regionally, for example, along the entire continental margin of the western South Atlantic, high aragonite contents (up to 50% by weight of the total sediments) are related to abundance maxima of pteropods (Fig. 4.13).

Another new approach to quantify the contribution of the main carbonate producers accurately was introduced by Frenz et al. (2005), who combined grain-size data and SEM observations. A characteristic minimum in the carbonate silt (CS)-size distributions subdivides the calcareous inventory into carbonate made of planktic foraminifers (>8 μm, equivalent sinking diameter (ESD) using sedigraph measurements) and coccoliths (<8 μm, ESD). The spatial distribution of foraminifer and coccolith carbonate in the South Atlantic surface sediments confirms the findings by Baumann et al. (2003) that foraminifers prefer mesotrophic regions, whereas coccoliths are dominant in oligotrophic areas (Fig. 4.14). While the fine silt peak is stable over water depths, coarse silt (mainly foraminifer fragments) decreases with increasing water depth. This indicates a higher dissolution susceptibility of foraminifers compared to coccoliths and can be used as a proxy for - carbonate dissolution. The fine silt (8–2 μm) can be dissected into Gaussian distributions. These represent individual coccolith species and can be used for quantification of their carbonate contribution on species level.

More detailed information about coccolith compositions in the South Atlantic surface sediments can be derived from a study by Böckel et al.

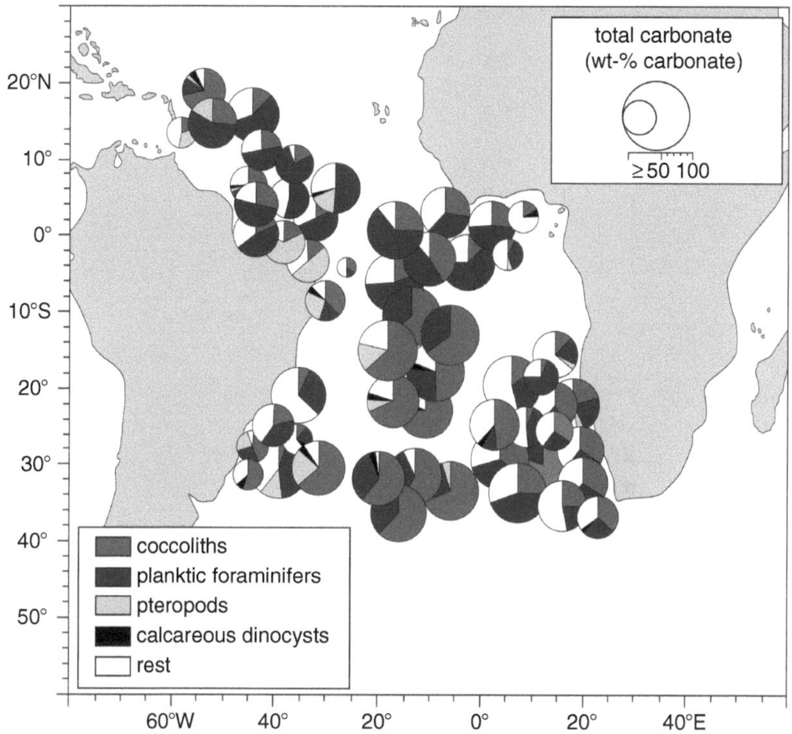

Figure 4.13 Quantitative contribution of the various calcareous plankton groups in South Atlantic surface sediments (redrawn and modified from Baumann et al., 2003).

(2006). In general, *Emiliania huxleyi* is the most abundant species. However, the taxa from the lower photic zone, composed of *Florisphaera profunda* and *Gladiolithus flabellatus*, often dominate the assemblages between 20°NL and 30°SL. If *E. huxleyi* is excluded, *Calcidiscus leptoporus* and *F. profunda* become the most abundant species, each dominating discrete oceanographic regimes (Fig. 4.15). While *F. profunda* is very abundant in the sediments underneath warmer, stratified surface waters with a deep nutricline, *C. leptoporus* is encountered in high-productivity environments. Furthermore, the results of a canonical correspondence analysis reveal affinities of *Gephyrocapsa* spp., *Helicosphaera* spp. and *Coccolithus pelagicus* for intermediate to higher nutrient conditions in a well-mixed upper water column. In contrast, *G. flabellatus* seems to be associated with high temperatures and salinities under low-nutrient conditions. Based on the relative abundances of *C. leptoporus*, *F. profunda*, *G. flabellatus*, *Helicosphaera* spp., *Umbilicosphaera foliosa*, *Umbilicosphaera sibogae* and a group of subordinate subtropical species, six surface-sediment assemblages have been identified which reflect the distribution and characteristics of the overlying surface waters (Fig. 4.15). Their

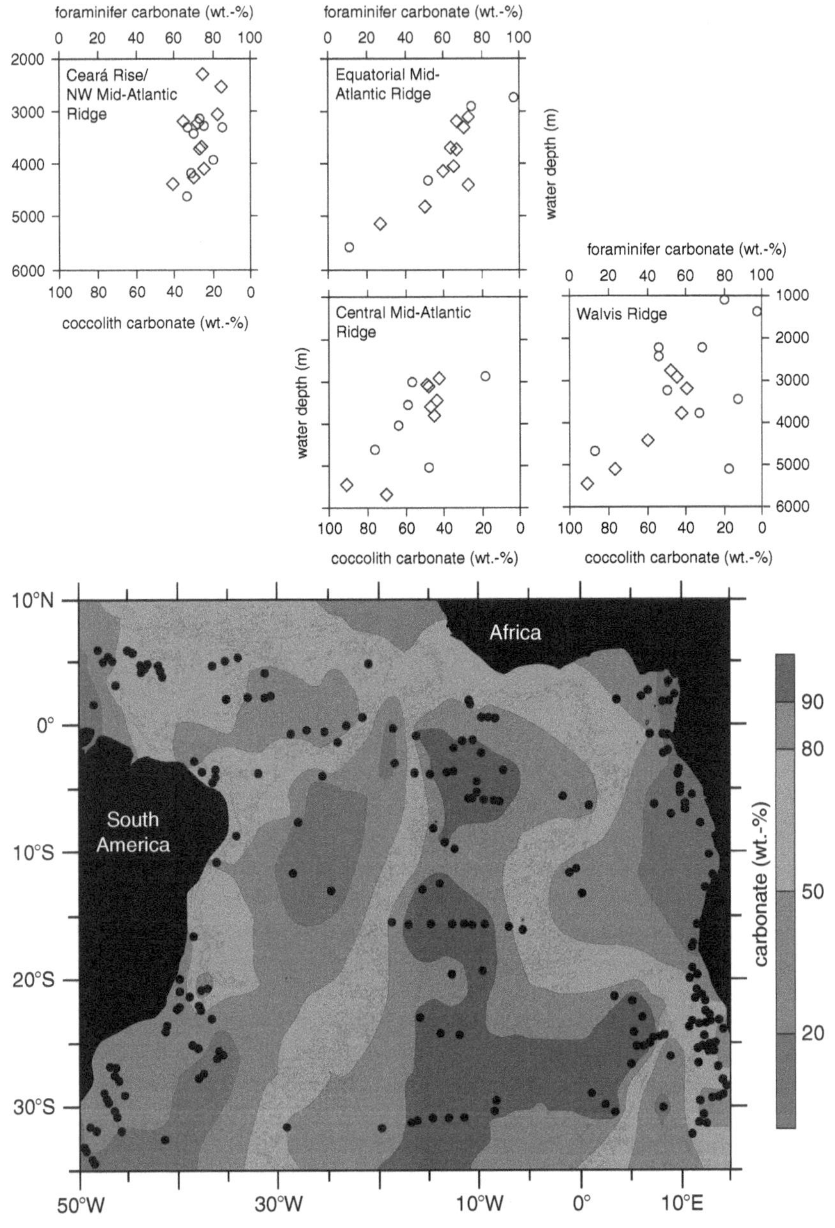

Figure 4.14 Contribution of foraminifers and coccoliths to bulk the pelagic carbonate contents in the South Atlantic surface sediments (redrawn and slightly modified from Frenz and Henrich, 2007). (A multi-colour version of this figure is on the included CD-ROM.)

distribution appears to be mainly a function of the relative position of the nutricline and thermocline in the overlying photic zone.

2.1.3. Opal

2.1.3.1. Parameters controlling biogenic opal production The largest part of the opal flux in modern oceans is a consequence of precipitation of dissolved orthosilicic acid by diatoms (Buesseler, 1998; Goldman, 1993; Nelson et al., 1995). The concentration of silicic acid inside diatom cells exceeds that of seawater by 2–3 orders of magnitude, which is maintained by active transport of silicic acid across the cell wall by multiple types of transporter proteins (Martin-Jézéquel et al., 2000). With such high cellular concentrations, diatoms manage the precipitation of their valves despite the fact that surface waters in modern oceans are highly undersaturated with respect to opal. Another consequence is that opal formation by diatoms is decoupled from carbon and nitrogen metabolism and continues to occur even in cells suffering from slow growth due to limitation of major nutrients, iron, temperature or light (Claquin et al., 2002). Iron deficiency, for example, decreases the growth rate of diatoms and thereby nitrogen and carbon uptake, resulting in more heavily silicified frustules as long as there is adequate silicate. Experimental results also show that diatoms build thicker frustules when limited by temperature or light (Durbin, 1977; Taylor, 1985).

The specific rate of silicon uptake by diatoms and the specific cell division rate both depend on, and increase with, a rising extracellular $Si(OH)_4$ concentration (Hildebrand et al., 1997; Martin-Jézéquel et al., 2000). Nevertheless, many diatoms can maintain cell division rates very close to their maximum rates at an extracellular $Si(OH)_4$ concentration that clearly limits silicon uptake (Nelson and Dortch, 1996). Diatoms do this by producing thinner frustules, thus decreasing their cellular silicic acid content and Si:C elemental ratio (Brzezinski et al., 1990).

Measurements of diatom abundances indicate that they tend to dominate whenever conditions become optimal for phytoplankton growth (Guillard and Kilham, 1978). These diatom-dominated situations include spring blooms, (coastal) upwelling blooms, equatorial divergences, river plumes, macro-tidal coastal ecosystems, ice edge blooms and transient open ocean blooms triggered by wind-mixing events, decay of ocean eddies and atmospheric dust inputs (Ragueneau et al., 2000). These hydrodynamic singularities tend to favour the growth of large phytoplankton cells such as diatoms and dinoflagellates (Legendre and Le Févre, 1989). The generally low surface/volume ratio of large cells leads to a requirement for a nutrient-rich habitat. In contrast, the higher surface/volume ratio of small coccolithophorids and silicoflagellates, and the extremely small ratio of picophytoplankton allow for a more efficient exploitation of low-nutrient concentrations (Chisholm, 1992).

Most of the radiolarians are unrelated to diatoms and it is questioned whether their flux is linked to primary production in the surface waters of modern oceans at all (Falkowski et al., 2003; Ragueneau et al., 2000). The temporal pattern of total radiolarian fluxes in the Gulf of Alaska, however, parallels that of siliceous phytoplankton groups, suggesting that radiolarians as a whole can also be considered as productivity indicators (Takahashi, 1997). At least some species appear to be related to water masses characterised by high productivity (Nigrini and Caulet, 1992). In particular, the small nassellarians are mainly feeding on primary producers (Swanberg and Eide, 1992).

2.1.3.2. Biogenic opal production rates The rate of biogenic opal production (by diatoms) has been measured for various provinces of the modern oceans by means of isotopic tracers (Nelson et al., 1995; Ragueneau et al., 2000). The global rate of biogenic silica production has been estimated to be between 2.0 and 2.8 × 10^{14} mole of silicon per square metre and year, equivalent to a mean production rate of 0.6–0.8 mol Si m^{-2} a^{-1} or 1.5–2.1 mmol Si m^{-2} d^{-1} (Nelson et al., 1995). Although the data obtained under different oceanographic conditions are consistent with this global estimate, they indicate intensive spatial and temporal variability (e.g. Nelson and Brzezinski, 1997).

Geochemists stress that the ratio between the silicon flux and the carbon flux (in short Si:C-flux rates) is an important parameter in opal production (Fischer et al., 2002; Nelson et al., 2002; Ragueneau et al., 2002). In most regions of the modern oceans, where surface waters are at least temporarily silicon-depleted, this ratio is generally low, ranging from ~0.02 in the North Atlantic to ~0.08 in the North Pacific (Fig. 4.16). Within the Southern Ocean, in contrast, the ratio between the Si and the C flux during production is significantly higher, increasing with latitude from ~0.1 in the

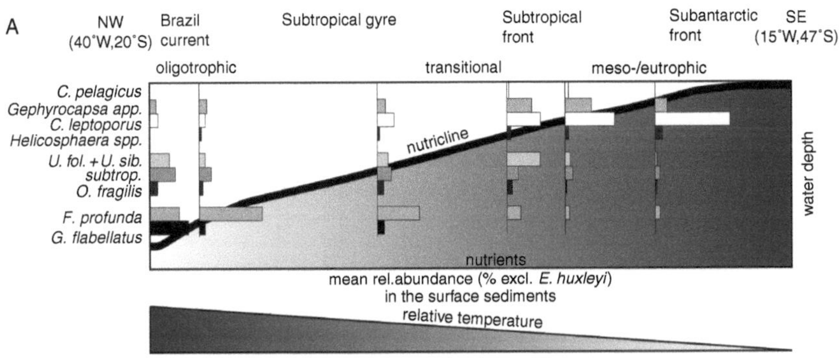

Figure 4.15 (Continued)

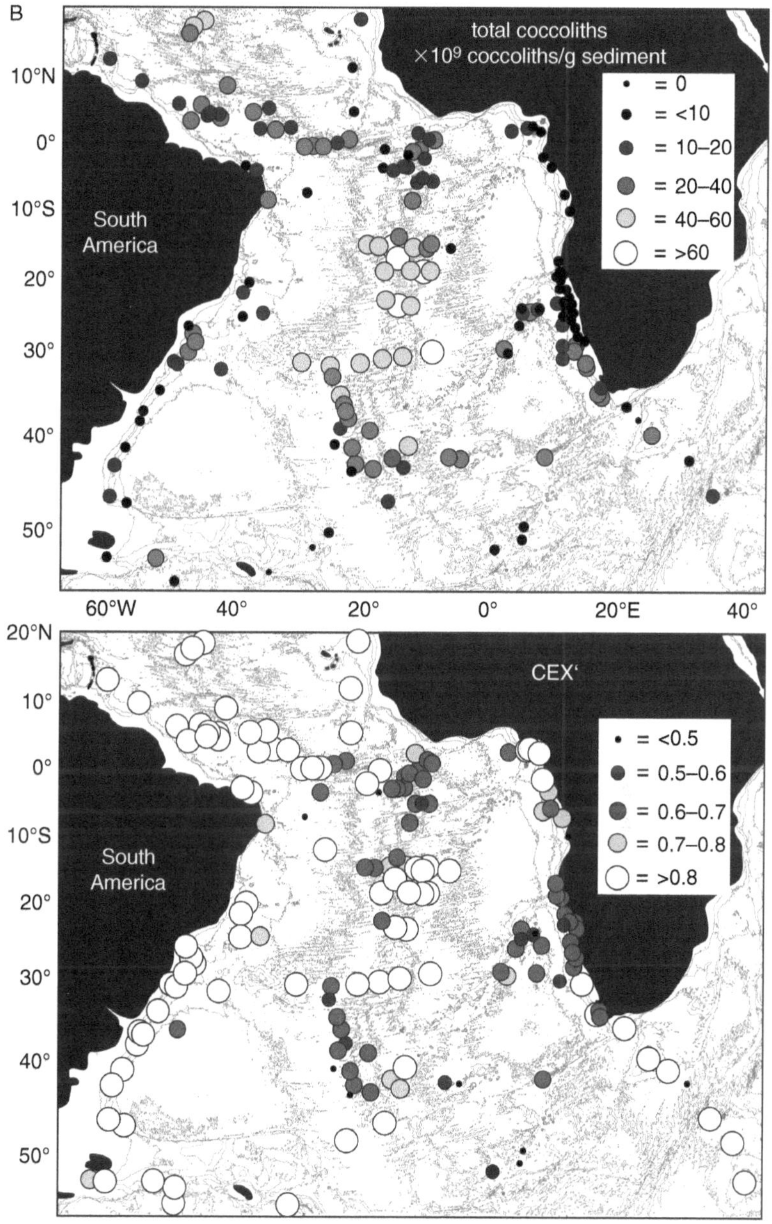

Figure 4.15 Coccolith assemblages in Atlantic surface sediments (redrawn and slightly modified from Böckel et al., 2006). (A) Absolute abundances of total coccoliths and CEX dissolution index; lower numbers point towards stronger dissolution. (B) Ocean-floor abundance patterns of coccoliths based on *E. huxleyi*-free counts reflect the properties of the surface waters. The distribution pattern seems to be mainly a function of the relative position of the nutricline and thermocline.

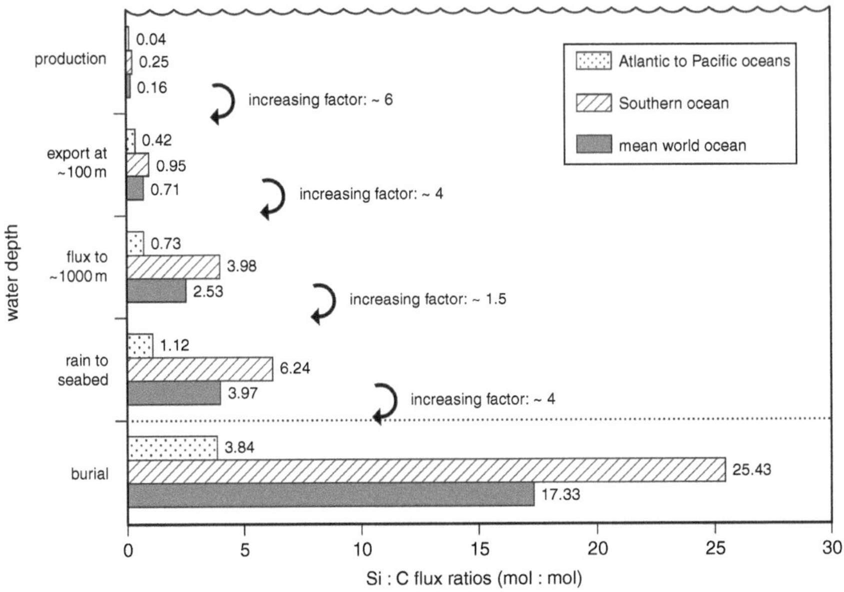

Figure 4.16 Mean annual ratio between the Si and the C flux in the water column of modern oceans and its regional variability (modified from Ragueneau et al., 2002). Note the strong increase of the ratio within the upper 1000 m of the water column and during burial.

Northern Antarctic Circumpolar Current (NACC) to ~0.4 in the Southern Antarctic Circumpolar Current (SACC).

Waters upwelled in the Pacific have a higher Si:N ratio than waters of the Atlantic, which is caused by silica enrichment along the way of the conveyor belt (see Chapter 11). In the Southern Ocean, the North–South trend of increasing ratios between the Si and the C fluxes during production reflects the silicic acid gradient across the APF (Morrison et al., 2001; Nelson et al., 2002; Pondaven et al., 2000a; Quéguiner and Brzezinski, 2002).

Based on the annual flux ratios of biogenic silica and particulate organic carbon in several biogeochemical provinces of the modern ocean, Ragueneau et al. (2002) provided an interpretation of the complex interplay of factors influencing opal production: in the Atlantic, the low silicic acid availability limits the contribution of diatoms to total production and leads to the production of diatom frustules that are weakly silicified. Both factors commonly yield the characteristic low ratio between the Si and C fluxes (~0.02) during uptake. Although silicon limitation occurs in the Pacific, its effects are weaker here and may be counterbalanced by Fe limitation, resulting in more heavily silicified frustules and an intermediate ratio of the Si and C fluxes (~0.04–0.08). In the Southern Ocean, finally, especially south of the APF, all factors (low temperature, reduced light intensity, short photoperiod, micronutrient limitation and sufficient availability of silicic acid) combine to produce highly silicified diatoms. There is an increasing abundance of large diatoms in silicic acid-rich waters south of the APF, whereas poorly silicified, smaller diatoms, being limited by the lower silicic acid availability, occur north of it. South of the APF region, Fe is a limiting factor, and the light regime is also unfavourable (Pondaven et al., 2000b), which results in enhanced ratios of the Si and C fluxes during production.

Thus, the ratio between the Si and C fluxes in the Southern Ocean is nearly six times higher than anywhere else, during production. It makes that most of the global accumulation of biogenic opal occurs around Antarctica, constituting the so-called opal belt (e.g. DeMaster, 1981) despite the only moderate primary productivity (compare Figs. 4.6 and 4.9). The formation of the opal belt also benefits from the low contribution of other sedimentary components to the deep-sea sedimentation and, beyond that, documents the high preservation potential of biogenic opal within that region, as described below (Pondaven et al., 2000a; Schlüter et al., 1998).

2.1.3.3. Biogenic opal from surface waters Seasonality strongly influences the export of biogenic opal. A general rise of opal fluxes with seasonality is observed (Berger and Wefer, 1990). This is ascribed to the central role that diatoms play in the export of organic matter from the euphotic zone and the ability of it to respond rapidly to changes in light and nutrient availability as driven by physical factors (Legendre et al., 1986). There is evidence of two annual diatom-export events at a wide range of locations (Kemp et al., 2000). The first of these events commonly follows the termination of the spring bloom (Alldredge and Gotschalk, 1989), whereas the second event results from the breakdown of the thermal stratification that typically occurs in autumn, and that leads to sedimentation maxima (Smetacek, 2000). The latter, called "fall dump", is made up of specific large diatoms (>50 μm) that are adapted to low-light conditions and that are able to migrate vertically between the euphotic zone and nutrient-rich deeper layers by regulating their buoyancy (Villareal et al., 1999).

Ragueneau et al. (2000) found, based on sediment-trap measurements, that opal export fluxes in the open ocean vary spatially by more than a factor of 100 on annual time scales. The highest annual export fluxes are observed in the Southern Ocean, which is mainly caused by the excessive contribution of well-silicified diatoms to phytoplankton production, as reflected in the high Si:C ratio (see above). Apart from coastal environments, opal fluxes peak along the opal belt of the Permanently Open Ocean Zone and the Polar Front Zone (Fig. 4.6). They are high even within the Marginal Winter Sea ice Zone (Fischer et al., 2002). Seasonally, ice-covered zones such as the Weddell Sea and the Ross Gyre, however, are characterised by lower opal fluxes, due to their shorter photoperiod (Fischer et al., 1988; Nelson et al., 2002).

With the exception of upwelling regions, annual opal export fluxes generally decrease rapidly towards lower latitudes, to the same degree as the carbonate flux increases. The two main reasons are the rising water temperatures, favouring opal dissolution (Lawson et al., 1978; Van Cappellen et al., 2002), and the more steady production, restraining export compared to areas with greater seasonal fluctuation (Berger and Wefer, 1990). Governed by low primary production, the central oligotrophic gyres are characterised by much lower opal export values than the mesotrophic and eutrophic areas. Upwelling regions show intermediate to high opal export fluxes, due to high primary production rates and, in cases of HNLC areas such as the Equatorial Pacific, iron limitation.

As can be seen from the patterns outlined above, opal-export fluxes show a strong correlation with the organic-matter-export fluxes (Fig. 4.17) despite, in particular, the more southern position of the circum-Antarctic

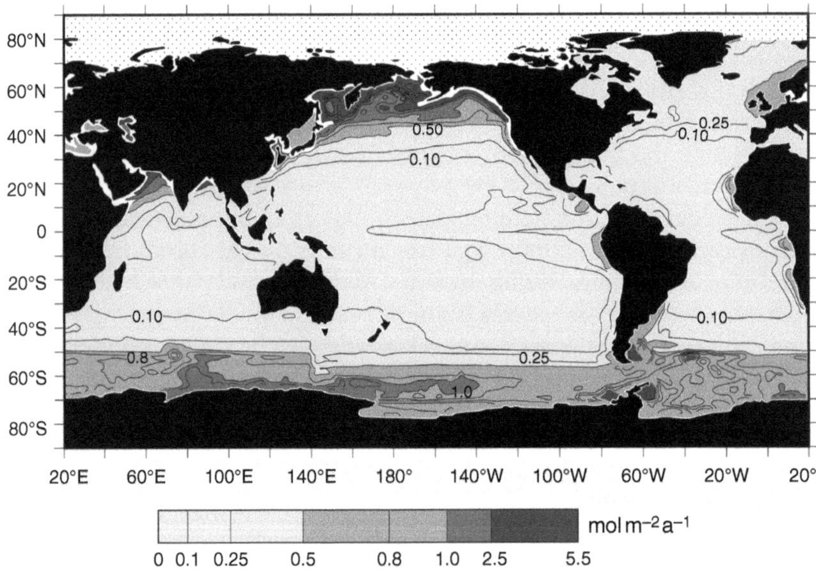

Figure 4.17 Map of the annual opal export in modern oceans (redrawn and modified from Sarmiento and Gruber, 2006). The Si flux is calculated by combining organic-matter export (based on satellite colour data) with an estimate of the opal/POC export ratio (data from Dunne et al., 2005). Note the strong correlation with the organic-matter export for many regions of the world (see Fig. 4.10). The highest opal fluxes occur in the Southern Ocean (40% of the global total), and the northern and equatorial Pacific (33% of the global total). The dotted signature outlines missing data fields. (A multi-colour version of this figure is on the included CD-ROM.)

opal-export belt and the small opal export in the northern North Atlantic (Sarmiento and Gruber, 2006).

Sarmiento and Gruber (2006) emphasised the striking difference between the cycling of silicon and organic carbon in the surface ocean, with a much lower global mean export ratio of organic carbon (0.23) than the mean export ratio of biogenic opal (0.50). The more efficient decomposition of organic matter (~75%) relative to dissolution of opal (~50%) in surface waters persists all the way down the water column and into the sediment.

2.2. Preservation: Patterns and proxies

2.2.1. Organic matter

2.2.1.1. Organic matter degradation The size continuum of organic matter in the oceans is divided for analytical reasons, by filtration, into dissolved organic matter (DOM) and particulate organic matter (POM). While DOM is transported by ocean circulation and mixing, POM is prone to pelagic settling. Therefore, POM plays the dominant role in export of organic matter from surface water and finally determines the organic carbon content of the sedimentary record. Heterotrophic biotic processes of biodegradation result in an intensive oxygen consumption and a dramatic decrease of the concentrations in both DOM and POM with depth. The twilight zone, immediately below the photic zone down to about 1000-m water depth, is a key layer for organic matter degradation (Boyd et al., 1999; Louanchi and Najjar, 2000; Tréguer et al., 2003). Inorganic nutrients such as nitrate and phosphate are regenerated from the organic matter. The nutrients are released together with dissolved CO_2 into subsurface waters.

Except for autotrophs, essentially all major plankton groups present in the euphotic zone occur in the twilight zone, albeit with lower abundances. Bacteria are responsible for the most heterotrophic respiration of organic carbon in the twilight zone, similar to their role in the surface ocean (Harris et al., 2001; Nagata et al., 2000). The bacteria begin to make way for the archaea. Below approximately 500 m, the archaea equal the heterotrophic bacteria in numbers and biomass (Karner et al., 2001). Because the degradation of organic matter increases with its residence time in the twilight zone, the settling velocity of the exported particulate material largely determines the rate of remineralisation. Because it does not sink, DOM is mostly consumed in the upper twilight zone (Carlson et al., 1994). Most of the POM also sinks too slowly to avoid remineralisation (~100 m per day).

Only a few types of exported particles sink fast enough to reach the deep sea (see Table 4.1). Among fast-sinking particles are phytoplankton aggregates (mainly diatoms) and large faecal pellets (such as those of salps, pteropods and krill) and houses of twilight zone appendicularians (Fortier et al., 1994), which may accumulate with several hundreds up to more than thousand metres per day (Fig. 4.12). These particles arrive in deep waters only a few days after their production in the upper water column and are important vehicles for the transfer of organic matter into the deep ocean (Conte et al., 2001; Lochte et al., 2003). Skeletons of calcite, aragonite and opal significantly increase the density of settling particles and hence their sinking velocity (Armstrong et al., 2002). Additional ballast from lithogenic particles scavenged from the water by the organic particles also increases the sinking velocity (Ittekkot, 1993).

The food quality of organic particles being exported is another factor determining degradation and flux in the water column (Lochte et al., 2003). Particles accumulating after phytoplankton blooms are composed of relatively fresh organic materials from senescent algal cells and are remineralised rapidly within days (Turley and Lochte, 1990). Under balanced-state conditions, in contrast, the food web recycles most material in the upper water column, and the small loss into deeper waters is of a rather resistant quality.

2.2.1.2. Transfer fluxes and accumulation rates of organic matter Below the twilight zone, biodegradation is clearly less important. Sediment-trap measurements and geochemical models indicate that most of the exported POC is remineralised in the upper 1000 m of the water column (Fig. 4.18). According to the inverse model of Schlitzer et al. (2003), about 4–10% of

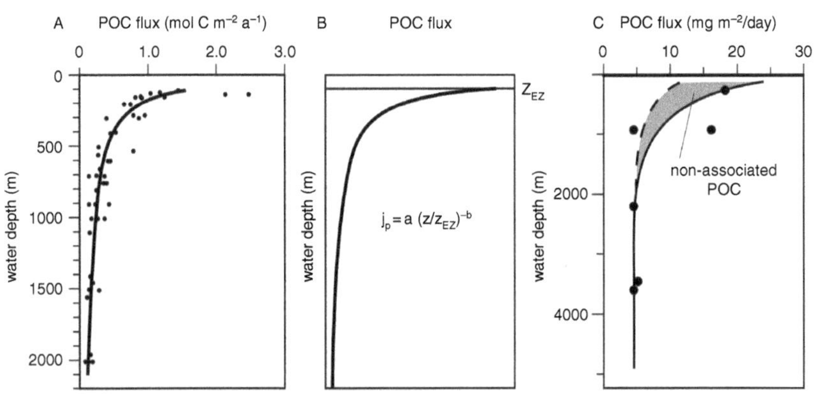

Figure 4.18 Particulate organic-matter (POC) flux in modern oceans. (A) Estimates from sediment-trap measurements (redrawn from Martin et al., 1987). (B) Exponential relationship to describe the water-depth dependence of the POC flux (redrawn from Schlitzer et al., 2003); *a*, export production at the base of the euphotic zone (z_{EZ}); *b*, dimensionless scaling factor determining the shape of particle flux profile and thus, the depth of remineralisation. (C) Disappearance of non-associated POC within the upper 2000 m of the water column, exemplified for the equatorial Pacific (modified from Armstrong et al., 2002); filled circles, measured POC fluxes; solid line, fitted total fluxes; dashed line, fitted ballast-associated fluxes.

the exported material in modern oceans arrives at 2000 m, and about the same proportion arrives at the sea floor (3–9%). A power law function is commonly used to describe the particle flux profile below the photic zone (Fig. 4.18B).

Sediment-trap measurements in several parts of the modern oceans all show the characteristic sharp drop-off of the organic carbon fluxes with increasing depth, implying that most of the sinking organic particles are either remineralised or broken down into non-sinking material within the upper few hundreds of metres of the water column (Berelson, 2001; Martin et al., 1987). Surprisingly, perhaps, the significant regional differences in export flux at the base of the euphotic zone (Fig. 4.11) contrast with a clearly less regional variability at greater oceanic depths (Lochte et al., 2003). For the Atlantic, Antia et al. (2001) found variations by a factor of 24 in shallow sediment traps at 200–1000 m (carbon flux amounts to 0.5–11.65 g m^{-2} a^{-1}), whereas deep-moored traps record differences by a factor of 4 between sites (carbon flux is 0.5–2.4 g m^{-2} a^{-1}). Thus, the regional differences in flux seasonality and composition of accumulating particles are no longer visible below 3000 m. Both the varying particle sinking rates and the different length of mid-water food chains blur the temporal variability that is still seen in subsurface waters (Antia et al., 2001).

Since inorganic ballast, namely, carbonate, opal and lithogenic materials, accelerates deep-water fluxes of POM, Armstrong et al. (2002) differentiated between two types of organic matter in sinking particles: one that is quantitatively associated with ballast minerals and likely protected from mineralisation (protected POC), and one that is non-associated (excess POC) and has virtually disappeared by 2000 m (Fig. 4.18C). Interestingly, different types of ballast appear to funnel different associated organic fractions, with carbonate having the largest protected fraction and opal the smallest (Francois et al., 2002; Klaas and Archer, 2002). A consequence of this is a regional variation in deep-ocean organic-matter fluxes, depending on the type of ballast produced or available in the photic zone.

It is important to note, however, that there are systematic discrepancies between geochemical model results and sediment-trap measurements (Usbeck et al., 2003). Results from ^{230}Th calibrations suggest that sediment traps tend to underestimate vertical particle fluxes at shallow water depths, whereas radionuclide calibrations indicate over-trapping as well as under-trapping for the deep traps (Buesseler, 1991; Scholten et al., 2001).

The reconstruction of ancient organic-matter fluxes from the sedimentary record is even more difficult. Each method that relies on the absolute accumulation rate of organic carbon and any other proxy indicator to determine changes in particle fluxes is sensitive to errors introduced (1) by sediment focusing, that is, the lateral advection of particles by deep-sea currents and (2) by variable rates of preservation of the proxy indicator. Deep-sea sediments thickened by laterally supplied materials have been preferentially cored during research cruises because of their expected high time resolution and complete records. When the accumulation rates of any sedimentary component from these sediment cores are used for mass-balance calculations (e.g. palaeo-export fluxes), huge overestimations may be the result.

It is possible to correct for sediment focusing and for variable rates of preservation by normalising concentrations of proxy indicators to ^{230}Th (e.g. Scholten et al., 2001; Suman and Bacon, 1989; Wefer et al., 1999; Yu et al., 2001). The flux of ^{230}Th to the sea floor is nearly constant and equal to the rate at which it is produced from ^{234}U dissolved in seawater. The high particle reactivity of ^{230}Th causes a very short residence time of only 5–10 years in the water column, preventing significant basin-to-basin transport. When lateral sediment redistribution has occurred, the flux of ^{230}Th into the sediments deviates from the expected value because, together with the redistributed particles, the adsorbed ^{230}Th is also redistributed. In case of sediment focusing, the flux of ^{230}Th exceeds the value expected from the production rate in the water column. The opposite is true when sediment winnowed at a location. The main disadvantage of the ^{230}Th method is its limited applicability to, at best, the last 300,000 years. A similar and promising proxy candidate to evaluate sediment redistribution further back in time is the measurement of stable ^{3}He retained in interplanetary dust particles within marine sediments (Marcantonio et al., 1996, 2001). If the ^{3}He-flux from cosmic dust has remained constant over geological time, sediment mass accumulation rates can be determined from the ^{3}He concentration in sediments.

2.2.1.3. *Organic matter deposition on the sea floor* Because of the flux characteristics outlined above, the water depth clearly controls how much organic carbon is deposited at the deep-sea floor (Armstrong et al., 2002; Berger et al., 1987; Martin et al., 1987; Fig. 4.18). While large fractions of the exported organic matter reaches the bottom of the shelves or the upper continental slopes, most of the exported material is remineralised by the time the material reaches the deep-sea floor of the open ocean. As calculated for the Southern Atlantic by Schlitzer et al. (2003), the rain rates of particulate organic carbon to the sea floor amount to 10–70% of the export flux along the south–west African and South American shelves, whereas the rain rates are much smaller and generally below 10% of the export in the open ocean, or even less than 5% (Fig. 4.19). Consequently, the overall pattern of bottom POC fluxes differs markedly from the map of export fluxes from the euphotic zone shown in Fig. 4.10 (compare Schlitzer et al., 2003).

When considering the organic matter deposition on the sea floor, it is crucial to regard the time frame. On shorter time scales, like seasonal cycles, a coherence between primary production and benthic fluxes is commonly

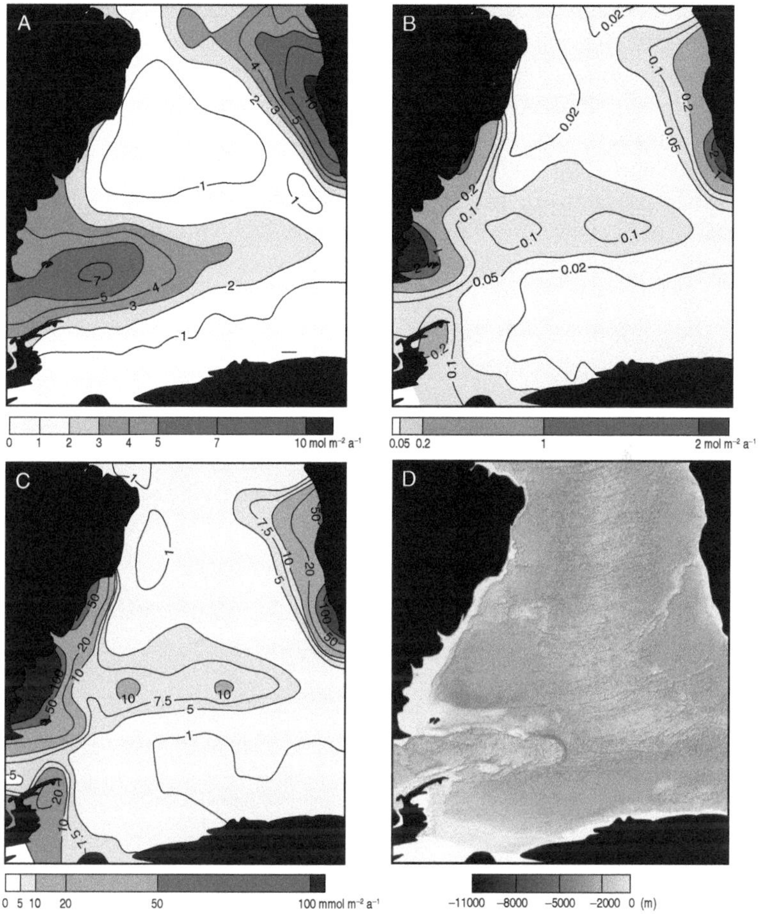

Figure 4.19 Particulate organic-carbon (POC) fluxes in the South Atlantic illustrating the influence of export production and water depth on burial of organic matter in deep-sea sediments (redrawn and slightly modified from Schlitzer et al., 2003). (A) Export flux of POC at 133 m depth. (B) POC flux to the sea floor (rain rate). (C) Accumulation flux of POC in the sediment (burial rate) as calculated by inverse modelling from oxygen, nutrient and carbon distribution. (D) Bathymetry of the South Atlantic illustrating the extent of shelf and deep-sea areas (etopo1 data after Amante and Eakins, 2009). Note that the pronounced high-productivity band centred in the south-west Atlantic and stretching eastwards is clearly reflected in the bottom fluxes on the broad shelf of South America but only weekly in the deep sea. (A multi-colour version of this figure is on the included CD-ROM.)

obliterated, due to the different speeds with which both systems function (Lochte et al., 2003). On longer time scales, however, a correlation between surface water productivity and benthic fluxes has been proved in several well-studied regions of the modern ocean (Jahnke, 1996; Pfannkuche and

Lochte, 2000; Schlüter and Sauter, 2000). Notable exceptions of this general pattern are found only in regions close to continental margins which receive organic matter from productive shelf areas (Hensen et al., 2000; Witte and Pfannkuche, 2000). In some cases, particulate organic carbon of terrestrial origin dominates the organic matter buried in ocean-margin sediments (e.g. Lyle et al., 1992; Villaneuva et al., 1997), even in regions that are far from the mouth of any major rivers. Supply of such terrestrial material from the shelf mostly occurs in the bottom turbidity layer (see Faugères and Mulder, 2011, this volume).

*2.2.1.4. **Degradation of organic matter at the water–sediment interface*** The water–sediment interface acts as a final filtre that determines in which form and quantity the accumulating organic matter is preserved in the deeper layers of the sediment (Fig. 4.20) (Lochte et al., 2003). This critical zone comprises (1) the lower few metres of water above the sediment with an increased concentration of resuspended sediment particles

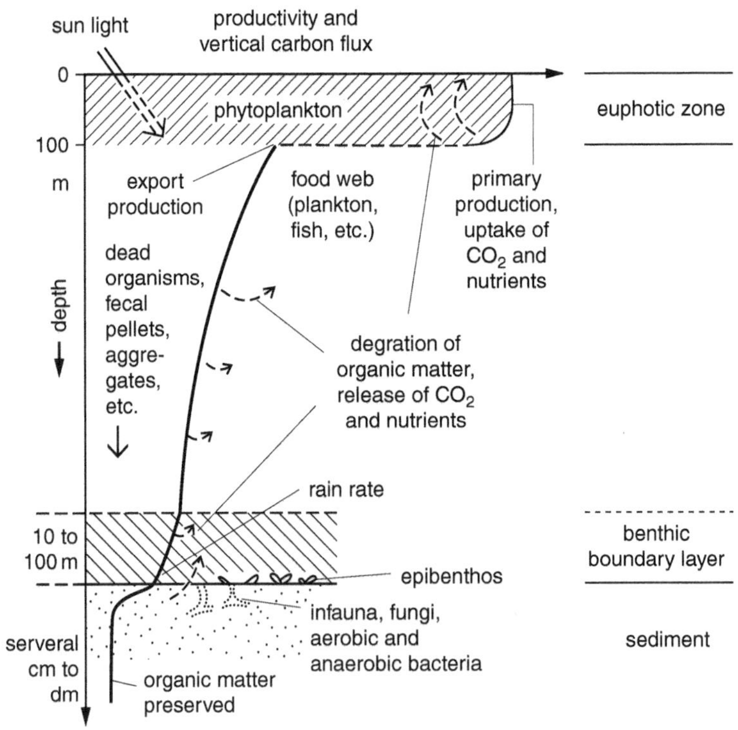

Figure 4.20 Scheme illustrating the main levels of organic-matter remineralisation and processes controlling the vertical organic-carbon flux from the zone of primary production to burial within the sediment (redrawn and modified from Einsele, 2000).

and elevated bacterial activity, which is called the "benthic boundary layer" (Ritzrau, 1996) and (2) the upper few centimetres of the sediment with a high activity of bacteria and archaea (Boetius et al., 2000; Ritzrau et al., 2001b). The biological activity in this zone is intimately related to the residence time of the organic matter and controlled by the influx and the composition of the organic material reaching the deep-sea floor (bentho-pelagic coupling). In accordance with this, bioturbation is positively related to the level of food input (e.g. Shimmield and Jahnke, 1995).

The biotic communities populating the deep-sea floor are well adapted to a low food supply of poor quality and possess specific capabilities to deal with this material (Boetius and Lochte, 1996). At times of high sedimentation rates (e.g. after the phytoplankton spring bloom), the accumulating organic material can form a fluffy layer of organic-rich particles on the sediment surface, changing the chemical and biological conditions in this layer (e.g. Thiel et al., 1989). Because of the high biological activity and changing physico-chemical conditions at the water–sediment interface, the accumulating organic material (and related proxy indicators) suffers major alterations. The highest degradation rates have been proved for relatively "fresh" organic matter that has settled quickly to the sea floor in regions with a large vertical flux. If the particles arrive slowly, benthic remineralisation does not follow directly the temporal rhythm of sedimentation, but the signal is spread out in time (Lochte et al., 2003).

Since most modern present-day oceans are well oxygenated, the organic compounds arriving at the water–sediment interface are primarily remineralised via aerobic respiration using oxygen (Emerson et al., 1985). In areas with high sedimentation rates of organic matter, the increased respiration in the sediments cannot be balanced by diffusion of oxygen and pore water irrigation from the overlying water column. In these sediments, oxygen becomes depleted and the pathways of microbial degradation processes change to anaerobic respiration. Denitrification, manganese reduction, iron reduction and sulphate reduction can be involved; this happens step by step, determined by the relative amounts of the free energy released by each reaction (Boetius et al., 2000; Ferdelman et al., 1999; see also Hesse and Schacht, 2011, this volume). Such conditions occur in the deep-sea sediments adjacent to intense upwelling systems at continental margins (Lochte et al., 2003). For instance, the upwelling systems off Namibia, Peru and Chile are characterised by extensive anoxic zones in the sediments, locally even extending into the water column. Anoxic conditions are also found in the upper sediment layers of the Arabian Sea, where the oxygen minimum zone of the water column impinges on the continental margins in regions where high organic matter fluxes from monsoonal upwelling occur. Furthermore, anoxic layers may also develop when turbidites cover the sea floor and prevent oxygen exchange with the overlying water column (Lochte et al., 2003).

2.2.1.5. Burial of organic matter in marine sediments Deep-sea sediments permanently store the organic matter that has survived the degradation in both the upper water column and at the water–sediment interface. The accumulation of this small residual fraction is generally correlated—over longer periods of time—with the fluxes to the sea floor, with ~10% of the bottom-reaching flux actually being accumulated (Bender and Heggie, 1984; Emerson et al., 1985). According to Schlitzer et al. (2003), the global mean preservation efficiency (accumulation rate relative to export flux) of organic carbon is 0.88%, which is within the range of 0.3–3% originally proposed by Berger (1989).

In addition to the main factors summarised above, that is, (1) the biological productivity and export in surface water, and (2) the preservation rate of organic matter during its transfer through the water column and its burial, the content of organic carbon preserved in marine sediments mainly depends on (3) the sedimentation rate of inorganic materials. All the processes related to these controlling factors are highly interrelated, as discussed in detail by Einsele (2000), Harris (2005), Katz (2005), Tyson (2005) and Sarmiento and Gruber (2006) from different methodological points of view.

The sedimentation rate of biogenic carbon, opal and siliciclastic particles has been described within this context as a two-edged sword (Einsele, 2000; Tyson, 2001, 2005). At very low sedimentation rates and under oxic bottom water conditions, the residence time of organic matter in the zone of benthic life is long (hundreds to thousands of years) and biodegradation is very intensive. With increasing dilution by inorganic materials, the residence time is reduced and the benthic community is more and more hindered to decompose the organic material near the sedimentary surface. In addition, infauna begins to feed in deeper sediments, because some high-quality food is still available. Burrowing activity of (larger) organisms facilitates rapid incorporation of carbon into the sediment, where it more effectively depletes the pore water and thus may reduce degradation (Lochte et al., 2003). This enhanced burial efficiency results in increasing organic carbon contents within buried sediments. Finally, if a critical threshold value of dilution by inorganic materials is exceeded, the reverse effect occurs: increasing sedimentation rates cause largely a dilution of the organic matter flux, and decreasing organic carbon contents within the sediments are the result (Fig. 4.21). Under anoxic bottom water conditions, much higher proportions of organic carbon are, principally, buried (Canfield, 1994; Hartnett et al., 1998).

Additional factors, partly related again to the water depth, are known to influence the burial rate of organic carbon in sediments. These include first of all (1) the nature of the oxidant to which the organic materials are exposed, with oxygen being particularly effective at degrading organic matter, and redox oscillation also appearing to be particularly efficient and (2) the association of organic matter with minerals, with smectite appearing

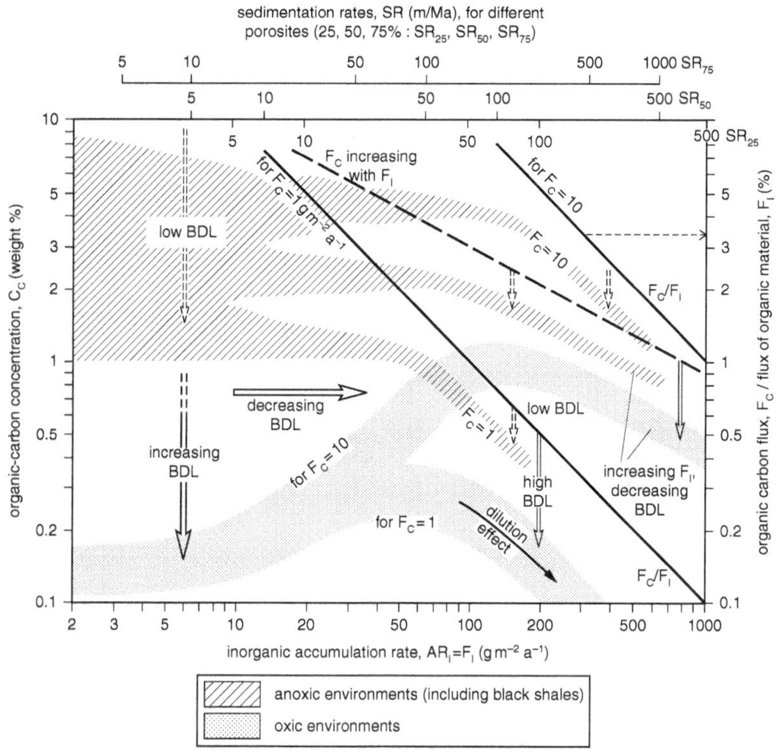

Figure 4.21 Conceptual scheme illustrating the preservation of organic matter in marine sediments as a function of the organic-carbon flux (F_C) to the water–sediment interface, flux of inorganic materials (F_I) to the sea floor and benthic degradation loss (BDL) (redrawn and modified from Einsele, 2000).

to be particularly effective in protecting organic matter from degradation (Hedges et al., 1999; Sarmiento and Gruber, 2006). The oxygen-exposure time usually rises with distance offshore as the sedimentation rate decreases and oxygen-penetration depth increases. The coating thickness of organic matter on mineral surfaces is important as well and can probably be related to specific oceanic regimes, that is, continental-margin, deep-ocean, upwelling and anoxic-basin environments (Hedges and Keil, 1995). Over time, many empirical relationships have been found between organic matter preservation and primary production, rain rate to the sediments, sedimentation rate, organic carbon degradation rate and bottom water oxygen concentration (cf. Harnett and Devol, 2003).

Although bioturbation has a major impact on sediment diagenesis, its influence on the organic matter decomposition is less well known. Large benthic organisms, the proportion of which in respect to the total biomass decreases from the more productive continental slopes towards the mid-oceanic regions, seem to be particularly important in mixing sediments and organic matter (Heip et al., 2001).

2.2.1.6. Proxies to reconstruct organic productivity from the sedimentary record Given that the present-day relationship between export flux and burial rate of organic carbon is comparable to that in the Cenozoic and perhaps the late Mesozoic (see Chapter 3), it should be possible to estimate past changes in surface water productivity from the accumulation rate of organic carbon. This approach requires that two key parameters do not vary greatly in space and time (Lochte et al., 2003): (1) the preservation of POM as it sinks through the water column and as it settles on the sea floor and (2) the remineralisation of organic materials after reaching the sea floor.

Both in their quantitative and marker signals, benthic fluxes and their accumulation in the sediments harbour proxy variables (specific sediment properties) that are used to reconstruct palaeoproductivity scenarios (Sarnthein et al., 1992; Wefer et al., 1999). Productivity is defined in this context as the flux of organic carbon (C_{org}/area/time), which means that most of these productivity proxies measure export production rather than primary production. A number of promising methods have been proposed (see Lochte et al., 2003; Wefer et al., 1999, for an overview). Many of these proxies are currently calibrated in order to test the basic assumptions and to resolve the limitations inherent to these tools.

The organic carbon content of sediments may provide a first rough estimate. To minimise the high potential error in this proxy, several algorithms for POC regeneration as a function of water depth and sinking speed have been developed (e.g. Armstrong et al., 2002; Dunne et al., 2005). Furthermore, an algorithm relating carbon preservation to sediment accumulation rate has been derived empirically (Mueller and Suess, 1979). Substantial uncertainty arises, however, from factors controlling organic matter preservation at the water–sediment interface during early diagenesis, including bottom-water oxygen concentration (compare Hesse and Schacht, 2011, this volume).

The equation relating productivity to C_{org} content (Berger et al., 1989) is $PPa/PPb = (C_{orga}/C_{orgb})q$, where $q = 0.6$–0.8 and C_{orgb} is the Holocene standard value. A complicating factor is the lateral transport of organic material down the continental slope. Much of the POM in regions close to continental margins can be terrestrial and old, redeposited from shelves by density currents.

Alternatively, specific organic biomarker compounds, known to be produced by marine phytoplankton, are used as proxies for palaeoproductivity studies of the organic carbon. Biomarkers may be specific to selected groups of phytoplankton (e.g. alkenones from coccolithophorids, dinosterol from dinoflagellates, brassicasterol from diatoms, isorenieratene from green sulphur bacteria), or they may include the full suite of the pigment transformation products of chlorophyll, known as chlorins. Use of these biomarkers eliminates the error introduced by the input of terrestrial organic matter, and supplies additional information about environmental conditions and the dominant groups of phytoplankton preserved in the sediments (e.g. Schubert et al., 1998).

Identification of biomarkers derived from green sulphur bacteria (Chlorobiaceae) or their diagenetic alteration products in sediments is indicative of anoxygenic photosynthesis. Green sulphur bacteria are typical of environments in which hydrogen sulphide extends into the photic zone, where it serves as the electron donor required for anoxygenic photosynthesis (Grice et al., 1996). They show that anoxic conditions were not confined to the lower parts of the water column or an oxygen minimum zone at mid-water depth but, in fact, extended into the photic zone, resulting in an almost completely euxinic water column. The presence of these biomarkers provides unequivocal evidence for photic zone euxinic (PZE) conditions in the past. Such conditions occurred, at least temporarily, during oceanic anoxic events (OAEs) in the Cretaceous and prevailed during the Permian–Triassic superanoxic event (Grice et al., 2005; Wagner et al., 2004; Weissert, 2011, this volume).

The use of biomarkers is sensitive to errors caused by sediment focusing and by variable rates of preservation of the biomarkers (Lochte et al., 2003). Local losses of biomarkers due to biological action or changes in redox conditions are another major problem. No method has been proposed until now to correct for such variations in the preservation of biomarkers. To obtain reliable results, palaeoproductivity reconstructions should generally rely on a multi-proxy approach (Lochte et al., 2003).

2.2.2. Carbonate

Carbonate dissolution is a complex process controlled by hydrostatic pressure, differential pCO_2 and CO_3^{2-} concentrations of intermediate- and deep-water masses, as well as size, mineralogy, crystallography and texture of calcareous biogenic materials (Henrich and Wefer, 1986). Various approaches have been used to assess the influence of these factors and to reconstruct the aragonite and calcite lysocline and compensation depth levels in various parts of the oceans (see review by Berger, 1979). At the lysocline, the rate of dissolution increases dramatically. Further below, there occurs a depth termed the carbonate compensation depth (CCD), below which the rate of supply of calcite equals the rate of dissolution, such that no calcite is deposited.

In addition, there is an ongoing discussion regarding the magnitude and intensity of the supralysoclinal dissolution at various levels of the oceans. Supralysoclinal dissolution is induced by degradation of organic matter in the water column and close to the sediment–water interface (Archer and Maier-Reimer, 1994; Jahnke et al., 1994). Milliman et al. (1999) suggested

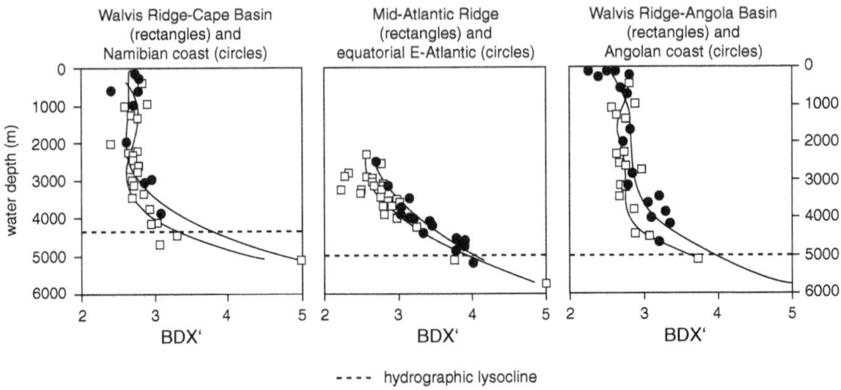

Figure 4.22 Determination of calcite preservation along several transects through the eastern South Atlantic as indicated by the BDX′ dissolution proxy (*Globigerina bulloides* index) in surface sediments. The broken line shows the depth level at which the hydrographic lysocline is recorded. Note the much shallower position of the sedimentary lysocline in all profiles, for example, where the BDX′ value increases above three (modified from Henrich et al., 2003).

that as much as 60–80% of the biogenic carbonate might be dissolved by biological processes above the hydrographic lysocline (Fig. 4.22).

In order to test and compare the applicability of the dissolution proxies most commonly used, Dittert et al. (1999) have carried out a dissolution experiment in the South Atlantic. The state of carbonate preservation in surface sediments was determined using various dissolution proxies, and the results were compared with the state of carbonate saturation indicated by hydrographic data. The set of conventional dissolution proxies tested included (1) *for bulk sediment*, the carbonate content, the weight percentages of the sand fraction (Berger et al., 1982; Wu and Berger, 1991) and the carbon rain ratio (Berger and Keir, 1984) and (2) *for microfossils*, the planktic foraminiferal concentration, the fragmentation index (Fi) of planktic foraminifers (Berger, 1973), the radiolarian/planktic foraminiferal ratio (r/pf) (Diester-Haass and Rothe, 1987), the benthic/planktic foraminiferal ratio (bf/pf) (Diester-Haass and Rothe, 1987; Parker and Berger, 1971), the percentage of dissolution-sensitive planktic foraminifer taxa (FDX, Berger, 1973) and the foraminiferal dissolution index defined according to the variable dissolution resistance of species (Berger, 1979) (Fig. 4.23).

The basic tenet underlying these bulk sediment and microfossil proxies is the progressive breakdown of carbonate particles during dissolution, leading to (1) a decrease in carbonate content and in the sand fraction grain size, as well as to an increase in fragmentation; (2) planktic foraminifer species assemblages being modified by preferentially removing surface dwellers with thin porous walls and enriching deep-dwelling, thick-walled, more

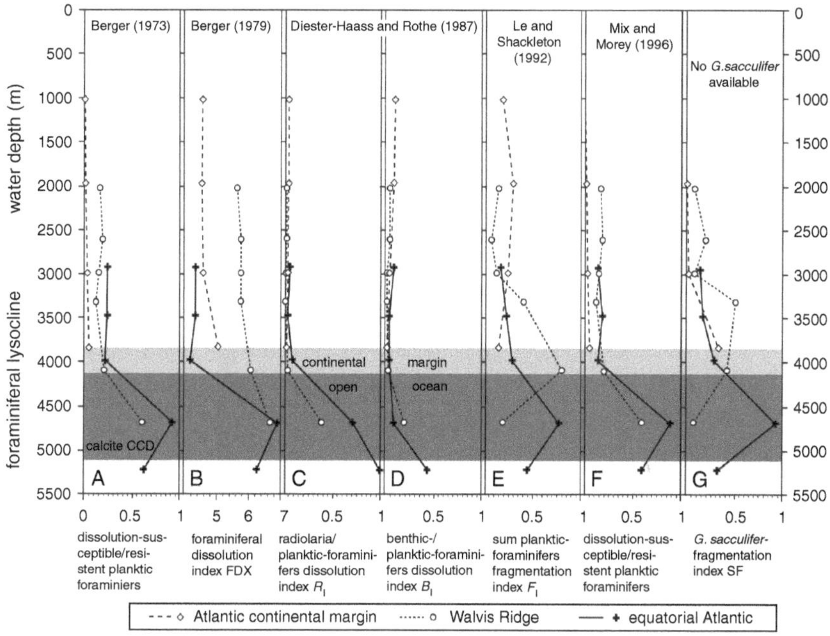

Figure 4.23 Conventional dissolution proxies (redrawn and slightly modified from Dittert et al. 1999). Foraminiferal dissolution parameters allow to distinguish the area above the calcite lysocline from the area below, irrespective of whether they are derived from the open ocean or from continental-margin transects. Increase in dissolution can be obtained at the top of the transition zone; towards the CCD, the parameters are disturbed due to a deteriorated database.

robust shells and (3) the more soluble planktic foraminifers being preferentially dissolved, enriching the more robust compounds, for example, benthic foraminifers, calcareous nannofossils and radiolarians.

However, as clearly pointed out in the assessment by Henrich et al. (2003), conventional proxies produce, in many cases, incorrect and misleading results, due to the following reasons: (1) bulk sediment parameters may be affected by contamination with non-carbonate material (e.g. input of aeolian dust, river supply of terrigenous fine-grained sediments), downslope resuspension and lateral advection or by winnowing of sediment by bottom currents and (2) the microfossil parameters may respond to changes in ecology through time. As a consequence, reconstructions of dissolution levels differ considerably depending on the proxies used. Henrich et al. (2003) emphasise that it should always be assured that the proxies under consideration were not influenced by the other factors discussed above. In addition, when studying fossil records, the assumption that other factors did not significantly vary through time should be tested and verified.

In order to avoid these discrepancies, semi-quantitative scanning electron microscope (SEM) and light microscope dissolution proxies were developed for planktic foraminifers (Henrich, 1989; Volbers and Henrich, 2002) and pteropods (Gerhardt and Henrich, 2001). SEM preservation studies have also been carried out on the spinose foraminifer *G. bulloides* (Dittert and Henrich, 2000; Volbers and Henrich, 2002), which reveals a high-ranking dissolution susceptibility, a consequence of its more porous test structure (Thunell and Honjo, 1981). *G. bulloides* is prevalent throughout the South Atlantic, with particularly high present-day abundances in the productive surface waters of the eastern South Atlantic (Kemle-von Mücke and Oberhänsli, 1999). Figure 4.24A presents a graphical illustration of the BDX′, with six stages of preservation (decreasing preservation from 0 to 5; for details on the analytical procedure and calculation of the BDX′, see Volbers and Henrich, 2002). The high precision of the BDX′ to reconstruct foraminiferal carbonate preservation in the modern South Atlantic and its power as a palaeoceanographic tool in the fossil record have been tested on South Atlantic surface sediments and LGM sediment samples (Volbers and Henrich, 2002, 2004) as well as for a Pliocene to Quaternary record from the Ceara Rise (Frenz et al., 2006).

Establishing reliable aragonite dissolution proxies is even much more difficult than for calcite. All conventional approaches, like simple fragmentation indices or the ratio to other groups, often fail to produce reliable results (Gerhardt et al., 2000). The main reason for this failure is the high fragility of certain pteropod species (compare Henrich et al., 2003 for a detailed discussion of other potential errors). These highly fragile species produce a large number of fragments by simple mechanical breakdown even in perfectly preserved sediment samples that have not undergone any dissolution. During incipient dissolution, these delicate fragments are preferentially dissolved, thus decreasing the fragmentation index in a sample. As a solution to this problem, Gerhardt and Henrich (2001) established a dissolution index for various species of the more robust pteropod genus *Limacina* by distinguishing characteristic features of progressive structural breakdown of tests under the light microscope. The *Limacina inflata* dissolution index (LDX) turned out to be the most practicable and useful in this group. Figure 4.24B presents a graphical illustration of the six steps of progressive dissolution of *L. inflata*. As dissolution proceeds, tests become milky and lustreless/opaque, they display additional damage and finally break down.

In the modern South Atlantic, the intermediate- and deep-water masses of the southern origin (AAIW, AABW) display weak to strong CO_3^{2-} undersaturation (CO_3^{2-} < 90 mol l^{-1}). North Atlantic Deep Water (NADW), in contrast, is oversaturated with respect to CO_3^{2-} (CO_3^{2-} > 110 mol l^{-1}: Broecker and Peng, 1982). Therefore, dissolution of carbonate increases in AAIW and AABW water masses, whereas NADW

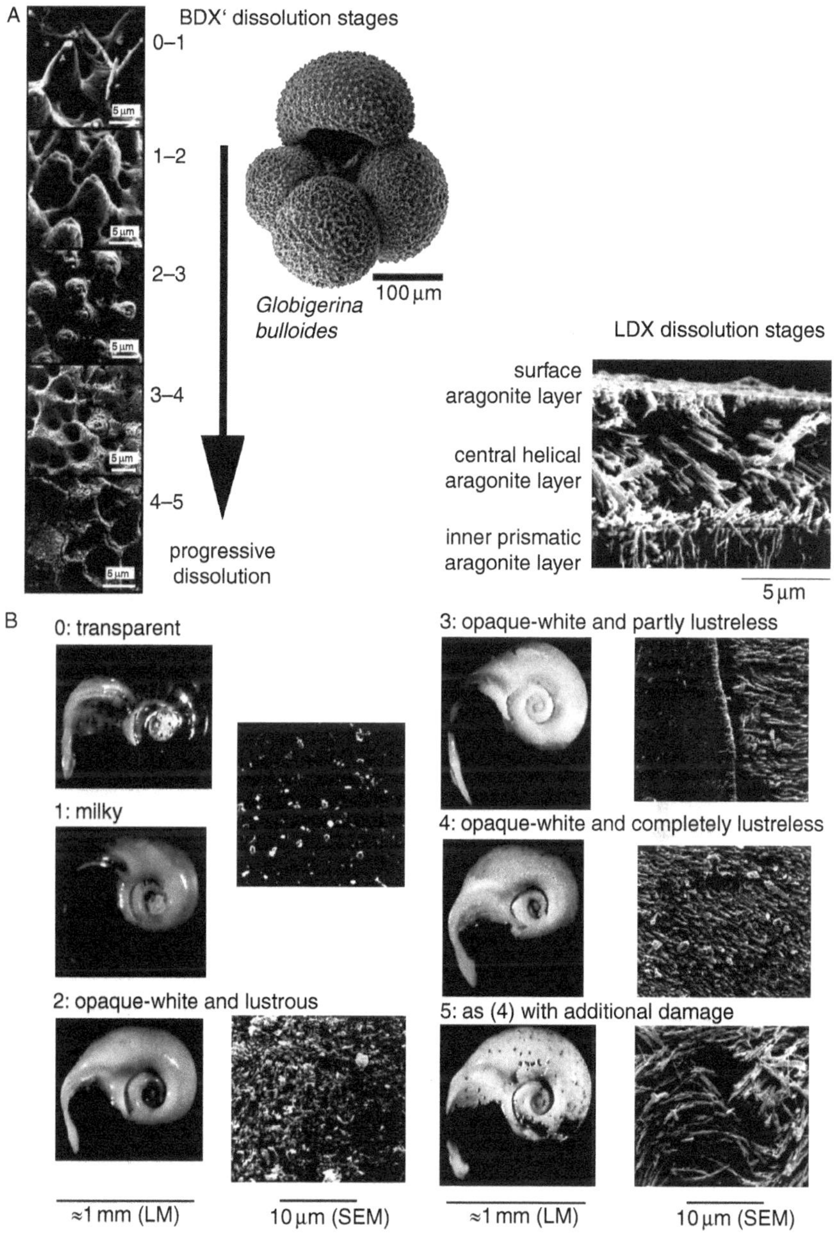

Figure 4.24 Graphical illustration of carbonate dissolution stages (redrawn and slightly modified from Volbers and Henrich, 2002; Gerhardt and Henrich, 2001). (A) *Globigerina bulloides* index (BDX′), a calcite dissolution proxy. (B) *Limacina inflata* index (LDX), an aragonite dissolution proxy.

fosters enhanced carbonate preservation. The sedimentary calcite lysocline, that is, the prominent horizon below which carbonate dissolution increases significantly, is mainly tied to the NADW/AABW boundary (Volbers and Henrich, 2002). The distribution of AABW in the South Atlantic is controlled by submarine barriers such as the MAR and the Walvis Ridge. As only small quantities of AABW are able to cross the sills of the Romanche Fracture Zone and the Walvis Passage, the eastern Guinea and Angola Basins are predominantly filled with NADW (Shannon and Chapman, 1991; Whitworth et al., 1999).

Figure 4.25 displays the results of a calibration of the BDX and LDX values in South Atlantic surface sediments, plotted against the carbonate ion concentration of the surface-, intermediate- and deep-water masses in a cross-section through the basins of the western South Atlantic. The BDX′ results are plotted in the upper graph. They reflect the hydrography very well, accurately displaying the state of carbonate saturation in the modern South Atlantic. In the AAIW, incipient calcite dissolution is indicated north of 25°SL. Overall good preservation in the NADW is recorded by BDX′ values of 2. The calcite lysocline is encountered at a water depth of 3500 m at 35°SL. It gradually increases to a depth of 3900 m at 25°SL and coincides with the LNADW/LCDW transition. Near the equator, the lysocline is uplifted into the LNADW. This might be explained by additional supralysoclinal dissolution due to increased organic productivity in the equatorial upwelling zone. The lower graph displays the results of the aragonite LDX proxy. Excellent preservation corresponds to a high CO_3^{2-} concentration within the SACW. Within the low CO_3^{2-} concentration core of the AAIW and CDW, at first a very good preservation is observed until a transition to moderate preservation indicates an upper aragonite lysocline at a water depth of approximately 750 m. Below this lysocline, a good preservation is observable again within the UNADW until the lower aragonite lysocline is encountered at a water depth of 2500 m. The aragonite compensation depth is recorded at a water depth of 3400 m as indicated by the LDX failure in all samples below this depth level.

Additional evidence on carbonate preservation can be derived from grain-size studies. Intensified fragmentation during dissolution will result in a progressive transfer of particles from coarser to finer size fractions, which may be assessed by analysing grain-size characteristics of the CS fraction. The potential to deduce the state of carbonate preservation from the grain size was independently developed by Gröger et al. (2003) and Stuut et al. (2002), using different size fractions and methods. Gröger et al. (2003) measured calcareous silt-size spectra in the Quaternary of Ocean Drilling Programme sites in the equatorial Atlantic at the Ceara Rise with a Micromeritics SediGraph 5100. They show by comparison with planktic foraminifer fragmentation proxies that, with progressive dissolution, the overall calcareous silt content decreases, and that the values for mean and

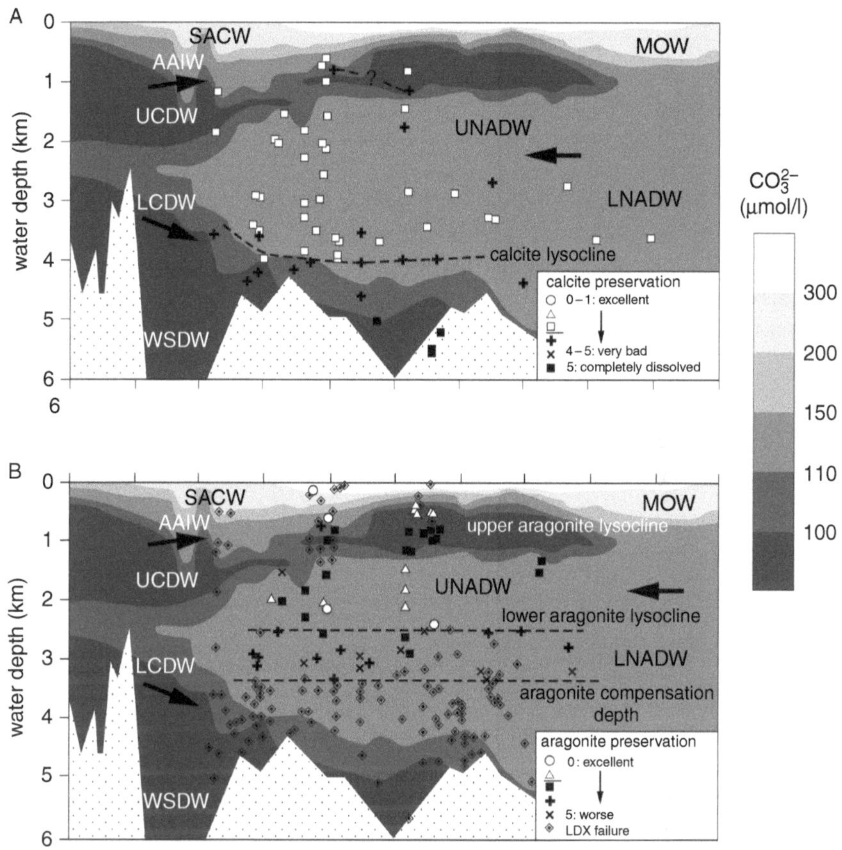

Figure 4.25 Results of a calibration of carbonate-dissolution proxies in South Atlantic surface sediments (redrawn and slightly modified from Henrich et al., 2003). Present-day water-mass distribution: SACW, South Atlantic Central Water; AAIW, Antarctic Intermediate Water; UCDW, Upper Circumpolar Deep Water; LCDW, Lower Circumpolar Deep Water; WSDW, Weddell Sea Deep Water; MOW, Mediterranean Overflow Water; UNADW, Upper North Atlantic Deep Water; LNADW, Lower North Atlantic Deep Water. (A) BDX′ values, a calcite dissolution proxy. (B) LDX values, an aragonite-dissolution proxy. (A multi-colour version of this figure is on the included CD-ROM.)

modal grain size in the coarse silt fraction decrease consistently. At the same time, the proportion of calcareous clay increases, indicating a continuous transfer of particles from coarser to finer grain sizes. Stuut et al. (2002) used a laser-particle sizer for measurements of grain-size distribution in Quaternary calcareous sediments from the Walvis Ridge. They defined a new dissolution index as the log-ratio of two coarse size fractions, size fraction A (25–90 μm) and size fraction B (>90 μm). Similar to Gröger et al. (2003), they related the decrease in the coarsest mode to progressive carbonate dissolution. In addition, Frenz and Henrich (2007) evaluated recently, in a calibration study on surface samples from the equatorial MAR ranging from 2746 to 5159 m water depth, how the CS grain size responds to carbonate dissolution. Figure 4.26 displays the results, indicating a clear shift in the amplitude and grain size of the coarse CS fraction (>8 μm, mainly consisting of foraminifer fragments), including a decrease with increasing water depth. This effect is amplified when the lysocline water depth is reached at ~4100–4200 m. The fine CS fraction (<8 μm, mainly consisting of coccoliths), however, increases in amplitude only, while grain size is rather stable, indicating a lower susceptibility of coccoliths to carbonate dissolution than foraminifers. These silt grain-size dissolution proxies have been successfully used to reconstruct the Miocene carbonate preservation record from the Walvis Ridge (Kastanja and Henrich, 2007). Here, significant decreases of preservation at 17.3, 16.3, 13.8, 13.0, 11.6 and 10.4 Ma coincide with Miocene glacial events (Mi-events), suggesting that

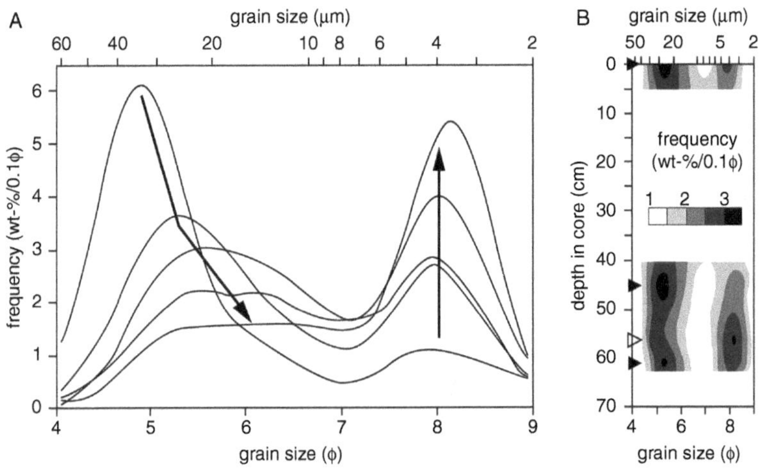

Figure 4.26 The response of CS grain size to carbonate dissolution. (A) Grain-size distributions of surface samples from the equatorial MAR province ranging from 2746 to 5159 m water depth. Amplitude and grain size of the coarse CS fraction (>8 μm, mainly consisting of foraminifer fragments) decreases with increasing water depth. This effect is amplified when the lysocline water depth is reached at approximately 4100–4200 m. The fine CS fraction (<8 μm, mainly consisting of coccoliths), however, increases in amplitude only, while grain size is rather stable, indicating a lower susceptibility of coccoliths to carbonate dissolution than foraminifers. (B) The same effect downcore in an example from the equatorial MAR province (3161 m water depth). Grain-size distributions are shown along sample depths (triangles) and interpolated in between. Samples above and below the LGM interval (filled triangles) reveal very similar CS grain-size distributions with a pronounced coarse fraction (foraminifer fragments). During the LGM interval (open triangle), the fine (coccolith) fraction became more pronounced due to increased dissolution in response to a shallower lysocline.

dissolution was caused by an increase in Southern Component Water (SCW) influence, which occurred as a response to the intensification of the Antarctic ice-sheet development. The SCW acted as the major deep-water source during the global Miocene cooling. At 10.4 Ma, a change to overall better preservation points to a weakening of the SCW influence, as a response to the initiation or strengthening of NADW.

2.2.3. Opal

2.2.3.1. Biogenic opal dissolution Opal dissolution is a major point of interest in recent studies (e.g. Fischer et al., 2002; Nelson et al., 2002; Pondaven et al., 2000a; Ragueneau et al., 2000, 2002) because biogenic silica in marine sediments is thought to be a significant palaeoproductivity proxy (e.g. Bareille et al., 1998; Diekmann et al., 2003; Mortlock et al., 1991; Romero et al., 1999).

Since seawater is undersaturated with respect to silica, the silicic acid availability controls not only biogenic production but also the dissolution of opal skeletons in surface waters, during settling at all depths in the water column and at the water–sediment interface during early diagenesis (Ragueneau et al., 2000, 2002; Sarmiento and Gruber, 2006). By comparing measured dissolution rates with production rates in several marine habitats, Nelson et al. (1995) found that on average 50% of the silica produced in the euphotic zone dissolves in the upper 50–100 m of the water column. There is, however, a wide regional range, as sediment-trap studies show, which can be explained in most cases by the temperature sensitivity of the dissolution rate. While in the warm waters of the equatorial Pacific, for example, almost 90% of the opal dissolves in the euphotic zone, it is only 10–30% in the cold waters of the Pacific and Atlantic sector of the Southern Ocean (Ragueneau et al., 2002). Thus, differences in the intensity of biogenic silica dissolution in surface waters are at least as important as differences in silica production rates for the control of the spatial pattern of opal efflux from the upper ocean (Ragueneau et al., 2000).

Two levels are particularly important, as most of the biogenic silica is recycled there along with organic matter (compare Fig. 4.20). The first main level is located within the upper 100 m, where 15–90% (mean 50%) of the opal produced in surface waters is dissolved and 65–85% of organic carbon is remineralised (Berger et al., 1989; Nelson et al., 1995; Suess, 1980; Tréguer et al., 1995, Van Cappellen et al., 2002). The second main level comprises the water–sediment interface, in particular the top few centimetres of the sediment column, where 60–95% of the opal and 90–99% of organic carbon rain escape burial (Emerson and Hedges, 1988; Ragueneau et al., 2000). During particle settling through the deeper water column, opal dissolution and organic matter remineralisation are less intense (Fig. 4.27).

Opal is recycled slower than organic matter, leading to a stepwise increase in the ratio between the Si and C fluxes with depth in the water

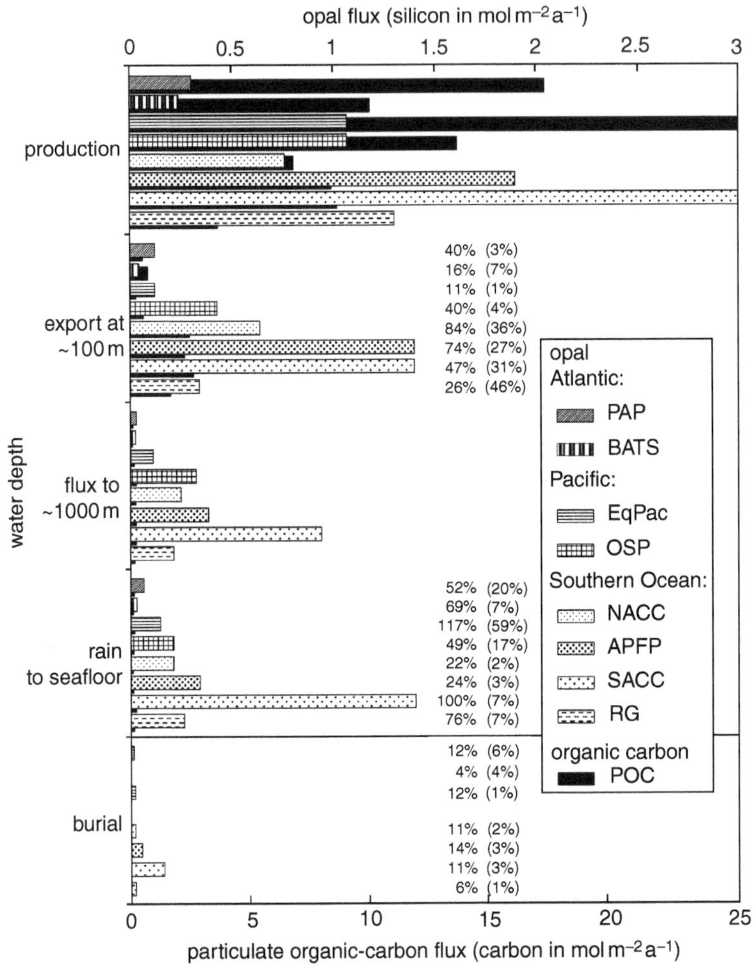

Figure 4.27 Mean annual opal fluxes against the background of particulate organic-carbon (POC) fluxes for several regions of the modern ocean (modified from Ragueneau et al., 2002). Note the more efficient decomposition of organic matter (~75%) compared with the dissolution of opal (~50%) in the surface water, which persists all the way down the water column. Export ratios, rain ratios and burial ratios are shown for both opal and organic matter (in brackets). Exemplified provinces are the North Atlantic (Porcupine Abyssal Plain, PAP), the oligotrophic Atlantic (Bermuda Atlantic Time-Series Study, BATS), the equatorial Pacific (EqPac), the North Pacific (Ocean Station Papa, OSP), the Pacific sector of the Northern Antarctic Circumpolar Current (NACC), the Pacific sector of the Antarctic Polar Front (APFP), the Pacific sector of the Southern Antarctic Circumpolar Current (SACC) and the Antarctic Ross Gyre (RG).

column and at the water–sediment interface (see Fig. 4.16). Despite large spatial and reservoir-dependent variability, Ragueneau et al. (2002) found that the order of increase of this ratio remains comparable for all open ocean provinces studied in the Atlantic, the Pacific and the Southern Ocean. Significant regional variation only exists at the depth of export, displaying higher ratios for the Atlantic and the Pacific ($\sim$10) than for the Southern Ocean ($\sim$3.5), which is attributed by them to the high organic carbon export efficiency in the Southern Ocean (compare Fig. 4.27).

2.2.3.2. Mechanisms controlling biogenic opal dissolution Up till now, the main processes influencing opal dissolution are not well quantified. A compilation by Nelson et al. (1995) and Ragueneau et al. (2000) designates the following four main points:

(1) Temperature plays an important role, showing an approximately 20-fold difference in the specific dissolution rate of opal between subequatorial and polar environments, as outlined above. Since the fraction of opal dissolution is much larger in warm waters than in cold waters, the diminishing opal dissolution rate with increasing water depth can be explained by decreasing temperatures (cf. Bidle et al., 2002).
(2) The ambient silicic acid concentration is essential for opal dissolution. It varies with the seasons and is, on larger scales, controlled by nutrient fractionation between ocean basins along the way of the conveyor belt (see Chapter 1). There is not a simple linear correlation between the degree of silica undersaturation and the dissolution kinetics of biogenic opal (Van Capellen et al., 2002): below a critical level of undersaturation, corresponding to $Si(OH)_4$ observed in deep oceans, the dissolution rate rises exponentially with increasing undersaturation. In contrast, incorporation of trace elements (in particular, aluminium) within the opaline matrix can decrease opal solubility significantly (Van Beusekom et al., 1997).
(3) The extent of dissolution is dependent upon the taxa, since morphology, chemical composition and residence time in the water column can vary substantially (Johnson, 1976; Takahashi, 1991). Because of taxon-specific dissolution, species diversity decreases with depth and thicker shelled forms increase in importance. Extensive dissolution within the water column can eliminate most, if not all, of the settling population of dissolution-susceptible taxa.
(4) Any process that destroys the organic coatings from the surfaces of diatom and radiolarian skeletons and hence exposes silica directly to sea water increases the dissolution rate. The breakdown of organic coating by bacterial action accelerates the rate of opal dissolution on diatom particles by more than an order of magnitude (Bidle et al., 2002).

2.2.3.3. Transfer of opal through the water column We outlined above that on average only half of the opal produced in the upper 100 m of the ocean survives dissolution and sinks (Nelson et al., 1995; Van Cappellen et al., 2002). Below the euphotic zone, however, opal fluxes do not severely decrease with water depth (Fig. 4.27). Consequently, opal rain rates to the sea floor are comparable to fluxes at the base of the twilight zone and opal preservation patterns of deep-sea sediments are more or less independent of water depth (Ragueneau et al., 2000, 2002). Only minor differences in dissolution rates exist between the various ocean basins. Notwithstanding this general rule, a distinct decrease in opal fluxes can be observed where ubiquitous silica undersaturation in the water column occurs, such as in the Greenland Sea (e.g. Von Bodungen et al., 1995).

Clear indications for ongoing dissolution in the water column with increasing depth are, for example, differences in the relative abundances of strongly and weakly silicified diatom taxa between trap levels, and a general increase in shell breakage of radiolarians with depth. Phaeodarian radiolarians, which are common in surface waters, dissolve almost completely during settling through the water column, whereas the signal of polycystine shells is relatively well recorded in deep-sea sediments (Abelmann, 1992; Takahashi, 1983).

At lower latitudes, a slight increase in opal flux in intermediate waters occurs at some sites and can be explained by sinking of deep-living radiolarians. Abelmann (1992) documented a marked lateral transport of radiolarians during the summer in the Southern Ocean. In the eastern Atlantic Ocean (Treppke et al., 1996), higher flux values for diatoms and silicoflagellates in the near-bottom sediment traps were observed.

2.2.3.4. Biogenic opal deposition on the sea floor Opal deposition rates on the sea floor either are measured by means of sediment traps moored just above the benthic nepheloid layer or are reconstructed from the difference between opal burial rates and pore water regenerative fluxes. Both methods provide data with good agreement for the few sites where such a comparison is possible (see Ragueneau et al., 2000). An exception is the Southern Ocean, where opal fluxes from surface waters and accumulation rates differ significantly because of intense lateral advection and sediment focusing (Abelmann and Gersonde, 1991, Dezileau et al., 2000, Diekmann et al., 2003, Lucchi and Rebesco, 2007, Rebesco et al., 2007).

On average, 25% of the biogenic opal that escapes from the surface waters and is transferred through the deep waters reaches the sea floor (Tréguer et al., 1995). Diatom skeletons prevail among the accumulating particles, while the contribution of radiolarian shells to the total flux is usually not more than 10% (Honjo, 1996). For many oceanic environments, however, Ragueneau et al. (2002) showed that the opal rain rate is substantially higher, reaching 50% in two regions of the Atlantic and some sectors of the Southern Ocean (Fig. 4.27). Even almost 100% of the exported opal reaches the sea floor in two regions of the Pacific. Ragueneau et al. (2002) found no obvious pattern in the variation from location to location.

Opal rain rates can be immense, since up to 90% of the opal skeletons produced in the upper ocean can be exported and because opal fluxes do not severely decrease with depth in open ocean regions. Opal deposition on the sea floor peaks under conditions of high silicic acid availability, as in the Pacific. The circum-Antarctic opal belt is a further example (Fig. 4.6). In other cases, only a small amount of the biogenic opal debris, including diatoms and radiolarians, or even almost nothing is deposited as sediment on the ocean floor. This occurs under conditions of low silicic acid availability, as known from the modern Atlantic.

Because of differential dissolution and many other processes such as lateral advection and zooplankton grazing, assemblages of opal shells found in marine sediments may differ substantially from the assemblages living in the surface waters (Leventer, 2003). Larger robust shells are much more likely to survive passage through the water column and burial in surface sediments than small, thin shells. In general, silicoflagellates and diatoms tend to dissolve well before radiolarians. Despite the impact of these factors, the distribution of diatoms in surface sediments is clearly related to surface water properties (e.g. Armand et al., 2005; Crosta et al., 2005; Cunningham and Leventer, 1998; Leventer, 1992; Romero et al., 2005; Zielinski and Gersonde, 1997).

2.2.3.5. Recycling of opal at the water–sediment interface and opal burial Most opal dissolution occurs at the water–sediment interface when settled skeletal elements are exposed to the deep-ocean water over a long time, and if the water is highly undersaturated with respect to opal. An average of 80% of the biogenic opal arriving at the sea floor dissolves within the uppermost layers of sediment prior to final burial (Ragueneau et al., 2002). This loss locally decreases to less than 70% in regions where the opal rain is highest, such as in the Bering Sea and the Southern Ocean.

An important aspect is that silicic acid concentrations in pore water rapidly increase within the first centimetres below the sea floor (Fig. 4.28). The general interpretation to account for such an increase is that opal undergoes dissolution during its progressive burial in the sediments, thereby enriching the pore waters in silicic acid (McManus et al., 1995). All profiles hitherto documented for various parts of the modern ocean show an increase to an asymptotic value between ~200 and ~800 mmol m^{-3} (Gallinari et al., 2002; Rabouille et al., 1997). The fact that these pore water concentrations are all higher than the concentrations in the water column implies a continuous backflow of silicic acid from the sediments into the highly undersaturated water column.

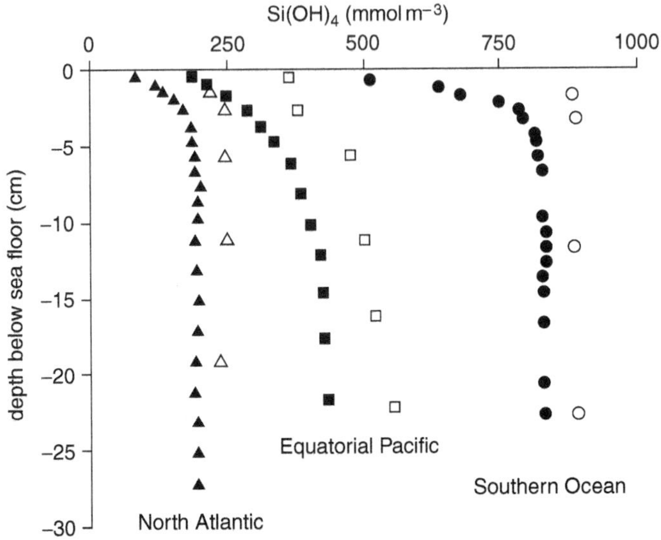

Figure 4.28 Pore-water $Si(OH)_4$ concentrations within the first centimetres below the sea floor (filled symbols) for several regions of the modern ocean (redrawn and slightly modified from Ragueneau et al., 2000). Also shown are apparent solubilities determined experimentally with flow-through reactors (open symbols). Note that the silicic acid concentrations in the pore water generally approach towards an asymptotic value that is well below that of saturation with opal, even in sediments that are rich in opal, such as in the Southern Ocean.

The high dissolution efficiency at the water–sediment interface and the relatively modest variation (60–96%) around the mean indicate that the dissolution of biogenic opal and the related flux of silicic acid out of the sediment is determined primarily by the pattern of opal rain to the sediment (Gallinari et al., 2002; Rabouille et al., 1997; Ragueneau et al., 2002). The higher the opal rain rate, the more the rise in asymptotic silicic acid concentration (Fig. 4.28).

Ragueneau et al. (2000) found an opal preservation discontinuity, which occurs at an opal rain rate of about 2 mmol m^{-2}•per day (Fig. 4.29). Above this threshold, the opal burial efficiency is high (up to 30%), whereas it is low (only about 30%) below this threshold. This separation of the preservation efficiency into two distinct groups occurs at an opal rain rate that corresponds approximately to a theoretical upper limit of the silicic acid flux out of the sediment, which is defined by the thermodynamic properties of opal and the physical properties of the sediments (see Dixit and Van Cappellen, 2003; Sarmiento and Gruber, 2006). If the opal rain reaching the sea floor exceeds the threshold silicic acid flux out of the sediments, any excess rain above this flux leads inevitably to burial. If the opal rain falls under conditions below this threshold value, it becomes difficult for

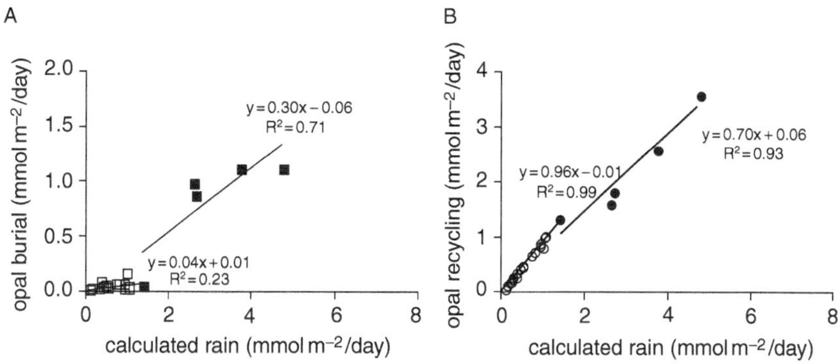

Figure 4.29 Opal preservation efficiencies (A) and recycling efficiencies (B) for conditions of low opal rain rates (open symbols) and high opal rain rates (filled symbols) during shallow burial from a wide range of locations within the oceans (redrawn and slightly modified from Ragueneau et al., 2000). (A) Burial rate of opal, plotted against the rain rate of opal arriving at the water–sediment interface. Areas with a low opal rain rate, which are representative of most of the deep ocean, typically have about 5% preservation efficiency, whereas the high rain-rate areas, which occur generally at high latitudes in the Bering Sea and the Southern Ocean, have about 30% preservation efficiency. (B) Flux of silicic acid from the sediments back to the water column, plotted against the rain rate.

preservation of opal to occur, unless the asymptotic concentration or some other properties of the sediments change (Fig. 4.29).

A set of additional factors may influence opal burial, including the temperature and silicic acid concentration of the bottom waters, the rate of bioturbation, the gross accumulation rate and the availability of specific trace elements (e.g. McManus et al., 1995; Ragueneau et al., 2000, 2001). With respect to the latter, aluminium is known to be of particular importance (Van Beusekom et al., 1997; Van Cappellen and Qiu, 1997). Its incorporation in the silica skeletons, during biomineralisation or after death of the organisms, significantly enhances the preservation efficiency of biogenic opal. Aluminium dissolved in pore water can also be incorporated in biogenic opal particles at the water–sediment interface during early diagenesis, causing chemical changes that further modify their solubility. In addition, aluminium may reprecipitate as amorphous Al–Si coatings or as authigenic clay minerals, which decreases the dissolution rate constant of opal with age, as the number of reactive sites on the diatom surface becomes more and more blocked (Van Cappellen et al., 2002). The biogeochemical processes of these models have, however, not yet been well confirmed (see Sarmiento and Gruber, 2006 for details). Some specific diagenetic aspects are highlighted by Hesse and Schacht (2011, this volume).

More reliable is the observation that the opal preservation potential exponentially rises with the gross sedimentation rate (Ragueneau et al.,

2000). At average deep-sea sedimentation rates of 1–4 cm ka^{-1}, only 1–5% of the biogenic opal arriving at the sea floor is buried within the sediment, whereas the high accumulation rates of siliceous oozes beneath the APF (10–30 cm ka^{-1}) permit burial of more than 80% of the deposited opal. A possible explanation for a link of the opal burial to the sedimentation rate is that a reactive zone of high dissolution exists within the top few centimetres of the sediments, causing the burial rate to increase if opal spends less time in this reactive zone due to a high sedimentation rate.

In analogy to the sedimentation rate effect, enhanced bioturbation, such as caused by seasonal phytodetritus deposition, may dramatically increase opal burial as well (Boudreau, 1996). On time scales of months, surface sediments are transported to depths of 3–6 cm (McClintic et al., 2008), where the concentrations of dissolved silica in the pore water are close to the solubilities of fresh diatoms, which should reduce the rate and probability of opal dissolution (Gallinari et al., 2008). These effects of age-dependent mixing are still largely unexplored, although bioturbation is known to be a key process in early diagenesis, particularly in determining the fate of biogenic opal (compare Boudreau, 1996; Uchman and Wetzel, 2011, this volume).

2.2.3.6. Biogenic opal as productivity proxy The abundance of opaline shells of diatoms and radiolarians on the deep-sea floor commonly correlates with overall patterns of surface water productivity, and the assemblages in the sediment are not very different from the planktic biocoenosis (Berger, 1976). The opal content is usually high in sediments below high-productivity regions and low elsewhere. Because of these observations that were made already relatively long ago, the biogenic opal content in deep-sea sediments has become a rough qualitative proxy for surface productivity, although there are major differences between the patterns of the opal and the organic carbon export from the euphotic zone that remain to be better explored (see above).

In several areas of the ocean, indeed, diatom abundance and assemblage composition excellently reflect surface water productivity. At such sites, the opal content is strongly correlated with the marine organic carbon values and other palaeoproductivity proxies such as the barium content (e.g. Bonn et al., 1998; Moreno et al., 2002). In other cases, however, there is no indication that biogenic opal may serve as a palaeoproductivity proxy (e.g. Lange and Berger, 1993).

The major limiting factor of the opal proxy is the regional and temporal variation of both the silicate supply into the photic zone and the opal dissolution in the water column and after deposition on the sea floor (Section 2.2.3). Furthermore, regulation of diatom growth by iron availability has been proved in the open ocean (see Section 2.1.3). Iron-limited conditions lead to more heavily silicified and thus faster-sinking diatoms. Increased sinking rates can result in more efficient carbon export and in preferred preservation of the more robust diatom frustules under such iron-deficient regimes, despite an overall low total productivity (Hutchins and Bruland, 1998). Consequently, downcore variations in the opal accumulation may also reflect a changing supply of iron to the surface waters rather than varying productivity (Wefer et al., 1999).

These limitations are particularly severe in HNLC regions. Opal sediment records of the Southern Ocean, for example, do not clearly show a glacial increase in export production in high-latitude surface waters (Dezileau et al., 2003; Frank et al., 2000; Mortlock et al., 1991), which has been invoked as a mechanism responsible for the glacial reduction in atmospheric concentrations of carbon dioxide, as deduced from ice core measurements (e.g. Anderson et al., 2002; Moore et al., 2000). A possible explanation of this misfit is that a higher iron input into the ocean occurred during glacials, which was introduced by strong aeolian dust supply (Kumar et al., 1995) or increased detrital particulate loads in upwelled ocean waters (Latimer and Filippelli, 2001). As a consequence, diatoms produced thinner valves that were more susceptible to dissolution (e.g. Wefer et al., 1999).

The accuracy of the opal proxy has been much improved by cross-correlating diatom species assemblages from the sedimentary record (e.g. Abrantes, 2000; Sancetta, 1992) with fluxes of diatom shells through the water column (Nelson et al., 2002; Ragueneau et al., 2000, 2002) and the fluxes of organic carbon. A general complication is, however, that differences in silica preservation often exceed the variations in productivity (Nelson et al., 1995). The effect of preservation in several areas of the open ocean may thus obscure the record of the most productive season, and no evidence of original variability need therefore be preserved in the sedimentary record (Romero et al., 1999; Sancetta, 1992). In contrast, diatoms in surface sediments underneath some coastal upwelling systems often record the high-productivity season because moderately robust species dominating during this season are preferentially preserved (e.g. Brodie and Kemp, 1994; Lange et al., 1998).

Combining biogenic opal with other proxies of export production is suggested to be the best way to achieve a coherent reconstruction of the past surface ocean history (Lochte et al., 2003; Romero et al., 1999; Wefer et al., 1999). These proxies include organic carbon burial rates, organic biomarker accumulation rates, the fluxes of benthic and planktic foraminifers, barium-accumulation rates, radionuclide ratios and redox-sensitive trace elements.

3. History and Evolution of Ancient Pelagic Factories

The production of pelagic sediments is critically dependent upon key groups of marine protists, plants and animals that form mineralised skeletons in large quantities in surface waters of the ocean. The biologically catalysed processes of calcification and silicification are controlled by specific metabolic pathways that have repeatedly coevolved in several groups of marine organisms during earth history (Falkowski and Raven, 1997). These homologous functional groups of organisms are related through similar biogeochemical processes rather than phylogenetic affiliation and are consequently named biogeochemical guilds (Falkowski et al., 2003, 2004a; Iglesias-Rodríguez et al., 2002).

Modern eukaryotic phytoplankton came to ecological importance not before the Mesozoic (e.g. Falkowski et al., 2004b; Knoll and Lipps, 1993). The sedimentary record shows, however, that phytoplankton continuously drives pelagic food webs and export production in the upper ocean since more than 2.7 Ga ago (Cavalier-Smith, 2006; Rosing and Frei, 2004; Summons et al., 1999). Cyanobacteria are the oldest clade of three basic evolutionary lineages of this diverse group of organisms (Delwiche, 1999). This prokaryotic oxygenic phytoplankton developed the metabolic system that utilises the energy of visible light to oxidise water and simultaneously reduce CO_2 to organic carbon. Cyanobacteria initiated a major global environmental change in earth history by releasing oxygen into the ocean/atmosphere system.

All other oxygen-producing phytoplankton in the ocean is eukaryotic. Green and red algae evolved more than 1.5 Ga ago in the Palaeoproterozoic oceans (Fig. 4.30) (Javaux et al., 2001). Synthesising different types of chlorophyll as a secondary pigment, they are at the origin of two major evolutionary clades, namely, the green and the red lineage (Delwiche, 1999; Falkowski et al., 2003). Both clades have their roots in a common ancestor interpreted to be the endosymbiotic enslavement of a cyanobacterium into a heterotrophic host cell. Subsequently, several independent secondary endosymbioses were formed during the phylogenetic evolution of phototrophic marine organisms in earth history. The open question, why dominant phytoplankton in modern oceans have overwhelmingly red plastids, as raised by Falkowski et al. (2004b), is intrinsically tied with the search for environmental controls of deep-sea sedimentation in earth history.

The green lineage, which is numerically dominated by the green algae, played a major role in oceanic food webs during the Neoproterozoic and Palaeozoic until the end-Permian extinction and gave rise to all higher (land) plants (Knoll and Lipps, 1993). Prasinophytes and chlorophytes essentially constituted marine phytoplankton (Butterfield, 1997; Martin, 1996; Martin et al., 2008; Mendelson, 1993). The red lineage, including the red algae and seaweeds, has risen to ecological prominence since the Triassic radiation (Falkowski et al., 2004b). It is represented by several modern phytoplankton clades, of which the diatoms, dinoflagellates, haptophytes (including the coccolithophorids) and the chrysophytes (including silicoflagellates) are the most important for pelagic sedimentation (Fig. 4.30, Chapter 1).

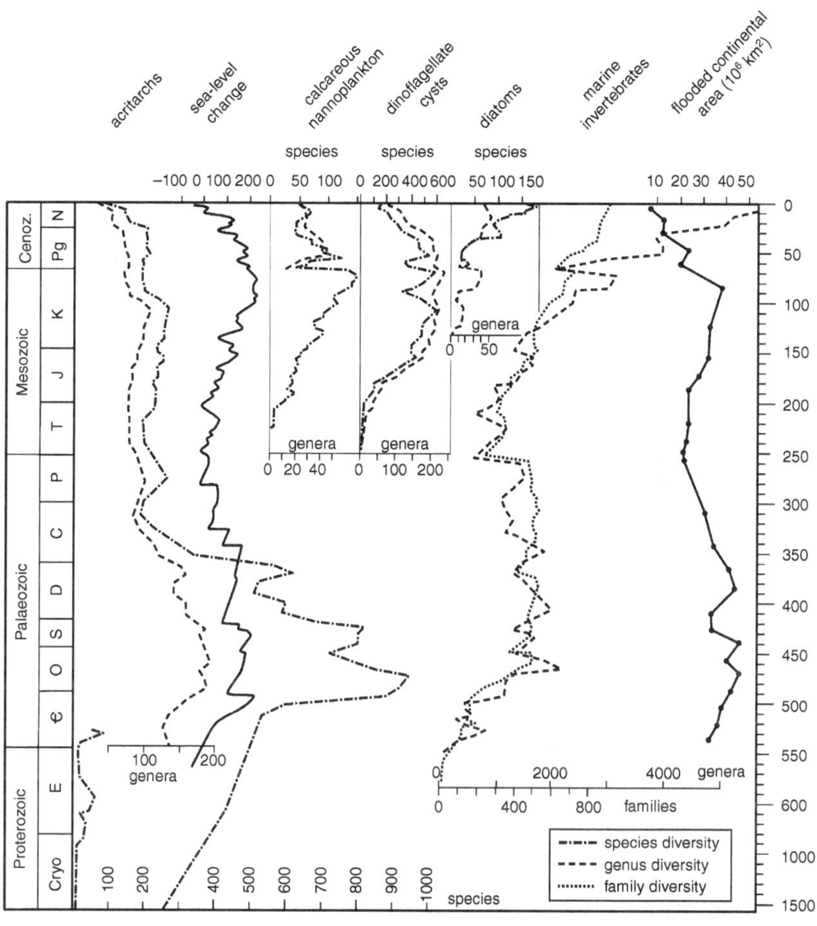

Figure 4.30 Comparison of eukaryotic phytoplankton diversity curves with sea-level change (Haq et al., 1987; Vail et al., 1977), flooded continental areas (Ronov, 1994) and marine genus-level and family-level invertebrate diversity (Sepkowski, 1997); redrawn and modified from Katz et al. (2004). Phytoplankton species-level diversity is from published studies (calcareous nannofossils: Bown et al., 2004; dinoflagellates: Stover et al., 1996; diatoms: Spencer-Cervato, 1999; acritarchs: Knoll, 1994; R.A. MacRae, unpublished data). Phytoplankton genus-level diversity is compiled from publicly available databases (calcareous nannofossils and diatoms: Spencer-Cervato, 1999; dinoflagellates and acritarchs: R.A. MacRae, unpublished data). All records are adjusted to the Berggren et al. (1995) (Cenozoic), Gradstein et al. (1995) (Mesozoic) and GSA (http://rock.geosociety.org/science/timescale/timescl.htm) (Palaeozoic) timescales. Interpretation of acritarch evolution is based on the Proterozoic and Early Cambrian record of acritarch species (Knoll, 1994) and on literature compilation that may include taxonomic synonyms (R.A. MacRae, unpublished data). Note change in vertical scale at 600 Ma.

3.1. Plankton evolution through time

3.1.1. Archaean–Proterozoic

During the Archaean, when no ozone layer existed, ultraviolet radiation probably largely excluded photosynthetic plankton from the upper photic zone of the marine water column. Phototrophic organisms were concentrated in stratified microbial mats (preserved as stromatolites) where iron-impregnated mineral grains or snow overlying thin ice facilitated enough protection (Cavalier-Smith, 2006). Marine plankton evolved between 2.5 and 2.1 Ga when the photic zone was fully colonised by cyanobacteria and protobacteria. This occurred during later stages of the Great Oxidation Event, a time span during which the ocean/atmosphere system became noticeably oxygenated. Atmospheric oxygen rose to $>1\%$ of modern levels and the ozone layer was formed (Gauthier-Lafaye and Weber, 2003; Holland, 2006).

Cyanobacteria appear to have dominated the primary production in Proterozoic oceans, followed by eukaryotic algae, which became important elements of marine ecosystems not before the end of the era (Anbar and Knoll, 2002; Butterfield, 2007; Knoll, 1989, 1992; Lipps, 1993; Summons et al., 1999; Tappan, 1980). Proterozoic organic-walled microproblematica called "acritarchs" largely represent, like those in Palaeozoic rocks, the remains of a thick-walled resting stages, or cysts, in the life cycle of planktic eukaryotic algae (e.g. Knoll, 1994; Mendelson, 1993; Vidal and Moczydlowska-Vidal, 1997). They occur abundantly in rocks younger than 1.5 Ga (Fig. 4.30). While a number of acritarchs are interpreted to be dinoflagellate cysts, many others can be assigned to the green algae or even golden-brown algae (Colbath and Grenfell, 1995; Moldowan and Talyzina, 1998). Butterfield (2007) emphasises, however, that Proterozoic acritarch assemblages may include, in some cases, vegetative remains of organisms that were heterotrophic rather than photosynthetic, and benthic rather than planktic. Biomarker molecules indicative of cyanobacteria and eukaryotes (hopanes, 2α-methylhopanes, terpanes and steranes), which are preserved in organic-rich sediments, add complementary information to the Proterozoic fossil record (e.g. Dutkiewicz et al., 2007; Schwark and Empt, 2006; Summons and Walter, 1990; Summons et al., 1999).

3.1.2. Palaeozoic

Although we do not clearly know which specific organisms mediated the primary production during the Palaeozoic, fossils especially in shales indicate that eukaryotic phytoplankton with morphological features similar to members of the green evolutionary lineage were abundant and diverse in the surface water of the oceans (Falkowski et al., 2004b; Tyson, 1995). Eukaryotic phytoplankton diversified in step with mesozooplankton and marine invertebrates during the Cambrian and Ordovician periods (Knoll and Lipps, 1993; Tappan, 1980). Acritarchs in the late Ediacaran and earliest Cambrian are represented by a default biota of unornamented sphaeromorphs, followed by a distinctive radiation of small, rapidly evolving ornamented forms during the Tommotian (Butterfield, 1997, 2003; Knoll et al., 2006). Several genera of morphologically distinct prasinophyte algae appeared at this time (Talyzina and Moczydlowska, 2000). A greater diversification of organic-walled phytoplankton followed during the Ordovician, with the expansion of baltisphaerid, veryhachid and other acritarch groups (Colbath and Grenfell, 1995; Knoll and Lipps, 1993). In response to an episode of extinction at the end of the Ordovician, acritarchs diversified rapidly again during the beginning of the Silurian. Towards the end of the Devonian, the acritarch diversity decreased again sharply, reflecting the Frasnian–Famennian extinction (Mendelson, 1993). Biomarkers from eukaryotic cell membranes imply a fundamental change in the green algae assemblage from more primitive, mainly C_{29}-sterane-producing algae, to modern C_{28}-sterane-producing algae at the Devonian/Carboniferous transition (Fig. 4.31; Schwark and Empt, 2006). This shift in the C_{28}/C_{29}-sterane ratio coincides with the appearance of euspondyle and metaspondyle dasycladales in the Tournaisian/Mississippian.

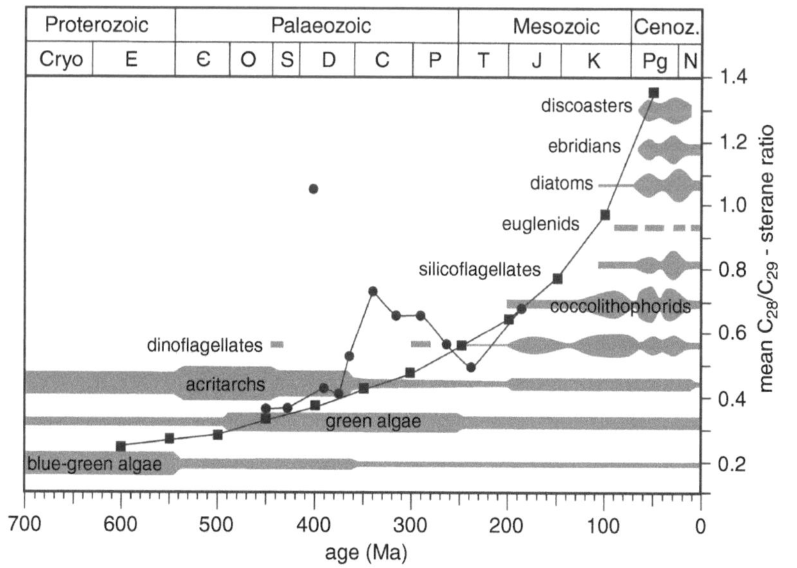

Figure 4.31 Geological distribution of important groups of phytoplankton compared to the mean C_{28}/C_{29} sterane ratio of marine sedimentary rocks and crude oils (redrawn and slightly modified form Schwark and Empt, 2006). The range of the phytoplankton groups has been adapted from Tappan and Loeblich, 1973. Filled circles, 500 rock samples averaged over stages (from Schwark and Empt, 2006) and filled squares, 400 oil samples averaged over 50 Ma steps (from Grantham and Wakefield, 1988).

Heterotrophic protists that developed skeletons since the Cambrian are radiolarians and foraminifers. While foraminifers lived exclusively benthic until the Early Jurassic (Culver, 1993), radiolarians were certainly present in planktic ecosystems at least by the Middle Cambrian, even though with very few species (Knoll and Lipps, 1993; Nazarov and Ormiston, 1986). The early radiolarians already produced siliceous skeletons and became considerably more numerous and more complex during the Ordovician, rising to sedimentological importance (De Wever et al., 2001). Cold- and warm water radiolarians are distinguishable since the Cambrian, and a deep-living radiolarian fauna had developed by at least the Silurian (Casey, 1993). The biodiversity of radiolarian families increased constantly throughout the Palaeozoic and decreased only slightly towards the end of the Permian (De Wever et al., 2006). Their skeletons are the principal components of many siliceous pelagic rocks (as discussed below).

During the Palaeozoic and through the Jurassic, the plankton food chain was governed by radiolarians, feeding on a variety of non-skeletal organisms (Baumgartner, 1987; Knoll and Lipps, 1993). Assuming overall low phytoplankton densities in mostly oligotrophic conditions during the Palaeozoic (Martin, 1995; Racki and Cordey, 2000), radiolarians may have thrived either in highly productive shallow waters, in oceanic gyres (probably with symbiotic algae), or in deeper oceanic sub-photic zones as detritivores and bacterivores (Afanasieva et al., 2005a,b; Casey, 1993; Vishnevskaya and Kostyuchenko, 2000).

Although acritarchs and radiolarians were particularly abundant during the Palaeozoic, further microplankton elements such as chrysophytes and tintinnids are documented only sporadically, indicating that they, too, may have had a more extensive but unpreserved record. Much better preserved dacryoconarids (styliolinids and nowakiids), which lived as nektoplankton, possibly in huge swarms, are important constituents of many deep-sea carbonates formed during most of the Devonian (Walliser and Reitner, 1999). Based on the presence of phytoplanktic acritarchs and radiolarians, as well as the additional planktic elements, Knoll and Lipps (1993) concluded that a complex pelagic ecosystem was fully established. Like the modern oceans, the Palaeozoic seas were inhabited beyond any doubt by cyanobacteria and abundant soft-bodied larger zooplankton organisms that have not been preserved as fossils.

The Permian/Triassic crisis was one of the most important in the evolution of marine organisms and included a major perturbation in the planktic ecosystems. Acritarchs and radiolarians both suffered severe extinctions (Casey, 1993; Knoll and Lipps, 1993; Kozur, 1998; Vishnevskaya, 1997), which occurred progressively, as documented particularly well for the radiolarians (De Wever et al., 2003, 2006). The collapse of the pelagic opal factory is recorded in a long-lasting radiolarite gap (Isozaki, 1997; Kakuwa, 1996; Kozur, 1998). The destructive impact of a volcanic winter, invoked by Kozur (1998), probably enhanced by positive feedbacks such as an expanding superanoxia accompanied by a decline of nutrients and productivity breakdown, was particularly effective in oceanic surface waters (Grice et al., 2005; Hallam and Wignall, 1997; Isozaki, 1997; Martin, 1998). Whatever the reasons, the Permian–Triassic transition is marked by a tremendous post-crisis radiolarian diversification (Casey, 1993).

3.1.3. Mesozoic

During the Mesozoic, stepwise evolutionary innovations among marine phytoplankton caused fundamental changes in deep-sea sedimentation (Mesozoic plankton revolution). Dinoflagellates, coccolithophores, diatoms and silicoflagellates, which are all members of the red evolutionary lineage, came to largely displace green eukaryotic algae in the photic environments of open-marine ecosystems (Falkowski et al., 2004a,b). In time with zooplankton, mainly represented by radiolarians, planktic foraminifers and tintinnids, they either diversified from pre-existing forms into a wider variety of new types or appeared and radiated for the first time (De Wever et al., 2001; Knoll and Lipps, 1993; Moldowan and Jacobson, 2000).

Radiolarians, the microplankton group with a nearly continuous Palaeozoic sedimentary record, are mostly characterised by the tremendous diversification of spumellarians and nassellarians in the Early–Mid Triassic (De Wever et al., 2003, 2006). From the Cretaceous onwards, they produced a number of new families many of which included symbiotrophes (Casey, 1993). Dinoflagellate cysts may be part of Late Proterozoic and Palaeozoic acritarch assemblages, but dinoflagellates first become diverse and abundant components of Late Triassic plankton associations (Edwards, 1993). Their diversity expanded and retracted several times during the Mesozoic, with species numbers reaching a maximum in Late Cretaceous (Santonian) times. Silicoflagellates first appeared in the Early Cretaceous and were abundant in the Late Cretaceous (McCartney, 1993). Their contribution to deep-sea siliceous oozes, however, remained relatively unimportant on a global scale.

Coccolithophores and associated nannoplankton first appear in the fossil record in Upper Triassic sediments (Bown, 2005; Bown et al., 1992, 2004; Roth, 1987, 1989). The oldest known nannofossils are relatively large and heavily calcified nannoliths in Carnian sediments from the Southern Alps. They are probably related to dinoflagellates and other planktic groups rather than to coccolithophores (Erba, 2006). First, tiny coccoliths, which show an extremely simple morphology, appeared in the Norian (Bown, 1998; Janofske, 1992). Triassic assemblages are of low diversity and only one coccolith species survived the Triassic–Jurassic extinction. Subsequent speciation during the Pliensbachian–Toarcian is one of the most important diversifications events in the history of calcareous nannoplankton (Erba, 2004; Mattioli and Erba, 1999). Turnover rates reached a remarkable peak in the early Tithonian, when both extinction and speciation rates were high

(Roth, 1987). This event marks the beginning of the replacement of Jurassic by Cretaceous nannoplankton (Bornemann et al., 2003; Roth, 1989). The Cretaceous diversity record, in contrast, is one of relatively uniform increase punctuated by short periods of turnover and decline (Siesser, 1993). Accelerated rates of origination occurred synchronously at low and high latitudes. As suggested by Bown et al. (2004), this pattern indicates long-term stability and widespread oligotrophy in photic zone environments under conditions of a greenhouse-mode climate (cf. Weissert, 2011, this volume). Coccolithophores became the dominant nannofossil group, and the sizes of coccoliths increased significantly. The cosmopolitan genus *Watznaueria* produced abundant coccoliths since the Middle Jurassic, which often dominate nannofossil assemblages and might be considered the *E. huxleyi* equivalent of the Mesozoic oceans (Erba, 2006). The highest global diversity of coccolithophorids in the evolutionary history of this group coincides with the widespread distribution of nannofossil chalks during the Campanian–Maastrichtian.

Radiating from a benthic ancestor, planktic foraminifers evolved during the Early Jurassic (Toarcian), probably as a result of the early Toarcian OAE (Hart et al., 2003). They first spread in areas of the northern Tethyan margin and the epicontinental seas of Europe (Culver, 1993). The palaeogeographic restriction of Jurassic early representatives and the shape of their test seem to indicate a meroplanktic mode of life, that is, benthic in the early stage and planktic in the late stage (Simmons et al., 1997). Tests of most Jurassic planktic foraminifers appear to be primarily composed of aragonite. Nonetheless, they contribute to the pelagic rain onto the deep-sea floor since this time.

From the Cretaceous onwards, radiolarians have had to share their herbivorous and symbiotrophic niches with radiating planktic foraminifers (Anderson, 1996; Casey et al., 1983). By the Hauterivian, the first true globigerinids developed, typifying the holoplanktic foraminiferal fauna of the Cretaceous oceans. Shells were then composed of calcite. Early Cretaceous forms, pre-hedbergellids and schackoinids, lived at or near the sea surface and were geographically widespread throughout the low- and mid-latitudes (BouDagher-Fadel et al., 1997). Many Cretaceous species inhabiting the photic zone were already able to house algal symbionts (Casey, 1993). During latest Early Aptian times, planktic foraminifers appear to have begun colonising deeper waters (BouDagher-Fadel, 1996). The macroperforate hedbergellids and their descents are likely to have been more capable of feeding in cooler and less oxygenated oceanic waters at subsurface depths than were their microperforate ancestors (Banner and Desai, 1988). Concomitant with this habitation of deeper environments is a more widespread and cosmopolitan distribution.

In general, there seems to be a strong relationship between evolution of planktic foraminifers and OAEs (Leckie et al., 2002). The habitation of deeper waters by planktic foraminifers and the occupation of the oceans worldwide may be related to the cessation of the Aptian OAEs and the rising sea level in the Albian (BouDagher-Fadel et al., 1997). An anoxic or dysoxic water column in many parts of the oceans, as suggested by Bralower et al. (1994), would have prevented the earlier dispersal of hedbergellids. Significant eustatic rises in sea level, as documented by Haq et al. (1988), caused a worldwide overstepping of the Early Cretaceous basin margins and correspond with the spread of deeper environments into continental areas such as Europe. This events lead to a turnover in the planktic foraminiferal record. Species diversity in planktic foraminifers culminated during the Late Cretaceous.

The oldest known diatoms are of Early Jurassic age (Toarcian), although they must have a much longer evolutional history (Barron, 1993; Medlin et al., 2000). The unknown ancestors probably lacked a well-developed test. The high susceptibility to dissolution of the siliceous frustules justifies why diatoms do not have an unequivocal older fossil record. Planktic diatoms are well represented during the Cretaceous and underwent a major evolutionary radiation in the middle of the Cretaceous (Harwood and Nikolaev, 1995; van den Hoek et al., 1998; Jousé, 1978; Kooistra et al., 2007), probably along with the ascent of terrestrial grasses that accelerated the silicate weathering cycle (Falkowski et al., 2004b; Katz et al., 2004).

With the transition to the Palaeogene, reorganisation occurred in the oceanic plankton and nekton communities. Some microplankton groups, in particular those inhabiting surface waters (nannoplankton and many planktic foraminifers), show rapid and extreme extinctions at the end of the Cretaceous. Deeper-water benthic species and organisms tolerant of variable conditions, such as cyst-forming species, survived this event much better (Knoll and Lipps, 1993). Both dinoflagellates and many diatoms produce cysts that may rest on the sea floor and enable survival during ecological change or short-lived catastrophic events. According to Barron (1993), diatom generic extinction (23%) at the Cretaceous/Palaeogene boundary is very low compared to generic extinction rates of coccolithophorids (73%), radiolarians (85%) and planktic foraminifers (92%). The dramatic reduction of the phytoplankton persisted for at least several hundreds of thousands of years after this event.

3.1.4. Cenozoic

Plankton groups rapidly radiated after the end-Cretaceous extinction event during the Palaeocene to reestablish a pelagic factory ecologically comparable to the one that had been broken down, although now composed of new species and families (Knoll and Lipps, 1993). The ecological success of marine phytoplankton derived from the red evolutionary lineage became manifest during the Cenozoic (Falkowski et al., 2004b).

Generally, major characteristics of the microplankton record can be attributed to the significant global plate-tectonic and climatic changes, especially the shifted patterns of productivity and circulation of water masses in the oceans (e.g. Lawver and Gahagan, 2003; Mackensen, 2004; Thomas et al., 2003). Most crucial for plankton evolution were changes in the structure and properties of oceanic water masses in the Palaeocene and at the end of the Eocene. The general cooling of high latitudes from the Miocene onwards provoked that pole-to-equator thermal gradients increased and oceanic frontal systems were strengthened (Barron, 2003; Bickert and Henrich, 2011, this volume).

Bown et al. (2004) have exemplified the striking correlation between phytoplankton diversity and long-term climate trends during the Cenozoic for coccolithophorids and associated calcareous nannoplankton (cf. Bickert and Henrich, 2011, this volume; Thierstein and Young, 2004). The overall pattern of diversity decline since the Eocene suggests that onset and evolution of the icehouse-mode climate resulted in a lowered diversity at temperate and high latitudes, and in global diversity minima during the Oligocene and Pliocene-to-Holocene. Conversely, the marked decline in coccolithophorid diversity went along with an exponential rise in marine diatom diversity, in particular since the early Miocene, reflecting an expansion of habitats favourable to diatom diversification by changes such as the formation of cool high-latitude water masses (Bown et al., 2004; Spencer-Cervato, 1999). Diatoms became the dominant group of phytoplankton. Synchronously, increased provincialism evolved since the Eocene–Oligocene, so that diatom assemblages of high and low latitudes can be more clearly distinguished (Barron, 1993).

The ascent of diatoms to dominance, in turn, has exerted some influence on radiolarian evolution, resulting from competition for dissolved silica in seawater. The number of radiolarian families began to diminish, and the average weight of radiolarian skeletons decreased during the Cenozoic (Casey, 1993; DeWever et al., 2001). Moreover, there is also an Oligocene-to-Pleistocene trend towards more finely silicified diatom frustules (Barron, 1993; Sancetta, 1999). While Late Cretaceous to Eocene diatom assemblages are dominated by relatively heavy silicified genera, these robust forms began to decline since the early Oligocene, and became gradually replaced by more finely silicified genera. Even more finely silicified and elongated forms dominate modern coastal upwelling assemblages in temperate regions. Despite a high primary production, the frustules of these delicate forms dissolve rapidly upon death, and they hardly contribute to deep-sea sedimentation. Barron (1993) assumed that it is an advantage for modern diatoms to be content with relatively small amounts of silica in order to exploit the high level of nutrients initially available during seasonal upwelling and to reproduce rapidly.

In particular, the Eocene–Oligocene transition has long been recognised as the most profound biologic, climatic and oceanographic change during the Cenozoic (e.g. Diester-Haass and Zahn, 1996; Knoll and Lipps, 1993; Mackensen, 2004; Salamy and Zachos, 1999; Zachos et al., 1993). Changes of microplankton across the Eocene/Oligocene boundary are well documented by planktic foraminifers and coccolithophorids (e.g. Boersma and Premoli Silva, 1991; Bown, 2005; Culver, 1993; Hallock et al., 1991).

The Eocene was a period of pronounced speciation, resulting in diverse, morphologically complex foraminiferal faunas characterised by marked latitudinal provinciality, as well as strong vertical segregation of species. The formation of high-latitude, cold-water masses, separated by distinctive oceanic fronts, began in the late Eocene Southern Ocean (Cervato and Burckle, 2003; Lazarus and Caulet, 1994; Nelson and Cooke, 2001). The emergence of a sub-Antarctic water-mass at this time introduced new, cold-water, photic-zone habitats. Eocene cooling gave rise to new environments in which nannoplankton could not flourish, or became competitively excluded, and provincialism expressed as diversity-loss at high latitudes (Aubry, 1992; Bown et al., 2004). The siliceous, zooplanktic radiolarians formed an Antarctic province in the late Eocene, with cool-water, cosmopolitan elements and true Antarctic endemic forms becoming increasingly common for the first time (Lazarus and Caulet, 1994). Most strikingly of all, the siliceous, phytoplanktic diatoms displayed a remarkably rapid diversity increase in the late Eocene (Spencer-Cervato, 1999).

This contrasts with the Oligocene, an epoch characterised by low diversity, low morphological complexity and reduced provinciality of foraminiferal faunas as a result of increased homogeneity of surface water masses and low rates of speciation. Other plankton groups that show decreasing diversity patterns during the Eocene–Oligocene transition are coccolithophorids, dinoflagellates and silicoflagellates.

The coccolithophorid diversity maximum in the Late Palaeocene–Early Eocene corresponds to a climate optimum and maximum dispersal of warm-water taxa (Pospichal and Wise, 1990). The diversity decline through the Eocene closely parallels the climatic deterioration marked by diminishing surface- and deep-water temperatures (Zachos et al., 2001). The Cenozoic coccolithophorid diversity minimum occurred during the Oligocene glacial maximum, which was characterised by the initial development of a semi-permanent ice sheet on Antarctica (Ivany et al., 2006).

During the Neogene, polar regions began to cool markedly, altering oceanographic and atmospheric circulation patterns and culminating in extensive ice sheets, at first in the Antarctic (since about 15 Ma) and then closely succeeded by an initial ice build-up in the Arctic (at about 12.6 Ma) (see Bickert and Henrich, 2011, this volume). Again, plankton and other groups changed in response, despite shifting biogeographic distributions. Diatoms, in particular, radiated significantly in the cold, nutrient-rich surface waters of high latitudes, significantly switching the pelagic factory to opal production, for example, culminating in widespread opal deposition in the Norwegian-Greenland Sea and the North Atlantic (Bohrmann et al., 1990; and reference therein). The modern pelagic diatom flora began to evolve during the Late Oligocene; the specific diatom assemblages characterising the North Pacific, the equatorial region and the Southern Ocean are found since the Late Miocene (Round et al., 1996; Sancetta, 1995, 1999). Diversity of coccolithophores and associated nannoplankton declined until the end of the Early Miocene, increased slightly in the Middle Miocene and finally declined again until the Pleistocene (Bown et al., 2004; Siesser, 1993). By the Neogene, diatoms were three or four times as diverse as the nannoplankton (Bown, 2005). The last major radiolarian reorganisation occurred at the Oligocene–Miocene transition (Casey, 1993).

In general, both phytoplankton and zooplankton became increasingly provincial during Cenozoic (Knoll and Lipps, 1993). Warm-water plankton was increasingly restricted to equatorial latitudes, while new, cool-water plankton developed at high northern and southern latitudes. While climate and oceanography changed significantly in response to the general cooling at high latitudes since the Eocene, a major, even more rapid change went on in the Pleistocene as glaciations waxed and waned (see Bickert and Henrich, 2011, this volume). The pelagic carbonate and opal factories rapidly migrated in response to shifting thermal and oceanographic conditions.

3.2. Start-up and growth of the biological pump

Achaean oceans are considered to have been mostly anoxic, sulphate-poor and iron-rich. The oxidation state of the ocean/atmosphere system increased stepwise from 2400 to 2200 Ma ago (Canfield et al., 2000; Des Marais et al., 1992; Farquhar et al., 2002; Holland, 1994, 2006) and from 800 to 580 Ma ago (Canfield and Teske, 1996; Derry et al., 1992), basically driven by photosynthetic biota (Fig. 4.32). This two-staged, progressive ventilation implies a stratified ocean with unique sulphidic chemistry for much of the Proterozoic (Canfield, 1998). Based on the iron chemistry of shales, Shen et al. (2003, 2008) argued that euxinic deep waters persisted beneath the oxidised surface waters in Proterozoic oceans. Strong palaeodepth gradients of sulphur isotopes indicate that the anoxic bottom waters were sulphidic, whereas the oxic surface waters contained only modest amounts of sulphate. Subsequent ventilation of the deep ocean after the Cryogenian global glaciations (snowball Earth), ending 635 Ma ago as inferred from shifts in Fe speciation, gave way for Ediacaran macro-biota to colonise the deep-sea floors. During the Phanerozoic, the oceans were well oxygenated or at least suboxic, and rich in sulphate as today (Holland, 2006). The oceanic surface waters have been oxygenated throughout, but anoxic conditions returned for long time spans of the Palaeozoic,

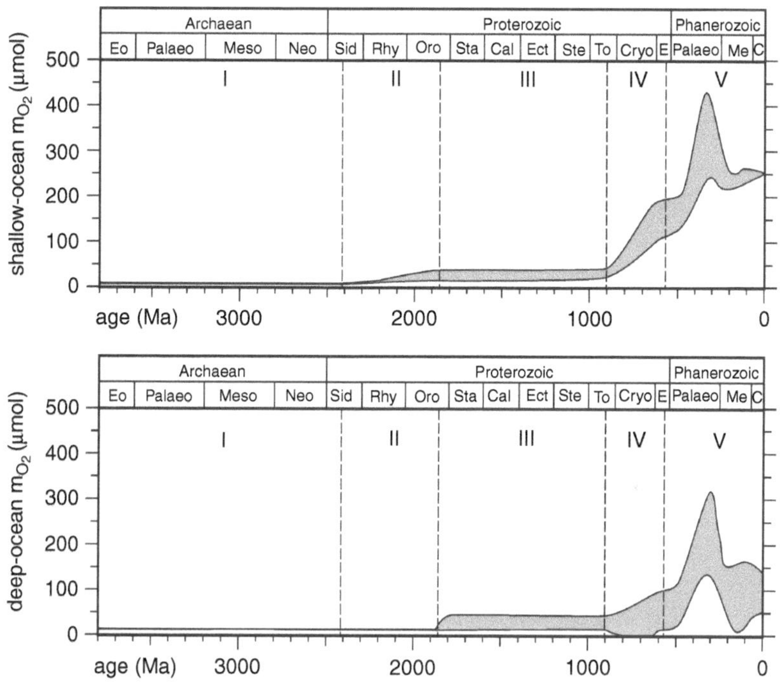

Figure 4.32 Tentative evolution of the oxygen concentration in (A) shallow and (B) deep oceans (redrawn and slightly modified from Holland, 2006). Roman numbers refer to the major evolutionary steps outlined by Holland (2006).

characterised by greenhouse climates limiting the supply of downwelling O_2 relative to demand from sinking organic matter (Berry and Wilde, 1978; Saltzman, 2005). Widespread deposition of organic carbon-rich sediments still occurred for short phases during the Mesozoic (OAEs) and required mechanisms that involve some combination of elevated biologic productivity and enhanced organic-carbon preservation under oxygen-depleted water masses in deep oceans (Kump et al., 2005).

In Proterozoic–Palaeozoic oceans, both cyanobacteria (prokaryotes) and green algae (eukaryotes) already mediated substantial primary production in surface waters (Falkowski et al., 2004a,b; Katz et al., 2004). Neoproterozoic representatives of the eukaryotic phytoplankton became adapted to different environmental regimes; the green lineages proliferated in oligotrophic open oceans underlain by anoxic waters, whereas the red lineages differentiated in coastal waters where both fully oxygenated waters and nutrients were more abundant (Anbar and Knoll, 2002).

For the Proterozoic, there is conflicting evidence concerning the efficiency of the biological pump, which has been operated by cyanobacteria-dominated primary producers. Upon death, the prokaryotic plankton left small particles that (presumably) settled very slowly through the water column, giving way to extensive microbial degradation in the upper parts of the water column. There is no evidence of zooplankton faecal pellets or macroscopic marine snow, the major modern vehicles that accelerate vertical transport of organic matter (see Chapter 2). Neither calcareous nor siliceous tests were available for providing the ballast required for rapid particle sinking. The carbon-isotopic composition of lipids in Proterozoic sediments supports such slow particle settling and indicates only limited export of organic matter from surface to deep waters (Logan et al., 1995). Thus, the Proterozoic biological pump would have relied mainly on the availability of mineral ballast from aeolian and riverine supply, if not downward mixing and advection of particulate and dissolved organic material played a significant role.

Despite these arguments, several lines of evidence indicate that the Proterozoic biological pump was already capable of effectively extracting carbon from the surface water and sequestering it at depth and within sediments (see Hotinski et al., 2004). These include comparable organic-carbon contents of Archaean, Proterozoic and Phanerozoic basinal shales (Strauss et al., 1992a) and a substantial fractionation between shallow-water bulk carbonate and organic carbon $\delta^{13}C$ reservoirs of $\sim$25‰ persisting since the Archaean [$\sim$3.5 Ga] (Schidlowski et al., 1983; Strauss et al., 1992b). Moreover, primary precipitates and cements in Palaeoproterozoic and Cambrian carbonate platforms record ancient carbon-isotope gradients of $\sim$0.5‰ between shallow- and deep-water environments (Hotinski et al., 2004; Surge et al., 1997). Those authors argue that this small but significant isotopic gradient is evidence of substantial biological pumping in pelagic environments: transport of elements from surface waters to deep waters via organic uptake, sinking and remineralisation (e.g. Broecker and Peng, 1982; see Chapter 1). These results indicate that both the Proterozoic prokaryotic and the Palaeozoic eukaryotic phytoplankton were already a significant element of the global carbon and nutrient cycles, which affected—if not regulated—the isotopic composition of both surface and deep ocean waters (Hotinski et al., 2004).

Steady-state considerations imply that the fraction of the total carbon input into the marine system buried as organic carbon has thus always been comparable to that being buried today, of which 17% is buried in the deep ocean (Berner, 1982). Hotinski et al. (2004) presume that the proportion of marine organic matter embedded in deep-sea sediments during the Precambrian was even larger, as there was likely no important supply of terrestrial organic carbon to marine environments.

The major shift in the source of marine primary productivity, from predominately cyanobacteria in the Proterozoic to predominately eukaryotic algae in the Phanerozoic was intimately linked to the rise of the eumetazoans, in particular to the introduction of herbivorous mesozooplankton (Butterfield, 2007; Summons et al., 1999). The expansion of larger phytoplankton and the improved availability of transport vehicles such as fast-sinking faecal pellets, in turn, had profound effects on biogeochemical cycling (Fig. 4.33). Logan et al. (1995) traced the consequences of the increased export flux of organic matter and the spatial extension of microbial remineralisation to the deep sea. Extensive recycling of organic matter in surface water of Proterozoic oceans meant that most

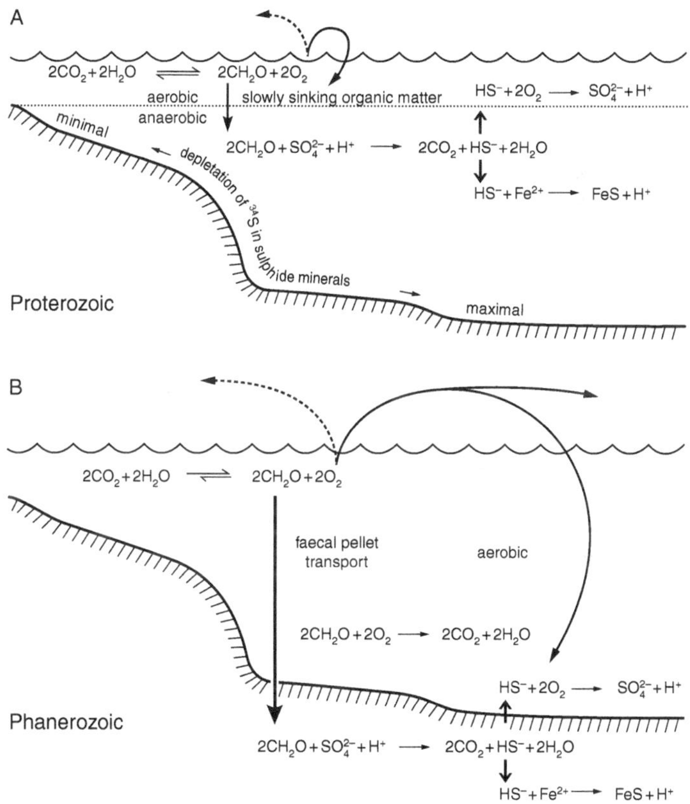

Figure 4.33 Contrasting biogeochemical cycles in Proterozoic and Phanerozoic oceans according to Logan et al. (1995; redrawn and slightly modified). (A) Proterozoic. Because of the slow settling of organic matter, demands by heterotrophic organisms for O_2 and other electron acceptors in surface waters were high. Irrespective of the extent of O_2 leaking to the atmosphere, production of sulphide by planktic sulphate-reducing bacteria was likely important. Together with the extensive remineralisation of organic matter itself, the sulphide will have inhibited transport of O_2 to deeper waters. (B) Phanerozoic. The rapid removal of organic matter from surface waters reduces the demand for O_2 and stimulates its transport throughout the water column. The increased export flux of organic carbon shifts microbial remineralisation, as well as the associated production of sulphide, to the deep sea and into the sediments.

photosynthetically produced oxygen was consumed where it was generated. Due to the ongoing escape of oxygen towards the atmosphere, supplies of oxidisable organic material would exceed those of oxygen (Logan et al., 1995). As a consequence, these authors argue that sulphate-reducing bacteria presumably mediated a significant portion of the remineralisation of organic matter. The production of sulphide and the consumption of oxygen near the ocean surface would have formed a "biological redox buffer" that inhibited transport of oxygen to the deep sea (negative feedback mechanism).

Based on the geochemical characteristics of hydrocarbons (preservation of algal-lipid skeletons), Logan et al. (1995) found that export of organic matter significantly increased with the transition to the Phanerozoic, which narrowed the consumption of oxygen in surface waters and finally destroyed the "biological redox buffer". Rising concentrations of oxygen at the sea surface, in turn, stimulated its transport to the deep sea. The final stages of organic-matter remineralisation by microbial heterotrophs and the associated production of sulphide obviously shifted from the upper to the lower water column, relocating the biological redox buffer to the sediments. Widespread ventilation of the deep sea and colonisation of the sea floor by heterotrophs were the most fundamental knock-on effects of this reorganisation (Butterfield, 2007; Logan et al., 1995).

The transition to pre-modern marine productivity of the Phanerozoic occurred by a stepwise introduction of additional trophic levels, particularly during the Ediacaran, the Early Cambrian and the Early Ordovician (see Butterfield, 1997, 2007). Standing biomass would have increased dramatically with this rapid expansion of pelagic food webs (Kerr and Dickie, 2001; Payne and Finnegan, 2006). Since large amounts of suspended marine biomass are susceptible to collapse, despite dynamic stability, Butterfield (2007) reasons that the biogeochemical perturbations associated with Phanerozoic mass extinctions derive largely from the reversion of the marine biomass spectrum to former (pre-Ordovician) conditions, including a temporary return of primary productivity to smaller (less exportable), possibly cyanobacteria-dominated, phytoplankton.

Biomass, metabolic rates and physical activity (such as predation) of the marine biosphere generally increased through geological time (e.g. Allmon and Ross, 2001; Bambach, 1993; Martin, 1995, 2001; Vermeij, 1987, 1995), which has been ascribed to increasing food availability; it is tightly related to the evolution of primary producers (e.g. Bambach, 1999; Martin, 1996, 2002). Both quantity and quality of food available for zooplankton have increased since the Neoproterozoic because of the evolving nutrient content of eukaryotic phytoplankton (Martin et al., 2008). Thus, phytoplankton produced organic matter with decreasing C:P and N:P ratios during geological time (Falkowski et al., 2004b; Quigg et al., 2003).

The evolution of phytoplankton stoichiometry and the change of ecological hegemony from green to red eukaryotic phytoplankton during the

Phanerozoic appears related to tectonic cycles of supercontinent rifting and reassembly (Wilson cycles) with associated changes in continental flooding and ocean redox chemistry (Falkowski et al., 2004b; Katz et al., 2004, 2005). The redox state, in particular, has controlled the availability of trace elements that play vital roles in mediating critical biochemical reactions in all phytoplankton (Quigg et al., 2003). Expanded oxygen-minimum zones and long-lived anoxia in deep water occurred through much of the time from the Ediacaran to the Triassic (e.g. Arthur and Sageman, 1994; Isozaki, 1997; Shen et al., 2003; Wignall and Twitchett, 2002) and enhanced the release of Fe (and Mn, P, NH_4^+) from bottom sediments. Such episodic anoxic conditions are thought to have promoted the dominance of green phytoplankton lineages (chlorophyll-b bearing), which use, for example, iron as a coenzyme in their photosynthetic pathways (Katz et al., 2004). As deep-ocean oxygenation became increasingly permanent through the Mesozoic, the Mo (and Cd, Cu, Zn, NO_3^-) availability increased. This allowed the red phytoplankton lineages (chlorophyll-c bearing), which preferentially incorporate molybdenum in their photosynthetic apparatuses, to expand (Anbar and Knoll, 2002; Falkowski et al., 2004b). Thus, dinoflagellates and coccolithophorids drove the biological pump from the Jurassic to the Palaeogene, followed by the diatoms during the Neogene (Martin, 1995, 1996). The short-lived (<0.5 Ma) OAEs of the Jurassic and Cretaceous had, however, only minor impacts on the evolution and ecologic significance of eukaryotic phytoplankton (Bown et al., 2004).

The modern phytoplankton more efficiently exported organic matter (e.g. Bambach, 1993). Substantial amounts of organic carbon were stored on the passive continental margins of the broadening Atlantic and on flooded continental interiors. The circum-Atlantic sediments have not yet been recycled through subduction during the current Wilson cycle, thereby providing long-term storage of large amounts of isotopically light organic carbon (Katz et al., 2004). This plankton-mediated increase in organic-carbon burial may account for as much as half of the long-term increase in $\delta^{13}C$ values recorded in organic carbon ($\delta^{13}C_{org}$) (Hayes et al., 1999) and marine carbonates ($\delta^{13}C_{carb}$) (Katz et al., 2005) from the Jurassic to the Miocene. In addition, the increased storage of organic carbon contributed to a gradual depletion in CO_2 from the ocean/atmosphere system and a simultaneous increase in the oxidation state (Katz et al., 2005).

3.3. Pelagic opal oozes and chert deposits: Growth of the silicate pump

Based on a thorough review of chert petrologies, Maliva et al. (2005) concluded that non-biogenic silica deposition was the characteristic mode of the Archaean and Proterozoic opal formation. Dissolved silica must have been removed from seawater by chemical processes, albeit perhaps through local microbial mediation (Siever, 1992). Palaeoproterozoic cherts associated with iron formations appear to have formed largely by direct silica-gel precipitation at or just below the sea floor (Klein, 2005). Some cherts are composed of silica types similar to those in Phanerozoic siliceous hydrothermal deposits (e.g. Konhauser et al., 2001), indicating that basinal, and perhaps global, oceanic silica concentrations were higher during the Palaeoproterozoic era than later (see Chapter 3.5). In Mesoproterozoic–Neoproterozoic strata, opal formation was largely restricted to peritidal environments. These early-diagenetic cherts typically occur as nodules or discontinuous beds within carbonate deposits, showing analogous depositional textures. They formed primarily by carbonate replacement with subsidiary direct silica precipitation (Màliva et al., 1989, 2005; and references therein).

With the radiation of the radiolarians during the Cambrian to Ordovician, the oldest known unequivocal pelagic sediments were deposited in the form of siliceous oozes (Maliva et al., 1989). These radiolarites were formed in offshore shelf and basinal environments, as radiolarians achieved high population densities predominantly in relatively low-energy subtidal environments in the pelagic water column (Casey, 1993; De Wever et al., 2001). Furongian–Tremadocian successions of ribbon-banded cherts and siliceous shales from Kazakhstan are so far the oldest-known radiolarian-ooze accumulations. They are associated with remnants of oceanic crust and are presumably deposited in deep-sea settings of equatorial palaeolatitudes (Tolmacheva et al., 2001).

With the beginning of biologically controlled precipitation of opal by radiolarians and benthic siliceous sponges, the global silica cycle fundamentally changed. Sedimented skeletons became a principal sink for oceanic silica (Maliva et al., 1989; Siever, 1992). The silica concentration of marine waters thus decreased, largely preventing silica supersaturation and early-diagenetic precipitation from marginal marine waters, as typically occurred during the late Proterozoic. Hence, increasing volumes of chert were deposited in open-shelf and deep-sea settings during the Early and Middle Ordovician (Kidder and Mumma, 2003).

From the Ordovician to the Jurassic, skeleton-forming plankton was essentially represented by radiolarians (Knoll and Lipps, 1993; Maliva et al., 1989). The major depositional environment of this radiolarian ooze was the deep sea, leading to the widespread formation of bedded cherts along continental margins and within open oceanic settings (De Wever et al., 1994; Hein and Parrish, 1987; Hesse, 1989; Jenkyns and Winterer, 1982; Jones and Murchey, 1986; Kidder and Erwin, 2001). The absence of calcareous fossils in many radiolarian cherts and their shale intercalations suggest that radiolarian ooze accumulated below the calcite compensation depth (CCD). Boss and Wilkinson (1991) argued that the CCD has been deepened, indeed, below mid-ocean ridge crests (>2.5 km water depth) not before the Late Carboniferous, and that significant pelagic carbonate accumulation on oceanic crust occurred only after the Late Jurassic, apparently due to deeply submerged mid-ocean ridges during the Permian–Jurassic time span (Fig. 4.35; see next chapter for details). Some Palaeozoic radiolarian cherts have, however, been documented from shallow-water environments, ranging from water depths of a few hundreds of metres to intertidal (e.g. Molina et al., 1999). The pelagic character of these deposits is a consequence of the distance from the continent but not necessarily of depositional depth.

Baumgartner (1987) considered radiolarian chert as the normal pelagic sediment during the Ordovician–Jurassic time span in far-offshore open oceanic settings, where dilution by terrigenous material and resuspension of shallow-water carbonates can be excluded. Consequently, radiolarites may directly overlie oceanic basalts in ophiolite terranes of pre-Late Jurassic age (see Boss and Wilkinson, 1991). This is due to the absence of (substantial) calcareous plankton production, which is one major reason for a much shallower CCD of the pre-Jurassic oceans in comparison to modern oceans (Bosellini and Winterer, 1975).

Increased deposition and preservation rates of radiolarian ooze are mostly considered as related to conditions of increased oceanic fertility. In analogy to modern oceans, the intensity of cold-water upwelling is generally seen as the primary factor influencing the bioproductivity and radiolarite deposition (e.g. Baldauf and Barron, 1990; De Wever and Baudin, 1996; De Wever et al., 1994; Dell'agnese and Clark, 1994; Hein and Parrish, 1987; Murchey and Jones, 1992; Seibold and Berger, 1993; Weissert, 1979). As an example, the equatorial convergence zones were a likely setting for many Tethyan and circum-Pacific Mesozoic radiolarites (Baumgartner, 1987). Abrupt changes in tropical upwelling and periodically enhanced radiolarian production, in response to Milankovitch climatic fluctuations, are suggested for ribbon-type radiolarite deposits in the central equatorial Pacific (Ogg et al., 1992). Consequently, De Wever et al. (1994) compared the radiolarite basins of the Tethyan Jurassic with the modern Owen and Somalia Basins, which are both regions of active monsoonal upwelling and blanked by pelagic opal ooze (see also Weissert, 2011, this volume).

Actualistic models of localised upwelling may be of limited applicability for environmental interpretations, however, since the modern biogeochemical cycle is representative only for the overall silica-depleted post-Eocene oceanic ecosystems (Racki and Cordey, 2000). This basic objection is emphasised by examples of widespread and continuous biosiliceous accumulation within the palaeo-Pacific (cf. Cordey and Schiarizza, 1993; Isozaki et al., 1990; Jones and Murchey, 1986). Racki and Cordey (2000) argued that the marine silica budget seems to have been less tightly linked to phosphorus and nitrogen cycles during the Palaeozoic–Mesozoic than in modern settings (see Siever, 1992). Moreover, a global productivity peak caused by an increase in the circulation rate cannot result, in itself, in increased opal burial over several million years: due to the short residence time of dissolved silica and other major nutrients ($<10^5$ a; e.g. Tréguer et al., 1995), the resources would be rapidly depleted. Long-lasting fertile conditions and widespread siliceous plankton blooms are causally tied to large-scale changes in global nutrient cycling, including (1) an elevated supply of nutrients to the oceans, (2) an effective redistribution of nutrients within the global ocean or (3) a combination of both (e.g. Farrell et al., 1995; Föllmi et al., 1993, 1994; Grant and Dickens, 2002). The addition of biolimiting elements to the ocean may be driven by a highly intensified flux of nutrients from riverine and aeolian sources, usually related to higher weathering rates in uplifted continental areas or to climate change. In addition, large-scale volcano-hydrothermal activity during major plate-boundary reconfigurations has been taken into consideration in some interpretations of widespread biosiliceous deep-sea records (see Racki and Cordey, 2000).

Productivity and sediment preservation as they occur nowadays (related to upwelling sites and CCD depth) depend on the meridian thermohaline oceanic circulation that is quite recent and closely related to the emergence and closure of the Panama isthmus (see Bickert and Henrich, 2011, this volume). The impact of a latitudinal thermohaline circulation is difficult to evaluate, and so is consequently the relation between bioproductivity and pelagic sediment deposition. Because the configuration of the continents was different from today, the thermohaline circulation was quite different, particularly in the pre-Jurassic oceans (e.g. Winguth and Maier-Reimer, 2005). It is even likely that there was no thermohaline circulation at all during the warmest periods of the Earth's history. Circulation model experiments suggest for the mid-Cretaceous, which is well known for its OAEs, an unsteady inactive thermohaline circulation, and the relationships between the thermohaline circulation and biogeochemical cycles were obviously different from those in the present-day ocean (Misumi and Yamanaka, 2008).

Bedded cherts are often composed of layers of radiolarites regularly alternating with intervals of siliceous shales, for example, the so-called ribbon radiolarites (Fig. 4.34). Such deposits usually result from hemipelagic settling of radiolarian tests through the water column. The regular sedimentary alternation arose from relative variations in the ratio of skeletal opal accumulation and the supply of detrital particles. Apart from such autochthonous deposits, several other origins of radiolarian-rich cherts have been described, including redeposition of siliceous tests by turbidity flows and contourite currents (e.g. De Wever et al., 2001; Hesse, 1990; Jenkyns and Winterer, 1982; Jones and Murchey, 1986; Knauth, 1994; Sarnthein and Faugères, 1993; Weissert, 1981a,b). In addition, early-diagenetic

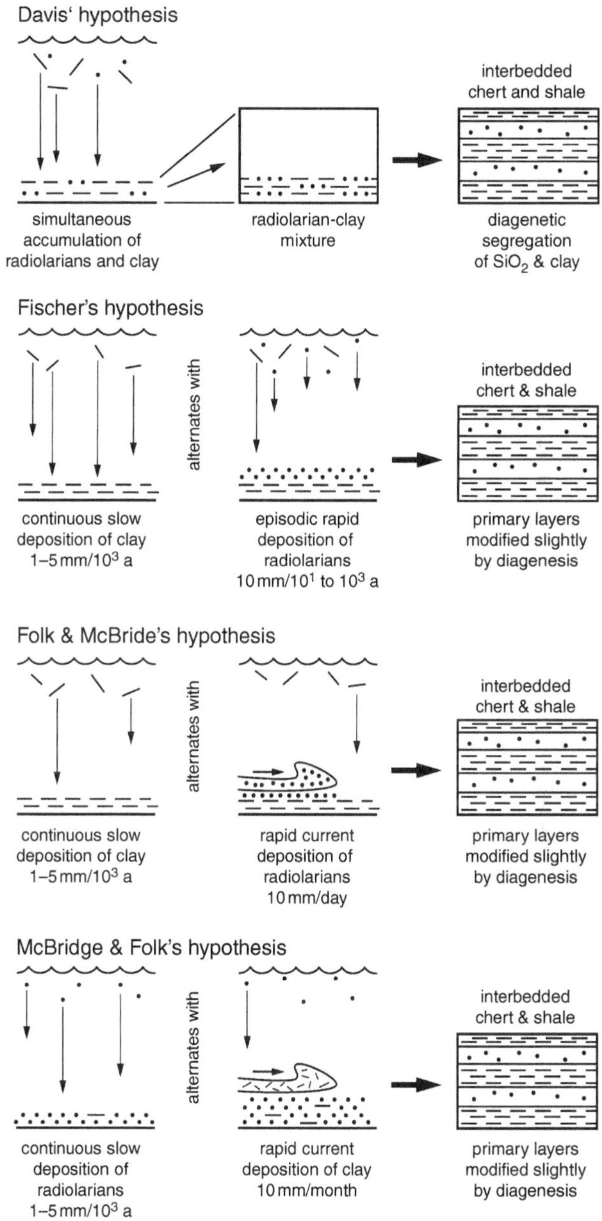

Figure 4.34 Hypotheses for the various origins of chert/shale rhythms (redrawn and slightly modified from McBridge and Folk, 1979).

segregation of silica from initially sub-homogeneous siliceous ooze has usually enhanced the lamination (Hesse, 1990; Hesse and Schacht, 2011, this volume). A common constituent of many radiolarian cherts are siliceous sponge spicules (Maliva et al., 1989) although pure spiculitic cherts, resulting mainly form benthic opal production, are more typical of platform, shelf and slope settings. Most nodular cherts, which are a common constituent of many shallow-water to upper continental-slope carbonates from the Ordovician to the Cretaceous, originated as sponge spicules.

With the rise of diatoms, beginning in the Late Cretaceous (Campanian), the composition of pelagic siliceous oozes started to change. In the Eocene, diatom populations had already displaced radiolarians as the overwhelmingly dominant opal-producing plankton, initiating a further fundamental change in the global marine-silica cycle (Maliva et al., 1989; Siever, 1991). The proliferation of diatoms stepwise reduced the concentration of dissolved silica in ocean water to its present low level and necessarily resulted in proportionally less silica removal by radiolarians and sponges (Fig. 4.36). Radiolarians remained quantitatively important only in a few regions, such as the peri-polar and equatorial belts, and the western sides of continents (Casey, 1993; Lisitzin, 1985). Shallow-water to upper-bathyal spiculitic oozes declined rapidly in abundance, and shelf environments ceased to be major sites of chert formation at all (Maldonado et al., 1999). Silicoflagellates were only locally rock-forming in the Cenozoic (Distanov and Glezer, 1973).

The early Eocene was a peculiar time span, with extraordinarily enhanced opal accumulation (e.g. Gibson and Towe, 1971; Moore, 2008a,b; Muttoni and Kent, 2007; Riech and Von Rad, 1979). During this time span, chert, opaline claystone, porcellanite and related rock types accumulated not only in a broad belt across the northern Atlantic (between ~20°N and ~50°N), but also in the equatorial Pacific and in circum-Mediterranean regions. The main chert layers can be traced as an acoustic horizon (Horizon Ac) virtually over the entire latitudinal extent of the western and eastern basins of the North Atlantic, although it corresponds in some areas to an erosional unconformity associated with volcanogenic debris (Gartner, 1970; Norris et al., 2001; Tucholke and Mountain, 1979). Sparse and poorly preserved diatom and radiolarian faunas are a puzzling characteristic of these cherts (Sanfilippo and Nigrini, 1998; Weaver and Wise, 1974).

The widespread occurrence of Eocene deep-sea cherts suggests that even opal deposits, including those with a recognisable amount of diatoms, may not simply mirror a high biosiliceous productivity in surface water of latitudinally focused upwelling zones (Muttoni and Kent, 2007). McGowran (1989) suggests an initial build-up of large amounts of dissolved silica from intensified weathering on continents—perhaps enhanced by silica from volcanism in the North Atlantic igneous province—in a warm ocean with sluggish circulation of the early Eocene climatic optimum. During the subsequent mid-Eocene global cooling, accelerated oceanic circulation and consequent upwelling made this silica reservoir available for increased opal production by diatoms and radiolarians (silica burp model), triggering widespread oceanic deposition of biosiliceous oozes.

The main interval of opal formation coincides, however, more closely with the early Eocene climatic optimum than with the subsequent global cooling in the mid-Eocene. Muttoni and Kent (2007) suggest that basin-to-basin fractionation by the thermohaline circulation could have further raised the high concentration of seawater with dissolved silica, especially in the North Atlantic, due to a reversed operating conveyor belt in the early Eocene.

Since the Oligocene, the occurrence of diatomites and radiolarites more clearly precedes the formation and the distribution of siliceous oozes (Maliva et al., 1989). Bedded deep-sea cherts and porcellanites underlie regions of former high primary productivity, especially zones of equatorial upwelling (Moore et al., 2004; Parés and Moore, 2005; Seibold and Berger, 1993). In modern oceans, opal sedimentation is virtually entirely biogenic, although hydrothermal alteration of basalts may have locally contributed to high rates of dissolved silica input into the ocean water.

Chert becomes less abundant with decreasing age in Pliocene–Pleistocene sediments, because insufficient time has passed for its formation. The opal-A of diatom and radiolarian tests is highly instable and is successively transformed into opal-CT and then into quartz (see Hesse and Schacht, 2011, this volume). These diagenetic processes are mainly controlled by time and host-rock lithology, together with temperature and pore-water chemistry (e.g. Hesse, 1990; Kastner et al., 1977). It explains why opal-CT (porcellanites, tripoli) dominates in the Cenozoic, whereas quartz (chert) mainly occurs in Mesozoic and older rocks.

3.4. Pelagic calcareous oozes and limestone deposits: Growth of the carbonate pump

Pelagic limestones do not occur in the Precambrian since evolutionary innovation gave organisms the ability to precipitate carbonate skeletons not before the time of the Ediacaran/Cambrian boundary (Wood et al., 2002). Prior to this development, there has been no significant biogenically driven production of calcite and aragonite. Carbonate deposition was restricted to heterogeneous nucleation and crystal growth on organic and inorganic surfaces in warm shallow-water environments (Grotzinger and James, 2000).

Biogenically-controlled carbonate precipitation became significant initially in early Palaeozoic shallow-water environments with the activities of benthic calcifiers, whereas the existence of a pelagic carbonate factory is enigmatic. The origin of lime mud constituting most of the Palaeozoic deep-sea carbonates remains unclear: organisms capable of producing pelagic micrite are virtually unknown from the Palaeozoic. It might well be that nannoplankton existed in the oceans, but it did not produce mineralised parts that were archived in deep-sea sediments (Erba, 2006). Tucker (1990a) suggested that the lime mud of Palaeozoic deep-sea carbonates could well be derived from the breakdown of macroskeletons, rather than from micro- and nannoplankton. Devonian pelagic coquinas consisting of dacryoconarids (styliolinids and/or nowakiids), which may be formed by exceptional blooming or catastrophic mortality of these planktic organisms (Walliser and Reitner, 1999), are a potential source of such calcareous deep-sea muds. Clear indications of bottom-current activity, however, which may account for skeletal breakdown, are documented from specific periods only (Hüneke, 2006; Hüneke and Stow, 2008).

Hay (2004) argued that Palaeozoic deep-sea carbonates mainly preserve fine-grained carbonate detritus winnowed from shallow-water carbonate platforms. If true, they would represent periplatform carbonates rather than pelagic deposits (see Henrich and Hüneke, 2011, this volume). An allochthonous origin seems to be irrelevant, however, for areas with a widespread record of mud-supported cephalopod limestones or nodular limestones and limited shallow-water carbonate production such as the eastern Moroccan Anti-Atlas (e.g. Wendt et al., 1984).

Besides the largely unknown carbonate-producing plankton, accumulation of pelagic calcareous oozes during the Palaeozoic was probably severely restricted for reasons of preservation. Boss and Wilkinson (1991) concluded that deposition of extensive biogenic shallow-water carbonates likely helped keeping up the CCD between ~1- and 2 km depth, thereby preventing significant carbonate-ooze accumulation in deep-sea settings (Fig. 4.35). The CCD appears to have deepened to below mid-ocean ridge crests (>2.5 km water depth) not before the Late Carboniferous, when calcareous oozes presumably began to accumulate on the oceanic crust (Mackenzie and Morse, 1992; Martin, 1995). An alternative model of Ridgwell (2005), mainly relying on data of atmospheric CO_2 proxy records, sea-level change and Ca^{2+} concentration, suggests a constantly greater depth for the Palaeozoic CCD (or more exactly the less deep calcite-saturation horizon) but a similar significant deepening of the CCD during Late Palaeozoic (see below for details; Fig. 4.38).

Low preservation rates are also presumed for pelagic carbonates of outer-shelf and slope settings. Like in modern oceans, the CCD probably shoaled towards the margins of continental shelves, where rain rates of particulate organic carbon are often quite high and dissolution is enhanced by the decay of organic matter (Emerson and Archer, 1990; see also the detailed discussion in Chapter 2). Based on this scenario, Martin (1995) assumed that large quantities of Cambro-Devonian records of calcareous plankton have

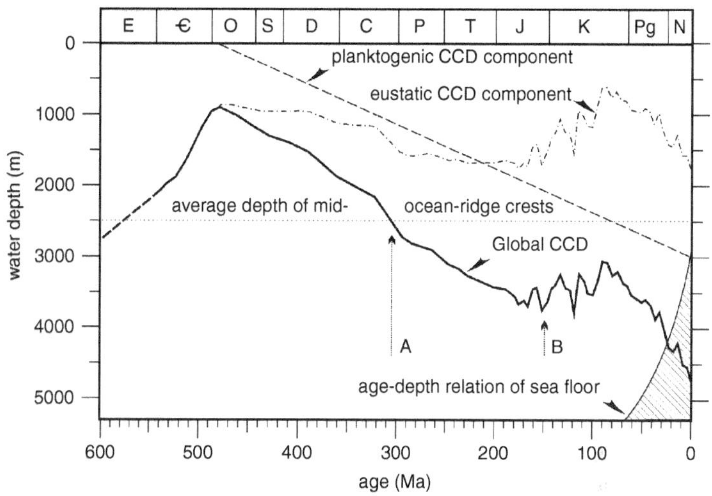

Figure 4.35 Model for the temporal variation of the carbonate compensation depth (CCD) during the Phanerozoic (redrawn and modified from Boss and Wilkinson, 1991), constrained by a minimum CCD for carbonate accumulation at 2.5 km water depth determined from sea-floor age/depth data (Sclater et al., 1980), close correspondence between sea level (Haq et al., 1987) and average CCD (Van Andel, 1975) for the Cretaceous–Neogene, and general absence of pelagic carbonate in Cambrian to Jurassic ophiolitic sedimentary complexes (see Boss and Wilkinson, 1991). The proposed global CCD (heavy solid line) was calculated from the hypothetical depth related to eustatic (dot-and-dashed line) and planktogenic (dashed line) controls on deep-sea carbonate accumulation. The eustatic component is based on the presumption of a larger oceanic carbonate production during continental emergence, as calculated from the sea levels of Haq et al. (1987) and Vail et al. (1977), and constrained to a depth of <2.5 km; the planktogenic component was calculated assuming a linear increase in carbonate productivity since ~500 Ma. The model implicates sub-equally important, but temporarily diverging, influences of the eustatic and evolutionary control on open-ocean carbonate production and preservation. The global CCD fell below the level of mid-ocean ridge crests at ~300 Ma (A), giving a maximum age for pelagic carbonate accumulation on oceanic crust. The slower production of oceanic crust, as inferred from a low sea level (first-order variation) between 300 and 150 Ma, resulted in still limited areas of the sea floor above the CCD and a potentially greater rarity of pelagic carbonate accumulation during this time span as well. Significant amounts of pelagic carbonates became preserved only after ~150 Ma (B).

probably been dissolved, leaving behind non-calcareous taxa such as radiolarians, acritarchs and graptolites. The composition of Phanerozoic ophiolite suites, which document that pelagic carbonate accumulation was comparatively rare in Palaeozoic deep-sea sediments, seems to support this view (Boss and Wilkinson, 1991).

The Mesozoic deep-sea record reveals the formation of carbonate ooze of unequivocal pelagic origin since the Jurassic (e.g. Bown et al., 2004; Hart et al., 2003; Knoll, 2003; Martin, 1995; Roth, 1987, 1989). During the Late

Jurassic and earliest Cretaceous, coccolithophores and related nannoplankton, together with calpionellids, were the most important carbonate producers of deep-sea oozes, while planktic foraminifers substantially contributed to pelagic rain not until the Early Cretaceous (Bornemann et al., 2003; Erba, 1989; Reháková and Michalík, 1997; Roth, 1989; Weissert and Erba, 2004). A significant but stepwise increase in pelagic sediment production and accumulation rates was the result for both oceanic and epicontinental settings.

The first pelagic carbonates consisting of coccolithophores and associated nannofossils have been described from the mid-late Jurassic, for example, the oldest oceanic limestones of Bathonian–Callovian age drilled at the North Atlantic continental margin off Florida and the stone bands of the British Kimmeridge Clay Formation (e.g. Gallois and Medd, 1979; Sheridan and Gradstein, 1983).

Enhanced sedimentation of pelagic carbonates has been reported from numerous low-latitude sections and drill cores of Kimmeridgian to Tithonian Age, particularly from the Tethys and the central Atlantic Ocean. Most striking is the shift from radiolarians-dominated siliceous sediments (e.g. the Rosso ad Aptici and the Ruhpolding Radiolarite) to nannofossils-dominated carbonates (e.g. the Maiolica Formation, the Biancone facies and the Malm-Aptychen Limestone) that have accumulated in the western and northern Tethys (Baumgartner, 1987; Lackschewitz et al., 1991; Schlager and Schöllnberger, 1974; Weissert, 1979, 1981a). This change corresponds with an evolutionary turnover of coccolithophorids at the Jurassic/Cretaceous boundary and is associated with substantial faunal changes in other plankton groups, including a global spread of holoplanktic globigerinid foraminifers, calpionellids and high rates of diversification in radiolarians (Bown et al., 2004; Erba, 1989; Roth, 1986, 1989). Roth (1986) argued that nannoplankton originated in shelf environments and epicontinental seas migrated into oceanic habitats only later (see Baumgartner, 1987).

Deep-sea records of the central Atlantic Ocean reveal two intervals of increase in the nannofossil carbonate record (Bornemann et al., 2003). A first significant maximum occurred in the middle and late Tithonian, probably caused by mass occurrences of more heavily calcified taxa (mainly nannolith assemblages). The second carbonate maximum in the late Berriasian was induced by a rise in absolute abundances of nannofossils, with highly diverse coccolith assemblages. This second maximum was amplified by an overall increase in the sedimentation rate, which reached for the first time values similar to coccolith accumulation rates in modern oceans.

During the Cenomanian and Turonian, nannoplankton increased again in abundance and diversity, and huge amounts of pelagic carbonate were deposited in the open ocean, on the deeply flooded shelves and in epicontinental seas (Erba, 2006). These Late Cretaceous chalks, which consist largely of coccoliths, are far more widespread than pelagic carbonates of any other period. Within the warm epicontinental seas, pelagic chalk accumulated at depths of ~100–500 m, locally above the storm-wave base, enabled by the exceptional high sea level (Haq et al., 1988; Kennedy, 1987; Kennedy and Garrisson, 1975; Lasseur et al., 2009; Muller et al., 2008; Quine and Bosence, 1991; Robaszynski et al., 1982; Surlyk, 1997).

The ecological proliferation of planktic calcifiers during the Late Jurassic and Cretaceous shifted the main global setting of limestone depositional from shallow to deep marine (Berger and Winterer, 1974; Boss and Wilkinson, 1991; Riding, 1993). This shift is considered as a mid-Mesozoic revolution (Erba, 2006; Ridgwell, 2005) since the powerful start-up of the pelagic carbonate factories fundamentally changed the mode of the marine carbonate cycle with wide-ranging consequences for both the oceanic geochemistry and the climate.

The step-wise expansion of pelagic carbonate sedimentation into oceanic deep-sea settings was accompanied by ongoing deepening of the global CCD throughout the Phanerozoic, along with increased plate spreading since the Late Jurassic (~150 Ma), that is, lifting of mid-ocean ridge crests (Fig. 4.35) (Boss and Wilkinson, 1991). In particular, the widespread increase of calcareous plankton production during the Late Jurassic and Cretaceous gave rise to a rapid and large deepening of the CCD of probably more than 1000 m (Winterer and Bosellini, 1981). On the other hand, the increase in pelagic carbonate production and accumulation during the Tithonian was closely related to major changes in palaeoenvironmental conditions (e.g. Weissert and Channell, 1989). With the opening of the central American gateway, to give an alternative view, a circum-global equatorial current system developed (see Ross and Scotese 1988). As a result, upwelling in the Mediterranean Tethys and the Atlantic ceased, while it could prevail or even intensify in other areas of the Tethys (De Wever, 1989).

3.5. Controls of secular changes in pelagic sedimentation and feedbacks

There is growing certainty that long-term patterns of pelagic sedimentation mirror the evolution of seawater chemistry and the interaction of global factors that modulate its changes during earth history (e.g. Erba, 2006; Eriksson et al., 2007; Knoll, 2003; Marchitto et al., 2005; Martin, 1995, 1996; Ridgwell, 2005; Riding, 1993; Sandberg, 1983; Stanley, 2006; Tucker, 1990b). Among these factors, both biotic causes such as evolutionary innovations or competition and abiotic causes, such as pCO_2, temperature, sea-floor spreading, sea level, climate and productivity of oceanic water masses, play a critical role. Ongoing discussions about the mutual interrelations among these factors makes it clear that we are just beginning to understand its impacts on the velocity and mode of evolution, the productivity of sediment-producing plankton and on the extent of sediment preservation. This is particularly true for control and feedback mechanisms during the Proterozoic and Palaeozoic.

The metabolic processes by which pelagic organisms precipitate skeletal minerals, as well as the forms and functions of the skeletons they create have been fashioned by natural selection through geological time. Major evolutionary changes trigger skeletal biomineralisation and thus are key events in the history of pelagic factories (cf. Knoll, 2003). As a long-term consequence, the production mode of ancient pelagic factories is intimately related to the composition of seawater, regardless of the fact that organisms may exert considerable control over their biomineralisation. Siever (1991, 1992), Maliva et al. (1989, 2005), Martin (1995, 1996), Anbar and Knoll (2002), Tyrrell and Zeebe (2004), Demicco et al. (2005), Holland (2004), Ridgwell (2005) and Stanley (2006) gave useful reviews of the geological history of ocean water and the available data, revealing that marine chemistry has changed significantly even during the younger earth history. Moreover, regulation modes of the ocean chemistry have been different from today (Zeebe and Westbroek, 2003), which has important implications for the interpretation of the geological record based on proxies.

Changes in the concentration of silicic acid in seawater through the Phanerozoic have affected the ability of organisms to secrete siliceous skeletons and vice versa (Fig. 4.36). The evolving silica budget may be thought of as a combination of two trends: (1) a fluctuating, but generally diminishing flux of exogenous silica (Gibbs et al., 1999) and (2) increasing biogenic removal of dissolved silica (Maliva et al., 1989, 2005). Major causes for readjustments in the global silica balance were related to increased weathering rates, in particular in uplifted areas, or were favoured by evolutionary factors such as grass vegetation on land; presumably, the readjustments were also related to peaks in magmatism and hydrothermal influx (e.g. Bown, 2005; François et al., 1993; Hardie, 1996; Siever, 1992; Steinberg, 1981).

Because silica is a minor component of seawater, biogenic extraction of opal—in particular by organisms of the pelagic opal factory—has exerted an important negative feedback on the ocean chemistry (Maliva et al., 1989, 2005; Stanley, 2006). Mainly, the proliferation of radiolarians early in the Palaeozoic significantly reduced the concentration of silica in the oceans, resulting in a transition from abiological silica deposition, characteristic of the Archaean and Proterozoic oceans, to the predominantly biologically controlled silica deposition of the Phanerozoic (Fig. 4.36). The late Mesozoic radiation of diatoms further reduced the concentration of silica, preventing siliceous sponges from forming reefs, and shifting the main depositional setting of opal from the shallow shelf to the deep shelf and oceanic environments (Maldonado et al., 1999).

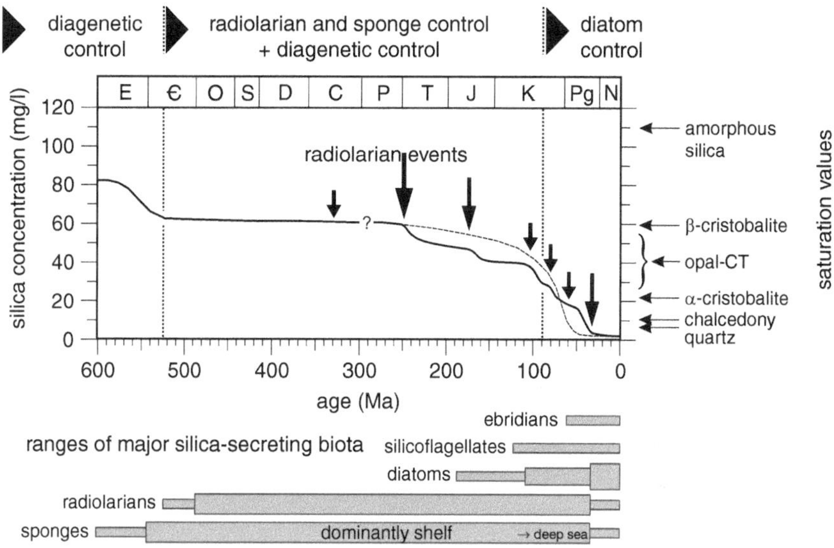

Figure 4.36 Evolution of the marine silica concentration from the latest Proterozoic to the present day (redrawn and modified from Racki and Cordey, 2000; Siever, 1991). Supposed silica concentrations of the ocean are shown against the background of saturation values for relevant silica minerals, the evolutionary ranges of major silica-secreting biota and the temporal trend in the regulation of oceanic SiO_2 precipitation. Instead of a gradual secular trend (broken line) as supposed by Siever (1991), a step-wise decrease in oceanic silica levels (solid line) is proposed by Racki and Cordey (2000), driven mainly by major radiolarian events (arrows); compiled from Berger (1991), Schubert et al. (1997) and Vishnevskaya (1997).

The significantly increasing utilisation of amorphous silica by radiolarians (and siliceous sponges) since the Ordovician enhanced the Si-output of the oceans. During the Palaeozoic, silica levels were lowered, possibly only a little bit higher than the most disordered opal-CT saturation levels (Siever, 1991). Within the silica-saturated greenhouse habitats (and because the overall robust radiolarian tests dissolve much more slowly), the pelagic opal production was linked less directly to the primary productivity than in modern oceans (e.g. Racki and Cordey 2000). Although controlled by severe nitrate and iron limitation (Berger, 1991; Berger et al., 1989), radiolarian productivity and silica accumulation from the Cambrian to the Jurassic would have been more independent of nutrient recycling and upwelling zones than today (Siever, 1991).

The final relevant biotic event was the mid-Cretaceous expansion of diatoms, which drastically lowered the concentration of dissolved silica to its present-day low level (Maliva et al., 1989). In modern oceans, the vast bulk of deep and surface waters are at least 1 order of magnitude undersaturated with respect to quartz, and close to two orders of magnitude undersaturated with respect to biogenic opal (Siever, 1991). Diatoms extract today all but a small fraction of the silica supplied to the ocean by rivers and mid-ocean ridge hydrothermal activity. Radiolarians are quantitatively important only in the Pacific equatorial belt, whereas siliceous sponges and silicoflagellates are considered unimportant on a global scale. The apparently relative independence of recent opal-mineralising plankton from the dissolved silica pool, as found by De Wever et al. (1994), probably reveals their progressive physiological specialisation towards reduced biomineralisation during the evolutionary history (Racki and Cordey, 2000). Consequently, the present biogeochemical cycle is representative only for the overall silica-depleted post-Eocene oceanic ecosystems, which broadly arose from the expansion of diatoms—the group extremely efficient in silica precipitation and closely linking the silica budget with phosphorus and nitrogen cycles.

Among the major constituents of seawater, Magnesium (Mg^{2+}) and Calcium (Ca^{2+}) were subjected to strong fluctuations during the Phanerozoic that influenced the pelagic carbonate production and preservation. The ratio of Mg^{2+} to Ca^{2+} and the absolute concentration of Ca^{2+} in seawater have not only controlled primarily the precipitation of non-skeletal carbonate (ooids, early marine cements) but also played important roles in biocalcification (Demicco et al., 2005; Hardie, 1996; Stanley and Hardie, 1998). As a response, the Phanerozoic oceans fluctuated between "aragonite" and "calcite sea" states, as discovered by Sandberg (1983, 1985), corresponding to times of high and low Mg^{2+}/Ca^{2+} ratios, respectively (Fig. 4.37). The observed secular fluctuation is attributed to the net effect of changes in the Ca-rich brine flux along mid-ocean ridges (Demicco et al., 2005; Hardie, 1996), which is not necessarily highlighted by mantle-plume activity, as evidenced from large igneous provinces (e.g. Courtillot and Renne, 2003; Ernst and Buchan, 2002). Anyhow, fluctuating Mg^{2+}/Ca^{2+} ratios have been confirmed by studies of fluid inclusions and bromine percentages in marine evaporites (Horita et al., 2002; Lowenstein et al., 2001, 2003; Siemann, 2003). Moreover, petrographic and chemical analyses indicate that the mineralogy of lime muds has also adhered to this pattern (Lasemi and Sandberg, 2000).

Organisms that are highly productive biomineralisers have responded to changes in the cationic composition of seawater in ways that mirror patterns for non-skeletal carbonates (Stanley and Hardie, 1998). With regard to pelagic sedimentation, the low-Mg calcite-producing coccolithophorids, which rose to ecological dominance in the Late Jurassic to Oligocene calcite ocean (Calcite II), are commonly cited as a striking example (e.g. Erba, 2006; Stanley, 2006; Young et al., 1994). They formed massive coccolith chalks in the warm shallow seas of the Late Cretaceous, after the Mg^{2+}/Ca^{2+} ratio of seawater had reached a very low value (high Ca^{2+} concentration). As the Mg^{2+}/Ca^{2+} ratio of seawater increased during the Cenozoic (decreasing Ca^{2+} concentration), individual coccoliths, on average, became

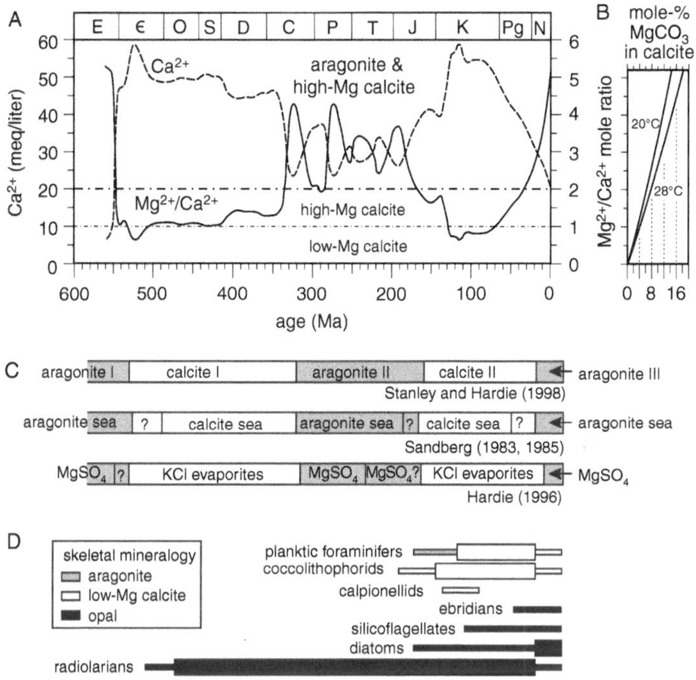

Figure 4.37 Evolution of calcium concentrations (Ca^{2+}) and magnesium/calcium ratios (Mg^{2+}/Ca^{2+}) in seawater influencing the secular variation of mineralogies for non-skeletal marine carbonates and evaporites, and mineralogy of important hypercalcifying marine taxa (redrawn and modified from Stanley, 2006). (A) Nucleation fields with respect to the Mg^{2+}/Ca^{2+} molar ratio of seawater for low-Mg calcite, high-Mg calcite and aragonite. (B) Incorporation of Mg^{2+} in non-skeletal calcite as a function of the ambient Mg^{2+}/Ca^{2+} ratio at two given temperatures (Füchtbauer and Hardie, 1976, 1980). (C) Temporal oscillations in the geological record between calcite and aragonite non-skeletal carbonates and the temporal oscillations between KCl and $MgSO_4$ marine evaporites which correlate with them (Hardie, 1996; Sandberg, 1983, 1985; Stanley and Hardie, 1998). Both fluctuations are predicted from estimates of global rates of mid-ocean-ridge accretion and the effects of these rates on seawater chemistry (Hardie, 1996). (D) Temporal distribution of primary-carbonate mineralogy of important planktic organisms producing pelagic sediment (Stanley and Hardie, 1998).

less massive and encrusted cells less thickly. By the Pliocene, when the modern aragonite ocean (Aragonite III) was established, prominent taxa such as the genus *Discoaster* secreted only narrow-rayed coccoliths that covered not more than 25% of the cell surface. This example also illustrates that changes in ocean chemistry mainly correlate with the initial evolutionary appearance and the proliferation of skeletal-forming clades but not necessarily with their last appearance, as discussed in more detail by Knoll (2003).

Based on experimental results, Stanley (2006) has evaluated the factors that govern both the mineralogy of skeletons and the rate at which skeletons form. Obviously, the high Mg^{2+}/Ca^{2+} ratio (low Ca^{2+}concentration) in modern seawater apparently limits population growth for the large majority of modern coccolithophore species, which mainly belong to old clades with Cretaceous roots (Calcite II). Those species fail to respond to nitrate, phosphate or iron fertilisation and are confined to oligotrophic waters, which serve them as a place of refuge. The species *E. huxleyi*, evolved fairly recently during Marine Isotope Stage 8 is, in contrast, a (atypical) well-adapted coccolithophore species, being saturated with calcium in modern low-calcium oceans (Aragonite III) and blooming prolifically.

As well as the concentration of Ca^{2+}, changes in concentration of CO_3^{2-} in the ocean water influenced the pelagic carbonate factories during the Earth's history and, in addition, the preservation potential of pelagic carbonate oozes. The carbonate ion (CO_3^{2-}) is one of the carbon compounds collectively termed dissolved inorganic carbon (DIC), comprising the total sum of $CO_{2(aq)}$ $(+H_2CO_3)$ + HCO_3^- + CO_3^{2-}, that all are related to one another and to calcium carbonate ($CaCO_3$) by aqueous carbonate equilibrium reactions (see Schneider et al., 2006; Zeebe and Wolf-Gladrow, 2001).

Zeebe and Westbroek (2003) have conceptualised the evolution in the global carbonate cycling over geological time as the following three distinct ocean modes (Table 4.2).

(1) *The Strangelove ocean mode*: The geochemistry-ruled Precambrian mode of $CaCO_3$ cycling drove a carbonate ocean in which biogenic precipitation of $CaCO_3$ was essentially absent. It was characterised by high-supersaturation with respect to the solid phase of $CaCO_3$ and generally, inorganic (or partly biologically mediated) formation of carbonates, comprising sea-floor encrustations, crystal fans and thick cement beds (Grotzinger and James, 2000). Progressively, younger Precambrian rocks show a decreasing abundance of such inorganically precipitated carbonates, which most likely reflects a progressive decline in the degree of ocean oversaturation (Grotzinger and Kasting, 1993).

(2) *The Neritan ocean mode*: With the beginning of biomineralisation in the Cambrian, biologically controlled carbonate precipitation in neritic environments became significant. The dominant mode of Ca^{2+} and CO_3^{2-} removal from seawater was biogenic-carbonate deposition mainly accomplished by benthic calcifying organisms in shallow water. The saturation state of the Neritan ocean has been highly susceptible to changes in the population or ecological success of shallow-water calcifiers. Thus, secular oscillation in sea level controlling the extent of shallow-water carbonate seas, in oceanic cation concentrations (Fig. 4.37), in weathering rates and in global temperatures—all

Table 4.2 Principal ocean modes in global carbonate cycling over geological time as conceptualised by Zeebe and Westbroek (2003)

Mode	Time	Main control	Characteristics
Cretan ocean	Jurassic to Holocene	Biologically controlled carbonate cycling: planktic carbonate production predominates	Significant stabilisation of the marine $CaCO_3$ saturation state
Mid-Mesozoic revolution: proliferation of calcareous plankton			
Neritan ocean	Cambrian to Triassic (short returns in early Palaeogene?)	Biologically controlled carbonate cycling: benthic shallow-water carbonate production predominates	Extreme variability of the marine $CaCO_3$ saturation state in response to perturbations
Cambrian revolution: phylogenetic achievement of biocalcification			
Strangelove ocean	Archaean to Proterozoic (short returns in Permian to Triassic?)	Geochemically controlled carbonate cycling: inorganic (at most partly biologically mediated) formation of carbonates in shallow water	High supersaturation with respect to the solid phase of $CaCO_3$

drove perturbations of carbonate deposition and ocean chemistry that could only be weakly buffered (Ridgwell and Zeebe, 2005).

(3) *The Cretan ocean mode*: The Mesozoic proliferation of calcareous plankton finally led to the establishment of a substantive deep-sea carbonate sink that introduced a new stabilising mechanism to the Earth system, namely, the carbonate compensation (see Broecker and Peng, 1987). This deep-sea carbonate buffer (negative feedback mechanism) critically regulates the ocean chemistry and the global carbon cycle as a whole until today (see Chapter 2.2.2). Large and rapid shifts back into Neritic- or even Strangelove ocean modes may have occurred for rather short time spans in the aftermaths of catastrophic events such as the Cretaceous/Tertiary bolide impact or mantle-plume activity (see discussion in Caldeira and Rampino, 1993; Courtillot and Renne, 2003; Kump, 1991).

Based on the successive carbonate-cycling modes outlined above (Zeebe and Westbroek, 2003), Ridgwell (2005) modelled the impact of variations in first-order boundary conditions on the saturation state and the carbonate chemistry of the global ocean. The model behaviour of the Phanerozoic ocean confirms a profound change, particularly in the regulation of the ocean saturation state ($\Omega_{calcite}$) that arose from the establishment of a robust planktic carbonate factory in the Mesozoic (Fig. 4.38). The model of carbon cycling between the atmosphere, the ocean and the sediments suggests that surface saturation ($\Omega_{calcite}$) of the ocean during the earlier Phanerozoic underwent substantial oscillations between 5.8 and 10.7 (Fig. 4.38A). Directly related, the depth of the calcite saturation horizon (CSH) shows a pronounced excursion during the Permian and Triassic, when it dropped below 6000 m, with the deep sea bathed in oversaturated water (Fig. 4.38B). With the onset of an effective accumulation of pelagic carbonate in the deep sea since the Jurassic, the oceanic surface saturation ($\Omega_{calcite}$) obtained stability, with deviations not more than ± 0.4 from the modern value of 4.8. The broad picture of the predicted CSH depth resembles the CCD reconstruction by Boss and Wilkinson (1991). A comparison should be made with care, however, because the CCD need not *a priori* have a fixed relationship with the depth of the CSH (Sigman et al., 1998). Clearly, more feasible are the CCD reconstructions in modern ocean basins since the Jurassic, which are based on the distribution in time of calcareous and non-calcareous deep-sea sediments while the age and depth of the oceanic basement are known (Fig. 4.39) (e.g. Berger and Rad, 1972; Lyle, 2003; Peterson and Backman, 1990; Sclater et al., 1977; Van Andel, 1975).

Ocean-surface pH, which responds primarily to changes in atmospheric pCO_2 and the oceanic concentration of Ca^{2+} according to the model of Ridgwell (2005), underwent substantial change during the Phanerozoic with a secular oscillation of up to ± 0.4 superimposed on an underlying increase of ~ 0.7 pH units from ~ 7.5 to a pre-industrial value of ~ 8.18 (Fig. 4.38C). For the Cenozoic, an alternative pH trend based on $\delta^{11}B$ proxy reconstructions by Pearson and Palmer (2000) suggests a more pronounced alkalification of 0.8 pH units and extensive secular oscillations. For a more detailed review of the changing ocean chemistry during younger earth history, the reader is referred to Tyrrell and Zeebe (2004) and Holland (2004).

There are also good arguments that fluctuations in the partial pressure of CO_2 (pCO_2) in ocean water have exerted a direct control on marine plankton evolution and productivity, in particular on calcareous phytoplankton, since carbon dioxide is involved in biochemical reactions of both photosynthesis and biocalcification (Erba, 2006). Experimental results have shown, indeed, that seawater CO_2 availability and pCO_2 are crucial for calcification in living coccolithophores (Riebesell et al., 2000; Rost and Riebesell, 2004). Increased CO_2 levels result in decreased calcification and a lower ratio of calcification to particulate organic carbon production (calcite/POC). Coccolith secretion probably represents a strategy to reduce the

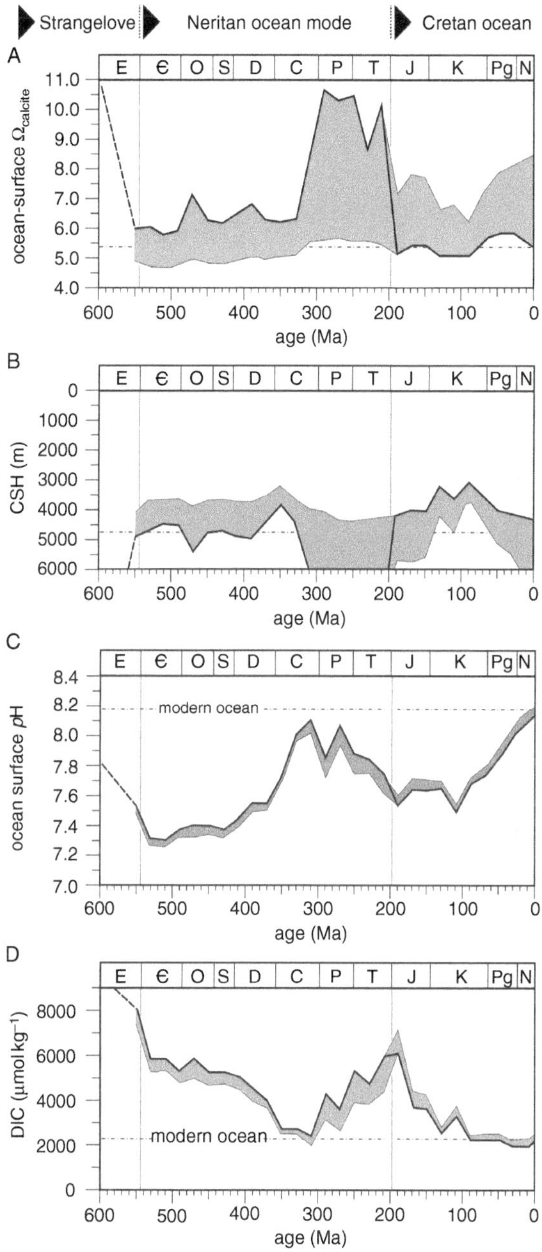

Figure 4.38 Evolution of ocean chemistry during the Phanerozoic based on marine carbonate-cycling modes (redrawn and slightly modified from Zeebe and Westbroek, 2003). (A) Mean (area-weighted) surface saturation state with respect to calcite ($\Omega_{calcite}$). Note that the precipitation of $CaCO_3$ from seawater is thermodynamically

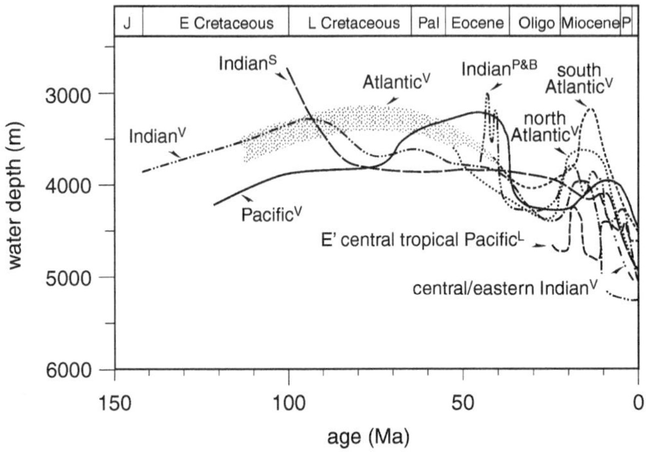

Figure 4.39 History of the calcite compensation depth (CCD) in modern ocean basins since the Late Jurassic; summarised from (V) van Andel (1975), (S) Sclater et al. (1977), (P&B) Peterson and Backman (1990) and (L) Lyle (2003).

energy cost of photosynthesis by producing—directly within the cell—the calcification waste product CO_2 (Paasche, 1962; Young, 1994). Consequently, under low CO_2 levels of the ocean/atmosphere system, coccolith mineralisation might represent an evolutionary strategy in order to maintain photosynthesis, whereas excess CO_2 would make calcification more dispensable to coccolithophore life, because surface waters are already (over-) saturated with carbon dioxide. In the latter case, smaller and less calcified coccoliths become dominant.

Based on the experimental results and available palaeontological data, Erba (2006) interpreted the Mesozoic coccolithophorid record as closely

favourable when $\Omega > 1$. During the Phanerozoic, the surface of the ocean was everywhere more than saturated with respect to the solid carbonate phase, and $CaCO_3$ precipitation took place under direct metabolic control of living organisms. (B) Depth in the ocean of the calcite saturation horizon (CSH) at which $\Omega = 1.0$ occurs. The greater the depth in the ocean, the more likely the ambient environment is under saturated (i.e. $\Omega < 1$) and $CaCO_3$ will tend to dissolve. The greater depth of the calcite lysocline (not shown) is geochemically defined at $\Omega = 0.8$, a value which marks a distinct increase in the dissolution rate (Broecker, 2003; Milliman et al., 1999). At the calcite compensation depth (CCD), which is still deeper, dissolution becomes sufficiently rapid for the dissolution flux back to the ocean to exactly balance the rain flux of calcite to the sediments. (C) Mean ocean-surface pH. (D) Mean (volume-weighted) dissolved inorganic-carbon (DIC) concentration. In all the panels, the grey band represents the difference between Cretan- and Neritan ocean behaviour (the uncertainty due to CO_2 and/or weathering is not shown). Vertical dotted lines indicate the phases of Cambrian and mid-Mesozoic revolutions.

interrelated with fluctuating pCO_2: low CO_2 resulted in increased rates of nannoplankton speciation and calcification, whereas a drastic rise in CO_2 caused reduction in nannofossil palaeofluxes. The Late Triassic appearance of coccolithophorids, the major diversification in the Early Jurassic and the turnover and nannofossil speciation in the late Tithonian—all correlate with low CO_2 levels in the ocean/atmosphere system and are interpreted as a compensation of calcareous nannoplankton in order to sustain growth (Erba, 2004; Weissert and Erba, 2004). In contrast, extinction events and major reductions in biogenic fluxes seem to be correlatable with episodes of increased volcanic activity, which are regarded as a source of excess atmospheric carbon dioxide. The drastic decline in nannofossil carbonate production during the Toarcian OAE, for example, was contemporaneous with the formation of the large Karoo–Farrar igneous province (e.g. Erba, 2004). Furthermore, downturns in biogenic carbonate production in the Valanginian and the Aptian, which occurred in both pelagic and neritic environments, are attributed to excess atmospheric CO_2 during episodes of increased volcanic activity (Weissert and Erba, 2004).

In addition to the factors highlighted above, a number of boundary conditions have influenced the evolution of pelagic sediment production and preservation during the earth's history, as among many others climate, sea level, nutrient budget and redox state of the ocean. For these factors, the interrelationship with pelagic sedimentation is, however, not as evident and straightforward as for ocean-water chemistry and pCO_2 (e.g. Bown et al., 2004; Erba, 2006; Falkowski et al., 2004b; Katz et al., 2004; Martin, 1995, 1996; Martin et al., 2008). In particular, the various feedback mechanisms involved are hardy to verify for the Phanerozoic oceans, which experienced evolving seawater compositions clearly different from today.

Closely related to the partial pressure of CO_2 (pCO_2) is the climate, which seems to be a less direct environmental factor in controlling the mode and intensity of pelagic sediment production. The systematic relationships are difficult to quantify, since climatic forcing of the biogenically mediated $CaCO_3$ precipitation provokes short- and long-term responses. The carbonate accumulation represents the principal long-term mechanism of CO_2 removal from the ocean. The $CaCO_3$ precipitation reaction, however, reduces the total sum of dissolved carbon species (DIC). The remaining carbon, in turn, is repartitioned in favour of $CO_{2(aq)}$, that raises ambient pCO_2 and pH in the surface ocean water (Ridgwell and Zeebe, 2005). Thus, the short-term effect is that enhanced pelagic (and benthic) carbonate production drives an increase in ocean pCO_2 and, with it, an increase in the atmospheric CO_2 concentration. Conversely, enhanced dissolution of deep-sea carbonate oozes drives a pCO_2 decrease. This is crucial for climatic interpretations of pelagic archives and is discussed in more detail for the Cretaceous greenhouse world by Weissert (2011, this volume) and for the Neogene icehouse world by Bickert and Henrich (2011, this volume).

Tappan (1982, 1986) and Martin (1995, 1996) have recognised an increase in nutrient levels of ocean surface waters throughout the Phanerozoic and suggested that this secular trend explains many aspects of the history of marine plankton. The surface nutrient levels may have been substantially lower from the Cambrian to the Devonian than those typical of modern oligotrophic waters, thereby restraining the development of large populations of calcareous nannoplankton and the accumulation of substantial calcareous oozes. Support comes from the secular trend in the carbon-isotope composition of marine carbonates ($\delta^{13}C_{carb}$), which is a proxy for the global organic carbon production and burial (see Goddéris and Joachimski, 2004). Long time spans of stability in $\delta^{13}C_{carb}$ from the Cambrian to the Devonian (warm climate) are interpreted to reflect negative feedbacks on productivity in a nitrogen-limited ocean in which anoxic-euxinic conditions led to increased denitrification (Saltzman, 2005). Nitrogen limitation may have persisted over long time spans, particularly if associated with a limited availability of the trace metals (particularly Fe and Mo) that are essential for N-fixing bacteria (Anbar and Knoll, 2002; Falkowski, 1997). In contrast, cool time spans that ventilated the oceans (Late Cambrian, Late Ordovician to Silurian, Late Devonian to Early Carboniferous) switched the ultimate limiting nutrient to P and allowed for large positive $\delta^{13}C_{carb}$ excursions, which signal episodic organic carbon burial that could be sustained by positive feedbacks between productivity and anoxia (Saltzman, 2005). During the Carboniferous–Permian, the nutrient levels and plankton productivity had increased to intermediate concentrations via glaciation, sea-level fall, and enhanced deep-ocean overturn and continental weathering (Martin, 1996).

Following the end-Permian extinctions, plankton re-expanded in response to both eustatic sea-level rise and increased nutrient levels (Martin, 1995, 1996). As rain rates of organic matter and $CaCO_3$ increased from the Triassic to the Cretaceous, bioturbation in deep-sea sediments became intensified (see Uchman and Wetzel, 2011, this volume), thereby recycling nutrients back to the surface waters, reinforcing productivity and calcareous-ooze formation. Enhanced rates of ocean circulation, continental erosion and bioturbation further raised the nutrient levels and productivity in the Neogene, as indicated by the expansion of diatoms. Martin et al. (2008) argued that rising nutrient levels and marine productivity may have fuelled a secular increase in marine biomass and diversity through the Phanerozoic.

As an immanent element of climate, upper-ocean turbulence and its influence on the temporal availability of nutrients have governed resource competition between modern phytoplankton groups on geological scales (Falkowski et al., 2004b; Katz et al., 2004). The different strategies in resource acquisition of coccolithophorids, dinoflagellates and diatoms are crucial for understanding the competition among these pelagic sediment producers during the Mesozoic and Cenozoic (Tozzi et al., 2004). Because planktic diatoms have evolved a nutrient-storage vacuole, which allows hoarding of inorganic nutrients frequently supplied to surface water in pulses (Raven, 1997), they can overcome light-dependent nutrient uptake in mixing systems and deprive competing taxa of these essential resources. Diatoms thereby dominate the pelagic sediment production when times of water-column stability are punctuated by high turbulence, such as storm events (see Fig. 4.8), whereas coccolithophores and dinoflagellates dominate when the water column is more permanently stratified (e.g. Brown and Yoder, 1994; Gibson, 2000; Hulburt, 1970; Iglesias-Rodríguez et al., 2002; Smetacek, 1999; Thomas et al., 1997).

Margalef (1994) recognised these fundamental differences in physiology and proposed that competition among the three major pelagic sediment producers could be related to upper-ocean turbulence and the supply of nutrients (Fig. 4.40). Moreover, the differences fit the classical r and K paradigm (MacArthur and Connel, 1966), where r-strategists (e.g. diatoms) dominate in high-mixing environments and K-strategists (e.g. coccolithophores) dominate during oligotrophic conditions. Tozzi et al. (2004) numerically explored how nutrient uptake by phytoplankton can interact with turbulence to influence the relative abundance of diatoms and

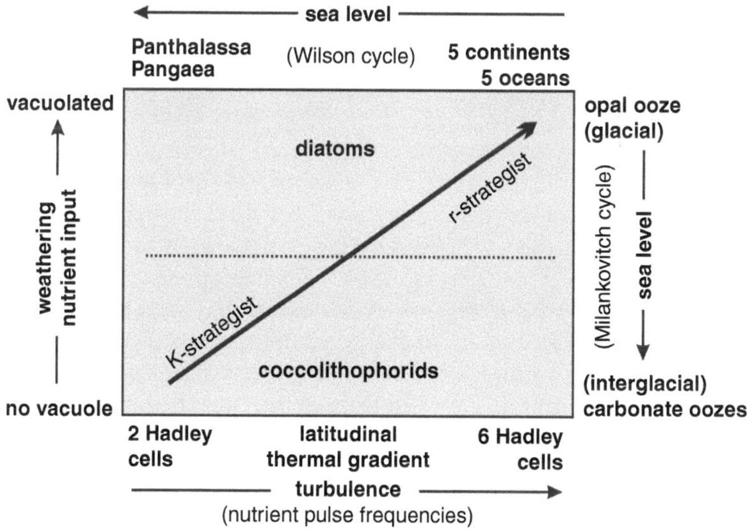

Figure 4.40 Extension of the "Margalef's mandala" (Margalef, 1997) to geological time scales (redrawn and slightly modified from Katz et al., 2004; Tozzi et al., 2004), showing that increased turbulence and subsequently improved availability of nutrients lead to a shift from K-strategist to r-strategist species. Turbulence and nutrient concentration in the ocean reflect the latitudinal thermal gradient that is a function of radiative forcing, continental configuration and sea level.

coccolithophorids. These model simulations qualitatively mimic field observations in modern oceans, as discussed in Chapter 1.

The so-called Margalef mandala has been extended over geological time (Falkowski et al., 2004b; Katz et al., 2004; Prauss, 2000; Tozzi et al., 2004) to explain the shift from carbonate- to opal-dominated pelagic factories during the Cenozoic (Fig. 4.40). The occurrence of widespread stable, subtropical and tropical oligotrophic water masses favoured the diversification of the broadly K-strategist calcareous nannoplankton and dinoflagellates for much of the Mesozoic and into the early Palaeogene (Bown, 2005). The Mesozoic was relatively warm, resulting in a thoroughly mixed atmosphere with nearly uniform temperatures over the surface of the Earth (Huber et al., 1995). The atmospheric heat transport towards the poles decreased the latitudinal thermal gradients; global winds and ocean circulation were both sluggish. Following the onset of permanent Antarctic glaciation during the Eocene/Oligocene transition, the atmospheric circulation changed dramatically (Crowley and North, 1991). A more intense thermohaline circulation, higher wind speeds and a decreased upper-ocean stability were the consequences (Barron et al., 1995; Chandler et al., 1992), favouring the r-strategist diatoms at the expense of the other phytoplankton groups. Diatoms became the most competitive of the phytoplanktic sediment producers, flourishing in all environments but particularly in those with high and pulsed nutrient inputs (Tozzi et al., 2004). The substantial increase in global siliceous-plankton export production, in turn, contributed to a prominent deepening of the CCD, of more than 1 km near the Eocene/Oligocene boundary (Fig. 4.39) (Harrison, 2000).

The Pleistocene deep-sea record reveals a periodicity of opal/calcite deposition corresponding to glacial/interglacial alternations, which may also be described with the nutrient uptake model of Tozzi et al. (2004) see (Fig. 4.40). Falkowski (2002) suggested that such alternations in pelagic sediment production were triggered by the upper-ocean turbulence linked to the Milankovitch cycle (Berger, 1988). Glacials appear characterised by higher wind speeds and a stronger thermal contrast between the equator and the pole, favouring diatoms more than coccolithophorids (Gargett, 1991). During interglacials, a more intense stratification, weaker winds and a less outspoken thermal contrast between the equator and the poles would tend to reduce upper-ocean mixing and favour coccolithophorids (Tozzi et al., 2004). In addition, the fluxes of silica and iron into the ocean have been shown to have been much larger during glacial maxima (Froelich et al., 1992; Harrison, 2000) and therefore they co-occurred with phases of higher turbulence. At least silica availability has undoubtedly stimulated the opal factory during the glacials.

The CCD, which remained relatively shallow during much of the middle Miocene-Pliocene at least in the Pacific (Rea and Leinen, 1985), deepened rapidly at the beginning of the Pleistocene, coinciding with the long-term drop in eustatic sea level (Fig. 4.39). Since then, it has fluctuated rapidly, possibly in relation to the onset of the northern hemisphere glacial cycles (see Bickert and Henrich, 2011, this volume).

ACKNOWLEDGEMENTS

We gratefully acknowledge T. Mulder for his helpful comments to improve the manuscript and P.O. Baumgartner for critically reading the section on ancient pelagic sediments. Tom van Loon suggested editorial changes that are highly appreciated. We especially thank D. Lau and H. Sengpiehl for creatively redrafting most of the figures.

REFERENCES

Abelmann, A., 1992. Radiolarian taxa from Southern Ocean traps (Atlantic sector). Polar Biol. 12, 373–385.

Abelmann, A., Gersonde, R., 1991. Biosiliceous particle flux in the Southern Ocean. Mar. Chem. 35, 503–536.

Abelmann, A., Gowing, M.M., 1996. Horizontal and vertical distribution pattern of living radiolarians along a transect from the Southern-Ocean to the South Atlantic subtropical region. Deep Sea Res. I 43, 361–382.

Abelmann, A., Gowing, M.M., 1997. Spatial distribution pattern of living polycystine radiolarian taxa—baseline study for paleoenvironmental reconstructions in the Southern Ocean (Atlantic sector). Mar. Micropaleontol. 30, 3–28.

Abelmann, A., Nimmergut, A., 2005. Radiolarians in the Sea of Okhotsk and their ecological implication for paleoenvironmental reconstructions. Deep Sea Res. II 52, 2302–2331.

Abrantes, F., 2000. 2,00, 000 yr diatom records from Atlantic upwelling sites reveal maximum productivity during LGM and a shift in phytoplankton community structure at 1,85, 000 yr. Earth Planet. Sci. Lett. 176, 7–16.

Afanasieva, M.S., Agarkov, Y.V., Amon, E.O., 2005a. Stages in the evolution of radiolarians (Polycystina) in the Phanerozoic. Paleontol. J. 6, 590–605.

Afanasieva, M.S., Amon, E.O., Agarkov, Y.V., 2005b. Radiolarians in the geological record. Paleontol. J. 39 (Suppl. 3), 135–392.

Alldredge, A.L., 1979. The chemical composition of macroscopic aggregates in two neritic seas. Limnol. Oceanogr. 24, 855–866.

Alldredge, A.L., Gotschalk, C.C., 1988. In situ settling behaviour of marine snow. Limnol. Oceanogr. 33, 339–351.

Alldredge, A.L., Gotschalk, C.C., 1989. Direct observations of the mass flocculation of diatom blooms: characteristics, settling velocities and formation of diatom aggregates. Deep Sea Res. 36, 159–171.

Alldredge, A.L., Silver, M.W., 1988. Characteristics, dynamics and significance of marine snow. Progr. Oceanogr. 20, 41–82.

Alldredge, A.L., Passow, U., Logan, B.E., 1993. The abundance and significance of a class of large, transparent organic particles in the ocean. Deep Sea Res. I 40, 1131–1140.

Allmon, W.D., Ross, R.M., 2001. Nutrients and evolution in the marine realm. In: Allmon, W.D., Bottjer, D.J. (Eds.), Evolutionary Paleoecology: The Ecological Context of Macroevolutionary Change. Columbia University Press, New York, pp. 105–148.

Almogi-Labin, A., 1982. Distribution of euthecosomatous pteropods in plankton tows, trap and surface sediment of Bermuda and Barbados (Northern Atlantic). In: Hemleben, C.

(Ed.), Cytologic and Ecologic Aspects of Morphogenesis, N. Jb. Geol. Paläont., 164, 92–95.

Amante, C., Eakins, B.W., 2009. ETOPO1 1 arc-minute global relief model: procedures, data sources and analysis. NOAA Technical Memorandum NESDIS NGDC-24, pp. 1–19.

Anbar, A.D., Knoll, A.H., 2002. Proterozoic ocean chemistry and evolution: a bioinorganic bridge. Science 297, 1137–1142.

Anderson, O.R., 1996. The physiological ecology of planktic sarcodines with applications to paleoecology: patterns in space and time. J. Eukaryot. Microbiol. 43, 261–274.

Anderson, R.F., Chase, Z., Fleisher, M.Q., Sachs, J., 2002. The Southern Ocean's biological pump during the Last Glacial Maximum. Deep Sea Res. II 49, 1909–1938.

Antia, A.N., Maaßen, J., Herman, P., Voß, M., Scholten, J., Groom, S., et al., 2001. Spatial and temporal variability of particle flux at the N.W. European continental margin. Deep Sea Res. 48, 3083–3106.

Antoine, D., Andre, J.M., Morel, A., 1996. Oceanic primary production 2. Estimation at global-scale from satellite (coastal zone color scanner) chlorophyll. Global Biogeochem. Cycles 10, 57–69.

Archer, D.E., Maier-Reimer, E., 1994. Effect of deep-sea sedimentary calcite preservation on atmospheric CO_2 concentration. Nature 367, 260–264.

Armand, L.K., Crosta, X., Romero, O., Pichon, J.J., 2005. The biogeography of major diatom taxa in Southern Ocean sediments: 1. sea ice related species. Palaeogeogr. Palaeoclimatol. Palaeoecol. 223, 93–126.

Armstrong, R.A., Jahnke, R.A., 2001. Decoupling surface production from remineralisation and benthic deposition: the role of mineral ballasts. US JGOFS News 11, 1–2.

Armstrong, R.A., Lee, C., Hedges, J.I., Honjo, S., Wakeham, S.G., 2002. A new, mechanistic model for organic carbon fluxes in the ocean: a quantitative role for the association of POC with ballast minerals. Deep Sea Res. II 49, 219–236.

Arrigo, K.R., Worthen, D., Schnell, A., Lizotte, M.P., 1998. Primary production in Southern Ocean waters. J. Geophys. Res. 103, 15587–15600.

Arthur, M., Sageman, B., 1994. Marine black shales: depositional mechanisms and environments of ancient deposits. Annu. Rev. Earth Planet. Sci. 22, 499–551.

Asper, V.L., 1987. Measuring the flux and sinking speed of marine snow aggregates. Deep-Sea Res. 34, 1–17.

Asper, V.L., Deuser, W.G., Knauer, G.A., Lohrenz, S.E., 1992. Rapid coupling of sinking particle fluxes between surface and deep ocean waters. Nature 357, 670–672.

Aubry, M.-P., 1992. Late Paleogene calcareous nannoplankton evolution: a tale of climatic deterioration. In: Prothero, D.R., Berggren, W.A. (Eds.), Eocene–Oligocene Climatic and Biotic Evolution. Princeton University Press, Princeton, pp. 272–309.

Bacon, M.P., 1987. Oceanic flux of Chernobyl fallout. Nature 329, 767–768.

Baldauf, J.G., Barron, J.A., 1990. Evolution of biosiliceous sedimentation patterns—Eocene through Quaternary: paleoceanographic response to polar cooling. In: Bleil, U., Thiede, J. (Eds.), Geological History of the Polar Oceans: Arctic Versus Antarctic. Kluwer Academic Publishing, Dordrecht, pp. 575–607.

Bambach, R.K., 1993. Seafood through time: changes in biomass, energetics, and productivity in the marine ecosystem. Paleobiology 19, 372–397.

Bambach, R.K., 1999. Energetics in the global marine fauna: a connection between terrestrial diversification and change in the marine biosphere. GeoBios 32, 131–144.

Banner, F.T., Desai, D., 1988. A review and revision of the Jurassic-Early Cretaceous Globigerinina, with especial reference to the Aptian assemblages of Speeton (North Yorkshire, England). J. Micropaleont. 7, 143–185.

Bareille, G., Labracherie, M., Bertrand, P., Labeyrie, L., Lavaux, G., Dignan, M., 1998. Glacial–interglacial changes in the accumulation rates of major biogenic components in southern Indian Ocean sediments. J. Mar. Syst. 17, 527–539.

Barker, S., Elderfield, H., 2002. Foraminiferal calcification response to glacial–interglacial changes in atmospheric CO_2. Science 297, 833–836.

Barron, J.A., 1993. Diatoms. In: Lipps, J.H. (Ed.), Fossil Prokaryotes and Protists. Blackwell, Cambridge, pp. 155–168.

Barron, J.A., 2003. Planktic marine diatom record of the past 18 m.y.: appearances and extinctions in the Pacific and Southern Oceans. Diatom Res. 18, 203–224.

Barron, E.J., Fawcett, P.J., Peterson, W.H., Pollard, D., Thompson, S.L., 1995. A simulation of mid-Cretaceous climate. Paleoceanography 10, 953–962.

Baumann, K.-H., Böckel, B., Donner, B., Gerhardt, S., Henrich, R., Vink, A., et al., 2003. Contribution of calcareaous plankton groups to the carbonate budget of South Atlantic surface sediments. In: Wefer, G., Mulitza, S., Rathmeyer, E. (Eds.), The South Atlantic in the Quaternary: Reconstruction of Material Budgets and Current Systems. Springer-Verlag, Berlin, pp. 81–99.

Baumgartner, P.O., 1987. Age and genesis of Tethyan Jurassic radiolarites. Eclogae Geol. Helv. 80, 831–879.

Bé, A.W.H., Gilmer, R.W., 1977. A zoographic and taxonomic review of euthecosomatous Pteropoda. In: Ramsay, A.T.S. (Ed.), Oceanic Micropalaeontology vol. 1. Academic Press, London, pp. 733–808.

Bé, A.W.H., Hutson, W.H., 1977. Ecology of planktonic foraminfera and biogeographic patterns of life and fossil assemblages in the Indian Ocean. Micropaleontology 23, 369–414.

Bé, A.W.H., Tolderlund, D.S., 1971. Distribution and ecology of living planktonic foraminifera in surface waters of the Atlantic and Indian Oceans. In: Funnell, B.M., Riedel, W. R. (Eds.), The Micropaleontology of the Oceans. Cambridge University Press, Cambridge, pp. 105–149.

Behrenfeld, M.J., Falkowski, P.G., 1997. Photosynthetic rates derived from satellite-based chlorophyll concentration. Limnol. Oceanogr. 42, 1–20.

Bender, M.L., Heggie, D.T., 1984. Fate of organic carbon reaching the sea floor: a status report. Geochim. Cosmochim. Acta 48, 977–986.

Berelson, W.M., 2001. The flux of particulate organic carbon into the ocean interior: a comparison of four U.S. JGOFS regional studies. Oceanography 14, 59–67.

Berger, W.H., 1970. Biogenic deep-sea sediments: fractionation by deep-sea circulation. Geol. Soc. Am. Bull. 18, 1385–1402.

Berger, W.H., 1973. Deep-sea carbonates: Pleistocene dissolution cycles. J. Foraminiferal Res. 3, 187–195.

Berger, W.H., 1976. Biogenous deep-sea sediments: production, preservation and interpretation. In: Riley, J.P., Chester, R. (Eds.), Chemical Oceanography vol. 5. Academic Press, London, pp. 265–389.

Berger, W.H., 1979. Preservation of foraminifera. In: Lipps, J.H., Berger, W.H., Buzas, M. A., Douglas, R.G., Ross, C.A. (Eds.), Preservation of Foraminifera, Cushman Foundation Spec. Publ., 13, 105–155.

Berger, W.H., 1985. CO_2 increase and climate prediction: clues from deep-sea carbonates. Episodes 8, 163–168.

Berger, A., 1988. Milankovitch theory and climate. Rev. Geophys. 26, 624–657.

Berger, W.H., 1989. Appendix. Global maps of ocean productivity. In: Berger, W.H., Smetacek, V.S., Wefer, G. (Eds.), Productivity of the Ocean: Present and Past. Wiley, Chichester, pp. 429–455.

Berger, W.H., 1991. Produktivität des Ozeans aus geologischer Sicht: Denkmodelle und Beispiele. Z. Dtsch. Geol. Ges. 42, 149–178.

Berger, W.H., Keir, R., 1984. Glacial–Holocene changes in atmospheric CO2 and the deep-sea record. In: Broecker, W.S., Hansen, J., Takahasi, T. (Eds.), Climate Processes and Climate Sensitivity. American Geophysical Union, Washington, D.C., pp. 337–351.

Berger, W.H., Rad, U., 1972. Cretaceous and Cenozoic sediments from the Atlantic Ocean. In: Hayes, D.E., Pimm, A.C. (Eds.), Initial Reports, Deep Sea Drilling Project vol. 14. U.S. Government Printing Office, Washington, pp. 787–954.
Berger, W.H., Wefer, G., 1990. Export production: seasonality and intermittency, and paleoceanographic implications. Palaeogeogr. Palaeoclimatol. Palaeoecol. 89, 245–254.
Berger, W.H., Winterer, E.L., 1974. Plate stratigraphy and the fluctuating carbonate line. Int. Assoc. Sedimentol. Spec. Publ. 1, 11–48.
Berger, W.H., Bonneau, M.-C., Parker, F.L., 1982. Foraminifera on the deep-sea floor: lysocline and dissolution rate. Oceanol. Acta 5, 249–258.
Berger, W., Fischer, K., Lai, C., Wu, G., 1987. Ocean productivity and organic carbon flux. Part I. Overview and maps of primary production and export production. University of California, SIO Technical Rep. Ref. Ser. California, pp. 87–30.
Berger, W.H., Smetacek, V.S., Wefer, G., 1989. Ocean productivity and paleoproductivity—an overview. In: Berger, W.H., Smetacek, V.S., Wefer, G. (Eds.), Productivity of the Ocean: Present and Past. Wiley, New York, pp. 1–34.
Berggren, W.A., Kent, D.V., Swisher, C.C., Aubry, M.P., 1995. A revised Cenozoic geochronology and chronostratigraphy. In: Berggren, W.A., Kent, D.V., Hardenbol, J. (Eds.), Geochronology, Time Scales and Global Stratigraphic Correlations: A Unified Temporal Framework for an Historical Geology, Soc. Econ. Paleontol. Mineral. Spec. Publ., vol. 54. 129–212.
Berner, R.A., 1982. Burial of organic carbon and pyrite sulfur in the modern ocean: its geochemical and environmental significance. Am. J. Sci. 282, 451–473.
Berner, R.A., Honjo, S., 1981. Pelagic sedimentation of aragonite: its geochemical significance. Science 211, 940–942.
Berry, W.B.N., Wilde, P., 1978. Progressive ventilation of the oceans—an explanation for the distribution of the Lower Paleozoic black shales. Am. J. Sci. 278, 257–275.
Betzer, P.R., Byrbe, R.H., Acker, J.G., Lewis, C.S., Jolley, R.R., Feely, R.A., 1984. The oceanic carbonate system: a reassessment of biogenic controls. Science 226, 1074–1077.
Bickert, T., Henrich, R., 2011. Climate records of deep-sea sediments: towards the Cenozoic Ice House. In: Hüneke, H., Mulder, T. (Eds.), Deep-Sea Sediments. Developments in Sedimentology, Elsevier, Amsterdam, vol. 63, pp. 793–823.
Bidle, K.D., Manganelli, M., Azam, F., 2002. Regulation of oceanic silicon and carbon preservation by temperature control on bacteria. Science 298, 1980–1984.
Bienfang, P.K., 1980. Herbivore diet affects fecal pellet settling. Can. J. Fish Aquat. Sci. 37, 1352–1357.
Bijma, J., Spero, H.J., Lea, D.W., 1999. Reassessing foraminiferal stable isotope geochemistry: impact of the oceanic carbonate system (experimental results). In: Fischer, G., Wefer, G. (Eds.), Use of Proxies in Paleoceanography. Springer, Berlin, pp. 489–512.
Billett, D.S.M., Lampitt, R.S., Rice, A.L., Mantoura, R.F.C., 1983. Seasonal sedimentation of phytoplankton to the deep-sea benthos. Nature 302, 520–522.
Böckel, B., Baumann, K.-H., Henrich, R., Kinkel, H., 2006. Distribution of coccoliths in South Atlantic and Southern Ocean surface-sediments in relation to environmental gradients. Deep Sea Res. I 53, 1073–1099.
Boersma, A., Premoli Silva, I., 1991. Distribution of Paleogene planktonic foraminifera: analogies with the recent. Palaeogeogr. Palaeoclimatol. Palaeoecol. 83, 29–48.
Boetius, A., Lochte, K., 1996. Effect of organic enrichments on hydrolytic potentials and growth of bacteria in deep-sea sediments. Mar. Ecol. Prog. Ser. 140, 239–250.
Boetius, A., Ferdelman, T., Lochte, K., 2000. Bacterial activity in sediments of the deep Arabian Sea in relation to vertical flux. Deep Sea Res. II 47, 2835–2875.
Bohrmann, G., Henrich, R., Thiede, J., 1990. Miocene to Quaternary palaeoceanography in the northern North-Atlantic—variability in carbonate and biogenic opal

accumulation. In: Bleil, U., Thiede, J. (Eds.), Geological History of the Polar Oceans: Arctic versus Antarctic, NATO Adv. Sci. Inst. Ser. C, vol. 308. 647–675.
Boltovskoy, D., 1998. Classification and distribution of South Atlantic recent polycystine radiolaria. Paleont. Electr. 1, 1–116.
Bonn, W.J., Gingele, F.X., Grobe, H., Mackensen, A., Fütterer, D.K., 1998. Palaeoproductivity at the Antarctic continental margin: opal and barium records for the last 400 ka. Palaeogeogr. Palaeoclimatol. Palaeoecol. 139, 195–211.
Bornemann, A., Aschwer, U., Mutterlose, J., 2003. The impact of calcareous nannofossils on the pelagic carbonate accumulation across the Jurassic–Cretaceous boundary. Palaeogeogr. Palaeoclimatol. Palaeoecol. 199, 187–228.
Bosellini, A., Winterer, E.L., 1975. Pelagic limestone and radiolarite of the Tethyan Mesozoic: a genetic model. Geology 3, 279–282.
Boss, S.K., Wilkinson, B.H., 1991. Planktogenic eustatic control on cratonic oceanic carbonate accumulation. J. Geol. 99, 497–513.
BouDagher-Fadel, M.K., 1996. Re-evaluation of the phylogeny of Early Cretaceous planktonic foraminifera (Praehedbergellidae). Cretaceous Res. 17, 767–771.
BouDagher-Fadel, M.K., Banner, F.T., Whittaker, J.E., 1997. The Early Cretaceous families. In: BouDagher-Fadel, M.K., Banner, F.T., Whittaker, J.E. (Eds.), The Early Evolutionary History of Planktonic Foraminifera. Chapman and Hall, London, pp. 53–56.
Boudreau, B.P., 1996. Diagenetic Models and Their Implementation. Springer, Berlin pp. 1–414.
Bown, P.R., 1998. Calcareous Nannofossil Biostratigraphy. Kluwer Academic Publishers, Cambridge 315pp.
Bown, P.R., 2005. Calcareous nannoplankton evolution: a tale of two oceans. Micropaleontology 51, 299–308.
Bown, P.R., Burnett, J.A., Gallagher, L.T., 1992. Calcareous nannoplankton evolution. Mem. Sci. Geol. 43, 1–17.
Bown, P., Lees, J.A., Young, J.R., 2004. Calcareous nannoplankton evolution and diversity through time. In: Thierstein, H.R., Young, J.R. (Eds.), Coccolithophores—From Molecular Processes to Global Impact. Springer, Berlin, pp. 481–508.
Boyd, P.W., Sherry, N.D., Berges, J.A., Bishop, J.K.B., Calvert, S.E., Charrette, M.A., et al., 1999. Transformations of biogenic particulates from the pelagic to deep ocean realm. Deep Sea Res. II 46, 2761–2792.
Boyd, P.W., Crossley, A.C., Di Tullio, G.R., Griffiths, F.B., Hutchins, D.A., Queguiner, B., et al., 2002. Control of phytoplankton growth by iron supply and irradiance in the Subantarctic Southern Ocean: experimental results from the SAZ project. J. Geophys. Res. 106, 31585–31595.
Bralower, T.J., Arthur, M.A., Leckie, R.M., Sliter, W.V., Allard, D.J., Schlanger, S.O., 1994. Timing and paleoceanography of oceanic dysoxia/anoxia in the late Barremian to early Aptian (early Cretaceous). Palaios 9, 335–369.
Brand, L.E., 1994. Physiological ecology of marine coccolithophores. In: Winter, A., Siesser, W.G. (Eds.), Coccolithophores. Cambridge University Press, Cambridge, pp. 39–49.
Brodie, I., Kemp, A.E.S., 1994. Variation in biogenic and detrital fluxes and formation of laminae in late Quaternary sediments from the Peruvian coastal upwelling zone. Mar. Geol. 116, 385–398.
Broecker, W.S., 2003. The oceanic $CaCO_3$ cycle. In: Holland, H.D., Turekian, K.K. (Eds.), Treatise on Geochemistry. Elsevier, Amsterdam, pp. 529–549.
Broecker, W.S., Peng, T.H., 1982. Tracers in the Sea. Eldigio Press, Columbia University, Palisades, NY 690pp.
Broecker, W.S., Peng, T.H., 1987. The role of $CaCO_3$ compensation in the glacial to interglacial atmospheric CO_2 change. Global Biogeochem. Cycles 1, 15–29.

Broerse, A.T.C., Ziveri, P., Van Hinte, J.E., Honjo, S., 2000. Coccolithophore export production, species composition, and coccolith-$CaCO_3$ fluxes in the NE Atlantic (34°N 21°W and 48°N 21°W). Deep Sea Res. II 47, 1877–1905.
Brown, C.W., Yoder, J.A., 1994. Coccolithophorid blooms in the global ocean. J. Geophys. Res. (Oceans) 99, 7467–7482.
Bruland, K.W., Silver, M.W., 1981. Sinking rates of fecal pellets from gelatinous zooplankton (salps, pteropods, doliolids). Mar. Biol. 63, 295–300.
Brzezinski, M.A., Olson, R.J., Chisholm, S.W., 1990. Silicon availability and cell-cycle progression in marine diatoms. Mar. Ecol. Prog. Ser. 67, 83–96.
Buesseler, K.O., 1991. Do upper-ocean sediment traps provide an accurate record of particle flux. Nature 353, 420–423.
Buesseler, K.O., 1998. The decoupling of production and particle export in the surface ocean. Global Biogeochem. Cycles 12, 297–310.
Burckle, L.H., Stanton, D., 1975. Distribution of displaced diatoms in the Argentina Basin. Nova Hedwig. Beih. 53, 283–292.
Burckle, L.H., Robinson, D., Cooke, D., 1982. Reappraisal of sea-ice distribution in Atlantic and Pacific sectors of the Southern Ocean at 18,000 yr BP. Nature 299, 435–437.
Butterfield, N.J., 1997. Plankton ecology and the Proterozoic–Phanerozoic transition. Paleobiology 23, 247–262.
Butterfield, N.J., 2003. Exceptional fossil preservation and the Cambrian explosion. Integr. Comp. Biol. 43, 166–177.
Butterfield, N.J., 2007. Macroevolution and macroecology through deep time. Palaeontology 50, 41–55.
Caldeira, K., Rampino, M.R., 1993. Aftermath of the end-Cretaceous mass extinction—possible biogeochemical stabilization of the carbon-cycle and climate. Paleoceanography 8, 5–525.
Canfield, D.E., 1994. Factors influencing organic carbon preservation in marine sediments. Chem. Geol. 114, 315–329.
Canfield, D.E., 1998. A new model for Proterozoic ocean chemistry. Nature 396, 450–453.
Canfield, D.E., Teske, A., 1996. Late Proterozoic rise in atmospheric oxygen concentration inferred from phylogenetic and sulphur-isotope studies. Nature 382, 127–132.
Canfield, D.E., Habicht, K.S., Thamdrup, B., 2000. The Archean sulfur cycle and the early history of atmospheric oxygen. Science 288, 658–661.
Carlson, C.A., Ducklow, H.W., Michaels, A.F., 1994. Annual flux of dissolved organic carbon from the euphotic zone in the northwestern Sargasso Sea. Nature 37, 405–408.
Casey, R.E., 1993. Radiolaria. In: Lipps, J.H. (Ed.), Fossil Prokaryotes and Protists. Blackwell, Cambridge, pp. 249–284.
Casey, R.E., Wigley, C.R., Perezguzman, A.M., 1983. Biogeographic and ecologic perspective on polycystine radiolarian evolution. Paleobiology 9, 363–376.
Cavalier-Smith, T., 2006. Cell evolution and Earth history: stasis and revolution. Philos. Trans. R. Soc. B 361, 969–1006.
Cervato, C., Burckle, L., 2003. Patter of first and last appearance in diatoms: oceanic circulation and the position of polar fronts during the Cenozoic. Paleoceanography 18, 1055. doi:10.1029/2002PA000805.
Chandler, M.A., Rind, D., Ruedy, R., 1992. Pangean climate during the Early Jurassic: GMC simulations and the sedimentary record of paleoclimate. Geol. Soc. Am. Bull. 104, 543–559.
Chisholm, S.W., 1992. Phytoplankton size. In: Falkowski, P.G., Woodhead, A.D. (Eds.), Primary Productivity and Biogeochemical Cycles in the Sea. Plenum, New York, pp. 213–238.

Claquin, P., Martin-Jézéquel, V., Kromkamp, J.C., Veldhuis, M.J.W., Kraay, G.W., 2002. Uncoupling of silicon compared to carbon and nitrogen metabolism, and role of the cell cycle, in continuous cultures of *Thalassiosira pseudonana* (Bacillariophyceae) under light, nitrogen and phosphorus control. J. Phycol. 38, 922–930.

Codispoti, L.A., 1983. On nutrient variability and sediments in upwelling areas. In: Suess, E., Thiede, J. (Eds.), Coastal Upwelling, Part A. Plenum, New York, pp. 125–145.

Colbath, G.K., Grenfell, H.R., 1995. Review of biological affinities of Paleozoic acid-resistant, organic-walled eukaryotic algal microfossils (including acritarchs). Rev. Palaeobot. Palynol. 86, 287–314.

Conte, M.H., Ralph, N., Ross, E.H., 2001. Seasonal and interannual variability in deep-ocean particle fluxes at the Oceanic Flux Program (OFP)/Bermuda Atlantic Time Series (BATS) site in the western Sargasso Sea near Bermuda. Deep Sea Res. II 48, 1471–1505.

Cordey, F., Schiarizza, P., 1993. A long-lived Panthalassic remnant: the Bridge River accretionary complex, Canadian Cordillera. Geology 21, 263–266.

Courtillot, V.E., Renne, P.R., 2003. On the ages of flood basalt events. C. R. Geoscience 335, 113–140.

Crosta, X., Romero, O., Armand, L.K., Pichon, J.-J., 2005. The biogeography of major diatom taxa in Southern Ocean sediments: 2. open-ocean related species. Palaeogeogr. Palaeoclimatol. Palaeoecol. 223, 66–92.

Crowley, T., North, G., 1991. Paleoclimatology. Oxford University Press, New York 339pp.

Culver, S.J., 1993. Foraminifera. In: Lipps, J.H. (Ed.), Fossil Prokaryotes and Protists. Blackwell, Cambridge, pp. 203–248.

Cunningham, W.L., Leventer, A., 1998. Diatom assemblages in surface sediments of the Ross Sea: relationship to present oceanographic conditions. Antarct. Sci. 10, 134–146.

Deibel, D., 1990. Still-water sinking velocity of fecal material from the pelagic tunicate *Dolioletta gegenbauri*. Mar. Ecol. Prog. Ser. 62, 55–60.

De Baar, H.J.W., De Jong, J.T.M., Bakker, D.C.E., Löscher, B.M., Veth, C., Bathmann, U., et al., 1995. Importance of iron for plankton blooms and carbon dioxide drawdown in the Southern Ocean. Nature 373, 412–414.

De Wever, P., 1989. Radiolarians, radiolarites, and Mesozoic paleogeography of the circum-Mediterranean Alpine Belts. In: Hein, J.R., Obradovic, J. (Eds.), Siliceous Deposits of the Tethys and Pacific Regions. Springer, Berlin, pp. 31–49.

De Wever, P., Baudin, F., 1996. Paleogeography of radiolarite and organic-rich deposits in Mesozoic Tethys. Geol. Rundsch. 85, 310–326.

De Wever, P., Azéma, J., Fourcade, E., 1994. Radiolaires et radiolarites: production primaire, diagenèse et paléogéographie. Bull. Cent. Rech. Explor. Prod. Elf-Aquitaine 18, 315–379.

De Wever, P., Dumitrica, P., Caulet, J.P., Nigrini, C., Caridroit, M., 2001. Radiolarians in the Sedimentary Record. Gordon and Breach, London, pp.1–533.

De Wever, P., O'Dogherty, L., Caridroit, M., Dumitrica, P., Guex, J., Nigrini, C., et al., 2003. Diversity of radiolarian families through time. Bull. Soc. Géol. France 174, 453–469.

De Wever, P., O'Dogherty, L., Gorican, S., 2006. The plankton turnover at the Permo-Triassic boundary, emphasis on radiolarians. Eclogae Geol. Helv. 99, S49–S62.

Dell'agnese, D.J., Clark, D.L., 1994. Siliceous microfossils from the warm Late Cretaceous and Early Cenozoic Arctic Ocean. J. Paleontol. 68, 31–47.

Delwiche, C.F., 1999. Tracing the thread of plastid diversity through the tapestry of life. Am. Nat. 154, 164–177.

DeMaster, D.J., 1981. The supply and accumulation of silica in the marine environment. Geochim. Cosmochim. Acta 45, 1715–1732.

Demicco, R.V., Lowenstein, T.K., Hardie, L.A., Spencer, R.J., 2005. Model of seawater composition for the Phanerozoic. Geology 33, 877–880.

Derry, L.A., Kaufman, A.J., Jacobsen, S.B., 1992. Sedimentary cycling and environmental change in the late Proterozoic: evidence from stable and radiogenic isotopes. Geochim. Cosmochim. Acta 56, 1317–1329.

Des Marais, D.J., Strauss, H., Summons, R.E., Hayes, J.M., 1992. Carbon isotope evidence for the stepwise oxidation of the Proterozoic environment. Nature 359, 605–609.

Deuser, W.G., Ross, E.H., 1980. Seasonal changes in the flux of organic carbon to the deep Sargasso Sea. Nature 283, 364–365.

Deuser, W.G., Muller-Karger, F.E., Evans, R.H., Brown, O.B., Esaias, W.E., Feldman, G. C., 1990. Surface ocean color and deep-ocean carbon flux: how close a connection. Deep Sea Res. 37, 1331–1343.

Dezileau, L., Bareille, G., Reyss, J.L., Lemoine, F., 2000. Evidence for strong sediment redistribution by bottom currents along the southeast Indian ridge. Deep Sea Res. I 47, 1899–1936.

Dezileau, L., Reyss, J.L., Lemoine, F., 2003. Late Quaternary changes in biogenic opal fluxes in the Southern Indian Ocean. Mar. Geol. 202, 143–158.

Dickson, R.R., Brown, J., 1994. The production of North Atlantic Deep Water: sources, rates, and pathways. J. Geophys. Res. 99 (C6), 12319–12341.

Diekmann, B., 2007. Sedimentary patterns in the late Quaternary Southern Ocean. Deep Sea Res. II 54, 2350–2366.

Diekmann, B., Fütterer, D.K., Grobe, H., Hillenbrand, C.D., Kuhn, G., Michels, K., et al., 2003. Terrigenous sediment supply in the polar to temperate South Atlantic: land–ocean links of environmental changes during the late Quaternary. In: Wefer, G., Mulitza, S., Ratmeyer, V. (Eds.), The South Atlantic in the Late Quaternary: Reconstruction of Material Budget and Current Systems. Springer, Berlin, pp. 375–399.

Diester-Haass, L., Rothe, P., 1987. Plio-Pleistocene sedimentation on the Walvis Ridge, Southeast Atlantic (DSDP Leg 75, Site 532)—influence of surface currents, carbonate dissolution and climate. Mar. Geol. 77, 53–85.

Diester-Haass, L., Zahn, R., 1996. Eocene–Oligocene transition in the Southern Ocean: history of water mass circulation and biological productivity. Geology 24, 163–166.

Dilling, L., Alldredge, A.L., 1993. Can chaetognath fecal pellets contribute to carbon flux? Mar. Ecol. Prog. Ser. 92, 51–58.

Distanov, U.G., Glezer, Z.I., 1973. Silicoflagellitite, a new type of opaline siliceous rock. Litol. Polezhn. Iskop. 9, 226–227 (in Russian).

Dittert, N., Henrich, R., 2000. SEM ultrastructure of a spinose planktonic foraminifer, *Globigerina bulloides* (d'Orbigny, 1826), and its paleoceano-graphic power. Deep Sea Res. I 47, 603–620.

Dittert, N., Baumann, K.-H., Bickert, T., Henrich, R., Huber, R., Kinkel, H., et al., 1999. Carbonate dissolution in the deep sea: methods, quantification and paleoceanographic application. In: Fischer, G., Wefer, G. (Eds.), Use of Proxies in Paleoceanography. Springer, Berlin, pp. 255–284.

Dittert, N., 1998. Late Quaternary planktic foraminifera assemblages in the South Atlantic Ocean: quantitative determination and preservational aspects. Berichte FB Geowiss. Univ. Bremen 126, 1–64.

Dixit, S., Van Cappellen, P., 2003. Predicting benthic fluxes of silicic acid from deep-sea sediments. J. Geophys. Res. Oceans 108, 3334. doi:10.1029/2002JC001309.

Dugdale, R.C., Wilkerson, F.P., 1998. Understanding the eastern equatorial Pacific as a continuous new production system regulating on silicate. Nature 391, 270–273.

Dugdale, R.C., Wilkerson, F.P., Minas, H.J., 1995. The role of a silicate pump in driving new production. Deep Sea Res. 42, 697–719.

Dunne, J.P., Murray, J.W., Aufdenkampe, A., Blain, S., Rodier, M., 1999. Silica: nitrogen coupling in the equatorial Pacific upwelling zone. Global Biogeochem. Cycles 13, 715–726.
Dunne, J.P., Armstrong, R.A., Gnanadesikan, A., Sarmiento, J.L., 2005. Empirical and mechanistic models for the particle export ratio. Global Biogeochem. Cycles 19, GB4026. doi:10.1029/2004GB002390.
Durbin, E.G., 1977. Studies of the autoecology of the marine diatom *Thalassiosira nordenskioeldii*. II. The influence of cell size on growth rate, and carbon, nitrogen, chlorophyll a and silica content. J. Phycol. 13, 150–155.
Dutkiewicz, A., George, S.C., Mossman, D.J., Ridley, J., Volk, H., 2007. Oil and its biomarkers associated with the Palaeoproterozoic Oklo natural fission reactors, Gabon. Chem. Geol. 244, 130–154.
Edwards, L.E., 1993. Dinoflagellates. In: Lipps, J.H. (Ed.), Fossil Prokaryotes and Protists. Blackwell, Cambridge, pp. 105–127.
Einsele, G., 2000. Sedimentary Basins: Evolution, Facies, and Sediment Budget, second ed. Springer, Berlin 792pp.
Emerson, S., Archer, D., 1990. Calcium carbonate preservation in the ocean. Philos. Trans. R. Soc. Lond. A 331, 29–40.
Emerson, S., Hedges, J.I., 1988. Processes controlling the organic carbon content of open ocean sediments. Paleoceanography 3, 611–634.
Emerson, S., Fischer, K., Reimers, C., Heggie, D., 1985. Organic carbon dynamics and preservation in deep-sea sediments. Deep Sea Res. 32, 1–21.
Erba, E., 1989. Upper Jurassic to Lower Cretaceous *Nannoconus* distribution in some sections from northern and central Italy. Mem. Sci. Geol. 41, 255–261.
Erba, E., 2004. Calcareous nannofossils and Mesozoic oceanic anoxic events. Mar. Micropaleontol. 52, 85–106.
Erba, E., 2006. The first 150 million years history of calcareous nannoplankton: biosphere–geosphere interactions. Palaeogeogr. Palaeoclimatol. Palaeoecol. 232, 237–250.
Eriksson, P.G., Banerjee, S., Catuneanu, O., Sarkar, S., Bumby, A.J., Mtimkulu, M.N., 2007. Prime controls on Archaean–Palaeoproterozoic sedimentation: change over time. Gondwana Res. 12, 550–559.
Ernst, R.E., Buchan, K.L., 2002. Maximum size and distribution in time and space of mantle plumes: evidence from large igneous provinces. In: Condie, K.C., Abbot, D., Des Marais, D.J. (Eds.), Superplume Events in Earth's History: Causes and Effects, *J. Geodynamics* 34, Elsevier, Amsterdam, 309–342 (Erratum, 34, 711–714 (2002)).
Fabry, V.J., Deuser, W.G., 1991. Aragonite and magnesian calcite fluxes to the deep Sargasso Sea. Deep Sea Res. 38, 713–728.
Fabry, V.J., Deuser, W.G., 1992. Seasonal changes in the isotopic compositions and sinking fluxes of euthecosomatous pteropod shells in the Sargasso Sea. Paleoceanography 7, 195–213.
Falkowski, P.G., 1997. Evolution of the nitrogen cycle and its influence on the biological sequestration of CO_2 in the ocean. Nature 387, 272–275.
Falkowski, P.G., 2002. The ocean's invisible forest—marine phytoplankton play a critical role in regulating the earth's climate. Could they also be used to combat global warming. Sci. Am. 287, 54–61.
Falkowski, P.G., Raven, J.A., 1997. Aquatic Photosynthesis. Blackwell Science, Oxford 375pp.
Falkowski, P., Barber, R., Smetacek, V., 1998. Biogeochemical controls and feedbacks on ocean primary productivity. Science 281, 200–206.
Falkowski, P., Laws, E.A., Barbar, R.T., Murray, J.W., 2003. Phytoplankton and their role in primary, new and export production. In: Fasham, M.J.R. (Ed.), Ocean

Biogeochemistry. The Role of the Ocean Carbon Cycle in Global Change. Springer, Berlin, pp. 99–121.
Falkowski, P.G., Katz, M.E., Knoll, A.H., Quigg, A., Raven, J.A., Schofield, O., et al., 2004a. The evolution of modern eukaryotic phytoplankton. Science 305, 354–360.
Falkowski, P., Schofield, O., Katz, M.E., Van de Schootbrugge, B., Knoll, A.H., 2004b. Why is the Land Green and the Ocean Red. In: Thierstein, H.R., Young, J.R. (Eds.), Coccolithophores—From Molecular Processes to Global Impact. Springer, Berlin, pp. 429–454.
Farquhar, J., Wing, B.A., McKeegan, K.D., Harris, J.W., Cartigny, P., Thiemens, M.H., 2002. Mass-independent sulfur of inclusions in diamond and sulfur recycling on early Earth. Science 298, 2369–2372.
Farrell, J.W., Raffi, I., Janecek, T.R., Murray, D.W., Levitan, M., Delaney, M., et al., 1995. Late Neogene sedimentation patterns in the eastern equatorial Pacific. Proc. ODP Sci. Results 138, 717–756.
Faugères, J.C., Mulder, T., 2011. Contour currents and contourite drifts. In: Hüneke, H., Mulder, T. (Eds.), Deep-Sea Sediments. Developments in Sedimentology, Elsevier, Amsterdam, vol. 63, pp. 149–214.
Feely, R.A., Sabine, C.L., Schlitzer, R., Bullister, J.L., Mecking, S., Greeley, D., 2004. Oxygen utilization and organic carbon remineralization in the upper water column of the Pacific Ocean. J. Oceanogr. 60, 45–52.
Ferdelman, T., Fossing, H., Neumann, K., Schulz, H.D., 1999. Sulfate reduction in surface sediments of the South-East Atlantic continental margin between 15′38′S and 27°57′S (Angola and Namibia). Limnol. Oceanogr. 44, 650–661.
Field, C., Behrenfeld, M., Randerson, J., Falkowski, P., 1998. Primary production of the biosphere: integrating terrestrial and oceanic components. Science 281, 237–240.
Finkel, Z.V., Irwin, A.J., Schofield, O., 2004. Resource limitation alters the 3/4 size scaling of metabolic rates in phytoplankton. Mar. Ecol. Prog. Ser. 273, 269–279.
Fischer, G., Futterer, D., Gersonde, R., Honjo, S., Ostermann, D., Wefer, G., 1988. Seasonal variations of particle flux in the Weddell Sea and its relation to ice cover. Nature 335, 426–428.
Fischer, G., Gersonde, R., Wefer, G., 2002. Organic carbon, biogenic silica and diatom fluxes in the marginal winter sea ice zone in the Polar Front Region: variations and differences in composition. Deep Sea Res. II 49, 1721–1745.
Foldvik, A., Gammelrød, T., 1988. Notes on Southern Ocean hydrography, sea-ice and bottom water formation. Palaeogeogr. Palaeoclimatol. Palaeoecol. 67, 3–17.
Föllmi, K.B., Weissert, H., Lini, A., 1993. Nonlinearities in phosphogenesis and phosphorus–carbon coupling and their implications for global change. In: Wollast, R., Mackenzie, F.T., Chou, L. (Eds.), Interactions of C, N, P and S Biogeochemical Cycles and Global Change, NATO Adv. Sci. Inst. Ser., 14, 447–474.
Föllmi, K.B., Weissert, H., Bisping, M., Funk, H.P., 1994. Phosphogenesis, carbon-isotope stratigraphy, and carbonate-platform evolution along the Lower Cretaceous northern Tethyan margin. Geol. Soc. Am. Bull. 106, 729–746.
Fortier, L., Le Fèvre, J., Legendre, L., 1994. Export of biogenic carbon to fish and to the deep ocean: the role of large planktonic microphages. J. Plankton Res. 16, 809–839.
Fowler, S.W., Knauer, G.A., 1986. Role of large particles in the transport of elements and organic compounds through the oceanic water column. Prog. Oceanogr. 16, 147–194.
Fowler, S.W., Buat-Menard, P., Yokoyama, Y., Ballestra, S., Holm, E., Nguyen, H.V., 1987. Rapid removal of Chernobyl fallout from Mediterranean surface waters by biological activity. Nature 329, 56–58.
Fowler, S.W., Small, L.F., 1972. Sinking rates of euphausiid fecal pellets. Limnol. Oceanogr. 17, 293–296.

Francois, R., Honjo, S., Krishfield, R., Manganini, S., 2002. Factors controlling the flux of organic carbon to the bathypelagic zone of the ocean. Global Biogeochem. Cycles 16, 1087.
François, L.M., Walker, J.C.G., Opdyke, B.N., 1993. The history of global weathering and the chemical evolution of the ocean–atmosphere system. In: Takahashi, R., Jeanloz, R., Dubie, D. (Eds.), Evolution of the Earth and Planets, AGU Geophys. Monogr., 74, 143–159.
Frank, M., Gersonde, R., Rutger van der Loeff, M., Bohrmann, G., Nürnberg, C.C., Kubik, P.W., et al., 2000. Similar glacial and interglacial export bioproductivity in the Atlantic sector of the Southern Ocean: multiproxy evidence and implications for glacial atmospheric CO_2. In: Paleoceanography, 15, 642–658.
Frenz, M., Henrich, R., 2007. Carbonate dissolution revealed by silt grain-size distribution: comparison of Holocene and Last Glacial Maximum sediments from the pelagic South Atlantic. Sedimentology 54, 391–404.
Frenz, M., Baumann, K.-H., Boeckel, B., Höppner, R., Henrich, R., 2005. Carbonates from the Mid-Atlantic Ridge: quantification of foraminifer and coccolith carbonate in the South Atlantic surface sediments means of carbonate grain-size distributions. J. Sediment. Res. 75, 468–479.
Frenz, M., Henrich, R., Zychla, B., 2006. Carbonate preservation patterns at the Ceara Rise—evidence for the Pliocene super conveyor. Mar. Geol. 232, 173–180.
Froelich, P.N., Blanc, V., Mortlock, R.A., Chillrud, S.N., Dunstan, W., Udomkit, A., et al., 1992. River fluxes of dissolved silica to the ocean were higher during glacials: Ge/Si in diatoms, rivers, and oceans. Paleoceanography 7, 739–767.
Füchtbauer, H., Hardie, L.A., 1976. Experimentally determined homogeneous distribution coefficients for precipitated magnesian calcites: application to marine carbonate cements. Geol. Soc. Am. Abstr. Prog. 8, 877.
Füchtbauer, H., Hardie, L.A., 1980. Comparison of Experimental and Natural Magnesian Calcites. International Association of Sedimentology, Bochum pp.167–169.
Gallinari, M., Ragueneau, O., Corrin, L., DeMaster, D.J., Treguer, P., 2002. The importance of water column processes on the dissolution properties of biogenic silica in deep-sea sediments I. Solubility. Geochim. Cosmochim. Acta 66, 2701–2717.
Gallinari, M., Ragueneau, O., DeMaster, D.J., Hartnett, H., Rickert, D., Thomas, C., 2008. Influence of seasonal phytodetritus deposition on biogenic silica dissolution in marine sediments—potential effects on preservation. Deep Sea Res. II 55, 2451–2464.
Gallois, R.W., Medd, A.W., 1979. Coccolith-rich marker bands in the English Kimmeridge Clay. Geol. Mag. 116, 247–260.
Gargett, A.E., 1991. Physical processes and the maintenance of nutrient-rich euphotic zones. Limnol. Oceanogr. 36, 1527–1545.
Garrison, T.S., 2007. Oceanography: An Invitation to Marine Science, sixth ed. Thomson Brooks/Cole, Belmont 588pp.
Gartner, S., 1970. Sea-floor spreading, carbonate dissolution level, and the nature of Horizon A. Science 169, 1077–1079.
Gauthier-Lafaye, F., Weber, F., 2003. Natural nuclear fission reactors: time constraints for occurrence, and their relation to uranium and manganese deposits and to the evolution of the atmosphere. Precambrian Res. 120, 81–100.
Gerhardt, S., Henrich, R., 2001. Shell preservation of *Limacina inflata* (pteropoda) in surface sediments from the central and South Atlantic ocean: a new proxy to determine the aragonite saturation state of water masses. Deep Sea Res. I 48, 2051–2071.
Gerhardt, S., Groth, H., Rühlemann, C., Henrich, R., 2000. Aragonite preservation in late Quaternary sediment cores on the Brazilian continental slope: implications for intermediate water circulation. Int. J. Earth Sci. 88, 607–618.
Gersonde, R., Crosta, X., Abelmann, A., Armand, L., 2005. Sea-surface temperature and sea-ice distribution of the Southern Ocean at the EPILOG Last Glacial Maximum—a

circum-Antarctic view based on siliceous microfossil records. Quat. Sci. Rev. 24, 869–896.
Gervais, F., Riebesell, U., Gorbunov, M.Y., 2002. Changes in primary producticity and chlorophyll in response to iron fertilization in the Southern Polar Frontal Zone. Limnol. Oceanogr. 47, 1324–1335.
Gibbs, M.T., Bluth, G.J.S., Fawcett, P.J., Kump, L.R., 1999. Global chemical erosion over the last 250 my: variations due to changes in paleogeography, paleoclimate, and paleogeology. Am. J. Sci. 299, 611–651.
Gibson, C.H., 2000. Laboratory and ocean studies of phytoplankton response to fossil turbulence. Dyn. Atmos. Oceans 31, 295–306.
Gibson, T.G., Towe, K.M., 1971. Eocene volcanism and the origin of Horizon A. Science 172, 152–154.
Gilmer, R.W., 1974. Some aspects of feeding in thecosomatous pteropod molluscs. J. Mar. Biol. Ecol. 15, 127–144.
Godderis, Y., Joachimski, M.M., 2004. Global change in the Late Devonian: modelling the Frasnian-Famennian short-term carbon isotope excursions. Palaeogeogr. Palaeoclimatol. Palaeoecol. 202, 309–329.
Goldman, J.C., 1993. Potential role of large oceanic diatoms in new primary production. Deep Sea Res. 40, 159–168.
Gorsky, G., Fisher, N.S., Fowler, S.W., 1984. Biogenic debris from the pelagic tunicate, *Oikopleura dioica*, and its role in the vertical transport of a transuranium element. Estuar. Coast Shelf Sci. 18, 13–23.
Gowing, M.M., 1986. Trophic ecology of phaeodarian radiolarians and vertical flux of radiolarians in the upper 2000 m of the North Pacific central gyre. Deep Sea Res. 33, 655–674.
Gradstein, F.M., Agterberg, F.P., Ogg, J.G., Hardenbol, H., Van Veen, P., 1995. A Triassic, Jurassic, and Cretaceous time scale. In: Berggren, W.A., Kent, D.V., Hardenbol, J. (Eds.), Geochronology, Time Scales and Global Stratigraphic Correlations: A Unified Temporal Framework for an Historical Geology, Soc. Sediment. Geol. Spec. Publ., vol. 54. 95–126.
Graf, A., 1989. Benthic–pelagic coupling in a deep-sea benthic community. Nature 341, 437–439.
Grant, K.M., Dickens, G.R., 2002. Coupled productivity and carbon isotope records in the southwest Pacific Ocean during the late Miocene–early Pliocene biogenic bloom. Palaeogeogr. Palaeoclimatol. Palaeoecol. 187, 61–82.
Grantham, P.J., Wakefield, L.L., 1988. Variations in the sterane carbon number distributions of marine source rock derived crude oils through geological time. Org. Geochem. 12, 61–73.
Grice, K., Schaeffer, P., Schwark, L., Maxwell, J.R., 1996. Molecular indicators of palaeoenvironmental conditions in an immature Permian shale (Kupferschiefer, Lower Rhine Basin, north-west Germany) from free and S-bound lipids. Org. Geochem. 25, 131–147.
Grice, K., Twitchett, R.J., Alexander, R., Foster, C.B., Looy, C., 2005. A potential biomarker for the Permian–Triassic ecological crisis. Earth Planet. Sci. Lett. 236, 315–321.
Gröger, M., 2002. Deep water circulation in the western equatorial Atlantic: inferences from carbonate preservation studies and silt grain-size analysis. Grain-size composition of Quaternary South Atlantic sediments and its paleoceanographic significance. Berichte FB Geowiss. Univ. Bremen 195, 1–95.
Gröger, M., Henrich, R., Bickert, T., 2003. Glacial–interglacial variability in lower North Atlantic Deep Water: interference from silt grain size analysis and carbonate preservation in the western equatorial Atlantic. Mar. Geol. 201, 321–332.

Grotzinger, J.P., James, N.P., 2000. Precambrian carbonates: evolution of understanding. In: Grotzinger, J.P., James, N.P. (Eds.), Carbonate Sedimentation and Diagenesis in the Evolving Precambrian World, Soc. Econ. Paleontol. Mineral. Spec. Pub., vol. 67. 3–22.

Grotzinger, J.P., Kasting, J.F., 1993. New constraints on Precambrian ocean composition. J. Geol. 101, 235–243.

Guillard, R.R.L., Kilham, P., 1978. The ecology of marine planktonic diatoms. In: Werner, D. (Ed.), The Biology of Diatoms. University of California Press, Berkeley, pp. 372–469.

Hallam, A., Wignall, P., 1997. Mass Extinctions and Their Aftermath. Oxford University Press, Oxford 328pp.

Hallock, P., Premoli Silva, I., Boersma, A., 1991. Similarities between planktonic and larger foraminiferal evolutionary trends through Paleogene paleoceanographic changes. Palaeogeogr. Palaeoclimatol. Palaeoecol. 83, 49–64.

Haq, B.U., Hardenbol, J., Vail, P.R., 1987. Chronology of fluctuating sea levels since the Triassic (250 million years ago to present). Science 235, 1156–1167.

Haq, B.U., Hardenbol, J., Vail, P.R., 1988. Mesozoic and Cenozoic chronostratigraphy and cycles of sea-level change. In: Wilgus, C.K., Hastings, B.S., Kendall, C.G.S.C., Posamentier, H.W., Ross, C.A., VanWagoner, J.C. (Eds.), Sea-Level Changes: An Integrated Approach, Soc. Econ. Paleontol. Mineral. Spec. Publ., vol. 42. 71–108.

Hardie, L.A., 1996. Secular variation in seawater chemistry: an explanation for the coupled secular variation in the mineralogies of marine limestones and potash evaporites over the past 600 m.y. Geology 24, 279–283.

Harnett, H.E., Devol, A.H., 2003. Role of a strong oxygen-deficient zone in the preservation and degradation of organic matter: a carbon budget for the continental margins of northwest Mexico and Washington State. Geochim. Cosmochim. Acta 67, 247–264.

Harris, N.B., 2005. The deposition of organic-carbon-rich sediments: models, mechanisms, and consequences—introduction. In: Harris, N.B. (Ed.), The deposition of organic-carbon-rich sediments: models, mechanisms, and consequences, Soc. Sediment. Geol. Spec. Publ., vol. 82. 1–6.

Harris, J.R.W., Stutt, E.D., Turley, C.M., 2001. Carbon flux in the northwest Mediterranean estimated from microbial production. Deep Sea Res. I 48, 231–248.

Harrison, K.G., 2000. Role of increased marine silica input on paleo-pCO_2 levels. Paleoceanography 15, 292–298.

Hart, M.B., Hylton, M.D., Oxford, M.J., Price, G.D., Hudson, W., Smart, C.W., 2003. The search for the origin of the planktic foraminifera. J. Geol. Soc. 160, 341–343.

Hartnett, H.E., Keil, R.G., Hedges, J.I., Devol, A.H., 1998. Influence of oxygen exposure time on organic carbon preservation in continental margin sediments. Nature 391, 572–574.

Harwood, D.M., Nikolaev, V.A., 1995. Cretaceous diatoms: morphology, taxonomy, biostratigraphy. In: Blome, C.D., Whalen, P.M., Katherine, R. (Eds.), Siliceous Microfossils. Paleontological Society, Lawrence, KS, pp. 81–106.

Hay, W.W., 2004. Carbonate fluxes and calcareous nannoplankton. In: Thierstein, H.R., Young, J.R. (Eds.), Coccolithophores—From Molecular Processes to Global Impact. Springer, Berlin, pp. 509–528.

Hayes, J.M., Strauss, H., Kaufman, A.J., 1999. The abundance of ^{13}C in marine organic matter and isotopic fractionation in the global biogeochemical cycle of carbon during the past 800 Ma. Chem. Geol. 161, 103–125.

Hebbeln, D., Wefer, G., 1991. Effects of ice coverage and ice-rafted material on sedimentation in the Fram Strait. Nature 350, 409–411.

Hedges, J.I., Keil, R.G., 1995. Sedimentary organic matter preservation: an assessment and speculative synthesis. Mar. Chem. 49, 81–115.

Hedges, J.I., Hu, F.S., Devol, A.H., Hartnett, H.E., Tsamakis, E., Keil, R.G., 1999. Sedimentary organic matter preservation: a test for selective degradation under oxic conditions. Am. J. Sci. 299, 529–555.

Hein, J.R., Parrish, J.R., 1987. Distribution of siliceous deposits in time and space. In: Hein, J.R. (Ed.), Siliceous Sedimentary Rock—Hosted Ores and Petroleum. Van Nostrand Reinhold, New York, pp. 10–57.

Heip, C.H.R., Duineveld, G., Flach, E., Graf, G., Helder, W., Herman, P.M.J., et al., 2001. The role of the benthic biota in sedimentary metabolism and sediment-water exchange processes in the Goban Spur area (NE Atlantic). Deep Sea Res. II 48, 3223–3243.

Hemleben, C., Spindler, M., Anderson, O.R., 1989. Modern Planktonic Foraminifera. Springer, Berlin 363pp.

Henrich, R., 1989. Glacial/interglacial cycles in the Norwegian Sea: sedimentology, paleoceanography, and evolution of late Pliocene to Quaternary northern hemisphere climate. Proc. ODP Sci. Results 104, 189–232.

Henrich, R., Hüneke, H., 2011. Hemipelagic advection and periplatform sedimentation. In: Hüneke, H., Mulder, T. (Eds.), Deep-Sea Sediments. Developments in Sedimentology, Elsevier, Amsterdam , vol. 63, pp. 353–396.

Henrich, R., Wefer, G., 1986. Dissolution of biogenic carbonates: effects of skeletal structure. Mar. Geol. 71, 341–362.

Henrich, R., Freiwald, A., Schäfer, P., Wehrmann, A., Kühlmann, D., Brachert, T., et al., 1995. Controls on modern carbonate sedimentation on warm-temperate to Arctic coasts, shelves and seamounts in the northern hemisphere: implications for fossil counterparts. Facies 32, 71–109.

Henrich, R., Baumann, K.H., Gerhardt, S., Gröger, M., Volbers, A., 2003. Carbonate preservation in deep and intermediate water masses in the South Atlantic: evaluation and geological record (a review). In: Wefer, G., Mulitza, S., Ratmeyer, V. (Eds.), The South Atlantic in the Late Quaternary: Reconstruction of Material Budget and Current Systems. Springer, Berlin, pp. 645–670.

Hensen, C., Zabel, M., Schulz, H.D., 2000. A comparison of benthic nutrient flues from deep-sea sediments off Namibia and Argentina. Deep Sea Res. II 47, 2029–2050.

Herman, Y., 1978. Pteropods. In: Haq, B., Boersma, A. (Eds.), Introduction to Marine Micropaleontology. Elsevier, New York, pp. 151–159.

Hesse, R., 1989. Silica diagenesis: origin of inorganic and replacement cherts. Earth-Sci. Rev. 26, 253–284.

Hesse, R., 1990. Origin of chert I. Diagenesis of biogenic siliceous sediments. In: McIlreath, I.A., Morrow, D.W. (Eds.), Diagenesis Geoscience. Geological Association of Canada, St. John's, Newfoundland, Canada, Reprint Ser. 4, pp. 227–251 (reprinted from Geosci. Canada 15(3), 1988).

Hesse, R., Schacht, U., 2011. Early diagenesis of deep-sea sediments. In: Hüneke, H., Mulder, T. (Eds.), Deep-Sea Sediments. Developments in Sedimentology, Elsevier, Amsterdam, vol. 63, pp. 557–714.

Hildebrand, M., Volcani, B.E., Gassmann, W., Schroeder, J.I., 1997. A gene family of silicon transporters. Nature 385, 688–689.

Holland, H.D., 1994. Early Proterozoic atmospheric change. In: Bengtson, S. (Ed.), Early Life on Earth. Nobel Symposium vol. 84. Columbia University Press, New York, NY, pp. 237–244.

Holland, H.D., 2004. The geologic history of seawater. In: Elderfield, H. (Ed.), The Oceans and Marine Geochemistry. Treatise on Geochemistry vol. 6. Elsevier, Amsterdam, pp. 583–625.

Holland, H.D., 2006. The oxygenation of the atmosphere and oceans. Phil. Trans. R. Soc. B 361, 903–915.

Hollister, C.D., McCave, I.N., 1984. Sedimentation under deep sea storms. Nature 309, 220–225.

Honjo, S., 1976. Coccoliths: production, transportation and sedimentation. Mar. Micropaleontol. 1, 65–79.

Honjo, S., 1996. Fluxes of particles to the interior of the open oceans. In: Ittekkot, P., Schäfer, P., Honjo, S., Depetris, P.J. (Eds.), Particle Flux in the Ocean. John Wiley & Sons, New York, pp. 91–154.

Honjo, S., Manganini, S.J., 1993. Annual biogenic particle fluxes to the interior of the North Atlantic Ocean studied at 34°N 21°W and 48°N 21°W. Deep Sea Res. 40, 587–607.

Honjo, S., Roman, M.R., 1978. Marine copepod fecal pellets: production, preservation, and sedimentation. J. Mar. Res. 36, 45–57.

Honjo, S., Dymond, J., Collier, R., Manganini, S.J., 1995. Export production of particles to the interior of the equatorial Pacific Ocean during the 1992 EqPac experiment. Deep Sea Res. II 42, 831–870.

Honjo, S., Francois, R., Manganini, S., Dymond, J., Collier, R., 2000. Particle fluxes to the interior of the Southern Ocean in the Western Pacific sector along 170°W. Deep-Sea Res. II (47), 3521–3548.

Horita, J., Zimmermann, H., Holland, H.D., 2002. Chemical evolution of seawater during the Phanerozoic: implications from the record of marine evaporites. Geochim. Cosmochim. Acta 66, 3733–3756.

Horner, R.A., Syvertsen, E.E., Thomas, D.P., Lange, C., 1988. Proposed terminology and reporting units for sea-ice algal assemblages. Polar Biol. 8, 249–253.

Hotinski, R.M., Kump, L.R., Arthur, M.A., 2004. The effectiveness of the Paleoproterozoic biological pump: a $\delta^{13}C$ gradient from platform carbonates of the Pethei Group (Great Slave Lake Supergroup, NWT). Geol. Soc. Am. Bull. 116, 539–554.

Huber, B.T., Hodell, D.A., Hamilton, C.P., 1995. Middle-late Cretaceous climate of the southern high latitudes: stable isotopic evidence for minimal equator-to-pole thermal gradients. Geol. Soc. Am. Bull. 107, 1164–1191.

Hulburt, E.M., 1970. Competition for nutrients by marine phytoplankton in oceanic, coastal, and estuarine regions. Ecology 51, 475–484.

Hüneke, H., 2006. Erosion and deposition from bottom currents during the Givetian and Frasnian: response to intensified oceanic circulation between Gondwana and Laurussia. Palaeogeogr. Palaeoclimatol. Palaeoecol. 234, 146–167.

Hüneke, H., Stow, D., 2008. Identification of ancient contourites: problems and palaeoceanographic significance. In: Rebesco, M., Camerlenghi, A. (Eds.), Contourites. Developments in Sedimentology, 60, Elsevier, Amsterdam, pp. 323–344.

Hutchins, D.A., Bruland, K.W., 1998. Iron limited diatom growth and Si:N uptake ratios in a coastal upwelling regime. Nature 393, 561–564.

Iglesias-Rodríguez, M.D., Brown, C.W., Doney, S.C., Kleypas, J., Kolber, D., Kolber, Z., et al., 2002. Representing key phytoplankton functional groups in ocean carbon cycle models: coccolithophorids. Global Biogeochem. Cycles 16, 1100. doi:10.1029/2001GB001454.

Isozaki, Y., 1997. Permo-Triassic superanoxia and stratified superocean: records from lost deep sea. Science 276, 235–238.

Isozaki, Y., Maruyama, S., Furuoka, F., 1990. Accreted materials in Japan. Tectonophysics 181, 179–205.

Itaki, T., 2003. Depth-related radiolarian assemblage in the water-column and surface sediments of the Japan Sea. Mar. Micropaleontol. 47, 253–270.

Ittekkot, V., 1993. The abiotically driven biological pump in the ocean and short-term fluctuations in atmospheric CO_2 contents. Global Planet. Change 8, 17–25.

Ivany, L.C., Van Simaeys, S., Domack, E.W., Samson, S.D., 2006. Evidence for an earliest Oligocene ice sheet on the Antarctic Peninsula. Geology 34, 377–380.

Jahnke, R.A., 1996. The global ocean flux of particulate organic carbon: areal distribution and magnitude. Global Biogeochem. Cycles 10, 71–88.

Jahnke, R.A., Craven, D.B., Gaillard, J.-F., 1994. The influence of organic matter diagenesis on $CaCO_3$ dissolution at the deep-sea floor. Geochim. Cosmochim. Acta 58, 2799–2809.

Janofske, D., 1992. Kalkiges Nannoplankton, insbesondere kalkige Dinoflagellaten-Zysten der alpinen Ober-Trias: Taxonomie, Biostratigraphie und Bedeutung für die Phylogenie der Peridiniales. Berliner Geowiss. Abh. E 4, 1–73.

Javaux, E.J., Knoll, A.H., Walter, M.R., 2001. Morphological and ecological complexity in early eukaryotic ecosystems. Nature 412, 66–69.

Jenkins, W.J., 1987. ^3H and ^3He in the Beta triangle: observations of gyre ventilation and oxygen utilization rates. J. Phys. Oceanogr. 17, 763–783.

Jenkyns, H.C., Winterer, E.L., 1982. Palaeoceanography of Mesozoic ribbon radiolarites. Earth Planet. Sci. Lett. 60, 351–375.

Johnson, T.C., 1976. Controls on the preservation of biogenic opal in sediments of the Eastern Tropical Pacific. Science 192, 887–890.

Jones, D.L., Murchey, B.L., 1986. Geologic significance of Paleozoic and Mesozoic radiolarian chert. Annu. Rev. Earth Planet. Sci. 14, 455–492.

Jousé, A.P., 1978. Diatom biostratigraphy on the generic level. Micropaleontology 24, 316–326.

Kakuwa, Y., 1996. Permian–Triassic mass extinction event recorded in bedded chert sequence in southwest Japan. Palaeogeogr. Palaeoclimatol. Palaeoecol. 121, 35–51.

Kalberer, M., Fischer, G., Pätzold, J., Donner, B., Segl, M., Wefer, G., 1993. Seasonal sedimentation and stable isotope records of pteropods off Cap Blanc. Mar. Geol. 113, 305–320.

Kamykowski, D., Zentara, S.J., 1989. Circumpolar plant nutrient covariation in the Southern Ocean: patterns and processes. Mar. Ecol. Prog. Ser. 58, 101–111.

Karner, M.B., DeLong, E.F., Karl, D.M., 2001. Archaeal dominance in the mesopelagic zone of the Pacific Ocean. Nature 409, 507–510.

Kastanja, M.M., Henrich, R., 2007. Grain size variations in pelagic carbonate oozes from the Walvis Ridge–SE Atlantic (ODP Site 1265): a Miocene record of carbonate sedimentation and preservation. Mar. Geol. 237, 97–108.

Kastner, M., Keene, J.B., Gieskes, J.M., 1977. Diagenesis of siliceous oozes. I. Chemical controls on the rate of opal-A to opal-CT transformation—an experimental study. Geochim. Cosmochim. Acta 41, 1041–1059.

Katz, B.J., 2005. Controlling factors on source-rock development—a review of productivity, preservation, and sedimentation rate. In: Harris, N.B. (Ed.), The Deposition of Organic-Carbon-Rich Sediments: Models, Mechanisms, and Consequences, Soc. Sediment. Geol. Spec. Publ., vol. 82. 7–16.

Katz, M.E., Finkel, Z.V., Grzebyk, D., Knoll, A.H., Falkowski, P.G., 2004. Evolutionary trajectories and biogeochemical impacts of marine eukaryotic phytoplankton. Annu. Rev. Ecol. Evol. Syst. 35, 523–556.

Katz, M.E., Wright, J.D., Miller, K.G., Cramer, B.S., Fennel, K., 2005. Biological overprint of the geological carbon cycle. Mar. Geol. 217, 323–338.

Keller, M.D., Bellows, W.K., Guillard, R.R.L., 1989. Dimethyl sulfide production in marine phytoplankton. Am. Chem. Soc. Symp. Ser. 393, 167–182.

Kemle-von Mücke, S., Oberhänsli, H., 1999. The distribution of living planktic foraminifera in relation to Southeast Atlantic oceanography. In: Fischer, G., Wefer, G. (Eds.), Use of Proxies in Paleoceanography. Examples from the South Atlantic. Springer, Berlin, pp. 91–115.

Kemp, A.E.S., Lange, C.B., Pike, J., 2000. The "fall dump"—a new perspective on the role of a "shade flora" in the annual cycle of diatom production and export flux. Deep Sea Res. II 47, 2129–2154.

Kennedy, W.J., 1987. Late Cretaceous and Early Paleocene Chalk Group sedimentation in the Greater Ekofisk area, North Sea Central Graben. Bull. Cent. Rech. Explor. Prod. Elf-Aquitaine 11, 91–126.

Kennedy, W.J., Garrisson, R.E., 1975. Morphology and genesis of nodular chalk and hardgrounds in the Upper Cretaceous of southern England. Sedimentology 22, 311–386.

Kerr, S.R., Dickie, L.M., 2001. The Biomass Spectrum: A Predator-Prey Theory of Aquatic Production. Columbia University Press, New York, NY 320pp.

Kidder, D.L., Erwin, D.H., 2001. Secular distribution of biogenic silica through the Phanerozoic: comparison of silica-replaced fossils and bedded cherts at the series level. J. Geol. 109, 509–522.

Kidder, D.L., Mumma, S.A., 2003. Silica-replaced oolites, bedded shelf cherts and Paleozoic changes in the silica cycle. Sediment. Geol. 162, 159–166.

Kiørboe, T., Hansen, J.L.S., 1993. Phytoplankton aggregate formation: observations of patterns and mechanisms of cell sticking and the significance of exopolymeric material. J. Plankton Res. 15, 993–1018.

Klaas, C., Archer, D.A., 2002. Association of sinking organic matter with various types of mineral ballast in the deep sea: implications for the rain ratio. Global Biogeochem. Cycles 16, 1116. doi:10.1029/2001GB001765.

Klein, C., 2005. Some Precambrian banded iron-formations (BIFs) from around the world: their age, geologic setting, mineralogy, metamorphism, geochemistry, and origins. Am. Mineral. 90, 1473–1499.

Kling, S.A., Boltovskoy, D., 1995. Radiolarian vertical distribution across the southern California Current. Deep Sea Res. I 42, 191–231.

Knauth, L.P., 1994. Petrogenesis of chert. Rev. Min. 29, 233–258.

Knoll, A.H., 1989. Evolution and extinction in the marine realm—some constraints imposed by phytoplankton. Philos. Trans. R. Soc. Lond. B 325, 279–290.

Knoll, A.H., 1992. The early evolution of eukaryotes: a geological perspective. Science 256, 622–627.

Knoll, A.H., 1994. Proterozoic and Early Cambrian protists: evidence for accelerating evolutionary tempo. Proc. Natl. Acad. Sci. USA 91, 6743–6750.

Knoll, A.H., 2003. Biomineralization and evolutionary history. Rev. Miner. Geochem. 54, 329–356.

Knoll, A.H., Lipps, J.H., 1993. Evolutionary history of prokaryotes and protists. In: Lipps, J.H. (Ed.), Fossil Prokaryotes and Protists. Blackwell, Boston, pp. 19–30.

Knoll, A.H., Javaux, E.J., Hewitt, D., Cohen, P., 2006. Eukaryotic organisms in Proterozoic oceans. Philos. Trans. R. Soc. Lond. B 361, 1023–1038.

Konhauser, K.O., Phoenix, V.R., Bottrell, S.H., Adams, D.G., Head, I.M., 2001. Microbial-silica interactions in Icelandic hot spring sinter: possible analogues for some Precambrian siliceous stromatolites. Sedimentology 48, 415–433.

Kooistra, W.H.C., Gersonde, R., Medlin, L.K., Mann, D.G., 2007. The origin and evolution of the diatoms: Their adaptation to a planktonic existence. In: Falkowski, P.G., Knoll, A.H. (Eds.), Evolution of Primary Producers in the Sea, Elsevier, Amsterdam, pp. 207–249.

Kozur, H.W., 1998. Some aspects of the Permian–Triassic boundary (PTB) and of the possible causes for the biotic crisis around this boundary. Palaeogeogr. Palaeoclimatol. Palaeoecol. 143, 227–272.

Kumar, N., Anderson, R.F., Mortlock, R.A., Froelich, P.N., Kubik, P., Dittrich-Hannen, B., et al., 1995. Increased biological productivity and export production in the glacial Southern Ocean. Nature 378, 675–680.

Kump, L.R., 1991. Interpreting carbon-isotope excursions: Strangelove oceans. Geology 19, 299–302.

Kump, L.R., Pavlov, A., Arthur, M.A., 2005. Massive release of hydrogen sulfide to the surface ocean and atmosphere during intervals of oceanic anoxia. Geology 33, 397–400.

Lackschewitz, K.S., Grützmacher, U., Henrich, R., 1991. Paleoceanography and rotational block faulting in the Jurassic carbonate series of the Chiemgau Alps (Bavaria). Facies 24, 1–24.

Lampitt, R.S., 1985. Evidence for the seasonal deposition of detritus to the deep-sea floor and its subsequent resuspension. Deep-Sea Res. 32, 885–897.

Lampitt, R.S., Antia, A.N., 1997. Particle flux in deep seas: regional characteristics and temporal variability. Deep Sea Res. 44, 1377–1403.

Lange, C.B., Berger, W.H., 1993. Diatom productivity and preservation in the western equatorial Pacific: the Quaternary record. In: Berger, W.H., Kroenek, L.W., Mayer, L. A. (Eds.), Proceedings ODP Science Results 130. Texas University, College Station, pp. 509–523.

Lange, C.B., Romero, O.E., Wefer, G., Gabric, A.J., 1998. Offshore influence of coastal upwelling off Mauritania, NW Africa, as recorded by diatoms in sediment traps at 2195 m water depth. Deep Sea Res. I 45, 985–1013.

Lasemi, Z., Sandberg, P.A., 2000. Linked cyclic changes in Phanerozoic ocean atmosphere composition and original mineralogy of micritic limestones. Geol. Soc. Am. Abstr. Progr. 32, A68.

Lasseur, E., Guillocheau, F., Robin, C., Hanot, F., Vaslet, D., Coueffe, R., et al., 2009. A relative water-depth model for the Normandy Chalk (Cenomanian-Middle Coniacian, Paris Basin, France) based on facies patterns of metre-scale cycles. Sediment. Geol. 213, 1–26.

Latimer, J.C., Filippelli, G.M., 2001. Terrigenous input and paleoproductivity in the Southern Ocean. Paleoceanography 16, 1–17.

Laws, E.A., Falkowski, P.G., Smith, W.O., Ducklow, H., McCarthy, J.J., 2000. Temperature effects on export production in the open ocean. Global Biogeochem. Cycles 14, 1231–1246.

Lawson, D.S., Hurd, D.C., Pankratz, H.S., 1978. Silica dissolution rates of decomposing assemblages at various temperatures. Am. J. Sci. 278, 1373–1393.

Lawver, L.A., Gahagan, L.M., 2003. Evolution of Cenozoic seaways in the circum-Antarctic region—Palaeogeogr. Palaeoclimatol. Palaeoecol. 198, 11–37.

Lazarus, D., Caulet, J.P., 1994. Cenozoic Southern Ocean reconstructions from sedimentologic, radiolarian, and other microfossil data. In: Kennett, J.P., Warnke, D.A. (Eds.), The Antarctic Paleoenvironment: A Perspective on Global Change, Pt. 2. Antarctic Res. Ser., vol. 60. 145–174.

Le, J., Shackleton, N.J., 1992. Carbonate dissolution fluctuations in the western equatorial Pacific during the Late Quaternary. Paleoceanography 7, 21–42.

Leckie, R.M., Bralower, T.J., Cashman, R., 2002. Oceanic anoxic events and plankton evolution: biotic response to tectonic forcing during the mid-Cretaceous. Paleoceanography 17, 1041. doi:10.1029/2001PA000623.

Legendre, L., Le Févre, J., 1989. Hydrodynamical singularities as controls of recycled versus export production in oceans. In: Berger, W.H., Smetacek, V.S., Wefer, G. (Eds.), Productivity of the Ocean: Present and Past. Wiley, S. Bernard, Dahlem Konferenzen, pp. 49–63.

Legendre, L., Demers, S., Lefaivre, D., 1986. Biological production at marine ergoclines. In: Nihoul, J.C.J. (Ed.), Marine Interface Ecohydrodynamics. Elsevier, Amsterdam, pp. 1–29.

Leventer, A., 1992. Modern distribution of diatoms in sediments from the George V Coast, Antarctica. Mar. Micropaleontol. 19, 315–332.

Leventer, A., 2003. Particulate flux from sea ice in polar waters. In: Thomas, D.N., Dieckmann, G.S. (Eds.), Sea Ice: An Introduction to its Physics, Biology, Chemistry, and Geology. Blackwell Science, Amsterdam, pp. 303–332.

Lipps, J.H., 1993. Introduction to fossil prokaryotes and protists. In: Lipps, J.H. (Ed.), Fossil Prokaryotes and Protists. Blackwell, Cambridge, pp. 1–10.

Lisitzin, A.P., 1972. Sedimentation in the World Ocean. Soc. Econ. Paleontol. Mineral. Spec. Publ. vol. 17, Tulsa, 218pp.

Lisitzin, A.P., 1985. The silica cycle during the last ice age. Palaeogeogr. Palaeoclimatol. Palaeoecol. 50, 241–270.

Lochte, K., Anderson, R., Francois, R., Jahnke, R., Shimmield, G., Vetrov, A., 2003. Benthic processes and the burial of carbon. In: Fasham, M.J.R. (Ed.), Ocean Biogeochemistry. Springer, Heidelberg, pp. 195–216.

Logan, G.A., Hayes, J.M., Hieshima, G.B., Summons, R.E., 1995. Terminal Proterozoic reorganization of biogeochemical cycles. Nature 376, 53–56.

Longhurst, A., Sathyendranath, S., Platt, T., Caverhill, C., 1995. An estimate of global primary production in the ocean from satellite radiometer data. J. Plankton Res. 17, 01245–01271.

Louanchi, F., Najjar, R.G., 2000. A global monthly mean climatology of phosphate, nitrate and silicate in the upper ocean: spring–summer production and shallow re-mineralization. Global Biogeochem. Cycles 14, 957–977.

Lowenstein, T.K., Timofeeff, M.N., Brennan, S.T., Hardie, L.A., Demicco, R.V., 2001. Oscillations in Phanerozoic seawater chemistry: evidence from fluid inclusions. Science 294, 1086–1088.

Lowenstein, T.K., Hardie, L.A., Timofeeff, M.N., Demicco, R.V., 2003. Secular variation in seawater chemistry and the origin of calcium chloride basinal brines. Geology 31, 857–860.

Lucchi, R.G., Rebesco, M., 2007. Glacial contourites on the Antarctic Peninsula margin: insight for palaeoenvironmental and palaeoclimatic conditions. In: Viana, A.R., Rebesco, M. (Eds.), Economic and Palaeoceanographic Significance of Contourite Deposits, Geol. Soc. London, Spec. Publ., vol. 276. 111–128.

Lyle, M., 2003. Neogene carbonate burial in the Pacific Ocean. Paleoceanography 18 (1059) doi:10.1029/2002PA000777.

Lyle, M., Zahn, R., Prahl, F.G., Dymond, J., Collier, R., Pisias, N.G., et al., 1992. Paleoproductivity and carbon burial across the California Current: the MULTITRACERS transect, 42°N. Paleoceanography 7, 251–272.

MacArthur, R.H., Connel, J.H., 1966. The Biology of Population. John Wiley, New York 200pp.

Mackensen, A., 2004. Changing Southern Ocean palaeocirculation and effects on global climate. Antarct. Sci. 16, 369–386.

Mackenzie, F.T., Morse, J.W., 1992. Sedimentary carbonates through Phanerozoic time. Geochim. Cosmochim. Acta 56, 3281–3295.

Madin, L.P., 1982. Production, composition and sedimentation of salp fecal pellets in oceanic waters. Mar. Biol. 67, 39–45.

Maldonado, M., Carmona, M.C., Uriz, M.J., Cruzado, A., 1999. Decline in Mesozoic reef-building sponges explained by silicon limitation. Nature 401, 785–788.

Maliva, R.G., Knoll, A.H., Siever, R., 1989. Secular change in chert distribution: a reflection of evolving biological participation in the silica cycle. Palaios 4, 519–532.

Maliva, R.G., Knoll, A.H., Simonson, B.M., 2005. Secular change in the Precambrian silica cycle: insights from chert petrology. Geol. Soc. Am. Bull. 117, 835–845.

Marcantonio, F., Anderson, R.F., Stute, M., Kumar, N., Schlosser, P., Mix, A., 1996. Extraterrestrial ^{3}He as a tracer of marine sediment transport and accumulation. Nature 383, 705–707.

Marcantonio, F., Anderson, R.F., Higgins, S., Stute, M., Schlosser, P., Kubik, P., 2001. Sediment focusing in the central equatorial Pacific Ocean. Paleoceanography 16, 260–267.
Marchitto, T.M., Lynch-Stieglitz, J., Hemming, S.R., 2005. Deep Pacific $CaCO_3$ compensation and glacial–interglacial atmospheric CO_2. Earth Planet. Sci. Lett. 231, 317–336.
Margalef, R., 1994. Dynamic aspects of diversity. J. Veg. Sci. 5, 451–456.
Margalef, R., 1997. Our biosphere. In: Kinne, O. (Ed.), Excellence in Ecology, Book 10. International Ecology Institute, Oldendorf/Luhe, p. 194.
Martin, R.E., 1995. Cyclic and secular variation in microfossil biomineralization: clues to the biogeochemical evolution of Phanerozoic oceans. Global Planet. Change 11, 1–23.
Martin, R.E., 1996. Secular increase in nutrient levels through the Phanerozoic: implications for productivity, biomass, and diversity of the marine biosphere. Palaios 11, 209–219.
Martin, R.E., 1998. Catastrophic fluctuations in nutrient levels as an agent of mass extinction: upward scaling of ecological processes. In: McKinney, M.L., Drake, J.A. (Eds.), Biodiversity Dynamics. Turnover of Populations, Taxa, and Communities. Columbia University Press, New York, pp. 405–429.
Martin, R.E., 2001. Marine plankton. In: Briggs, D.E.G., Crowther, P.R. (Eds.), Paleobiology II: A Synthesis. Blackwell, Oxford, UK, pp. 309–312.
Martin, R.E., 2002. Cyclic and secular trends in preservation through geologic time: implications for the evolution of biogeochemical cycles. In: De Renzi, M., Alonso, M. V.P., Belinchón, M., Peñalver, E., Montoya, P., Márquez-Aliaga, A. (Eds.), Current Topics on Taphonomy and Fossilization—Taphos: International Conference on Taphonomy. Ayuntamiento de Valencia, Valencia, Spain, pp. 67–76.
Martin, J.H., Fitzwater, S.E., 1988. Iron deficiency limits phytoplankton growth in the North-East Pacific subarctic. Nature 331, 341–343.
Martin, J.H., Knauer, G.A., Karl, D.M., Broenkow, W.W., 1987. VERTEX—carbon cycling in the northeast Pacific. Deep Sea Res. A 34, 267–285.
Martin, J.H., Fitzwater, S.E., Gordon, R.M., Hunter, C.N., Tanner, S.J., 1993. Iron, primary production and carbon–nitrogen flux studies during the JGOFS North Atlantic Bloom Experiment. Deep Sea Res. II 40, 115–134.
Martin, R.E., Quigg, A., Podkovyrov, V., 2008. Marine biodiversification in response to evolving phytoplankton stoichiometry. Palaeogeogr. Palaeoclimatol. Palaeoecol. 258, 277–291.
Martin-Jézéquel, V., Hildebrand, M., Brzezinski, M.A., 2000. Silicon metabolism in diatoms: implications for growth. J. Phycol. 36, 821–840.
Mattioli, E., Erba, E., 1999. Synthesis of calcareous nannofossil events in Tethyan Lower and Middle Jurassic successions. Riv. Ital. Paleontol. Stratigr. 105, 343–376.
McBridge, E.F., Folk, R.L., 1979. Features and origin of Italian Jurassic radiolarites deposited on continental crust. J. Sediment. Petrol. 49, 837–868.
McCartney, K., 1993. Silicoflagellates. In: Lipps, J.H. (Ed.), Fossil Prokaryotes and Protists. Blackwell, Cambridge, pp. 143–154.
McClintic, M.A., DeMaster, D.J., Thomas, C.J., Smith, C.R., 2008. Testing the FOODBANCS hypothesis: seasonal variations in near-bottom particle flux, bioturbation intensity, and deposit feeding based on ^{234}Th measurements. Deep Sea Res. II 55, 2425–2437. doi:10.1016/j.dsr2.2008.06.003.
McGillicuddy, D.J., Robinson, A.R., Siegel, D.A., Jannasch, H.W., Johnson, R., Dickey, T.D., et al., 1998. Influence of mesoscale eddies on new production in the Sargasso Sea. Nature 394, 263–266.
McGillicuddy, D.J., Anderson, L.A., Doney, S.C., Maltrud, M.E., 2003. Eddy-driven sources and sinks of nutrients in the upper ocean: results from a 0.1° resolution model of the North Atlantic. Global Biogeochem. Cycles 17, doi:10.1029/2002GB001987.
McGowran, B., 1989. Silica burp in the Eocene ocean. Geology 17, 857–860.

McIntyre, A., Bé, A.W.H., 1967. Modern Coccolithophoridae of the Atlantic Ocean I. Placoliths and cyrtoliths. Deep-Sea Res. 14, 561–579.
McManus, J., Berelson, W.M., Hammond, D.E., Kilgore, T.E., DeMaster, D.J., Ragueneau, O., et al., 1995. Early diagenesis of biogenic silica: dissolution rates, kinetics, and paleoceanographic implications. Deep Sea Res. II 42, 871–903.
Medlin, L.K., Kooistra, W.C.H.F., Schmid, A.-M.M., 2000. A review of the evolution of the diatoms—a total approach using molecules, morphology and geology. In: Witkowski, A., Sieminska, J. (Eds.), The Origin and Early Evolution of the Diatoms: Fossil, Molecular and Biogeographical Approaches, Polish Acad. Sci., Krakow, pp. 13–35.
Mendelson, C.V., 1993. Acritarchs and prasinophytes. In: Lipps, J.H. (Ed.), Fossil Prokaryotes and Protists. Blackwell, Cambridge, pp. 77–104.
Michaels, A.F., Silver, M.W., 1988. Primary production, sinking fluxes and the microbial food web. Deep Sea Res. 35, 473–490.
Milliman, J.D., 1993. Production and accumulation of calcium carbonate in the ocean: budget of a non-steady state. Global Biogeochem. Cycles 7, 927–957.
Milliman, J.D., Droxler, A.W., 1996. Neritic and pelagic carbonate sedimentation in the marine environment: ignorance is not bliss. Geol. Rundsch. 85, 496–504.
Milliman, J.D., Troy, P.J., Balch, W.M., Adams, A.K., Li, Y.H., Mackenzie, F.T., 1999. Biologically mediated dissolution of calcium carbonate above the chemical lysocline. Deep Sea Res. I 46, 1653–1669.
Misumi, K., Yamanaka, Y., 2008. Ocean anoxic events in the mid-Cretaceous simulated by a 3-D biogeochemical general circulation model. Cretaceous Res. 29, 893–900.
Mix, A.C., Morey, A.E., 1996. Climate feedback and Pleistocene variations in the Atlantic south equatorial current. In: Wefer, G., Berger, W., Siedler, W.H., Webb, D.J. (Eds.), The South Atlantic: Present and Past Circulation. Springer, Berlin, pp. 503–525.
Moldowan, J.M., Jacobson, S.R., 2000. Chemical signals for early evolution of major taxa: biosignatures and taxon-specific biomarkers. Int. Geol. Rev. 42, 805–812.
Moldowan, J.M., Talyzina, N.M., 1998. Biogeochemical evidence for dinoflagellate ancestors in the Early Cambrian. Science 281, 1168–1170.
Molina, J.M., O'Dogherty, L., Sandoval, J., Vera, J.A., 1999. Jurassic radiolarites in a Tethyan continental margin (Subbetic, southern Spain): palaeobathymetric and biostratigraphic considerations. Palaeogeogr. Palaeoclimatol. Palaeoecol. 150, 309–330.
Molina-Cruz, A., Welling, L., Caudillo-Bohorquez, A., 1999. Radiolarian distribution in the water column, southern Gulf of California, and its implication in thanatocoenose constitution. Mar. Micropaleont. 37, 149–171.
Moore, T.C., 2008a. Chert in the Pacific: biogenic silica and hydrothermal circulation. Palaeogeogr. Palaeoclimatol. Palaeoecol. 261, 87–99.
Moore, T.C., 2008b. Biogenic silica and chert in the Pacific Ocean. Geology 36, 975–978.
Moore, J.K., Abbott, M.R., Richman, J.G., Nelson, D.M., 2000. The Southern Ocean at the last glacial maximum: a strong sink for atmospheric carbon dioxide. Global Biogeochem. Cycles 14, 455–475.
Moore, J.K., Doney, S.C., Glover, D.M., Fung, I.Y., 2002. Iron cycling and nutrient-limitation patterns in surface waters of the World Ocean. Deep Sea Res. II 49, 463–507.
Moore, T.C., Backman, J., Raffi, I., Nigrini, C., Sanfilippo, A., Pälike, H., et al., 2004. The Paleogene tropical Pacific: clues to circulation, productivity, and plate motion. In: Paleoceanography, 19, PA3013. doi:10.1029/2003PA000998.
Morel, F.M.M., Rueter, J.G., Price, N.M., 1991. Iron nutrition of phytoplankton and its possible importance in the ecology of ocean regions with high nutrient and low biomass. Oceanography 4, 56–61.
Moreno, A., Nave, S., Kuhlmann, H., Canals, M., Targarona, J., Freudenthal, T., et al., 2002. Productivity response in the North Canary Basin to climate changes during the last 250000 yr: a multi-proxy approach. Earth Planet. Sci. Lett. 196, 147–159.

Morrison, J.M., Gaurin, S., Codispoti, L.A., Takahashi, T., Millero, F.J., Gardner, W.D., et al., 2001. Seasonal evolution of the hydrographic properties in the Antarctic Circumpolar Current at 1701°W during 1997–1998. Deep Sea Res. II 48, 3943–3972.
Morse, J.W., Mackenzie, J.T., 1990. Geochemistry of Sedimentary Carbonates. Developments in Sedimentology48, Elsevier, Amsterdam pp.1–696.
Mortlock, R.A., Charles, C.D., Froelich, P.N., Zibello, M.A., Saltzman, J., Hays, J.D., et al., 1991. Evidence for lower productivity in the Antarctic Ocean during the last glaciation. Nature 351, 220–223.
Mueller, P.J., Suess, E., 1979. Productivity, sedimentation rate, and sedimentary organic matter in the oceans—I. Organic carbon preservation. Deep Sea Res. 26 A, 1347–1362.
Mulitza, S., Donner, B., Fischer, G., Paul, A., Pätzold, J., Rühlemannn, C., et al., 2003. The South Atlatic oxygen isotope record of planktic foraminifera. In: Wefer, G., Mulitza, S., Ratmeyer, V. (Eds.), The South Atlantic in the Late Quaternary: Reconstruction of Material Budget and Current Systems. Springer, Berlin, pp. 121–142.
Muller, R.D., Sdrolias, M., Gaina, C., Steinberger, B., Heine, C., 2008. Long-term sea-level fluctuations driven by ocean basin dynamics. Science 319, 1357–1362.
Murchey, B.L., Jones, D.L., 1992. A mid-Permian chert event; widespread deposition of biogenic siliceous sediments in coastal, island arc and oceanic basins. Palaeogeogr. Palaeoclimatol. Palaeoecol. 96, 161–174.
Muttoni, G., Kent, D.V., 2007. Widespread formation of cherts during the early Eocene climate optimum. Palaeogeogr. Palaeoclimatol. Palaeoecol. 253, 348–362.
Nagata, T., Fukuda, H., Fukuda, R., Koike, I., 2000. Bacterioplankton distribution and production in deep Pacific waters: large-scale geographic variations and possible coupling with sinking particle fluxes. Limnol. Oceanogr. 45, 426–435.
Nazarov, B.B., Ormiston, A.R., 1986. Trends in the development of Paleozoic radiolaria. Mar. Micropaleontol. 11, 3–32.
Nelson, D.M., Brzezinski, M.A., 1997. Diatom growth and productivity in an oligotrophic mid-ocean gyre: a 3-yr record from the Sargasso Sea near Bermuda. Limnol. Oceanogr. 42, 473–486.
Nelson, C.S., Cooke, P.J., 2001. History of oceanic front development in the New Zealand sector of the Southern Ocean during the Cenozoic—a synthesis. N. Z. J. Geol. Geophys. 44, 537–555.
Nelson, D.M., Dortch, Q., 1996. Silicic acid depletion and silicon limitation in the plume of the Mississippi River: evidence from kinetic studies in spring and summer. Mar. Ecol. Prog. Ser. 136, 163–178.
Nelson, D.M., Tréguer, P., Brzezinski, M.A., Leynaert, A., Quéguiner, B., 1995. Production and dissolution of biogenic silica in the ocean: revised global estimates, comparison with regional data and relationship to biogenic sedimentation. Global Biogeochem. Cycles 9, 359–372.
Nelson, D.M., Anderson, R.F., Barber, R.T., Brzezinski, M.A., Buesseler, K.O., Chase, Z., et al., 2002. Vertical budgets for organic carbon and biogenic silica in the Pacific sector of the Southern Ocean 1996–1998. Deep Sea Res. II 49, 1645–1674.
Nigrini, C., Caulet, J.P., 1992. Late Neogene radiolarian assemblage's characteristic of Indo-Pacific areas of upwelling. Micropaleontology 38, 139–164.
Norris, R.D., Klaus, A., Kroon, D., 2001. Mid-Eocene deep water, the Late Paleocene Thermal Maximum and continental slope mass wasting during the Cretaceous–Paleogene impact. In: Kroon, D., Norris, R.D., Klaus, A. (Eds.), Western North Atlantic Paleogene and Cretaceous Paleoceanography, Geol. Soc. London Spec. Publ., vol. 183. 23–48.
Oberhänsli, H., Benier, C., Meinecke, G., Schmidt, H., Schneider, R., Wefer, G., 1992. Planktonic foraminfera as tracers of ocean currents in the eastern South Atlantic. Paleoceanography 7, 607–632.

Ogg, J.G., Karl, S.M., Behl, R.J., 1992. Jurassic through Early Cretaceous sedimentation history of the central equatorial Pacific and of Sites 800 and 801. Proc. ODP Sci. Results 129, 571–613.

Oschlies, A., 2002. Can eddies make ocean deserts bloom. Global Biogeochem. Cycles 16, 1106. doi:10.1029/2001GB001830.

Paasche, E., 1962. Coccolith formation. Nature 193, 1094–1095.

Paffenhöfer, G.A., Knowles, S.C., 1979. Ecological implications of fecal pellet size, production and consumption by copepods. J. Mar. Res. 37, 35–49.

Parés, J.M., Moore, T.C., 2005. New evidence for the Hawaiian hotspot plume motion since the Eocene. Earth Planet. Sci. Lett. 237, 951–959.

Parker, F.L., Berger, W.H., 1971. Faunal and solution pattern of planktonic foraminifera in surface sediments of the south Pacific. Deep Sea Res. 18, 73–107.

Passow, U., 2000. Formation of transparent exopolymer particles TEP, from dissolved precursor material. Mar. Ecol. Prog. Ser. 192, 1–11.

Passow, U., Shipe, R.F., Murray, A., Pak, D.K., Brzezinski, M.A., Alldredge, A.L., 2001. The origin of transparent exopolymer particles (TEP) and their role in the sedimentation of particulate matter. Cont. Shelf Res. 21, 327–346.

Payne, J.L., Finnegan, S., 2006. Controls on marine animal biomass through geological time. Geobiology 4, 1–10.

Pearson, P.N., Palmer, M.R., 2000. Atmospheric carbon dioxide concentrations over the past 60 million years. Nature 406, 695–699.

Peinert, R., Bodungen, B.V., Smetacek, V.S., 1989. Food web structure and loss rate. In: Berger, W.H., Smetacek, V.S., Wefer, G. (Eds.), Productivity of the Ocean: Present and Past. Wiley & Sons, Chichester, pp. 35–48.

Peterson, L.C., Backman, J., 1990. Late Cenozoic carbonate accumulation and the history of the carbonate compensation depth in the western equatorial Indian Ocean. Proc. ODP Sci. Res. 115, 467–489.

Pfannkuche, O., Lochte, K., 2000. The biogeochemistry of the deep Arabian Sea: overview. Deep Sea Res. II 47, 2615–2628.

Pickett-Heaps, J., Schmid, A.M.M., Edgar, L.A., 1990. The cell biology of diatom valve formation. Prog. Phycol. Res. 7, 1–168.

Pollock, D.E., 1997. The role of diatoms, dissolved silicate and Antarctic glaciation in glacial/interglacial climatic change: a hypothesis. Global Planet. Change 14, 113–125.

Pondaven, P., Ragueneau, O., Tréguer, P., Hauvespre, A., Dezileau, L., Reyss, J.-L., 2000a. Resolving the 'opal paradox' in the Southern Ocean. Nature 405, 168–172.

Pondaven, P., Ruiz-Pino, D., Fravalo, C., Tréguer, P., Jeandel, C., 2000b. Interannual variability of Si and N cycles at the time series station KERFIX between 1990 and 1995—a 1-D modelling study. Deep Sea Res. I 47, 223–257.

Pospichal, J.J., Wise, S.W., 1990. Paleocene to middle Eocene calcareous nannofossils of ODP Sites 689 and 690, Maud Rise, Weddell Sea. Proc. ODP Sci. Res. 113, 613–638.

Prauss, M., 2000. The oceanographic and climatic interpretation of marine palynomorphs phytoplankton distribution from Mesozoic, Cenozoic and Recent sections. Göttinger Arb. Geol. Paläontol. 76, 1–235.

Quéguiner, B., Brzezinski, M.A., 2002. Biogenic silica production rates and particulate matter distribution in the Polar Frontal Zone of the Southern Ocean (Atlantic sector) during austral spring 1992. In: Deep Sea Res. II, 49, 1765–1786.

Quigg, A., Finkel, Z.V., Irwin, A.J., Rosenthal, Y., Ho, T.-Y., 2003. The evolutionary inheritance of elemental stoichiometry in marine phytoplankton. Nature 425, 291–294.

Quine, M., Bosence, D.W.J., 1991. Stratal geometries, facies and sea-floor erosion in Upper Cretaceous Chalk, Normandy, France. Sedimentology 38, 1113–1152.

Rabouille, C., Gaillard, J.-F., Tréguer, P., Vincendeau, M.-A., 1997. Biogenic silica recycling in surficial sediments across the Polar Front of the Southern Ocean (Indian sector). Deep Sea Res. II 44, 1151–1176.
Racki, G., Cordey, F., 2000. Radiolarian palaeoecology and radiolarites: is the present the key to the past. Earth-Sci. Rev. 52, 83–120.
Ragueneau, O., Treéguer, P., Leynaert, A., Anderson, R.F., Brzezinski, M.A., DeMaster, D.J., et al., 2000. A review of the Si cycle in the modern ocean: recent progress and missing gaps in the application of biogenic opal as a paleoproductivity proxy. Global Planet. Change 26, 317–365.
Ragueneau, O., Gallinari, M., Corrin, L., Grandel, S., Hall, P., Hauvespre, A., et al., 2001. The benthic silica cycle in the Northeast Atlantic: annual mass balance, seasonality and importance of non-steady-state processes for the early diagenesis of biogenic opal in deep-sea sediments. Prog. Oceanogr. 50, 171–200.
Ragueneau, O., Dittert, N., Pondaven, P., Tréguer, P., Corrin, L., 2002. Si/C decoupling in the world ocean: is the Southern Ocean different. Deep Sea Res. II 49, 3127–3154.
Ravelo, A.C., Fairbanks, R.G., Philander, S.G.H., 1990. Reconstructing tropical Atlantic hydrography using planktic foraminifera and an ocean model. Paleoceanography 5, 409–431.
Raven, J.A., 1997. The vacuole: A cost-benefit analysis. Adv. Bot. Res. 25, 59–86.
Rea, D.K., Leinen, M., 1985. Neogene history of the calcite compensation depth and lysocline in the South-Pacific Ocean. Nature 316, 805–807.
Rebesco, M., Camerlenghi, A., Volpi, V., Neagu, C., Accettella, D., Lindberg, B., et al., 2007. Interaction of processes and importance of contourites: insights from the detailed morphology of sediment Drift 7, Antarctica. In: Viana, A.R., Rebesco, M. (Eds.), Economic and Palaeoceanographic Significance of Contourite Deposits, Geol. Soc. London Spec. Publ., vol. 276. 95–110.
Reháková, D., Michalík, J., 1997. Evolution and distribution of calpionellids—the most characteristic constituents of Lower Cretaceous Tethyan microplankton. Cretaceous Res. 18, 493–504.
Rhein, M., Schott, F., Fischer, J., Send, U., Stramma, L., 1996. The deep water regime in the equatorial Atlantic. In: Wefer, G., Berger, W., Siedler, W.H., Webb, D.J. (Eds.), The South Atlantic—Present and Past Circulation. Springer, Berlin, pp. 261–271.
Ridgwell, A.T., 2005. A Mid Mesozoic Revolution in the regulation of ocean chemistry. Mar. Geol. 217, 339–357.
Ridgwell, A.T., Zeebe, R.E., 2005. The role of the global carbonate cycle in the regulation and evolution of the Earth system. Earth Planet. Sci. Lett. 234, 299–315.
Riding, R., 1993. Phanerozoic patterns of marine $CaCO_3$ precipitation. Naturwissenschaften 80, 513–516.
Riebesell, U., Zondervan, I., Rost, B., Tortell, P.D., Zeebe, R.E., Morel, F.M.M., 2000. Reduced calcification of marine plankton in response to increased atmospheric CO_2. Nature 407, 364–367.
Riech, V., Von Rad, U., 1979. Eocene porcellanites and Early Cretaceous cherts from the western North Atlantic. Init. Rep. Deep Sea Drill. Proj. 43, 437–455.
Ritzrau, W., 1996. Microbial activity in the benthic boundary layer (BBL): small scale distribution and its relationship to the hydrodynamic regime. J. Sea Res. 36, 171–180.
Ritzrau, W., Graf, G., Scheltz, A., Queisser, W., 2001a. Benthic-pelagic coupling and carbon dynamics in the northern North Atlantic. In: Schäfer, P., Ritzrau, W., Schlüter, M., Thiede, J. (Eds.), The Northern North Atlantic: A Changing Environment. Springer, Berlin, pp. 207–224.
Ritzrau, W., Graf, G., Schlüter, M., 2001b. Exchange processes across the sediment water interface. In: Schäfer, P., Ritzrau, W., Schlüter, M., Thiede, J. (Eds.), The Northern North Atlantic: A Chinging Environment. Springer, Berlin, pp. 199–206.

Robaszynski, F., Alcayde, G., Amedro, F., Badillet, G., Damotte, R., Foucher, J.C., et al., 1982. Le Turonien de la région-type; Saumurois et Touraine, stratigraphie, biozonations, sédimentologie. Bull. Cent. Rech. Explor. Prod. Elf-Aquitaine 6, 119–225.

Romero, E., Lange, C.B., Fischer, G., Treppke, U.F., Wefer, G., 1999. Variability in export production documented by downward fluxes and species composition of marine planktic diatoms: observations from the tropical and equatorial Atlantic. In: Fischer, G., Wefer, G. (Eds.), Use of Proxies in Paleoceanography, Examples from the South Atlantic. Springer, Berlin, pp. 365–392.

Romero, O.E., Armand, L.K., Crosta, X., Pichon, J.-J., 2005. The biogeography of major diatom taxa in Southern Ocean sediments: 3. tropical/subtropical species. Palaeogeogr. Palaeoclimatol. Palaeoecol. 223, 49–65.

Ronov, A.B., 1994. Phanerozoic transgressions and regressions on the continents: a quantitative approach based on areas flooded by the sea and areas of marine and continental deposition. Am. J. Sci. 294, 777–801.

Rosing, M., Frei, R., 2004. U-rich Archaean sea-floor sediments from Greenland—indications of >3700 Ma oxygenic photosynthesis. Earth Planet. Sci. Lett. 217, 237–244.

Ross, M., Scotese, C.R., 1988. A hierarchical tectonic model of the Gulf of Mexico and Caribbean region. Tectonophysics 155, 139–168.

Rost, B., Riebesell, U., 2004. Coccolithophores and the biological pump: responses to environmental changes. In: Thierstein, H.R., Young, J.R. (Eds.), Coccolithophores—From Molecular Processes to Global Impact. Springer, Berlin, pp. 99–125.

Roth, P.H., 1986. Mesozoic palaeoceanography of the North Atlantic and Tethys oceans. In: Summerhayes, C.P., Shackleton, N.J. (Eds.), North Atlantic Palaeoceanography, Geol. Soc. Spec. Publ. London, vol. 21. 299–320.

Roth, P.H., 1987. Mesozoic calcareous nannofossil evolution: relation to paleoceanographic events. Paleoceanography 2, 601–611.

Roth, P.H., 1989. Ocean circulation and calcareous nannoplankton evolution during the Jurassic and Cretaceous. Palaeogeogr. Palaeoclimatol. Palaeoecol. 74, 111–126.

Roth, P.H., 1994. Distribution of coccoliths in oceanic sediments. In: Winter, A., Siesser, W.G. (Eds.), Coccolithophores. Cambridge University Press, Cambridge, pp. 199–218.

Round, F.E., Crawford, R.M., Mann, D.G., 1996. The Diatoms: Biology and Morphology of the Genera. Cambridge University Press, Cambridge 747pp.

Salamy, K.A., Zachos, J.C., 1999. Latest Eocene–Early Oligocene climate change and Southern Ocean fertility: inferences from sediment accumulation and stable isotope data. Palaeogeogr. Palaeoclimatol. Palaeoecol. 145, 61–77.

Saltzman, M.R., 2005. Phosphorus, nitrogen, and the redox evolution of the Paleozoic oceans. Geology 33, 573–576.

Samtleben, C., Bickert, T., 1990. Coccoliths in sediment traps from the Norwegian Sea. Mar. Micropaleontol. 16, 39–65.

Sancetta, C., 1992. Primary production in the glacial North Atlantic and North Pacific Oceans. Nature 360, 249–251.

Sancetta, C., 1995. Diatoms in the Gulf of California—seasonal flux patterns and the sediment record for the last 15,000 years. Paleoceanography 10, 67–84.

Sancetta, C., 1999. Diatoms and marine paleogeography. In: Stoermer, E.F., Smol, S.P. (Eds.), The Diatoms: Applications for the Environmental and Earth Sciences. Cambridge University Press, Cambridge pp. 374–386.

Sandberg, P.A., 1983. An oscillating trend in Phanerozoic non-skeletal carbonate mineralogy. Nature 305, 19–22.

Sandberg, P.A., 1985. Nonskeletal aragonite and pCO_2 in the Phanerozoic and Proterozoic. Am. Geophys. Union Monogr. 32, 585–594.

Sanfilippo, A., Nigrini, C., 1998. Upper Paleocene–Lower Eocene deep-sea radiolarian stratigraphy and the Paleocene/Eocene series boundary. In: Aubry, M.-P., Lucas, S.G., Berggren, W.A. (Eds.), Late Paleocene–Early Eocene Climatic and Biotic Events in the Marine and Terrestrial Records. Columbia University Press, New York, pp. 244–276.

Sarmiento, J.L., Gruber, N., 2006. Ocean Biogeochemical Dynamics. Princeton University Press, Princeton 503pp.

Sarnthein, M., Faugères, J.-C., 1993. Radiolarian contourites record Eocene AABW circulation in the equatorial East Atlantic. Sediment. Geol. 82, 145–155.

Sarnthein, M., Pflaumann, U., Ross, R., Tiedemann, R., Winn, K., 1992. Transfer functions to reconstruct ocean paleoproductivity: a comparison. In: Summerhayes, C.P., Prell, W.L., Emeis, K.C. (Eds.), Upwelling Systems: Evolution since the Early Miocene, Geol. Soc. London Spec. Publ., vol. 64. 411–427.

Schidlowski, M., Hayes, J.M., Kaplan, I.P., 1983. Isotopic inferences of ancient biochemistries: carbon, sulphur, hydrogen, and nitrogen. In: Schopf, J.W. (Ed.), Earth's Earliest Biosphere: Its Origin and Evolution. Cambridge University Press, Cambridge, pp. 149–186.

Schlager, W., Schöllnberger, W., 1974. Das Prinzip stratigraphischer Wenden in der Schichtfolge der Nördlichen Kalkalpen. Mitt. Geol. Ges. Wien 66/67, 165–196.

Schlitzer, R., 2000. Applying the adjoint method for global biogeochemical modeling. In: Kasibhatla, P., Heimann, M., Hartley, D., Mahowald, N., Prinn, R., Rayner, P. (Eds.), Inverse Methods in Global Biogeochemical Cycles, Amer. Geophys. Union Monogr. Ser., vol. 114. 107–124 Boulder.

Schlitzer, R., 2002. Carbon export fluxes in the Southern Ocean: results from inverse modeling and comparison with satellite based estimates. Deep Sea Res. II 49, 1623–1644.

Schlitzer, R., 2004. Export production in the North Pacific derived from dissolved oxygen, nutrient and carbon data. J. Oceanogr. 60, 53–62.

Schlitzer, R., Usbeck, R., Fischer, G., 2003. Inverse modeling of particulate organic carbon fluxes in the South Atlantic. In: Wefer, G., Mulitza, S., Rathmeyer, V. (Eds.), The South Atlantic in the Late Quaternary—Reconstruction of Material Budget and Current Systems. Springer, Berlin, pp. 1–19.

Schlüter, M., Sauter, E.J., 2000. Spatial budget of organic carbon flux to the seafloor of the northern North Atlantic (60°N–80°N). Global Biogeochem. Cycles 14, 329–340.

Schlüter, M., Rutgers van der Loeff, M.M., Holby, O., Kuhn, G., 1998. Silica cycle in surface sediments of the South Atlantic. Deep Sea Res. I 45, 1085–1109.

Schmitz, W.J., McCartney, M.S., 1993. On the North Atlantic circulation. Rev. Geophys. 31, 29–49.

Schneider, R.R., Schulz, H.D., Hensen, C., 2006. Marine carbonates: their formation and destruction. In: Schulz, H.D., Zabel, M. (Eds.), Marine Geochemistry. Springer, Berlin, pp. 311–338.

Scholten, J.C., Fietzke, J., Rutgers van der Loeff, M.M., Mangini, A., Koeve, W., Stoffers, P., et al., 2001. Trapping efficiencies of sediment traps from the deep eastern North Atlantic: the ^{230}Th calibration. Deep Sea Res. 48, 2383–2408.

Schröder-Ritzrau, A., Andruleit, H., Jensen, S., Samtleben, C., Schäfer, P., Matthiessen, J., et al., 2001. Distribution, export and alteration of fossilizable plankton in the Nordic Seas. In: Schäfer, P., Ritzrau, W., Schlüter, M., Thiede, J. (Eds.), The Northern North Atlantic. A Changing Environment. Springer, Berlin, pp. 81–104.

Schubert, J.K., Kidder, D.L., Erwin, D.H., 1997. Silica-replaced fossils through Phanerozoic. Geology 25, 1031–1034.

Schubert, C.J., Villaneuva, J., Calvert, S.E., Cowie, G.L., Von Rad, U., Schulz, H., et al., 1998. Stable phytoplankton community structure in the Arabian Sea over the past 200,000 years. Nature 394, 563–566.

Schuster, S., Herndl, G.J., 1995. Formation and significance of transparent exopolymeric particles in the northern Adriatic Sea. Mar. Ecol. Prog. Ser. 124, 227–236.
Schwark, L., Empt, P., 2006. Sterane biomarkers as indicators of Palaeozoic algal evolution and extinction events. Palaeogeogr. Palaeoclimatol. Palaeoecol. 240, 225–236.
Sclater, J.G., Abbott, D., Thiede, J., 1977. Paleobathymetry and sediments of the Indian Ocean. In: Heirtzler, J.R. (Ed.), Indian Ocean Geology and Biostratigraphy. American Geophysical Union, Washington, pp. 25–60.
Sclater, J.G., Jaupart, C., Galson, D., 1980. The heat flow through oceanic and continental crust and the heat loss of the Earth. Rev. Geophys. Space Phys. 18, 269–311.
Seibold, E., Berger, W.H., 1993. The Sea Floor. An Introduction to Marine Geology, second ed. Springer, Heidelberg 356pp.
Sepkowski, J., 1997. Biodiversity: past, present, and future. J. Paleontol. 71, 533–539.
Shanks, A.L., Trent, J.D., 1980. Marine snow: sinking rates and potential role in vertical flux. Deep-Sea Res. 27, 137–144.
Shannon, L.V., Chapman, P., 1991. Evidence of Antarctic bottom water in the Angola Basin at 32°S. Deep Sea Res. 38, 1299–1304.
Shen, Y., Knoll, A.H., Walter, M.R., 2003. Evidence for low sulphate and anoxia in a mid-Proterozoic marine basin. Nature 423, 632–635.
Shen, Y., Zhang, T.G., Hoffman, P.F., 2008. On the coevolution of Ediacaran oceans and animals. Proc. Natl. Acad. Sci. USA 105, 7376–7381.
Sheridan, R.E., Gradstein, F.M., 1983. Site 534: Blake-Bahama Basin. In: Sheridan, R.E., Gradstein, R.M. (Eds.), Initial Reports of the Deep Sea Drilling Project vol. 76. U.S. Government Printing Office, Washington D.C, pp. 141–340.
Shimmield, G.B., Jahnke, R.A., 1995. Particle flux and its conversion to the sediment record: open ocean upwelling systems. In: Summerhayes, C.P., Emeis, K.-C., Angel, M.V., Smith, R.L., Zeitzschel, B. (Eds.), Upwelling in the Ocean: Modern Processes and Ancient Records. John Wiley & Sons, New York, pp. 171–192.
Siemann, M.G., 2003. Extensive and rapid changes in seawater chemistry during the Phanerozoic: evidence from Br contents in basal halite. Terra Nova 15, 243–248.
Siesser, W.G., 1993. Calcareous nannoplankton. In: Lipps, J.H. (Ed.), Fossil Prokaryotes and Protists. Blackwell, Boston, pp. 169–201.
Siever, R., 1991. Silica in the oceans: biological–geochemical interplay. In: Schneider, S.H., Boston, P.J. (Eds.), Scientists on Gaia. MIT Press, Cambridge, pp. 287–295.
Siever, R., 1992. The silica cycle in the Precambrian. Geochim. Cosmochim. Acta 56, pp. 3265–3272.
Sigman, D.M., McCorkle, D.C., Martin, W.R., 1998. The calcite lysocline as a constraint on glacial/interglacial low-latitude production changes. Global Biogeochem. Cycles 12, 409–427.
Silver, M.W., Alldredge, A.L., 1981. Bathypelagic marine snow: deep-sea algal and detrital community. J. Mar. Res. 39, 501–530.
Silver, M.W., Bruland, K.W., 1981. Differential feeding and fecal pellet composition of salps and pteropods, and the possible origin of the deep-water flora and olive-green 'cells'. Mar. Biol. 62, 263–273.
Simmons, M.D., BouDagher-Fadel, M.K., Banner, F.T., Whittaker, J.E., 1997. The Jurassic Favusellacea, the earliest Globigerina. In: BouDagher-Fadel, M.K., Banner, F.T., Whittaker, J.E. (Eds.), The Early Evolutionary History of Planktonic Foraminifera. Chapman & Hall, London, pp. 17–52.
Small, L.F., Fowler, S.W., Ünlü, M.Y., 1979. Sinking rates of natural copepod fecal pellets. Mar. Biol. 51, 233–241.
Smayda, T.J., 1971. Normal and accelerated sinking of phytoplankton in the sea. Mar. Geol. 11, 105–122.

Smetacek, V., 1991. Die Primärproduktion der marinen Plankton–Algen. In: Hempel, G. (Ed.), Biologie der Meere, Spektrum der Wissenschaft, 125, 33–44.
Smetacek, V., 1999. Diatoms and the ocean carbon cycle. Protist 150, 25–32.
Smetacek, V., 2000. The giant diatom dump. Nature 406, 574–575.
Spencer-Cervato, C., 1999. The Cenozoic deep sea microfossil record: explorations of the DSDP/ODP sample set using the Neptune database. Palaeontol. Electron. 2, 1–270.
Sprengel, C., Baumann, K.H., Henderiks, J., Henrich, R., Neuer, S., 2002. Modern coccolithophore and carbonate sedimentation along a productivity gradient in the Canary Island region: seasonal export production and surface accumulation rates. Deep Sea Res. II 49, 3577–3598.
Stanley, S.M., 2006. Influence of seawater chemistry on biomineralization throughout Phanerozoic time: paleontological and experimental evidence. Palaeogeogr. Palaeoclimatol. Palaeoecol. 232, 214–236.
Stanley, S.M., Hardie, L.A., 1998. Secular oscillations in carbonate mineralogy of reef-building and sediment-producing organisms driven by tectonically forced shifts in seawater chemistry. Palaeogeogr. Palaeoclimatol. Palaeoecol. 144, 3–19.
Steinberg, M., 1981. Biosiliceous sedimentation, radiolarite periods and silica budget fluctuations. Oceanol. Acta 4 (Suppl.), 149–154.
Stoermer, E.F., Smol, J.P., 2001. The Diatoms: Applications for the Environmental and Earth Sciences. Cambridge University Press, Cambridge 482pp.
Stover, L.E., Brinkhuis, H., Damassa, S.P., de Verteuil, L., Helby, R.J., Monteil, E., et al., 1996. Mesozoic-Tertiary dinoflagellates, acritarchs and prasinophytes. In: Jansonius, J., McGregor, D.C. (Eds.), Palynology: Principles and Applications, Amer. Assoc. Strat. Palynol. Found., vol. 2. 641–750 Washington.
Strauss, H., Des Marais, D.J., Hayes, J.M., Summons, R.E., 1992a. Concentrations of organic carbon and maturities and elemental compositions of kerogens. In: Schopf, J. W., Klein, C. (Eds.), The Proterozoic Biosphere: A Multidisciplinary Study. Cambridge University Press, Cambridge, pp. 95–101.
Strauss, H., Des Marais, D.J., Hayes, J.M., Summons, R.E., 1992b. The carbon-isotopic record. In: Schopf, J.W., Klein, C. (Eds.), The Proterozoic Biosphere: A Multidisciplinary Study. Cambridge University Press, Cambridge, pp. 117–129.
Stuut, J.B., Prins, M.A., Jansen, H.F., 2002. Fast reconnaissance of carbonate dissolution based on the size distribution of calcareous ooze on Walvis Ridge. Mar. Geol. 190, 581–589.
Suess, E., 1980. Particulate organic carbon flux in the oceans—surface productivity and oxygen utilization. Nature 288, 260–263.
Suman, D.O., Bacon, M.P., 1989. Variations in Holocene sedimentation in the North American basin determined from ^{230}Th measurements. Deep Sea Res. 36, 869–878.
Summons, R.E., Walter, M.R., 1990. Molecular fossils and microfossils of prokaryotes and protists from Proterozoic sediments. Am. J. Sci. 290A, 212–244.
Summons, R.E., Jahnke, L.L., Hope, J.M., Logan, G.A., 1999. 2-methylhopanoids as biomarkers for cyanobacterial oxygenic photosynthesis. Nature 400, 554–557.
Surge, D.M., Savarese, M., Dodd, J.R., Lohmann, K.C., 1997. Carbon isotopic evidence for photosynthesis in Early Cambrian oceans. Geology 25, 503–506.
Surlyk, F., 1997. A cool water carbonate ramp with bryozoan mounds: late Cretaceous–Danian of the Danish Basin. In: James, N.P., Clarke, J.A.D. (Eds.), Cool-Water Carbonates, Soc. Econ. Paleontol. Mineral. Spec. Publ., vol. 56. 293–308 Tulsa.
Suzuki, N., Kato, K., 1953. Studies on suspended materials: marine snow in the sea. I. Sources of marine snow. Bull. Fac. Fish. Hokkaido Univ. 4, 132–135.
Swanberg, N.R., Eide, L.K., 1992. The radiolarian fauna at the ice edge in the Greenland Sea during summer 1988. J. Mar. Res. 50, 297–320.

Taguchi, S., 1982. Seasonal study of fecal pellets and discarded houses of appendicularia in a subtropical inlet, Kaneohe Bay, Hawaii. Est. Coast Shelf Sci. 14, 545–555.
Takahashi, K., 1983. Radiolaria: sinking population, standing stock, and population rate. Mar. Micropaleontol. 8, 171–181.
Takahashi, K., 1991. Radiolaria: flux, ecology, taxonomy in the Pacific and Atlantic. In: Honjo, S. (Ed.), Ocean Biocoenosis Series 3. Woods Hole Oceanographic Institution, Washington, p. 303.
Takahashi, K., 1997. Siliceous microplankton fluxes in eastern subarctic Pacific, 1982–1986. J. Oceanogr. 53, 455–466.
Takeda, S., 1998. Influence of iron availability on nutrient consumption ratio of diatoms in oceanic waters. Nature 393, 774–777.
Talyzina, N.M., Moczydlowska, M., 2000. Morphological and ultrastructural studies of some acritarchs from the Lower Cambrian Lukati Formation. Estonia. Rev. Palaeobot. Palynol. 112, 1–21.
Tappan, H., 1980. The Paleobiology of Plant Protists. Freeman, San Francisco 1028pp.
Tappan, H., 1982. Extinction or survival: selectivity and causes of Phanerozoic crises. In: Silver, L.T., Schultz, P.H. (Eds.), Geological Implications of Impacts of Large Asteroids and Comets on the Earth, Geol. Soc. Amer. Spec. Pap., vol. 190. 265–276.
Tappan, H., 1986. Phytoplankton: below the salt at the global table. J. Paleontol. 60, 545–554.
Tappan, H., Loeblich Jr., A.R., 1973. Evolution of the oceanic plankton. Earth-Sci. Rev. 9, 207–240.
Taylor, N.J., 1985. Silica incorporation in the diatom *Coscinodiscus granii* as affected by light intensity. Br. Phycol. J. 20, 365–374.
Thiel, H., Pfannkuche, O., Schriever, G., Lochte, K., Gooday, A.J., Hemleben, C., et al., 1989. Phytodetritus on the deep-sea floor in a central oceanic region of the Northeast Atlantic. Biol. Oceanogr. 6, 203–239.
Thierstein, H.R., Young, J.R. (Eds.), 2004. Coccolithophores—From Molecular Processes to Global Impact. Springer, Berlin 565pp.
Thomas, W.H., Tynan, C.T., Gibson, C.H., 1997. Turbulence–phytoplankton interrelationship. Prog. Phycol. Res. 12, 283–324.
Thomas, D.J., Bralower, T.J., Jones, C.E., 2003. Neodymium isotopic reconstruction of late Paleocene–early Eocene thermohaline circulation. Earth Planet. Sci. Lett. 209, 309–322.
Thunell, R.C., Honjo, S., 1981. Calcite dissolution changes in the South Atlantic Ocean: response to initiation of northern hemisphere glaciation. Paleoceanography 4, 565–583.
Tolmacheva, T.J., Danelian, T., Popov, L.E., 2001. Evidence for 15 m.y. of continuous deep-sea biogenic siliceous sedimentation in early Paleozoic oceans. Geology 29, 755–758.
Tozzi, S., Schofield, O., Falkowski, P., 2004. Historical climate change and ocean turbulence as selective agents for two key phytoplankton functional groups. Mar. Ecol. Prog. Ser. 274, 123–132.
Tréguer, P., Nelson, D.M., Van Bennekom, A.J., De Master, D.J., Leynaert, A., Quéguiner, B., 1995. The silica balance in the world ocean: a re-estimate. Science 268, 375–379.
Tréguer, P., Legendre, L., Rivkin, R.T., Ragueneau, O., Dittert, N., 2003. Water column biogeochemistry below the euphotic zone. In: Fasham, M.J.R. (Ed.), Ocean Biogeochemistry. The Role of the Ocean Carbon Cycle in Global Change. Springer, Berlin, pp. 145–156.
Treppke, U.F., Lange, C.B., Wefer, G., 1996. Vertical fluxes of diatoms and silicoflagellates in the eastern equatorial Atlantic, and their contribution to the sedimentary record. Mar. Micropaleontol. 28, 73–96.

Tsuda, A., Saito, H., Machida, R.J., Shimode, S., 2003. A mesoscale iron enrichment in the western Subarctic Pacific induces a large centric diatom bloom. Science 300, 958–961.

Tucholke, B.E., Mountain, G.S., 1979. Seismic stratigraphy, lithostratigraphy and paleosedimentation patterns in the North Atlantic basin. In: Talwani, M., Hay, W., Ryan, W.B. F. (Eds.), Deep Drilling Results in the Atlantic Ocean: Continental Margins and Paleoenvironment vol. 3. American Geophysical Union, Washington, DC, pp. 58–86.

Tucker, M.E., 1990a. Carbonate depositional systems II: deeper-water facies of pelagic and resedimented limestones. In: Tucker, M.E., Wright, V.P. (Eds.), Carbonate Sedimentology. Blackwell, Oxford, pp. 228–283.

Tucker, M.E., 1990b. The geological record of carbonate rocks. In: Tucker, M.E., Wright, V.P. (Eds.), Carbonate Sedimentology. Blackwell, Oxford, pp. 401–421.

Turley, C.M., Lochte, K., 1990. Microbial response to the input of fresh detritus to the deep-sea bed. Palaeogeogr. Palaeoclimatol. Palaeoecol. 89, 3–23.

Turner, J.T., 1977. Sinking rates of fecal pellets from the marine copepod *Pontella meadii*. Mar. Biol. 40, 249–259.

Turner, J.T., 2002. Zooplankton fecal pellets, marine snow and sinking phytoplankton blooms. Aquat. Microb. Ecol. 27, 57–102.

Turner, J.T., Ferrante, J.G., 1979. Zooplankton fecal pellets in aquatic ecosystems. Bioscience 29, 670–677.

Tyrrell, T., Zeebe, R.E., 2004. History of carbonate ion concentration over the last 100 million years. Geochim. Cosmochim. Acta 68, 3521–3530.

Tyson, R.V., 1995. Sedimentary Organic Matter. Chapman & Hall, London 615pp.

Tyson, R.V., 2001. Sedimentation rate, dilution, preservation and total organic carbon: Some results of a modelling study. Org. Geochem. 32, 333–339.

Tyson, R.V., 2005. The "productivity versus preservation" controversy: cause, flaws, and resolution. In: Harris, N.B. (Ed.), The Deposition of Organic-Carbon-Rich Sediments: Models, Mechanisms, and Consequences, Soc. Sediment. Geol. Spec. Publ., 82, 7–16.

Uchman, A., Wetzel, A., 2011. Deep-sea ichnology: the relationships between depositional environment and endobenthic organisms. In: Hüneke, H., Mulder, T. (Eds.), Deep-Sea Sediments. Developments in Sedimentology, Elsevier, Amsterdam, vol. 63, pp. 517–556.

Usbeck, R., Schlitzer, R., Fischer, G., Wefer, G., 2003. Particle fluxes in the ocean: comparison of sediment trap data with results from inverse modelling. J. Mar. Syst. 39, 167–183.

Vail, P.R., Mitchum, R.M., Thompson, M.S., 1977. Seismic stratigraphy and global changes of sea level, part 4: global cycles of relative changes of sea level. In: Payton, C. E. (Ed.), Seismic stratigraphy and global changes of sea level, Amer. Assoc. Petrol. Geol. Mem., vol. 26. 83–97 Tulsa.

Van Andel, T.H., 1975. Mesozoic/Cenozoic calcite compensation depth and the global distribution of calcareous sediments. Earth Planet. Sci. Lett. 26, 187–194.

Van Beusekom, J.E.E., Van Bennekom, A.J., Tréguer, P., Morvan, J., 1997. Aluminium and silicic acid in water and sediments of the Enderby and Crozet basins. Deep Sea Res. II 44, 987–1004.

Van Cappellen, P., Qiu, L., 1997. Biogenic silica dissolution in sediments of the Southern Ocean: I. Solubility. Deep Sea Res. 44, 1109–1128.

Van Cappellen, P., Dixit, S., Van Beusekom, J., 2002. Biogenic silica dissolution in the oceans: Reconciling experimental and field-based dissolution rates. Glob. Biogeochem. Cycles 16, 1075, doi:10.1029/2001GB001431.

van den Hoek, C., Mann, D.G., Jahns, H.M., 1998. Algae, An Introduction to Phycology. Cambridge University Press, Cambridge 640pp.

Van der Loeff, M.M., Berger, G.W., 1991. Scavenging and particle flux: seasonal and regional variations in the Southern Ocean. Mar. Chem. 35, 553–567.

Van der Spoel, S., 1972. Pteropoda Thecosomata. Conseil International Pour Exploration De La Mer. Zooplankton Sheet 140–142, 1–12.

Vermeij, G.J., 1987. Evolution and Escalation: An Ecological History of Life. Princeton University Press, Princeton, NJ 527pp.

Vermeij, G.J., 1995. Economics, volcanoes, and Phanerozoic revolutions. Paleobiology 21, 125–152.

Vidal, G., Moczydlowska-Vidal, M., 1997. Biodiversity, speciation, and extinction trends of Proterozoic and Cambrian phytoplankton. Paleobiology 23, 230–246.

Villaneuva, J., Grimalt, J.O., Cortijo, E., Vidal, L., Labeyrie, L.D., 1997. A biomarker approach to the organic matter deposited in the North Atlantic during the last climatic cycle. Geochim. Cosmochim. Acta 61, 4633–4646.

Villareal, T.A., Pilskaln, C., Bzrezinski, M., Lipschultz, F., Dennet, M., Gardner, G.B., 1999. Upward transport of oceanic nitrate by migrating diatom mats. Nature 397, 423–433.

Vishnevskaya, V.S., 1997. Development of Palaeozoic–Mesozoic radiolaria in the northwestern Pacific rim. Mar. Micropaleontol. 30, 79–95.

Vishnevskaya, V.S., Kostyuchenko, A.S., 2000. The evolution of radiolarian biodiversity. Paleontol. J. 34, 124–130.

Volbers, A., Henrich, R., 2002. Present water mass calcium carbonate corrosiveness in the eastern South Atlantic inferred from ultrastructural breakdown of *Globigerina bulloides* in surface sediments. Mar. Geol. 186, 471–486.

Volbers, A., Henrich, R., 2004. Calcium carbonate corrosiveness in the South Atlantic during the Last Glacial Maximum as inferred from changes in the preservation of *Globigerina bulloides*: a proxy to determine deep water circulation pattern. Mar. Geol. 204, 43–57.

Von Bodungen, B., Antia, A., Bauerfeind, E., Haupt, O., Peeken, I., Peinert, R., et al., 1995. Pelagic processes and vertical flux of particles: an overview of a long-term comparative study in the Norwegian Sea and Greenland Sea. Geol. Rundsch. 84, 11–27.

Wagner, T., Damste, J.S.S., Hofmann, P., Beckmann, B., 2004. Euxinia and primary production in Late Cretaceous eastern equatorial Atlantic surface waters fostered orbitally driven formation of marine black shales. Paleoceanography 19, PA3009. doi:10.1029/2003PA000898.

Walliser, O.H., Reitner, J., 1999. Coquinas of pelagic fossils in pelagic facies: allodapic or autochthonous. N. Jb. Geol. Paläont. Abh. 214, 111–128.

Weaver, F.M., Wise Jr., S.W., 1974. Opaline sediments of the southeastern coastal plain and Horizon A: biogenic origin. Science 184, 899–901.

Wefer, G., Fischer, G., 1993. Seasonal patterns of vertical particle flux in equatorial and coastal upwelling areas of the eastern Atlantic. Deep Sea Res. I 40, 1613–1645.

Wefer, G., Suess, E., Balzer, W., Liebezeit, G., Müller, P.J., Ungerer, C.A., et al., 1982. Fluxes of biogenic components from sediment trap deployment in circumpolar waters of the Drake Passage. Nature 299, 145–147.

Wefer, G., Berger, W.H., Bickert, T., Donner, B., Fischer, G., Kemle-von Mücke, S., et al., 1996. Late Quaternary surface circulation of the South Atlantic: The stable isotope record and implications for heat transport and productivity. In: Wefer, G., Berger, W.H., Siedler, G., Webb, D.J. (Eds.), The South Atlantic: Present and Past Circulation. Springer, Berlin, Heidelberg, pp. 461–502.

Wefer, G., Berger, W.H., Bijma, J., Fischer, G., 1999. Clues to ocean history: a brief overview of proxies. In: Fischer, G., Wefer, G. (Eds.), Use of Proxies in Paleoceanography: Examples from the South Atlantic. Springer, Berlin, pp. 1–68.

Weissert, H., 1979. Die Paläoozeanographie der südwestlichen Tethys in der Unterkreide. Mitt. Geol. Inst. Eidg. Tech. Hochschule Univ. Zürich 226, 1–174.

Weissert, H., 1981a. Depositional processes in an ancient pelagic environment: the Lower Cretaceous Maiolica of the Southern Alps. Eclogae Geol. Helv. 74, 339–352.

Weissert, H., 1981b. The environment of deposition of black shales in the Early Cretaceous: an ongoing controversy. In: Warme, J.E., Douglas, R.G., Winterer, E.L. (Eds.), The Deep Sea Drilling Project: A Decade of Progress, Soc. Econ. Paleontol. Mineral. Spec. Publ., Vol. 32. 547–560.

Weissert, H., 2011. Mesozoic pelagic sediments—archives for ocean and climate history during green-house conditions. In: Hüneke, H., Mulder, T. (Eds.), Deep-Sea Sediments. Developments in Sedimentology, Elsevier, Amsterdam, vol. 63, pp. 765–792.

Weissert, H., Channell, J.E.T., 1989. Tethyan carbonate carbon isotope stratigraphy across the Jurassic–Cretaceous boundary: an indicator of decelerated global carbon cycling. Paleoceanography 4, 483–494.

Weissert, H., Erba, E., 2004. Volcanism, CO_2 and palaeoclimate: a Late Jurassic–Early Cretaceous carbon and oxygen isotope record. J. Geol. Soc. London 161, 695–702.

Wendt, J., Aigner, T., Neugebauer, J., 1984. Cephalopod limestone deposition on a shallow pelagic ridge: the Tafilalt Platform (Upper Devonian, eastern Anti-Atlas, Marocco). Sedimentology 31, 601–625.

Westbroek, P., Buddemeier, B., Coleman, M., Kok, D.J., Fautin, D., Stal, L., 1994. Strategies for the study of climate forcing by calcification. In: Doumenge, F. (Ed.), Past and present biomineralization processes. Considerations about the carbonate cycle, Bull. Inst. Océanogr. Monaco, Vol. 13. 37–60.

Westbroek, P., Brown, C.W., van Bleijswijk, J., Brownlee, C., Brummer, G.J., Conte, M., et al., 1993. A model system approach to biological climate forcing. The example of *Emiliania huxleyi*. Global Planet. Change 8, 27–46.

Wheeler, A.J., Stadnitskaia, A., 2011. Benthic deep-sea carbonates: reefs and seeps. In: Hüneke, H., Mulder, T. (Eds.), Deep-Sea Sediments. Developments in Sedimentology, Elsevier, Amsterdam, vol. 63, pp. 397–455.

Whitworth, T. III, Warren, B.A., Nowlin, J., Rutz, S.B., Pillsbury, R.D., Moore, M.I., 1999. On the deep western-boundary current in the Southwest Pacific Basin. Prog. Oceanogr. 43, 1–54.

Wignall, P., Twitchett, R., 2002. Permian–Triassic sedimentology of Jameson Land, East Greenland: incised submarine channels in an anoxic basin. J. Geol. Soc. London 159, 691–703.

Winguth, A.M.E., Maier-Reimer, E., 2005. Causes of the marine productivity and oxygen changes associated with the Permian-Triassic boundary: A reevaluation with ocean general circulation models. Mar. Geol. 217, 283–304.

Winter, A., Jordan, R.W., Roth, P.H., 1994. Biogeography of living coccolithophores in ocean waters. In: Winter, A., Siesser, W.G. (Eds.), Coccolithophores. Cambridge University Press, Cambridge, pp. 161–177.

Winterer, E.L., Bosellini, A., 1981. Subsidence and sedimentation on Jurassic passive continental margin, Southern Alps, Italy. Bull. Am. Assoc. Petrol. Geol. 65, 394–421.

Witte, U., Pfannkuche, O., 2000. High rates of benthic carbon remineralisation in the abyssal Arabian Sea. Deep Sea Res. II 47, 2785–2804.

Wollast, R., 1994. The relative importance of biomineralization and dissolution of $CaCO_3$ in the global carbon cycle. In: Dovmenge, F. (Ed.), Past and Present Biomineralization Processes: Considerations about the Carbonate Cycle, Bull. Inst. Oceanogr. Monaco, 16–36.

Wood, R.A., Grotzinger, J.P., Dickson, J.A.D., 2002. Proterozoic modular biomineralized metazoan from the Nama Group, Namibia. Science 296, 2383–2386.

Worthington, L.V., 1976. On the North Atlantic circulation. Johns Hopkins Oceanogr. Studies 6, 110pp.

Wu, G., Berger, W.H., 1991. Pleistocene $\partial^{18}O$ record from Ontong Java Plateau: effects of winnowing and dissolution. Mar. Geol. 96, 193–209.

Yamanaka, Y., Tajika, E., 1996. The role of the vertical fluxes of particulate organic matter and calcite in the oceanic carbon cycle: studies using an ocean biogeochemical general circulation model. Global Biogeochem. Cycles 10, 361–382.

Yoon, W.D., Kim, S.K., Han, K.N., 2001. Morphology and sinking velocities of fecal pellets of copepod, molluscan, euphausiid, and salp taxa in the northeastern tropical Atlantic. Mar. Biol. 139, 923–928.

Young, J.R., 1994. Functions of coccoliths. In: Winter, A., Siesser, W.G. (Eds.), Coccolithophores. Cambridge University Press, Cambridge, pp. 63–82.

Young, J.R., Ziveri, P., 2000. Calculation of coccolith volume and its use in calbration of carbonate flux estimates. Deep Sea Res. II 47, 1679–1700.

Young, J.R., Bown, P.B., Burnett, J.A., 1994. Paleontological perspectives. In: Green, J.C., Leadbeater, B.S.C. (Eds.), The Haptophyte Algae, Syst. Assoc. Special, vol. 51. 379–392.

Youngbluth, M.J., Bailey, T.G., Davoll, P.J., Jacoby, C.A., Blades-Eckelbarger, P.I., Griswold, C.A., 1989. Fecal pellet production and diel migratory behavior by the euphausiid *Meganyctiphanes norvegica* effect benthic-pelagic coupling. Deep-Sea Res. 36, 1491–1501.

Yu, E.F., Francois, R., Bacon, M.P., Honjo, S., Fleer, A.P., Manganini, S.J., et al., 2001. Trapping efficiency of bottom-tethered sediment traps estimated from the intercepted fluxes of ^{230}Th and ^{231}Pa. Deep Sea Res. I 48, 865–889.

Zachos, J.C., Lohmann, K.C., Walker, J.C.G., Wise, S.W., 1993. Abrupt climate change and transient climates during the Palaeogene—a marine perspective. J. Geol. 101, 191–213.

Zachos, J.C., Pagani, M., Sloan, L., Thomas, E., Billups, K., 2001. Trends, rhythms, and aberrations in global climate 65 Ma to present. Science 292, pp. 686–693.

Zeebe, R.E., Westbroek, P., 2003. A simple model for the $CaCO_3$ saturation state of the ocean: the "Strangelove", the "Neritan", and the "Cretan" Ocean. Geochem. Geophys. Geosyst. 4, 1104. doi:10.1029/2003GC000538.

Zeebe, R.E., Wolf-Gladrow, D., 2001. CO_2 in Seawater: Equilibrium, Kinetics, Isotopes. Oceanographic Series, Elsevier, New York vol. 65, p. 360.

Zielinski, U., Gersonde, R., 1997. Diatom distribution in Southern Ocean surface sediments (Atlantic sector): implications for paleoenvironmental reconstructions. Palaeogeogr. Palaeoclimatol. Palaeoecol. 129, 213–250.

Ziveri, P., Thunell, R.C., 2000. Coccolithophore export production in Guaymas Basin, Gulf of California: response to climate forcing. Deep-Sea Res. II (47), 2073–2100.

CHAPTER 5

Hemipelagic Advection and Periplatform Sedimentation

Rüdiger Henrich*[,1] *and* Heiko Hüneke[‡]

Contents

1. Introduction

About 90% of the sediment load generated by weathering and erosion on land, some 26 gigatons per year, is deposited at the ocean margins. Carried towards the sea by fluvial, glacial or aeolian transport, the sediments' seaward journey forms a chain of consecutive resedimentation processes acting on time scales from minutes to thousands of years. Major filters in the source-to-sink, that is land-to-ocean, sediment transfer are the continental

* Department of Sedimentology/Paleoceanography at the Faculty of Geosciences - University of Bremen, Klagenfurter Straße, Bremen, Germany
‡ Institute of Geography and Geology, University of Greifswald, Greifswald, Germany
1 Corresponding author.
E-mail address: henrich@uni-bremen.de

Developments in Sedimentology, Volume 63
ISSN 0070-4571, DOI: 10.1016/S0070-4571(11)63005-7

shelves. Shelves may be regarded as conduits for sediments to the continental margins. Commonly, high-energy conditions resulting from tides and storms dominate in near coastal shallow shelf seas. Coastal sediment transport is characterized by the injection of sand delivered by river mouths (i.e. "river mouth bypassing": Swift and Thorne, 1991) to the longshore drift, which results from obliquely breaking wavefronts inundating the coastal zone, whereas mud (i.e. silt and clay) is exported via turbid plumes and hyperpycnal flows, spreading obliquely across the shelf under the combined influence of buoyancy, shelf current and Coriolis forces (see Mulder, 2011, this volume, Section 1.5.3). Many such plumes end on the mid-shelf as a well-defined front contributing to the mid-shelf sink, which forms thick and extensive veneers of the mid-shelf silt depocentres. However, a certain amount of mud always escapes from the continental margin (Nitrouer and Wright, 1994).

Another possibility for sediment to bypass the shelf is by long-distance transport in the form of sediment-laden plumes. Examples of such extensive high-buoyancy fluxes are the plumes from Arctic rivers, which extend up to 500 km offshore well into the Arctic ocean basin, as well as the huge Amazon brackish plume, of which the jet reaches a mid-shelf position 200 km offshore, where it is turned towards the NW, parallel to the shelf contours, by the strong offshore North Brazilian Current (see the chapters in Nitrouer and Kuehl, 1995). The sea level acts as a prominent control on shelf-sediment dynamics on glacial/interglacial to millennium time scales. During lowstands, like the Last Glacial Maximum, the sea-level position was close to shelf break, inducing erosion on major parts of the exposed shelves as well as deposition of shelf-edge deltas (e.g. Winn et al., 1995); during highstands, like under modern conditions, starvation of the sediment supply to the outer shelf often goes parallel with highstand coastal sediment trapping. Significant secondary sediment availability is often related to intervals of sea-level rise and fall associated with intense reworking activity (e.g. palimpsest sediments and ravinement).

To conclude, a significant volume of shelf sediments is stored during highstands, at least temporarily, on the shelf, whereas sediment export to the continental slope is highest during lowstands and transgressions. Thus, sediment bypassing is common during lowstands, when fluvial material is delivered to the shelf edge and/or dust is supplied directly towards the continental slope. In addition, direct sediment drainage by canyons cutting deeply back into a shelf, and strong outer-shelf currents lead to rapid export of shelf sediments. The material crossing the shelf break is generally transported either in suspension (nepheloid layer; see Salon et al., 2008) or as gravity-driven density flows or currents. In addition to sea-level control, sediment bypassing may also be strongly affected by climatic changes in the hinterland, which in turn influence the sediment dynamics on the shelves and continental slopes, for example by variations in wind strength, thus controlling sediment input, shifts in precipitation intensity via surface runoff as well as changes in current and upwelling intensity (see examples underneath in the section on hemipelagic advection).

In addition to hemipelagic advection, predominately by terrigenous sediments, open-oceanic pelagic sediments may be diluted by neritic particles derived from shallow-water carbonate platforms (Schlager and James, 1978). Compositional variations of the periplatform oozes mainly reflect changes in the depositional conditions on the shallow-water platforms, because the vast quantity of excess bank-top production and the limited storage volume in the shallow-water realm may make large volumes of sediment available for off-bank export. It is obvious that also here, like in case of shelf-sediment export, the impact of sea-level fluctuations and climatic changes is very high and will be reflected by variations in the composition and accumulation patterns of periplatform sediments (see detailed discussion in Section 3 on periplatform sediments underneath).

2. Hemipelagic Advection

2.1. Contribution of fluvial supply to pelagic and hemipelagic sediments

In many regions of the ocean, river supply is by far the most important contributor of terrigenous material. The amount of sediment supplied by rivers has been shown to be proportional to the surface area of the drainage basin (Milliman, 1995; Milliman and Meade, 1983). It also depends on uplift rates in the drainage basin (Hovius, 1998) and on the climate, as this parameter determines the nature of the soil cover (Mulder and Syvitski, 1996). Figure 5.1 presents a recent compilation of annual river-discharge rates by Walsh and Nittrouer (2009). The highest discharges are recorded from the huge Asian rivers, in the catchment areas of which prominent collision zones (e.g. Himalayas) coincide with intense chemical weathering under tropical monsoonal climates. High rates are also registered from rivers with huge drainage basins in the tropical climate belt, like the Amazon and the Orinoco, whereas generally much lower rates characterize the discharge from the tectonically stable African craton (Summerfield and Hulton, 1994). In the recent past, damming has caused drastic changes in the sediment discharge by rivers (see Meade, 1996; Syvitski, 2003).

Important aspects are the timing and nature of the sediment discharge, and, in particular, the grain-size distribution and water/sediment ratio (Orton and Reading, 1993). The concentration of suspended sediment load is important and is generally larger for medium-sized rivers rather than for large rivers because of the dilution of sediment by the higher water discharge of the latter (see Mulder and Syvitski, 1995). Discharged into the river mouth, the material may be further dispersed on the shelf and

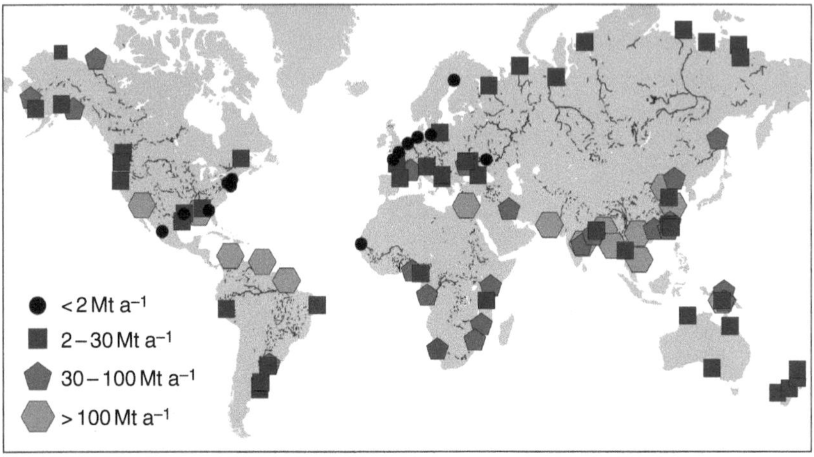

Figure 5.1 Magnitude of sediment discharge at large river mouths. Data are adopted from Milliman and Syvitski (1992) and Hovius (1998) (redrawn and slightly modified after Walsh and Nittrouer, 2009).

transported to the continental margin via surface plumes (hypopycnal flows), or as a dilute-suspension bottom layer or a sediment gravity flow (hyperpycnal plumes) (see Mulder, 2011, this volume, Sections 1.5.3 and 1.6). Flocculation is a key process limiting transport distances; if coupled with estuarine circulation, it results in rapid deposition close to the river mouth (Geyer et al., 2004). In general, most sediment is deposited at or near the seafloor where rivers meet the coastal ocean. These sediments can subsequently move across and along continental margins in the bottom boundary layer. They can experience multiple episodes of transport and re-deposition until their ultimate site of accumulation is reached. Finally, they are available to be pirated by contour currents (Rebesco and Camerlenghi, 2008; see also Faugères and Mulder, 2011, this volume).

Diffusive transport (by waves and tides) and advective (by currents) transport are assumed to control the mud distribution on the margins, and gravity-driven flows are thought to be important for moving sediment down the continental slope (McCave, 1972; Mulder, 2011, this volume).

Consequently, mud is not equally distributed over the shelf but is rather transferred by a wide variety of dispersal systems and deposited at different locations. Recently, Walsh and Nittrouer (2009) developed a new classification comprising five end members of fine-grained fluvial sediment dispersal systems (Fig. 5.2), for example the estuarine accumulation-dominated (EAD), the canyon-captured (CC), the proximal accumulation-dominated (PAD), the marine dispersal-dominated (MDD) and the subaqueous delta-clinoform (SDC) systems. These divisions are based on the dominant mode and pattern of sedimentation for each type. The main characteristics of each

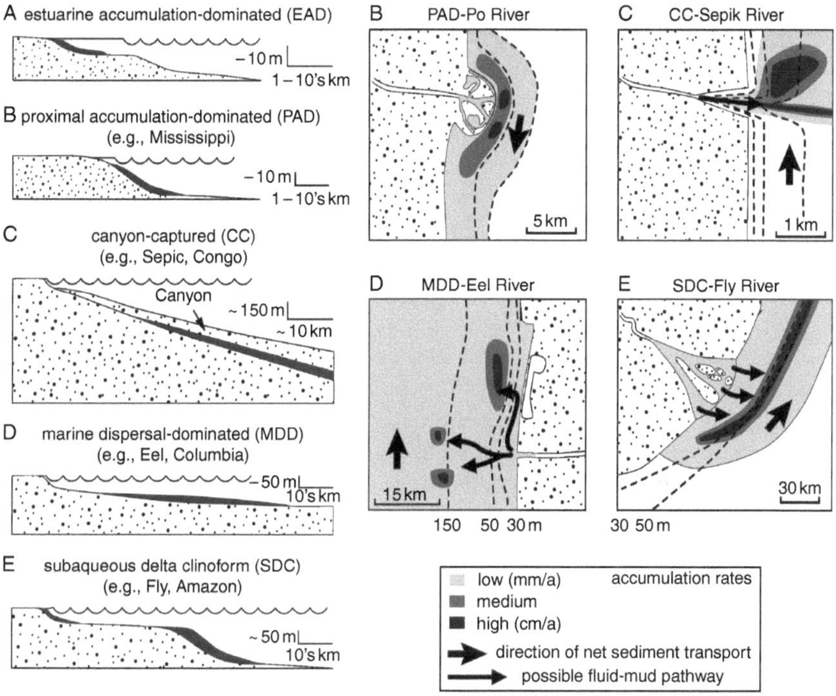

Figure 5.2 Major types of marine river-sediment dispersal systems (left side) and representative examples of each type (except EAD, right side) (redrawn and slightly modified after Walsh and Nittrouer, 2009). Examples for the PAD (B), CC (C), MDD (D) and SDC (E) are the Po, Sepik, Eel and Fly rivers, respectively (right). These example systems do not reflect end-members. The spatial distribution of sediment-accumulation rates are shown in the figures for each example system. Locations of maximum fine-grained sediment accumulation in each system type (left) and examples (right) are shown in grey. (A multi-colour version of this figure is on the included CD-ROM.)

type are summarized below, with particular emphasis on its potential as a source for the supply of fine-grained sediments to the continental slope and the deep sea.

(1) EAD systems are characterized by flocculation and estuarine circulation. Hence, sediments generally are rapidly deposited and mainly stored within the estuary, whereas only a small portion might escape. As a result, EAD systems are relatively inefficient in contributing mud to the continental margin and to the deep-sea realm.
(2) Flocculation and rapid deposition of mud are also the dominant features of PAD systems. However, since the estuaries are already filled up in these systems, mud is transferred to the inner shelf, where it is mainly deposited because of prevalent insignificant tidal and current regimes.

As a consequence, PAD systems also hinder the transfer of mud to the continental slope and deep sea, and may be dispersed further offshore.

(3) In contrast to EAD and PAD systems, mud is directly supplied into a canyon at CC river mouths and the majority passes through the canyon as sediment gravity flows (turbidity currents, debris flows) into the deep sea (cf. Mulder, 2011, this volume, Section 3.3.2). Excellent examples of such submarine canyon fine-grained sediment transfer come from the Sepik, Eel, Congo and Ganges–Brahmaputra rivers.

(4) MDD systems, characterized by bottom-boundary-layer transport and sediment gravity flows, develop where rivers discharge their sediment load into an oceanographic environment with moderate to large waves and/or currents. As a consequence, MDD systems may efficiently disperse most river sediments on the continental margin.

(5) SDC systems, typical features of broad, low-gradient margins, are characterized by a "tide-dominated delta" (triangular-shaped) as a result of strong tidal currents which induce current regimes carrying the fine-sediment load farther away from the shoreline. Examples of this type are the Amazon and Ganges–Brahmaputra systems.

In conclusion, frequency, size and distribution of the various dispersal systems strongly depend on shelf geometry and oceanography. Hence, any change in this configuration due to sea-level oscillations will immediately affect the configuration of these dispersal systems, as recorded by many studies for the last postglacial sea-level rise and summarized by Walsh and Nittrouer (2009). During sea-level lowstand, when rivers debouched at the modern shelf break, CC systems must have been considerably more common than today (Milliman and Syvitski, 1992). As the sea level rose rapidly after the Last Glacial Maximum, most rivers systems stored their sediment loads in estuarine valleys crossing the shelf (i.e. they were EAD systems). When the sea-level rise slowed around 8–6 ka BP, the storage capacity of some estuaries was rapidly reached, and deltas began to form around the world (Warne and Stanley, 1995).

2.2. Contribution of terrigenous dust to hemipelagic and pelagic sediments

In arid and semi-arid tropical and subtropical ocean provinces, dust supply is by far the most important terrigenous contribution to pelagic and hemipelagic sedimentation, with the most important source regions being the Sahara–Sahel in northern Africa and the Gobi–Taklamakan in central Asia (Fig. 5.3) (Harrison et al., 2001).

Figure 5.4, reproduced from Goudie and Middleton (2001), clearly displays three main dust source regions in northern Africa based on Total Ozone Mapping Spectrometer (TOMS) data, for example (1) the Bodélé Depression between Tibesti and Lake Chad, (2) a large swathe of parts of Mauritania, Mali

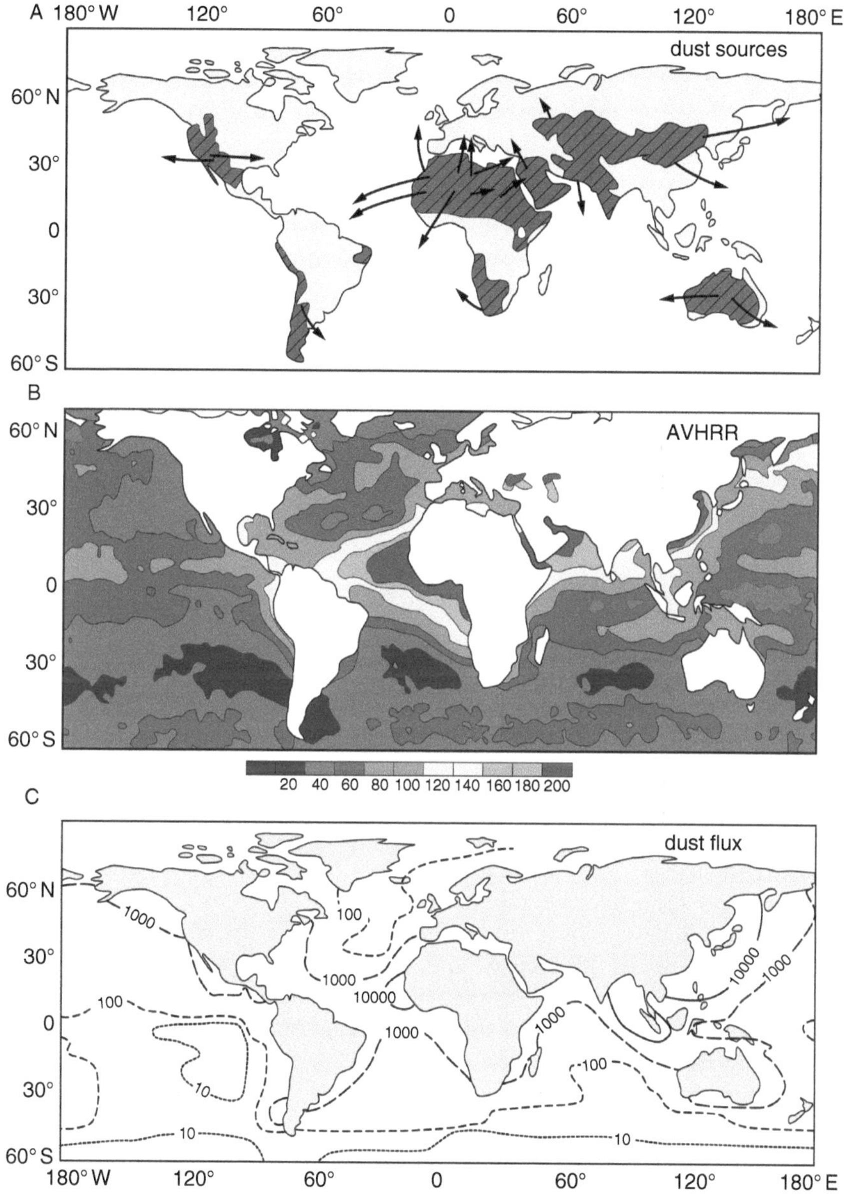

Figure 5.3 Present-day locations of dust sources, transport paths and deposition zones. (A) Present-day dust source regions and wind trajectories reconstructed from observations of dust storms (redrawn and slightly modified after Livingstone and Warren, 1996). (B) Zones of high atmospheric dust concentrations, inferred from the mean annual equivalent aerosol optical depth ($\times 1000$) as measured by an Advanced Very High Resolution Radiometer (AVHRR). (C) Global fluxes (mg m^{-2} a^{-1}) of mineral aerosols to the ocean (redrawn and slightly modified after Harrison et al., 2001). (A multi-colour version of this figure is on the included CD-ROM.)

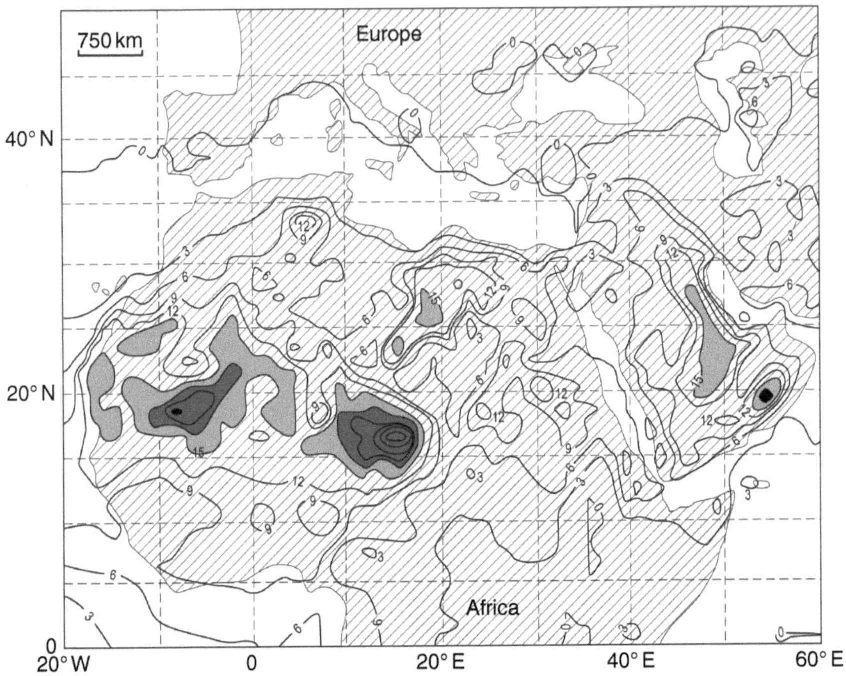

Figure 5.4 Annual mean Aerosol Index for the Sahara, derived from TOMS (redrawn and slightly modified after Goudie and Middleton, 2001).

and southern Algeria, and (3) the Nubian Desert in Egypt and northern Sudan. The importance of Bodélé as a dust source results from two main factors: (1) extreme dryness (Faya Largeau receiving an average annual rainfall of just 17 mm) and (2) extensive winnowing of silty alluvium by wadi streams draining the Tibesti Massif, and of silty deposits from an expanded Lake Chad during early Holocene and Pleistocene pluvials (Goudie and Middleton, 2001). Dust from the second important source (Mali, Mauritania and Algeria) originates from an area of low relief bounded on the North and East by uplands. Wadis draining the upland may have transported silt-rich alluvium into the area, thus providing the source for the dust.

Dust-raising winds vary from small-scale convective vortices, dust devils of the order of metres, to gust fronts associated with thunderstorm events and haboobs, which can be up to 1000 km wide (Middleton, 1986a,b). The formation of a deep thermally mixed layer over the southern Sahara results, for example in dust-laden air being carried to elevations of 3–5 km and incorporated into the African easterly jet (Karyampudi et al., 1999). Saharan dust is regularly transported from its source areas along three main transport paths: westward over the North Atlantic Ocean, northward across the Mediterranean to Europe, sometimes as far north as Scandinavia and along easterly trajectories across the eastern Mediterranean (see references in Goudie and Middleton, 2001). Dust transported within the jet across the Atlantic reaches the Caribbean and the Amazon Basin within 5–7 days (Glaccum and Prospero, 1980; Prospero and Lamb, 2003; Prospero et al., 1970, 1981).

Schütz et al. (1981) have modelled the annual mass budget of dust transported from the Sahara over the Atlantic (Fig. 5.5). A depositional rate of up to 20 cm per 1000 years occurs over the first 2000 km, whereas—when most of the mass of dust plume has fallen out at distances of over 2000 km—a zone of comparatively low accumulation rates (1–2 cm per 1000 years) occurs.

In contrast, the lofting of dust-laden air into the upper troposphere over central Asia is associated with the passage of successive cold fronts.

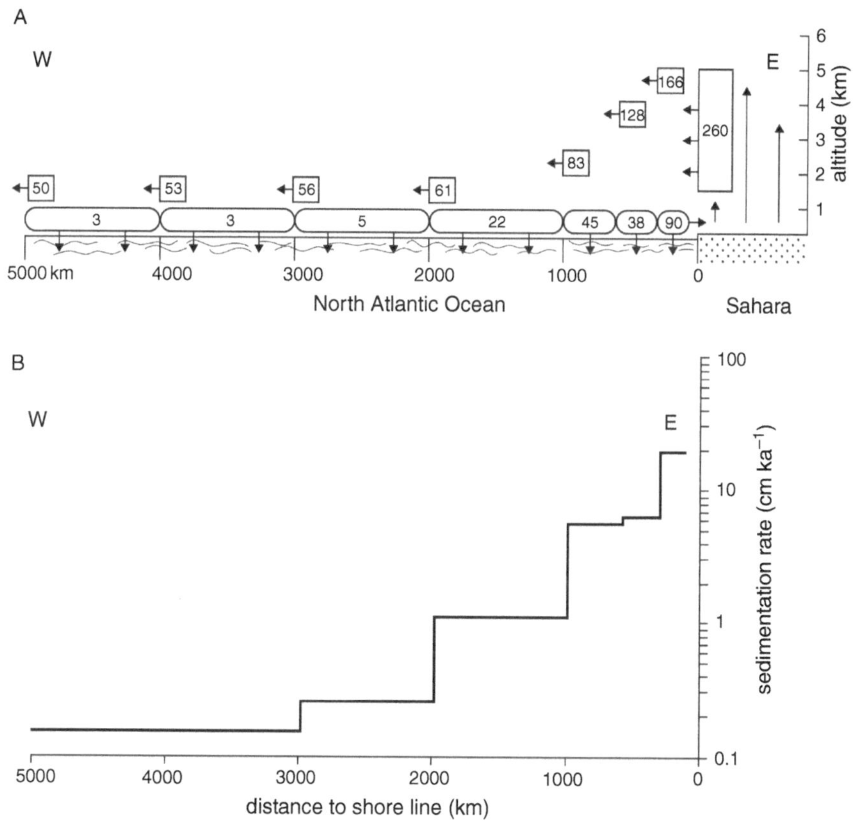

Figure 5.5 Schütz et al.'s (1981) model of dust transport and accumulation over the North Atlantic offshore from the Sahara (redrawn and slightly modified after Goudie and Middleton, 2001). (A) Annual mass budget at 15–24°N (10^6 t a^{-1}). (B) Sedimentation rate of dust in deep-sea sediments versus distance from the sediment source (cm ka^{-1}).

Long-distance transport of this dust by the westerly jet results in it crossing the Pacific in 12–18 days (Merrill et al., 1994). With respect to supply rates, the Sahara is by far the most important source region, export estimates being 0.6–0.7 Pg a^{-1}, whereas Asian dust transport rates into the North Pacific are of the order of two orders of magnitude lower (see the excellent review of export rates from various source regions by Harrison et al., 2001).

Mineralogical studies in regions downwind of major deserts have shown that mineral dust is generally a major component of ocean sediments (Leinen et al., 1994; Rea, 1994; Sarnthein and Koopman, 1980). In the central North Pacific, for example, between 75% and 98% of the ocean-floor sediments are of aeolian origin (Blank et al., 1985). Dust consists of a mixture of minerals, including quartz, clay minerals, calcite, gypsum and iron oxides, occurring either as individual mineral grains or as pure or mixed-mineral aggregates. Saharan dust is richer in iron and consequently darker in colour than Asian dust. Sarnthein et al. (1982) also distinguished between northern and southern source areas. Dust from the South Sahara and Sahel south of 20–25°N is less rich in carbonate but richer in kaolinite and montmorillonite, whereas carbonate contents are higher in the North and Central Sahara (up to 20–50%) and the dominant clay minerals are illite, chlorite, palygorskite and montmorillonite. Moreover, it is possible to distinguish Saharan dust from Sahelian dust by its mineralogy and colour even after it has been transported across the Atlantic to Barbados (Caquineau et al., 1998; Carlson and Prospero, 1972). The size of the dust particles ranges from up to 50 μm in diameter close to source regions to about 0.1–20 μm (median diameter of 1.5–3 μm) in far distal areas after long-distance transport in the atmosphere. Dust may efficiently alter the chemistry of the atmosphere and thus contribute to changes of regional and global climates. Modern dust is usually highly alkaline, because of the high carbonate content of most arid and semi-arid soils, and thus has a neutralizing effect on rainwater acidity. In addition, considerable amounts of nitrate and sulphate can be adsorbed on the surface of dust particles (Li-Jones and Prospero, 1998; Savoie and Prospero, 1989); the effectiveness of dust in scavenging nitrogen or sulphur from the atmosphere increases with distance from the major dust sources. Another aspect is that the removal of nitrogen and sulphur compounds from the atmosphere reduces the overall aerosol loading of the atmosphere, and consequently affects the irradiation balance. In addition, mineral dust may provide an important source of nutrients, particularly Si, Fe and P, for marine ecosystems. In many regions, dust inputs provide the sole or main source of key nutrients (Chadwick et al., 1999). Most spectacular and recently intensively discussed in this respect is the effect of iron fertilization. The input of iron-rich dust to low-chlorophyll regions could have stimulated productivity in these regions and thus have contributed to the export of carbon to the deep ocean, providing an important process for millennial-scale regulation of the atmospheric pCO_2;

it may have been a contributory cause of the low pCO_2 during the last glacial maximum (LGM) (Falkowski et al., 1998; Martin, 1990; Martin et al., 1994; Oba and Pedersen, 1999; Prentice and Sarnthein, 1993).

During the LGM much of the world experienced increased dust deposition, from Africa to the Atlantic (three to five times higher than present-day values), in the North Pacific (one to two times present-day rates) as well as less pronounced increases in the Arabian Sea (60–80%) and in the polar regions (see Harrison et al., 2001). However, there were also regions where dust fluxes were less than today, for example in the Gulf of Guinea, in the tropical Pacific and off northwestern South America (Rea, 1994). Three factors have been invoked to explain why the LGM atmospheric dust loadings and dust depositional rates were higher: higher wind velocities, a reduced intensity of the hydrological cycle and expansion of dust source areas. An increase in dust supply due to an expansion of source areas could result from: (1) extensive fine-grained outwash deposition along the margins of the ice sheets at the northern hemisphere (Junge, 1979); (2) the increased land area particularly in the tropics and subtropics, with the exposure of the continental shelves resulting from the lowered sea level (Ono and Naruse, 1997) and (3) the reduced vegetation cover and/or soil moisture (Genthon, 1992; Joussaume, 1990; Petit et al., 1981). In marine sediments, increased wind strengths have been inferred from increases in the median grain size of aeolian material (Clemens and Prell, 1990; Rea and Leinen, 1988; Sarnthein and Koopman, 1980), wind-transported materials including diatoms deflated from desiccated lakes (Hooghiemstra, 1989; Stabell, 1989) and from a shoaling of thermocline depths off northern Africa (Molfino and McIntyre, 1990; Ruddiman, 1997) and in the equatorial Pacific (Andreasen and Ravelo, 1997).

In conclusion, the hyper-arid core region of the Sahara considerably expanded during glacials due to increased aridity and overall higher wind speeds, while dust export to the ocean significantly increased (Sarnthein et al., 1982), huge erg systems widely expanded on the shelves (Lancaster et al., 2002) and deposition of aeolian sand turbidites occurred on the NW African slope (Sarnthein and Diester-Haass, 1977). A major outcome from studies of the sedimentary archives of two NW African canyon systems, that is the Timiris Canyon off Mauritania, and the Dakar Canyon off Senegal, clearly revealed a close correlation of climatically controlled shifts in shelf and slope sediment dynamics and the triggering of turbidity currents in the canyons (Henrich et al., 2009). The records of both canyons (Henrich et al., 2010; Pierau et al., 2010) indicate a high frequency of siliciclastic turbidity currents during deglaciation-related sea-level rise (Fig. 5.6), obviously induced by remobilization of huge aeolian dunes fields that had expanded close to the shelf edge during glacial exposure (Lancaster et al., 2002). Further, a high frequency of turbidites deposited during glacials is recorded in the canyons. In addition, turbidites were also deposited, though sporadically, during intermediate sea-level rises in the late phase of Heinrich events

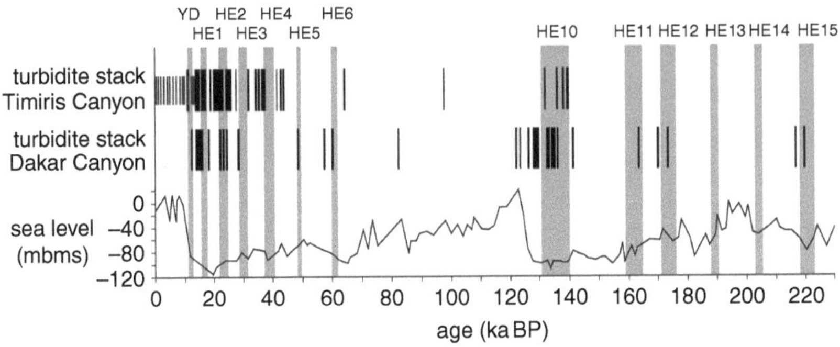

Figure 5.6 Comparison of turbidite activity in the Timiris Canyon and the Dakar Canyon. Stack plot of all turbidite events recorded in these canyon systems compared to a global sea-level curve (Siddall et al., 2003). White space between the intervals with turbidite events represents continuous hemipelagic background sedimentation (redrawn and slightly modified after Henrich et al., 2009). Mbms, metres below mean sea level.

(Pierau et al., 2010). These seem also to be related to an increased dust supply and remobilization of dunes. In the Timiris Canyon, the youngest and cyclic turbidite events (every 900 years) are recorded over the entire Holocene in core 8509 from the upper Timiris Canyon. They originated in the tributaries after severe droughts with a high dust supply, as evidenced by a shelf record displaying 900 year spaced distinct short-lasting dust events (Hanebuth and Henrich, 2009).

Analysis of DSDP and ODP cores offshore the Sahara provides hints on the long-term record of dust deposition in the Atlantic. Aeolian activity dates back to the Early Cretaceous (Lever and McCave, 1983), and aeolian dust is persistently present in Neogene sediments (Sarnthein et al., 1982). However, aeolian activity appears to become much more pronounced in the late Tertiary as evidenced by a maximum of aeolian accumulation rates and a coarsening of grain size in the latest Miocene (between 6 and 5 Ma) and subsequent elevated dust fluxes during the Late Pliocene and Quaternary, in particular during the last 2.5 Ma (Stein, 1985). Interesting in this respect is that, from about 2.5 Ma, the great inland lakes of the Sahara began to dry out, which closely matches in time a strengthening of northern hemisphere glaciation (cf. Bickert and Henrich, 2011, this volume).

2.3. Contribution of glaciomarine material to hemipelagic and pelagic sediments

Sea ice, continental ice sheets and their outlet glaciers influence the pelagic realm and atmospheric circulation fundamentally. It is therefore of crucial importance to understand their main controls in modern and ancient systems. At the northern hemisphere, sea ice is nowadays formed predominately on the shelves of the Arctic Ocean, especially in the East Siberian, the Kara and the Laptev Seas as well as on the Alaskan shelf. The newly formed ice is collected and transported by the two main ice-drift streams, the Beaufort Gyre and the Transpolar Drift (Fig. 5.7). Sediment-rich sea ice occurs mostly within the Beaufort Gyre (Osterkamp and Gosink, 1984; Reimnitz and Kempema, 1987) and within those parts of the Transpolar drift which include a high proportion of multiyear ice from the Siberian shelves (Pfirman et al., 1989). The sediment-rich ice leaves the Arctic Ocean through the Fram Strait and is transported further southwards within the East Greenland Current. In contrast, the ice on the eastern side,

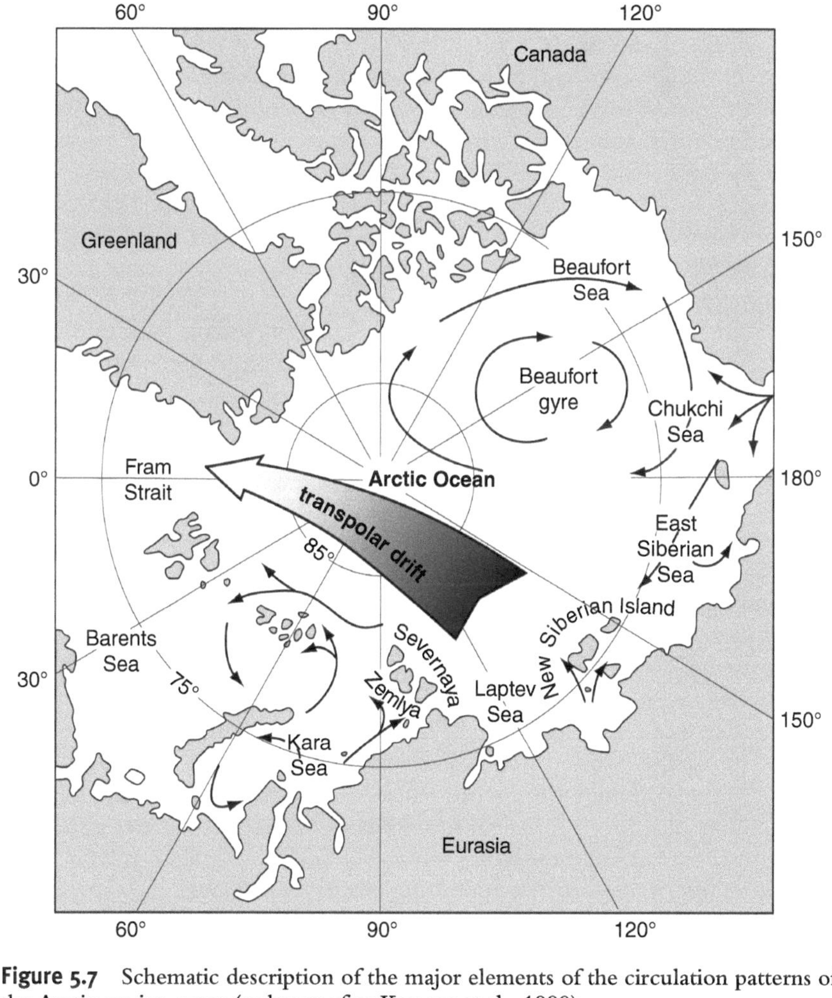

Figure 5.7 Schematic description of the major elements of the circulation patterns of the Arctic sea ice cover (redrawn after Kassens et al., 1999).

over the Yermark Plateau north of Svalbard, and the ice west of Franz Josephs Land is much thinner and cleaner (Pfirman et al., 1989).

There has been much discussion about the predominant sediment type in sea ice, and how sediments derived from sea ice and iceberg may be distinguished (see Elverhøi and Henrich, 1996, for a detailed discussion). Mechanisms incorporating sediment into sea ice include: aeolian transport (Burgeois et al., 1985); in shallow shelf regions: stirring up, re-suspension and freeze-in of fine-grained terrigenous sediments (Barnes et al., 1982; Gilbert, 1990; Osterkamp and Gosink, 1984; Reimnitz et al., 1987) in addition to freezing of the entire water column and anchor-ice formation (Barnes et al., 1982; Osterkamp and Gosink, 1984; Reimnitz et al., 1987; Kempema et al., 1989) and finally in near coastal areas: river discharge onto sea ice and rock fall from cliffs. In general, sediments in sea ice and on the sea floor beneath the ice cover in the Arctic Ocean predominately consist of silt and clay, indicating that stirring up suspension load in shallow shelf seas during strong storms is currently the most important formation process (Wollenburg, 1993). A large amount of sea-ice sediment is transported over long distances and released at the ice margins. In the Fram Strait, sea-ice sediment is thus released, contributing up to 35% of total sediment flux (Hebbeln and Wefer, 1991), while flux rates under the dense pack-ice fields in the Arctic Ocean are extraordinarily low (Elverhøi et al., 1989; Wollenburg, 1993).

Sediment transport and release by icebergs is strongly dependent on the type of glacier or continental ice stream. Under present-day conditions, subpolar glaciers are the most prominent glacier type in the circum-Arctic region as well as on the Antarctic Peninsula (Boulton, 1972; Domack, 1988; Dowdeswell et al., 1998; Lawson, 1982; Powell and Molnia, 1989). Subpolar glaciers are characterized by a variety of patterns (Fig. 5.8):

- They reveal a seasonally changing wet and dry base due to alternating surface melting and/or rainfall during summer, and freezing during winter. Such glaciers develop a meltwater discharge at their base during summer, and incorporate huge amounts of sediment into the basal parts of the ice during winter freezing. As a result, they are extraordinarily dirty with a thick, sediment-rich basal ice package.
- In contact with the sea, they develop steep tidewater ice margins with submarine meltwater discharges and overflow, interflow and, in some cases, underflow suspension load plumes (Powell and Molnia, 1989). Suspension load may be distributed at the surface over several tens of kilometres offshore and several hundreds of kilometres parallel to the ice margin (Pfirman and Solheim, 1989).
- Another important feature of tidewater glaciers is their tendency to surge, a process during which huge parts of glacier ice become unstable, take off and deliver huge ice volumes in the form of icebergs to the sea (Solheim and Pfirman, 1985).

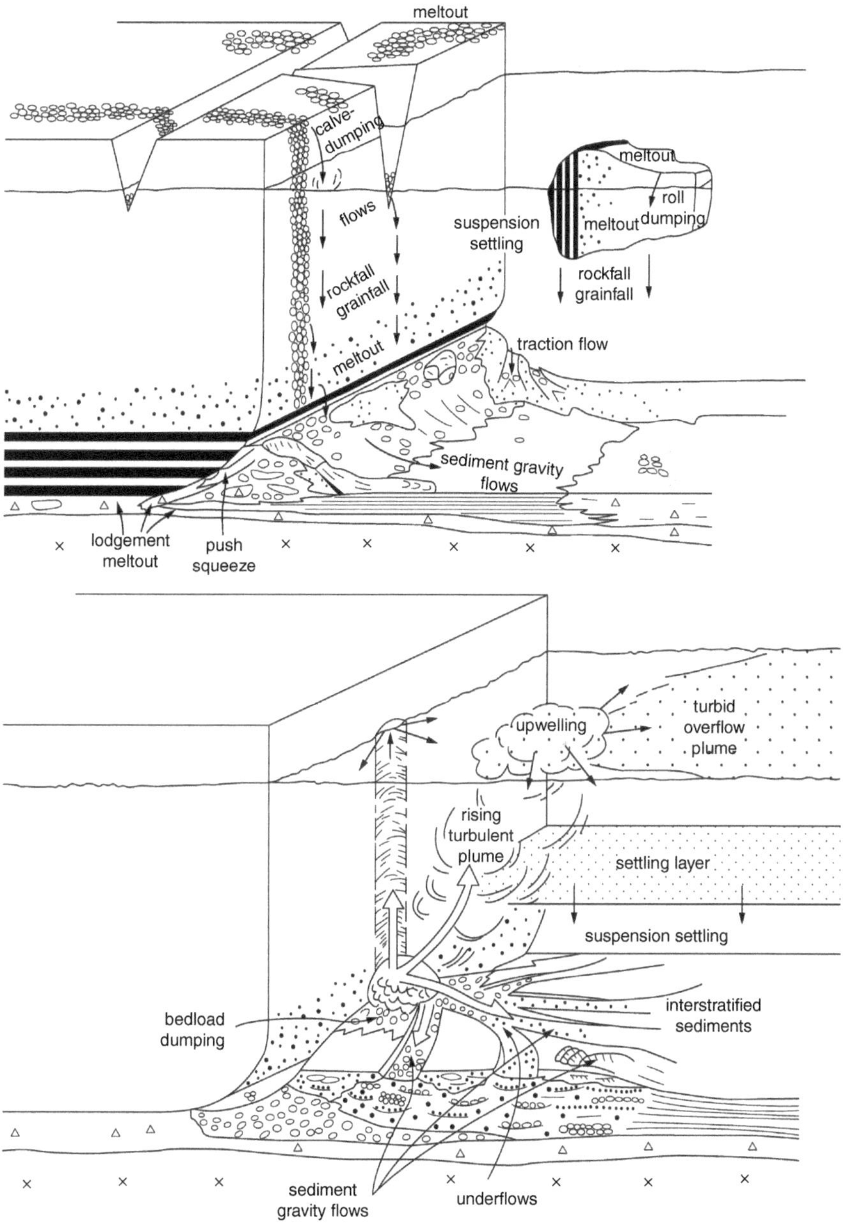

Figure 5.8 Processes and lithofacies at temperate and subpolar tidewater fronts (redrawn and slightly modified after Powell and Molnia, 1989).

As a result, icebergs release sediments of all grain sizes, sorting classes and composition. In most cases, the iceberg sediments contain large amounts of terrigenous coarse material; this material is called "ice-rafted debris" (IRD; see Ruddiman, 1977) in sedimentological investigations. Previous studies of high-latitude marine sediments have shown that coarse terrigenous particles (>63 μm) can be interpreted as IRD if other transport mechanisms (e.g. gravity flows, boundary currents) can reasonably be ruled out (see references in Hebbeln et al., 1998). Sediments transported on sea-ice floes generally contain less than 10% material >63 μm (Pfirman et al., 1989; Wollenburg, 1993). In contrast, icebergs release also significant quantities of sand, gravel and boulders (Dowdeswell and Dowdeswell, 1989). Therefore, high contents of terrigenous material >63 μm can generally be interpreted as indicators of increased iceberg drift. However, variable grain-size spectra have been used in IRD studies, for example in the Nordic seas. IRD studies are either based on the >500 μm fraction (Bischof, 1990, 1994; Bischof et al., 1991; Hebbeln et al., 1994; Spielhagen, 1991) or on the 125–500 μm fraction (Henrich et al., 1989, 1995; Baumann et al., 1995).

Because of the shifts in dominant sediment discharge at tidewater ice margins, a typical offshore sediment distribution may be observed. Proximal areas tend to be characterized by a high discharge of meltwater suspension mixed with sporadic coarse sediment from icebergs (IRD), and are thus characterized by a high accumulation rate of "dropstone" mud. Farther offshore, the influence of suspension load decreases and intermediate amounts of IRD-rich diamicton are deposited. In distal, far offshore regions, a small amount of "dropstone" mud is deposited (cf. Brodzikowski and Van Loon, 1991).

Quantitative studies of glaciomarine-influenced hemipelagic and pelagic sediments from the Nordic seas have shown that their IRD content can be correlated to the onshore glacial history of the Fennoscandian and the Svalbard/Barents Sea ice sheets (Baumann et al., 1995; Mangerud et al., 1998; Fig. 5.9). Large amounts of IRD in the sediments coincide with the extension of the ice sheets over the continental shelves. In addition, qualitative studies of IRD composition have been carried out in order to reconstruct drift paths of icebergs by identifying tracer lithologies with clearly defined source areas like, for example, coal, Cretaceous chalk or typical assemblages of igneous and metamorphic rocks (see references in Hebbeln et al., 1998).

In the Antarctic region, the situation is completely different. There, a typical high-relief margin has developed due to strong isostatic subsidence and glacial over-deepening by variations in the volume of the huge Antarctic ice cap. Extensive ice-shelf environments developed under extreme polar aridity, resulting in generally low sedimentation rates on the shelf and slope because of very restricted meltwater production and supply of IRD (Fütterer et al., 1988). Hence, terrigenous sedimentation is dominated by

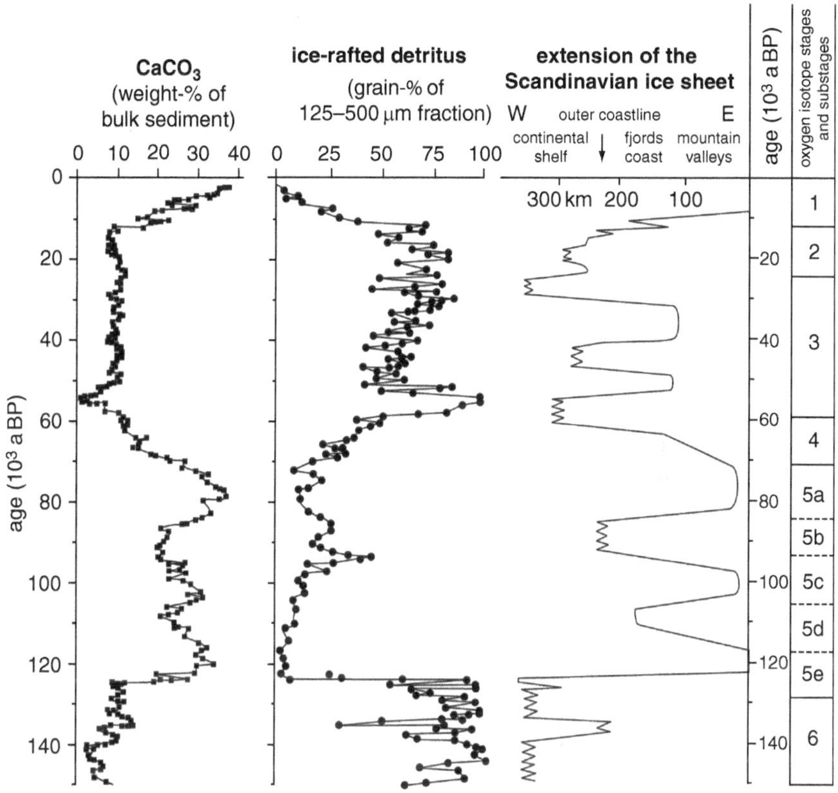

Figure 5.9 Comparison of the glaciation curve of Scandinavia with IRD and the carbonate content in sediments from the Norwegian Greenland Sea for the last two glacial–interglacial cycles (redrawn and slightly modified after Baumann et al., 1995).

silt- and clay-sized detritus with small proportions of gravel and sand delivered by icebergs. Calved icebergs predominately drift with the Antarctic coastal current, hence high IRD concentrations are restricted to the shelves, whereas IRD contents generally are low on the slopes and further offshore, for example less than 15% by weight (Diekmann and Kuhn, 1999). Diekmann et al. (2000) presented a detailed summary of the present-day IRD distribution around Antarctica, displaying the following spatial distribution: (1) large IRD fluxes are recorded from the southeastern Weddell Sea, where prominent ice streams drain the East Antarctic ice sheet; (2) moderate IRD fluxes are found off Donning Maud Land and in the Bellingshausen Sea off the Antarctic peninsula; (3) farther offshore, in the pelagic realm, small IRD fluxes occur along the Polar front, whereas moderate IRD input is observed in the area of warm surface waters to the North; (4) unusually large IRD fluxes appear east of the South Sandwich

Islands and near Bouvet Island. Part of the ice-rafted material might feed the contourite system under control of the strong Antarctic Circumpolar Current (ACC) (see Faugères and Mulder, 2011, this volume).

At glacial/interglacial time scales, much larger interglacial IRD fluxes are registered in the Weddell Sea (Grobe and Mackensen, 1992) (Fig. 5.10), whereas the maximum fluxes in the Bellingshausen Sea off the Antarctic peninsula and the northern Scotia Sea are recorded as short-term spikes in glacials and coincide with the glacial terminations (Diekmann et al., 2000). Farther offshore in the southeastern Atlantic, IRD maxima were registered only during glacials. To conclude, the spatial and temporal distribution of IRD along Antarctica and in the Scotia Sea reflects the calving rates of Antarctic and Patagonian ice masses, while the IRD abundances are related to the presence of cold surface waters (Diekmann et al., 2003). On glacial/interglacial time scales, East Antarctic ice-sheet dynamics follows 100-ka cycles with the highest iceberg discharge during interglacial sea-level high-stands, when floating ice shelves developed, whereas the subpolar ice masses on the Antarctic Peninsula and in Patagonia showed the highest iceberg discharges during the ends of glacials because of rapid disintegration of the far advanced ice masses during these time spans.

During the last decade, numerous studies in the North and South Atlantic used a variety of proxies, including IRD. These studies have revealed that climatic oscillations were faster than Milankovitch frequencies that are manifested by Dansgaard–Oeschger cycles and the so-called *Heinrich events* (H) (Bond et al., 1992). The picture of giant iceberg armadas discharging from the rapidly collapsing Laurentide ice sheet and crossing the Atlantic Ocean has been recognized as an important scenario inducing variable climatic feedback in the ocean/atmosphere system (Bond and Lotti, 1995). Grousset et al. (2001) found that during H1 and H2 about four to six major discharges occurred roughly at a century time scale, each IRD pulse comprising a "precursor" IRD from Europe/Iceland as the source area, followed by Laurentide-derived IRD. The inter-hemispheric cause-and-effect relationships of these short-term cycles and events are starting now to become better understood; for example, during Heinrich events, the trade-wind strength increased, resulting in an increased dust supply to the tropical and subtropical Atlantic (Jullien et al., 2007). There is now common consensus that ocean/ice-sheet interaction plays an essential role, and that continental climatic oscillations probably have a strong imprint on deep-sea records. However, there is ongoing discussion whether the main ice sheets in the northern and southern hemispheres responded simultaneously, and which were the actual forcing mechanisms. One of these debates is related to the bipolar see–saw hypothesis (Broecker, 1998, 2000), which builds on the millennial-scale asynchrony of Antarctic and Greenland climate records during the last glacial. High-latitude and/or equatorial sea-surface perturbations as well as oscillations of the

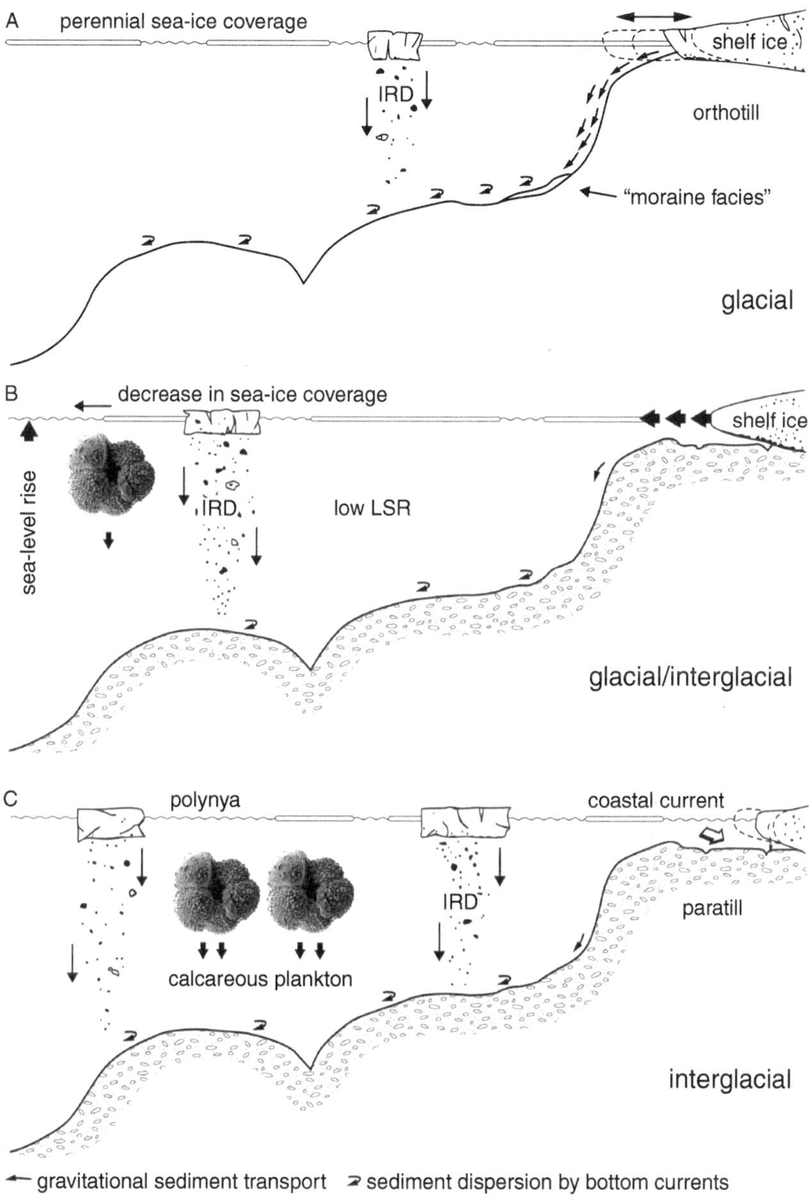

Figure 5.10 Environmental reconstruction of glacial–interglacial facies evolution at the Antarctic margin. (A) Glacial setting characterized by reduced biogenic productivity, deposition of fine-grained sediment by downslope currents and a shallow CCD at about 2000 m. (B) Transitional glacial to interglacial setting showing increased calving of icebergs from rapidly floating ice shelves during the end of glacials, followed by high silica production in an early phase of the next interglacial. (C) Interglacial setting showing a high biogenic calcareous productivity at the outer slope and a deep CCD at about 3000 m (redrawn and modified after Grobe, 1987). LSR, linear sedimentation rate; IRD, ice-rafted debris.

deep-ocean conveyor belt are proposed as driving forces (Seidov and Maslin, 2001). The bipolar see–saw hypothesis predicts that, at the very least, the surface of the South Atlantic should respond abruptly in an anti-phase fashion to changes in the North. Recently, Barker et al. (2009) have presented the first clear evidence of this rapid opposite response, obtained from a temperature record in the South Atlantic; Severinghaus (2009) gave simultaneously an intriguing explanation what could be a possible chain of events causing such a see–saw response during the last deglaciation.

2.4. Quantification and provenience of hemipelagic sediment

In pelagic and hemipelagic environments, sediments are affected and formed by a wide variety of transport processes, including fluvial and aeolian supply, ice rafting, wave and storm currents, littoral and geostrophic currents as well as submarine mass movements. According to the specific marine setting, the importance of the various processes varies, which is reflected by differences in sorting and typical modes of grain-size distribution. After removing the organic, carbonate and opal fraction, grain-size spectra of the suite of remaining terrigenous particles were frequently used to decipher the variable input sources and transport modes.

In general, three basic types of current-influenced sediments can be distinguished: (1) sediments deposited by low-energy processes; (2) sediments representing remnants of erosional processes (residual sediments) and (3) well-sorted sediments (Höppner and Henrich, 1997). According to this model, the first type of sediment is characterized by a positive (i.e. fine) skewness due to the lack of coarse grains. They occur where current velocities slowed down enough to allow even fine particles to settle. In contrast, the fine fraction of residual sediments is lacking due to the winnowing effect of strong currents. These deposits display a negative (i.e. coarse) skewness. Both accumulation and residual deposits are rather poorly sorted. The third type, which is intermediate between the first two, is well sorted, that is it displays a narrow range of grain sizes: both coarse and fine components are absent, due to frequent re-suspension/deposition cycles, which ideally results in a very good sorting and a symmetrical frequency distribution. Hence, for the interpretation of grain-size distributions the symmetry or "shape" matters more than mere "size". On the other hand, size is an indicative parameter to trace energy regimes and the distance to the source. For example, waves and strong currents leave behind coarser sediments than lower-energy regimes. Good progress regarding the influence of current strength on deep-sea sediments was achieved by McCave and co-workers, who investigated drift sediments in the north-east Atlantic (see Faugères and Mulder, 2011, this volume). They found the terrigenous silt to be bimodal, showing a prominent minimum at about 10 μm (Bianchi and McCave, 2000; Robinson and McCave, 1994; McCave et al., 1995a,b).

They interpreted the terrigenous fine fraction (< 10 μm) as having settled in flocculated form, thus being harder to be remobilised by currents due to its cohesive behaviour. Therefore, it was concluded that size distribution is not sensitive to current sorting. Finally, these authors referred to the terrigenous 63–10 μm fraction as "sortable silt" and the considered its mean (modal) grain size indicative for the current strength. A large number of publications manifested the applicability of this tool for the reconstruction of deep-ocean currents in the context of climate variability (e.g. Bianchi and McCave, 1999; Bianchi et al., 2001; Diekmann et al., 2003; Gröger et al., 2003; McCave et al., 1995a,b; McCave, 1997; Revel et al., 1996; Yokokawa and Franz, 2002).

Because of the complexity of the various sedimentary processes and their potential interference, the applicability of the various methodological approaches has to be tested and calibrated for each region. For example, Rea and Hovan (1995) deduced from grain-size distributions in a set of North Pacific surface samples that a clear aeolian signal (prominent peak at 2 μm, well-sorted) appeared to be more prominent farther downwind in the central North Pacific, in contrast to more hemipelagic sediments (even size distribution, poor sorting) close to the continental margins. Similarly, Joseph et al. (1998), who studied sediments from Bermuda Rise and Blake Outer Ridge, distinguished between turbidites, hemipelagic and pelagic intercalations, and drift sediments on the basis of grain-size distributions and magnetic fabrics of the sediments, determining grain alignment due to random settling (pelagic) versus transport along the sea floor (drift sediments and turbidites). A useful tool to distinguish between aeolian and fluvial sediment supply and current sorting is the so-called *Koopmann index* (Koopmann, 1981), which utilizes the relationship between the modal grain size of the terrigenous fraction (> 6 μm) and the proportion of this fraction relative to the bulk sediment. By this method, well-sorted aeolian dust is identified by an increase in the modal grain size with increasing proportion of coarse silt, whereas high amounts of clay and silt are indicative for fluvial material; moderate modal grain sizes characterize sediments that experienced removal of the fines due to winnowing by currents. Koopmann's concept was successfully applied to trace the climatic record of dust versus fluvial supply off NW Africa back into the Miocene (Tiedemann et al., 1989) and of Holocene sediments in the northwestern Indian Ocean (Sirocko and Sarnthein, 1989). Lamy et al. (1998) developed another approach. They used a combination of silt grain-size analysis and clay-mineralogical results to trace the spatial and temporal variation in aeolian versus fluvial sediment input along the Chilean continental margin. The grain-size distribution there appears to be controlled by the primary grain size of the source rocks and the mode of entrainment, for example aeolian in the subtropical North versus fluvial farther South.

Another important terrigenous proxy in deep-sea sediments is constituted by clay minerals, since their composition reflects the prevailing weathering regimes in the continental source regions; these depend on the intensity of physical versus chemical weathering regimes and pedogenesis in the different climate belts (Chamley, 1989). Clay-mineralogical studies have been widely applied as an important palaeoceanographic tool (see Chamley, 1997). Excellent examples of the usefulness of this method come from the South Atlantic (Biscaye, 1965; Petschik et al., 1996). Figure 5.11 provides a

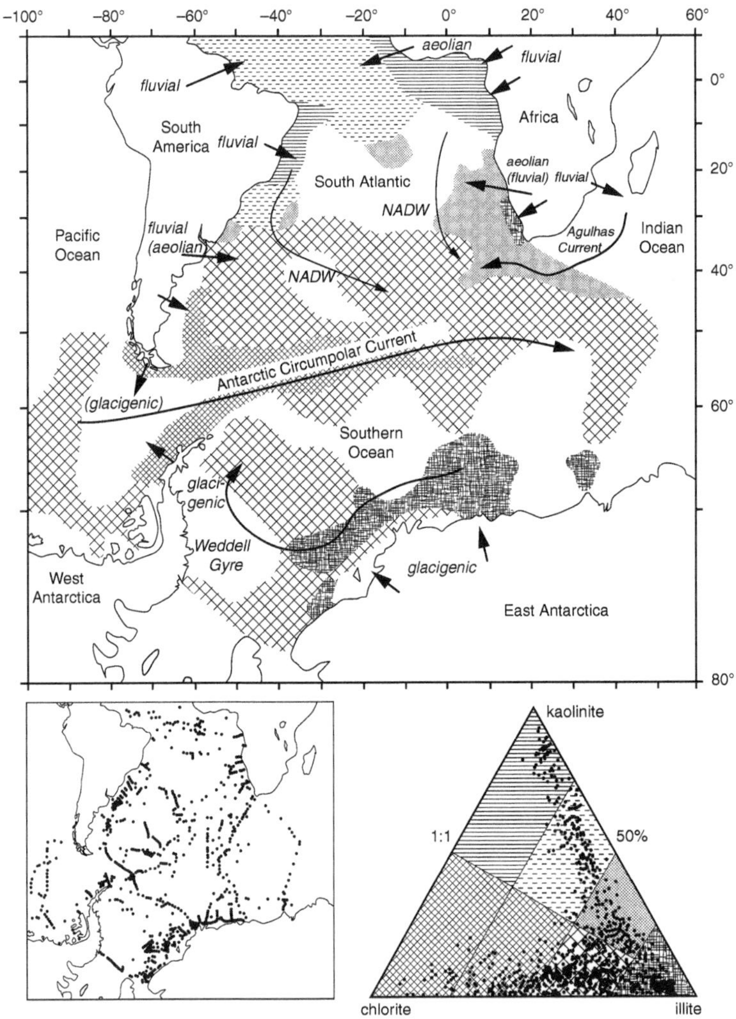

Figure 5.11 Present-day clay-mineral provinces and modes of terrigenous clay supply and dispersal. The lower left map shows the distribution of samples from the sedimentary surface. The ternary concentration diagram visualizes the clay-mineral assemblages (redrawn and slightly modified after Diekmann et al., 2003).

compilation by Diekmann et al. (2003) of South Atlantic clay-mineral provinces in surface sediments. Kaolinite- and chlorite-bearing assemblages are recognized as contrasting end members of low and high latitudes, respectively. Provinces dominated by kaolinite or mixed kaolinite/illite assemblages appear north of the ACC, whereas mixed illite/chlorite assemblages are restricted to the South. In more detail, the following regional features are recognized: (1) a kaolinite-dominated province off tropical Africa and off Brazil, mainly supplied by the tropical Congo and Niger rivers as well as by dust from the Sahel Zone (i.e. from fossil and modern tropical soils) on the African side, and from fluvial suspensions draining the deeply weathered Santos/Sao Paulo Plateaus, which are covered by lateritic soils; (2) an illite-dominated province passing offshore into an illite/kaolinite assemblage of southern Africa, illite being mainly derived from deserts and semi-arid regions and carried by the trade winds and less by fluvial discharge; (3) a chlorite province under the ACC, originating from the Andean mobile belts and from the Antarctic Peninsula. The various clay-mineral assemblages of South Atlantic surface sediments are further dispersed along the pathways of the main deep-water masses, providing a useful palaeoceanographic tool. High kaolinite/chlorite ratios in interglacial intervals document the inflow of kaolinite-bearing suspensions entrained in the filaments of the Agulhas Current retroflection and within the NADW, whereas low kaolinite/chlorite ratios indicate that the region was mainly bathed by ACC water masses. Hence, the operation of the warm-route conveyor mode is evident for interglacials, implying southward injection of relatively warm saline NADW in the ACC, whereas the cold-route conveyor mode dominated by cold southern source water masses during glacials (Diekmann et al., 2003).

Different input processes and transport phenomena may, however, often have affected the sediments, which makes identification of the individual processes and their relative contribution very complicated. A solution to this problem is offered by the concept of end-member modelling (Weltje, 1997). This algorithm provides a mathematical tool to decompose particle-size distributions into a limited number of subpopulations. This unmixing technique has been successfully applied to a variety of sedimentary environments, for instance (1) to discriminate proximal and distal aeolian input and fluvial discharge in the Arabian Sea (Prins and Weltje, 1999; Prins et al., 2000), in the eastern South Atlantic (Stuut et al., 2002) and off NW Africa (Holz et al., 2004, 2007; Fig. 5.12); (2) to trace iceberg discharge and current sorting in North Atlantic sediments (Prins et al., 2001, 2002) and (3) to decipher the sediment discharge onto the Argentine margin by the Rio de la Plata, to detect current winnowing by vigorous currents like the Malvinas Current, and to quantify fine-grained hemipelagic settling of river suspensions along the Brazil Malvinas Confluence (Frenz et al., 2003).

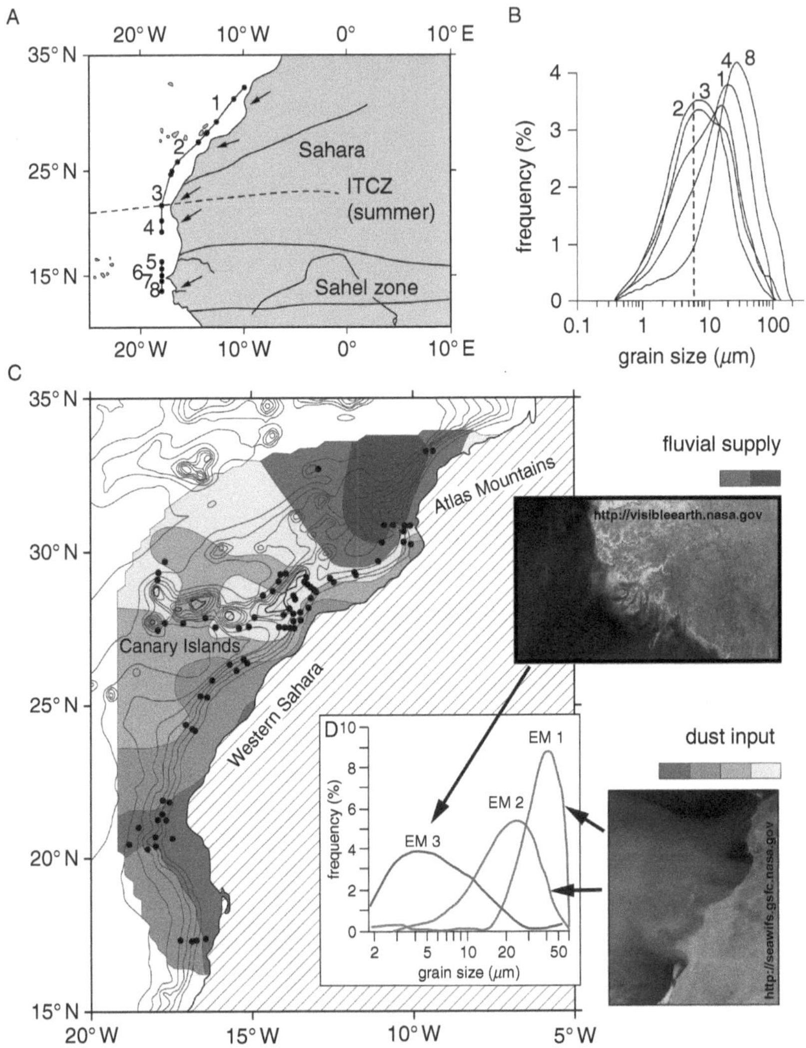

Figure 5.12 Grain-size distributions of present-day aeolian dust collected off the northwest African coast with a dust sampler on board of the RV Meteor (redrawn and modified after Holz et al., 2004). (A) Studied transects. The dotted line indicates the summer position of the Intertropical Convergence Zone (ITCZ). The arrows highlight the prevailing surface-wind directions. (B) The exemplified grain-size distributions are the result of collecting dust fallout along the illustrated transects. (C–D) Spatial variation of modelled end-member contributions of the carbonate-free silt fraction of surface sediments offshore northwest Africa. Seabed sample locations indicated by black dots. Isopleths were calculated using the gridding method of kriging. Contour map displaying sediment contribution by coarse-grained aeolian dust (black signature—end-member EM 1) and fine-grained aeolian dust (grey colour shades—end-member EM 2) and fluvially discharged mud (dotted signature—end-member EM 3). (A multi-colour version of this figure is on the included CD-ROM.)

3. Periplatform Carbonates

3.1. Off-bank transport of shallow-water carbonates

In the vicinity of shallow-water carbonate platforms, pelagic sediments gradually merge into carbonate oozes that include a growing proportion of shallow-water carbonate debris and mud. Schlager and James (1978) established the term "periplatform ooze" for this type of deep-sea deposit. A variable but often considerable amount of shallow-water components is supplied from the carbonate-platform tops and margins to the adjacent deep-ocean water column, mixing with oceanic planktic components (Heath and Mullins, 1984).

Off-bank transport of coarse-grained material from carbonate platforms occurs mainly during the passage of storms along open leeward bank margins, as documented for the Bahamas (Hine et al., 1981; Wilber et al., 1993). Tidal currents and normal wave action cause only minor sand movement, but relatively weak currents are able to winnow fine-grained shallow-water sediment and transport it to the slope. Such off-bank transport of carbonate-platform mud is achieved mainly by wind- and tide-driven advection (Heath and Mullins, 1984; Pilskaln et al., 1989). It can be accelerated under cold-front conditions and the vertical transport of these bank-derived components through the adjacent deep-ocean water column is recognized to be temporarily forced by "density cascading" (see Mulder, 2011, this volume, Section 1.5.3; Wilson and Roberts, 1992, 1993, 1995).

Both bank-derived and open-ocean materials collectively rain down through the water column, at least through it lower part, and settle onto the deep-sea floor. Thus, periplatform deposits generally miss any traction features, derived from bed-load transport, and other characteristics of internal organization, as they are attributed to deposits of many density flows, which travel horizontally near the sea floor (see Mulder, 2011, this volume). Despite (seasonal) variable supply of bank-derived and planktic components, periplatform sedimentation is a rather continuous (quasi-steady) process similar to hemipelagic sedimentation, contrasting with the event character of most density-flow depositional processes. Due to the variable supply of bank-derived and plankton materials periplatform carbonates may, nevertheless, show gradual changes from mud- to grain-supported textures and systematic grain-size trends, such as the coarsening upwards documented by Westphal (1998) from the Late Pliocene of the western Great Bahama Bank.

The most distinct attribute of periplatform sediments is that they have two input sources (Boardman and Neumann, 1984; Boardman et al., 1986; Reijmer et al., 1988; Schlager and James, 1978). In present-day oceans, shallow-water carbonate platforms produce and supply neritic material of mainly metastable mineralogy (aragonite and high-Mg calcite), while the open ocean contributes planktic material of essentially stable mineralogy (low-Mg calcite). This compositional property and its variation have been well studied for periplatform carbonates of the Bahamas and the Florida shelf (Lantzsch et al., 2007; Turpin et al., 2008; Westphal, 1998). It was found that the major part of *bank-derived* sediment occurs as aragonite needles (~5 μm in length), which are delivered to the deep-sea floor in the form of small (<62 μm) aggregates (Neumann and Land, 1975). The aragonite needles appear to be the end-products of the natural disintegration of calcareous green algae or are precipitations from oversaturated seawater within the shallow bank-top lagoons (Gischler and Zingeler, 2002; Milliman et al., 1993; Shinn et al., 1989; Stockman et al., 1967). High-magnesium calcite (HMC) mainly occurs in the form of coralline algal remains and skeletal fragments of benthic foraminifers or as cements. Additional shallow-water constitutes of variable mineralogy are cortoids, peloids, ostracods, echinoderms, bryozoans, gastropods, scleractinian corals, acritarchs and dinoflagellate cysts. It was also found that low-magnesium calcite (LMC) of *open-ocean* origin is principally derived from coccolithophorids and planktic foraminifers. Aragonite shells from pteropods are not found to contribute significantly to the Bahamian periplatform muds.

Both the shallow-water platform and the open ocean can supply sand- and mud-sized particles, though in variable proportions (Table 5.1). In particular the amount of bank-derived calcareous sand depends on the frequency and intensity of hydrodynamic events mediating the off-bank flux. The composition of periplatform sediments is, therefore, not a straightforward mirror of platform-top sedimentation, as usually fine-grained material is exported to the slopes more or less permanently, while coarser grains are left behind on the platform top (Lantzsch et al., 2007).

The unique character of modern periplatform sediments, consisting of two mineralogically distinct assemblages of biogenic carbonate constituents, is challenged by their diagenetic overprint that extensively obliterates primary signals, as found for fine-grained calcareous rhythmites from many Palaeozoic–Mesozoic periplatform settings. Such rhythmites would potentially hold important information on palaeoenvironmental conditions (e.g. Munnecke and Westphal, 2005). It is unclear, however, where the aragonite constituents did come from, especially during times of calcite seas (see Hüneke and Henrich, 2011, this volume, Fig. 37) and how much dissolved aragonite was lost to the sea water during early marine diagenesis. A conspicuous characteristic of the calcareous rhythmites is the intercalation of two rock types that have undergone two entirely different diagenetic pathways. This differential diagenesis has not only the potential to distort primary environmental signals significantly, but also to mimic primary signals (see Hesse and Schacht, 2011, this volume; Westphal, 2006).

Pilskaln et al. (1989) found that during fair-weather, storm-free periods, no significant lateral advection of bank-derived carbonate components

Table 5.1 Vertical periplatform carbonate flux, mineralogy and flux constituents at 500 m water depth between the Little Bahama Bank and the northern Great Bahama Bank (from Pilskaln et al., 1989).

Fraction (carbonate flux)	Mineralogy	Particle types	Provenance
>1 mm 4.48 mg m^{-2} d^{-1}	LMC 45% A 55%	Planktic foraminifers Pteropods	Predominantly open-ocean (planktic) material
63 μm–1 mm 12.14 mg m^{-2} d^{-1}	LMC 56% HMC 16%	Planktic foraminifers Benthic foraminifers; extremely rare red algae and echinoderm fragments	
	A 28%	Pteropods Zooplankton faecal pellets (containing coccoliths, diatom skeletons, other fragments, organic material)	
<63 μm 30.99 mg m^{-2} d^{-1}	LMC 46% HMC 20%	Coccolith Fragments of benthic foraminifers, spicules from tunicates and soft corals, echinoderm skeletal elements, encrusting red-algae fragments	Predominantly bank-derived (neritic) material
	A 34%	Aggregates of aragonite needles (from calcareous green algae) and green-algae skeletal fragments; extremely rare pteropod-shell fragments	

occurs in the upper water column around the Bahama Platform. In a case study to quantify the off-bank transport, a sediment trap was moored at 500 m in the Northwest Providence Cannel, which is a 200–2500 m deep inter-platform setting between the Little Bahama Bank and the northern Great Bahama Bank; it represents an extensive region of periplatform sedimentation (Table 5.1). The low vertical flux measured for shallow-water components clearly contrasts with the high sedimentation rates calculated for Holocene periplatform sedimentation within this area, where sedimentation rates are of the order of 9.4 cm 10^{-3} years (Boardman and Neumann, 1984). In addition, flux constituents differ significantly from those of the underlying sediments (see Pilskaln et al., 1989 for details).

The sediment-trap data imply that flux and deposition of bank-derived materials in the periplatform environment are variable on a short-term scale. A relatively minor proportion of bank-derived components is deposited during fair-weather conditions, but most of it is delivered during the passage of the frequent, low-amplitude seasonal storms and the occasional hurricanes (Pilskaln et al., 1989). In addition, density cascading can explain the major discrepancy between the calculated high accumulation rates close to the Bahamas platforms and the low flux rates measured by Pilskaln et al. (1989) in the upper water column. The process of density cascading suggests that fine-grained sediment is entrained by high-density waters and transported to the adjacent ocean (Wilson and Roberts, 1995). These sediment-charged hyperpycnal flows are regularly generated across the Bahama and Florida platforms in response to winter cold fronts and summer periods of intense evaporation (Wilson and Roberts, 1992). Once released from platform areas, hyperpycnal waters, together with entrained sediments, sink to their compensation level in the adjacent ocean (Fig. 5.13). Initially, the confined flow is attached to the slope and sinks along a preferential

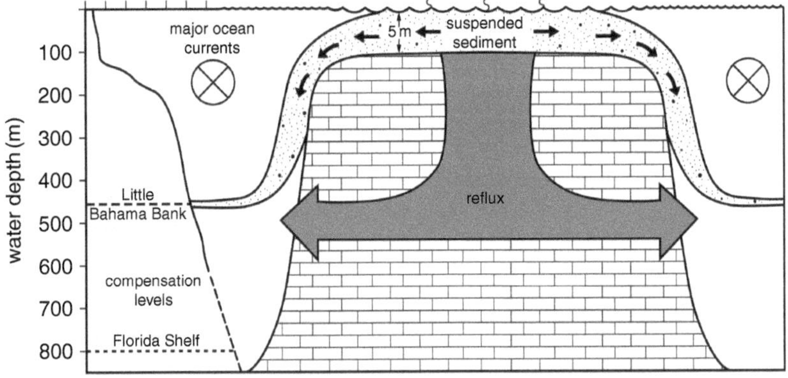

Figure 5.13 Schematic cross-section of a shallow-water carbonate platform, showing off-bank density flows attached to the slope, followed by slope separation, giving way to periplatform settling (redrawn and slightly modified after Wilson and Roberts, 1993). The density profile for deep-ocean water adjacent to the Bahama Banks exemplifies the compensation levels for hyperpycnal waters at the Little Bahama Bank (illustrated case) and the Florida Shelf (28th February 1990, 26.55°N, 76.75°W).

bathymetric pathway, followed by slope separation and creation of an extensive nepheloid layer, approximately along their isopycnal levels (McCave, 1983). Under these circumstances, coarse-grained sediments are quickly deposited, but fine-grained sediments may become partially trapped at the density interface (Fig. 5.14). If the compensation level for density cascading exceeds the depth at the base of the platform slope, however, erosion results, carving characteristic "plunge pools" and a "bump-and-trough" morphology into muds draping the carbonate slope apron (Wilber et al., 1993). Plunge pools are also related to hydraulic jumps in turbidity currents arriving at the slope-base break.

Density cascading also explains how periplatform muds accumulate at rapid rates along the western margin of the Great Bahama Bank and the southwestern Florida shelf in the presence of strong surface currents operating in the Straits of Florida (Brooks and Holmes, 1990). It accelerates the vertical flux of shallow-water mud to depth at which it cannot be remobilized by ocean-surface currents (Wilson and Roberts, 1995). On

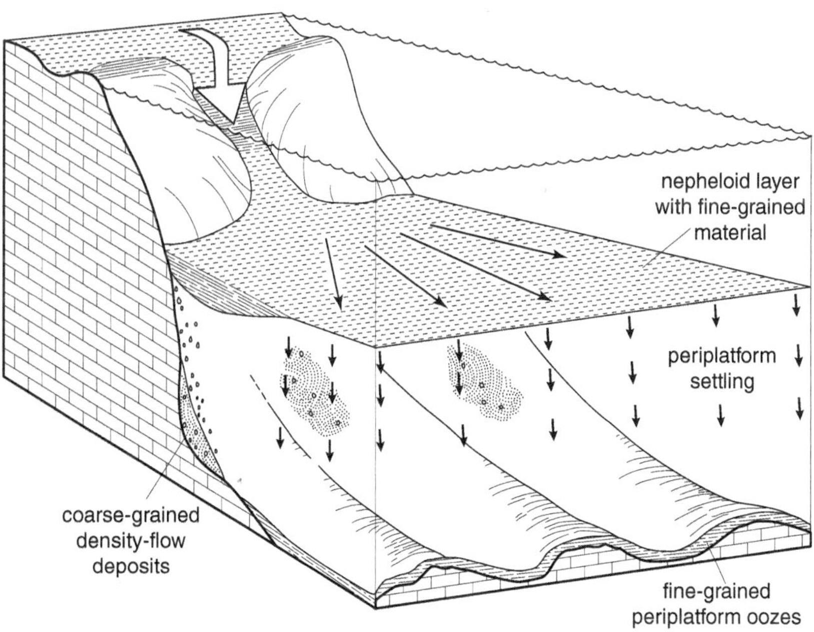

Figure 5.14 Block diagram of a density cascading event at the margin of a shallow-water carbonate platform, illustrating periplatform sedimentation (redrawn and slightly modified after Wilson and Roberts, 1995). Initially, the flow is attached to the slope and sinks along a preferential bathymetric pathway, followed by slope separation and production of an extensive nepheloid layer and finally, vertical settling of benthic shoal-water biogenic materials over slope and basin settings. Note that the effects of ocean currents and the Coriolis force are not taken into consideration.

the other hand, density cascades may pass into contour currents. If the density force is weak, the density cascade becomes diffuse and may run-in bank-parallel deep-ocean currents (e.g. Faugères and Mulder, 2011, this volume; Mullins et al., 1980). Where density instabilities are strong, flow confinements may induce self-sustaining sediment erosion and transport of coarse material. This transformation from a thermohaline flow to a flow in which sediment contributes significantly to excess-density driving force, results in the development of a density flow (see Mulder, 2011, this volume).

3.2. Controls of compositional variation in periplatform oozes

Compositional variations in periplatform oozes mainly reflect changes in the depositional conditions on the shallow-water platforms, because the vast quantity of excess bank-top production and the limited storage volume in the shallow-water realm may provide large volumes of sediment available for off-bank export (Mullins et al., 1984; Neumann and Land, 1975; Roth and Reijmer, 2004). Therefore, the impact of both sea-level fluctuations and climatic changes on the accumulation patterns of periplatform sediments is high.

The highstand-shedding model is widely applied for flat-topped carbonate platforms to explain cyclic variations in the sedimentation rate and the composition of periplatform sediments adjacent to carbonate platforms (Droxler and Schlager, 1985; Reijmer et al., 1988; Schlager et al., 1994). This implies that a depositional system sheds, in principle, most sediment into the adjacent basin during highstands of the sea level. This off-bank export includes periplatform deposition, but also other types of gravity-controlled sedimentation processes (see Mulder, 2011, this volume). During highstands, the platform top is flooded, enlarging the sediment-production area, whereas the top is exposed during lowstands, confining the production area to a narrow fringe of reefs and limiting the export of carbonate sediments to deeper waters (Dravis, 1996). Platform tops rapidly lithify during sea-level falls when the sea floor is winnowed due to lowering of the wave base or sediments become exposed to fresh water (Schlager et al., 1994).

Highstand shedding is most pronounced on low-latitude, rimmed platforms, whereas it is reduced on carbonate ramps (Schlager et al., 1994; Turpin et al., 2008). In cool-water carbonate systems, the reverse may occur, termed "lowstand shedding", which is a basic principle of siliciclastic systems (e.g. Kindler et al., 2006). In the Neogene, highstand shedding has been observed around all rimmed carbonate platforms studied to date (e.g. Grammer et al., 1993; Lantzsch et al., 2007; Reijmer and Andresen, 2007; Rendle-Bühring and Reijmer, 2005; Rendle et al., 2000). Several detailed investigations strongly suggest highstand shedding of ancient platforms, too (e.g. Cadjenovic et al., 2008; Mawson and Tucker, 2009; Schlager et al., 1994). This basic model explains most aspects of compositional variation in periplatform oozes, including the carbonate content, carbonate mineralogy, organic content and grain-size. Important additional controls are the topography of the platform, the leeward-versus-windward orientation of the periplatform setting with respect to the shallow-water platform source and proximality–distality trends.

The impact of climatically controlled sea-level fluctuations on the sedimentation patterns present in periplatform sediments has been highlighted for the northern slope of the Little Bahama Bank by Lantzsch et al. (2007). The studied Pleistocene–Holocene record typifies a mid-slope depositional environment of an isolated (peritidal) carbonate platform and consists of periplatform carbonates, except for coarser intervals with cemented debris that are mainly supplied by turbidity and debris flows. A 26-m-thick succession of carbonate oozes reveals largely platform-controlled deposition processes for the past 375,000 years (Lantzsch et al., 2007): (1) interglacial average aragonite accumulation rates (11 g cm^{-2} ka^{-1}) are distinctly higher than during glacial intervals (3 g cm^{-2} ka^{-1}). Under conditions of a high sea-level, a large production area for aragonite existed and excess sediment could be supplied to the open ocean. Glacial lowstands caused a decrease in size of the production area and therefore resulted in decreasing aragonite accumulation rates in periplatform settings. (2) The organic content was influenced by interglacial/glacial-related sea-level fluctuations similar to the aragonite content. During phases of widespread bank-top flooding, organic production on the platform increased, because of the larger production area, leading to the highest export rates of organic material to the open ocean during highstands. During glacial phases of bank-top exposure, the organic production was restricted, resulting in a decreased export of organic carbon from the platform. (3) The grain sizes in deposits of glacial stages are considerably larger than those found for interglacial stages, suggesting that the contribution of fine material from the platform top decreased during glacial lowstands (decreased input of aragonite needles). These patterns agree with the highstand-shedding model proposed by Droxler and Schlager (1985), Reijmer et al. (1988) and Schlager et al. (1994).

The compositional variation of periplatform sediments also depends on the orientation of the carbonate platform with respect to the prevailing wind direction and, accordingly, its type of margin setting (accretionary/bypass/erosional). Reverse correlations in the grain-size distribution patterns have been recognized by Rendle-Bühring and Reijmer (2005) for the accretionary leeward (western) margin and the erosional windward (eastern) margin of Great Bahama Bank in response to glacial/interglacial environmental changes (Fig. 5.15). The leeward-margin periplatform sediments are generally finer during interglacials and coarser during glacials, as described from many other periplatform settings in the Caribbean and the western North Atlantic (e.g. Grammer et al., 1993; Lantzsch et al., 2007; Rendle et al., 2000; Westphal, 1998). The windward margin, however, shows the reverse

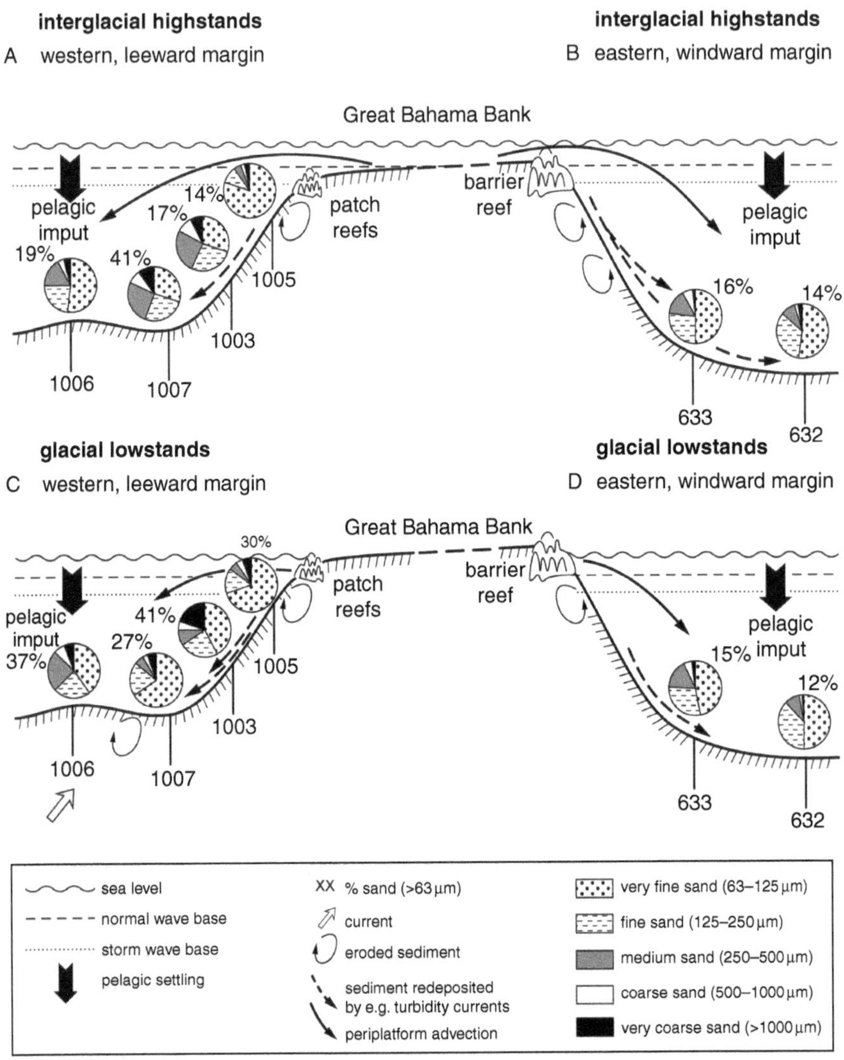

Figure 5.15 Sediment dynamics in periplatform settings on windward and leeward platform margins for the Great Bahama Bank (redrawn and slightly modified after Rendle-Bühring and Reijmer, 2005). (A–B) Interglacial highstand. High sediment production and input from the platform top give way to the formation of thick, fine-grained highstand wedges on the slightly inclined, more stable, leeward slopes (A). Eroded coarser sediments are redeposited at the toe-of-slope. On the steeper, windward margin (B), supplied sediments are rapidly eroded due to slope instability, and transferred to the deep sea by sediment gravity flows. (C–D) Glacial lowstand. Reduced sediment production and input, in combination with a lower wave base, result in sediment-starved slopes. Upper leeward-slope sediments are redeposited in the basin (C), where strengthened currents lead to winnowing and/or non-deposition of fines. The windward-margin upper slopes are more stable (D), limiting redeposition of slope sediments by sediment gravity flows to the deep sea.

pattern, with coarser sediments (sands, especially medium to very coarse) during interglacials and generally finer grained deposits (low sand content) during glacials. Rendle-Bühring and Reijmer (2005) conclude that different sedimentary processes are responsible for the grain-size distribution patterns in the opposing slope settings of the Great Bahama Bank. The *leeward* margin is controlled by the productivity-export pattern of the platform, whereas the *windward* margin is subjected to gravity-induced resedimentation processes on the slope, as indicated by the increased occurrence of turbidites during interglacials (Fig. 5.15). The reverse trend according to platform orientation can also be due to winnowing by hydrodynamic processes (drift, swell).

The type of slope developed at opposing platform margins, that is accretionary for the leeward margin and bypassing and/or erosional for the windward margin (Schlager and Camber, 1986; Schlager and Ginsburg, 1981) is controlled by changes in slope angle, which in turn is determined by the grain size of the periplatform sediments supplied from the shallow-water carbonate factory (Rendle-Bühring and Reijmer, 2005).

3.3. Trends of proximality and distality

Proximality and distality trends in compositional variation with respect to shallow-water platform sources have been illustrated along two carbonate margin-to-basin transects of the Northern Nicaragua Rise by Reijmer and Andresen (2007). The analysis of periplatform sediments and related density-flow deposits in the surrounding of Pedro Bank revealed the presence of characteristic depositional environments since the mid-Pleistocene (last 300 ka). The spatial distribution in mineralogy and grain size along the leeward, downcurrent margin of the platform shows a distinct facies pattern (Reijmer and Andresen, 2007):

(1) During sea-level highstands (interglacials) the fine-sediment fraction (<63 μm) dominates periplatform carbonates along the entire leeward transect that was studied, without any spatial variations. The fine-grained materials are mainly of shallow-water origin, primarily aragonite needles. This dominance holds for distances up to 40 km from the margin. Farther downslope, aragonite is still the most abundant mineral, but the pelagic carbonate-mineral content (LMC) increases. Within the subordinate coarse-fraction classes (>63 μm), the very fine sand fraction dominates only at proximal sites (<20 km), due to the influence of fine neritic sediments shed off-bank. More distal sites (>20 km) show a more bimodal distribution pattern in the coarse grain sizes, with maximum percentages within the very fine and medium sand fraction. This indicates a mixed bank-derived/open-ocean signal.
(2) During sea-level lowstands (glacials), a twofold partition in the spatial distribution of the periplatform carbonates is evident. A proximal

environment (<28 km) with enhanced coarse-fraction percentages gives way to a distal environment (>28 km) with a strong dominance of the fine fraction (>90%). The increased coarse-fraction content in proximal areas is the result of various interacting processes: (i) a lower input of fine neritic sediments; (ii) increased current winnowing and (iii) reworking on the upper slope during a lowered sea level, and the export of this material to the base-of-slope. In the subordinate coarse-fraction classes, a similar distribution pattern is found at all studied locations, showing a bimodal grain-size distribution. This indicates a low shallow-water content within the coarse-fraction, and similar open-ocean influence on the sedimentation in the leeward, down-current platform vicinity during glacials (Reijmer and Andresen, 2007).

Thus, periplatform carbonates contain a wealth of information on shallow-water and oceanic conditions, and they record processes occurring in the water column through which they settle. As summarized above, present-day carbonate-platform slopes offer the possibility of studying the interplay between allochthonous and pelagic sedimentation, and submarine cementations.

ACKNOWLEDGMENTS

We gratefully acknowledge Thierry Mulder for his critical comments to improve the manuscript. Tom van Loon suggested editorial changes that are highly appreciated. We especially thank Dagmar Lau and Heiko Sengpiehl for carefully redrafting most of the figures.

REFERENCES

Andreasen, D.J., Ravelo, A.C., 1997. Tropical Pacific Ocean thermocline depth reconstructions for the last glacial maximum. Paleoceanography 12, 395–414.

Barker, S., Diz, P., Vautravers, M.J., Pilke, J., Knorr, G., Hall, I.R., et al., 2009. Interhemispheric Atlantic see-saw response during the last deglaciation. Nature 457, 1097–1102.

Barnes, P.W., Reimnitz, E., Fox, D., 1982. Ice rafting of fine-grained sediment, a sorting and transport mechanism, Beaufort Sea, Alaska. J. Sediment. Petrol. 52, 493–502.

Baumann, K.-H., Lackschewitz, K., Mangerud, J., Spielhagen, R.F., Wolf-Welling, T.C. W., Henrich, R., et al., 1995. Reflection of the Scandinavian ice sheet fluctuations in Norwegian Sea sediments. Quat. Res. 43, 185–197.

Bianchi, G.G., McCave, I.N., 1999. Holocene periodicity in North Atlantic climate and deep-ocean flow south of Iceland. Nature 397, 515–517.

Bianchi, G.G., McCave, I.N., 2000. Hydrography and sedimentation under the deep western boundary current on Björn and Gardar Drifts, Iceland Basin. Mar. Geol. 165, 137–169.

Bianchi, G.G., Vautravers, M., Shackleton, N.J., 2001. Deep flow variability under apparently stable North Atlantic Deep Water production during the last interglacial of the subtropical NW Atlantic. Paleoceanography 16, 306–316.

Bickert, T., Henrich, R., 2011. Climate records of deep-sea sediments: towards the Ice House. In: Hüneke, H., Mulder, T. (Eds.), Deep-Sea Sediments. Developments in Sedimentology, Elsevier, Amsterdam, vol. 63, pp. 793–823.
Biscaye, P.E., 1965. Mineralogy and sedimentation of recent deep-sea clay in the Atlantic Ocean and adjacent seas and oceans. Geol. Soc. Am. Bull. 76, 803–832.
Bischof, J.F., 1990. Dropstones in the Norwegian-Greenland Sea—indications of a Late Quaternary circulation pattern? In: Bleil, U., Thiede, J. (Eds.), Geological History of the Polar Oceans: Arctic versus Antarctic, NATO Adv. Sci. Inst. Ser. C, 308, 499–518.
Bischof, J.F., 1994. The decay of the Barents ice sheet as documented in Nordic Seas ice-rafted debris. Mar. Geol. 117, 35–55.
Bischof, J., Koch, J., Kubisch, M., Spielhagen, R.F., Thiede, J., 1991. Nordic Sea surface ice drift reconstructions: evidence from ice rafted coal fragments during oxygen isotope stage 6. In: Dowdeswell, J.A., Scourse, J.D. (Eds.), Glaciomarine Environments: Processes and Sediments, Geol. Soc. London, 53, 235–251. Spec. Publ.
Blank, M., Leinen, M., Prospero, J.M., 1985. Major Asian aeolian inputs indicated by the mineralogy of aerosols and sediments in the western North Pacific. Nature 314, 84–86.
Boardman, M.R., Neumann, A.C., 1984. Sources of periplatform carbonates: northwest Providence Channel Bahamas. J. Sediment. Petrol. 54, 1110–1123.
Boardman, M.R., Neumann, A.C., Baker, P.A., Dunlin, L.A., Kenter, R.J., Hunter, G.E., et al., 1986. Banktop responses to Quaternary fluctuations in sea level recorded in periplatform sediments. Geology 14, 28–31.
Bond, G.C., Lotti, R., 1995. Iceberg discharges into the North Atlantic on millennial time scales during the last glaciation. Science 267, 1005–1010.
Bond, G., Heinrich, H., Broecker, W., Labeyrie, L., McManus, J., Andrews, J., et al., 1992. Evidence for massive discharges of icebergs into the North Atlantic ocean during the last glacial period. Nature 360, 245–249.
Boulton, G.S., 1972. Modern Arctic glaciers as depositional models for former ice sheets. J. Geol. Soc. London 128, 361–393.
Brodzikowski, K., Van Loon, A.J., 1991. Glacigenic Sediments. Developments in Sedimentology, Elsevier, Amsterdam, 49, 674 pp.
Broecker, W.S., 1998. Paleocean circulation during the last deglaciation: a bipolar see-saw? Paleoceanography 13, 119–121.
Broecker, W.S., 2000. Was a change in thermohaline circulation responsible for the Little Ice Age? Proc. Natl. Acad. Sci. Am. 97, 1339–1342.
Brooks, G.R., Holmes, C.W., 1990. Modern configuration of the southwest Florida carbonate slope – development by shelf margin progradation. Mar. Geol. 94, 301–315.
Burgeois, J.C., Koerner, R.M., Alt, B.T., 1985. Airborne pollen: a unique air mass tracer, its influx to the Canadian High Arctic. Ann. Glaciol. 7, 109–116.
Cadjenovic, D., Kilibarda, Z., Radulovic, N., 2008. Late Triassic to Late Jurassic evolution of the Adriatic Carbonate Platform and Budva Basin, Southern Montenegro. Sediment. Geol. 204, 1–17.
Caquineau, S., Gaudichet, A., Gomes, L., Magonthier, M.-C., Chatenet, B., 1998. Saharan dust: clay ratio as a relevant tracer to assess the origin of soil-derived aerosols. Geophys. Res. Lett. 27, 983–986.
Carlson, T.N., Prospero, J.M., 1972. Large-scale movement of Saharan air outbreaks over the northern equatorial Atlantic. J. Appl. Meteorol. 11, 283–297.
Chadwick, O.A., Derry, L.A., Vitousek, P.M., Huebert, B.J., Hedin, L.O., 1999. Changing sources of nutrients during four million years of ecosystem development. Nature 397, 491–497.
Chamley, H., 1989. Clay Sedimentology. Springer, Berlin, 623 pp.

Chamley, H., 1997. Clay mineral sedimentation in the ocean. In: Paquet, H., Clauer, N. (Eds.), Soils and Sediments: Mineralogy and Geochemistry. Springer, Berlin, pp. 269–302.

Clemens, S.C., Prell, W.L., 1990. Late Pleistocene variability of Arabian Sea summer monsoon winds and continental aridity: eolian records from the lithogenic component of deep-sea sediments. Paleoceanography 5, 109–145.

Diekmann, B., Kuhn, G., 1999. Provenance and dispersal of glacial-marine surface sediments in the Weddell Sea and adjoining areas, Antarctica: ice-rafting versus current transport. Mar. Geol. 158, 209–231.

Diekmann, B., Kuhn, G., Rachold, V., Abelmann, A., Brathauer, U., Fütterer, D.K., et al., 2000. Terrigenous supply in the Scotia Sea (Southern Ocean): response to Late Quaternary ice dynamics in Patagonia and on the Antarctic Peninsula. Palaeogeogr. Palaeoclimatol. Palaeoecol. 162, 357–387.

Diekmann, B., Fütterer, D.K., Grobe, H., Hillenbrand, C.D., Kuhn, G., Michels, K., et al., 2003. Terrigenous supply in the polar to temperate South Atlantic: land-ocean links of environmental changes during the Quaternary. In: Wefer, G., Mulitza, S., Ratmeyer, V. (Eds.), The South Atlantic in the Late Quaternary: Reconstruction of Material Budget and Current Systems. Springer, Berlin, pp. 375–399.

Domack, E.W., 1988. Biogenic facies in the Antarctic glaciomarine environment: basis for a polar glaciomarine summary. Palaeogeogr. Palaeoclimatol. Palaeoecol. 63, 357–372.

Dowdeswell, J.A., Dowdeswell, E.K., 1989. Debris in icebergs and rates of glaciomarine sedimentation: observations from Spitsbergen and a simple model. J. Geol. 97, 221–231.

Dowdeswell, J.A., Elverhøi, A., Spielhagen, R., 1998. Glaciomarine sedimentary processes and facies on the polar North Atlantic margins. In: Elverhøi, A., Dowdeswell, J., Funder, S., Mangerud, J., Stein, R. (Eds.), Glacial and Oceanic History of the Polar North Atlantic Margins, Quat. Sci. Rev., 17, 243–272.

Dravis, J.J., 1996. Rapidity of freshwater calcite cementation; implications for carbonate diagenesis and sequence stratigraphy. Sediment. Geol. 107, 1–10.

Droxler, A., Schlager, W., 1985. Glacial versus interglacial sedimentation rates and turbidite frequency in the Bahamas. Geology 13, 799–802.

Elverhøi, A., Henrich, R., 1996. Glaciomarine environments: ancient glaciomarine environments. In: Menzies, J. (Ed.), Past Glacial Environments—Sediments, Forms, and Techniques. Butterworth-Heinemann, Oxford, pp. 179–212.

Elverhøi, A., Pfirman, S.L., Solheim, A., Larssen, B.B., 1989. Glaciomarine sedimentation in epicontinental seas exemplified by the Northern Barents Sea. Mar. Geol. 85, 225–250.

Falkowski, P.G., Barber, R.T., Smetacek, V., 1998. Biogeochemical controls and feedbacks on ocean primary production. Science 281, 200–206.

Faugères, J.C., Mulder, T., 2011. Contour currents and contourite drifts. In: Hüneke, H., Mulder, T. (Eds.), Deep-Sea Sediments. Developments in Sedimentology, Elsevier, Amsterdam, vol. 63, pp. 149–214.

Frenz, M., Höppner, R., Stuut, J.-B.W., Wagner, T., Henrich, R., 2003. Surface sediment bulk geochemistry and grain-size composition related to the oceanic circulation along the South American continental margin in the Southwest Atlantic. In: Mulitza, S., Ratmeyer, V., Wefer, G. (Eds.), The South Atlantic in the Late Quaternary: Reconstruction of Material Budget and Current Systems. Springer, Berlin, pp. 347–373.

Fütterer, D., Grobe, H., Grüning, S., 1988. Quaternary sediment patterns in the Weddell Sea: relations and environmental conditions. Paleoceanography 3, 351–361.

Genthon, C., 1992. Simulations of desert dust and sea-salt aerosols in Antarctica with a general circulation model of the atmosphere. Tellus B 44, 371–389.

Geyer, W.R., Hill, P.S., Kineke, G.C., 2004. The transport, transformation and dispersal of sediment by buoyant coastal flows. Cont. Shelf Res. 24, 927–949.

Gilbert, R., 1990. Rafting in glaciomarine environments. In: Glaciomarine Environments and Processes, Geol. Soc. London, 53, 105–120. Spec. Publ.
Gischler, E., Zingeler, D., 2002. The origin of carbonate mud in isolated carbonate platforms of Belize Central America. Int. J. Earth Sci. 91, 1054–1070.
Glaccum, R.A., Prospero, J.M., 1980. Saharan aerosols over the tropical North Atlantic – Mineralogy. Mar. Geol. 37, 295–321.
Goudie, A.S., Middleton, N.J., 2001. Saharan dust storms: nature and consequences. Earth. Sci. Rev. 56, 179–204.
Grammer, G.M., Ginsburg, R.N., Harris, P.M., 1993. Timing of deposition, diagenesis, and failure of steep carbonate slopes in response to a high-amplitude/high-frequency fluctuation in sea level, Tongue of the Ocean, Bahamas. In: Loucks, R.G., Sarg, J.F. (Eds.), Carbonate Sequence Stratigraphy: Recent Developments and Applications, Am. Assoc. Petrol. Geol. Mem., 57, 107–131.
Grobe, H., 1987. Facies classification of glacio-marine sediments in the Antarctic. Facies 17, 99–109.
Grobe, H., Mackensen, A., 1992. Late Quaternary climate cycles as recorded in sediments from the Antarctic continental margin. In: Kennett, J.P., Warnke, D.A. (Eds.), The Antarctic Paleoenvironment: A Perspective on Global Change, Part One, Antarct. Res. Ser. (Am. Geophy. Union), 56, 349–376.
Gröger, M., Henrich, R., Bickert, T., 2003. Glacial–interglacial variability in lower North Atlantic Deep Water: interference from silt grain size analysis and carbonate preservation in the western equatorial Atlantic. Mar. Geol. 201, 321–332.
Grousset, F.E., Cortijo, E., Huon, S., Herve, L., Richter, T., Burdloff, D., et al., 2001. Zooming in on Heinrich layers. Paleoceanography 16, 240–259.
Hanebuth, T.J.J., Henrich, R., 2009. Recurrent decadal-scale dust events over Holocene western Africa and their control on canyon turbidite activity (Mauritania). Quat. Sci. Rev. 28, 261–270.
Harrison, S.P., Kohfeld, K.E., Roelandt, C., Claquin, T., 2001. The role of dust in climate changes today, at the Last Glacial Maximum and in the future. Earth Sci. Rev. 54, 43–80.
Heath, K.C., Mullins, H.T., 1984. Open-ocean, off-bank transport of fine grained carbonate sediment in the northern Bahamas. In: Stow, D.A.B., Piper, D.J.W. (Eds.), Fine-Grained Sediments: Deep Water Processes and Facies. Blackwell, London, pp. 199–208.
Hebbeln, D., Wefer, G., 1991. Effects of ice coverage and ice-rafted material on sedimentation in the Fram Strait. Nature 350, 409–411.
Hebbeln, D., Dokken, T., Andersen, E.S., Hald, M., Elverhøi, A., 1994. Moisture supply for northern ice-sheet growth during the Last Glacial Maximum. Nature 370, 357–360.
Hebbeln, D., Henrich, R., Baumann, K.-H., 1998. The polar North Atlantic—Paleoceanography of the last interglacial/glacial cycle. In: Elverhøi, A., Dowdeswell, J., Funder, S., Mangerud, J., Stein, R. (Eds.), Glacial and Oceanic History of the Polar North Atlantic Margins, Quat. Sci. Rev. 17, 125–153.
Henrich, R., Kassens, H., Vogelsang, E., Thiede, J., 1989. Sedimentary facies of glacial-interglacial cycles in the Norwegian Sea during the last 350 ka. Mar. Geol. 86, 283–319.
Henrich, R., Wagner, T., Goldtschmidt, P., Michels, K., 1995. Depositional regimes in the Norwegian-Greenland Sea: the last two glacial to interglacial transitions. Geol. Rundsch. 84, 28–48.
Henrich, R., Hanebuth, T.J.J., Cherubini, Y., Krastel, S., Pierau, R., Zühlsdorff, C., 2009. Climate induced trubidity current activity in NW-African canyon systems. In: Mosher, D.C., Shipp, R.C., Moscardelli, L., Chaytor, J., Baxter, C., Lee, H., Urgeles, R. (Eds.), Submarine Mass Movements and their Consequences. Advances in Natural and Technological Hazards Research, Vol. 28, Springer, Austin, Texas, pp. 447–461.

Henrich, R., Cherubini, Y., Meggers, H., 2010. Climate and sea level induced turbiditic activity in a canyon system offshore the hyperarid Western Sahara: the Timiris Canyon. Mar. Geol. 275, 178–198.

Hesse, R., Schacht, U., 2011. Early diagenesis of deep-sea sediments. In: Hüneke, H., Mulder, T. (Eds.), Deep-Sea Sediments. Developments in Sedimentology, Elsevier, Amsterdam, vol. 63, pp. 557–714.

Hine, A.C., Wilber, R.J., Bane, J.M., Neumann, A.C., Lorenson, K.R., 1981. Offbank transport of carbonate sands along open, leeward bank margins; northern Bahamas. Mar. Geol. 42, 327–348.

Holz, C., Stuut, J.B., Henrich, R., 2004. Terrigenous sedimentation processes along the continental margin off NW-Africa: implications from grain-size analyses of seabed sediments. Sedimentology 51, 1145–1154.

Holz, C., Stuut, J.B., Henrich, R., Meggers, H., 2007. Variability in terrigenous sedimentation processes off Northwest Africa and its relation to climatic changes: interferences from grain-size distributions of a Holocene marine record. Sediment. Geol. 202, 499–508.

Hooghiemstra, H., 1989. Variations of the N.W. African trade wind regime during the past 140, 000 years: changes in pollen flux evidenced by marine sediment records. In: Leinen, M., Sarnthein, M. (Eds.), Palaeoclimatology and Palaeometeorology: Modern and Past Patterns of Global Atmospheric Transport. Kluwer Academic Publishing, Dordrecht, pp. 733–770.

Höppner, R., Henrich, R., 1997. Kornsortierungsprozesse am Argentinschen Kontinentalhang anhand von Siltkorn-Analysen. Zentralbl. Geol. Palaontol. I 7–9, 897–905.

Hovius, N., 1998. Controls on sediment supply by large rivers. Soc. Econ. Paleontol. Mineral. 42, 47–69. Spec. Publ.

Hüneke, H., Henrich, R., 2011. Pelagic sedimentation in modern and ancient oceans. In: Hüneke, H., Mulder, T. (Eds.), Deep-Sea Sediments. Developments in Sedimentology, Elsevier, Amsterdam, vol. 63, pp. 215–351.

Joseph, L.H., Rea, D.K., Pluijm, B.A., 1998. Use of grain size and magnetic fabric analyses to distinguish among depositional environments. Paleoceanography 13, 491–501.

Joussaume, S., 1990. Three-dimensional simulation of the atmospheric cycle of desert dust particles using a general circulation model. J. Geophys. Res. 95, 1909–1941.

Jullien, E., Grousset, F., Malaizé, B., Duprat, J., Sanchez-Goni, M., Eynaud, F., et al., 2007. Low-latitude "dusty events" vs. high-latitude "icy Heinrich events". Quat. Res. 68, 379–386.

Junge, C., 1979. The importance of mineral dust as an atmospheric constituent. In: Morales, C. (Ed.), Saharan Dust: Mobilization, Transport, Deposition; Papers and Recommendations. Scientific Committee on Problems of the Environment (SCOPE) of the International Council of Scientific Unions.Wiley, Chichester, p. 297.

Karyampudi, V.M., Palm, S.P., Reagen, J.A., Fang, H., Grant, W.B., Hoff, R.M., et al., 1999. Validation of the Saharan dust plume conceptual model using lidar, Meteosat, and ECMWF data. Bull. Am. Meteorol. Soc. 80, 1045–1075.

Kassens, H., Bauch, H.A., Dmitrenko, I., Eicken, H., Hubberten, H.W., Melles, M., et al., 1999. Land–Ocean Systems in the Siberian Arctic: Dynamics and History. Springer, New York, 711 pp.

Kempema, E.M., Reimnitz, E., Barnes, P.W., 1989. Sea-ice entrainment and rafting in the Arctic. J. Sediment. Petrol. 59, 308–317.

Kindler, P., Ruchonnet, C., White, T., 2006. The Southern Marion Platform (Marion Plateau, NE Australia) during the early Pliocene: a lowstand-producing, temperate-water carbonate factory. In: Pedley, H.M., Carannante, G. (Eds.), Cool-Water Carbonates: Depositional Systems and Palaeoenvironmental Controls, Geol. Soc. London, 255, 269–282. Spec. Publ.

Koopmann, B., 1981. Sedimentation von Saharastaub im subtropischen Nordatlantik während der letzten 25 000 Jahre. Meteor Forschungsergebnisse C35, 23–59.

Lamy, F., Hebbeln, D., Wefer, G., 1998. Terrigeous sediment supply along the Chilean margin: modern regional patterns of texture and composition. Geol. Rundsch. 87, 477–494.

Lancaster, N., Kocurek, G., Singhvi, A., Pandey, V., Deynoux, M., Ghienne, J.F., et al., 2002. Late Pleistocene and Holocene dune activity and wind regimes in the western Sahara Desert of Mauritania. Geology 30, 991–994.

Lantzsch, H., Roth, S., Reijmer, J.J.G., Kinkel, H., 2007. Sea-level related resedimentation processes on the northern slope of Little Bahama Bank (Middle Pleistocene to Holocene). Sedimentology 54, 1307–1322.

Lawson, D.E., 1982. Mobilisation, movement and deposition of active subaerial sediment flow, Matanuska Glacier Alaska. J. Geol. 90, 279–300.

Leinen, M., Prospero, J.M., Arnold, E., Blank, M., 1994. Mineralogy of aeolian dust reaching the North Pacific Ocean: 1. Sampling and analysis. J. Geophys. Res. 99, 21017–21023.

Lever, A., McCave, I.N., 1983. Eolian components in Cretaceous and Tertiary North Atlantic sediments. J. Sediment. Petrol. 53, 811–832.

Li-Jones, X., Prospero, J.M., 1998. Variations in the size distribution of non-sea-salt sulfate aerosol in the marine boundary layer at Barbados: impact of African dust. J. Geophys. Res. 103, 16073–16084.

Livingstone, I., Warren, A., 1996. Aeolian Geomorphology: An Introduction. Longmann, London, 211 pp.

Mangerud, J., Dokken, T., Hebbeln, D., Heggen, B., Ingolfsson, O., Landvik, J., et al., 1998. Fluctuations of the Svalbard-Barents Sea ice sheet during the last 150 000 years. In: Elverhøi, A., Dowdeswell, J., Funder, S., Mangerud, J., Stein, R. (Eds.), Glacial and oceanic history of the polar North Atlantic margins, Quat. Sci. Rev. 17, 11–42.

Martin, J.H., 1990. Glacial–interglacial CO_2 change: the iron hypothesis. Paleoceanography 5, 1–13.

Martin, J.H., Coale, K.H., Johnson, K.S., 1994. Testing the iron hypothesis in ecosystems of the equatorial pacific-ocean. Nature 371, 123–129.

Mawson, M., Tucker, M., 2009. High-frequency cyclicity (Milankovitch and millennial-scale) in slope-apron carbonates: Zechstein (Upper Permian), North-East England. Sedimentology 56, 1905–1936.

McCave, I.N., 1972. Transport and escape of fine-grained sediment from shelf areas. In: Swift, D.J.P., Duane, D., Pilkey, O.H. (Eds.), Shelf Sediment Transport. Dowden Hutchinson and Ross, Stroudsburg, PA, pp. 225–248.

McCave, I.N., 1983. Particulate size spectra and origin of nepheloid layers of the Nova Scotian continental rise. J. Geophys. Res. 88, 7647–7666.

McCave, I.N., 1997. Recent sedimentation beneath the Deep Western Boundary Current of northern New Zealand. Deep Sea Res. I 44, 1203–1237.

McCave, I.N., Manighetti, B., Beveridge, N.A.S., 1995a. Circulation in the glacial North Atlantic inferred from grain-size measurements. Nature 374, 149–152.

McCave, I.N., Manighetti, B., Robinson, S.G., 1995b. Sortable silt and fine grained sediment size/composition slicing: patterns for paleocurrent speed and paleoceanography. Paleoceanography 10, 593–610.

Meade, R.H., 1996. River-sediment inputs to major deltas. In: Milliman, J.D., Haq, B.U. (Eds.), Sea-Level Rise and Coastal Subsidence. Kluwer Academic Publishers, Dordrecht, pp. 63–85.

Merrill, J.T., Arnold, E., Leinen, M., Weaver, C., 1994. Mineralogy of aeolian dust reaching the North Pacific Ocean: 2. Relationship mineral assemblages atmospheric transport patterns. J. Geophys. Res. 99, 21025–21032.

Middleton, N.J., 1986a. Dust storms in the Middle East. J. Arid Environ. 10, 83–96.

Middleton, N.J., 1986b. A geography of dust storms in south-west Asia. J. Climatol. 6, 183–196.
Milliman, J.D., 1995. Sediment discharge to the ocean from small mountainous rivers: the New Guinea example. Geo Mar. Lett. 15, 127–133.
Milliman, J.D., Meade, R.H., 1983. World-wide delivery of river sediment to the oceans. J. Geol. 91, 1–21.
Milliman, J.D., Syvitski, J.P.M., 1992. Geomorphic tectonic control of sediment discharge to the ocean – the importance of small mountainous rivers. J. Geol. 100, 525–544.
Milliman, J.D., Freile, D., Steinen, R.P., Wilber, R.J., 1993. Great Bahama Bank aragonitic muds: mostly inorganically precipitated, mostly exported. J. Sediment. Petrol. 63, 589–595.
Molfino, B., McIntyre, A., 1990. Precessional forcing of nutricline dynamics in the equatorial Atlantic. Science 249, 766–769.
Mulder, T., 2011. Gravity processes on continental slope, rise and abyssal plains. In: Hüneke, H., Mulder, T. (Eds.), Deep-Sea Sediments. Developments in Sedimentology, Elsevier, Amsterdam, vol. 63, pp. 25–148.
Mulder, T., Syvitski, J.P.M., 1995. Turbidity currents generated at river mouths during exceptional discharges to the world oceans. J. Geol. 103, 285–299.
Mulder, T., Syvitski, J.P.M., 1996. Climatic and morphologic relationships of rivers: implications of sea level fluctuations on river loads. J. Geol. 104, 509–523.
Mullins, H.T., Neumann, A.C., Wilber, R.J., Hine, A.C., Chinburg, S.J., 1980. Carbonate sediment drifts in northern straits of Florida. Am. Assoc. Petrol. Geol. Bull. 64, 1701–1717.
Mullins, H.T., Heath, K.C., Vanburen, H.M., Newton, C.R., 1984. Anatomy of a modern open-ocean carbonate slope—northern Little Bahama Bank. Sedimentology 31, 141–168.
Munnecke, A., Westphal, H., 2005. Variations in primary aragonite, calcite, and clay in fine-grained calcareous rhythmites of Cambrian to Jurassic age—an environmental archive? Facies 51, 592–607.
Neumann, A.C., Land, L.S., 1975. Lime mud deposition and calcareous algae in the Bight of Abaco, Bahamas: a budget. J. Sediment. Petrol. 45, 763–786.
Nitrouer, C.A., Kuehl, S.A. (Eds.), 1995. Geological Significance of Sediment Transport and Accumulation on the Amazon Continental Shelf. Mar. Geol. 125, 175–401.
Nitrouer, C.A., Wright, L.D., 1994. Transport of particles across continental shelves. Rev. Geophys. 32, 85–113.
Oba, T., Pedersen, T.F., 1999. Paleoclimatic significance of eolian carbonates supplied to the Japan Sea during the last glacial maximum. Paleoceanography 14, 34–41.
Ono, Y., Naruse, T., 1997. Snowline elevation and eolian dust flux in the Japanese islands during isotope stages 2 and 4. Quat. Int. 37, 45–54.
Orton, G.J., Reading, H.G., 1993. Variability of deltaic processes in terms of sediment supply, with particular emphasis on grain size. Sedimentology 40, 475–512.
Osterkamp, T.E., Gosink, J.P., 1984. Observations and anlyses of sediment-laden sesa ice. In: Barnes, P.W., Schell, D.M., Reimnitz, E. (Eds.), The Alaskan Beaufort Sea: Ecosystems and Environments. Academic Press, London, pp. 73–93.
Petit, J.R., Briat, M., Royer, A., 1981. Ice age aerosol content from East Antarctic ice core samples and past wind strength. Nature 293, 391–394.
Petschik, R., Kuhn, G., Gingele, G., 1996. Clay mineral distribution in surface sediments of the South Atlantic: sources, transport, and relation to oceanography. Mar. Geol. 139, 203–229.
Pfirman, S.L., Solheim, A., 1989. Subglacial meltwater discharge in the open-marine tidewater glacier environment; observations from Nordaustlandet, Svalbard archipelago. Mar. Geol. 86, 265–281

Pfirman, S., Wollenburg, I., Thiede, J., 1989. Lithogenic sediment on Arctic pack ice: potential aeolian flux and contribution to deep sea sediments. In: Leinen, M., Sarnthein, M. (Eds.), Paleoclimatology and Paleometeorology: Modern and Past Patterns of Global Atmospheric Transport, NATO Adv. Sci. Inst. Ser. C, 282, 463–493.

Pierau, R., Hanebuth, T.J.J., Krastel, S., Henrich, R., 2010. Late Quaternary climatic and sea level changes recorded in turbidite activity, Dakar Canyon, NW Africa. Quat. Res. 73, 385–392.

Pilskaln, C.H., Neumann, A.C., Bane, J.M., 1989. Periplatform carbonate flux in the northern Bahamas. Deep Sea Res. A 36, 1391–1406.

Powell, R.D., Molnia, B.F., 1989. Glaciomarine sedimentary processes, facies and morphology of the south-southeast Alaska shelf and fords. Mar. Geol. 85, 359–390.

Prentice, I.C., Sarnthein, M., 1993. Self-regulatory processes in the biosphere in the face of climate change. In: Eddy, J., Oeschger, H. (Eds.), Global Changes in the Perspective of the Past. Wiley, Chichester, pp. 29–38.

Prins, M., Weltje, G., 1999. End-member modelling of siliciclastic grain-size distributions: the late Quaternary record of eolian and fluvial sediment supply to the Arabian Sea and its paleoclimatic significance. In: Harbaugh, J., Watney, W.L., Rankey, E., Slingerland, R., Goldstein, R., Franseen, E. (Eds.), Numerical Experiments in Stratigraphy: Recent Advances in Stratigraphic and Sedimentologic Computer Simulations, Soc. Econ. Paleontol. Mineral. 62, 91–111. Spec. Publ.

Prins, M.A., Postma, G., Weltje, G.J., 2000. Controls on terrigenous sediment supply to the Arabian Sea during the late Quaternary: the Makran continental slope. Mar. Geol. 169, 351–371.

Prins, M.A., Troelstra, S.R., Kruk, R.W., Van der Borg, K., De Jong, A.J., Weltje, G.J., 2001. The Late Quaternary sediment record from Reykjanes Ridge, North Atlantic. Radiocarbon 43, 939–947.

Prins, M.A., Bouwer, L.M., Beets, C.J., Troelstra, S.R., Weltje, G.J., Kruk, R.W., et al., 2002. Ocean circulation and iceberg discharge in the glacial North Atlantic: inferences from unmixing of sediment size distribution. Geology 30, 555–558.

Prospero, J.M., Lamb, P.J., 2003. African droughts and dust transport to the Caribbean: climate change implications. Science 302, 1024–1027.

Prospero, J.M., Bonatti, E., Schubert, C., Carlson, T.N., 1970. Dust in the Caribbean atmosphere traced to an African dust storm. Earth Planet. Sci. Lett. 9, 287–293.

Prospero, J.M., Glaccum, R.A., Nees, R.T., 1981. Atmospheric transport of soil dust from Africa to South America. Nature 289, 570–572.

Rea, D.K., 1994. The paleoclimatic record provided by eolian deposition in the deep sea: the geologic history of wind. Rev. Geophys. 32, 159–195.

Rea, D.K., Hovan, S.A., 1995. Grain size distribution and depositional processes of the mineral component of abyssal sediments: lessons from the North Pacific. Paleoceanography 10, 251–258.

Rea, D.K., Leinen, M., 1988. Asian aridity and the zonal westerlies: Late Pleistocene and Holocene record of eolian deposition in the Northwest Pacific Ocean. Palaeogeogr. Palaeoclimatol. Palaeoecol. 66, 1–8.

Rebesco, M., Camerlenghi, A. (Eds.), 2008. Contourites. Developments in Sedimentology, Vol. 60. Elsevier, Amsterdam, 663 pp.

Reijmer, J.J.G., Andresen, N., 2007. Mineralogy and grain size variations along two carbonate margin-to-basin transects (Pedro Bank, Northern Nicaragua Rise). Sediment. Geol. 198, 327–350.

Reijmer, J.J.G., Schlager, W., Droxler, A.W., 1988. Site 632: Pliocene–Pleistocene sedimentation cycles in a Bahamian Basin. In: Austin, J.A.J., Schlager, W. (Eds.), Proc. ODP Sci. Res. Vol. 166, 213–220. (College Station, Texas).

Reimnitz, E., Kempema, M., 1987. Field observations of slush ice generated during freeze-up in Arctic coastal waters. Mar. Geol. 77, 219–231.

Reimnitz, E., Kempema, M., Barnes, P.W., 1987. Anchor ice, seabed freezing, and sediment dynamics in shallow Arctic Seas. J. Geophys. Res. 92C, 14671–14678.

Rendle, R.H., Reijmer, J.J.G., Kroon, D., Henderson, G.M., 2000. Mineralogy and sedimentology of the Pleistocene to Holocene on the leeward margin of Great Bahama Bank. In: Swart, P., Eberli, G.P., Malone, M., Sarg, J.F. (Eds.), Proc. ODP Sci. Res. Vol. 166, 61–76. (College Station, TX).

Rendle-Bühring, R.H., Reijmer, J.J.G., 2005. Controls on grain-size patterns in periplatform carbonates: marginal setting versus glacio-eustacy. Sediment. Geol. 175, 99–113.

Revel, M., Cremer, M., Grousset, F.E., Labeyrie, L., 1996. Grain-size and Sr-Nd isotopes as tracers of paleo-bottom current strength, Northeast Stlantic Ocean. Mar. Geol. 131, 233–249.

Robinson, S.G., McCave, I.N., 1994. Orbital forcing of bottom-current enhanced sedimentation on Feni Drift, NE Atlantic, during the mid-Pleistocene. Paleoceanography 9, 943–972.

Roth, S., Reijmer, J.J.G., 2004. Holocene Atlantic climate variations deduced from carbonate periplatform sediments (leeward margin, Great Bahama Bank). Paleoceanography 19, PA1003. doi:10.1029/2003PA000885.

Ruddiman, W.F., 1977. Late Quaternary deposition of ice-rafted sand in the subpolar North Atlantic (lat 40° to 65°N). Geol. Soc. Am. Bull. 88, 1813–1827.

Ruddiman, W.F., 1997. Tropical Atlantic terrigenous fluxes since 25, 000 yrs B.P. Mar. Geol. 136, 189–207.

Salon, S., Crise, A., Van Loon, A.J., 2008. Dynamics of the bottom boundary layer. In: Rebesco, M., Camerlenghi, A. (Eds.), Contourites. Developments in Sedimentology, Vol. 60. Elsevier, Amsterdam, pp. 83–97.

Sarnthein, M., Diester-Haass, L., 1977. Eolian-sand turbidites. J. Sediment. Petrol. 47, 868–890.

Sarnthein, M., Koopman, B., 1980. Late Quaternary deep-sea record of Northwest African dust supply and wind circulation. Palaeoecol. Afr. 12, 239–253.

Sarnthein, M., Thiede, J., Pflaumann, U., Erlenkeuser, H., Fütterer, D.K., Koopmann, B., et al., 1982. Atmospheric and oceanic circulation patterns off Northwest Africa during the past 25 million years. In: Von Rad, U., Hinz, K., Sarnthein, M., Seibold, E. (Eds.), Geology of the Northwest African Continental Margin. Springer, Berlin, pp. 545–604.

Savoie, D.L., Prospero, J.M., 1989. Comparison of oceanic and continental sources of non-sea-salt sulfate over the Pacific Ocean. Nature 339, 685–687.

Schlager, W., Camber, O., 1986. Submarine slope angles, drowning unconformities and self-erosion of limestone escarpments. Geology 14, 762–765.

Schlager, W., Ginsburg, R.N., 1981. Bahama carbonate platforms—the deep and the past. Mar. Geol. 44, 1–24.

Schlager, W., James, N.P., 1978. Low-magnesium calcite limestones forming at the deep-sea floor, Tongue of the Ocean, Bahamas. Sedimentology 25, 675–702.

Schlager, W., Reijmer, J.J.G., Droxler, A., 1994. Highstand shedding of carbonate platforms. J. Sediment. Res. 64, 270–281.

Schütz, L., Jaenicke, R., Pietrek, H., 1981. Saharan dust transport over the North Atlantic Ocean. In: Péwé, T.L. (Ed.), Desert Dust, Geol. Soc. Am. Spec. Pap. GSA, Boulder, 186, 87–100.

Seidov, D., Maslin, M., 2001. Atlantic Ocean heat piracy and the bipolar climate see-saw during Heinrich and Dansgaard-Oeschger events. J. Quat. Sci. 16, 321–328.

Severinghaus, J., 2009. Southern see-saw seen. Nature 457, 1093–1094.

Shinn, E.A., Steinen, R.P., Lidz, B.H., Swart, P.K., 1989. Whitings, a sedimentologic dilemma. J. Sediment. Petrol. 59, 147–161.

Siddall, M., Rohling, E.J., Almogi-Labin, A., Hemleben, C., Meischner, D., Schmelzer, I., et al., 2003. Sea-level fluctuations during the last glacial cycle. Nature 423, 853–858.

Sirocko, F., Sarnthein, M., 1989. Wind-borne deposits in the northwestern Indian Ocean: record of Holocene sediments versus modern satellite data. In: Sarnthein, M. (Ed.), Paleoclimatology and Paleometeorology: Modern and Past Patterns of Global Atmospheric Transport. Kluwer Acad. Publ, Dordrecht, pp. 401–433.

Solheim, A., Pfirman, S., 1985. Sea-floor morphology outside a grounded surging glacier—Brsvellbreen, Svalbard. Mar. Geol. 65, 127–143.

Spielhagen, R.F., 1991. Die Eisdrift in der Franstraße während der letzten 200.000 Jahre. Geomar Rep. 4, 1–133.

Stabell, B., 1989. Deflation and humidity during the past 700 000 years in NW Africa from the marine record. In: Leinen, M., Sarnthein, M. (Eds.), Palaeoclimatology and Palaeometeorology: Modern and Past Patterns of Global Atmospheric Transport. Kluwer Academic Publishing, Dordrecht, pp. 771–777.

Stein, R., 1985. The post-Eocene sediment record of DSDP site 366: implications for African climate and plate tectonic drift. Geol. Soc. Am. Mem. 163, 305–315.

Stockman, K.W., Ginsberg, R.N., Shinn, E.A., 1967. The production of lime mud by algae in South Florida. J. Sediment. Petrol. 37, 633–648.

Stuut, J.B., Prins, M.A., Schneider, R.R., Weltje, G.J., Jansen, J.H.F., Postma, G., 2002. A 300 kyr record of aridity and wind strength in southwestern Africa: inferences from grain-size distributions of sediments on Walvis Ridge, SE Atlantic. Mar. Geol. 180, 221–234.

Summerfield, M.A., Hulton, N.J., 1994. Natural controls of fluvial denudation rates in major world drainage basins. J. Geophys. Res. B 99, 13871–13883.

Swift, D.J.P., Thorne, J.A., 1991. Sedimentation on continental margins, I: a general model for shelf sedimentation. In: Swift, D.J.P., Oertel, G.F., Tillman, R.W., Thorne, J.A. (Eds.), Shelf Sand and Sandstone Bodies, Int. Ass. Sediment., 14, 3–31. Spec. Publ.

Syvitski, J.P.M., 2003. Sediment fluxes and rates of sedimentation. In: Middleton, G.V. (Ed.), Encyclopedia of Sediments and Sedimentary Rocks. Kluwer Academic Publishers, Dordrecht, pp. 600–606.

Tiedemann, R., Sarnthein, M., Stein, R., 1989. Climatic changes in the western Sahara: aeolo-marine sediment record of the last 8 million years (Sites 657-661). In: Ruddiman, W., Sarnthein, M. (Eds.), Proc. ODP Sci. Res. Vol. 108, 241–261.

Turpin, M., Emmanuel, L., Renard, M., 2008. Nature and origin of carbonate particles along a transect on the western margin of Great Bahama Bank (Middle Miocene): sedimentary processes and depositional model. Bull. Soc. Geol. France 179, 231–244.

Walsh, J.P., Nittrouer, C.A., 2009. Understanding fine-grained river-sediment dispersal on continental margins. Mar. Geol. 263, 34–45.

Warne, A.G., Stanley, D.J., 1995. Sea-level change as critical factor in development of basin margin sequences: new evidence from late Quaternary record. J. Coast. Res. 17, 231–240. Spec. Issue.

Weltje, G., 1997. End-member modelling of compositional data: numerical-statistical algorithms for solving the explicit mixing problem. J. Math. Geol. 29, 503–549.

Westphal, H., 1998. Carbonate platform slopes—a record of changing conditions. Lecture Notes on Earth Science, Vol. 75. Springer, Berlin, 179 pp.

Westphal, H., 2006. Limestone–marl alternations as environmental archives and the role of early diagenesis: a critical review. Int. J. Earth Sci. (Geol. Rundsch.) 95, 947–961.

Wilber, R.J., Whitehead, J., Halley, R.B., Milliman, J.D., 1993. Carbonate-periplatform sedimentation by density flows: a mechanism for rapid off-bank and vertical transport of shallow-water fines: Comment. Geology 21, 667–668.

Wilson, P.A., Roberts, H.H., 1992. Carbonate-periplatform sedimentation by density flows: a mechanism for rapid off-bank and vertical transport of shallow-water fines. Geology 20, 713–716.

Wilson, P.A., Roberts, H.H., 1993. Carbonate-periplatform sedimentation by density flows: a mechanism for rapid off-bank and vertical transport of shallow-water fines: Reply. Geology 21, 668–669.
Wilson, P.A., Roberts, H.H., 1995. Density cascading: off-shelf sediment transport, evidence and implications, Bahama Bank. J. Sediment. Res. A65, 45–56.
Winn, R.D., Roberts, H.H., Kohl, B., Fillon, R.H., Bouma, A.H., Constans, R.E., 1995. Latest Quaternary deposition on the outer shelf, northern Gulf of Mexico: facies and sequence stratigraphy from Main Pass Block 303 core. Geol. Soc. Am. Bull. 107, 851–866.
Wollenburg, I., 1993. Sedimenttransport durch das arktische Meereis—Die rezente lithogene und biogene Materialfracht. Ber. Polarforsch. 127, 1–159.
Yokokawa, M., Franz, S.O., 2002. Changes in grain size and magnetic fabric at Blake-Bahama Outer Ridge during the late Pleistocene (marine isotope stages 8–10). Mar. Geol. 189, 123–144.

CHAPTER 6

Benthic Deep-Sea Carbonates: Reefs and Seeps

A.J. Wheeler* *and* A. Stadnitskaia†

Contents

1. Introduction

A common assumption is that carbonates and their lithified counterparts—limestones and dolomites—are indicative of biosedimentary accumulations in tropic shelf seas. Although carbonate accumulation rates are an

* School of Biological, Earth & Environmental Sciences and Environmental Research Institute, University College Cork, Cork, Ireland
† Department of Marine Organic Biogeochemistry, Royal Netherlands Institute for Sea Research (Royal NIOZ), Texel, The Netherlands

Developments in Sedimentology, Volume 63
ISSN 0070-4571, DOI: 10.1016/S0070-4571(11)63006-9

order of magnitude higher in the shallow-water tropics than in other settings (see for a comparison: Langer, 2008; Lindberg and Mienert, 2005; Vecsei, 2004), thereby generating thicker limestone successions, it is not justified to assume that carbonates are uncommon elsewhere. In fact, carbonates are formed across most latitudes (e.g. cool-water carbonates) and they are not restricted to shallow water but are also found in all but the deepest abyssal and hadal settings.

In modern marine environments, the vast majority of carbonates are of biological origin (Milliman, 1974; Sellwood, 1978; Tucker and Wright, 1990). Their distribution is directly linked to ecological parameters, that is, temperature, salinity, availability of light, nutrient and food accessibility, substrate, and the presence/absence of siliciclastics. These parameters regulate essential conditions for the growth of organisms secreting calcium carbonate. In terms of areal extent, deep-water carbonates cover a far larger percentage of the global ocean seafloor than shallow-water tropic carbonates, and they account for nearly half of the deep-water seafloor (Fig. 6.1). The vast majority of this accumulation is foraminiferal and coccolithophorid ooze, and to a lesser extent, pteropod ooze (see Section 2). Accumulations tend to occur beyond the influence of diluting terrigenous sediment supply and at water depths above the carbonate compensation

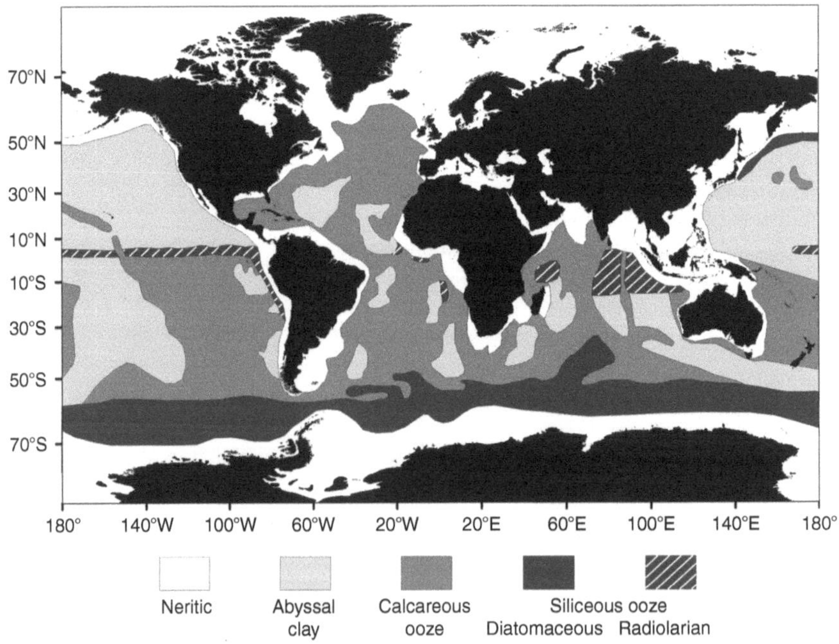

Figure 6.1 Global distribution of marine sediments showing the extent of deep-sea carbonates (modified from Thurman and Burton, 2001).

depth (CCD), below which carbonates go into solution (see Section 2.2). Equatorial and high-latitude deep-water sediments tend to be dominated by siliceous radiolarian ooze or diatomites.

But what are carbonates? Quite simply, they are deposits or accumulations that contain more than 50% carbonate. In modern marine settings, this invariably means calcium carbonate ($CaCO_3$), although many ancient carbonates have become significantly enriched in magnesium to remineralise as dolomites, $CaMg(CO_3)_2$. Other forms of naturally occurring marine carbonates exist but not in significant concentrations or accumulations and usually as minor components in other deposits, such as magnesite ($MgCO_3$) in evaporites, rhodochrosite ($MnCO_3$) in manganese-rich sediments, ankerite ($Ca(Mg,Fe,Mn)(CO_3)_2$) in iron-rich sediments and siderite ($FeCO_3$) as cements and concretions. Calcium carbonate takes two crystalline habits, the rhombohedral calcite and the orthorhombic aragonite, both of which are commonly found. Calcite may be enriched in magnesium to form high-magnesium calcite (containing more than 4% $MgCO_3$), although this is rare in deep-water settings (Morse, 2003). Aragonites are less stable than calcite and are therefore less likely to be preserved in ancient deposits, where they will revert to calcite: a process known as inversion.

A distinction is made between carbonate sediments where the carbonate is predominantly in the form of biological remains (see Sections 2 and 3) and authigenic carbonate accumulations, which form by the chemical precipitation of carbonates from seawater under certain conditions (see Section 4). Both occur in deep-water settings and are the foci of this chapter. Reference is often made to primary and secondary limestones, where primary limestones are those deposits which were originally deposited as carbonates, and secondary limestones are those that have become enriched in carbonates through diagenetic processes. Due to the relatively labile nature of carbonates, carbonate diagenetic processes are rife and complex. The formation of secondary deep-water carbonates due to carbonate enrichment through cementation and concretionary processes is not considered in this chapter. For an overview of carbonate diagenesis, the reader is directed to Bathurst (1975), Tucker and Wright (1990), Morse (2003) and also see Chapter 9.

To form an accumulation of carbonate, quite simply the input of carbonate must exceed, by more than 50%, the input of other non-carbonate sedimentary components. This can occur in three ways:

- if the carbonate production is high;
- if the input of non-carbonate components is low;
- if a combination of the above conditions exists.

Carbonate formation due solely to a high carbonate input occurs where a high degree of biological productivity by calcifying organisms exists, or if the rates of authigenic-carbonate formation are high. In deep water, this is typified by the formation of deep-sea reefs or by association with cold seeps.

Sessile calcareous reefal organisms may accumulate significant deposits of carbonate remains composed not only of their own remains but also of the calcareous remains of other organisms living in the reef habitat. Deep-sea cold seeps fuel authigenic-carbonate formation through microbial consumption of escaping methane. Both of these scenarios tend to occur on the continental margin in environments otherwise dominated by terrigenous sediment input. These two scenarios form the focus of this chapter and are explored fully in Sections 3 and 4.

In deeper waters, beyond the continental margins and onto the abyssal plain, the low input of terrigenous sediment results, possibly enhanced by high surface productivity, in the accumulation of foraminiferal, coccolithophorid and pteropod oozes (see Section 2). A more detailed account of pelagic sedimentation patterns can be found in Chapter 4.

2. Carbonate Bentho–Pelagic Coupling

Extensive accumulations of deep-water carbonates across the global seafloor (Fig. 6.1) are dependent upon pelagic carbonate productivity in the surfacial ocean waters. Large concentrations of calcareous plankton live in the epipelagic zone (down to a depth of 200 m), where they feed and reproduce, with some organisms migrating down as deep as 900 m during daylight hours. When the organisms die, their tests sink down to the seafloor, where they accumulate as calcareous oozes.

2.1. Calcareous ooze

The main contributors to calcareous oozes include the planktonic foraminifers, coccolithophores and pteropods (Fig. 6.2). As the main contributors to deep-sea carbonates, these are briefly reviewed here, although a more thorough appraisal can be found in Chapter 4. The foraminifers (Fig. 6.2A) are unicellular protists ranging in size from 1 to 30 μm and are a diverse group which can be roughly divided into spinose (Globigerinidae) and non-spinose (Globorotaliidae) forms. The distribution of species is controlled by the availability of light and food, with increasing species diversity and decreasing test size from the tropics to higher latitudes. Planktonic foraminifers are about twice as abundant in the Pacific as in the Atlantic, although foraminiferal sedimentation rates in the latter are higher (Kennish, 2001). In areas of strong benthic currents, sediment winnowing produces "foram sands" (see Chapter 3). The ability of foraminifers to preserve oxygen-isotope ratios in equilibrium with seawater within their tests has made them especially useful for the determination of paleoceanic temperatures.

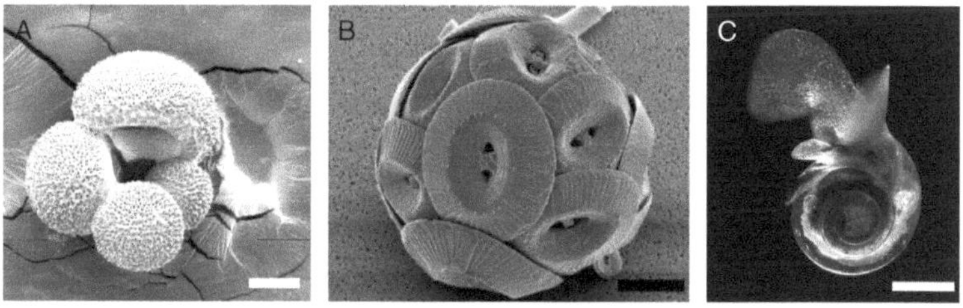

Figure 6.2 Examples of pelagic carbonate plankton. (A) The common planktonic foraminifers *Globigerina bulloides* as found in marine sediments, the live *Globigerina* is covered by long spines that are not preserved in the fossil. The scale bar is approx. 100 μm. (Image courtesy of the U.S. National Oceanic and Atmospheric Administration). (B) The coccolithosphere *Coccolithus pelagicus* from the mid-Atlantic. The scale bar is approx. 1 μm. (Image courtesy of Richard Lampitt and Jeremy Young, The Natural History Museum, London). (C) The pteropod *Limacina helicina*, swimming. The scale bar is approx. 1 cm. (Image courtesy and copyright of Russ Hopcroft, University of Alaska, Fairbanks). (A multi-colour version of this figure is on the included CD-ROM.)

Coccolithophores are nannoplanktonic calcifying algae that secrete skeletal edifices known as spheres (10–50 μm across); they are built of discoidal structures called "coccoliths" measuring 2–10 μm across (Fig. 6.2B). On death, the spheres tend to disarticulate into individual coccoliths. Like foraminifers, coccolithophores are most abundant in the tropics and decrease in abundance towards higher latitudes. Coccolith oozes are more common at temperate latitudes.

Pteropods (or sea butterflies) are pelagic gastropods of the order Opisthobranchia of which the suborder Thecosomata secretes a thin aragonitic shell 0.3–10 mm long (Fig. 6.2C). As aragonite is more susceptible to solution, pteropod oozes are only found on seamounts and mid-ocean ridges. They are more common in the tropics and other areas of high biological productivity.

Calcareous oozes cover $\sim 0.14 \times 10^9$ km^2 of the world's seafloor and accumulate at an average rate of about 1–3 mm a^{-1} (Kennett, 1982). The distribution of the deep-sea oozes is strongly dictated by surface productivity. The total $CaCO_3$ flux from the surface ocean is shown graphically in Fig. 6.3, although not all of this carbonate will accumulate on the seafloor. It is estimated that between 20% and 60% of the pelagic marine carbonate production comes from coccoliths (Broerse, 2000; Milliman, 1993; Wollast, 1994; Ziveri et al., 1999) with blooms in coccolithophores detectable from space (Fig. 6.4). Contributions to surface ocean $CaCO_3$ flux from foraminifers are less but still significant, accounting for a further 21.3% (Langer, 2008; Langer et al., 1997). The average planktonic foraminiferal global production rate is estimated at approximately 3.5 g m^{-2} a^{-1} or 1.2 billion tons per year (Langer, 2008; Langer et al., 1997). The contributions from pteropods are significantly smaller, although locally significant for deep-water carbonate accumulation.

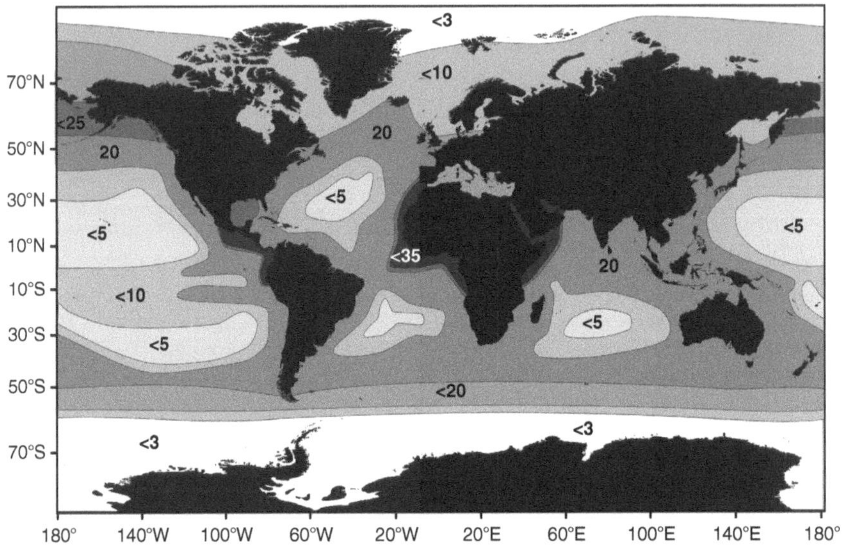

Figure 6.3 Flux of calcium carbonate from surface water (in g $CaCO_3$ m^{-2} a^{-1}) (modified from Milliman, 1993).

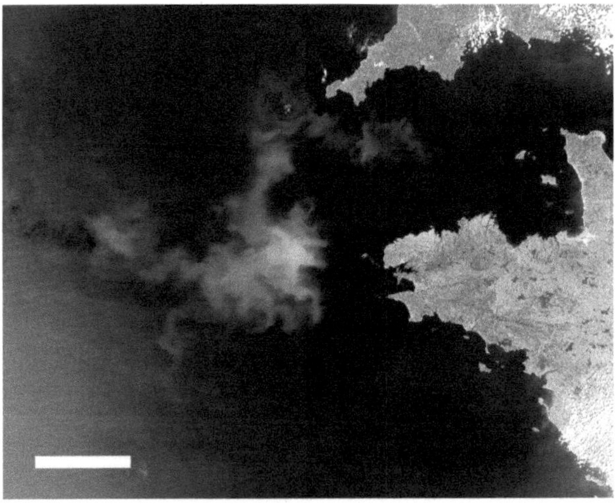

Figure 6.4 A coccolithophore bloom offshore of Brittany, extending across the shelf break and over the Porcupine Abyssal Plain. The bloom had been developing for a few days before this image was taken on June, 15, 2004 by the Moderate Resolution Imaging Spectroradiometer (MODIS) on NASA's Aqua satellite. The scale bar is approx. 100 km. (Image courtesy of Jacques Descloitres, MODIS Rapid Response Team, NASA/GSFC). (A multi-colour version of this figure is on the included CD-ROM.)

The pelagic input of $CaCO_3$ from primary producers at the surface is by far the most significant contribution of carbonate to deep-marine sediments. Additional contributions come from a range of benthic calcifying organisms, most significantly benthic calcareous foraminifers, various crustaceans, echinoderms and molluscs, and framework-building cold-water corals. Although the contribution to the deep-sea carbonates from benthic organisms is limited in comparison with pelagic carbonate imports, this balance may be reversed where benthic organisms form biological reefs (sensu Roberts et al., 2009; see Section 3)

2.2. Controls on calcareous ooze distribution: Calcite and aragonite dissolution

The distribution of calcareous oozes differs between oceans (Fig. 6.1; see also Chapter 4). The Atlantic is calcareous rich, while the Pacific is calcareous poor, whereas intermediate concentrations exist in the Indian Ocean. This distribution is largely a result of differential preservation on the seafloor as opposed to differences in primary productivity (see Chapters 3 and 4).

Figure 6.5 shows the depth at which seawater becomes saturated in aragonite and calcite. Below these depths, aragonite and calcite go into dissolution. Aragonite (of which pteropods are composed) and calcite (of which foraminifers and coccoliths are composed) are not preserved on the seafloor, thereby defining the depth limit for calcareous oozes. The aragonite and calcite saturation depths in the Pacific are considerably less than in the Atlantic, thus explaining the relative distribution of calcareous sediments in both oceans. As most deep waters in the Pacific are undersaturated in $CaCO_3$, $CaCO_3$ in the deepest basins undergoes permanent dissolution.

The average depth limit for the accumulation of calcareous ooze (the calcite saturation depth) in the South Pacific is approximately 3000 m, rising to less than 1000 m in the North Pacific. In the Atlantic, it is considerably deeper, at approximately 4500 m, and in the Indian Ocean, it is on average 3500 m. The accumulation of pteropod ooze is limited by the aragonite saturation depth, which is approximately 3000 m in the Atlantic but less than 1000 m in large parts of the Pacific.

The aragonite and calcite saturation depths are defined by temperature, pressure and the partial pressure of CO_2 (pCO_2), with the dissolution of $CaCO_3$ governed by the equation $CaCO_3 + H_2O + CO_2 <====> Ca^{2+} + 2HCO_3^-$

The more CO_2 can be held in solution, the more $CaCO_3$ will be dissolved. Since more CO_2 can be held in solution at lower temperatures and higher pressures, $CaCO_3$ is more soluble in deeper waters.

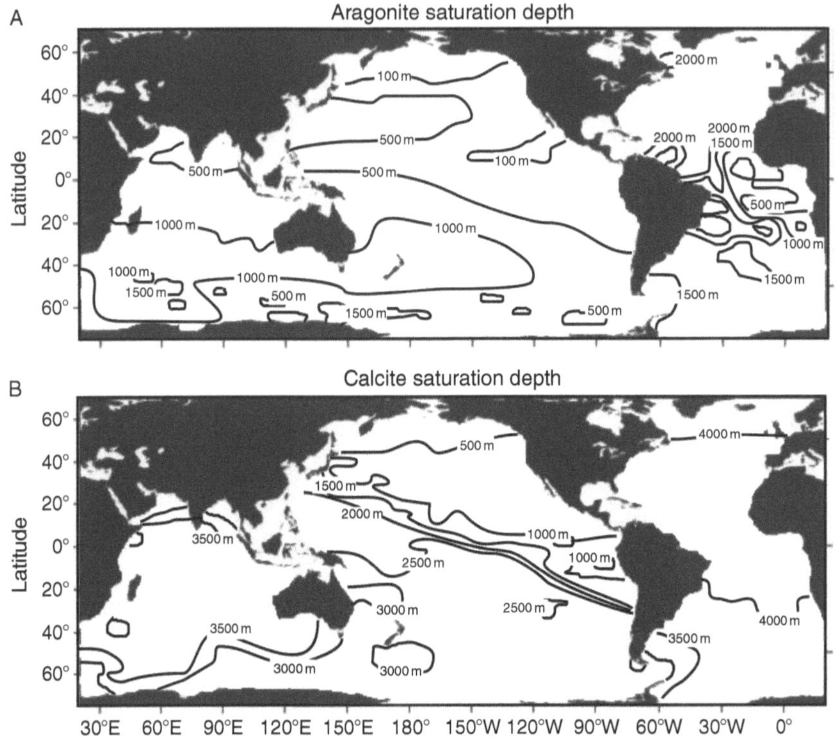

Figure 6.5 The distribution of (A) the aragonite and (B) the calcite saturation depth ($\Omega = 1$) in the oceans. The level at which aragonite and calcite are in thermodynamic equilibrium is known as the saturation depth. When the degree of saturation (Ω) exceeds 1, seawater is supersaturated with aragonite and calcite; in contrast, seawater is undersaturated with respect to these minerals when $\Omega < 1$ (modified from Feely et al., 2004).

3. Calcareous Aphotic Reefs

Despite the extensive formation of pelagic calcareous oozes and their undeniable dominance as the global deep-water carbonates, higher rates of deep-water carbonate accumulation exist in limited areas due to the processes controlling the formation of benthic reefs (see Section 3.1) and of authigenic carbonate (see Section 4). Reef-forming organisms are sessile and generate microhabitats that attract other organisms. This has three consequences which result in enhanced carbonate accumulation rates. In the first place, aphotic reefs tend to consist of dense agglomerations of sessile organism which, if calcareous, result in high local carbonate fluxes. Secondly, the reef microhabitat attracts other organisms, some of which are also calcareous; and thirdly, once a reef is established, organisms tend to grow on the remains of the former reef, resulting in the sustained accumulation of carbonates at the reef site. Perhaps more significantly, aphotic reefs are often "biodiversity hotspots".

With improved access to deep water over the past decade, calcareous aphotic reefs have become the focus of recent research. However, we have known about the existence of these reefs for some time. Early studies noted the presence of cold-water corals in shallow high-latitude waters (e.g. Gunnerus, 1768; Pontoppidan, 1755) with deep-water samples being dredged in the mid-nineteenth century (see Cairns, 2001a and references therein). Samples from NE Atlantic deep-sea coral reefs were discussed in a series of presentations at the Zoological Society of London as early as the 1870s (Duncan, 1870, 1873, 1878) with further notable early studies including those of Joubin (1922) and Teichert (1958). The advent of echo sounders allowed the topography of the deep-water reefs to be appreciated (Stetson et al., 1962), and the structure of the reefs was finally observed for the first time a little later, during Alvin submersible dives off South Carolina and the Florida Straits (U.S.A.) in the 1960–1970s (Milliman et al., 1967; Neumann et al., 1977).

In deep water, two carbonate reef contexts are significant: those offered by cold-water corals and those offered by deep-sea oysters. As a footnote, the term "reef" has many definitions, and in this context, we use the definition of Roberts et al. (2006, 2009), which defines biogenic reefs as long-lived structures, the growth of which is balanced by (bio)erosion, forming local topographic highs that alter hydrodynamic and sedimentary regimes and form structural habitats for many other species.

Aphotic reef deposits (both coral and oyster) are also valuable as archives of environmental and climate change (see Roberts et al., 2009, and references therein; Wisshak et al., 2009a).

3.1. Cold-water corals

3.1.1. Cold-water corals, reefs and mounds

There are over 5100 coral species of coral, of which more than 65% are aphotic: the so-called cold-water (or deep-sea) coral species (Cairns, 2007; Roberts et al., 2009). These include both calcareous and non-calcareous forms. Of the calcareous forms, the following species are the most significant framework-forming cold-water corals capable of creating biogenic reefs: *Enallopsammia profunda, Goniocorella dumosa, Lophelia pertusa, Madrepora oculata, Oculina varicosa* and *Solenosmilia variabilis*.

Figure 6.6 shows the global distribution of the main cold-water coral-reef sites discovered to date. This map is inevitably an underestimate of the distribution, as new reef and mound sites are being discovered on a continual basis as we explore more and more of the deep-sea seafloor. Nor does

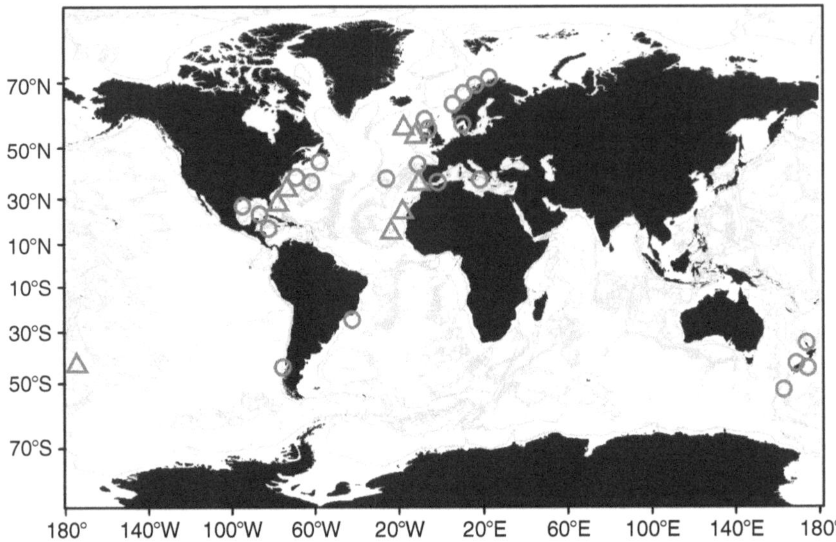

Figure 6.6 Distribution of the main cold-water coral-reef (O) and coral carbonate-mound (Δ) sites (modified from Roberts et al., 2009).

the map include individual occurrences of calcareous cold-water corals, and there is the distinct possibility that small cold-water coral reefs are also overlooked. The map shows a concentration of reefs in the North Atlantic and off New Zealand. The paucity of reefs in the Pacific and Indian Oceans is possibly due in part to the shallower aragonite saturation depth, although individual colonies can be recorded (Roberts et al., 2009). There is some bias in the distribution due to exploration efforts, although this does not account for the pattern shown.

Figure 6.6 also shows the known distribution of coral carbonate mounds (many of which also support contemporary coral reefs). Their distribution on continental margins is mainly symptomatic of the ecological tolerances of the coral-reef organisms. Coral carbonate mounds are distinct from cold-water coral reefs, as their definition relies on their stratigraphic properties. Coral carbonate mounds are topographic structures formed through *successive* periods of reef formation and hemipelagic sedimentation (Roberts et al., 2006, 2009). They are biogeological constructions consisting of alternating layers of reef deposits and intervening hemipelagic sedimentation. Their topography and environmental setting make them ideal locations for renewed reef formation over tens of thousands to millions of years, punctuated by periods of reef die-back and erosion. It should be noted that "coral bank" is also a valid term but that this term simply refers to a topographic high, whether a reef or a coral carbonate mound, the development of which has been influenced by the presence of cold-water corals.

The known distribution of coral carbonate mounds is more restricted than that of cold-water coral reefs as multiple reef-formation episodes are required to form these structures. The number of mounds is also an underestimate as seismic data, and in many cases, drilling or coring, are required to verify whether the bank is a single reef or has formed through successive periods of reef development—a coral carbonate mound. This bias is further compounded by differences in exploration activity between the oceans. Nevertheless, there appears to be a genuine tendency for coral carbonate mound development on both sides of the North Atlantic. This implies that parts of the North Atlantic have provd repeatedly favourable for cold-water coral-reef development over millennia. This may be because the aragonite saturation depth (Fig. 6.5A) is deepest in the North Atlantic. With climatically induced fluctuations in the aragonite saturation depth occurring through time, it may be that the North Atlantic is one of a few deep-water settings where cold-water coral-reef formation is repeatably viable.

Cold-water coral reefs come in a range of shapes and sizes. Typically, reefs are tens to hundreds of metres across, raising several to tens of metres above the seafloor. As reefs may colonise existing bedrock highs and, of course, form "live" tops to coral carbonate mounds, the cold-water corals tend to be denser and better developed on mound summits and upstream flanks where food (principally particulate organic matter, POM) is more abundant. In areas where current speeds are high enough to inhibit the growth of a coral framework, coral reefs may prosper in the lee side of coral carbonate mounds, for example, on the Giant Mound, northern Porcupine Bank, offshore of Ireland (Dorschel et al., 2009).

Coral carbonate mounds by contrast are often hundreds of metres to several kilometres across, rising tens to hundreds of metres above the seafloor. The largest known examples, from the Logachev mound province on the eastern Rockall Bank, offshore Ireland, are up to 380 m high, covering over 5 km at the base (de Haas et al., 2009; Mienis et al., 2006). Both cold-water reefs and coral carbonate mounds tend to occur in clusters of several mounds. Clusters of coral carbonate mounds are often called "mound provinces" and may extend tens of kilometres along continental margins.

Groups of colonies tend to spread out from the centre of the reef and grow preferentially upstream, forming an ovoid reef plan aligned with the prevailing current direction (Fig. 6.7). In many instances, the morphology of the colonised substrate may dictate that an "inherited" reef morphology develops (Wheeler et al., 2007). Examples include the Sula Ridge, offshore Norway, which has developed on the boulder-strewn levees of iceberg ploughmarks (Freiwald et al., 1999). Coral carbonate mounds, due to their long development histories, tend to exhibit "developed" morphologies aligned to the dominant current regime. Conical examples are rare with most forming ridge-shaped morphologies. Joining of reefs and mounds is common, with double-ridged or irregularly shaped accumulations being common.

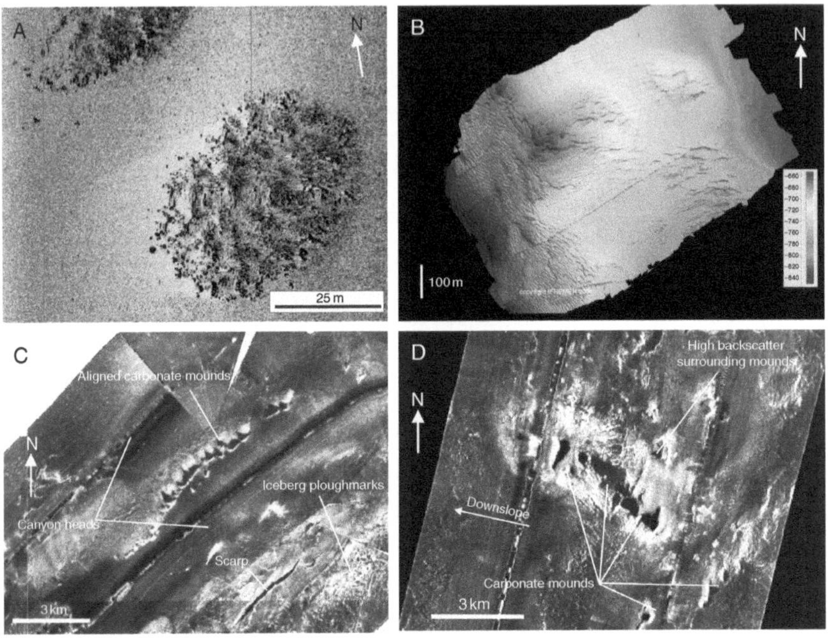

Figure 6.7 Examples of cold-water coral reefs and mounds. (A) An example of a small ovoid reef in the Darwin Mounds, northern Rockall Trough, Scotland (modified from Wheeler et al., 2008). (B) ROV-MBES microbathymetry image of a Logachev mound, Rockall Bank, offshore Ireland, showing a ridged morphology aligned to the dominant current direction with a coral cover forming terraces (modified from Wheeler et al., 2007). (C) Alignment of coral carbonate mounds (the Hedge Mounds) on a topographic high on the Porcupine Bank, offshore Ireland (modified from Wheeler et al., 2007). (D) Complex, multi-summited mound (the Giant Mound) of the Porcupine Bank, offshore Ireland (modified from Wheeler et al., 2007). (A multi-colour version of this figure is on the included CD-ROM.)

3.1.2. Mound-initiation theories (cold-seep versus environmental controls)

The size of coral carbonate mounds has provoked two conflicting theories for the initiation and development of cold-water coral reefs and coral carbonate mounds. One theory, the hydraulic theory, proposes that cold-water coral reefs benefit from hydrocarbon seepage, whereas the alternative theory simply sees cold-water corals as thriving under favourable hydrodynamic and environmental conditions (Fig. 6.8).

The hydraulic theory, first proposed by Hovland (1990), draws a link between cold seeps (see Section 4) and cold-water coral reefs. The theory states that cold-water coral reefs and coral carbonate mounds grow in areas of enhanced organic productivity, due to elevated phosphorous, nitrate, sulphur and carbon fluxes to the seafloor through sub-sea fluid expulsion. Chemosynthetic bacterial breakdown of expelled compounds form the basis

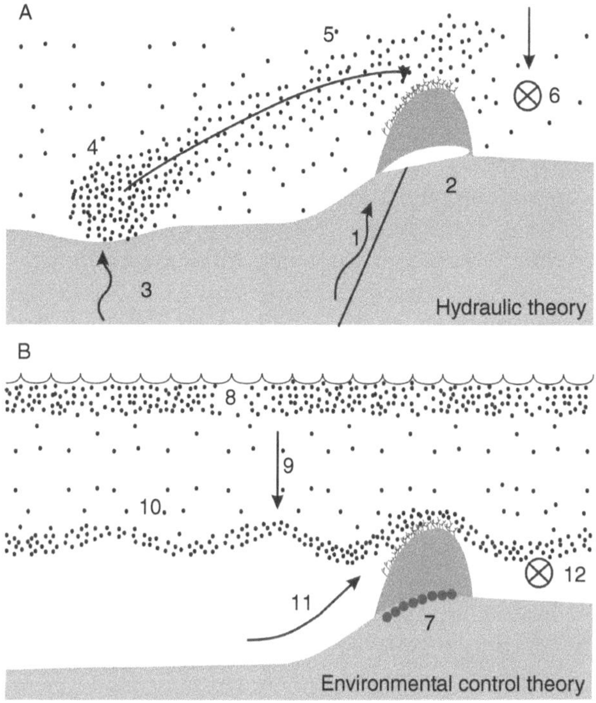

Figure 6.8 Schematic diagram summarising the hydraulic and environmental control theories for mound initiation and development. (A) Hydraulic theory. (1) Gas seepage may lead to the development of (2) methane-derived authigenic carbonates that may act as a substratum for coral settlement. (3) Gas seepage at the seabed provides a source of food, increasing (4) biomass along the food chain. (5) Cold-water corals are supported by this food chain. (6) Along-slope and hemipelagic sedimentation provides a mineral sediment infill of the reef structure. (B) Environmental-control theory. (7) Erosion of the seafloor by strong currents generates suitable coarse-grained substrata for coral settlement. (8) Surface primary productivity underpins the food chain. (9) This settles through the water column as marine "snow" and may (10) become concentrated at water-mass boundaries and transported to the coral reef, possibly assisted by internal waves. (11) Benthic hydrodynamics helps to enhance the food flux and prevents coral polyp smothering by deposition of fine sediments. (12) Along-slope and hemipelagic sedimentation provides a mineral sediment infill of the reef structure (from Roberts et al., 2009).

of a food chain from which cold-water corals benefit. In addition, microbially mediated authigenic-carbonate formation (see Section 4.4) may provide an appropriate hard substratum for cold-water corals to grow on. Interstitial authigenic-carbonate formation may contribute more carbonate to cement the reef deposits, adding further stability to the growing carbonate structures. Hovland et al. (1994) gave further support to the theory by noting that "giant" coral carbonate mounds in the Hovland mound province (northern Porcupine Seabight, offshore Ireland) directly overlie faults connected to hydrocarbon reserves, but this seismic interpretation was later disputed by Bailey et al. (2003). Henriet et al. (1998) suggested that the hydrocarbon source "fuelling" the Hovland mounds may have come from decaying gas hydrates.

Numerous observations of cold-water coral reefs growing in close proximity to or even in pockmarks have been noted and are fully explored in Hovland (2008): on the Norwegian Shelf (Hovland, 2005; Hovland and Mortensen, 1999; Hovland et al., 1998; Lindberg et al., 2007), offshore Ireland (Naeth et al., 2006; Van Rooij et al., 2009), Darwin Mounds, offshore UK (Huvenne et al., 2009; Masson et al., 2003; Wheeler et al., 2008), Gulf of Cadiz (Foubert et al., 2008; Wienberg et al., 2009), offshore Brazil (Sumida et al., 2004) and in the Gulf of Mexico (Moore and Bullis, 1960; Rezak et al., 1986). However, despite the coincidence of some cold-water reefs with hydrocarbon seeps, there are many more examples where no evidence of seepage can be found.

Even where cold-water corals and seeps coexist, no evidence has been found that there is any fundamental relationship between the seep and the cold-water corals. The $\delta^{13}C$ values for authigenic carbonates at seep sites typically vary between $-38‰$ and $-5‰$ PDB (Boetius and Suess, 2004; Suess et al., 1999) and in organic tissue from methane-seep macrobenthos averaging $-43‰$ PDB (Levin and Mendoza, 2007). $\delta^{13}C$ values from cold-water coral frameworks and organic tissue suggest a surface productivity source for the metabolised carbon with $\delta^{13}C$ values at $-2.0‰$ and $-9.9‰$ PDB and $-20‰$ PDB, respectively (Blamart et al., 2005; Duineveld et al., 2004; Lutringer et al., 2005; Mikkelsen et al., 1982; Mortensen and Rapp, 1998; Spiro et al., 2000). Similarly, in terms of fauna assemblages, Cordes et al. (2008) found little similarity between the faunal assemblages of cold-seep fauna and nearby cold-water corals in the Gulf of Mexico, suggesting that the food chains were not significantly interconnected. No examples of cold-seep fauna coexisting with cold-water corals have been found, with a possible exception of small *Beggiatoa* mats near Norwegian *L. pertusa* reefs (Hovland, 2005, 2008; Hovland and Mortensen, 1999).

One exception where cold-water coral reefs can benefit from cold-seep activity is in the formation of hard substrate for corals to colonise. This can take the form of authigenic carbonates or pockmark gravel lags. Three examples of coral directly colonising pockmarks are known: the Darwin Mounds in the northern Rockall Trough, UK (Huvenne et al., 2009; Masson et al., 2003; Wheeler et al., 2008), the Kristin field, Norwegian Shelf (Hovland, 2005) and offshore Brazil (Sumida et al., 2004). It therefore appears that a hard substrate under the appropriate environmental conditions is all that is required to initiate cold-water coral colonisation that may eventually lead to reef and mound development. In fact, most cold-water corals seem to be able to establish themselves on a variety of natural hard substrates, including rock outcrops, boulders and gravel (e.g. Dorschel et al., 2009) as well as man-made structures such as submarine cables (Duncan, 1877) and oil platforms (Bell and Smith, 1999; Gass and Roberts, 2006; Roberts, 2002).

Temperature and salinity tolerances for the main framework-forming cold-water corals vary between 4 and 13 °C and 35–37 ppt, respectively (Roberts et al., 2006; Rogers, 1999; Taviani et al., 2006), with *Oculina* reefs existing in often warmer waters: from 10 to 30 °C (Roberts et al., 2009). Dullo et al. (2008) noted that cold-water coral reefs in the NE Atlantic occur only within a narrow water-density envelope (27.35–27.65 kg m^{-3}), which may have a control on the dispersal of coral larvae. Importantly, currents need to be strong enough to ensure that coral polyps are not smothered by fine-grained sediment, but their velocity should not be more than 100 cm s^{-1}, as higher velocities can topple colonies (Frederiksen et al., 1992). Cold-water corals are diverse feeders, feeding on POM and live copepods and other zooplankton (Duineveld et al., 2004; Freiwald, 2002; Kiriakoulakis et al., 2006; Mortensen, 2001). Cold-water corals often occur at water-mass boundaries in areas of high surface productivity where POM becomes concentrated in intermediate nepheloid layers. Water-mass boundaries on continental margins may also be subjected to internal waves which locally intensify benthic hydrodynamic conditions and nutrient mixing, enhancing food supply from benthic zooplankton (Frederiksen et al., 1992; Mienis et al., 2007). Similarly, cold-water corals may benefit from enhanced food supply due to Taylor Columns and current intensification around banks (Dorschel et al., 2007a; White et al., 2005, 2007) or through the seasonal trapping of a "dome" on nutrient-rich water on banks in the spring, which seasonally drains to enrich deeper water where corals thrive (White et al., 1998; White et al., 2005). Thiem et al. (2006) also demonstrated that coral communities on the Norwegian shelf break also benefit from oceanographic enhanced food supply generated by an interaction between Atlantic along-slope currents and Ekmann transport induced by atmospheric low-pressure systems.

3.1.3. Mound development models

The model for the development of cold-water coral reefs from initial coral larvae settlement was presented by Squires (1964) from uplifted cold-water coral exposures from Palliser Bay, New Zealand (Fig. 6.9). The model envisages the spread of coral colonies over the available substrate to form a coral thicket which eventually defines the footprint of the coral reef or mound. As the coral colonies grow, the thicket increases in density and attracts a number of other organisms to the microhabitat that has been created (Messing et al., 1990; Mullins et al., 1981). Typical carbonate contributors to the coral reef sediments other than corals include molluscs, serpulids, brachiopods, bryozoans, barnacles, echinoderms, ostracods and

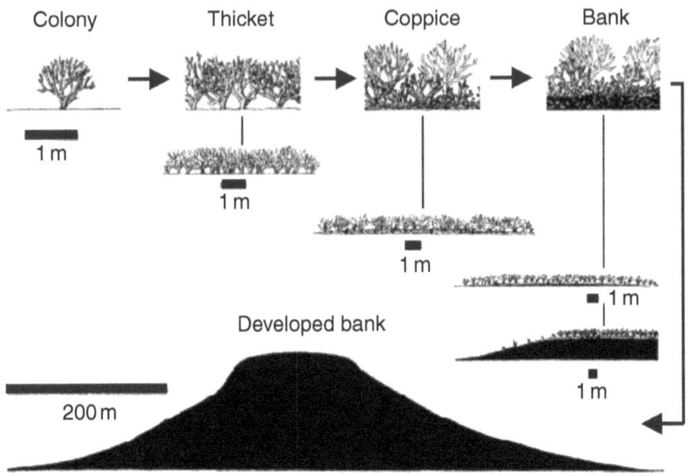

Figure 6.9 Model of cold-water coral-reef development, showing stages from (A) coral colony to (B) thicket to (C) coppice and eventual to (D) mound (bank) development (modified from Squires, 1964).

benthic foraminifers (De Mol et al., 2002). The coral thicket passes to the coral coppice stage, so that the habitat becomes vertically zoned with live coral colonies at the surface, and other organisms, including bioeroders, at a lower level. It is at this stage that the coppice becomes a reef.

The role of bioeroders is crucial. As the coral colony grows, only the upper branches support coral polyps with the base and lower branches susceptible to bioerosion by sponges and fungi (Beuck and Freiwald, 2005). These bioeroders bore into the coral skeleton, structurally weakening it until it collapses. In this way, the corals contribute coral fragments to the accumulating carbonate sediments at the base of the reef to form packstones. Further breakdown of the coral fragments leads to the generation of a micritic component. At the same time, calcareous components from other dead coral-habitat dwellers may also be added to the sediment.

As well as coral bioerosion products, the reef sediment also contains significant components of hemipelagic components. This may, in many cases, greatly exceed the amount of reef-generated carbonates and is an important contributor to the reef-development processes. This sediment contains few fines but typically includes a mixture of sands and foraminifers. The contribution of sand to the coral-reef sediment may exceed 50%, which means that in some cases, cold-water coral-reef sediments may not, strictly speaking, be carbonates, although they are still calcareous.

As the corals grow, they present an obstacle to the flow of benthic boundary-layer waters (see Faugères and Mulder, 2011, this volume) and effectively increase the frictional drag component (Mullins et al., 1981).

This has the effect of slowing down the flow, thereby reducing the sediment-carrying capacity of the water and stimulating sedimentation at the reef site. In addition, the obstacle of the coral colonies creates eddies in the water flow, which tends to deflect sediment transport into the centre of the reef. Roberts et al. (2009) and Wheeler et al. (submitted) show the change in sand-ripple morphology from moderate flow regime cuspate to lower flow regime straight-crested ripples as they pass into the Moira Mounds. The diffraction of sand-ripple crests around coral colonies can also be seen. The ability of the coral reef to baffle sediment has important implications for the development of the reef. In the first place, it buries the lower levels of the coral colonies, thereby providing some protection to the coral colony from bioerosional toppling and also offering physical support. If the sedimentation rates do not outstrip the coral growth rates, the corals do not seem to be adversely affected by the sedimentation at their base. In fact, the input of sediment into the reef helps to elevate the reef higher up into the water column, where current velocities and consequently food supply, are enhanced. Wheeler et al. (2008) showed for the Darwin Mounds that, as current velocities increased, bedforms changed from rippled sands to sediment waves and that a corresponding increase in mound height and downstream reef-plan elongation (from circular to ovoid) occurred.

As well as trapping sediment and slowing down the water through the coral framework, corals also trap POM. Mienis et al. (2007) note that tidal variations in current velocity also help to resuspend sediment, including POM, within the coral reef.

The continued development and expansion of reefs may cause them to coalesce over a large area (Wheeler et al., 2010). Analysis of seismic data suggests that many coral carbonate mounds are initiated over a wide base (Colman et al., 2005; De Mol et al., 2002, 2005, 2007; Huvenne et al., 2003; Mienis et al., 2006).

As cold-water coral reefs are sensitive to climatic variations, the reef may eventually die due to a number of factors. As climate changes, the depth at which water masses occur may change and the coral reef may find itself existing in a water mass, the properties of which are not conducive to coral growth. Such an occurrence was noted by Rüggeberg et al. (2007) as a probable cause of the demise of coral reefs on the Propellor Mound, offshore Ireland. Alternatively, background sedimentation rates may outstrip coral growth rates, resulting in a burial of the reef. These alternatives actually often work together; for example, during a change from interglacial to glacial conditions in the NE Atlantic, water masses may become less stratified, the sea level may drop and fluvioglacial sediment supply and glaciomarine sedimentation (especially IRD) may increase over the shelf.

If the conditions improve, however, the raised topography of the former reef as well as the availability of appropriate substrates for colonisation (exposed dropstones, hardgrounds or former coral frameworks) make the site ideal for renewed coral-reef development. In this way, coral carbonate mounds develop through successive phases of reef development and intervening hemipelagic sedimentation and/or erosion. The possibility that glacial hemipelagic sediment draped over reef deposits may slump off the steep topography of coral carbonate mounds was mentioned by Dorschel et al. (2005). Such a mechanism would also expose substrates fit for renewed colonisation.

Numerous, broadly similar, models have been published for coral-mound development (De Mol et al., 2002, 2005; Dorschel et al., 2005; Henriet et al., 2003; Huvenne et al., 2005; Kenyon et al., 2003; Roberts et al., 2006; Roberts et al., 2009; Rüggeberg et al., 2007); the model of Roberts et al. (2006) is given as an example (Fig. 6.10). De Mol et al. (2005), Henriet et al. (2003) and Huvenne et al. (2005) proposed that coral carbonate mounds may undergo an initial period of rapid accumulation, the so-called "active" phase, followed by a period of "slow-down" and eventual "retirement" where the mound may become buried. The "booster" and "slow-down" phases are recorded in the Challenger Mound (Kano et al., 2007) with initial growth from 2.70 to 1.67 Ma ago at a rate of 12.6 cm ka^{-1}, followed by an erosion phase coinciding with the change in dominant periodicity of glacial response from 41 to 100 ka (Lisiecki and Raymo, 2007). The upper-mound subsequence above the unconformity recorded renewed reef development at a lower rate of 2.3 cm ka^{-1}. Henriet et al. (2003) postulated that the booster phase may be due to the influence of hydrocarbon seepage (see Section 3.1.2), although there is no evidence to support the role of hydrocarbons in the development of the Challenger Mound (Roberts et al., 2009; Williams et al., 2006). Alternatively, the rapid growth phases reflect ideal climatic condition for renewed coral growth prior to 1 Ma, whereas more extreme oscillations in climate from 1 Ma to present (Lisiecki and Raymo, 2007) may have been less conducive to reef development.

Very few studies deal with the diagenetic processes altering cold-water coral-reef deposits. Noé et al. (2006) note several diagenetic phases with coral carbonate mound sediments. Compaction and de-watering are followed by the growth of calcite microspar and forms the main cementation processes. This initially occurs on the rims of foraminiferal tests but then spreads into all pore spaces. Due to the low permeability of the sediment, lithification can be patchy. The extent of neomorphism and dissolution of aragonite components varies from site to site. The formation of hardgrounds due to prolonged seabed exposure under a rigorous current regime is common to many coral carbonate mounds (de Haas et al., 2009; Paull et al., 2000; Wheeler et al., 2005) and adds stability to the coral mounds (de Haas et al., 2009) as well as potential settlement sites for future coral-reef development. Foubert et al. (2008) note significant post-depositional alteration of carbonates in the Moroccan coral carbonate mound successions.

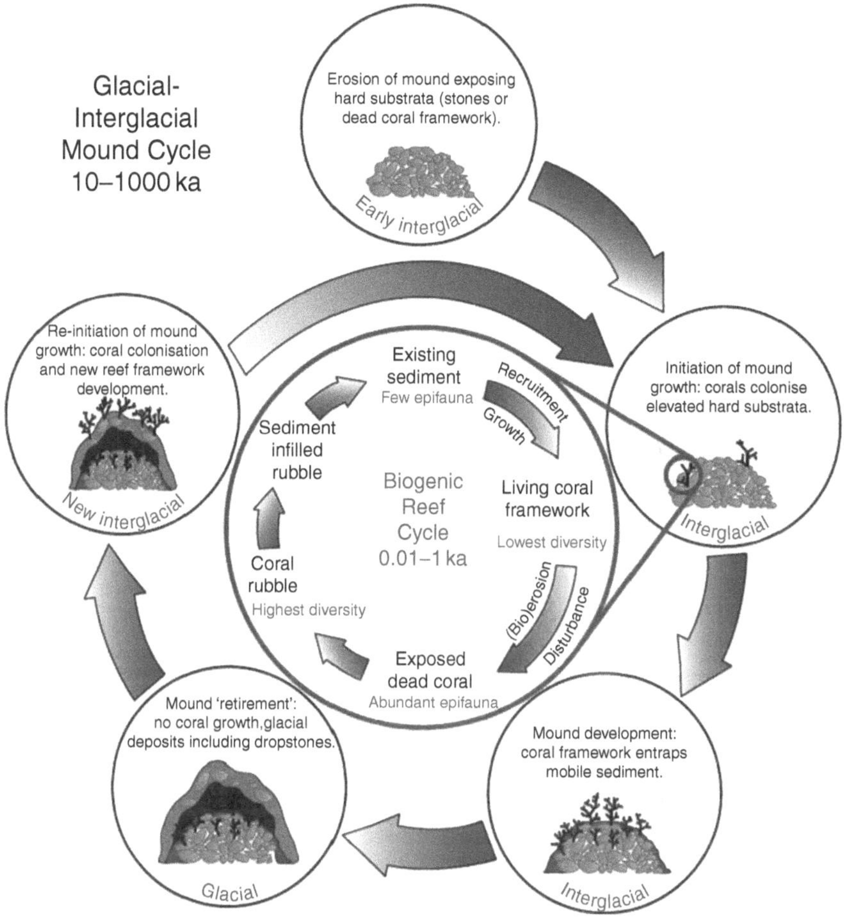

Figure 6.10 Model of coral carbonate mound development. Outer circle: cyclic stages of carbonate-mound growth from initiation, development, retirement and recolonisation. Inner circle: smaller-scale cycle of reef microhabitats, succession and faunal diversity (from Roberts et al., 2006). (A multi-colour version of this figure is on the included CD-ROM.)

3.1.4. Accumulation rates

Cold-water coral-reef and coral carbonate mound accumulation rates are spatially variable. Nevertheless, the accumulation rates are very high. Reed et al. (2005) record accumulation rates for *Oculina varicosa* reefs offshore Florida as 16 mm a^{-1}. For *L. pertusa* reefs, lower rates of 5 mm a^{-1} and 4.3 mm a^{-1} have been calculated for the Fugløy reef and Sula Ridge, respectively, offshore Norway (Freiwald et al., 1999; Lindberg et al., 2007). Although considerably lower than for the *Oculina varicosa* reefs, these rates are still impressively high if compared with rates of calcareous-ooze formation (1–3 mm a^{-1}:

Kennett, 1982). Nevertheless, much lower reef-accumulation rates of 0.067–0.5 mm a^{-1} are calculated for the coral carbonate mounds offshore Ireland (Dorschel et al., 2007b; Frank et al., 2005). These rates are time-averaged and include periods of erosion and non-deposition, and intercalated hemipelagic sedimentary deposition. The fact that the coral carbonate mounds form positive features means that their accumulation rates outstrip that of the surrounding seabed.

With respect to the carbonate flux, Lindberg and Mienert (2005) calculated rates of 54–188 g m^{-2} a^{-1} from *L. pertusa* reefs, offshore Norway. Comparable flux rates were calculated by Titschack et al. (2009), viz. 173 g m^{-2} a^{-1} for the Challenger Mound (offshore Ireland) during its active growth phase (2.70 to 1.67 Ma). Since 1 Ma ago, the Challenger Mound carbonate flux was reduced to 57 g m^{-2} a^{-1}. This is, however, still considerably higher than calculated rates for the Propeller Mound, offshore Ireland, which is in its "retirement phase" (0.277–6.160 g m^{-2} a^{-1}: Dorschel et al., 2007b). Titschack et al. (2009) estimate the carbonate production for the entire Belgica mound province as 3.2×10^3g m^{-2} a^{-1} (for the 2.70–1.67 Ma time interval) and 1.0×10^3 g m^{-2} a^{-1} (for the last million years), whilst Lindberg and Mienert (2005) estimate that cold-water corals may account for $<1\%$ of the total marine calcium-carbonate production.

3.2. Deep-sea oysters

Recently discovered and locally significant accumulation centres for deep-sea carbonates are formed by the deep-sea pycnodontine oyster *Neopycnodonte zibrowii* (Fig. 6.11). They grow on steep rock faces and have therefore been difficult to sample by traditional sampling. Consequently, we know very little about them, and recent discoveries have caused much interest. Until recently, pycnodontine oysters were in fact known only from the fossil record, where they were once widespread. It is suggested that modern *Neopycnodonte zibrowii* is surviving as a relict member of this ancient fauna in deep-sea refugia (Wisshak et al., 2009b). With the increased usage of remotely operated vehicles (ROVs) as survey platforms, such inaccessible habitats are now becoming the focus of more targeted research.

One notable and well-studied site exists between 420 and 500 m water depth under overhangs in the Faial Channel, Azores (Wisshak et al., 2009a,b). Here, *Neopycnodonte zibrowii* can grow up to 30 cm across, and is often found grown one on top of the other to form multi-specimen stacks. Accumulations may contain several hundreds of individuals (Wisshak et al., 2009a,b). Individuals can live for several centuries; an age of 370–470 year has been found for one individual based on AMS ^{14}C ages (Wisshak et al., 2009a).

Figure 6.11 *Neopycnodonte zibrowii* oyster bed growing under an overhanging outcrop of volcanic bedrock at 500-m water depth in the Azores. The image is approximately 1 m across (from Wisshak et al., 2009a). (A multi-colour version of this figure is on the included CD-ROM.)

Similar occurrences of *Neopycnodonte zibrowii* have also been found on the Ostrea Spur and Guilvinec Canyon, Bay of Biscay, NE Atlantic, where they also occur typically under rock overhangs and on steep to vertical faces on the flank of a submarine canyon at between 540 and 586 m water depth (Van Rooij et al., 2007; submitted). Nine banks were observed, with the oysters forming dense accumulations with up to 100 individuals per m^2 including live and dead (articulated and recently disarticulated) individuals.

Elsewhere, *Neopycnodonte zibrowii* has been found on the El Idrissi Bank in the Alboran Sea, western Mediterranean at 390–490 m water depth (Gofas et al., 2007), offshore Sicily, Italy (Taviani et al., 2006), Gorringe Bank, off Portugal (Auzende and Shipboard Party, 1984) and south of Madeira, Portugal (Hoernle and Shipboard party, 2001).

4. Cold Seeps and Related Carbonates

In addition to deep-sea carbonates formed from the remains of calcareous organisms, carbonates can also form as a result of chemical reactions in sediments. In this case the trigger for carbonate precipitation is methane-based microbial metabolism. Where there is an outflow of hydrocarbons from the seafloor, these carbonates display different morphologies and can be found in the shapes of crusts, pavements, chimneys and concretions; these are the focus of this section. The seepage of methane from the seafloor, and hence the transfer of carbon from the geological reservoir to the ocean reservoir, is a significant component of the carbon biogeochemical cycle (see Section 4.1). The carbonates that are formed during this processes record changes in this cycle (see Section 5.2).

4.1. Methane and the earth system: A small molecule with a big impact

Methane (CH_4) is the smallest hydrocarbon molecule and consists of one carbon atom (prefix meth-) covalently bound with four hydrogen atoms in a tetrahedron. Despite its structural simplicity, methane is an important greenhouse gas (after water vapour and CO_2), playing a significant role in global climate change and atmospheric chemistry. Over a 20-year emission period, 1 ton of CH_4 will have the same climate impact as 72 tons of CO_2 (Intergovernmental Panel on Climate Change, 2007). A rapid global release of methane from its subsurface reservoirs would have a major impact on the world's climate, a scenario which we think may have happened in the past at the end of the Paleogene (Dickens et al., 1995; Nisbet et al., 2009; Zachos et al., 2005).The observed chemical lifetime (or residence time) of methane in today's atmosphere is 12 years and is over twenty times more effective in trapping heat than the same amount of CO_2. These natural abilities make methane a more hazardous greenhouse gas than carbon dioxide; hence the need to understand its growth rate and future projection.

Continental margins are important areas for the discharge of greenhouse gases and they are sensitive to, and key players in, climate change. Methane escaping from the seafloor plays an important part in the carbon biogeochemical cycle. This is one of the reasons why deep-sea methane/hydrocarbon seepage and their associated carbonate deposits have been of key interest to research during recent years. Methane is buoyant and migrates upwards, making gas seepage into the hydrosphere a natural and dynamic process. Migration is strongly associated with geological and tectonic factors, which can either prohibit fluid passage, forming secondary hydrocarbon accumulations in the subsurface, or induce fluid transport, determining migration modes and mechanisms.

As well as forming carbonates (see Section 4.4), the continuous upward migration of methane-rich fluids may also form submarine gas-hydrate accumulations in the subsurface, storing considerable quantities of methane (Bouriak and Akhmetzhanov, 1998; Ginsburg and Soloviev, 1997; Ivanov et al., 1998; Soloviev and Ginsburg, 1997). Gas hydrates are ice-like crystalline solids composed of a mixture of water and natural gas, frequently methane (see Chapter 9). They are stable under specific pressure and temperature conditions (Henriet and Mienert, 1998; Kvenvolden et al., 1993, and references therein), filling the pore spaces of sediments. Accordingly, in areas of gas seepage, gaseous compounds, particularly methane, may be present in a dissolved form, as a gas phase and as a solid phase (in hydrate form), depending on gas composition, temperature and pressure.

The estimated amount of natural gas contained in the world's gas-hydrate accumulations is three orders of magnitude larger than the amount of methane currently present in the atmosphere.

The spatial distribution of methane accumulations within the marine realms is still not fully revealed. This means that, although we know that methane venting at the seafloor is a potentially important component of the global carbon cycle, we cannot reliably estimate the geological flux of methane to the ocean and the atmosphere. In addition, the regulation of methane transport, fluxes and turnover in marine sediments is also uncertain. This lack of information and understanding leads to uncertainties in the development of future climate scenarios which, through feedback, will also influence future methane release. Having said this, it should be mentioned that in the past couple of decades, significant advances have given us a better understanding of the interactions between seep fluids, methane, microorganisms, minerals and the environment in general.

Given the significance of methane seeps, understanding the past activity of methane seepage and its interactions with the Earth climate system is of crucial importance. One potential source of information about seepage activity is constituted by seep deposits, that is, seep-associated carbonates.

4.2. Submarine cold seepage

Submarine hydrothermal vents with their associated black smokers are well known to the public and have been intensely studied for several decades. However, recent geological exploration of the ocean floor has led to numerous discoveries of a different type of seep, where submarine fluids are discharged at the same temperature as the surrounding seawater. These vents are called "cold seeps" and include everything from small-scale gas bubbling at the seafloor (Hovland and Judd, 1988) to catastrophic bursts of fluids from the subsurface (Ivanov et al., 1998). These seeps are often associated with authigenic (microbially-mediated) carbonate formation (see also Chapter 9) and are concentrated along, and within, continental margins. Since the 1980s, the exploration of cold seeps has received prominence in science, as it was shown that degassing and defluidisation at continental margins are significant contributors to the Earth climate and biogeochemical cycles.

Cold seepage often results from the upward migration of fluids from the subsurface through zones of weakness, that is, fault systems, cracks and fissures, affecting the morphology of the seafloor. Fluid escape at the seafloor can result in the formation of (1) pockmarks, that is, negative topographic features formed due to fluid escape (Hovland and Judd, 1988), (2) large fissure eruptions of sediments from the subsurface (e.g. Ivanov et al., 1998) and/or (3) mud volcanoes (Fig. 6.12).

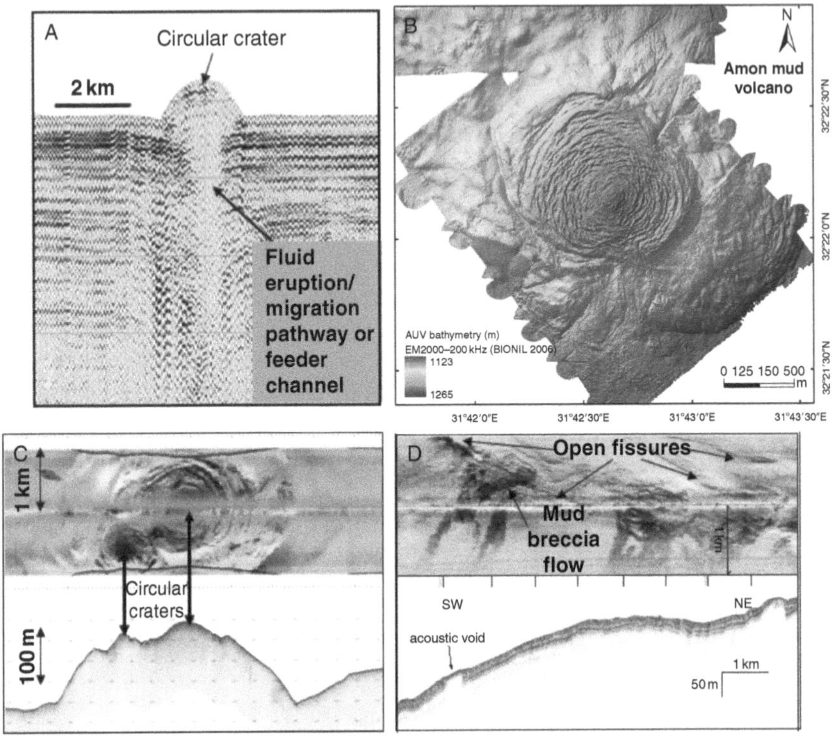

Figure 6.12 Deep-sea cold seep structures. (A) Fragment of a seismic line across the Kovalevsky Mud Volcano, Central Black Sea showing the subsurface gas-migration pathway. (Kenyon et al., 2002). (B) AUV-acquired multibeam bathymetry image of the Amon mud volcano, 1120-m water depth, Nile Deep Sea Fan, Egypt (Dupré et al., 2008). (C) Fragment of an OREtech deep-towed sidescan sonar across the Yuma Mud Volcano from the Gulf of Cadiz (NE Atlantic; Pinheiro et al., 2003). (D) Fragment of MAK-1 deep-towed sidescan sonar across the Odessa Mud Volcano from the Sorokin Trough, showing a fissure eruption of mud breccia (NE Black Sea). See http://ioc.unesco.org/ttr/ (Ivanov et al., 1998). (A multi-colour version of this figure is on the included CD-ROM.)

4.2.1. Mud volcanoes

Where substantial, concentrated discharges of sub-seafloor fluids occur, mud volcanoes are formed and, as a major morphological feature where significant concentration of microbially mediated carbonates are found, they are given special attention here. Mud volcanoes are kilometre-scale, low-temperature seepage-related geomorphological features and some of the most remarkable indications of fluid venting (Ivanov et al., 1998). They are similar to magmatic volcanoes in the sense that their eruptions are powerful, but instead of lava, mud volcanoes expel a complex mixture of products, including hydrocarbon gases (e.g. methane and wet hydrocarbons), hydrogen sulphide, carbon dioxide, petroleum, pore water and mud.

Hydrocarbon gases (especially methane) and petroleum often represent one of the main constituents of the emitted fluid and, on-land, mud-volcano eruptions are often accompanied by vigorous burning.

Mud volcanoes are often elevated above the seafloor as dome-like structures, reaching several kilometres in diameter and hundreds of metres in height (Fig. 6.12A and B; Ivanov et al., 1998, and references therein). However, mud volcanoes may also form a caldera-like depression of a few kilometres in diameter. This type of mud volcanoes results from the collapse of the seafloor following explosive eruptions of large volumes of ejected sedimentary gas- and water-saturated material (Woodside et al., 1997). On continental margins, mud volcanoes and seeps can significantly alter the topography and induce continental-slope instability. In shallower waters, mud volcanoes sometimes form islands and banks.

Mud volcanoes develop as a result of strong lateral or vertical compression of the Earth's crust, which allows deep-lying sediments to move upwards and be transported to the seafloor. Through the feeder channel of a mud volcano (Fig. 6.12A), large volumes of clastic material, saturated with gas and other fluids, are also transported in a clay/silt fluidised matrix and deposited on the seafloor (Ivanov et al., 1998). This emitted sedimentary material is called "argille scaliose", "diapiric mélange" or more frequently, "mud breccia" (Akhmanov, 1996; Cita et al., 1981) and is a complex mixture of a matrix and rock fragments, mechanically incorporated into the eruption deposit by the powerful upward transport of fluids (Akhmanov, 1996; Akhmanov and Woodside, 1998). These fragments of rocks represent sediments potentially deposited millions of years ago, buried in the subsurface and brought to the seafloor due to the mud-volcano activity. Mud breccias, rock clasts and matrix contain important information regarding the composition and genesis of sediments in the subsurface, their maturity and hydrocarbon potential of the area (Akhmanov, 1996; Ovsyannikov et al., 2003). Accordingly, mud volcanoes can be considered as "free bore-holes", providing us material from great depths below the seafloor, perhaps up to 20 km deep, where drilling has not been possible yet (Kholodov, 2002b).

Comprehensive investigations of numerous mud volcanic/seepage provinces have provided overwhelming evidence for the role of hydrocarbon gases in their formation. Mud volcanoes and seeps, as general indicators of hydrocarbon migration, provide a view of the hydrocarbon potential of the deeper sediments (Drozd et al., 1981; Gubkin, 1932; Guliev and Feizullayev, 1997; Ivanov et al., 1998; Jones and Drozd, 1979, 1983; Kholodov, 2002a,b; Mousseau and Williams, 1979; Richers and Jones, 1986). The connection between seepage and hydrocarbon reservoirs in the deep subsurface was advocated by Link (1952) who stated that "oil and gas seeps gave the first clues to most oil-producing regions. Many great oil fields are the direct result of seepage drilling". Mud volcanoes and gas-venting areas commonly occur in petroliferous regions (Guliev and Feizullayev, 1997), and the hydrocarbon gases emitted are often thermogenic (i.e. formed by the breakdown of organic matter in the subsurface under the influence of subsurface burial temperature and pressure conditions). The presence of a seep thus indicates the subsurface presence of a mature source of gas, the existence of migration pathways as well as the absence of a proper seal.

An alternative origin of methane is microbial decomposition of shallowly buried organic matter under anoxic conditions, that is, methanogenesis (producing microbial or biogenic gas). Biogenic gas sources are more diffuse, so often lead to the formation of pockmarks. Biogenic and thermal methane can be readily identified by differences in their ^{13}C signature (Rice and Claypool, 1981; Sackett, 1978; Stahl et al., 1979) and both can be incorporated in authigenic carbonates.

4.2.2. Distribution and occurrence of cold-seepage and mud-volcano provinces

Cold seeps and mud volcanoes are most common in areas of recent tectonic activity, that is, within the mobile tectonic belts of the Alpine-Himalayas, Pacific and Central-Asia and in areas with thick (5–20 km) sedimentary successions. Numerous field studies in the last decades have revealed that spontaneous discharge of subsurface fluids is a worldwide phenomenon, and such discharges are now known to be a common feature on active and passive continental margins (see Table 6.1). The areas of known seeps on the European and north African margins are shown in Fig. 6.13.

4.3. Seafloor methane habitats and the marine microbial methane filter

The deep-water methane-seepage environment is commonly non-thermophilic, oxygen-limited, light-free, sometimes brine-impacted, sulphate-containing and sulphide- and methane/(hydrocarbons)-rich. When methane-saturated fluids percolate through subsurface sediments and reach the seafloor, the sedimentary geochemistry of the seepage-affected area is radically altered. The local environments become oxygen-depleted and exhibit a range of temperature, salinity, pH and oxidation/reduction-potential (Eh) conditions. The internal fluidodynamics, including the intensity and partial pressure of migrating fluid, the mechanisms of fluid transport (e.g. focused fluid flow vs. diffusion) and the formation of gas hydrates affect the sedimentary porosity and permeability. The ascending methane-rich fluids form the basis for abundant, yet poorly understood, microbial communities and a diversity of seep-associated chemosynthetic fauna (Fig. 6.14; Lanoil et al., 2001; Olu et al., 1996, 1997). The myriad of microbes thriving in such chemotrophy-based habitats are generally aerobic and anaerobic methane-, sulphide- or sulphur-oxidising and heterotrophic prokaryotes. The main factors determining the biogeochemical pathways

Table 6.1 Locations of well-studied mud volcanoes

Locations		Locations	
Active margins	References	*Passive margins*	References
Makran accretionary prism	Von Rad et al. (1996),White (1982)	Black Sea	Ginsburg and Soploviev (1998), Ivanov et al. (1989, 1996, 1998), Michaelis et al. (2002)
Barbados	Lallemant et al. (1990), Le Pichon et al. (1990)	Vøring Plato	Bouriak et al. (2000)
Aleutian subduction zone	Suess et al. (1998), Wallmann et al. (1997)	Nile deep-sea fan	Loncke et al. (2002), Mascle et al. (2001)
Cascadia margin	Cragg et al. (1996), Han and Suess (1989)	Gulf of Mexico	Roberts and Aharon (1994)
Nankai Trough	Taira et al. (1992)	Blake Outer Ridge	Galimov and Kvenvolden (1983), Paull et al. (1999)
Chilean margin	Sellanes et al. (2004)	Norwegian margin of the Barents Sea	Pimenov et al. (1999), Vogt et al. (1999)
Peruvian margin	Kvenvolden et al. (1990)		
Eastern Mediterranean	Ivanov et al. (1996), Limonov et al. (1996), Premoli Silva et al. (1996), Volgin and Woodside (1996)		
Gulf of Cadiz	Gardner (2001), Kenyon et al. (2000, 2001, 2002), Pinheiro et al. (2003)		
Anaximander Seamounts	Woodside et al. (1998), Zitter (2004)		

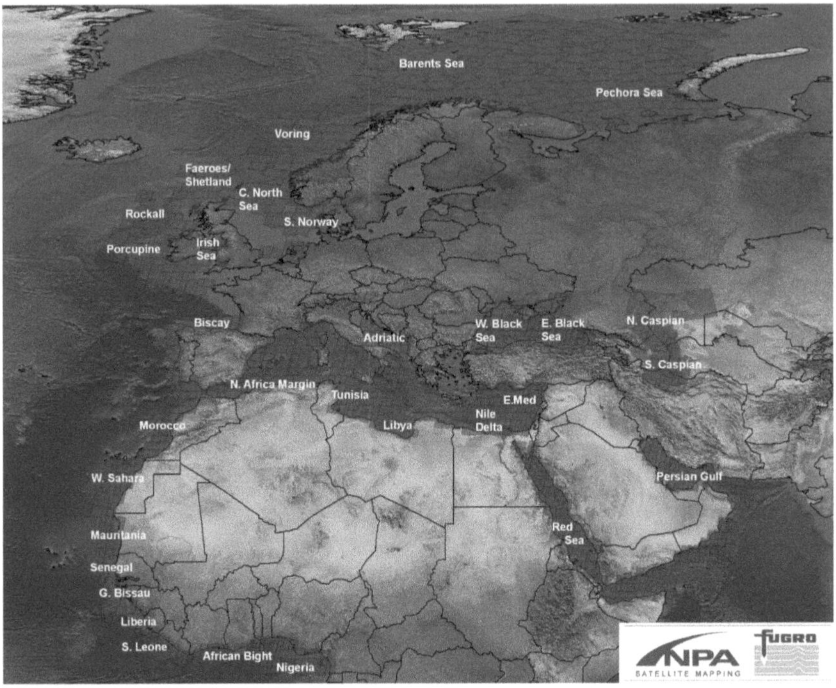

Figure 6.13 Fugro NPA's most recent SAR seepage coverage for Europe and NW Africa, showing the widespread distribution of seepage hotspots on continental shelves and margins, and in deep-sea basins. Fugro NPA's seep database is continually updated (see http://fugro-npa.com; reproduced with permission by Fugro NPA).

and microbial mutual benefits in the various seepage environments are still largely unknown.

Compared to continents, the contribution of the oceans to the global methane flux to the atmosphere is small (3–5%; Intergovernmental Panel on Climate Change, 1994). This is due either to temporarily subsurface methane sequestration by sub-seafloor gas hydrates or to microbial methane-consumption processes in marine sediments or in the water column. The heterogeneous distribution of migrated fluids in the subsurface and the complex dynamics of seepage systems induce local sharp changes in the sediment chemistry. This is one of the main controls governing the spatial distribution of microbial communities in the subsurface and the development of chemosynthetic ecosystems at the seafloor. The main factors determining the biogeochemical pathways, from which the microbial communities' benefit in the array of seepage environments, are poorly constrained. However, it is recognised that cold-seep microbial communities represent the dominant biological filter of our planet, being principally responsible for the

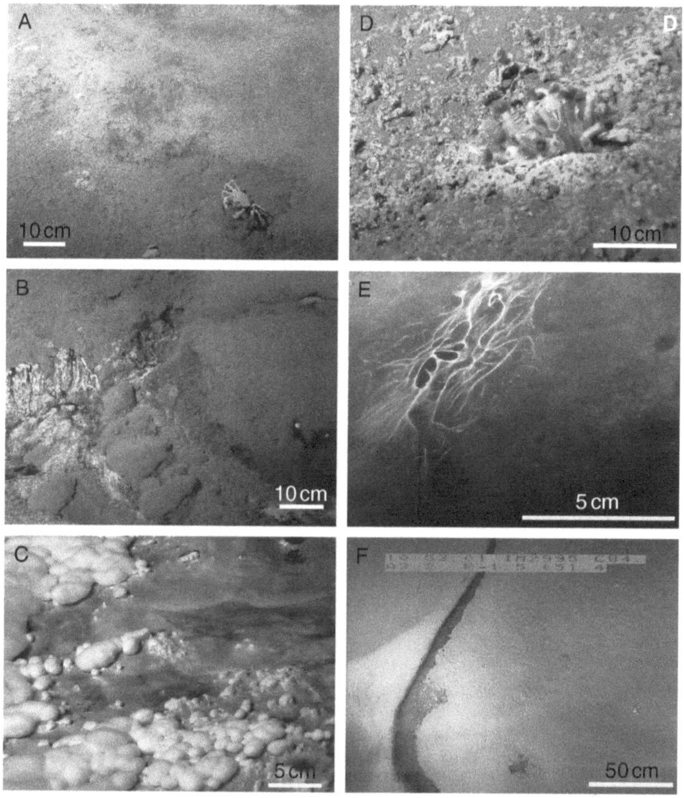

Figure 6.14 Cold-seep habitats and related chemosynthetic ecosystems at 3-km water depth in the Eastern Mediterranean. Seafloor images taken by the Ifremer manned-submersible Nautile during the Nautinil cruise (2003) of the RV L'Atalante (Images copyright of Ifremer, Brest, France) (Huguen et al., 2009). (A) Brine-impacted cold-seepage food web. A chemosynthetic crab thrives in environments rich in methane, sulphide and emitting brine solutions. The white crab is eating worms living in the sediments. (B) Microbial mat systems in the brine-impacted seepage. The cushion-like mat is formed by bacteria closely related to the neutrophilic Fe(II)-oxidising betaproteobacterium *Leptothrix ochracea*. The white mat is composed of filamentous sulphur produced by the sulphide-oxidising *epsilonproteobacterium Candidatus Arcobacter sulfidicus* (see for details: Omoregie et al., 2008). (C) Filamentous bacterial mat covering the surface of methane- and brine-saturated mud volcanic deposits. (D) Methane-derived carbonates used as an anchor for sessile seep-associated vestmentifera tube worms. These tube worms are dependent on sulphide availability and sulphide-oxidising symbionts. (E) Chemosynthetic sulphide-oxidising filamentous proteobacteria *Beggiatoa* growing heterotrophically in the presence of oxygen. *Beggiatoa* is one of the largest prokaryotes. (F) Methane-saturated brine lake (left) with orange microbial mats covering the brine lake margin (right). (A multi-colour version of this figure is on the included CD-ROM.)

biogeochemical breakdown of greenhouse gases and for the cycling of carbon, sulphur, trace elements and some toxic chemicals.

Only recently has the microbial anaerobic oxidation of methane (AOM) been recognised as the key biogeochemical process that regulates the Earth's climate and as an important methane sink in marine environments (Reeburg, 2007, and references therein). Valentine and Reeburgh (2000) estimated that AOM removes 5–20% of the total methane flux from the ocean every day. Preliminary estimates made by Hinrichs and Boetius (2002) for marine sediments which are not affected by methane seepage, suggest that AOM could affect 300 Tg of methane per year. Taking into account the fact that AOM rates at methane-seepage areas are much higher, and that the total area covered by seeps at the ocean floor is still uncertain, the removal of methane by AOM is potentially considerably larger.

In marine environments, AOM occurs at the base of the sulphate-reducing zone where upwardly migrating methane encounters sulphate derived from the sea and moving downwards through the sediments. Nearly all methane from the sediments is removed before sulphate becomes depleted, as was first reported by Martens and Berner (1974) and later by Barnes and Goldberg (1976). AOM is a sulphate-dependent process. It engages electron transfers from the methane to the sulphate and results in the formation of sulphide and bicarbonate in equimolar quantities:

$$CH_4 + SO_4^{2-} \rightarrow HCO_3^- + HS^- + H_2O. \qquad (6.1)$$

Using inhibition experiments, Hoehler et al. (1994) showed that a consortium of methanogenic archaea and sulphate-reducing bacteria could mediate anaerobic methanotrophy. It was hypothesised that the use of hydrogen as an electron donor by the sulphate-reducing bacteria results in a low partial pressure of hydrogen, thereby creating thermodynamically favourable conditions for methanogenic archaea to act as methane-oxidisers (Alperin and Reeburgh, 1985; Hoehler and Alperin, 1996; Hoehler et al., 1994; Reeburgh, 1976; Zender and Brock, 1979). Therefore, AOM was postulated as a net two-step reaction that can be expressed through the following formula:

$$CH_4 + 2H_2O \rightarrow CO_2 + 4H_2(\text{methanogenic archaea}), \qquad (6.2)$$

$$SO_4^{2-} + 4H_2 + H^+ \rightarrow HS^- + 4H_2O\,(\text{sulfate} - \text{reducing bacteria}). \qquad (6.3)$$

Following this study, alternative microbial mechanisms for this process were later suggested by Valentine and Reeburgh (2000).

The first visual evidence for structured archaeal-bacterial AOM symbiosis was found by Boetius et al. (2000) using fluorescence for *in situ* hybridisation (FISH) with specific 16S rRNA-targeted oligonucleotide probes. These AOM-mediating microbes represent a complex syntrophic consortium of *an*aerobic *me*thane-oxidising archaea (ANME: Boetius et al., 2000; Hinrichs et al., 1999; Knittel et al., 2005) and diverse branches of the sulphate-reducing *Deltaproteobacteria* (Boetius et al., 2000; Knittel and Boetius, 2009 and references therein). Phylogenetic analyses of ribosomal RNA gene sequences have revealed three distinct lineages amongst the Euryarchaeota capable of anaerobic methanotrophy: ANME-1, ANME-2 and ANME-3 (Boetius et al., 2000; Hinrichs et al., 1999; Knittel et al., 2005; Niemann et al., 2006, Orphan et al., 2001a,b). Archaea belonging to the ANME-1 cluster do not contain any cultured relatives (Hinrichs et al., 1999; Knittel et al., 2003, 2005; Orphan et al., 2002). The ANME-2 archaeal guild are affiliated to the cultured members of the methanogenic *Methanosarcinales* (Boetius et al., 2000; Knittel et al., 2003, 2005; Orphan et al., 2001a,b). The ANME-3 archaea are more closely related to the cultivated *Methanococcoides* spp. than ANME-1 and ANME-2 (Knittel et al., 2005). In vitro experiments indicated that ANME-1- and ANME-2-dominated communities use methane and sulphate in a stoichiometry of about 1:1 (Nauhaus et al., 2005).

Follow-up studies of seabed fluid flow suggest that the AOM consortium of archaea/sulphate-reducing bacteria coupling can vary in composition (e.g. Hinrichs and Boetius, 2002; Michaelis et al., 2002; Orphan et al., 2001a; Pernthaler et al., 2008). In addition, Joye et al. (2004) reported an absence of correlation between AOM and sulphate-reduction rates, and Orphan et al. (2001a, 2002) showed the existence of bacteria-free archaeal cells. Pernthaler et al. (2008) showed that sometimes the diversity of the ANME's bacterial partners is not restricted to sulphate-reducing bacteria and can be represented by members of *Alphaproteobacteria* and *Betaproteobacteria*. These indicate that occasionally methanotrophic archaea can establish AOM independently from sulphate-reducing bacteria and there is a larger assortment of oxidants in seepage environments than was considered previously. The latter was recently shown by Beal et al. (2009) who reported alternative electron acceptors (oxidants) for marine AOM. Incubation experiments on methane-seep sediments from the Eel River Basin in California revealed that anaerobic methane-consuming microorganisms are also capable of oxidising methane using manganese (birnessite) and iron (ferrihydrate) instead of sulphate, following:

$$CH_4 + 4MnO_2 + 7H^+ \rightarrow HCO_3^- + 4Mn^{2+} + 5H_2, \qquad (6.4)$$

$$CH_4 + 8Fe(OH_3) + 15H^+ \rightarrow HCO_3^- + 8Fe^{2+} + 21H_2O. \qquad (6.5)$$

This finding revealed a clear shift in the AOM-involved bacterial community from known sulphate-reducing bacteria to putative metal-reducing microorganisms, designating an important link between AOM and metal-reduction processes (Beal et al., 2009).

Despite these findings, no such methane-oxidising systems have been cultured in isolation so far. This limits a clear perception of the mechanisms by which methane is being oxidised in nature and thus hampers the understanding of the processes responsible for net sources and sinks of methane in the ocean.

4.4. Microbial carbonate genesis

Microbial metabolic behaviour can chemically affect local environments, generating conditions favourable for carbonate precipitation (see Chapter 9). This is known as microbially induced carbonate formation (Castanier et al., 1999). Carbonate rocks are ubiquitous and known already from Precambrian times, which make them unique geological archives of the biogeological evolution of the Earth (see Chapter 4).

The ability of living microorganisms to mediate carbonate precipitation is an important geobiological process. Nevertheless, little is known about the role and coupling of microorganisms and the formation of individual carbonate minerals, including aragonite, calcite and dolomite. The microbial interposition in the nucleation of the carbonate also remains obscure. Sedimentary chemistry determines the initial growth of microbial communities, symbioses and later the environment-specific microbial ecosystem. Through diverse metabolic pathways engaged to the carbon, nitrogen and sulphur cycles, and via complex direct and indirect interactions between concurrent or successive microbial metabolisms, microorganisms eventually create an environment that is favourable for carbonate precipitation (c.f. Castanier et al., 1999; Dupraz et al., 2009, and references therein).

Precipitation of carbonate minerals can be theoretically regulated by a number of factors:

(1) temperature, salinity, evaporation processes, substrate availability and the presence/absence of siliciclastics;
(2) carbonate alkalinity ($[HCO_3^-] + 2 \times CO_3^{2-}]$);
(3) the presence of free calcium;
(4) the state of saturation, which is a logarithmic function of the ion activity product ($[Ca^{2+}] \times [CO_3^{2-}]$) and the solubility product (K_{sp}) of a given carbonate mineral (Stumm and Morgan, 1996); and
(5) the quality and availability of organic matter.

The role of microbes in carbonate precipitation has been known for a long time. The first assumption on microbial involvement in carbonate genesis was put forward by Murray and Irvine (1889/1890) and later by Murray and Hjort (1912). These authors demonstrated carbonate formation being due to the chemical interaction between ammonium carbonate, resulting from the decomposition of nitrogenous organic material, and calcium sulphate derived from the water column:

$$(NH_4)2CO_3 + CaSO_4 \rightarrow CaCO_3 + (NH_4)2SO_4. \qquad (6.6)$$

Solid evidence for the microbial involvement in carbonate formation was given by Nadson (1928). His incubation experiments resulted in the inference that carbonate precipitation is more dependent on proper environmental conditions than on a specific group of bacteria. As to the role of microorganisms in carbonate precipitation, Nadson pioneered the importance of bacterial sulphate reduction, especially in the formation of dolomite:

$$CaSO_4 + CH_3COOH \rightarrow CaCO_3 + H_2S + H_2O + CO_2 \qquad (6.7)$$

or

$$CaMg(SO_4)_2 + 2CH_3COOH \rightarrow CaMg(CO_3)_2 + 2H_2S + 2H_2O + 2CO_2. \qquad (6.8)$$

Laboratory experiments performed by Deelman (1975) demonstrated the precipitation of calcium carbonate (aragonite) through the activity of sulphate-reducing bacteria. Later, Gautier (1985) and Paull et al. (1992) showed that the aerobic oxidation of organic matter is of minor importance for carbonate formation, whereas the sulphate-reduction process is believed to be one of the important processes accountable for carbonate precipitation, supporting the hypothesis of Nadson (1928).

4.4.1. Methane-derived carbonates

In environments that are not affected by methane, carbonate precipitation can be mediated in two ways: (1) autotrophically via oxygenic and anoxygenic photosynthesis and non-methylotrophic methanogenesis, and (2) heterotrophically, involving the nitrogen and sulphur cycles. In methane-saturated sediments, the production of bicarbonate ions via AOM (Eq. (6.1)) leads to an increase of alkalinity, inducing carbonate precipitation (Eq. (6.9)). The biogeochemistry of AOM is not well constrained and we are only just beginning to understand the environmental controls of the processes and the process itself. Still, the lack of possibilities of carrying out culture experiments with anaerobic methanotrophs leaves a gap in the knowledge on the biogeochemical binding of two phenomena: AOM and associated carbonate genesis.

By focusing on methane-related carbonate precipitation, Ritger et al. (1987) suggested that anaerobic methanotrophy contributes significantly to alkalinity production in sediments (Eq. (6.1)) that thereby causes carbonate precipitation. Chemical analyses showed that these carbonates are significantly depleted in ^{13}C (up to approx. -69‰: Campbell et al., 2002) compared to the "normal" pelagic carbonates (Aloisi et al., 2000, 2002; Paull et al., 1992; Peckmann et al., 1999a,b, 2001, 2002; Stakes et al., 1999; Thiel et al., 1999; Von Rad et al., 1996). The very low $\delta^{13}C$ values of these authigenic carbonates suggest that methane is the major carbon source oxidised as CO_2 by methanotrophic archaea (Eq. (6.2)). In other words, the outcome of the three suggested AOM pathways (Eqs. (6.1), (6.4) and (6.5)) is the production of methane-derived ^{13}C-depleted bicarbonate and the consequent formation of ^{13}C-depleted carbonates (Eq. (6.9)). This is the first indication that AOM plays a role in carbonate precipitation.

$$C{*}H_4 + SO_4^{2-} + Ca^{2+} \rightarrow CaC{*}O_3 \downarrow + H_2S + H_2O, \qquad (6.9)$$

where * indicates the methane-derived carbon which will be partially incorporated into the resulting carbonate.

The occurrence of neoformed diagenetic carbonates is a widespread phenomenon in cold-seep settings (Aloisi et al., 2000, 2002; Campbell, 2006; Michaelis et al., 2002; Peckmann and Thiel, 2004; Peckmann et al., 1999a,b, 2001; Ritger et al., 1987; Roberts and Aharon, 1994; Von Rad et al., 1996). Where they formed as a result of syntrophic interactions between methanotrophic archaea and sulphate-reducing bacteria.

From a biogeochemical perspective, the diagenesis of methane-derived carbonate is principally different from that occurring in regular marine settings. The presence of ANME-1, ANME-2 and ANME-3 has been found up till now to be restricted to anoxic, methane-rich environments, and sulphate-reducing bacteria are, for the most part, obligate anaerobes (Madigan et al., 2009). Methane-derived carbonates are therefore precipitated exclusively within the sediments under oxygen-limited conditions (Stadnitskaia et al., 2005). The oxygenated seawater is an obstacle for carbonate precipitation since aerobic methane oxidation results in acidity increase, which stimulates carbonate dissolution:

$$CH_4 + 2O_2 \rightarrow CO_2 + 2H_2O. \qquad (6.10)$$

Methane-derived, ^{13}C-depleted carbonates have been documented globally from the Neoproterozoic to recent times (Jiang et al., 2003; Peckmann and Goedert, 2005, and references therein). This gives robust evidence that AOM has been an important metabolic process throughout geological history.

4.4.2. Molecular tracers of AOM-mediating microorganisms: Lipid biomarkers

Lipid biomarkers are organic components with a specific structure that can be related to a particular source organism. The strength of the lipid-biomarker approach is that, by simply determining the chemical structure of a compound, it is possible to define an organism to the level of kingdoms, and in some cases, orders. For instance, the most distinctive characteristic of Archaea is the presence of ether core membrane lipids composed of isoprenoidal units linked to a glycerol moiety (Comita et al., 1984; De Rosa et al., 1983, 1991; Hoefs et al., 1997; Hopmans et al., 2000; Langworthy, 1985; Langworthy et al., 1982; Pancost et al., 2000, 2001a,b; Schouten et al., 1998; Stadnitskaia et al., 2003, 2008a). In contrast, membrane lipids from bacteria contain ester-bound to glycerol fatty acids. These differences are often used as chemotaxonomical markers for the archaeal and bacterial domains and their specific groups.

The application of compound-specific carbon-isotope studies to archaeal lipids derived from seepage-associated sediments and carbonates revealed that these biomarkers are significantly ^{13}C-depleted. This proved that archaea use methane-derived carbon as their carbon source (Aloisi et al., 2002; Elvert et al., 2000; Hinrichs et al., 1999, 2000; Pancost et al., 2000, 2001a,b; Schouten et al., 2003; Stadnitskaia et al., 2003, 2005, 2008b; Teske et al., 2002; Thiel et al., 1999, 2001; Zhang et al., 2002, 2003). Lipids of methanotrophic archaea often display $\delta^{13}C$ values around $-100‰$ and lower (see Fig. 6.15 for structures).

The distribution of archaeal lipids in methane-related carbonates and seepage sediments can also provide evidence for establishing which archaeal group is dominant in the environment, and it can also reveal a shift in the archaeal community structure during the development of the carbonate (Stadnitskaia et al., 2008b). Archaea in the ANME-II group are thought to produce mainly pentamethylicosane (I: see Fig. 6.15 for structures) together with archaeol (III) and hydroxyarchaeols (IV, V) as their membrane lipids, with concentrations of hydroxyarchaeols exceeding those of archaeol (Blumenberg et al., 2004). Glycerol tetraethers (VI, VII, VIII, IX) and pentamethylicosenes (II) have been proposed as the majority of membrane lipids in archaea of the ANME-1 group, but occur in ANME-2 group in small amounts (Blumenberg et al., 2004; Elvert et al., 2005).

The co-occurrence of ^{13}C-depleted lipids from uncharacterised sulphate-reducing bacteria (X–XIV: Pancost et al., 2001a,b; Stadnitskaia et al., 2005; Werne et al., 2002) and from the *Desulfosarcina/Desulfococcus* species (XV: Elvert et al., 2003) marks a close metabolic association between sulphate-reducing bacteria and methanotrophic archaea. Consequently, as ^{13}C-depletion of carbonates indicates the incorporation of methane-derived carbon into the carbonate structure, ^{13}C-depletion of archaeal and bacterial lipids from the same carbonate is strong microbiological evidence for AOM

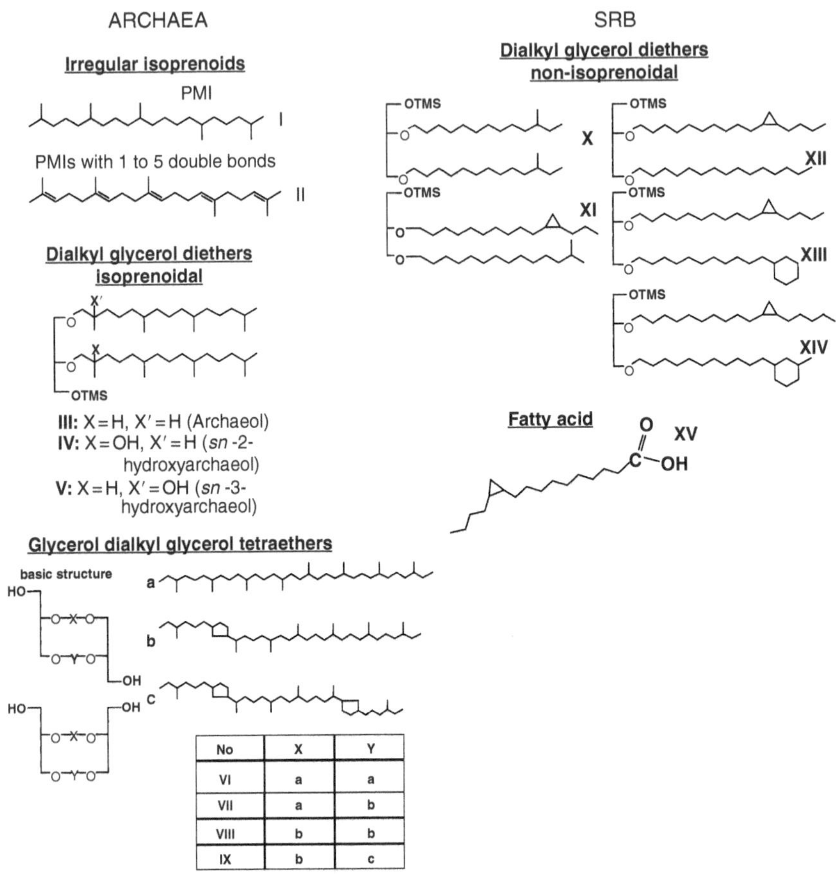

No	X	Y
VI	a	a
VII	a	b
VIII	b	b
IX	b	c

Figure 6.15 Chemical structures of archaeal and sulphate-reducing bacteria lipid biomarkers often found in methane-derived carbonates and gas-saturated sediments.

(Boetius et al., 2000; Hinrichs et al., 1999; Orphan et al., 2001a) as the main process inducing carbonate precipitation.

4.4.3. Environmental controls

Methane-related carbonate precipitation is an exclusively microbially induced process. The mechanism controlling the preferential development of seepage-related anaerobic methanotrophs is still uncertain. In methane-rich microenvironments, the carbonate alkalinity level is controlled by AOM rates. Recently, it was shown that changes in the methane partial pressure and *in situ* temperature could influence AOM rates and the AOM community structure (Nauhaus et al., 2002, 2006; Treude, 2003). Archaea of the ANME-1 and ANME-2 groups can coexist (Nauhaus et al., 2006).

Nevertheless, archaea from the ANME-2 group were found to have a preferential niche in habitats of elevated methane partial pressures (Blumenberg et al., 2004). The methane partial pressure is an important environmental factor, since it controls the mode of methane transport, its solubility in the fluid mixture and degassing.

Methane-related carbonate genesis does not seem to be controlled by temperature and salinity. The direct evidence that AOM-prokaryotes induce and govern carbonate precipitation has only recently been shown and proved (Michaelis et al., 2002; Stadnitskaia et al., 2005). Contemporary applications of lipid biomarkers and 16S rDNA sequences to living methanotrophic mats and modern carbonates revealed diverse associations of methane-oxidising ANMEs and sulphate-reducing bacteria, which to some extent, are restrained by the intensity of the methane flux (e.g. Blumenberg et al., 2004; Michaelis et al., 2002; Stadnitskaia et al., 2005). This suggests that certain methanotrophic communities are endemic to certain seepage environments. Nevertheless, unresolved issues remain concerning our understanding of the biogeochemical factors that drive carbonate precipitation and physical interactions among methane-dependent microbes.

Modern and ancient seep-related carbonates display different shapes and morphologies, characteristic carbonate fabrics and mineralogies, as well as stable-carbon and oxygen-isotope compositions (Aloisi et al., 2000, 2002; Hovland et al., 1987; Peckmann and Thiel, 2004; Peckmann et al., 1999a,b, 2001, 2002; Roberts and Aharon, 1994; Stakes et al., 1999). Seep-related carbonates can appear as expansive fractured pavements, carbonate towers and "mushrooms" rising from the seafloor (Fig. 6.16) as well as in the form of newly formed small fragile carbonate crusts and concretions within the sediment. The impressive diversity of authigenic carbonate structures is symptomatic of the variability in their formation histories.

Localised environments developed by AOM-mediated prokaryotes at seeps allow carbonate precipitation within the sediments even below the CCD (Greinert et al., 2002). Such natural attributes of seep-related carbonates assign them as exceptional geological archives facilitating continued advances in our understanding of the diversity of microbes capable of turning gaseous methane into limestone/dolostone and providing records of geosphere/biosphere coupling through geological time.

5. Past and Future

The global distribution of deep-sea carbonates (Fig. 6.1) is determined, with the exception of authigenic carbonates that form within the sediments, by the depth of the carbonate and aragonite saturation

Figure 6.16 The diversity of seep carbonates from the Nile deep-sea fan seepage area, Eastern Mediterranean. Both the carbonate samples and the images were obtained by the remotely operated vehicle QUEST-4000 (MARUM, Germany) during the Bionil cruise (R/V Meteor, M-70b; Images copyright of MARUM, University of Bremen, Germany). (A) Carbonate tower outcropping at the seafloor. (B) "slate" carbonate crust partially outcropping at the seafloor, surrounded by fields of white and brownish-orange dead clams. (C) Laminated carbonate dome. (D) Sampling of one of the chimneys of the carbonate tower (A) with the QUEST-4000. (E) The chimney, forming a sphere-shaped piece, being hollow inside, with two open conduits. (F) Carbonate sample from a carbonate outcrop in the central part of a pockmark. (A multi-colour version of this figure is on the included CD-ROM.)

depth (see Section 1). This depth is controlled by a number of factors, including pressure, temperature and pCO_2. Due to gas exchange across the ocean surfaces, pCO_2 in the oceans is in equilibrium with pCO_2 in the atmosphere (albeit slightly offset in time), reflecting the influence of global "greenhouse climate" conditions. Therefore, the presence or absence of extensive deep-sea carbonates, and shallowing occurrences thereof, reflects the prevailing climatic regime on a geological timescale: absent or scarce carbonates in greenhouse climates and potentially abundant carbonates in cold climates (see Chapter 4).

5.1. Deep-sea biogenic carbonates

The paleontological record of cold-water corals is reviewed by Roberts et al. (2009) (see also Cairns, 2001b; Squires, 1958). Scleractinians evolved following the Permian/Triassic mass extinction 251.4 Ma ago. By the Cretaceous, Dendrophylliidae and Oculinidae cold-water corals had developed thickets and mounds (Roberts et al., 2009) with early occurrences of *Petrophyllia* deep-water coral thickets outcropping near Lamy, New Mexico, USA of Mid-Cretaceous (99–89 Ma) age (Coates and Kauffman, 1973).

The distribution of cold-water coral fossil outcrops on land is patchy, requiring significant uplift to make the deep-water deposits emerge. On Seymour Island, Antarctica, Early Paleocene (66.5–62 Ma) *Madrepora* coral thickets are preserved in the Sobral Formation (Filkorn, 1994). *Madrepora* is, at present, a cosmopolitan reef-forming coral with concentrations in the North Atlantic and Caribbean (Roberts et al., 2009). The earlier recorded cold-water corals in the Pacific are from Late Eocene (40–34 Ma) deposits on Eua Island, Tonga (Wells, 1977). Compressive forces in the Tonga Trench have uplifted oceanic basalts and overlying deep-water deposits including limestones. The cold-water corals recorded by Wells (1977) are not *in situ* but preserved at the base of turbiditic ash deposits that slid down the side of the uplifted island. Wells (1977) noted a variety of cold-water corals from these deposits although, unfortunately, the original coral outcrop has now been reclaimed by the jungle.

Compressive tectonics affecting the Mediterranean Basin has also uplifted deep-water deposits containing cold-water corals. *Oculina* coral banks of Tortonian age (12–7 Ma) are found in northern Calabria, Italy (Mastandrea et al., 2002); not far away, on the other side of the Messina Strait, in NE Sicily, Plio-Pleistocene (2.5 Ma) outcrops contain deep-water *L. pertusa*, *M. oculata* and *D. dianthus* fossils as well as other scleractinians and isidiid octocorals (Di Geronimo et al., 2005). At La Montagna, Messina, coral rubble is developed into a 21-m tall mound structure dominated by *L. pertusa*, *M. oculata* and *Desmophyllum cristagalli*, with subordinate *Enallopsammia scillae* and *Dendrophyllia cornigera* cold-water coral fossils (Titschack et al., 2008; Fig. 6.17).

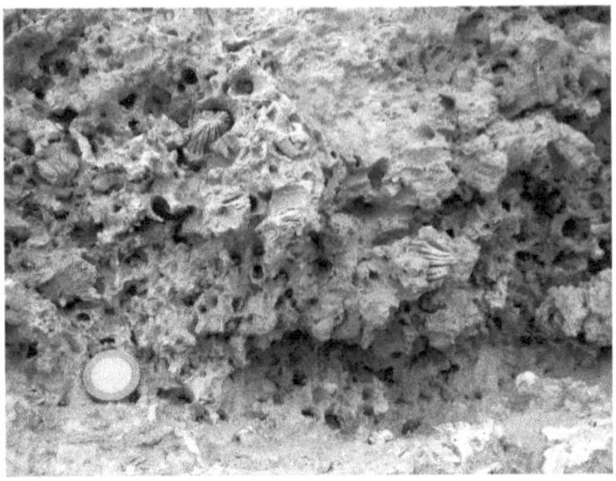

Figure 6.17 A close-up of Plio-Pleistocene cold-water coral deposits showing large *Desmophyllum cristagalli* from La Montagna, Messina, Sicily. Coin for scale is 2 cm. (A multi-colour version of this figure is on the included CD-ROM.)

The palaeontology of New Zealand cold-water corals is well documented by Squires (1958). Squires's (1964) seminal paper on cold-water coral reef development was based on observations of *Lophelia parvisepta* coral thickets at Lake Ferry, Palliser Bay, New Zealand of Early Pliocene (5–3.5 Ma) age (see Section 3.1.3). These developed in 500–700-m water depth (Devereux, 1967). These outcrops have also disappeared, this time due to coastal erosion. However, Late Miocene (12–5 Ma) *Lophelia parvisepta* corals also outcrop in the cut bank of the Pahaoa River near Hinakura, Wairarapa (Squires, 1964).

The deep-sea oyster *Pycnodonte zibrowii* is also preserved in the fossil record, with Middle Danian (Paleocene) deposits found at Faxe in Denmark, although these may have grown in a shallower setting than contemporary finds, possibly between 100 and 300 m (Bernecker and Weidlich, 1990, 2005).

5.2. Ancient methane-derived carbonates: using fossil lipid biomarkers and fossil rRNA gene sequences

Methane-derived carbonates occur within the seafloor and form irrespective of climate changes and the depth of the CCD. Their geological history extends arguably back to the Precambrian (Jiang et al., 2003). Whether this process has remained unchanged through geological time is debatable. Methane-derived carbonates represent complex habitats holding specific groups of microorganisms and so-called interspecies, the interrelationships of which remain largely unknown. In seepage environments, anaerobic cycling of carbon is commonly accomplished by coupled mutualistic interrelationships, as no single type of anaerobic microorganism is capable of consuming complex organic molecules (Canfield, 2005).

To address the role of AOM processes in the formation of ancient carbonate fabrics, a combination of organic and inorganic geochemical techniques and molecular ecological methods has been applied to a fossil methane-related carbonate crust (Stadnitskaia et al., 2008b). The idea was to develop the most effective scientific strategy for the reconstruction of fossil methane-consuming microbial communities and associated "carbonate formation history". Concentrations of microbial lipids, their stable-carbon isotope composition, sequences of fossil 16S rRNA genes of anaerobic methanotrophic archaea in combination with mineralogical and stable-carbon and oxygen-isotope data from bulk carbonate were obtained for seven horizons of a crust from the Gulf of Cadiz, NE Atlantic (Fig. 6.18).

The data revealed AOM as the main process that induced the formation of the carbonate crust. All data show changes in the AOM archaeal community structure related to two main phases of the carbonate-crust development. In the upper part of the crust, archaeal lipid biomarkers and fossil 16S rRNA gene sequence data indicate that ANME-2 was the dominant archaeal guild in the AOM community, whereas ANME-1 members dominated the AOM community during the precipitation of the lower part of the carbonate. This indicates that methane concentrations were initially high and then dropped during crust growth. This is in agreement with the stable-oxygen isotopic data and with the observed changes in mineralogy, signifying a downward accretion of the crust below the sediment–water interface.

In addition, depth profiles of $\delta^{13}C$ values of fossil archaeal and sulphate-reducing bacteria lipids (Fig. 6.18) show rather constant values along the crust. This suggests that, despite apparent changes in methane-flux velocities entailed by the changes in the archaeal community structure, the metabolic pathways for methane consumption of both ANMEs remained similar during the formation of the crust.

Comparative analysis of ANME sequences from the fossil carbonate crust with those previously discovered in the present-day flourishing methane-related biomes (Knittel et al., 2005; Michaelis et al., 2002; Stadnitskaia et al., 2005) revealed close phylogenetic similarities within the ANME-1 archaeal guild (Fig. 6.19). Hence, identified ancient AOM communities seemed to have similar ecological preference as the present-day anaerobic methanotrophs, at least those affiliated with the ANME-1 guild. The phylogenetic proximity of fossil and currently active ANME-1 signifies that these anaerobic methanotrophs can maintain their environmental niche for several thousands of years.

5.3. Deep-sea carbonates of the future

The future occurrence of deep-sea biogenic carbonates is dependent on variations in the depth of the CCD. There are indications that contemporary changes in global climate may be causing a shallowing of the CCD

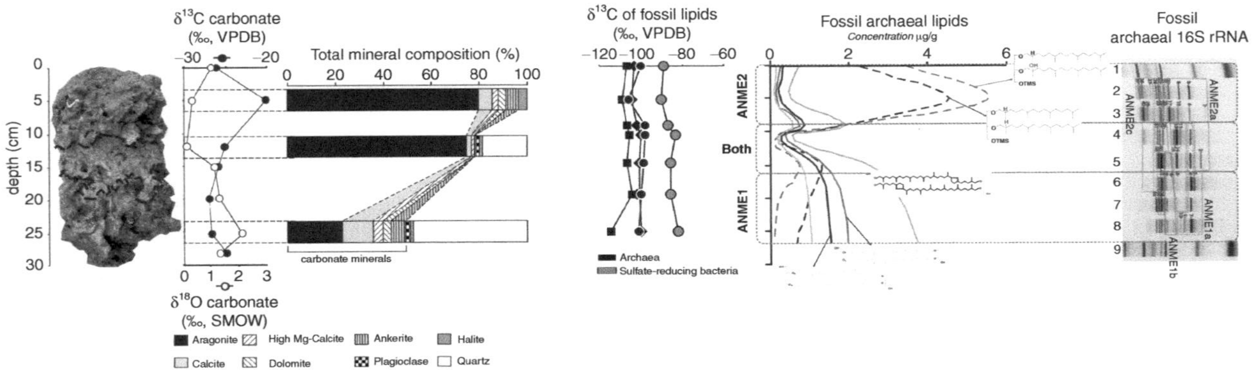

Figure 6.18 Stable-carbon and oxygen-isotope profiles, mineralogical composition, $\delta^{13}C$ values (in ‰, VPDB) of archaeal and sulphate-reducing bacteria lipid biomarkers, concentration profiles (μg/g of dry sediment) of archaeal lipids and DGGE analysis of PCR products obtained with primers specific for the 16S rRNA encoding genes of anaerobic methanotrophs (ANMEs; modified from Stadnitskaia et al., 2008b).

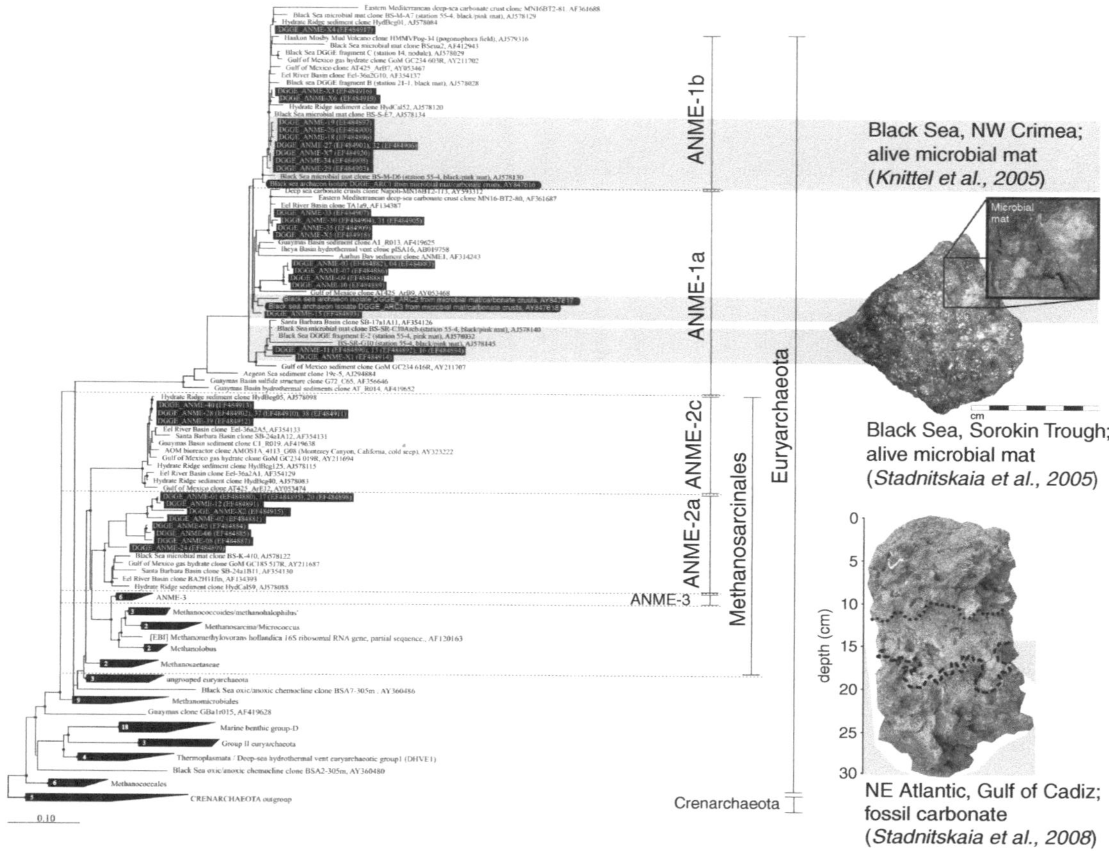

Figure 6.19 Phylogenetic tree of 16S rRNA gene sequences showing the affiliation of anaerobic methane oxidising archaea. Fragments are highlighted in black squares using white letters referring to the different sequences encountered in the fossil carbonate. ANME sequences found in the carbonate crusts and related microbial mats in the Black Sea seeps (Knittel et al., 2005; Michaelis et al., 2002; Stadnitskaia et al., 2005) are indicated within the coloured bar. The scale bar represents 10% sequence divergence (modified from Stadnitskaia et al., 2008b). (A multi-colour version of this figure is on the included CD-ROM.)

through the process of ocean acidification stimulated by increased absorption of CO_2 by the ocean from the atmosphere (Wei et al., 2009). Shallowing of the CCD during times of rapid global warming has occurred in the past, most notably at the Paleocene–Eocene Thermal Maximum (Zachos et al., 2005). Not only did this event affect the preservation potential of deep-sea carbonate oozes, but it was also accompanied by a mass extinction arising from the dramatic environmental changes occurring at this time.

Stanley and Fautin's (2001) "naked coral" hypothesis implies that coral calcification was a response to major shifts in seawater chemistry, with corals surviving as non-calcified organisms during acidified oceanic conditions (and that they were correspondingly poorly preserved in the fossil record). This hypothesis partly explains the sudden appearance of highly evolved calcified scleractinians 14 Ma after the Permian–Triassic extinction (Stanley and Fautin, 2001), possibly from the soft-bodied descendants of the scleractiniamorphs (Ezaki, 2000; Wendt, 1990). Future anthropogenically induced increases in atmospheric pCO_2 may cause a shallowing of the aragonite saturation depth, heralding the potential future demise of cold-water corals (Guinotte et al., 2006). However, this may not necessarily mean the extinction of cold-water scleractinia but perhaps, in the worst case scenario at least, their continued evolution as non-calcified forms.

The future of seep-related carbonates is perhaps more promising as methane will continue to be expelled from the seafloor as it has done so for hundreds of millions of years. Microbially mediated authigenic carbonate formation will continue to occur, and, as these processes occur in the sediment, ocean acidification will have little to no effect.

REFERENCES

Akhmanov, G.G., 1996. Lithology of mud breccia clasts from the Mediterranean Ridge. Mar. Geol. 132, 151–164.

Akhmanov, G.G., Woodside, J.M., 1998. Mud volcanic samples in the content of the Mediterranean Ridge mud diapiric belt. In: Robertson, A.H.F., Emeis, K.-C., Richter, C., Camerlenghi, A. (Eds.), Proc. ODP, vol. 160. Scientific Results, College Station, TX, pp. 597–606.

Aloisi, G., Pierre, C., Rouchyb, J.-M., Foucher, J.-P., Woodside, J., the MEDINAUT Scientific Party, 2000. Methane-related authigenic carbonates of eastern Mediterranean Sea mud volcanoes and their possible relation to gas hydrate destabilization. Earth Planet Sci. Lett. 184, 321–338.

Aloisi, G., Bouloubassi, I., Heijs, S.K., Pancost, R.D., Pierre, C., Sinninghe Damsté, J.S., et al., 2002. CH_4-consuming microorganisms and the formation of carbonate crusts at cold seeps. Earth Planet Sci. Lett. 203, 195–203.

Alperin, M.J., Reeburgh, W.S., 1985. Inhibition experiments on anoxic marine sediments suggests methane is not directly oxidized by sulfate-reducing bacteria. In: Book Monograph: 7th International Symposium on Environmental Biogeochemistry, 6.

Auzende, J.-M., Shipboard Party, 1984. Géologie du Banc de Gorringe, Campagne CYA-GORII. Résultats des campagnes à la mer 27. Publications du Centre National pour l'Exploration des Oceans (C.N.E.X.O), Brest, 65 pp.

Bailey, W., Shannon, P.M., Walsh, J.J., Unnithan, V., 2003. The spatial distributions of faults and deep sea carbonate mounds in the Porcupine Basin, offshore Ireland. Mar. Petrol. Geol. 20, 509–522.

Barnes, R.O., Goldberg, E.D., 1976. Methane production and consumption in anoxic marine sediments. Geology 4, 297–300.

Bathurst, R.G.C., 1975. Carbonate Sediments and their Diagenesis, second ed. Elsevier, Amsterdam Developments in edimentology, No. 12, 658 pp.

Beal, E.J., House, C.H., Orphan, V.J., 2009. Manganese- and iron-dependent marine methane oxidation. Science 325, 184–187.

Bell, N., Smith, J., 1999. Coral growing on North Sea oil rigs. Nature 402, 601.

Bernecker, M., Weidlich, O., 1990. The Danian (Paleocene) coral limestone of Fakse, Denmark: a model for ancient aphotic, azooxanthellate coral mounds. Facies 22, 103–138.

Bernecker, M., Weidlich, O., 2005. Azooxanthellate corals in the Late Maastrichtian–Early Paleocene of the Danish basin: bryozoan and coral mounds in a boreal shelf setting. In: Freiwald, A., Roberts, M. (Eds.), Cold-Water Corals and Ecosystems. Springer, Berlin, pp. 3–26.

Beuck, L., Freiwald, A., 2005. Bioerosion patterns in a deep-water *Lophelia pertusa* (Scleractinia) thicket (Propeller Mound, northern Porcupine Seabight). In: Freiwald, A., Roberts, J.M. (Eds.), Cold-Water Corals and Ecosystems. Springer, Berlin, pp. 915–936.

Blamart, D., Rollion-Bard, C., Cuif, J.-P., Juillet-Leclerc, A., Lutringer, A., van Weering, T.C.E., et al., 2005. C and O isotopes in a deep-sea coral (*Lophelia pertusa*) related to skeletal microstructure. In: Freiwald, A., Roberts, J.M. (Eds.), Cold-Water Corals and Ecosystems. Springer, Berlin, pp. 1005–1020.

Blumenberg, M., Seifert, R., Reitner, J., Pape, T., Michaelis, W., 2004. Membrane lipid patterns typify distinct anaerobic methanotrophic consortia. PNAS 101, 11111–11116.

Boetius, A., Suess, E., 2004. Hydrate Ridge: a natural laboratory for the study of microbial life fueled by methane from near-surface gas hydrates. Chem. Geol. 205, 291–310.

Boetius, A., Ravenschlag, K., Schubert, C.J., Rickert, D., Widdel, F., Gieske, A., et al., 2000. A marine anaerobic consortium apparently mediating anaerobic oxidation of methane. Nature 407, 623–626.

Bouriak, S.V., Akhmetzhanov, A.M., 1998. Origin of gas hydrate accumulations on the continental slope of the Crimea from geophysical studies. In: Henriet, J.-P., Mienert, J. (Eds.), Gas Hydrates: Relevance to World Margin Stability and Climatic Change. Special Publications 137, Geological Society, London, pp. 215–222.

Bouriak, S., Vanneste, M., Saoutkine, A., 2000. Inferred gas hydrates and clay diapirs near the Storrega slide on the southern edge of the Vøring Plateau, offshore Norway. Mar. Geol. 163, 125–148.

Broerse, A.T.C., 2000. Coccolithophore export production in selected ocean environments: seasonality, biogeography and carbonate production. PhD Dissertation, Vrije Universiteit, Amsterdam,185 pp.

Cairns, S.D., 2001a. A brief history of taxonomic research on azooxanthellate Scleractinia (Cnidaria: Anthozoa). Bull. Biol. Soc. Wash. 10, 191–203.

Cairns, S.D., 2001b. A generic revision and phylogenetic analysis of the Dendrophylliidae (Cnidaria: Scleractinia). Smithson. Contrib. Zool. 615, 1–76.

Cairns, S.D., 2007. Deep-water corals: an overview with special reference to diversity and distribution of deep-water scleractinian corals. Bull. Mar. Sci. 81, 311–322.

Campbell, K.A., 2006. Hydrocarbon seep and hydrothermal vent paleoenvironments and paleontology: past developments and future research directions. Palaeogeogr. Palaeoclimatol. Palaeoecol. 232, 362–407.

Campbell, K.A., Farmer, J.D., Des Marais, D., 2002. Ancient hydrocarbon seeps from the Mesozoic convergent margin of California: carbonate geochemistry, fluids and paleoenvironments. Geofluids 2, 63–94.

Canfield, D.E., 2005. The early history of atmospheric oxygen: homage to Robert M. Garrels. Annu. Rev. Earth Planet. Sci. 33, 1–36.

Castanier, S., Le Métayer-Levrel, G., Perthuisot, J.-P., 1999. Ca-carbonates precipitation and limestone genesis—the microbiogeologist point of view. Sed. Geol. 126, 9–23.

Cita, M.B., Ryan, W.B.F., Paggi, L., 1981. Prometheus mud breccia. An example of shale diapirism in the Western Mediterrranean Ridge. Annales Géologiques des Pays Héllé-niques 30, 543–569.

Coates, A.G., Kauffman, E.G., 1973. Stratigraphy, paleontology and paleoenvironment of a Cretaceous coral thicket, Lamy, New Mexico. J. Paleontol. 47, 953–968.

Colman, J.G., Gordon, D.M., Lane, A.P., Forde, M.J., Fitzpatrick, J.J., 2005. Carbonate mounds off Mauritania, Northwest Africa: status of deep-water corals and implications for management of fishing and oil exploration activities. In: Freiwald, A., Roberts, J.M. (Eds.), Cold-Water Corals and Ecosystems. Springer, Berlin, pp. 417–441.

Comita, P.B., Gagosian, R.B., Pang, H., Costello, C.E., 1984. Structural elucidation of a unique macrocyclic membrane lipid from a new, extremely thermophilic, deep-sea hydrothermal vent archaebacterium, *Methanococcus jannaschii*. J. Biol. Chem., 259, 15234–15241.

Cordes, E.E., McGinley, M.P., Podowski, E.L., Becker, E.L., Lessard-Pilon, S., Viada, S.T., et al., 2008. Coral communities of the deep Gulf of Mexico. Deep Sea Res. Part I 55, 777–787.

Cragg, B.A., Parkes, R.J., Fry, J.C., Weightman, A.J., Rochelle, P.A., Maxwell, J.R., 1996. Bacterial populations and processes in sediments containing gas hydrates (ODP Leg 146: Cascadia Margin). Earth Planet. Sci. Lett. 139, 497–507.

de Haas, H., Mienis, F., Frank, N., Richter, T.O., Steinacher, R., de Stigter, H., et al., 2009. Morphology and sedimentology of (clustered) cold-water coral mounds at the south Rockall Trough margins, NE Atlantic Ocean. Facies 55, 1–26.

De Mol, B., Van Rensbergen, P., Pillen, S., Van Herreweghe, K., Van Rooij, D., McDonnell, A., et al., 2002. Large deep-water coral banks in the Porcupine Basin, southwest of Ireland. Mar. Geol. 188, 193–231.

De Mol, B., Henriet, J.P., Canals, M., 2005. Development of coral banks in Porcupine Seabight: do they have Mediterranean ancestors? In: Freiwald, A., Roberts, J.M. (Eds.), Cold-Water Corals and Ecosystems. Springer, Berlin, pp. 515–533.

De Mol, B., Kozachenko, M., Wheeler, A., Alvares, H., Henriet, J.P., Olu-Le Roy, K., 2007. Thérèse Mound: a case study of coral bank development in the Belgica Mound Province, Porcupine Seabight. Int. J. Earth Sci. 96, 103–120.

De Rosa, M., Gambacorta, A., Nicolaus, B., Chappe, B., Albrecht, P., 1983. Isoprenoid ethers: backbone of complex lipids of the archaebacterium Sulfolobus solfataricus. Biochim. Biophys. Acta 753, 249–256.

De Rosa, M., Trincone, A., Nicolaus, B., Gambacorta, A., 1991. Archaebacteria: lipids, membrane structure, and adaptation to environmental stresses. In: di Prisco, (Ed.), Life Under Extreme Conditions. Springer-Verlag, Berlin, pp. 61–87.

Deelman, J.C., 1975. Two mechanisms of microbial carbonate precipitation. Naturwissenschaften 62, 484–485.

Devereux, I., 1967. Oxygen isotope palaeotemperature measurements on two Tertiary deep-water coral thickets from Wairarapa, New Zealand. Palaeogeogr. Palaeoclimatol. Palaeoecol. 3, 447–456.

Dickens, G.R., O'Neil, J.R., Rea, D.K., Owen, R.M., 1995. Dissociation of oceanic methane hydrates as a cause of carbon-isotope excursion at the end of the Paleocene. Paleoceanography 10, 965–971.
Di Geronimo, I., Messina, C., Rosso, A., Sanfilippo, R., Sciuto, F., Vertino, A., 2005. Enhanced biodiversity in the deep: early Pleistocene coral communities from southern Italy. In: Freiwald, A., Roberts, J.M. (Eds.), Cold-Water Corals and Ecosystems. Springer, Berlin, pp. 61–86.
Dorschel, B., Hebbeln, D., Rüggeberg, A., Dullo, W.-C., Freiwald, A., 2005. Growth and erosion of a cold-water coral covered carbonate mound in the Northeast Atlantic during the Late Pleistocene and Holocene. Earth Planet. Sci. Lett. 233, 33–44.
Dorschel, B., Hebbeln, D., Foubert, A., White, M., Wheeler, A.J., 2007a. Hydrodynamics and cold-water coral facies distribution related to recent sedimentary processes at Galway Mound west of Ireland. Mar. Geol. 244, 184–196.
Dorschel, B., Hebbeln, D., Rüggeberg, A., Dullo, W.-C., 2007b. Carbonate budget of a cold-water coral carbonate mound: propeller Mound, Porcupine Seabight. Int. J. Earth Sci. 96, 73–83.
Dorschel, B., Wheeler, A.J., Huvenne, V.A.I., de Haas, H., 2009. Cold-water coral mounds in an erosive environmental setting: TOBI side-scan sonar data and ROV video footage from the northwest Porcupine Bank, NE Atlantic. Mar. Geol. 264, 218–229.
Drozd, R.J., Pazdersky, G.J., Jones, V.T., Weismann, T.J., 1981. Use of compositional indicators in prediction of petroleum production potential (abs.). American Chemical Society Meeting, Atlanta, GA, 29th March–3rd April.
Duineveld, G.C.A., Lavaleye, M.S.S., Berghuis, E.M., 2004. Particle flux and food supply to a seamount cold-water coral community (Galicia Bank, NW Spain). Mar. Ecol. Prog. Ser. 277, 13–23.
Dullo, W.-C., Flögel, S., Rüggeberg, A., 2008. Cold-water coral growth in relation to the hydrography of the Celtic and Nordic European continental margin. Mar. Ecol. Prog. Ser. 371, 165–176.
Duncan, P.M., 1870. On the Madreporaria dredged up in the expeditions of H.M.S. 'Porcupine'. Proc. R. Soc. Lond. 18, 289–301.
Duncan, P.M., 1873. A description of the Madreporaria dredged up during the expeditions of H.M.S. Porcupine in 1869 and 1870. Part 1. Trans. Zool. Soc. Lond. 8, 303–344.
Duncan, P.M., 1877. On the rapidity of growth and variability of some Madreporaria on an Atlantic Cable, with remarks upon the rate of accumulation of foraminiferal deposits. Proc. R. Soc. Lond. 26, 133–137.
Duncan, P.M., 1878. A description of the Madreporaria dredged up during the expeditions of H.M.S. Porcupine in 1869 and 1870. Part 2. Trans. Zool. Soc. Lond. 10, 235–249.
Dupraz, C., Reid, R.P., Braissant, O., Decho, A.W., Norman, R.S., Visscher, P.T., 2009. Processes of carbonate precipitation in modern microbial mats. Earth Sci. Rev. 96, 141–162.
Dupré, S., Buffet, G., Mascle, J., Foucher, J.-P., Gauger, S., Boetius, A., et al., 2008. High resolution mapping of large gas emitting mud volcanoes on the Eygyptian continental margin (Nile Deep Sea Fan) by AUV surveys. Marine Geophys. Res. 29, 275–290.
Elvert, M., Suess, E., Greinert, J., Whiticar, M.J., 2000. Archaea mediating anaerobic methane oxidation in deep-sea sediments at cold seeps of the eastern Aleutian subduction zone. Org. Geochem. 31, 1175–1187.
Elvert, M., Boetius, A., Knittel, K., Jørgensen, B.B., 2003. Characterization of specific membrane fatty acids as chemotaxonomic markers for sulfate-reducing bacteria involved in anaerobic oxidation of methane. Geomicrobiol. J. 20, 403–419.
Elvert, M., Hopmans, E.C., Treude, T., Boetius, A., Suess, E., 2005. Spatial variations of methanotrophic consortia at cold methane seeps: implications from a high-resolution molecular and isotopic approach. Geobiology 3, 195–209.

Ezaki, Y., 2000. Palaeoecological and phylogenetic implications of a new scleractiniamorph genus from Permian sponge reefs, south China. Palaeontology 43, 199–217.

Faugères, J.-C., Mulder, T., 2011. Contour currents and contourite drifts. In: Hüneke, H., Mulder, T. (Eds.), Deep-Sea Sediments. Developments in Sedimentology, Elsevier, Amsterdam, vol. 63, pp. 149–214.

Feely, R.A., Sabine, C.L., Lee, K., Berelson, W., Kleypas, J., Fabry, V.J., et al., 2004. Impact of anthropogenic CO_2 on the $CaCO_3$ system in the oceans. Science 305, 362–366.

Filkorn, H.F., 1994. Fossil scleractinian corals from James Ross Basin, Antarctica. Antarct. Res. Ser. 65, 1–96.

Foubert, A., Depreiter, D., Beck, T., Maignien, L., Pannemans, B., Frank, N., et al., 2008. Carbonate mounds in a mud volcano province off north-west Morocco: key to processes and controls. Mar. Geol. 248, 74–96.

Frank, N., Lutringer, A., Paterne, M., Blamart, D., Henriet, J.-P., van Rooij, D., et al., 2005. Deep-water corals of the northeastern Atlantic margin: carbonate mound evolution and upper intermediate water ventilation during the Holocene. In: Freiwald, A., Roberts, J.M. (Eds.), Cold-Water Corals and Ecosystems. Springer, Berlin, pp. 113–133.

Frederiksen, R., Jensen, A., Westerberg, H., 1992. The distribution of the scleractinian coral *Lophelia pertusa* around the Faroe Islands and the relation to internal tidal mixing. Sarsia 77, 157–171.

Freiwald, A., 2002. Reef-forming cold-water corals. In: Wefer, G., Billett, D., Hebbeln, D., Jørgensen, B.B., Schlüter, M., van Weering, T.C.E. (Eds.), Ocean Margin Systems. Springer, Berlin, pp. 365–386.

Freiwald, A., Wilson, J.B., Henrich, R., 1999. Grounding Pleistocene icebergs shape recent deep-water coral reefs. Sed. Geol. 125, 1–8.

Galimov, E.B., Kvenvolden, K.A., 1983. Concentrations and carbon isotopic compositions of CH_4 and $C0_2$ in gas from sediments of the Blake outer ridge, deep sea drilling project leg 76. Initial Reports of the Deep Sea Drilling Project, vol. 76,. 393–407.

Gardner, J.M., 2001. Mud volcanoes revealed and sampled on the Western Moroccan continental margin. Geophys. Res. Lett. 28, 339–342.

Gass, S.E., Roberts, J.M., 2006. The occurrence of the cold-water coral *Lophelia pertusa* (Scleractinia) on oil and gas platforms in the North Sea: colony growth, recruitment and environmental controls on distribution. Mar. Pollut. Bull. 52, 549–559.

Gautier, D.L., 1985. Interpretation of early diagenesis in ancient marine sediments. In: Gautier, D.L., Kharaka, Y.K., Surdam, R.C. (Eds.), Relationship of Organic Matter and Mineral Diagenesis. SEPM, Tulsa, Oklahoma, USA, pp. 6–78 Short Course 17.

Ginsburg, G., Soloviev, V., 1997. Methane migration within the submarine gas-hydrate stability zone under deep-water conditions. Mar. Geol. 137, 49–57.

Ginsburg, G.D., Soploviev, V.A., 1998. Submarine Gas Hydrates. VNII Okeangeologia, St. Petersburg.

Gofas, S., Freiwald, A., López Correa, M., Remia, A., Salas, C., Taviani, M., et al., 2007. Oyster beds in the deep sea. In: Jordaens, K., van Houtte, N., van Goethem, J., Backeljau, T. (Eds.), World Congress of Malacology (Antwerp, Belgium, 15–20 July 2007), 80 (Abstracts).

Greinert, J., Bohrmann, G., Elvert, M., 2002. Stromatolitic fabric of authigenic carbonate crusts: result of anaerobic methane oxidation at cold seeps in 4,850 m water depth. Int. J. Earth Sci. (Geol. Rundsch) 91, 698–711.

Gubkin, I.M., 1932. Uchenie o Nefti. Moscow, USSR (in Russian).

Guinotte, J.M., Orr, J., Cairns, S., Freiwald, A., Morgan, L., George, R., 2006. Will human-induced changes in seawater chemistry alter the distribution of deep-sea scleractinian corals? Front. Ecol. Environ. 4, 141–146.

Guliev, I.S., Feizullayev, A.A., 1997. All About Mud Volcanoes. Nafta-Press, Baku.

Gunnerus, J.E., 1768. Om nogle Norske coraller. Kongel. Norske Vidensk. Selsk. Skr. 4, 38–73.
Han, M.W., Suess, E., 1989. Subduction induced pore fluid venting and the formation of authigenic carbonates along the Cascadia continental margin: implications for the global Ca cycle. Palaeogeogr. Palaeoclimatol. Palaeoecol. 71, 97–118.
Henriet, J.-P., Mienert, J. (Eds.), 1998. Gas Hydrates. Relevance to World Margin Stability and Climatic Change. Geological Society, London Special Publications 137, 338 pp.
Henriet, J.-P., De Mol, B., Pillen, S., Vanneste, M., Rooij, D.V., Versteeg, W., et al., 1998. Gas hydrate crystals may help build reefs. Nature 391, 648–649.
Henriet, J.-P., Guidard, S., ODP "Proposal 573" Team, 2003. Carbonate mounds as a possible example for microbial activity in geological processes. In: Wefer, G., Billett, D., Hebbeln, D., Jørgensen, B.B., Schlüter, M., van Weering, T.C.E. (Eds.), Ocean Margin Systems. Springer, Heidelberg, pp. 439–456.
Hinrichs, K.U., Boetius, A.B., 2002. The anaerobic oxidation of methane: new insights in microbial ecology and biogeochemistry. In: Wefer, G., Billett, D., Hebbeln, D., Jørgensen, B.B., Schlüter, M., van Weering, T. (Eds.), Ocean Margin Systems. Springer-Verlag, Heidelberg, pp. 457–477.
Hinrichs, K.-U., Hayes, J.M., Sylva, S.P., Brewer, P.G., De Long, E.F., 1999. Methane-consuming archaebacteria in marine sediments. Nature 398, 802–806.
Hinrichs, K.-U., Summons, R.E., Orphan, V., Sylva, S.P., Hayes, J.M., 2000. Molecular and isotopic analysis of anaerobic methane-oxidizing communities in marine sediments. Org. Geochem. 31, 1685–1701.
Hoefs, M.J.L., Schouten, S., De Leeuw, J.W., King, L.L., Wakeham, S.G., Sinninghe Damsté, J.S., 1997. Ether lipids of planktonic archaea in the marine water column. Appl. Environ. Microbiol. 63, 3090–3096.
Hoehler, T.M., Alperin, M.J., 1996. Anaerobic methane oxidation by a methanogen-sulfate reducer consortium: geochemical evidence and biochemical considerations. In: Lidstrom, M.E., Tabita, R.F. (Eds.), Microbial Growth on C-1 Compounds. Kluwer, Dordrecht, The Netherlands, pp. 326–333.
Hoehler, T.M., Alperin, M.J., Albert, D.B., Martens, C.S., 1994. Field and laboratory studies of methane oxidation in an anoxic marine sediment: evidence for a methanogen-sulfate reducer consortium. Glob. Biogeochem. Cycles 8, 451–463.
Hoernle, K., Shipboard party, 2001. METEOR Cruise No. M51, Leg 1, VULKOSO—Vulkanismus Ostatlantik-Alboran. GEOMAR, Bremen 94 pp.
Hopmans, E.C., Schouten, S., Pancost, R.D., van der Meer, M.T.J., Sinninghe Damsté, J. S., 2000. Analysis of intact tetraether lipids in archaeal cell material and sediments by high performance liquid chromatography/atmospheric pressure chemical ionization mass spectrometry. Rapid Commun. Mass Spectrom. 14, 585–589.
Hovland, M., 1990. Do carbonate reefs form due to fluid seepage? Terra Nova 2, 8–18.
Hovland, M., 2005. Pockmark-associated coral reefs at the Kristin field off Mid-Norway. In: Freiwald, A., Roberts, J.M. (Eds.), Cold-Water Corals and Ecosystems. Springer, Berlin, pp. 623–632.
Hovland, M., 2008. Deep Water Coral Reefs: Unique Biodiversity Hot-Spots. Springer-Praxis, Chichester.
Hovland, M., Judd, A., 1988. Seabed Pockmarks and Seepages. Impact on Geology, Biology and the Marine Environment. Graham and Trotman, London 293 pp.
Hovland, M., Mortensen, P., 1999. Norske korallrev og prosesser i havbunnen. John Grieg Forlag, Bergen, Norway 160 pp.
Hovland, M., Talbot, M., Qvale, H., Olaussen, S., Aasberg, L., 1987. Methane-related carbonate cements in pockmarks of the North Sea. J. Sediment. Petrol. 57, 881–892.
Hovland, M., Croker, P.F., Martin, M., 1994. Fault-associated seabed mounds (carbonate knolls?) off western Ireland and north-west Australia. Mar. Petrol. Geol. 11, 232–246.

Hovland, M., Mortensen, P., Brattegard, T., Strass, P., Rokeongen, K., 1998. Ahermatypic coral banks off Mid-Norway: evidence for a link with seepage of light hydrocarbons. Palaios 13, 189–200.
Huguen, C., Foucher, J.P., Mascle, J., Ondréas, H., Thouement, M., Gontharet, S., et al., 2009. Menes caldera, a highly active site of brine seepage in the Eastern Mediterranean sea: "In situ" observations from the NAUTINIL expedition (2003). Mar. Geol. 261, 138–152.
Huvenne, V.A.I., De Mol, B., Henriet, J.P., 2003. A 3D seismic study of the morphology and spatial distribution of buried coral banks in the Porcupine Basin, SW of Ireland. Mar. Geol. 198, 5–26.
Huvenne, V.A.I., Beyer, A., de Haas, H., Dekindt, K., Henriet, J.P., Kozachenko, M., et al., 2005. The seabed appearance of different coral bank provinces in the Porcupine Seabight, NE Atlantic: results from sidescan sonar and ROV seabed mapping. In: Freiwald, A., Roberts, J.M. (Eds.), Cold-Water Corals and Ecosystems. Springer, Berlin, pp. 535–569.
Huvenne, V.A.I., Masson, D.G., Wheeler, A.J., 2009. Sediment dynamics of a sandy contourite: the sedimentary context of the Darwin cold-water coral mounds, Northern Rockall Trough. Int. J. Earth Sci. 98, 865–884.
Intergovernmental Panel on Climate Change, 2007. IPCC Fourth Assessment Report: Climate Change 2007 (AR4). Geneva, Switzerland.
Intergovernmental Panel on Climate Change, 1994. Radiative Forcing of Climate Change and Evaluation of the IPCC IS 92 Emission Scenarios. Cambridge Univ. Press, New York.
Ivanov, M.K., Konyukhov, A.I., Kul'nitskii, L.M., Musatov, A.A., 1989. Gryazevie vulkani v glubokovodnoy chasti Chernogo moria. Vestn. Mosc. Univ. ser. Geol. 3, 48–54 in Russian.
Ivanov, M.K., Limonov, A., van Weering, T., 1996. Comparative characteristic of the Black Sea and Mediterranean Ridge mud volcanoes. Mar. Geol. 132, 253–271.
Ivanov, M.K., Limonov, A.F., Woodside, J.M., 1998. Extensive deep fluid flux through the sea floor on the Crimean continental margin (Black Sea). In: Henriet, J.-P., Mienert, J. (Eds.), Gas Hydrates. Relevance to World Margin Stability and Climatic Change. Geological Society, London, pp. 195–213 Special Publications, 137.
Jiang, G., Kennedy, M.J., Christie-Blick, N., 2003. Stable isotopic evidence for methane seeps in Neoproterozoic postglacial cap carbonates. Nature 426, 822–826.
Jones, V.T., Drozd, R.J., 1979. Predictions of oil or gas potential by near-surface geochemistry. AAPG Bull. 63, 699.
Jones, V.T., Drozd, R.J., 1983. Prediction of oil or gas potential by near-surface geochemistry. AAPG Bull. 67, 932–952.
Joubin, M.L., 1922. Les coraux de mer profonde nuisibles aux chalutiers. Office Scientifique et Technique des Pêches Maritimes. Notes et Mémoires 18, 5–16.
Joye, S.B., Boetius, A., Orcutt, B.N., Montoya, J.P., Schulz, H.N., Erickson, M.J., et al., 2004. The anaerobic oxidation of methane and sulfate reduction in sediments from Gulf of Mexico cold seeps. Chem. Geol. 205, 219–238.
Kano, A., Ferdelman, T.G., Williams, T., Henriet, J.P., Ishikawa, T., Kawagoe, N., et al., 2007. Age constraints on the origin and growth history of a deep-water coral mound in NE Atlantic drilled during Integrated Ocean Drilling Program Expedition 307. Geology 35, 1051–1054.
Kennett, J.P., 1982. Marine Geology. Prentice Hall, Englewood Cliffs, NJ.
Kennish, M., 2001. Practical Handbook of Marine Science, third ed. CRC Press, Boca Raton, FL 857 pp.
Kenyon, N., Ivanov, M.K., Akhmetzhanov, A., Akhmanov, G., 2000. Multidisciplinary study of geological processes on the North East Atlantic and Western Mediterranean margins. Preliminary results of geological and geophysical investigations during TTR-9

cruise of R/V Professor Logachev, June–July, 1999, Intergovernmental Oceanographic Commission Technical Series 56, 56 pp.
Kenyon, N., Ivanov, M.K., Akhmetzhanov, A., Akhmanov, G. (Eds.), 2001. Geological processes in the Mediterranean and Black Seas and North East Atlantic. Preliminary results of investigations during the TTR-11 cruise of R/V Professor Logachev, July–September 2001, In: Intergovernmental Oceanographic Commission Technical Series 62, 63 pp.
Kenyon, N.H., Ivanov, M.K., Akhmetzhanov, A.M., Akhmanov, G.G. (Eds.), 2002. Geological Processes in the Mediterranean and Black Seas and North East Atlantic: Preliminary Results of Investigations during the TTR-11 cruise of RV Professor Logachev July–September, 2001, In: Intergovernmental Oceanographic Commission Technical Series 62, 113 pp.
Kenyon, N.H., Akhmetzhanov, A.M., Wheeler, A.J., van Weering, T.C.E., de Haas, H., Ivanov, M.K., 2003. Giant carbonate mud mounds in the southern Rockall Trough. Mar. Geol. 195, 5–30.
Kholodov, V.N., 2002a. Mud volcanoes, their distribution regularities and genesis: communication1. Mud volcanic provinces and morphology of mud volcanoes. Lithol. Miner. Resour. 37 (3), 197–209.
Kholodov, V.N., 2002b. Mud volcanoes: distribution regularities and genesis (Communication 2. Geological-geochemical peculiarities and formation model). Lithol. Miner. Resour. 37 (4), 193–309.
Kiriakoulakis, K., Fisher, E., Wolff, G.A., Freiwald, A., Grehan, A., Roberts, J.M., 2006. Lipids and nitrogen isotopes of two deep-water corals from the North-East Atlantic: initial results and implications for their nutrition. In: Freiwald, A., Roberts, J.M. (Eds.), Cold-Water Corals and Ecosystems. Springer, Berlin, pp. 715–729.
Knittel, K., Boetius, A., Eilers, A.L.H., Lochte, K., Linke, O.P.P., 2003. Activity, distribution, and diversity of sulfate reducers and other bacteria in sediments above gas hydrate (Cascadia Margin, Oregon). Geomicrobiol. J. 20, 269–294.
Knittel, K., Lösekann, T., Boetius, A., Kort, R., Amann, R., 2005. Diversity and distribution of methanotrophic archaea at cold seeps. Appl. Environ. Microbiol. 71, 467–479.
Kvenvolden, K.A., Frank, T.J., Golan-Bac, M., 1990. Hydrocarbon gases in Tertiary and Quaternary sediments offshore Peru-results and comparisons. In: Suess, E., von Huene, R. et al., Proc. ODP, Sci. Results, 112. Ocean Drilling Program, College Station, TX, pp. 505–516.
Kvenvolden, K.A., Ginsburg, G., Soloviev, V., 1993. Worldwide distribution of subaquatic gas hydrates. Geo. Mar. Lett. 13, 32–40.
Lallemant, S.J.C., Henry, P., Le Pichon, X., Foucher, J.-P., 1990. Detailed structure and possible fluid paths at the toe of the Barbados accretionary wedge (ODP Leg 110 area). Geology 18 (9), 854–857.
Langer, M., 2008. Assessing the contribution of foraminiferan protists to global ocean carbonate production. J. Eukaryot. Microbiol. 55, 163–169.
Langer, M.R., Silk, M.T., Lipps, J.H., 1997. Global ocean carbonate and carbon dioxide production: the role of reef foraminifera. J. Foramin. Res. 27 (4), 271–277.
Langworthy, T.A., 1985. Lipids of archaebacteria. In: Woese, C.R., Wolfe, R.S. (Eds.), The BacteriaVol. VIII. Academic Press, New York, pp. 459–497.
Langworthy, T.A., Tornabene, T.G., Holzer, G., 1982. Lipids of archaebacteria. In Zbl. Bakt. Hyg., I. Abt. Orig. C3, 228–244.
Lanoil, B., Sassen, R., La Duc, M.T., Sweet, S.T., Nealson, K.H., 2001. Bacteria and Archaea physically associated with Gulf of Mexico gas hydrates. Appl. Environ. Microbiol. 67, 5143–5153.
Le Pichon, X., Henry, P., Lallemant, S., 1990. Water flow in the Barbados accretionary complex. J. Geophys. Res. 95, 8945–8967.

Levin, L.A., Mendoza, G.F., 2007. Community structure and nutrition of deep methane-seep macrobenthos from the North Pacific (Aleutian) Margin and the Gulf of Mexico (Florida Escarpment). Mar. Ecol. 28, 131–151.

Limonov, A.F., Woodside, J.M., Cita, M.B., Ivanov, M.K., 1996. The Mediterranean Ridge and related mud diapirism; a background. Mar. Geol. 132, 7–19.

Lindberg, B., Mienert, J., 2005. Postglacial carbonate production by cold-water corals on the Norwegian Shelf and their role in the global carbonate budget. Geology 33, 537–540.

Lindberg, B., Berndt, C., Mienert, J., 2007. The Fugloy Reef at 70°N; acoustic signature, geologic, geomorphologic and oceanographic setting. Int. J. Earth Sci. 96, 201–213.

Link, W.K., 1952. Significance of oil and gas seeps in World oil exploration. AAPG Bull. 36, 1505–1541.

Lisiecki, L.E., Raymo, M.E., 2007. Plio-Pleistocene climate evolution: trends and transitions in glacial cycle dynamics. Q. Sci. Rev. 26, 56–69.

Loncke, L., Gaullier, V., Bellaiche, G., Mascle, J., 2002. Recent depositional patterns of the Nile deep-sea fan from echo-character mapping. AAPG Bull. 86, 1165–1186.

Lutringer, A., Blamart, D., Frank, N., Labeyrie, L., 2005. Paleotemperatures from deep-sea corals: scale effects. In: Freiwald, A., Roberts, J.M. (Eds.), Cold-Water Corals and Ecosystems. Springer, Berlin, pp. 1081–1096.

Madigan, M.T., Martinko, J.M., Dunlap, P.V., Clark, D.P., 2009. Brock Biology of Microorganisms, twelfth ed. Pearson Benjamin Cummings, San Francisco 596 pp.

Martens, C.S., Berner, R.A., 1974. Methane production in the interstitial waters of sulfate-depleted marine sediments. Science 185, 1167–1169.

Mascle, J., Zitter, T., Bellaiche, G., Droz, L., Gaullier, V., Loncke, L., et al., 2001. The Nile deep sea fan: preliminary results from a swath bathymetry survey. Mar. Petrol. Geol. 18 (4), 471–477.

Masson, D.G., Bett, B.J., Billett, D.S.M., Jacobs, C.L., Wheeler, A.J., Wynn, R.B., 2003. The origin of deep-water, coral-topped mounds in the northern Rockall Trough, Northeast Atlantic. Mar. Geol. 194, 159–180.

Mastandrea, A., Muto, F., Neri, C., Papazzoni, C.A., Perri, E., Russo, F., 2002. Deep-water coral banks: an example from the 'Calcare di Mendicino' (Upper Miocene, northern Calabria, Italy). Facies 47, 27–42.

Messing, C.G., Neuman, A.C., Lang, J.C., 1990. Biozonation of deep-water lithoherms and associated hardgrounds in the northeastern Straits of Florida. Palaios 5, 15–33.

Michaelis, W., Seifert, R., Nauhaus, K., Treude, T., Thiel, V., Blumenberg, M., et al., 2002. Microbial reefs in the Black Sea fueled by anaerobic oxidation of methane. Science 297, 1013–1016.

Mienis, F., van Weering, T., de Haas, H., de Stigter, H., Huvenne, V., Wheeler, A., 2006. Carbonate mound development at the SW Rockall Trough margin based on high resolution TOBI and seismic recording. Mar. Geol. 233, 1–19.

Mienis, F., de Stigter, H.C., White, H., Duineveld, G., de Haas, H., van Weering, T.C.E., 2007. Hydrodynamic controls on cold-water coral growth and carbonate mound development at the SW and SE Rockall Trough Margin, NE Atlantic Ocean. Deep Sea Res. Part I 54, 1655–1674.

Mikkelsen, N., Erlenkeuser, H., Killingley, J.S., Berger, W.H., 1982. Norwegian corals: radiocarbon and stable isotopes in *Lophelia pertusa*. Boreas 11, 163–171.

Milliman, J.D., 1974. Marine Carbonates. Recent Sedimentary Carbonates Part 1. Springer-Verlag, New York 375 pp.

Milliman, J.D., 1993. Production and accumulation of calcium carbonate in the ocean: budget of a non steady state. Glob. Biogeochem. Cycles 7, 927–957.

Milliman, J.D., Manheim, F.T., Pratt, R.M., Zarudzki, E.F.K., 1967. ALVIN Dives on the Continental Margin Off the Southeastern United States July 2–13, 1967. Woods Hole Oceanographic Institution, Massachusetts, USA 64 pp.

Moore, D.R., Bullis, H.R., 1960. A deep-water coral reef in the Gulf of Mexico. Bull. Mar. Sci. 10, 125–128.
Morse, J.W., 2003. Formation and diagenesis of carbonate sediments. In: Mackenzie, F.T. (Ed.), Treatise on Geochemistry 7. Elsevier, Amsterdam, pp. 67–86.
Mortensen, P.B., 2001. Aquarium observations on the deep-water coral *Lophelia pertusa* (L., 1758) (Scleractinia) and selected associated inverterbrates. Ophelia 54, 83–104.
Mortensen, P.B., Rapp, H.T., 1998. Oxygen and carbon isotope ratios related to growth line patterns in skeletons of *Lophelia pertusa* (L) (Anthozoa, Scleractinia): implications for determining of linear extension rates. Sarsia 83, 433–446.
Mousseau, R.J., Williams, J.C., 1979. Dissolved hydrocarbons in coastal waters of North America (abstract). AAPG Bull. 63, 699–700.
Mullins, H.T., Newton, C.R., Heath, K., Vanburen, H.M., 1981. Modern deep-water coral mounds north of Little Bahama Bank: criteria for recognition of deep-water coral bioherms in the rock record. J. Sed. Petrol. 51, 999–1013.
Murray, Sir John, Hjort, J., 1912. The Depths of the Ocean. London, Macmillan 821 pp.
Murray, J., Irvine, R., 1889. On coral reefs and other carbonate of lime formations in modern seas. Proc. Roy. Soc. Lond. A17, 79–109.
Nadson, G.A., 1928. Beitrag zur Kenntniss der bakteriogenen Kalkablagerung. Arch. Hydrobiol. 191, 154–164.
Naeth, J., di Primio, R., Horsfield, B., Schaefer, R.G., Shannon, P.M., Bailey, W.R., et al., 2006. Hydrocarbon seepage and carbonate mound formation: a basin modelling study from the Porcupine Basin (offshore Ireland). J. Pet. Geol. 28, 147–166.
Nauhaus, K., Boetius, A., Krüger, M., Widdel, F., 2002. In vitro demonstration of anaerobic oxidation of methane coupled to sulfate reduction in sediment from a marine gas hydrate area. Environ. Microbiol. 4, 296–306.
Nauhaus, K., Treude, T., Boetius, A., Krüger, M., 2005. Environmental regulation of the anaerobic oxidation of methane: a comparison of ANME-I and ANME-II communities. Environ. Microbiol. 7, 98–106.
Neumann, A.C., Kofoed, J.W., Keller, G.H., 1977. Lithoherms in the Straits of Florida. Geology 5, 4–10.
Niemann, H., Lösekann, T., de Beer, D., Elvert, M., Nadalig, T., Knittel, K., et al., 2006. Novel microbial communities of the Haakon Mosby mud volcano and their role as a methane sink. Nature 443, 854–858.
Nisbet, E.G., Jones, S.M., Maclennan, J., Eagles, G., Moed, J., Warwick, N., et al., 2009. Kick-starting ancient warming. Nat. Geosci. 2, 156–159.
Noé, S., Titschack, J., Freiwald, A., Dullo, C., 2006. From sediment to rock: diagenetic processes of hardground formation in deep-water carbonate mounds of the NE Atlantic. Facies 52, 183–203.
Olu, K., Sibuet, M., Harmegnies, F., Foucher, J.-P., Fiala-Medioni, A., 1996. Spatial distribution of diverse cold seep communities living in various diapiric structures of the southern Barbados prism. Prog. Oceanogr. 38, 347–376.
Olu, K., Lance, S., Sibuet, M., Henry, P., Fiala-Medioni, A., Dinet, A., 1997. Cold seep communities as indicator of fluid expulsion patterns through mud volcanoes seaward of the Barbados accretionary prism. Deep-Sea Res. 44, 811–841.
Omoregie, E.O., Mastalerz, V., de Lange, G., Straub, K.L., Kappler, A., Røy, H., et al., 2008. Biogeochemistry and community composition of iron- and sulfur- precipitating microbial mats at the Chefren Mud Volcano (Nile Deep Sea Fan, Eastern Mediterranean). Appl. Environ. Microbiol. 74, 3198–3216.
Orphan, V.J., House, C.H., Hinrichs, K.-U., McKeegan, K.D., DeLong, E.F., 2001a. Methane-consuming archaea revealed by directly coupled isotopic and phylogenic analysis. Science 293, 484–487.

Orphan, V.J., Hinrichs, K.-U., Ussler III, W., Paull, C.K., Taylor, L.T., Sylva, S.P., et al., 2001b. Comparative analysis of methane-oxidizing archaea and sulfate-reducing bacteria in anoxic marine sediments. Appl. Environ. Microbiol. 67, 1922–1934.

Orphan, V.J., House, C.H., Hinrichs, K.-U., McKeegan, K.D., DeLong, E.F., 2002. Multiple archael groups mediate methane oxidation in anoxic cold seep sediments. Proc. Natl. Acad. Sci. 99, 7663–7668.

Ovsyannikov, D.O., Sadekov, A.Y., Kozlova, E.V., 2003. Rock fragments from mud volcanic deposits of the Gulf of Cadiz: an insight into the Eocene–Pliocene sedimentary succession of the basin. Mar. Geol. 195, 211–221.

Pancost, R.D., Sinninghe Damsté, J.S., De Lint, S., van der Maarel, M.J.E.C., Gottschal, J. C., The Medinaut shipboard scientific party, 2000. Biomarker evidence for widespread anaerobic methane oxidation in Mediterranean sediments by a consortium of methanogenic archaea and bacteria. Appl. Environ. Microbiol. 66, 1126–1132.

Pancost, R.D., Bouloubassi, I., Aloisi, G., Sinninghe Damsté, J.S., the Medinaut shipboard scientific party, 2001a. Three series of non-isoprenoidal dialkyl glycerol diethers in cold-seeps carbonate crusts. Org. Geochem. 32, 695–707.

Pancost, R.D., Hopmans, E.C., Sinninghe Damsté, J.S., the Medinaut shipboard scientific party, 2001b. Archaeal lipids in Mediterranean cold-seeps: molecular proxies for anaerobic methane oxidation. Geochim. Cosmochim. Acta 65, 1611–1627.

Paull, C.K., Chanton, J.P., Neumann, A.C., Coston, J.A., Martens, C.S., Showers, W., 1992. Indicators of methane-derived carbonates and chemosynthetic organic carbon deposits: examples from the Florida Escarpment. Palaios 7, 361–376.

Paull, C.K., Lorenson, T.D., Boroworski, W.S., Ussler, W., Olsen, K., Rodriguez, N.M., et al., 1999. Isotopic composition of CH_4, CO_2 species, and sedimentary organic matter within ODP Leg 164 sample from the Blake Ridge: gas source implications. Proc. ODP Sci. Results 164, 67–78.

Paull, C.K., Neumann, A.C., am Ende, B.A., Ussler III, W., Rodriguez, N.M., 2000. Lithoherms on the Florida-Hatteras slope. Mar. Geol. 166, 83–101.

Peckmann, J., Goedert, J.L., 2005. Geobiology of ancient and modern methane-seeps. Palaeogeogr. Palaeoclimatol. Palaeoecol. 227, 1–6.

Peckmann, J., Thiel, V., 2004. Carbon cycling at ancient methane-seeps. Chem. Geol. 205, 443–467.

Peckmann, J., Thiel, V., Michaelis, W., Clari, P., Gaillard, C., Martire, L., et al., 1999a. Cold seep deposits of Beauvoisin (Oxfordian; southeastern France) and Marmorito (Miocene; northern Italy): microbially induced authigenic carbonates. Int. J. Earth Sci. 88, 60–76.

Peckmann, J., Walliser, O.H., Reigel, W., Reitner, J., 1999b. Signatures of hydrocarbon venting in a middle Devonian carbonate mound (Hollard Mound) at the Hamar Laghdat (Antiatlas, Morocco). Facies 40, 281–296.

Peckmann, J., Reimer, A., Luth, U., Luth, C., Hansen, B.T., Heinicke, C., et al., 2001. Methane-derived carbonates and authigenic pyrite from the northwestern Black Sea. Mar. Geol. 177, 129–150.

Peckmann, J., Goedert, J.L., Thiel, V., Michaelis, W., Reitner, J., 2002. A comprihensive approach to the study of methane-seep deposits from the Lincoln Creek Formation, western Washington State, USA. Sedimentology 49, 855–873.

Pernthaler, A., Dekas, A.E., Brown, C.T., Goffredi, S.K., Embaye, T., Orphan, V.J., 2008. Diverse syntrophic partnerships from deep-sea methane vents revealed by direct cell capture and metagenomics. PNAS 105, 7052–7057.

Pimenov, N., Savvichev, A., Rusanov, I., Lein, A., Egorov, A., Gebruk, A., et al., 1999. Microbial processes of carbon cycle as the base of food chain of Haakon Mosby mud volcano benthic community. Geomar. Lett. 19, 89–96.

Pinheiro, L.M., Ivanov, M.K., Sautkin, A., Akhmanov, G., Magalhães, V.H., Volkonskaya, A., et al., 2003. Mud volcanism in the Gulf of Cadiz: results from the TTR-10 cruise. Mar. Geol. 195, 131–151.
Pontoppidan, E., 1755. The Natural History of Norway. A. Linde, London.
Premoli Silva, I., Erba, E., Spezzaferri, S., Cita, M.B., 1996. Age variation in the source of the diapiric mud breccia along and across the axis of the Mediterranean Ridge accretionary complex. Mar. Geol. 132, 175–202.
Reeburg, W.S., 2007. Oceanic methane biogeochemistry. Chem. Rev. 107, 486–513.
Reeburgh, W.S., 1976. Methane consumption in Cariaco Trench waters and sediments. Earth Planet. Sci. Lett. 28, 337–344.
Reed, J.K., Pomponi, S.A., Weaver, D., Paull, C.K., Wright, A.E., 2005. Deep-water sinkholes and bioherms of South Florida and the Pourtalès Terrace—habitat and fauna. Bull. Mar. Sci. 77, 267–296.
Rezak, R., Bright, T.J., McGail, D.W., 1986. Reefs and Banks of the Northern Gulf of Mexico: Their Geological, Biological and Physical Dynamics. John Wiley & Sons, New York 259 pp.
Richers, D.M., Jones, V.T., 1986. The 1983 Landsat soil-gas geochemical survey of Patrick Draw area, Wyoming. AAPG Bull. 70, 869–887.
Rice, D.D., Claypool, G.E., 1981. Generation, accumulation and resource potential of biogenic gas. AAPG Bulletin 65, 5–21.
Ritger, S., Carson, B., Suess, E., 1987. Methane-derived authigenic carbonates formed by subduction-induced pore-water expulsion along the Oregon/Washington margin. Geol. Soc. Am. Bull. 98, 147–156.
Roberts, J.M., 2002. The occurrence of the coral *Lophelia pertusa* and other conspicuous epifauna around an oil platform in the North Sea. J. Soc. Underwater Technol. 25, 83–91.
Roberts, H.H., Aharon, P., 1994. Hydrocarbon-derived carbonate buildups of the northern Gulf of Mexico continental slope: a review of submersible investigations. Geo. Mar. Lett. 14, 135–148.
Roberts, J.M., Wheeler, A.J., Freiwald, A., 2006. Reefs of the deep: the biology and geology of cold-water ecosystems. Science 312, 543–547.
Roberts, J.M., Wheeler, A.J., Cairns, S., Freiwald, A., 2009. Cold-Water Corals: the Biology and Geology of Deep-Sea Coral Habitats. Cambridge University Press, Cambridge.
Rogers, A.D., 1999. The biology of *Lophelia pertusa* (Linnaeus 1758) and other deep-water reef-forming corals and impacts from human activities. Int. Rev. Hydrobiol. 84, 315–406.
Rüggeberg, A., Dullo, C., Dorschel, B., Hebbeln, D., 2007. Environmental changes and growth history of a cold-water carbonate mound (Propeller Mound, Porcupine Seabight). Int. J. Earth Sci. 96, 57–72.
Schouten, S., Hoefs, M.J.L., Koopmans, M.P., Bosch, H.-J., Sinninghe Damsté, J.S., 1998. Structural characterization, occurrence and fate of archaeal ether-bound acyclic and cyclic biphytanes and corresponding diols in sediments. Org. Geochem. 29, 1305–1319.
Sackett, W.M., 1978. Carbon and hydrogen isotope effects during the thermocatalic production of hydrocarbons in laboratory simulation experiments. Geochimica et Cosmochimica Acta 42, 571–580.
Schouten, S., Wakeham, S.G., Hopmans, E.C., Sinninghe Damsté, J.S., 2003. Biogeochemical evidence that thermophilic Archaea mediate the anaerobic oxidation of methane. Appl. Environ. Microbiol. 69, 1680–1686.
Sellanes, J., Quiroga, E., Gallardo, V.A., 2004. First direct evidence of methane seepage and associated chemosynthetic communities in the bathyal zone of Chile. J. Mar. Biol. Ass. U.K. 84, 1065–1066.

Sellwood, B.W., 1978. Shallow-water carbonate environments. In: Reading, H.G. (Ed.), Sedimentary Environments and Facies. Blackwell Science Publications, London, pp. 259–313.
Soloviev, V., Ginsburg, G., 1997. Water segregation in the course of gas hydrate formation and accumulation in submarine gas-seepage fields. Mar. Geol. 137, 59–68.
Spiro, B., Roberts, M., Gage, J., Chenery, S., 2000. $^{18}O/^{16}O$ and $^{13}C/^{12}C$ in an ahermatypic deep-water coral *Lophelia pertusa* from the North Atlantic: a case of disequilibrium isotope fractionation. Rapid Commun. Mass Spectrom. 14, 1332–1336.
Squires, D.F., 1958. The Cretaceous and Tertiary corals of New Zealand. Paleontol. Bull. 29, 1–107.
Squires, D.F., 1964. Fossil coral thickets in Wairarapa, New Zealand. J. Paleontol. 38, 904–916.
Stadnitskaia, A., Baas, M., Ivanov, M.K., Van Weering, T.C.E., Sinninghe Damsté, J.S., 2003. Novel archaeal macrocyclic diether core membrane lipids in a methane-derived carbonate crust from a mud volcano in the Sorokin Trough, NE Black Sea. Archaea 1, 165–173.
Stadnitskaia, A., Muyzer, T.G., Abbas, B., Coolen, M.J.L., Hopmans, E.C., Baas, M., et al., 2005. Biomarker and 16S rDNA evidence for anaerobic oxidation of methane and related carbonate precipitation in deep-sea mud volcanoes of the Sorokin Trough, Black Sea. Mar. Geol. 217, 67–96.
Stadnitskaia, A., Bouloubassi, I., Elvert, M., Hinrichs, K.-U., Sinninghe Damsté, J.S., 2008a. Extended hydroxy archaeol, a novel lipid biomarker for anaerobic methanotrophy in cold seepage habitats. Org. Geochem. 39, 1007–1014.
Stadnitskaia, A., Nadezhkin, D., Abbas, B., Blinova, V., Ivanov, M.K., Sinninghe Damsté, J. S., 2008b. Carbonate formation by anaerobic oxidation of methane: evidence from lipid biomarker and fossil 16S rDNA. Geochim. Cosmochim. Acta 72, 1824–1836.
Stakes, D.S., Orange, D., Paduan, J.B., Salamy, K.A., Maher, N., 1999. Cold-seeps and authigenic carbonate formation in Monterey Bay, California. Mar. Geol. 159, 93–109.
Stahl, W.J.,, 1979. Carbon isotopes in petroleum geochemistry. In: Jäger, E., Hunziker, J.C. (Eds.), Lectures in isotope geologySpringer-Verlag, pp. 274–282.
Stanley, G.D., Fautin, D.G., 2001. The origins of modern corals. Science 291, 1913–1914.
Stetson, T.R., Squires, D.F., Pratt, R.M., 1962. Coral banks occurring in deep water on the Blake Plateau. Am. Mus. Novit. 2114, 1–39.
Suess, E., Bohrmann, G., von Huene, R., Linke, P., Wallmann, K., Lammers, S., et al., 1998. Fluid venting in the Aleutian subduction zone. J. Geophys. Res. 103 (B2), 2597–2614.
Stumm, W., Morgan, J.J., 1996. Aquatic Chemistry, Chemical Equilibria and Rates in Natural Waters. 3rd ed John Wiley & Sons, Inc., New York, p. 1022.
Suess, E., Torres, M.E., Bohrmann, G., Collier, R.W., Greinert, J., Linke, P., et al., 1999. Gas hydrate destabilization: enhanced dewatering, benthic material turnover and large methane plumes at the Cascadia convergent margin. Earth Planet. Sci. Lett. 170, 1–16.
Sumida, P.Y.G., Yoshinaga, M.Y., Madureira, L., Hovland, M., 2004. Seabed pockmarks associated with deepwater corals off SE Brazilian continental slope, Santos Basin. Mar. Geol. 207, 159–167.
Taira, A., Hill, I., Firth, J., Berner, U., Brückmann, W., Byrne, T., et al., 1992. Sediment deformation and hydrogeology of the Nankai accretionary prism: synthesis of shipboard results of ODP Leg 131. Earth Planet. Sci. Lett. 109, 431–450.
Taviani, M., Freiwald, A., Zibrowius, H., 2006. Deep coral growth in the Mediterranean Sea, an overview. In: Freiwald, A., Roberts, M. (Eds.), Cold-Water Corals and Ecosystems. Springer, Berlin, pp. 137–156.
Teichert, C., 1958. Cold- and deep-water coral banks. Bull. Am. Assoc. Petrol. Geol. 42, 1064–1082.

Teske, A., Hinrichs, K.-U., Edgcomb, V., De Vera Gomez, A., Kysela, D., Sylva, S.P., et al., 2002. Microbial diversity of hydrothermal sediments in the Guaymas basin: evience for anaerobic methanotrophic communities. Appl. Environ. Microbiol. 68, 1994–2007.

Thiel, V., Peckmann, J., Seifert, R., Wehrung, P., Reitner, J., Michaelis, W., 1999. Highly isotopically depleted isoprenoids: molecular markers for ancient methane venting. Geochim. Cosmochim. Acta 63, 3959–3966.

Thiel, V., Peckman, J., Richnow, H.H., Luth, U., Reitner, J., Michaelis, W., 2001. Molecular signals for anaerobic methane oxidation in Black Sea seep carbonates and microbial mat. Mar. Chem. 73, 97–112.

Thiem, Ø., Ravagnan, E., Fosså, J.H., Bernsten, J., 2006. Food supply mechanisms for cold-water corals along a continental shelf. J. Mar. Syst. 60, 207–219.

Thurman, H.V., Burton, E.A., 2001. Introductory Oceanography, ninth ed. Prentice Hall, Englewood Cliffs.

Titschack, J., Vertino, A., Pino, P., Rosso, A., 2008. Plio-Pleistocene cold-water coral deposits at La Montagna (Messina, Sicily) submarine dune or cold-water coral mound. In: Abstract 26th Regional Meeting of the International Association of Sedimentology (IAS), Bochum, Germany.

Titschack, J., Thierens, M., Dorschel, B., Schulbert, C., Freiwald, A., Kano, A., et al., 2009. Carbonate budget of a cold-water coral mound (Challenger Mound, IODP Exp.307). Mar. Geol. 259, 36–46.

Treude, T., 2003. Anaerobic Oxidation of Methane in Marine Sediments. Ph.D. Thesis, Bremen University, Bremen, Germany.

Tucker, M.E., Wright, V.P., 1990. Carbonate Sedimentology. Blackwell Science, Oxford 482 pp.

Valentine, D.L., Reeburgh, W.S., 2000. New perspectives on anaerobic methane oxidation. Environ. Microbiol. 2, 477–484.

Van Rooij, D., Huvenne, V., Le Guilloux, E., Foubert, A., Wheeler, A., Staelens, P., et al., 2007. Deep-water oyster cliffs at La Chapelle Bank (Celtic Margin). Geophys. Res. Abstr. 9, 08811.

Van Rooij, D., Huvenne, V.A.I., Foubert, A., Blamart, D., Staelens, P., Henriet, J.-P., et al., 2009. The Enya mounds: a lost mound-drift competition. Int. J. Earth Sci. 98, 849–863.

Van Rooij, D., De Mol, L., Le Guilloux, E., Wisshak, M., Huvenne, V.A.I., Moeremans, R, Henriet, J.-P., 2010. Environmental setting of deep-water oysters in the Bay of Biscay. Deep-Sea Res. I 57, 1561–1572. doi:10.1016/j.dsr.2010.09.002.

Vecsei, A., 2004. A new estimate of global reefal carbonate production including the fore-reefs. Glob. Planet. Change 43, 1–18.

Vogt, P.R., Gardner, J., Crane, K., 1999. The Norwegian-Barents-Svalbard (NBS) continental margin: introducing a natural laboratory of mass wasting, hydrates, and ascent of sediment, pore water, and methane. Geo. Mar. Lett. 19, 2–21.

Volgin, A.V., Woodside, J.M., 1996. Sidescan sonar images of mud volcanoes from the Mediterranean Ridge; possible causes of variations in backscatter intensity. Mar. Geol. 132, 39–53.

Von Rad, U., Rösch, H., Berner, U., Geyh, M., Marchig, V., Schulz, H., 1996. Authigenic carbonates derived from oxidized methane vented from Makran accretionary prism off Pakistan. Mar. Geol. 136, 55–77.

Wallmann, K., Linke, P., Suess, E., Bohrmann, G., Sahling, H., Schlüter, M., et al., 1997. Quantifying fluid flow, solute mixing, and biogeochemical turnover at cold vents of the eastern Aleutian subduction zone. Geochim. Cosmochim. Acta 61, 5209–5219.

Wei, G., McCulloch, M.T., Mortimer, G., Deng, W., Xie, L., 2009. Evidence for ocean acidification in the Great Barrier Reef of Australia. Geochem. Cosmochim. Acta 73, 2332–2346.

Wells, J.W., 1977. Eocene Corals from Eua, Tonga. Geological Survey Professional Paper G640, 1–13 and 17–18.

Wendt, J., 1990. The first aragonitic rugose coral. J. Paleontol. 64, 335–340.

Werne, J.P., Baas, M., Sinninghe Damsté, J.S., 2002. Molecular isotopic tracing of carbon flow and trophic relationships in a methane-supported benthic microbial community. Limnol. Oceanogr. 47, 1694–1701.

Wheeler, A.J., Beck, T., Thiede, J., Klages, M., Grehan, A., Monteys, F.X., et al., 2005. Deep-water coral mounds on the Porcupine Bank, Irish margin: preliminary results from Polarstern ARK-XIX/3a ROV cruise. In: Freiwald, A., Roberts, J.M. (Eds.), Cold-Water Corals and Ecosystems. Springer, Berlin, pp. 393–402.

Wheeler, A.J., Beyer, A., Freiwald, A., de Haas, H., Huvenne, V.A.I., Kozachenko, M., et al., 2007. Morphology and Environment of Deep-water Coral Mounds on the NW European Margin. Int. J. Earth Sci. 96, 37–56.

Wheeler, A.J., Kozachenko, M., Masson, D.G., Huvenne, V.A.I., 2008. The influence of benthic sediment transport on cold-water coral bank morphology and growth: the example of the Darwin Mounds, NE Atlantic. Sedimentology 55, 1875–1887.

Wheeler, A.J., Kozachenko, M., Henry, L.-A., de Haas, H., Huvenne, V.A.I., Masson, D. G., Olu-Le Roy, K., 2010. The Moira Mounds, small cold-water coral banks in the Porcupine Seabight, NE Atlantic: an early stage growth phase for future coral carbonate mounds? Mar. Geol 55, doi: 10.1016/j.margeo.2010.08.006.

White, R.S., 1982. Deformation of the Makran accretionary sediment prism in the Gulf of Oman (north-west Indian Ocean). In: Leggett, J.K. (Ed.), Trench-Forearc Geology: Sedimentation and Tectonics on Modern and Ancient Active Margins. Geological Society, London, pp. 351–372 Special Publication 10.

White, M., Mohn, C., Orren, M.J., 1998. Nutrient distributions across the Porcupine Bank. ICES J. Mar. Sci. 55, 1082–1094.

White, M., Mohn, C., de Stigter, H., Mottram, G., 2005. Deep-water coral development as a function of hydrodynamics and surface productivity around the submarine banks of the Rockall Trough, NE Atlantic. In: Freiwald, A., Roberts, J.M. (Eds.), Cold-Water Corals and Ecosystems. Springer, Berlin, pp. 503–514.

White, M., Bashmachnikov, I., Arístegui, J., Martins, A., 2007. Physical processes and seamount productivity. In: Pitcher, T.J., Morato, T., Hart, P.J.B., Clark, M.R., Haggan, N., Santos, R.S. (Eds.), Seamounts: Ecology, Fisheries and Conservation. Blackwell Publishing, Oxford, pp. 65–84.

Wienberg, C., Hebbeln, D., Fink, H.G., Mienis, F., Dorschel, B., Vertino, A., et al., 2009. Scleractinian cold-water corals in the Gulf of Cádiz—First clues about their spatial and temporal distribution. Deep-Sea Res. I 56, 1873–1893.

Williams, T., Kano, A., Ferdelman, T., Henriet, J.-P., Abe, K., Andres, M.S., et al., 2006. Cold-water coral mounds revealed. EOS Trans. Am. Geophys. Union 87, 525–526.

Wisshak, M., López Correa, M., Gofas, S., Salas, C., Taviani, M., Jakobsen, J., et al., 2009a. Shell architecture, element composition, and stable isotope signature of the giant deep-sea oyster *Neopycnodonte zibrowii* sp. n. from the NE Atlantic. Deep-Sea Res. I 56, 374–407.

Wisshak, M., Neumann, C., Jakobsen, J., Freiwald, A., 2009b. The 'living-fossil community' of the cyrtocrinoid *Cyathidium foresti* and the deep-sea oyster *Neopycnodonte zibrowii* (Azores Archipelago). Palaeontol. Palaeoclimatol. Palaeoecol. 271, 77–83.

Wollast, R., 1994. The relative importance of biomineralization and dissolution of $CaCO_3$ in the global carbon cycle. In: Doumenge, F. (Ed.), Past and Present Biomineralization Processes. Considerations about the Carbonate Cycle, 13–36 Bulletin de l'Institute océanographique, Monaco 13.

Woodside, J.M., Ivanov, M.K., Limonov, A.F., 1997. Neotectonics and fluid flow through seafloor sediments in the Eastern Mediterranean and Black Seas: Part II. Black Sea.

Preliminary results of geological and geophysical investigations during the ANAXPROBE/TTR-6 cruise of R/V Gelendzhik, July–August 1996, 226 Intergovernmental Oceanographic Commission Technical Series 48.
Woodside, J.M., Ivanov, M.K., Limonov, A.F., Shipboard scientists of the ANAXIPROBE expeditions, 1998. Henriet, J.-P., Mienert, J. (Eds.), Gas Hydrates. Relevance to World Margin Stability and Climatic Change. Geological Society, London, pp. 177–193 Special Publications 137.
Zachos, J.C., Rohl, U., Schellenberg, S.A., Sluijs, A., Hodell, D.A., Kelly, D.C., et al., 2005. Rapid acidification of the ocean during the Paleocene–Eocene thermal maximum. Science 308, 1611–1616.
Zender, A.J.B., Brock, T.D., 1979. Methane formation and methane oxidation by methanogenic bacteria. J. Bacteriol. 137, 420–432.
Zhang, C.L., Li, Y., Wall, J.D., Larsen, L., Sassen, R., Huang, Y., et al., 2002. Lipid and carbon isotopic evidence of methane-oxidizing and sulfate-reducing bacteria in association with gas hydrates from the Gulf of Mexico. Geology 30, 239–242.
Zhang, C.L., Pancost, R.D., Sassen, R., Qian, Y., Macko, S.A., 2003. Archaeal lipid biomarkers and isotopic evidence of anaerobic methane oxidation associated with gas hydrates in the Gulf of Mexico. Org. Geochem. 34, 827–836.
Zitter, T., 2004. Mud Volcanism and Fluid Emissions in Eastern Mediterranean Neotectonic Zones. Thesis. Free University of Amsterdam, The Netherlands.
Ziveri, P., Young, J.R., Van Hinte, J.E., 1999. Coccolithophore export production and accumulation rates. GeoRes. Forum 5, 41–56.
Knittel, K., Boetius, A., 2009. Anaerobic oxidation of methane: progress with an unknown process. Ann. Rev. Microbiol. 63, 311–334.

CHAPTER 7

Volcaniclastic Processes and Deposits in the Deep-Sea

Steven N. Carey*,[1] *and* Jean-Luc Schneider†

Contents

* Graduate School of Oceanography, University of Rhode Island, Narragansett, Rhode Island, USA
† Université Bordeaux 1, Observatoire Aquitain des Sciences de l'Univers, CNRS-UMR EPOC, Talence Cedex, France
[1] Corresponding author.
E-mail address: scarey@gso.uri.edu

Developments in Sedimentology, Volume 63
ISSN 0070-4571, DOI: 10.1016/S0070-4571(11)63007-0

1. Introduction

Volcanic activity at ocean-spreading centres, large oceanic islands, and especially subduction zones, leads to several types of eruptive and non-eruptive processes that generate large volumes of volcaniclastic sediments, and results in the deposition of fragmental material in the deep-sea. The contribution of these products may reach as high as 5–10% of the total clastic flux on the Earth's surface (Fisher and Schmincke, 1994). Volcaniclastic material forms an important part of sedimentary successions, particularly along active margins of both cordilleran and island-arc environments, and around large volcanic oceanic islands. They record valuable information about sedimentary processes and geodynamic setting through their chemical composition and facies architecture. As a result of their often isochronous character over large areas, they also form important stratigraphic markers and event signals. Moreover, marine volcaniclastic formations can be the host of ore-mineral deposits with economic interest. Finally, volcaniclastic deposits allow stratigraphic reconstruction of past volcanic activity on both local and regional scales, and are thus a key for certain aspects of volcanic-hazards assessment.

This chapter presents an overview of the large diversity of volcaniclastic processes and deposits that can be recognized in deep-marine environments (i.e. beyond the shelf/slope break). In many cases, the origin of volcaniclastic deposits in the deep-sea are linked to volcanic and sedimentary processes that occur sub-aerially, along the littoral zone, and on the shelf. The various aspects of their origin, recognition, and interpretation will be emphasized from the individual depositional interval to the volcaniclastic apron scale.

2. Volcaniclastic Materials: The Evidence of Volcanic Activity

Clastic marine sediments typically reflect many aspects of their sedimentary sources. More than other clastic particles, however, the distinctive morphologies and chemical compositions of volcaniclastic particles allow for the determination of processes that lead to their generation. The terminology for volcaniclastic material used herein is from Fisher (1961, 1966) and Schmid (1981) (Fig. 7.1). The term 'volcaniclastic' is defined to include the entire spectrum of clastic material composed in part, or entirely, of fragments of volcanic composition and origin, formed by any particle-forming mechanism (e.g. pyroclastic, hydroclastic, epiclastic, autoclastic), transported by any mechanism, deposited in any physiographic environment, including

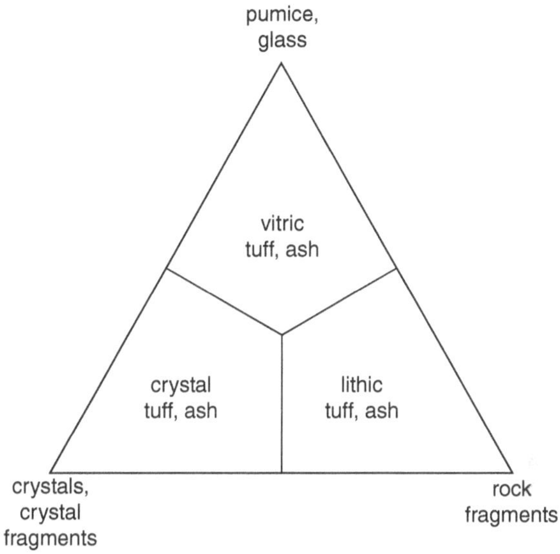

Figure 7.1 Subdivision of volcaniclastic deposits based on their fragmental composition. Reproduced from Schmid (1981) with permission of the Geological Society of America.

marine settings, or mixed with any other volcaniclastic type or with any non-volcanic fragment types in any proportions (tuffites).

2.1. Sub-aerial and sub-marine formation of volcanic particles

Important types of volcaniclastic particles are pyroclastic, hydroclastic, autoclastic, alloclastic, and epiclastic (Fisher and Smith, 1991; Schneider and Fisher, 1996). They are mainly distinguished on the basis of morphological characteristics (Heiken and Wohletz, 1985). Pyroclasts form from gas-rich magma as pressure is reduced during eruptions. When magma ascends towards the surface, dissolved gases, such as water and carbon dioxide, come out of solution to form bubbles that eventually burst and fragment the magma into small pieces (explosive eruption). The concentration and release of gas can vary substantially depending on magma viscosity, volatile content, and the ability of a separate gas phase to move relative to the magma. Low-viscosity basaltic magmas commonly fragment by fire fountaining (sustained discharge of magma spray) or Strombolian activity (intermittent bursting of large gas bubbles that are able to rise faster than the magma-ascent rate). In both of these processes, the magmatic-gas content may be increased by the collection of exsolved bubbles at the top of a magma chamber or by the coalescence of bubbles rising through a slowly ascending column of magma (a.o. Head and Wilson, 2003). Products of this type of activity include a wide range of particle sizes from aerodynamically shaped bombs and scoria clasts to glassy ash-sized material including microscoria (small vesicle-rich particles), hyaloclastites (blocky glassy particles), limu o Pele (quenched bubble-wall fragments), and Pele's hair (thin, hair-like fragments of glass). The common feature of this material is that they either contain abundant vesicles (remnant gas bubbles) or wholly consist of bubble walls from exsolved magmatic gases (Fig. 7.2). In addition, crystals that were present in the magma prior to eruption are included within glassy fragments, or liberated to form separate particles. Non-juvenile clasts consist of fragments of wall rock that become incorporated in the eruptive mixture.

Viscous magmas of andesitic, dacitic, and rhyolitic composition are commonly fragmented more energetically by exsolution of dissolved gases. This is attributed to the generally higher dissolved gas contents of such magmas (Wallace and Anderson, 2000) and the ability of gas-bubble pressure to build to high levels prior to fragmentation due to the viscous confining forces (Sparks, 1978). In these magmas, gas bubbles cannot rise independently of the magma, and exsolution triggers a rapid rise in the viscosity of the remaining melt phase. Fragmentation occurs when the excess bubble pressure exceeds the tensile strength of the magma foam. Such explosive eruptions may be sustained (Plinian or sub-Plinian style) or short-lived releases of excess pressure when a vent or conduit becomes blocked (Vulcanian style). Pumice or quenched magmatic foam is produced dominantly by Plinian to Vulcanian eruptions, often in very large volumes during single events (up to 10^3 km^3). Its density can vary significantly from

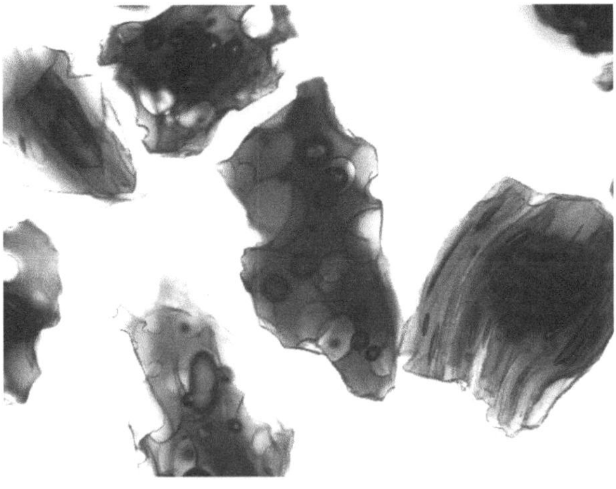

Figure 7.2 Highly vesicular basaltic glass grains from the 1996 Vatnajökull explosive eruption in Iceland. Grains are approximately 125 μm in diameter.

<0.5 to >1.5 g cm^{-3} and thus may either float or sink in water. Eventually, all pumice will sink due to the adsorption of water through small pores in vesicles walls (Whitham and Sparks, 1986), but some may stay buoyant for periods up to several years. The smaller products of highly explosive activity of viscous magma include micropumice, bubble-wall shards, crystals with adhering glass, and lithic fragments (Fig. 7.3).

Pyroclastic particles deposited in marine environments can be produced by both sub-aerial and sub-marine eruptions. Sub-aerial eruptions can produce enormous quantities of pyroclasts that can eventually be deposited in the deep-sea following a variety of transport processes, such as widespread fallout or sediment gravity flows. In contrast, the importance of sub-marine explosive activity is probably underestimated because of the scarcity of direct observations. The presence of pumice rafts, gas bubbling on the sea surface, local sea-water colour changes (Hedervari, 1984; McClelland et al., 1989), and seismic activity (Talandier, 1989) are all direct indicators of sub-marine volcanic explosions. The nature of sub-marine explosions depends upon the water depth (hydrostatic pressure), composition, and volatile content of the ascending magma (Burnham, 1983; Cas, 1992; Head and Wilson, 2003; McBirney, 1963; Stix, 1991; White et al., 2003). Gas expansion and resulting sub-marine volcanic explosions are theoretically possible down to a depth of 3000 m, which corresponds to the critical point of water. Head and Wilson (2003) suggest that explosions could be possible at depths in excess of 3000 m and that a wide range of pyroclastic deposits can be expected. They call upon various methods of volatile concentration within the magma in order to achieve gas contents high enough to drive

Figure 7.3 Highly vesicular dacitic glass shards from the 1991 explosive eruption of the Hudson volcano in Chile. Scale bar is 100 µm.

explosive volcanism at such great depths. The styles of sub-marine explosive activity are dependent on the magma composition, volatile content, discharge rate, and water depth. Potential sub-marine explosive eruption styles include Plinian (sustained: Kokelaar and Busby, 1992; Kano, 2003), Vulcanian, Strombolian, and fire fountaining (Head and Wilson, 2003; Mueller and White, 1992). Deposits of fragmented rhyolitic pumice and basaltic scoria that erupted sub-aqueously have been found in several places in the ocean, at depths between 1500 and 2000 m (Cashman and Fiske, 1991; Fiske et al., 1998; Kato, 1987). In addition, basaltic scoria and vesicular shards have been discovered at depths up to 3800 m (Clague et al., 2003a; Eissen et al., 2003; Fouquet et al., 1998; Gill et al., 1990; Hékinian et al., 2000), providing evidence for deep-water explosive activity. However, it appears that many sub-aqueous explosions resulting from primary gas exsolution preferentially occur at shallower depths, above the suggested maximum volatile fragmentation depth (VFD) (Fisher and Schmincke, 1984).

Hydroclasts form by interaction of magma with external non-juvenile water that produces chilled glass particles by either explosive (hydroclast) or non-explosive (hyaloclast) means (Batiza and White, 2000; Heiken and Wohletz, 1985; Wohletz, 1983). Sub-marine explosive eruptions triggered by the interaction of magma and external water are also controlled by water depth, as fragmentation relies on the expansion associated with the conversion of sea-water to vapour. Kokelaar (1986) has summarized the complexities of these processes and emphasized the potential feedback mechanisms between magma/water interactions and degassing of dissolved volatile components. An important fragmentation process is contact-surface explosivity. It is driven by the cyclic formation and collapse of a vapour layer at the point of contact between water and hot magma. The collapse of the vapour leads to fragmentation and exposure of a new hot surface to initiate a new cycle. As the cycle time is quite short, of the order of microseconds, the process can lead to a sustained explosion if sufficient fuel (magma) and coolant (water) are available in the right proportions (Wohletz, 2003). Experimental studies suggest that initiation of the process requires an external energy trigger that may be provided by gas exsolution or the fluid dynamics of magma discharge (Zimanowski and Buttner, 2003). Most explosive eruptions of this type are thought to occur in water depths less than 1500 m. In relatively shallow water, copious amounts of hydroclastic material can be produced by Surtseyan-style activity such as the recent sub-marine eruptions of Kavachi volcano, Solomon Islands, in 2000. Particles produced by explosive magma/water interactions typically consist of blocky, curviplanar, or mossy glassy particles with some remnant vesicles (Fig. 7.4) (Wohletz, 1983).

Explosive sub-marine fragmentation of magma due to interaction with external water may also occur by bulk interaction steam explosivity (Kokelaar, 1986). In this process, flowing magma locally traps water,

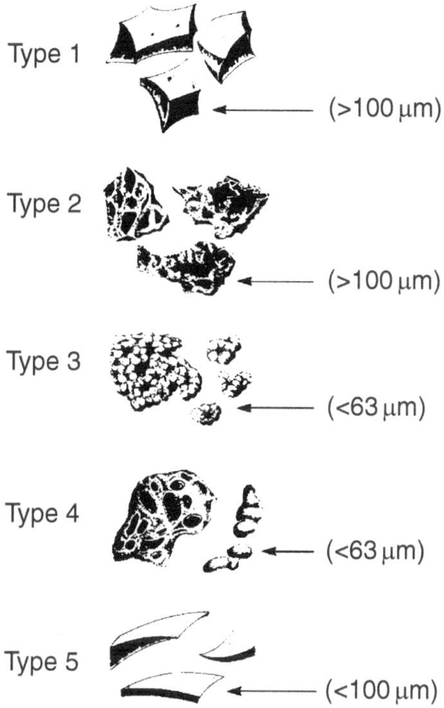

Figure 7.4 Major types of volcanic particles are formed by the interaction of magma with water. Reproduced from Wohletz (1983) with permission from Elsevier.

which is then vaporized by heating. The rapid expansion of the trapped water, as it converts to vapour, fragments the magma in a short-lived release. The water pressure limits the depth at which magma fragments by this process and it is unlikely to be important below ~1500 m.

Non-explosive interactions of magma and water can also generate significant quantities of hyaloclastites by cooling–contraction granulation (Carlisle, 1963; Fisher and Schmincke, 1984; Honnorez and Kirst, 1975; Pichler, 1965). Hyaloclastites form not only during sub-marine effusion of lava but also when lava flows into the sea (Fig. 7.5) (Jones and Nelson, 1970; Moore et al., 1973). Tidal formation of hyaloclastites is enhanced for low-viscosity lava flows, but progression of more viscous flows can be disturbed by phreatic explosions that form rootless littoral cones. Hyaloclastite deltas can develop when lava flows prograde on their coastal hyaloclastite deposits. The sequential evolution of these volcaniclastic deposits can record sea-level fluctuations because hydroclastic fragmentation of the moving lava flows mainly occurs along the shoreline (Fig. 7.5) (Jones and Nelson, 1970). Other hyaloclastite deposits consist of basaltic material and form in association with relatively deep-water eruptions on seamounts

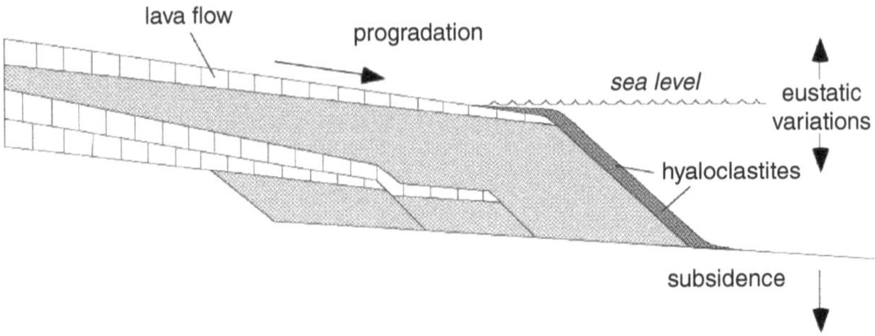

Figure 7.5 Construction of a hyaloclastite delta as a sub-aerial lava flow enters the sea. Modified from Schneider (2000).

(Batiza et al., 1984) and along mid-ocean ridges (Batiza and White, 2000). Hyalocastites of more silicic composition have been recognized in uplifted successions and from some sub-glacial eruptions, but there are few examples of recent deposits from the deep-sea. In addition, the intrusion of magma into water-saturated sediments can result in the formation of hyaloclastite particles in peperites (Busby-Spera and White, 1987; Skilling et al., 2002). Batiza and White (2000) have suggested that the presence of pelagic sediment surrounding deep-sea vents might facilitate the formation of hyaloclastites because of the potential for enhanced mixing and formation of vapour bubbles from heated sea-water.

Autoclastic fragments form by mechanical friction or gravity crumbling of moving viscous lava flows (Kano et al., 1991) and growing domes. Alloclastic fragments form by disruption of pre-existing volcanic rocks by igneous processes beneath the Earth's surface, with or without intrusion of fresh magma.

All of the above clastic types are formed contemporaneously with volcanic activity. They also may be reworked by various dispersal agents and are then termed 'reworked pyroclastic' or 'reworked hydroclastic'.

Epiclastic particles are lithic clasts and crystals derived from weathering and erosion of any type of pre-existing rock. Epiclasts are considered to be volcaniclastic particles if the pre-existing rocks are volcanic. Pyroclasts and hydroclasts are the dominant volcaniclastic particle types that occur in marine settings.

2.2. Significance of volcaniclastic deposits

Volcaniclastic particles, by their morphological characteristics, reveal important information about fragmentation processes and the environment of formation. Pyroclastic particles provide evidence of explosive volcanism and the occurrence of volatile-rich magmas. Hydroclastic particles demonstrate the importance of phreatic and phreatomagmatic explosions related magma/water interactions, or of thermal quenching of the magma or lava flows. In contrast, the chemical composition of the particles reveals the petrological nature of the source magmas and gives insight into the broader geodynamic setting of magma generation such as mid-ocean ridge spreading centres, backarc spreading centres, subduction zones, or hot spots. Moreover, precise geochronological techniques can be applied to these deposits in order to accurately assess the timing and duration of volcanic activity and to correlate volcaniclastic deposits over large areas. The facies characteristics of volcaniclastic accumulations allow reconstruction of key sedimentary processes, mainly gravity-driven, and of interactions and flow transformations (Fisher, 1983) during sub-marine transport. Normal marine sedimentary processes, for example, bottom currents and slope instabilities, can also affect previously deposited volcaniclastic materials and play important roles in the final architectural configuration of deep-water successions.

The high preservation potential in the deep-marine environment means that volcaniclastic deposits are particularly useful for the reconstruction of the history of volcanic activity at source areas along subduction-related basins, oceanic spreading centres, and around oceanic volcanic islands and seamounts.

3. Transport and Deposition of Volcaniclastics to the Deep-Sea

The majority of deep-sea volcaniclastics does not form in close proximity to their final depositional site. In many cases, the particles are formed either sub-aerially or in shallow water and are transported to the deep-sea by two primary processes: fallout or some type of sediment gravity flow. The latter represents the most important mechanism in terms of the overall flux of volcaniclastics to the deep-sea.

3.1. Gravitational settling: fallout

The most effective mechanism for widespread dispersal of volcaniclastic material to deep-sea sediments is a large sub-aerial explosive eruption that generates an atmospherically transported ash plume. Injection of volcanic plumes into the stratosphere favours widespread dispersion of tephra to very large areas of the ocean, like ash fallout during the 1991 eruption of Mt. Pinatubo (Philippines) over the South China Sea (Wiesner et al., 2004). Settling of ash onto the sea surface and eventually to the sea-floor can produce isochronous deposits of fine-grained volcanic ash over tens of thousands of square kilometres (Kennett, 1981). Although these deposits have been interpreted as forming by passive fallout of particles through the water column, Carey (1997) demonstrated that the fallout of ash on the water surface triggers the formation of vertical gravity currents that accelerate the settling of ash. The resulting layers are typically several to tens of centimetres thick, depending on the distance to the source and magnitude of the eruption. Basal contacts with the underlying sediment are usually sharp, and the upper contacts are often bioturbated by burrowing organisms (Fig. 7.6). Thickness and grain size of tephra layers are controlled by the volume of ejected particles, their initial size range, eruptive column height, and prevailing wind strength (Fisher and Schmincke, 1984). In many cases, the layers are normally graded with a concentration of crystals at their base. In distal areas, usually more than 50 km from the source, the principal components are bubble-wall glass shards with minor crystals and lithic components (Fig. 7.3). The distribution patterns of some marine tephra layers indicate that advection by ocean currents can impact the final shape of the depositional area (Ninkovich and Shackleton, 1975).

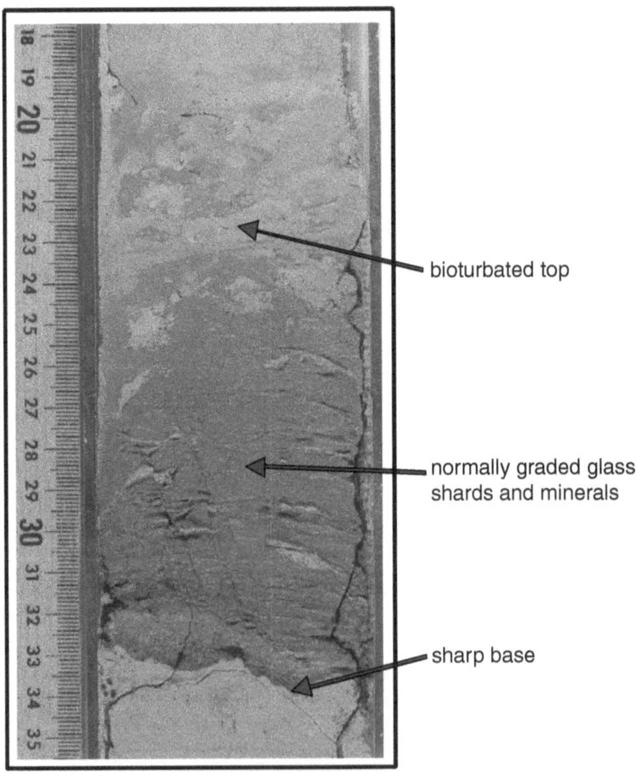

Figure 7.6 Typical volcanic-ash fall layer from ODP Site 999 in the western Caribbean Sea (S. Carey). (A multi-colour version of this figure is on the included CD-ROM.)

Widespread volcanic ash, or tephra layers, in the deep-sea are very useful tools for correlation and dating of volcanic events, determination of eruption frequency, and estimation of erupted volumes (Schmincke and Van den Bogaard, 1991). Keller et al. (1978), for instance, demonstrated that the eruptive chronology of Mediterranean volcanism could be reconstructed from the deep-sea sedimentary record. An excellent example of a widespread marine tephra layer was produced by the 75,000-year BP eruption of the Toba caldera on the island of Sumatra in Indonesia (Buhring et al., 2000; Ninkovich et al., 1978). The eruption discharged over 1000 km^3 of silicic magma and resulted in the deposition of volcanic ash to the west and northeast of Sumatra (Fig. 7.7). A minimum area of $>1,000,000$ km^2 was affected by the fallout and the layer represents an important chronostratigraphic horizon in Indian Ocean deep-sea sediments.

Another method for the dispersal of tephra to deep-sea sediments is by either pumice or ice rafting. Pumice rafts form during shallow sub-marine or sub-aerial explosive volcanism when large quantities of low-density pumice are discharged into the ocean. Pumice may float for periods up to several years and be transported thousands of kilometres by surface currents. For example, following the 1883 explosive eruption of the Krakatau volcano in Indonesia, floating pumice was able to drift all the way across the Indian Ocean and reach the west coast of Africa (Simkin and Fiske, 1983). During transport, some of the pumice may become water-logged and sink

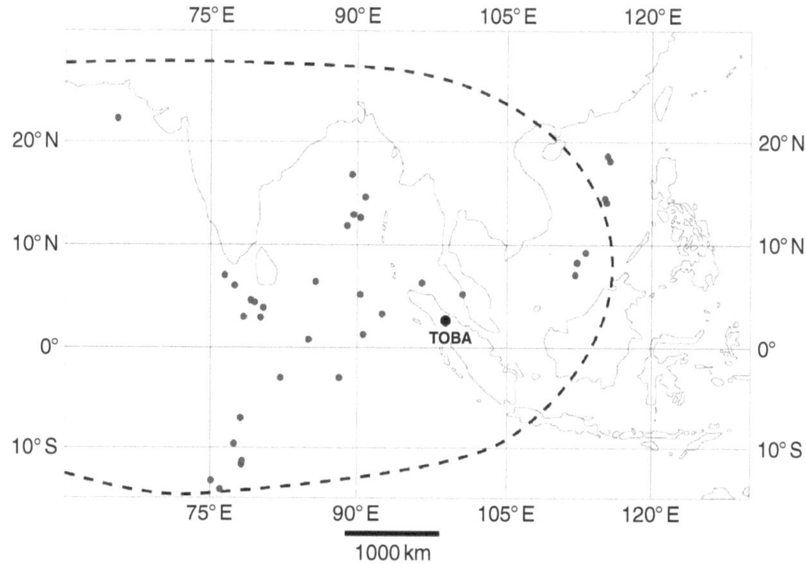

Figure 7.7 Distribution of the Toba ash layer in marine sediments (dashed line) from the Indian Ocean and South China Sea (modified from Buhring et al., 2000).

into the deep-sea. In high latitudes, volcanic-ash fallout may occur on sea ice. Drifting and subsequent melting of the ice can result in the dispersal of tephra to the deep-sea over distances of hundreds of kilometres (Lacasse et al., 1996).

Fallout of tephra may also occur from sub-marine eruption plumes that are generated by explosive eruption of silicic, gas-rich magma. Tephra layers, resulting from this type of activity, typically display bimodal grain-size distributions because of similar settling velocities of water-logged pumice and lithics, and of absence of impact fragmentation of pumice during emplacement (Cashman and Fiske, 1991).

3.2. Sediment gravity flows

3.2.1. Volcaniclastic turbidites

In the sea, the main agents for transport of volcaniclastic material are sediment gravity flows, in particular turbidity currents (Fisher, 1984; Schneider, 2000; Mulder, 2011, this volume, Chapter 2.1). The source material for volcaniclastic turbidity currents can be tremendously variable (Schneider et al., 2001). Virtually all of the fragmentation processes described above can supply material for turbidites that may occur contemporaneously with volcanic activity. These include both sub-aerial and sub-marine explosive eruptions, and the non-explosive quench granulation that typically occurs in deeper water. Such processes lead to primary volcaniclastic turbidites that can be initiated by a number of different syn-eruptive events. The requirements for the formation of turbidity currents include rapid accumulation of volcaniclastic debris, a sloping surface, mixing of sediment with sea-water, and a trigger mechanism to start sediment movement. Turbidity currents are often generated when sub-aerially produced pyroclastics are discharged into the sea by pyroclastic flows (Mandeville et al., 1996), block-and-ash flows (Hart et al., 2004; Picard et al., 2006), lahars, or jökulhlaups (Maria et al., 2000). Pyroclastic flows are non-turbulent to turbulent poly-dispersions of pyroclastic particles sustained by gas that flow rapidly under the influence of gravity (Carey, 1991; Cas and Wright, 1987; Fisher and Schmincke, 1984; Schneider, 1997; Wilson and Houghton, 2000).

In the case of gas-supported particle flows such as pyroclastic flows and block-and-ash flows, the formation of turbidites requires the replacement of interstitial high-temperature gas with water. This mixing and replacement must occur for both sub-aerially and sub-aqueously generated high-temperature flows. Direct observations indicate that, when hot pyroclastic flows enter water, phreatic secondary explosions related to rapid heat transfer between hot pyroclastic material and water can occur (Legros and Druitt, 2000; Mandeville et al., 1996; Sigurdsson et al., 1991; Walker, 1979; Young et al., 1997). Evidence of such hydroclastic processes has been recognized in ancient formations (Wright and Coward, 1977). Littoral explosions favour the mixing of pyroclasts with sea-water and subsequent replacement of interstitial gas by the water (Freundt, 2003; Freundt and Schmincke, 1998; Legros and Druitt, 2000; Walker, 1979). Flow transformations also occur during the entrance of pyroclastic flows into the sea (Freundt, 2003; Schneider et al., 2004). The entrance into the water of block-and-ash flows is accompanied by elutriation of fine-grained particles (Schneider et al., 2004), and when volcaniclastic debris flows (e.g. lahars or jökulhlaups) enter the sea, the flows are typically diluted and transformed into a turbidity current (Lacasse et al., 1996; Maria et al., 2000).

Pyroclastic flows can also originate from sub-marine eruptions (Fiske and Matsuda, 1964; Kokelaar and Busby, 1992). Schneider et al. (1992) have described deposits in the Lower Carboniferous of southern Vosges (NE France) and in the Upper Cretaceous of Anatolia (Turkey) that correspond to welded ignimbrites inter-bedded in marine formations, and that were emplaced under sub-aqueous conditions. These sub-marine ignimbrites display lateral variations that grade from proximal coarse welded breccia deposits to distal lapilli tuffs, consisting of two depositional intervals. The base is normally graded, they are partly welded in their median part, and often columnar jointed. The upper parts of the successions are planar-laminated fine ash tuffs, likely emplaced by dilute volcaniclastic turbidity currents.

Another important source of material for turbidity currents is the reworking of unconsolidated volcaniclastic material previously deposited in the marine environment. The resulting deposits are termed 'secondary volcaniclastic turbidites'. In this case, the remobilization, transport, and emplacement mechanisms are purely sedimentary. Volcano-tectonic activity and associated seismicity are the main triggering processes of secondary volcaniclastic turbidity currents.

Sub-aerial and sub-marine erosion of consolidated volcanic formations can also generate epiclastic particles that can be transported to the deep-sea by turbidity currents. The contribution of epiclastic volcaniclastics to deep-sea sediments has not been thoroughly studied in many volcanic environments, but may contribute a significant volume flux in some areas.

*3.2.1.1. **Primary turbidites*** The sequences of many volcaniclastic turbidites are analogous to those of classical turbidites and their thicknesses range from a decimetre to several metres (Wright and Mutti, 1981). Commonly, the sequences display two distinct divisions (Fig. 7.8). The base is rich in coarse material, generally normally graded, and can contain an important fraction of non-juvenile, even non-volcanic material. Differences exist in grading depending on the density of pumiceous components (e.g. symmetric grading; Cole and DeCelles, 1991; Cole and Stanley, 1994; Schneider and Fisher, 1996) (Mulder, 2011, this volume, Chapter 2.2.5). Fluid-escape

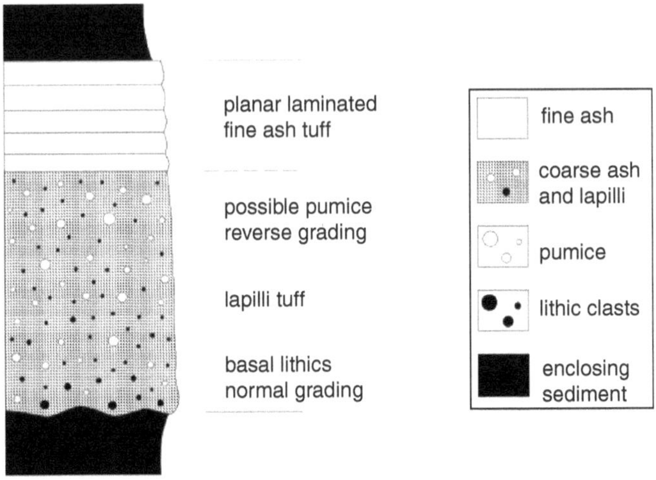

Figure 7.8 Subdivisions of a typical primary volcaniclastic turbidite deposit. Modified from Schneider (2000).

structures can be present in the lower part of the deposits that are similar to those observed in deposits related to fluidized flows (Lowe, 1982). Often the base of the deposits contains rip-up clasts of underlying sediment, illustrating the erosive nature of the turbidites.

The upper division is constituted of fine-grained material (ash ± clay) with planar lamination. Current-ripple structures can be present, but less abundant than in classical turbidites. The top of the deposit grades upwards into overlying sediment and is sometimes bioturbated. The enrichment in vitric, or glassy particles in the top of the turbidites is related to elutriation that occurs from the dense base and head of the flow during transport (Sparks and Wilson, 1983). Consequently, the base of the deposit is enriched in coarser and denser particles.

Identification of primary volcaniclastic turbidites requires a clear link between the deposit and a syn-eruptive event, such as pyroclastic-flow discharge into the sea. The relatively homogeneous chemical composition of vitric particles within a single turbidite argues in favour of a syn-eruptive origin, suggesting that the pyroclasts were emitted during a single event. Mixing of primary, homogenous volcaniclastic material with epiclastic particles from the substratum supports the existence of mixing and erosion during transport. Thus, primary volcaniclastic turbidites can contain significant proportions of non-volcanic material (Sigurdsson et al., 1980; Whitham, 1989). In the case of turbidites derived from jökulhlaups, the volcaniclastic particles will be essentially reworked hydroclasts that were formed contemporaneously with a sub-glacial eruption or remobilized from previous deposits on jökulhlaup paths (Maria et al., 2000).

3.2.1.2. Secondary turbidites The syn-eruptive volcaniclastic deposits accumulated in the littoral zone or on the shelf by various eruptive processes can be subsequently reworked by exclusively sedimentary processes (Fig. 7.9). Reworking occurs in conjunction with volcanic eruptions or during non-volcanic periods. Redistribution of unconsolidated deposits occurs mainly by sedimentary gravity flows but also by bottom currents (Faugères and Mulder, 2011, this volume, Chapter 3). Remobilization can also result from the weathering and erosion of pre-existing rocks in emerged volcanic domains, producing epiclastic material.

Depositional intervals of secondary volcaniclastic turbidites are very similar to those of primary ones, but the particles may display a large compositional spectrum because of mixing. Individual sequences can be much thicker than primary volcaniclastic turbidites, because the volume of the remobilized material is independent of the erupted volumes during single eruptions. Secondary turbidites are not necessarily contemporaneous with eruptive phases; some can be coeval to volcanic activity because remobilization is induced by primary volcaniclastic deposit overload, or contemporaneous volcano-seismic activity. Reworking can occur for an extended period following initial sedimentary accumulation. Large-scale mass-wasting processes of volcanic material have been recognized on sub-marine volcaniclastic aprons, island arcs, and continental active margins.

An example of volcaniclastic reworking is given by the Miocene Obispo Formation exposed along the south-central coast of California (Schneider and Fisher, 1996). In this formation, about 250–450 km^3 of rhyodacitic pyroclastic material that was likely erupted from a caldera complex was deposited in a marine environment. Two distinct facies were recognized in these pyroclastic deposits. (1) A bedded facies that is composed of decimetres to meter thick turbiditic sequences. The sequences have a coarse-grained symmetrically graded base (with centimetre-size pumice fragments) that grades upwards into a planar- and ripple-laminated division with a very

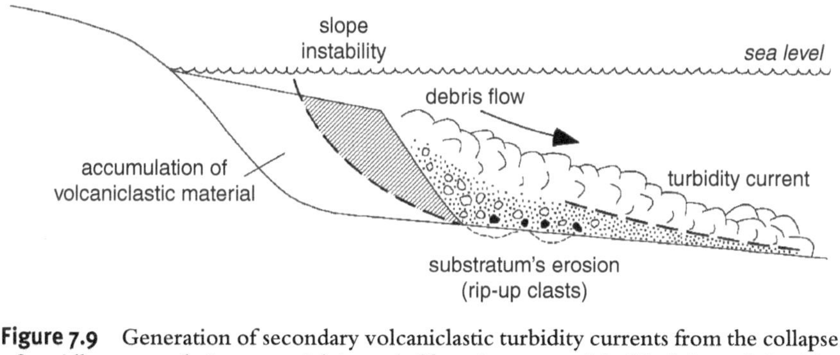

Figure 7.9 Generation of secondary volcaniclastic turbidity currents from the collapse of rapidly accumulating material in a shelf environment. Modified from Schneider (2000).

fine bioturbated ash tuff on top. (2) A more proximal massive facies that forms thick plurimetric coarse deposits that contain packets of disorganized bedded tuff. Spectacular deformation structures (slump structures, brecciated layers, large rip-up clasts pulled out from the underlying bedded sediments and folded within the massive tuff) are exhibited beneath or within the base of the massive tuff, and indicate that the moving material was water-saturated and poorly indurated. The characteristics of the massive facies strongly suggest reworking of the turbiditic deposits of the bedded facies.

3.2.2. Debris flows

Debris flows are a common method of sediment transport in many volcanic environments adjacent to marine basins (e.g. Ballance and Gregory, 1991; Cas et al., 1981; Mulder, 2011, this volume, Chapter 2.1.3.5). An important difference between volcaniclastic and non-volcaniclastic debris flows is that the former are usually poorer in clay-size material compared to the latter (Smith, 1986). This can affect the particle support mechanisms by reducing the effectiveness of the cohesive matrix strength in volcaniclastic debris flows.

Deposition of debris flows occurs predominantly by en masse freezing of the moving flow as the shear force can no longer overcome the flow strength (Mulder, 2011, this volume, Chapters 2.2.3 and 2.3.4). Cas et al. (1981) described a thick succession of deep-water volcaniclastic deposits that resulted from turbidity currents, grain flows, debris flows, and avalanches being derived from an emergent explosive volcanic centre. Massive units up to tens of metres thick, consisting of quartzo-feldspathic material with little grading, are likely to represent deposition from cold debris flows that formed by rapid influx of pyroclastics into the marine environment from a sub-aerial source.

In the marine environment, debris flows can have significant mobility. To the west of the island of Hierro (Canary Islands) is a volcaniclastic debris-flow deposit (Canary flow) that extends for about 500 km to the Madeira Abyssal Plain (Masson, 1996). The flow deposit is correlated to a catastrophic failure of the northern flank of Hierro island that produced the El Golfo debris avalanche. Such deep-sea deposits thus provide an important record of catastrophic volcanic island collapse that occurs in shallow water and may have significant volcanic-hazard implications.

3.2.3. Sub-marine pyroclastic flows

A unique type of high-concentration volcaniclastic gravity-flow deposit found in marine environments is produced by sub-marine pyroclastic flows. These are deposited at high temperature with an interstitial gas phase instead of water (Yamada, 1984). They can be generated by entrance of pyroclastic flows into the sea from the collapse of a sub-aerial eruption, as happened during the 1883 Krakatau eruption in the Sunda Strait of Indonesia or the recent eruptions of the Soufrière Hills volcano (Hart et al.,

2004). In the case of the Krakatau eruption, the pyroclastic flows decoupled between a denser base that continued their path sub-aqueously along the sea-floor and emplaced at high temperature (Mandeville et al., 1994), and a less dense diluted and turbulent top (surge) that travelled sub-aerially over the sea surface (Carey et al., 1996; Mandeville et al., 1996; Sigurdsson et al., 1991). Several examples of sub-marine pyroclastic flows, some welded, have been documented from uplifted successions (Cole and Decelles, 1991; Howells et al., 1979; Kano, 1990; Schneider et al., 1992). The deposits exhibit characteristics such as thick massive units (tens of metres), an abundance of juvenile vesicular clasts, compositional homogeneity, and evidence for high-temperature emplacement, similar to their sub-aerial counterparts. Cas and Wright (1991) argue that the majority of true sub-marine pyroclastic flows is limited to shallow-water environments, and that the mixing of hot flows with water would severely limit transport to deep water. However, sub-marine pyroclastic flows can also form from deep sub-marine eruption of volatile-rich magma associated with high magma discharge. Kokelaar and Busby (1992) proposed that emplacement of high-temperature pyroclastic flows can occur without significant mixing of ambient sea-water if the flows have sufficient thickness, velocity, and abundance of fine ash.

3.2.4. Sub-marine volcanic debris avalanches

Gravity instabilities are important processes that occur on sub-marine slopes, both along passive and active margins (cf. Mulder, 2011, this volume; Le Friant, 2001; Shaller, 1991). The process begins with a large-scale rock or sediment slide that rapidly transforms into a fast moving debris avalanche. Volcanic debris avalanches (VDAs) are large-volume ($>10^6$ m^3) rapid granular gravity flows of rocky masses that result from large-scale destabilizations and collapse of the flanks of volcanic edifices. They are not necessarily related to eruptive episodes. Debris-avalanche generation accompanies the evolution of most volcanoes, particularly stratovolcanoes, and these events limit their vertical growth. These processes are well known on volcaniclastic aprons around sub-aerial volcanoes (Glicken, 1991; Schneider, 1997). During the last decade, numerous oceanographic surveys have also revealed the importance of debris-avalanche sedimentation in sub-marine settings around volcanic islands, seamounts, and volcanic arcs as well (Deplus et al., 2001; Holcomb and Searle, 1991; Krastel et al., 2001; Labazuy, 1996; Le Friant, 2001; Le Friant et al., 2003; Moore and Mark, 1992; Moore et al., 1989, 1994, 1995; Takahashi et al., 2002).

One of the most intriguing aspects of VDAs is their abnormally high mobility based on their effective friction coefficient, $\mu = H/L$, where H is the initial fall height and L the runout length. Among all VDAs recognized on Earth, sub-marine VDAs are the most voluminous and the most mobile, with effective friction coefficients of $0.03 < H/L < 0.13$ (Shaller, 1991). The volumes of the resulting deposits can reach several thousand cubic kilometres. One of the largest documented examples is the Alika debris-avalanche deposit off the island of Hawaii, with a volume of about 2000 km^3 (Lipman et al., 1988).

The destabilization that leads to an initial rockslide is frequently co-seismic on volcanic islands like in Hawaii (Moore et al., 1994, Takahashi et al., 2002). Large-scale rotational landslides can also affect oceanic island flanks, but these movements are typically slow and incremental. Their thickness can be up to 10 km for a length of over 100 km. Parts of these large landslides can collapse to form sub-marine debris avalanches. Remnant scarps from collapse events can be found in both the sub-aerial and sub-marine environment (Lipman et al., 1988), and often exhibit a horse-shoe shaped outline. In addition to the ability for moving large quantities of sediment to the deep-sea, an important aspect of VDAs is the potential for the generation of large-scale tsunamis (Ward, 2001).

During transport, debris avalanches undergo a dynamic disintegration that leads to a progressive grain-size reduction (Pollet and Schneider, 2004) and to spreading out of the material. Sub-marine transport is favoured by interstitial water that saturates the moving mass, and efficiently contributes to the reduction of the internal friction. The presence of water also induces formation of turbidity currents in distal areas by dilution of the fine-grained particles of the avalanche with sea-water (Garcia, 1996). Strong erosion of the sediments of the substratum occurs at the base of most debris avalanches, and leads to sediment ingestion. Consequently, deposits often display a larger volume than the missing volume associated with the initial rockslide scarp in the VDA source area (Deplus et al., 2001). In some instances, disintegration does not occur, and slump deposits have been recognized, as around Réunion Island in the Indian Ocean (Labazuy, 1996).

Deposits from VDAs are chaotic and usually display large, internally fractured blocks within a finer-grained matrix (Glicken, 1991). Deposits form chaotic fan-shaped lobes. The thickness ranges from 0.05 up to 2 km, like around Hawaii (Lipman et al. 1988). The presence of large blocks leads to a characteristic hummocky surface of the deposits. This feature often appears better developed in distal parts of the deposits. This morphology supports the existence of dynamic disintegration during transport. VDA depositional fans can develop by successive events like off the shores of Réunion Island (Cochonat et al., 1990; Labazuy, 1996) and in the Lesser Antilles (Deplus et al., 2001). In proximal areas, slumped benches can be observed, whereas VDAs deposits spread more distally. The latter contain large blocks that can reach up to several kilometres in size. These chaotic deposits are poorly sorted, and blocks float within a silty to sandy matrix, as revealed by direct observations of the sea-floor (Labazuy, 1996; Yokose and

Lipman, 2004). Block/matrix-volume ratios allow the distinction between the block and matrix facies as for sub-aerial VDAs deposits (Glicken, 1991).

The characteristics of sub-marine VDAs deposits can be recognized in the modern oceans through oceanographic survey techniques. Swath bathymetry and side-scan sonar data allow recognition of the hummocky topography of these deposits (Fig. 7.10). Transects using a 3.5-kHz sediment echosounder reveal hummocky surfaces exhibiting a characteristic hyperbolic acoustic facies. On seismic lines, VDAs deposits can be identified by characteristic seismic-reflection facies (Deplus et al., 2001). Proximal deposits display highly chaotic facies with incoherent areas containing few internal reflections that may be bordered by more continuous reflections. However, distal deposits are represented by more continuous reflectors, although chaotic facies remain. The hummocky and/or chaotic sequences of reflections thin with distance from the source, and onlap or downlap underlying reflectors (Leslie et al., 2002). Block reflectivity is very high in

Figure 7.10 Side-scan sonar image of the sea-floor off of Oahu and Molokai, showing characteristic hummocky topography associated with a debris avalanche. Reproduced from Moore et al. (1989) with permission from the American Geophysical Union.

comparison to that of the matrix on side-scan sonar data (Cochonat et al., 1990; Labazuy, 1996). Remotely operated vehicles (Mulder et al., 2011, this volume, Chapter 1.3.4) also allow interesting detail observations of sub-marine VDAs deposits, and the distinction between various depositional facies (Yokose and Lipman, 2004).

Some large-scale volcaniclastic gravity-flow deposits in uplifted marine successions have been interpreted as resulting from the entrance of sub-aerial VDA into the sea. The Parnell Grits, recognized in Miocene formations of New Zealand (0.06 km^3) (Ballance and Gregory, 1991), correspond to a succession of very poorly sorted breccia layers that contain large rip-up blocks (up to 90 m long). These units were produced by flows that travelled at least 40 km from source. Because of their considerable momentum, these debris avalanches were able to strongly erode the substratum. Ballance and Gregory (1991) suggest that entrance of VDAs into the sea was accompanied by water ingestion and transformation of the avalanche into other types of high-concentration gravity flows such as debris flows and high-concentration grain flows. The ingestion of water by sub-aerially generated VDAs is an important process for generating a spectrum of flow types in the marine environment (Schneider et al., 2004).

3.3. Reworking of sub-marine volcaniclastic deposits

Sediment gravity flows are the principal dispersal agents of volcaniclastic material in many sub-marine settings, although other processes can lead to the transport and re-deposition of particles. Once emplaced, volcaniclastic deposits are subjected to all of the effects of normal sedimentary processes that occur in the deep-sea. Processes such as secondary gravity flows (Mulder, 2011, this volume, Chapter 2.1.3), bottom currents, and hemi-pelagic sedimentation (Hüneke and Heinrich, 2011, this volume, Chapter 6), induce transformations of primary volcaniclastic deposits (Stow et al., 1998). Some of them involve down-slope reworking, transport, and redeposition away from the original eruption site or location of primary sedimentation (Busby-Spera, 1985; Carey and Sigurdsson, 1984; Fisher, 1984). Consequently, they are not favourable for the preservation of volcaniclastic deposits in marine sedimentary successions. Moreover, all these processes produce a strong facies convergence among volcaniclastic products of distinct origin.

For instance, bottom currents (Faugères and Mulder, 2011, this volume, Chapter 3) can remobilize primary volcaniclastic deposits and mix them with other deep-sea sediments (Bahk et al., 2005). Moreover, bottom currents form a variety of distinctive sedimentary structures during reworking such as current ripples or crag-and-tail structures (Wright, 2001; Faugères and Mulder, 2011, this volume, Chapter 3.3.4). Bioturbation can mix tephra layers with non-volcaniclastic sediment, and also leads to structureless deposits. In highly oxygenated areas, the biological activity does not favour the preservation of thin volcanic-ash layers. However, anoxic marine basins provide the ideal environments for the preservation of extremely detailed records of volcaniclastic deposition (a.o. Scasso, 2001).

Sediment fluxes during post-primary emplacement and post-eruptive processes are difficult to establish. However, it seems reasonable to consider that syn-eruptive sedimentation and episodic post-eruptive gravity flows involve a high flux, whereas post-eruptive transport by traction load is continuous and generates a low flux (Wright, 1996).

4. Volcaniclastic Contribution to Marine Sedimentation

In marine basins adjacent to active margins and oceanic volcanic islands (Carey, 2000; Carey and Sigurdsson, 1984; Marsaglia, 1995; Smith and Landis, 1995), a strong interaction between volcanic activity and marine sedimentation occurs (Carey, 2000; Cas and Wright, 1987; Fisher, 1984; Orton, 1996; Schneider, 2000; Suthern, 1985). Volcaniclastic deposition occurs intermittently within the background non-volcanic sedimentation of the local environment or, at a larger scale, in the sedimentary basin. Consequently, most volcaniclastic deposits appear as event intervals in the sedimentary record. The facies evolution is influenced by these interactions, and by two important factors: (1) sediment-flux rate (related to the magmatic productivities of the source region), and (2) flow-transformation processes that can affect volcaniclastic gravity flows (Fisher, 1983), the most important dispersal mechanisms in the marine environment.

The volcanic influence on sedimentation in the deep-sea can be direct or indirect, and the sedimentary record reflects temporal changes in clastic supply rates and pathways. The architecture of the deposits in volcaniclastic successions is largely influenced by the location of the volcanic source(s) and the nature of the eruptive activity. In general, volcanism and its related deposits are highly discontinuous through time (Cas and Wright, 1987). Direct volcaniclastic sedimentation is contemporaneous with eruptive periods (syn-eruptive periods). In marine settings, important deposits result from ash fallout that is influenced by dominant winds. A large part of the sub-marine volcaniclastic deposits in volcanic areas is related to pyroclastic flows, which can be sub-aerial or sub-marine in origin (Schneider et al., 1992, 2001). Volcaniclastic deposits that are directly related to sub-marine eruptions generally occur in the vicinity of their volcanic sources and their dispersal may be more restricted. Indirect volcaniclastic-sedimentation results from the reworking of previously deposited volcanic material. It occurs both during syn-eruptive and inter-eruptive periods.

Over the past several decades, the number of publications devoted to sub-marine volcaniclastic processes has increased dramatically. It is now possible to get a much clearer picture of the type, abundance and depositional processes of volcaniclastic deposits in deep-sea environments surrounding mid-ocean ridges, oceanic hotspot islands, and convergent margins (island arc and continental subduction zones). In Section 5, we discuss examples from these environments with the objective of building a set of distinguishing characteristics for the identification of their deposits in both the modern ocean and older, uplifted marine sections.

5. Volcaniclastic Sedimentation in Various Deep-Sea Environments

In the first part of this chapter, we presented individual volcanic and volcaniclastic-sedimentation processes in a generic framework. In this section, we consider how the style of volcanism and configuration of volcanic centres can influence volcaniclastic successions by examining the types of volcaniclastic deposits found in modern deep-sea environments adjacent to mid-ocean ridges, large oceanic islands such as Hawaii, and subduction zones. The discussion will be supplemented by evidence from uplifted marine successions where the volcanic environment can be inferred and the three-dimensional architecture of deposits is often better revealed. In addition, the observations of deposits will be related to recent theoretical models for the nature of sub-marine eruptions in both shallow- and deep-water environments.

5.1. Volcaniclastic deposits related to mid-ocean ridge volcanism

Mid-ocean ridges represent the sites where roughly three quarters of the Earth's annual volcanic output take place. They are almost wholly sub-marine with depths usually over 1500 m. Volcanism is concentrated in the vicinity of the ridge crest in an area referred to as the 'neovolcanic zone'. The shape of the ridge crest varies considerably around the world and is related to the rate at which the plates are spreading apart. Slow-spreading ridges (1–4 cm per year), such as the Mid-Atlantic Ridge, consist of a well-defined axial valley, usually 8–20 km wide, surrounded on each side by steep, fault-bounded ridges that rise up to 1–2 km above the axial valley floor. Intermediate spreading centres (4–8 cm per year), such as the Juan de Fuca Ridge in the northwest Pacific, have a smaller axial valley at the ridge crest that varies from 1 to 5 km in width and is flanked by fault scarps of 50–1000 m high. In contrast, fast-spreading ridges (8–16 cm per year), such as the East Pacific Rise, have no large axial valley at the ridge crest. Instead, a very narrow depression, usually only 5–40 m deep and 40–250 m wide, is the main expression of the neovolcanic zone. These neovolcanic zones are areas of potential accumulation of volcaniclastic deposits.

In general, the majority of volcanic activity along mid-ocean ridges is effusive and occurs along fissures, with emplacement of tholeiitic basalt as pillow, sheet, and lobate lava flows (Perfit and Chadwick, 1998). Fragmentation of glassy pillow crusts by thermal quenching can lead to the formation of basaltic hyaloclastites (Honnorez and Kirst, 1975) or breccias (Carlisle, 1963) that are commonly closely associated with pillow flows. Hyaloclastite often forms the matrix between lava tubes and bedded deposits, and can also be reworked by gravity flows and then redeposited within topographic lows. Significant quantities of hyaloclastites have been identified from drilling of 13-million years old ocean crust at DSDP site 396B, east of the Mid-Atlantic Ridge (Schmincke et al., 1978). The origin of these thick units (90 m) had been attributed to spallation of pillow rinds and granulation of lavas. Head and Wilson (2003) have recently suggested that they may instead be due to localized Hawaiian-style eruptions with emplacement of material by sediment gravity flows formed by collapsing mixtures of hyaloclastites and water.

Compared with other volcanic environments, mid-ocean ridge volcanism produces a relatively narrow range of magmatic compositions. Mid-ocean ridge basalt (MORB) is the most common product over a large part of the ridge system, with lesser amounts of enriched E-MORB, a variety of MORB with elevated levels of incompatible elements and distinct isotopic signatures, or more alkalic basalt produced in areas that are influenced by mantle plumes. Because of the low volatile contents of most MORB magmas, it has generally been assumed that explosive volcanism driven by primary gas exsolution would occur only very rarely in the deep-water ridge environment. However, recent studies have identified several examples of basaltic volcaniclastics that are likely linked to deep sub-marine explosive eruptions along mid-ocean ridges (e.g. Clague et al., 2003a; Eissen et al., 2003; Fouquet et al., 1998; Hékinian et al., 2000).

Along the mid-Atlantic ridge, south of the Azores, Fouquet et al. (1998) and Eissen et al. (2003) documented the existence of basaltic volcaniclastics along three ridge segments that span a depth range of 500–1700 m (Fig. 7.11). The deposits contain highly vesicular glass, Pele's hair, curved bubbles walls, and dense, low vesicularity sideromelane clasts. Most of the units consist of mm- to cm-scale, poorly sorted beds with relatively sharp contacts. Outcrop thicknesses range from up to 400 m along the shallowest ridge segment to only a few metres in the deepest segment. With increasing water depth, the observed areal distribution of the volcaniclastics also decreased from 65 to 2 km^2. Eissen et al. (2003) suggest that the highly vesicular volcaniclastics were produced by deep-water explosive eruptions (Fig. 7.14). Fragmentation of the magma was the result of the release of

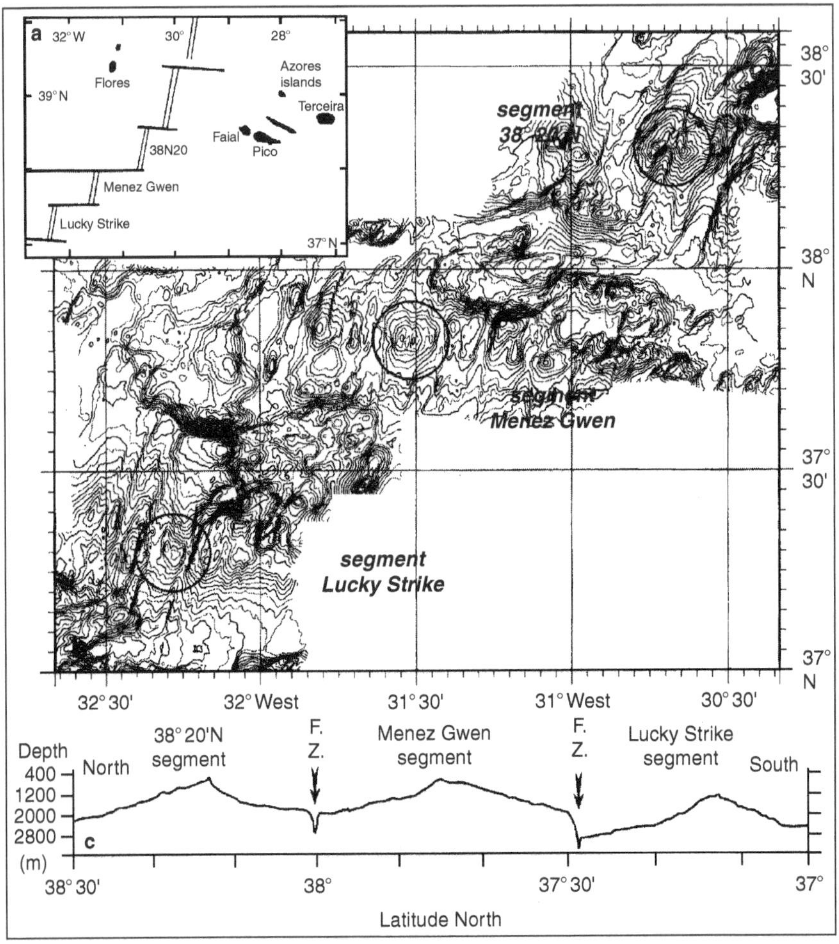

Figure 7.11 Bathymetry of the Mid-Atlantic Ridge, south of the Azores Islands, where explosively derived volcaniclastics were recovered. Reproduced from Eissen et al. (2003) with permission from Elsevier.

CO_2 and H_2O from enriched E-MORBs that had retained most of their juvenile gas content. As disruption of the magma occurred by primary gas-bubble bursting, additional fragmentation and modification of particles was caused by bulk magma/water and surface magma/water interactions. Deposition of material from the explosive eruptions occurred by fallout from a convective plume developed over the vent and from sediment gravity flows generated by collapse of dense portions of the eruption column (Fig. 7.12). Inter-beds of calcareous sediment within the bedded volcaniclastics indicate that volcanism occurred over an extended period of time and numerous individual eruptive events.

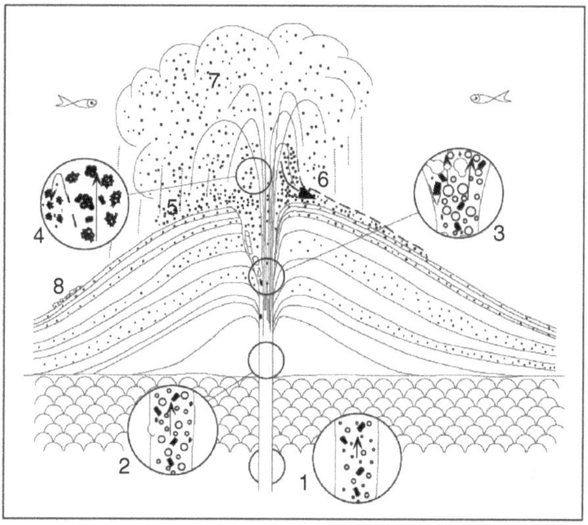

Figure 7.12 Model of a submarine explosive eruption (modified from Eissen et al., 2003). (1) Magma rises and begins to exsolve gas; (2) bubbles expand as pressure is decreased and some lithic material is incorporated; (3) bubbles coalesce and more incorporation of lithic material take place; (4) fragmentation of bubble-rich magma; (5) fallout of fragmented clasts through the water column; (6) generation of sediment gravity flows; (7) finer particles dispersed above the vent; (8) reworking of material by gravity flows.

Head and Wilson (2003) re-interpreted the origin of the volcaniclastics described by Fouquet et al. (1998) and Eissen et al. (2003) by suggesting that additional gas accumulation played a role in creating the conditions necessary for the deep sub-marine explosive eruptions. Deposits found along the shallower parts of the ridge were attributed to Hawaiian fire-fountaining activity caused by the accumulation of CO_2-rich foam at the top of the magmatic reservoirs. Explosive eruptions in the deeper sections of the ridge were attributed to sporadic Strombolian-style activity triggered by a high rise rate and coalescence of bubbles.

Highly vesicular volcaniclastic deposits have also been recovered along the mid-Atlantic ridge at depths from 1500 to 1700 m near 34°50′N (Hékinian et al., 2000). Vesicular hyaloclastite deposits up to 15 m thick with crude graded bedded were found along the crest of a volcanic crater (Fig. 7.13). The composition of the glassy shards corresponds to the most enriched alkali basalts found in this ridge area. Hékinian et al. (2000) suggest that formation of the highly vesicular particles was also by deep explosive eruptions. However, it was proposed that the highly explosive nature of the eruptions required some mechanism to concentrate the volatile content of the magma. They suggested that gas bubbles accumulated within a magma

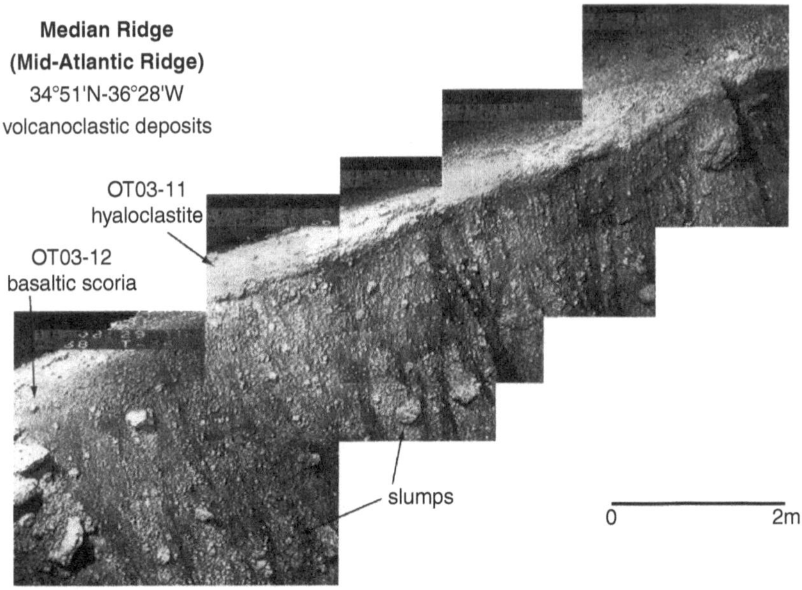

Figure 7.13 Hyaloclastite deposit overlying a pyroclastic deposit along the Mid-Atlantic Ridge, south of the Azores, at a depth of 1600 m. Reproduced from Hékinian et al., 2000 with permission from Elsevier.

column that was blocked at the surface by some type of plug. An explosive event occurred when a sufficiently large pocket of gas had accumulated and raised the pressure in excess of the strength of the conduit blockage.

Evidence of even deeper explosive fragmentation of magma along mid-ocean ridges was presented by Clague et al. (2003a) for the Gorda Rise in the northeast Pacific Ocean. Submersible dives to depths of 3800 m recovered thin volcaniclastic deposits with abundant limu o Pele fragments and Pele's hair (Fig. 7.14). This is an interesting occurrence because the origin of limu o Pele had previously been attributed to the incorporation and vaporization of external water by lava flows, leading to explosive fragmentation of bubbles (Hon et al., 1988). However, the Gorda Rise deposits are below the critical point of sea-water, so its expansion to a vapour can not be called upon as a fragmentation mechanism in this situation. Instead, Clague et al. (2003a) suggest that the formation of the particles was the result of explosive eruptions of Strombolian style. Excess CO_2 accumulated and coalesced into large gas pockets at the top of the magma reservoirs and was then released as a slug flow. Material was ejected to heights of perhaps only a few hundred metres above the vent (e.g. Head and Wilson, 2003) and then settled in the surrounding area as fallout or perhaps transported by gravity flows generated by collapse of the eruption plumes. The relatively low height of the

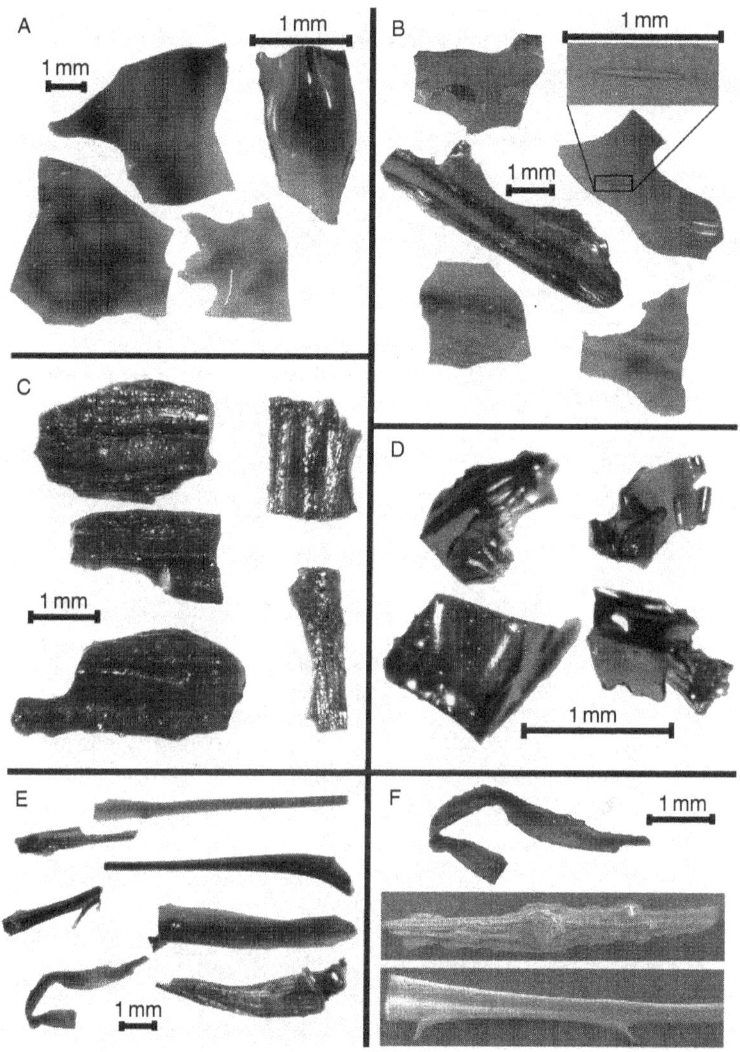

Figure 7.14 Glassy basaltic clasts recovered from the Gorda Ridge. Reproduced from Clague et al. (2003a,b) with permission from the American Geophysical Union. (A, B) Limu o Pele; (C) small spatter fragments; (D) limu o Pele; (E, F) ribbon spatter and Pele's hair.

eruption columns and drag by the surrounding water significantly limits the dispersal area from these types of deep sub-marine eruptions (White et al., 2003).

Based on the recent studies of volcaniclastic deposits along deep mid-ocean ridges, they appear to consist predominantly of localized bedded units with thickness up to several hundred metres. They form either by deep sub-marine explosive (Strombolian and Hawaiian) eruptions or thermal contraction/granulation of lava flows. Deposition occurs by fallout in the water, emplacement from eruption-fed gravity flows, or local sloughing off of lava flow margins. The morphology and vesicularity of the particles, along with the outcrop characteristics, are key to interpreting the nature of the source-fragmentation mechanisms. With decreasing water depth, other processes of fragmentation such as bulk interaction steam explosivity and contact-surface steam explosivity are likely to become more important, although their relative roles have yet to be thoroughly documented.

5.2. Volcaniclastic deposits related to seamounts

Seamounts are relatively small sub-marine volcanoes that develop adjacent to mid-oceanic ridges, on the oceanic crust in intra-plate settings, and also in extensional forearc basins along island arcs. Only limited studies have been conducted on seamounts, and our knowledge of these volcanic structures is quite restricted. However, many volcanic and volcaniclastic processes and facies on seamounts appear to be very similar to those that occur along mid-ocean ridges (Batiza and White, 2000; Batiza et al., 1984; Staudigel and Schmincke, 1984). Pillowed lava flows are common, as are hyaloclastite gravity flows, because of the presence of steep slopes on seamounts (Lonsdale and Batiza, 1980).

Hyaloclastites deposits on some Pacific seamounts at depths from 1240 to 2500 m have been studied in detail (Batiza et al., 1984; Smith and Batiza, 1989). The deposits contain blocky, sliver, and fluidal basaltic glass shards, similar to limu o Pele. Most beds are thin (< 1 m), finely laminated, normally graded, and exhibit imbrication of elongated shards. The origin of the beds was originally attributed to sediment gravity flows generated by sub-marine fire fountaining. Shard formation was thought to have occurred primarily by cooling–contraction granulation. It is now believed that flows are related to the shock granulation and spalling of thin lava flows. Batiza and White (2000) note that hyaloclastite flows are virtually absent from the fast-spreading East Pacific Rise, despite the fact that high magma-discharge rates would seem to favour their formation in such an environment. They suggest that the presence of pelagic sediment, typically found on the crest of slow spreading ridges and seamounts may be an important factor in hyaloclastite formation. Lava flows that move over unconsolidated sediment can entrain the material and heat the pore water until it vaporizes and causes fragmentation of the magma by bubble bursting and thermal-shock granulation. Intrusion of hypabyssal lava into wet marine sediment can also lead to the formation of hydroclastic breccias called 'peperites' (Busby-Spera and White, 1987).

As with mid-ocean ridges, the occurrence of deep-water explosive volcanism at seamounts, such as Hawaiian fire fountaining, Strombolian, and Vulcanian eruptions, is theoretically possible if there is sufficient magmatic-gas content (Head and Wilson, 2003). Clague et al. (2003b) have described basaltic volcaniclastic deposits from the summit of the Loihi seamount at depths between 1058 and 1160 m. These relatively thin units consist of mudstones to less consolidated layers of black sand and gravel. The morphologies of glassy fragments include dense angular pieces, concoidal tachylite fragments, limu o Pele, and Pele's hair. Also included are a variety of lava fragments, hydrothermally altered basalt, and hydroclastic pyrite. The abundance of limu o Pele and other pyroclastic fragments suggest sub-marine explosive eruptions of Strombolian and/or Hawaiian fire-fountaining style. Deposition of the juvenile fragments is likely to have occurred by fallout through the water column from eruption columns several hundred metres high, and by the generation of sediment gravity flows by collapse dense particle/water mixtures. The dispersal of material was sufficient to reach a large area of Loihi's summit region. Subsequent reworking of the unconsolidated material resulted in the formation of current-rippled deposits on the sea-floor.

In addition to sub-marine explosive eruptions driven by magmatic-gas exsolution, the lithic and mixed nature of some of the Loihi deposits suggest that occurrence of phreatic and phreatomagmatic eruptions as well. Clague et al. (2003b) propose that caldera-summit collapses led to access of sea-water to hot rocks, resulting in steam-driven explosions. This activity fragmented pre-existing lava flows and hydrothermally altered formations.

In general, the volcaniclastic deposits of seamounts are restricted in their distribution and show some similarities to those produced at mid-ocean ridges. However, unlike at the ridge environment, volcaniclastic material generated at seamounts may be transported to deeper water by secondary sediment gravity flows and other mass-wasting processes that are likely to be more active on steep-sided seamounts (Wright, 2001).

5.3. Volcaniclastic sedimentation around large oceanic islands

Some of the largest volcanoes on Earth are associated with a high magma flux from deep-seated mantle plumes. Examples include the shield volcanoes and calderas in the Hawaiian and Canary islands. These enormous edifices, about 90% of which are sub-marine, are an important source of volcaniclastic material to the deep-sea. Information about their structure and growth has been derived largely from studies of uplifted successions (a.o. Staudigel and Schmincke, 1984) coupled with deep drilling into sediment aprons on their flanks (a.o. Schmincke et al., 1995). Studies of the Canary Islands and, in particular, Gran Canaria have led to detailed models for the generation and transport of volcaniclastics in the marine environment during major ocean-island construction (Schmidt and Schmincke, 2000). Unlike mid-ocean ridges and small seamounts, large ocean islands are capable of producing widespread volcaniclastic deposits up to hundreds of kilometres from the source, and in water depths of several thousand metres.

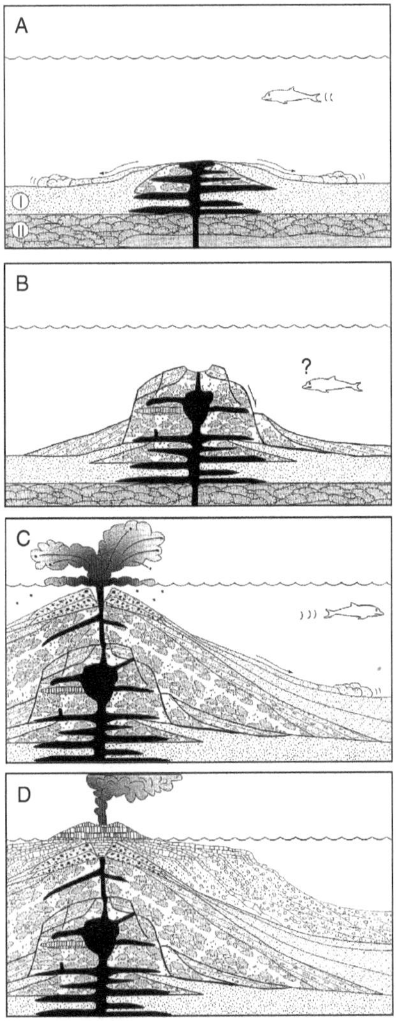

Figure 7.15 Evolution and growth of an oceanic island. Reproduced from Schmidt and Schmincke (2000) with permission from Elsevier. (A) Beginning of island growth, intrusion of dikes and extrusion of pillow lavas; (B) deep-water stage dominated by intrusive and extrusive (pillows and sheet flows); (C) activity become explosive with more fragmental material being produced; (D) sub-aerial emergence of the island and formation of lava deltas.

The nature and abundance of volcaniclastics derived by these types of volcanoes changes with time as they grow from the sea-floor to sub-aerial islands. Their evolution has been broken down into three main stages of growth (Fig. 7.15). All of them begin with a deep-water stage (1000–5000 m) where volcanism penetrates existing ocean crust and sediments. Effusive activity dominates this stage with the formation of thick pillow lava successions and associated intrusive rocks. The amount of volcaniclastics is limited to the local formation of hyaloclastites by quench granulation. Explosive eruptions may also occur and their importance in the deep-water evolution of oceanic island may not yet be appreciated. Magmas associated with hotspot oceanic islands are typically more enriched in volatile components such as water and carbon dioxide, thus potentially providing the necessary gas contents for deep-water pyroclastic activity.

During island growth, the slopes become gradually steeper and reworking is enhanced by mass-wasting processes (pillow breccia, slumps, debris-flow deposits) (Mulder, 2011, this volume, Chapter 2.1.3) that can transport clastic material over a larger area. A second stage of evolution occurs at intermediate and shallow-water depth ($\sim<1000$ m, Fig. 7.17). As hydrostatic pressure decreases, the vesicularity of lavas and volcaniclastic products increases significantly. Sub-marine explosive eruptions become more common, being driven by exsolution of magmatic gas, contact-surface steam explosivity, and bulk interaction steam explosivity. These generate increasing quantities of volcaniclastic material that accumulates by fallout and primary sediment gravity flows near the eruptive vents. However, the steep slopes of the growing edifice are conducive to the further mobilization of material by sediment gravity flows. These flows feed the growth of a laterally expanding volcaniclastic apron at the base of the seamount. Thus an important aspect of island growth is both the shallowing of the eruptive vents and the lateral spreading of a clastic apron of material in the deep-water environment.

The third stage of island growth occurs when volcanism is able to build a sub-aerial edifice and establish vents that isolate magma from interactions with water (Fig. 7.15). Sub-aerial volcanism can produce lava flows that enter the sea from land and produce lava-delta successions. Clastic material is formed by the brecciation of lava flows and localized surface steam explosivity. Steep-fronted lava deltas are susceptible to mass-wasting processes and clastic material can be transported to deeper water by turbidity currents and debris flows. Sub-aerial volcanism may also contribute volcaniclastic material to the deep-sea through fallout of tephra from explosive eruptions, discharge of pyroclastic flows into the sea, entrance of lahars into the sea, and the fluvial supply of eroded sub-aerial volcanics.

An important process in the evolution of oceanic islands is the large-scale displacement of the island's flanks by slumping and/or debris avalanches (Moore et al., 1994). Slumps are gradual, intermittent movements of the island flanks that can affect successions up to 10 km thick and 100 km wide. Displacement of several metres may occur in association with earthquakes. Debris avalanches, however, are catastrophic failures of the island flanks that send large volumes of material downslope at high velocity. They may be up to 2 km thick and travel several hundred kilometres into the deep-sea, producing impressive deposits of the order of hundreds of cubic kilometres. Mixing of the debris avalanche with water as it moves downslope can trigger the secondary generation of debris flows and turbidity currents that can travel up to 600 km from source (Masson, 1996). Studies of the Hawaiian and Canary islands suggest that debris avalanches and slumps are more common during the mature stages of island growth, but may also occur during the sub-marine stages. These events play an important role in the growth of deep-water volcaniclastic aprons surrounding large ocean islands.

Volcaniclastic aprons were first recognized by Menard (1956) from seismic surveys around oceanic volcanic islands. Aprons often correspond to the coalescence of volcaniclastic fans that form during several sedimentation periods. Except for fine-grained tephra fallout that can be dispersed over very large distances, only a small proportion of the volcaniclastic products are transported sub-aqueously away from these volcaniclastic aprons. Kilometre-thick volcaniclastic aprons typically represent more than 90% of the rock volume of volcanic islands (Carey 2000; Orton 1996; Schmincke et al., 1995).

A combination of seismic profiling and ODP drilling on the flanks of Gran Canaria (Canary Islands) has provided a detailed look at the nature and evolution of volcaniclastic aprons (Schmincke et al., 1995). Gran Canaria is a mature oceanic island that began its sub-marine growth stage more than 14 million years ago. The volcaniclastic apron can be sub-divided into three facies based on the seismic signature and lithology of the successions (Fig. 7.16). Closest to the island is the seismically chaotic flank facies. It is characterised by rough topography, discontinuous reflectors, and consists mostly of pillow breccias and hyaloclastite flows. The succession is draped over a core of pillow basalts and intrusives and extends into relatively deep water. The formation of this facies is attributed to the shallow sub-marine to emergent phase of island growth. Satellite sub-marine volcanic cones can also be present with related lava flows. Because of volcano-seismic activity and the accumulation of large volumes of volcaniclastic deposits on the proximal apron, slope-instability structures are common. They are revealed by the presence of scarps and coarse debris-flow deposits in the bottom of canyons (Gamberi, 2001) and on slopes (Wright, 1996).

Overlying the flank facies is a more seismically coherent succession referred to as the slope facies (Fig. 7.16). It consists of slumps, discontinuous bedding, and massive sediment gravity flows. This facies extends further into deep water than the flank facies and represents the more distal

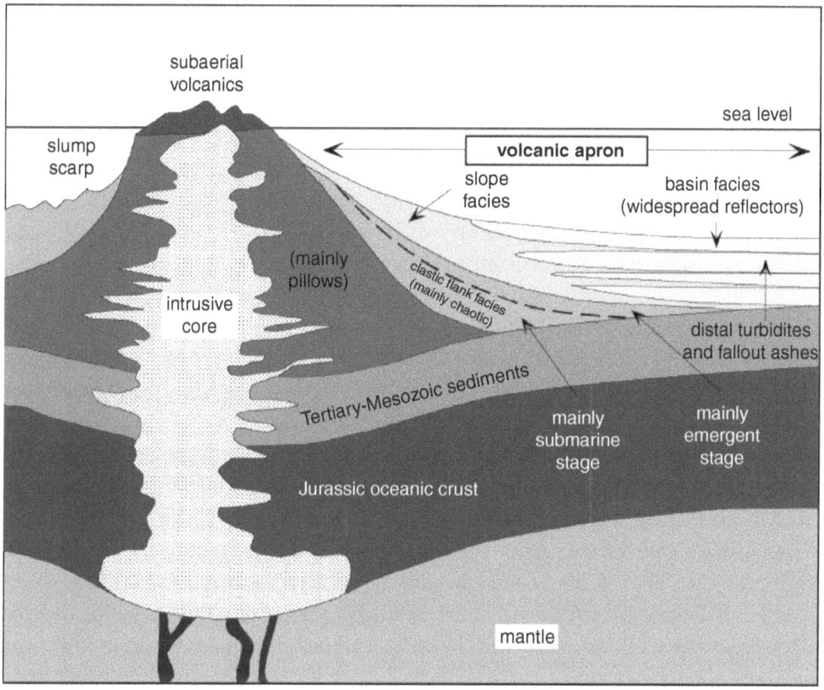

Figure 7.16 Facies relationships of an oceanic island, modified from Schmincke et al. (1995). Volcaniclastic sediments are concentrated in the flank, slope, and basin facies.

equivalents of sediment gravity flows that were also generated during the shallow sub-marine and emergent phases of island growth.

The slope facies inter-fingers distally with a more widespread basin facies (Fig. 7.16). This facies exhibits well-defined seismic reflectors and extends up to several hundred kilometres from the source. Depositional units include ash falls, debris flows, and turbidites inter-bedded with background pelagic and hemipelagic sediment. The basin facies represents a major repository of volcaniclastics that have been generated during the long-term evolution of oceanic islands.

5.4. Volcaniclastic sedimentation in subduction-zone settings

Island arcs and convergent continental margins are also important suppliers of volcaniclastic sediment to the deep-sea. They are, perhaps, the most complex and diverse source of this type of material in the world's oceans, owing to several factors. First, subduction-zone volcanism produces a much larger spectrum of magma compositions than mid-ocean ridges and large oceanic islands. In particular, more evolved and volatile-rich magmas are significantly more abundant in subduction zones. The higher volatile contents and more viscous nature of the magmas favour fragmentation by explosive eruptions and the generation of a substantially higher proportion of volcaniclastic material. Secondly, the configuration of subduction zones is such that there are multiple sources (i.e. volcanoes) of volcaniclastic material arranged in a curvilinear fashion adjacent to back- and forearc basins, and trenches. These deep-water traps for volcaniclastic sediment thus receive material from many different volcanic centres over linear distances of hundreds of kilometres. It has been estimated that oceanic island arcs are composed of more than 90% fragmental material (Garcia, 1978).

The sources of volcaniclastic sediment in subduction-zone (island-arc) environments are remarkably diverse and summarized in Fig. 7.17. Primary sources include highly explosive eruptions of sub-aerial volcanoes that generate ash fallout and pyroclastic-flow discharge into the sea. This material may be transported directly to the deep-sea by passive settling or a variety of sediment gravity flows. Alternatively, the material may be temporarily deposited in the shallow-water environment and then remobilized by currents or sediment gravity flows for subsequent deposition in deep water (Fig. 7.17). The common occurrence of explosive volcanism in subduction zones produces sub-aerial deposits that are susceptible to erosion and reworking (Cas and Wright, 1987), eventually making their way into the marine environment by complex paths that may also involve temporary staging in shallow water.

Subduction zones are also the sites of significant sub-marine volcanism, as most island-arc volcanoes undergo an evolution that involves a deep sub-marine stage, growth into intermediate to shallow water, and eventual emergence to form a sub-aerial edifice. It is likely that island arcs may be the sites of more abundant deep to intermediate depth explosive eruptions than other volcanic environments owing to the volatile-rich nature of the magmas. Recent detailed exploration of sub-marine arc volcanoes in the Kermadec and Izu-Bonin island arcs has shown a great abundance of geomorphic features such as craters, calderas, and pyroclastic deposits indicative of explosive activity (Fiske et al., 2001; Wright et al., 2003; Yuasa and Kano, 2003). Such sub-marine eruptions are likely to be more energetic than those at mid-ocean ridges and oceanic islands because of the higher magmatic volatile content, more viscous nature of the magmas, and larger individual volumes of magma that are involved in subduction-zone eruptions.

An important component of sub-marine volcanism in subduction zones is the production of abundant pumice during explosive events. The nature and dynamics of sub-marine explosive pumice-forming eruptions (pyroclastic) is complex and still not well understood. The presence of water and the potential for the formation of steam results in a three-phase system (gas, particles, and water) that has novel sedimentation and transport behaviours (Cashman and Fiske, 1991; Fiske et al., 2001; Kano, 2003). Explosive

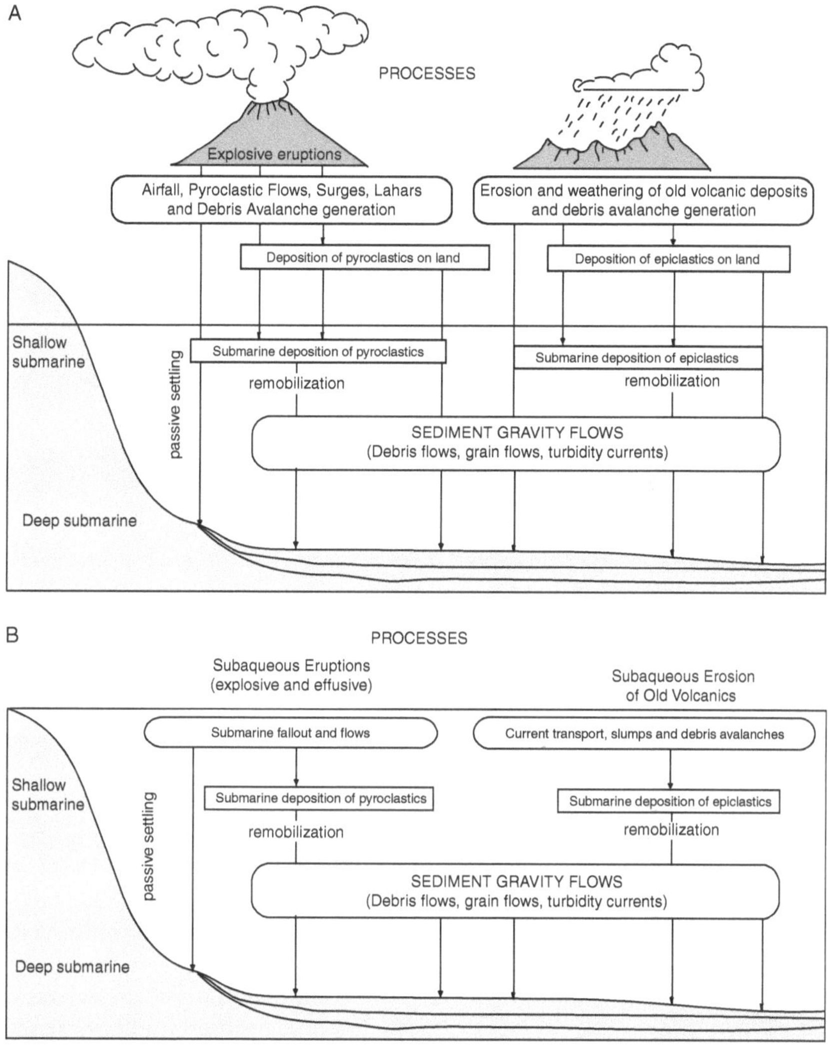

Figure 7.17 Sources and transport processes in the sub-aerial and submarine environments of an active island arc. Reproduced from Carey (2000) with permission from Elsevier.

eruptions of silicic magma will form sub-marine eruption columns that eject hot pumice and gas into sea-water (Fig. 7.18). If the mass flux is very high, a buoyant mixture of pumice, steam and hot water will rise and potentially reach the surface, where it spreads out to form a mushroom-shaped mixture of water and particles. Most pumice is less dense than sea-water, but will eventually become water-logged by condensation of internal steam and

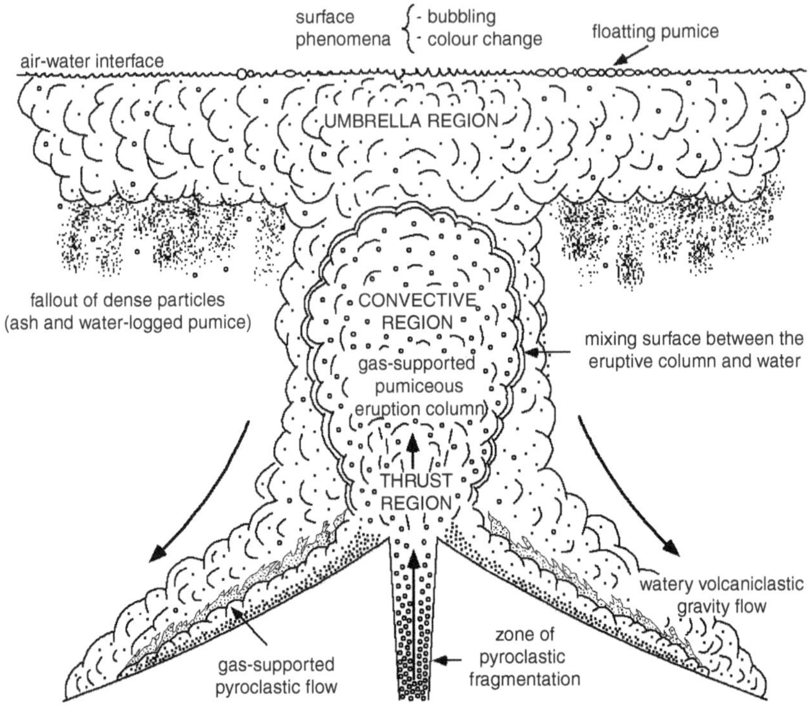

Figure 7.18 Model of a submarine pumice-producing eruption column. Modified from Schneider (2000).

absorption of water (Manville et al., 1998; Whitham and Sparks, 1986). In particular, hot pumice can become negatively buoyant very quickly upon exposure to water. Thus, pumice that initially rises in the water column will eventually settle back to the sea-floor, but at a much slower rate than in air. The timescale for the transition to negative buoyancy is proportional to the size of the pumice (Manville et al., 1998) and may thus result in reverse size grading of sub-marine fallout deposits.

If the pyroclastic mass flux is low and the sub-marine eruption column is able to entrain enough water, the steam phase may recondense and the bulk density will increase significantly. This will trigger collapse of the column to form syn-eruptive sediment gravity flows. These are likely to be water-dominated flows (Allen and Stewart, 2003; Kano, 2003), although some collapses might sufficiently shield the collapsing mixture from further entrainment of sea-water such that hot, gas-dominated sub-marine pyroclastic flows may be formed (a.o. Kokelaar and Busby, 1992). Even when parts of the column collapse, significant amounts of pumice and ash may still rise convectively and then begin to fallout once water-saturation occurs (Fig. 7.18).

This may result in a hybrid depositional mode of fallout being synchronously mixed with collapsing sediment gravity flows (Fiske et al., 2001).

Existing evidence for sub-marine pyroclastic eruptions that produce pumice suggest that they can occur at least to depths down to 1500 m, if the magmatic volatile contents are high enough. These events are potentially important for deep-sea volcaniclastic sedimentation as they can produce large volumes of pumice and ash during single events. Fiske et al. (2001) estimate that a sub-marine eruption of the Myojin Knoll caldera in the Izu-Bonin arc produced more than 40 km^3 of pumiceous tephra, and Wright et al. (2003) suggest that the Healy caldera in the southern Kermadec arc discharged 10–15 km^3 of pumice at water depths of about 1000 m.

Subduction zones are also the sites of sub-marine Hawaiian, Strombolian, and Vulcanian eruptions, as has been proposed for mid-ocean ridges and large oceanic islands Head and Wilson (2003). It is likely that phreatomagmatic eruptions may also be more common at subduction zones, because of the potential feedback between fragmentation by primary gas release and other hydroclastic fragmentation mechanisms (Kokelaar, 1986).

Recent marine geological surveys have highlighted an additional important source of volcaniclastic material at subduction zones: sub-aerial and

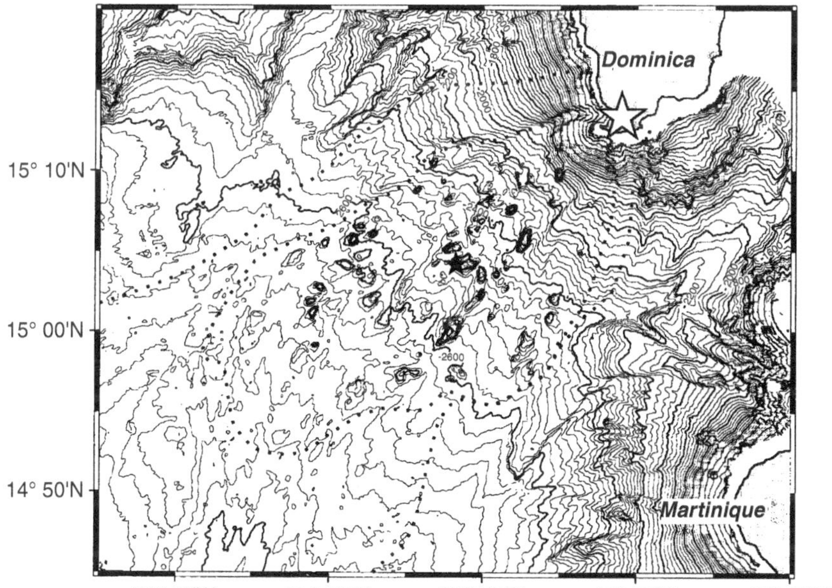

Figure 7.19 Bathymetric map of the sea-floor off the coast of Dominica (Lesser Antilles) showing distinctive hummocky topography produced by a submarine debris avalanche. Reproduced from Deplus et al. (2001) with permission from Elsevier.

sub-marine debris avalanches (Fig. 7.19). In the Lesser Antilles, there is abundant evidence for collapse of island edifices and flanks that travelled into the deep backarc Grenada Basin (Boudon et al., 1984; Deplus et al., 2001; Wadge and Isaacs, 1988). It appears that island-arc volcanoes, like

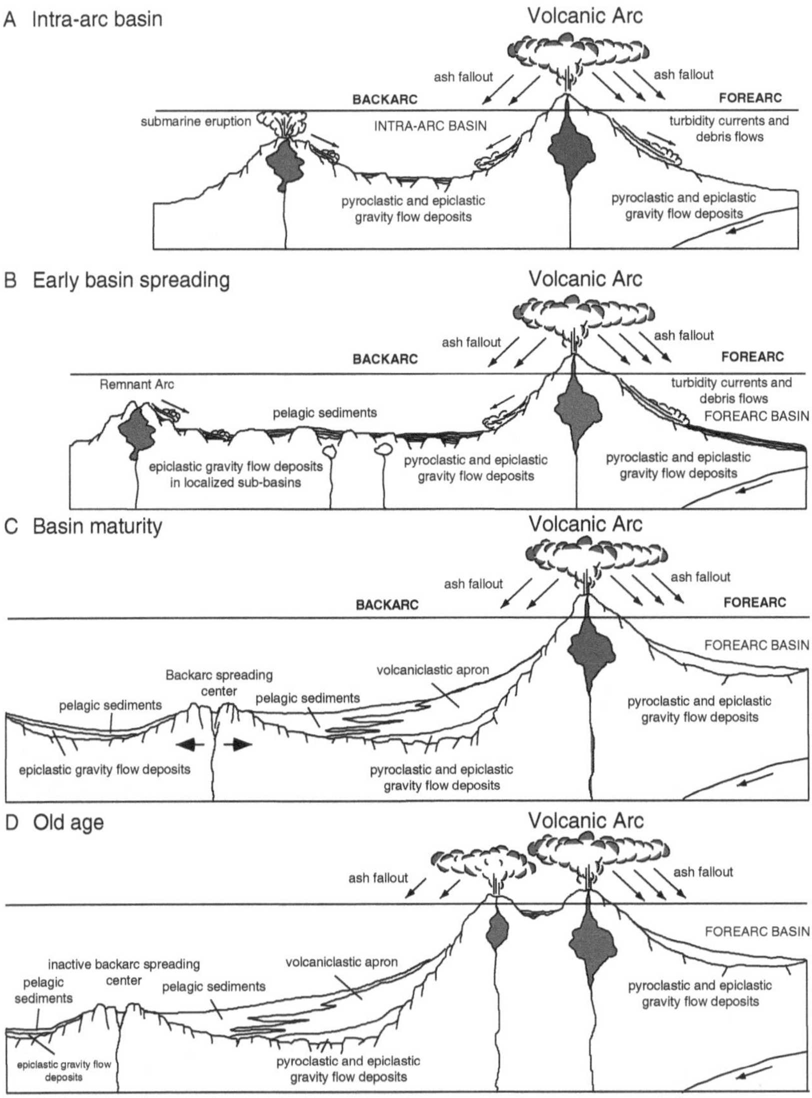

Figure 7.20 Sources and processes of volcaniclastic sedimentation during the evolution of an oceanic island arc with backarc basin formation. Reproduced from Carey (2000) with permission from Elsevier.

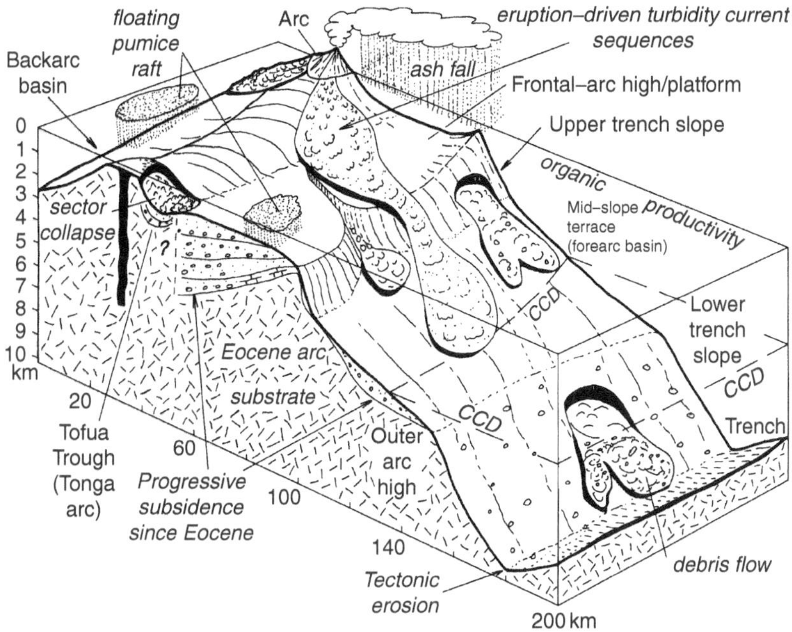

Figure 7.21 Processes of volcaniclastic sedimentation in a forearc environment of an oceanic island arc. Reproduced from Underwood et al. (1995) with permission from the American Geophysical Union.

other large oceanic islands, are also susceptible to these catastrophic failures during their growth and evolution.

Volcaniclastic sedimentation in subduction zones (island arcs) can be considered within different structural elements of a typical oceanic subduction zone (Fig. 7.20). The forearc area is structurally complex with small basins and discontinuous slopes owing to uplift and tectonism associated with subduction. Volcaniclastic sediment is supplied to the forearc from sub-aerial and sub-marine eruptions along the volcanic front (Fig. 7.21). Depositional mechanisms include fallout from sub-aerial eruptions, syn-eruptive sediment gravity flows, and debris-flow generation by trench-slope failure. In the Izu-Bonin forearc area, drilled at ODP Site 792 (1798 m water depth), deposits include decimetre-scale layers of vitric sand and silt, meter-scale massive to bedded vitric sandstones, and heterogeneous massive to graded conglomerates.

Many oceanic island arcs undergo progressive evolution during which time the volcanic front repeatedly rifts, creating a series of backarc basins. Early rifts that are in close proximity to the volcanic front can experience extremely high fluxes of volcaniclastic sediment. The Sumisu rift in the Izu-Bonin arc is an example of a small arc basin with accumulation rates of volcaniclastic sediments as high as 1000 m per million years. ODP drilling

(Site 790) in this area recovered spectacular, thick (15–30 m) pumiceous gravels that grade upwards into sands and silts, forming a series of distinct fining-upwards sequences. The origin of these thick pumice sequences is attributed to large-scale sub-marine explosive eruptions from nearby sub-marine calderas. During these eruptions, high-concentration turbidity currents and debris flows were likely generated from collapse of the sub-marine eruption column and rapid mixing with sea-water. Distribution of pumice may also have occurred by pumice rafting of material ascended to the sea surface.

Small intra-arc basins continue to evolve to young marginal, or backarc basins (Fig. 7.20). Their structure is also complex owing to the migration of volcanism back to the main volcanic-arc front. They typically exhibit a horst-and-graben configuration leading to many potential depositional sites. Successions in these types of basins (e.g. ODP Site 838 in the Tofua arc of the SW Pacific) include basal units of polymict vitric gravels, sandstones, and siltstones with inter-beds of hemipelagic ooze. These are linked to localized sources of epiclastic material associated with mass-wasting processes and are transported by sediment gravity flows. Overlying the basal successions are pumiceous gravels (up to 10 m thick), vitric sands, and silts. The pyroclastic nature of these units indicates that they are derived from explosive volcanism of intra-basinal seamounts, supplying large volumes of pumiceous material to nearby depositional sites. The uppermost units in these types of arc environments consist of relatively fine-grained ash layers that represent fallout from more distal eruptions (sub-aerial arc) and local redistribution of pyroclastic material by turbidity currents. A general fining-upwards trend reflects the continuous growth of the backarc basin and migration of primary volcanism back to the main volcanic front of the arc.

Mature backarc basins contain thick volcaniclastic successions that consist of material derived from the sub-aerial/sub-marine volcanic front, intra-basinal seamounts, and a backarc spreading centre (Fig. 7.20). Such successions may contain records up to tens of million years in duration and reflect the dynamic evolution of the volcanic front and the backarc basin. In the Parece-Vela Basin in the western Pacific, the sedimentary succession illustrates the changing nature of volcaniclastic-sediment sources and transport mechanisms. At DSDP Site 450, the succession begins with pillow basalts intruded into wet sediments, formed during the early stages of basin rifting. Overlying the basalts is 250 m of vitric tuffs and volcaniclastic conglomerates that accumulated at a rate as high as 100 m per million year. These represent fallout from sub-aerial eruptions, high- and low-concentration syn-eruptive turbidites and debris flows, and redeposition of epiclastic material. The primary source of material during this stage was the main volcanic front of the arc. Above this interval, the volcaniclastics become finer grained and less abundant, being inter-bedded more prominently with hemipelagic sediment. Ash-fallout layers are more abundant within this interval. Deposition during this interval reflects the progressive movement of the site away from the volcanic arc and the main sources of volcaniclastic sediment. Finally, the succession is capped by tens of metres of pelagic clay, where the volcaniclastic input is no longer significant.

A schematic representation of the evolution of an oceanic island arc and its associated volcaniclastic sedimentation is shown in Fig. 7.20 (from Carey, 2000). The figure simplifies the complexity of the volcaniclastic-sediment sources, but attempts to show spatial relationships between sources and depositional sites on a large scale. In this model, there are important differences between the forearc and backarc sedimentation histories. The forearc is a more long-lived feature of an island arc and many of the depositional sites retain their relative proximity to volcanic sources. Consequently, the sedimentary records of forearcs may contain a longer depositional interval of arc activity compared to backarc basins (Fig. 7.20). In contrast, the backarc area is subject to a temporal evolution that involves progressive basin widening and eventually isolation from the active volcanic arc. Backarc basin successions thus record changes in the relative importance of different arc sources as a function of time and changes in the position of depositional sites relative to those sources. An important sedimentary feature of backarc basins, recognized early on by Karig (1971), is the wedge of volcaniclastics that develops within the basin at the base of the volcanic arc. This wedge of sediments records the volcanic history of the arc during the period when the backarc basin is active, prior to rifting of the volcanic line (Fig. 7.20).

6. Importance of Volcaniclastic Aprons in the Deep-Sea

6.1. Comparison of volcaniclastic aprons

Two of the most significant deep-water volcaniclastic depositional areas are found adjacent to large oceanic islands, such as Hawaii, and in backarc basins bordering active oceanic island arcs. Menard (1956) recognized the volumetric importance of volcaniclastic aprons in relation to large oceanic islands, but it is clear that island arcs create substantial deep-water accumulations as well. Facies associations on sub-marine volcaniclastic aprons are complex and correspond to primary volcanic and volcaniclastic, reworked volcaniclastic, and volcanic epiclastic and non-volcanic products. Interstratified non-volcanic, marine sediments (often fossiliferous) are very useful to establish age controls of the apron pile. The architecture of the facies associations is governed by magmatic productivity, and the nature of the transport and sedimentation processes. Non-volcanic processes, such as erosion, that occur sub-aerially on the volcanic source also play a role in sub-marine apron sedimentation (Fisher and Schmincke, 1994; Schneider et al., 1998). The term 'volcaniclastic apron' is used generally for successions in both environments, but there are likely to be important differences between oceanic islands and island arcs. First, oceanic islands are considerably larger than individual arc volcanoes and thus a single oceanic island will likely feed a larger deep-water apron. However, the size of individual eruptions at subduction zones can be much larger than at oceanic islands, and thus individual primary volcaniclastic-sedimentation events recorded in arc aprons may greatly surpass those at oceanic islands. Second, volcaniclastic aprons in backarc basins are generally more elongated in geometry and fed by multiple sources compared to oceanic island aprons that are more circular in shape. Third, volcanic arcs are characterised by more explosive eruptions, and thus the sediments are more highly fragmented and finer grained than those of oceanic islands. And finally, the diversity in magmatic compositions, and thus products, is significantly greater in subduction zones than in oceanic islands.

Despite these differences, there are features that both share in common. In the proximal parts of the aprons, the steep slopes display a seismically chaotic facies. These slopes are dominated by coarse deposits, and are generally incised by numerous gullies and canyons. Volcaniclastic turbidites related to the discharge of volcaniclastic gravity flows or triggered from delta slopes are inter-calated in the sedimentary succession (Carey and Sigurdsson, 1984; Ollier et al., 1998; Schmincke et al., 1995). Proximal debris-avalanche deposits are related to sub-aerial collapse events, but can also originate on the apron slopes. Numerous primary volcanic and volcaniclastic deposits are frequent in proximal volcaniclastic aprons. More distally, usually over 1500 m in depth, the facies become more organized and stratified. Volcaniclastic turbidites fed from sub-marine canyons and fallout deposits dominate. They are inter-stratified with the normal marine sediments that form the background sedimentation. Distal debris-avalanche deposits are also present in these distal deep-water parts of the aprons. The mean grain size of the deposits in distal aprons is finer than in proximal areas.

6.2. Dynamics of volcaniclastic aprons

The geometry of volcaniclastic aprons both results from, and influences the sedimentation processes. Influx of sediment creates and modifies the topography of the apron surface. Structural deformation related to magma ascent may cause rapid local subsidence or uplift, and steep slopes are created as a consequence of rapid accumulation of large volumes of volcaniclastic material. Because of the high heat flow in the proximal areas, hydrothermal

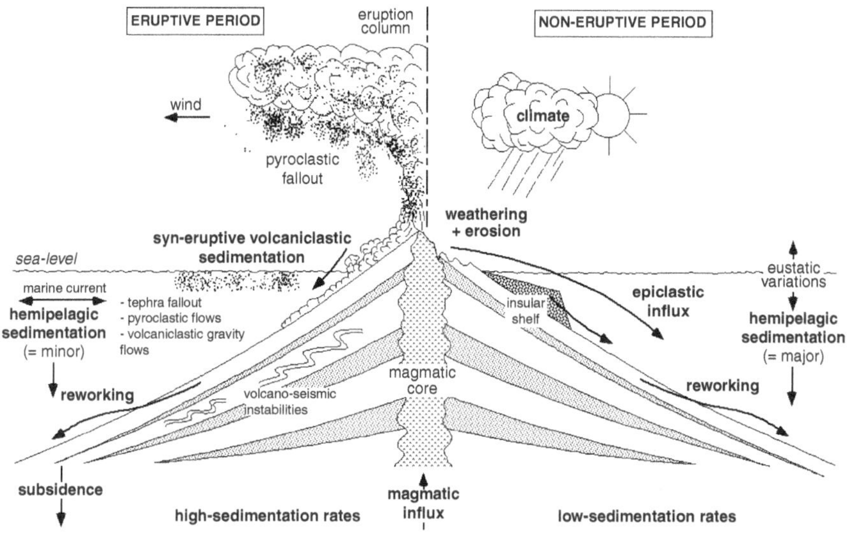

Figure 7.22 Sources and processes of sedimentation on marine volcaniclastic aprons during eruptive and non-eruptive periods. Modified from Schneider (2000).

alteration affects the deposits, and volcanic-related earthquakes may cause their destabilization, and favour slope-failure events.

Volcaniclastic apron evolution (Schneider, 1998) is controlled by alternation of growth (=aggradation) and destabilization (=erosion). Vertical aggradation results from accumulation of volcaniclastic material during eruptive periods (mainly primary volcanic and volcaniclastic products) (Fig. 7.22). A minor flux of pelagic sediments also contributes to the growth phases. High sedimentation rates are the cause of instabilities on the steep slopes of the proximal aprons and favour potential failure. During non-eruptive periods, epiclastic materials produced by weathering and erosion of sub-aerial volcanic areas are transferred toward the apron (Fig. 7.22), but their sedimentation rates are generally low (Schneider et al., 1998).

Sub-marine erosion of volcaniclastic aprons mainly results from large-scale mass-wasting events triggered by volcano-tectonic activity. These instabilities occur preferentially in the proximal parts of the aprons. Large volumes of volcaniclastic material are then transferred towards deeper and distal domains of the aprons. A part of this material can also reach deeper areas and abyssal plains that constitute the ultimate sedimentation environments for some volcaniclastic particles. Remobilization by bottom currents seems to play a minor role on volcaniclastic aprons.

Various factors control volcaniclastic apron dynamics; the most important is magmatic productivity (Fig. 7.22). Sedimentation rates on the apron reflect the volcanic influx, and their estimate is a good way to reconstruct the volcanic activity of the source. During periods of strong volcanic activity, the sedimentation rate is high (Gamberi, 2001; Schmincke et al., 1995; Schneider et al., 1998). Subsidence seems to play only a minor role, and is related to the effects on the elastic lithosphere by the load of the growing volcanic centre and of its sub-marine apron. This is particularly evident for oceanic islands such as Hawaii where a prominent moat is developed at the base of the island. During non-eruptive periods, sub-aerial areas are subject to normal weathering and erosion. Epiclastic products are transferred toward the sub-marine apron and more distal sub-marine environments. Because of the elasticity of the lithosphere, an isostatic rebound occurs, generating uplift of the volcaniclastic apron. Sea-level fluctuations also impact sedimentation on deep-sea aprons. High-level stands are favourable to the storage of volcaniclastic sediments on the shelf, whereas sea-level lowstands are accompanied by important sedimentary remobilization and sediment movement towards the deeper parts of the aprons.

From all these observations, it seems that subsidence only plays a minor role in the evolution of sub-marine volcaniclastic aprons. The total space (accommodation) available for storing volcaniclastic sediments on aprons is mainly furnished by the thickness of the water column. During periods of strong volcanic activity, and consequently of high sedimentation rates, aprons growth mainly vertically and also laterally by transfer of sediments by the contemporaneous gravity flows.

7. Economic Aspects of Sub-Marine Volcaniclastic Deposits

7.1. Sub-marine volcaniclastic deposits as targets of oil exploration

Volcaniclastic formations are usually considered as poor targets for oil exploration, because heat flows are high in volcanic areas and vitreous particles suffer strong alteration into clay and zeolite minerals. Because of the latter reason, primary porosity and permeability is lost within the deposits (Davies et al., 1979; Mathiesen and McPherson, 1991; Seeman and Scherer, 1984; Surdam and Boles, 1979). However, recent studies suggest that formation of a secondary porosity is common within volcaniclastic formations (Hawlander, 1990; Mathiesen and McPherson, 1991). Moreover, heat flows are high only in the vicinity of volcanic edifices and magmatic bodies. In deep-marine basins located distally from volcanic sources, volcaniclastic deposits can accumulate as thick successions without a high heat flow. In this regard, the morphology and grain size of the volcaniclastic particles could make these deposits potentially interesting as hydrocarbon reservoirs (Imbert, 2011, this volume, Chapter 10.4.2.4).

Some examples of oil reservoirs associated with marine volcaniclastic are known (Bouysse and Mascle, 1994). In the Lesser Antilles, oil and gas indices and shows had been identified in the Paleogene Saba Bank volcaniclastic. Interesting targets for exploration include backarc marine basins that receive large influxes of volcaniclastic sands.

7.2. Ore resources related to sub-marine volcaniclastic deposits

Many of the economically important mineral deposits in volcanic environments are associated with intrusive rocks that form at depths of several kilometres beneath the surface (White and Herrington, 2000). For example, porphyry copper, molybdenum, and gold deposits are found in ancient volcanic-arc terranes in close proximity to calcalkaline intrusive bodies. However, an important class of volcanic mineral deposits that are commonly associated with shallower levels of emplacement and that are found within volcaniclastic sediments are volcanically hosted massive sulphide deposits (VHMS). These deposits make significant contributions to the world's supplies of copper, lead, silver, and gold. The two principal types are copper- and zinc-bearing (CZ) and zinc-, lead- and copper-bearing (ZLC). They are associated with both volcanic arcs (Kuroko-type) and oceanic spreading centres (Cyprus-type). Kuroko-type deposits may consist of either CZ or ZLC assemblages, with CZ deposits typically being associated with more felsic rocks. Cyprus-type deposits are almost exclusively associated with mafic rock compositions.

The origin of VHMS deposits is linked to circulation and discharge of metal-bearing hydrothermal fluids on the sea-floor. Convective circulation of sea-water is driven by the heat from crustal magma bodies. These magma bodies also contribute some of the metals and gases to the hydrothermal fluids. Precipitation of minerals occurs on the sea-floor in the form of chimneys consisting of minerals such as pyrite, chalcopyrite, pyrrhotite, and sphalerite, as hot fluids mix with colder, more oxidized sea-water. Minerals also form just below the surface within the hydrothermal plumbing system. Because of the porosity of volcaniclastic sediments they can be important sites of mineralization caused by the circulation of hydrothermal fluids.

VHMS deposits formed at oceanic spreading centres are commonly hosted within a succession of pillow lavas, breccias, and dykes. In general, the proportion of volcaniclastic sediment is low in these environments and thus the linkage of VHMS with sediments is marginal. However, in volcanic-arc environments where volcaniclastic deposits are much more abundant, VHMS deposits are found in close association with large volumes of sediment. For example, a VHMS deposit in the Archean Sturgeon Lake Complex (Ontario, Canada) is hosted within a thick succession of felsic subaqueously erupted pyroclastic flows (Hudak et al., 2003), and Kessek and

Busby (2003) describe thick sub-marine pyroclastic-flow deposits, erupted at depths up to 1.4 km, that host an important Ordovician VHMS deposit in the Bald Mountain area of Maine (USA). In the Iberian Pyrite Belt (SW Iberian Peninsula), the largest massive sulphide ore deposits on Earth developed during the Late Devonian and Early Carboniferous in relatively shallow-marine environments (~150 m); they are related to felsic volcanic and volcaniclastic deposits (Sáez et al., 1996).

Studies of ancient VHMS deposits are useful for determining the three-dimensional relationships of mineral deposits to enclosing volcaniclastic sediments and magma bodies. However, the exact sub-marine environment

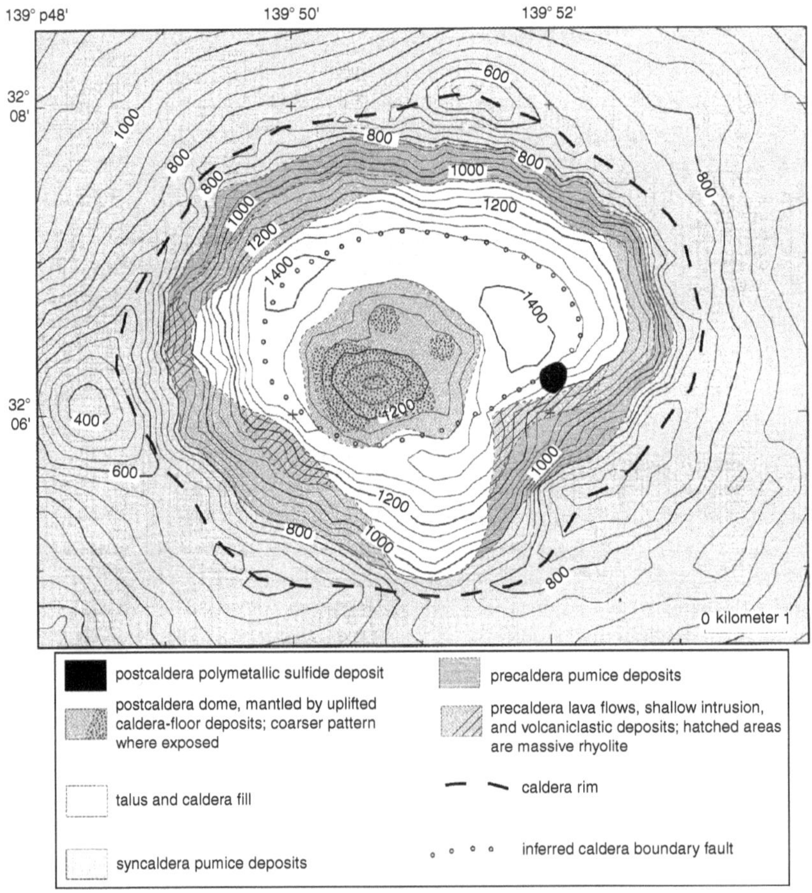

Figure 7.23 Bathymetric map of the sub-marine caldera of the Myojin Knoll volcano in the Izu-Bonin island arc of the western Pacific. A polymetallic sulphide deposit is located along the eastern wall of the caldera. Reproduced from Fiske et al. (2001) with permission from the Geological Society of America.

of formation is often difficult to define precisely. Formation of VHMS deposits in the modern oceans has been studied extensively at mid-ocean ridges, but only recently have these deposits been discovered and examined in subduction zones. One of the best examples of a modern Kuroko-type VHMS deposit was discovered within the Myojin Knoll sub-marine caldera (Iizasa et al., 1999). A polymetallic sulphide deposit about 400 × 400 m in size occurs on the eastern floor of the caldera (Fig. 7.23). It is rich in gold and silver and is actively forming from circulation of hydrothermal fluids near the caldera wall. Sub-marine arc calderas and their associated volcaniclastic deposits provide many of the necessary requirements for the formation of Kuroko-style VHMS deposits (Ohmoto, 1996). Ring faults establish pathways for sea-water to circulate into the crust, where sub-caldera magma bodies deliver the heat for convective flow. In addition, the common occurrence of sub-marine dome formation following caldera subsidence provides high localized thermal gradients and sources of metals and gas for circulating fluids. VHMS deposits have also been recognized at sub-marine volcanoes in the Kermadec arc of the SW Pacific (Wright et al., 1998).

8. Sub-marine Volcaniclastic Deposits as Tools for Natural-Hazard Assessment

Currently, there is an increasing interest in using sub-marine volcaniclastic deposits to better understand the history of volcanic activity in different environments. These data are useful for volcanic and other natural hazards assessment. The analysis of thick volcaniclastic deposits in well-documented sedimentary successions is one of the best ways to study the evolution of eruptive activity of volcanic islands and littoral volcanic centres over large time periods. Indeed, preservation of the 'volcaniclastic signal' is better on the sea-floor than in sub-aerial environments that suffer rapid erosion.

Reconstruction of remote volcanic activity can be made by the analysis of widespread tephra layers. Ash layers in the deep-sea, particularly those present on bathymetric highs where oceanic sedimentation rates are the lowest, are generally not affected by erosive processes that can occur on the distal slopes of sub-marine volcaniclastic aprons. Volumes of individual tephra layers can be determined from their thicknesses and correlations between sediment cores. Individual volcanic sources can be recognized from geochemical analysis of the tephra particles (a.o. Sarna-Wojicki et al., 1987) and the dating of events can be constrained by biostratigraphic analysis of the inter-bedded hemipelagic and pelagic sediments. Recurrence patterns of explosive volcanic activity can be reconstructed from these types of data. For example, Paterne et al. (1990) used marine tephrochronology in the central Mediterranean area to recognise a 23 ka periodicity of Italian explosive volcanism during the last 190 ka.

A better knowledge of the volcanic activity is important, particularly in shallower areas where volcanic eruptions can induce effects on the dynamics of the water column. The discharge of volcaniclastic gravity flows into the sea has a high tsunamigenic potential. Such processes constitute major hazards in coastal areas that are exposed to entry of debris avalanches (Le Friant et al., 2003; McCoy and Heiken, 2000; Ward, 2001) and of pyroclastic flows (Carey et al., 2000) into the sea. Numerical modelling has been carried out to better understand tsunami-wave propagation related to these processes (Mangeney et al., 2000).

The study of volcaniclastic successions also allows reconstruction of volcanic activity over longer time intervals. With an accurate volcano-stratigraphic timescale of past activity, it is possible to establish the timing of major episodes of volcanic activity. Tephra layers recovered by ODP drilling in the western Caribbean defined two major episodes of Central American explosive volcanism during the late Eocene and early Miocene (Fig. 7.24) (Sigurdsson et al., 2000). On somewhat smaller timescales, three successive flank collapse events have been identified for Mount Pelée (Martinique island, Lesser Antilles) from sub-marine volcanic debris-avalanche deposits with recurrence intervals of about 25 ka (Deplus et al., 2001; Le Friant, 2001). Consequently, 25 ka is the average time for volcanic edifice reconstruction by magmatic influx in this area.

Sub-marine seismic networks allow the detection of sub-marine volcanic activity, which is often related to sub-marine earthquakes (Talandier, 1989). Such monitoring is important because sub-marine earthquakes and eruptions may trigger sub-marine flank collapse, having tsunamigenic potential.

9. Conclusions

Volcanic activity in and around the world's oceans generates large quantities of fragmental material that are transported and deposited in the deep-sea. The production of particles occurs by sub-aerial and sub-marine explosive volcanism driven by the exsolution of dissolved magmatic gases or interactions of water and magma. In addition, the erosion of existing volcanic formations contributes to the flux of volcaniclastic sedimentation. The production of sub-marine volcaniclastics is strongly depth-dependent, with highest rates occurring in water depths less than 500 m. However, increasing evidence suggests that particle formation from explosive activity can take place as deep as 3800 m. A large diversity of particle types, density, and chemical composition are generated by volcanism in different tectonic environments. In ancient marine deposits, the particle characteristics thus

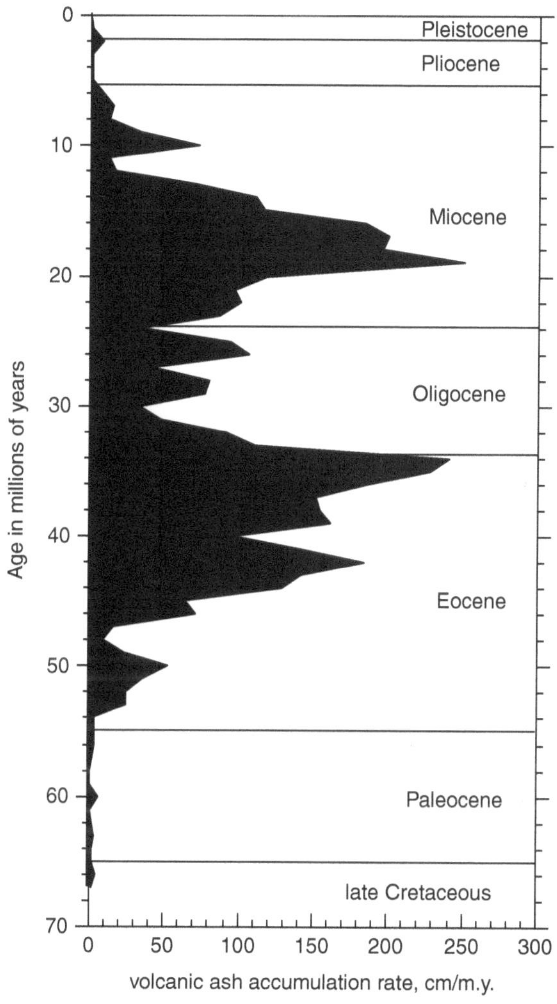

Figure 7.24 Record of the volcanic-ash accumulation rate at ODP site 999 in the western Caribbean Sea. Modified from Sigurdsson et al. (2000).

contain important information about the source volcanism, transport processes, and geodynamic environment.

Primary transport mechanisms for volcaniclastic sediments include passive settling through the water column from volcanic plumes, downslope movement by sediment gravity flows, and current redeposition. Sediment gravity flows are responsible for the majority of volcaniclastic-sediment transport to the deep-sea and can vary substantially in sediment concentration. The importance of debris avalanches from large oceanic islands and subduction-zone volcanoes in supplying large quantities of sediment to deep-sea basins has now been recognized. These events may result in a spectrum of sediment gravity flows, such as debris flows and turbidity currents, being generated during single events. There is now also a greater appreciation for the frequency of large-volume explosive sub-marine eruptions, as shown by the discovery of numerous sub-marine calderas in oceanic island arcs. Eruptions of this type are capable of producing tens to hundreds of km^3 of pumiceous debris that is distributed by hot and cold sediment gravity flows.

The long-term growth and evolution of volcanic islands in the marine environment is associated with the formation of volcaniclastic aprons that serve as repositories for large volumes of volcaniclastic sediment. Apron successions record the variation in volcanic activity of the source region and have a complex facies architecture as a result of multiple sediment sources. Around oceanic islands, volcaniclastic aprons extend up to several hundred kilometres and contain significant amounts of basaltic fragmental material. In oceanic island arcs, aprons accumulate in backarc basins and are more wedge-shaped in geometry. They contain a much larger diversity of particle compositions, owing to the larger range of source magmas in arcs and because the degree of fragmentation is typically greater.

Many important polymetallic mineral deposits, such as VHMS are closely associated with volcaniclastic sediments. In particular, recent discoveries of VHMS deposits in sub-marine arc calderas foreshadows significant new areas of exploration. Ancient volcaniclastic aprons may turn out to be areas of potential interest for hydrocarbon exploration as the physical properties and geothermal gradients in part of these sub-marine systems are conducive to hydrocarbon production.

Unlike the sub-aerial environment where volcanic terranes are often subject to enhanced rates of erosion and degradation, the deep-sea environment favours the preservation of event stratigraphy. Volcaniclastic sediments are thus important recorders of eruptive activity and can provide fundamental information about the frequency, duration, and nature of volcanism in a variety of tectonic environments. As such, they are increasingly being used as tools for volcanic-hazard assessment.

REFERENCES

Allen, S.R., Stewart, A.L., 2003. Products of explosive subaqueous felsic eruptions based on examples from the Hellenic Island arc, Greece. In: White, J.D.L., Smellie, J., Clague, D. (Eds.), Explosive Subaqueous Volcanism, AGU Geophysical Monograph 140, 285–298.

Bahk, J.J., Lee, S.H., Yoo, H.S., Back, G.G., Chough, S.K., 2005. Late Quaternary sedimentary processes and variations in bottom-current activity in the Ulleung Interplain Gap, East Sea (Korea). Mar. Geol. 217, 119–142.

Ballance, P.F., Gregory, M.R., 1991. Parnell Grits—Large subaqueous volcaniclastic gravity flows with multiple particle-support mechanisms. In: Fisher, R.V., Smith, G.A. (Eds.), Sedimentation in Volcanic Settings, SEPM Spec. Publ. 45, 189–200.

Batiza, R., White, J.D.L., 2000. Submarine lavas and hyaloclastite. In: Sigurdsson, H., Houghton, B., McNutt, S.R., Rymer, H., Stix, J. (Eds.), Encyclopedia of Volcanoes. Academic Press, San Diego, pp. 361–381.

Batiza, R., Fornari, D.J., Vanko, D.A., Lonsdale, P., 1984. Craters, calderas and hyaloclastites on young pacific seamounts. J. Geophys. Res. 89, 8371–8390.

Boudon, G., Semet, M., Vincent, P., 1984. Flank failure-directed blast eruption at Soufriere, Guadeloupe, French West Indies: a 3,000 yr old Mt. St. Helens? Geology 12, 350–353.

Bouysse, P., Mascle, A., 1994. Sedimentary basins and petroleum plays around the French Antilles. In: Mascle, A. (Ed.), Hydrocarbon and Petroleum Geology of France, EAPG Spec. Publ. 4, 431–443.

Buhring, C., Sarnthein, M., Leg 184 Shipboard Scientific Party, 2000. Toba ash layers in the South China Sea: evidence for contrasting wind directions during eruption ca. 74 ka. Geology 28, 275–278.

Burnham, C.W., 1983. Deep submarine pyroclastic eruptions. Econ. Geol. Monogr. 5, 142–148.

Busby-Spera, C.J., 1985. A sand rich submarine fan in the lower Mesozoic Mineral King Caldera complex, Sierra Nevada, California. J. Sediment. Petrol. 55, 376–391.

Busby-Spera, C.J., White, J.D.L., 1987. Variation in peperite textures associated with differing host-sediment properties. Bull. Volcanol. 49, 765–775.

Carey, S.N., 1991. Transport and deposition of tephra by pyroclastic flows and surges. In: Fisher, R.V., Smith, G.A. (Eds.), Sedimentation in Volcanic Settings, SEPM Spec. Publ. 45, 39–57.

Carey, S.N., 1997. Influence of convective sedimentation on the formation of widespread tephra fall layers in the deep sea. Geology 25, 839–842.

Carey, S.N., 2000. Volcaniclastic sedimentation around island arcs. In: Sigurdsson, H., Houghton, B., McNutt, S.R., Rymer, H., Stix, J. (Eds.), Encyclopedia of Volcanoes. Academic Press, San Diego, pp. 627–642.

Carey, S.N., Sigurdsson, H., 1984. A model of volcanogenic sedimentation in marginal basins. In: Kokelaar, B.P., Howells, M.F. (Eds.), Marginal Basin Geology: Volcanic and Associated Sedimentary and Tectonic Processes in Modern and Ancient Marginal Basins, Geol. Soc. Lond. Spec. Publ. 16, 37–58.

Carey, S.N., Sigurdsson, H., Mandeville, C., Bronto, S., 1996. Pyroclastic flows and surges over water, an example from the 1883 Krakatau eruption. Bull. Volcanol. 57, 493–511.

Carey, S., Sigurdsson, H., Mandeville, C., Bronto, S., 2000. Volcanic hazards from pyroclastic flow discharge into the sea: examples from the 1883 eruption of Krakatau, Indonesia. Geol. Soc. Am. Spec. Pap. 345, 1–14.

Carlisle, D., 1963. Pillow breccias and their aquagene tuffs, Quadra Island, British Columbia. J. Geol. 71, 48–71.

Cas, R.A.F., 1992. Submarine volcanism: eruption style, products and relevance to understanding the host-rock successions to volcanic-hosted massive sulfide deposits. Econ. Geol. 87, 511–554.

Cas, R.A.F., Wright, J.V., 1987. Volcanic Successions—Modern and Ancient. Allen & Unwin, London, 528 pp.

Cas, R.A.F., Wright, J.V., 1991. Subaqueous pyroclastic flows and ignimbrites: an assessment. Bull. Volcanol. 53, 357–380.

Cas, R.A.F., Powell, C., Fergusson, C., Jones, J., Roots, W., Fergusson, J., 1981. The lower Devonian Kowmung volcaniclastics: a deep-water succession of mass-flow origin, northeastern Lahlan Fold Belt, N.S.W.. J. Geol. Soc. Aust. 28, 271–288.

Cashman, K.V., Fiske, R.S., 1991. Fallout of pyroclastic debris from submarine volcanic eruptions. Science 253, 275–280.

Clague, D., Davis, A., Dixon, J., 2003a. Submarine strombolian eruptions on the Gorda Mid-Ocean Ridge. In: White, J.D.L., Smellie, J., Clague, D. (Eds.), Explosive Subaqueous Volcanism, AGU Geophys. Monogr. 140, 111–128.

Clague, D., Batiza, R., Head, J., Davis, A., 2003b. Pyroclastic and hydroclastic deposits on Loihi Seamount, Hawaii. In: White, J.D.L., Smellie, J., Clague, D. (Eds.), Explosive Subaqueous Volcanism, AGU Geophys. Monogr. 140, 73–95.

Cochonat, P., Lénat, J.-F., Bachèlery, P., Boivin, P., Cornaglia, B., Deniel, C., et al., 1990. Importance des dépôts gravitaires dans la mise en place d'un système volcano-sédimentaire sous-marin (volcan de la Fournaise, Ile de la Réunion). C.R. Acad. Sci. Paris 311 (II), 679–686.

Cole, R.B., DeCelles, P.G., 1991. Subaerial to submarine transitions in early Miocene pyroclastic flow deposits, southern San Joaquin basin, California. Geol. Soc. Am. Bull. 103, 221–235.

Cole, R.B., Stanley, R.G., 1994. Sedimentology of subaquatic volcaniclastic sediment gravity flows in the Neogene Santa Maria Basin, California. Sedimentology 41, 37–54.

Davies, D.K., Almon, W.R., Bonis, S.B., Hunter, B.E., 1979. Deposition and diagenesis of Tertiary-Holocene volcaniclastics, Guatemala. In: Scholle, P.A., Schluger, P.R. (Eds.), Aspects of Diagenesis, SEPM Spec. Publ. 26, 281–306.

Deplus, C., Le Friant, A., Boudon, G., Komorowski, J.-C., Villemant, B., Harford, C., et al., 2001. Submarine evidence for large-scale debris avalanches in the Lesser Antilles Arc. Earth Planet. Sci. Lett. 192, 145–157.

Eissen, J.-P., Fouquet, Y., Hardy, D., Ondréas, H., Eissen, J.-P., Fouquet, Y., et al., 2003. Recent MORB volcaniclastic explosive deposits formed between 500 and 1750 m.b.s.l. on the axis of the Mid-Atlantic Ridge, South of the Azores. In: White, J.D.L., Smellie, J.L., Clague, D.A. (Eds.), Explosive Subaqueous Volcanism, AGU Geophys. Monogr. 140, 143–166.

Faugères, J.-C., Mulder, T., 2011. Contour currents and contourite drifts. In: Hüneke, H., Mulder, T. (Eds.), Deep-Sea Sediments. Developments in Sedimentology, Elsevier Amsterdam, vol. 63, pp. 149–214.

Fisher, R.V., 1961. Proposed classification of volcaniclastic sediments and rocks. Geol. Soc. Am. Bull. 72, 1409–1414.

Fisher, R.V., 1966. Rocks composed of volcanic fragments and their classification. Earth Sci. Rev. 1, 287–298.

Fisher, R.V., 1983. Flow transformations in sediment gravity flows. Geology 11, 273–274.

Fisher, R.V., 1984. Submarine volcaniclastic rocks. In: Kokelaar, B.P., Howells, M.F. (Eds.), Marginal Basin Geology: Volcanic and Associated Sedimentary and Tectonic Processes in Modern and Ancient Marginal Basins, Geol. Soc. Lond. Spec. Publ. 16, 5–27.

Fisher, R.V., Schmincke, H.U., 1984. Pyroclastic Rocks. Springer Verlag, Berlin 472 pp.

Fisher, R.V., Schmincke, H.U., 1994. Volcaniclastic sediment transport and deposition. In: Pye, K. (Ed.), Sediment Transport and Depositional Processes. Blackwell Scientific Publications, pp. 351–388.

Fisher, R.V., Smith, G.A., 1991. Volcanism, tectonics and sedimentation. In: Fisher, R.V., Smith, G.A. (Eds.), Sedimentation in Volcanic Settings, SEPM Spec. Publ. 45, 1–5.

Fiske, R.S., Matsuda, T., 1964. Submarine equivalents of ash flows in the Tokiwa Formation, Japan. Am. J. Sci. 262, 76–106.

Fiske, R.S., Cashman, K.V., Shibata, A., Watanabe, K., 1998. Tephra dispersal from Myojinsho, Japan, during its shallow submarine eruption of 1952–1953. Bull. Volcanol. 59, 262–275.

Fiske, R.S., Naka, J., Iizasa, K., Yuasa, M., Klaus, A., 2001. Submarine silicic caldera at the front of the Izu-Bonin Arc, Japan, voluminous seafloor eruption of rhyolite pumice. Geol. Soc. Am. Bull. 113, 813–824.

Fouquet, Y., Eissen, J.P., Ondréas, H., Barriga, F., Batiza, R., Danyushevsky, L., 1998. Extensive volcaniclastic deposits at the Mid-Atlantic Ridge axis: results of deep-water basaltic explosive activity. Terra Nova 10, 280–286.

Freundt, A., 2003. Entrance of hot pyroclastic flows into the sea: experimental observations. Bull. Volcanol. 65, 144–164.

Freundt, A., Schmincke, H.-U., 1998. Emplacement of ash layers related to high-grade ignimbrite P1 in the sea around Gran Canaria. In: Weaver, P.P.E., Schmincke, H.-U., Firth, J.V., Duffield, W. (Eds.), Gran Canaria and Madeira Abyssal Plain, Proc. ODP, Sci. Res. 157, 201–218.

Gamberi, F., 2001. Volcanic facies associations in a modern volcaniclastic apron (Lipari and Vulcano offshore, Aeolian Island Arc). Bull. Volcanol. 63, 264–273.

Garcia, M.O., 1978. Criteria for the identification of ancient volcani arcs. Earth Sci. Rev. 14, 147–165.

Garcia, M.O., 1996. Turbidites from slope failure on Hawaiian volcanoes. In: McGuire, W. J., Jones, A.P., Neuberg, J. (Eds.), Volcano Instability on the Earth and Other Planets, Geol. Soc. Lond. Spec. Publ. 110, 281–294.

Gill, J., Torssander, P., Lapierre, H., Taylor, R., Kaiho, K., Koyama, M., et al., 1990. Explosive deep-water basalt in the Sumisu backarc rift. Science 248, 1214–1217.

Glicken, H., 1991. Sedimentary architecture of large volcanic-debris avalanches. In: Fisher, R.V., Smith, G.A. (Eds.), Sedimentation in Volcanic Settings, SEPM Spec. Publ. 45, 99–106.

Hart, K., Carey, S., Sigurdsson, H., Sparks, R.S.J., Robertson, R.E.A., 2004. Discharge of pyroclastic flows into the sea during the 1996–1998 eruptions of the Soufrière Hills volcano, Montserrat. Bull. Volcanol. 66, 599–614.

Hawlander, H.M., 1990. Diagenesis and reservoir potential of volcanogenic sandstones—Cretaceous of the Surat Basin, Australia. Sediment. Geol. 66, 181–195.

Head III, J.W., Wilson, L., 2003. Deep submarine pyroclastic eruptions: theory and predicted landforms and deposits. J. Volcanol. Geotherm. Res. 121, 155–193.

Hedervari, P., 1984. Catalog of submarine volcanoes and hydrological phenomena associated with volcanic events—1500 B.C. to December 31, 1899. U.S. Department of Commerce World Data Center of a Solid Earth Geophysics, Report SE-36, 62 pp.

Heiken, G., Wohletz, K., 1985. Volcanic Ash. University of California Press, Berkeley 246 pp.

Hékinian, R., Pineau, F., Shilobreeva, S., Bideau, D., Gracia, E., Javoy, M., 2000. Deep sea explosive activity on the Mid-Atlantic Ridge near 34°50'N: magma composition, vesicularity and volatile content. J. Volcanol. Geotherm. Res. 98, 49–77.

Holcomb, R.T., Searle, R.C., 1991. Large landslides from oceanic volcanoes. Mar. Geotech. 10, 19–32.

Hon, K., Heliker, C., Kjargaard, J., 1988. Limu o Pele: a new kind of hyaloclastitic tephra from Kilauea volcano, Hawaii. Geol. Soc. Am. Abstr. 20, A112–A113.

Honnorez, J., Kirst, P., 1975. Submarine basaltic volcanism: morphometric parameters for discriminating hyaloclastites from hyalotuffs. Bull. Volcanol. 39, 1–25.

Howells, M.F., Leveridge, B.E., Addison, R., Evans, C., Nutt, M., 1979. The Capel Curig Volcanic Formation, Snowdonia, North Wales: variations in ash-flow tuffs related to emlacement environment. In: Harris, A., Holland, C., Leake, B. (Eds.), The Caledonides of the British Isles reviewed, Geol. Soc. Lond. Spec. Publ. 8, 611–618.

Hudak, G., Morton, R., Franklin, J., Petersen, D., 2003. Morphology, distribution, and estimated volumes for intracaldera tuffs associated with volcanic-hosted massive sulphide deposits in the Archean Stugeon Lake Caldera complex, Northwestern Ontario. In: White, J.D.L., Smellie, J., Clague, D. (Eds.), Explosive Subaqueous Volcanism, AGU Geophys. Monogr. 140, 345–359.

Hüneke, H., Henrich, R., 2011. Pelagic sedimentation in modern and ancient oceans. In: Hüneke, H., Mulder, T. (Eds.), Deep-Sea Sediments. Developments in Sedimentology, Elsevier Amsterdam, vol. 63, pp. 215–351.

Iizasa, K., Fiske, R.S., Ishizuka, O., Yusa, M., Hashimoto, J., Ishibashi, J., et al., 1999. A Kuroko-type polymetallic deposit in a submarine silicic caldera. Science 283, 975–977.

Imbert, P., 2011. Industrial application of deep-sea sediments. In: Hüneke, H., Mulder, T. (Eds.), Deep-Sea Sediments. Developments in Sedimentology, Elsevier, Amsterdam, vol. 63, pp. 715–764.

Jones, J.G., Nelson, P.H.H., 1970. The flow of basalt lava from air into water—its structural expression and stratigraphic significance. Geol. Mag. 107, 13–19.

Kano, K., 1990. An ash-flow tuff emplaced in shallow water, Early Miocene Koura Formation, southwest Japan. J. Volcanol. Geotherm. Res. 40, 1–9.

Kano, K., 2003. Subaqueous pumice eruptions and their products: a review. In: White, J.D. L., Smellie, J., Clague, D. (Eds.), Explosive Subaqueous Volcanism, AGU Geophys. Monogr. 140, 213–229.

Kano, K., Takeuchi, K., Yamamoto, T., Hoshizumi, H., 1991. Subaqueous rhyolite block lavas in the Miocene Ushikiri Formation, Shimane Peninsula, SW Japan. J. Volcanol. Geotherm. Res. 46, 241–253.

Karig, D., 1971. Origin and development of marginal basins in the Western Pacific. J. Geophys. Res. 76, 2542–2561.

Kato, Y., 1987. Woody pumice generated with submarine eruption. J. Geol. Soc. Japan 77, 193–206.

Keller, J., Ryan, W.B.F., Ninkovich, D., Altherr, R., 1978. Explosive volcanic activity in the Mediterranean over the past 200,000 years as recorded by deep-sea sediments. Geol. Soc. Am. Bull. 89, 591–604.

Kennett, J., 1981. Marine tephrochronology. In: Emiliani, C. (Ed.), The Sea. Wiley, New York, pp. 1373–1436.

Kessek, L., Busby, C., 2003. Analysis of VHMS-hosting ignimbrites erupted at bathyal water depths (Ordovician Bald Mountain Sequence, N. Maine). In: White, J.D.L., Smellie, J., Clague, D. (Eds.), Explosive Subaqueous Volcanism, AGU Geophys. Monogr. 140, 361–379.

Kokelaar, P., 1986. Magma-water interactions in subaqueous and emergent basaltic volcanism. Bull. Volcanol. 48, 275–289.

Kokelaar, P., Busby, C., 1992. Subaqueous explosive eruption and welding of pyroclastic deposits. Science 257, 196–201.

Krastel, S., Schmincke, H.-U., Jacobs, C.L., Rihm, R., Le Bas, T.P., Alibès, B., 2001. Submarine landslides around the Canary Islands. J. Geophys. Res. 106 (B3), 3977–3997.

Labazuy, P., 1996. Recurrent landslides events on the submarine flank of Piton de la Fournaise volcano (Reunion Island). In: McGuire, W.J., Jones, A.P., Neuberg, J. (Eds.), Volcano Instability on the Earth and Other Planets, Geol. Soc. Lond. Spec. Publ. 110, 295–306.

Lacasse, C., Sigurdsson, H., Carey, S., Paterne, M., Guichard, F., 1996. North Atlantic deep-sea sedimentation of late Quaternary tephra from the Iceland hotspot. Mar. Geol. 129, 207–235.

Le Friant, A., 2001. Les déstabilisations de flanc des volcans actifs de l'arc des Petites Antilles: origines et conséquences. Ph.D. thesis Univ. Paris 7, 378 pp.

Le Friant, A., Boudon, G., Deplus, C., Villemant, B., 2003. Large scale flank collapse events during the activity of Montagne Pelée, Martinique, Lesser Antilles. J. Geophys. Res. 108, 2055. doi:10.1029/2001JB001624.

Legros, F., Druitt, T.H., 2000. On the emplacement of ignimbrite in shallow-marine environments. J. Volcanol. Geotherm. Res. 95, 9–22.

Leslie, S.C., Moore, G.F., Morgan, J.K., Hills, D.J., 2002. Stratigraphy of the Frontal Hawaiian Moat: implications for sedimentary processes at the leading edge of an oceanic hotspot trace. Mar. Geol. 184, 143–162.

Lipman, P.W., Normark, W.R., Moore, J.G., Wilson, J.B., Gutmacher, C.E., 1988. The giant submarine Alika debris slide, Mauna Loa, Hawaii. J. Geophys. Res. 93 (B5), 4279–4299.

Lonsdale, P., Batiza, R., 1980. Hyaloclastite and lava flows on young seamounts examined with a submersible. Geol. Soc. Am. Bull. 91, 545–554.

Lowe, D.R., 1982. Sediment gravity flows: II, depositional models with special reference to the deposits of high-density turbidity currents. J. Sediment. Petrol. 52, 279–297.

Mandeville, C.W., Carey, S.N., Sigurdsson, H., 1994. Paleomagnetic evidence for high temperature emplacement of the 1883 subaqueous pyroclastic flows from Krakatau volcano, Indonesia. J. Geophys. Res. 99, 9487–9504.

Mandeville, C.W., Carey, S.N., Sigurdsson, H., 1996. Sedimentology of the Krakatau 1883 submarine pyroclastic deposits. Bull. Volcanol. 57, 512–529.

Mangeney, A., Heinrich, P., Roche, R., Boudon, G., Cheminée, J.-L., 2000. Modeling of debris avalanche and generated water waves: application to real and potential events in Monteserrat. Phys. Chem. Earth A 25 (9–11), 741–745.

Manville, V., White, J.D.L., Houghton, B.F., Wilson, C.J.N., 1998. The saturation behaviour of pumice and some sedimentological implications. Sediment. Geol. 119, 5–16.

Maria, A., Carey, S., Sigurdsson, H., Kincaid, C., Helgadóttir, G., 2000. Source and dispersal of jökulhlaup sediments discharged to the sea following the 1996 Vatnajökull eruption. Geol. Soc. Am. Bull. 112, 1507–1521.

Marsaglia, K.M., 1995. Interarc and backarc basins. In: Busby, C.J., Ingersoll, R.V. (Eds.), Tectonics of Sedimentary Basins. Blackwell Science, pp. 299–329.

Masson, D.G., 1996. Catastrophic collapse of the volcanic island of Hierro 15 ka ago and the history of landsides in the Canary Island. Geology 24, 231–234.

Mathiesen, M.E., McPherson, J.G., 1991. Volcaniclastic deposits: implications for hydrocarbon exploration. In: Fisher, R.V., Smith, G.A. (Eds.), Sedimentation in Volcanic Settings, SEPM Spec. Publ. 45, 27–36.

McBirney, A.R., 1963. Factors governing the nature of submarine volcanism. Bull. Volcanol. 26, 455–469.

McClelland, L., Simkin, T., Summers, M., Nielsen, E., Stein, T.C., 1989. Global Volcanism 1975–1985. Prentice Hall, Englewood Cliffs/American Geophysical Union, 657 pp.

McCoy, F.W., Heiken, G., 2000. Tsunami generated by the Late Bronze Age eruption of Thera (Santorini), Greece. Pure Appl. Geophys. 157, 1227–1256.

Menard, H.W., 1956. Archipelagic aprons. Bull. Am. Assoc. Petrol. Geol. 40, 2195–2210.

Moore, J.G., Mark, R.K., 1992. Morphology of the island of Hawaii. GSA Today 2, 257–259.

Moore, J.G., Phillips, R.L., Grigg, R.W., Peterson, D.W., Swanson, D.A., 1973. Flow of lava into the sea, 1969–1971, Kilauea volcano, Hawaii. Geol. Soc. Am. Bull. 84, 537–546.

Moore, J.G., Clague, D.A., Holcomb, R.T., Lipman, P.W., Normark, W.R., Torresan, M. E., 1989. Prodigious submarine landslides on the Hawaiian Ridge. J. Geophys. Res. 94 (B12), 17465–17484.

Moore, J.G., Normark, W.R., Holcomb, R.T., 1994. Giant Hawaiian landslides. Annu. Rev. Earth Planet. Sci. 22, 119–144.

Moore, J.G., Bryan, W.B., Beeson, M.H., Normark, W.R., 1995. Giant blocks in the South Kona landslide, Hawaii. Geology 23, 125–128.

Mulder, T., 2011. Gravity processes on continental slope, rise and abyssal plains. In: Hüneke, H., Mulder, T. (Eds.), Deep-Sea Sediments. Developments in Sedimentology, Elsevier, Amsterdam, vol. 63, pp. 25–148.

Mulder, T., Hüneke, H., van Loon, A.J., 2011. Progress in Deep-Sea sedimentology. In: Hüneke, H., Mulder, T. (Eds.), Deep-Sea Sediments. Developments in Sedimentology, Elsevier Amsterdam, vol. 63, pp. 1–24.
Mueller, W., White, J.D.L., 1992. Felsic fire-fountaining beneath Archean seas: pyroclastic deposits of the 2730 Ma Hunter Mine Group, Quebec, Canada. J. Volcanol. Geotherm. Res. 54, 117–134.
Ninkovich, D., Shackleton, N., 1975. Distribution, stratigraphy, position and age of ash layer 'L' in the Panama Basin regions. Earth Planet. Sci. Lett. 27, 20–34.
Ninkovich, D., Sparks, R.S.J., Ledbetter, M.J., 1978. The exceptional magnitude and intensity of the Toba eruption, Sumatra: an example of the use of deep-sea tephra layers as a geological tool. Bull. Volcanol. 41, 286–298.
Ohmoto, H., 1996. Formation of volcanogenic massive sulfide deposits: the Kuroko perspective. Ore Geology Rev. 10, 135–177.
Ollier, G., Cochonat, P., Lénat, J.-F., Labazuy, P., 1998. Deep-sea volcaniclastic sedimentary systems: example of La Fournaise volcano, Réunion Island, Indian Ocean. Sedimentology 45, 293–330.
Orton, G.J., 1996. Volcanic environments. In: Reading, H.G. (Ed.), Sedimentary Environments: Processes, Facies and Stratigraphy. 3rd ed. Blackwell Science, pp. 485–567.
Paterne, M., Labeyrie, J., Guichard, F., Mazaud, A., Maitre, F., 1990. Fluctuations of the Campanian explosive volcanic activity (South Italy) during the past 190,000 years, as determined by marine tephrochronology. Earth Planet. Sci. Lett. 98, 166–174.
Picard, M., Schneider, J.-L., Boudon, G., 2006. Contrasted sedimentary processes along a convergent margin: the Lesser Antilles arc system. Geo-Mar. Lett. 26, 397–410.
Pichler, H., 1965. Acid hyaloclastites. Bull. Volcanol. 28, 293–310.
Perfit, M., Chadwick, W., 1998. Magmatism at Mid-Ocean Ridges Contraints from Volcanological and Geochemical Invesgations In: Faulting and Magmatism at Mit-Ocean Ridgs. Geophysical Monograph 106, 59–115, American Geophysical Union Washington, DC.
Pollet, N., Schneider, J.-L., 2004. Rock mass dynamic disintegration processes of the Flims Holocene sturzstrom (Swiss Alps). Earth Planet. Sci. Lett. 221, 433–448.
Sáez, R., Almodóvar, G.R., Pascual, E., 1996. Geological constraints on massive sulphide genesis in the Iberian Pyrite Belt. Ore Geol. Rev. 11, 429–451.
Sarna-Wojicki, A.M., Morrison, S.D., Meyer, C.E., Hillhouse, J.W., 1987. Correlation of Upper Cenozoic tephra layers between sediments of the Western United States and Eastern Pacific Ocean and comparison with biostratigraphic and magnetostratigraphic age data. Geol. Soc. Am. Bull. 98, 207–223.
Scasso, R., 2001. High frequency of explosive volcanic eruptions in a Late Jurassic volcanic arc: the Ameghino formation in the Antarctic Peninsula. J. Sedimant. Res. 71, 101–106.
Schmid, R., 1981. Descriptive nomenclature and classification of pyroclastic deposits and fragments: recommendations of the IUGS Subcommission on the systematics of igneous rocks. Geology 9, 41–43.
Schmidt, R., Schmincke, H.-U., 2000. Seamounts and island building. In: Sigurdsson, H., Houghton, B., McNutt, S.R., Rymer, H., Stix, J. (Eds.), Encyclopedia of Volcanoes. Academic Press, San Diego, pp. 383–402.
Schmincke, H.-U., Van den Bogaard, P., 1991. Tephra layers and tephra events. In: Einsele, G., Ricken, W., Seilacher, A. (Eds.), Cycles and Events in Stratigraphy. Springer Verlag, Heidelberg, pp. 329–429.
Schmincke, H.-U., Robinson, P., Ohnmacht, W., Flower, M., 1978. Basaltic hyaloclastites from hole 396B, DSDP Leg 46. In: Dmitriev, L., Heirtzler, J., Aguilar, R., Cambon, P., Dick, H.J.B., Dungan, M. et al., Init. Repts. D.S.D.P., 341–348.
Schmincke, H.-U., Weaver, P.P.E., Firth, J.V., Baraza, J., Bristow, J.F., Brunner, C., 1995. Gran Canaria and Madeira Abyssal Plain. Proc. ODP, Init. Repts. 157, 843 pp.

Schneider, J.-L., 1997. Ecoulements pyroclastiques et avalanches de débris en domaine volcanique. In: Ildefonse, B., Allain, C., Coussot, P. (Eds.), Des grands Écoulements Naturels à la Dynamique du Tas de Sable. Introduction aux Suspensions en Géologie et en Physique. Cemagref Editions, 51–70, Paris.
Schneider, J.-L., 1998. Du volcan au sédiment: les formations volcanoclastiques. Habilitation à Diriger des Recherches, Univ. Lille 1, 444 pp.
Schneider, J.-L., 2000. Volcaniclastic sedimentation in submarine settings: products and processes. In: Leyrit, H., Montenat, C. (Eds.), Volcaniclastic Rocks, from Magmas to Sediments. Gordon & Breach Science Publishers, Amsterdam, pp. 175–192.
Schneider, J.-L., Fisher, R.V., 1996. Obispo Formation, California: remobilized pyroclastic material. U.S. Geol. Surv. Bull. 1995 (O), O1–O21.
Schneider, J.-L., Fourquin, C., Paicheler, J.C., 1992. Two examples of subaqueously welded ash-flow tuffs: the Visean of southern Vosges (France) and the Upper Cretaceous of northern Anatolia (Turkey). J. Volcanol. Geotherm. Res. 49, 365–383.
Schneider, J.-L., Brunner, C.A., Kuttner, S., 293-313, 1998. Epiclastic sedimentation during the upper Miocene-lower Pliocene volcanic hiatus of Gran Canaria: evidence from Sites 953 and 954. In: Weaver, P.P.E., Schmincke, H.-U., Firth, J.V., Duffield, W. (Eds.), Gran Canaria and Madeira Abyssal Plain, Proc. ODP, Sci. Res., 157, 293–313.
Schneider, J.-L., Le Ruyet, A., Chanier, F., Buret, C., Ferrière, J., Proust, J.-N., et al., 2001. Primary or secondary distal volcaniclastic turbidites: how to make a distinction An example from the Miocene of New Zealand (Mahia Peninsula, North Island). Sediment. Geol. 145, 1–22.
Schneider, J.-L., Pérez Torrado, F.J., Gimeno Torrente, D., Wassmer, P., Cabrera Santana, M.C., Carracedo, J.C., 2004. Sedimentary signatures of the entrance of coarse-grained volcaniclastic flows into the sea: the example of the breccia units of the Las Palmas Detritic Formation (Mio-Pliocene, Gran Canaria, eastern Atlantic, Spain). J. Volcanol. Geotherm. Res. 138, 295–323.
Seeman, U., Scherer, M., 1984. Volcaniclastics as potential hydrocarbon reservoirs. Clay Miner. 9, 457–470.
Shaller, P.J., 1991. Analysis and Implications of Large Martian and Terrestrial Landslides. Ph.D. thesis CalTech, Pasadena, 586 pp.
Sigurdsson, H., Sparks, R.S.J., Carey, S.N., Huang, T.C., 1980. Volcanogenic sedimentation in the Lesser Antilles Arc. J. Geol. 88, 523–540.
Sigurdsson, H., Carey, S., Mandeville, C., Bronto, S., 1991. Pyroclastic flows of the 1883 Krakatau eruption. EOS Trans. 72 (36), 380–381 377+.
Sigurdsson, H., Kelley, S., Leckie, M., Carey, S., Bralower, T., King, J., 2000. History of circum-Caribbean explosive volcanism: ^{40}Ar/^{39}Ar dating of tephra layers. In: Leckie, M., Sigurdsson, H., Acton, G., Draper, G. (Eds.), Proc. ODP, Sci. Res. 165, 299–314.
Simkin, T., Fiske, R.S., 1983. Krakatau 1883—The Volcanic Eruption and Its Effects. Smithsonian Institution Press, Washington, 464 pp.
Skilling, I.P., White, J.D.L., McPhie, J., 2002. Peperite: a review of magma-sediment mingling. J. Volcanol. Geotherm. Res. 114, 1–17.
Smith, G.A., 1986. Coarse-grained nonmarine volcaniclastic sediment: terminology ad depositional process. Geol. Soc. Am. Bull. 97, 1–10.
Smith, T.L., Batiza, R., 1989. New field and laboratory evidence for the origin of hyaloclastite flows on seamount summits. Bull. Volcanol. 51, 96–114.
Smith, G.A., Landis, C.A., 1995. Intra-arc basins. In: Busby, C.J., Ingersoll, R.V. (Eds.), Tectonics of Sedimentary Basins. Blackwell Science, pp. 263–298.
Sparks, R.S.J., 1978. The dynamics of bubble formation and growth in magmas—a review and analysis. J. Volcanol. Geotherm. Res. 3, 1–37.
Sparks, R.S.J., Wilson, C.J.N., 1983. Flow-head deposits in ash turbidites. Geology 11, 348–351.

Staudigel, H., Schmincke, H.-U., 1984. The Pliocene seamount series of La Palma/Canary Islands. J. Geophys. Res. 89 (B13), 11195–11215.
Stix, J., 1991. Subaqueous, intermediate to silicic-composition explosive volcanism: a review. Earth Sci. Rev. 31, 21–53.
Stow, D.A.V., Taira, A., Ogawa, Y., Soh, W., Tanigushi, H., Pickering, K.V., 1998. Volcaniclastic sediments, process interaction and depositional setting of the Mio-Pliocene Miura Group, SE Japan. Sediment. Geol. 115, 351–381.
Surdam, R.C., Boles, J.R., 1979. Diagenesis of volcanic sandstones: a review. In: Scholle, P. A., Schluger, P.R. (Eds.), Aspects of Diagenesis, SEPM Spec. Publ. 26, 227–242.
Suthern, R.J., 1985. Facies analysis of volcaniclastic sediments: a review. In: Brenchley, P.J., Williams, B.P. (Eds.), Sedimentology, Recent Developments and Applied Aspects, Geol. Soc. Lond. Spec. Publ. 18, 123–146.
Takahashi, E., Lipman, P.W., Garcia, M.O., Naka, J., Aramaki, S., 2002. Hawaiian volcanoes: deep underwater perspectives. AGU Geophys. Monogr. 128, 418.
Talandier, J., 1989. Submarine volcanic activity: detection, monitoring and interpretation. EOS Trans. 70, 561–568.
Underwood, M., Ballance, P., Clift, P., Hiscott, R., Marsaglia, K., Pickering, K., et al., 1995. Sedimentation in forearc basins, trenches, and collision zones of the western Pacific: a summary of results from the Ocean Drilling Program. In: Taylor, B., Natland, J. (Eds.), Active Margins and Marginal Basins of the Western Pacific, Am. Geophys. Union Monogr. 88, 315–354.
Wadge, G., Isaacs, M., 1988. Mapping the volcanic hazards from Soufriere Hills volcano, Montserrat, West Indies using an image processor. J. Geol. Soc. London 145, 541–551.
Walker, G.P.L., 1979. A volcanic ash generated by explosions where ignimbrite entered the sea. Nature 281, 642–646.
Wallace, P., Anderson, A.T., 2000. Volatiles in magmas. In: Sigurdsson, H., Houghton, B., McNutt, S.R., Rymer, H., Stix, J. (Eds.), Encyclopedia of Volcanoes. Academic Press, San Diego, pp. 149–169.
Ward, S.N., 2001. Landslide tsunami. J. Geophys. Res. 106, 11201–11215.
White, J.D.L., Smellie, J.L., Clague, D.A., 2003. Introduction: a deductive outline and topical overview of subaqueous explosive volcanism. In: White, J.D.L., Smellie, J., Clague, D. (Eds.), Explosive Subaqueous Volcanism, AGU Geophys. Monogr. 140, 1–23.
White, N.C., Herrington, R., 2000. Mineral deposits associated with volcanism. In: Sigurdsson, H., Houghton, B., McNutt, S.R., Rymer, H., Stix, J. (Eds.), Encyclopedia of Volcanoes. Academic Press, San Diego, pp. 897–913.
Whitham, A.G., 1989. The behaviour of subaerially produced pyroclastic flows in a subaqueous environment: evidence from the Roseau eruption, Dominica, West Indies. Mar. Geol. 86, 27–40.
Whitham, A.G., Sparks, R.S.J., 1986. Pumice. Bull. Volcanol. 48, 209–223.
Wiesner, M.G., Wetzel, A., Catane, S.G., Listanco, E.L., Mirabueno, H.T., 2004. Grain size, areal thickness distribution and controls on sedimentation of the 1991 Mount Pinatubo tephra layer in the South China Sea. Bull. Volcanol. 66, 226–242.
Wilson, C.J.N., Houghton, B.F., 2000. Pyroclastic transport and deposition. In: Sigurdsson, H., Houghton, B., McNutt, S.R., Rymer, H., Stix, J. (Eds.), Encyclopedia of Volcanoes. Academic Press, San Diego, pp. 545–554.
Wohletz, K.H., 1983. Mechanisms of hydrovolcanic pyroclast formation: grain size, scanning electron microscopy and experimental results. J. Volcanol. Geotherm. Res. 17, 31–63.
Wohletz, K.H., 2003. Water/magma interactions: physical considerations for deep submarine environment. In: White, J.D.L., Smellie, J., Clague, D. (Eds.), Explosive Subaqueous Volcanism, AGU Geophys. Monogr. 140, 25–50.

Wright, I.C., 1996. Volcaniclastic processes on modern submarine arc stratovolcanoes: sidescan and photographic evidence from the Rumble IV and V volcanoes, southern Kermadec Arc (SW Pacific). Mar. Geol. 136, 21–39.

Wright, I.C., 2001. In situ modification of modern submarine hyaloclastic/pyroclastic deposits by oceanic currents: an example from the Southern Kermadec arc (SW Pacific). Mar. Geol. 172, 287–307.

Wright, J.V., Coward, M.P., 1977. Rootless vents in welded ash-flow tuffs from northern Snowdonia, North Wales, indicating deposition in a shallow water environment. Geol. Mag. 114, 133–140.

Wright, J.V., Mutti, E., 1981. The Dali Ash, Island of Rhodes, Greece: a problem interpreting submarine volcanogenic sediments. Bull. Volcanol. 44, 153–167.

Wright, I.C., De Ronde, C.E.J., Faure, K., Gamble, J.A., 1998. Discovery of hydrothermal sulfide mineralization from southern Kermadec arc volcanoes (SW Pacific). Earth Planet. Sci. Lett. 164, 335–343.

Wright, I.C., Gamble, J.A., Shane, P., 2003. Submarine silicic volcanism of the Healy caldera, southern Kermadec arc (SW Pacific) I—volcanology and eruption mechanisms. Bull. Volcanol. 65, 15–29.

Yamada, E., 1984. Subaqueous pyroclastic flows, their development and their deposits. In: Kokelaar, B.P., Howells, M.F. (Eds.), Marginal Basin Geology, Volcanic and Associated Sedimentary and Tectonic Processes in Modern and Ancient Marginal Basins, Geol. Soc. Lond. Spec. Publ. 16, 29–35.

Yokose, H., Lipman, P.W., 2004. Emplacement mechanisms of the South Kona slide complex, Hawaii Island: sampling and observations by remotely operated vehicle Kaiko. Bull. Volcanol. 66, 569–584.

Young, S., Sparks, S., Robertson, R., Lynch, L., Aspinall, W., 1997. Eruption of Soufriere Hills Volcano in Montserrat continues. EOS Trans. 78 (38), 408–409 401+.

Yuasa, M., Kano, K., 2003. Submarine silicic calderas on the Northern Shichito-Iwojima Ridge, Izu-Ogasawara (Bonin) Arc, Western Pacific. In: White, J.D.L., Smellie, J., Clague, D. (Eds.), Explosive Subaqueous Volcanism, AGU Geophys. Monogr. 140, 231–243.

Zimanowski, B., Buttner, R., 2003. Phreatomagmatic explosions in subaqueous volcanism. In: White, J.D.L., Smellie, J., Clague, D. (Eds.), Explosive Subaqueous Volcanism, AGU Geophys. Monogr. 140, 51–60.

CHAPTER 8

Deep-Sea Ichnology: The Relationships Between Depositional Environment and Endobenthic Organisms

A. Uchman*[,1] *and* A. Wetzel[‡]

Contents

* Institute of Geological Sciences, Jagiellonian University, Kraków, Oleandry 2a, Poland
‡ Geologisch-Paläontologisches Institut der Universität, Basel, Switzerland
1 Corresponding author.
E-mail address: alfred.uchman@uj.edu.pl

Developments in Sedimentology, Volume 63
ISSN 0070-4571, DOI: 10.1016/S0070-4571(11)63008-2

1. Introduction

Animals living on and within the sea floor disturb the primary sedimentary structures and produce a new fabric, the so-called ichnofabric. This process is called bioturbation. As a rule, deep-sea sediments continuously accumulating in an oxygenated setting are completely bioturbated. Identifiable bioturbational structures of recurrent shape are called 'trace fossils'. Because of their biogenic origin, both unidentifiable bioturbational structures and trace fossils are essential components of ichnofabrics and carry information about the *in situ* (palaeo)ecological conditions. Therefore, they are a valuable source of information in various analyses of deep-sea sediments.

For a long time, deep-sea ichnology developed a little behind shallow-water ichnology, mostly because the deep sea remained a mysterious environment until the fifties of the twentieth century. In the first half of nineteenth century, however, trace fossils from flysch deposits were described as algae (fucoids) without recognition of their deep-sea origin (e.g. Fischer-Ooster, 1858). After the true nature of trace fossils had been recognized (Nathorst, 1881), they were increasingly studied (Abel, 1935; Richter, 1928; Seilacher, 1953, 1955). Their potential as ecological indicators drastically increased when modern environments were studied in addition. Advanced actuo-ichnologic studies of modern traces, i.e. *lebensspuren*, and bioturbation started in shallow-water settings (Frey, 1970; Richter, 1924, 1938; Schäfer, 1956), then also in the deep sea (Reineck, 1973; Ekdale, 1977; Gaillard, 1988; Wetzel, 1981; Ohta, 1984). At first, the deep-sea trace fossils were related to parameters like the energy of the environment (turbiditic vs. non-turbiditic) and indirectly to bathymetry (Seilacher, 1964, 1967, 1974). Later, other parameters were taken into account such as sedimentation rate, particle flux, organic-carbon content, pore-water chemistry, etc. (e.g. Wetzel, 1991). Furthermore, the understanding of the interrelationship in benthic communities and their dependence on ecologic parameters has improved greatly (Lopez et al., 1989; Gage and Tyler, 1991). Concomitantly, fossil deep-sea deposits were interpreted ecologically based on their trace-fossils content (Ekdale, 1985; Ekdale and Mason, 1988; Leszczyński 1991a,b, 1993b; Uchman, 1991b, 1995a, 2004b; Wetzel and Uchman, 1998a,b, 2001). Since the 1990s, the analysis of ichnofabrics is applied also to deep-sea sediments taking into account all aspects of fabric formation.

It is a purpose of this contribution to provide an overview of the ichnologic imprint on deep-sea sediments and the-state-of-the-art regarding its interpretation, as well as to provide examples of the ichnological record of some deep-sea processes and phenomena.

2. The Deep-Sea Floor as Habitat

For a long time, the habitat of the deep sea was considered as being a dark, very stable setting (time-stability hypothesis: Sanders, 1968). Discoveries during the past half century show, however, that the abyss is a dynamic and variable environment in many aspects, such as food supply, oxygenation or substrate consistency (e.g. Gage and Tyler, 1991).

To evaluate the activity of organisms on and within the deep-sea floor, a tranquil setting is considered for reference; there, the benthic fauna is fuelled only by the vertical flux of organic matter (see Hüneke and Henrich, 2011, this volume). However, the primary productivity within the photic zone of the ocean fluctuates significantly throughout the year (e.g. Antoine et al., 1996). When settling down, especially the fresh (labile) components are degraded (Wakeham et al., 1997). Therefore, the proportion of the export production reaching the sediment surface exponentially decreases with increasing water depth (Suess, 1980). Consequently, the benthic biomass decreases with increasing water depth (Rowe, 1983). If, however, the vertical particle flux is supplemented by lateral transport, the benthic biomass is higher than expected where lateral particle supply becomes accessible to the benthic fauna (Lavaleye et al., 2002). The lateral flux is caused by various bottom currents and by turbidity currents (see Mulder, 2011, this volume; Faugères and Mulder, 2011, this volume). The latter can transport a large amount of plant debris that is rich in cellulose, which is not edible by macro-benthic animals, but that becomes so after break-down by microbes (Gooday and Turley, 1990).

Exposed on the sediment surface, the organic matter is decomposed. Therefore, the amount of C_{org} within the sediment is strongly affected by the sedimentation rate (Müller and Suess, 1979; see Hüneke and Henrich, 2011, this volume). The arrival of surplus organic matter on the sea floor leads to decreasing oxygenation of the pore water within a few weeks (e.g. Balzer et al., 1987; Gehlen et al., 1997; Soetaert et al., 1998). This signal can be directly received by the endobenthic animals as all macro-organisms depend on the availability of oxygen. Therefore, observations in the recent and in fossil deposits have been used to establish relationships between oxygenation, endofauna and the trace fossils produced (see below).

Most of the deep-sea floor is covered by sediments, which are soft or soupy on the surface and become stiffer with increasing burial depth due to de-watering through consolidation and diagenetic processes. There are also places where a rocky or firm bottom appears due to erosion, for instance in submarine canyons. The substrate consistency is an important habitat factor that even can influence the composition of the fauna. In turn, trace fossils can be used to evaluate substrate consistency (see below).

The influence of temperature on the deep-sea organisms is less known. In the recent, temperatures in the deep sea are quite uniform and low, but in the geological past, like the Early Eocene, temperatures were higher than 10 °C (e.g. Brass et al., 1982; Shackleton, 1986). Such changes influence, oxygenation, development of microbes and turn-over of organic matter (e.g. Thomas et al., 2000). The trace-fossil diversity in the Phanerozoic deep-sea deposits appears to be partly related to temperature changes (Uchman, 2004a). This is also valid for shallow-marine trace fossils and burrowing organisms (Goldring et al., 2004). The intensity of bioturbation probably is affected in a similar way.

3. Bioturbation

To quantify burrowing activity, bioturbation has been described as a diffusion-like process comprising an infinite number of small random steps (e.g. DeMaster et al., 1991). Within the so-called surface mixed layer, the sediment is churned by bioturbation so effectively that even short-lived radio-isotopes exhibit a roughly constant concentration; below the mixed layer, the radiotracer concentration decreases exponentially (e.g. Thomson et al., 2000) (Fig. 8.1). For stable oceanic environments, the mean thickness of the mixed layer has been estimated to be in the range of 9.8 ± 4.5 cm (Boudreau, 1998). The thickness of the surface mixed layer has been related to the flux of organic matter to the sea floor (Trauth et al., 1997). Therefore, it is not surprising that very high mixing rates have been found in areas receiving very large amounts of benthic food (e.g. Legeleux et al., 1994).

However, the mixed-layer concept only addresses the turn-over of sedimentary particles, but not the resulting bioturbational fabric, that is, ichnofabric. The chance of a *lebensspur* to become preserved as trace fossil increases with burrowing depth of its producer (e.g. Werner and Wetzel, 1982). Consequently, the interval in between—with low rates of bioturbation of deeply penetrating organisms where preservational potential of trace fossils is highest—has been introduced as 'transitional layer' (Berger, 1982). Inversely, as the population density/biomass decreases with depth in sediment (e.g. Gage and Tyler, 1991), the time required to bioturbate a sedimentary layer increases, in particular because the endofauna is often patchily distributed (e.g. Jumars and Ekman, 1983).

The thickness of the mixed layer in the modern deep-sea floor of the Atlantic and Pacific ranges from 3 to 8 cm, and the transitional layer is 20–35 cm thick (Berger et al., 1979; Ekdale and Berger 1978; Ekdale et al., 1984). The upper part of the transitional layer (5 cm) displays intensive bioturbation of a so-called 'lumpy nature' (Berger et al., 1979).

This purely descriptive concept has been refined for the bioturbational texture by the tiering concept that takes into account the ecologic conditions changing with depth in the sediment. The sea floor is vertically split up into

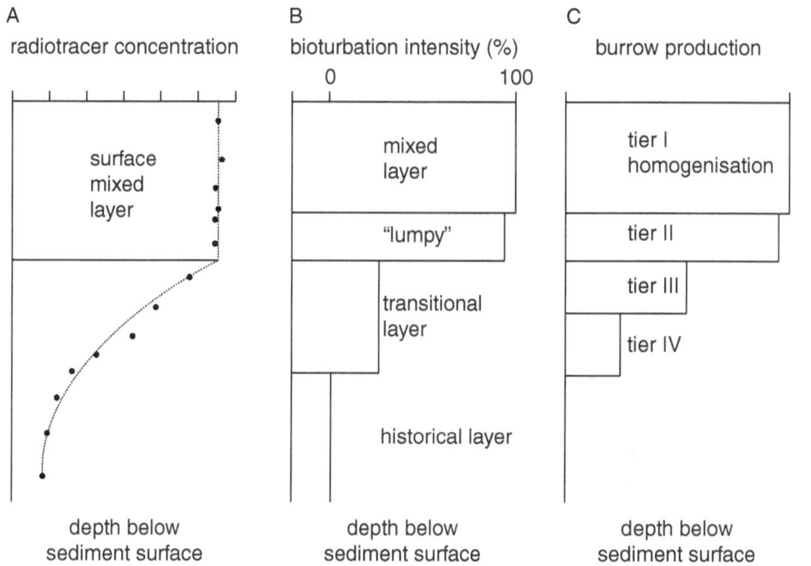

Figure 8.1 Subdivision of the bioturbated zone. (A) Concentration profiles of short-lived radiotracers suggesting a homogeneously mixed layer and no mixing below. (B) Bioturbation intensity within a multi-layer model using sediment particles as tracers. (C) Subdivision into tiers based on the cross-cutting relationships; traces cross-cutting each other belong to one tier, and traces of deeper tier cross-cut those of shallower ones. In fact, burrow production appears too low to affect radiotracer distribution, but dominate the fabric.

different habitat intervals that are occupied by various infaunal organisms (Jumars and Ekman, 1983; Jumars et al., 1990; Gage and Tyler, 1991). The vertical zonation of animals is documented by tiered burrows ('Gefügestockwerke': Reineck et al., 1967), later called tiers (Ausich and Bottjer, 1982). Each tier is defined as a distinct depth interval containing co-occurring traces, which intersect each other (Fig. 8.1) (Wetzel, 1981, 1984; Gaillard, 1984; Bromley and Ekdale, 1986). In the fossil record, at a distinct level the youngest burrow cross-cuts all pre-existing traces, whereas the oldest bioturbation structure is penetrated by all others. Studying cross-cutting relationships, a number of tiers can be established. The number of tiers within the bioturbated zone depends on the fractionation of ecospace. Commonly three to five tiers can be distinguished in aerobic deep-marine sediments (Wetzel, 1981; Bromley and Ekdale, 1986). The tiering pattern is diversified in different types of deep-sea deposits (Uchman, 1991a; Wetzel, 1991), and delicate near-surface burrows show a tiering (Leszczyński, 1991a). The cross-cutting relationships can result from vertical migration of one infaunal community keeping pace with the sedimentary surface during accumulation (Bromley and Ekdale, 1986; Miller, 1991a) or from the juxtaposition of burrows of replacing communities (Wetzel and Uchman, 2001; Miller et al., 2004; Savary et al., 2004). Erosion can significantly disturb the tiering pattern and preservation of tiers (e.g. Baldwin and McCave, 1999).

Using the tiering concept for environmental analysis, steady-state conditions are often invoked. However, at a closer look, this is not always likely: in slowly accumulating Holocene sediments the deepest tier may be as deep as 50 cm (Werner and Wetzel, 1982, p. 280); the development of such an ichnofabric reflecting steady-state conditions would need steady conditions for more than 5000 years, a requirement seldom met as can be deduced from the stable-isotope record of climate change (e.g. Seibold and Berger, 1996). It is therefore necessary to evaluate whether the benthic fauna (and the related trace fossils) of the deeper tiers are associated with present-day conditions or whether they represent responses to past conditions.

4. Trace Fossils

When sediment is bioturbated, two general types of bioturbational sedimentary structures can be distinguished:

- *Biodeformational structures*, which have no distinct outlines and do not display a recurrent geometry that allows their classification; biodeformational structures are formed within soft to soupy sediment (Wetzel and Uchman, 1998a) and destroy pre-existing physical structures (Fig. 8.2);
- *Trace fossils*, which exhibit sharp outlines and possess a characteristic recurrent geometry that allows their classification in terms of ichnotaxonomy (e.g. Häntzschel, 1975).

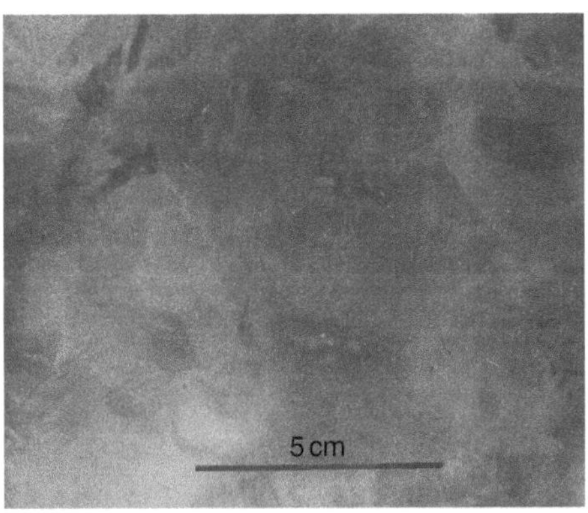

Figure 8.2 Biodeformational structures in modern deep-sea sediments off NW Africa (X-ray radiograph negative; Core 12329 Institute of Geosciences Kiel. For details, see Wetzel, 1981). (A multi-colour version of this figure is on the included CD-ROM.)

In the fossil record, two main behavioural and hence, often differently preserved groups of trace fossils are distinguished. Trace fossils that are produced on, or very close to, the sediment surface become preserved only when they are scoured and cast by turbidity currents. For such burrows Książkiewicz (1954) introduced the term 'pre-depositional' because they were produced prior to turbidite sedimentation. The trace fossils penetrating previously deposited beds were called 'post-depositional' (see also Seilacher, 1962; Leszczyński, 1993a). In fact, the 'post-depositional' suite consists as a rule of deeply penetrating trace fossils, whereas the 'pre-depositional' suite comprises some large trace fossils, but especially highly organised, often delicate burrows for which Fuchs (1895) coined the term 'graphoglyptids'. The pre-depositional trace fossils are mostly mud dwellers (Kern, 1980).

In most cases, the producers of deep-sea trace fossils are unknown, and commonly they are referred to as 'worms' of unknown taxonomic affinity. More can be said about their ethology. Ethological interpretation of fossils is based on the analysis of their functional morphology. The basic ethological categories were introduced by Seilacher (1953) and new ones were added later (see Pemberton et al., 2001 for a review). In the deep sea, the most common categories are

- *pascichnia*: mostly horizontal burrows produced by vagile deposit-feeding organisms;
- *fodinichnia*: more or less stationary structures produced during deposit feeding;
- *agrichnia*: delicate, shallow, regular burrow systems mainly in the form of various meanders, spirals, rosettes or nets, produced for trapping or farming of microbes or other very small organisms; trace fossils of this category are called 'graphoglyptids' (Fuchs, 1895; Seilacher, 1977a);
- *chemichnia*: more or less stationary structures produced by organisms feeding on chemosymbiotic microbes; the organisms maintain a connection to oxygenated waters, but penetrate into anoxic sediments rich in sulphides or nitrates that are necessary for the microbe feeding;
- *domichnia*: open burrows used as dwelling structures; intergradations to fodinichnia are common;
- *repichnia*: structures recording moving of organisms.

Some of the deep-sea trace fossils display a very complex morphology pointing to more than one ethological activity. This is caused by a complex behaviour of the trace-maker, commonly while adapting to changing environmental conditions. There are more guesses than proofs in this matter (for a discussion, see Miller, 2003).

Many publications address deep-sea trace fossils and describe them systematically. For the Palaeozoic, these are Delgado (1910), Pfeiffer (1968), Osgood (1970), Chamberlain (1971), Pickerill (1980), Benton (1982a), Stepanek and Geyer (1989), Crimes and Crossley (1991), Orr (1995, 1996, and references therein), and for the Mesozoic and Cenozoic these are, among others, Sacco (1888), Squinabol (1890), Azpeitia Moros (1933), Crimes (1973, 1977), Książkiewicz (1977), Crimes et al. (1981), McCann and Pickerill (1988), Leszczyński and Seilacher (1991), Miller (1991a, 1993), Uchman (1995a, 1998, 1999, 2001), and Buatois et al. (2001, and references therein).

Ichnotaxonomy is important in keeping order in the dramatically increasing quantities of new information, because its rules provide a codified discipline in description and provide the basis for comparison. Researchers increasingly realize the need for objective morphological criteria (ichnotaxobases) to classify trace fossils properly. In the following, some common trace fossils are described and figured; first, the post-depositional (Figs. 8.3 and 8.4) and then the pre-depositional trace fossils (Figs. 8.5 and 8.6).

4.1. Post-depositional trace fossils

Chondrites (Fig. 8.3A) is a three-dimensional, regularly branching tunnel system consisting of an open connection to the surface and numerous tunnels that ramify at angles of 30–60° to form a dendritic pattern. Interpreted as deep-tier chemichnion (for a discussion, see Fu, 1991; Uchman, 1999).

Dictyodora (Fig. 8.3C) is a meandering to looping deep-tier pascichnion composed of a basal sub-cylindrical string and a dorsal crest. *Dictyodora* is known from Palaeozoic deep-sea sediments. For a discussion, see Benton and Trewin (1980).

Nereites (Fig. 8.3B) is a winding to regularly meandering, more or less horizontal pascichnion, consisting of a median back-filled tunnel enveloped by an even to lobate zone of reworked sediment. Synonyms of *Nereites* are post-depositional *Helminthoida* and probably pre-depositional *Spirophycus* (Uchman, 1995a). Modern *Nereites* are produced in relation to the redox boundary, about 1–2 cm above the Mn-stained horizon (Wetzel, 2002).

Ophiomorpha (Figs. 8.3D and 8.5E) is represented by variable oriented, branched cylinders, locally with a granulated wall. In post-Tithonian deep-sea sediments, *Ophiomorpha annulata* and *Ophiomorpha rudis* occur. *Ophiomorpha nodosa* is typical of shallow-marine sediments. Interpreted to be produced by crustaceans. For a discussion, see Tchoumatchenco and Uchman (2001) and Uchman (2001, 2009).

Palaeophycus is a branched or unbranched, smooth or ornamented, lined, essentially cylindrical, predominantly horizontal burrow of variable diameter; fill is passive, typically structureless and of the same lithology as the host rock. Ascribed to pascichnia or domichnia. For a discussion, see Pemberton and Frey (1982) and Keighley and Pickerill (1995).

Phycosiphon (Fig. 8.3E and F) consists of compound antler-like spreiten which are formed by small U-shaped inclined parts. A spreiten is a laminated biogenic structure composed of closely spaced successive tunnel walls as burrow is shifted laterally through the sediment. The width of a single

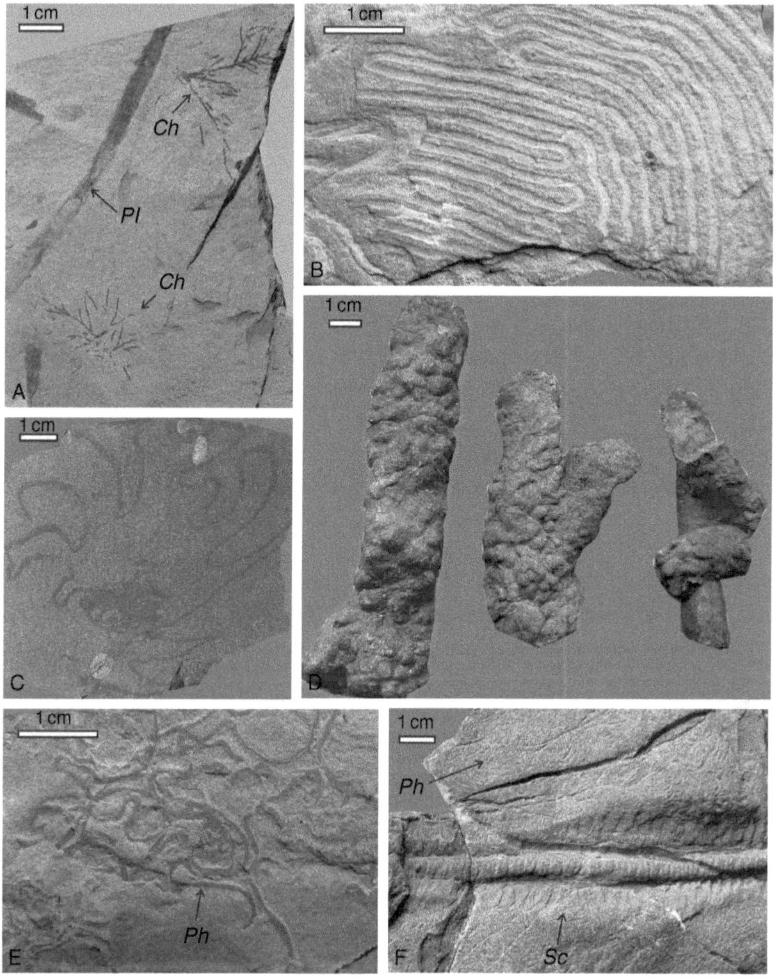

Figure 8.3 Some post-depositional trace fossils. (A) *Chondrites intricatus* (Brongniart) (Ch) and *Planolites beverleyensis* Billings (Pl). INGUJ193P1, endichnial full reliefs in turbiditic marlstone, parting surface, Krzeczkowa, Holownia Siliceous Marls (Turonian–Lower Santonian), Skole Nappe, Polish Carpathians. (B) *Nereites irregularis* Schafhäutl. INGUJ143P6, weathered endichnial full relief in a turbiditic sandstone/siltstone, top view, Lubomierz (Paleocene), Magura Nappe, Polish Carpathians. (C) *Dictyodora liebeana* Geinitz in horizontal section. INGUJ191P31, turbiditic mudstone, Kulm facies, Olšovec quarry, Moravia, Czech Republic. (D) *Ophiomorpha rudis* (Książkiewicz). INGUJ144P18a (left), b, c (right), full reliefs extracted from sandy mudstone at the top of thick-bedded turbidites, Wierchomla Wielka quarry, Magura Formation (Eocene), Polish Carpathians. In the right, tunnels with granulated wall, formed in an unstable substrate, are cross-cut by a smooth unwalled burrow formed in a stiff ground. (E) *Phycosiphon incertum* Fischer-Ooster (Ph). INGUJ192P12, endichnial full relief on a parting surface at the top of a turbiditic bed, Cisna Beds (Maastrichtian–Paleocene), Cisna, Dukla Unit, Polish Carpathians. (F) *Scolicia prisca* Quatrefages (Sc). INGUJ144P16, weathered endichnial full relief at the top of a turbiditic sandstone, Bystrica Formation (Eocene), Magura Nappe, Polish Carpathians. The sculptured fabric on the surface is caused by *Phycosiphon incertum* (Ph).

spreite is 1–5 mm, and the diameter of the marginal tube is 0.5–1 mm. The marginal tubes show a thin lining. Interpreted as fodinichnion. For a discussion, see Wetzel and Bromley (1994).

Planolites (Fig. 8.3A) consists of rarely branched, straight or gently curved, inclined or horizontally oriented, cylindrical tunnels. Differences in diameter, penetration depth and sediment infill may allow distinguishing different types (Wetzel, 1981). Active filling of the burrow is indicated by grain-size differences in relation to the surrounding sediment and relatively close packing of the infill; interpeted as pascichnion. For a discussion, see Pemberton and Frey (1982) and Keighley and Pickerill (1995).

Scolicia (Figs. 8.3F and 8.4A) is a large, bilaterally symmetrical, subcylindrical pascichnion filled with meniscate lamellae, more or less divided into two concave sets. The basal face shows faint trilobate morphology and the upper face a slight axial groove. On traverse cross-sections, the concentric structure of the bilobate lamellae surrounds an eccentric axis that is interpreted as a drain fill. *Scolicia* is produced by *Spatangus*-like irregular echinoids (for a discussion, see Uchman, 1995a).

Skolithos consists of straight, unbranched vertical shafts more or less perpendicular to the bedding. Its diameter is uniform and commonly ranges from 2 to10 mm. For a discussion, see Alpert (1974) and Schlirf and Uchman (2005).

Teichichnus is a vertical, blade-like spreiten consisting of several closely spaced, horizontal or inclined, longitudinally nested individual tubular burrows adjoining single parent trunks; burrows within the spreite are displaced upward or downward. For a discussion, see Fillion and Pickerill (1990).

Thalassinoides (Fig. 8.4B) is a system consisting of horizontally branched tunnels connected to the surface by more or less vertical shafts; swellings may occur at points of branching. A typical asymmetrical (eccentric) fill structure often results from active filling by the burrowing organism or collapse of the burrow walls. The tubes are 5–20 mm in diameter and smooth-walled. Interpreted to be produced by crustaceans. For a discussion, see, among others, Ekdale (1992).

Trichichnus (Fig. 8.4C) is a thread-like and rarely branching cylindrical burrow commonly exceeding 20 cm in length. Its fill is usually pyritized. For a discussion, see McBride and Picard (1991) and Uchman (1995a, 1999).

Zoophycos (Fig. 8.4D) is a spreiten structure composed of distinct levels or coilings and/or lobes. The regular fill structure appears as crescentic en-echelons in vertical sections. Open (collapsed) or actively filled marginal tubes indicate lower or higher oxygen content, respectively, in the respiration water (Wetzel and Werner, 1981). For details see, among others, Bromley and Hanken, (2003).

4.2. Pre-depositional trace fossils

Most of the pre-depositional forms are represented by graphoglyptids, which are interpreted as agrichnia. They are common on the soles of turbiditic sandstones, but never abundant. They are preserved mostly as semi-reliefs

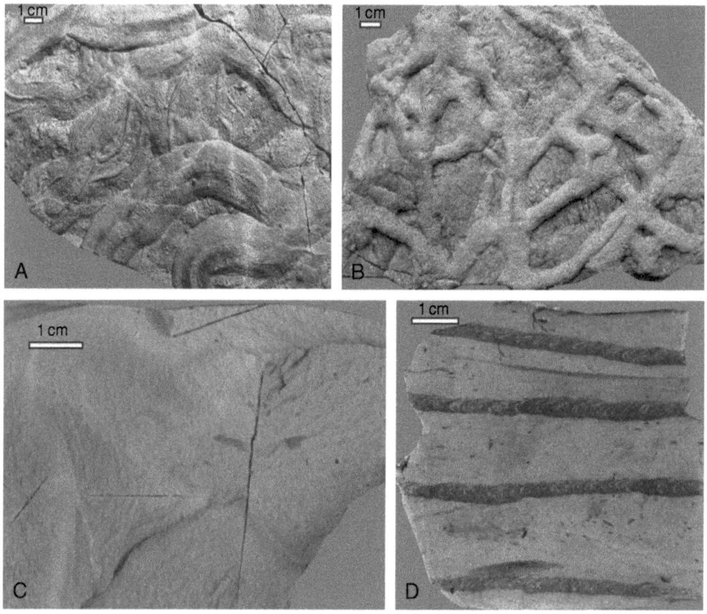

Figure 8.4 Some other post-depositional trace fossils from the Polish Carpathians. (A) *Scolicia plana* Książkiewicz, INGUJ144P10b, hypichnial full relief in a turbiditic sandstone bed, Konina, Beloveža Formation (Eocene). (B) *Thalassinoides* isp., INGUJ144P12, hypichnial full relief in a turbiditic sandstone bed, Żleźnikowa Wielka, Żeleźnikowa Formation (Eocene), Magura Nappe. (C) *Trichichnus linearis* Frey, INGUJ193P2, endichnial full relief in turbiditic marlstone, parting surface, Krzeczkowa, Holownia Siliceous Marls (Turonian–Lower Santonian), Skole Nappe. (D) *Zoophycos* isp., INGUJ194P1, in cross-section, full relief in a hemipelagic marlstone, Węglówka, Węglówka Marls (Senonian), Subsilesian unit.

resulting from scouring and casting of shallow burrow systems. Their tubes are mostly less than 2 mm wide. For a discussion of graphoglyptids, the reader is referred to Seilacher (1977a), Miller (1991b) and Uchman (1995a, 1998, 2003). The first six trace fossils described below belong to the graphoglyptids:

Cosmorhaphe (Fig. 8.5A) is a hypichnial, semicircular ridge forming first- and second-order, regular meanders; that means a narrowly meandering tube (i.e. secondary meanders) forms a wide first-order meander.

Helminthorhaphe (Fig. 8.5B) is a hypichnial, semicircular ridge forming only first-order, regular meanders.

Lorenzinia (Fig. 8.5C) is a hypichnial structure composed of semicircular bars and knobs radially arranged around a central empty area.

Megagrapton (Fig. 8.5D) is composed of hypichnial semicircular ridges, forming an irregular net.

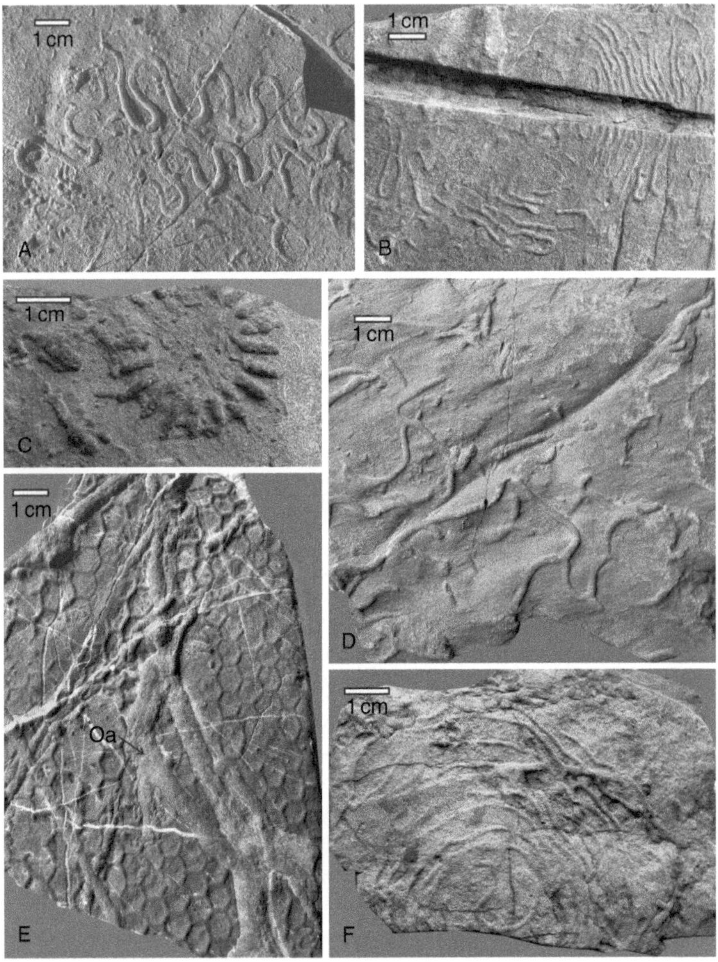

Figure 8.5 Some pre-depositional (an exception in E) trace fossils from the, Polish Carpathians, mainly graphoglyptids, preserved in hypichnial semi-reliefs. (A) *Cosmorhaphe sinuosa* Azpeitia Moros, INGUJ144P17, Słopnice, Beloveža Formation (Eocene), Magura Nappe. (B) *Helminthorhaphe flexuosa* Uchman, INGUJTF UJ 49, Poręba Wielka, Inoceramian Beds (Senonian-Paleocene), Magura Nappe. (C) *Lorenzinia carpathica* (Zuber), TF UJ 447, Lipnica Mała—Gubernasówka, Variegated Shale (Paleocene), Magura Nappe. (D) *Megagrapton submontanum* Azpeitia Moros, INGUJ144P11b, Piorunka, Beloveža Formation (Eocene), Magura Nappe. (E) *Paleodictyon majus* Meneghini in Peruzzi cross-cut by post-depositional *O. annulata* Książkiewicz (Oa), INGUJ144P13, Uhryń, Beloveža Formation (Eocene), Magura Nappe. (F) *Spirorhaphe involuta* (De Stefani), INGUJTF UJ 2603A, Znamirowice, Ciężkowice Sandstone (Eocene), Silesian Nappe.

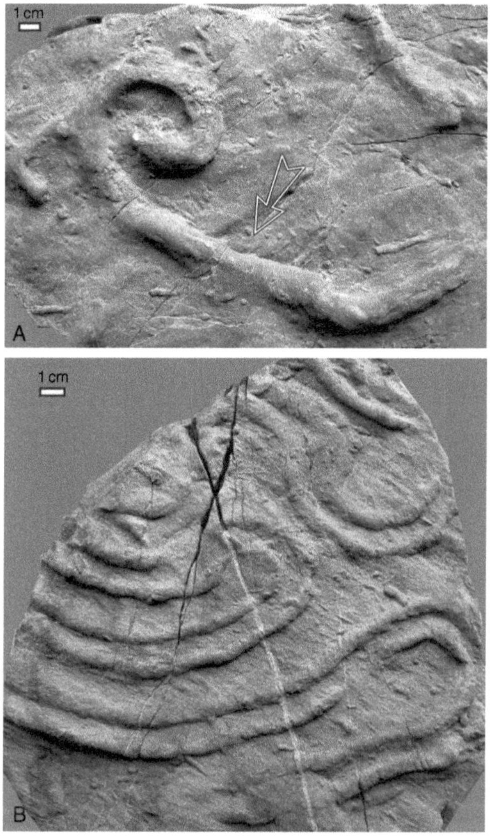

Figure 8.6 Some pre-depositional non-graphoglyptid trace fossils preserved in hypichnial semi-reliefs, Polish Carpathians. (A) *Spirophycus bicornis* (Heer), TF UJ 204, Zubrzyca Górna, Moniaków, Beloveža Formation (Eocene), Silesian Nappe. Note the asymmetry of the semi-relief. The arrow points to the approximate direction of the flow. (B) *Scolicia strozzii* (Savi and Meneghini), INGUJ144P14, Łabowiec, Beloveža Formation (Eocene), Magura Nappe.

Paleodictyon (Fig. 8.5E) is composed of hypichnial semicircular ridges forming a regular hexagonal net. The meshes of the net are 1–50 mm wide.

Spirorhaphe (Fig. 8.5F) is a hypichnial semicircular ridge forming a spiral with a loop in the centre.

Spirophycus (Fig. 8.6A) is a horizontal ridge bent at one end in a spiral. This non-graphoglyptid trace fossil is probably a preservational variant of *Nereites*, i.e. scoured and cast. For a discussion, see Książkiewicz (1977) and Uchman (1998).

Scolicia strozzii (= *Taphrhelminthopsis*) (Fig. 8.6B) is a bilobate hypichnial smooth ridge, with a central furrow. At least in Late Jurassic and younger sediments, it is interpreted as a washed-out and cast shallow irregular echinoid burrow, which in full relief would be preserved as *Scolicia* (Uchman, 1995a).

5. Interpretation of Trace Fossils and Ichnofabrics

The biogenic sedimentary structures can be used to decipher the environmental conditions by looking at a specific trace fossil or at the whole ichnofabric as such. The combination of both attempts will provide maximum information.

5.1. Trace-fossil boundaries

Besides the taxonomy of the trace fossils, the boundary of the burrows stores information about the sediment consistency. The consistency of the substrate can be estimated independently by measuring the porosity and the burrow geometry (Wetzel and Aigner, 1986). Direct observations allow to distinguish five categories (Wetzel and Uchman, 1998a):

- *Loose ground*: tubes within loose grains of sandy substrates are stabilized by a mucus lining (e.g. *Skolithos*) or by a pellet-armed wall (e.g. *Ophiomorpha*);
- *Soup ground*: animals 'swim' through the sediment producing biodeformational structures or turbichnia (e.g. Leszczyński, 2004) or assemble a 'floating pile foundation' in suspension-rich settings (Wetzel and Bromley, 1996); thin-walled tubes are not stable and can show a characteristic 'mantle and swirl' appearance in cross-section (Lobza and Schieber, 1999);
- *Soft ground*: thin-walled (e.g. *Palaeophycus*) or pelleted tubes (e.g. *Ophiomorpha*) are stable;
- *Stiff ground*: unlined tunnels are stable (e.g. *Thalassinoides* or *Zoophycos*);
- *Firm ground*: animals use body appendages to burrow (e.g. the producers of *Rhizocorallium* or *Thalassinoides*) and burrow walls often exhibit a sculptured ornamentation.

The formation of some burrows requires special consistency of sediment. *Zoophycos* is formed in soft to stiff sediment that has been ascribed to a drop in the sedimentation rate (Olivero, 1996; Olivero and Gaillard, 1996; Savary et al., 2004). *O. annulata* and *O. rudis* can penetrate substrates of different consistency. In loose sands, for instance, when penetrating thick turbidites, it displays a distinct granulated wall preventing the burrow from collapsing (Fig. 8.3D). When produced along bedding planes, its lower part is smooth, without a wall because the underlying mud was already stiff

(Fig. 8.3D), but the roof is locally reinforced by a granulated wall preventing collapse (Kern and Warme, 1974). In firm sediments, there is no wall, but scratch marks of the trace-maker can be seen (Uchman, 2001, 2009).

5.2. Ichnofacies

To distinguish between the different trace-fossil associations characterizing shallow- and deep-marine deposits, so-called molasse-type deposits in shallow foreland basins and turbidite-dominated, so-called flysch deposits, in deep basins (i.e. trenches) in front of orogens, the ichnofacies model (Fig. 8.7) was introduced by Seilacher (1964, 1967) and was later refined by Frey and Seilacher (1980) and Frey et al. (1990). At the beginning, the resolution of this model for the deep sea was low; flysch deposits were grouped into the *Nereites* ichnofacies, which is characterized by graphoglyptids and meandering trace fossils (Seilacher, 1967). Muddy deposits below the upper offshore zone were positioned in the *Zoophycos* ichnofacies, which is typified by the ichnogenera *Zoophycos*, *Phycosiphon* and *Chondrites*. Traditionally, this ichnofacies is related to slope deposits (Seilacher, 1967; Frey and Seilacher, 1980; Frey at al., 1990). As the principal ecologic factors (organic-matter input, sedimentation rate, grain size, oxygenation) vary significantly, a metric calibration of the various ichnofacies as suggested by Frey and Pemberton (1985) is of very limited precision (Wetzel, 1983).

The ichnofacies are considered to some extent as taphofacies, depending on the preservation of the trace fossils (e.g. Bromley and Asgaard, 1991). For instance, graphoglyptids characterizing the *Nereites* ichnofacies require delicate

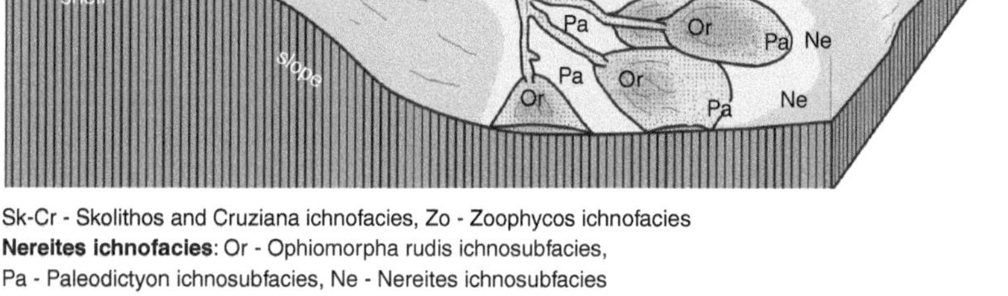

Sk-Cr - Skolithos and Cruziana ichnofacies, Zo - Zoophycos ichnofacies
Nereites ichnofacies: Or - Ophiomorpha rudis ichnosubfacies,
Pa - Paleodictyon ichnosubfacies, Ne - Nereites ichnosubfacies

Figure 8.7 Ichnofacies from shelf to deep sea (modified after Uchman, 2007a).

scouring and casting to be preserved. Photographs of the modern deep-sea floor, however, document graphoglyptid lebensspuren (e.g. Gaillard, 1991), also in places where their preservation seems almost impossible due to lack of such processes (Ekdale, 1985; Miller, 1991b). This is probably true for many pelagites preserving trace fossils typical of the *Zoophycos* ichnofacies.

With time, the *Nereites* ichnofacies has been subdivided into the *Ophiomorpha rudis* ichnosubfacies for thick-bedded sandstones in channels and proximal lobes in the turbiditic systems (Uchman, 2001, 2009), the *Paleodictyon* ichnosubfacies for more sandy 'normal' flysch, and the *Nereites* ichnosubfacies for mud-rich distal flysch (Seilacher, 1974). Roughly, the *Ophiomorpha rudis–Paleodictyon–Nereites* ichnosubfacies may express a bathymetric trend from inner to outer fan. In addition, within sandy flysch deposits, patterned trace fossils are frequent (e.g. *Paleodictyon*, *Desmograpton*) and muddy deposits contain relatively more meandering forms. However, the diversity and the composition of the trace-fossil associations are affected by the lithological variability and other ecologic factors (Uchman, 2001).

The *Nereites* ichnofacies occurs within a wide bathymetric range. The *Paleodictyon* ichnosubfacies occurs in an Eocene turbiditic succession a few tens of metres below tempestites in the Sinop Basin, Turkey (Uchman et al., 2004). This ichnosubfacies occurs even in tempestite-bearing, deep intrashelf troughs reported from the Late Cretaceous shelf of Tanzania (Ernst and Zander, 1993; Gierlowski-Kordesch and Ernst, 1987). However, it is an open question as to whether these cases should be regarded as some odd exceptions or as an indication for a habitat with such conditions. The ichnoassemblage of the Early Cretaceous Kamchia Formation, Bulgaria, contains a mixture of forms typical of both the *Nereites* ichnofacies (*Squamodictyon*) and the *Cruziana* ichnofacies (*Curvolithus*, *Gyrochorte*). Probably, sediments of the Kamchia Formation were deposited in an offshore or deeper setting with storm sand layers and background marly sedimentation. It is possible that storm currents transported producers of typical traces from the shelf into the deeper sea (Uchman and Tchoumatchenco, 2003).

In general, co-occurrence of deep-sea and shelf trace fossils can be caused by the transport of trace-makers by storm and other currents from the shelf to the deep sea. The occurrence of 'shallow-water' trace fossils, mostly *Ophiomorpha* and *Thalassinoides*, in deep-sea sediments has thus been explained (Crimes, 1977; Föllmi and Grimm, 1990; Wetzel, 1984). However, most of the producers of so-called shallow-water trace fossils, such as *Ophiomorpha*, appear to be well adapted to the deep-sea environment, working there for a long time in deep tiers. Size analysis of *Ophiomorpha rudis* suggests that all age groups of the trace-maker occur in the deep sea, and ichnotaxonomic analysis shows distinct differences on the ichnospecies level between the deep 'shallow-water' forms and true shelf trace fossils (Uchman, 1995a). Thus, most of the deep 'shallow-water' forms belong probably to the deep-sea residents.

5.3. Ichnoassemblages

The composition and diversity of deep-sea ichnoassemblages can record palaeoenvironmental and ecological parameters. Ekdale (1985) distinguished between K-selected ichnotaxa and r-selected ichnotaxa. K-selected ichnotaxa are produced by animals adapted to a stable environment with low or moderate ecological stress, which usually reproduce slowly. Their taxonomic diversity is high but their abundance is relatively low. Graphoglyptids (category agrichnia) are the best example. Tracemakers of r-selected trace fossils are adapted to instability and high environmental stress. The morphology of these trace fossils is commonly simple and their diversity is low.

In turbiditic sediments (see Section 2.5 in Chapter 2), K-selected forms are roughly mostly pre-depositional trace fossils, whereas r-selected forms are mostly among the post-depositional trace fossils (Tunis and Uchman, 1996b). The ratio between r-selected to K-selected ichnotaxa can vary from formation to formation in one basin in response to environmental changes such as oxygenation, sedimentation rate, grain size, etc. (Tunis and Uchman, 1996a; Uchman, 1991b, 1992, 2004b). Commonly, ichnoassemblages rich in graphoglyptids occur in red to green thin- to medium-bedded turbiditic deposits, related to moderate oligotrophic conditions (Miller, 1991b; Tunis and Uchman, 1996b).

Not all turbidity currents bring enough oxygen allowing colonization by infauna if there is a general oxygen deficiency in the basin. Many turbidites are not colonized because of this reason (Leszczyński, 1991b; Uchman, 1991b, 1992). The trace-fossil diversity is low if oxygenation drops below a threshold value. However, the inverse can also occur: when a slowly accumulating sediment is very well oxygenated and if only little organic matter is available for the burrowing organisms, diversity decreases. This is well demonstrated by the relationship between trace-fossil diversity and sediment colour: the highest diversity was found in grey sediments, and the lowest in red or black (Leszczyński, 1993b). The diversity can also be lowered by too frequent disturbances by turbidites (Leszczyński, 2003).

5.4. Ichnofabrics

5.4.1. Pelagites

Trace fossils are suitable to evaluate several ecologic parameters of pelagic and hemipelagic sediments that accumulate rather continuously (see Hüneke and Henrich, 2011, this volume). If the steady accumulation, however, is interrupted by the deposition of turbidites, the previous tiering structure can become fossilized, depending on the erosional power of the currents (see Wetzel and Aigner, 1986). Such frozen tiers are useful to calibrate ichnofabrics. The development of an ichnofabric is affected by many interrelated environmental factors, especially organic-matter supply,

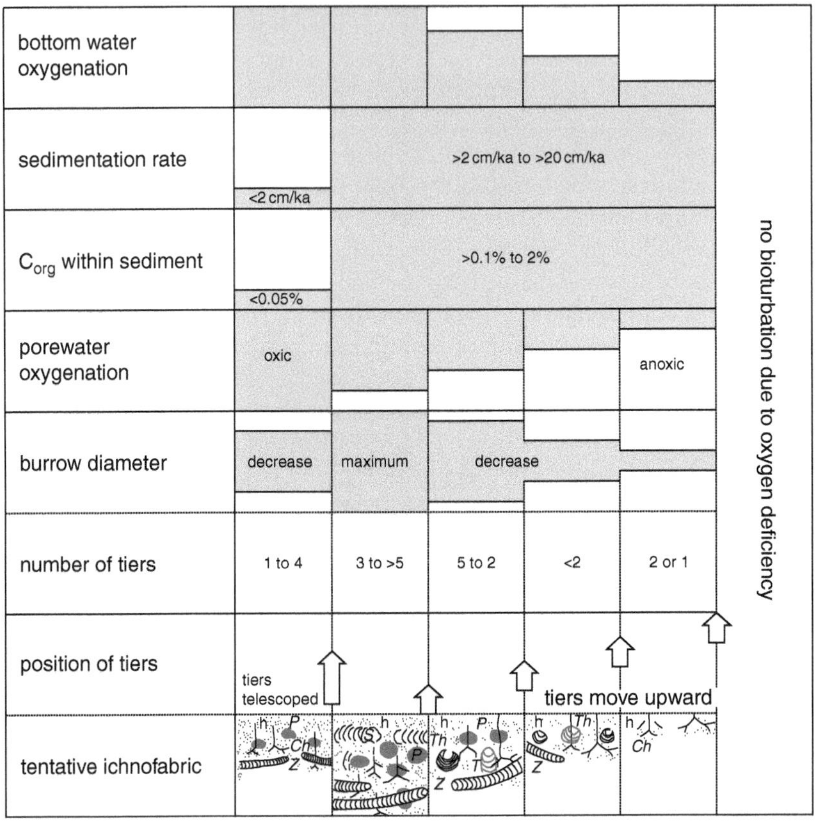

Figure 8.8 The response of burrowing organisms in deep-marine hemipelagic settings to changing environmental factors based on observations off NW Africa (modified after Wetzel, 1991). *h*, homogeneous surface layer; *C*, *Chondrites*, *P*, *Planolites*, *S*, *Scolicia*, *T*, *Teichichnus*, *Th*, *Thalassinoides*, *Z*, *Zoophycos*.

sedimentation rate and oxygenation (Fig. 8.8). As these can be only partially unravelled by ichnofabric analysis, the potential effects are discussed first, and then two standard sediment types having oxic or anoxic pore water, respectively, are outlined.

The *organic-matter content* of the deposits depends on the sedimentation rate (Müller and Suess, 1979) and affects, in turn, the oxygenation of the pore water (see Hesse and Schacht, 2011, this volume). These factors cannot be sharply distinguished and directly evaluated. There is, however, one exception. If the availability of benthic food is very high, biodeformational structures normally >2 cm in diameter dominate the bioturbate texture. It appears that the endobenthic animals burrow without distinct behavioural specialization and, hence, no trace fossils are produced (Wetzel, 1981, 1991). In modern sediments off NW Africa, biodeformational structures dominate if C_{org} content is >2%; below this value, trace fossils are present. Some trace fossils can indicate food fluctuation. For instance, the trace-maker of *Zoophycos* collects food mostly from the sedimentary surface and can keep it for periods of temporary oligotrophy (Miller and D'Alberto, 2001; Löwemark et al., 2004a).

The *sedimentation rate* controls the burial of organic matter. Therefore, the trophic level of sediments cannot be evaluated from trace fossils for a wide range of sedimentation rates (<3 to >30 cm/ka). Such deposits are completely bioturbated if fully oxygenated. Nonetheless, trends can be found (Fig. 8.8). With increasing sedimentation rate, the vertical extension of tiers may increase, and deeply penetrating burrows such as *Thalassinoides* and *Zoophycos* may become dominant in deep tiers. In intermediate tiers, patchy bioturbation by the producers of *Phycosiphon* can occur; furthermore, *Chondrites* and *Teichichnus* may be present. In the case of retarded sediment input, the penetration depth tends to decrease. At a drastically reduced sediment input, little organic matter is buried and the substrate starts to stiffen. In this case, burrows are less compacted, often sharply walled (stiff ground), and passively infilled, and they show claw-sculptured ornamentation indicating firm ground (Savrda, 1995).

Regarding *oxygenation*, three different situations are commonly distinguished with respect to the oxygen content of the bottom/pore water (Rhoads and Boyer, 1982): aerobic/oxic (>1.0 ml O_2/l), dysaerobic/dysoxic (1.0–0.3–0.1 ml O_2/l) and anaerobic/anoxic (<0.3–0.1 ml O_2/l). With respect to bioturbation, it is useful to distinguish within the aerobic setting between (1) completely oxic sediments—having also well-oxygenated pore water—characterized by brownish sediments and (2) partially oxic sediments, which become oxygen deficient and, hence, greyish/green within the bioturbated zone. Under anaerobic conditions, black shales form as a rule (e.g., Weissert, 2011, this volume).

5.4.2. Oxic deposits

Oxic deposits accumulate very slowly, mainly in the deep sea and, hence, they are low in organic matter (e.g. Mangini et al., 2001). Such sediments are pale, brownish or reddish coloured, as a result of the oxidized state of iron. Food is restricted, at least deeper within sediment, but the near-surface benthic food level may be increased after plankton blooms. The sediment is completely bioturbated and the number of tiers is low, often <3. Below a thin mixed layer, small-sized burrows dominate and become preserved (Fig. 8.8). The penetration depth of burrows is generally low (up to 10–15 cm). The substrate is soft to stiff. Fossil examples are variegated shales from the Carpathians (Leszczyński and Uchman, 1993; Mikuláš et al., 2009).

The ichnofabric is formed commonly by *Palaeophycus*, *Planolites*, *Phycosiphon*, *Nereites*, *Asterosoma*, *Teichichnus*, *Thalassinoides*, *Chondrites* and *Zoophycos*. Intercalated turbidites may preserve a pre-depositional graphoglyptid fauna. Although the environment is not oxygen-limited, trace fossils occur that have an open connection to the sea floor. It is not yet clear whether this is to enhance respiration water supply or to catch food from the surface during bloom times (Wetzel and Uchman, 1998a).

5.4.3. Partially oxic deposits

Partially oxic sediments accumulate under well-oxygenated bottom water, but burial of organic matter leads to the consumption of oxygen within the bioturbated zone and to greyish to green colours, as iron is in a reduced state. Under such conditions, even slight changes in the oxygen content of the bottom water can affect the burrowing fauna and the resultant ichnofabric considerably (Bromley and Ekdale, 1984; Savrda and Bottjer, 1986; Wetzel, 1983).

Based on many observations, some general trends have been found (Fig. 8.8). The deviations between the various observations are due to the different habitats studied rather than that one of the concurrent observations is in error (for details see, for instance, Wetzel, 1991; Wetzel and Uchman 1998a). In the following, only changes related to decreasing oxygenation are outlined; the development for an increasing oxygen content is just the reverse. When the oxygen content in the bottom water decreases, the penetration depth, the burrow diameter, the trace-fossil diversity and the number of tiers also decrease, while burrows with an open connection to the surface become more frequent (Wetzel, 1983, 1991). The tiers show a clear order of disappearance, starting with the shallowest (Bromley and Ekdale, 1984).

Ichnological models for the fossil record differ to some degree from observations of modern benthic communities (Wheatcroft, 1989). The various ichnological models may also seem to differ significantly from each other, but in fact the differences are small. The benthic-community succession paradigm (Rhoads and Boyer, 1982) differs mainly from ichnological observations (Bromley and Ekdale, 1984; Savrda and Bottjer, 1986) with respect to the first colonization by surface grazers versus colonizing by the producers of *Chondrites* or *Thalassinoides*. At a closer look, it is a matter of preservation potential rather than of missing uniformitarism: burrows produced by organisms living at, or close to, the sedimentary surface during re-oxygenation events have a very low preservation potential. These sediments are rich in organic matter and soft or even soupy (e.g. Wetzel, 1990). Thus, animals burrowing near the surface produce biodeformational structures that may be compressed to pseudo-lamination during compaction. During further re-oxygenation, such traces will not be preserved due to subsequent overprint by trace fossils of deep tiers (Behl and Kennett, 1996). Alternatively, during persisting low-oxygen conditions, trace-fossil producers, for instance those that make *Chondrites*, may penetrate the biodeformational structures. The absence of lamination has rarely been recognized as a phase of previous bioturbation.

Across the zone with minimum oxygen within the Indian Ocean, however, neither the maximum burrow size nor the maximum penetration depth of the burrows correlates to bottom-water oxygenation (Meadows et al., 2000; Smith et al., 2000), but the ichnofabric was studied only on fresh core surfaces that in fact provide not such detailed information of the ichnofabric as X-ray radiographs do (Werner, 1967; Wetzel, 1981, 2010). Therefore, it is not clear if the burrows have been produced under the observed low-oxygen situation. Furthermore, when oxygenation deteriorates in modern environments, the decrease in penetration depth and in burrow diameter may differ for the various morphological types of burrows (Wetzel, 1981, 1983). One can distinguish between burrows without a permanent connection to the sea floor (like *Planolites*, *Scolicia*), burrows with a single tube being connected to the sea floor (*Chondrites*, *Thalassinoides* and J-type *Zoophycos*) and burrows with U-shaped tubes (like *Arenicolites*, *Diplocraterion*, *Teichichnus*, *Thalassinoides*) with multiple openings or with one simple U-type form (*Zoophycos*).

5.4.4. Bottom-current affected settings

The effects of deep-sea bottom currents on the habitat have been studied mainly in modern settings that are influenced by contour currents (e.g., Faugères and Mulder, 2011, this volume). In the recent, such areas have been investigated in great detail (e.g. Nowell and Hollister, 1985; Stow et al., 2002 and references therein). The mean flow velocity of contour currents is in the range of 10–30 cm/s (McCave et al., 1980). In fact, the current speed can fluctuate within a few days to several weeks from quiescence to benthic storms (>70 cm/s: Richardson et al., 1981). Contour currents may therefore rework the sea floor and take-up particles that are settling down. Thus, such currents can carry a considerable amount of fine grains including particulate organic matter (e.g. McCave, 1985; Thomsen et al., 2002) and supply food to the deep-marine biota (e.g. Thistle et al., 1985). Therefore, the benthos exhibits a higher population density and biomass than in adjacent tranquil settings (e.g. Thistle et al., 1985).

At the contour-current affected HEBBLE (*H*igh *E*nergy *B*enthic *B*oundary *L*ayer *E*xperiment) site off Nova Scotia, the macro- and meiofauna is on average extremely abundant for that depth and comparable to sites 2000 m shallower (Thistle et al., 1985). Due to the repeated physical disturbance, the faunal parameters resemble those obtained in recolonization experiments (Thistle et al., 1991). With respect to behavioural groups, ordinary suspension feeders are not abundant, probably as their filter apparatus can easily be plugged when the suspension concentration is very high for some time (Thistle et al., 1991). Taxa, however, that passively extract drifting particles by maintaining a relief on the sea floor exhibit a pronounced abundance (Aller and Aller, 1986). Similarly, in many organism groups, taxa that can enter the sediment occur more abundantly than at tranquil sites (Thistle et al., 1985, p. 103).

The high-trophic mode of bioturbation in contourite settings results in an enhanced rate of burrow production, a distinct depth of bioturbation and a specialized behaviour, as observations of the modern sea floor suggest. In contrast to numerous biological studies, there are only few detailed investigations of bioturbation and ichnofabrics addressing contour current settings, such as those by Fu and Werner (1994, 2000), Baldwin and McCave (1999), Löwemark et al. (2004b), and Wetzel et al. (2008). They mainly exhibit a clear relationship between ichnofabrics and grain size; the effects of the other environmental factors are not unequivocal. Therefore, a clear ichnological criterion recording the contourite mode of bioturbation is not available yet.

5.4.5. Turbidites

Turbidites, like other event beds, are colonized mostly from the top. Turbidity currents and other gravity-induced flows can transport trace-makers from shallower to deeper waters. These trace-makers can survive the transportation and produce burrows (Crimes, 1977; Föllmi and Grimm, 1990), but this mechanism seems to be overestimated in interpretations of the ichnological record (see end of Section 5.2).

In fact, turbiditic/pelagic or turbiditic/hemipelagic couplets consist of rapidly deposited turbidites and slowly deposited background sediment (see Mulder, 2011, this volume). Basically, two intervals can be distinguished in the couplets, a spotty and an elite layer (Uchman, 1999). They correspond roughly to the mixed layer and the transitional layer, respectively, in modern deep-sea sediments (Ekdale and Berger, 1978; Berger et al., 1979; Ekdale et al., 1984; Bromley, 1990, 1996, and references therein). The characteristic spotty layer is entirely bioturbated and it occupies the uppermost part of a turbidite bed, commonly corresponding to Bouma's Td–Te intervals (Fig. 8.9; Uchman, 1999; see also Section 2.5 in Chapter 2). Oval spots of different colour and sharpness of contours are visible against the indistinctly mottled background. The spots are cross-sections of trace fossils, commonly *Planolites* or *Thalassinoides*. In some layers, the colour contrast is so low that the layer seems to be structureless. This picture can change in thin-bedded flysch, if some deep-tier trace fossils of the so-called multi-layer colonizers *sensu* Uchman (1995b), such as *Chondrites* and *Ophiomorpha*, penetrate from the overlying bed to the spotty layer of the underlying bed (Rajchel and Uchman, 1998). In cross-section, they are visible as last-generation spots, with very distinct margins, strong colour contrast, and commonly different lithology from the host rock. The lithology of the spotty layer differs from that of the underlying sediment. It is more fine-grained and shows a different colour or tint. In sediments deposited below the CCD, it is free of carbonate, at least in the upper part. Above the CCD, the $CaCO_3$ concentration in the spotty layer can be higher than in the

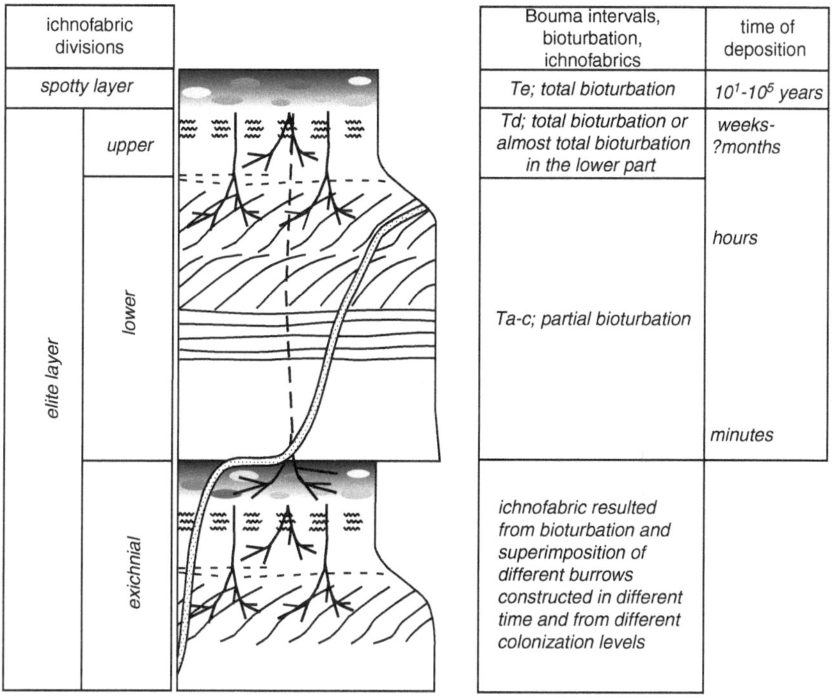

Figure 8.9 Model of ichnofabrics in turbidite/hemipelagite couplets (modified after Uchman, 1999).

underlying siliciclastic deposits if the supply of pelagic carbonate particles is significant. In oligotrophic, well-oxygenated environments, the spotty layer is lighter than the underlying sediments. In eutrophic environments, the spotty layer is commonly darker than the sediments underneath (Wetzel and Uchman, 1998a). Some trace fossils of the spotty layer are preserved as pre-depositional forms on soles of overlying turbidites. They belong to the so-called background ichnofauna (Leszczyński, 1993a), which includes foremost graphoglyptids, which in cross-sections are hardly recognizable.

The spotty layer is sometimes mistaken for purely pelagic or hemipelagic sediment, but in fact it represents the mixed layer in modern deep-sea sediments, which in turbiditic deposits is composed not only of pelagic grains but, due to bioturbation, also of some turbiditic sediments (Uchman 1995a, 1999). Pelagic sediment grains are not only piped down into the turbiditic sediment, but turbiditic sediment is also conveyed upwards to the deep-sea floor that is covered mostly with pelagic particles. This can be deduced from the fact that, on the modern deep-sea floor, differently coloured sediment of the transitional layer is conveyed to the sea floor by burrowers (Ekdale et al., 1984).

The spotty layer cannot be entirely considered as a fossil mixed layer, especially with respect to thickness. The formation of the spotty layer in turbiditic environments is probably a dynamic process. At first, a thin mixed layer is formed at the top of the sediment after deposition of a turbidite. Trace fossils produced during this phase display very low lithological contrast and do not have sharp margins, because the sediments are saturated with water. At the same time, pelagic particles accumulate. As the rate of biological reworking exceeds the rate of the accumulation, the pelagic material becomes totally mixed. It is piped down in burrows within the turbiditic sediment. The lower boundary of the spotty layer can be either distinct (because of the lithological contrast) or gradational (due to intense bioturbation by producers of *Scolicia* or *Ophiomorpha*).

The most eye-catching trace fossils (elite *sensu* Ekdale and Bromley, 1991) occur below the spotty layer in the elite layer. In most marly turbidites, deep-tier elite *Chondrites*, *Planolites*, *Nereites*, *Phycosiphon* and *Scolicia* occur. To estimate the depth of burrowing, the distance to the top of the elite layer is a useful measure. It is not clear whether the equivalent of the 'lumpy' layer *sensu* Berger et al. (1979) is in the lower part of the spotty layer (at least in some units) or in the upper elite layer (Uchman, 1999).

The ichnofabric analysis is useful to reconstruct the complex colonization history of a single layer in turbiditic/hemipelagic sediments. It appears that the cross-cutting relationships between trace fossils result not from accretion of sediment and upward shifting of tiers as in (hemi)pelagic sediments, but from sequential colonization in time (Wetzel and Uchman, 2001). Freshly deposited turbiditic sediments contain relatively well-oxygenated pore water and a relatively high amount of food. At first, the sediment is colonized by small opportunistic deposit feeders producing *Phycosiphon*, rather than by the larger sediment-feeding producers of *Nereites* (Fig. 8.10). Both burrows penetrate the sediment horizontally and have no connection to the sea floor, benefiting from the oxygenated pore water. When oxygen and food decreases to the level that horizontal mining becomes inefficient, the sediment is colonized by stationary chemosymbionts producing *Chondrites* utilizing the gradient between reducing pore water below and oxygenated bottom water on the sea-floor. During the long periods between the deposition of successive turbidites, commonly hundreds or thousands of years, the slowly deposited background sediment is colonized by graphoglyptids. There are also trace-makers independent of this succession, such as echinoids producing *Scolicia*, and the multi-layer colonizers *sensu* Uchman (1995b) disturbing the ichnofabric to a varying degree. Such a scenario of sequential colonization, elaborated on the basis of Paleogene flysch, is also applicable to older formations, including at last part of the Upper Palaeozoic (Mikuláš et al., 2002).

In Late Cretaceous calciturbidites of Italy, crustacean feeders are present that colonized the calcarenitic lower part, and produced various structures.

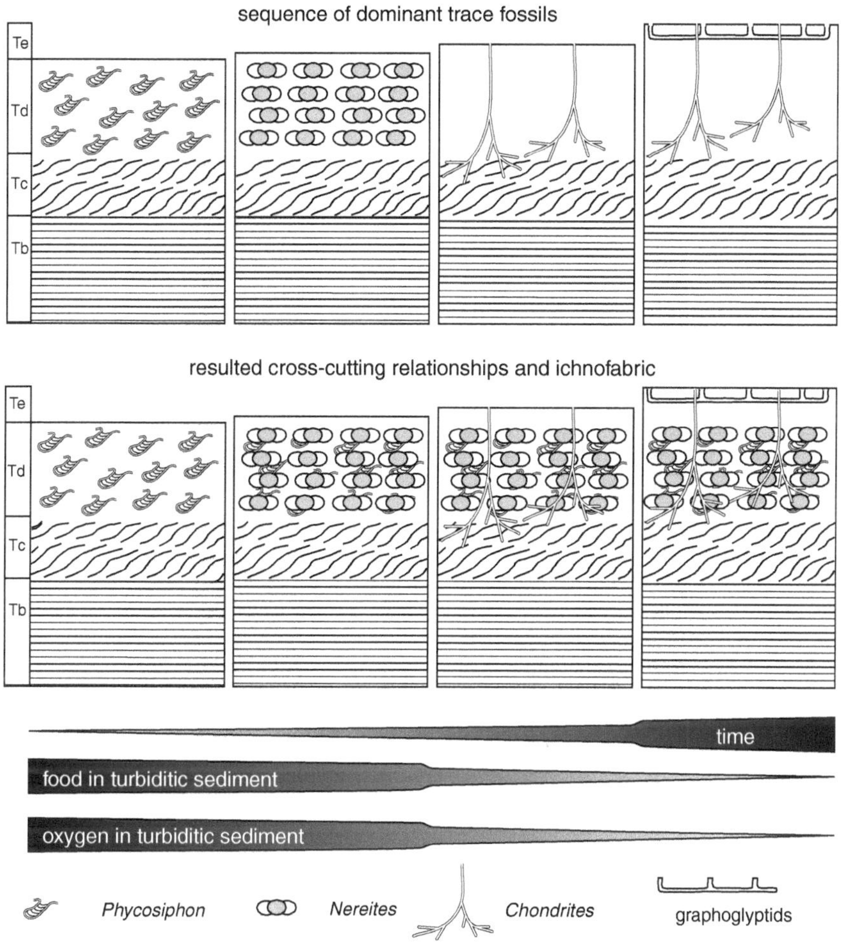

Figure 8.10 Model of sequential colonization in turbidite/hemipelagite couplets according to Wetzel and Uchman (2001), modified after Uchman (2007a). Tb, Tc, Td, Te indicate the Bouma intervals (see Section 2.5 in Chapter 2).

Burrows of subsequent colonizers display a calcilutitic mantle and a calcarenitic filling, and burrows of latter colonizers (*Chondrites*, *Planolites*) show only calcilutitic filling. The latter colonizers are believed to work during prolonged setting of calcareous mud (Miller et al., 2004).

Trace-fossil preservation and diversity in siliciclastic and carbonate (marly) Cretaceous–Paleogene flysch deposits are different (Uchman, 1999, 2007b). Generally, trace fossils in marly turbidites are more often of post-depositional than of pre-depositional origin. Graphoglyptids are relatively rare. This is probably a result of preservation conditions and benthic food level. For preservation of pre-depositional trace fossils, particularly graphoglyptids, delicate erosion is necessary, which scours their burrow systems before casting. Grains of the Upper Jurassic and younger marls consist mostly of microfossil remnants, which behave hydraulically like silt or even clay particles. As a consequence, the turbidity currents of calcareous mud were not able to cause erosion similar to those of siliciclastic mud. Thus, marly turbidites mask a shallow-tier ichnofauna. On the other hand, the marly flysch displays features of eutrophic environments, in which graphoglyptids are less abundant than in oligotrophic environments (Uchman, 2003). Only a few trace fossils apparently are characteristic of either siliciclastic or marly sediment, respectively. Most probably, *Cladichnus fischeri* is restricted to marly sediments. *Nereites irregularis* is very common in marly turbidites; however, it occurs also in carbonate-free sediments, such as in the variegated shales of the Upper Paleocene–Lower Eocene in the Carpathians (Uchman, 1999). The long settling time of carbonate suspension, which can last months, may lead to such a special ecological situation that can make trace-producers active in calciturbidites (Miller et al., 2004).

In some beds or packages of beds, the ichnofabric is dominated by one or two trace fossils, such as *Scolicia*, which record the opportunistic invasion of their trace-maker to the sea-floor. This phenomenon is related to the bulldozing effect *sensu* Thayer (1979), who concluded that relatively large burrowers prevent or reduce colonization of the substrate by immobile suspension feeders. Thus, the term 'bulldozing effect' may be expanded to include the above-described phenomenon in deep-sea flysch environments (Uchman 1995a).

5.5. Trace fossils as indicators of some sedimentological parameters

Trace fossils are particular sedimentary structures reflecting not only behaviour of their trace-makers but also contemporaneous and later physical processes in the environment. One of them is erosion by bottom or turbidity currents (see Section 2.5 in Chapter 2). It is not uncommon that in thin-bedded turbidites no evidences of erosion are found at the base, except for graphoglyptids or other trace fossils that require scouring for preservation. If the tiering pattern is recognized and calibrated, the depth of erosion below a turbidite can be estimated by the presence of trace fossils from a given tier preserved as semi-relief on the sole of the turbidite (Wetzel and Aigner, 1986).

Some pre-depositional trace fossils display characteristic asymmetric scouring, which in the case of lacking better evidence can be used for the reconstruction of the current direction of the mass flow. The steeper slopes of ridges are always up-current, similarly to flute casts (Fig. 8.6A). The orientation of some trace fossils, like *Paleodictyon* meshes, can also indicate the palaeocurrent direction (Crimes and Crossley, 1980). Many taphonomic features of graphoglyptids, including fluting, point to bottom current scouring acting long before their casting on the base of turbiditic bed (Monaco, 2008).

6. Evolutionary Aspects

Trace-makers are, just like other organisms, influenced by evolutionary changes. The ichnological record of the changes and their impact on sediments is intriguing. The problem was outlined by Seilacher (1962), but the first attempts to analyse the evolution of deep-sea trace-fossil communities were made later (Seilacher, 1974, 1977b, 16 formations) and Crimes (1974), followed by McCann (1990; 34 formations), Orr (2001; 50 Ordovician–Carboniferous formations), and Uchman (2004a; 151 Phanerozoic flysch formations) (Fig. 8.11).

A sharp increase in the number of ichnogenera in Cretaceous–Neogene flysch was described by Crimes (1974), but he emphasized that 'more

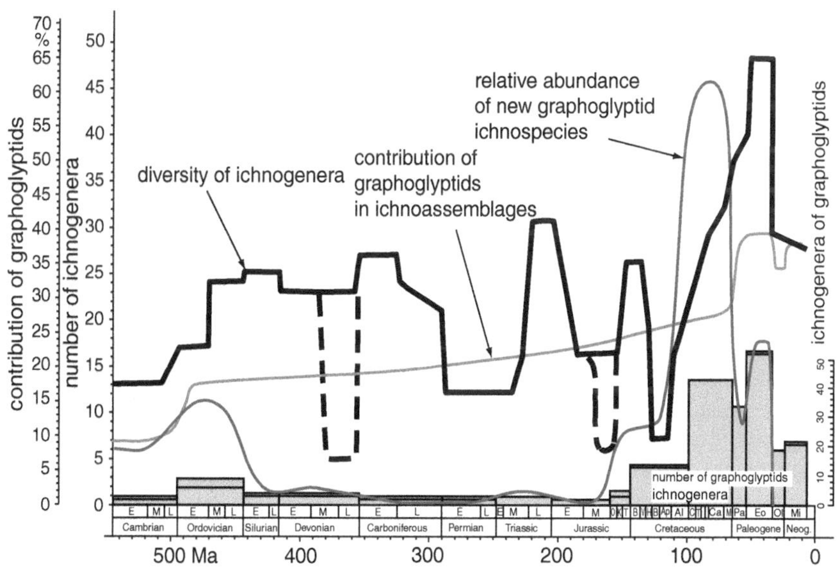

Figure 8.11 Diversity curve of deep-sea ichnogenera (the dashed line shows some less well-documented portions of the curve). The curves show the contribution of graphoglyptids in ichnoassemblages and the relative abundance of new graphoglyptid ichnospecies, whereas the chart shows the number of graphoglyptid ichnogenera in the Phanerozoic (upper portions of the columns express ichnogenera tentatively included in the graphoglyptids). Based on Uchman (2003, 2004a) with complementary data by Wetzel et al. (2007).

detailed studies' were needed. An increasing diversity of deep-sea trace fossils during the Phanerozoic, with a rapid acceleration in the Cretaceous, was found by Seilacher (1974, 1976), who subsequently mentioned a gradual increase in diversity of 'flysch ichnocoenoses' through the Phanerozoic (Seilacher, 1977b), but then came back to his previous idea and proposed a 'mid-Cretaceous diversity burst' (Seilacher, 1978). These views were tested by McCann (1990), who concluded that the increase of diversity was neither gradual during most of the Phanerozoic nor rapid in the Cretaceous, but who also drew attention to a methodological problem. For instance, the Jurassic–Tertiary flysch of the Polish Carpathians has been taken as one formation, although it includes several different units deposited in a number of basins (Książkiewicz, 1977). According to Orr (2001), the diversity of deep-sea trace-fossil assemblages shows a distinct increase from the Cambrian to the Ordovician, and later differences are considered as 'not obvious'.

The diversity versus time graph (Fig. 8.11) (Uchman, 2004a) shows that the diversity of deep-sea trace fossils markedly changed through the Phanerozoic, with peaks in the Ordovician–Early Silurian and Early Carboniferous, a low in the Permian–older Late Jurassic, a peak in the Tithonian–Aptian, a low in the Albian, and the maximum peak in the Eocene. The contribution of graphoglyptids rose gradually up to the end of the Mesozoic, with a peak in the Paleocene–Eocene and a low in the Oligocene. The diversity changes were influenced mostly by competition for food, bottom-water temperatures, sediment oxygenation and also, indirectly, by changes of the number of flysch deposits (i.e. suitable deep-sea habitats). There is no clear influence of the major biotic crises, such as those at the Ordovician/Silurian, Frasnian/Famennian, Triassic/Jurassic and Cretaceous/Tertiary boundaries (for details, see Uchman et al., 2005), on the diversity of deep-water trace fossils, except for the lower-rank Eocene/Oligocene crisis. However, the Ordovician/Silurian, Cretaceous/Tertiary and Paleocene/Eocene crises influenced the graphoglyptid ichnodiversity and their relative abundance (Uchman, 2003). The decrease of diversity after the Late Carboniferous was probably caused by the Gondwana glaciations and then reinforced by the Permian/Triassic mass extinction. The recovery took a long time, and the diversity remained low until the end of the Jurassic, although the newest data from Oman (Wetzel et al., 2007) show that the diversity in the Tethyan refuges can have been high again already in the Late Triassic.

The diversity of graphoglyptids (Uchman, 2003) was low from the Cambrian to the Middle Jurassic, with a low peak in the Ordovician followed by a drop in the Silurian. In the Late Jurassic, after the late Palaeozoic low, it increased gradually and markedly in the Late Cretaceous, probably during the Turonian. It dropped again in the Paleocene, and subsequently increased to a maximum in the Eocene. The Oligocene was marked by a sharp decrease in graphoglyptid diversity and frequency, followed by an increase in the Miocene (Uchman, 2003). The Late

Cretaceous radiation of graphoglyptids may be correlated with global changes in the circulation of waters that were rich in plankton and organic matter. The ichnologic radiation probably was delayed in the Early Cretaceous by the widespread anoxia at that time. The Eocene (total number of ichnotaxa) or Paleocene (total number of ichnotaxa peak in the diversity of graphoglyptids) and the Eocene peak in the frequency of graphoglyptids are related to the common occurrence of oceanic oligotrophy in the Late Paleocene and Early Eocene (Uchman, 2003). Graphoglyptids display an increase in complexity with time (Seilacher, 1977a), showing a distinct acceleration in the infaunal colonization during the Late Cretaceous, when the farming activity of their trace-makers became a common feeding strategy.

Furthermore, individual ichnotaxa changed their environmental range through time. *Zoophycos*, in general, migrated to the deep sea during the Jurassic (Bottjer et al., 1988; Olivero, 2003), but occurs rarely in Miocene shelf sediments (Pervesler and Uchman, 2004). *Ophiomorpha* expanded its range, since *O. rudis* (Książkiewicz 1977) and *O. annulata* (Książkiewicz 1977) occur in flysch sediments since the Tithonian and document the invasion of large burrowing crustaceans to the deep sea, together with echinoids producing *Scolicia* (Tchoumatchenco and Uchman, 2001). Also cross-cutting relationships between burrows can change through geological time. As an example, Goldring et al. (1991) described *Phycosiphon* (which they referred to as *Anconichnus*) in a shallow-tier position in the muddy parts of Jurassic tempestites, and these are cross-cut by *Chondrites*, *Palaeophycus*, *Rhizocorallium*, *Teichichnus* and *Thalassinoides*. In contrast, Wetzel (1981) described *Phycosiphon* (which he referred to as *Helminthopsis*) from an intermediate tier position in modern sediments off NW Africa. A complex tiering pattern in the transition layer occurred in the deep sea after the Middle Ordovician (Orr, 2003).

The evolutionary changes were so rapid that some ichnospecies, for example those belonging to the Palaeozoic *Dictyodora* (Benton, 1982b), display narrow stratigraphic ranges and can be used in stratigraphy (Uchman, 2004a). As shown by their relationship with tuff layers, beds that contain characteristic trace fossils occupy the same stratigraphic position and can be used for delineation of ichnostratigraphic units within a basin (Pieńkowski and Westwalewicz-Mogilska, 1986).

The *Nereites* ichnofacies has time constraints. It can be distinguished since the Ordovician, when graphoglyptids colonized the deep sea (Orr, 2001). Before, in the Cambrian, graphoglyptids occurred only in shelf sediments. A specific trace-fossil association related to shallow burrowers and surface structures associated with microbial mats occurs in the Vendian–Cambrian (Buatois and Mángano, 2003). The *Ophiomorpha rudis* ichnosubfacies can be distinguished since the Tithonian, when its index taxon first occurred in flysch deposits (Tchoumatchenco and Uchman, 2001; Uchman, 2009).

7. Perspective

Trace fossils are still underestimated in sedimentological analyses. For instance, detailed divisions of turbiditic facies, like those by Pickering et al. (1989), still insufficiently take ichnofabrics into account. Furthermore, there is a great gap between deep-sea palaeoichnology and neoichnology. The world of modern endobenthic burrowing animals is only little known and their interrelationships to the fluctuating environmental factors are rarely investigated. Complex neoichnological research in different deep-sea environments is needed in order to recognize and identify the trace-makers, their burrowing and overall behaviour and distribution. Probably, laboratory experiments using different deep-sea invertebrates and simulation of sedimentary processes will gain significant advances in this matter. Special expeditions using submersible vehicles should be undertaken. Proper employment of robots for resin casting, for example, would be a significant forward step in ichnological research in the deep sea.

Other achievements are expected in the further recognition of trace fossils and ichnofabrics, especially in palaeoenvironmentally well-recognized facies tracts, and by their comparisons based on standardized data. Ichnological data of deep-sea sediments from some periods are underrepresented. This is especially true for the Cambrian, Devonian and the Late Carboniferous–Jurassic, which have been studied only in a few places (Uchman, 2004a).

Acknowledgements

A. U.'s research benefited financially from the Alexander von Humboldt Foundation, the Jagiellonian University (DS funds) and the Komitet Badań Naukowych (Poland) as a 2004–2005 research project. A. W.'s research was supported by the Swiss National Science Foundation (grant nos. PLPJ 041522, 2100-052256.97, 200021-112128) and the University of Basel.

References

Abel, O., 1935. Vorzeitliche Lebensspuren. Gustav Fischer, Jena 644 pp.

Aller, J.Y., Aller, R.C., 1986. Evidence for localized enhancement of biological activity with tube and burrow structures in deep-sea sediments at the HEBBLE site, western North Atlantic. Deep-Sea Res. 33, 755–790.

Alpert, S.P., 1974. Systematic review of the genus *Skolithos*. J. Paleont. 49, 509–521.

Antoine, D., André, J.-M., Morel, A., 1996. Oceanic primary production 2. Estimation at global scale from satellite (coastal zone color scanner) chlorophyll. Global Biogeochem. Cy. 10, 57–69.

Ausich, W.I., Bottjer, D., 1982. Tiering in suspension-feeding communities on soft substrata through the Phanerozoic. Science 216, 173–174.

Azpeitia Moros, F., 1933. Datos para el estudio paleontólogico del flysch de la Costa Cantábrica y de algunos otros puntos de España. Bol. Inst. Geol. Min. Esp. 53, 1–65.
Baldwin, C.T., McCave, I.N., 1999. Bioturbation in an active deep-sea area: implications for models of trace fossil tiering. Palaios 14, 375–388.
Balzer, W., Erlenkeuser, H., Hartmann, M., Müller, P.J., Pollehne, F., 1987. Diagenesis and exchange processes at the benthic boundary layer. In: Rumohr, J., Walger, E., Zeitschel, B. (Eds.), Seawater-Sediment Interactions in Coastal Waters. Lecture Notes on Coastal and Estuarine Studies 13. Springer, Berlin, pp. 111–161.
Behl, J.R., Kennett, J.P., 1996. Brief interstadial events in the Santa Barbara Basin, NE Pacific, during the past 670 kyr. Nature 379, 243–246.
Benton, M.J., 1982a. Trace fossils from Lower Palaeozoic ocean–floor sediments of the Southern Uplands of Scotland. Trans. R. Soc. Edinb., Earth Sci. 73, 67–87.
Benton, M.J., 1982b. *Dictyodora* and associated trace fossils from the Palaeozoic of Thuringia. Lethaia 15, 115–132.
Benton, M.J., Trewin, N.J., 1980. *Dictyodora* from the Silurian of Peeblesshire, Scotland. Palaeontology 23, 501–513.
Berger, W.H., 1982. The benthic interface of deep-sea carbonates: a three-tiered sequence controlled by depth of deposition. In: Fanning, K.A., Manheim, F.T. (Eds.), The Dynamic Environment of the Ocean Floor. Heath and Company, Lexington, MA, pp. 95–114.
Berger, W.H., Ekdale, A.A., Bryant, P.F., 1979. Selective preservation of burrows in deep-sea carbonates. Mar. Geol. 32, 205–230.
Bottjer, D.J., Droser, M.L., Jablonski, D., 1988. Palaeoenvironmental trends in the history of trace fossils. Nature 333, 252–255.
Boudreau, B.P., 1998. Mean mixed depth of sediments: the where-fore and the why. Limnol. Oceanogr. 43, 524–526.
Brass, G.W., Southam, J.R., Peterson, W.H., 1982. Warm saline bottom water in the ancient ocean. Nature 296, 620–623.
Bromley, R.G., 1990. Trace Fossils: Biology and Taphonomy. Unwin Hyman, London.
Bromley, R.G., 1996. Trace Fossils. Biology, Taphonomy and Applications, 2nd ed. Chapman and Hall, London.
Bromley, R.G., Asgaard, U., 1991. Ichnofacies: a mixture of taphofacies and biofacies. Lethaia 24, 153–163.
Bromley, R.G., Ekdale, A.A., 1984. *Chondrites*: a trace fossil indicator of anoxia in sediments. Science 224, 872–874.
Bromley, R.G., Ekdale, A.A., 1986. Composite ichnofabric and tiering of burrows. Geol. Mag. 123, 49–65.
Bromley, R.G., Hanken, N.-M., 2003. Structure and function of large, lobed *Zoophycos*, Pliocene of Rhodes, Greece. Palaeogeogr. Palaeoclimatol. Palaeoecol. 192, 79–100.
Buatois, L.A., Mángano, M.G., 2003. Early colonization of the deep sea: ichnologic evidence of deep-marine benthic ecology from the Early Cambrian of northwest Argentina. Palaios 18, 572–581.
Buatois, L.A., Mángano, M.G., Sylvester, Z., 2001. A diverse deep-marine ichnofauna from the Eocene Tarcau Sandstone of the eastern Carpathians, Romania. Ichnos 8, 23–62.
Chamberlain, C.K., 1971. Morphology and ethology of trace fossils from the Ouachita Mountains, southeast Oklahoma. J. Paleont. 45, 212–246.
Crimes, T.P., 1973. From limestones to distal turbidites: a facies and trace fossil analysis in the Zumaya flysch (Paleocene-Eocene), North Spain. Sedimentology 20, 105–131.
Crimes, T.P., 1974. Colonisation of the early ocean floor. Nature 248, 328–330.
Crimes, T.P., 1977. Trace fossils of an Eocene deep-sea fan, northern Spain. In: Crimes, T.P., Harper, J.C. (Eds.), Trace fossils 2, Geol. J. Spec., 71–90, Issue 9.

Crimes, P.T., Crossley, J., 1980. Inter-turbidite current orientation from trace fossils with an example from the Silurian flysch of Wales. J. Sediment. Petrol. 50, 821–830.
Crimes, T.P., Crossley, J.D., 1991. A diverse ichnofauna from Silurian flysch of the Aberystwyth Grits Formation, Wales. Geol. J. 26, 27–64.
Crimes, T.P., Goldring, R., Homewood, P., Van Stuijvenberg, J., Winkler, W., 1981. Trace fossil assemblages of deep-sea fan deposits, Gurnigel and Schlieren flysch (Cretaceous–Eocene). Eclogae. Geol. Helv. 74, 953–995.
Delgado, J.F.N., 1910. Terrains palézoïques du Portugal. Étude sur les fossiles des schichtes à néréites de San Domingo et des schichtes à néréite et à graptolites de Barrancos. Com. Serv. Geol. Port. 56, 1–68.
DeMaster, D.J., Brewster, D.C., McKee, B.A., Nittrouer, C.A., 1991. Rates of particle scavenging, sediment reworking, and longitudinal ripple formation at the HEBBLE site based on measurements of ^{234}Th and ^{210}Pb. Mar. Geol. 99, 423–444.
Ekdale, A.A., 1977. Abyssal trace fossils in worldwide deep sea drilling project cores. In: Crimes, T.P., Harper, J.C. (Eds.), Trace Fossils 2, Geol. J. Spec., 163–182 Issue 9.
Ekdale, A.A., 1985. Paleoecology of the marine endobenthos. Palaeogeogr. Palaeoclimatol. Palaeoecol. 50, 63–81.
Ekdale, A.A., 1992. Muckraking and mudslinging: the joys of deposit-feeding. In: Maples, C.G., West, R.R. (Eds.), Trace Fossils, Paleont. Soc., Short Course in Paleont. 5, 145–171.
Ekdale, A.A., Berger, W.H., 1978. Deep-sea ichnofacies: modern organism traces on and in pelagic carbonates of the western equatorial Pacific. Palaeogeogr. Palaeoclimatol. Palaeoecol. 23, 268–278.
Ekdale, A.A., Bromley, R.G., 1991. Analysis of composite ichnofabrics: an example in uppermost Cretaceous chalk of Denmark. Palaios 6, 232–249.
Ekdale, A.A., Mason, T.R., 1988. Characteristic trace-fossil association in oxygen-poor sedimentary environments. Geology 16, 720–723.
Ekdale, A.A., Muller, L.N., Novak, M.T., 1984. Quantitative ichnology of modern pelagic deposits in the abyssal Atlantic. Palaeogeogr. Palaeoclimatol. Palaeoecol. 45, 189–223.
Ernst, G., Zander, J., 1993. Stratigraphy, facies development, and trace fossils of the Upper Cretaceous of southern Tanzania (Kilwa District). In: Abbate, E., Sagri, M., Sassi, F.P. (Eds.), Geology and Mineral Resources and Surrounding Regions, Relazioni e Monografie Agrarie Suptropicali e Tropicali, Nuova Serie 113, 259–278.
Faugères, J.-C., Mulder, T., 2011. Contour currents and contourite drifts. In: Hüneke, H., Mulder, T. (Eds.), Deep-Sea Sediments. Developments in Sedimentology, Elsevier, Amsterdam, vol. 63, pp. 149–214.
Fillion, D., Pickerill, R.K., 1990. Ichnology of the Upper Cambrian to Lower Ordovician Bell Island and Wabana groups of eastern Newfoundland, Canada. Palaeontogr. Can. 7, 1–119.
Föllmi, K.B., Grimm, K.A., 1990. Doomed pioneers: gravity-flow deposition and bioturbation in marine oxygen-deficient environments. Geology 18, 1069–1072.
Frey, R.W., 1970. Environmental significance of recent marine lebensspuren near Beaufort, North Carolina. J. Paleont. 44, 89–90.
Frey, R.W., Pemberton, G.S., 1985. Biogenic structures in outcrops and cores. I. Approach to ichnology. Bull. Canad. Petrol. Geol. 33, 72–115.
Frey, R.W., Seilacher, A., 1980. Uniformity in marine invertebrate ichnology. Lethaia 13, 183–207.
Frey, R.W., Pemberton, G.S., Saunders, T.D.A., 1990. Ichnofacies and bathymetry: a passive relationship. J. Paleont. 54, 155–158.
Fu, S., 1991. Funktion, Verhalten und Einteilung fucoider und lophoctenoider Lebensspuren. Cour. Forsch. Inst. Senck. 135, 1–79.

Fu, S., Werner, F., 1994. Distribution and composition of biogenic structures on the Iceland-Faeroe Ridge: relation to different environments. Palaios 9, 92–101.

Fu, S., Werner, F., 2000. Distribution, ecology and taphonomy of the organism trace *Scolicia*, in northeast Atlantic deep-sea sediments. Palaeogeogr. Palaeoclimatol. Palaeoecol. 156, 289–300.

Fuchs, T., 1895. Studien über Fucoiden und Hieroglyphen. Denkschr. Kais. Akad. Wiss. Wien, math.-nat. Kl. 62, 369–448.

Gage, J.D., Tyler, P.A., 1991. Deep-sea biology: a natural history of organisms at the deep-sea floor. Cambridge Univ., Press, Cambridge.

Gaillard, C., 1984. Bioturbation des sédiments pélagiques du Crétacé inférieur dans le Bassin Vocontien (Chaînes Subalpines méridionales, France). Geobios. Spec. Mem. 8, 205–214.

Gaillard, C., 1988. Bioturbation récente au large de la Nouvelle-Calédonie. Premiers résultats de la campagne Biocal. Oceanogr. Acta 11, 389–399.

Gaillard, C., 1991. Recent organism traces and ichnofacies on the deep-sea floor off New Caledonia, southwestern Pacific. Palaios 6, 302–315.

Gehlen, M., Rabouville, C., Ezat, U., Guidi-Guilvard, L.D., 1997. Drastic changes in deep-sea sediment porewater composition induced by episodic input of organic matter. Limnol. Oceanogr. 42, 980–986.

Gierlowski-Kordesch, E., Ernst, F., 1987. A flysch trace fossil assemblage from the Upper Cretaceous shelf of Tanzania. In: Mathies, G., Schandelmeir, H. (Eds.), Current Research in African Earth Sciences. Extended Abstracts. 14th Colloquium on African Geology, Berlin (West), 18–22 August 1987. A.A. Balkema, Rotterdam, pp. 217–221.

Goldring, R., Pollard, J.E., Taylor, A.M., 1991. *Anconichnus horizontalis*: a pervasive ichnofabric-forming trace fossil in the post-Paleozoic offshore siliciclastic facies. Palaios 6, 250–263.

Goldring, R., Cadée, G.C., D'Alessandro, A., de Gibert, J.M., Jenkins, R., Pollard, J.E., 2004. Climatic control on trace fossil distribution in the marine realm. In: McIlroy, D. (Ed.), The Application of Ichnology to Palaeoenvironmental and Stratigraphic Analysis, Geol. Soc. Lond., Spec. Publ. 228, 77–93.

Gooday, A.J., Turley, C.M., 1990. Responses of benthic organisms to inputs of organic material to the ocean floor: a review. Phil. Trans. R. Soc. Lond. A331, 119–138.

Häntzschel, W., 1975. Trace fossils and problematica. In: Teichert, C. (Ed.), Treatise on Invertebrate Paleontology, Part W, Miscellanea, Supplement 1. Geological Society of America and University of Kansas Press, Boulder, Colo., and Lawrence, Kans, pp. W1–W269.

Hesse, R., Schacht, U., 2011. Early diagenesis of deep-sea sediments. In: Hüneke, H., Mulder, T. (Eds.), Deep-Sea Sediments. Developments in Sedimentology, Elsevier, Amsterdam, vol. 63, pp. 557–714.

Hüneke, H., Henrich, R., 2011. Pelagic sedimentation in modern and ancient oceans. In: Hüneke, H., Mulder, T. (Eds.), Deep-Sea Sediments. Developments in Sedimentology, Elsevier, Amsterdam, vol. 63, pp. 215–351.

Jumars, P., Ekman, J.E., 1983. Spatial structure within deep-sea benthic communities. In: Rowe, G.T. (Ed.), Deep-Sea Biology. The Sea. Vol. 8. Wiley, New York, pp. 399–451.

Jumars, P.A., Mayer, L.M., Deming, J.W., Barros, J.A., Wheatcroft, R.A., 1990. Deep-sea deposit-feeding strategies suggested by environmental and feeding constraints. Phil. Trans. R. Soc. Lond. A331, 85–101.

Keighley, D.G., Pickerill, R.K., 1995. The ichnotaxa *Palaeophycus* and *Planolites*: historical perspectives and recommendations. Ichnos 3, 301–309.

Kern, J.P., 1980. Origin of trace fossils in Polish Carpathians flysch. Lethaia 13, 347–723.

Kern, P.J., Warme, J.E., 1974. Trace fossils and bathymetry of the Upper Cretaceous Point Loma Formation, San Diego, California. Geol. Soc. Am. Bull. 85, 893–900.

Książkiewicz, M., 1954. Uziarnienie frakcjonalne i laminowane we fliszu karpackim (Graded and laminated bedding in the Carpathian Flysch). Rocznik Pol. Tow. Geol. 22, 399–471.
Książkiewicz, M., 1977. Trace fossils in the flysch of the Polish Carpathians. Palaeont. Polon. 36, 1–208.
Lavaleye, M.S.S., Duineveld, G.C.A., Berghuis, E.M., Kok, A., Witbaard, R., 2002. A comparison between the megafauna communities on the N.W. Iberian and Celtic continental margins—effects of coastal upwelling. Prog. Oceanogr. 52, 459–476.
Legeleux, F., Reyss, J.-L., Schmidt, S., 1994. Particle mixing rates in sediments of the northeast tropical Atlantic from $^{210}Pb_{xs}$, ^{137}Cs, $^{238}Th_{xs}$ and $^{234}Th_{xs}$ downcore distribution. Earth Planet. Sci. Lett. 128, 545–562.
Leszczyński, S., 1991a. Trace-fossil tiering in flysch sediments: examples from the Guipúzcoan flysch (Cretaceous–Paleogene), northern Spain. Palaeogeogr. Palaeoclimatol. Palaeoecol. 88, 167–184.
Leszczyński, S., 1991b. Oxygen-related controls on predepositional ichnofacies in turbidites, Guipúzcoan flysch (Albian-lower Eocene), northern Spain. Palaios 6, 271–280.
Leszczyński, S., 1993a. A generalized model for the development of ichnocoenoses in flysch deposits. Ichnos 2, 137–146.
Leszczyński, S., 1993b. Ichnocoenosis versus sediment colour in upper Albian to lower Eocene turbidites, Guipúzcoan province, northern Spain. Palaeogeogr. Palaeoclimatol. Palaeoecol. 100, 251–265.
Leszczyński, S., 2003. Bioturbation structures in the Holovnia Siliceous Marls (Turonian–Lower Santonian) in Rybotycze (Polish Carpathians). Annu. Soc. Geol. Polon. 73, 103–122.
Leszczyński, S., 2004. Bioturbation structures of the Kropivnik Fucoid Marls (Campanian–lower Maastrichtian) of the Huwniki–Rybotycze area (Polish Carpathians). Geol. Quart. 48, 35–60.
Leszczyński, S., Seilacher, A., 1991. Ichnocoenoses of a turbidite sole. Ichnos 1, 293–303.
Leszczyński, S., Uchman, A., 1993. Biogenic structures of organic-poor sediments: examples from the Paleogene variegated shales, Polish Outer Carpathians. Ichnos 2, 267–275.
Lobza, V., Schieber, J., 1999. Biogenic sedimentary structures produced by worms in soupy, soft muds: observations from the Chattanooga Shale (Upper Devonian) and experiments. J. Sediment. Res. 69, 1041–1049.
Lopez, G., Taghon, G., Levinton, J.E. (Eds.), 1989. Ecology of Marine Deposit Feeders. Springer, Berlin 322pp.
Löwemark, L., Lin, I.-T., Wang, C.-H., Huh, C.-A., Wei, K.-Y., Chen, C.-W., 2004a. Ethology of the *Zoophycos*-producer: arguments against the gardening model from delta-13 C_{org} evidences of the spreiten material. Terr. Atmos. Ocean. Sci. 15, 713–725.
Löwemark, L., Schönfeld, J., Werner, F., Schäfer, P., 2004b. Trace fossils as a paleoceanographic tool: evidence from Late Quaternary sediments of the southwestern Iberian margin. Mar. Geol. 204, 27–41.
Mangini, A., Jung, M., Laukenmann, S., 2001. What do we learn from peaks of uranium and of manganese in deep sea sediments. Mar. Geol. 177, 63–78.
McBride, E.F., Picard, M.D., 1991. Facies implications of *Trichichnus* and *Chondrites* in turbidites and hemipelagites, Marnoso-arenacea Formation (Miocene), northern Apennines, Italy. Palaios 6, 281–290.
McCann, T., 1990. Distribution of Ordovician–Silurian ichnofossil assemblages in Wales—implications for Phanerozoic ichnofaunas. Lethaia 23, 243–255.
McCann, T., Pickerill, R.K., 1988. Flysch trace fossils from the Cretaceous Kodiak Formation of Alaska. J. Paleont. 62, 330–347.
McCave, I.N., 1985. Properties of suspended sediment over the HEBBLE area on the Nova Scotia rise. Mar. Geol. 66, 169–188.

McCave, I.N., Lonsdale, P.F., Hollister, C.D., Gardner, W.D., 1980. Sediment transport over the Hatton and Gardar contourite drifts. J. Sed. Petrol. 50, 1049–1062.
Meadows, A., Meadows, P.S., West, F.J.C., Murray, J.M.H., 2000. Bioturbation, geochemistry and geotechnics of sediments affected by the oxygen minimum zone on the Oman continental slope and abyssal plain, Arabian Sea. Deep-Sea Res. II 47, 259–280.
Mikuláš, R., Lechotský, T., Bábek, O., 2002. Lower Carboniferous ichnofabrics of the Culm facies: a case study of the Moravice Formation (Moravia and Silesia, Czech Republic). Geol. Carpathica 53, 141–148.
Mikuláš, R., Skupien, P., Bubík, M., Vašíček, Z., 2009. Ichnology of the Cretaceous Oceanic Red Beds (Outer Western Carpathians, Czech Republic. Geol. Carpathica 60, 233–250.
Miller III, W., 1991a. Intrastratal trace fossil zonation, Cretaceous flysch in northern California. Ichnos 1, 161–171.
Miller III, W., 1991b. Paleoecology of graphoglyptids. Ichnos 1, 305–312.
Miller III, W., 1993. Trace fossil zonation in Cretaceous turbidite facies, northern California. Ichnos 3, 11–28.
Miller III, W., 2003. Paleobiology of complex trace fossils. Palaeogeogr. Palaeoclimatol. Palaeoecol. 192, 3–14.
Miller III, W., D'Alberto, L., 2001. Paleoethologic implications of *Zoophycos* from Late Cretaceous and Paleocene limestones of the Venetian Prealps, northeastern Italy. Palaeogeogr. Palaeoclimatol. Palaeoecol. 166, 237–247.
Miller III, W., Stefani, C., Grandesso, P., 2004. Alternation of ecologic regimes in a deep-marine carbonate basin: calciturbidite trace fossils from the Cretaceous Scaglia Rossa, northeastern Italy. Palaeogeogr. Palaeoclimatol. Palaleoecol. 204, 317–330.
Monaco, P., 2008. Taphonomic features of *Paleodictyon* and other graphoglyptid trace fossils in Oligo-Miocene thin-bedded turbidites, Northern Apennines, Italy. Palaios 23, 667–682.
Müller, P.J., Suess, E., 1979. Productivity, sedimentation rate and sedimentary organic matter in the oceans. I. Organic carbon preservation. Deep-Sea Res. 26, 1347–1362.
Mulder, T., 2011. Gravity processes on continental slope, rise and abyssal plains. In: Hüneke, H., Mulder, T. (Eds.), Deep-Sea Sediments. Developments in Sedimentology, Elsevier, Amsterdam, vol. 63, pp. 25–148.
Nathorst, A.G., 1881. Om spar af några evertebrerade djur m. m. och deras paleontologiska betydelse. Kong. Sv. Vatens. – Akad. Handl. 18, 1–104.
Nowell, A.R.M., Hollister, C.D. (Eds.), 1985. Deep Ocean Sediment Transport. Preliminary Results of the High Energy Benthic Boundary Layer Experiment. Elsevier, Amsterdam 418pp.
Ohta, S., 1984. Star-shaped feeding traces produced by echiurian worms on the deep-sea floor of the Bay of Bengal. Deep-Sea Res. 31, 1415–1432.
Olivero, D., 1996. *Zoophycos* distribution and sequence stratigraphy. Examples from the Jurassic and Cretaceous deposits of southeastern France. Palaeogeogr. Palaeoclimatol. Palaeoecol. 123, 273–287.
Olivero, D., 2003. Early Jurassic to Late Cretaceous evolution of *Zoophycos* in French Subalpine Basin (southeast France). Palaeogeogr. Palaeoclimatol. Palaeoecol. 192, 59–78.
Olivero, D., Gaillard, C., 1996. Paleoecology of Jurassic *Zoophycos* from south-eastern France. Ichnos 4, 249–260.
Orr, P.J., 1995. A deep-marine ichnofaunal assemblage from Llandovery strata of the Welsh Basin, west Wales. UK. Geol. Mag. 132, 267–285.
Orr, P.J., 1996. The ichnofauna of the Skiddaw Group (early Ordovician) of the Lake District, England. Geol. Mag. 133, 193–216.
Orr, P.J., 2001. Colonization of the deep-marine environment during the early Phanerozoic: the ichnofaunal record. Geol. J. 36, 265–278.

Orr, P.J., 2003. Ecospace utilization in early Phanerozoic deep-marine environments: deep bioturbation in the Blakely Sandstone (Middle Ordovician), Arkansas, USA. Lethaia 36, 97–106.
Orr, P.J., Benton, M., Trewin, N.H., 1996. Deep marine trace fossil assemblages from the Lower Carboniferous of Menorca, Balearic Islands, western Mediterranean. Geol. J. 31, 235–258.
Osgood, R.G., 1970. Trace fossils of the Cincinnati Area. Palaeontogr. Am. 6, 193–235.
Pemberton, G.S., Frey, R.W., 1982. Trace fossil nomenclature and the *Planolites–Palaeophycus* dilemma. J. Paleont. 56, 843–881.
Pemberton, G.S., Spila, M., Pulham, A.J., Saunders, T., MacEachern, J.A., Robbins, D., et al., 2001. Ichnology and sedimentology of shallow to marginal marine systems: Ben Nevis and Avalon Reservoirs, Jeanne d'Arc Basin. Geol. Ass. Can., Short Course Notes, vol. 15. 343 pp.
Pervesler, P., Uchman, A., 2004. Ichnofossils from the type area of the Grund Formation (Miocene, Lower Badenian) in northern Lower Austria (Molasse Basin). Geol. Carpathica 55, 103–110.
Pfeiffer, H., 1968. Die Spurenfossilien des Kulms (Dinants) und Devons der Frankenwälder Querzone (Thüringen). Jahrb. Geol. 2, 651–717.
Pickerill, R.K., 1980. Phanerozoic flysch trace fossil diversity - observations based on an Ordovician flysch ichnofauna from the Aroostook-Metapedia Carbonate Belt of northern New Brunswick. Can. J. Earth Sci. 17, 1259–1270.
Pickering, K.T., Hiscott, R.N., Hein, F.J., 1989. Deep Marine Environments: Clastic Sedimentation and Tectonics. London, Unwin Hyman 416pp.
Pieńkowski, G., Westwalewicz-Mogilska, E., 1986. Trace fossils from the Podhale Flysch Basin, Poland - an example of ecologically based lithocorrelation. Lethaia 19, 53–65.
Rajchel, J., Uchman, A., 1998. Ichnological analysis of an Eocene mixed marly-siliciclastic flysch deposits in the Nienadowa Marls Member, Skole Unit, Polish Flysch Carpathians. Annu. Soc. Geol. Pol. 68, 61–74.
Reineck, H.-E., 1973. Schichtung und Wühlgefüge in Grundproben vor der ostafrikanischen Küste. "Meteor" Forsch.-Ergebn. C16, 67–81.
Reineck, H.-E., Gutmann, W.F., Hertweck, G., 1967. Das Schlickgebiet südlich Helgoland als Beispiel rezenter Schelfablagerungen. Senck. Leth. 48, 219–275.
Rhoads, D.C., Boyer, L.F., 1982. The effects of marine benthos on physical properties of sediments: a successional perspective. In: McCall, P.L., Tevesz, M.J.S. (Eds.), Animal–Sediment Relations. Plenum, New York, pp. 3–52.
Richardson, M.J., Wimbush, M., Mayer, L., 1981. Exceptionally strong near-bottom flows on the continental rise of Nova Scotia. Science 213, 887–888.
Richter, R., 1924. Flachseebeobachtungen zur Paläontologie und Geologie, VII-XI. Senckenbergiana 6, 119–165.
Richter, R., 1928. Die fossilen Fährten und Bauten der Würmer, ein Überblick über ihre biologischen Grundformen und deren geologische Bedeutung. Paläont. Zeitschr. 9, 193–241.
Richter, R., 1938. Spuren von Schnecken–versteinert? Natur und Volk 68, 337–342.
Rowe, G.T., 1983. Biomass and production of the deep-sea macrobenthos. In: Rowe, G.T. (Ed.), The Sea. Vol. 8. Wiley and Sons, New York, pp. 97–121.
Sacco, F., 1888. Note di paleoicnologia italiana. Atti Soc. Ital. Sci. Nat. 31, 151–192.
Sanders, H.L., 1968. Marine benthic diversity: a comparative study. Am. Nat. 100, 243–282.
Savary, B., Olivero, D., Gaillard, C., 2004. Calciturbidite dynamics and endobenthic colonization. Example from a Late Barremian (Early Cretaceous) succession of southeastern France. Palaeogeogr. Palaeoclimatol. Palaeoecol. 211, 221–239.
Savrda, C.E., 1995. Ichnologic applications in paleoceanographic, paleoclimatic, and sea-level studies. Palaios 10, 565–577.

Savrda, C.E., Bottjer, D.J., 1986. Trace-fossil model for reconstruction of paleo-oxygenation in bottom waters. Geology 14, 3–6.
Schäfer, W., 1956. Wirkungen der Benthos-Organismen auf den jungen Schichtverband. Senck. Leth. 37, 183–263.
Schlirf, M., Uchman, A., 2005. Revision of the ichnogenus *Sabellarifex* Richter, 1921 and its relationship to *Skolithos* Haldeman, 1840 and *Polykladichnus* Fürsich, 1981. J. Syst. Palaeont. 3, 115–131.
Seibold, E., Berger, W.H., 1996. The Sea Floor, 3rd ed. Springer, Berlin, 356 pp.
Seilacher, A., 1953. Studien zur Palichnologie. I. Über die Methoden der Palichnologie. Neues Jb. Geol. Paläont. Abh. 96, 421–452.
Seilacher, A., 1955. Spuren und Fazies im Unterkambrium. In: Schindewolf, O.H., Seilacher, A. (Eds.), Beiträge zur Kenntnis des Kambriums in der Salt Range (Pakistan), Akad. Wiss. Liter. Mainz. Abh. math.-nat. Kl. 10, 373–399.
Seilacher, A., 1962. Paleontological studies of turbidite sedimentation and erosion. J. Geol. 70, 227–234.
Seilacher, A., 1964. Biogenic sedimentary structures. In: Imbrie, J., Newell, N. (Eds.), Approaches to Paleoecology. Wiley and Sons, New York, pp. 296–316.
Seilacher, A., 1967. Bathymetry of trace fossils. Mar. Geol. 5, 413–428.
Seilacher, A., 1974. Flysch trace fossils: evolution of behavioural diversity in the deep-sea. Neues Jb. Geol. Paläont. Mh. 1974, 223–245.
Seilacher, A., 1976. Evolution von Spuren-Vergesellschaftungen. Zentrabl. Geol. Paläont., Teil 2, 396–402.
Seilacher, A., 1977a. Pattern analysis of *Paleodictyon* and related trace fossils. In: Crimes, T.P., Harper, J.C. (Eds.), Trace Fossils 2, Geol. J., Spec., 289–334 Issue 9.
Seilacher, A., 1977b. Evolution of trace fossil communities in the deep sea. In: Hallam, A. (Ed.), Patterns of Evolution. Elsevier, Amsterdam, pp. 359–376.
Seilacher, A., 1978. Evolution of trace fossil communities in the deep sea. Neues Jb. Geol. Paläont. Abh. 157, 251–255.
Shackleton, N.J., 1986. Paleogene stable isotope events. Palaeogeogr. Palaeoclimatol. Palaeoecol. 57, 91–102.
Smith, C.R., Levin, L.A., Hoover, D.J., McMurtry, G., Gage, J.D., 2000. Variations in bioturbation across the oxygen minimum zone in the northwest Arabian Sea. Deep-Sea Res. II 47, 227–257.
Soetaert, K., Herman, P.M.J., Middleburg, J.J., Heip, C., 1998. Assessing organic matter mineralization, degradability and mixing rate in an ocean margin sediment (Northeast Atlantic) by diagenetic modeling. J. Mar. Res. 56, 519–534.
Squinabol, S., 1890. Alghae and Pseudoalghae fossili italiane. Atti Soc. Ling. Sci. Nat. Geogr. 1 (29–49), 166–199.
Stepanek, J., Geyer, G., 1989. Spurenfossilien aus dem Kulm (Unterkarbon) des Frankenwaldes. Beringeria 1, 1–55.
Stow, D.A.V., Pudsey, C.J., Howe, J.A., Faugères, J.-C., Viana, A.R. (Eds.), 2002. Deep-Water Contourite Systems: Modern Drifts and Ancient Series, Seismic and Sedimentary Characteristics, Geol. Soc. London Mem. 22, 464 pp.
Suess, E., 1980. Particulate organic carbon flux in the oceans–surface productivity and oxygen utilization. Nature 288, 260–263.
Tchoumatchenco, P., Uchman, A., 2001. The oldest deep-sea *Ophiomorpha* and *Scolicia* and associated trace fossils from the Upper Jurassic–Lower Cretaceous deep-water turbidite deposits of SW Bulgaria. Palaeogeogr. Palaeoclimatol. Palaeoecol. 169, 85–99.
Thayer, C.W., 1979. Biological bulldozers and the evolution of marine benthic communities. Nature 203, 458–461.

Thistle, D., Yingst, J.Y., Fauchald, K., 1985. A deep-sea benthic community exposed to strong near-bottom currents on the Scotian Rise (western Atlantic). Mar. Geol. 66, 91–112.

Thistle, D., Ertman, S.C., Fauchald, K., 1991. The fauna of the HEBBLE site: patterns in standing stock and sediment-dynamic effects. Mar. Geol. 99, 413–422.

Thomas, E., Zachos, J.C., Bralower, T.J., 2000. Deep-sea environments on a warm earth: latest Paleocene-early Eocene. In: Huber, B.T., MacLeod, K.G., Wing, S.L. (Eds.), Warm Climates in Earth History. Cambridge University Press, Cambridge, pp. 132–160.

Thomsen, L., Van Weering, T., Gust, G., 2002. Processes in the benthic boundary layer at the Iberian continental margin and their implication for carbon mineralization. Progr. Oceanogr. 52, 315–329.

Thomson, J., Brown, L., Nixona, S., Cook, G.T., MacKenzie, A.B., 2000. Bioturbation and Holocene sediment accumulation fluxes in the north-east Atlantic Ocean (Benthic Boundary Layer experiment sites). Mar. Geol. 169, 21–39.

Trauth, M., Sarnthein, M., Arnold, M., 1997. Bioturbational mixing depth and carbon flux at the seafloor. Paleoceanography 12, 517–526.

Tunis, G., Uchman, A., 1996a. Trace fossil and facies changes in the Upper Cretaceous–Middle Eocene flysch deposits of the Julian Prealps (Italy and Slovenia): consequences of regional and world-wide changes. Ichnos 4, 169–190.

Tunis, G., Uchman, A., 1996b. Ichnology of the Eocene flysch deposits in the Istria peninsula, Croatia and Slovenia. Ichnos 5, 1–22.

Uchman, A., 1991a. Diverse tiering patterns in Paleogene flysch trace fossils of the Magura nappe, Carpathians Mountains, Poland. Ichnos 1, 287–292.

Uchman, A., 1991b. Trace fossils from stress environments in Cretaceous–Paleogene flysch of the Polish Outer Carpathians. Annu. Soc. Geol. Pol. 61, 207–220.

Uchman, A., 1992. An opportunistic trace fossil assemblage from the flysch of the Inoceramian beds (Campanian-Palaeocene), Bystrica Zone of the Magura Nappe, Carpathians, Poland. Cret. Res. 13, 539–547.

Uchman, A., 1995a. Taxonomy and palaeoecology of flysch trace fossils: The Marnoso-arenacea Formation and associated facies (Miocene, Northern Apennines, Italy). Beringeria 15, 3–115.

Uchman, A., 1995b. Tiering patterns of trace fossils in Paleogene flysch deposits of the Carpathians, Poland. Geobios Mem. Spec. 18, 389–394.

Uchman, A., 1998. Taxonomy and ethology of flysch trace fossils: a revision of the Marian Książkiewicz collection and studies of complementary material. Annu. Soc. Geol. Polon. 68, 105–218.

Uchman, A., 1999. Ichnology of the Rhenodanubian Flysch (Lower Cretaceous–Eocene) in Austria and Germany. Beringeria 25, 65–171.

Uchman, A., 2001. Eocene flysch trace fossils from the Hecho Group of the Pyrenees, northern Spain. Beringeria 28, 3–41.

Uchman, A., 2003. Trends in diversity, frequency and complexity of graphoglyptid trace fossils: evolutionary and palaeoenvironmental aspects. Palaeogeogr. Palaeoclimatol. Palaeoecol. 192, 123–142.

Uchman, A., 2004a. Phanerozoic history of deep-sea trace fossils. In: McIlroy, D. (Ed.), The Application of Ichnology to Palaeoenvironmental and Stratigraphic Analysis, Geol. Soc. Lond., Spec. Publ. 228, 25–139.

Uchman, A., 2004b. Deep-sea trace fossils controlled by palaeo-oxygenation and deposition: an example from the Lower Cretaceous dark flysch deposits of the Silesian Unit, Carpathians, Poland. Fossils Strata 51, 39–57.

Uchman, A., 2007a. Deep-sea ichnology: development of major concepts. In: Miller, W. (Ed.), Trace Fossils: Concepts, Problems, Prospects. Elsevier, Amsterdam, pp. 248–267.

Uchman, A., 2007b. Deep-sea trace fossils from the mixed carbonate-siliciclastic flysch of the Monte Antola Formation (Late Campanian-Maastrichtian), North Apennines, Italy. Cret. Res. 28, 980–1004.
Uchman, A., 2009. The *Ophiomorpha rudis* ichnosubfacies of the *Nereites* ichnofacies: characteristics and constraints. Palaeogeogr. Palaeoclimat. Palaeoecol. 276, 107–119.
Uchman, A., Tchoumatchenco, P., 2003. A mixed assemblage of deep-sea and shelf trace fossils from the Lower Cretaceous (Valanginian) Kamchia Formation in the Troyan region, central Fore-Balkan, Bulgaria. Annu. Soc. Geol. Pol. 73, 27–34.
Uchman, A., Janbu, N.E., Nemec, W., 2004. Trace fossils in the Cretaceous–Eocene flysch of the Sinop-Boyabat Basin, Central Pontides, Turkey. Annu. Soc. Geol. Pol. 74, 197–235.
Uchman, A., Bubík, M., Mikuláš, R., 2005. The ichnological record across the Cretaceous/ Tertiary boundary in turbiditic sediments at Uzgruň (Moravia, Czech Republic). Geol. Carpathica 56, 57–65.
Von Fischer-Ooster, C., 1858. Die fossilen Fucoiden der Schweizer Alpen, nebst Erörterung über deren geologisches Alter. Huber u. Comp., Bern 74pp.
Wakeham, S.G., Lee, C., Hedges, J.I., Hernes, P.J., Peterson, M.I., 1997. Molecular indicators of diagenetic status in marine organic matter. Geoch. Cosmoch. Acta 61, 5363–5368.
Weissert, H., 2011. Mesozoic pelagic sediments – archives for ocean and climate history during green-house conditions. In: Hüneke, H., Mulder, T. (Eds.), Deep-Sea Sediments. Developments in Sedimentology, Elsevier, Amsterdam, vol. 63, pp. 765–792.
Werner, F., 1967. Röntgen-Radiographie zur Untersuchung von Sedimentstrukturen. Umschau 16, 532.
Werner, F., Wetzel, A., 1982. Interpretation of biogenic structures in oceanic sediments. Bull. Inst. Géol. Bassin d'Aquitaine 31, 275–288.
Wetzel, A., 1981. Ökologische und stratigraphische Bedeutung biogener Gefüge in quartären Sedimenten am NW-afrikanischen Kontinentalrand. "Meteor" Forsch. Ergebn C34, 1–47.
Wetzel, A., 1983. Biogenic sedimentary structures in a modern upwelling region: NW African continental margin. In: Thiede, J., Suess, E. (Eds.), Coastal Upwelling and Its Sediment Record, Part B, Sedimentary Records of Ancient Coastal Upwelling. Plenum, New York, pp. 123–144.
Wetzel, A., 1984. Bioturbation in deep-sea fine-grained sediments: influence of sediment texture, turbidite frequency and rates of environmental changes. In: Stow, D.A.V., Piper, D.J.W. (Eds.), Fine Grained Sediments: Deep-Water Processes and Facies, Geol. Soc. Spec. Publ. 15, 597–608.
Wetzel, A., 1990. Interrelationships between porosity and other geotechnical properties of slowly deposited, fine-grained marine surface sediments. Mar. Geol. 92, 105–113.
Wetzel, A., 1991. Ecologic interpretation of deep-sea trace fossil communities. Palaeogeogr. Palaeoclimatol. Palaeoecol. 85, 47–69.
Wetzel, A., 2002. Modern *Nereites* in the South China Sea—ecological associations with redox conditions in the sediment. Palaios 17, 507–515.
Wetzel, A., 2010. Deep-sea ichnology: Observations in modern sediments to interpret fossil counterparts. Acta Geol. Pol. 60, 125–138.
Wetzel, A., Aigner, T., 1986. Stratigraphic completeness: tiered trace fossils provide a measuring stick. Geology 14, 234–237.
Wetzel, A., Bromley, R.G., 1994. *Phycosiphon incertum* revisited: *Anconichnus horizontalis* is its junior subjective synonym. J. Paleont. 68, 1396–1402.
Wetzel, A., Bromley, R.G., 1996. The ichnotaxon *Tasselia ordamensis* and its junior synonym *Caudichnus annulatus*. J. Paleont. 70, 523–526.

Wetzel, A., Uchman, A., 1998a. Biogenic sedimentary structures in mudrocks—an overview. In: Schieber, J., Zimmerle, W., Sethi, P. (Eds.), Shales and Mudrocks I. Schweizerbart, Stuttgart, pp. 351–369.

Wetzel, A., Uchman, A., 1998b. Trophic level in the deep-sea recorded by ichnofabrics: an example from Palaeogene flysch in the Carpathians. Palaios 13, 533–546.

Wetzel, A., Uchman, A., 2001. Sequential colonization of muddy turbidites: examples from Eocene Beloveza Formation, Carpathians, Poland. Palaeogeogr. Palaeoclimat. Palaeoecol. 168, 171–186.

Wetzel, A., Werner, F., 1981. Morphology and ecological significance of *Zoophycos* in deep-sea sediments off NW Africa. Palaeogeogr. Palaeoclimat. Palaeoecol. 32, 185–212.

Wetzel, A., Blechschmidt, I., Uchman, A., Matter, A., 2007. A highly diverse ichnofauna in late Triassic deep-sea fan deposits of Oman. Palaios 22, (in press).

Wetzel, A., Werner, F., Stow, D.A.V., 2008. Bioturbation and biogenic sedimentary structures in contourites. In: Rebesco, M., Camerlenghi, A. (Eds.), Contourites. Dev. Sediment. Elsevier, Amsterdam, pp. 183–202.

Wheatcroft, R.A., 1989. Comment on "Characteristic trace-fossil associations in oxygen-poor sedimentary environments" Geology 17, 674.

CHAPTER 9

Early Diagenesis of Deep-Sea Sediments

Reinhard Hesse*[,1] *and* Ulrike Schacht[†]

Contents

* Earth and Planetary Sciences, McGill University, Montreal, Quebec, Canada
† Australian School of Petroleum, The University of Adelaide, Adelaide, SA, Australia
[1] Corresponding author.
E-mail address: Reinhard.Hesse@mcgill.ca

Developments in Sedimentology, Volume 63
ISSN 0070-4571, DOI: 10.1016/S0070-4571(11)63009-4

1. Introduction

1.1. Deep-sea sediments: Their water depth, diagenetic significance and reactivity

Deep-sea sediments cover about two thirds of the earth's surface, but vary considerably in thickness and facies from the continental margins to the deep-ocean basins. There is no consensus on the minimum water depth required for sediments to be called "deep", but for practical purposes, a depth of 500 m on the upper continental slope is probably a good upper limit. This excludes sediments with shallow-water affinities on the uppermost slope that may be deposited during sea-level low-stands or shed from carbonate platforms and reef slopes during high stands. It is also beyond the reach of storms affecting bottom sediments. It intentionally excludes pelagic carbonates such as chalks deposited on the continental shelves during sea-level highstands as well as some black shales because shelves are the shallow-water portions of the continental margins, although in exceptional situations such as shelf troughs formed by ice streams they may reach 1000 m water depth. Diagenesis, and in particular early diagenesis, is an important aspect of deep-sea sediments as it determines the geochemical cycling and the residence time of important elements such as silicium, phosphorus or nitrogen in the ocean. Diagenetic products may record significant paleoceanographic and paleoclimatic signals.

The diagenesis of deep-sea sediments depends very much on the reactivity of the sediment constituents. Sediments rich in organic matter or volcanic components may undergo rapid diagenetic change even during shallow burial and only shortly after deposition or while still exposed to bottom water on the ocean floor. Organic matter, biogenic silica in the form of opal-A and volcanic glass are the most reactive sediment constituents. Sediments lacking organic matter such as the brown abyssal clay may not react diagenetically for long periods of time. Also the low-magnesium calcite of many planktonic micro-organisms is diagenetically fairly inert. As a rule, diagenetic change is sluggish in the majority of deep-sea sediments because of the uniformly low temperatures on the deep ocean floor and moderate to low geothermal gradients characterizing large parts of the ocean basins and continental margins. A distinct exception is the mid-ocean ridge (MOR) province with its high geothermal gradients. However, it is free of sediment on its crest, except where the ridge meets continental margins, and carries only a thin sediment cover on its flanks. In many deep-sea sediments of the open ocean there are consequently few if any newly formed minerals at shallow burial depth during early diagenesis and only modest changes occur during intermediate levels of diagenesis. Towards the continental margins, the concentration of organic matter increases progressively from the low levels in ocean basins raising the reactivity of the sediment and leading to pronounced diagenetic reactions.

The general strategy for this review is to gain first insight in the diagenesis of the various classes of deep-sea sediments by looking at typical pore-water profiles, and to depict the mineral reactions that are associated with changes in the pore-water chemistry. In doing so, we shall proceed from the simplest systems encountered in pelagic sediments on relatively young ocean crust with pore-water profiles lacking chemical gradients to the most highly reactive organic-matter rich sediments on the continental margins at the oxygen-minimum zone. A special section is devoted to highly reactive volcanogenic sediments.

1.2. Significance of pore-water studies for tracing diagenetic processes and reactions

A powerful tool used to monitor the beginning of changes in the sediment composition during early diagenesis is pore-water analysis. Even minor changes affecting the solids that may be too small to be detected with common analytical techniques may leave a large enough signal in the pore-water composition to be picked up by routine hydrochemical analysis. Variations in the chemical composition of pore waters may point to the locale where reactions in the sediment are presently occurring, because diffusion generally is too slow to wipe out the signals from ongoing reactions. Diffusion, on the other hand, helps to smooth pore-water profiles and to suppress noise. Water is the principal transport medium for dissolved species that are carried by advection or diffusion from the sites of dissolution to the sites of precipitation. Where water is absent because mineral cements or non-aqueous fluids such as hydrocarbons fill the pores, diagenetic reactions may be inhibited or drastically slowed down. The presence of water in the sediment pores is thus a prerequisite for diagenetic reactions to proceed. By advection and diffusion, water also facilitates exchange between the hydrosphere and deeper parts of the sediment column in sedimentary basins. Last not least, pore-water chemistry can be used to assess thermodynamic mineral-solution equilibria in diagenetic environments. To summarize, pore waters are much more sensitive to changes in the diagenetic environment than the solids, and pore-water analyses are therefore the preferred analytical tool at least for the study of early diagenesis and are still very useful at intermediate levels of diagenesis.

Pore-water profiles are characteristic of specific sediment types only if they are reaction-controlled. In advection- and/or diffusion-dominated profiles, fluid-chemistry related to the nature of the host sediment is overprinted by ions derived from external sources. Since this chapter is subdivided into sections based on sediment-type, advection and diffusion-dominated profiles that do not fit this classification scheme have been advanced to the beginning of the section on pelagic sediments, although some of the profiles will be discussed in more detail with specific pelagic sediment types at a later stage.

1.3. Boundary between early and intermediate diagenesis

The boundary between early and intermediate diagenesis is conveniently placed at about 75 °C, where most bacterial reactions cease and the temperature is high enough to provide the activation energies required to trigger thermocatalytic reactions. The existing database from deep-sea drill cores essentially covers the realm of early diagenesis; some deeper holes also penetrate middle diagenetic sediments. Sediments having undergone advanced diagenesis (>150 °C; Hesse, 1990a) are generally beyond the limits of conventional deep-water drilling from scientific research vessels except in geothermal areas. They will, however, become future drilling targets with the new Japanese riser drill-ship Chikyu.

Much of what is known about the early diagenesis of deep-sea sediments is the fruit of 40 years of deep-sea drilling during the Deep-Sea Drilling Project (DSDP, 1968–1983), the Ocean Drilling Program (ODP, 1985–2003) and the Integrated Ocean Drilling Program (IODP, since 2003). The uppermost tens of meters have been intensely studied with piston corers that, like the French Calypso corer operated on the RV Marion Dufresne 2, can penetrate up to almost 70 m of sediment.

2. Pelagic Sediments: Characteristics and Lithology-Independent Pore-Water Profiles

2.1. Deposition of pelagic sediments

Pelagic sediments are the deposits of the open ocean that accumulate on the ocean floor protected from terrestrial influence (see Hüneke and Henrich, 2011, this volume). They are not necessarily deep but are usually located at great distance from the continents. They have a lack of detrital terrigenous components, a generally low sedimentation rate, and low to moderate organic-matter concentrations in common. Pelagic sediments at a water depth above the calcite compensation level (CCL) are composed mostly of biogenic constituents, predominantly carbonates or mixed carbonate/siliceous components. At great water depths below the compensation level, they consist of siliceous tests and skeletal elements.

In areas of low surface productivity outside of upwelling zones, very little biogenic material reaches the ocean floor. In these barren areas under the large circulation gyres north and south of the equator, brown abyssal clay forms the residual sediment after carbonate dissolution. The brown clay is characterized by the lowest sedimentation rates on earth with typical values less than 5 m per million years (or <5 μm per year). Biogenic pelagic sediments may have rates as high as 200 m per million years (Scholle et al., 1983). Hemipelagic sediments that are characteristic of the continental margins have higher concentrations of terrigenous components, mostly clay, due to their proximity to the continents, in contrast to the true or eupelagic sediments of the open ocean (see Henrich and Hüneke, 2011, this volume).

2.2. Advection-dominated pore-water profiles

Pelagic sediments on ocean floors younger than $\sim$80 Ma with a thickness of less than 200 m commonly have advection-dominated pore-water profiles that lack chemical gradients. They are more or less straight-line vertical profiles essentially of seawater composition regardless of the host-sediment lithology (Fig. 9.1), and they result from the downward advection of seawater. They occur in the vast areas of intake of seawater into the MOR hydrothermal convection system and exit as supercritical fluids in the "black smokers" at the ridge axis and their cooler off-axis siblings, the "white smokers". Advection in the intake areas with downward directed flow apparently is fast enough to wipe out gradients caused by reactions.

Secondary and tertiary hydrothermal convection cells extend off-axis from the ridge crest to crustal ages of up to 80 Ma. In still older crust, convection stops because the pores in the pillow basalt become closed due to sediment infilling and cementation. The overlying sediments also undergo cementation and compaction, thus sealing off the upper crust from the downwelling fluids from above.

2.3. Diffusion-dominated pore-water profiles

Where the sediment thickness exceeds 200 m, the number of drill sites displaying pore-water gradients increases significantly. For a thickness of more than 500 m, sites without gradients do not exist (Fig. 9.2). The sites with gradients are either diffusion-dominated or reaction-dominated. Diffusion-dominated sites are characterized by negative linear correlations between the concentration changes for calcium (Δ[Ca]) and magnesium (Δ[Mg]) and between calcium concentration and oxygen-isotope ratios (Figs. 9.3 and 9.4). The negative linear correlation between the concentration changes Δ[Ca] and Δ[Mg] is due to hydrolysis reactions of

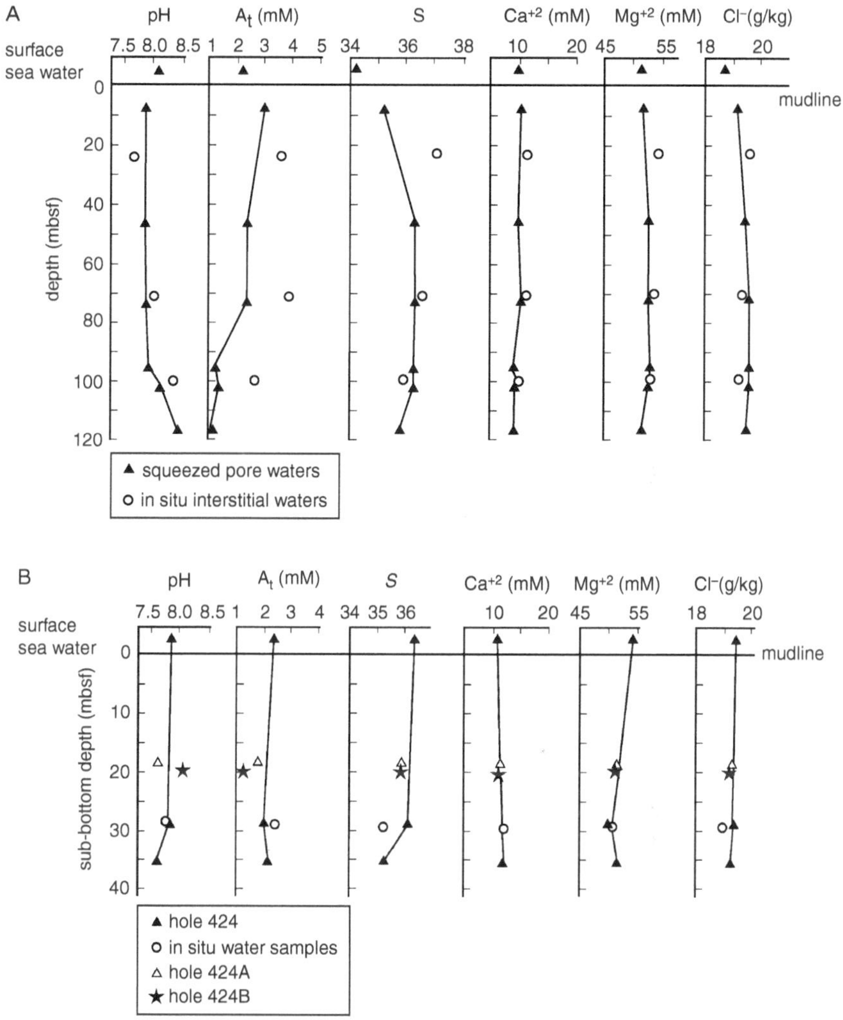

Figure 9.1 Sites 420 and 424 (redrawn and modified from Humphris and Hallman, 1980). (A) Advection-dominated pore-water profiles with straight-line sea-water composition in pelagic sediments on East Pacific Rise DSDP Site 420. (B) Site 424.

aluminosilicates in the basaltic crust that release Mg, Ca, hydroxyl-ions and silicic acid as main dissolved species. The Mg and part of the OH-ions are removed by the precipitation of Mg-bearing secondary minerals such as brucite [$Mg_3(OH)_6$] and saponite [$R_{(x-y)/m}{}^{m+} \cdot nH_2O(Mg_{3-y}R_y{}^{3+})(OH)_2(Si_{4-x}Al_x)O_{10}$ with a trivalent cation $R_y{}^{3+}$ substituting in part for Mg in the octahedral sheet]. Excess Mg supplied by downward

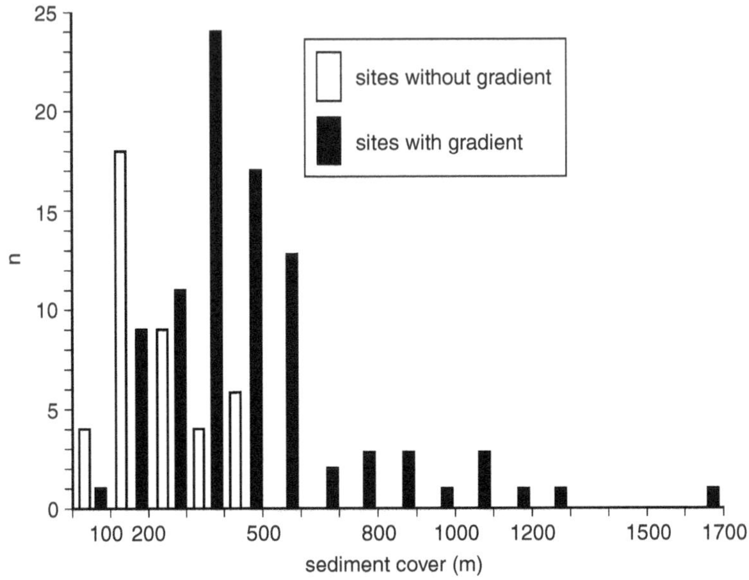

Figure 9.2 Frequency distribution of drill sites with and without concentration gradients for major pore-water cations in pelagic (and hemipelagic) sediments as a function of sediment thickness on basaltic crust (based on DSDP legs 1–85, data for legs 1–53, compiled by McDuff, 1981, redrafted from Hesse 1990a, fig. 5). *n*, number of drill sites.

diffusion is also removed with hydroxyl ions, whereas the Ca that is set free, for example by the breakdown of feldspars, diffuses away. Upward diffusion of Ca is driven by the low calcium concentration of seawater (11 mM), which is about 1/5 of that of magnesium (55 mM).

Where hydrolysis reaction rates are low, Mg uptake may exceed Ca release. In this case, the excess upward diffusion of Ca^{2+} is compensated by a downward diffusional flux of seawater-supplied Na^+ that is reflected by a distinct Na^+ gradient. Diffusion-dominated profiles with their coupled changes for [Ca] and [Mg] have been successfully modelled using diffusion coefficients that vary with depth (McDuff, 1978; McDuff and Gieskes, 1976). A strong argument in favour of diffusion-controlled coupled depth changes in [Ca] and [Mg] is the correlation with a downward decrease in the oxygen-isotope ratios seen in many deep-sea drill-hole profiles (Fig. 9.4).

The $\delta^{18}O$ decrease, which amounts to 4–5‰ down to a few hundreds of meters subsurface depth, can best be explained by preferential removal of the heavy-oxygen isotope in alteration reactions of mafic minerals of the basaltic crust and volcanogenic sediments involving the precipitation of phyllosilicates (see Section 11; Gieskes and Lawrence, 1981; Lawrence et al., 1975). A small upward $\delta^{18}O$ increase (probably not more than 0.5‰) could be due to the withdrawal of isotopically light water from the ocean and long-term storage in

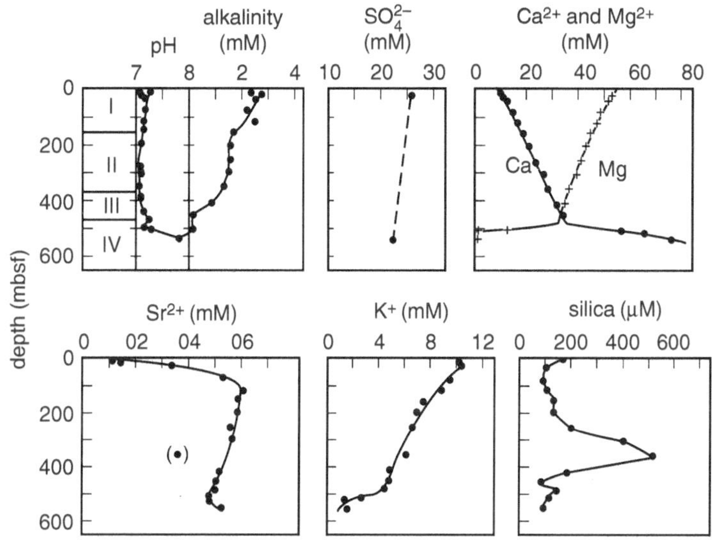

Figure 9.3 Diffusion-dominated pore-water profiles with negative linear correlation between Ca^{2+} and Mg^{2+}, DSDP Site 528 on the Walvis Ridge, South Atlantic. Dissolved Mg drops rapidly to zero below the first basalt; at the same level Ca increases rapidly. Dissolved Sr shows a typical trend for pelagic carbonates with a maximum around 100 m, resulting from carbonate recrystallisation. Lithology: I, calcareous ooze; II, nanno-fossil ooze and chalk; III, chalk with interbedded turbidites; IV, interbedded sedimentary rocks and basalt (redrawn and modified from Gieskes, 1983).

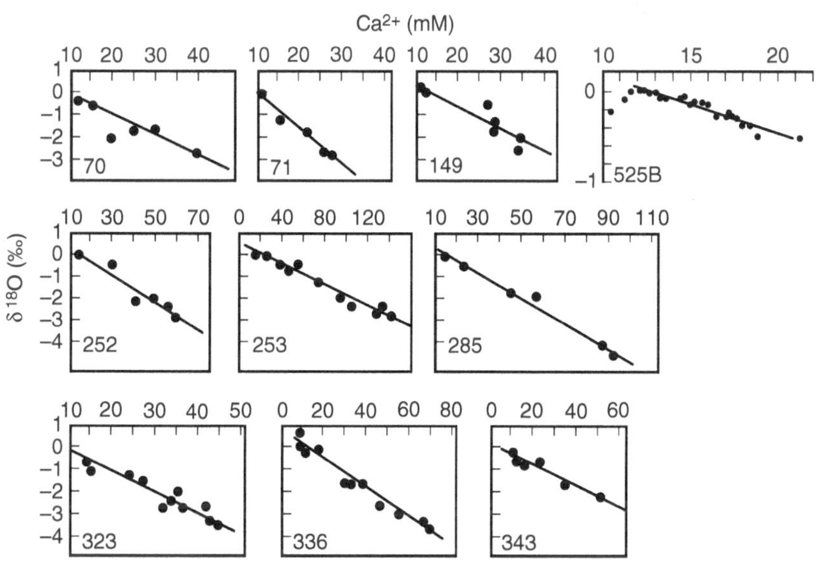

Figure 9.4 Negative linear correlation between Ca^{2+} and $\delta^{18}O$ in diffusion-dominated pore-water profiles of the DSDP (redrawn and modified from Gieskes, 1983; data from Lawrence and Gieskes, 1981).

continental ice sheets since the Miocene (Bath and Shackleton, 1984), which would make seawater progressively heavier towards the present, reaching the Vienna Standard Mean Ocean Water (VSMOW) value of 0‰ only at the terminations of the Pleistocene glaciations.

3. Brown Abyssal Clay

A good example of pelagic sediment with diffusion-controlled pore-water profiles is the brown abyssal clay. Its occurrence at great water depth below the CCL implies that it is commonly encountered in older parts of ocean basins where the effects of downward seawater advection in the crust do no longer prevail. The brown abyssal clay is virtually free of reactive organic carbon (C_{org}). A few tenths of a percent (or less) of highly refractive organic matter may be present, according to analyses of many Pacific DSDP drill holes (Fig. 9.5).

In the absence of reactive organic carbon, which is the main electron donor, iron and manganese remain in their oxidized, high-valence states in the sediment, which thus retains its brown colour during burial. More deeply buried abyssal brown clay may change its colour to red, which is the colour of red shale, the ancient equivalent of brown abyssal clay in mountain belts. This change is due to the conversion of iron-hydroxides to hematite. Although freely dissolved oxygen, which is supplied by downward diffusion, may disappear completely from the brown clay at deeper burial levels, the pore water does not become reducing enough to change the Fe^{3+} to Fe^{2+}, preserving the brown or red colour.

Brown clay forms below the CCL. It consists of the sediment particles that remain after dissolution of the calcareous components, that is eolian quartz, clay minerals, volcanic-ash particles, subordinate siliceous tests and authigenic precipitates (including zeolites, limonite and manganese oxides); hence its very low accumulation rates. The bulk of the sediment is eolian dust. Red clays in the North Pacific, which receive their non-authigenic components from the deserts of central Asia, have higher sedimentation rates (0.5–6 m per million years) than those of the South Pacific.

4. Biogenic Siliceous Sediments

4.1. Deposition of biogenic siliceous oozes and scope of the review of silica diagenesis

Siliceous ooze accumulates on the deep-sea floor under zones of increased surface bioproduction, where the dilution by biogenic pelagic carbonate is suppressed below the CCL. Increased primary production is encountered in

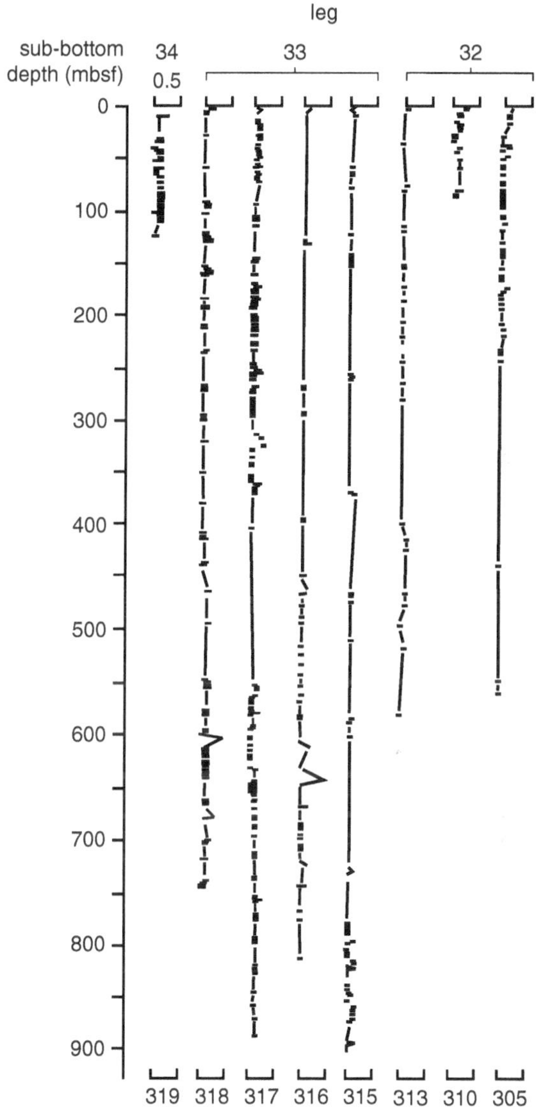

Figure 9.5 Concentration of (non-reactive) organic matter in brown abyssal clay.

zones of oceanic upwelling, where nutrients released by dissolution of solid particles at depth are returned to the surface. In the Pacific Ocean, regions of upwelling correspond to five major belts of silica accumulation on the sea floor (Fig. 9.6): (1) an equatorial belt north of the equator formed under the upwelling zone associated with the equatorial divergence; (2) the sub-Antarctic belt related to upwelling north of the Antarctic convergence;

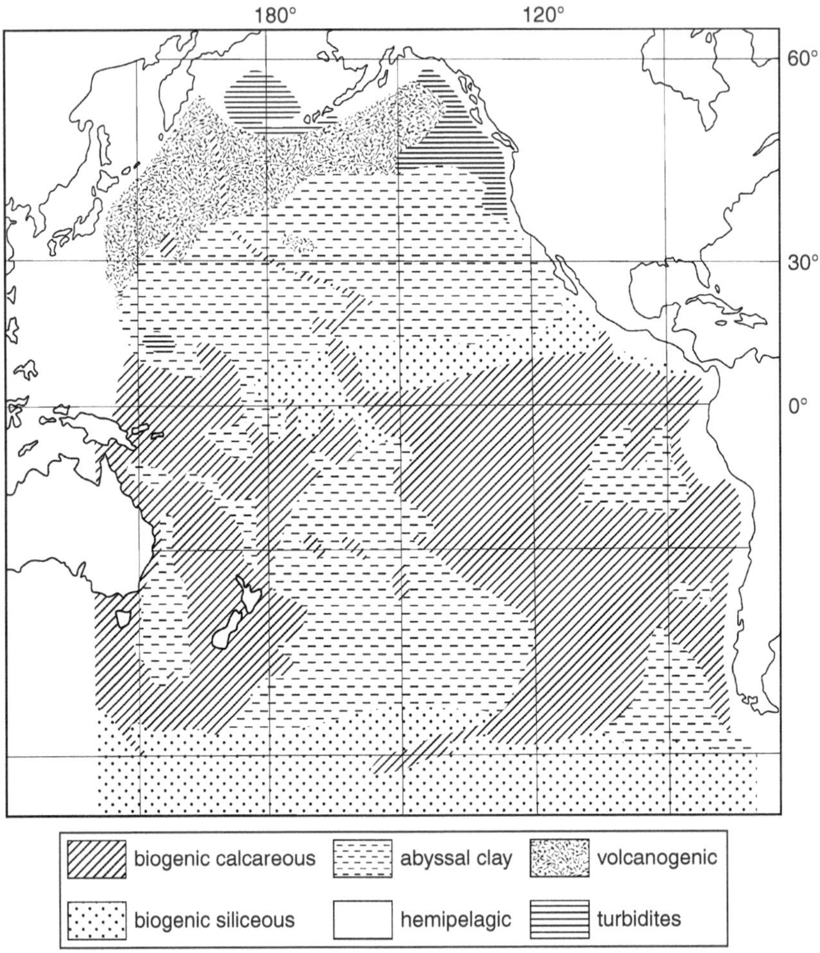

Figure 9.6 Major belts of silica deposition on the floor of the Pacific Ocean. Note that the belt of siliceous sediments south of the Aleutian and Kurile-Kamtschatka island arcs are mixed with volcanogenic material and shown as volcanogenic sediments (redrafted from Hesse, 1990b, fig. 2).

(3) the sub-Arctic belt south of the Arctic convergence; (4) the Gulf of California affected by continental west-coast/marginal-sea upwelling; and (5) the Okhotsk and Bering Seas (marginal-sea upwelling driven by offshore winds).

In the equatorial and low-latitude areas of high primary production, mixed calcareous/siliceous oozes are deposited, because in the surface waters of the tropical seas calcareous micro- and nannoplankton thrives together with the silica producers. The boundary of these mixed calcareous/siliceous oozes with the equatorial silica belt immediately to the north marks the CCL intersecting the seafloor. Present-day pelagic silica producers include diatoms, radiolarians, silicoflagellates and siliceous sponges which build their frustules, shells or skeletal elements of opal-A, the least stable solid silica phase (see below).

The present review of the early diagenesis of biogenic siliceous oozes covers a wide range of topics from pore-water chemistry to burial-diagenetic stages, characteristics of the solid silica phases, nature of the conversion mechanisms and their interpretation in terms of Ostwald processes, crystallographic structural changes of opal-CT and quartz during progressive burial, rate-controlling factors of the diagenetic reactions, the formation of bedded chert and the physical diagenesis of biogenic siliceous deep-sea sediments.

4.2. Reaction-controlled pore-water profiles of biogenic siliceous sediments

Biogenic siliceous sediments are an example of pelagic sediments with reaction-controlled pore-water profiles, at least for the main dissolved species. However, where silica deposition occurs within reach of the convection cells of the MOR hydrothermal circulation system, reaction-controlled silica gradients are wiped out. The profiles show consistent increases in dissolved silica with subbottom depth. Values exceeding 1 mM at a few hundred meters below the seafloor are not uncommon (Fig. 9.7). These dissolved-silica profiles reflect the progressive dissolution in the subsurface of opal-A.

In many drill holes, the gradual downward increase in dissolved silica is followed by an abrupt decrease below a certain depth (Fig. 9.7), consistent with reprecipitation as opal-CT (see below). At greater subsurface depth, a second, less pronounced maximum may be present in holes with sufficient core depth to have penetrated the opal-CT to quartz conversion level.

4.3. Burial-diagenetic stages of siliceous sediments based on solid silica phases versus detrital content

Opal-A, the amorphous, thermodynamically least stable form of silica that makes up the biogenic silica particles, dissolves easily and is reprecipitated as opal-CT. This is still a metastable phase which in a further reaction step corresponding to the second dissolved silica maximum is converted into α-quartz, the ultimate stable silica phase under earth surface and near-surface conditions. Quartz does generally not precipitate directly from silica-supersaturated fluids. It is preceded by the less well-ordered metastable phases opal-A′ (non-biogenic) and/or cristobalite/tridymite (opal-CT), which illustrates the significance of reaction kinetics for diagenetic processes. During burial, biogenic siliceous sediments as a rule thus go through three diagenetic stages: the opal-A stage of unlithified siliceous

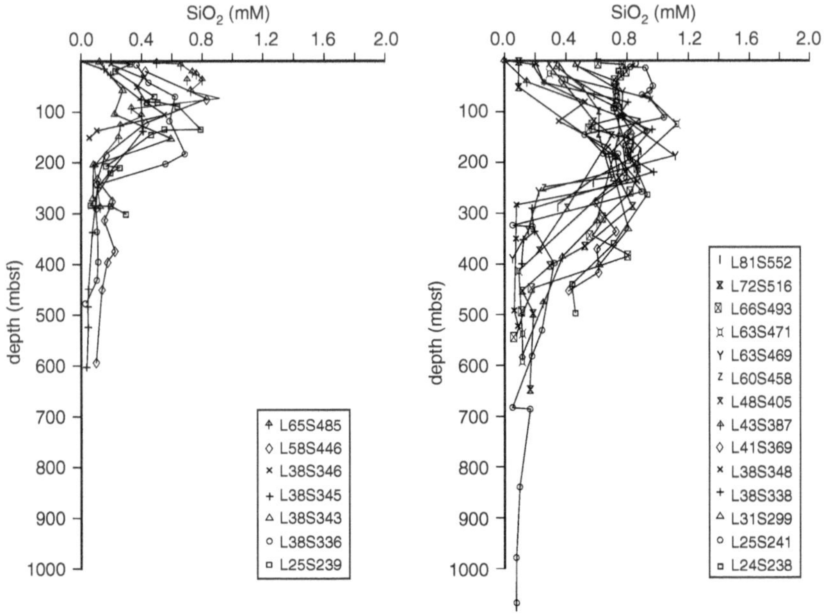

Figure 9.7 Trend of downward increasing dissolved-silica concentrations in pore waters from DSDP drill holes. L, DSDP leg; S, DSDP Site. (redrafted from Hesse, 1990b, fig. 10).

ooze, the opal-CT stage of porcelanite and the microquartz or quartz stage of chert *sensu stricto* (quartz chert).

There is a dichotomy in the nomenclature of the rock products of the various stages of silica diagenesis. Calvert (1971) suggested a nomenclature of siliceous rocks strictly based on mineralogy, according to which porcelanite consists predominantly of opal-CT and chert predominantly of (micro-)quartz, a usage largely followed by the DSDP and ODP. Originally, however, the distinction between the rock types, porcelanite and chert, was based on the ratio of the detrital (clay) minerals to silica minerals, because this is what determines the macroscopic appearance of the rocks in outcrop; it thus was a suitable basis for field descriptions (Bramlette, 1946). In this scheme, porcelanite refers to a porous rock with dull or matte lustre, similar to unglazed porcelain. Chert denotes a dense vitreous, hard and brittle rock. In the Miocene Monterey Formation of California, which consists of deep-marine siliceous sediments exposed on land, rocks that are richest in silica (pure diatomites) appear macroscopically as cherts although they may still consist entirely of opal-CT. Diatomaceous shale, on the other hand, has the appearance of porcelanite, even after having reached the quartz stage. Isaacs et al. (1983) argue that the distinction between porcelanite and chert should be based primarily on detrital content.

4.4. Equilibrium solubilities and characteristics of the solid silica phases in biogenic siliceous sediments

X-ray-amorphous silica named opal-A (Jones and Segnit, 1971) is mineralogically similar to precious opal, showing only a broad hump near 4 Å in its X-ray diffraction pattern (Fig. 9.8), like that of silica glass. It is highly porous and has a substantial water content (4–9% in precious opal, 10–14% in biogenic opal-A). Opal-A dissolves easily in seawater, which is highly under-saturated with respect to opaline silica everywhere in the modern oceans. Actual concentrations of dissolved silica in the ocean range from a few μM up to 150–250 μM. They are thus 10–1000 times lower than the equilibrium solubility of amorphous silica, which is between 1 and 2 mM at 25 °C and a pH $\leq$ 9.

The equilibrium solubility of silica depends not only on temperature (Fig. 9.9A) but also on a number of other factors, notably surface area and particle size of the solid silica phase (Fig. 9.9B), and to a lesser extent pressure (Fig. 9.9C). It increases significantly at pH values >9–10 because, in addition to the dissolved undissociated orthosilicic or monomeric acid, the first and second dissociation steps of silica produce $H_3SiO_4^-$ and $H_2SiO_4^{2-}$ ions that raise the solubility (Fig. 9.9D; Williams and Crerar, 1985; Williams et al., 1985). The ageing of biogenic siliceous tests affects solubility through changes in surface area and crystallite size in the shell wall (Hurd and Theyer, 1975). Equilibrium solubilities of the other common low-temperature silica polymorphs, opal-CT and α-quartz, are 1 and 2 orders of magnitude lower than those of opal-A, respectively (Fig. 9.9A).

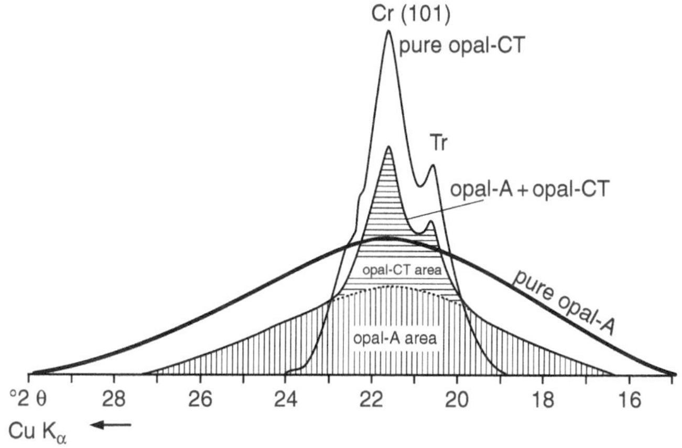

Figure 9.8 XRD pattern of pure opal-A, pure opal-CT and a mixture of opal-A and opal-CT (from Von Rad et al., 1978).

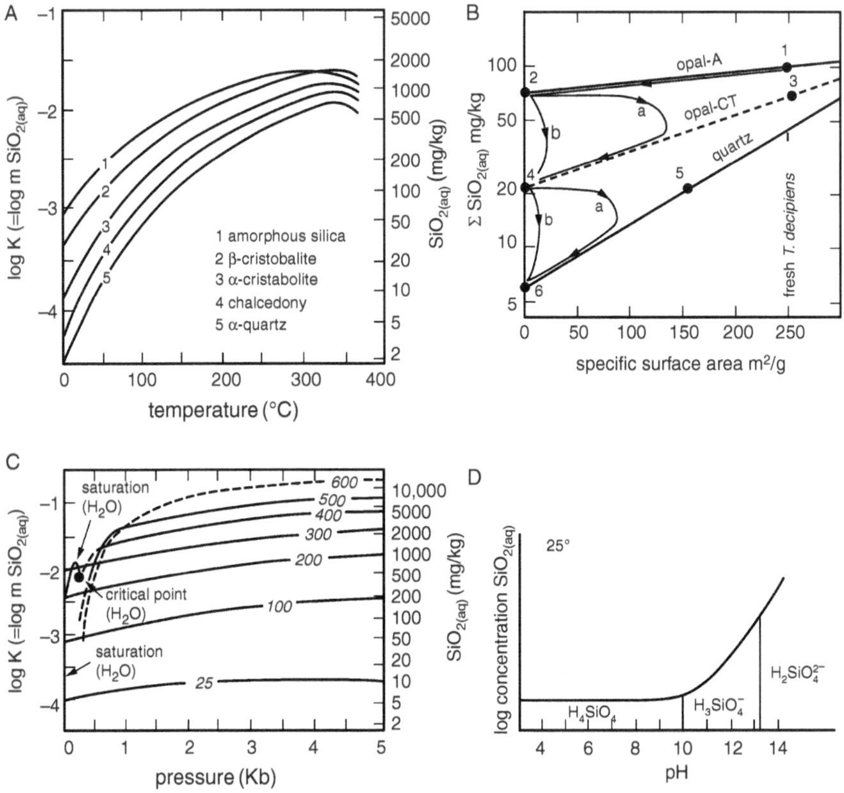

Figure 9.9 Equilibrium solubility of various silica phases as a function of (A) temperature, (B) specific surface area, (C) pressure (for quartz at temperatures between 25 and 600 °C), and (D) pH; redrawn and modified (A, C, D) redrafted from Williams and Crerar (1985), (B) redrafted from Williams et al. (1985).

In supersaturated solutions, dissolved-silica polymerizes, forming first oligomers (dimers, tetramers and ring structures), and later higher molecular-weight polymers as siloxane (Si–O–Si) bonds develop through combination of silanol (≡Si–OH) groups:

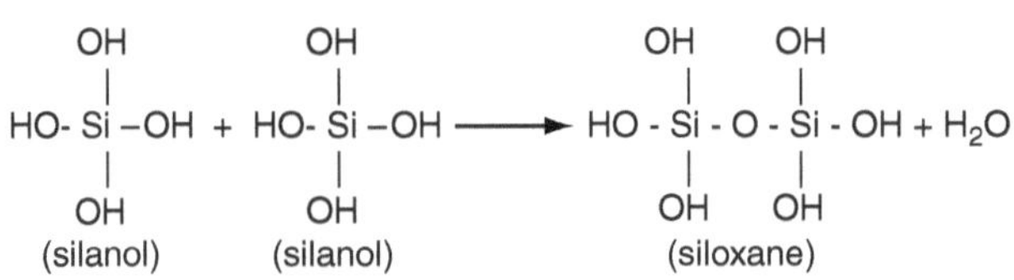

When supersaturation persists, high molecular-weight polymers (with molecular weights up to 10,000) can form. Such polymers have colloidal dimensions (more than 50 Å) and may remain suspended as sols as long as

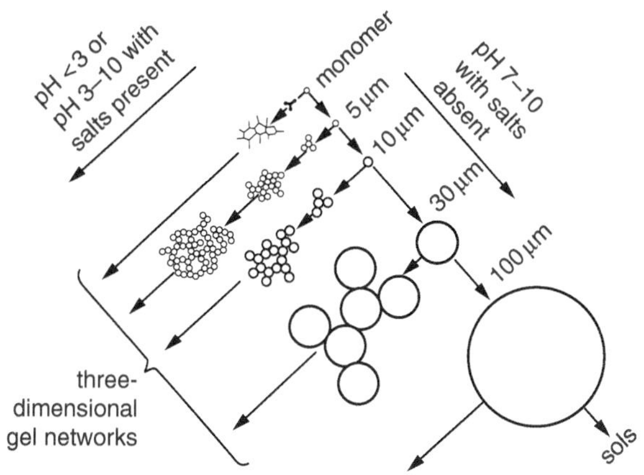

Figure 9.10 Silica sols and gels. Polymerization behaviour of dissolved silica as function of pH and ionic strength of the solution (redrawn and modified from Williams and Crerar, 1985).

the pH remains relatively high and the salinity low. Otherwise they will form cross-links with neighbours and coagulate into gels (Fig. 9.10). Silica polymers display a negative surface charge, down to a point of zero charge (PZC, i.e. the pH at which the residual surface charge disappears) as low as 2±0.5 (Parks, 1965). Silica colloids thus repel each other unless the surface charge is neutralized by other ions in solution such as metal hydroxides. The hydroxide most commonly used for silica precipitation in industrial applications, $Mg(OH)_2$ (Iler, 1979), is also thought to be instrumental in the nucleation of opal-CT (see below).

Opal-CT denotes a modification of opal which has structural characteristics of both α-cristobalite and α-tridymite (Jones and Segnit, 1971). Opal-CT thus is the low-temperature form (α-form) of cristobalite/tridymite formerly called lussatite (Mallard, 1890). On X-ray powder diagrams, the main diffraction peak of opal-CT is a doublet at 4.1 and 4.3 Å (Figs. 9.11B–I). Wise et al. (1972) were the first to observe the occurrence of small opal-CT spheres, named opal-CT lepispheres (Weaver and Wise, 1972), under the scanning electron microscope (SEM) in deep-sea drilling samples (Fig. 9.12A). These spheres are intergrowths of tiny cristobalite-tridymite plates consisting of opal-CT blades 2–5 µm long, 0.05–0.10 µm thick and displaying ragged or rounded edges (Fig. 9.12B). Opal-CT plates in "juvenile" lepispheres reveal regular intergrowth, typically showing dihedral angles between adjacent plates of 70–71° characteristic of (304)- and (106)-twinning laws of macroscopic crystals of tridymite (Flörke et al., 1976). The diameter of individual lepispheres usually does not exceed the length of the blades (~5 µm) of which they are composed. Larger lepispheres are almost invariably aggregate forms (Fig. 9.12C; see also figs. 1C, D, 3A, B in Carver, 1980).

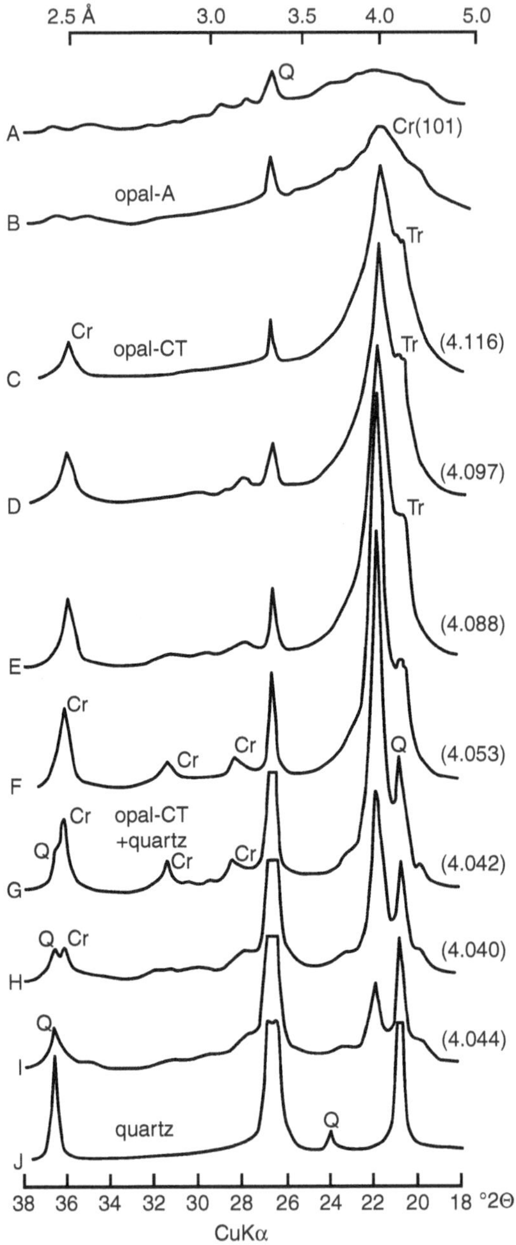

Figure 9.11 Authigenic SiO_2. (A) X-ray diffractogram for opal-A (broad hump at about 4 Å) and detrital quartz. (B through I) Doublet peaks at 4.1 and 4.3 Å for cristobalite (Cr) and tridymite (Tr), respectively, in diatomaceous shale in the Monterey Formation of California. Note peak sharpening and shift in the position of the (101) cristobalite diffraction (from 4.116 to 4.04 Å) with progressive burial. (G) Appearance of authigenic quartz (modified from fig. 5 in Murata and Larson, 1975).

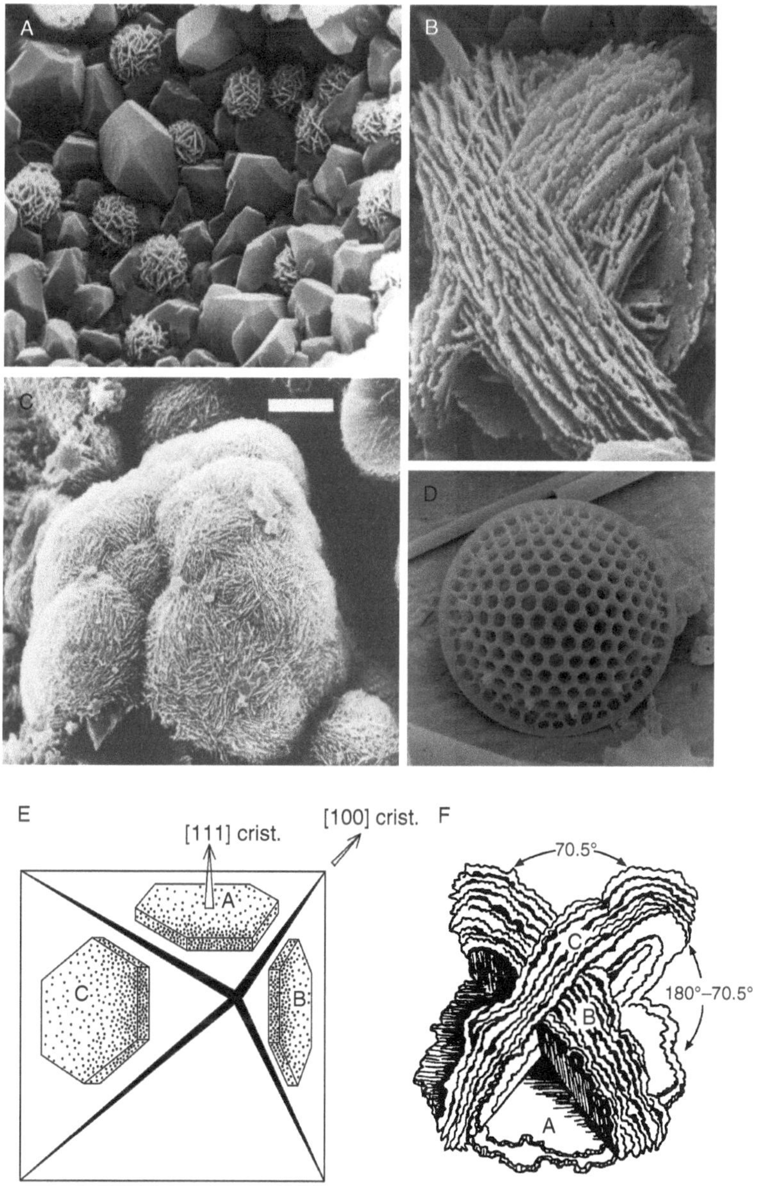

Figure 9.12 Opal-CT (figures (E) and (F) redrafted from Flörke et al., 1976; SEM photos of (A), (B), (D) courtesy U. von Rad). (A) Small opal-CT lepispheres (2–3 μm in diameter) growing on euhedral calcite in cavity of foraminifera in partially silicified Maastrichtian chalk (DSDP Leg 14, Site 144, Core 3, Section 2, 103–104 cm). (B) Opal-CT blades (~1 μm in length) of juvenile lepisphere displaying twinning angle of 70 °C. (C) Composite lepisphere 50 μm in diameter that resulted from coalescence of numerous smaller individual lepispheres (DSDP 12-117A, core catcher sample 2, from Flörke et al., 1976; reprinted with permission from Springer Verlag). (D) Sieve structure of diatom frustule (20 μm in diameter). (E) Faces of a cristobalite octahedron. (F) Schematic drawing showing the intersection angle of the faces of a cristobalite octahedron corresponding to the example in (B).

4.5. Nature of the conversion mechanism of opal-A to opal-CT: Dissolution/reprecipitation

Only a small percentage of the large amount of biogenic silica (opal-A) tests produced by planktonic organisms in the surface ocean reaches the sea floor, and only a fraction of this escapes dissolution during the first meters of burial. Even this very small proportion of tests of solution-resistant species will ultimately undergo dissolution at greater subbottom depths. The effects of progressive dissolution have been documented by systematic SEM studies of diatom oozes (e.g. Hein et al., 1978). Breakage of partially dissolved diatom valves accompanies and enhances dissolution (Fig. 9.13), culminating in the complete destruction of the tests. The progressive dissolution of solid silica particles in the subsurface is reflected in dissolved-silica profiles from DSDP

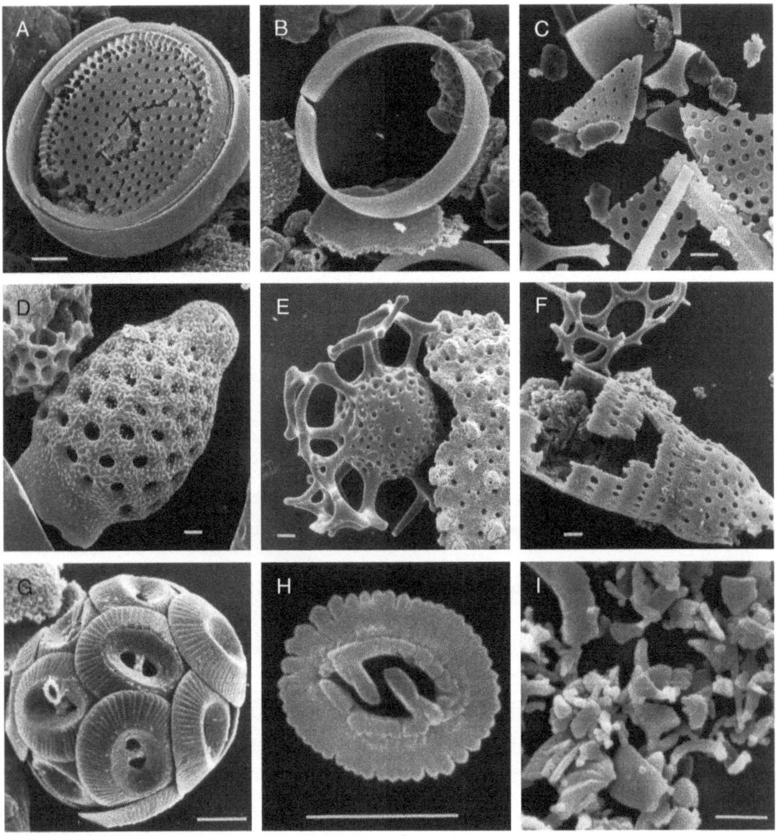

Figure 9.13 SEM photographs showing progressive breakage and dissolution (courtesy D. Fütterer, modified from fig. 16 in Fütterer, 2006; Reproduced with kind permission from Springer). (A–C) Diatom valves. (D–F) Radiolarian tests. (G–I) Coccolithophores and coccoliths. Scale bar is 1 μm.

holes with their characteristic downward increases (Fig. 9.7). The abrupt decrease that follows corresponds to the reprecipitation of silica as opal-CT.

Continued dissolution of opal-A in sediment during burial is the result of slowly increasing temperature and pressure. Siliceous tests of diatoms and radiolarians have large specific surface areas ranging from several tens to 450 m^2/g (Kastner et al., 1977), compared with 0.1 m^2/g of crushed quartz in the 5–3 φ (25–75 μm) size range (Van Lier et al., 1960). The sieve structure of the porous test walls of these organisms (Fig. 9.12D) is only partly responsible for the large specific surface areas. It is the size of the small opal-A domains (2.5–4.5 Å diameter) in the tests which causes the high specific surface areas (Hurd et al., 1979). The surface area can significantly affect solubility: with a large surface-area/volume ratio of a substance, small changes in pressure and temperature may markedly increase solubility. Suppression of dissolution-inhibiting factors such as the removal of protective coatings of organic matter may further enhance solubility. This explains the continuation of opal-A dissolution during burial, even though up to 99% of the opal originally produced in surface seawater may have already been dissolved during settling and initial burial.

Dissolution during sediment burial occurs, in contrast to the earlier dissolution, in a more or less closed system, in which concentration levels may reach supersaturation before opal-A dissolution has gone to completion. In this case, dissolution will be interrupted by precipitation of a less soluble non-biogenic opal-A phase, designated opal-A′, which forms overgrowths on partially dissolved siliceous tests (Hein et al., 1978). The crystallite size calculated from X-ray diffraction data for opal-A′ is larger (20–27 Å) than for biogenic opal-A (12–16 Å). In individual DSDP holes, opal-A′ overgrowths have been found to occur only over a narrow stratigraphic range of a few meters, indicating that the overgrowths redissolve shortly after formation together with the remaining opal-A (Hein et al., 1978). Their presence would explain the oscillating fluctuations seen in the concentration-depth profiles of dissolved silica (Fig. 9.7).

The discovery by Wise et al. (1972), Weaver and Wise (1972) and Berger and von Rad (1972) of opal-CT lepispheres with the euhedral crystal shapes of cristobalite/tridymite blades provided proof that the recrystallisation of siliceous oozes to porcelanite occurs through a dissolution/reprecipitation mechanism. Lepispheres develop only where crystallization takes place in open pore spaces such as the cavities of microfossils (Figs. 9.12A and B). More commonly, a densely felted mass of opal-CT forms which may impregnate the sediment and/or replace other mineral phases. The latter process may involve pseudomorphic replacement of opal-A by opal-CT in radiolarian tests, which perfectly preserves the shape of the shell (Fig. 9.14) but is nevertheless a dissolution/reprecipitation process. It probably proceeds on a "crystal"-by-"crystal" scale with local precipitation immediately following dissolution. A matrix of organic matter, which is

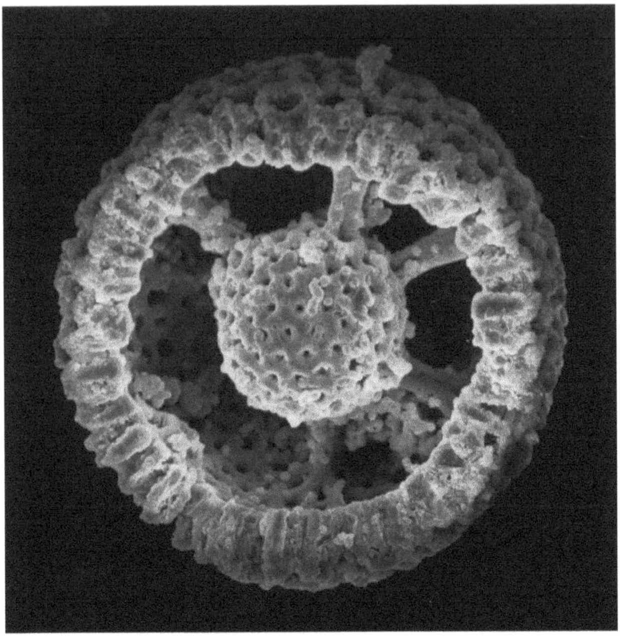

Figure 9.14 Pseudomorphic replacement of opal-A of a radiolarian test by opal-CT (DSDP 41-366-23-, 42–44 cm; Riech and Von Rad, 1979, reprinted with permission of the American Geophysical Union; SEM photo courtesy U. von Rad). Test wall is ~20 μm thick.

not or only partially affected by dissolution, may serve as a template which helps preserve the original shape of the shell. The process may be similar to the mechanism of silicification of ooids (Hesse, 1987) or to the petrifaction of wood. Where radiolarians occur embedded in lutitic pelagic limestone, a test replaced by opal-CT may also be simply a cast of the former opal-A shell.

4.6. Opal-CT to quartz conversion

The opal-CT to quartz conversion was originally perceived as a solid-state reaction based on hydrothermal experiments by Ernst and Calvert (1969). However, Stein and Kirkpatrick (1976) examined the reaction products of these experiments under the SEM and found mainly quartz fibres, much larger than the original grains of crushed porcelanite. Short but thick quartz crystals also appeared in the longer runs. The re-examination thus showed that the conversion of opal-CT to quartz in the experiments had occurred by a dissolution-precipitation mechanism. This is in line with the pore-water profiles from some deep-sea drill holes that show a second, deeper dissolved-silica maximum at greater depth, most likely corresponding to the dissolution of opal-CT and the subsequent precipitation of quartz.

4.7. Diagenetic silica phase conversions as examples of Ostwald processes

The recrystallisation of very fine-grained opal-A to somewhat coarser grained opal-A′ is an example of Ostwald ripening; the dissolution of opal-A′ and the reprecipitation as opal-CT are an example of Ostwald's step rule. Recognition of an intermediate stage of inorganically precipitated opal-A′ in the opal-A to opal-CT transition (Hein et al., 1978) demonstrates that the conversion is not a single-step process but involves a series of dissolution and reprecipitation reactions. This reaction series is predicted by the model, based on surface-area effects (Williams et al., 1985) (Fig. 9.9B), which illustrates the effects of reaction kinetics on phase changes. It is equally applicable to the opal-CT to quartz conversion.

In the hypothetical SiO_2–H_2O system of Williams et al. (1985), opal-A of a diatom species with a specific surface area, say of 250 m^2/g, that is that of the radiolarian species *Thalassiosira decipiens* (Fig. 9.9B), will have a solubility of about 1.5 mM. If the dissolution of the frustules of this species is fast relative to nucleation and growth of new silica phases, the solution will soon become supersaturated with respect to opal-A of a lower surface area, that is opal-A′, which will then precipitate, and the solution will evolve along the pathway from point 1 to point 2 in Fig. 9.9B. This is the process called "Ostwald ripening", that is grain-coarsening of material belonging to the same phase.

Near point 2, the effect of surface area on opal-A solubility becomes negligible and opal-CT, the solid silica phase with the next lower solubility, shall precipitate. The metastable opal-CT is a classical example for Ostwald's step rule, which states that the conversion of an unstable to a stable mineral phase at the low temperatures near the earth's surface (here opal-A to quartz) may require one or more intermediate metastable phases. This is a consequence of the crystallization kinetics (Morse and Casey, 1988).

Reaction kinetics explain why opal-CT is required as an intermediate metastable phase in the sequence of diagenetic silica conversions. Quartz cannot form directly from the dissolution of opal-A, at least not from an equilibrium solution, because the solubility in equilibrium with opal-A of any specific surface area is too high. At such high silica concentrations, the faces of any embryonic quartz crystal would be crowded by polymerized silica that has not had the time to be properly fitted into the crystal lattice. Quartz growth will be blocked and a less well-ordered phase, opal-CT, forms instead. Only when the "equilibrium solubility" of this phase has been lowered sufficiently through Ostwald ripening, will quartz crystallization become possible.

The effects of reaction kinetics are displayed in an informative way by the Williams et al. (1985) model (Fig. 9.9B). Depending on the mutual relationship between the rate of dissolution of opal-A and the rates of nucleation and growth of opal-CT, the evolution of the solution will follow pathways from point 2 to 4 either along curve "a" (high nucleation rate) or curve "b" (low-nucleation rate). In the first case, rapid nucleation (relative to growth) leads to a relatively large specific surface area, because the newly formed crystals are small and numerous. Only when the silica removal rate exceeds the dissolution rate of the remaining opal-A, will silica concentration drop, and will the surface area of the newly formed opal-CT decrease (curve 2–4a). If, on the other hand, the growth rate of opal-CT exceeds the nucleation rate early in the process, the fluid should evolve along pathway 2–4b. The opal-CT to quartz conversion discussed in a subsequent section follows analogous pathways.

Lowering of the equilibrium concentration of dissolved silica through Ostwald ripening of the opal-CT phase also counterbalances a solubility increase with rising temperature during burial. The solubility of a small surface-area opal-CT at 110 °C is only slightly higher (2.5 mM vs. ~2.2 mM) than that of the large surface-area opal-CT at 50 °C and supersaturated only 1.5 times with respect to chalcedony or cryptocrystalline quartz at 110 °C (Table 9.1). Since the growth of a new, more highly ordered silica phase such as chalcedony or quartz is favoured by low supersaturation, the ordering process during burial of opal-CT ultimately facilitates the precipitation of quartz by lowering the equilibrium solubility (or preventing a solubility rise with increasing temperature).

Lowering the equilibrium solubility in the silica maturation process also enhances silica transport by diffusion, as the precipitating silica phase with the lower solubility will generate a concentration gradient towards the dissolving, less mature phase with its higher solubility (Landmesser, 1993). Because of the higher density of chalcedony compared to opal-CT and opal-A, considerable addition of silica by diffusion is required, if the volume of the solid silica phases is to remain constant in the maturation process.

4.8. Crystallographic structural changes of opal-CT and quartz in the porcelanite and quartz-chert stages

4.8.1. Sharpening and shift of the opal-CT diffraction peaks

Despite the important finding of a dissolution-to-precipitation step in the opal-CT to quartz conversion (Stein and Kirkpatrick, 1976), the results of a high-precision X-ray diffraction study of Murata and Larson (1975) still left the involvement of low-temperature solid-state processes in the

Table 9.1 Solubility of β-cristobalite as approximation for opal-CT solubility (A), temperatures for the opal-A to opal-CT conversion (B), temperatures for the opal-CT to quartz conversion (C) and densities and refractive indices for the various silica phases (D)

(A) Opal-CT		with high specific surface area at 50 °C:			~2.2 mM
		with high specific surface area at 110 °C:			~4.8 mM
		with low specific surface area at 110 °C:			2.5 mM
chalcedony or cryptocrystalline quartz at 110 °C:					~1.7 mM
(B) Formation location	Age (Ma)	Temp (°C)	Method	Remarks	Reference
Bering Sea DSDP sites 184, 185		35–51	Downhole logging	500–600 mbsf	Hein et al. (1978)
Monterey Fm. Temblor Range, CA		41–56	$\delta^{18}O$	Interstitial water With $\delta^{18}O$ = 0‰	Murata et al. (1977)
		50		Geotherm. grad. 700 m subbottom	Murata et al. (1977)
Monterey Fm. Santa Maria Valley	4–10	2–33	$\delta^{18}O$		Pisciotto (1981) Behl (1992), Matheney and Knauth (1993)
(C) Monterey Fm.		55–110	$\delta^{18}O$		Murata et al. (1977)
Temblor range			Heat flow		Murata and Larson (1975)
Monterey Fm.		30–61	$\delta^{18}O$		Pisciotto (1981)
Santa Maria valley					Behl (1992)
(D) Silica phase		Density (g/cm^3)		Refractive index	
Opal-A		<2.2 (maximum)		1.40–1.45	Füchtbauer (1988)
Opal-CT		2.0–2.3		1.45–1.49	
Chalcedony		2.6		1.53–1.54	

reaction as a possibility. In siliceous deep-marine sediments of the Monterey Formation in the Temblor Range in California, the main diffraction peak of opal-CT (4.1 Å), which is the (101) diffraction of α-cristobalite, undergoes a distinct shift from 4.11 to 4.04 Å with increasing burial (Fig. 9.11B–I). This decrease in the d-spacing is accompanied by a progressive sharpening of the peak and a gradual disappearance of the α-tridymite (0001) peak (at 4.32–4.26 Å), which is finally replaced by the quartz (100) peak at the opal-CT to quartz transition (Fig. 9.11G–J). The shift in the d-spacing of the (101) peak in the Temblor Range succession proceeds continuously over the entire burial range of the opal-CT zone, although it is relatively rapid at the top and bottom of the opal-CT zone and very slow in the middle (Fig. 9.15).

In contrast, Pisciotto (1981) found for the Santa Maria Valley, that from the middle of the opal-CT zone the shift becomes more rapid with increasing burial depth. The bulk density of the sediments, on the other hand, does not show any systematic increase over this range of burial depths (730–2030 m below reference level) (Fig. 9.15). Oxygen-isotope data from the same burial-depths display a similar trend: no systematic change within the opal-CT zone (average value of $\delta^{18}O = +29.4‰$ relative to VSMOW), but a significant decrease by about 5‰ within a short distance below the opal-CT/quartz boundary. Murata et al. (1977) concluded that the depth-dependent structural ordering of cristobalite, reflected by the changes of the d(101) spacing, is mostly an internal solid-state adjustment within the opal-CT stage that does not substantially change the amount of silica per unit volume of rock. The abrupt change in density and isotopic ratios at the opal-CT/quartz stage boundary, however, indicates complete dissolution and reprecipitation.

However, "neither the structure of the mineral opal-CT nor the mineralogic significance of the d-spacing of opal-CT is" sufficiently well understood to interpret the changes in the structure of opal-CT as solid-state ordering (Isaacs et al., 1983). For instance, the sharpening of the (101)-cristobalite peak with increasing burial may reflect a growing crystallite size of the opal-CT, which in turn may result from gradual dissolution of smaller crystals and redeposition of the dissolved silica on larger ones by an Ostwald ripening process. The oxygen-isotope composition may in fact change very little within the opal-CT stage because the isotopic ratio of growing larger crystallites may be locally inherited from dissolving smaller ones. The density may also change very little because the morphology of the opal-CT blades remains essentially unchanged.

Effects will be seen only after a significant portion of the rock has been converted to the new phase. These considerations would make the problem amenable to interpretations based entirely on dissolution/reprecipitation reactions and would eliminate the need for solid-state reactions, which are prohibitively slow at the low temperatures of diagenesis.

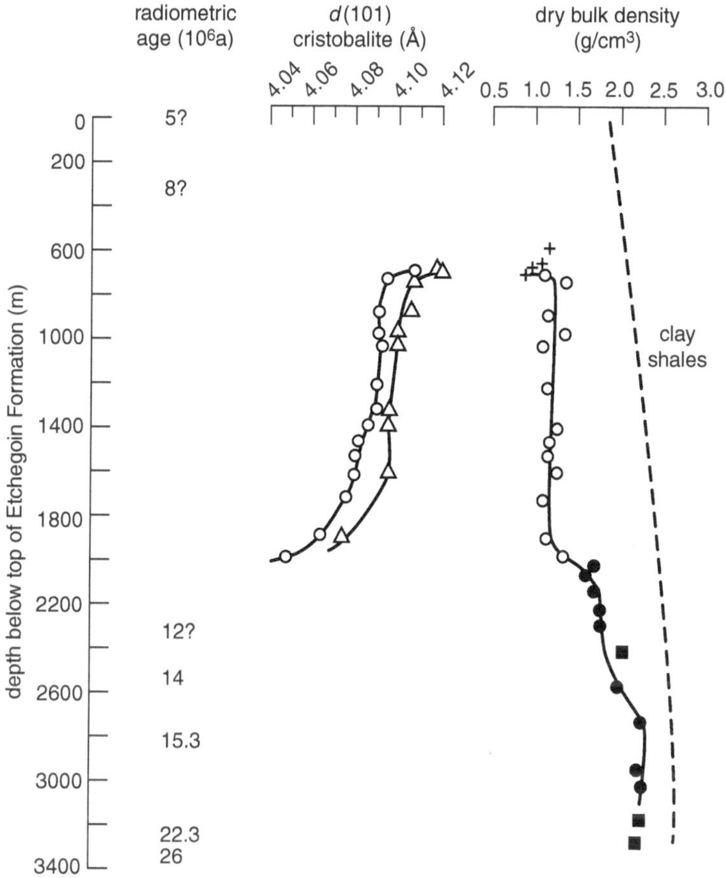

Figure 9.15 Shift in the $d(101)$ spacing of opal-CT in porcelanite (circles) and cherts (triangles) with burial in the Monterey Formation section of Chico Martinez Creek, California. Density data: crosses, diatomaceous shale; open circles, opal-CT porcelanite; filled circles, quartz chert; filled squares, shale; dashed line, density of normally compacting shale for comparison (redrawn and modified from Murata and Larson, 1975; fig. 6).

4.8.2. Crystallinity index of quartz

Crystallographic changes of the solid silica phases during progressive diagenesis continue in the quartz stage, as demonstrated by the improvement of quartz crystallinity with burial, comparable to the sharpening of the (101) cristobalite peak in the opal-CT stage. The quartz crystallinity index of Murata and Norman (1976) is based on the quintuplet XRD-reflection in the high-angle region between 67 and 69° 2Θ (Fig. 9.16A). The (212) peak at 67.74° 2Θ measures the effects of the recrystallisation of cryptocrystalline (crystallite size <1 μm) to microcrystalline authigenic

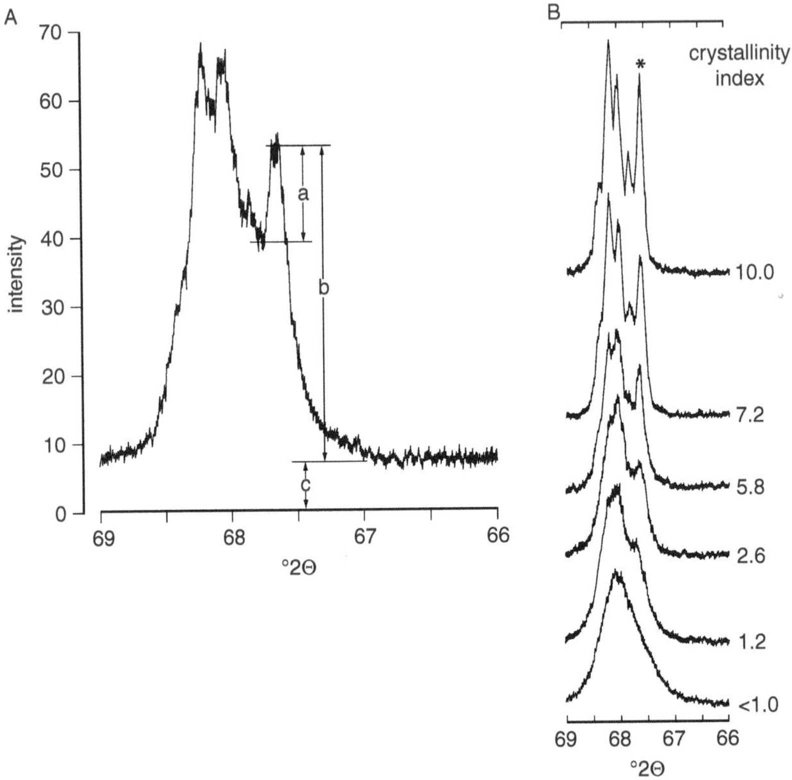

Figure 9.16 Quartz crystallinity determined from diffractograms. (A) Quartz-crystallinity index defined as $CI = 10Fa/b$, where a is the height of the (212) reflection above the shoulder of the quintuplet peak at about 68° 2θ on the high-angle side, while b is the total height of the (212) reflection above the base line (c), and F is a scaling factor that varies with instrumental settings and adjusts the crystallinity index to a scale of 0–10. Determination of the quartz crystallinity index requires high-precision XRD measurements at slow scanning speed (0.25°2θ per min). (B) Tracings of diffractograms showing increasing quartz crystallinity from bottom to top. Asterisk marks the (212) peak at 67.74°2θ used for determination of the index (redrawn and modified from Murata and Norman, 1976; fig. 1).

quartz during diagenesis and low-grade metamorphism, most probably the increase in crystal size.

The crystallinity of authigenic quartz is characteristically poor (e.g. 2.0–3.2 in the Monterey Formation), even in quartz-cherts s.s. Cretaceous quartz cherts in the West Pacific at about 400 m subsurface depth have crystallinities generally less than 1 (using a scaling factor of 1.36; Pisciotto, 1980). Franciscan cherts show considerably better crystallinities (Fig. 9.16B), but reach high values (above 8.0) only in metamorphosed rocks (Murata and Norman, 1976).

4.9. Rate-controlling factors of the opal-A to opal-CT conversion in deep-sea environments

4.9.1. Temperature and time

Temperature is the dominant rate-determining factor for the diagenetic silica conversions, as illustrated by the experiments of Kastner et al. (1977) for the opal-A to opal-CT transformation (described below), and the experiments of Ernst and Calvert (1969) for the opal-CT to quartz conversion (see Section 4.6). The temperature control is based on the Arrhenius equation for the rate constant, k, of a chemical reaction:

$$k = A\exp\left(\frac{-E_a}{RT}\right)$$

where E_a is the activation energy, T is the temperature, R is the gas constant, and A is a frequency factor. During the diagenesis of deep-sea pelagic sediments, the effect of pressure is generally negligible compared with the effect of temperature and other factors. The pressure is generally hydrostatic in these environments and varies little over the burial intervals of a few hundred meters where the first silica reactions take place.

The actual *in situ* temperatures for the opal-A to opal-CT conversion range from 35 to 56 °C (Table 9.1B). A considerably lower temperature range of 2–33 °C was estimated for the top of the opal-CT zone (the depth where opal-CT exceeds 5 wt%) in the Monterey Formation of the Santa Maria Valley in California (Behl, 1992; Matheney and Knauth, 1993; Pisciotto, 1981). The last-mentioned authors used a method of stepwise fluorination, which removes the internal water prior to isotopic analysis, whereas the higher temperatures of the earlier studies were based on measurements uncorrected for the internal water content.

In pelagic sediments of the modern ocean away from MORs, the conversion occurs at shallow subsurface depths of a few hundred meters, where temperatures rarely exceed 20–30 °C. At these low temperatures, the reactions are very slow and the opal-A to opal-CT conversion takes tens of millions of years to go to completion (Fig. 9.17). The conversion can be traced regionally on seismic-reflection profiles because the transition from siliceous ooze to porcelanite is a major regional lithification event, that in the Northwest Pacific, for example, is marked by the occurrence of an "opaque" seismic reflector in the subsurface below an upper transparent layer, which first appears at the meridian of Hawaii (at approx. 155°W; Fig. 9.18) in mid-Tertiary siliceous sediments (15–20 Ma), which rest on 90 Ma Late Cretaceous ocean crust. The reflector increases in thickness westwards from less than 20 m to 200–300 m in the West Pacific, where it represents the combined thicknesses of siliceous rocks and pelagic limestones (see, e.g. DSDP legs 7, 16, 17, 20, 32, 55, 61, 62, 89). The opaque

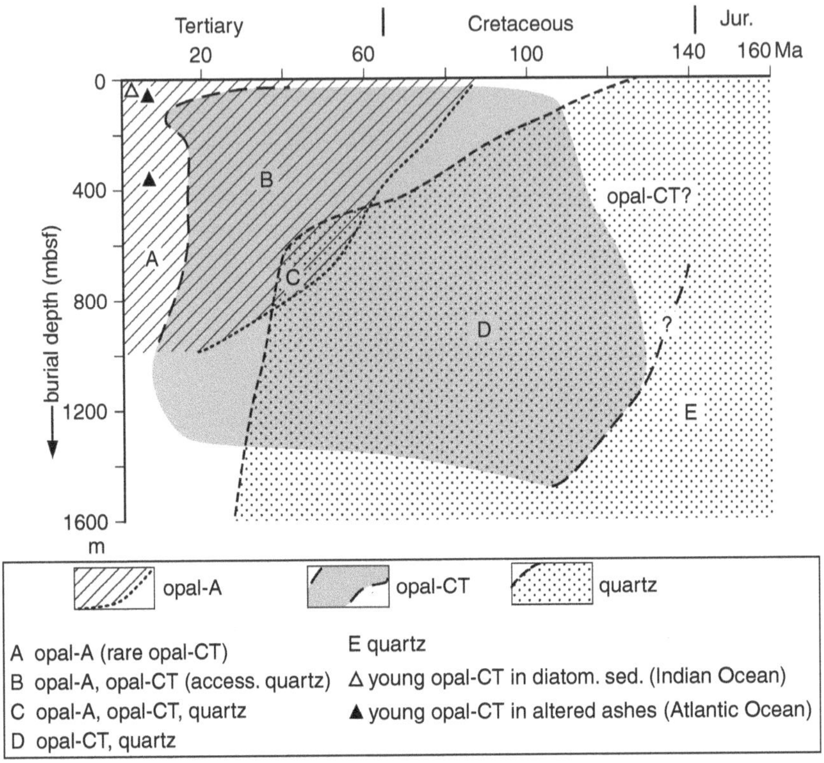

Figure 9.17 Burial depth/age fields of the various silica modifications based on DSDP results (modified from Riech and Von Rad, 1979). For discussion, see text.

layer can be deciphered on seismic profiles as a distinct interval because it is underlain by older, relatively poorly consolidated brown abyssal clay that produces a second lower, seismically transparent layer. The upper opaque layer on North Pacific seismic profiles west of Hawaii thus records a diagenetic history that needed at least 15 million years before it produced bulk physical changes. This illustrates the importance of the factor time, which plays a major role at the low temperatures of diagenesis.

Exceptionally young opal-CT has been observed by Weaver and Wise (1973) and Bohrmann et al. (1994) in Pliocene sediments in the Indian Ocean on the Kerguelen Plateau (<5 Ma old) and the Southwest Indian Ridge (<4.2 Ma), respectively. The former authors speculated that it occurred adjacent to igneous dikes and sills injected close to the sediment/water interface. Bohrmann et al. (1994) reported an even younger (<0.43 Ma old), well-cemented opal-CT layer from the Maud Rise off Antarctica. Oxygen-isotope analyses of four examples of anomalously shallow (5.5–14 m subbottom depth) opal-CT occurrences from the Southern Ocean

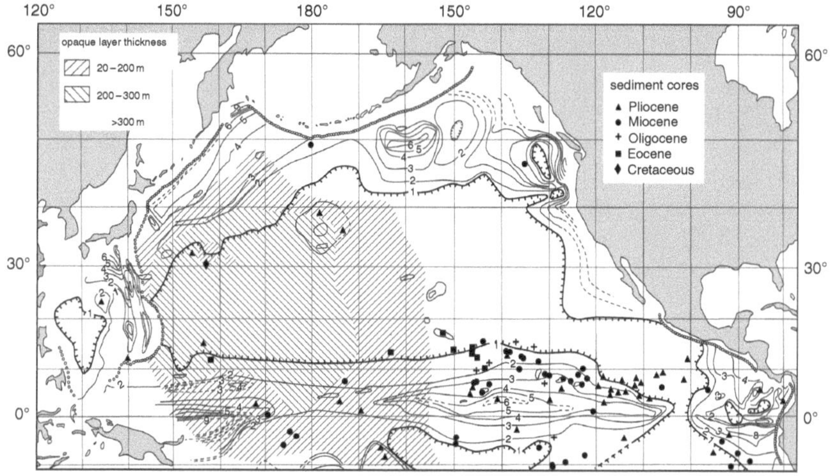

Figure 9.18 Thickness of the "opaque" seismic reflector in West Pacific indicating the westward growth in thickness of the porcelanite/pelagic limestone succession due to continuing lithification of opal-A rich sediments with age and burial (redrafted from fig. 2 in Wyllie, 1971, after Ewing et al., 1968). Isopachs show the thickness of the upper transparent layer with contour intervals of approximately 100 m.

suggest that the opal-CT formed at temperatures between 0 and 4 °C in line with their present burial depth (Botz and Bohrmann, 1991). Bohrmann et al. (1994) attributed these anomalously shallow and young opal-CT occurrences to the extreme purity of the siliceous host rocks in which they were found.

4.9.2. Host-rock lithology

In addition to temperature and time, host-rock lithology may promote or retard the diagenesis of siliceous sediments. Since Bramlette's (1946) classic work on the Monterey Formation, numerous exceptions from the general maturation sequence siliceous ooze → opal-CT porcelanite → quartz chert have been recognized. For example, opal-CT porcelanite and/or quartz-chert beds occur in the Monterey Fm. within the zone of non-recrystallized diatomaceous mudstone and diatomite, and (quartz-)chert layers occur within the porcelanite zone showing that burial depth and temperature are not the only factors controlling the maturation sequence. In other cases, the transition to quartz seems to have occurred directly from opal-A without an intermediate opal-CT stage (e.g. Lancelot, 1973). As observed during early legs of the DSDP (e.g. Keene, 1975; Lancelot, 1973; Von Rad and Rösch, 1972, 1974), the lithology of the host sediment has a distinct influence on silica conversions. For example, in many clayey sedimentary successions, silica maturation has reached the opal-CT stage of porcelanite, whereas quartz chert has formed in accompanying calcareous sediments of the same age and thermal maturation level.

Experiments by Kastner et al. (1977) clarified the role of host-rock lithology and associated solution chemistry in the opal-A to opal-CT transition. Diatom frustules and radiolarian tests were made to react with sea water or distilled water at 25 and 150 °C in the presence of either calcareous ooze or montmorillonitic clay for one day and for 1–3 or 6 months, with the result that precipitation of opal-CT lepispheres was observed after 2–3 months, but only in the experiments with sea water in the presence of calcite. Precipitation of lepispheres occurred preferentially on the surfaces of foraminifers. In each of the experiments in which opal-CT crystallization occurred, a lowering of the Mg^{2+} and $(OH)^-$ concentrations at a 1:2 ratio was observed, parallel with the decrease in the concentration of dissolved silica. In experiments with montmorillonitic clays, dissolution of siliceous tests took place, but opal-CT did not precipitate. However, small amounts of an Mg-rich clay formed.

The results highlighted the significance of a magnesium-hydroxide compound (MHC), which attracts silanol groups with their highly negative surface charge and causes them to precipitate and nucleate as opal-CT lepispheres. Kent and Kastner (1985) postulated that the MHC is the mineral sepiolite. Opal-CT precipitation apparently is more rapid than sepiolite growth once the nuclei have formed. The role of calcium carbonate, which also dissolves in the process, is thought to resupply alkalinity consumed in the formation of the nuclei. Seawater supplies the required Mg^{2+}. When the (OH)-groups have been exhausted, the rate of nucleus formation drops and the existing opal-CT embryos grow to well-developed lepispheres, provided continued opal-A dissolution supplies the required dissolved silica. In experiments with montmorillonitic clays, the clay minerals competed for the Mg^{2+} and thus prevented the formation of MHC nuclei. Consequently no opal-CT formed in these experiments, including the 6-month runs.

4.9.3. Opal-A dissolution and opal-CT precipitation as rate-limiting factors

Extending these hydrothermal experiments, Kastner and Gieskes (1983) determined the rate-limiting steps in the dissolution/reprecipitation process. The overall reaction rate is a function of the relative rates of dissolution, nucleation and growth of the silica phases involved. Which one of these processes will be the rate-limiting factor depends on the specific chemical and physical conditions.

Amorphous silica of varying specific surface area in aqueous solutions of 0.03 M $MgCl_2$ and 0.03 M $NaHCO_3$ was heated to temperatures of 50, 75, 100, 125 and 150 °C for periods of 1–30 days (Fig. 9.19). In all experiments, Mg^{2+} and $(OH)^-$ concentrations decreased at the 1:2 ratio previously established for the process, and opal-CT lepispheres or embryonic

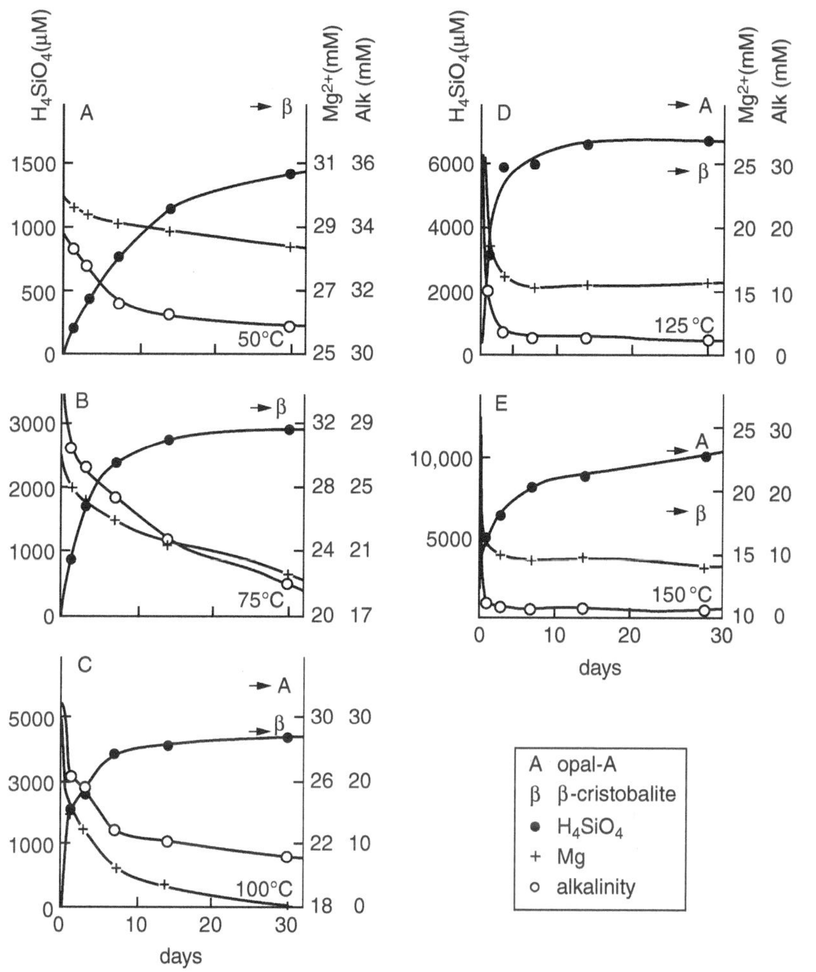

Figure 9.19 Opal-A dissolution as the rate-limiting factor in the opal-A to opal-CT conversion (redrafted from fig. 1 in Kastner and Gieskes, 1983). For interpretation, see text.

lepispheres formed, confirming that the MHC serves to nucleate the opal-CT. In experiments with low surface-area silica (i.e. Eocene radiolarians with a specific surface area of 5.3 m^2/g), the opal-A dissolution rate turned out to be the rate-controlling step at the lower temperatures (50, 75 and 100 °C). The concentration of dissolved silica remained below the equilibrium solubility of β-cristobalite, the silica phase which most closely approximates a poorly ordered opal-CT and for which solubility data are available. Nucleus formation and growth of opal-CT apparently kept pace with opal-A dissolution at these temperatures.

At 125 and 150 °C, however, opal-CT precipitation became the rate-limiting step after a few days, while the concentration of dissolved-silica rose above the equilibrium solubility of β-cristobalite. In experiments with higher-surface area Ludox silica (with a specific surface area of 61.5 m^2/g), opal-CT precipitation was the rate-limiting process at all temperatures: β-cristobalite solubility was rapidly exceeded, but opal-A equilibrium solubility concentrations were not attained (Fig. 9.20). The results of these experiments also clearly illustrate the dominant role of temperature as rate-controlling factor in the conversion of opal-A to opal-CT.

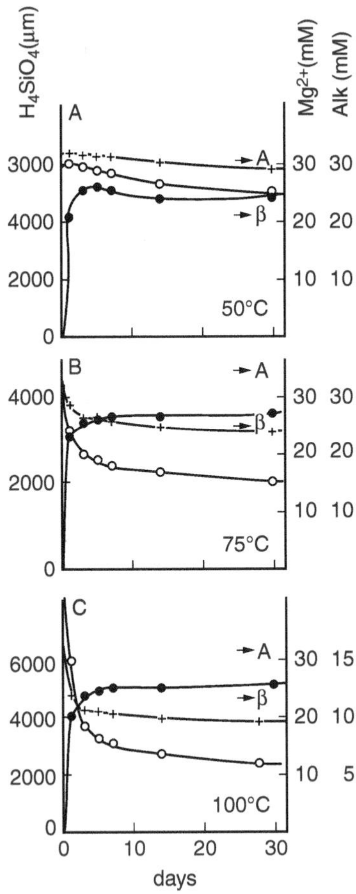

Figure 9.20 Opal-CT precipitation as the rate-limiting factor in the opal-A to opal-CT conversion (redrafted from fig. 2 in Kastner and Gieskes, 1983).

4.10. Rate-controlling factors of the opal-CT to quartz conversion

4.10.1. Temperature

Temperature estimates for the opal-CT to quartz transformation based on oxygen-isotope data and heat-flow considerations for the Monterey Formation range from 55 to 110 °C (Table 9.1C) (Murata and Larson, 1975; Murata et al., 1977). The actual temperature at which the conversion occurs at a particular locality depends in a complex way of many factors, including the earlier maturation history of the opal-CT. Pisciotto's (1981) temperature estimates for the base of the opal-CT zone in the Monterey Fm. (35–61 °C) are consistently lower than those of earlier authors, but they are supported by the results of Behl (1992), who mentions values as low as 30–40 °C.

4.10.2. Host-rock lithology

The effects of host-rock lithology on the second silica conversion reaction are opposite to those of the first reaction. Isaacs (1982) observed that the first opal-CT that appears in the Monterey Formation in a progressive burial sequence, has a d(101) spacing which varies inversely with the detrital mineral content of the host sediment (Fig. 9.21). This observation confirms the results of Murata and Larson (1975) that, at any given burial depth, the opal-CT d-spacing in (detritus-rich) porcelanite is 0.004–0.015 Å smaller than in associated (detritus-poor) chert (Fig. 9.15). This indicates retardation of opal-CT nucleation in clayey sediments because of the competition for Mg^{2+} between the MHC and clay minerals, consistent with the results of Kastner et al. (1977). Growth of opal-CT is delayed until the silica concentration reaches a metastable equilibrium with the lower-surface-area opal-CT (of a *d*-spacing of 4.08 Å or less). Such reduced silica concentrations in pore waters of clay-rich host sediments can be accomplished most easily by adsorption of silica on detrital minerals (Siever and Woodford, 1973) and by the neoformation of silica-rich clay minerals. A prerequisite is that their nucleation and growth rates are faster than those of the silica phases that could potentially form. These and other possible scenarios are discussed by Williams et al. (1985).

The detrital mineral content, while retarding the opal-A to opal-CT conversion, enhances the opal-CT to quartz conversion. In sediments with more than 70% detrital minerals, the initial opal-CT has a d(101) spacing of 4.08 Å or less (Fig. 9.21). In these sediments, the conversion to quartz should occur more rapidly than in sediments with less than 30% detrital minerals with an initial d(101) spacing of 4.11 Å, because quartz precipitation does not start before the d-spacing of opal-CT is reduced to below 4.07 Å by Ostwald ripening. An abbreviated opal-CT stage in clayey host sediment appears at first glance, however, to be at variance with the earlier observation (i.e. Lancelot, 1973) that the silica maturation sequence is

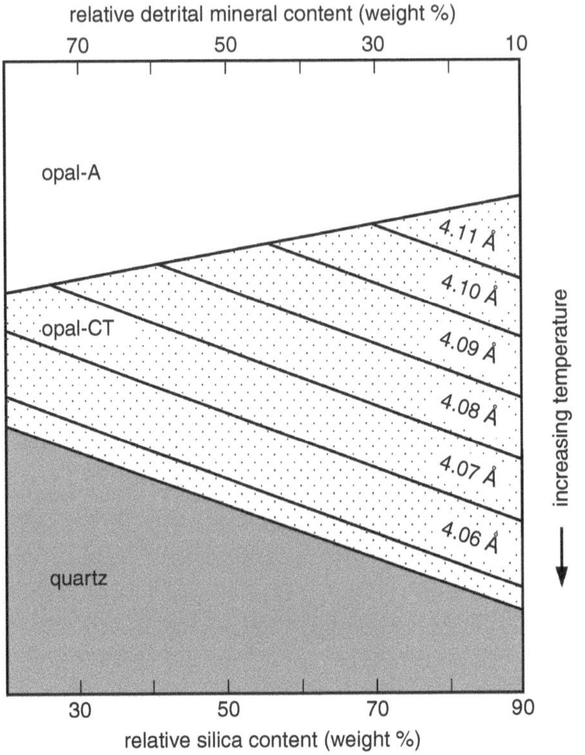

Figure 9.21 XRD *d*-spacing for first appearance of opal-CT and opal-CT—quartz conversion as a function of the detrital content (redrafted from Isaacs, 1982).

accelerated in calcareous host rocks, leading to the relatively early formation of quartz chert. The dénouement is that the initial retardation of the opal-A to opal-CT conversion in clayey sediments delays the entire silica maturation sequence to an extent that cannot be compensated for by later acceleration, due to an abbreviated opal-CT stage. Although the opal-CT stage may in fact last longer in carbonates than in clay-rich siliceous sediments, this may, contrary to Lancelot's (1973) assumption, be outweighed by a considerably earlier initiation of the entire conversion sequence in carbonates.

The field relationship between the detrital mineral content and opal-CT transformation rates reported by Isaacs (1982) from the Santa Barbara coast is not observed everywhere in the Monterey Formation. Pisciotto (1981), for example, found lower temperatures not only for the first appearance of opal-CT but also for quartz in the Santa Barbara Basin to be associated with a lower detritus content.

The role of organic matter in the silica conversion reactions is that, in organic-matter-rich sediments, the rate of silica diagenesis is reduced

(Hinman, 1987). Organic acids released by organic-matter maturation dissociate and lower the pH, thereby reducing the carbonate alkalinity. This, in turn, would slow down the conversions by decreasing the silica-polymerization rate. The initial opal-CT formed in organic-matter-rich sediments should have a lower d(101) spacing because of a slower nucleation rate under decreased carbonate alkalinity. In so far as organic-matter content may be positively correlated with the detrital mineral content, Hinman's (1987) results are in accordance with Isaac's findings. However, Hinman (1987) did not find evidence that an initially lower d(101) spacing of opal-CT in organic-matter- and detritus-rich sediments also accelerates the subsequent transition to quartz. On the contrary, quartz formation in the Santa Maria Basin seemed to be slowed down because it occurred at higher temperatures in porcelanite than in chert.

4.11. Absence of opal-CT as an intermediate metastable phase

It seems unlikely that the opal-CT stage can be bypassed entirely in the normal diagenetic silica maturation process, except under special conditions. Rare exceptions from Ostwald's step rule seem to be restricted to micro-environments in sediments such as the cavities of foraminifers (e.g. Keene, 1975) or radiolarians in calcareous or siliceous pelagic sediments. These are often filled with chalcedony, whereas the host rocks are calcareous opal-CT rocks or diatomaceous (opal-A) shales (Isaacs, 1982). This indicates anomalously early quartz (i.e. chalcedony) precipitation. Precipitation of chalcedony occurs in spherical open cavities indicating an absence of compaction due to limited overburden weight and very shallow burial. This chalcedony does not appear to have had an opal-CT precursor in the cavities, because elsewhere in the host-rock opal-CT is still present without any sign that the conversion to quartz has started. Most likely, the chalcedony formed directly from opal-A, possibly because the concentration of dissolved silica in the micro-environment of the cavities, for some unknown reason, remained below the β-cristobalite equilibrium solubility during the early stage of opal-A dissolution. Among the mechanisms discussed by Williams et al. (1985) to maintain low silica concentrations in pelagic carbonates is the early precipitation of calcium zeolites, but no specific studies have been devoted as yet to this problem.

Direct opal-A to quartz conversion is possible also at elevated temperatures. An interesting example is the occurrence of volcanic dykes and sills in the sediment-covered spreading ridge of the Guaymas Basin in the Gulf of California. The silica modifications observed there depend on the distance from the sills (Kastner and Siever, 1983). In the vicinity of basalt or dolerite sills 30 m thick or more, the only silica phase present is authigenic quartz. At greater vertical distance from the sills, opal-A or opal-CT are found but no quartz. However, the absence of opal-A and presence of quartz extends

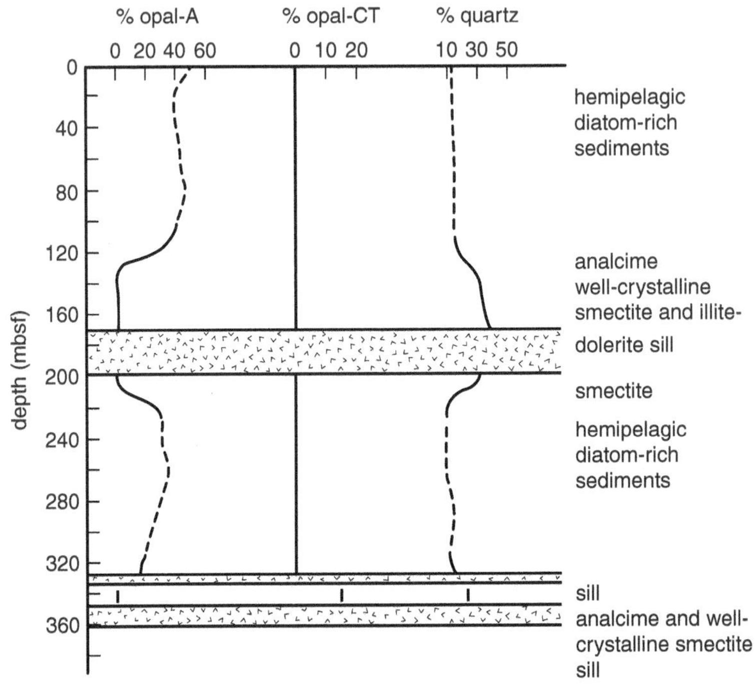

Figure 9.22 Distribution of silica phases as a function of distance from igneous sills on sedimented spreading centre in the Guaymas Basin, Gulf of California (DSDP Site 481; redrawn and modified from Kastner and Siever, 1983; fig. 3). Quartz between 0 and 120 mbsf is detrital in origin.

to considerably greater distances (up to 45 m in DSDP Site 481) above than below sills (Fig. 9.22), indicating that hydrothermal convection initiated by the intrusion operates basically above the sills. The amount of quartz found in the opal-A-free zone is less than expected from opal-A dissolution. Much of the silica dissolved from opal-A is carried away by convection; this reduces the silica activity below opal-CT solubility, facilitating direct precipitation of quartz (Fig. 9.22). Being only a few tens of thousands of years old, the quartz- and opal-CT bearing sediments of the Guaymas Basin spreading centre are among the youngest biogenic siliceous sediments known to have undergone silica conversions.

4.12. Time/burial-depth distribution of silica phases in deep-sea diagenetic environments

The age/burial-depth plot of Fig. 9.17 summarizes the distribution of the various silica phases in oceanic sedimentary successions on the basis of DSDP data (Riech and Von Rad, 1979). Opal-A has been found in sediments as old as

85 Ma (Late Cretaceous), but only at shallow subsurface depths. The oldest opal-A reported is an occurrence of very well preserved radiolarian tests that are overgrown by opal-A′ in Lower Berriasian claystone from the outer trench slope of the Marianas trench at 6300 m water depth (Ogawa and Kawata, 1998). With increasing subsurface depth and temperature, the maximum survival time is shortened significantly. For example, opal-A is normally absent at 1000 m subsurface depth in sediments older than about 20 Ma.

Opal-CT takes a minimum of about 10 million years to first appear. Exceptionally young and shallow opal-CT occurrences in the Southern Ocean that formed at very low temperatures are difficult to explain. The youngest example of opal-CT formation in the Guaymas Basin, which is only a few tens of thousands of years old (Kastner and Siever, 1983), is clearly related to elevated temperatures on a sedimented ridge. On the other hand, opal-CT may still persist in sediments 100–120 Ma old. Little new opal-CT forms, however, in pre-Tertiary sediments (older than 65 Ma), because generally no source material (opal-A) is left in those sediments. No opal-CT has been found in pre-Cretaceous sediments (older than 144 Ma).

There is considerable overlap on the age/depth graph of Fig. 9.17 between the fields for the three main silica phases. The quartz field, for example, overlaps major parts of the opal-CT field and to a lesser extent even the opal-A field, illustrating how local variations in heat flow, pore-fluid chemistry, host-rock lithology and other factors can affect the rate of the diagenetic conversions. Quartz cherts need a minimum of 30–40 million years to form at burial depths of 500 m or more, and considerably longer at shallower depths. Genuine quartz cherts are, consequently, rare in Cenozoic pelagic sediments, but become predominant in Lower Cretaceous and older siliceous rocks. They are the exclusive lithology of siliceous sediments in Palaeozoic and Precambrian sedimentary successions.

4.13. Diagenetic formation of bedded chert?

Many ancient bedded chert formations for which geological and sedimentological evidence suggests a deep-water origin, appear to have formed from siliceous pelagic sediments that were redeposited by turbidity currents (e.g. Imoto and Saito, 1973; McBride and Folk, 1979; Nisbet and Price, 1974; Robertson, 1977) or ocean bottom currents as inferred from characteristic primary sedimentary structures and grain fabrics (characterized by current-parallel alignment of sponge spicules and radiolarians).

Alternative interpretations that have been suggested as explanations for bedded cherts such as (1) variations in surface-water productivity, (2) variations in the input rate of the terrigenous component, and (3) diagenetic segregation of silica into chert-rich beds and silica-poor claystones from initially homogeneous siliceous muds, do not adequately explain the sharp lithological boundaries between the chert and intercalated shale layers.

These should have been obliterated by the burrowing activity of benthic organisms if the sediments accumulated by a slow pelagic rain of particles. Nearly all "bedded cherts" contain shale intercalations, often red or green in colour, that comprise 5–40% of the succession.

Well-bedded radiolarian cherts abound in the Mesozoic, particularly the Jurassic, when the "ribbon cherts" of the Tethyan realm and California formed, whereas equivalents of these ribbon cherts have rarely been observed in DSDP drill-cores or in Tertiary chert-bearing formations exposed on land (e.g. Iijima and Utada, 1983) with the exception of the Monterey Formation (Bramlette, 1946). Jenkyns and Winterer (1982) ascribed the Jurassic radiolarite/shale rhythms to cyclic variations in surface-water productivity (coupled with turbidity-current and bottom-water activity) in small ocean basins or marginal seas that caused periodic increases in radiolarian abundance.

The difference in bedding style between the Mesozoic ribbon cherts and younger siliceous deposits might, however, be related to the difference in the diagenetic behaviour between radiolarians and diatoms that led to different diagenetic histories of these rocks. Jurassic (and older) ribbon cherts consist predominantly of radiolarians (see Hüneke and Henrich, 2011, this volume, section 3). Cenozoic siliceous sediments are largely diatomites reflecting the evolutionary success of the diatoms, which only appeared in the Jurassic (Harper and Knoll, 1975) and since then have become the most efficient silica extractors in oceans. The early diagenetic stabilization of radiolarians by chalcedony without an intermediate opal-CT phase (cf. Section 4.11) might be the key for the occurrence of Jurassic radiolarites as ribbon cherts, which appear to be rare in younger, diatom-dominated siliceous sediments. Early fixation of silica as chalcedony would reduce its mobility during early diagenesis, and help to preserve the original bedding. Recognition of symmetrical geochemical changes (from the centre of beds upwards to the top and downwards to the base) and symmetrical grading of radiolarian abundance (Sano, 1983; Steinberg et al., 1983) supports the diagenetic origin (or modification) of bedding in some ribbon cherts that was originally proposed by Davis (1918) (see also Jones and Murchey, 1986).

In summary, there seems to be general agreement that the bedding in radiolarian cherts is not due to a single process (e.g. Baltuck, 1983; Hein and Karl, 1983; Jenkyns and Winterer, 1982). Diagenesis may play an important role among the processes involved.

4.14. Physical diagenesis of biogenic siliceous deep-sea sediments

Treatment of the diagenesis of siliceous deep-sea sediments would be incomplete without consideration of the physical diagenesis in cherts. A stepwise compaction history is characteristic of siliceous sediments during burial, which contrasts with the more gradual compaction history of argillaceous sediments

(Fig. 9.15). According to Tada (1991), a sharp porosity reduction from 75% to 35% over a few tens of meters is associated with the opal-A to opal-CT conversion. Differential dewatering and compaction in the course of the conversion has been suggested as the cause for the development of giant polygonal ridges up to 50 m high in Oligocene-Miocene sedimentary successions of the Norwegian Sea that have been imaged in 3D-seismic profiles. The ridges are matched by "touch-downs" from an overlying layer so that large, up to 2 km wide, sediment cushions or hummocks are enclosed by the ridges and touch-downs (Davies, 2005; Davies et al., 1999).

Other physical diagenetic features characteristic of chert formations are dykes, diapirs and breccias; these have been well described in the literature (e.g. Snyder et al., 1983; Taliaferro, 1934). They attest to differential compaction and variable rates of silica recrystallisation and cementation. In a given succession of siliceous sediments, some layers may already be lithified or semi-lithified while others are not. Shaking during an earthquake may rupture the lithified layers and cause injection of unlithified sediment into the fractures from adjacent uncemented layers, particularly if these had been overpressurized. The opal-A to opal-CT and the opal-CT to quartz conversions are dehydration reactions. They may generate excess pore pressure and cause hydraulic fracturing, particularly if assisted by earthquakes. Dykes in Cretaceous radiolarian porcelanites from the West Pacific east of the Marianas deep-sea trench/island arc (Fig. 9.23) probably reflect seismic activity associated with the Marianas subduction zone. Similar mechanisms of dyke and vein formation are invoked by Steinitz (1970) and Snyder et al. (1983) for examples in Israel and California. On the basis of 3-D seismics and drill holes, Davies et al. (2006) suggested that overpressure development associated with dewatering during the opal-A to opal-CT conversion in biogenic siliceous sediments of several North Atlantic basins may have triggered giant sand injections from intercalated sand layers above the conversion front.

5. Biogenic Pelagic Carbonates

Above the CCL, the tests of foraminifers and coccoliths, which are the major carbonate constituents of pelagic sediments, dilute the biogenic siliceous oozes to the extent that either mixed carbonate/siliceous oozes or pelagic carbonate oozes with low silica content accumulate. Besides chert-nodule formation, the main early diagenetic reaction in these fine-grained biogenic carbonate sediments is recrystallisation of the delicate tests of the calcareous organisms involving (pressure) solution and reprecipitation. Dissolution occurs because of their large reactive surface area. The result is chalk, a partially lithified fine-grained carbonate that ultimately is turned into lutitic limestone. Above the aragonite compensation level

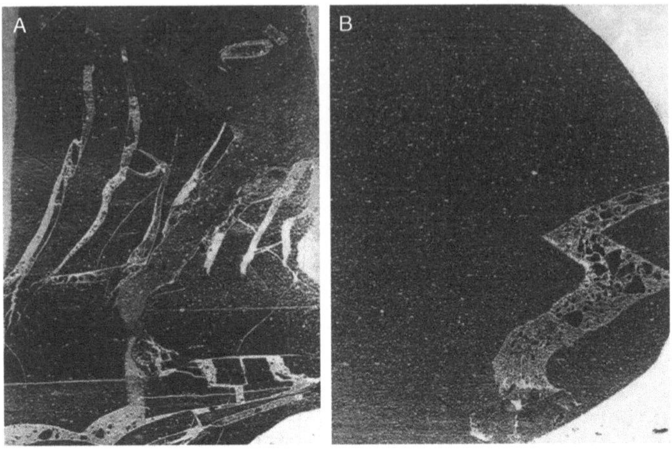

Figure 9.23 Samples from the Marianas subduction zone in West Pacific (DSDP Leg 20, Site 198A), (A) Multiple fractures filled with opal-CT sediment and fragments that were cemented by microquartz and chalcedony. Light-coloured margins and ends of the smaller fractures consist almost entirely of cement because these parts of the fractures were too small to be invaded by the mobilized sediment. Thin section view is ~2.5 cm high. (sample 5cc#4) (fig. 21 from Hesse, 1990b, reproduced with kind permission of the Geological Association of Canada). (B) Folded chert dyke in Early Cretaceous radiolarian porcelanite (sample 5cc#1). The folding is due to differential compaction. The folded dyke behaved as a more competent layer, while the surrounding host sediment was still undergoing compaction. Abundant porcelanite fragments in the microdyke are cemented by microquartz and chalcedony. The pure chalcedonic cement at the dyke margin was probably precipitated into pore space that was generated by shrinkage of the host sediment during ongoing compaction. Thin section view is ~2 cm high.

(ACL), which is 2–3 km shallower than the CCL, the shells of aragonite-precipitating organisms such as pteropods (planktonic gastropods) can also form a major component of pelagic sediments.

Some pelagic carbonates display distinct bedding rhythms—caused by periodic variations in the concentration of terrigenous fine-grained particles, mostly clay minerals—that are reflected in colour variations and that are related to Milankovitch paleoclimate cycles (e.g. Fischer et al., 2009). However, rhythmic limestone/marlstone alternations may also be accentuated or even originate during burial diagenesis due to differential dissolution, cementation and compaction.

5.1. Dissolved Sr and Sr-isotope anomalies in pore waters

The lithification event associated with carbonate dissolution and reprecipitation is reflected by a distinct strontium anomaly in the pore waters at a hundred to a few hundreds of meters subsurface depth (Figs. 9.3 and 9.24).

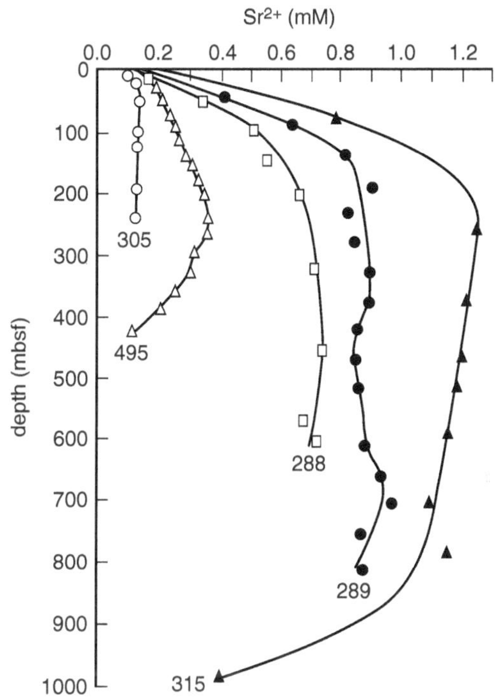

Figure 9.24 Typical dissolved Sr^{2+} concentration profiles in pelagic carbonates, Pacific drill sites of the DSDP (redrafted from Baker et al., 1982).

The Sr^{2+} concentration of biogenic calcite is three to five times higher than that of the inorganic calcite reprecipitated from solution (Baker et al., 1982; Elderfield et al., 1982). Thus a distinct signal of dissolved Sr in pore water is associated with the calcareous ooze/chalk boundary where the Sr curve has its maximum. From this maximum, dissolved Sr is transported by diffusion both upwards towards the sediment/water interface and downwards towards a sink deeper in the sediment column (volcanogenic sediment) or the basaltic oceanic crust. The ooze/chalk boundary may be viewed as a diagenetic front that moves upwards through the subsiding sediment column during burial (Gieskes et al., 1986). At the ooze/chalk boundary, the recrystallisation rate appears to be at its maximum; lower in the chalk, recrystallisation is slowed down considerably. The precipitation of calcite frequently occurs as overgrowths on single-crystal skeletal elements of the coccoliths, especially discoasters (Fig. 9.25).

The rate of carbonate recrystallisation in pelagic sediments depends also on the sedimentation rate (Gieskes and Johnston, 1984), which controls the amount of organic matter that is buried. The rate of HCO_3^- production

Figure 9.25 Calcite overgrowth with well-developed crystal faces on discoaster (10 μm in diameter). Early Pliocene carbonate turbidite, DSDP Site 199, Caroline Abyssal Plain (fig. 21, from Hesse, 1990a, reproduced with kind permission of the Geological Association of Canada).

during early diagenesis, which in turn controls calcite dissolution, depends on the organic-matter concentration. The Sr maximum occurs thus at strongly variable subsurface depths. The solid phase celestine ($SrSO_4$) is a minor diagenetic mineral in marine carbonates (Hanor, 2000). Where the ooze/chalk transition occurs in the sulfate-reduction zone, dissolved Sr^{2+} may be removed by celestine (Baker and Bloomer, 1988). It is also produced in the water column by acantharians, marine protozoans, both in their skeletons and their cysts, but it reaches the seafloor only at shallow to intermediate water depths.

The diffusive transport of Sr in pelagic carbonates is illustrated by the strontium-isotope ratios. Sr-isotope curves for the pore waters of pelagic carbonates can be divided into two parts (Fig. 9.26): an upper part above the Sr-maximum, where $^{87}Sr/^{86}Sr$ ratios are generally lower than the contemporaneous sea-water curve of Palmer and Elderfield (1985) and Veizer et al. (1999), and a lower part below the maximum, where they tend to be higher than or equal to paleo-seawater ratios for coeval sediments. Because seawater has become increasingly heavier in strontium-isotope ratios during the last 100 Ma towards today, the Sr-maximum is a source of relatively light Sr isotopes for the section above it and of relatively heavy isotopes for the section below it. Transport is by both upward and downward diffusion from the maximum, where the isotopic composition is closest to that of seawater of the age of the host sediment.

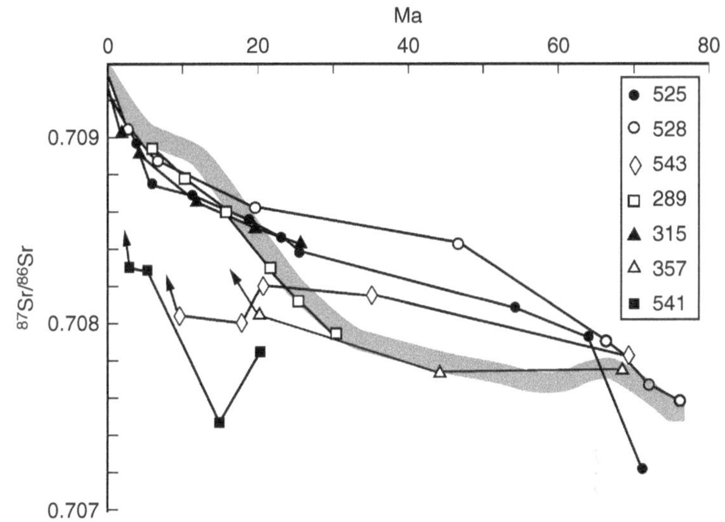

Figure 9.26 Sr-isotope ratios versus sediment age in pore waters from pelagic carbonates (DSDP drill sites) and Sr-isotope curve of seawater (stippled) for the last 80 Ma (redrawn and modified from Gieskes et al., 1986). DSDP Site 289 on the Ontong-Java Plateau in the Southwest Pacific differs from other carbonate sites in that the Sr-isotopes composition of the pore water below the ooze/chalk boundary is very close to the paleoseawater curve. In this site, maximum dissolved Sr-concentrations are reached at about 400 m subbottom, and below this depth no significant decrease takes place. This suggests that the downward diffusive flux in this thick chalk-limestone section is very small or nil—in line with the isotope data, which also suggest that isotopic equilibrium exists in the chalk section of this hole between the pore waters and their host sediments. In carbonate-poor sediments of sites 541 and 543 near the toe of the Lesser Antilles island-arc slope, Sr-isotopic ratios of the pore waters in the upper part of the holes fall much below the paleoseawater curve and are due to volcanic-ash alteration (see Section 11).

5.2. Diagenesis of rhythmic calcareous ooze/marl alternations

The principal question concerning the origin of rhythmic limestone/marl (stone) alternations is whether their rhythmicity reflects external causes (such as cyclical climate or/and relative sea-level changes, productivity or deep-water dissolution cycles that cause periodic variations in the supply rate of either fine-grained terrigenous particles or biogenic carbonate) or whether it reflects the effects of burial diagenesis, or a combination of some or all of these factors? Ever since Ricken (1986) proposed diagenetic bedding as a model for the origin of limestone/marlstone alternations, the redistribution of carbonate by dissolution in the marlstone and reprecipitation in the limestone during diagenesis has become a popular mechanism to explain the origin of rhythmic carbonates. Westphal et al. (2004), who studied a Lower Cretaceous

(Valanginian) rhythmic limestone/marlstone succession from the Blake-Bahama Basin in the West Atlantic, came to the conclusion that diagenesis was probably a major factor in developing the rhythmic character of the sediments, however, not in the sense of the simple calcite export-import mechanism between the marlstone and limestone envisaged by the original diagenetic-bedding hypothesis. Coccoliths are equally well preserved in the two lithologies making differential dissolution of calcitic marl components unlikely.

As the sediments may have been deposited above the ACL, fine-grained aragonite supplied from the Bahama Banks could have been the constituent that dissolved and supplied the extra carbonate for cementation of the limestones. Aragonite dissolution is facilitated by bacterially mediated organic-matter oxidation that slightly lowers the pH to a value that is sufficient for dissolution of the more soluble aragonite but not of calcite. Calcite can in fact precipitate due to the increased alkalinity while aragonite dissolves. Aragonitic fossils in the limestone layers were neomorphosed to calcite. High strontium concentrations averaging 9 mM support an aragonite precursor, whereas calcite dissolution typically leads to Sr enrichment in the 4.5 mM range. Environmental indicators such as palynomorphs and cross-plots of elements from geochemical analyses showed changes incoherent with the lithological alternations and did not provide conclusive evidence for external factors operating on land or in the sea that could have controlled the development of the rhythms. Diagenesis must have taken place early, as demonstrated by the absence of compaction features in the limestones (Fig. 9.27). Munnecke et al. (2001) developed a quantitative model to estimate the original composition of the sediment before diagenesis. In the case of the Blake-Bahama-Basin rhythmites, the apparently high original aragonite concentrations (of the order of 40%) required for cementation of the limestones remain an unsolved problem because of the distance from the Bahama Banks, whereas precursor aragonite exceeding 40% is not uncommon in shallow-water rhythmic carbonate successions and may reach levels as high as 70%.

Phanerozoic limestone/marlstone alternations are most abundant in areas above the ACL adjacent to potentially aragonite-producing shelf areas and carbonate banks, independent of their occurrence in aragonite or calcite seas, as shown by a literature search of Westphal and Munnecke (2003). The findings underline the involvement of aragonite in the differential diagenesis of these alternations. The explosion of carbonate-producing planktonic micro- and nanno-organisms since the Cretaceous has lowered the CCL to abyssal depths and extended the deposition of pelagic calcareous ooze/marl successions into the deep sea. These undergo different pathways of diagenesis not or only subordinately involving aragonite as a precursor carbonate mineral and displaying compaction both in the limestone and intercalated slightly cemented marl.

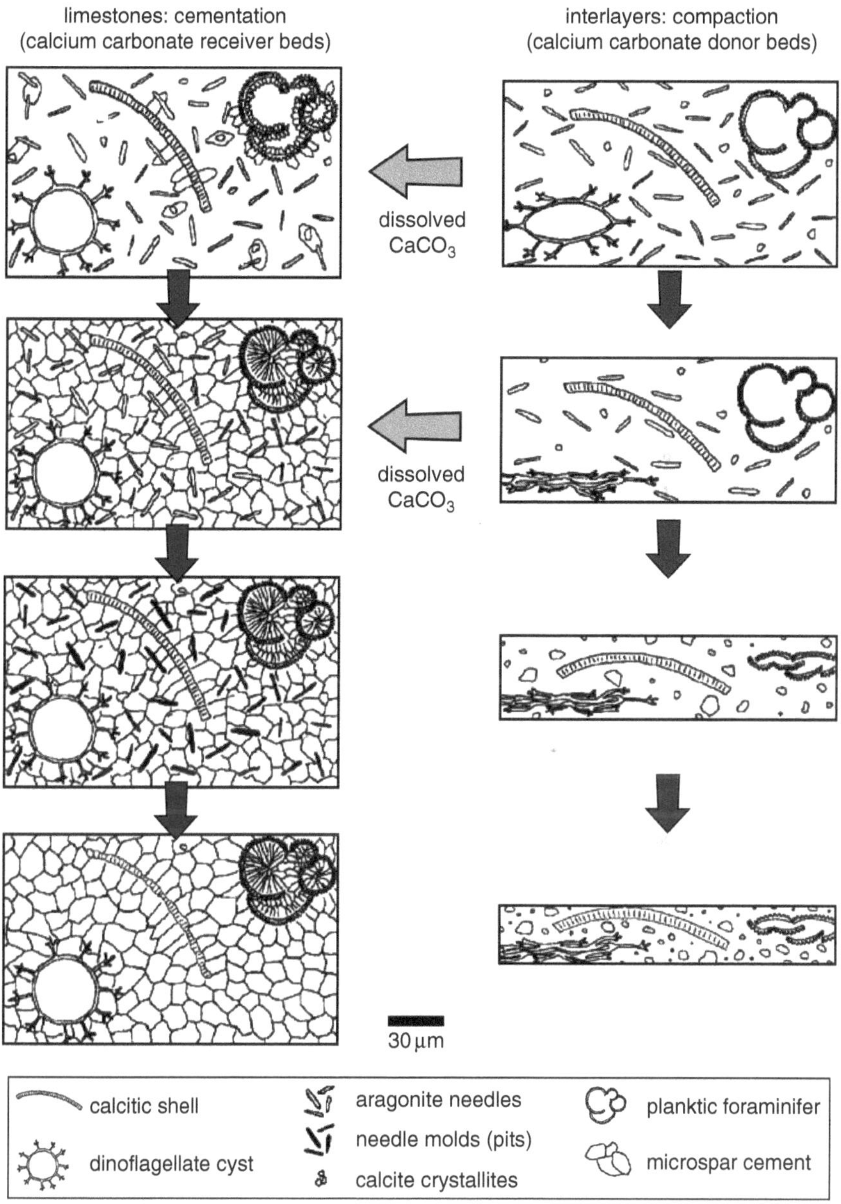

Figure 9.27 Differential diagenesis in rhythmic carbonate successions (modified from fig. 3 in Westphal, 2006, with permission of the author). Whereas the limestone beds are cemented early and thus mechanically stabilized, the interlayers loose carbonate by aragonite dissolution and undergo compaction.

Rhythmic pelagic/hemipelagic carbonate successions have been widely used for high-resolution chronostratigraphy throughout the Mesozoic-Cenozoic, based on the assumption that they are caused by orbital forcing. Unless lithological parameters can be identified that were not affected by diagenesis, the obtained frequencies may have been shifted or biased diagenetically and may not reflect paleoclimatic changes caused by orbital insolation changes (Westphal, 2006).

Finally, it should be mentioned that early-diagenetic carbonate cementation in deep-sea pelagic carbonates may start close to the sea floor, similar to shallow-water carbonate hardgrounds, as has been convincingly demonstrated for the eastern Mediterranean Sea involving mostly high-magnesium calcite cements that in part form nodules (Miliman and Müller, 1973, 1977; Müller and Fabricius, 1974).

6. Hemipelagic Sediments

With decreasing distance to the continents, a progressively higher influx of terrigenous components leads to the deposition of hemipelagic sediments on the continental margins. They cover the rise and slope and parts of the oceanic crust closest to the margins, where they may alternate with intercalated gravity-flow deposits of varying thickness (see Henrich and Hüneke, 2011, this volume; Mulder, 2011, this volume). The organic-matter concentration in hemipelagic sediments tends to correlate with sedimentation rate (Fig. 9.28) (Heath et al., 1977; Müller and Suess, 1979). Organic matter is a main driver of early diagenesis in hemipelagic deep-sea sediments.

6.1. Early diagenetic organic-matter oxidation

Organic matter, the reduced carbon compound, is the strongest reductant in freshly deposited sediments. The organic matter dispersed in the sediment provides an energy source for sediment-dwelling organisms, foremost bacteria, to maintain their metabolism through oxidation reactions. Because of their small size in the less than micrometer range, bacteria, the main representatives of the prokaryotes, possess a large surface-area/mass ratio that favours the exchange of dissolved substances with the pore water through the cell walls. Bacterially mediated organic-matter oxidation reactions are among the first and most efficient mechanisms to alter the chemical composition of pore waters, starting immediately after deposition and leading to remineralization. The sub-seafloor biosphere is the largest prokaryotic habitat on Earth, although metabolic rates of these communities are among the slowest (Parkes et al., 2005). Parkes et al. (1994) provided the first proof of the presence of bacteria in

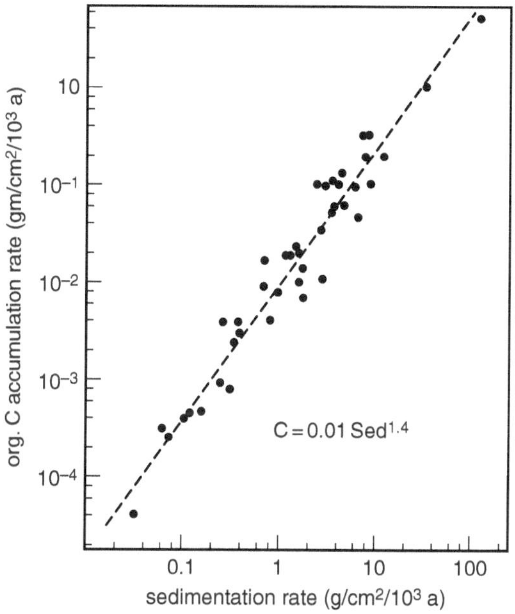

Figure 9.28 Correlation of the organic-carbon mass-accumulation rate and the sedimentation rate for open-marine pelagic and hemipelagic environments (redrawn and modified from Heath et al., 1977).

deep (>500 m) subsurface burial environments, thus extending the range of the biosphere to considerable sub-seafloor depths.

Bacteria are energy opportunists. Microbial organic-matter decomposition and conversion into carbon dioxide and methane is accomplished by at least five different oxidant-specific populations of bacteria (aerobic bacteria, nitrate reducers, sulphate reducers, carbonate reducers or methanogens and fermenters) that follow one another during burial in the order of decreasing energy efficiency of their metabolic reactions (Table 9.2) (Froelich et al., 1979). They give rise to the well-established vertical sequence of organic-matter decomposition zones with increasing burial depth (Fig. 9.29; Claypool and Kaplan, 1974; Curtis, 1978) comprising (1) the oxidation zone, (2) the nitrate-reduction zone, (3) the sulphate-reduction (SR) zone, (4) the carbonate-reduction zone and (5) the fermentation zone. At temperatures above 75 °C, bacterial activity ceases and thermocatalytic reactions start, constituting (6) the thermocatalytic decarboxylation zone. The oxidation zone corresponds to what is called "oxic diagenesis", the nitrate-reduction zone to "suboxic diagenesis", and the top of the SR zone to the top of "anoxic diagenesis". The lower boundary of zone (5) defines the lower boundary of "early diagenesis".

Table 9.2 Stages of organic-matter oxidation

C:N:P = 106:16:1 = Redfield ratio for primary organic matter[a]	$\Delta G^{o\prime}$ of metabolic processes coupled with oxidation reactions (in kcal mol^{-1} glucose)
(1) Oxidation by freely dissolved O_2 (aerobic respiration)	
$(CH_2O)_{106}(NH_3)_{16}(H_3PO_4) + 138O_2 \rightarrow 106CO_2 + 16HNO_3 + H_3PO_4 + 122H_2O$	−763
(2) Nitrate reduction	
$(CH_2O)_{106}(NH_3)_{16}(H_3PO_4) + 94.4HNO_3 \rightarrow 106CO_2 + 42.2N_2 + 16NH_3 + H_3PO_4 + 177.2H_2O$	−724
$(CH_2O)_{106}(NH_3)_{16}(H_3PO_4) + 84.8\ HNO_3 \rightarrow 106CO_2 + 42.4N_2 + 16NH_3 + H_3PO_4 + 148.4H_2O$	−658
Manganese reduction	
$(CH_2O)_{106}(NH_3)_{16}(H_3PO_4) + 236MnO_2 + 472H^+ \rightarrow 236\ Mn^{2+} + 106CO_2 + 8N_2 + H_3PO_4 + 366H_2O$	−748 (birnessite) −698 (pyrolusite)
Iron reduction	
$(CH_2O)_{106}(NH_3)_{16}(H_3PO_4) + 212\ Fe_2O_3$ (or 424 FeOOH) $+ 848\ H^+ \rightarrow 424Fe^+ + 106CO_2 + 16NH_3 + H_3PO_4 + 530H_2O$ (or $742H_2O$)	−337 (hematite) −318 (limonitic goethite)
(3) Sulphate reduction	
$(CH_2O)_{106}(NH_3)_{16}(H_3PO_4) + 53(SO_4)^{2-} \rightarrow 106CO_2 + 16NH_3 + 53S^{2-} + H_3PO_4 + 106H_2O$	−91
(4) Carbonate reduction	
$(CH_2O)_{106}(NH_3)_{16}(H_3PO_4) \rightarrow 53CO_2 + 53CH_4 + 16NH_3 + H_3PO_4$	−84

[a] The organic matter involved in reactions (1)–(4) does not necessarily have this ratio.

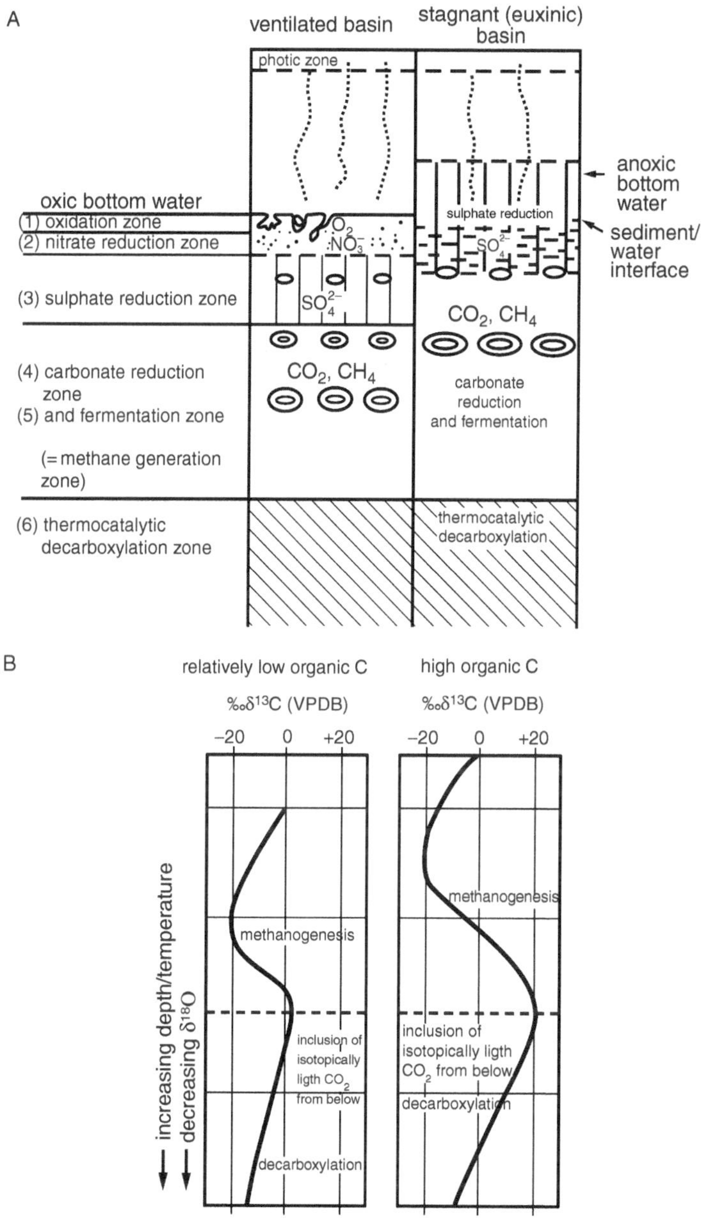

Figure 9.29 Stages of organic-matter oxidation in anoxic sediments in ventilated (or intermediate sedimentation-rate) basins and stagnant (or high-sedimentation-rate) basins (redrafted from Hesse et al., 2004). (A) In stagnant basins, the organic-matter decomposition zones are raised to shallower burial depths and zones (1)–(3) occur above the sediment/water interface compared with well-aerated basins (modified from Claypool and Kaplan, 1974). (B) Schematic $\delta^{13}C$ profiles for dissolved carbonate in the two types of basins (modified from Mozley and Burns, 1993).

6.1.1. The oxidation zone

In the oxidation zone, freely dissolved oxygen is available from sea water trapped in the sediment during deposition (connate water) or supplied by diffusion from overlying bottom water. Microbial degradation of organic molecules leads to a loss of functional groups, particularly hydrocarbon chains and carboxyl groups. The smaller molecules (e.g. acetate, lactate, propionate, isobutyrate) are partly converted to carbon dioxide according to an overall reaction of the form:

$$CH_2O + O_2 \rightarrow CO_2 + H_2O$$

where CH_2O is shorthand used to denote the bulk composition of organic matter.

Aerobic bacteria are the most efficient users of oxygen. Use of oxygen by aerobic respirators may lead to temporary or local oxygen depletion, particularly in regions of rapid sedimentation where replenishment from the overlying bottom water by diffusion does not keep pace with consumption. Aerobic organisms will then inevitably cease to exist, but facultative aerobes which can switch from an aerobic to an anaerobic mode of respiration may still be present.

The oxidation zone, if present, is easily recognized by its brownish-yellowish sediment colour. In organic-matter-rich sediments it may be only a few centimetres or millimetres thick, in contrast to the carbon-free brown abyssal clay in which the oxidation zone may be tens to hundreds of meters thick. In stagnant or euxinic basins, the oxidation zone is absent. The amount of dead animal or plant matter left undestroyed in the sediment after passage through the oxidation zone is a function of its residence time in this zone.

6.1.2. The nitrate-reduction zone

The nitrate-reduction zone commences where the concentration of dissolved oxygen drops below about 0.5 $cm^3/10^3$ cm^3 H_2O (Devol, 1978), which corresponds to the oxygen level at which most benthic macroorganisms disappear from the sediment (Rhoads and Morse, 1971). Dysaerobic bacteria, which live at dissolved oxygen levels between 1.0 and 0.1 $cm^3/10^3$ cm^3 H_2O, are characteristic of this environment. Concentration of nitrate tends to increase from the ambient bottom water values of 0.03–0.04 mM (mM = millimoles/10^3 cm^3) to a maximum in the oxidation zone where ammonia released from organic-matter decomposition is oxidized (Froelich et al., 1979). From this maximum, nitrate decreases downwards and disappears at the base of the nitrate-reduction zone. Nitrate thus becomes a tracer of both oxic and suboxic diagenesis as it is both generated and consumed in these zones. The contribution of denitrification to the total sediment organic-matter degradation has been estimated at 7–11% (Middelburg et al., 1996). The change from positive to negative electrochemical potential characterizes the lower boundary of the nitrate-reduction zone, below which anaerobic bacteria appear.

When the electrochemical potential is lowered sufficiently, oxides and hydroxides of manganese and iron, which at higher oxidation potential have a very low solubility, will be reduced and become soluble (Fig. 9.30; Froelich et al., 1979) thus serving as important sources of oxidant (electron acceptor) in addition to the nitrate of the pore water. Manganese is reduced nearly simultaneously with nitrate, consistent with comparable amounts of free energy for both reactions (Emerson and Hedges, 2003). The Mn^{2+} produced is either released by diffusion to the bottom seawater or reoxidized by oxygen. Iron reduction occurs after manganese reduction at somewhat greater depth, commonly overlapping with the SR zone, where it is grasped by the sulphide produced to form Fe-sulphides. In the upper SR zone, about 1/3 of the organic-matter oxidation is due to iron reduction and 2/3 to sulphate reduction, as shown by a recent study of Arctic fjord sediments (Finke et al., 2007). Both manganese and iron reduction are hydrogen-ion consuming reactions (Table 9.2), increasing the pH, which may be a prerequisite for the precipitation of early diagenetic carbonates in concretions (see Section 12.2).

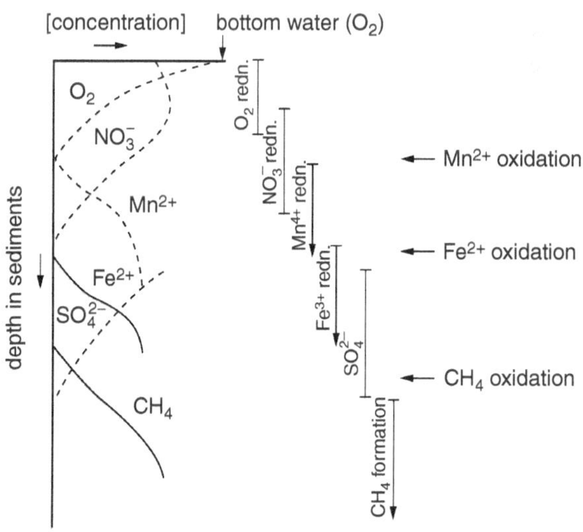

Figure 9.30 Schematic diagram showing the depth distribution of dissolved O_2, NO_3^-, Mn^{2+}, Fe^{2+}, SO_4^{2-} and CH_4 in oxic, suboxic and anoxic diagenetic environments (redrawn and modified from Emerson and Hedges, 2003).

6.1.3. The sulphate-reduction zone

In the anoxic SR zone below the nitrate-reduction zone, only a few bacterial species survive that can tolerate the toxic effects of hydrogen sulphide produced in this zone. The sulphate-reducing bacterium *Desulfovibrio desulfuricans* is the dominant species. The sulphate reducers oxidize relatively small organic molecules such as lactic acid and four-carbon dicarboxylic acids. These are produced by fermenting bacteria the symbiosis of which with the sulphate reducers is a prerequisite for the SR process. A simplified equation for the rather complex reaction is:

$$2CH_2O + SO_4{}^{2-} \rightarrow S^{2-} + 2H^+ + 2HCO_3{}^-.$$

As soluble ferrous iron usually becomes available in terrigenous sediments through reduction of ferric oxyhydroxides and oxides in the SR zone and the lower part of the nitrate-reduction zone, precipitation of metastable iron-monosulphides will occur instantaneously, which keeps dissolved sulphide concentrations low (<5 mM). The monosulphides are later converted to pyrite. In iron-poor marine carbonate oozes, the process will lead to elevated sulphide levels and produce sour gases.

The SR zone is considerably thicker than the nitrate-reduction zone because dissolved sulphate is about 3 orders of magnitude more abundant (27 mM) in oceanic bottom waters than nitrate. The combined thickness of the nitrate- and SR zones in organic-matter rich marine sediments is, however, often less than a few meters and may be as thin as 5 cm in certain regions (Finke et al., 2007; Murray et al., 1978; Reeburgh, 1983).

6.1.4. The carbonate-reduction zone

In each of the foregoing three oxidation steps, one of the main products of bacterial organic-matter decomposition is carbonic acid and its dissociated species bicarbonate, $HCO_3{}^-$, or carbonate, $CO_3{}^{2-}$. Below the SR zone, carbonate itself becomes one of the main oxidants for further bacterial organic-matter oxidation and leads to the production of methane in the carbonate-reduction (or methane-generation) zone.

Bacterial carbonate reduction appears to require hydrogen as a reducing intermediate, which is also derived in the process of bacterial organic-matter degradation (Berner, 1980). An example is the oxidation of ethanol, which yields acetic acid and hydrogen according to the reaction:

$$CH_3CH_2OH + H_2O \rightarrow CH_3COOH + 2H_2.$$

The CO_2 generated in this and the previous zones (according to a reaction of the general type $CH_2O + H_2O \rightarrow CO_2 + 4H$, which is enzyme-catalyzed and produces atomic H) will then be reduced by methanogenic bacteria such as *Methanobacterium thermoautotrophicum* in the redox reaction:

$$8H + CO_2 \rightarrow CH_4 + 2H_2O.$$

The last two equations taken together show that the process of methane formation can be viewed as a disproportionation reaction in which the valence of carbon in the organic matter is simultaneously raised to a higher (+4 in carbon dioxide) and a lower valence state (−4 in methane) (Goldhaber and Kaplan, 1974). Whether a direct transformation of organic compounds into methane by bacteria is possible, is still an open question. Some recent results suggest that certain groups of bacteria can use CO_2 and H_2 for methanogenesis (Newberry et al., 2004). Although methanogenic bacteria exist over a wide temperature range (0–75 °C: Zeikus and Wolfe, 1972), the optimum for individual species spans only a few degrees. For example, the thermophylic *M. thermoautotrophicum* has its optimal temperature range between 65 and 70 °C.

Methane production and sulphate reduction are not mutually exclusive processes (Claypool and Kvenvolden, 1983), but they seem to be fairly well separated, because CH_4 levels in the SR zone are low, and significant CH_4 production seems to start only after the disappearance of more than 80%, if not complete depletion of the dissolved sulphate (Sansone and Martens, 1981). Possible reasons for the lack of significant overlap are (1) that free hydrogen may not be available for CO_2 reduction in the presence of sulphate-reducing bacteria and (2) that the methane which may be produced in the SR zone is immediately oxidized to CO_2. Oxidation also inhibits the buildup of CH_4 by upward diffusion to any substantial levels in the SR zone.

6.1.5. The fermentation zone

The fifth oxidation zone involves bacterial fermentation reactions in which the oxygen contained in organic compounds is transferred to oxidize organic matter, simultaneously yielding CO_2 and CH_4. This may not be a separate step, as has been discussed for bacterial carbonate reduction, which requires fermentation processes in symbiotic reactions, as stated above.

6.1.6. The thermocatalytic decarboxylation zone

Bacterial activity more or less ceases at temperatures above 75 °C. Only few thermophylic populations still exist at higher temperatures. Organic-matter oxidation, however, continues spontaneously at temperatures above 75 °C under abiogenic conditions once the activation energy has been reached. These thermocatalytic reactions include the decarboxylation of organic acids according to the general formula:

$$R - COOH \rightarrow RH + CO_2$$

where the organic acid group, COOH, is converted to CO_2, and R− is the organic radical. This process is probably very important in generating aggressive acidic pore waters at depths that may be responsible for the dissolution of carbonates and feldspars, and for the generation of secondary porosity in sandstones, but is beyond the scope of this review, which is restricted to the realm of early diagenesis.

The biochemical reactions involved in bacterial metabolism are subject of a wealth of recent geomicrobiological studies but are still poorly understood. Probably all of them include enzymatic reactions. It is known, for example, that methanogens can only use acetic acid or acetate and shorter hydrocarbon molecules for conversion into methane (Mechalas, 1974). Because acetic acid levels in the pore waters of marine sediments are generally relatively low, other strains of bacteria must be present to produce the intermediate substances which are then being used in the methane-generating process. Concentrations of acetic acid and other volatile fatty acids are expected to be low at greater subsurface depths, due to rapid consumption by bacteria. The existence of large bacterial populations in deeply buried sediments older than 10 Ma has been surprising. It has been attributed to the increased generation of acetate with increasing temperature during burial (Wellsbury et al., 1997). Similarly, sulphate-reducing bacteria use relatively short-chained organic acids which must be provided by symbiotic bacteria breaking down larger organic molecules. Thus different interacting bacterial populations must be present at any one time in the SR and methane-generation zones that are involved in a series of simultaneous and partially symbiotic reactions. With the conventional techniques of pore-water analyses, we only see the end products of these chain reactions.

6.2. Suboxic diagenesis: Reaction-controlled pore-water profiles and mineralization products in suboxic pelagic to hemipelagic environments

The organic-matter content of hemipelagic sediments on the continental rise and adjacent parts of the oceanic crust is generally still relatively low and does not allow the early-diagenetic oxidation reactions to proceed beyond the oxic to suboxic stage (Fig. 9.30). In these environments, the oxidant demand leads to depletion of the free-oxygen reservoir but not the nitrate reservoir. According to Müller and Mangini (1980), early diagenesis will remain in the suboxic stage for sedimentation rates of <40 m per million years. With higher rates, increased organic-matter content will eventually lead to exhaustion of the nitrate reservoir and establishment of anoxic conditions of the SR zone. Nitrate reduction is accompanied by Mn and Fe reduction that liberate the divalent Mn^{2+} and Fe^{2+} ions to the pore water and make them available for incorporation in ferromanganese nodules and crusts (see below) and complex Mn- and Fe-bearing carbonate concretions (see Section 12.7)

In the Madeira Abyssal Plain in the East Atlantic, the invasion of turbidity currents into an environment of abyssal brown-clay sedimentation caused fluctuations in organic-matter concentrations that are reflected in colour variations between the brown clays ($<0.3\%$ C_{org}) and the olive-grey marl turbidites (1–2% C_{org}). As the colours reveal, diagenesis in the brown layers is oxic, whereas in the turbidites it is anoxic. An oxidation front moved down into the organic-matter-rich marl turbidites from the overlying pelagic sediment as long as the sediment was in diffusive contact with sea water. It turned the upper part of the olive-green turbidites yellow-grey indicating suboxic diagenesis (Colley et al., 1984), as confirmed by pore-water data of Wilson et al. (1985). Intercalated turbidites and pelagic brown clay in the western North Atlantic Nares Abyssal Plain show the same kind of heterogeneous diagenesis (De Lange, 1986) as in the Madeira Abyssal Plain.

A suboxic zone of unusual thickness (of several meters) in the shallow subsurface has been postulated for shallow shelf environments off New Jersey during the Paleocene–Eocene thermal maximum that are characterized by gigantic biogenic magnetite (Schumann et al., 2008).

6.2.1. Ferromanganese nodules and crusts

Manganese-nodule fields are among the more spectacular features that have been discovered on the deep-sea floor (Fig. 9.31). They vary largely in size (from micronodules to more than meter size), shape (from spherical to ellipsoidal to tabular), surface texture (from smooth to botryoidal to

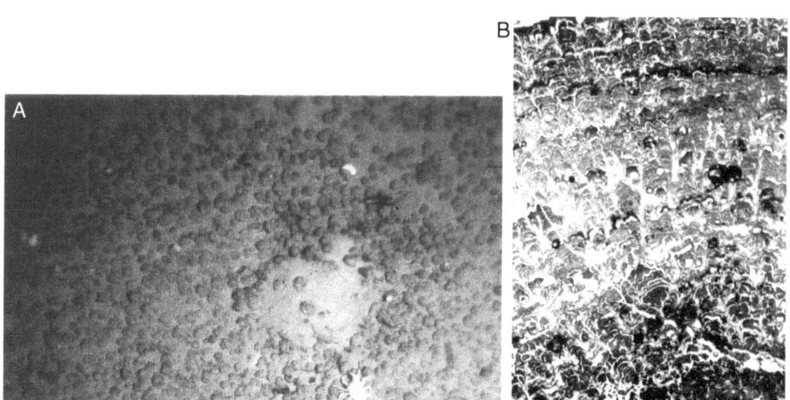

Figure 9.31 Ferromanganese nodule field within a sediment-filled channel from the top of the Fisher Ridge off Costa Rica, 3460 m water depth, East Pacific (from Frieling and Mrazek, 2005; with kind permission of the authors). (A) This underwater photo shows an area ($\sim$3 m^2) densely covered by tennis-ball size nodules of more or less uniform diameter (B) Section through a ferromanganese nodule with elevated manganese contents near the core and the margins (black stained).

rough) and abundance (Glasby, 2006). Micronodules are generally <1 mm in diameter and form around siliceous tests, calcareous microfossils or volcaniclastic fragments. Macronodules, in contrast, are >3–6 mm in diameter so that there is a size gap between 1 and 3 mm, which reflects the fact that the macronodules require a nucleus several mm in diameter. They precipitate around rock or bone fragments such as weathered volcanic rock, pumice, shark teeth or whale ear-bones, forming concentric laminae and layers of varying Mn and Fe concentration.

Ferromanganese nodules have attracted considerable economic interest because of the high metal concentrations of Co, Ni and Cu (up to 5% for these elements combined) in certain nodule provinces. They form in areas of low sedimentation rates mostly in deep water below the CCL, that is, in the realm of brown abyssal clay and biogenic siliceous ooze deposition, but also in pelagic carbonates close to the CCL, for example, in the Peru Basin in the East Pacific (Stackelberg, 2000) and in areas of strong bottom currents such as under the northward flowing Antarctic Bottom Water (AABW) in the SW Pacific. Low sedimentation rates are a prerequisite for Mn-nodule formation because of the required low organic-matter concentrations in sediment that facilitate suboxic diagenesis, but also because of the slow growth rates of the nodules. Growth occurs mainly at the sediment/water interface. There are three main sources of Mn in the nodules: sea water, early diagenesis and hydrothermal activity.

Three principal Mn-oxide minerals occur in the nodules and crusts: todorokite or 10 Å manganate (also called 10 Å manganite), birnessite or 7 Å manganate (7 Å manganite), and δMnO_2 or vernadite. The basic building block of most Mn-oxide minerals is the MnO_6 octahedron. The octahedra are combined by sharing edges and/or corners into a large variety of structures. In todorokite, triple chains of MnO_6 octahedra are arranged in frameworks that form large zeolite-like tunnels with dominantly square cross-section. The tunnels accommodate H_2O and/or cations such as K^+, Na^+, Ca^{2+}, Mg^{2+}, Ni^{2+}, Co^{2+}, Cu^{2+}, Zn^{2+}. Mn^{3+}, but also Ni^{2+} and Mg^{2+} substitute for Mn^{4+} in the octahedra to offset charges on the tunnel cations. Birnessite consists of randomly stacked MnO_6 octahedral sheets that alternate with layers of hydrated cations (e.g. Na^+, Ca^{2+}, Mg^{2+}). Because of the small grain size (in the 50–100 Å range) and poor crystallinity of birnessite, the crystal structure has been determined using synthetic birnessite-like phases (e.g. Post, 1999). In the synthetic precipitation of Na-birnessite, an expandable 10 Å phyllomanganate phase called "buserite" appears first; it collapses upon drying to 7 Å. This might also be a common precursor phase in natural Mn nodules before they dry out. It can be differentiated from todorokite by expansion of the interlayer spacing with dodecylammonium hydrochloride to 25 Å. Vernadite appears to be analogous to the synthetic phase δMnO_2. Manganese oxides also have a high adsorption capacity and can adsorb many of the cations mentioned above, but also anions (e.g. PO_4^{3-}). The diagenetic evolution from buserite to todorokite to δMnO_2 is a sequence of increasing degree of oxidation and dewatering.

The proportion of the diagenetically derived versus hydrogenous Mn and the type of diagenesis strongly affect the chemistry of the nodules (Glasby, 2006). Nodules with the highest metal concentrations of Co, Ni and Cu appear to form in areas of oxic diagenesis and have Mn/Fe ratios up to 5. Nodules formed in areas of suboxic diagenesis (such as in the Peru Basin) may have Mn/Fe ratios of up to 50, but lower concentrations of other metals. The metals enter the nodules via the buserite precursor phase, in which Ni^{2+}, Cu^{2+} and Co^{2+} substitute for the hydrated interlayer Na^+. Fe^{2+} cannot enter the buserite lattice because it would be oxidized by Mn^{4+} to insoluble FeOOH. In the belt of siliceous oozes in the NE Pacific between the Clarion and Clipperton fracture zones, the Fe^{2+} reacts with dissolved silica to form nontronite, an Fe-smectite ($R_{0.33}{}^{+} \cdot nH_2O$ $Fe_2{}^{3+}(OH)_2(Si_{3.67}Al_{0.33})O_{10}$; with R^+ being predominantly Na^+). This explains the very high Mn/Fe ratios in the nodules. The diagenetic component of the nodules in this belt increases from W to E and from N to S, as sedimentation rates and thus organic-carbon concentration increase into these directions. During suboxic diagenesis, Ni^{2+}, Cu^{2+} and Co^{2+} are released to the pore water and trapped in micronodules in the sediment.

The dominant form of Co in manganese nodules seems to be Co^{3+}, which has almost the same ionic radius of 0.53 Å as Fe^{3+} and Mn^{4+} and may substitute for these ions in either MnO_2 or FeOOH. Diagenetic processes also influence the surface texture of the nodules. Diagenetic nodules have botryoidal to rough surfaces, whereas hydrogenous nodules tend to have smooth surfaces. Hydrogenous nodules, which receive the elements for their authigenic components largely from seawater, have Mn/Fe ratios of about unity and Ni + Cu concentrations of $<1\%$. The concentration of dissolved Mn in the open ocean ranges from 0.1 to 3 nmol/kg (Bruland et al., 1994), which is still above the equilibrium concentration with respect to MnO_2 or MnOOH, which is possible due to the slow oxidation rate of Mn^{2+} in solution. The red-clay host sediments of the hydrogenous ferromanganese-nodule province of the SW Pacific have considerably lower Mn, Ni, Cu and Ba contents but higher Fe and Co contents than the siliceous-ooze host sediments of the nodule belt north of the equatorial Pacific between the Clarion and Clipperton fracture zones (Stoffers et al., 1981).

Which are the mechanisms that keep the nodules growing at the sediment surface and do they eventually get buried? It seems there is agreement among experts in the field that bioturbation by benthic megafauna is an essential process to keep the nodules at the sediment surface. However, as mentioned, if the growth rates of the nodules are lower than the local sedimentation rates, only little bioturbational support may be required and both mechanisms, bioturbation and low growth rates, may act together in maintaining the nodules at the surface. In areas of low-density endobenthonic communities such as under the AABW in the SW Pacific, the bottom current itself may be instrumental in preventing the nodules from getting buried (Glasby et al., 1983). However, nodules do get buried as studies in Peru Basin have shown (e.g. Stackelberg, 2000), where the combined abundance of nodules in the subsurface to a depth of 2.5 m is 1.25 times that at the surface. Buried Mn nodules have also been found in cores of the DSDP. The surface textures of buried nodules may show evidence for partial dissolution.

Co-rich ferromanganese crusts and pavements form on seamounts and current-swept submarine plateaus (such as the Blake Plateau in the West Atlantic) in water depths > 1000 m. According to their Mn/Fe ratios in the range of 1–2, they are hydrogenous deposits with δMnO_2 being the main Mn mineral. Their high Co (in the 0.5–2% range) and Ni and low Cu contents reflect the association with δMnO_2. To be of potential economic interest, a minimum Co concentration of > 0.8% and an average thickness of > 40 mm are required. Shallow-water ferromanganese concretions have been reported from a number of marginal marine basins such as the Baltic, the Black and Kara Seas and the Jervis Inlet, British Columbia. They consist mostly of todorokite and have low transition element metal concentrations.

Hydrothermal manganese deposits occur at MOR crests, in back-arc spreading centres and ocean island volcanoes. They represent late-stage hydrothermal activity at temperatures of 5–20 °C. Mineralogically they consist of todorokite and/or 7 Å manganate with high Mn/Fe ratios and low Co, Ni, Cu concentrations. Basal carbonates overlying the MOR basalts typically display a pinkish stain caused by the precipitation of Mn oxides, indicating that hydrothermally derived Mn-rich fluids that rise as plumes from the MOR affect the oldest deposited sediments on the ridge crest tens to hundreds of kilometres away from the median valley. About 90% of all Mn supplied to the ocean has a hydrothermal source; the remainder is river- and dust-borne (Glasby, 2006).

Modern methods of dating manganese nodules and determining their growth rates are based on the $^{230}Th/^{232}Th$, $^{231}Pa/^{230}Th$ and ^{10}Be methods. The half life of ^{10}Be is 1.5×10^6 years, which enables dating up to 15 Ma old samples using the accelerator mass spectrometer (AMS). Growth rates of deep-sea hydrogenous nodules are between a fraction of a mm per million years to a few mm per million years. These are among the slowest growing substances on Earth. Co-rich crusts also accumulate at slow rates of 1–2 mm per million years. Diagenetic nodules of the Peru Basin achieve higher rates of between 100 and 200 mm per million years, whereas shallow-water ferromanganese concretions have growth rates up to 0.2 mm per year, or up to 5 orders of magnitude higher than those of the slowest growing hydrogenous nodules. High-resolution dating has provided evidence for strong fluctuations in nodule growth rates, with high rates mainly correlating with interglacial stages and low rates with glacial stages. Their laminations, that reflect compositional variations, are in accordance with intermittent growth. In the North Pacific nodule field between the Clarion and Clipperton fracture zones, growth started at the Oligocene/Miocene boundary. Growth of crusts on the Magellan Seamounts in the West Pacific started in the Late Cretaceous. Interruptions of growth are common with major regional hiatuses occurring at the Cretaceous/Paleocene and Eocene/Oligocene boundaries (quoted after sources in Glasby, 2006).

6.2.2. Phosphorites

Early diagenesis plays an important role in the formation of phosphorites and phosphatic sediments, many of which occur in deep-water settings. Mineralization reactions forming phosphatic sediments are significant sinks for phosphorus. Some ancient deposits such as the Permian Phosphoria Formation of the western US are of astonishingly large size and represent significant phosphorus removal from the ocean compared to present annual river supply. Phosphorites typically occur in organic-matter-rich sediments (and their ancient equivalents in the form of black shales), siliceous oozes (or cherts), but also in glauconitic sediments (green sands and shales). Phosphorites associated with carbonates are rare and usually of low grade. These deposits are mostly thin-bedded. However, in the Phosphoria Formation, some extraordinary thick beds (1–>2 m) occur.

Giant phosphorite deposits, which are important economic phosphorous resources, occur throughout the geologic column and are exclusively marine. They include Precambrian occurrences in China, Cambrian deposits of Kazakhstan, Ordovician examples from the Russian Platform (Estonia, St. Petersburg region), the Permian Phosphoria Formation of Idaho, Montana, Wyoming, Utah and Colorado—which is the largest deposit of the world—Jurassic deposits in Mexico, Jurassic–Cretaceous deposits of the Russian Platform, Late Cretaceous–Paleocene occurrences in North Africa, Israel and Jordan, the Miocene Monterey Formation of California, the Plio–Pleistocene Bone Valley Formation of Florida, and modern outer-shelf and upper-slope regions off Peru and Namibia. Estimates for the Phosphoria Formation range from 7×10^{14} to 1.7×10^{15} kg P_2O_5. For comparison, the modern oceans contain 3×10^{14} kg dissolved P_2O_5 (or about 1/2–1/5 of that contained in the Phosphoria Fm.). The duration of the deposition of the Phosphoria Formation is estimated to be 15×10^6 years, which gives an average depositional rate of 10^8 kg per year. By comparison, the present total river supply to the world ocean is estimated at 4.5×10^9 kg per year. In other words, about 1/50 of the total annual river supply to the ocean would have been removed by deposition in the Phosphoria Sea of the Permian.

The predominant P-bearing mineral is carbonate-fluor apatite (CFA) or its cryptocrystalline form collophane, which has an approximate formula of $Ca_{10}(PO_4, CO_3)_6F_{2-3}$. Small amounts of Na^+ and Mg^{2+} substitute for Ca^{2+}

and some SO_4^{2-} for CO_3^{2-}. The cryptocrystalline form of carbonate-hydroxyl apatite is called dahllite. The mineral phosphorite $Ca_3(PO_4)_2$ does not seem to play a major role in natural phosphate deposits.

Modern phosphoritic sediments occur in areas of strong upwelling driven by offshore winds of the subtropical west coasts of continents. These areas are associated with a high primary productivity and burial of particulate organic matter under the oxygen-minimum zone on the upper slope and outer shelf. Kasakov (1937) was the first to recognize the association of phosphorite deposits with areas of upwelling and his study has been seminal for later work, although his suggestion that direct precipitation of apatite occurred in the water column is not shared by modern hypotheses of phosphogenesis (for a review, see Glenn et al., 1994). Phosphorite deposits are thus not chemical sediments in the strict sense. Their origin is considered predominantly diagenetic. Manheim et al. (1975), who studied modern deposits offshore Peru, showed that the oxidation of particulate organic matter settling through the water column caused the oxygen-minimum zone in the area to impinge on the sea-floor between 100 and 500 m. It correlates with a pH minimum due to intense CO_2 production. The P_2O_5 concentration in the water column (mainly as HPO_4^{2-} and $H_2PO_4^-$) in the oxygen-minimum zone reaches 0.02 μM (or 2 μg-at PO_4-P/L). Maxima of dissolved PO_4^{3-} in interstitial waters of DSDP and ODP drill sites of the region occur around 100 mbsf (discussed in Section 7.2) and are due to P release during organic-matter decomposition and the subsequent removal in authigenic apatite.

The diagenetic origin of phophorite deposits in areas of upwelling is widely accepted today. However, the subsurface depth of the formation of phosphorite deposits is most likely not associated with the PO_4^{3-} maximum in the pore waters of the drill cores, which reflects only minor apatite precipitation below. Precipitation of the phosphorite deposits occurs more likely within centimetres below the sediment/water interface (Glenn and Arthur, 1988) in more or less uncompacted, organic-carbon-rich muds. The sites of precipitation may have to remain in diffusive communication with the bottom water to enhance resupply of dissolved phosphate. The apatite content is a function of the residence time of the sediment close to the seafloor. Offshore Peru and Baja California, a spike of anomalously high dissolved PO_4^{3-} near the sediment/water interface is present, associated with the earliest precipitated CFA. Its origin is still debated. It may be related to a number of processes, including suboxic bacterial degration of organic matter, dissolution of fish debris, "iron-pumping" by which Fe-oxides are reduced and the adsorbed phosphate is released to the pore water, and other mechanisms. Shallow-precipitated authigenic apatite is in line with the formation of phosphoritic hardgrounds, concretions or as individual phosphate grains. It makes the exhumation of originally muddy sediments by bottom-current winnowing during sea-level fall and the concentration in granular phosphoritic beds more easily understandable. Granular deposits also include megadeposits such as the Cretaceous giant phosphorite sand waves and glauconitic greensands of Egypt (Glenn and Arthur, 1990). Glauconite formation is frequently associated with the desorption of phosphorous to the pore waters during reduction of FeOOH. The reduced iron becomes incorporated in glauconite.

Not all phosphorite occurrences are associated with upwelling and high primary productivity. On the eastern shelf of Australia, Pleistocene to Holocene phosphorite nodules occur on a continental margin that lacks prominent coastal upwelling and the geochemial characteristics associated with it. The organic-carbon concentration of the sediments is $<0.5\%$, compared to values off Peru that reach up to 20%. Very efficient organic-matter utilization by bacteria may in part compensate for the low concentrations, and low sedimentation rates and strong reworking may result in a long residence time of the sediment close to the seafloor.

6.2.3. Glauconite

Glauconite formation takes place in similar low-sedimentation-rate open-marine environments with limited terrigenous input that are characteristic of phosphorite deposition. The association of glauconitic sediments with phosphorite deposits is widespread. Odin and Matter (1981) introduced the new term "glaucony" for green grains consisting of mixed-layer clays of a ferric smectite and glauconitic mica. Glauconite deposits are known from all but the polar oceans. A few lacustrine occurrences have been discovered. Optimal conditions for glauconite formation appear to exist on the outer shelf/upper slope but many deep-water occurrences down to 2000 m water depth have been reported (e.g. Odin and Stephan, 1981), some with up to 35% glauconite as, for example, E of Patagonia. Not all of the deeper occurrences can be explained by redeposition or tectonic subsidence of the original site of formation.

Glauconite formation proceeds in the shallow subsurface (down to a few decimetres) under oxic to suboxic conditions and is essentially complete after about a million years (Odin and Fullagar, 1988). Suboxic conditions are required at least in the later stages of glauconite formation because of the incorporation of divalent iron in glauconite. The divalent iron is derived from the reduction of Fe-oxyhydroxides in the suboxic zone. This is reflected by the structural formula for glauconite:

$K_{(x+y)}(Fe^{3+}, Al, Mg, Fe^{2+})(OH)_{\sim 2}(Si_{4-x}Al_x)O_{10}$ with $Fe^{3+} \gg Al$ and $x = 0.2$–0.6; y (=sum of the dioctahedral cations) = 0.4–0.6 (with $Mg > Fe^{2+}$). Glauconite formation involves the alteration of suitable porous substrates of a large compositional variety (e.g. foraminifers, ostracodes, bryozoans, sponge spicules, fecal pellets, volcaniclastic debris and even quartz, feldspar and mica) by the uptake of potassium into smectitic clays from seawater. It is not clear whether the clays originally filled the pores and were subsequently altered or were precipitated as authigenic minerals in the glauconitization process. The predominantly rounded shape of the grains points to foraminifers as a widespread host for glauconite formation. Odin and Fullagar (1988) refer to the gap in iron content between illitic ($<10\%$ Fe_2O_3) and glauconitic minerals ($>15\%$ Fe_2O_3) as evidence that glauconite minerals do not form by the progressive Fe substitution for Al in octahedral sites of existing smectite. Instead they argue that the Fe^{3+} is fixed in the mineral structure of the precursor smectite at an early stage prior to the incorporation of K^+ (because the iron content shows little or no change from smectite to glauconitic mica) and that the uptake of Fe^{2+} at a late stage is a minor compositional shift. The precursor smectite in fact is a ferric montmorillonite, which is unstable. Together with kaolinite it is involved in the reaction: ferric montmorillonite + kaolinite + $K^+ \rightarrow$ Fe-beidellite – glauconitic mica mixed layer [where $Fe^{3+} \gg$ Al in Fe-beidellite $Na_{0.33}nH_2O(Fe^{3+}, Al_2)(OH)_2(Si_{3.67}Al_{0.33})O_{10}$]. This is a true equilibrium reaction that is reversible during weathering. The termination of glauconite formation during early diagenesis at shallow subsurface depth reflects the fact that the supply of K^+ by diffusion from overlying seawater is cut off.

6.3. Anoxic hemipelagic sediments

Approaching the upper, steeper part of the continental margin (the continental slope), sedimentation rates increase compared with the rise. Rates up to 500 m per million years and more are no exception, and initial organic-matter concentrations commonly exceed 2–3% organic carbon (C_{org}) because organic particles are degraded less during the shorter sinking time and distance. In addition to the terrestrial organic matter delivered with the terrigenous suspended sediments and by turbidity currents, marine organic matter is supplied at increasing rates by high biogenic surface production in areas of upwelling of nutrient-rich deeper waters. Where the oxygen-minimum zone of the water column that is associated with the settling organic debris intercepts the slope, organic-carbon concentrations may reach record levels of $>20\%$. The oxygen-minimum zone varies in depth but in many regions of the ocean is centered above 500 m water depth and may be as shallow as a 100 m. Rapid burial also enhances the initial preservation of organic matter.

In this environment, benthic organisms thrive and become bigger. Thus bioturbation penetrates deeper and is more effective making animal irrigation competitive with molecular diffusion. The role of diffusion in altering concentration anomalies is also diminished or suppressed, because sedimentation rates exceed diffusion rates. Under these circumstances, anoxic diagenesis leads to methane generation. In areas of the slope below 500 m water depth, pressure and temperature near the sea-floor are in the gas-hydrate stability field. In wide areas on the slope, gas hydrates are encountered because methane generation is sufficiently intense to permit hydrate formation. The presence of hydrates can be geochemically recognized by specific pore-water anomalies (see Section 7). Another trademark of anoxic sediments is the early-diagenetic precipitation of authigenic carbonates (see Section 12.2 and following).

6.4. Pore-water/depth profiles in anoxic sediments

The sequence of bacterial organic-matter decomposition reactions, outlined in the section on early-diagenetic organic-matter oxidation, is best illustrated by changes in the pore-water composition in rapidly deposited continental-margin sediments. During burial, the sediment passes rapidly through the (1) oxidation and (2) nitrate-reduction zones and then experiences anoxic diagenesis in the (3) SR, (4) carbonate-reduction, (5) fermentation and (6) thermocatalytic decarboxylation zones. The main chemical species released to the pore water from the microbial organic-matter breakdown and concomitant reduction of oxidants are the nutrients ΣCO_2 (including the species CO_2, H_2CO_3, HCO_3^- and CO_3^{2-}), phosphate, ammonia and sulphide. $O_{2(aq)}$, NO_3^-, SO_4^{2-} and CO_3^{2-} are the oxidants that are consumed in the process (Table 9.2). CO_2 production occurs in all six zones, CO_2 consumption from zone 4 downward. A distinction between the CO_2 released and the CO_2 consumed at different stages is possible due to the strong isotopic fractionation effects associated with the bacterial methane generation beginning in zone 4.

The carbon-isotopic values change through the six organic-matter decomposition zones during burial from negative to positive and back again to negative $\delta^{13}C$ values (Fig. 9.32). During the first three organic-matter degradation steps, negligible isotopic fractionation occurs. The CO_2 (or HCO_3^- and CO_3^{2-}) released has about the same isotopic composition as the parent marine organic matter, about −25‰ $\delta^{13}C$ (relative to the PDB standard). The pore-water CO_2 in the upper three zones therefore gradually approaches $\delta^{13}C$ values of −20‰ to −25‰ (Fig. 9.32, especially Site 174A). In the carbonate-reduction and fermentation zones, the disproportionation of organic matter into CH_4 and CO_2 is associated with a strong kinetic isotope-fractionation effect. The CO_2 with light carbon is preferentially reduced to CH_4 (Rosenfeld and Silverman, 1959). The C-isotopic composition of the CH_4 generated is about 70‰ lighter than the carbon of the parent material and may attain $\delta^{13}C$ values as negative as −90‰ to −100‰. Through a Rayleigh-distillation process, the residual CO_2 is progressively enriched in ^{13}C reaching positive δ-values (Fig. 9.32) as high as +15‰ to +25‰ (e.g. Curtis et al., 1972). As the dissolved carbonate becomes progressively heavier, so does the methane that is

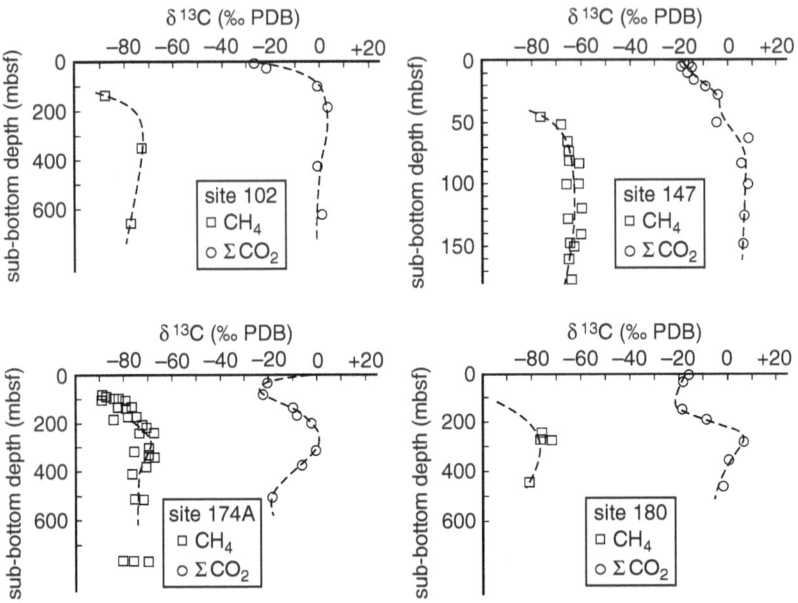

Figure 9.32 $\delta^{13}C$ depth trends in CH_4 and CO_2 in four DSDP drill sites (102, Blake-Bahama Outer Ridge; 147, Cariacou Trough; 174A, Astoria Fan; 180, Aleutian trench floor; redrawn and modified from Claypool and Kaplan, 1974).

produced from it at deeper levels. $\delta^{13}C$-values as high as +36‰ to +38‰ have been measured for CO_2 coexisting with methane as heavy as −41‰ in deeper parts of DSDP Sites 568 and 570 on the Guatemalan trench slope (Fig. 9.33C) (Claypool et al., 1985), whereas methane of biogenic origin is usually lighter than −55‰. $\delta^{13}C$-values heavier than −45‰ are characteristic of thermogenic gas (Schoell, 1983). However, methane produced from acetate is about 20‰ heavier than that produced from CO_2 (Jenden and Kaplan, 1986) and might be the source for anomalously heavy $\delta^{13}C$ values in these sites. There is no evidence for upward migration of thermocatalytic methane off Guatemala.

The $\delta^{13}C$ curves for both methane and dissolved carbonate show parallel trends (about 70 δ-units apart) with depth (Figs. 9.32 and 9.33C). In many cases, these curves are characterized by a decrease in $\delta^{13}C$ at greater sub-bottom depths, due to the release of relatively light carbon from the breakdown of organic matter by thermokatalytic decomposition reactions at temperatures exceeding 75 °C. These reactions are not associated with carbon-isotope fractionation, and release CO_2 with the same negative $\delta^{13}C$ values as the source organic-material.

In order to avoid the effects of gas hydrates on pore-water composition, which will be discussed in Section 7, the relatively shallow-water ODP drill

Site 724 on the slope of the Oman margin (ODP leg 117) drilled in 593 m water depth, which is outside the hydrate stability field, is used to show the effects of anoxic diagenesis on pore-water composition. Similar shallow anoxic sites are located on the West African margin (ODP leg 175).

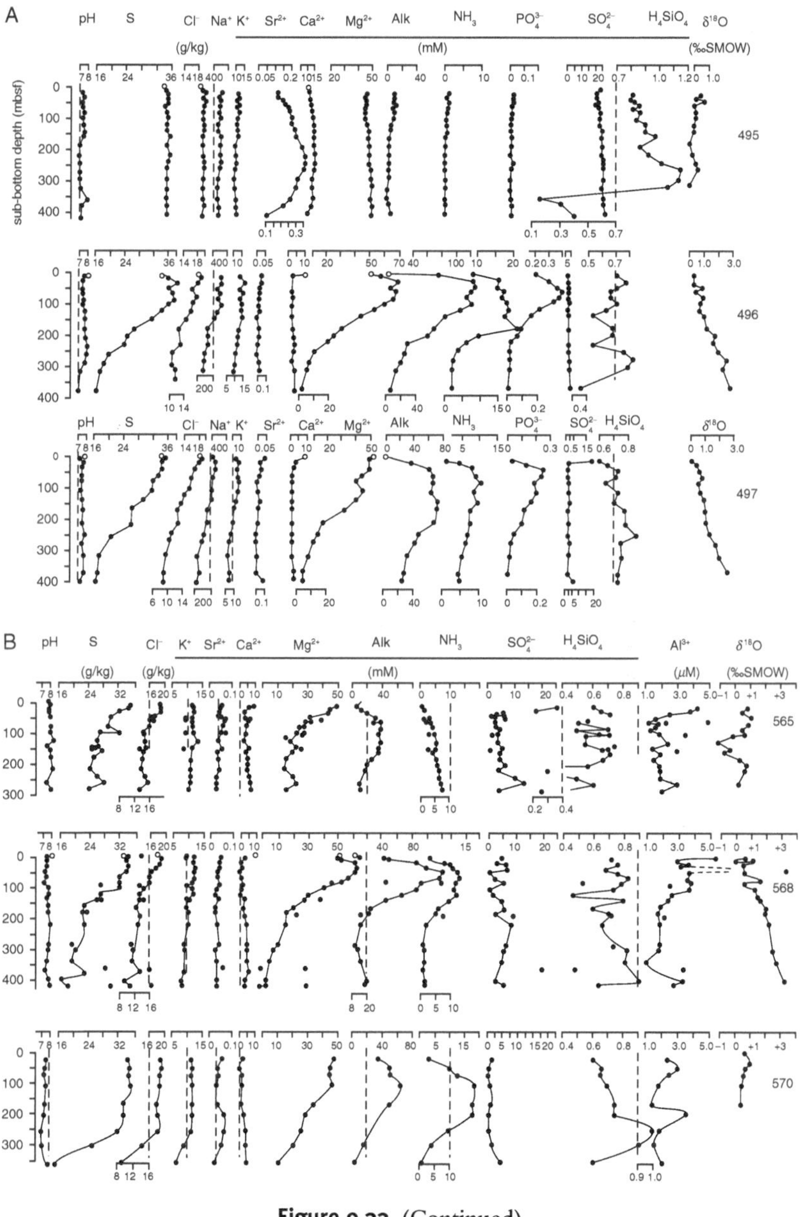

Figure 9.33 (Continued)

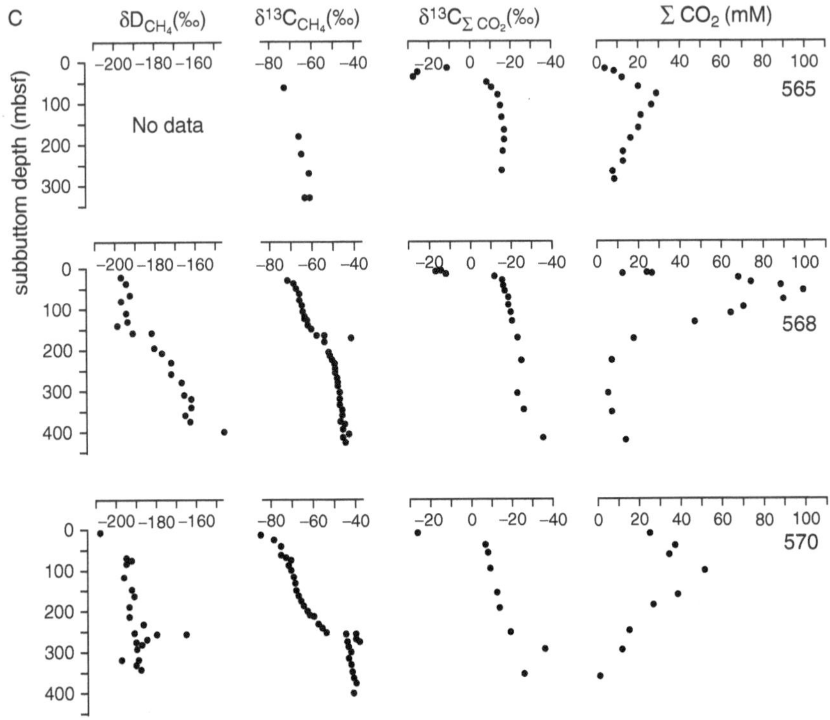

Figure 9.33 Pore-water chemistry and isotopic composition of DSDP drill-sites from the Middle America trench slope off Guatemala and Costa Rica. (A) Pore-water chemistry and isotopic composition of DSDP drill-sites 495, 496, 497 (DSDP Leg 67: Harrison et al., 1982) on the Middle America trench slope off Guatemala and oceanic crust of the Cocos plate (Site 495). Site 495 contains between 1% and 2% organic carbon and is intermediate between suboxic and anoxic diagenesis. It displays slight enrichments in ammonia, phosphate and alkalinity and a slight decrease in sulphate near the top of the sediment column. Although the changes are minor, they show that organic-matter decomposition reactions do in fact occur at this site. Note decrease in pore-water silica concentration at the opal-A to opal-CT transition. (B) Pore-water chemistry and isotopic composition of DSDP drill-sites 565, 568 and 570 (DSDP Leg 84: Hesse et al., 1985) on the Middle America trench slope off Guatemala. (C) Carbon- and hydrogen-isotopic ratios and total dissolved inorganic-carbon concentrations for Sites 565, 568, 570 (from Claypool et al., 1985). Squares for Sites 565 (continental slope off Nicoya Peninsula, Costa Rica) and 568 identify *in situ* water samples.

Sedimentation rates of Site 724 are intermediate to high, ranging from 60 to 120 m per million years in most of the hole (Shipboard Scientific Party, 1989). It does not display the typical decrease in chlorinity seen in hydrate-bearing sections (see Section 7). Below 80 mbsf, chloride values remain close to the bottom-water value. The maximum at about 50 mbsf is difficult to explain, as it exceeds considerably the maximum associated with

Pleistocene sea-level lowstands related to removal of fresh-water from the ocean and storage in continental ice sheets (McDuff, 1985; Schrag et al., 1996; see below). Depletion of sulphate occurs at about 50 mbsf. Reappearance of detectable although minor concentrations of sulphate at somewhat greater depth (Fig. 9.34) may be due to upward migration from a source below the drilled section (Pedersen and Shimmield, 1991). The depth of sulphate depletion depends on the organic-matter concentration but also on the nature of the organic matter. On the Peru continental margin (ODP leg 112), sulphate-depletion gradients are more than twice those on the Oman margin for comparable organic-matter concentrations, suggesting that the organic matter, which is dominantly marine in both regions, is more highly reactive in the former than in the latter (Pedersen and Shimmield, 1991). On the Peru margin, the biogenic component of the sediment is diatom-dominated; on the Oman margin, coccolithophorids

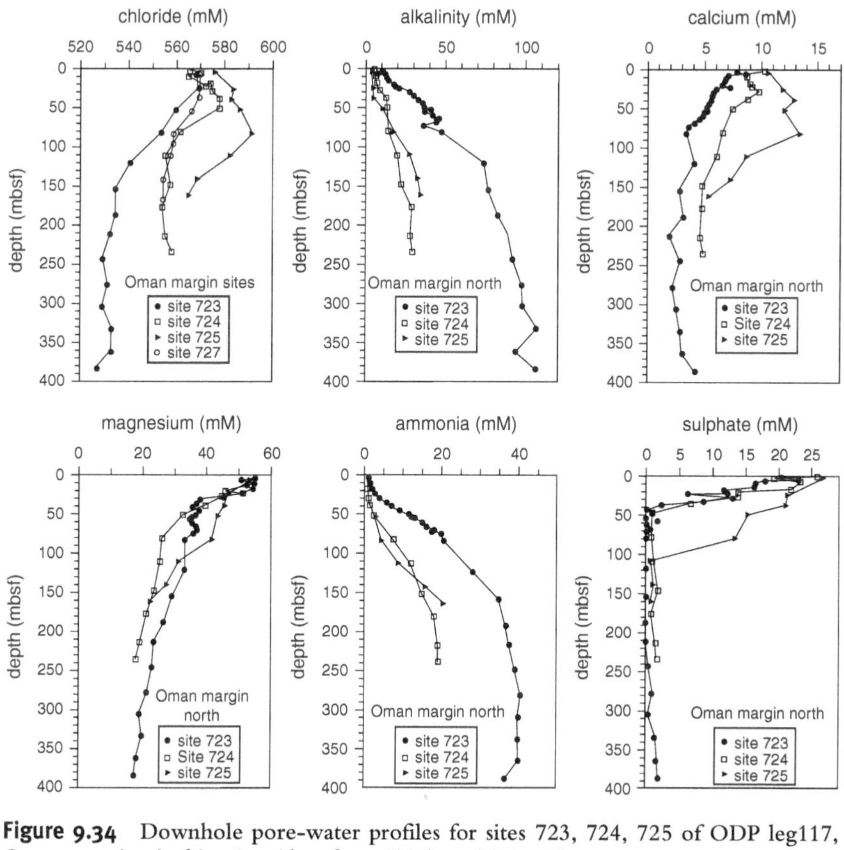

Figure 9.34 Downhole pore-water profiles for sites 723, 724, 725 of ODP leg117, Oman margin, Arabian Sea (data from Shipboard Scientific Party, 1989; Pedersen and Shimmield, 1991).

predominate. As will be shown for hydrate-bearing sediments, the gradients also depend on an upward methane flux and anaerobic methane oxidation (AMO) at the base of the SR zone.

6.5. Euxinic sediments

The present-day oceans are characterized by vigorous circulation driven by thermohaline convection that is largely initiated by the sinking of cold polar surface water which spreads out at depth through all ocean basins. The turnover rate leads to an ocean mixing time of approximately 1000 years. Under these conditions pelagic sediments are well oxidized because the supply of O_2 to the bottom waters is faster than its consumption by the oxidation of organic matter. At present, oxygen-free euxinic surface sediments are therefore restricted to semi-enclosed stagnant basins such as the Black Sea (below 200 m water depth), the bottom of deep troughs such as that of the Kurile-Kamchatka deep-sea trench or the Cariacou Trough in the southern Caribbean Sea, but include also some shallower regions where the water of the oxygen-minimum zone has become fully oxygen-depleted. They are characterized by finely laminated, organic-matter rich sediments that lack any trace of bioturbation. In contrast anoxic sediments that lost their oxygen during early diagenesis, sometimes called "gyttjas", usually show signs of bioturbation. The vertical position of the stages of organic-matter oxidation in euxinic as compared to ventilated basins is shown in Fig. 9.29.

Well-aerated bottom waters have not prevailed in the oceans at all times and, as recently as during the Plio-Pleistocene, sapropel layers in the eastern Mediterranean Sea record intervals of periodic stagnation. Ryan and Cita (1977) estimated that the up to two dozens layers comprise a cumulative time of about 40,000 years deposited since 5 Ma, that is, most were brief and lasted only a few thousands of years. Their organic-carbon (C_{org}) content varies between 1% and 30% (Böttcher et al., 1998) (average: 4%); their cumulative thickness in the eastern Mediterranean is 2 m on average. They occur over an area of 0.5×10^6 km^2 so that the total carbon removal amounts to 1.6×10^9 kg per year, compared to 1.2×10^9 kg per year in the pelagic sediments of all other ocean basins together during the same time. Ryan and Cita (1977) speculated that sapropel formation may have correlated with the end of glacial episodes when increased influx of fresh water into the basin from melting glaciers formed a less salty water lid in the basin that prevented turnover, causing periodic anoxia in the bottom water. However, the youngest sapropel layer S_1 has been deposited during the Holocene warm Climatic Optimum between 9 and 6 ka BP, suggesting that the role of ice caps controlling the initiation of sapropel formation should not be overestimated. Rohling and Hilgen (1991) established a close correlation between sapropel formation and orbitally forced climate variations that is supported by detailed analyses of cyclic geochemical variations in sapropel-bearing sediments sampled during ocean drilling in the eastern Mediterranean (Wehausen and Brumsack, 2000). Times of sapropel formation coincided closely with minima in the 21,000 year precessional cycle which occur at times when the perihelion occurs in the northern hemisphere summer, causing a maximum summer insolation and a minimum winter insolation. The resulting increased seasonal and land/sea temperature contrasts then enhanced the summer monsoonal circulation, which led to increased nutrient-rich fresh-water discharges of the Nile river into the eastern Mediterranean. These discharges alone, however, seem to be insufficient to have caused stagnation in the basin and complex additional atmospheric circulation changes in the Mediterranean region must be invoked (Rohling and Hilgen, 1991). The trends of the pore-water chemical profiles of the sapropel-bearing ODP drill holes are principally not different from those of sediments undergoing anoxic diagenesis, because diffusion smoothes out the differences between sapropel-bearing and sapropel-barren layers during early diagenesis, but in many Mediterranean drill sites they are modified by evaporite dissolution and brine advection from the underlying Messinian salt.

Numerous episodes of sapropel formation have been recorded during the warm climates of the Cretaceous anoxic oceanic events (AOEs) reported in the Proceedings of the ODP (see Weissert, 2011, this volume). The geochemical study of Wortmann et al. (1999) of Aptian black-green shale cycles in the deep-water Rhenodanubian Supergroup of the East Alps provided strong evidence based mainly on barium, silica and manganese distributions that the micro-laminated black muds were deposited under stagnant, anoxic conditions during low-productivity episodes, whereas the bioturbated hemipelagic green muds were most likely deposited in a suboxic environment during high productivity episodes. The cyclicity likely followed Milankovitch periodicities, but the precision of the dating methods precluded to constrain the studied age interval sufficiently. It was concluded that diagenesis had little effect on the distribution of the key elements with the exception of Mn.

In the Precambrian, the oceans were anoxic before the Great Oxidation Event that took place around 2.4 Ga (Sverjensky and Lee, 2010).

7. Gas-hydrate Bearing Sediments

7.1. Crystal chemistry, stability and evidence for the occurrence of natural-gas hydrates

Gas-hydrates are ice-like substances that consist of water and a gas. Their crystal structure contains polyhedra formed mostly by pentagonal H_2O-rings, in contrast to the hexagonal water-rings in normal ice that display a honeycomb pattern (Fig. 9.35). The polyhedra of structure-I hydrates are

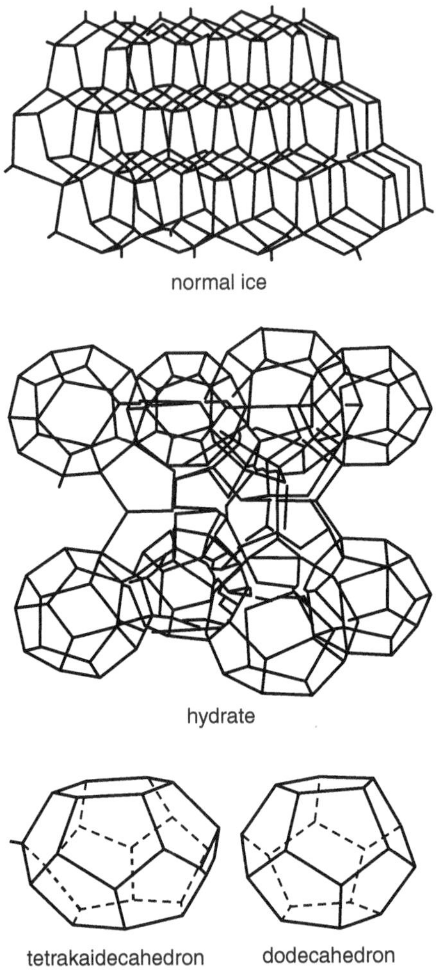

Figure 9.35 Structure of normal ice and hydrate structure I (redrafted from fig.1 in Hesse, 2003).

12-sided and 14-sided cages, called pentagonal dodecahedra and tetrakaidecahedra, 5.1 and 5.8 Å in diameter, respectively, which accommodate the gas molecules. From the cage-forming capacity of the hydrate structure, which is somewhat similar to that of zeolites, the synonym clathrate is derived. Most common among naturally occurring hydrates are CH_4 and CO_2 as guest molecules in the cages, but ethane (C_2H_6), hydrogen sulphide (H_2S) and nitrogen (N_2) do occur (e.g. Kastner et al., 1995; Swart et al., 2000). Higher hydrocarbons up to iso-butane fit into the cages of structure-II (Sloan, 1998). An ideal methane hydrate-I structure, in which all cavities are occupied, contains 46 water molecules, which form 8 cages, 2 dodecahedra and 6 tetrakaidecahedra, giving it the chemical formula $CH_4 \times 5\frac{3}{4}H_2O$. In natural hydrates, some of the cavities are not filled, making $CH_4 \times 6H_2O$ a good approximation for a methane hydrate at relatively high pressure.

The presence of the gas molecules confers a higher stability to the hydrate structure than normal ice, due to the formation of hydrogen bridges or van-der-Waals bonds between the gas molecules and the water molecules of the host structure, while no true chemical bonding is involved. The stability of the hydrates increases with increasing P, in contrast to normal ice, and is extended to +30 °C at 7–8 km water depth (~70–80 MPa, Fig. 9.36). Bottom-water temperatures on the present-day deep ocean floor are low (below 10 °C at 500 m) so that most of the area of the oceans below 500 m water depth is within the hydrate stability field. At the low temperatures of the Arctic Ocean, hydrates are stable at depths as shallow as 200 m.

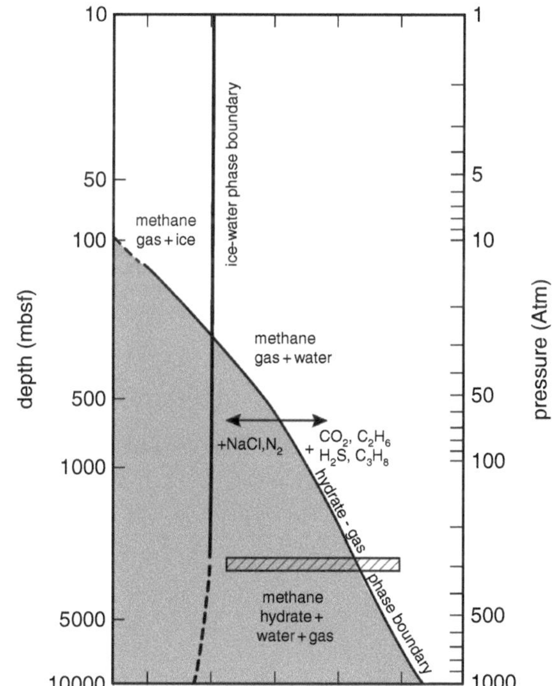

Figure 9.36 Stability field of methane hydrate (modified from Kvenvolden, 1998, fig. 3). The presence of CO_2, H_2S and higher hydrocarbons raises the stability of methane hydrate and the presence of N_2 or NaCl lowers it as indicated by arrows.

The area of the ocean floors where hydrates actually occur, however, is much smaller than the area of hydrate stability and restricted to regions of the continental slopes and rises where organic-matter-rich sediments abound, excluding the vast organic-matter-deprived areas of the subtropical gyres on the deep-sea floor oceanwards of the rises and trenches.

Thermodynamically, the pressure dependence of hydrate stability is the consequence of a negative change in molar volume, ΔV, during the formation of gas hydrate from water and gas (McIver, 1981), which is pressure-dependent:

$$CH_{4(aq)} + nH_2O_{(liq)} \leftrightarrow CH_4nH_2O_{(solid)}$$

$$\Delta V = nV_{hydrate} - V_{CH_4(aq)} - nV_{H_2O(liq)} = 22.68n - 34.5 - 18.02n\left(cm^3/mol\right)$$

At low pressure (100 kPa), $n = 7$ and $\Delta V = -1$ cm^3/mol. Through an increase in pressure, the occupancy of the cavities increases, approaching the ideal formula value of $n = 5.75$, for which $\Delta V = -8$ cm^3/mol.

Evidence for the occurrence of submarine gas-hydrate zones is provided by the bottom-simulating reflector (BSR) on seismic profiles, a strong, seafloor-parallel reflector at the hydrate base that separates the poorly reflecting, relatively transparent hydrate zone from normally reflective sediments below (Fig. 9.37). It shows a phase reversal indicative of the presence of free gas below the hydrate zone (Shipley et al., 1979). A BSR that has been found associated with the opal-A → opal-CT conversion

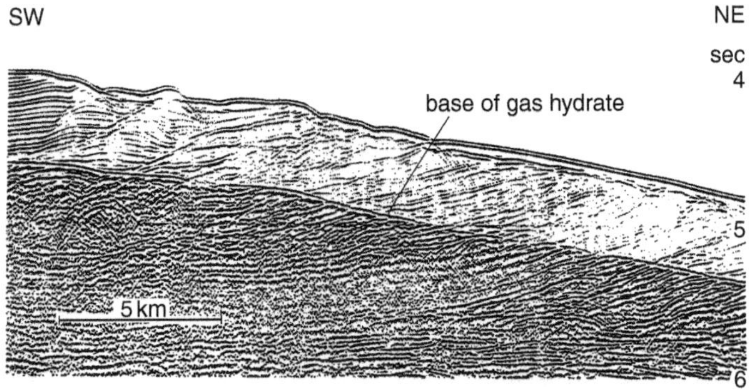

Figure 9.37 Multichannel seismic profile from the crest and the eastern flank of the Blake Ridge, showing the base of gas-hydrate marked by a strong bottom-simulating reflector (BSR) and lowered amplitudes caused by the presence of hydrates, which reduces the impedance contrast between different lithologies. Vertical scale: Two-way seismic travel time in seconds (redrafted from fig. 4 from Hesse, 2003; modified from Shipley et al., 1979).

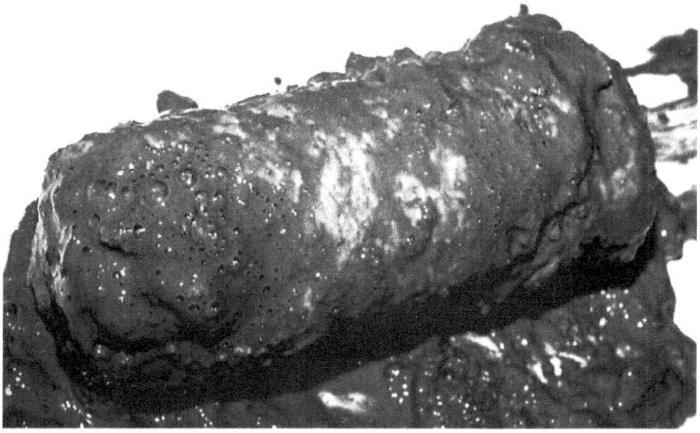

Figure 9.38 Massive hydrate. 14 cm long piece recovered from Site 997, Blake Ridge (fig. 5 from Hesse, 2003, reprinted with the permission of Elsevier).

front in the North Pacific (Hein et al., 1978), does not show this phase reversal. Where the BSR is absent due to a lack of free gas, hydrate occurrence cannot be predicted on a regional scale. On-the-spot detection of the presence of hydrate in drill cores is hampered by the fact that visual recognition is very difficult because in fine-grained sediment the very small hydrate crystals are generally highly dispersed and do not survive core retrieval. Massive hydrates or solidly hydrate-impregnated sediments, on the other hand, have only rarely been recovered during coring operations (Fig. 9.38). Low temperatures close to (and even below) 0 °C caused by endothermic hydrate melting, however, are one of the better indicators for the presence of hydrate. Temperature measurements (Paull et al., 1996) or infrared imaging (Riedel et al., 2006) are therefore routinely carried out in deep-sea drilling operations in potentially hydrate-bearing sediments during initial core inspection. Vigorous degassing is characteristic of drill cores with high hydrate concentrations. Confirmation of the actual presence of hydrate can be obtained from specific pore-water anomalies, as outlined in the following sections.

7.2. Pore-water profiles of gas-hydrate-bearing sediments

7.2.1. Carbonate alkalinity, ammonia and dissolved phosphate

Pore-water profiles of holes in the East Pacific off Guatemala, drilled in two parallel transects on the landward slope of the Middle America trench during DSDP legs 67 and 84, are representative for high sedimentation-rate continental margin settings bearing gas hydrates. They display chemical trends with unusual clarity. Drill Sites 496, 497 and 568 on the mid-slope in water depths between 2000 and 2400 m are characterized by extreme maxima for carbonate alkalinity, ammonia and dissolved phosphate between 50 and 200 m subbottom (Fig. 9.33A and B). The maxima occur in the carbonate-reduction and fermentation zones, that is below the SR zone, which is about 5 m thick in the region. Carbonate alkalinity exceeds 120 mM between 23 and 45 m subbottom in Site 496—the second highest value ever reported from deep-water sediments, only exceeded by values as high as 250 mM in trench-slope sediments off Peru (Kvenvolden and Kastner, 1990). The fact that the carbonate alkalinity maximum occurs within the carbonate-reduction (or methane-generation) zone rather than at its upper boundary, shows that carbonate/bicarbonate production continues in this zone due to various fermentation reactions. Initially, production is faster than consumption, although eventually consumption by methane generation and, as discussed later, by precipitation of carbonate concretions becomes dominant. The first microbiological study of gas-hydrate-bearing sediments carried out on the Cascadia margin (ODP Leg 146) has shown that bacterial processes are strongly affected by gas and fluid venting (Cragg et al., 1996). In particular, bacterial activity is significantly inhibited in H_2S hydrates, probably due to high concentrations of H_2S.

Sinks for dissolved ammonia and phosphate released by bacterial organic-matter decomposition are more difficult to identify than those for carbonate. Dissolved ammonia usually displays a maximum below the alkalinity maximum (21 mM at 175 mbsf in Site 496, Fig. 9.33A), while a phosphate maximum occurs in an intermediate position (0.4 mM at 56 mbsf in Site 496). The build-up of both ammonia and phosphate concentrations to their maxima in the methane-generation zone underlines the importance of continuing fermentation processes (e.g. deamination of proteins). There are no known ammonia-bearing minerals in anoxic sediments except the highly unstable struvite ($NH_4Mg(PO_4) \times 6H_2O$), the occurrence of which in modern marine sediments has yet to be demonstrated. The decrease of ammonia from its maximum in the methane-generation zone can be explained, in part, by downward diffusion (Lerman, 1977). In gas-hydrate-bearing drill-sites, the decrease is partially caused by dilution from hydrate water during sampling. However, the rapid drop generally seen in organic-matter-rich anoxic sediments requires an additional sink at greater depth, which is assumed to be ion exchange for K^+ in illitic clays. Ammonium ions are incorporated into interlayer positions of clay minerals with high layer charges, particularly so-called "expandable" illites, for example, in oxidized pelagic sediments of the Central Pacific (Müller, 1977). Similar reactions are likely to occur in anoxic sediments leading to the fixation of dissolved ammonia in crystal lattices of phyllosilicates that will then carry it to great burial depth, down to the realm of metamorphism (Itihara and Honma, 1979).

Phosphate diagenesis in rapidly deposited hemipelagic sediments of the continental margins is similar to that of ammonia, as the pore-water profiles indicate, but also different because authigenic-mineral phases incorporating phosphate exist, namely apatite, as discussed in Section 6.2.2. Vivianite is an iron-bearing phosphate [$Fe_3(PO_4)_2 \times 8H_2O$] that can give the sediment an amazing blue colour. Apatite precipitation is favoured over vivianite in the presence of fine-grained calcium carbonate as nuclei. In the absence of such nuclei, vivianite may form instead. Solid phosphate minerals are difficult to detect because they occur only in trace amounts (see, however, Section 6.2.2. on phosphorites). A significant portion of the phosphate fixation that is required to interpret the downward decreases in the profiles of dissolved phosphate may also be due to adsorption. In this context, dissolved fluoride profiles in Peru continental-margin sediments are of interest as they indicate diffusion of F^- into the sediment from the overlying bottom water (the concentration of which is about 70 μM) and uptake in authigenic carbonate fluorapatite, which is a significant sink for fluor (Froelich et al., 1983).

7.2.2. Pore-water chemical and isotopic anomalies associated with submarine gas-hydrate zones: Coupled chlorinity decrease and $\delta^{18}O$ increase

In hydrate-bearing sediments of trench-slope Sites 496 and 497 off Guatemala, a significant downward chlorinity decrease (to less than half of sea-water chlorinity) coupled with a major $\delta^{18}O$ increase was recognized for the first time (Fig. 9.33A and B) (Harrison et al., 1982; up to 3.3‰ at the bottom of nearby Site 568: Hesse et al., 1985). These coupled changes that subsequently have been found associated with many drilled hydrate-bearing sections (e.g. Gieskes et al. 1985; Jenden and Gieskes, 1983; Kvenvolden and Kastner, 1990) are now generally attributed to the release of fresh water by hydrate melting, either at the base of the hydrate zone or during the sampling process. The freshening is the result of the salt-exclusion effect (Hesse and Harrison, 1981; Ussler and Paull, 1995). Gas hydrates, like normal ice, do not incorporate dissolved sea salts in the crystal structure. The oxygen-isotope effect associated with the hydrate crystallisation is the result of solid/fluid isotope fractionation that causes preferential uptake of the heavy isotope ^{18}O in the solid phase and depletion in the fluid. Hydrate formation thus causes salt and isotope fractionation. As a consequence, the remaining pore fluid not involved in hydrate formation will be enriched in dissolved salts and light isotopes, the opposite of what is observed in submarine hydrate zones.

The dénouement for this apparent discrepancy lies in effects that overprint the salt-exclusion effect: burial compaction and diffusion. During burial and compaction, partial separation of the solid sediment particles from their surrounding pore fluids takes place: the solids are buried, whereas the liquids stay in place and thus in fact move upwards relative to the solids (in other words, an upward advective compaction flow occurs). In rapidly deposited anoxic sediments, hydrates form at shallow subsurface depth a few meters below the sediment surface where porosity is high, typically in the 80% range. During burial, the hydrate crystals carry fresh, isotopically heavy water in solid, ice-like form to greater depth. When the hydrate melts at the base of the hydrate zone, say at 400 mbsf, it releases the fresh, isotopically heavy water, which remixes with the remaining pore water. The porosity at this depth is much reduced by compaction, typically to 40% or half of the original volume. Mixing thus causes freshening and an increase in $\delta^{18}O$. Mixing occurs also as a sampling artefact when hydrate melts in samples taken from sediment above the hydrate base at shallower subsurface depths that have undergone less compaction. The degree of freshening and $\delta^{18}O$ increase is proportionately less. In this way the relatively smooth trends of down-hole chlorinity decrease and [related down-hole] $\delta^{18}O$ increase observed in many drill holes can be explained qualitatively (Fig. 9.33A and B, Sites 496, 497, 568).

The apparent absence of the salt-exclusion effect at the roof of the hydrate zone is principally the result of diffusion which, given enough time, will dissipate the chlorinity spike caused by hydrate crystallisation. In the same way the negative oxygen-isotope anomaly will disappear diffusively. A contributing factor is the generally low hydrate formation rate in areas of biogenic methane production. Apart from a notable exception (see below), the postulated chlorinity increase and $\delta^{18}O$ decrease at the top of the hydrate zone are therefore generally not found. Upward diffusion out of the hydrate zone and back to the sea floor is the major mechanism that produces the overall chlorinity deficit in hydrate zones; however, dilution can also be caused by fresh water migrating upwards from deep-seated sources where it is released by the dehydration of hydrous mineral phases (see below).

7.2.3. Preserved salt-exclusion effect at the roof of the hydrate zone on the Hydrate Ridge off Oregon

On the Hydrate Ridge off Oregon, a strong chlorinity increase at the top of the hydrate zone to up to 1100 mM Cl^- (or almost twice seawater chlorinity, with a single sample containing almost 1400 mM) has been detected (Suess et al., 2000) and documented during more recent ocean drilling (ODP leg 204) (Torres et al., 2004). The prerequisite for the salt-exclusion effect to be preserved in this locality is vigorous hydrate formation from methane carried upwards by fluids along fault zones in actively dewatering sediments of the accretionary wedge off the coast of Oregon (see Section 10). Methane venting is manifested in methane plumes in the water column at this and other vent sites including the Blake Ridge Diapir (e.g. Egeberg, 2000), the Gulf of Mexico (e.g. Brooks et al., 1987), the Arabian Sea (e.g. Von Rad et al., 2000) and many others.

Alternative mechanisms such as brine advection (see Section 8 on evaporite dissolution), hydrothermal processes (see Section 9 on hydrothermal activity), and buried Pleistocene connate seawater can also cause a chlorinity increase. The latter is associated with small Cl^- maxima of a few millimoles per 10^3 cm^3 in excess of the local bottom-water chlorinity. These pore-water Cl^- maxima record increased seawater chlorinity during glacial epochs (McDuff, 1985; Schrag et al., 1996, 2002). Diffusion did not have sufficient time to dissipate the effect. The small maxima that have been observed near the top of the hydrate zone at DSDP Mid-America trench slope Sites 496 and 497 (Harrison et al., 1982), 568 (Hesse et al., 1985), ODP Peru trench-slope Site 688 (Kastner et al., 1990), and ODP Site 997 on the Blake Ridge (Hesse et al., 2000) most likely reflect this connate-water effect rather than the salt-exclusion effect.

7.2.4. Coupled pore-water anomalies: Diagnostic tool for hydrate recognition

The two anomalies, the downward chlorinity decrease and $\delta^{18}O$ increase, if coupled, are strong evidence for the presence of hydrates and are one of the more commonly used diagnostic tools for hydrate recognition in drill cores. The same coupled anomalies can be generated, however, by dehydration reactions of clay minerals (Yeh and Savin, 1977) and gypsum (Fontes, 1965), but these start at higher temperatures (>50–60 °C: Weaver, 1989, Table 7-1) than hydrate dissociation and thus do not overlap with the hydrate stability field. Furthermore, smectite dewatering during the smectite-to-illite reaction causes not only ^{18}O enrichment of the water, but at the same time deuterium (D) depletion (Yeh, 1980) and can thus be differentiated from the effects of hydrate dissociation, in addition to a shift in alkali ions (see next section). Such waters can be imported from higher temperature regions by advection, but, again, their source should be recognisable due to the expected D depletion in clay reactions.

The coupled anomalies are sufficient indicators for the occurrence of hydrates; they are, however, not necessary. For example, hydrate occurrence associated with freshening of the pore waters without an oxygen-isotope ratio increase has been reported from DSDP Site 565 on the Mid-America Trench slope off Costa Rica (Hesse et al., 1985; repeat drilling at nearby Site 1041 confirmed the chlorinity reduction due to hydrate dissociation: Kimura et al., 1997). Here, a nearly continuous downward chlorinity decrease (Fig. 9.33B) is accompanied by a zone of negative $\delta^{18}O$ values at subbottom depths between 95 and 170 m (with a minimum δ-value of −1.26‰: Hesse et al., 1985), apparently caused by the alteration of volcanic glass. The heavy isotope ^{18}O is preferentially taken up by the clays and/or zeolites formed from the glass. The ensuing lowering of the pore-water $\delta^{18}O$ (Lawrence et al., 1975; Perry et al., 1976), superimposed on the hydrate effect, has obliterated the latter (Fig. 9.39). Hydrate-bearing ODP Site 859 near the Chile triple junction (Zheng et al., 1995) and ODP Sites 888–892 on the Cascadia margin (Kastner et al., 1995) may show the

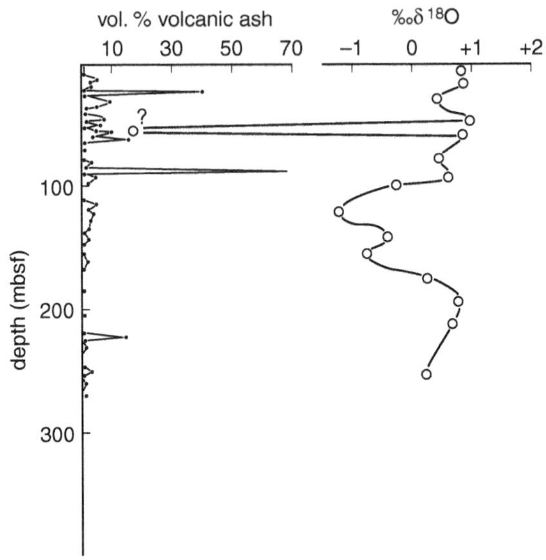

Figure 9.39 Concentration of volcanic ash and $\delta^{18}O$ of interstitial waters of Site 565, Middle America trench slope off Costa Rica (redrafted from fig. 11 from Hesse, 1990c). Note that negative $\delta^{18}O$ values appear below 100 mbsf, where the volcanic-glass content decreases generally below 5%, most likely due to diagenetic alteration.

same effect. Alternatively, advection of an isotopically light pore fluid could be responsible for the lowered values.

Freshening of the pore waters, on the other hand, can be caused by effects other than hydrate dissociation, such as meteoric-water influx (Manheim, 1967; Manheim and Paull, 1981), burial of brackish or fresh water during sea-level lowstands (Manheim and Schug, 1978), or the dehydration reactions already mentioned (illite-smectite, gypsum, opaline silica, etc.). Except for the latter group, these effects would be associated with decreasing oxygen-isotope ratios and can thus be differentiated from the hydrate-dissociation effect. Freshening of the pore waters unaccompanied by the heavy oxygen-isotope enrichment therefore can have different causes, including hydrate dissociation combined with an overprinting mechanism.

7.3. Chlorinity decrease as a tool to estimate hydrate concentrations: the diffusion-advection model

The degree of freshening of the pore waters in hydrate-bearing sediments provides a potential geochemical tool to calculate hydrate concentrations, if the effect of hydrate dissociation can be separated from other freshening mechanisms such as fresh or low-chlorinity water import by advection. This requires establishing the pore-water chlorinity before hydrate dissociation, which is possible using a pressure core-sampler that is pushed into the sediment ahead of the drill bit to collect interstitial water samples under *in situ* PT-conditions.

In ODP Site 997 in the Blake Ridge gas-hydrate field in the West Atlantic, one of the better studied submarine hydrate occurrences (DSDP legs 11, 76, ODP Leg 164), the top of the hydrate zone is postulated to occur at 24 mbsf based on pore-water data (Egeberg and Dickens, 1999; Hesse et al., 2000). The bottom occurs at 452 mbsf at the depth of the BSR (Fig. 9.40). For Site 997, shipboard samples squeezed in the laboratory show an approximately 10% chloride decrease from 558 mM Cl^- (local bottom-water chlorinity) to about 510 mM in the upper 200 m of the hydrate zone. *In situ* samples obtained from the interval from 50 to 150 mbsf display on average only half the chloride decrease of the corresponding shipboard samples, yet all values are lower than the local bottom-water chloride content (Fig. 9.41). This suggests that only about half of the freshening can be attributed to fresh-water release from hydrate melting; the remainder is due to advection of a low-chlorinity water and downward diffusion of Cl^- (see below). Advection is indicated by the more or less straight vertical chlorinity profile below the hydrate base at 452 mbsf, where the shipboard samples reach a plateau level of 506 mM (Fig. 9.40).

Hydrate abundance and distribution as quantified for Site 997 with a combined advection-diffusion model (Egeberg and Dickens, 1999) show that the continuous downward $\delta^{37}Cl$ decrease (Fig. 9.40) is the result of diffusive mixing of two isotopically distinct reservoirs: seawater and an isotopically depleted low-chlorinity water that is advected into the hydrate zone from below its base. The model approximates both the shipboard chlorinity measurements and the *in situ* values with smooth curves (Fig. 9.41).The crucial input to the model besides the *in situ* chlorinities are advection rates, which are obtained from fitting the model to the chlorine-isotope profile of Hesse et al. (2000) by trial and error (Hesse et al., 2001, 2006). The source of the advected low-chlorinity and isotopically light water is not known. Ages of the pore water of 55 Ma (in sediment that is 1.8–6 Ma old) determined with the aid of the radioactive isotope ^{129}I point to organic matter (from which the iodine is derived) of Paleocene/Eocene age (Fehn et al., 2000) deep in the sedimentary section (or of a mixture of such material of Cretaceous to Miocene age). At several kilometres subsurface depth, clay reactions such as the smectite-to-illite reaction (Ransom et al., 1995) could produce the required isotopic fractionation.

Although the chlorinity gradient disappears below the hydrate zone, so that downward chloride diffusion must stop below about 450 mbsf (Fig. 9.41A), the chlorine-isotope ratios continue to decrease below that depth, indicating continued isotope diffusion. For each of the two stable

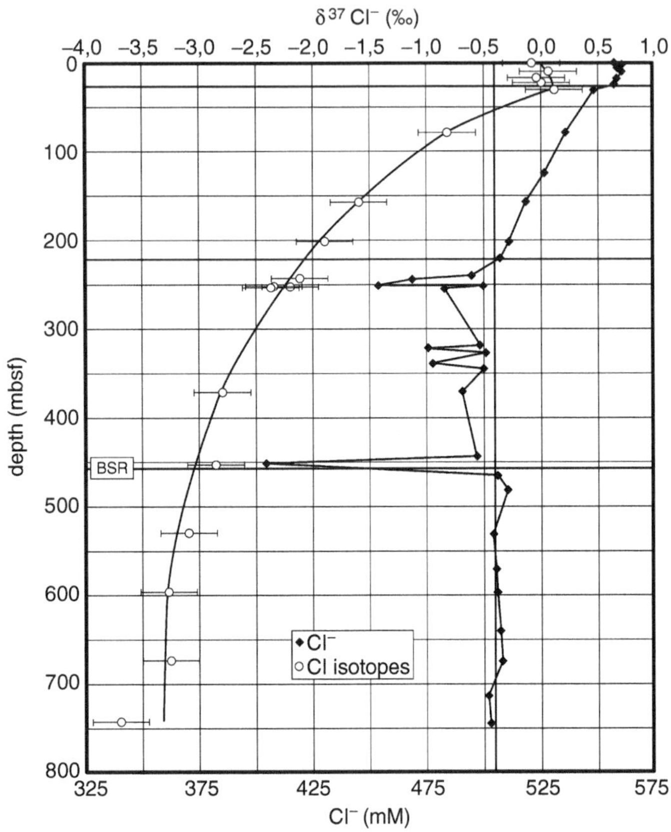

Figure 9.40 Downhole δ^{37}Cl and chlorinity profiles for Site 997 (modified from fig. 1 in Hesse et al., 2000). In the main hydrate zone between 220 and 452 mbsf, some pronounced low-chlorinity peaks (e.g. 390 mM at 330 mbsf and 405 mM at 451 mbsf) indicate layers of higher hydrate concentration, whereas the average concentration of 500 mM is only slightly less than the plateau level below, suggesting rather low average hydrate concentrations. From the Cl^- data set available for Site 997, only those samples have been included for which isotope measurements were made. The vertical line at 506 mM Cl^- represents the plateau value of the low-chlorinity water advected from below the base of the cored section. BSR, bottom-simulating reflector.

chlorine isotopes, a gradient, albeit small, is maintained (Fig. 9.42A) that facilitates ongoing isotope diffusion. A best fit of the advection–diffusion model to the chlorine-isotope curve is obtained for an advection rate of −0.18 mm per year (the minus sign referring to upward advection) for Site 997 (Fig. 9.42B).

Advection rates of −0.18 mm per year as determined for Site 997 are relatively high compared to rates that would result from compaction flow alone. The latter should not exceed sedimentation rates (about 125 m per

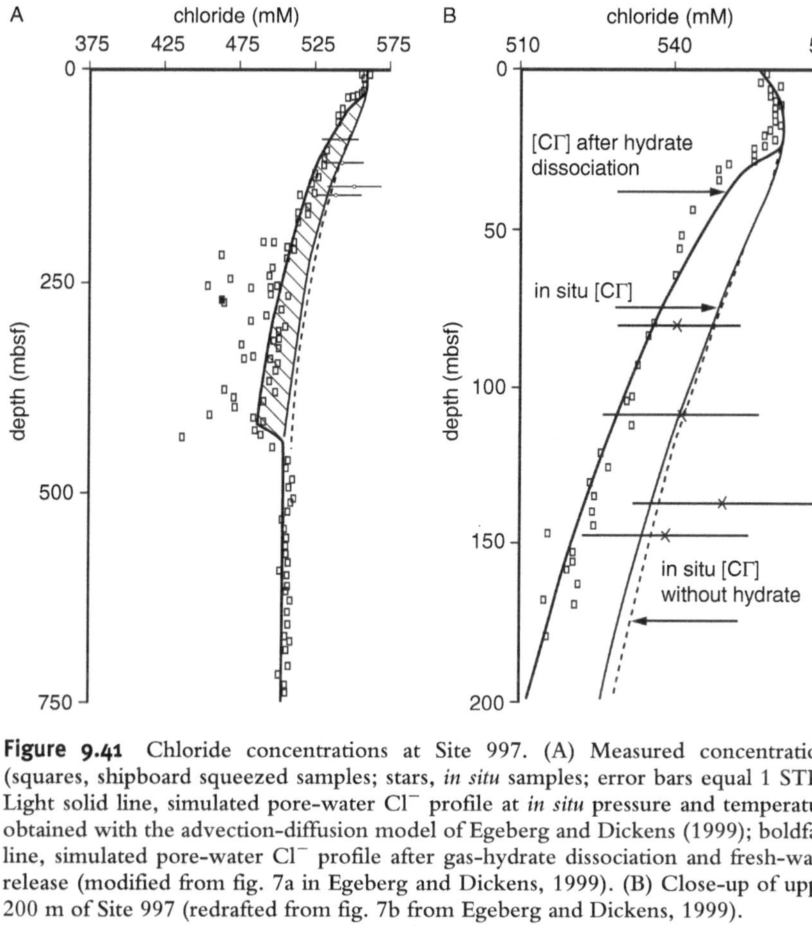

Figure 9.41 Chloride concentrations at Site 997. (A) Measured concentrations (squares, shipboard squeezed samples; stars, *in situ* samples; error bars equal 1 STD). Light solid line, simulated pore-water Cl^- profile at *in situ* pressure and temperature obtained with the advection-diffusion model of Egeberg and Dickens (1999); boldface line, simulated pore-water Cl^- profile after gas-hydrate dissociation and fresh-water release (modified from fig. 7a in Egeberg and Dickens, 1999). (B) Close-up of upper 200 m of Site 997 (redrafted from fig. 7b from Egeberg and Dickens, 1999).

million years or 0.12 mm per year on average for the last 6 Ma, and <0.05 mm per year for the last million years at the Blake Ridge drill sites: Paull et al., 1996, p. 291), if these equal the subsidence rates. Seismic profiles across the Blake Ridge (Fig. 9.43) show a pattern of deep-reaching subvertical faults that extend upwards through the hydrate zone and could have served as conduits for fluid flow from a deep-seated source. Since hydrate-impregnated horizons within the hydrate zone may act as a seal, the hydrate zone at the Blake Ridge may have the overall effect of a caprock focusing fluid flow along fault zones. Flow would be focused especially into the broad anticline-shaped hydrate zone at the crest of the ridge where advected solutes would then be dissipated by diffusion. The BSR at Blake Ridge only occurs under the ridge crest and disappears towards the flanks, where free

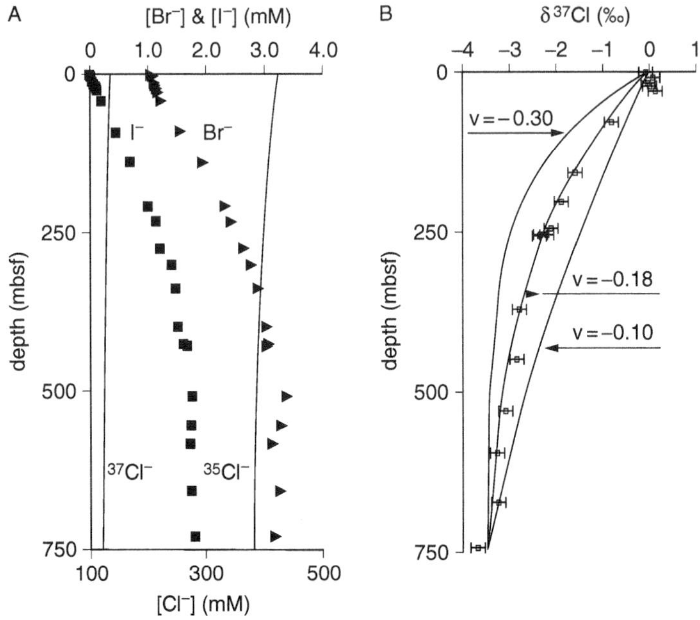

Figure 9.42 Measured and modelled isotope curves. (A) Measured ^{37}Cl, ^{35}Cl, Br^- and I^- concentration profiles (smoothed for ^{37}Cl, ^{35}Cl) (fig. 6 from Hesse et al., 2000). (B) Modelled Cl-isotope curve for ODP Site 997 using three different advection rates, a mobility ratio for diffusion of the light and heavy Cl isotopes of 1.0023 (Eggenkamp et al., 1994), and diffusion coefficients from Li and Gregory (1974) adjusted to varying pore-water viscosity (Out and Los, 1980) and sediment porosity (Paull et al., 1996) (fig. 5 from Hesse et al., 2000).

gas seems to be absent below the hydrate zone, supporting the idea of a gas cushion trapped underneath the hydrates.

Br^- and I^- concentrations in Site 997 interstitial waters show a large increase over seawater values: about three times for Br^- and 3 orders of magnitude for I^-,and remain more or less constant at these high levels below ~400 mbsf (Fig. 9.42A). These high concentrations necessitate advection of water enriched in these halogens from below, because the moderate organic-matter content of the host sediments (1.2% on average: Fig. 9.44A) is insufficient to generate the high concentrations by *in situ* bacterial organic-matter decomposition. The high acetate and dissolved organic-carbon contents below 350 mbsf (Egeberg and Barth, 1998) also require a diagenetically strongly altered fluid, the source of which would be below the base of the drilled succession. The fact that the acetate has not been used by methanogenic bacteria may seem surprising but may be due to the high interstitial concentrations of methane

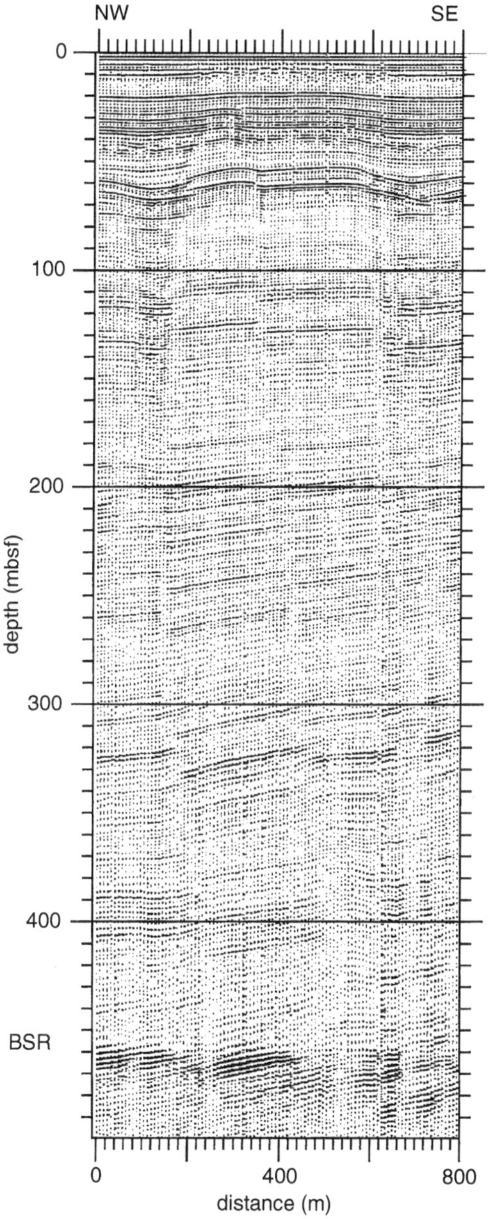

Figure 9.43 Deep-reaching faults on the Blake Ridge that penetrate the bottom-simulating reflector (BSR) at about 450 mbsf and show vertical displacements of up to 15 m. Fault spacing is tens to hundreds of meters (redrawn and modified from Wood and Ruppel, 2000; fig. 5).

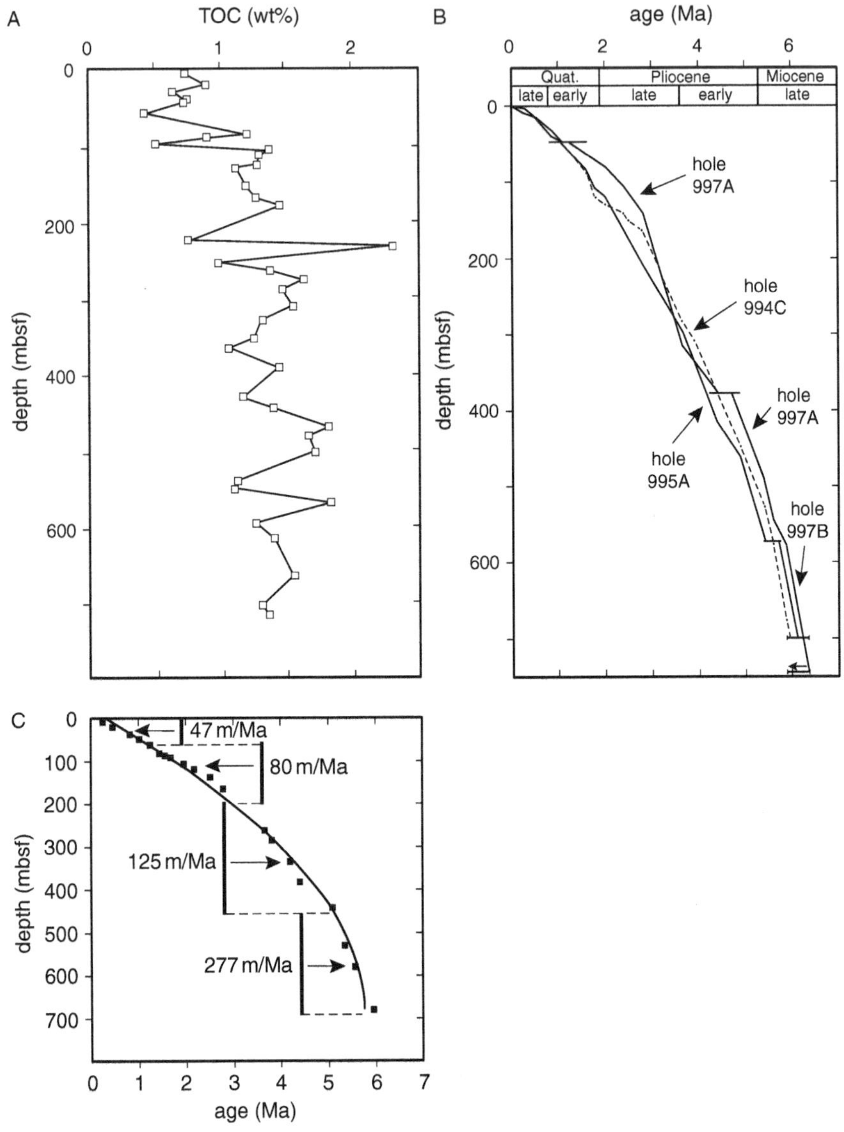

Figure 9.44 ODP Site 997. (A) Total organic carbon (TOC) (fig. 26 from Paull et al., 1996). (B) Age-depth relationship (fig. 13 from Paull et al., 1996). (C) Sedimentation rates (fig. 10 from Rodriguez et al., 2000).

at these depths, which have an inhibiting effect on microbial activity (W. Borowski, personal communication, 2000).

The extent of pore-water freshening caused by hydrate dissociation is obtained by subtraction of the modelled shipboard chlorinity curve from the *in situ* curve (Fig. 9.41). Results for Site 997 give low hydrate concentrations, on average 2.3% of the pore space in the hydrate zone between 24 and 452 mbsf (Fig. 9.45), in line with the small chlorinity decrease determined during previous DSDP drilling in the region (DSDP Sites 533 and 534: Claypool and Threlkeld, 1983; Jenden and Gieskes, 1983). Geophysical methods, which use the seismic velocity increase and amplitude reduction ("blanking") caused by the presence of hydrates (e.g. Lee et al., 1993), yield higher average values for the Blake Ridge drill sites, varying according to different methods and authors between 4% and >10% of the pore space (e.g. Holbrook et al., 1996; Lee, 2000; Tinivella and Ledolo, 2000).

The results show that simply assuming seawater chlorinity as a base line against which to measure the degree of dilution caused by hydrate dissociation, as has often been done in the past, is insufficient and would yield wrong results.

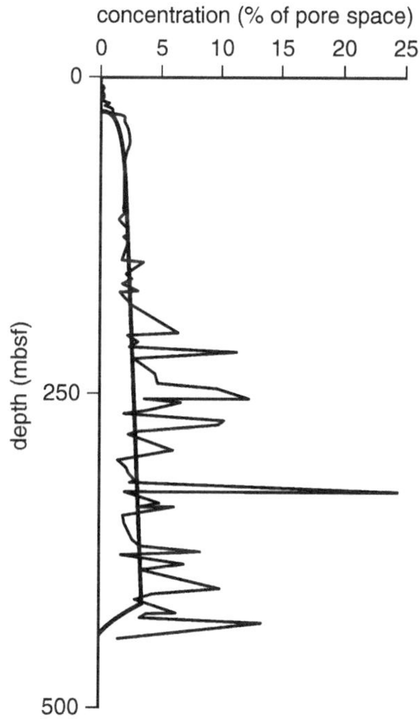

Figure 9.45 Calculated hydrate distribution for Site 997 (redrafted from fig. 8 from Egeberg and Dickens, 1999) yielding an average filling of the pore space by hydrate of 2.3%. The boldface line does not include isolated hydrate peaks obtained from low-chlorinity peaks, which would raise the average concentration to 3.8%.

7.4. Other geochemical anomalies associated with submarine hydrate zones

7.4.1. Dilution of major ion concentrations

Fresh-water release by hydrate dissociation affects all major ions in the same way as Cl^-. This dilution accentuates the downhole decrease of many ionic species below common mid-depth maxima imposed by other processes, such as dolomite and other authigenic carbonate precipitation (lowering of, e.g. $[Mg^{2+}]$, $[Ca^{2+}]$, $[CO_3^-]$), apatite precipitation (lowering of $[PO_4^{3-}]$), or ion exchange in clays (lowering, e.g. of $[NH_4^+]$).

The dilution effect of hydrate dissociation can be differentiated from the effects of fresh-water release by other reactions such as smectite dewatering, which occurs at greater depths below the base of the hydrate zone, by using chloride-normalised profiles. Smectite dewatering is associated with ion exchange, for example, K^+ uptake in exchange for Na^+ release. Whereas hydrate water would have no effects on the Cl^--normalised profiles of these two species, smectite dewatering would change them. In addition, hydrogen-isotope ratios would discriminate between the two reactions, as mentioned before. However, differentiating between dilution caused by hydrate dissociation and fresh-water release in the smectite-illite reaction with the aid of alkalis may be difficult in practise, because long advection distances may alter the effects of the reaction.

7.4.2. Sulphate gradient in the sulphate-reduction zone

In the SR zone above the Blake Ridge-Carolina Rise gas-hydrate field, linear gradients of pore-water sulphate reach values between -1.2 and -2.9 mM SO_4/m (corresponding to a depth of the sulphate/methane interface of between 20 and 10 mbsf). They have been interpreted to reflect a significant downward diffusive flux of sulphate driven by sulphate consumption at the base of the SR zone due to AMO (Fig. 9.46) (Borowski et al., 1996). Up to 35% of the total sulphate flux has been ascribed to this mechanism as an alternative to removal by bacteria within the SR zone. Since the stoichiometry of the reaction $CH_4 + SO_4^{2-} \rightarrow HCO_3^- + HS^- + H_2O$ requires mol-by-mol consumption of methane and sulphate, a significant upward methane flux (of up to 1.8×10^{-3} mmol cm^{-2} a^{-1}) must be involved. If conditions are conducive to hydrate formation, sulphate profiles are related to the presence of gas hydrate in underlying sediments by proxy (Borowski et al., 1999). The effects of AMO on SR within the sulphate/methane transition zone were studied in the upwelling region off Chile at stations between 800 and 3000 m water depth (Treude et al., 2005) with the shallowest station showing the highest rates.

Supporting evidence for SR by AMO comes from strongly lowered $\delta^{13}C$ values of the CO_2 reservoir ($=\Sigma CO_2$) at the base of the SR zone. At the Blake Ridge, $\delta^{13}C$ values as negative as -39‰ (Fig. 9.47A) have been measured

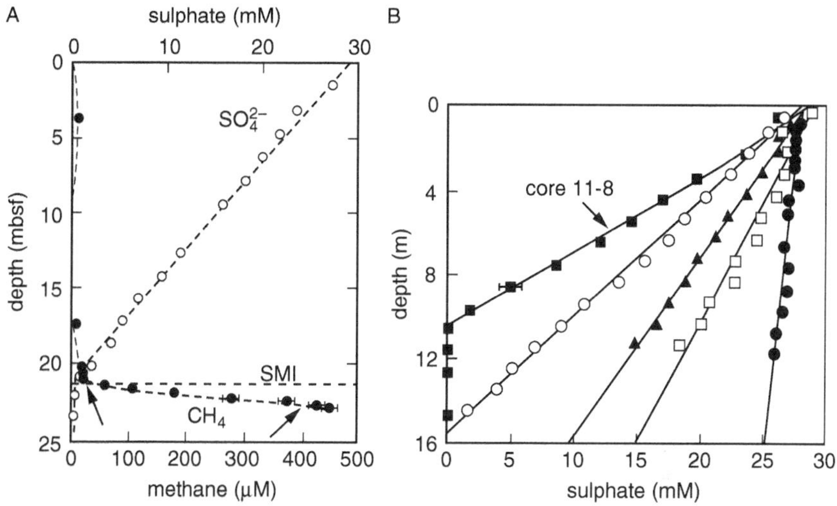

Figure 9.46 Relationship of sulphate gradients to anaerobic methane oxidation. (A) Measured sulphate and methane concentrations at ODP Site 995 (Paull et al., 1996) compared to model curve (dashed line) which assumes steady-state conditions and molecular diffusion as the only transport process (fig. 5A from Borowski et al., 2000). (B) Sulphate gradients for 5 piston cores from the Carolina Rise and Blake Ridge continental slope. SMI, sulphate/methane interface (redrafted from fig. 1 from Borowski et al., 1996).

(Borowski et al., 2000), which are more negative than the most negative values that could be derived by bacterial decomposition of marine organic matter in the SR zone (−30‰: Deines, 1980). Isotopically lighter terrestrial organic matter could also contribute to lower the $\delta^{13}C$ values in the pore water (Sackett and Thompson, 1963). Blake Ridge organic matter averages −21‰ $\delta^{13}C_{PDB}$, whereas the $\delta^{13}C_{CH_4}$ is lighter than −60‰, reaching values as low as −84‰ (Paull et al., 2000). From these numbers, the 35% contribution to the CO_2-pool stemming from methane oxidation has been derived.

7.4.3. Authigenic carbonates

Methane oxidation increases alkalinity and, if accompanied by reactions that maintain a high pH (such as Fe reduction) and consequently a high carbonate activity, induces carbonate precipitation. The sulphate/methane interface region above hydrate zones should therefore be a preferred site for the precipitation of authigenic carbonates, and these should carry a low-$\delta^{13}C$ isotopic signature characteristic of methane-derived carbonate, as found in organogenic deep-sea dolomite at Blake Ridge (Fig. 9.47B) (Pierre et al., 2000, Rodriguez et al., 2000).

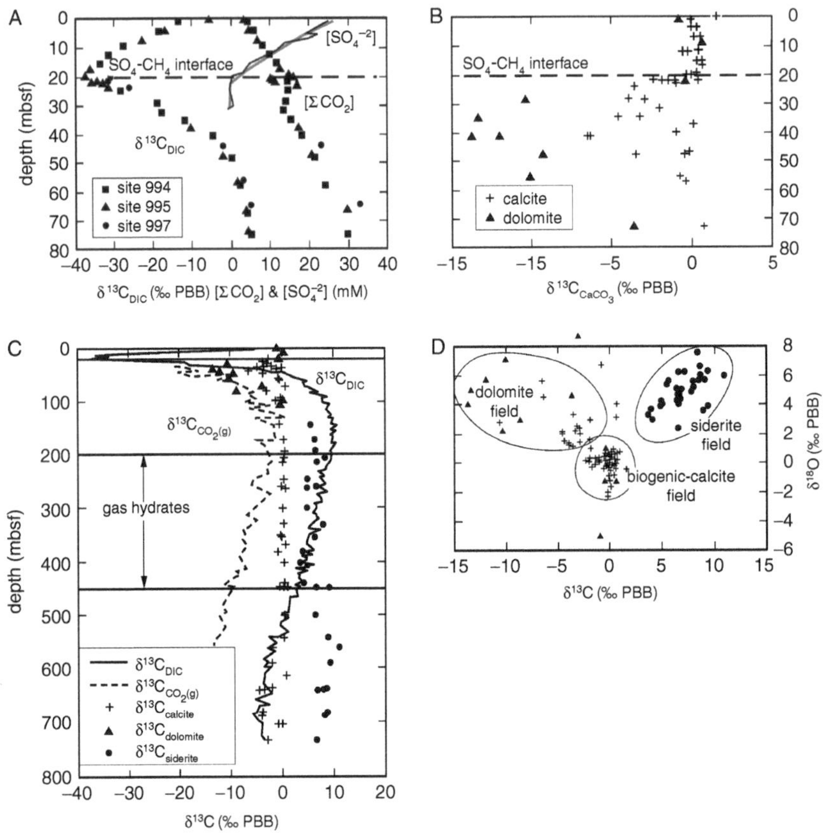

Figure 9.47 $\delta^{13}C$ values. (A) $\delta^{13}C$ measurements of total dissolved inorganic carbon (DIC) and concentrations of total dissolved carbon dioxide (ΣCO^2) and SO_4^{-2} for interstitial water samples from the upper 80 m of pooled ODP sites 994, 995 and 997, showing $\delta^{13}C$ values near −40‰ near the sulphate/methane interface (results for complete sites shown in (C)). (B) $\delta^{13}C$ values of authigenic carbonates precipitated from waters shown in (A). Note that authigenic dolomite in the SR zone shows a signature characteristic of the top of the SR zone and seems to have been precipitated very early, whereas dolomite in the methane-generation zone only started to form about 10 m below the sulphate/methane interface (figs. 8a,b from Rodriguez et al., 2000). (C) Combined $\delta^{13}C$ plots for interstitial waters and authigenic carbonates of pooled OPD Sites 994, 995 and 997. $\delta^{13}C_{DIC}$, carbon-isotope ratio in dissolved inorganic carbon; $\delta^{13}C_{CO_2(g)}$ = carbon-isotope ratio in CO_2 from the gas phase (fig. 7 from Rodriguez et al., 2000; note that these authors assume the top of the gas-hydrate zone to occur at 200 mbsf, whereas Egeberg and Dickens, 1999, and Hesse et al., 2000, placed it at 24 mbsf). (D) $\delta^{18}O$–$\delta^{13}C$ cross plots for the same authigenic carbonates as shown in (C) (symbols same as in (C); fig. 6 from Rodriguez et al., 2000).

8. Effects of Evaporite Dissolution on Pore-Water Chemistry

Halite dissolution in the vicinity of salt domes and evaporite layers is the main, although not the only source of high-salinity NaCl and (Ca, Na_2) Cl_2 brines which represent the high-salinity end member of pore-water profiles in terms of salinity variations, the counterpart to the low-salinity profiles. Typically they occur at greater depths in sedimentary basins that are encountered in deeper wells. However, increases in chlorinity at relatively shallow depth within the realm of early diagenesis have been reported from a number of oceanic drill sites of the DSDP and ODP in regions known to be underlain by evaporites, for example, the Mediterranean Sea (McDuff et al., 1978; Sayles et al., 1972), the Red Sea (Manheim et al., 1974), and Atlantic continental margins at the Blake Ridge Diapir in ODP Site 996 (Egeberg, 2000), off the Guyanas (Waterman et al., 1972), Namibia (Sotelo and Gieskes, 1978), Morocco (Couture et al., 1978; Gieskes et al., 1980), and the Milano Dome in ODP Sites 970A,B in the Eastern Mediterranean (De Lange and Brumsack, 1998). In some of these, the increase in chloride concentration is not matched by the sodium increase, for example, at Site 374 in the Balearic Basin of the Western Mediterranean (McDuff et al., 1978), indicating dissolution of other complex chlorides (Fig. 9.48). At this site, the rare magnesium-rich mineral lueneburgite [$Mg_3(PO_4)_2B_2O(OH)_4 \cdot 6H_2O$] has been detected (Müller and Fabricius, 1978).

9. Sediment-Covered Mid-Ocean Ridges: Hydrothermal Activity and Intrusion of Igneous Dykes and Sills

The pore-water chemistry in discharge areas of hydrothermal convection cells under MORs contrasts, not surprisingly, strikingly with the straight-line profiles of sea-water composition in the intake (recharge) areas that were presented at the beginning of this chapter. Hydrothermal activity, like evaporite dissolution, produces highly saline fluids. Circulating several kilometres deep in the convection cells of the oceanic crust, they have been heated to the critical temperature of water and reacted intensely with the rocks. Discharging hydrothermal solutions have been studied in the black and white smokers on the ridge crest (e.g. Edmond et al., 1979; Von Damm et al., 1985), but will not be discussed here because hydrothermal geochemistry other than in relation to early diagenesis is beyond the scope of the present review. On sediment-covered ridges, however, the ascending hot solutions or intruding igneous dykes and sills

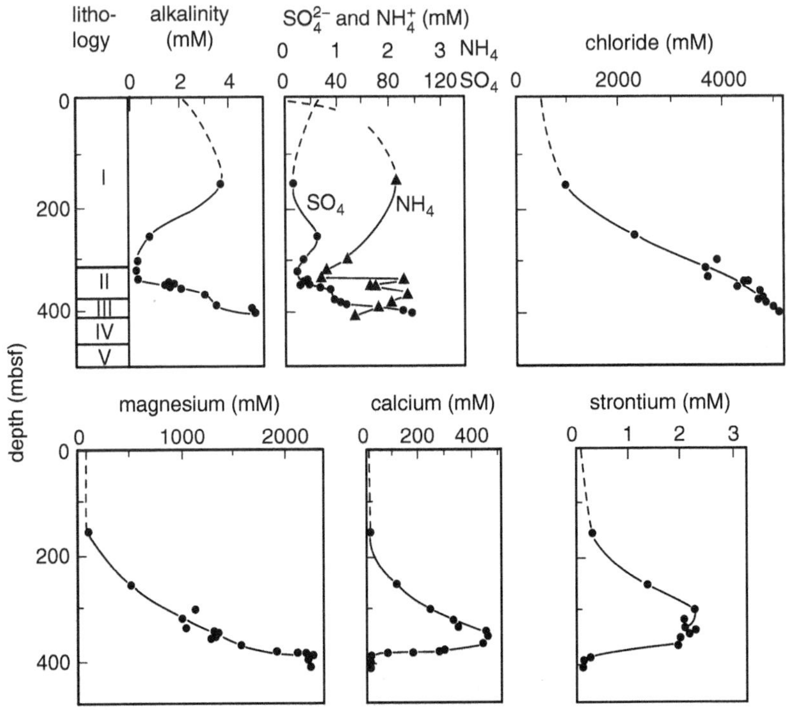

Figure 9.48 Interstitial water profiles for DSDP Site 374 in the Balearic Basin of the western Mediterranean Sea, indicating dissolution of evaporite minerals at 380 mbsf (redrawn and modified from Gieskes, 1983, after McDuff et al., 1978). Lithology: I, marls; II, nannofossil ooze; III, dolomitic marls; IV, gypsum, anhydrite; V, halite.

interact actively with the sediments during early diagenesis and cause complex pore-water profiles, which fall on the borderline between diagenesis and hydrothermal alteration.

Diagenetic effects of hydrothermal fluids in the Guaymas Basin of the Gulf of California are reflected by a distinct set of anomalous pore-water profiles (DSDP Site 477: Gieskes et al., 1982). The basin is located over a segment of the spreading ridge axis with a high heat-flow anomaly characterized by a geothermal gradient of 88 °C/100 m, so that early-diagenetic temperatures are surpassed at less than 100 mbsf. The measured bottom-hole temperature at 300 m sub-seafloor depth was 200 °C. A chloride increase with depth (Fig. 9.49), unlike that in evaporite dissolution sites, is related to water removal in hydration reactions. Hydrothermal alteration of the sediment releases alkali metals to the pore waters, causing distinct downward increases in Li^+, K^+ and Rb^+. The Ca^{2+} increase and Mg^{2+} decrease with depth are reminiscent of diffusion-controlled sites,

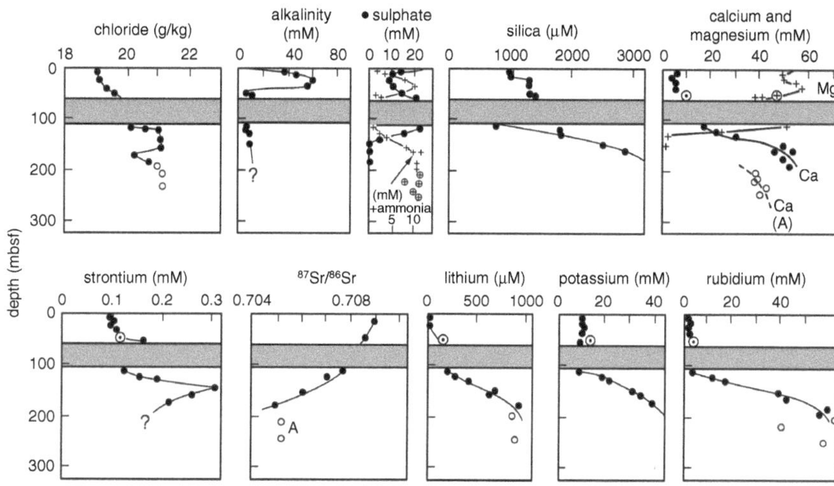

Figure 9.49 Hydrothermally influenced pore-water profiles of DSDP Site 477 (open circles: Site 477A), Guaymas Basin, Gulf of California Shaded bar: basaltic sill between 58 and 105.5 mbsf. (redrawn and modified from Gieskes, 1983; after Gieskes et al., 1982).

despite the very high sedimentation rates (>2000 m per million years). Ca^{2+} is probably released and Mg^{2+} taken up by the hydrothermally altered volcanic rocks of the basaltic crust and by volcaniclastic sediments in the same way as in diffusion-dominated sites. The Sr^{2+} maximum at 140 m subbottom depth may indicate Sr^{2+} removal deeper in the hole by basalt/seawater interaction at low rock/water ratios, as observed elsewhere (Menzies and Seyfried, 1979). This may also be the cause for the downward decreasing $^{87}Sr/^{86}Sr$ ratios. Dissolved-silica data (Fig. 9.22) and associated solid phases have already been discussed (Section 4.11).

10. Early Diagenesis in Active Margins Affected by Advective Lateral Fluid Flow

Sediments of subduction-zone complexes beneath modern trench slopes undergo active tectonic deformation leading to thrust faulting, early penetrative fracturing, development of "scaly clays", rehealing of the fractures by early-diagenetic cements or trapped clay matrix and a generally high degree of compaction. This tectonic setting is characterized by large-scale fluid expulsion from the imbricated wedges of thrust sheets (Moore and Vrolijk, 1992) that finds expression in stairway-shaped pore-water profiles (see below). Theoretically, dewatering may be a diffuse, trench-slope wide process; what has been found at most drilled margins, however, is upward flow focused along landward-dipping thrust planes and faults or permeable sediment layers. Extensive venting of fluids at the seafloor has been documented for these margins by authigenic carbonate precipitation including chemoherm formation (e.g. Han et al., 2004; Ritger et al., 1987), spectacular mud volcanism (e.g. Langseth et al., 1988), and dense benthic communities, the food chain of which starts with chemosynthetic bacteria which use methane for their metabolism (e.g. Von Rad et al., 2000).

In the pore-water profiles of active-margin drill sites, distinct step patterns have been observed that require rapid lateral advection along horizons of increased permeability or below seals, either fault zones or lower-porosity stratigraphic levels. The process must be fast enough to prevent the advected solutes from being diffused away. Ions that are not or only slightly involved in reactions (such as, e.g. Cl^-, Br^- and Li^+) and isotopes not undergoing fractionation in the drilled section are particularly suited to pin-point advective-flow horizons. Examples are the Cl^- profile at Site 683 (Fig. 9.50) from the Peru trench slope (Kastner et al., 1990) and the strontium-isotope profiles of Sites 888–891 (Kastner et al., 1995) from the Cascadia margin (Fig. 9.51). The low-chlorinity zone in the pore-water

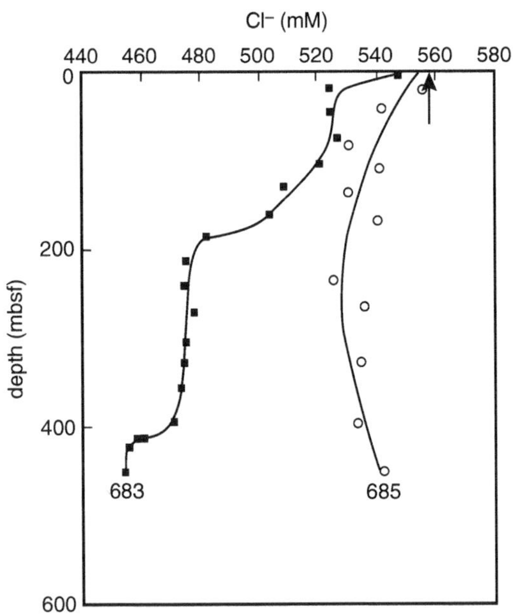

Figure 9.50 [Cl^-]-depth profile for ODP Site 683 (filled squares) at the Peru convergent margin (redrawn and modified from Kastner et al., 1990; fig. 25A), showing a stairway pattern indicative of advective injection of low-Cl^- fluids at three distinct depth levels. The arrow indicates bottom-water chlorinity.

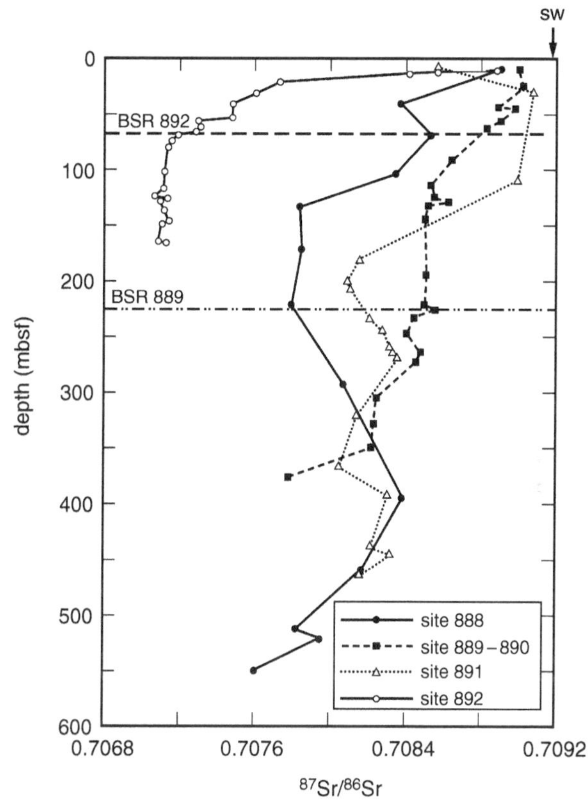

Figure 9.51 $^{87}Sr/^{86}Sr$-depth profiles for ODP Sites 888–891 at the Cascadia convergent margin (redrawn and modified from Kastner et al., 1995; fig. 5), showing a similar type of step curves as in Fig. 9.53, whereas Site 892 is indicative of advection from below, that is, from the décollement zone of the Cascadia imbricated wedge, which was not penetrated by the drill. The upper part of the profiles indicates diffusive mixing of the advected low $^{87}Sr/^{86}Sr$ fluid with seawater. BSR, bottom-simulating reflector.

profile of Site 808 at the toe of the Nankai Trough accretionary prism between 560 mbsf and the bottom of the cored section at 1290 mbsf, where [Cl^-] decreases to 440 mM, is also attributed to lateral advection from greater depth, as there is insufficient smectite for *in situ* fresh-water production by clay dewatering to cause a 20% lowering of the chlorinity (Kastner et al., 1993). The alternative of *in situ* smectite dewatering or dewatering during shipboard squeezing suggested by Fitts and Brown (1999) cannot explain the extent of freshening at this and probably other sites.

The chlorinity increase in the upper part of (hydrate-bearing) Site 808 (with an average geothermal gradient of 110 °C/km) is probably caused by hydration reactions similar to the ones inferred for the Guaymas Basin mentioned above, although the source of this saline water may be deeper-seated and it may also have been advected into this part of the drill site (Kastner et al., 1993). Advection of water with a chlorinity (~600 mM) in excess of seawater chlorinity from below into hydrate-bearing sediments characterises ODP Sites 859 and 860 in the accretionary wedge near the Chile triple junction (Froelich et al., 1995). The raised chlorinity probably has the same cause as that in the upper part of Site 808 and in the Guaymas Basin, that is, hydration reactions in the subsurface below the drilled succession due to the high geothermal gradients (in the 100 °C/km range) in the vicinity of the subducted zero-age crust of the Chile Ridge, an active spreading ridge. A special case high-lighting the role of lateral or oblique advection is the occurrence of CH_4–H_2S hydrate (with up to 10% H_2S of the released gas) that was found at shallow subsurface depth of <19 mbsf in Cascadia margin Site 892. Here, lateral advection, probably along a low-angle fault zone, appears to inject a methane-rich fluid from intermediate depths into the SR zone, causing methane oxidation coupled with sulphide production and formation of a mixed CH_4–H_2S hydrate (Kastner et al., 1995, p. 383).

In other sites, like 889/890 or 892 (Figs. 9.51 and 9.52), the chemical (Cl^-, Li^+) and isotope gradients ($\delta^{18}O$, $^{87}Sr/^{86}Sr$) indicate advection from below, suggesting that the source of the anomalous fluids is located in the décollement zone of these active-margin sites. At Site 888 off Victoria Island, British Columbia, and Sites 891 and 892 offshore central Oregon (Fig. 9.53), the ^{18}O-depleted water is advected by fluid flow, which is confined to distinct flow conduits in the imbricated prism of the Cascadia convergent margin, as suggested by the step curves. Other evidence for advection at the Cascadia margin sites besides the O- and Sr-isotope- and [Cl^-] and [Li^+] depth profiles comes from the presence of thermogenic (higher than C_1) hydrocarbons (Kastner et al., 1995). Other chemical species that have been used to demonstrate advection include boron [B^{3+}] and its isotope ^{11}B, which shows a strong increase in the lower half of ODP Site 1150 in the Japan trench slope that correlated with a [Cl^-] decrease (Deyhle and Kopf, 2002) and a [Sr^{2+}] increase (Fig. 9.54A) (Deyhle et al., 2004). The isotopes display a marked box-shaped downward increase of $\delta^{11}B$ in the fluid-flow-affected zone, and a decrease of $\delta^{37}Cl$ (though much less pronounced and less regular than for Blake Ridge Site 997) and $^{87}Sr/^{86}Sr$ (Fig. 9.54B) (Deyhle et al., 2004). Distinct increases in [B^{3+}] were also observed immediately below the décollement in Sites 1040 and 1043 at the toe of the Costa Rica trench slope (Kopf et al., 2000).

The master thrust between the upper and lower tectonic plates was first drilled during ODP Leg 110 on the Lesser Antilles active-margin traverse. Site 671 at the toe of the Lesser Antilles accretionary wedge off Barbados penetrated the sediments accreted to the (upper) Caribbean plate, the main décollement zone and the sediments of the incoming (lower) Atlantic plate.

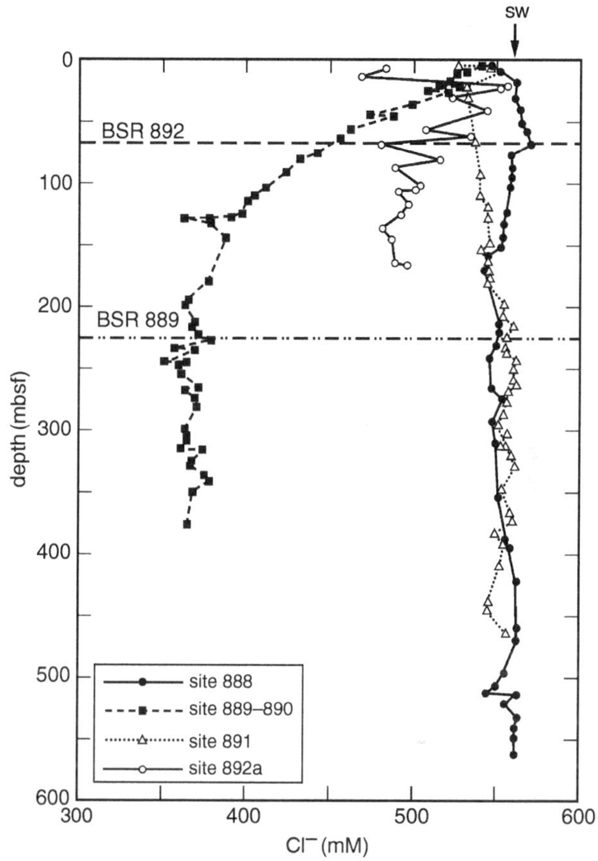

Figure 9.52 [Cl^-]-depth profiles for ODP drill sites 888–892 at the Cascadia convergent margin (redrawn and modified from Kastner et al., 1995; fig. 2). Site 889/890 is indicative of advective flow of a low-[Cl^-] fluid from below the drilled succession and diffusive mixing with seawater. The low-chlorinity peak at ~130 mbsf is most probably due to hydrate dissociation. [Cl^-] for Sites 888 and 891 behaves conservatively, except for a few narrow, non-distinct confined fluid-flow horizons, whereas the fluctuations in the upper part of Site 892A above the BSR seem to indicate hydrate dissociation (and/or lateral advection of a low-[Cl^-] fluid).

The pore fluids of the décollement zone and the closest fault zone immediately above it display a distinct methane anomaly with a concentration up to 0.5 mM, whereas [CH_4] is close to zero in the sediments of the upper plate (Fig. 9.55). Similar but less pronounced anomalies were observed in drill Sites 672 and 676. A permeable Eocene sand horizon below the décollement also shows elevated methane levels. The methane is thermogenic according to its isotopic composition and must come from a deep-seated

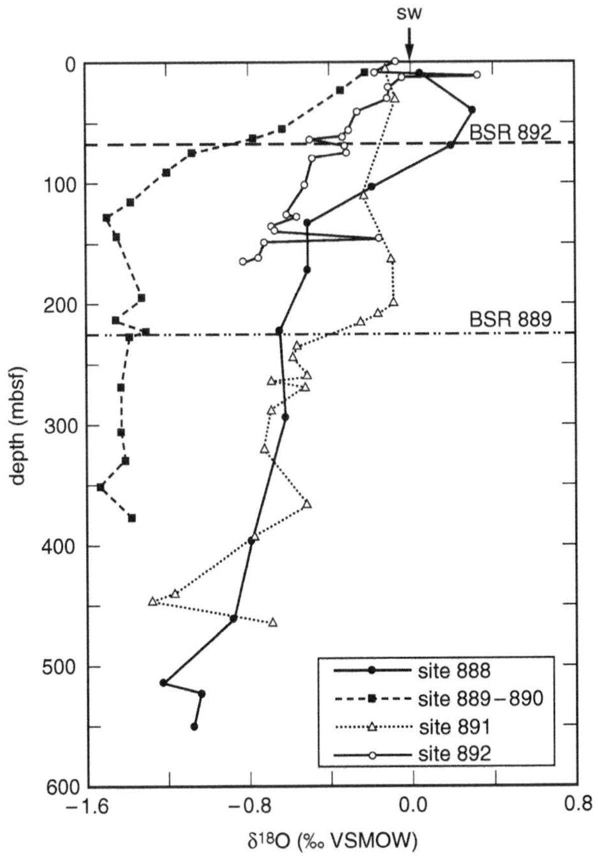

Figure 9.53 $\delta^{18}O$ depth profiles for Cascadia margin ODP drill sites (redrawn and modified from Kastner et al., 1995; fig. 7). The step curves for Sites 888 and 891 suggest lateral advection. The vertical part of Site 889/890 is probably due to vertical advection including slight positive excursions due to hydrate dissociation. The increase in the $\delta^{18}O$ values above ~110 mbsf reflects mixing with seawater, as does the entire curve of Site 892. The latter includes some positive and negative excursions, which may be due to hydrate dissociation (uppermost excursion) but are otherwise unexplained.

source tapped by the décollement zone, some 45–60 km to the West under more landward parts of the wedge (Moore et al., 1988).

Among the imbricated wedges hitherto studied by deep-sea drilling, only the interstitial waters from the Nankai Trough (ODP leg 131: Scientific Party, 1990) provide some evidence for diffuse migration of low-chlorinity water up-wedge below the décollement zone. The chloride minimum may indicate a past advection event now dissipated by diffusion. Davis et al. (1990) suggested a similar mechanism for the Cascadia margin based on heat-flow and multichannel seismic-reflection data.

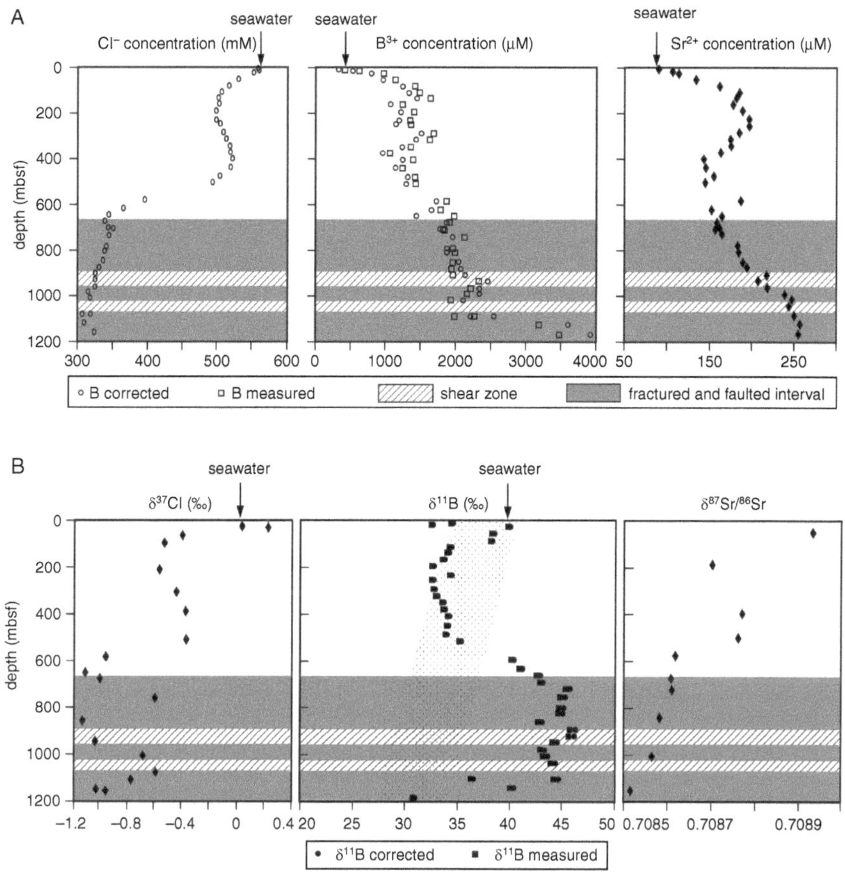

Figure 9.54 $[Cl^-]$, $[B^{3+}]$, $[Sr^{2+}]$ and related isotope depth profiles from ODP Japan trench slope Site 1150 (redrafted from figs. 2 and 3 from Deyhle et al., 2004). (A) $[Cl^-]$, $[B^{3+}]$ and $[Sr^{2+}]$ depth profiles. (B) Related isotope depth profiles ($\delta^{37}Cl$, $\delta^{11}B$ and $^{87}Sr/^{86}Sr$). The corrected *in situ* $\delta^{11}B$ values show only small deviations from the measured $\delta^{11}B$ values that lie within the analytical error of the method. The grey corridor represents the anticipated fractionation trend with depth.

11. Early Diagenesis of Volcanogenic Deep-Sea Sediments

Large volumes of tephra have been repeatedly released during explosive volcanic eruptions along the Pacific Ring of Fire (e.g. Kutterolf et al., 2007; see Carey and Schneider, 2011, this volume). These ashes form layers

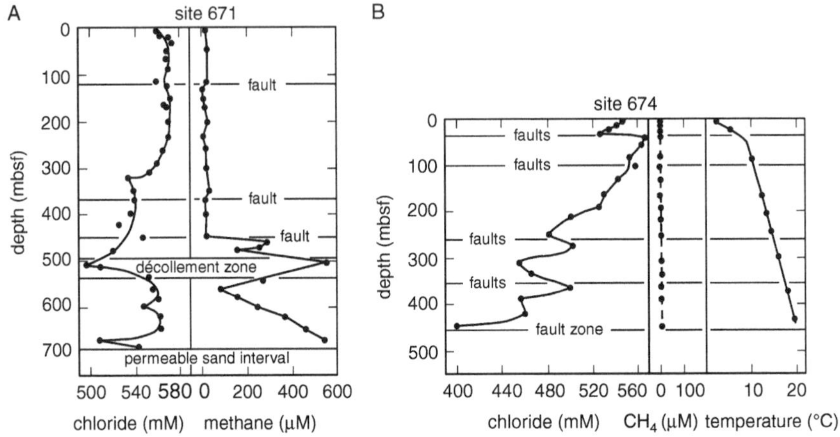

Figure 9.55 Methane and cloride anomalies at the décollement zone of the Lesse Antilles subduction zone. (A) Décollement zone between the upper Caribbean plate and the underthrusting Atlantic plate at the toe of the Lesser Antilles island-arc slope with methane and chloride anomalies (Site 671). Note that the methane anomaly is associated only with the décollement zone and a fault immediately adjacent to it as well as with an Eocene sand horizon below the décollement, but not with faults higher in the accretionary wedge (neither in hole 671 nor in hole 674). (B) Low-chloride waters, on the other hand, are also found in fault zones at greater distance from the décollement (e.g. Site 674, drilled up-slope 17 km west of Site 671). (From ODP Leg 110 Scientific Party, 1987.)

up to several centimetres thick in Pacific sediments. Straub and Schmincke (1998) evaluated the overall tephra input into the Pacific Ocean sediments by arc volcanism going back to the Mid-Miocene with the result that 10–13 km^3 of volcanoclastic sediments are produced per kilometre arc length in a million years. An extrapolation over the life-time of major Pacific volcanic arcs and hotspot chains, combined with a volume estimate of the distal tephra component, indicates a minimum of 9.3×10^6 km^3 of tephra ejecta, corresponding to 23% (by volume) of the Pacific oceanic sediments. In the following section, the significance of the alteration of volcanic matter for the chemical signatures of sediments and pore waters is reviewed. This includes the distribution of major and trace elements as well as isotopic data in pore waters that are influenced by such alteration.

11.1. Alteration of volcanic glasses in marine sediments

Electron microprobe (EMP) analyses of single glass shards taken from marine ash layers show a significant modification of the chemical composition of mafic glass shards. An important increase in the Fe_{alt}/Fe_{fresh} ratio (where Fe_{alt} is the iron content in the altered glass) was determined by

Schacht, 2005 while Mg_{alt}/Mg_{fresh}, and K_{alt}/K_{fresh} ratios increased only slightly and the concentrations of some elements (Ca, Na, Al) decreased significantly. Crovisier et al. (1983) described the release of Ca into solution and the addition of seawater Mg as the principal chemical exchange between basaltic glass and the leaching solution. The degree of alteration increases with age and results in the formation of a palagonitic rim followed by the formation of secondary minerals like smectites and zeolites (Bach et al., 2001). Palagonite is a yellow-brownish alteration product of glass.

In contrast, only few studies were performed investigating alteration rates and mechanisms of glasses of intermediate to felsic composition (e.g. Schacht et al., 2009; Wolff-Boenisch et al., 2004; Yokoyama and Banfield, 2002). EMP analyses of felsic shards, experimentally altered as well as taken from marine ash layers, show an active uptake of H_2O. Almost constant concentrations for network-forming elements such as Si and Al indicate only little replacement of Si and Al. Network-modifying elements (K, Na, Ca) show varying concentrations, indicating either losses or gains during alteration and the formation of a leached layer. Its thickness does not exceed a few micrometers. The stage of hydration can proceed until a saturation point of about 3 wt% of H_2O is reached. Further weakening and breakage of the Si–O–Si bonds of the glass structure results in a nearly complete dissolution of the glass. A palagonite-like interstage product is missing, resulting in the apparent unaltered appearance of marine felsic ash layers (e.g. Schacht et al., 2009).

11.2. Pore-water chemistry

11.2.1. Dissolved nutrients in pore waters of volcanogenic marine surface sediments

Schacht et al. (2008) investigated the chemical composition of pore waters related to marine ash-bearing sediment cores and thus to volcanic-ash alteration. The sediments were composed of terrigenous detrital material derived from the adjacent continent and contained several distinct ash layers of mafic to felsic composition. Biogenic opal and carbonate were only minor components. The terrigenous fraction was mainly composed of smectite and other clay minerals.

The composition of the pore fluids retrieved from these deep-marine surface sediments offshore Central America (cores A, B, C, D) show a systematic alkalinity increase (Fig. 9.56A) with subbottom depth and sulphate depletion (Fig. 9.56B) typical for rapidly accumulated slope sediments rich in organic carbon (Zuleger et al., 1996), as described in the section on early diagenesis of anoxic sediments.

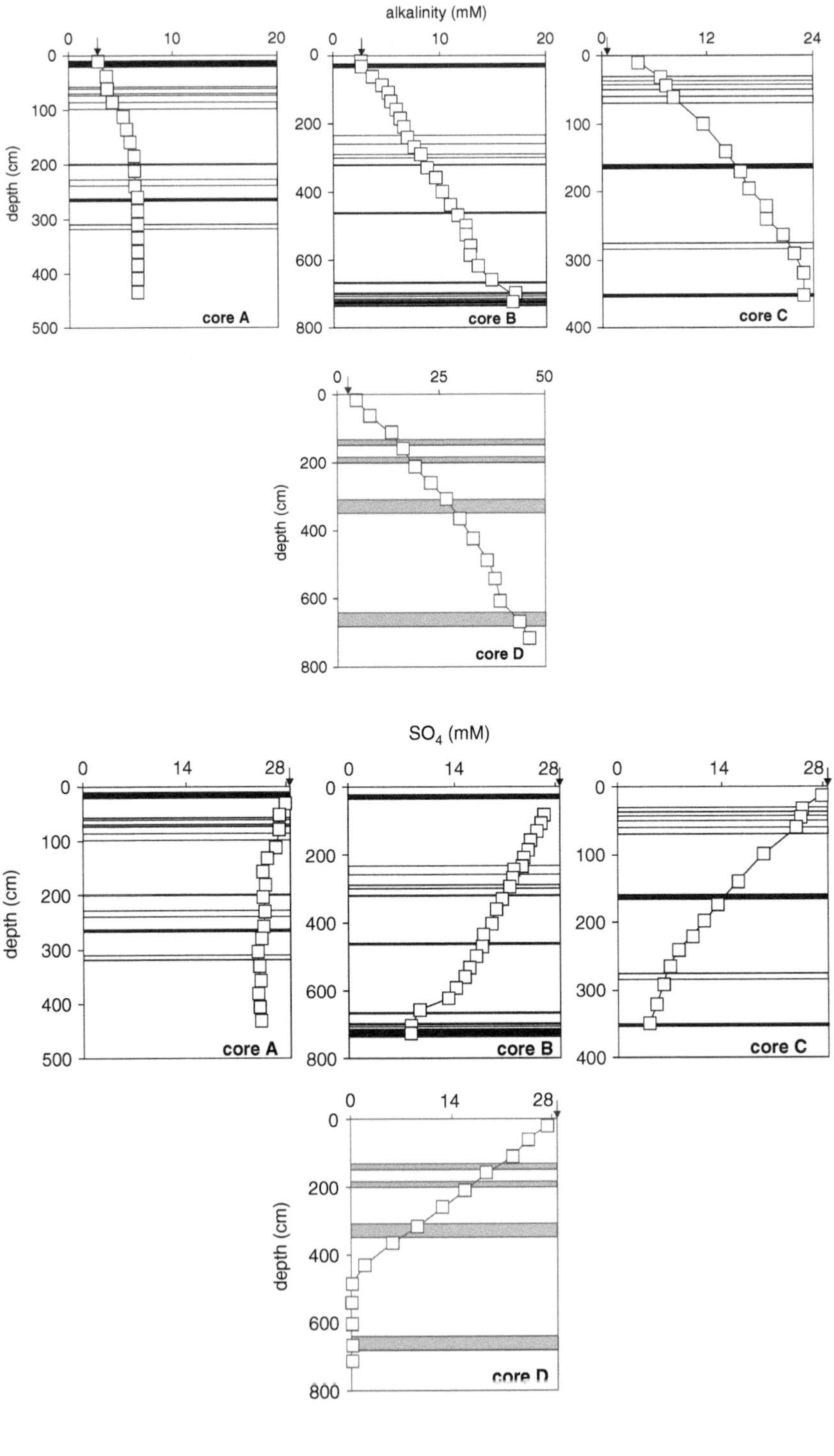
alkalinity (mM)
depth (cm)
core A
core B
core C
core D
SO4 (mM)
depth (cm)
core A
core B
core C
core D

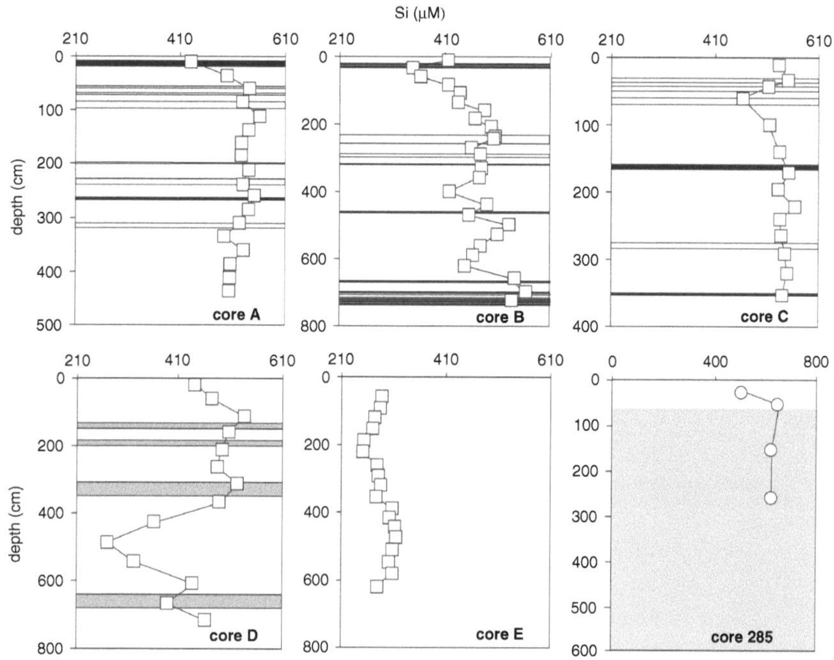

Figure 9.57 Silica pore-water profiles of ash-bearing and ash-free surface sediments (Schacht et al., 2008) and deeply buried ash-bearing sediments (Gieskes and Lawrence, 1981). Symbols as in Fig. 9.56.

11.2.2. Dissolved silica in pore waters of marine volcanogenic sediments

The concentration versus depth profiles of dissolved silica in pore waters of the shallow subsurface sediment cores referred to above (A, B, C, D) and one core of deeply buried sediments (DSDP Site 285) are remarkably similar to each other (Fig. 9.57) considering the differences in sedimentary biogenic-opal contents (Fig. 9.58). Dissolved-silica concentrations vary from approximately 300 μM near the sediment/seawater interface to about 500 μM at depth, but always stay far below their equilibrium solubility of ~1000 μM (Hurd, 1973). Even though the biogenic opal contents vary between the five ash-bearing cores, they are still low compared to other marine sediments. The consistently high dissolved-silica concentrations in pore waters of ash-bearing surface and deeply buried sediments described by

Figure 9.56 Pore-water profiles of ash-bearing and ash-free surface sediments (redrawn and modified from Schacht et al., 2008). Black bars indicate mafic ash layers, white bars felsic ash layers, and grey bars sediments enriched in volcanic material. Arrows mark the overlying seawater concentration of the element concerned.

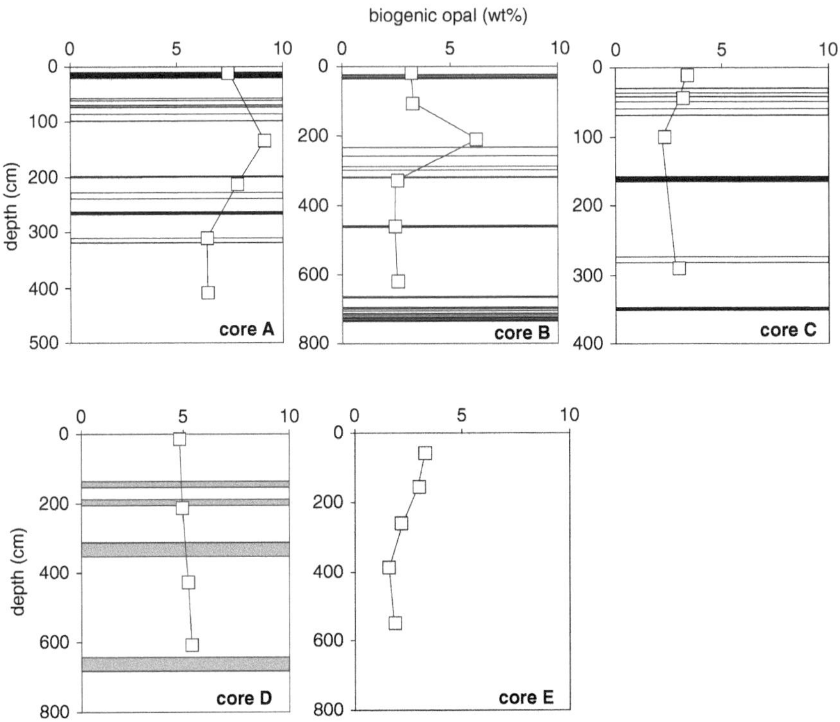

Figure 9.58 Contents of biogenic opal (wt%) in ash-bearing and ash-free surface sediments (redrawn and modified from Schacht et al., 2008). Symbols as in Fig. 9.56.

Schacht et al. (2008) and Gieskes and Lawrence (1981) therefore suggest an ongoing dissolution of volcanic glass, which generally has a significantly higher solubility than other silicates or aluminosilicate minerals.

The silica pore-water concentration of ash-free sediment core E (Fig. 9.57) is rather low, starting around concentrations of 280 μM at the very top of the gravity core. It is relatively stable with only a minor decrease with depth, reaching maximum values of 310 μM. The biogenic-opal content of core E is within the same range as described above for the ash-bearing cores (Fig. 9.58). This observation confirms that, concerning the cases presented here, the elevated silica pore-water concentrations of up to 600 μM in ash-bearing cores are probably caused by the alteration of volcanic ashes rather than by biogenic-opal dissolution.

Schacht et al. (2008) applied a transport reaction model to the silica and ash data of the marine sediment cores A, B, C and D to constrain the rate of dissolved-silica release during volcanic-glass alteration. The modelling showed that further silica release is inhibited when the dissolved silica concentration in the pore fluids approaches a saturation value of 450 μM. Hence ash dissolution occurs mainly in the upper sediment horizons, where the dissolved-silica concentrations are low and smaller than the saturation value. A good fit to the data was obtained applying the following kinetic rate law:

$$R_G = k_{GS} G \left(1 - \frac{Si}{Si_{max}}\right)$$

where R_G is the rate of glass dissolution, G is the glass concentration in the core, Si_{max} is the saturation concentration of dissolved silica (450 μM) and Si is the concentration of dissolved silica in the investigated pore fluids. The kinetic constant k_{GS} was determined as $3 \times 10^{-5}\ a^{-1}$ by fitting the model to the data.

11.2.3. Major elements in pore waters of volcanogenic deep-sea sediments

DSDP Site 285 in the South Fiji Basin was cored to a depth of 710 mbsf and is characterized by a high volcanic input below 62 m subbottom depth. The volcanic material with many glass shards in various stages of alteration has been described as intermediate to acidic in nature (Andrews et al., 1975). The pore waters of this deep-penetrating site are enriched in dissolved Ca and depleted in Mg (Gieskes and Lawrence, 1981) relative to bottom seawater (Fig. 9.59). The directions of element transfer between ash/glass and seawater inferred from these trends are similar to those observed in laboratory experiments at moderate to elevated temperatures (e.g. Berger

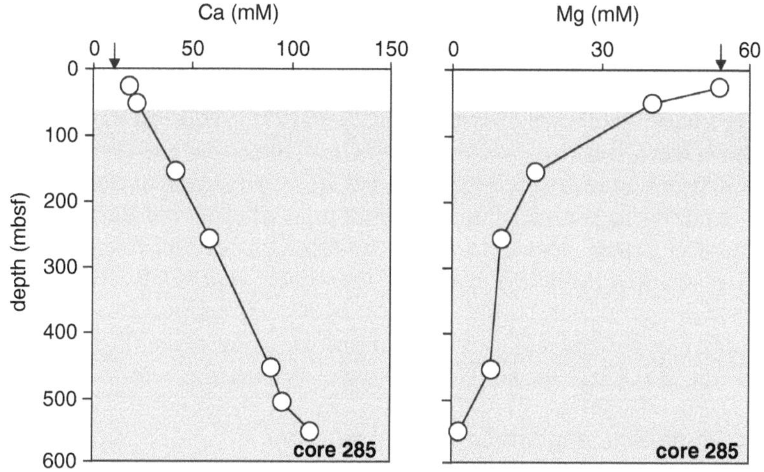

Figure 9.59 Ca and Mg pore-water profiles of deeply buried ash-bearing sediments (redrawn and modified Gieskes and Lawrence, 1981). Symbols as in Fig. 9.56.

et al., 1987; Schacht, 2005; Seyfried and Bischoff, 1979; Staudigel and Hart, 1983) and were thus ascribed to ash alteration.

11.2.4. Rare-earth elements in pore waters of marine volcanogenic sediments

Two marine sediment gravity cores were recovered during research cruise *RV Sonne* 173/3 along the Pacific margin offshore Nicaragua. Thick ash layers dominate the recovered sediments. Pore-water samples of these—one incoming plate (No. 15) and one slope site (No. 18)—sediment cores have been analysed by Schacht et al., 2010 for light rare-earth elements. The pore waters of core 18 are enriched in dissolved rare-earth-element (REE) contents (Fig. 9.60) compared to the overlying seawater composition (Table 9.4), demonstrating that REE are mobilised during early diagenesis. However, REE concentrations show no or only minor concentration maxima within ash layers. Core 18 represents the environment for continental slope sediments. In this situation, ash layers obviously provide only as little REE input into pore waters as described for the background sediments Schacht et al., 2010.

Core 15 represents the incoming-plate environment of mildly reducing sediments. Its REE-concentration-depth profile shows enriched REE contents in pore waters of background sediments compared to seawater (Fig. 9.60), but in contrast to core 18, the REE concentrations are especially high in pore waters of volcanic-ash layers (note scales in Fig. 9.60). The largest REE enrichment is observed for Ce, which reaches concentrations more than 100 times higher than the deep-ocean water value of 5.89 pmol kg^{-1} (Table 9.3).

During the microbial degradation of organic matter, metabolites such as ammonia and phosphate are released into the pore water (Fig. 9.61) to be transported through the sediment column via diffusion and burial advection (Schacht et al., 2010). The highest concentrations of dissolved REE occurred in the two prominent ash layers of core 15, where pronounced maxima in dissolved phosphate were observed (Fig. 9.61). On one hand, this suggests in all other sections of the two studied gravity cores, that the dissolved-REE concentrations were lowered when phosphate precipitation occurred. On the other hand, it suggests that high concentrations of dissolved REE in marine pore waters can only be maintained in the absence of phosphate-precipitation processes (marked by arrows in Fig. 9.61) (Schacht et al., 2010). The carbonate-fluoride-apatite, which is a common authigenic mineral in marine sediments, may be a major sink not only for dissolved phosphate (Delaney, 1998; Wallmann, 2003) but also for REE released from marine volcanic-ash layers.

11.2.5. Isotopic composition of pore waters

The above conclusion that volcanic matter and its alteration have a significant effect on pore-water chemistry is supported by studies of the $^{87}Sr/^{86}Sr$ ratio of dissolved strontium as well as by studies of the oxygen-isotopic

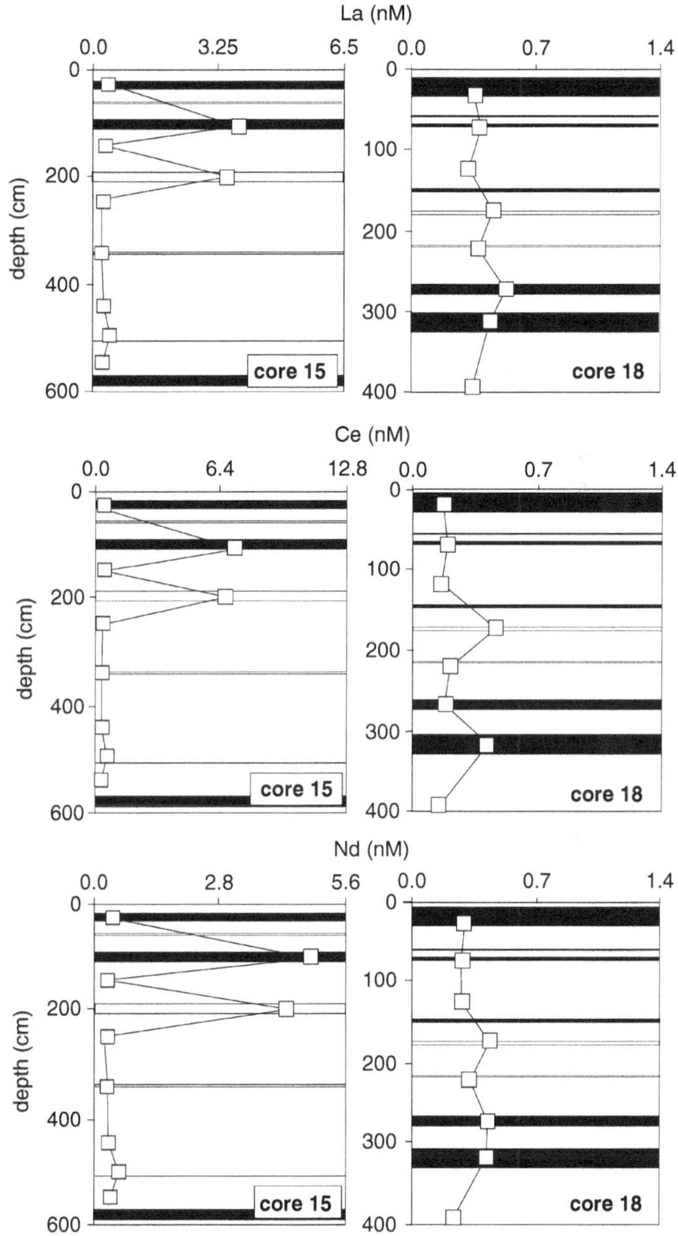

Figure 9.60 Pore-water profiles of selected REE in ash-bearing surface sediments (redrawn and modified from Schacht et al., 2010). Symbols as in Fig. 9.56.

Table 9.3 Rare-earth-element composition of North Pacific seawater for two different depth horizons according to Piepgras and Jacobsen (1992)

	La	Ce	Pr	Nd	Sm	Eu	Unit
Depth (m)	Average surface water North Pacific						
0–1000	25.7	5.66	No data	16.2	3.04	0.81	pmol/kg
Depth (m)	Average deep water North Pacific						
1000–6000	53.8	5.89	No data	36.2	6.84	1.79	pmol/kg

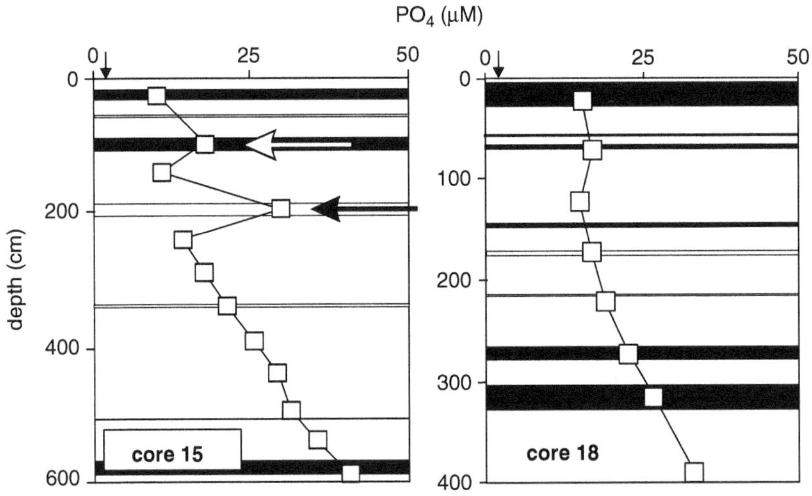

Figure 9.61 PO_4 pore-water profiles of ash-bearing surface sediments (redrawn and modified from Schacht et al., 2010). Symbols as in Fig. 9.56. Small red vertical arrows mark the overlying seawater concentration of the element concerned. Big red horizontal arrows mark horizons without phosphate precipitation.

composition of pore waters (Elderfield and Gieskes, 1982; Gieskes, 1983; Gieskes and Lawrence, 1981; Gieskes et al., 1984). This effect was already discussed in previous sections on biogenic pelagic sediments and diffusion-controlled pore-water profiles.

DSDP Sites 541 and 542 were drilled near the toe of the Lesser Antilles island-arc slope in the accretionary wedge seaward off the Barbados Ridge complex and Site 543 into the adjacent Atlantic oceanic crust. Penetration was up to 430 mbsf. As shown in the section on biogenic pelagic carbonates, mid-depth maxima in dissolved Sr are related to carbonate recrystallisation. Sr-isotope ratios above the maximum are lower than seawater values for the age of the host sediment, and below the maximum they are higher due to upward and downward diffusion from the Sr maximum, respectively.

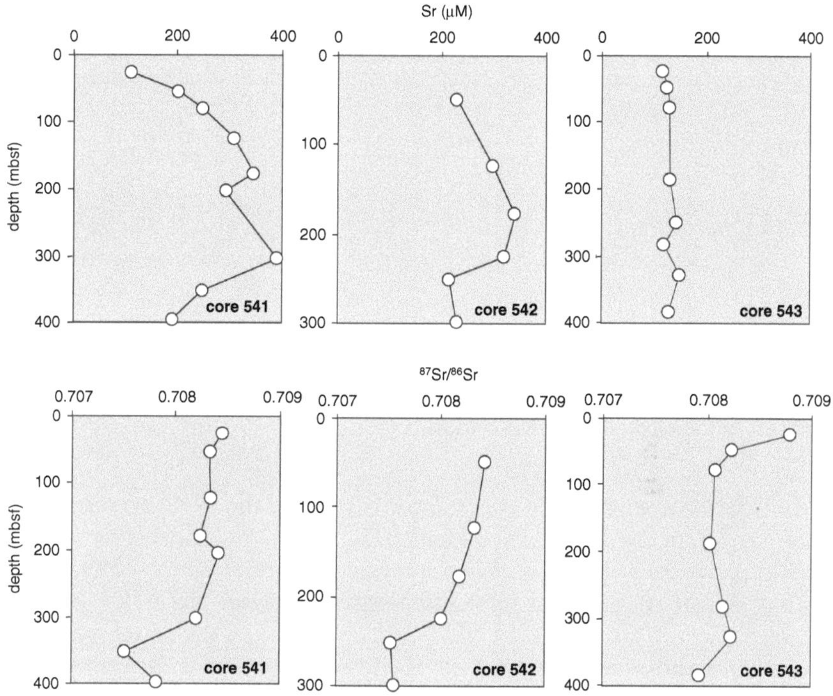

Figure 9.62 Sr and $^{87}Sr/^{86}Sr$ pore-water profiles of deeply buried ash-bearing sediments (redrawn and modified from Gieskes et al., 1984).

The pore-water depth profiles of all three carbonate-poor Sites 541, 542, 543 (Fig. 9.62) are marked by a relatively large and rapid decrease in the $^{87}Sr/^{86}Sr$ ratios lacking the upper part of downward increasing values. The Sr-isotope ratios of the pore waters in the upper part of the holes fall much below the paleo-seawater curve (Fig. 9.62B) and are due to volcanic-ash alteration (Gieskes et al., 1984). At Site 543, no significant increases in dissolved Sr occurred (Fig. 9.62A), so that it was possible to estimate the volcanic contribution to the $^{87}Sr/^{86}Sr$ ratio. For this the $^{87}Sr/^{86}Sr$ ratio was taken to be 0.708 at 100 mbsf (for contemporaneous seawater, $^{87}Sr/^{86}Sr$ is ~0.709). If volcanic material were to have an $^{87}Sr/^{86}Sr$ ratio of 0.703 before alteration, the contribution by the volcanic material to the $^{87}Sr/^{86}Sr$ ratio of dissolved strontium would be ~ 17%. This should be a minimum estimate because it ignores diffusion and assumes a relatively low amount of volcanic matter involved (Elderfield and Gieskes, 1982). In core 541 at ~150 mbsf, the $^{87}Sr/^{86}Sr$ ratio is ~0.7083. At this depth, the concentration of dissolved Sr is ~300 μM, and the contemporaneous seawater has an $^{87}Sr/^{86}Sr$ ratio of ~0.709. With these values, the calculated volcanic contribution to the observed $^{87}Sr/^{86}Sr$ ratio is at least 30%.

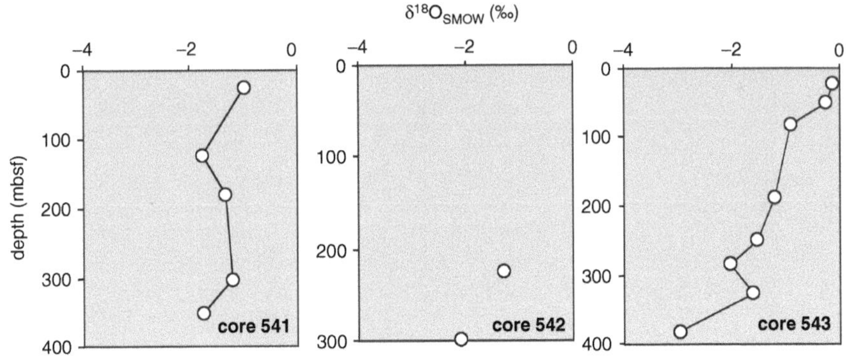

Figure 9.63 $\delta^{18}O$ pore-water profiles of deeply buried ash-bearing sediments (redrawn and modified from Gieskes et al., 1984).

In DSDP Sites 541–543, a substantial portion of the $\delta^{18}O$ decrease must have been generated in the sediments (Fig. 9.63), presumably as a result of volcanic-matter alteration. With a mean value of $\delta^{18}O$ of -1‰, about 5–15 m of ash are required in a closed system to cause the $\delta^{18}O$ decrease (Lawrence and Gieskes, 1981). This estimate could be increased when the possibility of diffusive exchange with the overlying ocean is also considered. However, this estimate would require alteration of about 20% of the sediments in the upper 150–200 m of the sediment column (Lawrence and Gieskes, 1981a,b).

11.3. Formation of zeolites in volcanogenic sediments

Synchronous decreases in $\delta^{18}O$ and Mg^{2+}, typical for diffusion-controlled pore-water profiles (see Section 2.3 on diffusion-controlled pore-water profiles using Ca^{2+} instead of Mg^{2+}) are related to volcanic-matter alteration processes involving the formation of zeolites and clay minerals. The crystallization of zeolites from volcanic material, particularly volcanic glass, may start on the sea floor. It proceeds through the formation of palagonite, a common zeolite precursor. In many pelagic drill sites, the pore waters display a downward decrease in K^+. This may be related to the formation of phillipsite [$(K,Ca)Al_3Si_5O_{16}\cdot 6H_2O$], the main potassium-bearing zeolite. Other common zeolites that form during early diagenesis in pelagic sediments such as clinoptilolite and analcite leave a less characteristic imprint on the pore-water composition. As Kastner and Stonecipher (1978) showed, phillipsite is most abundant in the youngest sediments and decreases with age and burial depth, opposite in behaviour to clinoptilolite. The transformation of zeolites into K-feldspar with progressively deeper burial is another sink for K^+, as are adsorption and ion exchange with clays. These processes lead into the realm of middle diagenesis. Feldspar precipitation in particular does not occur during early burial diagenesis.

11.4. Formation of smectites in volcanogenic sediments

The smectite concentration maximum of 70% and more in the <2 μm fraction of pelagic sediments in remote oceanic areas of the Central and South Pacific is mostly due to alteration of volcanic glass on the modern sea floor or beneath it in the sediment, although some of the smectite is supplied by erosion from oceanic volcanic islands. According to estimates by Griffin et al. (1968), about half the smectite in surface sediments of the Pacific and South Atlantic Oceans and two thirds in the Indian Ocean (Chester et al., 1974) are authigenic in origin. These authigenic smectites are mostly nontronitic (Fe-rich montmorillonitic smectites) and typically form from the alteration of basaltic volcanic matter. Smectites with lower Fe-contents and increased Mg- and Al-concentrations, on the other hand, are derived from non-basaltic parent material. Authigenic smectites have also been observed in the tests of siliceous organisms, where they form by reaction of silica with iron oxides (Chamley and Millot, 1972).

At 400–500 m subsurface depth in DSDP drill-holes 322 and 323 in the Bellinghausen Abyssal Plain of the Southeast Pacific, Mg-rich authigenic smectite intergrown with opal-CT formed in the same way from the dissolution of biogenic opal-A and volcanic glass (Kastner and Gieskes, 1976). Smectites in pelagic sediments usually lack interstratification with other clay minerals. Commonly they occur in paragenesis with the zeolite phillipsite with which they share a REE abundance pattern marked by a negative cerium (Ce) anomaly that is typical of sea-water derivation (Piper, 1974a,b). South Pacific smectites are characterized by this pattern, strongly supporting their early-diagenetic marine origin. Smectites in Tertiary sediments from the North Pacific, on the other hand, lack the seawater signature and have REE patterns similar to associated illites, for which a terrigenous-detrital origin has been suggested.

12. Early-Diagenetic Mineralization Reactions in Anoxic Deep-Water Sediments

The first mineralization reactions in freshly deposited organic-matter-rich deep-water sediments include the formation of iron sulphides, occasionally manganese sulphide, and various carbonates which are the main authigenic constituents of concretions. Carbonate concretions are a trademark of black-shale formations in the geological record, pointing at early-diagenetic organic-matter decomposition as the carbon source.

They conserve an amazing wealth of information concerning the chemical and isotopic composition of the pore waters and the physical growth conditions during early diagenesis. Mineralization reactions that produce volumetrically subordinate solid phases such as apatite are significant for shaping the pore-water depth profiles of the ions involved (i.e. phosphate) and have been discussed above (Section 7.2.1).

12.1. Early-diagenetic sulphide precipitation

In the presence of soluble ferrous iron, the sulphide produced in the SR zone immediately reacts to form metastable iron monosulphides such as mackinawite, greigite and amorphous FeS. Mackinawite is deficient in sulphur ($FeS_{0.90}$ to $FeS_{0.96}$), and up to 10% of the Fe is substituted for by Ni^{2+} and Co^{2+}. Greigite (Fe_3S_4) is a cubic mineral, probably with the structure of an inverse spinel. These intermediate phases are kinetically favoured at high supersaturation levels over the direct precipitation of pyrite, which has a much lower solubility product (2.4×10^{-28} as compared to 2.8×10^{-18} for mackinawite: Berner, 1967; Goldhaber and Kaplan, 1974). Pyrite originating from the transformation of these precursor phases displays a characteristic framboidal (raspberry-like) structure composed of tiny euhedral crystals of uniform size. This particular structure has been produced experimentally by reacting iron monosulphides with elemental sulphur (Berner, 1969; Farrand, 1970). Euhedral pyrite, which may overgrow the early framboidal pyrite, forms later at lower saturation levels (Berner, 1984).

These relationships are corroborated by the distribution of sulphur isotopes in pyrite-bearing carbonate concretions. A large isotopic fractionation is related to bacterial sulphate reduction, producing sulphide about 50‰ lighter than seawater sulphate (Goldhaber and Kaplan, 1980). Seawater sulphate at present has a $\delta^{34}S$ value of +20‰ relative to the CDT (Canyon Diablo Troilite) standard. Open-system conditions as in stagnant basins, where sulphate reduction starts above the sediment/water interface (Fig. 9.29), should supply light sulphide with $\delta^{34}S$ values in the range −20‰ to −30‰ to the sediment. Closed-system conditions, as during rapid sulphate reduction in organic-matter-rich sediments, on the other hand, should involve a Rayleigh distillation leading to increasingly heavier sulphides, as the light sulphur is preferentially withdrawn in the early stages. The sulphur-isotope composition of early-diagenetic pyrite may then provide information on the environment, in which anoxic, organic-matter-rich sediments were deposited.

In Cretaceous shales from the Western Interior Seaway of North America, two distinct sulphur-isotope trends are observed (Fig. 9.64). Samples with high organic-matter concentrations (4–10%) display a narrow range of very light sulphur-isotope compositions (pyrite $\delta^{34}S$ = −35‰ to −25‰) indicating open-system conditions, that is, sulphate reduction largely in the

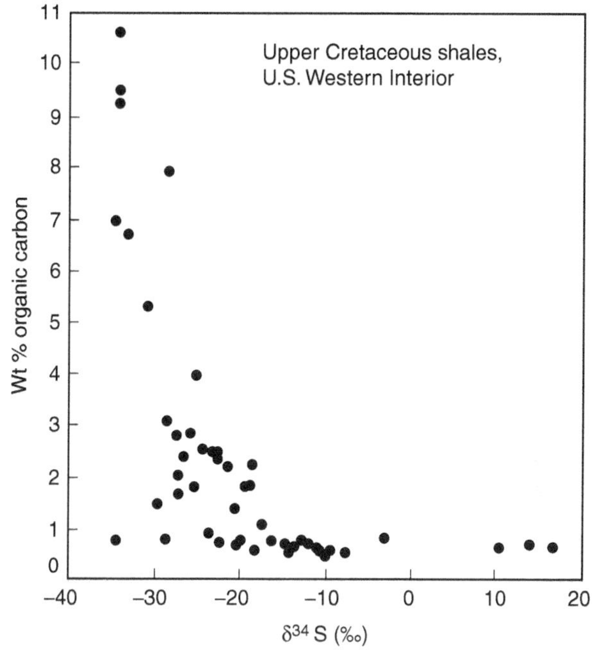

Figure 9.64 Percentage of organic carbon versus $\delta^{34}S$ of disseminated pyrite in Cretaceous shales from the Western Interior of North America (redrawn and modified from Gautier, 1985).

water column above the sediment surface. Samples with less than 3% organic carbon, on the other hand, show a wide range of $\delta^{34}S$ values (from −35‰ to +18‰), suggesting closed-system conditions, or a rapid change from initially open- to later closed-system conditions, probably under oxidizing bottom waters (Gautier, 1985; Schwarcz and Burnie, 1973). Wignall and Newton (1998) used the diameter of framboidal pyrite as an indicator of redox conditions. Tiny framboids with diameters of <5 μm and a narrow size range are indicative of pyrite precipitation in the water column under stagnant basin conditions, if certain precautions are taken into consideration.

Euhedral pyrite, which in part nucleated on the framboidal pyrite, becomes progressively heavier from the centre ($\delta^{34}S$ = −24‰ to −14.5‰) towards the concretion margins ($\delta^{34}S$ = −5.5‰ to −2.5‰; Fig. 9.65), reflecting the establishment of closed-system conditions when the resupply of light sulphate from sea water by diffusion ceased. Isotopically heavy, large euhedral pyrite crystals, which line the rims of these concretions, show displacive and sectorial growth, forming hopper shapes and occasionally cone-in-cone structures (Carstens, 1986).

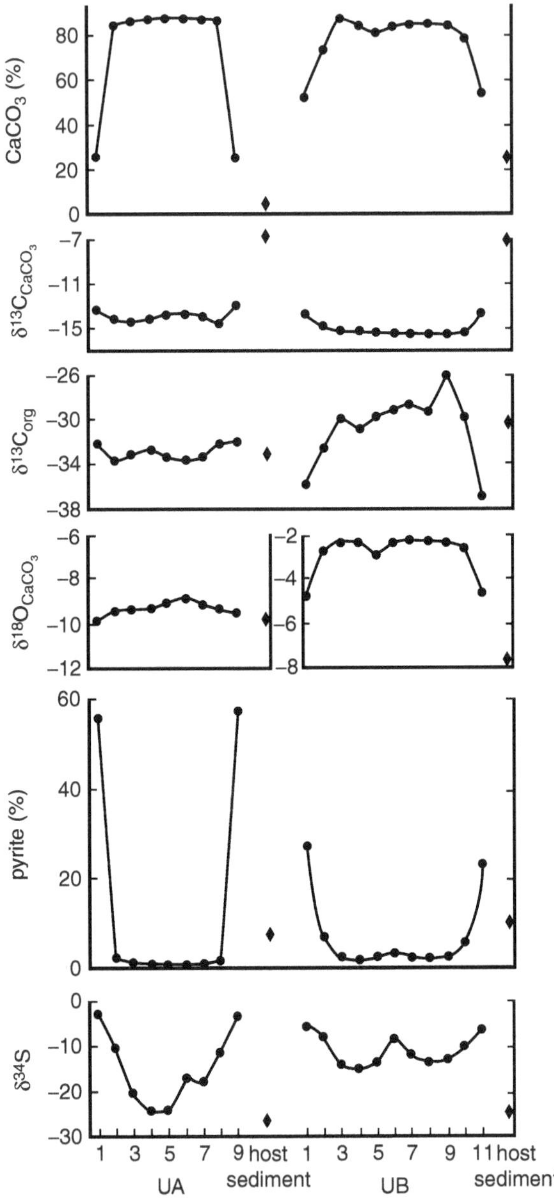

Figure 9.65 Centre-to-rim mineralogical and isotopic variations in Liassic (Early Jurassic) carbonate concretions from northern England (redrawn and modified from Coleman and Raiswell, 1981). Centre of concretion UA at sample 5, of concretion UB at sample 6.

The rare occurrence of authigenic manganese sulphide (alabandite) in modern anoxic sediments of the Baltic Sea (Baron and Debyser, 1957; Suess, 1979) has been ascribed to high dissolved Mn^{2+} concentrations that are typical for anoxic bottom water in local depressions of this restricted basin. Thus, alabandite precipitates instead of the usual iron sulphides.

12.2. Authigenic carbonates: Calcite and siderite

12.2.1. Chemical environment of authigenic carbonate precipitation and growth of calcite and siderite concretions

In organic matter-rich sediments, authigenic carbonates are commonly precipitated from the pore waters during early diagenesis. During the initial stages of organic-matter decomposition in the oxidation and nitrate-reduction zones, carbonate dissolves because the pH is low and buffered by the carbonate system. An increasing P_{CO_2} due to organic-matter oxidation in oxic environments will lower the pH and raise the bicarbonate rather than the carbonate activity. Prerequisite for carbonate precipitation is a relatively high pH, which is buffered by reactions other than those of the carbonate system, as in anoxic environments. The reduction of manganese and iron oxides and hydroxides in the nitrate- and SR zones raises the pH through alkalinity production according to reactions of the type:

$$Me_2O_3 + 3H_2O + 2e \rightarrow 2Me^{2+} + 6OH^-,$$

and

$$4MeOOH + H_2O + CH_2O \rightarrow 4Me^{2+} + CO_2 + 8OH^-$$

where Me^{2+} represents Mn^{2+} or Fe^{2+} ions. Hydrogen-ion consumption by ammonia production has the same effect. Under these conditions, an increased P_{CO_2} will increase carbonate activity sufficiently to cause supersaturation with respect to carbonates (Coleman, 1985; Suess, 1979). Additional carbonate alkalinity may be generated by anaerobic oxidation of methane and upward diffusion from the base of the SR zone (Borowski et al., 2000; Raiswell, 1987). In the carbonate-reduction and fermentation zones, hydrogen consumption during bacterial methane formation (in the presence of metal ions) according to a reaction of the type:

$$Me^{2+} + 2HCO_3^- + 8H^+ + 8e \rightarrow CH_4 + MeCO_3 + 3H_2O$$

will contribute alkalinity only if metals (Fe, Mn) are still available for reduction, and if the reaction is coupled with the two reactions above (Raiswell, 1987).

Under these circumstances, calcite, dolomite, siderite, rhodochrosite and more complex iron–manganese carbonates may form in organic-matter oxidation zones (2)–(5). Fe-carbonates (ferroan calcite, siderite) do normally not form in the SR zone in the presence of dissolved sulphide, which competes for the dissolved iron, unless the iron-reduction rate exceeds the SR rate (Pye et al., 1990). Authigenic calcite is among the earliest diagenetic minerals and may start to precipitate in the SR zone, together with or shortly after the first precipitation of Fe-monosulphides, if prior dissolution of carbonates in the oxidation and nitrate-reduction zones has raised the Ca activity sufficiently. The authigenic carbonates may be finely dispersed in the sediment as small crystals or may form concretions: the most conspicuous result of early-diagenetic carbonate precipitation.

12.2.2. Elemental and isotopic evidence

Chemical and isotopic compositions of carbonate concretions in black-shale successions reveal concentric centre-to-rim variations that record the oxidation processes of organic-matter and related pore-water trends discussed in previous sections. In the case of a concretion starting to form in the SR zone and continuing to grow in the carbonate-reduction and fermentation zones, an early calcite nucleus will be surrounded by a sideritic rim, as shown by examples from the Gammon Shale (Gautier, 1985; Gautier and Claypool, 1984). Oxygen isotopes at the centre are similar to those of calcite precipitated from modern sea water at temperatures above 15 °C ($\delta^{18}O$ values between −2‰ and −3‰) and become progressively lighter outward in the siderite rim (as low as −7‰) (Irwin et al., 1977). The calcite is associated with abundant framboidal pyrite, leaving little

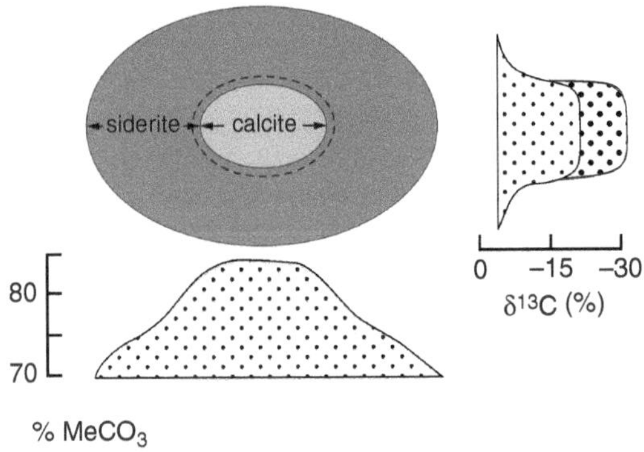

Figure 9.66 Schematic representation of the growth stages of calcite/siderite concretions from Cretaceous shales of the Western Interior Seaway of the USA (redrawn and modified from Gautier, 1985).

doubt that the iron-free calcite formed in the SR zone. Carbonate crystals in contact with the pyrite framboids radiate outward from the latter, indicating that the pyrite formed first. The calcite probably formed in the lower part of the SR zone, because the $\delta^{13}C$ values of -25 to -20‰ (Fig. 9.66) show the maximum effect of organic-matter oxidation. The carbon-isotope signature of calcite concretions formed in the SR zone may represent a mixture of carbonate derived from sulphate reduction, methane oxidation and from dissolution of marine calcium carbonate, and may span the range from 0 to -25‰ $\delta^{13}C$.

Uptake of iron into the carbonate phase during siderite precipitation, on the other hand, shows that the concretions continue to grow in the methane-generation and fermentation zones (4) and (5). In these zones, dissolved iron becomes available and the carbonate gets isotopically distinctly heavier because of the withdrawal of light carbon in methane. In the SR zone, ferrous iron concentrations in the presence of dissolved sulphide (at concentrations typically of the order of $10^{-3.5}$ M) are 10^{-16} M (for pH $= 7.5$ and Eh $= -0.245$ mV), far too low for iron carbonates to form. Siderite formation requires dissolved Fe^{2+} concentrations of at least 10^{-7} M (Curtis, 1967). Siderite concretions in continental terrigenous sediments tend to have higher $\delta^{13}C$ values than those precipitated in marine sediments, because of the lack or near-absence of sulphate in fresh-water environments. Consequently, because little organic matter is lost during sulphate reduction, the $\delta^{13}C$ is not lowered, and a higher proportion of organic matter is preserved for methanogenesis (Mozley and Wersin, 1992). Direct precipitation of siderite from the water column has been suggested for finely disseminated siderite in Plio-Pleistocene sideritic muds of the Black Sea when the basin had become a fresh-water lake undergoing evaporative concentration beyond roughly 1/12 of its original volume (Rajan et al., 1996). Because of the low sulphate concentrations in the pore waters of Black Sea sediments at this time, sulphate reduction was of minor importance. Anoxic diagenesis in this case was methanic (methane-dominated) and non-sulphidic (Berner 1981).

12.3. Organogenic dolomite ("deep-sea dolomite")

12.3.1. Deep-sea environment of modern and ancient authigenic dolomite in anoxic hemipelagic sediments

Besides calcite and siderite, dolomite and high-magnesium calcite are common authigenic carbonates in anoxic sediments. The diagenetic origin of organogenic dolomite in deep-water terrigenous muds and biogenic siliceous oozes is now firmly established based on the occurrence in deep-sea drill sites on the slopes of active and passive margins in several kilometres water depth (e.g. the California Borderland: Kelts and McKenzie, 1982; Pisciotto and Mahoney, 1981; the Middle America trench slope:

Matsumoto, 1983; Wada et al., 1981; the SW African continental margin: Pufahl and Wefer, 2002). Like other authigenic carbonates in mudrocks, the authigenic deep-sea dolomite occurs as scattered rhombs in unconsolidated argillaceous sediments (e.g. Lumsden, 1988) and their ancient equivalents (Landing et al., 1992) and as concretions or diagenetic beds.

An early-diagenetic origin and organic-matter source of the carbonate were first suggested by Bramlette (1946) and Spotts and Silverman (1966) for bedded and nodular dolomite in the Miocene Monterey Formation. The Monterey Formation has attracted much attention more recently as the main source rock for hydrocarbons in California (e.g. Garrison et al., 1984). Diatoms in these concretions may be preserved as opal-A when the surrounding siliceous sediments have reached the opal-CT or quartz stage, illustrating the efficiency of the carbonate cement in sealing off the concretions from the pore solutions. Monterey-type dolomite concretions as old as Ordovician have been studied in detail (Hesse et al., 2004). Direct precipitation of dolomite from aqueous solution is a less common mode of dolomite formation than the widespread replacement origin in shallow-water carbonates.

Mineralogical, chemical and isotope data for organogenic deep-water dolomite in the Middle Ordovician Cloridorme Formation of the Northern Appalachians in Quebec constrain the physical and chemical growth conditions of these concretions and diagenetic beds (Hesse et al., 2004). The host sediments of the dolomite concretions and diagenetic beds are turbidites with intercalated hemipelagic sediments (Enos, 1969a,b) which presently contain between 0.1% and 2.5% organic carbon.

12.3.2. Physical conditions of dolomite-concretion growth: Porosity at the time of precipitation

In concretions or diagenetic beds, the carbonates characteristically are pore-filling cements the decreasing concentration of which from centre (up to 85 or 90% by volume) to rim (as little as 25%) reflect the decreasing porosity at the time of precipitation (Hesse et al., 2004). For porosities estimated in this way, the term “minus-cement porosity” (Taylor, 1950) is commonly used. High minus-cement porosities, in the range of 70–90%, in the inner shells of many concretions attest to the early diagenetic initiation of concretion growth. The carbonate content of concretions is a good estimator of paleoporosity (and therefore burial depth) at the time of concretion growth under the following conditions.

(1) The carbonate crystals in the concretions did not displace detrital components. Cementation by micritic carbonate is not displacive, in contrast to slowly growing, low-nucleation rate cone-in-cone and related spherulitic calcite cements or calcium sulphates that create

their own growth space and cannot be used for porosity estimates (e.g. Al-Aasm et al., 1993).

(2) No carbonate recrystallisation and replacement have taken place. Preservation of the growth-related isotopic zonation of the concretions and diagenetic beds argues against recrystallisation (see also Section 12.3.5).

(3) The host-sediment was originally carbonate-free. If the host sediment contains detrital or biogenic carbonate and/or authigenic sulphide minerals precipitated prior to or during concretion growth, minus-cement porosities have to be corrected accordingly. For the Cloridorme dolomites, this correction was done by assuming that the discrete calcite identified in XRD (ranging from 0% to 39%) was detrital in origin. In thin-sections, non-ferroan calcite grains, like detrital silt-sized quartz, are on average twice as large as the authigenic sucrose ferroan dolomite crystals with a diameter of 10–15 μm. Higher calcite contents (>30%) are found in Cloridorme concretions or beds which grew in carbonate-rich mud turbidites (e.g. diagenetic bed β6-2: Fig. 9.67A).

(4) The fine-grained carbonate cement completely fills the pore space at the time of concretion growth. As far as the timing of complete filling of the pore space by authigenic cement is concerned, the model of overlapping cementation shells provides insight into the problem (see Section 12.4).

The carbonate concentration of the Cloridorme concretions and diagenetic beds (Table 9.4) generally shows a systematic centre-to-margin decrease (Fig. 9.67A).

Paleoporosities estimated from minus-cement porosities may be converted to burial depths using compaction curves for appropriate tectonic environments. For the Cloridorme Formation, comparison with compaction curves from deep-sea drilling sites (Hesse et al., 2004; Rocker, 1974; Shephard et al., 1982) suggests that the measured maximum minus-cement porosities (87%) correspond to a burial depth of <1 mbsf, where concretion growth started. Minus-cement porosities for the margins of most concretions and some diagenetic beds in the range of 80–70% suggest a depth of <25 mbsf, certainly not deeper than 75 mbsf, where concretion growth stopped. Considerations concerning the diffusive communication length (Section 12.3.7) suggest a depth limit for active concretion growth of ~10 mbsf. Field studies, however, provide evidence for concretion growth that has occurred at kilometre depth. For example, siderite concretions from the East Coast continental rise off the United States may have formed at burial depths of 800–1200 m, well beyond the early-diagenetic stage, in accordance with their isotopic signatures (DSDP Leg 93: Von Rad and Botz, 1987).

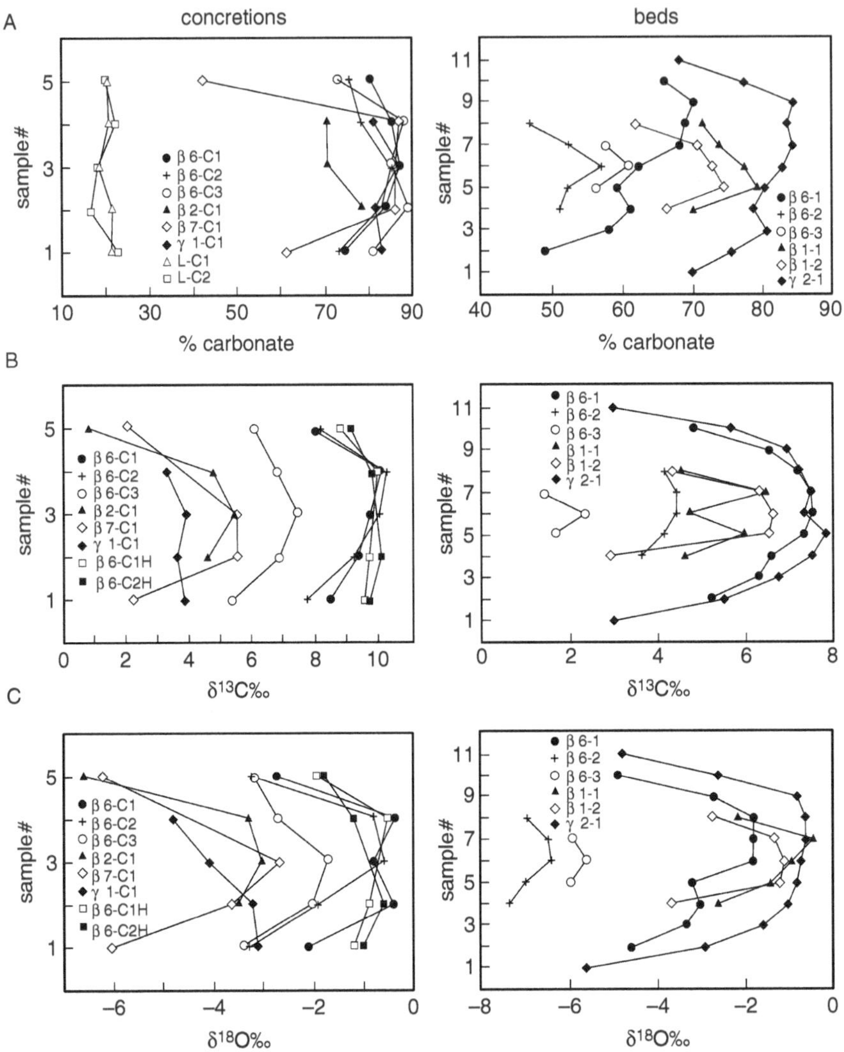

Figure 9.67 Carbonate content, $\delta^{13}C$ and $\delta^{18}O$ values across concretions and diagenetic beds of the Cloridorme and Levis Formations and the Cow Head Group. Sample numbers in ascending order across vertical traverses. H, horizontal traverse (redrawn and modified from Hesse et al., 2004; fig. 7). (A) The greater variability in carbonate content of the beds compared to the concretions indicates a longer duration of the growth of the former, reflecting the coalescence of individual concretions in a concretion horizon. Although the samples were collected from traverses across several diagenetic beds, random sampling of the beds apparently resulted in different growth stages being captured. A few traverses represent parts of beds that had completed most of their cementation early (β1, γ2); in other beds, growth started much later (β6). (B) Slight decreases in the $\delta^{13}C$ values of the central samples in diagenetic beds β1-1 and γ2-1 compared to adjacent samples probably reflect commencement of dolomite growth before peak methane generation. (C) The $\delta^{18}O$-profiles display similar comet-shaped decreases from centre to the margins of the concretions and beds as the carbonate- and $\delta^{13}C$-profiles reflecting decreases of the three parameters with increasing burial depth. For the horizontal traverses of the concretions the decreases are significantly smaller than for the vertical traverses.

Table 9.4 Minus-cement porosities, estimated burial depths and temperatures for concretions and diagenetic beds of the Ordovician Cloridorme and Levis Formations, Quebec Appalachians (from Hesse et al., 2004)

	Cloridorme Formation				Levis Formation	
	Concretions		Beds			
	Centre	Margin	Centre	Margin	Centre	Margin
Minus-cement porosity (%)	90	43	85	46	25	17
Estimated burial depth (m)		≤1–10 (max 75)				≤350–600
$\delta^{13}C$ (‰)	+10.2	+0.8	+7.7	+1.2	−3.2	−5.2
$\delta^{18}O$ (‰)	−0.4	−6.6	−0.5	−7.5	−4.1	−5.3
Estimated T (°C) (uncorrected)	17	43	17	48	32	38

Diagenetic beds of the Cloridorme Formation show significantly larger variations in minus-cement porosities than the concretions (Fig. 9.67A). This is to be expected, if diagenetic beds, also called "sheet concretions" (Pirrie and Marshall, 1991), result from the coalescence of concretions in concretion horizons during prolonged growth. Those parts of the beds originally located between concretions that have not yet coalesced become cemented much later and consequently have lower minus-cement porosities.

Concretion and bed growth in all examples discussed stopped a long time before compaction ceased. Independent evidence for substantial post-concretion compaction of the host sediment is obtained from the reduction of the distance between distinct depositional layers traced laterally from within a concretion into the not-carbonate cemented host rock between concretions (Fig. 9.68).

Other observations to determine the state of compaction at the time of cement precipitation include (1) the degree of preservation of the original shape of fossils and fecal pellets, and (2) the degree of preferred orientation of clay particles. Clay-fabric analyses of Carboniferous carbonate concretions from England (Oertel and Curtis, 1972) have shown that, in the centre of the concretions, clay-particle orientation is essentially random, as would be expected for a flocculated clay suspension with 80–90% water content in which carbonate precipitation occurred. Towards the concretion margins, a distinct trend emerges of increasingly preferred orientation of the basal planes of clay particles parallel to the bedding. This reflects the increasing effect of burial compaction on the host sediment during continuing

Figure 9.68 Dolomite concretion from the β_6-member of the Cloridorme Formation, wave-cut platform at Grande Vallée, Quebec, in a host sediment of black slate with intercalated centimetre-thin calcisiltite turbidites. The concretion margins are formed by two thin silt turbidites, which are 40 cm apart in the centre of the concretion and < 8 cm at the end where the concretion tapers out into the host sediment, indicating $> 80\%$ of post-cementation compaction. Stratigraphic top towards the right (fig. 3C from Hesse et al., 2004, reprinted with permission of the International Association of Sedimentologists).

concretion growth. In black shales of the Upper Devonian Rhinestreet Shale of western New York State, the fabric of randomly oriented clay particles extends into the immediate vicinity of large carbonate concretions along their equatorial plane, where it is preserved in the pressure shadow of the concretions at a short distance (<20 cm) in the host shale (Lash and Blood, 2004a).

12.3.3. Chemical environment of dolomite-concretion growth

Organogenic deep-sea dolomite displays a wide range in $\delta^{13}C$ values from -30 to $+30$‰, which has been interpreted as evidence for its formation in the SR, carbonate-reduction and fermentation zones (Arthur et al., 1983; Gautier, 1985; Hennessy and Knauth, 1985, Hesse et al., 2004). The carbon-isotope composition shows coherent differences between dolomite concretions with $\delta^{13}C$ values as heavy as +30‰, and calcite and siderite concretions that rarely exceed values of +3‰ and +7‰, respectively. These differences provide evidence that suggests that calcite and siderite concretions do not to form in sediments undergoing intense methanogenesis (Mozley and Burns, 1993). Dolomite, on the other hand, typically forms in the zone of methanogenesis at shallow burial depths (e.g. at ~3 mbsf in Kau Bay: Middelburg et al., 1990), that is, in sediments with rapid organic-matter decomposition. Although Baker and Burns (1985) suggested that isotopically very light dolomite may have been precipitated in the SR zone, such dolomite may form by methane oxidation. Dolomite, high-magnesium calcite and aragonite with $\delta^{13}C$ ranging from −32‰ to −66‰ that occur as concretions, crusts, chimneys and in chemoherms on the sea floor of the Oregon, Washington and Costa Rica continental slopes (Han et al., 2004; Ritger et al., 1987; Russell et al., 1967) are related to methane advection associated with active sediment dewatering in the accretionary prism above the Cascadia and Middle America Trench subduction zones (see Section 10). The carbonate ions derived from methane oxidation raise alkalinity sufficiently to cause carbonate precipitation.

The carbon- and oxygen-isotope profiles across both concretions and beds of the Chloridorme Formation show similar comet-like shapes of decreasing delta values from the centre to the margins (Fig. 9.67B and C). Positive $\delta^{13}C_{VPDB}$ values of the concretions and diagenetic beds (Table 9.4) indicate carbonate precipitation well within the methane-generation zone. The more or less uniform decrease of the $\delta^{13}C$ values of the concretions and beds from the centre to the rims by 2–5‰ is clear evidence that the beginning of dolomite precipitation with a few exceptions took place at or after peak methane generation (Fig. 9.67B). Combined with the porosity data (>80%), the maximum $\delta^{13}C$ values in the centre of some concretions (10‰) and beds (8‰) suggest that peak methane generation took place at very shallow subsurface depths, <1 mbsf in some cases (see also Raiswell, 1988), and <5–10 mbsf in the remainder. The SR zone overlying the methane-generation zone must have been correspondingly thin, perhaps as thin as a few centimetres (Murray et al., 1978) in the case of the concretion with a central porosity of 87%. The outward decrease in the $\delta^{13}C$ values reflects a contribution from thermocatalytically generated carbonate advected from greater depth which has a light isotope signature. This isotopically light carbonate must have migrated upwards for more than a kilometre in the sediment column by advection along faults and then have been dispersed by diffusion, because thermocatalytic reactions start at temperatures above 70 °C, which is well above the temperatures encountered by the Cloridorme concretions during their growth history (see below). Similar patterns of $\delta^{13}C$ shifts from core to rim (as well as $\delta^{18}O$ shifts, see below) were observed in dolomitic cements in the Cretaceous Mancos Shale (Klein et al., 1999), although the rims reached values as negative as almost −10‰.

In the Monterey Formation, dolomite concretions span a wider range in $\delta^{13}C$ values than associated bedded dolomite, probably indicating more rapid precipitation of the beds during a shorter burial interval compared to the concretions (Kushnir and Kastner, 1984). In the Cloridorme Formation, concretions and beds show the opposite relative behaviour. There is a slight difference between the top and bottom samples, both for individual concretions and beds of the Cloridorme Formation, suggesting that the majority of the concretions may have continued to grow longer at the bottom than at the top, while the reverse behaviour is true for some of the beds (fig. 9.10 in Hesse et al., 2004).

12.3.4. Temperature of precipitation

The $\delta^{18}O_{VPDB}$ profiles for the concretions and beds (Table 9.4) principally match the $\delta^{13}C_{VPDB}$ profiles in shape (Fig. 9.67B and C). Assuming a $\delta^{18}O$ value of −6‰ for Ordovician seawater (Popp et al., 1986; Railsback, 1990; Veizer, 1995; Veizer et al., 1997), the observed range of $\delta^{18}O$ values yields burial temperatures between 17 and 45–50 °C for the centre and rims, respectively (Fig. 9.69). A temperature range of 17–28 °C derived from the $\delta^{18}O$ values in the centre of the concretions and some of the beds would be consistent with warm and saline Ordovician deep bottom waters as

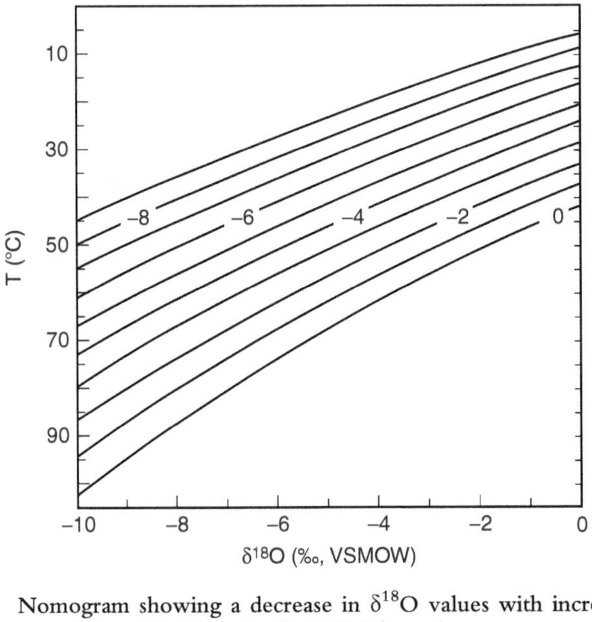

Figure 9.69 Nomogram showing a decrease in $\delta^{18}O$ values with increasing burial depth and temperature (modified from Land, 1983). The various curves represent temperature/$\delta^{18}O$ pairs for dolomite that precipitated from pore water of different $\delta^{18}O$ composition, ranging from 0‰ to −9‰ (redrafted from fig. 9 from Hesse et al., 2004).

determined elsewhere for continental-slope settings of the Trenton Group from within the same Ordovician foreland basin in Quebec and New York State (Railsback et al., 1990 in which the Cloridorme Formation was deposited). The warm saline bottom waters reflect a salinity-stratified ocean, in which temperature increased with water depth, and probably originated from evaporation in low-latitude marginal seas analogous to the modern Red Sea and Persian Gulf.

The average regional paleogeothermal gradient was 35 °C/km in the Cambro-Ordovician St. Lawrence Lowlands, which represent the passive-margin flank of the Taconian foreland basin (Yang and Hesse, 1993). With this gradient, the increase towards the concretion rims should not exceed 2–3 °C for most samples, to a maximum of 10 °C for the deeper grown parts of dolomite beds, corresponding to $\delta^{18}O$ values in the 0‰ to −2‰ range, which about half the samples adhere to (Fig. 9.67C). The remainder of the samples cannot be explained by temperature-dependent equilibrium fractionation (Fig. 9.69) from buried Ordovician seawater (connate water with a $\delta^{18}O$ value of −6‰) and require alternative mechanisms to accommodate 4–6‰ of the measured negative $\delta^{18}O$ values.

The discrepancy between isotopically derived paleoburial temperatures and burial-depth estimates based on compaction evidence (minus-cement porosity) is a frequently encountered problem in carbonate-concretion studies (e.g. Coleman and Raiswell, 1981). Influx of meteoric water was probably responsible for excessively negative $\delta^{18}O$ values of late-diagenetic calcite-fill in the septarian cracks of the Jurassic Oxford Clay concretions in England (Hudson, 1978), but can be excluded for deep-sea authigenic carbonates that have never been in contact with meteoric water before tectonic uplift such as the Cloridorme concretions. Alteration of volcanic glass to clay and zeolites that would shift the oxygen-isotope composition of the water from which the carbonate precipitated, towards more negative δ-values (Lawrence and Gieskes, 1981; Lawrence et al., 1979; Perry et al., 1976), is a likely source of light oxygen in the Cloridorme Formation, which contains abundant bentonites (Enos, 1969a) and volcanic rock fragments (Ko, 1985). It is difficult, however, to accomplish a required shift of −6‰ to −7‰ in the short burial-depth intervals (<25 mbsf) available. In the thin SR zone of the Cloridorme, the residence time of the organic matter would have been too short to allow for significant organic-matter degradation that could have shifted the oxygen isotopes towards lighter values (Sass et al., 1991). However, water from the thermocatalytic decarboxylation zone depleted not only in ^{13}C, as referred to above, but also in ^{18}O can be advected to shallower sub-bottom depths. Assuming that the $\delta^{18}O$ signature of Ordovician marine organic matter was not much different from its modern counterpart, organic-matter degradation by decarboxylation could have contributed to lower the $\delta^{18}O$ of the Cloridorme pore waters. Thus a number of mechanisms involving isotopically evolved pore water from which the dolomite precipitated at shallow subsurface depth, could all have the same effect of contributing to the negative oxygen-isotope anomaly. Anomalously heavy oxygen isotopes (with $\delta^{18}O$ values up to +8‰) in siderite, calcite and dolomite concretions of the present Atlantic continental slope off the USA (DSDP Sites 103, 104 and 533), on the other hand, may record the melting of former gas hydrates (Matsumoto, 1989).

An interesting sideline concerning advection is the possibility that clastic dykes may serve as pathways for fluid flow, as revealed by North Sea oil-field studies (Hurst et al., 2003; Jonk et al., 2005). Pressure release during pump tests showed that reservoirs at different stratigraphic levels were in hydrological communication, most likely due to the presence of dykes (Lonergan et al., 2001). For some of the Cloridorme concretions (i.e. β_7 and γ_1: Fig. 9.67), minus-cement porosity values suggest a very shallow origin (<5 mbsf), in conflict with the low isotopic values both for O and C, that require the presence of evolved pore waters. At the shallow depth of precipitation, the pore water would have sea-water isotopic signatures due to the effects of diffusion. Since the thick-bedded sandy channel-fill or turbidite-lobe facies associations of the β_7 and γ_1 members of the Cloridorme contain abundant clastic dikes, the anomalous pore waters could have been advected upward from greater subsurface depths by passing through uncemented sand dykes.

12.3.5. Recrystallisation of protodolomite

Recrystallisation at deeper burial levels and higher temperatures is another possibility to explain the unreasonably high isotopic temperatures that has been suggested for Devonian calcite (Dix and Mullins, 1987) and Cretaceous (Lawrence, 1991) and Miocene (Burns and Baker, 1987; Compton, 1988a) dolomite concretions. Recrystallisation is of principal interest beyond its role as an alternative explanation for unreasonably low oxygen-isotope ratios (Burns and Baker, 1987), because it touches on the question whether the commonly observed well ordered, near-stoichiometric dolomite in concretions results from a precursor protodolomite. The ferroan dolomite of the Cloridorme Formation is well-ordered and of near-stoichiometric composition (average of 51% $CaCO_3$ and 49% $MgCO_3$). It contains an average 2 mol% Fe (Hesse et al., 2004).

Evidence concerning the recrystallisation of carbonate concretions is controversial and appears to be in conflict with systematic centre-to-margin isotope variations displayed by many of the studied examples (e.g. Hesse et al., 2004; Mozley and Burns, 1993). To preserve the gradients across the concentric shells, recrystallisation would have to change the isotopic signatures selectively from the rims inward but leaving them unchanged in the centre, whereas recrystallisation is expected to homogenize the signatures. Furthermore, examples of recent organogenic dolomite in anoxic hemipelagic sediments from Gulf of California and Kau Bay, Indonesia, contain well-ordered dolomite unlikely to have recrystallized from a precursor primary protodolomite (Middelburg, 1990; Shimmield and Price, 1984). Van Lith et al. (2003) have documented the precipitation of near-stoichiometric, well-ordered dolomite in the presence of sulphate-reducing bacteria, leaving open the possibility that methanogens may have a similar effect.

However, examples for the recrystallisation of precursor protodolomite have been reported from Plio-Pleistocene concretions in the Bering Sea (Hein et al., 1979a) and also from the Santa Maria Basin of California. In California, the Sisquoc Formation contains very fine-grained, disordered dolomite, whereas dolomite in concretions of the older, underlying Monterey Formation is well ordered. Assuming that it also started as protodolomite would imply that it recrystallized in the course of burial (Compton, 1988b). Malone et al. (1994) have provided evidence for recrystallisation in the Monterey Formation. What is responsible for this difference in the primary precipitate between well ordered, near-stoichiometric dolomite in some cases and poorly ordered Ca-rich protodolomite in others remains uncertain (Mazzullo, 2000), although bacteria may play a crucial role, as mentioned.

12.3.6. Growth-limiting ions for dolomite concretions

Concretions will stop growing when the supply of one (or more) of the ions required for precipitation ceases. Among the ionic constituents of dolomite, carbonate is not a precipitation-limiting factor during early diagenesis because it is abundantly generated from internal sources by organic-matter decomposition. In an expanded SR zone, it is also supplied by the dissolution of solid carbonate particles. Of the two major cations of dolomite, Mg has limited internal sources (silicates), whereas calcium may have significant indigenous sources in the form of detrital terrigenous or biogenic calcite which, in the Cloridorme Formation, was introduced into the basin by turbidity currents (Enos, 1969b). However, dissolution of available detrital calcium carbonate probably did not contribute dissolved Ca^{2+} because of the limited thickness of the combined oxidation, nitrate-reduction and upper SR zones where carbonate dissolution may take place. Under these circumstances, Mg and Ca are supplied by diffusion and the molar ratio in sea water of about 5:1 (Veizer, 1983) initially determines the relative availability of Mg and Ca for authigenic carbonate formation, suggesting that Ca^{2+} and not Mg^{2+} would be the limiting ion for dolomite precipitation (Compton, 1988a).

12.3.7. Role of diffusion for supplying growth-limiting ions to concretions

Diffusional transport of Ca^{2+}, the most likely growth-limiting ion for dolomite growth, will decrease with continuing burial and will eventually cease due to the increasing degree of compaction which eliminates permeability. The distance over which diffusion may supply a chemical species required for precipitation is defined as the diffusive communication length $Z = \Phi D_s/\omega$ (Gieskes, 1975), where D_s is the sediment- diffusion coefficient, Φ is porosity, and ω the sedimentation rate. The sediment-diffusion coefficient is affected by tortuosity (Θ) according to $D_s = D/\Theta^2 = \Phi^2 D$ (Berner, 1980, p. 36), so that porosity (Φ) enters the equation for Z in the third power. With a diffusion coefficient for Ca^{2+} in water at 25 °C of 4.4×10^{-6} cm^2 s^{-1} (Li and Gregory, 1974), a sedimentation rate of 4×10^2 m per million years for the Cloridorme Formation (Hiscott et al., 1986), and a concretion-rim porosity of 45–75%, the diffusive communication length for Ca in the Cloridorme Formation will be on the order of 2–14 m. Diffusion drawing mineral constituents from a source volume of host sediment with a radius of ~10 m around the concretions is thus feasible. When the distance from the sea floor of ~10 m is exceeded, concretion growth stops because diffusion cannot keep pace with subsidence once a threshold level of compaction has been surpassed.

12.3.8. Controls on the mineralogy of marine carbonate concretions

Of the two hypotheses that have emerged from the literature, Scotchman's (1991) sedimentation-rate hypothesis attributes the authigenic concretion carbonate mineralogy to the sedimentation rate. Slaughter and Hill (1991) relate the mineralogy of the primary authigenic carbonates to the type of organic-matter being oxidized. Scotchman follows Curtis and Coleman's (1986) suggestion based on observations in the Kimmeridge Clay of England, where ferroan dolomite concretions are associated with the highest sedimentation rates, pyrite/ferroan calcite concretions with intermediate rates, and calcite concretions with low sedimentation rates of lime-mud deposition. Slaughter and Hill (1991) related organogenic dolomite formation to protein-rich marine organic matter, without discussing the conditions for the origin of calcite concretions. Mozley and Burns (1993) suggested that the role of the sedimentation rate in controlling concretion mineralogy may be indirect and may be related to the rate of organic-matter decomposition, which is strongly correlated with the sedimentation rate (Heath et al., 1977; Müller and Suess, 1979). This raises the question of which process associated with sedimentation rate and/or organic-matter decomposition rate will facilitate either dolomite or calcite precipitation.

A prerequisite for authigenic carbonate precipitation is raising the pH from the relatively low values in the oxidation zone in order to increase carbonate activity (at the expense of bicarbonate: Coleman, 1985). This is accomplished through metal (Fe^{3+}, Mn^{4+}) reduction as well as ammonia generation in the nitrate- and SR zones as mentioned before. Slaughter and Hill (1991) pointed out the role of NH_3 as the only base strong enough to raise the pH in the presence of acids. Ammonia production is most prolific from protein-rich marine organic matter.

A high pH promotes dolomite precipitation for several reasons. It favours formation of the neutral $MgCO_3{}^0$ ion pair, which has a low enough hydration energy to react at the dolomite surface, whereas Mg^{2+} has a high hydration energy (Dasent, 1982), which prevents its involvement in the dolomitization reaction. A high pH also removes hydrogen ions from the surface oxygens of dolomite or calcite, which at a lower pH are hydrogen-saturated and reduce the rate of crystallization. Increasing the ionic strength will have the same effects as raising the pH.

One of the main arguments favouring either calcite or dolomite as the primary authigenic carbonate mineral in a concretion is whether or not the concretion started to grow in the SR zone. Sulphate is a strong inhibitor of dolomite precipitation (Baker and Kastner, 1981) because it forms ion pairs with both Mg^{2+} and Ca^{2+} in seawater and marine pore waters: a stronger pair with Mg^{2+}, a weaker pair with Ca^{2+}. In the presence of phosphate, Ca^{2+} also forms a strong ion pair with $PO_4{}^{3-}$. These ion pairs are too strongly bonded to be broken at high-energy kink sites at the crystal surface. If $MgCO_3{}^0$, $CaCO_3{}^0$, $MgSO_4{}^0$ and $CaPO_4{}^-$ are the dominant ion pairs in the pore solution, as Slaughter and Hill (1991) assumed, $MgSO_4{}^0$ pairs, which are dominant in the SR zone, will block active crystallization (or dissolution) sites and prevent dolomite precipitation (or dissolution).

Because of the inhibitor effect of sulphate, pore waters in the SR zone may be supersaturated with respect to well-ordered dolomite by up to a thousand times (Compton, 1988a; Middelburg et al., 1990). Although isotopically very light dolomite ($\delta^{13}C$ as light as $-30‰$) may form in the SR zone (Baker and Burns, 1985), such dolomite is more likely to originate from AMO at the base of the SR zone (Borowski et al., 1996; Middelburg et al., 1990; Raiswell, 1987, 1988; Ritger et al., 1987; Rodriguez et al., 2000). Middelburg (1990), on the other hand, argued that dolomite may in fact start to form in the SR zone, based on isotopic evidence from sub-recent anoxic sediments of Kau Bay, Indonesia. This is possible if the degree of dolomite supersaturation of the pore water has been sufficient to overcome the effects of sulphate inhibition (Compton, 1988a; Middelburg et al., 1990).

In low-sedimentation-rate environments, the combined thickness of the oxidation, nitrate-reduction and SR zones will be large enough to allow sufficient dissolution of detrital or biogenic calcium carbonate to supply sufficient Ca^{2+} for the precipitation of a non-ferroan calcite cement in the lower part of the SR zone. Additional carbonate alkalinity may be generated by upward diffusion and AMO at the base of the SR zone, accompanied by pH-raising reactions, as mentioned.

In high- to very high-sedimentation-rate environments, in oxygen-minimum zones or stagnant basins, the SR zone may be very thin (<1 m; e.g. Nissenbaum et al., 1972; 0.05 m: Murray et al., 1978) not allowing sufficient calcium-carbonate dissolution for subsequent calcite precipitation in this zone. As soon as sulphate is depleted, ferroan dolomite may start to precipitate in the methane-generation zone. This may be the scenario which applies to the Cloridorme dolomites.

Dix and Mullins (1987) noted secular trends in the distribution of Phanerozoic carbonate concretions, relating dolomite as the primary authigenic carbonate mineral to times of sea-level highstands and claiming its absence in concretions prior to the Devonian. The results of Hesse et al. (2004), Landing et al. (1992) and Wright (1997) extend dolomite-precipitating periods in deep-water sediments back to the Ordovician and the Cambrian, respectively, when terrestrial vegetation did not exist (Arnold, 1969) and the carbonate of the concretions was derived solely from marine organic matter. Sea-level fluctuations, including short-duration eustatic changes, may well have affected concretion formation (as opposed to the diffuse dissemination of authigenic carbonates in sediment) by changing deep-water sedimentation rates, as discussed before.

12.4. Overlapping cementation shells (sediment-stabilization hypothesis)

The concentric isotopic and geochemical zonation of concretions has traditionally been used as an argument for a growth model that assumes addition of material to the concretion as successive concentric shells in an onion-skin or snow-ball fashion. Implicit in this model is the assumption that each previous shell is (more or less) completely cemented before a new one is added (e.g. Gautier, 1985). Mozley's (1989, 1996) groundbreaking observation of compositionally complexly zoned concretions, now called the "sediment-stabilization hypothesis", was originally based on ultraviolet-light and cathodoluminescence studies (see also Feistner, 1989). It suggests that concretions do not grow in a snowball-like fashion, but that the entire future volume of a concretion contains early formed cements the concentration of which is, however, higher in the centre of the concretion than at the rims. This conclusion was derived from the observation of different luminescence colours in successive generations of siderite crystals. The earliest crystals formed throughout the entire concretion body rather than only at the centre, where their concentration, however, was highest. The concentration decreased towards the margins, where they were widely scattered and isolated. Later overgrowth on the early crystals was much less voluminous in the centre than at the margins.

In view of this concept of concretion growth, many of the conventional interpretations and conclusions—and in particular those concerning the temperature and subsurface chemistry based on isotopic data and equilibrium fractionation models—require modification to include more complex mixing models. Hennessy and Knauth (1985) anticipated this kind of precipitation pattern in order to explain anomalous isotopic trends in concretion zonation. The concept of overlapping (or penetrating) cementation shells suggests that the pore space of the host sediment may be incompletely cemented in the early stages of concretion formation, but not towards the end, leaving the minus-cement porosity derived from authigenic carbonate concentration unaffected. Introduction of later cements into incompletely cemented inner concretion shells would raise the initial isotopic temperatures retroactively and thus in fact reduce the problem of anomalously high temperatures of precipitation. However, in many cases this increase will be insignificant, because the total temperature range for the growth of a concretion is low and may not exceed 5–10 °C.

An attractive aspect of the concept is that it helps to overcome the problem of a solidly cemented concretion centre becoming gravitationally unstable in the many cases where cementation starts very early during diagenesis, when the surrounding host sediment is still in a soupy state (therefore the name "sediment-stabilization hypothesis"). Tilted concretions (tilted with respect to bedding) in the Devonian Rhinestreet Shale of New York State (Lash and Blood, 2004b) may represent an example of more complete cementation at a very early stage, enabling destabilization (sinking and rotation) of the growing concretion within the uncompacted host sediment.

Deviations from concentric growth patterns revealed by non-concentric chemical and isotopic zonations (e.g. in Cretaceous concretions from New Zealand: Lawrence, 1991) are due to concretion growth around two or more centres and later coalescence of the (more or less concentric) seed concretions in a single ellipsoidal body.

12.5. Septarian concretions

Incomplete cementation in the early stages of concretion growth has also been inferred from septarian concretions that contain cracks that are characteristically widest in the centre and become narrower towards the concretion margins. These have been interpreted as shrinkage cracks (Raiswell, 1971; Gautier, 1982, 1985). The implication is that a small volume of pore space in the concretions (generally $<3\%$) was left uncemented initially. If so, porosity estimates from minus-cement porosity would be too high by the volume of the cracks. The shrinkage-crack hypothesis has been re-evaluated by Astin (1986) and Pratt (2001) in terms of deviatoric compaction stress and seismic-shock events, respectively. Support for Astin's hypothesis comes from the subvertical orientation of the cracks and their near-polygonal arrangement in plan view. If, on the other hand, they were shrinkage cracks due to dewatering, the less saline early-diagenetic pore water trapped in the concretions could still be drawn out osmotically by the more saline evolved water in the host sediment at later diagenetic stages.

Isotopic evidence for the "Moeraki boulders" of New Zealand, very large (up to 2 m in diameter) spherical septarian concretions, suggests that the first calcite cement in the septarian cracks was precipitated shortly after or during termination of concretion growth from continuously evolving pore waters of marine origin, whereas later septarian cements show the effects of meteoric-water inflow (Thyne and Boles, 1989). Similarly, cements in septarian concretions of the Jurassic Oxford Clay in England record the passage of at least four generations of pore fluids, the first one of marine origin (Hudson et al., 2001). Complex histories of pore-water evolution appear to be reflected in many septarian-concretion cements (Coniglio et al., 2000). Isotopic evidence from septarian concretions in the Lower Jurassic Whiteaves Formation of the Queen Charlotte Islands, British Columbia, shows that the first septarian cement, which is a fibrous calcite, was precipitated early at very shallow burial depth from marine-derived pore water under the influence of methane oxidation (Desrochers and Al-Aasm, 1993). Shallow burial would favour the traditional interpretation as shrinkage cracks over Astin's (1986) mechanism. Whewellite, a pinkish to wine-yellow, relatively rare calcium-oxalate-monohydrate mineral, has been detected as a late cement in septarian concretions (Lippmann, 1955; Pecora and Kerr, 1954). The origin of the oxalate is attributed to oxidation of organic matter.

12.6. Growth of concretion layers versus diffuse authigenic carbonate precipitation

An intriguing question is why authigenic carbonate occurs in concretion layers or beds at a given stratigraphic level rather than as diffusely distributed crystals throughout the host sediment. Carbonate concretions in Devonian and younger rocks are well known for their role as hosts of fossil fish and other petrifacts, pointing to a direct relationship between decaying organic matter, in particular to protein decomposition, and carbonate precipitation (e.g. Weeks, 1957).

Where no obvious seeding material is present, a small perturbation of the geochemical or physical environment along a particular horizon may focus carbonate precipitation. This may be a slight difference in porosity, permeability or the original organic-matter concentration or composition which may attract bacterial activity at distinct stratigraphic levels. Diffusion of solutes from the pore water in the surrounding sediment to the low-concentration region around the concretion where the dissolved species are consumed by bacterially mediated precipitation then favours continued concretion growth at the original site of precipitation rather than nucleation elsewhere, in a manner similar to calcite cementation in sandstones (Bjørkum and Walderhaug, 1990).

The micritic nature of carbonate crystals in concretions shows that reduced nucleation energy in the presence of already existing authigenic or detrital nuclei is not responsible for focused precipitation in concretions, because each tiny crystal needs an individual nucleus. A more realistic alternative is the hypothesis of clustering bacterial populations that has attracted much attention (e.g. de Craen et al., 1999; Vasconcelos and McKenzie, 1997). Bacteria consuming small organic molecules may be most abundant near active carbonate-precipitation sites where the products of microbial metabolism are removed in the solid carbonate phases, thus maintaining concentration gradients that enhance the diffusion of the reactants.

The growth of concretions in distinct layers and eventually their coalescence in diagenetic beds may, however, require drastic changes in sedimentation rates, typically encountered in alternating turbidite/hemipelagite successions. This may involve intermittent interruption of deposition and burial (Raiswell, 1987, 1988). Otherwise, authigenic dolomite rhombs will be widely disseminated in the sediment (e.g. Henderson et al., 1984; Landing et al., 1992; Lumsden, 1988). In this respect, the formation of concretionary carbonates appears similar to that of nodular chert bands (flint) in the Cretaceous chalk, which has been related to rhythmic sedimentation linked to sea-level changes (Clayton, 1986).

12.7. Complex authigenic carbonates

The occurrence of complex authigenic carbonates rich in Mn and Fe, which show transitions to end-member rhodochrosite and siderite, has been described only from a few DSDP drill sites (e.g. Hein et al., 1979a; Okada, 1980; Wada et al., 1981). The isotopic composition of similar concretions in Cretaceous black shales of the Western Alps (Tassé and Hesse, 1984) suggests the formation at somewhat greater burial depths in the methane-generation zone, which requires manganese mobilization at those depths. Secondary maxima in Mn-depth profiles between 100 and 200 mbsf at various DSDP drill sites show a close correlation with maxima in the dissolved-silica profiles and point to opal-A dissolution as a manganese source (Gieskes, 1981).

12.8. Authigenic barite concretions

Microcrystalline barite ($BaSO_4$) of 30 × 50 μm size is common in high-sedimentation-rate environments of the continental margins, and is thought to be precipitated in the water column in cavities within decaying biological debris (Dehairs et al., 1980). The distribution of marine barite in the water column and surface deep-sea sediment correlates with the organic carbon flux from the surface to the deep ocean, the carbon export production

(Paytan and Griffith, 2007). The preservation potential of this relatively labile "bio-barite" in deep-sea sediment is high above and within the SR zone. According to McManus et al. (1994), the $Ba/CaCO_3$ ratio is constant in the upper 0.5 mm in East Pacific continental-margin sediments and is thus apparently not affected by early diagenesis. The paleoceanographic significance of barite results from its potential to serve as a proxy for the organic-carbon flux and carbon-burial rate. Since the barite concentration is low (generally $< 1\%$), precise bulk accumulation rates are required to make use of this barite proxy for the purpose of determining changes in ocean export productivity, that is, the carbon flux to the deep ocean which is one of the most important parameters in the global carbon cycle, and which is directly linked to climate change.

Below the SR zone, barite dissolves remobilizing barium, which migrates upward in pore-water compaction stream and, being reprecipitated, forms a diagenetic barite front at the base of the SR zone that moves upwards with subsidence. Dissolved barium concentrations in deep-sea drill sites from the Pacific margins are contragredient with sulphate. Dissolved barium is almost quantitatively removed in the presence of sulphate (Fig. 9.70). Barite concretions form (1) potato-shaped microcrystalline nodules, (2) nodules with well-developed barite crystals commonly displaying a radial fabric and sometimes resembling the appearance of a hedgehock and (3) composite nodules resulting from the coalescence of spherulitic nuclei (Breheret and Brumsack, 2000). Septarian concretions (cf. Section 12.5) and cone-in-cone structures have been observed; the latter are commonly thought to be indicative of low-nucleation rates.

When sedimentation rates increase rapidly or where the base of the SR zone occurs at the sediment/water interface, the conditions for barite concretions or layers to form may be favourable (Torres et al., 1996). Breheret and Brumsack (2000), in contrast, assumed that early-diagenetic barite concretions formed in organic-matter-rich sediments when the sedimentation rate decreased or breaks in sedimentation occurred, so that concretion horizons reflect breaks in sedimentation when sulphate from overlying sea water could diffusively invade the sediment and precipitate barite. Sulphur- and oxygen-isotope evidence shows that such barite concretions do not require hydrothermal activity for their formation (Leśniak et al., 1999).

13. Early Diagenetic Clay-Mineral Formation

Apart from the alteration of volcanic material, the neoformation of clay minerals in pelagic environments at shallow burial levels has traditionally been considered subordinate (Kastner, 1981), although small amounts

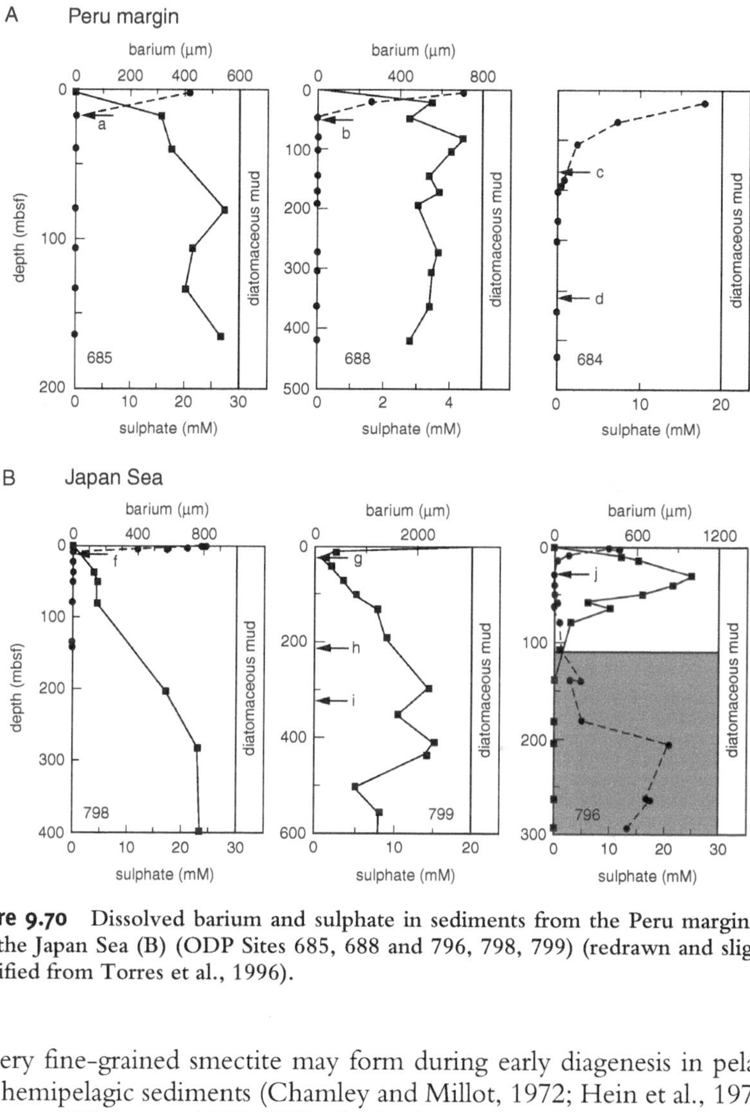

Figure 9.70 Dissolved barium and sulphate in sediments from the Peru margin (A) and the Japan Sea (B) (ODP Sites 685, 688 and 796, 798, 799) (redrawn and slightly modified from Torres et al., 1996).

of very fine-grained smectite may form during early diagenesis in pelagic and hemipelagic sediments (Chamley and Millot, 1972; Hein et al., 1979b; Johnson, 1976, pp. 1280–1282; Moberly et al., 1968). More recently, however, Emerson and Hedges (2003) postulated precipitation of authigenic aluminosilicates in marine sediments to be one of the most widespread diagenetic reactions. This would be supported by decreases of dissolved aluminium at shallow subsurface depths (Stoffyn-Egli, 1982).

Authigenic clay minerals may also form at deeper levels in the subsurface of high-sedimentation-rate regions (e.g. DSDP Sites 565 and 568, Fig. 9.33). Decreases in dissolved potassium in shallow pore waters of modern sediments, on the other hand, do not require the neoformation of clay minerals but can be explained by sorption and cation exchange (Hover et al., 2002). New smectite may also form deeper in the subsurface in many hemipelagic or terrigenous argillaceous sediments (Holzapfel and Chamley, 1986), but it has rarely been detected in deep-sea drilling samples (see, however, Kastner and Gieskes, 1976) because, if present, its existence is masked by the predominant detrital clay fraction (Heinemann and Füchtbauer, 1982). Special analytical techniques are required to demonstrate the presence of a small authigenic component against an overwhelming detrital background (De Lange and Rispens, 1986; Helm, 1985).

ACKNOWLEDGEMENTS

R. H. wishes to thank Bob Garrison for a critical review of the entire manuscript and helpful suggestions for improvement, Joris Gieskes for a thorough review of Sections 1–6, Radomir Petrovich for most detailed comments on an earlier version of Section 4 and J. Kallmaier for providing essential references. Tom van Loon suggested editorial changes that are most welcomed. D. Lau and H. Sengpiehl are thanked for redrafting most of the figures and co-editor Heiko Hüneke for his patience and encouragement getting the manuscript into its final shape. Financial support from NSERC (Natural Sciences and Engineering Research Council of Canada) that made writing of this review possible is gratefully acknowledged.

REFERENCES

Al-Aasm, I.S., Muir, I., Morad, S., 1993. Diagenetic conditions of fibrous calcite vein formation in black shales: petrographic, chemical and isotopic evidence. Bull. Can. Petrol. Geol. 41, 46–56.

Andrews, J.E., Packman, G., et al., 1975. Site report 285. Init. Rep. Deep Sea Drilling Proj. 30, 27–67.

Arnold, C.A., 1969. The fossil plant record. In: Tschudy, R.H., Scott, R.A. (Eds.), Aspects of Palynology. Wiley-Interscience, New York, pp. 127–143.

Arthur, M.A., Anderson, T.F., Kaplan, I.R., Veizer, J., Land, L.S. (Eds.), 1983. Stable Isotopes in Sedimentary Geology, In: Soc. Econ. Paleontologists Mineralogists, Short Course Lecture Notes 10, 151pp.

Astin, T.R., 1986. Septarian crack formation in carbonate concretions from shales and mudstones. Clay Miner. 21, 617–632.

Bach, W., Alt, J., et al., 2001. The geochemical consequences of late-state low-grade alteration of lower ocean crust at the SW Indian Ridge: results from ODP Hole 735B (Leg 175). Geochim. Cosmochim. Acta 65, 3267–3287.

Baker, P.A., Bloomer, S.H., 1988. The origin of celestite in deep-sea carbonate sediments. Geochim. Cosmochim. Acta 52, 335–339.

Baker, P.A., Burns, S.J., 1985. Occurrence and formation of dolomite in organic-rich continental margin sediments. Am. Assoc. Petrol. Geol. Bull. 69, 1917–1930.

Baker, P.A., Kastner, M., 1981. Constraints on the formation of sedimentary dolomite. Science 213, 214–216.

Baker, P.A., Gieskes, J.M., Elderfield, H., 1982. Diagenesis of carbonate in deep-sea sediments—evidence from Sr/Ca ratios and interstitial dissolved Sr^{2+} data. J. Sediment. Petrol. 52, 71–82.

Baltuck, M., 1983. Some sedimentary and diagenetic signatures in the formation of bedded radiolarite. In: Iijima, A., Hein, J.R., Siever, R. (Eds.), Siliceous Deposits in the Pacific Region. Dev. Sediment. Elsevier, Amsterdam, 36, pp. 299–315.
Baron, G., Debyser, I., 1957. Sur la présence dans des vases organiques de la mer Baltique de sulfure manganeux-hexagonal. Comptes Rendues Acad. Sci. Français 245, 1148–1150.
Bath, A., Shackleton, N., 1984. Oxygen and hydrogen isotope studies in squeezed pore waters, Deep Sea Drilling Project, Leg 74, hole 525B: evidence for mid-Miocene ocean isotopic change. In: Moore, T.C., Rabinowitz, P.D. (Eds.), Init. Rep. Deep Sea Drilling Proj. 74. U.S. Government Printing Office, Washington, DC, pp. 697–699.
Behl, R.J., 1992. Chertification in the Monterey Formation of California and deep sea sediments of the West Pacific. Unpubl. Ph.D. thesis, Univ. Calif. Santa Cruz, 302pp.
Berger, W.H., Von Rad, U., 1972. Cretaceous and Cenozoic sediments from the Atlantic Ocean. In: Hayes, D.E. (Ed.), Init. Rep. Deep Sea Drilling Proj. 14. U.S. Government Printing Office, Washington, DC, pp. 787–954.
Berger, G., Schott, J., Loubet, M., 1987. Fundamental processes controlling the first stage of alteration of a basalt glass by seawater: an experimental study between 200 and 320 °C. Earth Planet. Sci. Lett. 84, 431–445.
Berner, R.A., 1967. Thermodynamic stability of sedimentary iron sulphides. Am. J. Sci. 265, 773–785.
Berner, R.A., 1969. The synthesis of framboidal pyrite. Econ. Geol. 64, 383–384.
Berner, R.A., 1980. Early Diagenesis. A Theoretical Approach. Princeton Series in Geochemistry. Princeton University Press, Princeton, N.J., 241pp.
Berner, R.A., 1981. A new geochemical classification of sedimentary environments. J. Sediment. Petrol. 51, 359–365.
Berner, R.A., 1984. Sedimentary pyrite formation: an update. Geochim. Cosmochim. Acta 48, 605–615.
Bjørkum, P.A., Walderhaug, O., 1990. Geometrical arrangement of calcite cementation within shallow marine sandstones. Earth Sci. Rev. 29, 145–161.
Bohrmann, G., Abelmann, A., Gersonde, R., Hubberten, H., Kuhn, G., 1994. Pure siliceous ooze, a diagenetic environment for early chert formation. Geology 22, 207–210.
Borowski, W.S., Paull, C.K., Ussler III, W., 1996. Marine pore-water sulphate profiles indicate in situ methane flux from underlying gas hydrate. Geology 24, 655–658.
Borowski, W.S., Paull, C.K., Ussler III, W., 1999. Global and local variations of interstitial sulphate gradients in deep-water, continental margin sediments: sensitivity to underlying methane and gas hydrates. Mar. Geol. 159, 131–154.
Borowski, W.S., Hoehler, T.M., Alperin, M.J., Rodriguez, N.M., Paull, C.K., Borowski, W.S., et al., 2000. Significance of anaerobic methane oxidation in methane-rich sediments overlying the Blake Ridge gas hydrates. In: Paull, C.K., Matsumoto, R., Wallace, P.J., Dillon, W.P. (Eds.), Gas Hydrate Sampling on the Blake Ridge and Carolina Rise. Proc. Ocean Drill. Progr. Sci. Results 164. Ocean Drilling Program, College Station, TX, pp. 87–99.
Böttcher, M.E., Brumsack, H.-J., De Lange, G.J., 1998. Sulfate reduction and related stable isotope (^{34}S, ^{18}O) variations in interstitial waters from the eastern Mediterranean Sea. In: Robertson, A.H.F., Emeis, K.-C., Richter, C., Camerlenghi, A. (Eds.), Proc. Ocean Drill. Progr., Scient. Results 160. Ocean Drilling Program, College Station, TX, pp. 365–373.
Botz, R., Bohrmann, G., 1991. Low-temperature opal-CT precipiation in Antarctic deep-sea sediments: evidence from oxygen isotopes. Earth Planet. Sci. Lett. 107, 612–617.
Bramlette, M.N., 1946. The Monterey Formation of California and the origin of its siliceous rocks. U.S. Geol. Survey, Prof. Paper 212, 5pp.
Breheret, J.-G., Brumsack, H.J., 2000. Barite concretions as evidence of pauses of sedimentation in the Marnes Bleues Formation of Vocontian Basin (SE France). Sediment. Geol. 130, 205–228.

Brooks, J.M., Kennicut II, M.C., Bidigare, R., Wade, T.L., Powell, E.N., Enoux, G.J., et al., 1987. Hydrates, oil seepage, and chemosynthetic ecosystems on the Gulf of Mexico slope: an update. Eos 68 (489), 499.

Bruland, K.W., Orians, K.J., Cowen, J.P., 1994. Reactive trace metals in the stratified central North Pacific. Geochim. Cosmochim. Acta 58, 3171–3182.

Burns, S.J., Baker, P.A., 1987. A geochemical study of dolomite in the Monterey Formation, California. J. Sediment. Petrol. 57, 128–139.

Calvert, S.E., 1971. Composition and origin of North Atlantic deep sea cherts. Contrib. Mineral. Petrol. 33, 273–288.

Carstens, H., 1986. Displacive growth of authigenic pyrite. J. Sediment. Petrol. 56, 252–257.

Carver, R.E., 1980. Petrology of Paleocene-Eocene and Miocene opaline sediments, southeastern Atlantic plain. J. Sediment. Petrol. 50, 569–582.

Carey, S.N., Schneider, J.-L., 2011. Volcaniclastic processes and deposits in the deep-sea. In: Hüneke, H., Mulder, T. (Eds.), Deep-Sea Sediments. Developments in Sedimentology, Elsevier, Amsterdam, vol. 63, pp. 457–515.

Chamley, H., Millot, G., 1972. Neoformation de montmorillonite à partir de diatomées et de cendres dans les sédiments marins de Santorin (Mediterranée orientale). Compt. Rendus, Acad. Sci. Français, Ser. D 272, 1132–1134.

Chester, R., Stoner, J.H., Johnson, L.R., 1974. Montmorillonite in surface detritus. Nature 249, 335–336.

Claypool, G.E., Kaplan, I.R., 1974. The origin and distribution of methane in marine sediments. In: Kaplan, I. R. (Ed.), Natural Gases in Marine Sediments. Mar. Sci. 3, 99–140.

Claypool, G.E., Kvenvolden, K.A., 1983. Methane and other hydrocarbon gases in marine sediment. Annu. Rev. Earth Planet. Sci. 11, 299–327.

Claypool, G.E., Threlkeld, C.N., 1983. Anoxic diagenesis and methane generation in sediments of the Blake Outer Ridge, DSDP Site 533, Leg 76. In: Sheridan, R.E., Gradstein, F.M. (Eds.), Init. Rep. Deep Sea Drilling Proj. 76. U.S. Government Printing Office, Washington, DC, pp. 391–402.

Claypool, G.E., Threlkeld, C.N., Mankiewicz, P.N., Arthur, M.A., Anderson, T.F., 1985. Isotopic composition of interstitial fluids and origin of methane in slope sediment of the Middle America Trench, Deep Sea Drilling Project Leg 84. In: Von Huene, R., Aubouin, J. (Eds.), Init. Rep. Deep Sea Drilling Proj. 84, 683–691.

Clayton, C.J., 1986. The chemical environment of flint formation in Upper Cretaceous chalks. In: Sieveking, G., de, G., Hart, M.B. (Eds.), The Scientific Study of Flint and Chert. Cambridge University Press, pp. 43–54.

Coleman, M.L., 1985. Geochemistry of diagenetic non-silicate minerals: kinetic considerations. Phil. Trans. R. Soc. Lond. 315A, 39–56.

Coleman, M.L., Raiswell, R., 1981. Carbon, oxygen and sulfur isotope variations in concretions from the Upper Lias of N.E. England. Geochim. Cosmochim. Acta 45, 329–340.

Colley, S., Thomson, J., Wilson, T.R.S., Higgs, N.C., 1984. Post-depositional migration of elements during diagenesis in brown clay and turbidite sequences in the North East Atlantic. Geochim. Cosmochim. Acta 48, 1223–1235.

Compton, J.S., 1988a. Sediment composition and precipitation of dolomite and pyrite in the Neogene Monterey and Sisquoc Formations, Santa Maria Basin Area, California. In: Shukla, V., Baker, P.A. (Eds.), Sedimentology and Geochemistry of Dolostones, Soc. Econ. Paleontol. Mineral. Spec. Publ., vol. 43. 53–64.

Compton, J.S., 1988b. Degree of supersaturation and precipitation of organogenic dolomite. Geology 16, 318–321.

Coniglio, M., Myrow, P., White, T., 2000. Stable carbon and oxygen isotope evidence of Cretaceous sea-level fluctuations recorded in septarian concretions from Pueblo, Colorado, U.S.A. J. Sediment. Res. 70, 700–714.

Couture, R., Miller, R.S., Gieskes, J.M., 1978. Interstitial water and mineralogical studies, Leg 41. In: Lancelot, Y., Seibold, E. (Eds.), Init. Rep. Deep Sea Drilling Proj. 41. U.S. Government Printing Office, Washington, DC, pp. 907–914.

Cragg, B.A., Parkes, R.J., Fry, J.C., Weightman, A.J., Rochelle, P.A., Maxwell, J.R., 1996. Bacterial populations and processes in sediments containing gas hydrates (ODP Leg 146: Cascadia Margin). Earth Planet. Sci. Lett. 139, 497–507.

Crovisier, J.-L., Thomassin, J., et al., 1983. Experimental basalt-seawater glass interaction at 50 °C: study of early developed phases by electron microscopy and X-ray photoelectron spectrometry. Geochim. Cosmochim. Acta 47, 377–387.

Curtis, C.D., 1967. Diagenetic iron minerals in some British Carboniferous sediments. Geochim. Cosmochim. Acta 31, 2109–2123.

Curtis, C.D., 1978. Possible links between sandstone diagenesis and depth-related geochemical reactions occurring in enclosing mudstones. J. Geol. Soc. Lond. 135, 107–117.

Curtis, C.D., Coleman, M.L., 1986. Controls on the precipitation of early diagenetic calcite, dolomite, and siderite concretions in complex depositional sequences. In: Gautier, D. L. (Ed.), Roles of Organic Matter in Sediment Diagenesis. Soc. Econ. Paleontol. Mineral. Spec. Publ. 38, 23–33.

Curtis, C.D., Petrowski, C., Oertel, G., 1972. Stable carbon isotope ratios within carbonate concretions: a clue to place and time of formation. Nature 235, 98–100.

Dasent, W.E., 1982. Inorganic Energetics, second ed. Cambridge University Press, New York 185pp.

Davies, R., 2005. Differential compaction in sedimentary basins due to silica diagenesis: a case study. Geol. Soc. Am. Bull. 117, 1146–1155.

Davies, R., Cartwright, J., Rana, J., 1999. Giant hummocks in deep-water marine sediments: evidence for large-scale differential compaction and density inversion during early burial. Geology 27, 907–910.

Davies, R.J., Huuse, M., Hirst, P., Cartwright, J., Yang, Y., 2006. Giant clastic intrusions primed by silica diagenesis. Geology 34, 917–920.

Davis, E.F., 1918. The radiolarian cherts of the Franciscan group. Univ. California Publ. Bull. Dept. Geol. 11, 235–432.

Davis, E.E., Hyndman, R.D., Villinger, H., 1990. Rates of fluid expulsion across the northern Cascadia accretionary prism: constraints from new heat flow and multichannel seismic reflection data. J. Geophys. Res. 95, 8869–8889.

De Craen, M., Swennen, R., Keppens, E.M., MacAuley, C.I., Kiriakoulakis, K., 1999. Bacterially mediated formation of carbonate concretions in the Oligocene Boom Clay of northern Belgium. J. Sediment. Res. 69A, 1098–1106.

De Lange, G.J., 1986. Early diagenetic reactions in interbedded pelagic and turbidite sediments in the Nares Abyssal Plain (western North Atlantic): consequences for the composition of sediment and interstitial water. Geochim. Cosmochim. Acta 50, 2543–2561.

De Lange, G.J., Brumsack, H.-J., 1998. The occurrence of gas hydrates in Eastern Mediterranean mud dome structures as indicated by pore-water composition. In: Henriet, J.-P., Mienert, J. (Eds.), Gas hydrates: relevance to world margin stability and climate change, Geol. Soc. Lond. Spec. Publ., vol. 137. 167–175.

De Lange, G.J., Rispens, F.B., 1986. Indications of diagenetically induced precipitate of an Fe-Si mineral in sediments from the Nares Abyssal Plain, western North Atlantic. Mar. Geol. 73, 85–97.

Dehairs, F., Chesselet, R., Jedwab, J., 1980. Discrete suspended particles of barite and the barium cycle in the ocean. Earth Planet. Sci. Lett. 49, 528–550.

Deines, P., 1980. The isotopic composition of reduced organic carbon. In: Fritz, P., Fontes, J.C. (Eds.), Handbook of Environmental Isotope Geochemistry. The Terrestrial Environmentvol. 1. Elsevier, Amsterdam, pp. 329–406.

Delaney, M., 1998. Phosphorus accumulation in marine sediments and the marine phosphorus cycle. Global Biochem. Cycles 12 (4), 563–572.

Desrochers, A., Al-Aasm, I.S., 1993. The formation of septarian concretions in Queen Charlotte Islands, B.C. Evidence for microbially and hydrothermally mediated reactions at shallow burial depth. J. Sediment. Petrol. 63, 282–294.

Devol, A.H., 1978. Bacterial oxygen uptake kinetics as related to biological processes in oxygen-deficient zones of the oceans. Deep-Sea Res. 25, 137–146.

Deyhle, A., Kopf, A., 2002. Strong B enrichment and anomalous $\delta^{11}B$ in pore fluids from the Japan Trench forearc. Mar. Geol. 183, 1–15.

Deyhle, A., Kopf, A., Frape, S., Hesse, R., 2004. Evidence for fluid flow in the Japan Trench forearc using isotope geochemistry (Cl, Sr, B): results from ODP Site 1150. Island Arc 13, 258–270.

Dix, G.R., Mullins, H.T., 1987. Shallow, subsurface growth and burial alteration of Middle Devonian calcite concretions. J. Sediment. Petrol. 57, 140–152.

Edmond, J.M., Measures, C., McDuff, R.E., Chan, L.H., Collier, R., Grant, B., et al., 1979. Ridge crest hydrothermal acticity and the balances of the major and minor elements in the oceans: the Galapagos data. Earth Planet. Sci. Lett. 46, 1–18.

Egeberg, P.K., 2000. Hydrates associated with fluid flow above salt diapirs (Site 996). In: Paull, C.K., Matsumoto, R., Wallace, P.J., Dillon, W.P. (Eds.), Gas Hydrate Sampling on the Blake Ridge and Carolina Rise. Proc. Ocean Drill. Progr. Sci. Results 164. Ocean Drilling Program, College Station, TX, pp. 219–228.

Egeberg, P.K., Barth, T., 1998. Contribution of dissolved organic species to the carbon and energy budgets of hydrate bearing deep sea sediments. ODP Site 997 Blake Ridge. Chem. Geol. 149, 25–35.

Egeberg, P.K., Dickens, G.R., 1999. Thermodynamic and pore-water halogen constraints on gas hydrate distribution at ODP Site 997 (Blake Ridge). Chem. Geol. 153, 53–97.

Eggenkamp, H.G.M., Middelburg, J.J., Kreulen, R., 1994. Preferential diffusion of ^{35}Cl relative to ^{37}Cl in sediments of Kau Bay, Halmahera, Indonesia. Chem. Geol. 116, 317–325.

Elderfield, H., Gieskes, J.M., 1982. Sr isotopes in interstitial waters of marine sediments from deep sea drilling cores. Nature 300, 493–497.

Elderfield, H., Gieskes, J.M., Baker, P.A., Oldfield, R.K., Hawkesworth, C.J., Miller, R., 1982. $^{87}Sr/^{86}Sr$ and $^{18}O/^{16}O$ ratios, interstitial water chemistry and diagenesis in deep-sea carbonate sediments of the Ontong Java Plateau. Geochim. Cosmochim. Acta 46, 2259–2268.

Emerson, S., Hedges, J., 2003. Sediment diagenesis and benthic flux. Treatise Geochem. 6, 293–319.

Enos, P., 1969a. Anatomy of a flysch. J. Sediment. Petrol. 39, 680–723.

Enos, P., 1969b. Cloridorme Formation, Middle Ordovician Flysch, Northern Gaspé Peninsula, Quebec. Geol. Soc. Am. Spec. Paper 117, 61pp.

Ernst, W.G., Calvert, S.E., 1969. An experimental study of the recrystallization of porcellanite and its bearing on the origin of some bedded cherts. Am. J. Sci. 267A, 114–133.

Ewing, J., Ewing, M., Aitken, T., Ludwig, W.J., 1968. North Pacific sediment layers measured by seismic profiling. In: Knopoff, L., Drake, C.L., Hart, P.J. (Eds.), The Crust and Upper Mantle of the Pacific area, Am. Geophys. Union, Geophysical Monograph, vol. 12. 147–173.

Farrand, M., 1970. Framboidal sulphides precipitated synthetically. Miner. Deposita 5, 237–247.

Fehn, U., Snyder, G., Egeberg, P.K., 2000. Dating of pore waters with ^{129}I: relevance for the origin of marine gas hydrates. Science 289, 2332–2335.
Feistner, K.W.A., 1989. Petrographic examination and re-interpretation of concretionary carbonate horizons from the Kimmeridge Clay, Dorset. J. Geol. Soc. Lond. 146, 345–350.
Finke, N., Vandieken, V., Jørgensen, B.B., 2007. Acetate, lactate, propionate, and isobutyrate as electron donors for iron and sulfate reduction in Arctic marine sediments. Microbiol. Ecol. 59, 10–22.
Fischer, A.G., Hilgen, F.J., Garrison, R.G., 2009. Mediterranean contributions to cyclostratigraphy and astrochronoloy. Sedimentology 56, 63–94.
Fitts, T.G., Brown, K.M., 1999. Stress-induced smectite dehydration: ramifications for patterns of freshening and fluid expulsion in the N, Barbados accretionary wedge. Earth Planet. Sci. Lett. 172, 179–197.
Flörke, O.W., Hollmann, R., Von Rad, U., Rösch, H., 1976. Intergrowth and twinning in opal-CT lepispheres. Contrib. Mineral. Petrol. 58, 235–242.
Fontes, J.-C.h., 1965. Fractionnement isotopique dans l'eau de cristallisation du sulphate de calcium. Geol. Rundsch. 55, 172–178.
Frieling, D., Mrazek, J., 2005. Geochemische und strukturelle Untersuchungen an Fe-Mn-Knollen vom Fisherrücken/Costa Rica. In: Mrazek, J., Hüneke, H. (Eds). Tiefseefoschung in Greifswald - Geomarines Themenheft mit ausgewählten Beiträgen aus dem Nord-Atlantik, dem Südwest-Pazifik und dem Ost-Pazifik vor Mittelamerika. Greifswalder Geowissenschaftliche Beiträge 13, 163–175.
Froelich, P.N., Klinkhammer, G.P., Bender, M.L., Luedtke, N.A., Heath, G.R., Cullen, D., et al., 1979. Early oxidation of organic matter in pelagic sediments of the eastern equatorial Atlantic: suboxic diagenesis. Geochim. Cosmochim. Acta 43, 1075–1090.
Froelich, P.N., Kim, K.H., Jahnke, R., Burnett, W.C., Soutar, A., Deakin, M., 1983. Porewater fluoride in Peru continental margin sediments: uptake from seawater. Geochim. Cosmochim. Acta 47, 1605–1612.
Froelich, P.N., Kvenvolden, K.A., Torres, M.E., Waseda, A., Didyk, B.M., Lorenson, T.D., 1995. Geochemical evidence for gas hydrate in sediment near the Chile triple junction. In: Lewis, S.D., Behrmann, J.H., Musgrave, R.J., Cande, S.C. (Eds.), Chile Triple Junction. Proc. Ocean Drill Progr. Sci. Results 141. Ocean Drilling Program, College Station, TX, pp. 279–286.
Füchtbauer, H. (Ed.), 1988. Sediment-Petrologie II: Sedimente und Sedimentgesteine. Schweizerbart, Stuttgart, 1141pp.
Fütterer, D.K., 2006. The solid phase of marine sediments. In: Schulz, H.D., Zabel, M. (Eds.), Marine Geochemistry. Springer Verlag, Berlin-Heidelberg, pp. 1–25.
Garrison, R.E., Kastner, M., Zenger, D.H., 1984. Dolomites of the Monterey Formation and Organic-Rich Units. In: Soc. Econ. Paleontol. Mineral., Pacific Section, Spec. Publ. 41, 215pp.
Gautier, D.L., 1982. Siderite concretions: indicators of early diagenesis in the Gammon shale (Cretaceous). J. Sediment. Petrol. 52, 859–871.
Gautier, D.L., 1985. Interpretation of early diagenesis in ancient marine sediments. In: Gautier, D.L., Kharaka, Y.K., Surdam, R.C. (Eds.), Relationship of Organic Matter and Mineral Diagenesis, Soc. Econ. Paleontol. Mineral., 6–78, Short Course Lecture Notes 17.
Gautier, D.L., Claypool, G.E., 1984. Interpretation of methanic diagenesis in ancient sediments by analogy with processes in modern diagenetic environments. In: McDonald, D.A., Surdam, R.C. (Eds.), Clastic Diagenesis, Am. Assoc. Petroleum Geologists Mem. vol. 37. 111–123.

Gieskes, J.M., 1975. Chemistry of interstitial waters of marine sediments. Annu. Rev. Earth Planet. Sci. 3, 433–453.

Gieskes, J.M., 1981. Deep sea drilling interstitial water studies: Implications for chemical alteration of the ocean crust, Layers I and II. In: Warme, J.E., Douglas, R.G., Winterer, E.L. (Eds.), Soc. Econ. Paaleontologists Mineralogists Spec. Publ., vol. 32, 149–167.

Gieskes, J.M., 1983. The chemistry of interstitial waters of deep-sea sediments: interpretation of Deep Sea Drilling Data. In: Riley, J.P., Chester, R. (Eds.), Chem. Oceanogr. vol. 8. 221–269.

Gieskes, J.M., Elderfield, H., Lawrence, J.R., Lakind, J., 1984. Interstitial water studies, Leg 78A. In: Biju-Duval, B., Moore, J.C., et al. (Eds.), Init. Rep. Deep Sea Drilling Proj. 78. U.S. Government Printing Office, Washington, DC, pp. 377–384.

Gieskes, J.M., Johnston, K., 1984. Interstitial water studies, Leg 81. In: Roberts, D.G., Schnitker, D. (Eds.), Init. Rep. Deep Sea Drilling Proj. 81. U.S. Government Printing Office, Washington, DC, pp. 829–836.

Gieskes, J.M., Lawrence, J.R., 1981. Alteration of organic matter in deep-sea sediments: evidence from the chemical composition of interstitial waters from deep-sea drilling cores. Geochim. Cosmochim. Acta 45, 1687–1703.

Gieskes, J.M., Graham, D., Ellis, R., 1980. Interstitial-water studies, Deep Sea Drilling Project sites 415 and 416. In: Lancelot, Y., Winterer, E.L. (Eds.), Init. Rep. Deep Sea Drilling Proj. 50. U.S. Government Printing Office, Washington, DC, pp. 691–693.

Gieskes, J.M., Kastner, M., Einsele, G., Kelts, K., Niemitz, J., 1982. Hydrothermal activity in the Guaymas Basin, Gulf of California: a synthesis. In: Curray, J.R., Moore, D.G. (Eds.), Init. Rep. Deep Sea Drilling Proj. 64. U.S. Government Printing Office, Washington, DC, pp. 1159–1167.

Gieskes, J.M., Johnston, K., Boehm, M., 1985. Interstitial water studies, Leg 66. In: Von Huene, R., Aubouin, J. (Eds.), Init. Rep. Deep Sea Drilling Proj. 84. U.S. Government Printing Office, Washington, DC, pp. 961–967.

Gieskes, J.M., Elderfield, H., Palmer, M.R., 1986. Strontium and its isotopic composition in interstitial waters of marine carbonate sediments. Earth Planet. Sci. Lett. 77, 229–235.

Glasby, G.P., 2006. Manganese: predominant role of nodules and crusts. In: Schulz, H.D., Zabel, M. (Eds.), Marine Geochemistry. Springer Verlag, Berlin-Heidelberg, pp. 371–427.

Glasby, G.P., Stoffers, P., Sioulas, A., Thijssen, T., Friedrich, G., 1983. Manganese nodule formation in the Pacific Ocean: a general theory. Geo-Mar. Lett. 2, 47–53.

Glenn, C.R., Arthur, M.A., 1988. Petrology and major element geochemistry of Peru margin phosphorites and associated diagenetic minerals: authigenesis in modern organic-rich sediments. Mar. Geol. 80, 231–267.

Glenn, C.R., Arthur, M.A., 1990. Anatomy and origin of a Crecaeous phosphorite-greensand giant, Egypt. Sedimentology 37, 123–154.

Glenn, C.R., Föllmi, K.B., et al., 1994. Phosphorous and phosphorites: sedimentology and environments of formation. Eclogae Geol. Helv. 87, 747–788.

Goldhaber, M.B., Kaplan, I.R., 1974. The sulphur cycle. In: Goldberg, E.D. (Ed.), The Sea, Marine Chemistry, vol. 5. 569–655.

Goldhaber, M.B., Kaplan, I.R., 1980. Mechanisms of sulfur incorporation and isotope fractionation during early diagenesis in sediments of the Gulf of California. Mar. Chem. 9, 95–143.

Griffin, J.J., Windom, H., Goldberg, E.D., 1968. The distribution of clay minerals in the world ocean. Deep-Sea Res. 15, 433–459.

Han, X., Suess, E., Sahling, H., Wallmann, K., 2004. Fluid venting activity on the Costa Rica margin: new results from authigenic carbonates. Int. J. Earth Sci. (Geol.Rundsch.) 93, 596–611.

Hanor, J.S., 2000. Barite-celestine geochemistry and environments of formation. Rev. Mineral. Geochem. 40, 193–275.

Harper, H.E., Knoll, A.H., 1975. Silica, diatoms, and Cenozoic radiolarian evolution. Geology 3, 175–177.

Harrison, W.E., Hesse, R., Gieskes, J.M., 1982. Relationship between sedimentary facies and interstitial water chemistry in slope, trench and Cocos plate sites from the Mid-America Trench transect, active margin off Guatemala. Leg 67, Deep Sea Drilling Project. In: Von Huene, R., Aubouin, J. (Eds.), Init. Rep. Deep Sea Drilling Proj. 67. U.S. Government Printing Office, Washington, DC, pp. 603–614.

Heath, G.R., Moore, T.C., Dauphin, J.P., 1977. Organic carbon in deep sea sediments. In: Andersen, N.R., Malahoff, A. (Eds.), The Fate of Fossil Fuel CO_2 in the Oceans. Plenum Press, New York, pp. 605–625.

Hein, J.R., Karl, S.M., 1983. Comparisons between open-ocean and continental margin chert sequences. In: Iijima, A., Hein, J.R., Siever, R. (Eds.), Siliceous Deposits in the Pacific Region. Developments in Sedimentology, Elsevier, Amsterdam, 36, pp. 25–43.

Hein, J.R., Scholl, D.W., Barron, J.A., Jones, M.J., Miller, J., 1978. Diagenesis of late Cenozoic diatomaceous deposits and formation of the bottom simulating reflector in the southern Bering Sea. Sedimentology 25, 155–181.

Hein, J.R., O'Neil, J.R., Jones, M.G., 1979a. Origin of authigenic carbonates in sediment from the deep Bering Sea. Sedimentology 27, 577–591.

Hein, J.R., Yeh, H.W., Alexander, E., 1979b. Origin of iron-rich montmorillonite from manganese nodule belt of the north equatorial Pacific. Clays Clay Miner. 27, 185–194.

Heinemann, C., Füchtbauer, H., 1982. Insoluble residues of the fine-grained sediments from the trench transect south of Guateemala, Deep Sea Drilling Project, Leg 67. In: Aubouin, J., Von Huene, R. (Eds.), Initial Repts. Deep Sea Drilling Project 67. U.S. Government Printing Office, Washington, DC, pp. 497–506.

Helm, R., 1985. Mineralogy and diagenesis of slope sediments offshore Guatemala and Costa Rica, Deep Sea Drilling Project Leg 84. In: Von Huene, R., Aubouin, J. (Eds.), Init. Rep. Deep Sea Drilling Proj. 85. U.S. Government Printing Office, Washington, DC, pp. 571–594.

Henderson Jr., N.C., Crawford, G.A., Dunham, J.B., Osborne, R.H., 1984. Relationship of diagenetic dolomite to depositional facies in diatomaceous strata, Upper Sisquoc Fm., Casmalia Hills, Santa Barbara Co., California. In: Garrison, R.E., Kastner, M., Zenger, D.H. (Eds.), Dolomites of the Monterey Formation and Organic Rich Units, Soc. Econ. Paleontol. Mineral., Pacific Section, Spec. Publ. 41, 119–140.

Hennessy, J., Knauth, L.P., 1985. Isotopic variations in dolomite concretions from the Monterey Formation, California. J. Sediment. Petrol. 55, 120–130.

Henrich, R., Hüneke, H., 2011. Hemipelagic advection and periplatform sedimentation. In: Hüneke, H., Mulder, T. (Eds.), Deep-Sea Sediments. Developments in Sedimentology, Elsevier, Amsterdam, vol. 63, pp. 353–396.

Hesse, R., 1987. Selective and reversible carbonate-silica replacements in Lower Cretaceous carbonate bearing turbidites of the Eastern Alps. Sedimentology 34, 1055–1077.

Hesse, R., 1990a. Early Diagenetic Pore-Water/Sediment Interaction: Modern Offshore Basins. In: McIlreath, I.A., Morrow, D.W. (Eds.), Diagenesis. Geosci. Canada, Reprint Ser. 4, 277–316 [Originally published in Geosci. Canada 13(3), 1986.].

Hesse, R., 1990b. Origin of chert I. Diagenesis of biogenic siliceous sediments. In: McIlreath, I.A., Morrow, D.W. (Eds.), Diagenesis. Geosci. Canada, Reprint Ser. 4, 227–251 [Originally published in Geosci. Canada 15(3), 1988.].

Hesse, R., 1990c. Pore-water anomalies in gas-hydrate bearing sediments of the deeper continental margins: facts and problems. J. Inclusion Phenomena Molecular Recogn. Chem. 8, 117–138.

Hesse, R., 2003. Pore water anomalies of submarine gas-hydrate zones as tool to assess hydrate abundance and distribution in the subsurface: what have we learned in the past decade. Earth-Sci. Rev. 61, 149–179.

Hesse, R., Harrison, W.E., 1981. Gas-hydrates (clathrates) causing pore-water freshening and oxygen isotope fractionation in deep-water sedimentary sections of the terrigenous continental margins. Earth Planet. Sci. Lett. 55, 453–462.

Hesse, R., Lebel, J., Gieskes, J.M., 1985. Interstitial water chemistry of gas hydrate bearing sections on the Middle-America Trench Slope, Deep Sea Drilling Project, Leg 84. In: Von Huene, R., Aubouin, J. (Eds.), Init. Rep. Deep Sea Drilling Proj. 84. U.S. Government Printing Office, Washington, DC, pp. 727–737.

Hesse, R., Frape, S., Egeberg, P.K., Matsumoto, R., 2000. Stable isotope studies (Cl, O, and H) of interstitial waters from Site 997, Blake Ridge gas hydrate field, West Atlantic. In: Paull, C.K., Matsumoto, R., Wallace, P.J., Dillon, W.P. (Eds.), Gas hydrate Sampling on the Blake Ridge and Carolina Rise. Proc. Ocean Drill. Progr. Sci. Results 164. Ocean Drilling Program, College Station, TX, pp. 129–137.

Hesse, R., Egeberg, P.K., Frape, S.K., 2001. Modelling of chlorine stable isotope ratios of a submarine gas-hydrate field. In: Cidu, R. (Ed.), Water-Rock Interactionvol. 10. Balkema, Lisse, pp. 1525–1528.

Hesse, R., Shah, J., Islam, S., 2004. Physical and chemical growth conditions of Ordovician organogenic deep-water dolomite concretions: implications for the $\delta^{18}O$ of Early Palaeozoic sea water. Sedimentology 51, 601–625.

Hesse, R., Egeberg, P.K., Frape, S.K., 2006. Chloride stable isotope ratios as tracer for pore-advection rates in a submarine gas-hydrate field: implication for hydrate concentration. Geofluids 6, 1–7.

Hinman, N.W., 1987. Organic and inorganic chemical controls on the rates of silica diagenesis: a comparison of a natural system with experimental results. Unpublished Ph.D. thesis, Univeristy of California, San Diego, 381pp (Univ. Microfilms).

Hiscott, R.N., Pickering, K.T., Beeden, D.R., 1986. Progressive filling of a confined Middle Ordovician foreland basin associated with the Taconic Orogeny, Quebec, Canada. In: Allen, P.A., Homewood, P. (Eds.), Foreland Basins, Spec. Publ. Int. Assoc. Sediment., 8, 309–325.

Holbrook, W.S., Hoskins, H., Wood, W.T., Stephen, R.A., Lizzaralde, D., The Leg 164 Science Party, 1996. Methane gas-hydrate and free gas on the Blake Ridge from vertical seismic profiling. Science 273, 1840–1843.

Holzapfel, T., Chamley, H., 1986. Les smectites latées du domaine Atlantique depuis le Jurassique supérieur: gisement et signification. Clay Miner. 21, 133–148.

Hover, V.C., Walter, L.M., Peacor, D.R., 2002. K uptake by modern estuarine sediments during early marine diagenesis, Mississippi delta plain, Louisiana, U.S.A. J. Sediment. Res. 72, 775–792.

Hudson, J.D., 1978. Concretions, isotopes, and the diagenetic history of the Oxford Clay (Jurassic) of central England. Sedimentology 25, 339–370.

Hudson, J.D., Coleman, M.L., Barreiro, B.A., Hollingworth, N.T.J., 2001. Septarian concretions from the Oxford Clay (Jurassic, England, UK): involvement of original marine and multiple external pore fluids. Sedimentology 48, 507–531.

Hüneke, H., Henrich, R., 2011. Pelagic sedimentation in modern and ancient oceans. In: Hüneke, H., Mulder, T. (Eds.), Deep-Sea Sediments. Developments in Sedimentology, Elsevier, Amsterdam, vol. 63, pp. 215–351.

Humphris, S.E., Hallman, C.M., 1980. Chemistry of interstitial waters sampled during Leg 54. In: Rosendahl, B.R., Hekinian, R. (Eds.), Initial Reports of the Deep Sea Drilling Project 54U.S. Government Printing Office, Washington, DC, pp. 387–393.

Hurd, D.C., 1973. Interactions of biogenic opal, sediment and seawater in the central Equatorial Pacific. Geochim. Cosmochim. Acta 37, 2257–2282.

Hurd, D.C., Theyer, F., 1975. Changes in the physical and chemical properties of biogenic silica from the central Equatorial Pacific. 1. Solubility, specific surface area, and solution rate constants of acid-cleaned samples. Analyt. Methods Oceanogr. Adv. Chem. Ser. 147, 211–230.
Hurd, D.C., Wenkham, C., Pankratz, H.S., Fugate, J., 1979. Variable porosity in siliceous skeletons: determination and importance. Science 203, 1340–1343.
Hurst, A., Cartwright, J., Huuse, N., Jonk, R., Schwab, A., Duranti, D., et al., 2003. Significance of large-scale injectites as long-term fluid conduits: evidence from seismic data. Geofluids 3, 263–274.
Iijima, A., Utada, M., 1983. Recent developments in the sedimentology of siliceous deposits in Japan. In: Iijima, A., Hein, J.R., Siever, R. (Eds.), Siliceous Deposits in the Pacific Region. Dev. Sediment. Elsevier, Amsterdam, 36, pp. 45–64.
Iler, R.K., 1979. Chemistry of Silica. Wiley-Interscience, New York, 866 pp.
Imoto, N., Saito, Y., 1973. Scanning electron microscopy of chert. Natl. Sci. Museum Bull. 16, 397–400, Tokyo.
Irwin, H., Curtis, C.D., Coleman, M., 1977. Isotopic evidence for source of diagenetic carbonates formed during burial of organic-rich sediments. Nature 269, 209–213.
Isaacs, C.M., 1982. Influence of rock composition on kinetics of silica phase changes in the Monterey Formation, Santa Barbara area, California. Geology 10, 304–308.
Isaacs, C.M., Pisciotto, K.A., Garrison, R.E., 1983. Facies and diagenesis of the Miocene Monterey Formation, California: a summary. In: Iijima, A., Hein, J.R., Siever, R. (Eds.), Siliceous Deposits in the Pacific Region. Dev. Sediment. Elsevier, Amsterdam, 36, pp. 247–282.
Itihara, Y., Honma, H., 1979. Ammonium in biotite from metamorphic and granitic rocks of Japan. Geochim. Cosmochim. Acta 43, 503–509.
Jenden, P.D., Gieskes, J.M., 1983. Chemical and isotopic composition of interstitial water from Deep Sea Drilling Project sites 533 and 534. In: Sheridan, R.E., Gradstein, F.M. (Eds.), Init. Rep. Deep Sea Drilling Proj. 76. U.S. Government Printing Office, Washington, DC, pp. 453–461.
Jenden, P.D., Kaplan, I.R., 1986. Comparison of microbial gases from the Middle America Trench and Scripps Submarine Canyon: implications for the origin of natural gas. Appl. Geochem. 1, 631–646.
Jenkyns, H.C., Winterer, E.L., 1982. Palaeoceanography of Mesozoic ribbon radiolarites. Earth Planet. Sci. Lett. 60, 351–375.
Johnson, T.C., 1976. Biogenic opal preservation in pelagic sediments of a small area in the eastern tropical Pacific. Geol. Soc. Amer. Bull. 87, 1273–1282.
Jones, D.L., Murchey, B., 1986. Geologic significance of Paleozoic and Mesozoic radiolarian chert. Annu. Rev. Earth Planet. Sci. 14, 455–492.
Jones, J.B., Segnit, E.R., 1971. The nature of opal. I. Nomenclature and constituent phases. J. Geol. Soc. Aust. 18, 57–68.
Jonk, R., Hurst, A., Duranti, D., Mazzini, A., Fallick, A.E., 2005. Origin and timing of sand injection, petroleum migration, and diagenesis in Tertiary reservoirs, South Viking Graben, North Sea. AAPG Bull. 89, 329–357.
Kasakov, A.V., 1937. The phosphorite facies and the genesis of phosphorites. Trans. Sci. Inst. Fertil. Insecto-Fungicides 142, 95–113 [in Russian].
Kastner, M., 1981. Authigenic silicates in deep-sea sediments: formation and diagenesis. In: Emiliani, C. (Ed.), The Sea. The Oceanic Lithosphere, vol. 7. Wiley, New York, pp. 915–980.
Kastner, M., Gieskes, J.M., 1976. Interstitial water profiles and sites of diagenetic reactions, Leg 35, DSDP. Bellinghausen abyssal plain. Earth Planet. Sci. Lett. 33, 11–20.

Kastner, M., Gieskes, J.M., 1983. Opal-A to opal-CT transformation: a kinetic study. In: Iijima, A., Hein, J.R., Siever, R. (Eds.), Siliceous Deposits in the Pacific Region. Dev. Sediment. Elsevier, Amsterdam, 36, pp. 211–227.

Kastner, M., Siever, R., 1983. Siliceous sediments of the Guaymas Basin: the effect of high thermal gradients on diagenesis. J. Geol. 91, 629–641.

Kastner, M., Stonecipher, S.A., 1978. Zeolites in pelagic sediments of the Atlantic, Pacific and Indian Oceans. In: Sand, L.B., Mumpton, F.A. (Eds.), Natural Zeolites: Occurrence, Properties, Use. Pergamon Press, pp. 199–220.

Kastner, M., Keene, J.B., Gieskes, J.M., 1977. Diagenesis of siliceous oozes. I. Chemical controls on the rate of opal-A to opal-CT transformation—an experimental study. Geochim. Cosmochim. Acta 41, 1041–1059.

Kastner, M., Elderfield, H., Martin, J.B., Suess, E., Kvenvolden, K.A., Garrison, R.E., 1990. Diagenesis and interstitial water chemistry at the Peruvian continental margin - major constituents and strontium isotopes. In: Suess, E., Von Huene, R. (Eds.), Peru Continental Margin. Proc. Ocean Drill. Progr. Sci. Results, vol. 112. Ocean Drilling Program, College Station, TX, pp. 413–440.

Kastner, M., Elderfield, H., Jenkins, W.J., Gieskes, J.M., Gamo, T., 1993. Geochemical and isotopic evidence for fluid flow in the western Nankai subduction zone, Japan. In: Hill, I. A., Taira, A., Firth, J.V. (Eds.), Nankai Trough. Proc. Ocean Drill. Progr., Scient. Res. 131. Ocean Drilling Program, College Station, TX, pp. 397–413.

Kastner, M., Sample, J.C., Whiticar, M.J., Hovland, M., Cragg, B.A., Parkes, J.R., 1995. Geochemical evidence for fluid flow and diagenesis at the Cascadia convergent margin. In: Carson, B., Westbrook, B.K., Musgrave, R.J., Suess, E. (Eds.), Cascadia margin. Proc. Ocean Drill. Progr. Sci. Results 146, Pt. I, Ocean Drilling Program, College Station, TX, pp. 375–384.

Keene, J.B., 1975. Cherts and porcellanites from the North Pacific, DSDP, Leg 32. In: Moberly, R., Larsen, R. (Eds.), Init. Rep. Deep Sea Drilling Proj. 32. U.S. Government Printing Office, Washington, DC, pp. 429–507.

Kelts, K., McKenzie, J.A., 1982. Diagenetic dolomite formation in Quaternary anoxic diatomaceous muds of Deep Sea Drilling Project Leg 64, Gulf of California. In: Curry, J.R., Moore, D.G. (Eds.), Init. Rep. Deep Sea Drilling Proj. 64 (part 2). U.S. Government Printing Office, Washington, DC, pp. 553–569.

Kent, D.B., Kastner, M., 1985. Mg^{2+} removal in the system Mg^{2+}—amorphous SiO_2–H_2O by adsorption and Mg-hydroxysilicate precipitation. Geochim. Cosmochim. Acta 49, 1123–2236.

Kimura, G., Silver, E.E., Blum, P., Shipboard Scientific Party Leg 170, 1997. Costa Rica Accretionary Wedge. Proc. Ocean Drill. Progr. Init. Repts. vol. 170. Ocean Drilling Program, College Station, TX, 458pp.

Klein, J.S., Mozley, P., Campbell, A., Cole, R., 1999. Spatial distribution of carbon and oxygen isotopes in laterally extensive carbonate cemented layers: implications for mode of growth and subsurface identification. J. Sediment. Res. 69, 184–201.

Ko, J., 1985. Controls on graywacke petrology in Middle Ordovician Cloridorme Formation: tectonic setting of source areas versus diagenesis. Unpubl. M.Sc. thesis, McGill Univ., 235pp.

Kopf, A., Deyhle, A., Zuleger, E., 2000. Evidence for deep circulation and gas hydrate dissociation using boron and boron isotopes of pore fluids in forearc sediments from Costa Rica (ODP Leg 170). Mar. Geol. 167, 1–28.

Kushnir, J., Kastner, M., 1984. Two forms of dolomite occurrences in the Monterey Formation, California; concretions and layers - a comparative mineralogical, geochemical and isotopic study. In: Garrison, R.E., Kastner, M., Zenger, D.H. (Eds.), Dolomites of the Monterey Formation and Organic Rich Units, Soc. Econ. Paleontol. Mineral. Pacific Section, Spec. Publ. 41, 171–183.

Kutterolf, S., Schacht, U., Wehrmann, H., Freundt, A., Moerz, T., 2007. Onshore to offshore tephrostratigraphy and marine ash layer diagenesis in Central America. In: Alvarado, G., Buntschuh, J. (Eds.), Central America—Geology, Resources and Hazards, vol. 2. Balkema, Rotterdam, pp. 395–423.

Kvenvolden, K.A., Kastner, M., 1990. Gas hydrates of the Peruvian outer continental margin. In: Suess, E., Von Huene, R. (Eds.), Proc. ODP, Sci. Results 112, Ocean Drilling Program, College Station, TX, pp. 517–526.

Kvenvolden, K.A., 1998. A primer on the geological occurrence of gas hydrates. In: Henriet, J.-P., Mienert, J. (Eds.), Gas Hydrates: Relevance to World Margin Stability and Climate Change. Geol. Soc. London, Spec. Publ., 137, pp. 9–30.

Lancelot, Y., 1973. Chert and silica diagenesis in sediments from the central Pacific. In: Winterer, E.L., Ewing, J.I. (Eds.), Init. Rep. Deep Sea Drilling Proj. 17. U.S. Government Printing Office, Washington, DC, pp. 377–405.

Land, L.S., 1983. The application of stable isotopes to studies of the origin of dolomite and to problems of diagenesis of clastic sediments. In: Arthur, M.A., Anderson, T.F., Kaplan, I.R., Veizer, J., Land, L.S. (Eds.), Stable Isotopes in Sedimentary Geology, Soc. Econ. Paleontologists Mineralogists, Short Course Lecture Notes 10, 4-1–4-22.

Landing, E., Benus, A.P., Whitney, P.R., 1992. Early and early Middle Ordovician continental slope deposition: shale cycles and sandstones in the New York Promontory and Quebec Reentrant region. New York State Museum/Geol. Survey Bull. 474, 40pp.

Landmesser, M., 1993. Mobilisation durch Reifung: Anmerkungen zu einem geochemischen Akkumulationsprinzip. Ber. Deutsch. Mineral. Ges., Beih. Eur. J. Mineral 5, 38.

Langseth, M.G., Westbrook, G.K., Hobart, M.A., 1988. Geophysical survey of a mud volcano seaward of the Barbados accretionary complex. J. Geophys. Res. 93, 1049–1061.

Lash, G.G., Blood, D.R., 2004a. Origin of shale fabric by mechanical compaction of flocculated clay: evidence from the Upper Devonian Rhinestreet Shale, Western New York, USA. J. Sediment. Res. 74, 110–116.

Lash, G.G., Blood, D.R., 2004b. Geochemical and textural evidence for early (shallow) diagenetic growth of stratigraphically confined carbonate concretions, Upper Devonian Rhinestreet black shale, western New York. Chem. Geol. 206, 407–424.

Lawrence, M.J.F., 1991. Early diagenetic dolomite concretions in the Late Cretaceous Herring Formation, eastern Marlborough, New Zealand. Sediment. Geol. 75, 125–140.

Lawrence, J.R., Gieskes, J.M., 1981. Constraints on water transport and alteration in the oceanic crust from the isotopic composition of pore water. J. Geophys. Res. 86, 7924–7934.

Lawrence, J.R., Gieskes, J.M., Broecker, W.S., 1975. Oxygen isotope and cation composition of DSDP pore waters and the alteration of Layer II basalts. Earth Planet. Sci. Lett. 27, 1–10.

Lawrence, J.R., Drever, J.I., Anderson, T.F., Brueckner, H.K., 1979. Importance of alteration of volcanic material in the sediments of Deep Sea Drilling Site 323: chemistry, $^{18}O/^{16}O$ and $^{87}Sr/^{86}Sr$. Geochim. Cosmochim. Acta 43, 573–588.

Lee, M.W., 2000. Gas hydrate amounts estimated from acoustic logs at the Blake Ridge, Sites 994, 995, and 997. In: Paull, C.K., Matsumoto, R., Wallace, P.J., Dillon, W.P. (Eds.), Gas Hydrate Sampling on the Blake Ridge and Carolina Rise. Proc. Ocean Drill. Progr. Sci. Results 164, Ocean Drilling Program, College Station, TX, pp. 193–198.

Lee, M.W., Hutchinson, D.R., Dillon, W.P., Miller, J.J., Agena, W.F., Swift, B.A., 1993. Method of estimating the amount of in situ gas hydrates in deep marine sediments. Mar. Petrol. Geol. 10, 493–506.

Lerman, A., 1977. Migrational processes and chemical reactions in interstitial waters. In: Goldberg, E.D., McCave,, I.N., O'Brien,, J.J., Steele, J.H. (Eds.), The Sea 6, Marine Modeling, John Wiley and Sons, New York, 695–738.

Leśniak, P.M., Łącka, B., Hladikova, J., Zieliński, G., 1999. Origin of barite concretions in the West Carpathian flysch, Poland. Chem. Geol. 158, 155–163.
Li, Y.-H., Gregory, S., 1974. Diffusion of ions in sea water and in deep-sea sediments. Geochim. Cosmochim. Acta 38, 703–714.
Lippmann, F., 1955. Ton, Geoden und Minerale des Barrême von Hoheneggelsen. Geol. Rundsch. 43, 475–503.
Lonergan, L., Lee, N., Jolley, R., Johnson, H., 2001. Remobilization and injection in deepwater depositional systems. In: Fraser, S.I., Johnson, H.D., Fraser, A.J., Evans, A.M. (Conveners), Petroleum Geology of Deepwater Depositional Systems: Advances in Understanding 3D Architecture, Geol. Soc. London, Progr. Abstr. (unpaginated).
Lumsden, D.N., 1988. Characteristics of deep-marine dolomite. J. Sediment. Petrol. 58, 1023–1031.
Mallard, M.E., 1890. Sur la lussatite, nouvelle variété minérale cristallisée de silice. Bull. Soc. Fr. Minéral. Cristallogr. 13, 63–66.
Malone, M.J., Baker, P.J., Burns, S.J., 1994. Recrystallization of dolomite: evidence from the Monterey Formation (Miocene), California. Sedimentology 41, 1223–1239.
Manheim, F.T., 1967. Evidence for submarine discharge of water on the Atlantic continental slope of the southern United States, and suggestions for further search. Trans. NY Acad. Sci. Ser. II 29, 839–853.
Manheim, F.T., Paull, C.K., 1981. Patterns of groundwater salinity changes in a deep continental-oceanic transect off the southeastern Atlantic coast of the USA. J. Hydrol. 54, 95–105.
Manheim, F.T., Rowe, G.T., Jipa, D., 1975. Marine phosphorite formation off Peru. J. Sediment. Petrol. 45, 243–251.
Manheim, F.T., Schug, D.M., 1978. Interstitial waters of Black Sea cores. In: Ross, D.A., Neprochnov, Y.P. (Eds.), Init. Rep. Deep Sea Drilling Proj. 42(part 2). U.S. Government Printing Office, Washington, DC, pp. 637–651.
Manheim, F.T., Waterman, L.S., Woo, C.C., Sayles, F.L., 1974. Interstitial water studies on small core samples, Leg 23 (Red Sea). In: Whitmarsh, R.B., Ross, D.A. (Eds.), Init. Rep. Deep Sea Drilling Proj. 23. U.S. Government Printing Office, Washington, DC, pp. 955–967.
Matheney, R.K., Knauth, L.P., 1993. New isotopic temperature estimates for early silica diagenesis in bedded cherts. Geology 21, 519–522.
Matsumoto, R., 1983. Mineralogy and geochemistry of carbonate diagenesis of the Pliocene and Pleistocene hemipelagic mud on the Blake Outer Ridge, Site 533, Leg 76. In: Sheridan, R.E., Gradstein, F.M. (Eds.), Init. Rep. Deep Sea Drilling Proj. 76. U.S. Government Printing Office, Washington, DC, pp. 411–427.
Matsumoto, R., 1989. Isotopically heavy oxygen-containing sideraite derived from the decomposition of methane hydrate. Geology 17, 707–710.
Mazzullo, S., 2000. Organogenic dolomitization in peritidal to deep-sea sediments. J. Sediment. Res. 70A, 10–23.
McBride, E.F., Folk, R.L., 1979. Features and origin of Italian Jurassic radiolarites deposited on continental crust. J. Sediment. Petrol. 49, 837–868.
McDuff, R.E., 1978. Conservative behaviour of calcium and magnesium in interstitial waters of marine sediments: identification and interpretation. Unpubl. Ph.D. thesis, Scripps Institution of Oceanography, University of California, San Diego, 183pp.
McDuff, R.E., 1981. Major cation gradients in DSDP interstitial waters and the role of diffusive exchange between sea water and upper oceanic crust. Geochim. Cosmochim. Acta 45, 1705–1713.
McDuff, R.E., 1985. The chemistry of interstitial waters, Deep Sea Drilling Project Leg 86. Init. Rep. Deep Sea Drilling Proj. 86U.S. Government Printing Office, Washington, DC pp. 675–687.

McDuff, R.E., Gieskes, J.M., 1976. Calcium and magnesium profiles in DSDP interstitial waters: diffusion or reaction. Earth Planet. Sci. Lett. 33, 1–10.
McDuff, R.E., Gieskes, J.M., Lawrence, J.R., 1978. Interstitial water studies, Leg 42A. In: Hsü, K., Montadert, L. (Eds.), Init. Rep. Deep Sea Drilling Proj. 42 (part 1.) U.S. Government Printing Office, Washington, DC, pp. 561–568.
McIver, R.D., 1981. Gas hydrates. In: Meyer, R.G., Olson, J.C. (Eds.), Long Term Energy Resources. Pitman, Boston, MA, pp. 713–726.
McManus, J., Berelson, W.M., Klinkhammer, G.P., Kilgore, T.E., Hammond, D.E., 1994. Remobilization of barium in continental margin sediments. Geochim. Cosmochim. Acta 4899–4907.
Mechalas, B.J., 1974. Pathways and environmental requirements for biogenic gas production in the ocean: In: Kaplan, I.R. (Ed.), Natural Gases in Marine Sediments. Mar. Sci. 3, 11–26.
Menzies, M., Seyfried, W.E., 1979. Basalt—seawater interaction: trace element and strontium isotope variations in experimentally altered glassy basalt. Earth Planet. Sci. Lett. 44, 463–472.
Middelburg, J.B.M., 1990. Early diagenesis and authigenic mineral formation in anoxic sediments of Kau Bay, Indonesia. Geol. Ultraiectina 71, 177pp.
Middelburg, J.J., De Lange, G.J., Kreulen, R., 1990. Dolomite formation in anoxic sediments of Kau Bay, Indonesia. Geology 18, 399–402.
Middelburg, J.J., Soetaert, K., Herman, P.M.J., Heip, C.H.R., 1996. Denitrification in marine sediments: model study. Global Biogeochem. Cycles 10, 661–673.
Miliman, J.D., Müller, J., 1973. Precipitation and lithification of magnesian calcite in the deep-sea sediments of the eastern Mediterranean Sea. Sedimentology 20, 29–45.
Miliman, J.D., Müller, J., 1977. Characteristics and genesis of shallow-water and deep-sea limestones. In: Andersen, N.R., Malahoff, A. (Eds.), The Fate of Fossil CO_2 in the Oceans. Plenum Publ. Co, New York, pp. 655–672.
Moberly Jr., R., Kimura, H.S., McCoy Jr., F.W., 1968. Authigenic marine phyllosilicates near Hawaii. Geol. Soc. Am. Bull. 79, 1449–1460.
Moore, J.C., Vrolijk, P., 1992. Fluids in accretionary prisms. Rev. Geophys. 30, 113–135.
Moore, J.C., Mascle, A., Taylor, E., Andreieff, P., Alvarez, F., Barnes, R., et al., 1988. Tectonics and hydrogeology of the northern Barbados Ridge: results from Ocean Drilling Program Leg 110. Geol. Soc. Am. Bull. 100, 1578–1593.
Morse, J.W., Casey, W.H., 1988. Oswald processes and mineral paragenesis in sediments. Am. J. Sci. 288, 537–560.
Mozley, S.P., 1989. Complex compositional zonation in concretionary siderite: implications for geochemical studies. J. Sediment. Petrol. 59, 815–818.
Mozley, S.P., 1996. The internal structure of carbonate concretions in mudrocks: a critical evaluation of the conventional concentric model of concretion growth. Sed. Geol. 103, 85–91.
Mozley, P.S., Burns, S.J., 1993. Oxygen and carbon isotopic composition of marine carbonate concretions: an overview. J. Sediment. Petrol. 63, 73–83.
Mozley, P.S., Wersin, P., 1992. Isotopic composition of siderite—an indicator of depositional environment. Geology 20, 817–820.
Mulder, T., 2011. Gravity processes on continental slope. rise and abyssal plains. In: Hüneke, H., Mulder, T. (Eds.), Deep-Sea Sediments. Developments in Sedimentology, Elsevier, Amsterdam, Vol. 63. pp. 25–148.
Müller, P.J., 1977. C/N ratios in Pacific deep-sea sediments: effect of inorganic ammonium and organic nitrogen compounds sorbed by clays. Geochim. Cosmochim. Acta 41, 765–776.
Müller, J., Fabricius, F., 1974. Magnesium-calcite nodules in the Ionian deep sea: an actualistic model for the formation of some nodular limestones. In: Hsü, K.J., Jenkyns,

H.C. (Eds), Pelagic Sediments on Land and Under the Sea. Int. Assoc. Sedimentol. Spec. Publ. 1, 235–247.
Müller, J., Fabricius, F., 1978. Lüneburgite in Upper Miocene sediments of the eastern Mediterranean Sea. In: Hsü, K., Montadert, L. (Eds.), Init. Rep. Deep Sea Drilling Proj. 42U.S. Government Printing Office, Washington, DC, pp. 661–664.
Müller, P.J., Mangini, A., 1980. Organic carbon decomposition rates in sediments of the Pacific manganese nodule belt dated by ^{230}Th and ^{231}Pa. Earth Planet. Sci. Lett. 51, 94–114.
Müller, P.J., Suess, E., 1979. Productivity, sedimentation rate, and sedimentary organic matter in the oceans—I. Organic carbon preservation. Deep-Sea Res. 26A, 1347–1362.
Munnecke, A., Westphal, H., Elrick, M., Reijmer, J.J.G., 2001. The mineralogical composition of precursor sediments of calcareous rhythmites: a new approach. Int. J. Earth Sci. (Geol.Rundsch.) 90, 795–812.
Murata, K.J., Larson, R.R., 1975. Diagenesis of Miocene siliceous shales, Temblor Range, California. J. Res. US Geol. Surv. 3, 553–566.
Murata, K.J., Norman II, M.B., 1976. An index of crystallinity for quartz. Am. J. Sci. 276, 1120–1130.
Murata, K.J., Friedman, I., Gleason, J.D., 1977. Oxygen isotope relations between diagenetic silica minerals in Monterey Shale, Temblor Rangem, California. Am. J. Sci. 277, 259–272.
Murray, J.W., Grundmanis, V., Smethie, W.M., 1978. Interstitial water chemistry in the sediments of Saanich Inlet. Geochim. Cosmochim. Acta 42, 1011–1026.
Newberry, C.J., Webster, G., Cragg, B.A., Parkes, R.J., Weightman, A.J., Fry, J.C., 2004. Diversity of procaryotes and methanogenesis in deep subsurface sediments from the Nankai Trough, Ocean Drilling Program Leg 190. Environ. Microbiol. 6, 24–287.
Nisbet, E.G., Price, I., 1974. Siliceous turbidites: bedded cherts as redeposited ocean ridge-derived sediments. In: Hsu, K., Jenkyns, H.C. (Eds), Pelagic Sediments on Land and Under the Sea. Int. Assoc. Sediment. Spec. Publ. 1, 351–366.
Nissenbaum, A., Presley, B.J., Kaplan, I.R., 1972. Early diagenesis in a reducing fjord, Saanich Inlet. British Columbia—I. Chemical and isotopic changes in major components of seawater. Geochim. Cosmochim. Acta 36, 1007–1027.
Odin, G.S., Fullagar, P.D., 1988. Geological significance of the glaucony facies. In: Odin, G. S. (Ed.), Green marine clays. Oolitic Ironstone Facies, Verdine Facies, Glaucony Facies and Celadonite-Bearing Rock Facies—A Comparative Study. Dev. Sediment. Elsevier, Amsterdam, 45, pp. 295–332.
Odin, G.S., Matter, A., 1981. De glauconarium origine. Sedimentology 28, 611–641.
Odin, G.S., Stephan, J.F., 1981. The occurrence of deep water glaucony from the eastern Pacific: the result of in situ genesis or subsidence. In: Watkins, J.S., Moore, J.C. (Eds.), Init. Rep. Deep Sea Drilling Proj. 66. U.S. Government Printing Office, Washington, DC, pp. 419–428.
Oertel, G., Curtis, C.D., 1972. Claystone-ironstone concretion preserving fabrics due to progressive compaction. Geol. Soc. Am. Bull. 83, 2597–2606.
Ogawa, Y., Kawata, T., 1998. Preservation of biogenic opal-A in earliest Cretaceous radiolarian claystone from the Western Pacific. J. Sediment. Res. 68, 435–439.
Okada, H., 1980. Pebbles and carbonate nodules from Deep Sea Drilling Project Leg 56 cores. In: Honza, E. (Ed.), Init. Rep. Deep Sea Drilling Proj. 56-57. U.S. Government Printing Office, Washington, DC, pp. 1089–1105.
Out, D.J.P., Los, J.M., 1980. Viscosity of aqueous solutions of univalent electrolytes from 5 to 95 °C. J. Sol. Chem. 9, 19–35.
Palmer, M.R., Elderfield, H., 1985. Sr isotope composition of sea water over the past 75 Myr. Nature 314, 526–528.

Parkes, R.J., Cragg, B.A., Baie, S.J., Getliff, J.M., Goodman, K., Rochelle, P.A., et al., 1994. Deep bacterial biosphere in Pacific Ocean sediments. Nature 371, 410–413.
Parkes, R.J., Webster, G., Cragg, B.A., Weightman, A.J., Newberry, C.J., Ferdelman, T.G., et al., 2005. Deep sub-seafloor prokaryotes stimulated at interfaces over geological time. Nature 436, 390–394.
Parks, G.A., 1965. The isoelectric points of solid oxides, solid hydroxides, and aqueous complex systems. Chem. Rev. 65, 177–198.
Paull, C.K., Matsumoto, R., Wallace, P., et al., 1996. Gas hydrate sampling on the Blake Ridge and Carolina Rise. In: Proc. Ocean Drill. Progr. Init. Reports. 164. Ocean Drilling Program, College Station, TX, 623pp.
Paull, C.K., Matsumoto, R., Wallace, P.J., Dillon, W.P. (Eds.), 2000. Gas hydrate sampling on the Blake Ridge and Carolina Rise. In: Proc. Ocean Drill. Progr. Sci. Results, 164. Ocean Drilling Program, College Station, TX, 459pp.
Paytan, A., Griffith, E.M., 2007. Marine barite: recorder of variations in ocean export productivity. Deep-Sea Res. II Top. Stud. Oceanogr. 54, 687–705.
Pecora, W.T., Kerr, J.H., 1954. Whewellite from a septarian limestone concretion marine shale near Havre,Montana. Am. Mineral. 39, 208–214.
Pedersen, T.F., Shimmield, G.B., 1991. Interstitial water chemistry, leg 117: contrasts with the Peru margin. In: Prell, W.L., Niitsuma, N. (Eds.), Proc. ODP, Sci. Results, 117. Ocean Drilling Program, College Station, TX, pp. 499–513.
Perry, E.A., Gieskes, J.M., Lawrence, J.R., 1976. Mg, Ca and O^{18}/O^{16} exchange in the sediment-pore water system Hole 149, DSDP. Geochim. Cosmochim. Acta 40, 413–423.
Piepgras, D., Jacobsen, S., 1992. The behaviour of rare earth elements in seawater: precise determination of variations in the North Pacific water column. Geochim. Cosmochim. Acta 56, 1851–1862.
Pierre, C., Rouchy, J.M., Gaudichet, A., 2000. Diagenesis in the gas hydrate sediments of the Blake Ridge: mineralogy and stable isotope compositions of the carbonate and sulphide minerals. In: Paull, C.K., Matsumoto, R., Wallace, P.J., Dillon, W.P. (Eds.), Gas Hydrate Sampling on the Blake Ridge and Carolina Rise. Proc. Ocean Drill. Progr. Sci Results, 164. Ocean Drilling Program, College Station, TX, pp. 139–146.
Piper, D.Z., 1974a. Rare earth elements in ferromanganese nodules and other marine phases. Geochim. Cosmochim. Acta 38, 1007–1022.
Piper, D., 1974b. Rare earth elements in the sedimentary cycle: a summary. Chem. Geol. 14, 285–304.
Pirrie, D., Marshall, J.D., 1991. Field relationships and stable isotope geochemistry of concretions from James Ross Island,Antarctica. Sediment. Geol. 71, 137–150.
Pisciotto, K.A., 1980. Chert and porcellanite from Deep Sea Drilling Project Site 436, Northwest Pacific. In: Langseth, M., Okada, H. (Eds.), Init. Rep. Deep Sea Drilling Proj. 56/57. U.S. Government Printing Office, Washington, DC, pp. 1133–1142.
Pisciotto, K.A., 1981. Diagenetic trends in the siliceous facies of the Monterey Shale in the Santa Maria region, California. Sedimentology 28, 547–571.
Pisciotto, K.A., Mahoney, J.J., 1981. Isotopic survey of diagenetic carbonates, Deep Sea Drilling Project Leg 63. In: Yeats, R.S., Haq, B.U. (Eds.), Init. Rep. Deep Sea Drilling Proj. 63. U.S. Government Printing Office, Washington, DC, pp. 595–609.
Popp, B.N., Anderson, F.T., Sandberg, P.A., 1986. Brachiopods as indicators of original isotopic compositions in some Paleozoic limestones. Geol. Soc. Am. Bull. 97, 1262–1269.
Pratt, B.R., 2001. Septarian concretions: internal cracking caused by synsedimentary earthquakes. Sedimentology 48, 189–213.
Pufahl, P.K., Wefer, G., 2002. Data report: petrographic, cathodoluminescent, and compositional characteristics of organogenic dolomites from the southwest African margin.

In: Wefer, G., Berger, W.H., Richter, C. (Eds.), Proc. ODP, Sci. Results 175, 1-17 (CD-ROM). Ocean Drilling Program, College Station, TX.

Pye, K., Dickson, J.A.D., Schiavon, N., Coleman, M.L., Cox, M., 1990. Formation of siderite—Mg-calcite—iron sulphide concretions in intertidal marsh and sandflat sediments, north Norfolk, England. Sedimentology 37, 325–343.

Railsback, L.B., 1990. Influence of changing deep ocean circulation on the Phanerozoic oxygen isotope record. Geochim. Cosmochim. Acta 54, 1501–1509.

Railsback, L.B., Ackerly, S.C., Anderson, T.F., Cisne, J.L., 1990. Palaeontological and isotope evidence for warm saline deep waters in Ordovician oceans. Nature 343, 156–159.

Raiswell, R., 1971. The growth of Cambrian and Liassic concretions. Sedimentology 17, 147–172.

Raiswell, R., 1987. Non-steady state microbial diagenesis and the origin of concretions and nodular limestones. In: Marshall, J.D. (Ed.), Diagenesis of Sedimentary Sequences. Geol. Soc. Lond. Spec. Publ. 36, 41–54.

Raiswell, R., 1988. Chemical model for the origin of minor limestone-shale cycles by anaerobic methane oxidation. Geology 16, 641–644.

Rajan, S., Mackenzie, F.T., Glenn, C.R., 1996. A thermodynamic model for water column precipitation of siderite in the Plio-Pleistocene Black Sea. Am. J. Sci. 296, 506–548.

Ransom, B., Spivack, A.J., Kastner, M., 1995. Stable Cl isotopes in subduction-zone pore waters: implications for fluid-rock reactions and the cycling of chlorine. Geology 23, 715–718.

Reeburgh, W.S., 1983. Rates of biogeochemical processes in anoxic sediments. Annu. Rev. Earth Planet. Sci. 11, 269–298.

Rhoads, D.C., Morse, J.W., 1971. Evolutionary and ecologic significance of oxygen-deficient marine basins. Lethaia 4, 413–428.

Ricken, W., 1986. Diagenetic Bedding: A Model for Limestone-Marl Alternations. Springer Verlag, Berlin 210pp.

Riech, V., Von Rad, U., 1979. Silica diagenesis in the Atlantic Ocean: Diagenetic potential and transformations. In: Talwani, M., Hay, W., Ryan, W.B.F. (Eds.), Deep drilling results in the Atlantic Ocean: continental margins and paleoenvironment. Am. Geophys. Union, M. Ewing Ser. 3, 315–341.

Riedel, M., Collett, T.S., Malone, M.J., The IODP Expedition 311 Scientists, 2006. Stages of gas-hydrate evolution on the northertn Cascadia margin. Sci. Drill. 3, 18–24.

Ritger, S., Carson, B., Suess, E., 1987. Methane-derived authigenic carbonates formed by subduction-induced pore water expulsion along the Oregon/Washington margin. Geol. Soc. Am. Bull. 48, 147–156.

Robertson, A.H.F., 1977. The origin and diagenesis of cherts from Cyprus. Sedimentology 24, 11–30.

Rocker, K., 1974. Physical Properties Measurements and Test Procedures for Leg 27. Init. Rep. Deep Sea Drilling Proj. 27. U.S. Government Printing Office, Washington, DC.

Rodriguez, N.M., Paull, C.K., Borowski, W.S., 2000. Zonation of authigenic carbonates within gas-hydrate bearing sedimentary sections on the Blake Ridge: offshore southeastern North America. In: Paull, C.K., Matsumoto, R., Wallace, P.J., Dillon, W.P. (Eds.), Gas Hydrate Sampling on the Blake Ridge and Carolina Rise. Proc. Ocean Drill. Progr. Sci. Results 164. Ocean Drilling Program, College Station, TX, pp. 301–312.

Rohling, E.J., Hilgen, F.J., 1991. The eastern Mediterranean climate at times of sapropel formation: a review. Geol. Mijnb. 70, 253–264.

Rosenfeld, W.D., Silverman, S.R., 1959. Carbon isotope fractionation in bacterial production of methane. Science 130, 1658–1659.

Russell, K.L., Deffeyes, K.S., Fowler, G.A., Lloyd, R.M., 1967. Marine dolomite of unusual isotopic composition. Science 155, 189–191.

Ryan, W.B.F., Cita, M.B., 1977. Ignorance concerning episodes of ocean-wide stagnation. Mar. Geol. 23, 197–215.
Sackett, W.M., Thompson, R.R., 1963. Isotopic organic carbon compositions of recent continental shelf derived clastic sediments of the Eastern Gulf Coast, Gulf of Mexico. Am. Assoc. Petrol. Geol. Bull. 47, 525–531.
Sano, H., 1983. Bedded cherts associated wth greenstones in the Sawadai ad Shimantogawa Groups, Southwest Japan. In: Iijima, A., Hein, J.R., Siever, R. (Eds.), Siliceous Deposits in the Pacific Region. Dev. Sediment. Elsevier, Amsterdam, 36, pp. 427–440.
Sansone, F.J., Martens, C.S., 1981. Methane production from acetate and associated methane fluxes from anoxic coastal sediments. Science 211, 707–709.
Sass, E., Bein, A., Almogi-Labin, A., 1991. Oxygen-isotope composition of diagenetic calcite in organic-rich rocks: evidence for ^{18}O depletion in marine anaerobic pore water. Geology 19, 839–842.
Sayles, F.L., Waterman, L.S., Manheim, F.T., 1972. Interstitial water studies on small core samples from the Mediterranean Sea. In: Ryan, W.B.F., Hsü, K.J. (Eds.), Init. Rep. Deep Sea Drilling Proj. 13 U.S. Government Printing Office, Washington, DC, pp. 801–808.
Schacht, U., 2005. Alteration of volcanic glasses in marine sediments: Laboratory experiments and field studies. University of Kiel, PhD-Thesis.
Schacht, U., Wallmann, K., Kutterolf, S., Schmidt, M., 2008. Volcanogenic sediment-seawater interactions and the geochemistry of pore waters. Chem. Geol. 249, 321–338.
Schacht, U., Kutterolf, S., Wallmann, K., Schmidt, M., 2009. Alteration of volcanic glass in seawater: an experimental study. Int. J. Earth Sci. [under revision].
Schacht, U., Wallmann, K., Kutterolf, S., 2010. The influence of volcanic ash alteration on the REE composition of marine pore waters. J. Geochem. Exp. 106 (1-3), 176–187.
Schoell, M., 1983. Genetic characterization of natural gases. Am. Assoc. Petrol. Geol. Bull. 67, 2225–2238.
Scholle, P.A., Arthur, M.A., Ekdale, A.A., 1983. Pelagic environment: In: Scholle, P.A., Bebout, D.G., Moore, C.H. (Eds.), Carbonate Depositional Environments. Am. Assoc. Petrol. Geol. Mem. 33, 619–691.
Schrag, D.P., Hampt, G., Murray, D.W., 1996. Pore fluid constraints on the temperature and oxygen isotopic composition of the glacial ocean. Science 272, 1930–1932.
Schrag, D.P., Adkins, J.F., McIntyre, K., Alexander, J.L., Hodell, D., Charles, C.D., et al., 2002. The oxygen isotopic om[ositiion of seawater during the Last Glacial Maximum. Quatern. Sci. Rev. 21, 331–342.
Schumann, D., Raub, T.D., Kopp, R.E., Guerquin-Kern, J.-L., Wu, T.-D., Rouiller, I., et al., 2008. Gigantism in unique biogenic magnetite at the Paleocene-Eocene Thermal Maximum. Proc. NY Acad. Sci., doi/10.1073 pnas.0803634105.
Schwarcz, H.P., Burnie, S.W., 1973. Influence of sedimentary environments on sulfur isotope ratios in clastic rocks. A review. Mineralium Deposita 8, 264–277.
Scotchman, I.C., 1991. The geochemistry of concretions from the Kimmeridge Clay Formation of southern and eastern England. Sedimentology 38, 79–106.
Seyfried, J.W.E., Bischoff, J., 1979. Low temperature basalt alteration by seawater: an experimental study at 70 °C and 150 °C. Geochim. Cosmochim. Acta 43, 1937–1947.
Shephard, L.E., Bryant, W.R., Chiou, W.A., 1982. Geotechnical properties of Middle America Trench sediments, Deep Sea Drilling Project Leg 66. Init. Rep. Deep Sea Drilling Proj. 66. U.S. Government Printing Office, Washington, DC pp. 475–492.
Shimmield, G.B., Price, N.B., 1984. Recent dolomite in hemipelagic sediments off Baja California. In: Garrison, R.E., Kastner, M., Zenger, D.H. (Eds.), Dolomites of the Monterey Formation and organic rich units. Soc. Econ. Paleontol. Mineral. Pacific Section Spec. Publ. 41, 5–18.
Shipboard Scientific Party, 1989. Site 724. In: Prell, W.L., Niitsuma, N. (Eds.), 1989. Proc. ODP, Init. Repts. 117. Ocean Drilling Program, College Station, TX, pp. 385–417.

Shipboard Scientific Party, ODP Leg 110, 1987. In: Mascle, A., Moore, J.C. (Eds.), 1987. Proc. ODP, Init. Repts. 110. Ocean Drilling Program, College Station, TX.
Shipboard Scientific Party, ODP leg 131, 1990. In: Taira, A., Hill, I., Firth, J.V. (Eds.), 1990. Proc. ODP, Init. Repts. 131. Ocean Drilling Program, College Station, TX.
Shipley, T.H., Houston, M.H., Buffler, R.T., Shaub, F.J., McMillenik, J., Ladd, J.W., et al., 1979. Seismic evidence for widespread possible gas hydrate horizons on continental slopes and rises. Am. Assoc. Petrol. Geol. Bull. 63, 2204–2213.
Siever, R., Woodford, N., 1973. Sorption of silica by clay minerals. Geochim. Cosmochim. Acta 37, 1851–1880.
Slaughter, M., Hill, R.J., 1991. The influence of organic matter in organogenic dolomitization. J. Sediment. Petrol. 61, 296–303.
Sloan, E.D., 1998. Clathrate Hydrates of Natural Gases, second ed. M. Decker, Chemical Industries 39, New York 705pp.
Snyder, W.S., Brueckner, H.K., Schweikert, R.K., 1983. Deformation styles in the Monterey Formation and other siliceous rocks. In: Isaacs, C.M., Garrison, R. (Eds.), Symposium Volume on Monterey Oilfields, Soc. Econ. Mineral. Paleontol. Pacif. Section, 151–170.
Sotelo, V., Gieskes, J.M., 1978. Interstitial water studies, Leg 40: shipboard studies. In: Bolli, H.M., Ryan, W.B.F. (Eds.), Init. Rep. Deep Sea Drilling Proj. 40. U.S. Government Printing Office, Washington, DC, pp. 549–554.
Spotts, J.H., Silverman, S.R., 1966. Organic Dolomite from Point Fermin, California. Am. Mineral. 51, 1144–1155.
Stackelberg, Von, 2000. Manganese nodules of the Peru Basin. In: Cronan, D.S. (Ed.), Handbook of Marine Minerals. CRC Press, Boca Raton, FL, pp. 197–238.
Staudigel, H., Hart, S., 1983. Alteration of basaltic glass: mechanisms and significance for oceanic crust-seawater budget. Geochim. Cosmochim. Acta 47, 337–350.
Stein, C.L., Kirkpatrick, R.F., 1976. Experimental porcelanite recrystallization kinetics: a nucleation and growth model. J. Sediment. Petrol. 46, 430–435.
Steinberg, M., Bonno-Courtois, C., Tlig, S., 1983. Geochemical contribution to the understanding of beeded chert. In: Iijima, A., Hein, J.R., Siever, R. (Eds.), Siliceous Deposits in the Pacific Region. Dev. Sediment. Elsevier, Amsterdam, 36, pp. 193–210.
Steinitz, G., 1970. Chert "dyke" structures in Senonian chert beds, southern Negev, Israel. J. Sediment. Petrol. 40, 1241–1254.
Stoffers, P., Glasby, G.P., Thijssen, T., Shrivastava, P.C., Melguen, M., 1981. The geochemistry of co-existing manganese nodules, micronodules, sediments and pore waters from five areas in the equatorial and South-West Pacific. Chem. Erde 40, 273–297.
Stoffyn-Egli, P., 1982. Dissolved aluminium in interstitial waters of recent sediments from the North Atlantic Ocean. Geochim. Cosmochim. Acta 46, 1345–1352.
Straub, S., Schmincke, H., 1998. Evaluating the tephra input into Pacific Ocean sediments: distribution in space and time. Geol. Rundsch. 87, 461–476.
Suess, E., 1979. Mineral phases formed in anoxic sediments by microbial decomposition of organic matter. Geochim. Cosmochim. Acta 43, 339–352.
Suess, E., Bohrmann, G., Rickert, D., Eisenhauer, A., Whiticar, M.J., 2000. Vorkommen, Gefüge und Zusammensetzung von oberflächennahen Gashydraten vom Hydratrücken (Cascadia Akkretionszone). Deutsche Forschungsgmeinschaft, ODP Kolloquium Jena. p.77 [updated version of: occurrence, fabric and composition of hydrates in near-surface sediments of the Cascadia margin. Eos Trans. 89 (46), F511, 1999].
Sverjensky, D.A., Lee, N., 2010. The Great Oxidation Event and mineral diversification. Elements 6, 31–36.
Swart, P.K., Wortmann, U.G., Mitterer, R.M., Malone, M.J., Smart, P.L., Feary, D.A., et al., 2000. Hydrogen-sulphide rich hydrates and saline fluids in the continental margin of South Australia. Geology 28, 1039–1042.

Tada, R., 1991. Compaction and cementation in siliceous rocks and their possible effect on bedding enhancement. In: Einsele, G. (Ed.), Cycles and Events in Stratigraphy. Springer Verlag, Berlin-Heidelberg, pp. 480–491.

Taliaferro, N.L., 1934. Contraction phenomena in cherts. Geol. Soc. Am. Bull. 45, 189–232.

Tassé, N., Hesse, R., 1984. Origin and significance of complex authigenic carbonates in Cretaceous black shales of the Western Alps. J. Sediment. Petrol. 54, 1012–1027.

Taylor, J., 1950. Pore space reduction in sandstone. Am. Assoc. Petrol. Geol. Bull. 34, 701–716.

Thyne, G.D., Boles, J.R., 1989. Isotopic evidence for origin of the Moeraki septarian concretions, New Zealand. J. Sediment. Petrol. 59, 272–279.

Tinivella, U., Ledolo, E., 2000. The Blake Ridge bottom simulating reflector transect: tomographic velocity field and theoretical model to estimate methane hydrate quantities. In: Paull, C.K., Matsumoto, R., Wallace, P.J., Dillon, W.P. (Eds.), Gas Hydrate Sampling on the Blake Ridge and Carolina Rise. Proc. Ocean Drill. Progr. Sci. Res. 164. Ocean Drilling Program, College Station, TX, pp. 273–281.

Torres, M.E., Brumsack, H.J., Bohrmann, G., Emeis, K.C., 1996. Barite formation in continental margin sediments: a new look at barium remobilization in the zone of sulphate reduction and formation of heavy barites in diagenetic fronts. Chem. Geol. 127, 125–139.

Torres, M.E., Wallmann, K., Tréhu, A.M., Bohrmann, G., Borowski, W.S., Tomaru, H., 2004. Gas hydrate growth, methane transport, and chloride enrichment at the southern summit of Hydrate Ridge, Cascadia margin off Oregon. Earth Planet. Sci. Lett. 226, 225–241.

Treude, T., Niggemann, J., Kallmeyer, J., Wintersteller, P., Schubert, C.J., Boetius, A., et al., 2005. Anaerobic oxidation of methane and sulfate reduction along the Chilean continental margin. Geochim. Cosmochim. Acta 69, 2767–2779.

Ussler III, W., Paull, C.K., 1995. Effects of ion exclusion and isotopic fractionation on pore water geochemistry during gas hydrate formation and decomposition. Geo-Mar. Lett. 15, 37–44.

Van Lier, J.A., De Bruyn, P.A., Overbeek, J.T.G., 1960. The solubility of quartz. J. Phys. Chem. 64, 1675–1682.

Van Lith, Y., Warthmann, R., Vasconcelos, C., McKenzie, J.A., 2003. Microbial fossilization in carbonate sediments: a result of the bacterial surface involvement in dolomite precipitation. Sedimentology 50, 237–245.

Vasconcelos, C., McKenzie, J.A., 1997. Microbial mediation of modern dolomite precipitation and diagenesis under anoxic conditions (Lagoa Vermelha, Rio de Janeiro, Brazil). J. Sediment. Res. 67, 378–390.

Veizer, J., 1983. Chemical diagenesis of carbonates: theory and application of trace element technique. In: Arthur, M.A., Anderson, T.F., Kaplan, I.R., Veizer, J., Land, L.S. (Eds.), Stable Isotopes in Sedimentary Geology, Soc. Econ. Paleontologists Mineralogists, 3-1–3-100, Short Course Lecture Notes 10.

Veizer, J., 1995. Reply to the comment by L.S. Land on "Oxygen and carbon isotopic composition of Ordovician brachiopods: Implications for coeval seawate". Geochim. Cosmochim. Acta 59, 2845–2846.

Veizer, J., Bruckschen, P., Pawellek, F., Diener, A., Podlaha, O.G., Carden, G.A.F., et al., 1997. Oxygen isotopic evolution of Phanerozoic seawater. Palaeogeogr. Palaeoclimatol. Palaeoecol. 132, 159–172.

Veizer, J., Ala, D., Azmy, K., Bruckschen, P., Buhl, D., Bruhn, F., et al., 1999. $^{87}Sr/^{86}Sr$, $\delta^{13}C$ and $\delta^{18}O$ evolution of Phanerozoic seawater. Chem. Geol. 161, 59–88.

Von Damm, K.L., Edmond, J.M., Grant, B., Measures, C.I., Walden, B., Weiss, R.F., 1985. Chemistry of submarine hydrothermal solutions at 21°N, East Pacific Rise. Geochim. Cosmochim. Acta 49, 2197–2220.

Von Rad, U., Botz, R., 1987. Authigenic Fe-Mn carbonates in the Cretaceous and Tertiary continental rise sediments of Deep Sea Drilling Project Site 603 off the eastern U.S.A. In: Van Hinte, J.E., Wise, S.W. (Eds.), Init. Rep. Deep Sea Drilling Proj. 93. U.S. Government Printing Office, Washington, DC, pp. 1061–1071.

Von Rad, U., Rösch, H., 1972. Mineralogy and origin of clay minerals, silica and authigenic silicates in Leg 14 sediments. In: Hayes, D.E., Pimm, A.C. (Eds.), Init. Rep. Deep Sea Drilling Proj. 14. U.S. Government Printing Office, Washington, DC, pp. 727–751.

Von Rad, U., Rösch, H., 1974. Petrography and diagenesis of deep-sea cherts from the central Atlantic. In: Hsu, K., Jenkyns, H.C. (Eds.), Pelagic Sediments on Land and Under the Sea. Int. Assoc. Sediment. Spec. Publ. 1, 327–347.

Von Rad, U., Riech, V., Rösch, H., 1978. Silica diagenesis in continental margin sediments off Northwest Africa. In: Lancelot, Y., Seibold, E. (Eds.), Init. Rep. Deep Sea Drilling Proj. 41. U.S. Government Printing Office, Washington, DC, pp. 879–905.

Von Rad, U., Berner, U., Delisle, G., Doose-Polinski, H., Fechner, N., Linke, P., et al., 2000. Gas and fluid venting at the Makran accretionary wedge off Pakistan. Geo-Marine Lett, 20, 10–19.

Wada, H., Niitsuma, N., Nagasawa, K., Okada, H., 1981. Deep sea carbonate nodules from the Middle America trench area off Mexico, Deep Sea Drilling Project Leg 66. In: Watkins, J.S., Moore, J.C. (Eds.), Init. Rep. Deep Sea Drilling Proj. 66. U.S. Government Printing Office, Washington, DC, pp. 453–474.

Wallmann, K., 2003. Feedbacks between oceanic redox states and marine productivity: a model perspective focused on benthic phosphorus cycling. Global Biochem. Cycles 17, 1084.

Waterman, L.S., Sayles, F.L., Manheim, F.T., 1972. Interstitial water samples on small core samples, Leg 14. In: Hayes, D.E., Pimm, A.C. (Eds.), Init. Rep. Deep Sea Drilling Proj. 14. U.S. Government Printing Office, Washington, D.C, pp. 753–762.

Weaver, C.E., 1989. Clays, Muds, and Shales. Dev. Sediment. Elsevier, Amsterdam, 44, 819pp.

Weaver, F.M., Wise Jr., S.W., 1972. Ultramorphology of deep sea cristobalitic chert. Nat. Phys. Sci. 237, 56–57.

Weeks, L.G., 1957. Origin of carbonate concretions in shales, Magdalena Valley, Columbia. Geol. Soc. Am. Bull. 68, 95–102.

Wehausen, R., Brumsack, H.-J., 2000. Chemical cycles in Pliocene sapropel-bearing and sapropel-barren eastern Mediterranean sediments. Palaeogeogr. Palaeoclimatol. Palaeoecol. 158, 325–352.

Weissert, H., 2011. Mesozoic pelagic sediments – archives for ocean and climate history during green-house conditions. In: Hüneke, H., Mulder, T. (Eds.), Deep-Sea Sediments. Developments in Sedimentology, Elsevier, Amsterdam, vol. 63, pp. 765–792.

Wellsbury, P., Goodman, K., Barth, T., Cragg, B.A., Barnes, S.P., Parkes, R.J., 1997. Deep marine biosphere fuelled by inceasing organic matter availability during burial and heating. Nature 388, 573–576.

Westphal, H., 2006. Limestone-marl alternations as environmental archives and the role of early diagenesis: a critical review. Int. J. Earth Sci. (Geol.Rundsch.) 95, 947–961.

Westphal, H., Munnecke, A., 2003. Limestone-marl alternations: a warm-water phenomenon. Geology 31, 263–266.

Westphal, H., Munnecke, A., Pross, J., Herrle, J.O., 2004. Multiproxy approach to understanding the origin of Cretaceous pelagic limestone-marl alternations (DSDP Site 391, Blake- Bahama Basin). Sedimentology 51, 109–126.

Wignall, P.B., Newton, R., 1998. Pyrite framboidal diameter as a measure of oxygen deficiency in ancient mudrocks. Am. J. Sci. 298, 537–552.

Williams, L.A., Crerar, D.A., 1985. Silica diagenesis. II. General mechanisms. J. Sediment. Petrol. 55, 312–321.

Williams, L.A., Parks, G.A., Crerar, D.A., 1985. Silica diagenesis. I. Solubility controls. J. Sediment. Petrol. 55, 301–311.

Wilson, T.R.S., Thomson, J., Colley, S., Hydes, D.J., Higgs, N.C., Sorensen, J., 1985. Early organic diagenesis: the significance of progressive subsurface oxidation fronts in pelagic sediments. Geochim. Cosmochim. Acta 49, 811–822.

Wise Jr., S.W., Buie, B.F., Weaver, F.M., 1972. Chemically precipitated sedimentary cristobalite and the origin of chert. Eclogae Geol. Helvetiae 65, 157–163.

Wolff-Boenisch, D., Gíslason, S., et al., 2004. The dissolution rate of natural glasses as a function of their composition at pH 4 and 10.6, and temperatures from 25 and 74 °C. Geochim. Cosmochim. Acta 68, 4843–4858.

Wood, W.T., Ruppel, C., 2000. Seismic and thermal investigations of the Blake Ridge gas hydrate area: a synthesis. In: Paull, C.K., Matsumoto, R., Wallace, P.J., Dillon, W.P. (Eds.), Gas hydrate sampling on the Blake Ridge and Carolina Rise. Proc. Ocean Drill. Progr. Sci. Results 164. Ocean Drilling Program, College Station, TX, pp. 253–264.

Wortmann, U.G., Hesse, R., Zacher, W., 1999. Major element analysis of cyclic black shales: paleoceanographic implications for the Early Cretaceous deep western Tethys. Paleoceanography 14, 525–541.

Wright, D.T., 1997. An organogenic origin for widespread dolomite in the Cambrian Eilean Dubh Formation, northwestern Scotland. J. Sediment. Res. 67A, 54–64.

Wyllie, P.J., 1971. The Dynamic Earth—Textbook in Geosciences. Wiley-Interscience, New York 476pp.

Yang, C., Hesse, R., 1993. Diagenesis and anchimetamorphism in an overthrust belt, external domain of the Taconian Orogen, southern Canadian Appalachians - II. Paleogeothermal gradients derived from maturation of different types of organic matter. Organ. Geochem. 20, 381–403.

Yeh, H.W., 1980. D/H ratios and late stage dehydration of shales during burial. Geochim. Cosmochim. Acta 44, 341–352.

Yeh, H.W., Savin, S.M., 1977. Mechanism of burial metamorphism of argillaceous sediments. Geol. Soc. Am. Bull. 88, 1321–1330.

Yokoyama, T., Banfield, J., 2002. Direct determination of the rates of rhyolite dissolution and clay formation over 52, 000 years and comparison with laboratory measurements. Geochim. Cosmochim. Acta 66, 2665–2681.

Zeikus, J.G., Wolfe, R.S., 1972. Methanobacterium thermoautotrophicum sp. n., an anaerobic, autotrophic, extreme thermophile. J. Bacteriol. 109, 707–713.

Zheng, Y., Froelich, P.N., Torres, M.E., Dia, A.N., 1995. Stable isotopes ($^{18}O/^{16}O$) and $^{87}Sr/^{86}Sr$ ratios in pore fluids of the Chile triple junction accretionary prism: implications for diagenesis and fluid migration. In: Lewis, S.D., Behrmann, J.H., Musgrave, R.J., Cande, S.C. (Eds.), Chile Triple Junction. Proc. Ocean Drill. Progr. Sci. Results 141. Ocean Drilling Program, College Station, TX, pp. 313–320.

Zuleger, E., Gieskes, J., You, C., 1996. Interstitial water chemistry of sediments of Costa Rica accretionary complex off the Nicoya Peninsula. Geophys. Res. Lett. 23, 899–902.

CHAPTER 10

Industrial Application of Deep-Sea Sediments

Patrice Imbert

Contents

Total CSTJF, Avenue Larribau, 64000 Pau, France
E-mail address: patrice.imbert@total.com

Developments in Sedimentology, Volume 63
ISSN 0070-4571, DOI: 10.1016/S0070-4571(11)63010-0

1. Specificity of the Oil and Gas Industry Viewpoint

There is in principle no difference between industry and academic geology: in the specific domain of the sedimentology of deep-water deposits, a turbidite bed cropping out in the Alps or buried underneath 6 km of sediments remains a turbidite bed. Both will be described and understood much in the same way, using the same facies classification schemes.

However, economic constraints tend to focus the interest of the oil and gas industry on objects of economic interest, hydrocarbon reservoirs in particular; for that reason, sandy turbidites are much better known in the oil and gas industry than silty contourites, or than deep-water carbonates that provided so far much less economic discoveries. As a counterpart to this focus, that could be seen as a narrowing of the field of knowledge, the quest for hydrocarbons in particular has led to the development of specific investigation techniques that in turn provide a fantastic dataset for understanding deep-water geology at all scales.

A key word to describe the specificity of the oil and gas industry viewpoint on deep-water sediments could be 'predictivity': what is required at all times from the petroleum geologist or geophysicist is an accurate prediction of 'what is down there' based on fragmentary evidence. As the process goes on, and more data are collected, the required precision increases. At the early stages of exploration, one key issue is the presence of a reservoir; at the later stage of field appraisal, the problem is to predict if the discovered reservoir is 'big enough' to justify the investments needed to produce the accumulation. Finally, during the production phase, the heterogeneity of the reservoir has to be predicted in detail beyond well control in order to maximise the final recovery of the hydrocarbons.

1.1. What is 'deep' in the hydrocarbon industry?

In the hydrocarbon industry, 'deep water' has different meanings for drilling or production engineers on the one hand, for geoscientists on the other. Engineers define limits on water depth based on technological criteria which are irrelevant here. Geologists commonly set the limit between the 'shallow' and 'deep' domains based on a combination of biostratigraphic evidence (abundant planktonics), sedimentological criteria (predominance of gravity deposits, no reworking by wave or tide action) and wireline log character (dominant coarsening-up parasequences

(Van Wagoner et al., 1987) in shelf sediments vs. more erratic sand distribution with common sharp-based sandbodies in the deep domain). Geophysicists generally line up the successive positions of the shelf edge (when and where such a morphological feature is present and seismically observable) to define the boundary between the shallow and deep domains. These definitions may sound a bit loose and can sometimes be conflicting. The final goal, however, is not to decide about the water depth at which a hydrocarbon-bearing sandbody was deposited, but to define best its size, geometry and heterogeneity so that it can be produced in an optimal way.

1.2. Deep-sea sediments or turbidites?

There is admittedly much more in deep water than siliciclastic turbidites. However, most hydrocarbons discovered so far in deep-water sediments are actually trapped in turbidite reservoirs and the vast majority of them in siliciclastic turbidites and grain (hyperconcentrated) flow deposits (Mulder, 2011, this volume, Chapter 2.3 and 2.4). There are a few exceptions, which will be mentioned in the following sections: a few major accumulations (e.g. in the Campeche basin, Mexico) are trapped in carbonate breccias resedimented in deep water and some Brazilian turbidites are interpreted to have been winnowed by contour currents, thereby improving their reservoir characteristics. But on the whole, deep-water sediments are seen as 'turbidites' in the oil and gas industry.

1.3. Technological constraints

There is in theory no direct link between the depth at which sediments were deposited and the present water depth where they lie. However, a good part of the turbidite reservoirs discovered so far lie along present-day passive continental margins (Gulf of Mexico, Gulf of Guinea, North Atlantic margin, Brazilian margin), in a setting much similar to that in which they were initially deposited. The direct consequence of that is that exploration of these domains could start only after the technology of drilling safely in deep water was developed ('safely' specifically refers to preventing a hydrocarbon eruption). DSDP and ODP wells have been drilled in deep water for a long time, but were all purportedly emplaced so as to carefully avoid the possibility of meeting overpressured hydrocarbons. For technical purposes, drilling and production engineers commonly set the limits between the 'shallow', 'deep' and 'ultra-deep' domains at 200 m and 1500 m respectively.

2. Hydrocarbon Exploration and Production in Deep Water

2.1. From exploration to production

The process of exploration and production of oil and gas can be subdivided into three successive phases, each of which has distinct needs and uses different tools, or uses the same tools with different focuses.

The first phase, exploration, begins with the identification of a potentially prospective area. If the prospectivity is confirmed by subsequent studies, often based on seismic acquisition and interpretation, it ends with the drilling of a borehole to test the presence of hydrocarbons. This phase can last a few years.

Once hydrocarbons are discovered, a second phase is in most cases necessary to decide whether or not the accumulation can be produced in an economical way. This second phase, appraisal, often requires additional seismic acquisition, or improved processing, and interpretation. Most of the time, it requires the drilling of several additional wells. The appraisal phase ends when the accumulation is understood well enough for a development plan to be defined (i.e. for example, definition of the well pattern, surface installations, that will be necessary to actually produce the hydrocarbons and bring them to the market). This phase, like exploration, can last several years, occasionally several decades.

The production phase starts with the drilling of the actual production wells (usually distinct from the exploration and appraisal ones) and ends when the production of hydrocarbons stops being economical. The production of an oil or gas field can last years to decades.

2.2. Basin-scale exploration issues

The technical side of oil and gas exploration is guided by the concept of 'petroleum system'. The 'petroleum system' is the combination of a mature source rock, a migration pathway into a reservoir and a trap that can have accumulated hydrocarbons in commercial quantities and retained them until now. The source rock, the reservoir or both can result from deposition in deep water. In some cases, the trap itself can result from sediment deposition (geomorphologic traps, i.e. sand bodies with a 'bumpy' top providing a dead end for hydrocarbons migrating upwards). The first issues are regional, with questions like 'can there be a source rock or reservoirs in an unknown area far from well control?' After years of exploration and the drilling of many wells, the answer to such regional questions is well known (for instance, the first successful hydrocarbon well is a definite proof that there is a mature source rock in the basin), but initial attempts to solve such issues have to refer to already explored basins thought to be similar (e.g. for a continental margin, the conjugate margin at the time of basin opening—if drilled—or another segment of the target margin, with all the uncertainties related to this search for an analogue with extremely limited information).

A number of discoveries were made in the past ten years or so in the Gulf of Mexico and the Gulf of Guinea, highlighting the petroleum potential of mud-rich turbidite systems of divergent passive margins. Correlatively, other types of turbidite systems, although they had previously provided a number of discoveries, were somewhat shadowed by this success. A good and simple summary on the classification of turbidite systems, seen from the industry viewpoint, can be found in Reading and Richards (1994) (Mulder, 2011, this volume, Fig. 2.18 in Chapter 2.3). The main limitation of this scheme is probably its static character (see Section 2.2.5). Turbidite systems evolve through time in response to fluctuating boundary conditions, sediment supply in particular. It remains, however, the best synthetic classification for operational purposes.

The current focus of exploration and production, due to the discoveries mentioned above, tends to bias the understanding of turbidite systems in the hydrocarbon industry towards channel-levee systems of mature passive margins.

Figure 10.1 illustrates the type of problem hydrocarbon exploration has to address: can a 'seismic anomaly', like local high amplitude, be a hydrocarbon-filled reservoir? In most cases, answering the question requires a good deal of geological reasoning. In the case shown, part of the evidence comes from regional knowledge, like the presence of a source rock in the basin evidenced by hundreds of producing oil and gas fields. 'Evidence' can also be derived from reference to similar features calibrated elsewhere in the basin. For instance, the recognition of a channel-levee complex overlying a lobe is almost straightforward by reference to known examples (see details in figure caption and in Section 4.2).

2.2.1. Source rock

The very first issue addressed by exploration is the existence of a source rock in the basin. Source rocks are organic-rich sediments whose 'maturation' under the effect of temperature can produce hydrocarbons. Some stratigraphic intervals are known to make prolific source rocks, that is, source rocks that yielded high amounts of hydrocarbons. Several of them correspond to global transgressions during which organic productivity was high and diluting clastic influx low. The Cenomanian–Turonian shales worldwide, the Silurian 'graptolite shales' and the upper Jurassic Kimmeridge clay are well-known examples. All three correspond to relatively deep-water deposits. This special type of deep-water sediments, however, is rarely studied from the viewpoint of sedimentology, and the issue of source rocks is most of the time addressed by geochemical studies. One of the reasons is the regional

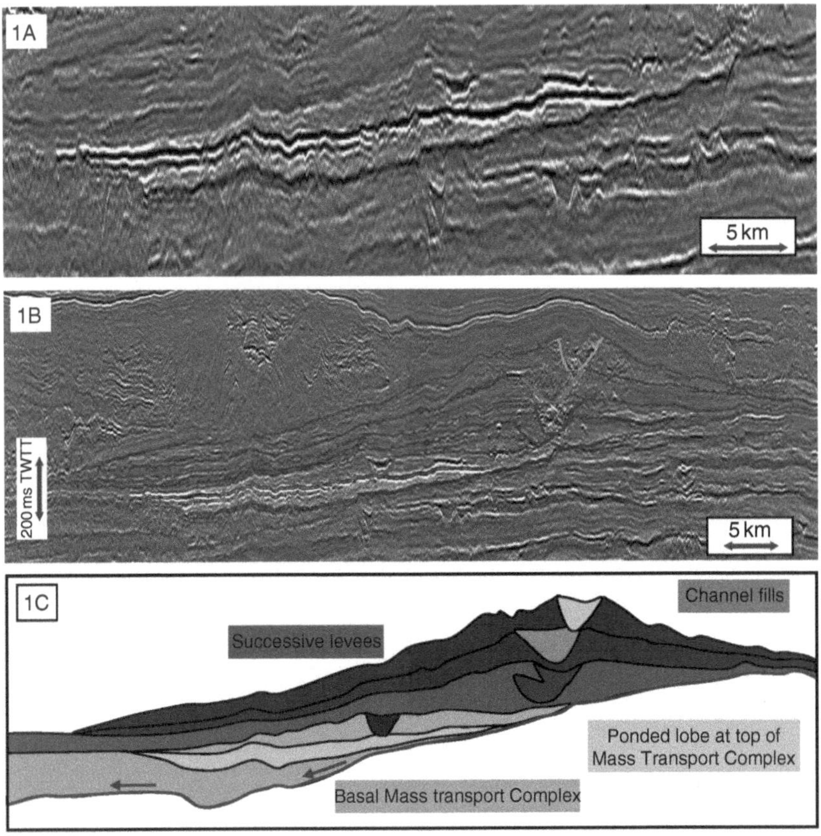

Figure 10.1 A hydrocarbon exploration issue. This figure shows how sedimentological knowledge can lead from the seismic observation of a bright amplitude (potentially indicative of hydrocarbons), to a consistent evaluation of its actual prospectivity. (A) An extract from a regional 2D line in the deep Gulf of Mexico showing a conspicuous strong amplitude in the middle of a less reflective background: could it be related to a gas-bearing reservoir? (B) An interpreted version of the same line, zoomed out to better see the context: the anomaly marks the beginning of a series of aggrading channel–levee complexes. (C) A line-drawing from the previous picking, highlighting the sedimentology of the interval. The final answer comes from a combination of sedimentological knowledge and other regional data. From this line only, it is possible to infer that the anomaly likely corresponds to a sand-rich turbidite lobe complex deposited at the top of a slumped mass and later overlain and locally eroded by successive episodes of channels (dominantly sand-filled) and levees (dominantly muddy). There is thus a high probability of having reservoir sands in the anomaly. As far as the trapping of hydrocarbons is concerned, the negative point is that the lobe sands are very likely eroded by the subsequent channels, providing a leakage pathway upwards for any incoming hydrocarbons. The anomaly remains undrilled to date, most likely due to the very high risk of having no hydrocarbon trap to the reservoir that lies down there. (A multi-colour version of this figure is on the included CD-ROM.)

character of source rocks in general, which does not necessitate a precise local description: in most cases, there is no such thing as a 'source rock body' with a limited size, like a 'reservoir body'. Large volumes are requested anyway to effectively source an economic hydrocarbon accumulation.

2.2.2. Reservoirs: How far into the basin?

This is a recurring concern in exploration, especially when dealing with undrilled areas: how far from the shelf break can significant amounts of sand be expected? The issue was addressed by some of the early DSDP/ODP drillings, for instance, DSDP leg 96 in the deep Gulf of Mexico: in a series of well bores along the channels of the Mississippi fan into the abyssal plain north of Cuba. This leg showed that the maximum amount of sand in the system was found in the lobes, in distal position (more than 500 km from the shelf break), while the channel-levee system was essentially muddy (Bouma et al., 1986).

At the regional exploration scale, there is often little concern about the details of the geometry of the reservoir: experience shows that predictions can be off by an order of magnitude (reservoir found 10 times thicker or thinner than anticipated). The issue is 'could sand ever reach this point or not?' And the answer, as shown by exploration failures or unanticipated successes, is far from obvious in many cases.

2.2.3. Seals and traps

Once hydrocarbons are formed in sufficient quantity in the source rock under the effect of temperature, they are expelled into the adjacent permeable sediments where they start moving ('migrating'). Migration is guided by pressure differences, and of course, by the geometry of the strata. Favourable geometries (e.g. an anticline) can make the hydrocarbons generated over a certain area to converge.

Seals are the impermeable layers that allow hydrocarbons migrating upward and laterally in a sedimentary basin to accumulate into commercial pools. Hemipelagites typically make good seals.

A trap is a configuration of reservoir and seal that constitutes a 'dead end' for migrating hydrocarbons. Many traps are structural (anticlines, for instance), but a trap can result from the combination of structural and sedimentary factors (e.g. a sinuous channel on a monocline), or even from sedimentary factors only. In the deep-water domain, classical examples can be found in the lower Eocene of the North Sea, where several gas fields are contained in 'sand heaps' resedimented at the foot of sandy lowstand deltas.

2.2.4. Method

Exploring an area consists in defining 'prospects', that is, objects that are worth drilling to check whether or not they correspond to hydrocarbon accumulations. One of the key elements to assess in the definition of a prospect is the reservoir that can be expected, in particular its size and geometry. This assessment most often has to be carried out on seismic sections (2D) or seismic volumes (3D). There are two types of approach to define the probability of finding a reservoir in a prospect, depending on the possibility to see sedimentary features on available seismic data.

When the seismic visibility is poor, all that can be done is to infer a likelihood of having sand deposited in the area, in particular based on regional knowledge, local palaeogeographic reconstructions and common-sense: sands get preferentially deposited in lows, not on highs. This lack of seismic visibility can result, *inter alia*, from a lack of acoustic impedance contrast between reservoir and non-reservoir lithologies, or from adverse structural conditions (sub-salt exploration, severe tectonic deformation).

When the seismic visibility is good, the task consists in identifying the sedimentary elements visible on seismic and reservoirs in particular. One frequent mistake is to believe that being able to recognise such reservoirs as channels and lobes ensures a good quality of interpretation. It is equally important to know other, non-reservoir types of sedimentary bodies, so as to be able to identify and avoid them. What is recommended is a 'check list' approach, simply meaning that a seismic interpreter must know as much as possible of what can be deposited in the deep domain. Since explorationists most of the time drill 'anomalies', a good knowledge of all that exists, reservoir or non-reservoir, may avoid costly mistakes.

Section 4 will focus on the various sedimentary features that have been recognised in deep water. Some are pervasive and have been described in virtually all deep basins; others appear to be restricted to specific settings. Knowing their existence and having seen them once in a textbook can save lots of efforts—and money.

2.2.5. Sequence stratigraphy of turbidite systems

Seismic stratigraphy (Vail et al., 1977) and its offspring, sequence stratigraphy (Posamentier and Vail, 1988), are two interpretation techniques developed to help predicting facies and reservoir distribution beyond well control in the hydrocarbon industry. These models add the time dimension to the depositional models like that of Reading and Richards (1994). According to sequence stratigraphy, and leaving aside details, turbidite sands are essentially deposited during periods of relative sea-level fall and low sea-level stand, that is, when sediments delivered by the rivers cannot be accommodated on the shelf (Vail et al., 1977) (see Chapter 2). The initial deposits of a sea-level cycle (starting by definition in the Vail model when the rate of sea-level fall is maximum) in the model are massive turbidite sands with a widespread extension in the basin, called the *basin-floor fan* (Van Wagoner et al., 1987). When the rate of sea-level fall decreases, the model predicts the deposition of channel-levee systems above the apex of the basin-floor fan, called the *slope fan* (Van Wagoner et al., 1987). The rest of the cycle was initially interpreted to be associated with sand starvation in the basin while coarse sediments were trapped on the shelf they built. After a few years, additional turbidite systems observed in the Gulf of Mexico within a cycle, but above the slope fan, were called the 'shingled turbidites' by reference to their seismic character and considered to correspond to local destabilisations at the front of the lowstand deltas (Vail, 1989, quoted as oral communication in Pacht, 1990).

A comparable model was proposed by Mutti (1985) (Mulder, 2011, this volume, Fig. 2.34 in Chapter 2.3.3.10) to account for the cycles observed in the style of turbidite system deposition in the South Pyrenees basin: massive, laterally continuous lobes associated with bypass facies in canyons and channels (Mutti's Type I systems interpreted as the first stage of deposition in a cycle), channel-fill deposits with small lobes attached (Type II systems, interpreted as Stage II), then thick piles of muddy sediments with occasional small channels (Type III systems, Stage III in the cycle).

Sequence stratigraphy of deep-sea turbidites was less developed anyway than for deposits on continental shelves, and exploration in the recent years has focused more on reservoir prediction according to seismic morphologies (e.g. channel recognition and facies prediction inside the fill) than on the actual timing of events, probably due in part to the widespread availability of 3D seismic.

2.3. Mature exploration

2.3.1. Successive phases of exploration in a basin

The first discoveries in a basin are usually made on conspicuous structures, the first to be imaged with the initial seismic grid. Success results in increased data acquisition, and the understanding of the basin progresses so that other geological concepts can be developed and tested. The first developments require the emplacement of heavy infrastructure (e.g. pipelines), which will later be available for subsequent discoveries. That way, exploration evolves continuously through time.

2.3.2. Case study: The evolution of exploration in the Gulf of Mexico over the past decades

After over a century of oil and gas exploration, most basins worldwide have been drilled extensively and many exploration wells target already recognised objectives, that is, are step-outs from existing discoveries. In that context, the quality of seismic data has often been fine-tuned so as to guarantee an optimal visibility of the target reservoirs. A good example is the exploration of the deep Gulf of Mexico: after initial successes in the 1980s, followed by the advances of technology in the subsequent years, the 1990s were the decade of amplitude chasing. The key to success in exploration was to identify on 3D seismic an amplitude anomaly whose downdip limit coincided with a contour line of the structural map of the target horizon: in this case, the seismic anomaly (impedance contrast) was inferred to be most likely due to the presence of a hydrocarbon accumulation limited by a horizontal fluid contact. Although there are interpretation pitfalls, this hypothesis proved to be a very valuable tool for exploration until all amplitudes were leased and drilled. The targets were essentially Pliocene and Pleistocene turbidite systems at that time; the success ratio of this period approached 60%, which is far above the normal success ratio in 'virgin' exploration, closer to 15%.

The improvements brought to seismic acquisition and processing during this first phase of amplitude-based exploration made it possible to image deeper horizons, revealing inverted depocentres deeper down the sedimentary series, in the Miocene in particular. Although no direct evidence of hydrocarbon was available in the absence of a good seismic contrast between oil sand, water sand and shale at this depth, inverted minibasins were drilled and yielded a number of significant discoveries from the late 1990s. As was the case in the first phase, exploration started by drilling the best imaged and biggest structures, moving on to more subtle and smaller traps as infrastructure became available.

The latest exploration target is the much deeper Palaeocene series of the abyssal plain. After a few exploration failures that helped provide a good age calibration of seismic, an interval initially thought from its seismic character to be massive shale or chalk was realised to consist of sand-rich turbidite lobes.

2.4. Appraisal: Focus on reservoirs

Appraisal starts once a discovery has been made, that is, once an exploration well has found enough hydrocarbons. At this stage, the knowledge is most of the time insufficient to decide whether the actual quantity discovered is sufficient to be economical. Even if the quantities are known to be economical, production investments (e.g. wells, platforms, pipelines) have to be dimensioned in the most cost-effective way, which is rarely possible with the limited information available at this early stage. Appraisal consists in acquiring more data (usually seismic and wells) and evaluating different scenarios of development until there is enough knowledge to guarantee a low risk of making the wrong decision. The geological aspects of appraisal concentrate on the reservoir, its geometry and heterogeneity (Fig. 10.2). Appraisal ends when the information gathered allows deciding safely on a development plan. In some favourable cases, the seismic image is precise

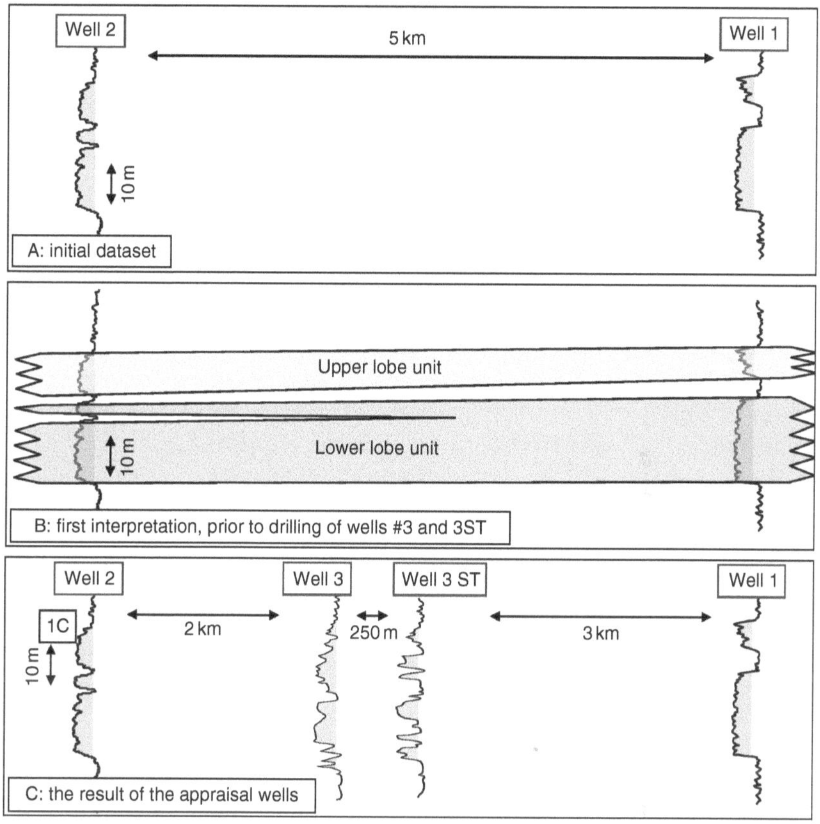

Figure 10.2 A field appraisal issue. Wells #1 and #2 were drilled independently through a set of turbidite lobes according to regional data and the available seismic. Well #2 was the discovery well of a small accumulation (not extending to well #1); the issue was to define reservoir geometry and size before starting a development. (A) The initial dataset, with the gamma-ray logs showing sand/shale distribution in the two wells. (B) The first interpretation based on the assumption that lobes made very laterally continuous sandbodies (see, for instance, Fig. 10.16 for reference). This assumption seemed to be confirmed by the similarity of the two wells. (C) adds to the previous set the gamma-ray logs of two intermediate wells, drilled 'for appraisal', that is, to test the geological hypothesis made previously before launching the actual development process. The architecture of the reservoir is clearly much less laterally continuous than anticipated. Well #3 was drilled first, and the correlation difficulties encountered led to the decision of drilling a 'sidetrack' well, that is, another leg from the same location to investigate the variability over a short distance. As shown on the panel, there was in that case a strong variability even over short distances. This, of course, led to a drastic revision of the 'sheet sand' model in this specific case. (A multi-colour version of this figure is on the included CD-ROM.)

enough to give direct clues on reservoir geometry, continuity and internal architecture. In many instances on the contrary, seismic resolution lies far below the level of detail needed, and reservoir estimate must go through a process of stochastic modelling, as discussed in Section 2.5.2.

2.4.1. What is a hydrocarbon reservoir?

In parallel with the classifications of turbidite systems and of turbidite facies, the oil and gas industry pays a strong attention to the understanding of reservoir bodies. In very crude terms, a reservoir is a geological 'bottle' containing hydrocarbons and water. What is addressed in this notion of reservoir is a geological body with fluid communication. A reservoir often consists of one architectural element, or a few of them connected by erosion surfaces or fault juxtaposition. Most of the time, reservoirs are approached as a connected collection of depositional elements and are characterised either as one entity or through the characterization of each element, depending on the precision of the data.

2.4.2. Reservoir properties

A reservoir is characterised not only by its geometry and size but also by the distribution of its internal heterogeneity (e.g. shale barriers between successive sand beds) and its petrophysical properties, porosity and permeability in particular. Porosity is usually measured by wireline logs with a good precision; permeability is more difficult to assess in the absence of a wireline measurement technique. Permeability can be approached at a small scale by direct measurement on core samples (a few centimetres to a few decimetres long), or can be evaluated at a much bigger scale by analysing results of well tests (see Section 3.2.5 for details).

2.4.3. The role of sedimentology during field appraisal

Exploration mostly dealt with predicting the presence of reservoir at a given location in the basin. Appraisal has to precise a lot the geometry and internal architecture/distribution of heterogeneity in the reservoir, based again on fragmentary evidence. Reservoir characteristics at this stage are interpreting from a combination of the data acquired locally (seismic, calibrated at the wells), regional knowledge (e.g. nearby wells, outcrops on the margins of the basin if appropriate) and inferences from better documented sedimentary systems, thought to be analogues of the investigated reservoir. Analogues can be developed fields in similar contexts, sedimentological analogues also often include shallow turbidite systems of the same type well imaged by seismic and outcrops showing similar sandbodies with a good visibility. The issue at this stage is to define the most appropriate analogue, which implies knowing a few of them at least.

2.5. Production: Optimising recovery and costs

2.5.1. The problem at issue

Once a commercial hydrocarbon accumulation has been evidenced by drilling comes the time of production. The industrial issue at this stage is to optimise the profitability of the process, that is, roughly to get as much hydrocarbons as possible from the field while minimising the costs. This is achieved through the drilling of a set of wells, whose pattern is defined so as to optimise the recovery. This optimisation is defined within the framework of a conceptual geological model. The geological issue at this stage is to get a description as precise as possible of the internal architecture of the reservoir, in particular of its heterogeneity. Sedimentary heterogeneity includes flow barriers or baffles (impermeable or less permeable layers or zones that will prevent hydrocarbons to flow from one side to the other) and drains (high permeability layers that will preferentially flow into a well penetrating them, potentially leaving behind hydrocarbons trapped in the background lithology). The well pattern to be used for production has to take into account this heterogeneity so as to optimise the economics of the project.

2.5.2. Geology seen through reservoir modelling

One essential tool in oil and gas production is the comparison of hydrocarbon production, as observed from the wells that penetrate the field, with what is expected from the geological model that geologists have in mind.

2.5.2.1. 3D reservoir modelling In most cases nowadays, the geological architecture of the field is represented as a 3D grid in which each cell contains the petrophysical properties expected from the geological model (Fig. 10.3A and B). A typical grid at the time of writing contains up to a few hundred thousand cells. A typical cell size would be 100 m × 100 m × 5 m. This size at present provides the best compromise between the precision needed (a geologist would always like a finer grid cell size to represent the concepts he has in mind) and the capabilities of currently available flow simulation software, as described in Section 2.5.2.2 (a flow simulation in this type of grid can be run overnight, making it possible to correct parameters rapidly).

2.5.2.2. Flow simulation Specific software can simulate the flow of hydrocarbons in the reservoir and define the production that would be achieved at each well for a given geological configuration and a given well pattern (Fig. 10.3C). Beyond sedimentology, many factors have to be taken into account, some geological like structure, faults and fractures, others related to fluids like viscosity, compressibility and others geomechanical like matrix compressibility, to cite a few. Simulation merely applies the laws of physics to the porosity–permeability grid previously defined. Flow simulation is typically run at several stages during the life of the field. The first

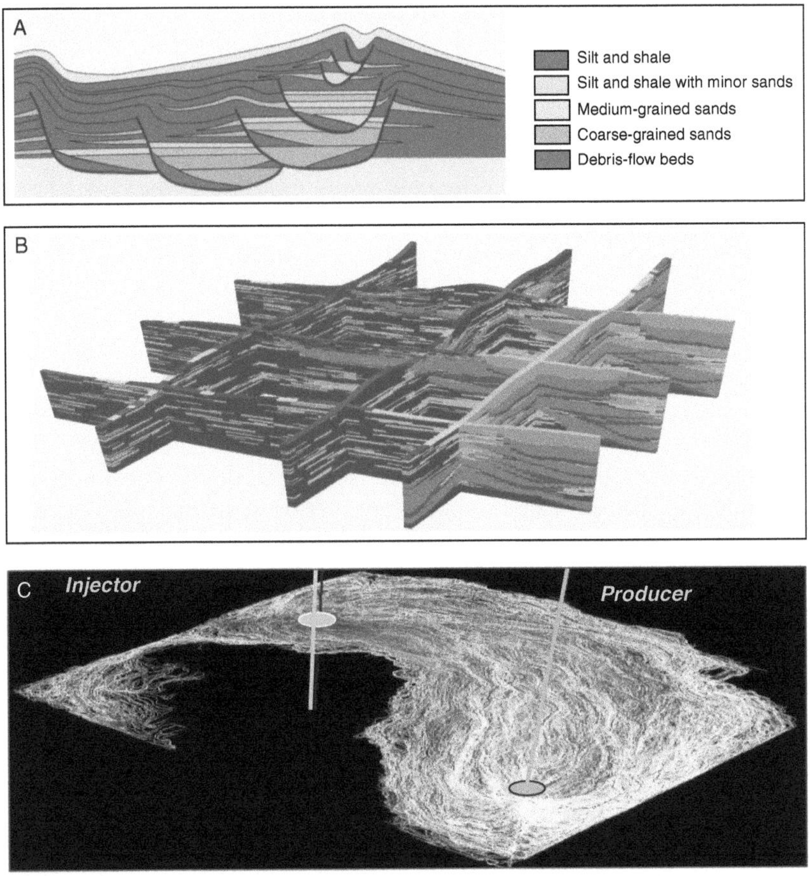

Figure 10.3 Reservoir sedimentology and production purposes ((B) and (C), R. Labourdette, unpublished material). Producing an oil or gas field implies an understanding of fluid behaviour so as to optimise the location of production wells, or to define the most effective way to recover hydrocarbons, for instance, by injecting water laterally so as to 'push' the hydrocarbons towards the producer wells. In order to predict fluid behaviour, it is necessary to convert geological concepts to 3D grids and to simulate fluid flow on these grids. (A) A classical concept of lithology distribution in a turbidite channel-fill, with successive episodes of infill, re-erosion and lateral migration. (B) A series of cross sections through a full 3D grid, again showing lithology distribution. Specific ranges of porosity and permeability can be ascribed to each facies (typically, orange and yellow facies will be porous and permeable, while green (levees) and purple (debris flows) will have low to very low permeabilities). (C) One type of flow simulation, with the 'threads' representing fluid flow lines expected in the permeability field defined in the previous step when a certain quantity of water is injected into the reservoir at the injector well. Analysing this sort of patterns allows to compare several locations for injection so as to eventually drill the one that provides the best recovery. (A multi-colour version of this figure is on the included CD-ROM.)

run can take place during the appraisal phase so as to define the best well pattern for production. The geological results of the production wells are used to update the model sooner or later in the process (in extreme cases where well results would be totally different from expectations, the full process may have to be re-run in order to modify the development scheme).

2.5.2.3. Hard data: tests and production history After the production starts, the production actually observed at the wells can be used as calibration points to modify the model (for instance, hydrocarbon flow of a certain level in a well may indicate that the communication with the neighbouring well is much less than expected initially, suggesting, for instance, the presence of a shale baffle between the two wells). Geology, through the geometry of the reservoir and its heterogeneity, is one of the parameters that may have to be revisited in this process.

3. Tools

The threefold process of exploration, appraisal and production is, of course, driven by economic considerations all along, but the previous sections aimed to show that a good understanding of the geology is a key to making sound decisions. As a result, specific techniques have been developed over the years to help define geological models, in particular reflection seismic and well data acquisition.

At the exploration stage, the main tools used are reflection seismic and regional well correlations. Until the late 1980s, 2D seismic (Mulder et al., 2011, this volume, Chapter 1) was the most generally used tool for exploration. Section 3.1 will focus on 3D seismic, but it must be pointed out that 2D seismic remains an excellent operational tool at basin scale, even for the sedimentologist. Several of the figures are taken from regional 2D surveys.

3.1. Seismic

3.1.1. Seismic visibility

The following sections will discuss the seismic character and geometry of turbidite systems. By its very principle, seismic can show geometry or character only when and where an impedance contrast exists between contrasted sediment bodies. In relatively shallow horizons (e.g. down to 3 s twtt[1] below seafloor), there is typically a good contrast between water-filled and

[1] Seismic sections are recorded in two-way transit time (twtt), that is, the time elapsed between the emission of the seismic signal at the source and its recording at the receiver. 'Two-way' refers to the fact that the signal travels down from the source to a reflector, then is reflected back up to the receiver. Typical seismic velocities in recent clastic sediments are reasonably close to 2000 ms^{-1}, so that a seismic time in seconds twtt is roughly equivalent to a depth in kilometres.

hydrocarbon-filled reservoirs, making it possible to directly 'see' the accumulations. The fluid-related contrast usually decreases with depth, so that only lithological contrasts become visible at greater depths (e.g. down to 5 s twtt below seafloor). At the same time, the contrast between sand/sandstone and mud/shale decreases and inverts with depth so that there is a zone of very little contrast between reservoirs and non-reservoirs. The images figured come from good contrast intervals, but exploration and production also deal with the others, of course.

3.1.2. Specificity of deep-water sediments

Sedimentary features are typically better seen on seismic in deep water than in shallow-water or continental deposits. Why is that? Essentially because deep-water sedimentary systems have lots of space to develop and can therefore build characteristic features like channels and levees on a scale largely exceeding seismic resolution. On the contrary, fluvio-deltaic deposits usually develop in the accommodation space created by a transgression, as happens nowadays along all continental margins. Even with the magnitude of the Holocene transgression, which is one of the largest recorded, the resulting deposits cannot exceed the accommodation created, that is, a thickness of about 100 m. In most cases, after compaction, the thickness of fluvial or deltaic sedimentary bodies in subsurface lies below seismic resolution. On the contrary, turbidite systems in deep water commonly show positive topography exceeding 500 m and make conspicuous features on seismic lines.

3.1.3. Three-dimensional seismic

The first commercial acquisition of 3D seismic took place in 1975 (Brown, 2004). Since that time, 3D seismic has become a routine tool for working at field scale and is becoming more and more often used for appraisal or even exploration purposes.

What is called '3D seismic' is actually a set of closely spaced seismic lines, most of the time acquired as parallel lines. The spacing of the lines is equal or close to the spacing between shot points (typically 12.5 or 25 m), so as to build a 3D block of the received signal (Fig. 10.4). A conventional seismic line is a representation of the signal received in two-way travel time, displayed along the acquisition path labelled in distance. In the same way, a 3D seismic block, or 'cube', represents the amplitude of the signal received in an (x, y, t) matrix where x and y are orthogonal coordinates in space and the two-way travel time of the signal from emission to reception. The standard processing applied to seismic lines (e.g. stack, filters, migration) is also applied to 3D blocks.

The key interest of this type of data is that the 3D block can be sampled in any direction. This means, for instance, reconstructing lines not only across the acquisition direction ('crosslines', as opposed to the acquired

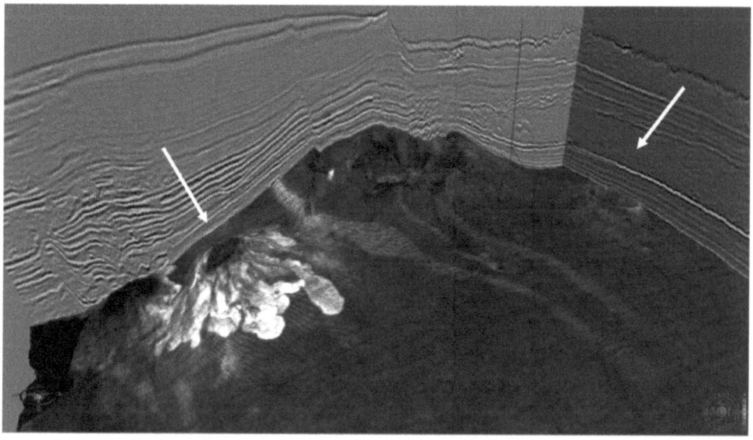

Figure 10.4 Mud volcano seen on a 3D seismic survey. (A) 3D seismic dataset is not much more than a set of very closely spaced lines, but this very high sampling is the key to the possibility of imaging the subsurface in 3D. This image displays two orthogonal seismic lines and the amplitude along a horizon, highlighting two small mud volcanoes (arrows) on the flank of a large anticline. Note the mud flows radiating from the vent of the volcano on the left-hand side. (A multi-colour version of this figure is on the included CD-ROM.)

'inlines') but also across or along any type of feature to help visualisation (e.g. across a channel, whatever its direction). Apart from these vertical sections, it is also possible to slice the volume horizontally, along a horizon or parallel to it, in these last cases making snapshots of the reflectivity of the sedimentary landscape at a certain geological time.

In parallel, the development of specific software for seismic interpretation has made the actual picking of horizons much easier, and makes it possible to image sedimentary systems buried beneath several kilometres of sediment almost as well as seafloor features (once again as long as the impedance contrast between the lithologies represented in the system, typically sand and shale, is sufficient).

3.2. Wells

Wells are primarily drilled to get to a target reservoir. In addition to this primary goal, they are an invaluable source of information, even when they fail to discover hydrocarbons. Rock samples are taken all along the borehole, and physical measurements are systematically recorded. All provide a wealth of information about the geology of the sediments penetrated.

Petroleum sedimentology had to specialise in the interpretation of data acquired along the well path, that is, on 1D data. In order to be able to forecast reservoir development away from a well, petroleum geologists have developed interpretation techniques relating well observations, vertical evolution, in particular, with reservoir continuity and size.

3.2.1. Rock samples

Contrary to surface geology, the hydrocarbon industry has to develop interpretations with a minimal amount of rock data, often without any at all. There are three types of samples commonly available from the boreholes: ditch cuttings, sidewall cores and full size cores. Cuttings are the chips of rock that are continuously brought to the surface as the drilling operation proceeds. They can and do provide highly valuable information while drilling on well site, but are often too small (a few millimetres), and their depth of provenance is too imprecise (a few metres) to allow any serious sedimentological study to be carried out. Cuttings are also the primary source of biostratigraphic data, that is, foraminifers, calcareous nannoplankton or palynomorphs (pollens, spores and other non-calcareous organisms). These are used for age-dating and environment definition.

When more precise information is needed on an interval after it has been drilled, it is possible to recover bigger samples from the borehole wall. The corresponding samples, called *sidewall cores*, are cylinders about 2.5 cm in diameter and 2.5 cm long. The conventional sampling process involves explosive to force a metallic cylinder into the borehole wall. Needless to say, the piece of rock recovered is usually disturbed by the process (fragmented grains, medium to poor preservation). A more recent procedure uses a little drill bit and recovers pristine samples perfectly usable for any geological purpose—at their small scale.

The best suited type of rock sample for sedimentological interpretation is a core. Due to the cost of coring, only hydrocarbon-bearing reservoir sections are cored in most cases, and many wells are not cored at all. The diameter of a core varies around 10 cm, which is largely enough for identifying sedimentary structures. Sedimentological core interpretation is usually carried out on slab sections, providing an excellent visibility of all structures in all lithologies, including silts and shales. From personal experience, there is nothing frustrating than working with cores compared to outcrops: although lateral continuity is missing, the 'quality of exposure' provided in shaly sections and the freshness of the contacts make sedimentological core interpretation a very pleasant experience.

3.2.2. The core description tool: Facies description and interpretation

The description and interpretation of turbidite facies from cores is one of the keys for predicting reservoir properties away from well control. Among the numerous facies classifications published, one of the most useful for the petroleum geologist is that proposed by Mutti (1992) (Mulder, 2011, this volume, Fig. 2.14 in Chapter 2.10). Although often referred to as 'the Mutti facies classification', it is actually proposed as a 'framework for a predictive classification of turbidite facies'. The key to the success lies in this predictive character. Mutti indicates that the scheme is derived from outcrop studies of mixed gravel/sand/mud systems (the Hecho Group, south Pyrenees and Miocene turbidite systems of the Apennines) and makes it clear that the scheme would have to be adapted to different source lithology. This caveat is often ignored anyway, and the 'framework' is most of the time used as a labelling tool. One of the keys of the classification is the identification of specific facies recording flow transformation and bypass zones, and consequently pointing to the location in the system.

Another key was given by Kneller and Branney (1995) (Mulder, 2011, this volume, Fig. 2.15 in Chapter 2.12). Kneller pointed out that deposition by turbidites was the result either of the time deceleration of a uniform flow, or of the space deceleration of a steady flow (or any combination thereof). The former case is more common in confined settings and will result in graded beds, the latter will rather characterise unconfined settings, such as lobes in open margin conditions. Although schematic, this can be taken as an educated first guess in the absence of a good seismic characterisation.

Taken together, Mutti's work on the downslope transformations of gravity flows and Kneller's approach of confinement from turbidite facies make it possible to predict the lateral and longitudinal evolution of turbidite beds rather than merely describe them. For instance, thick intervals of parallel laminations should not be *described* as 'base and top cut-out Bouma sequences restricted to their Tb term', but rather *understood* as deposits from a steady flow in an area of flow depletion (e.g. widening of the flow, or decrease in slope). In terms of prediction, what could be expected downflow is a thick interval of climbing ripples.

3.2.3. Wireline logs

Most wireline logs record physical parameters (resistivity, density, radioactivity and sonic velocity to mention the most commonly recorded) of the rocks surrounding the borehole. The typical sampling rate is about 15 cm (a half-foot), and resolution varies from a few tens of centimetres to a few metres. Higher resolution electric tools using multiple electrodes distributed around the borehole were initially developed to define the structural dip of strata in the well (by comparing the position of resistivity contrasts along several generatrices of the hole). They evolved in the late 1980s into real imaging tools with a resolution of 1 cm. Spectacular examples are obtained in slumped series, for instance (Fig. 10.5).

3.2.4. Sedimentological interpretation of logs: 'log patterns'

One challenge the industry has to address is the inference of the 3D geometrical characteristics of sedimentary bodies from 1D well data. For instance, what is the size of a reservoir drilled by a well, or how laterally

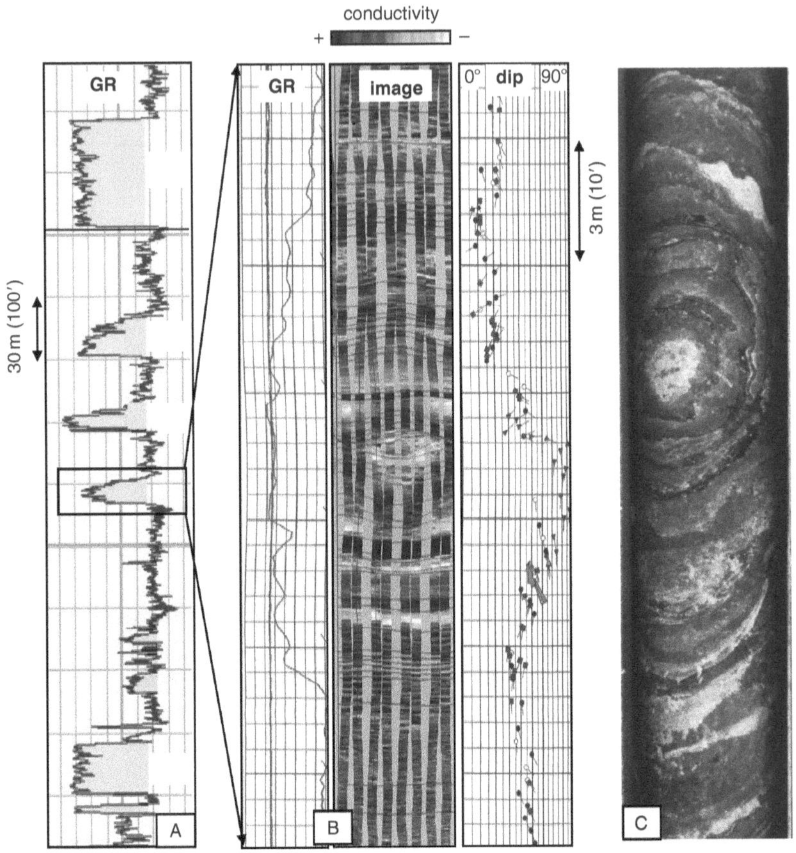

Figure 10.5 Example of an 'image log' recording from a slump fold in turbidite series. (A) A gamma-ray log in the series, showing the sharp-based character of all sandbodies but one (boxed) which looks more symmetrical. (B) A close-up on this specific interval, with again the gamma-ray log in the left track (red curve). The middle track is a representation of the resistivity around the borehole (measured by over 100 small electrodes applied onto the borehole wall). Flat layers appear as horizontal bands, while dipping layers show up as sine curves, the amplitude being directly related to the dip of the event. The right-hand track displays the dips calculated from this resistivity display. (C) For reference is a core photo from a slump fold (in a different stratigraphic interval and another part of the world) that can be used as a reading guide for the image log. The core diameter is 7.5 cm. (A multi-colour version of this figure is on the included CD-ROM.)

continuous is a shale barrier encountered in a reservoir. For that purpose, petroleum geologists are trying to define rules to interpret the nature of a sedimentary body from its log response along the well bore. While quantitative interpretation focuses on the combination of several log responses at each sample in depth, sedimentological interpretation relies on the vertical succession of log facies. In particular, the vertical evolution of sand content is in most cases readily available from the gamma-ray log (measurement of the natural radioactivity of the series penetrated by the borehole): sand normally has a low radioactivity compared to shales, so that the gamma-ray log provides a direct reading of the vertical evolution of the sand content. Even in the more difficult case of radioactive sands, other indicators often allow a good estimate of the sand/shale ratio. The first interpretations were proposed for deltaic sediments, where a 'coarsening-up' log pattern (radioactivity decreasing upwards) is interpreted to reflect the progradation of a mouth bar or shoreface, while a 'fining-up' pattern (radioactivity increasing upwards) can correspond to the infill of a channel or to a transgression. In deep-water series, very roughly, a 'fining-up pattern', over a few metres to a few tens of metres is often interpreted to correspond to channel deposits, while 'coarsening-up patterns' over a few metres are meant to indicate lobe deposits. It must be made clear that this is just a rule of thumb, and that many exceptions occur; in particular, fining-up patterns only indicate a progressive reduction of sand influx into the turbidite system, and can be recorded in the lobes as well as in the channels. The coarsening-up pattern, when present, still appears to be more frequently related to lobes. Erratic patterns with 'randomly' alternating packages of sand and shale also appear to be more frequent in lobes, while massive, blocky sands are more common in the channels of mud-rich systems (Fig. 10.6).

3.2.5. Pressure data

This is a typical hydrocarbon industry dataset: pressure measurements along a well, or between wells, define reservoir communication (more precisely give evidence on the absence of communication between two reservoir intervals when their pressures are not compatible). Specific logging tools can provide pressure measurements along a borehole (several tens of measurements can be made in a few hours along one or several reservoirs penetrated by a borehole), thereby indicating which parts are in vertical communication (common pressure regime) and which ones are isolated from the others.

Well testing is a more complex operation, and consists in producing hydrocarbons from the reservoir during a certain amount of time, typically ranging from a few hours to months, while recording the evolution of the downhole pressure and flow. Pressure recording goes on after the production phase of the test is finished; the way the pressure builds up during that second phase gives quite precise indications about the geometry of the connected reservoir. In particular, 'no-flow boundaries' (in geological terms, impermeable faults or turbidite channel margins for instance) can be detected, and their distance to the well estimated.

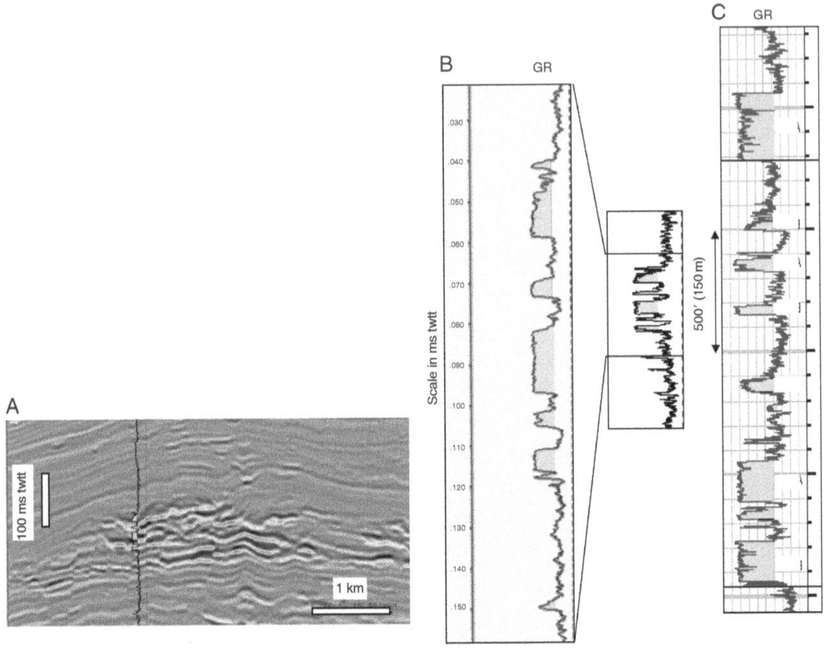

Figure 10.6 Log patterns in turbidite deposits. Beyond their use to define lithology, porosity and fluids in a borehole at any level, logs can be used as a sedimentological indicator, essentially using their vertical evolution. Although many variations exist, there is a general tendency for turbidite channels to be filled by overall fining and thinning-up successions. However, lobes may exhibit either a thickening and coarsening-up ('prograding') pattern or sometimes like channels a fining and thinning-up pattern. As a result, an upward increase of shale indicators (e.g. GR log) may be found in both channels and lobes, whereas an upward increase is usually characteristic of prograding medial to distal lobes. Levees often exhibit a fining-up character during the progressive build-up of the levee, typically remaining in a range of mixed lithologies (usually shaly sands to silty shales). The two examples illustrate log responses of a channel complex on the one hand ((A) seismic line, (B) close-up on the gamma-ray log) and of a pile of lobes on the other (C). The two examples come from several 1000 km apart and from different stratigraphic intervals. The message here is clear. There is no straightforward way of discriminating channels from lobes based on the vertical evolution of mud content only. (A multi-colour version of this figure is on the included CD-ROM.)

One last insight into the geology of the reservoir(s) is provided by the evolution of the hydrocarbon field during its commercial production. Parameters like the daily production of each well, sometimes of each reservoir within a well, or the evolution of downhole or wellhead pressures with time are monitored and recorded. All provide indications on communications between sand bodies, which may otherwise only be hypothesised from general geological understanding, with all the inherent uncertainty.

None of these techniques is specific to reservoirs deposited in deep water, but all of them contribute to the understanding of turbidite sand bodies as much as outcrop studies for instance. Issues like the lateral continuity of a foot-thick shale interbed between successive turbidite beds in a lobe are impossible to approach along the borehole (not enough sampling), nor can they be assessed from seismic (far below resolution). Only dynamic data can tell whether the shale interbed is continuous, and if discontinuous, what the proportions of 'holes' in it can be.

3.3. The use of analogues

The main purpose of well data analysis is to derive 3D parameters (lateral extension and continuity of reservoirs, heterogeneity distribution inside) from the observation along the borehole. While seismic often gives clues down to a certain scale, there is always a level of detail that is required beyond the reach of any tool, that is, some space for guesswork and uncertainty evaluation. As a result, all interpretations have to be based on a combination of the observation of local data on the one hand (seismic and well), and of concepts on the other. Concepts in turn have to be derived from better imaged systems thought to show the same characteristics as the studied system or reservoir. Such systems, considered as 'analogues', are often found at the outcrop. In the case of relatively recent turbidite systems, especially those deposited along mature passive margins and still lying in deep water, the shallow section of the system at issue is most of the time much better imaged seismically and often provides excellent clues on the behaviour of the turbidite system.

What is the limitation of analogues? The main danger is to rely on one single system (a well-described outcropping system, or a producing field with superb seismic imaging) as 'the analogue' and to forget that each turbidite system is unique. For instance, industry-academia consortia like the 'Brushy Canyon consortium', led by the University of Colorado at Boulder, CA, USA, have provided the industry with a wealth of data, of observations, and have deeply influenced the way turbidite systems are perceived (e.g. Gardner and Borer, 2000). The Brushy Canyon system (Permian of west Texas) is built of silty/sandy turbidites that reworked material of initial aeolian origin and deposited it at the foot of a carbonate platform. It could be very dangerous to draw turbidite reservoirs deposited in totally different contexts as 'little Brushy Canyons' without adapting. All companies have their favourite in-house analogues, often in foredeep basins, which should be treated with the same caveats, avoiding in particular copying directly geometries from the outcrop to the subsurface case.

More generally, turbidite systems can crop out only when they have been uplifted, often in mountain ranges. Many of the outcropping turbidite systems are the infill of the foredeep stage of the basin, that is, the remnants of relatively small systems deposited in confined basins with a very active sourcing, often involving cannibalism. On the contrary, most of the prolific (from a petroleum viewpoint) turbidite basins are the infill of passive continental margins by progradation of the continental slope. This problematic restriction to the actualism concept is discussed in Chapter 2.

4. Geology of Deep-Water Deposits Seen from the Hydrocarbon Industry Viewpoint

4.1. Method

4.1.1. Scales and hierarchy of turbidite deposits

A classification always has some degree of arbitrary in it. From biggest to smallest, it is usually convenient to distinguish turbidite systems (e.g. the Mississippi 'deep-sea fan'), sequences (e.g. the Pliocene lobe/channel-levee/hemipelagite set of Fig. 10.1), architectural elements (seismic-scale set of layers with similar origin and characteristics, e.g. one channel) and eventually bedsets and beds. One bed corresponds to the deposit of one individual resedimentation event, and can be comprised of one or a succession of facies, while a bedset (Campbell, 1967) is a succession of similar beds. In addition, erosion of one bed by the next may eventually make the identification of individual events difficult, at least locally (e.g. in a core). More detailed classifications of stratigraphic hierarchy can be useful in specific cases, especially when dealing with reservoir heterogeneity. For instance, reservoir studies in the Gulf of Guinea have distinguished a hierarchy of channel complex, channel storey, individual channel, each associated with one type of heterogeneity.

One key issue is that of seismic visibility and resolution. Geology can be interpreted with a reasonable degree of certainty down to the resolution scale of the available seismic. Below that scale, it is only possible to interpret by reference to concepts and analogues. Figures 10.7 and 10.8 show contrasting examples, one exhibits an exceptional seismic visibility and the other is more on the 'below average' side. Reservoir description will obviously be treated differently in the two cases.

4.1.1.1. Turbidite system: classification Turbidite systems, that is, the big scale, are an exploration issue. What is needed then is a prediction of the type of reservoirs they can contain, their overall sand content. The classification of Reading and Richards (1994) is a practical approach based on the comparison of a number of recent and ancient turbidite systems (Mulder, 2011, this volume, Fig. 2.18 in Chapter 2.3).

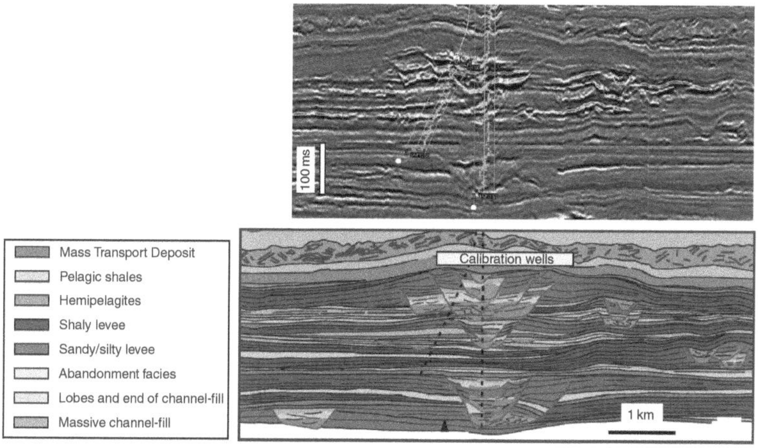

Figure 10.7 Example of high-resolution seismic and its interpretation. (A) A seismic line flattened below the base of the main reservoir so that it appears more or less as it was deposited. (B) A 'geoseismic section', that is, an interpretation of the seismic line calibrated on the wells that penetrate the structure. Seismic interpretation is heavily dependent on the quality of the survey used and also on the ratio between the resolution of the seismic dataset used and the size of sedimentary bodies, reservoirs in particular (image and interpretation by F. Temple). (A multi-colour version of this figure is on the included CD-ROM.)

4.1.1.2. Architectural elements Turbidite 'architectural elements' (initially defined by Miall, 1985 for fluvial sediments, (Mulder, 2011, this volume, Chapter 2.3.3), and transposed to turbidites by Clark and Pickering, 1996) or depositional elements (Mutti and Normark, 1987) could be defined as sedimentary units of relatively homogeneous properties at seismic scale (exploration or appraisal), or with a scale exceeding a few cells of the 3D reservoir model in the case of a producing field. Although the definition may sound a bit loose, a good identification of depositional elements of deep-water systems (turbidite or not) is the key to a good reservoir description, and deep-water elements will be described in detail in Section 4.2.

4.1.1.3. Facies Turbidite facies have been mentioned on the section about core description. The facies classification used must have a predictive character for exploration and initial appraisal, and must be related to reservoir properties when the issue moves from reservoir prediction to reservoir description. At this latter stage, the stress is put on the relationship between facies and petrophysical properties.

4.1.1.4. Reservoir properties The ultimate subject of interest for the oil and gas industry is porous network in which hydrocarbons have been accumulated and through which they will flow into the production wells.

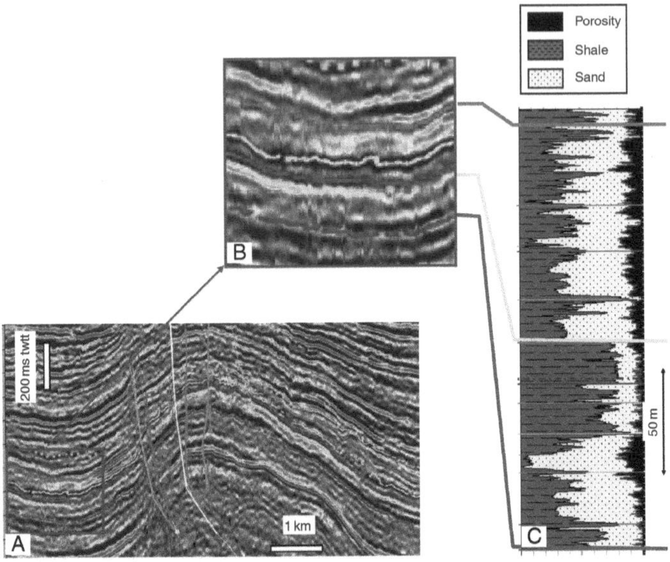

Figure 10.8 Example of low-resolution seismic. (A) The general seismic character on a turbidite field offshore Gabon. (B) A close-up on the reservoir series and (C) a lithological log of the reservoir itself (top reservoir is the magenta marker; base reservoir is the red pick). The log shows the vertical distribution of shale and sand in the series. In contrast with Fig. 10.7, this is an example of 3D seismic with a limited resolution. The factors limiting the resolution are both geological and geophysical: contrary to the previous case where sand and shale intervals were thick enough and acoustically well contrasted, turbidite reservoirs here are essentially thin beds (10 cm on average, i.e. well below any seismic resolution), rather cemented and show limited acoustic contrast with the background sedimentation. (A multi-colour version of this figure is on the included CD-ROM.)

All the geological features discussed previously are studied with the objective of better understanding the distribution of this porous network. A remote sensor that would measure in 3D the porosity and permeability in a field from the surface with enough detail would have a strongly negative influence on the number of jobs for geologists in the oil and gas industry. Fortunately for geologists, this tool will probably not exist for another few decades at least. Two parameters are essential to characterise a hydrocarbon accumulation: *porosity*, to define the quantity of hydrocarbons in place in the field, and *permeability*, to assess their capability to flow. In reservoir sands, porosity initially depends on the sorting of the grains for the most part and is, of course, progressively decreased by compaction and diagenesis during burial. Permeability on the contrary depends very strongly on the size of the sand grains. Reducing by a factor 2 the size of a porous network with a

3D photocopier (should the tool exist) would not change its porosity at all, but would decrease its permeability.

Turbidite sands are generally well sorted due to their mode of deposition, and to the fact that bioturbation (a mixer of grain sizes) is often limited in the sand beds, poor in nutrients and having been deposited very rapidly. As a consequence, there is little contrast in porosity between coarse-grained and fine-grained turbidites where the sediments have undergone limited diagenesis only (e.g. Tertiary systems along passive margins that make a good part of the hydrocarbon targets at present). On the contrary, the corresponding permeabilities will be much contrasted. As a result, the techniques developed to relate permeability to porosity often do not operate properly in this type of reservoir.

4.2. Architectural elements

When Mutti and Normark (1987, 1991) realised after looking at a number of turbidite systems that they were all different and that no single model could account for the variety observed, the authors recognised five major elements that appeared in many systems and proposed that all turbidite systems were made of a combination of these. Initially, the elements recognised were the following four: Large-scale erosional features (i.e. canyons), channels, lobes and overbank deposits (i.e. thick packages of fine-grained sediments); a fifth element, channel-lobe transition, was added in some of the publications of that period. This short list can be expanded into a 'check list' to be used as a reference for seismic interpretation in particular (Stow and Mayall, 2000, Mutti et al., 2009) (Mulder, 2011, this volume, Fig. 2.20 in Chapter 2.3.3). Expanding the list means including non-turbidite deposits, like hemipelagites, contourites or mass-transport deposits. It also means including some seismic-scale effects of post-burial remobilisation that can make bad pitfalls for seismic interpreters. Figure 10.9 shows a possible 'check list' for the geological interpretation of seismic, along with a first idea of the reservoir properties that can be expected from each type of element.

4.2.1. Resedimentation features

4.2.1.1. Canyons Canyons, or 'large-scale erosional features' (Mutti and Normark, 1987) are characterised by their dominantly erosional history (Mulder, 2011, this volume, Chapter 2.3.3.1). Their final preserved infill is usually complex, with a number of stacked erosion surfaces, making them unattractive exploration targets. Some canyon-fills are filled with hydrocarbons (the Chicontepec canyon in Mexico), but their production remains difficult due to the complexity of reservoir architecture (Fig. 10.10).

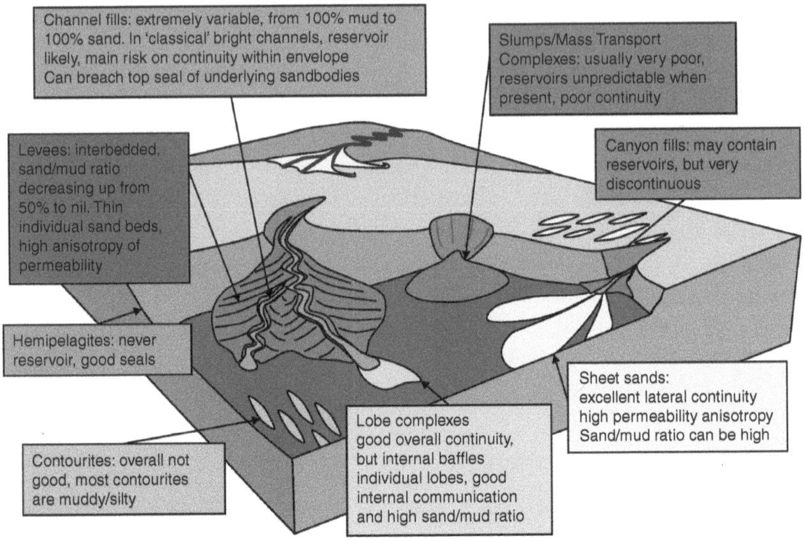

Figure 10.9 Reservoir properties of depositional elements. This schematic view summarises the reservoir characteristics that can be expected at the scale of exploration from the key architectural elements of turbidite systems. It is definitely not intended to solve production-scale problems, for which a higher degree of detail is requested, as well as a thorough analysis of the hierarchy of scales in the elements. (A multi-colour version of this figure is on the included CD-ROM.)

4.2.1.2. ***Channels*** Turbidite channels (Mulder, 2011, this volume, Chapter 2.3.3.5) are one of the favourite targets of oil and gas exploration: their geometry often makes them easy to recognise on 3D seismic images, ensuring a high success rate. They also have more isotropic reservoir properties than lobes, that in contrast are more layer-cake and therefore more difficult to produce (poorer recovery). At the same time, within one turbidite complex, channels preferentially develop in a shallower and more proximal position, making them the first targets of an exploration process that moved over the past decades from onshore to offshore to deep water. It is interesting to note that the oil and gas industry was a major contributor in the knowledge and understanding of turbidite channels, thanks to the quality of seismic data. Meandering channels in deep water were first described on the Amazon deep-sea fan, following DSDP leg 155 (Damuth et al., 1988; Flood and Damuth, 1987). However, deep-sea meanders were still considered for several years as characterising episodes of low sand supply, therefore unlikely to make reservoirs. The rationale behind that 'feeling' (rather than interpretation) was probably that sandy turbidity currents, capable of transferring sand to the deep domain were too 'sudden and strong' (Allen, 2001) to make nice and smooth meanders, implicitly associated with low-gradient, steady rivers flowing across

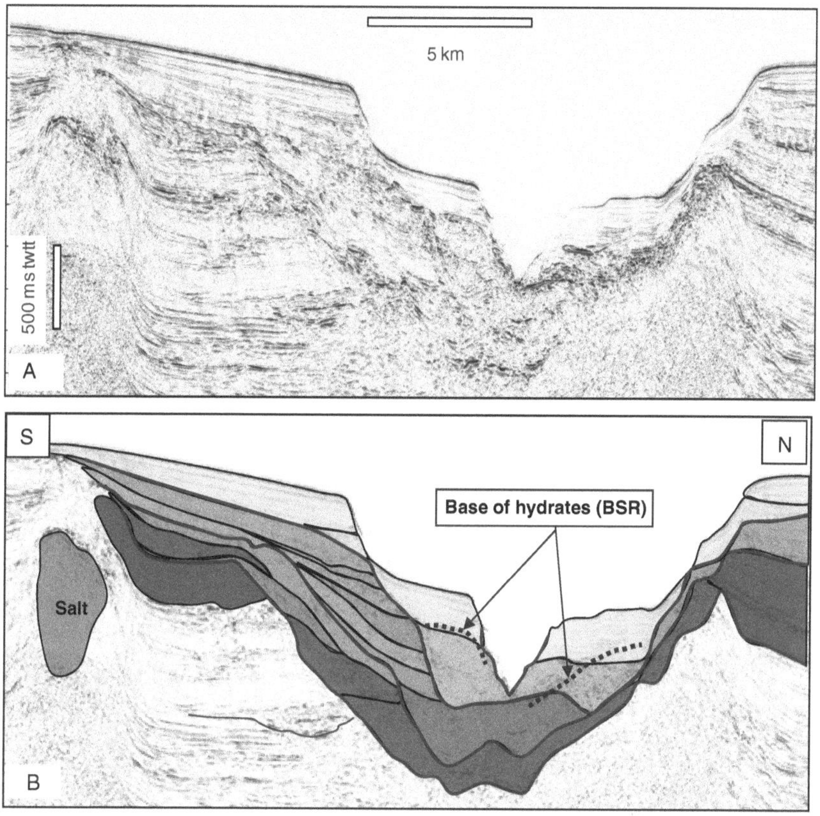

Figure 10.10 Sedimentary architecture of a canyon-fill. (A) This seismic line shot across the Zaire canyon offshore Angola illustrates the complex architecture of the sedimentary infill of a turbidite canyon. (B) Interpreted version highlighting the key elements of the canyon infill: the internal structure shows numerous phases of infill and re-erosion, with net deposition at the end of the process. Patches with a wiggle texture correspond to chaotic seismic facies, interpreted as mass-transport complexes; semi-transparent patches are dominantly draping packages corresponding to fine-grained sedimentation dominated by hemipelagites; while dotted intervals are packages of sediment wedging away from the axial thalweg of the canyon and interpreted as levee deposits confined inside the large-scale erosional morphology of the canyon (terraces). (A multi-colour version of this figure is on the included CD-ROM.)

mature or 'old' landscapes (Davis, 1889). Exploration successes in the Lower Congo basin, in particular with the discovery of the Girassol field offshore Angola showed that meandering turbidite channels could indeed be filled with sand and represent valid exploration targets (Kolla et al., 2001; also Fig. 10.11). Following these hydrocarbon discoveries, a number of outcropping sandy turbidite channels have been or are being reinterpreted as meandering, sometimes based on real evidence, sometimes in a more opportunistic way it seems.

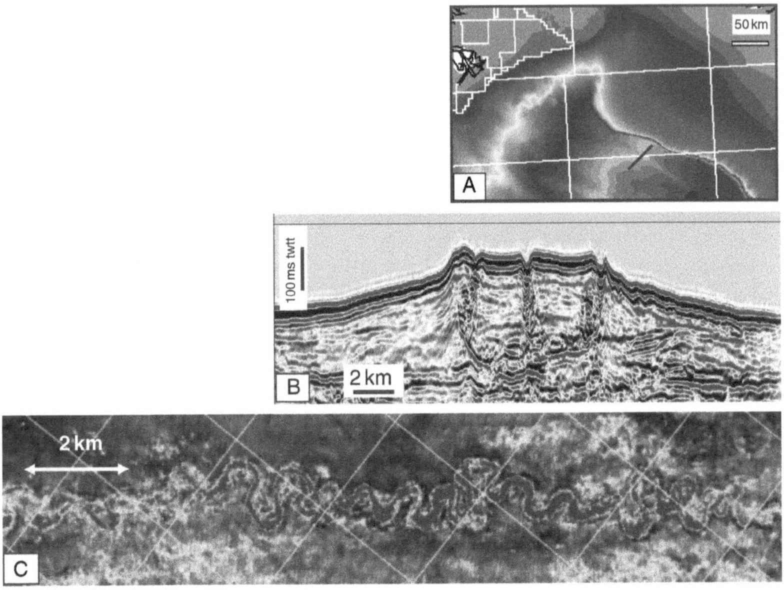

Figure 10.11 Meandering turbidite channels. (A) Bathymetric map of the seafloor of the north-eastern Gulf of Mexico showing a small ridge (dotted line) running parallel to the base of the Florida Escarpment. This ridge corresponds to a single meandering channel, illustrated by (B) (approximate location is shown by the little SW–NE trending black line). This figure highlights a classical pitfall on 2D seismic interpretation: a seismic line shot across a channel complex cuts the channel three times where it intersects a meander loop, giving the false feeling that it represents a branching point. There is no bifurcation here, as shown by the very linear character of the seafloor feature, and also by mapping from a seismic grid. (C) From Angola, and illustrates the typical character of meandering turbidite channels on 3D seismic. (A multi-colour version of this figure is on the included CD-ROM.)

4.2.1.3. Mass-transport deposits and complexes Thick intervals of en masse resedimentation can represent up to several tens of percentage of sedimentary series in deep water. The terms 'Mass-Transport Deposit' and 'Mass-Transport Complex' (Mulder, 2011, this volume, Chapter 2.3.3.9) were coined by P. Weimer (1989) to describe seismic intervals with a disorganised character without implying a specific depositional process. It is preferred to 'slump' or 'debris flow' bed commonly used: 'slump' conveys a notion of partial disorganisation only, preserving the original bedding of the series, while 'debris flow' implies a specific rheology (Mulder, 2011, this volume, Chapter 2.1.3.5). Mass-transport complexes can reach several hundred metres in thickness and extend over hundreds of square kilometres. While some exhibit distinct geometries, for instance, with imbricated thrusts at the front of the complex, others just appear as noisy packages, even within one complex. In bad cases, the geometry of mass-transport complexes can mimic that of reservoir elements.

Mass-transport deposits/complexes develop at all scales: small events are commonly part of the complex infill history of turbidite channels, while bigger events can cover a good part of a sedimentary basin (Figs. 10.12 and 10.13) (see Chapter 2).

4.2.1.4. ***Levees*** Turbidite levees (Mulder, 2011, this volume, Chapter 2.3.3.5) are normally made up of interbedded sands and shales, often with a low sand/shale ratio. In that respect, they usually make poor reservoirs and are rarely exploration targets in themselves. However, they often provide additional resources to their companion channel. Turbidite levees are very spectacular objects on seismic sections, and often represent the most conspicuous feature (Fig. 10.14). On large deep-sea fans, individual levee wedges can exceed 500 m in thickness and 50 km in width. They appear as wedges thinning away from a central axis, typically with high amplitude reflections.

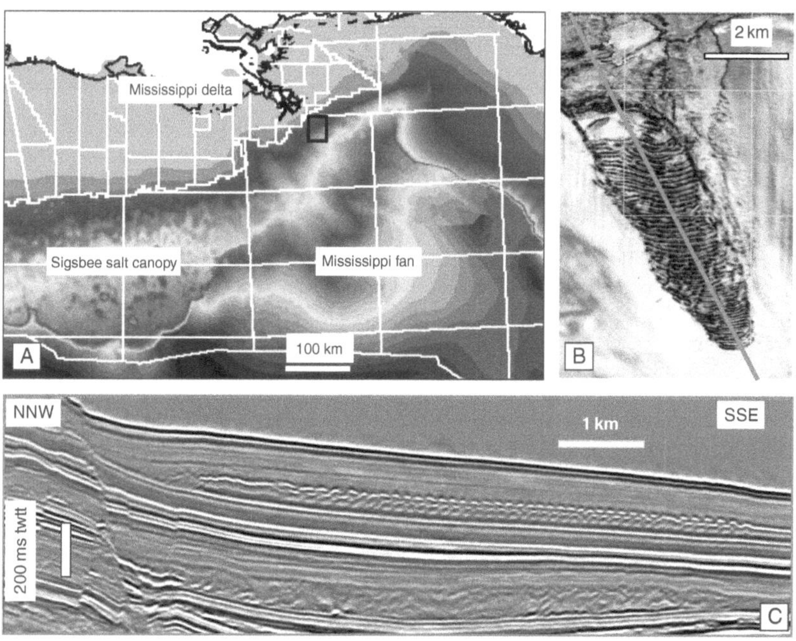

Figure 10.12 Small mass-transport complex from the Pliocene of the deep Gulf of Mexico. (A) Seafloor bathymetry map with the location of the MTC shown by the black square. (B) Map view (dip map from 3D seismic interpretation) of a small slump. (C) Cross section along the slump highlighting the extreme regularity and lateral continuity of the imbricated thrust folds. (A multi-colour version of this figure is on the included CD-ROM.)

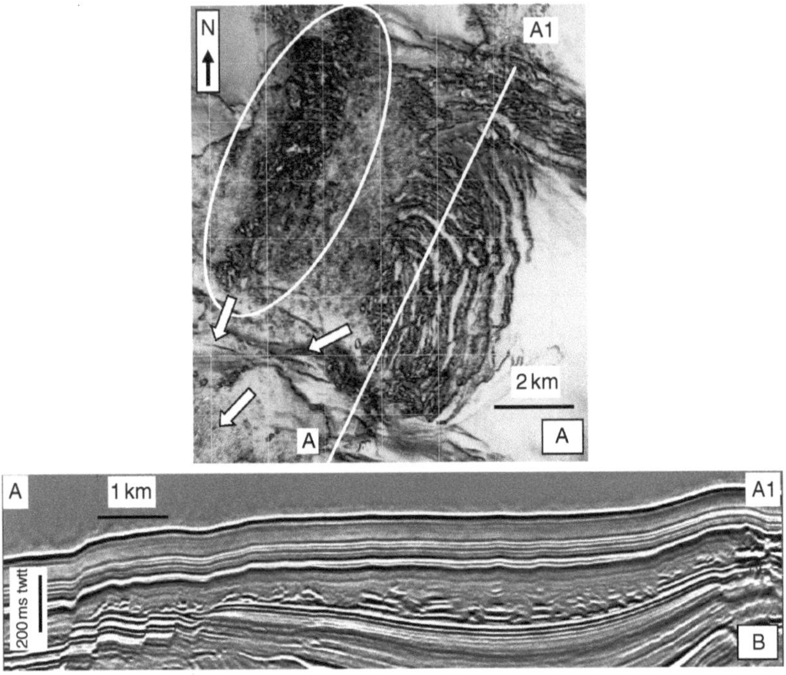

Figure 10.13 Proximal extensional area of a Mass-Transport Complex from the Pliocene of the Gulf of Mexico. Horizon slice of the seismic coherence cube through the complex on (A), cross section along the red line on (B). The most spectacular part of a mass-transport complex is often the distal end with regular imbricated thrusts, as shown in Fig. 10.12. This section and slice both illustrate the seldom imaged source area of an MTC, with incipient rafts that 'froze' before they could move any longer (right-hand part). The upper, left-hand part of the map (blue ellipse) shows the more typical shattered appearance of the body of another MTC coming from further upslope. Arrows denote the transport direction.

One interesting issue is that of sediment waves (Fig. 10.15; see discussion in Migeon, 2000; Migeon et al., 2001; Mulder, 2011, this volume, Fig.2.28 in Chapter 2.3.3.5). Sediment waves were initially described on seismic surveys and seafloor images, typically with a km-scale wavelength. They can develop on levees of turbidite channels or in association with contourite drifts (Faugères and Mulder, 2011, this volume, Fig. 3.20 in Chapter 3.4.2.3). Similar features have been described at the outcrop (e.g. Anderskouv et al., 2007), but seismic-scale sediment waves have not been reported so far from outcrop data. It would seem surprising that no such sediment wave crops out anywhere in the world. What is more likely is that their recognition would require either very large, well-oriented outcrops: such multi-kilometric sedimentary objects would require continuous outcrops of at least twice their wavelength to be identified unambiguously. Alternatively, they can be

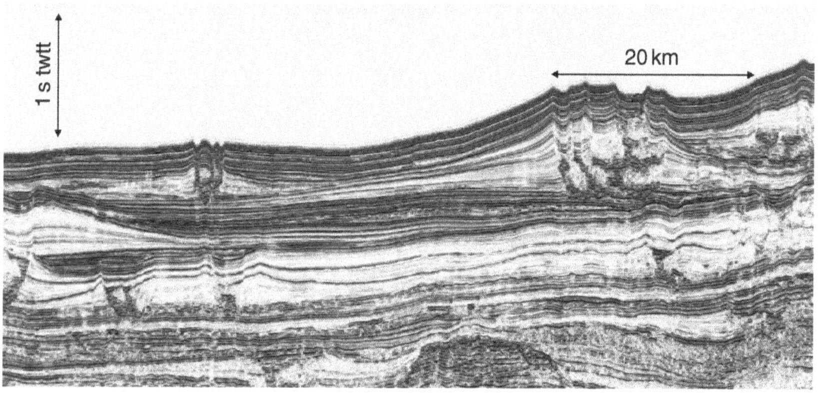

Figure 10.14 Turbidite levees and hemipelagites from the Indus fan. While Fig. 10.11 focused on the channel morphology, this one shows well the typical wedging of turbidite levee systems. The morphology observed on 2D varies according to the angle of the section with the axis of the channel: in the lower interval, for instance, the line is clearly almost parallel to the channel, which it cuts at several locations. The wedging indicates in a way the proximity of the channel to the line. The whole system is draped by some 200 ms of hemipelagites: note how individual reflectors can be followed with almost no change in thickness all along the line, including above the previously deposited channels and levees.

interpreted from discontinuous outcrops, but only by geologists familiar with the concept and specifically looking for them (Mulder, 2011, this volume, Figs. 2.26 and.2.27 in Chapter 2.3.3.5).

4.2.1.5. ***Lobes and sheet sands*** The term of 'turbidite lobe' covers a wide range of deposits and features from extremely laterally extensive 'sheet sands' to actually lobate features identified on seafloor surveys (Mulder, 2011, this volume, Chapter 2.3.3.10). They were second to channels in interest because they usually occur in a more distal position of the turbidite systems. More distal either implies deeper water and more technical constraints, or an older, deeper interval of a prograding margin, and even more technical constraints and difficulties again. Here again, like with channels, significant evolutions of understanding are currently taking place.

Based on fieldwork on foredeep turbidite systems, Mutti (1992) concluded that 'lobes' and 'sheet sands' were synonymous, meaning that turbidite lobes made very continuous and laterally extensive sand bodies. The archetype of turbidite lobes in that respect were those studied in the Apennines (e.g. the Contessa turbidite, one specific key bed that can be followed over more than 100 km along the basin), or the sandy packages developed in the Eocene Hecho Group of the South Pyrenees basin, where sand units several metres thick, as well as individual turbidite beds, can be traced over several tens of kilometres. Figure 10.16 shows an example of this

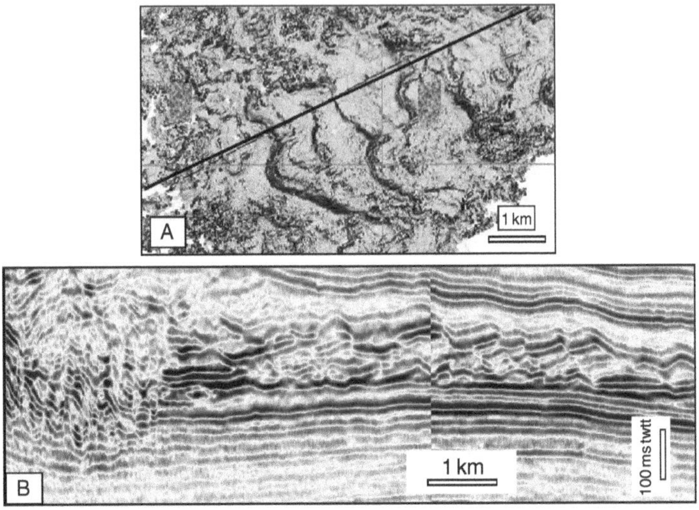

Figure 10.15 Sediment waves on turbidite levees. (A) Dip map of the top of the levee (undulated red marker on the section). (B) Section along the red line. Sediment waves are commonly observed on levees and can strongly alter the archetypal wedging character. Sediment waves generally have a 1–2-km spatial wavelength. They can be distinguished from collapse faults by the much lower angle of the planes of discontinuity (ca. 10°) and by the impossibility to actually correlate seismic vents from one 'compartment' to the next. Note that sediment waves can also be formed by other types of mud-laden currents, like contour currents. (For interpretation of the references to colour in this figure legend, the reader is referred to the Web version of this chapter).

type of continuity, as observed in the Annot sandstone (Palaeogene of the Alps, French–Italian border).

This concept would imply an excellent correlativity of turbidite beds and units between wells, which is seldom the case. Perfect peak-to-peak correlation between sand bodies in boreholes located several kilometres apart can indeed be found in some foredeep basins, but most drilled examples only show an overall similarity of sand packages at the scale of a few tens of metres, but poor bed-to-bed correlation inside those packages beyond a few kilometres, or less, as shown in Fig. 10.2. As a result, it is convenient to distinguish 'lobes' on the one hand, defined as relatively limited features with limited detail correlation between neighbouring wells, from 'sheet sands' on the other. In a very practical way, sheet sands would be packages of sand that extend beyond the limits of the field under study. The notion here is purely pragmatic and hydrocarbon production-minded: beds continuous over 25 km^2 may be considered as 'sheet sands' in the context of a small hydrocarbon field and 'lobes' in one covering a larger area.

Figure 10.16 Contrasting lobe styles at the outcrop and on seismic. (A) A photographic montage from the western Alps (Vallon du Lauzanier, Annot Sandstone Formation). The thumbnail in the lower, left-hand part is a 'copy and paste' of the black box on the crest, enlarged to correct for perspective effect and juxtaposed to the correlative series some 2 km away. Note the perfect bed-to-bed correlation. The white arrow approximately corresponds to 50 m. (B) and (C) Horizon slices from a turbidite lobe in the south Atlantic and show the 'braided' character of some subsurface turbidite lobes, difficult to reconcile with the layer-cake image provided by the outcrop. Subsurface studies also indicate that many turbidite lobe reservoirs do not show the continuity that should be expected from sheet sands. (A multi-colour version of this figure is on the included CD-ROM.)

*4.2.1.6. **Massive sand mounds*** Some of the early discoveries of the North Sea were made in 'sand heaps' that do not adequately fall in any of the above-mentioned categories. The Frigg field, for instance, discovered in 1971, is a seismic mound deposited beyond the shelf break of the Eocene North Sea (Héritier et al., 1979). A quick geometric reconstruction indicates that the water depth at the time when the reservoir was deposited did not exceed a few hundred metres.

Very soon after the 'deep water' origin of the field was established, geologists working in the area noticed the striking contrast between the turbidites they could see at the outcrop, characterised by a very flat top, and their mounded reservoirs. Several theories were proposed to account for that discrepancy. Enjolras et al. (1986), in a team including E. Mutti, proposed a 'contourite' interpretation to account for the fact that the sedimentary slopes observed at the top of the Frigg sand heap reached or locally exceeded 18° (Fig. 10.17) The rationale behind the interpretation was negative evidence:

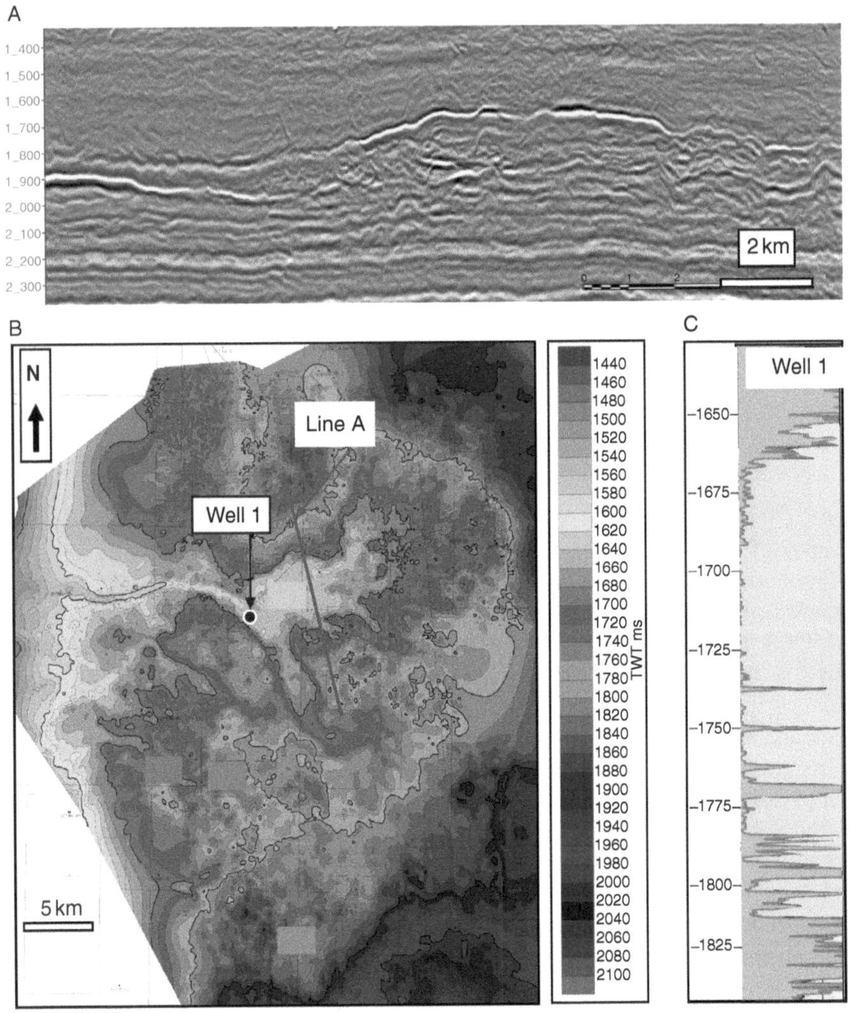

Figure 10.17 Deep-water Massive Sands. Deep-water massive sands are probably one of the less well-understood ingredient of deep-water systems, particularly well developed in the North Sea basin. (A) A seismic section through a massive sand mound. The top of the sand calibrates to the strong reflector in the upper middle part of the image. (B) A two-way time map of this top sand, highlighting the strong relief of the mound, with depositional dips locally exceeding 15°, well above what is normally observed on turbidite sands. (C) A lithological log from well 1 drilled through the sand mound. The track represents the respective percentages of sand (yellow) and shale (green) along the well trajectory. The interval 1665–1735 m is totally sandy, while some shaly interbeds develop below 1735 m. Seismic lines linking this mound to the coeval shelf indicate that it was deposited in water depths not exceeding 300 m, which probably explains in part the morphological difference with really deep-water turbidites. See discussion in text about the various interpretations proposed to explain their origin and specificity. (A multi-colour version of this figure is on the included CD-ROM.)

the sand heap had been deposited in 'deep water', that is, beyond the reach of shelf agents like fluvial, tide and wave influence. However, it did not match the 'turbidite' concept exemplified by foredeep basin classical series. What other depositional agent could have been active in this setting? Permanent, deep (contour *s.l.*) currents were the envisioned solution.

A few years later, additional observations made on the shelf coeval of the Frigg reservoirs made it clear that Frigg originated from a mass collapse of a very sandy shelf and had not been pushed a long way by bottom currents. Based on this observation and on core evidence, Shanmugam et al. (1995) came up with a different interpretation based on the very same observation that the morphologies of the North Sea do not match 'turbidite concept'. Shanmugam acknowledged the presence of the source of sand upslope of Frigg and proposed a non-turbulent depositional mechanism, which he called 'sandy debris flow' (frictional flows of Mulder and Alexander (2001) or grain flow of Nardin et al. (1979) to account for the observations. This interpretation seems closer to reality, but post-burial effects like sand injection and diapirism are known to have affected many massive sands of the same area (e.g. Dixon et al., 1995), and may play a role in the mounding. These effects are detailed in Section 4.2.3.

4.2.1.7. Resedimented carbonates There are far less known examples of hydrocarbon accumulations in resedimented carbonates, which means that carbonate turbidites are poorly known to the industry compared to their siliciclastic counterpart. The most prolific examples can be found in the Mexican part of the Gulf of Mexico, where huge oil accumulations are trapped in a series of deep-water carbonate breccias dated from the Jurassic to the Palaeocene - Eocene issue. The most prolific reservoir is the Cretaceous Tamabra Formation. The Tamabra Formation is comprised of allochthonous carbonate debris-flow breccias and carbonate turbidites deposited at the foot of the El Abra limestone platform (Magoon et al., 2001). About one third of 12×10^9 m^3 of oil are thought to be accumulated in the Tamabra Formation. In the same basin, resedimentation of the Yucatan platform in the Cretaceous and Palaeocene produced the carbonate breccia reservoirs of the giant Cantarell field (Grajales-Nishimura et al., 2000). About 60% of the oil production of Cantarell comes from the impact breccia associated with the Chicxulub crater (the impact that is interpreted to be implied in the environmental crisis at the Cretaceous/Palaeocene boundary) (Shepherd, 2009). Late Jurassic breccias in the same area make the reservoirs of the Chac field. Unfortunately, little detail information has been published on these fields.

4.2.2. The fine-grained continuum

4.2.2.1. Hemipelagites Hemipelagic shales (Hüneke and Heinrich, 2011, this volume, Chapter 4) make a good part of the series drilled by hydrocarbon wells searching for turbidite reservoirs. They also make most of the seals allowing migrating hydrocarbons to be directed towards these reservoirs, and ultimately trapping them there. Hemipelagites are normally characterised by draping reflectors on seismic sections and in most cases are easy to identify when thick enough (Fig. 10.14). As mentioned in the section on levees, hemipelagites act as a background sedimentation in which more episodic turbidites are intercalated. Intervals dominated by hemipelagites occur when the influx of turbidites is switched off by external factors (e.g. transgressive/highstand times in non-tectonic contexts) or in areas out of reach of turbidity currents.

*4.2.2.2. **Contourites*** As discussed in the 'massive sand mounds' section above, contourites (Faugères and Mulder, 2011, this volume, Chapter 3) became popular in the oil and gas industry when the role of deep permanent currents was invoked by Mutti and co-workers (Enjolras et al., 1986) to explain the mounded geometry of the deep-water sand bodies of the Tertiary of the North Sea. The rationale proposed was the following: turbidites from all outcropping examples lie flat and horizontal, within a fraction of a degree. The reservoirs of the Frigg field exhibit sedimentary dips commonly in excess of 10°, implying a different type of process. The other process known in deep water was deposition of reworking by permanent currents, hence the interpretation proposed.

Mutti (1992, personal communication) at that time suggested that turbidites were probably more common in foredeep basins, contourites making the bulk of passive margin deep-water sediments Although the role of contourites is now thought to be relatively low in making hydrocarbon reservoirs, contourite accumulations are known in hydrocarbon basins and deserve correct identification, be it only to avoid mistaking a contourite levee for a sandy lobe.

Permanent deep currents have been interpreted as a winnowing agent for turbidites that would otherwise have been laminated, for instance, in the Campos Basin, Brazil (Gonthier et al., 2003; Moraes et al., 2007; Mutti, 1992, pp. 19–24; Viana, 2001) or in the Gulf of Mexico (Shanmugam et al., 1993). Contourites have definitely been identified in hydrocarbon basins (e.g. Rigollet, 2001), but have not been formally identified so far as the principal agent responsible for the deposition of a commercial reservoir (with the possible exception of Brazil, but data from the national operator are scarce) (Viana et al., 2007).

*4.2.2.3. **Thin-bedded, muddy turbidites*** A close look at seismically well-imaged levee systems often reveals alternating episodes of 'real' levee construction with packages of reflectors wedging out away from the channel and of 'hemipelagic drape' with packages keeping their thickness all along the section. A phase-by-phase analysis of the seismic carried out on the Rhone turbidite system has shown an alternation of (1) episodes thinning out away from the shelf break and keeping an approximately constant thickness across the slope and (2) episodes thinning out laterally away from the channel and keeping a more or less constant thickness along lines parallel to channels. This is interpreted to reflect an alternation of phases during which most of the sediment is carried out into the basin through the turbidite channels (fan activity, with sand likely to be deposited in the distal part of the fan as well) and phases during which sediment is distributed within (subsurface, intermediate, or bottom) nepheloid layers extending from the shelf (e.g. during storms or fluvial floods).

On a different scale, it is interesting to note that the little final draping phase above the meandering channel of Fig. 10.11C can be traced into the 2-km-thick Mississippi turbidite system some 200 km to the SW. Hemipelagite or levee? Probably both, the clay particles that fell out from suspension to drape the channel (hemipelagite by definition) came in majority from the overbanks of the Mississippi channels (levee by definition).

4.2.3. Post-depositional effects

Deep-water sediments appear to be particularly prone to post-burial remobilisation, probably because they are often deposited as isolated porous bodies, without pressure communication with the surface. As a result, fluid pressure inside the reservoir builds up during burial until fluids escape by breaching the top seal, sometimes injecting sand into the surrounding shales. Other specific deep-water phenomena include the effects of gas hydrates (Wheeler and Stadnitskaia, 2011, this volume, Chapter 6.4.1). Figure 10.18 summarises the most common seismic-scale effects of post-burial remobilisation in deep-water sediments.

4.2.3.1. Injected sands The concept of sand injection was initially defined from outcrop observations in the nineteenth century (pseudo-dikes, Darwin, 1910). Sandstone dykes were then the subject of many academic studies (Parize, 1988; Parize et al., 2007), but remained largely unknown to petroleum geologists until difficult problems occurred during the appraisal of hydrocarbon discoveries in the North Sea. The Balder field, for instance, was discovered in 1969 but was not developed until 2001. The 30-year span between the discovery and the initial production is due in part to the extreme complexity of the reservoir, which is strongly affected by sand remobilisation and injection. Sand injections, sills and dykes, are commonly observed in the cores of certain basins: Lower Congo basin (Braccini and Penna, 2005, personal communication), Eocene of the North Sea (Dixon et al., 1995; Hurst and Cartwright, 2007), Cretaceous of the North Sea (unpublished work, also Jackson, 2007). Figure 10.19 shows a set of seismic anomalies initially thought to be a turbidite channel, based on 2D seismic interpretation. Subsequent acquisition of a 3D survey indicated that the anomalies actually develop as roughly circular complexes rather than linear channels,

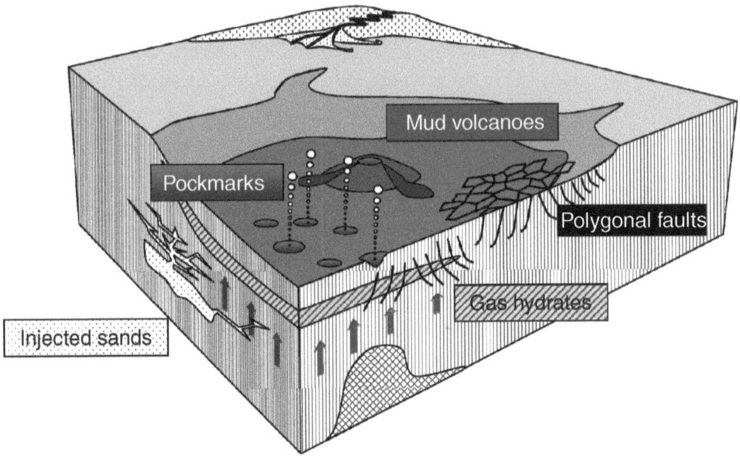

Figure 10.18 Post-depositional effects in deep water. The diagram schematically illustrates some of the most frequent effects of sediment remobilization by intra-formational fluid flow in deep water. Mud volcanoes and pockmark develop as surface features, sand (or mud) injection and diapirism occur at depth. Polygonal fault networks appear to develop in response to dewatering of very low-permeability series that cannot expel their compaction waters along the bedding. Gas hydrates are discussed extensively in Wheeler and Stadnitskaia (2011, this volume, Chapter 6.4.1) and Hesse and Schacht (2011, this volume, Chapter 9.7.1); their seismic response is shown in Fig. 10.21. (A multi-colour version of this figure is on the included CD-ROM.)

and a precise picking revealed a flower shape rooted in an underlying bulge. Drilling through the anomaly down to deeper objectives eventually revealed that it consisted of sands injected upwards from a deeper massive sandbody some time after its burial. Interestingly enough, the sand sill and dykes in this example are much thinner than suggested by their seismic response: the sands are injected into over-pressured shales and what makes the seismic reflection is actually the contrast between the low-impedance over-pressured shales and the higher-impedance normally pressured shales around the sills and dykes (Fig. 10.19).

Surprisingly enough, the abundance of sand injections appears to be a regional, rather than local, characteristic. While injected sands can be seen in most cores of the Eocene of the Viking Graben, or in the Lower Congo basin, they are apparently quite rare in the Gulf of Mexico.

4.2.3.2. Polygonal fault networks in thick shale series Once again, this specific seismic facies (often colloquially referred to as 'polygonal shales') could not be identified until 3D seismic revealed its structure. It consists of shaly series affected by closely spaced normal faults, with the faults dying out progressively in depth without any evidence of flattening on a decollement level, for instance (Fig. 10.20B and C). Horizontal slices through 3D volumes of seismic indicated that the faults were arranged into polygonal

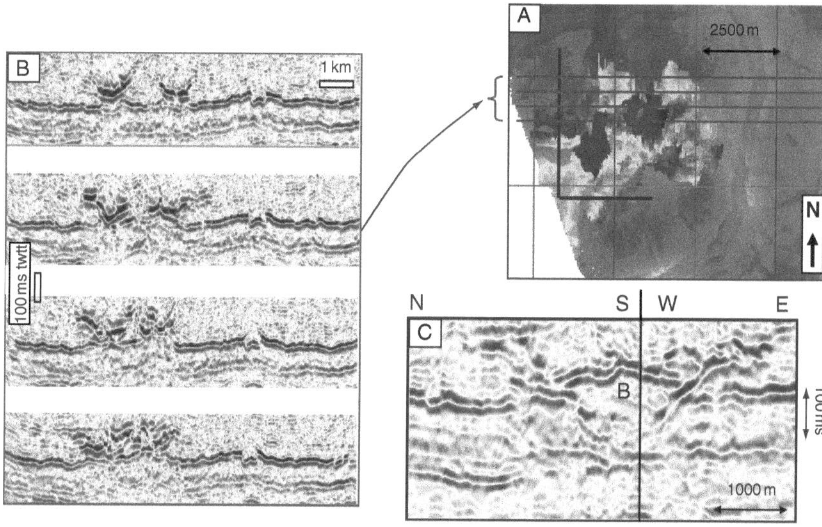

Figure 10.19 Sand injection in the Eocene of the North Sea. (A) A two-way time map of the blue reflection that appears around 1850 ms twtt on the four sections of (B) (location in black on (A)).The irregular patches correspond to the anomalies that 'float' above the more continuous marker below. (C) The composite line shown by the dotted 'L' shape on (A). The seismic anomaly shown by the four parallel lines was initially interpreted on 2D data as a channel-fill with poorly imaged levees. A 3D survey was acquired after drilling had proven that there was very little sand in the anomaly. It revealed that the interpreted 'channel' actually consisted of several isolated patches of anomalies. Mapping the anomalies showed a series of sheets 'floating' above and laterally to a deep bulge (B on C). Subsequent drilling confirmed that the bulges of the area are filled with sand and that the reflective sheets are sands injected upwards from their parent sandbody. (A multi-colour version of this figure is on the included CD-ROM.)

cells, typically a few hundred metres in diameter (Fig. 10.20A). They are interpreted (Cartwright, 1994) to correspond to the episodic dewatering of piles of sediment too impermeable to expel their compaction water along the bedding during mechanical compaction.

*4.2.3.3. **Hydrate-related effects*** Methane hydrates are an ice-like solid, a crystalline lattice where molecules of gas are trapped in a 'cage' of water molecules (Wheeler and Stadnitskaia, 2011, this volume, Chapter 6.4.1; Hesse and Schacht, 2011, this volume, Chapter 9.7.1). Hydrates are stable at low temperature and high pressure, for instance, below 500 m of water depth at temperatures of 5 °C. They were once made popular as a potential hydrocarbon resource (Kvenvolden, 1993). There is less optimism nowadays about their capacity to provide exploitable gas in the next few years. However, countries like Japan where proven hydrocarbon reserves are very small, but where the presence of hydrates has been shown by seismic, put a great deal of technical effort into finding a way of actually producing the gas contained in this source.

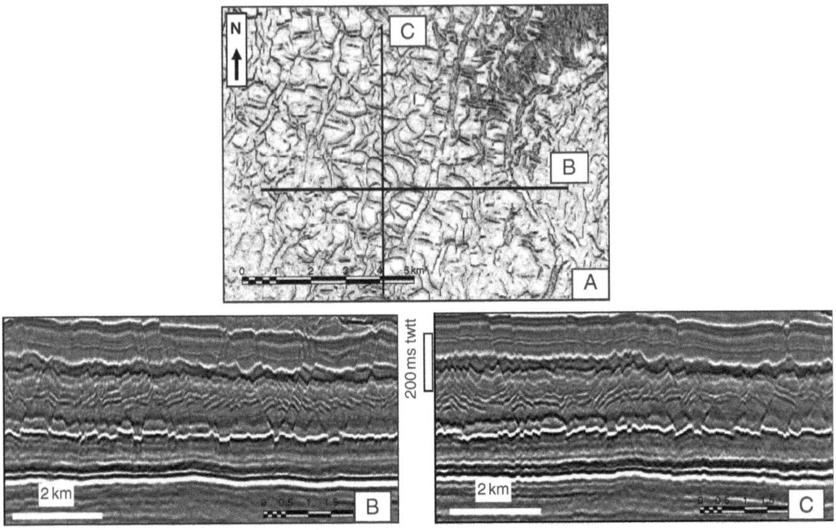

Figure 10.20 Polygonal fault network in low-permeability series. Map (A) and two orthogonal sections (B, C) from a 3D survey of the Carnarvon basin, Australia NW shelf. The sections are from the original cube (amplitude) and the map is a 'time slice' (horizontal section) at 3100 ms twtt through the corresponding 'coherency cube' (coherency is a seismic attribute highlighting discontinuities, like faults). The faults affect a 200-ms-thick (ca. 300 m) interval of carbonate ooze. Note how they die out at depth. They are interpreted to correspond to contraction by fluid escape rather than extension (Cartwright, 1994).

Hydrates are stable in a specific domain of (low) temperature and (high) pressure. Below the seafloor in deep-water settings, the stability domain typically extends down for a few hundred metres before the increase of temperature counterbalances the pressure increase and the hydrate dissociates into gas and water. The base of the stability domain in areas where gas and hydrates are present is often marked by a reflector parallel to the seafloor and secant to geological features called the 'bottom-simulating reflector' (commonly abbreviated into BSR; Fig. 10.21). Even in the absence of BSR, a contrast is often observed on seismic between the uppermost few hundreds of milliseconds, with well-behaved continuous reflections, and the deeper section much less organised. This may suggest an effect of freezing (by hydrates) in the upper part and 'thawing' below the stability zone once the sediment pile builds up.

4.2.3.4. Pockmarks and conical injections Pockmarks (Wheeler and Stadnitskaia, 2011, this volume, Chapter 6.4.2) are another good brain-teaser for seismic interpreters. V-shaped structures up to 100 m deep are commonly observed on seismic lines, at the seafloor or deeper, in several areas of the world (e.g. Gulf of Guinea, North Sea). They were typically understood as

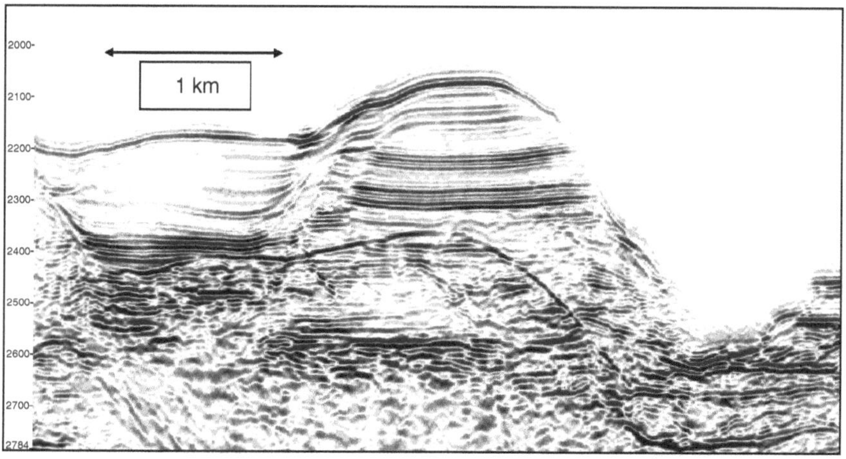

Figure 10.21 The 'bottom-simulating reflector' (BSR), base of the hydrate stability zone. This seismic section across the Zaire canyon showing the 'BSR' typically marking the base of the hydrate-bearing zone in deep water. The BSR is the strong blue reflection parallel to the seafloor some 250 ms twtt deeper. It is clearly secant to the stratification. The BSR corresponds to the contrast between a zone above where hydrates increase the acoustic impedance of the sediment and a zone below where free gas may be trapped, thereby decreasing the impedance with respect to the initial situation. As a result, it appears with a polarity opposite to that of the seafloor. (A multi-colour version of this figure is on the included CD-ROM.)

'submarine erosions' and intuitively related to channels of some sort until 3D seismic became available and revealed that they were actually circular in map view, that is, conical features (Fig. 10.22). Such structures, known as 'pockmarks', are known to correspond to fluid expulsion features (see a complete review in Hovland and Judd, 1988 and Judd and Hovland, 2007).

Other conical features, rather than eroding, appear to "push up" a keystone of sediment above the regional level, sometimes generating forced folds. They are interpreted as conical intrusions and can reach diameters of several kilometers and intrude thicknesses of several tens of meters (Shoulders and Cartwright, 2004, Cartwright et al., 2008). It had been known for years on 2D profiles and vaguely interpreted as an erosion surface probably corresponding to an emersion phase, until 3D seismic revealed that what appeared as V-shaped valleys on 2D sections were actually conical features, up to 200 m deep and 2 km in diameter. These were initially interpreted as pockmarks, but recent investigations on better quality seismic have shown that the series inside the cone match those outside, indicating that the features are injected cones rather than actual pockmarks.

4.2.3.5. Mud volcanoes More than a mere curiosity for the hydrocarbon industry, mud volcanoes (Wheeler and Stadnitskaia, 2011, this volume, Chapter 6.4.2) sometimes serve as providers of surface shows (e.g. around

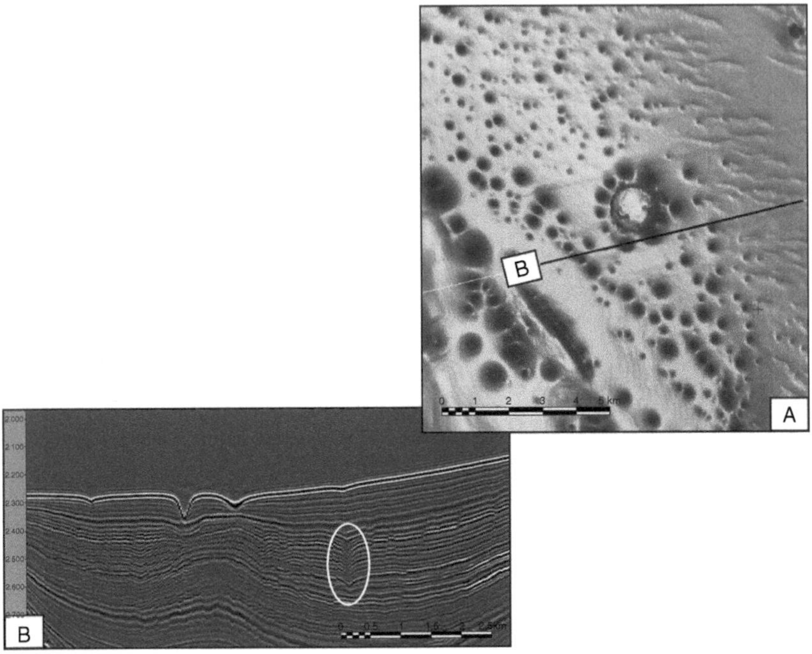

Figure 10.22 Pockmarks at the seafloor in the Gulf of Guinea (Angola). (A) Combination of seafloor bathymetry (colour scale) and dip to give a relief rendering. (B) Cross section along the red line on the map. Individual pockmarks appear as funnel-shaped depressions up to 1 km in diameter. In cross-sectional view, they erode previously deposited layers. Note the presence of fossil pockmarks (white ellipse). (For interpretation of the references to colour in this figure legend, the reader is referred to the Web version of this chapter).

Baku in Azerbaijan), and also make very effective pitfalls for the seismic interpreter who first comes across one on a seismic section. They can make, for instance, intriguing Christmas-tree shapes, difficult to interpret at first glance. Figure 10.4, used earlier to demonstrate the principle of 3D seismic, is a good illustration of a small mud volcano with its successive mudflows radiating from a central vent.

4.2.4. Oddities—volcanics

There are active volcanoes in deep water (Carey and Schneider, 2011, this volume, Chapter 6). More frequently in hydrocarbon basins, the deep-water realm can be within the reach of active volcanoes that shed their ejecta directly into deep water, or whose ejecta are resedimented in the deep domain. One famous example of deep-water volcaniclastic formation is the 'Balder tuff' that marks the limit between the Palaeocene and the Eocene in the North Sea. The Balder tuff appears as a strong seismic marker, very continuous regionally. It corresponds to ash falls from the break-up of the north Atlantic some 500–1000 km to the west. A few reservoirs are associated with the tuff interval, but with a 'normal' siliciclastic provenance.

Resedimentation of volcanic ash into the deep domain has been observed in the deep Gulf of Mexico, where Pleistocene tuffs were resedimented into a 'fan' fed by a channel. Age dating shows that the resedimented tuffs are coeval to a volcanic explosion that took place in the Yellowstone caldera at 2.1 Ma. The explosion ejected 2500 km^3 tephra into the atmosphere, known as the Huckleberry Ridge Tuff Formation (Smith and Siegel, 2000), enough for 50 km^3 of it to be carried down along the turbidite pathway of the time into the abyssal plain (unpublished work, related problems mentioned by Friedman, 2006).

4.3. Reservoir characterisation of architectural elements

Identifying architectural elements is just the first step of the interpretation. The next step consists in defining their reservoir properties in such a way that the appropriate decision can be made.

Reservoir elements (like channels and lobes) are characterised by their geometry and reservoir properties while non-reservoir ones (mass-transport complexes, hemipelagites), when significant, only need to be given a geometry. Very common elements like levees are often neglected initially because they do not make exploration targets in themselves but can become significant when the project proceeds as secondary contributors. It has been shown, for instance, that the proximal, relatively sandy levees associated with a turbidite channel could contribute significantly to hydrocarbon production. If the levees are sandy enough, injecting water into the levees could sweep channel hydrocarbons more effectively than injecting directly into the channel itself (Labourdette et al., 2007).

4.3.1. Reservoir characterization in exploration

At the stage of exploration, the key issue is to define the amount of hydrocarbons likely to be accumulated in a 'prospect'. The approach is often a broad-brush one, especially when seismic quality is not excellent: what is assessed is then an overall proportion of reservoir lithologies in a thick interval (e.g. 30% reservoir in a 150-m- thick interval). For turbidites, it is often difficult to go far beyond a 'channel-dominated' or 'lobe-dominated' classification.

More precisely, the in-place hydrocarbon volume is typically assessed multiplying the gross rock volume of the reservoir-bearing interval above the expected fluid contact by the proportion of reservoir in this volume (called the 'net to gross' ratio), the porosity of the reservoir (proportion of pore volume) and the proportion of oil or gas in the porosity. Due to the absence of hard data at this stage, each of these parameters is given a range of possible values. The range can be wide or narrow depending on the availability of data nearby, or on general knowledge. For instance, the porosity in Pliocene turbidites along a passive margin with a limited burial (<2000 m) can be expected high, for instance, in a 25–35% range. Other more precise parameters like permeability are usually not quantified at this stage, and economics are run using an overall recovery factor.

4.3.2. Reservoir characterization during appraisal

If the prospect is drilled, the exploration well will provide a calibration point for the evaluation of all parameters, and also for a much more precise identification of architectural elements. The vertical distribution of reservoirs will be known precisely, at well location at least, as well as the porosity of the penetrated interval. For instance, the expected 30% reservoir out of 150 m may become in the well four 50-m-thick channel complexes separated by 10-m-thick shales (of course, neither the sands nor the total thickness of the reservoir series needs to add up to what was predicted initially).

Other key reservoir parameters like permeability will most of the time be made available by coring or pressure testing. These data will allow to significantly reduce the uncertainty on most of the previous parameters. Most of the time, the reservoirs observed in the well will be redefined as isolated or combined architectural elements, for which a geometry can be guessed. In case of success, new seismic may be acquired, or existing data may be reprocessed to enhance their quality at reservoir level.

As drilling proceeds and provides new data, 3D reservoir models are normally built and populated with geometries, porosities and permeabilities observed in the wells, as described in Section 2.5.2 above. They are updated periodically as new data become available. Leaving the details aside, the approach consists in populating a 3D grid with architectural elements at a scale compatible with the dataset, then defining a distribution of reservoir properties for each element and filling each cell with those properties. The grid fill is governed by geostatistics so that it honours simultaneously the expected distribution, the expected rate of real and vertical variation and well data in the drilled cells of the model.

4.3.3. Reservoir characterisation during production

When production starts, the main difference with the previous stages is that reservoir-scale 'dynamic' data become available (dynamic in the sense, communications between reservoir intervals that could previously only be guessed become visible through production history). The evolution of pressures and flows with time allow to 'see' the whole reservoir and naturally lead to a redefinition of the 3D model. The normal process is to iterate between geological (and other) parameters and the actual observed production until the simulated production matches well observations.

ACKNOWLEDGEMENTS

TGS is gratefully acknowledged for allowing to publish material from their multi-client seismic database of the Gulf of Mexico (Figs. 10.1, 10.11, 10.12, 10.13).

Many colleagues at Total contributed to the examples shown in this book and are acknowledged here: Martine Bez, Jacqueline Camy-Peyret, Douba (Cheikh Ahmed Ould Ahmed Benan), Jean-Bernard Joubert, Richard Labourdette, François Temple.

REFERENCES

Allen, J.R.L., 2001. Principles of Physical Sedimentology. Blackburn, Caldwell, NJ 272pp.

Anderskouv, K., Damholt, T., Surlyk, F., 2007. Late Maastrichtian chalk mounds, Stevns Klint, Denmark—combined physical and biogenic structures. Sediment. Geol. 200, 57–72.

Bouma, A.H., Coleman, J.M., Meyer, A.W., 1986. Introduction, objectives and principal results of deep sea drilling project leg 96, in deep sea drilling project initial report, Bouma, A.H., 96, pp. 15–36, 39–421.

Braccini, E., Penna, E., 2005. Sand Injections in Angola Deep Offshore. SPWLA meeting presentation, Luanda.

Brown, A.R., 2004. Interpretation of Three-Dimensional Seismic Data, sixth ed. AAPG Memoir 42 and SEG Investigations in Geophysics,Tulsa, OK, USA, vol. 9. Chapter 1: Introduction, pp. 1–25.

Carey S.N., Schneider, J.-L., 2011. Volcaniclastic processes and deposits in the deep-sea. In: Hüneke, H., Mulder, T. (Eds.), Deep-Sea Sediments. Developments in Sedimentology, Elsevier, Amsterdam, vol. 63, pp. 457–515.

Clark, J.D., Pickering, K.T., 1996. Architectural elements and growth patterns of submarine channels: application to hydrocarbon exploration, AAPG Bull. Tulsa, OK, USA, vol. 80 (2), 194–220.

Campbell, C.V., 1967. Lamina, laminaset, bed and bedset. Sedimentology 8, 7–26.

Cartwright, J.A., 1994. Episodic basin-wide hydrofracturing of overpressured Early Cenozoic mudrock sequences in the North Sea Basin. Mar. Petrol. Geol. 11, 587–607.

Cartwright, J.A., James, D., Huuse, M., Vetel, W., Hurst, A., 2008. The geometry and emplacement of conical sandstone intrusions. J. Struct. Geol. 30 (7), 854–867.

Damuth, J.E., Flood, R.D., Kowsmann, R.O., Belderson, R.H., Gorini, M.A., 1988. Anatomy and growth pattern of Amazon Deep-Sea Fan as revealed by long-range side-scan sonar (GLORIA) and high-resolution seismic studies. AAPG Bull. Tulsa, OK, USA, 72, 885–911.

Darwin, C.R., 1910. Geological Observations on Coral Reefs, Volcanic Islands, and on South America: Being the Geology of the Voyage of the Beagle, Under the Command of Capt. FitzRoy, During the Years 1832-36. Ward Lock & Co, London, Melbourne & Toronto [first published Smith, Elder & Co., London, 1846]. Chapter 6, 423pp.

Davis, W.M., 1889. The rivers and valleys of Pennsylvania. Natl. Geogr. Mag. 1, 183–253.

Dixon, R.J., Schofield, K., Anderton, R., Reynolds, A.D., Alexander, R.W.S., Williams, M.C., Davies, K.G., 1995. Sandstone diapirism and clastic intrusion in the Tertiary sub-marine fans of the Bruce-Beryl Embayment, Quadrant 9, UKCS. In: Hartley, A.J., Prosser, D.J. (Eds.), Characterization of Deep Marine Clastic Systems. Geological Society, London, Special Publication 94, 77–94.

Enjolras, J.M., Gouadain, J., Mutti, E., Pizon, J., 1986. New turbidite model for the lower tertiary sands in the South Viking Graben. In: Habitat of Hydrocarbons on the Norwegian continental shelf. N.P.S. Graham & Trotman, London, pp. 171–178.

Faugères, J.-C., Mulder, T., 2011. Contour currents and contourite drifts. In: Hüneke, H., Mulder, T. (Eds.), Deep-Sea Sediments. Developments in Sedimentology, Elsevier, Amsterdam, vol. 63, pp. 149–214.

Flood, R.D., Damuth, J.E., 1987. Quantitative characteristics of sinuous distributary channels on the Amazon Deep-Sea Fan. Geol. Soc. Am. Bull. 98, 728–738.

Friedman, B., 2006. Origin form massive yellowstone volcanoes? ash deposits can be deceiving. AAPG Explorer. Tulsa, OK, USA, 28 Dec.

Gardner, M.H., Borer, J.M., 2000. Submarine channel architecture along a slope to basin profile, brushy canyon formation, West Texas. In: Bouma, A.H., Stone, C.G. (Eds.), Fine-Grained Turbidite Systems, AAPG Memoir 72/SEPM. Tulsa, OK, USA, Spec. Publ. No. 68, 195–211, Chapter 19.

Gonthier, E., Faugeres, J.C., Viana, A., Figueiredo, A., Anschutz, P., 2003. Upper Quaternary deposits on the Sao Tome deep-sea channel levee system (South Brazilian Basin): major turbidite versus contourite processes. Mar. Geol. 199 (1–2), 159–180.

Grajales-Nishimura, J.M., Cedillo-Pardo, E., Rosales-Domínguez, C., Morán-Zenteno, D.J., Alvarez, W., et al., 2000. Chicxulub impact: the origin of reservoir and seal facies in the southeastern Mexico oil fields. Geology 28 (4), 307–310.

Héritier, F.E., Lossel, P. and Wathne, E., 1979. Frigg field – large submarine-fan trap in lower Eocene rocks of North Sea Viking Graben. AAPG Bull. Tulsa, OK, USA, vol. 63 (11), 1999–2020.

Hesse, R., Schacht, U., 2011. Early diagenesis of deep-sea sediments. In: Hüneke, H., Mulder, T. (Eds.), Deep-Sea Sediments. Developments in Sedimentology, Elsevier, Amsterdam, vol. 63, pp. 557–714.

Hovland, M., Judd, A.G., 1988. Seabed Pockmarks and Seepages. Graham and Trotman, London, 293pp.

Hüneke, H., Henrich, R., 2011. Pelagic sedimentation in modern and ancient oceans. In: Hüneke, H., Mulder, T. (Eds.), Deep-Sea Sediments. Developments in Sedimentology, Elsevier, Amsterdam, vol. 63, pp. 215–351.

Hurst, A., Cartwright, J.A. (Eds.), 2007. Sand Injectites: Implications for Hydrocarbon Exploration, AAPG Memoir. Tulsa, OK, USA, vol. 87, 288.

Jackson, C.A.-L., 2007. The geometry, distribution, and development of clastic injections in slope systems: seismic examples from the Upper Cretaceous Kyrre Formation, Måløy slope, Norwegian margin. In: Hurst, A., Cartwright, J. (Eds.), Sand Injectites: Implications for Hydrocarbon Exploration and Production, AAPG Memoir. Tulsa, OK, USA, vol. 87, 37–48.

Judd, A.G., Hovland, M., 2007. Seabed fluid flow: the impact on geology, biology and the marine environment. Cambridge University Press, Cambridge, UK, 475p.

Kolla, V., Bourges, Ph., Urruty, J.-M., Safa, P., 2001. Evolution of deep-water Tertiary sinuous channels offshore Angola (west Africa) and implications for reservoir architecture, AAPG Bull. Tulsa, OK, USA, vol. 85 (8), 1373–1405.

Kneller, B.C., Branney, M.J., 1995. Sustained high-density turbidity currents and the deposition of thick massive beds. Sedimentology 42, 607–616.

Kvenvolden, K.A., 1993. Gas hydrates-geological perspective and global change. Rev. Geophys. 31, 173–187.

Labourdette, R., Crumeyrolle, P., Remacha, E., 2007. Characterisation of dynamic flow patterns in turbidite reservoirs using 3D outcrop analogues: example of the Eocene Morillo turbidite system (south-central Pyrenees, Spain). Mar. Petrol. Geol. 25 (3), 255–270.

Magoon, L.B., Hudson, T., Cook, H., 2001. Pimienta-Tamabra(!)—a giant supercharged petroleum system in the southern Gulf of Mexico, onshore and offshore Mexico. In: Bartolini, Claudio, Buffler, Richard T., Cantú-Chapa, Abelardo (Eds.), The Western Gulf of Mexico Basin: Tectonics, Sedimentary Basins, and Petroleum Systems, AAPG Memoir. Tulsa, OK, USA, vol. 75, 83–125, Chapter 4.

Miall, A.D., 1985. Architectural element analysis: a new method of facies analysis applied to fluvial deposits. Earth Sci. Rev. 22, 261–308.
Migeon, S., 2000. Dunes Géantes et levées Sédimentaires en domaine marin profond: Approches Morphologique, Sismique et Sédimentologique. Thèse, Université Bordeaux I, 288 p.
Migeon, S., Savoye, B., Zanella, E., Mulder, T., Faugères, J.-C., Weber, O., 2001. Detailed seismic-reflection and sedimentary study of turbidite sediment waves on the Var sedimentary Ridge (SE France): significance for sediment transport and deposition and for the mechanisms of sediment-waves construction. Mar. Petrol. Geol. 18, 179–208.
Moraes, M.A.S., Maciel, W.B., Braga, M.S.S., Viana, A.R., 2007. Bottom-current reworked Palaeocene–Eocene deep-water reservoirs of the Campos Basin, Brazil. In: Viana, A.R., Rebesco, M. (Eds.), Economic and Palaeoceanographic Significance of Contourite Deposits. Geological Society, London, Special Publications, 27, pp.81–94.
Mulder, T., 2011. Gravity processes on continental slope, rise and abyssal plains. In: Hüneke, H., Mulder, T. (Eds.), Deep-Sea Sediments. Developments in Sedimentology, Elsevier, Amsterdam, vol. 63, pp. 25–148.
Mulder, T., Alexander, J., 2001. The physical character of subaqueous sedimentary density currenys qand their deposits. Sedimentology, 48, 269–299.
Mulder, T., Hüneke, H., van Loon, A.J. 2011. Progress in Deep-Sea sedimentology. In: Hüneke, H., Mulder, T. (Eds.), Deep-Sea Sediments. Developments in Sedimentology, Elsevier, Amsterdam, vol. 63, pp. 1–24.
Mutti, E., 1985. Hecho turbidite system, Spain. In: Normark, W.R., Bouma, A.H., Barnes, N.E. (Eds.), Submarine Fans and Related Turbidite Systems. Springer-Verlag, New York, pp. 205–208.
Mutti, E., 1992. Turbidite Sandstones. Agip (Agenzia Generale Italiana Petroli), San Donato Milanese, 275 p.
Mutti, E., Bernoulli, D., Ricci Lucchi, F. ,Tinterri, R., 2009. Turbidites and turbidity currents from Alpine 'flysch' to the exploration of continental margins. Sedimentology 56, 267–318.
Mutti, E., Normark, W.R., 1987. Comparing examples of modern and ancient turbidite systems: problems and concepts. In: Legget, J.K., Zuffa, G.G. (Eds.), Deep Water Clastic Deposits: models and Case Histories. Graham and Trotman, London, pp. 1–38.
Mutti, E., Normark, W.R., 1991. An integrated approach to the study of turbidite systems. In: Weimer, P., Link, H., (Eds.), Seismic facies and sedimentary processes of submarine fans and turbidite systems. Graham and Trotman, London, pp. 1-38.
Nardin, T.R., Hein, F.J., Gorsline, D.S., Edwards, B.D., 1979. A review of mass movement processes, sediment and acoustic characteristics, and contrasts in slope and base-of-slope systems versus canyon-fan-basin floor systems. In: Doyle, L.J., Pilkey, O.H. (Eds.), Geology of Continental Slopes, SEPM, Special Publications, 27, pp.61–73.
Pacht, J.A., Bowen, B.E., Beard, J.H., Shaffer, L., 1990. Sequence-stratigraphy of Plio-Pleistocene depositional facies in the offshore Louisiana south additions. Gulf Coast Assoc. Geol. Soc. Trans. 4, 643–659.
Parize, O., 1988. Sills et dykes gréseux sédimentaires: Paléomorphologie, fracturation précoce, injection et compaction. Thése Doctorat Géologie, Ėcole Nationale Supérieure des Mines de Paris–Université Lille I: Mémoire des Sciences de la Terre, Ecole des Mines de Paris, n 7, 333pp.
Parize, O., Beaudoin, B., Eckert, S., Friès, G., Hadj-Hassen, F., Schneider, F., et al., 2007. The Vocontian Aptian and Albian syndepositional clastic sills and dikes: a field-based mechanical approach to predict and model the early fracturing of marly-limy sediments. In: Hurst, A., Cartwright, J. (Eds.), Sand Injectites: Implications for Hydrocarbon Exploration and Production, AAPG Tulsa, OK, USA, 87, 163–172.

Posamentier, H.W., Vail, P.R., 1988. Eustatic control on clastic deposition II-sequence and systems tracts models. In: Wilgus, C.K. et al., (Eds.), Sea level changes: an integrated approach, SEPM Special Publication, 42,125–154.
Reading, H.G., Richards, M., 1994. Turbidite systems in deep-water basin margins classified by grain size and feeder system. AAPG Bull. Tulsa, OK, USA, vol. 78 (5), 792–822.
Rigollet, C., 2001. Valorisation sédimentologique de l'information sismique: application au comblement des bassins profonds de la marge atlantique nord-européenne du Crétacé à l'Actuel. unpublished PhD thesis, Ecole des Mines de Paris, 507 pp.
Shanmugam, G., Spalding, T.D., Rofheart, D.H., 1993. Process sedimentology and reservoir quality of deep-marine bottom-current reworked sands (sandy contourites): an example from the Gulf of Mexico. AAPG Bull. Tulsa, OK, USA, vol. 77 (7), 1241–1259.
Shanmugam, G., Bloch, R.B., Mitchell, S.M., Beamish, G.W.J., Hodgkinson, R.J., Damuth, J.E., et al., 1995. Basin-floor fans in the north sea: sequence stratigraphic models vs. sedimentary facies. AAPG Bull. Tulsa, OK, USA, vol. 79 (4), 477–511.
Shepherd, M., 2009. Less common reservoir types. In Shepherd, M., Oil field production Geology, AAPG Memoir. Tulsa, OK, USA, vol. 91, 311–312.
Shoulders, S., Cartwright, J.A., 2004. Constraining the depth and timing of large-scale conical sandstone intrusions. Geology 32, 661–665.
Smith, R.B., Siegel, L.J., 2000. Windows into the Earth: The Geologic Story of Yellowstone and Grand Teton National Parks. Oxford University Press, New York, 256p.
Stow, D.A.V. Mayall, M., 2000. Deep-water sedimentary systems: New models for the 21st Century, in Marine and Petroleum Geology, vol. 17(2), 125–135.
Vail, P.R., Mitchum Jr. R.M., Thompson III, S., 1977. Seismic stratigraphy and global changes of sea level: part 3. relative changes of sea level from coastal onlap: section 2. application of seismic reflection configuration to stratigraphic interpretation. In: Payton, C.E. (Ed.), Seismic Stratigraphy-Applications to Hydrocarbon Exploration, AAPG Memoir. Tulsa, OK, USA, vol. 26, 63–81.
Van Wagoner, J.C., Mitchum Jr. R.M., Posamentier, H.W., Vail, P.R., 1987. Seismic stratigraphy interpretation using sequence stratigraphy: part 2: key definitions of sequence stratigraphy. In: Bally, A.W. (Ed.), Atlas of Seismic Stratigraphy, AAPG Studies in Geology No 27, Tulsa, OK, USA, vol. 1, 11–14.
Viana, A., 2001. Seismic expression of shallow- to deep-water contourites along the south-eastern Brazilian margin. Mar. Geophys. Res. 22, 509–521.
Viana, A.R., Almeida Jr. W., Nunes, M.C.V., Bulhões, E.M., 2007. The economic importance of contourites. In: Viana, A.R., Rebesco, M. (Eds.), Economic and Paleogeographic Significance of Contourite Deposits. Geological Society, London, Special Publications, 276, 1–23.
Weimer, P., 1989. Sequence stratigraphy of the Mississippi Fan (Plio-Pleistocene), Gulf of Mexico: Geo-Marine Letters, v. 9, 185–272.
Wheeler, A.J., Stadnitskaia, A., 2011. Benthic deep-sea carbonates: reefs and seeps. In: Hüneke, H., Mulder, T. (Eds.), Deep-Sea Sediments. Developments in Sedimentology, Elsevier, Amsterdam, vol. 63, pp. 397–455.

CHAPTER 11

Mesozoic Pelagic Sediments: Archives for Ocean and Climate History during Green-House Conditions

Helmut Weissert

Contents

1. Introduction

Pioneering studies of recent oceanic deposits and of deep-sea sediments in mountain ranges date back more than 100 years. Land-locked geologists at that time already could benefit from findings of sea-borne experts. A thorough understanding of sedimentation processes in modern oceans was needed before depositional conditions of fossil pelagic and other deep-sea sediments in oceans and mountain ranges could be reconstructed. In this study, I will trace the close interaction between marine- and land-based geologists since the late nineteenth century when deep-sea sediments

Department of Earth Sciences, ETH-Z, Zürich, Switzerland
E-mail address: helmut.weissert@erdw.ethz.ch

Developments in Sedimentology, Volume 63
ISSN 0070-4571, DOI: 10.1016/S0070-4571(11)63011-2

were sampled from all the major oceans during the Challenger expedition (1872–1876). Alpine Mesozoic sediments were identified as deep-sea sediments at the turn of the twentieth century. It took another 50 years until these sediments became recognised as important archives of basin evolution and fingerprints of Mesozoic oceanography. The development of a new research field, today known as 'palaeoceanography', was initiated through the international 'Deep Sea Drilling Project' (DSDP) and its successor projects, ODP and IODP (Ocean Drilling Project and Integrated Ocean Drilling Project, respectively). By using Mesozoic deep-sea sediments from the Alpine mountain range as an example, I will demonstrate how an integrated sedimentological, micropalaeontological and geochemical approach, combined with an 'earth-systems concept', resulted in a new understanding of the evolution of oceans, life and climate through time.

2. Oceans Explored

In his book 'Les grand navigateurs du XVIIIe siècle' (1879), Jules Verne tells us the fascinating story of the famous and courageous world explorers and their voyages across the wide oceans into unknown territories. At the time Jules Verne wrote this book, the golden age of the great explorers of the world was almost over. Undiscovered remained the deep abyss of the sea. Until the mid-nineteenth century, few expeditions chose the oceans as the prime target for their investigations. Little was known on water chemistry, on life in the deep oceans, and on current patterns, and even less about deep-sea sediments, when John Murray and Charles Thompson started their great expedition in 1872. Their British navy ship, HMS *Challenger*, brought them across all the major oceans, and when the researchers returned to England in 1876, they came back with water samples, with numerous measurements and with dredged sediments from 362 stations in all the oceans around the globe. The team combined information on water depth and temperatures with measurements of current intensity and with distribution maps of deep-sea sedimentary facies. Sediment samples were stored and studied at the University of Edinburgh. John Murray and Alphonse François Renard (1891) described the findings of the cruise as 'the greatest advance in the knowledge of our planet since the celebrated discoveries of the fifteenth and sixteenth centuries'. The results of these expeditions were published between 1885 and 1895 in a report consisting of 50 volumes (see Thomson and Murray, 1911).

Geologists studying sediments in mountain ranges learned about the distribution of sediments in modern oceans. The Challenger expedition showed how deep-sea sediments vary through oceans and through different water depths. Shell material was missing at the deepest sites. Aragonite shells disappeared at shallower water depths than calcitic remnants. Ocean chemistry seemed to control sediment types. The findings of the Challenger expedition made it possible that a first sound concept of deep-sea sedimentation could be formulated at the very end of the nineteenth century.

The *Challenger* results turned out to be of great importance for geology. Geologists had recognised long ago that thick marine sedimentary successions formed important parts of mountain ranges. These sediments were considered as deposits formed in a deep ocean. The American geologist Dana (see Dott, 1997) introduced the term 'geosyncline' for these disappeared ocean basins. Neumayr in Vienna proposed that the 'geosyncline sediments' of Alps and Himalaya were deposited in a 'central Mediterranean' extending from the North American continent through Alps and Mediterranean into the Himalayas (Neumayr, 1883). This ocean was later given the name 'Tethys' by Eduard Suess (1885). Of course, the main questions remained to be answered: what type of ocean was frozen into the mountain ranges? Could the sedimentary facies provide information on the oceanic environment? A better understanding of open-marine sedimentary facies became possible only with the findings of the Challenger expedition. The Austrian geologist Theodor Fuchs recognised as early as 1877 that aragonite shells were absent in the alpine Late Jurassic and Early Cretaceous Aptychus Limestones (Fuchs, 1877). Based on findings of the *Challenger,* he interpreted this absence as evidence for aragonite dissolution in deep water. He concluded that these limestones were formed in a bathyal environment (see also Jenkyns and Hsü, 1974). A few years later, Fuchs (1898) stated that any interpretation of alpine sediments is based on an understanding of sedimentation processes in the modern oceans. Fuchs can be considered as one of the founders of 'Alpine palaeoceanography': 'Es kann ja principiell gewiss keinem Zweifel unterliegen, dass das eigentliche Studienobjekt des Geologen das Festland ist und nicht das Meer, aber ebenso gewiss ist wohl auch, dass ein sehr grosser Theil des Festlandes im Meere gebildet wurde und dass eine Kenntnis und Beurtheilung der sedimentären Formationen nicht ohne genaue Kenntnis der heutigen Meeresablagerungen ein Ding der Unmöglichkeit ist' (Fuchs, 1898).

3. Deep-Sea Sediments: From Oceans to Mountain Ranges

With the findings of the Challenger expedition and with the new developments in alpine tectonics, the promising beginning for a new era in geological sciences was set at the turn to the twentieth century. Following the recommendation by Fuchs (1898), Alpine sediments were studied and compared with the sediments sampled during the Challenger expedition, which Murray and Renard (1891) had described in their report on 'The deep-sea deposits based on the specimens collected during the voyage of HMS Challenger in the years 1872–1876'. Haug (1900), Steinmann (1905) and Heim (1924) were among the Alpine geologists who used the findings of the Challenger expedition for their new interpretation of Alpine sediments. Steinmann (1905, 1925) interpreted Mesozoic radiolarian cherts as deep-sea deposits. With this interpretation he disagreed with John Murray, who could not see any similarities between sediments in mountain ranges and the deep-sea sediments he had collected during the Challenger expedition (see discussion in Jenkyns and Hsü, 1974).

Arnold Heim (1924) investigated an alpine Tethys shelf succession of Cretaceous age. He attempted to reconstruct the physical oceanography for the Early Cretaceous Tethys Ocean and he compared fine-grained Tethyan deposits with 'Challenger sediments'. He interpreted condensed sediments and phosphatized hardgrounds as products of strong shelf-current activity. With his remarkable study, Arnold Heim can be considered as one of the first physical palaeoceanographers. Heim also discussed the possible source of fine-grained, micritic carbonate in Cretaceous shelf deposits considered as chemical precipitates. This interpretation was falsified in later investigations of Cretaceous neritic and pelagic limestones in the Swiss Alps.

The occurrence of deep-sea sediments in mountain belts could best be explained with a mobilistic concept in geology. Many Alpine geologists therefore favoured the early continental-drift theory by Alfred Wegener. The rejection of Wegener's continental-drift theory left these geologists with no convincing model, which could have explained the occurrence of deep-sea sediments in the Alpine nappe pile. Therefore, it seems not surprising that only few authors supported the proposition by Steinmann and others that deep-sea sediments existed in the Alps (e.g. Tercier, 1939). Interest in a further investigation of these sediments remained small and authors continued to question the occurrence of deep-sea sediments in the Alps and in other mountain chains. Grunau (1947) proposed that all the Alpine radiolarite deposits, considered as deep-sea deposits by Steinmann (1905), were formed under shallow-water conditions. Jenkyns and Hsü (1974) expressed the opinion that part of the long-lasting controversy on deep-sea sediments was of semantic nature. Today, the term 'deep-sea sediment' is often replaced by the term 'pelagic sediment', a term that has no depth connotation. Pelagic sediments are defined as sediments with less than 25% particles >5 μm from neritic, continental or volcanic sources and their depth of deposition is debatable, normally over 200 m (cf. Hüneke and Henrich, 2011, this volume; Henrich and Hüneke, 2011, this volume).

The beginning of a new era in ocean and sediment research dates back into the middle of the twentieth century. World War II had a profound impact on marine research. The American Navy started to use sound transmission in search for hostile submarines in their anti-submarine warfare. This method was later developed into echo sounding and into marine seismics. Bathymetric mapping on a global scale became possible. The travel time of acoustic signals could be used for estimating the water depth. Maurice Ewing, a Lamont-based seismologist, combined this method with a seismic-reflection system, where hydrophones were linked with a low-frequency airgun as a sound source. Ewing was now able to investigate seafloor and sub-seafloor conditions in the worlds oceans and, together with Bruce Heezen, he was the first who detected the continuity of mid-ocean ridges (Ewing and Heezen, 1960). Marine-explosion seismology was further developed and new data on sediment thickness and crustal structure were acquired. The investigation of the modern seafloor culminated in the discovery of seafloor spreading (Dietz, 1961) and in the formulation of the theory of plate tectonics (e.g. Vine, 1966), which resulted in a major paradigm change in earth sciences.

New technical developments were not limited to geophysics. Mass-spectrometry was introduced as a new method in geochemistry. Harold Urey and his Chicago research group started the rapidly growing research field of stable-isotope geochemistry (Urey et al., 1951). Pelagic sediments were investigated for their isotope geochemical composition, and Cesare Emiliani (1954, 1955) presented the first palaeoclimate record based on oxygen-isotope analyses of benthic foraminifers in pelagic sediments from the Pacific Ocean.

The paradigm change in the earth sciences triggered by the discovery of seafloor spreading lead to the start of the Deep Sea Drilling Programme (DSDP). It started 1968 with the newly built drilling vessel, *Glomar Challenger.* The new international project revolutionised marine geology and the investigation of the history of the oceans and of global climate (see Berger, 1974; Hsü, 1992; Hsu and Jenkyns, 1974).

4. Pelagic Sediments—A New Field of Research for Sedimentologists and Stratigraphers

The volume 'Pelagic sediments: on land and under the sea' (Hsu and Jenkyns, 1974) marks an outstanding starting point of modern deep-sea sedimentology and the investigation of pelagic sediments and sedimentary rocks. The volume contains contributions on deep-sea sedimentation and the fluctuating carbonate line (Berger and Winterer, 1974), on the origin of pelagic nodular limestones (e.g. Jenkyns, 1974) and on the diagenesis of pelagic limestones (e.g. Schlanger and Douglas, 1974) and of chert (e.g. Calvert, 1974; Cecca et al., 1994), on benthic life and on the formation of manganese nodules (Wendt, 1974).

The questions addressed in that volume mirror many of the major research topics in sedimentology of pelagic or deep-sea deposits since the middle of the twentieth century. One of the basic questions in deep-sea sedimentology concerns the mineralogical composition and the biogenic source of rock-forming particles (Figs. 11.1–11.3). While pelagic limestones of Late Jurassic and younger age consist of calcareous nannoplankton and planktic foraminifera as major rock-forming constituents, pre-Late Jurassic pelagic carbonates contain significant amounts of micrite. Some authors interpret pelagic micrite as 'peri-platform ooze' formed as 'automicrite' on platforms and transported by currents from carbonate platforms into pelagic realms (e.g. Reuning et al., 2005). Considerable amounts of aragonitic peri-platform ooze are observed today in basinal sediments located at distances of 50–100 km from carbonate environments. ODP Site 1066, located west of the Bahamas carbonate platform, illustrates the importance of peri-platform

Figure 11.1 Pelagic sedimentary rocks, Jurassic-Early Cretaceous. (A) Rosso Ammonitico, nodular limestone and marlstone, Southern Alps, Breggia Gorge (Switzerland): Cyclic sedimentation with 21 ka precession cycle (Weedon, 1989). (B) Radiolarian chert, part of alpine ophiolite succession, Klosters (Eastern Switzerland). (C) Bioturbated radiolarian chert (S. Alps, Breggia Gorge, Switzerland), non-bioturbated reworked horizons were formed by bottom currents. (D) Chert layer, Maiolica Formation (S. Alps), completely silicified lower part and incomplete silicification of the upper part formed by current reworking of radiolarian ooze. (A multi-colour version of this figure is on the included CD-ROM.)

Figure 11.2 Early Cretaceous pelagic limestone, Maiolica Formation, S. Alps, N. Italy. (A) Maiolica Formation (Pusiano section, N. Italy; see Bersezio and Fornaciari, 1987). (B) Close-up of Maiolica Formation with white limestone beds and chert layers (S. Alps, N. Italy). (C) SEM picture, nannofossil limestone, Maiolica Formation, S. Alps. Length of white bar: 10 μm. (D) Slump fold, Maiolica Formation, S. Alps, N. Italy. (A multi-colour version of this figure is on the included CD-ROM.)

ooze as an ooze- or rock-forming constituent in sediments otherwise considered as pelagic sediments (Eberli et al., 1997). Westphal (2006) believes that many of the fine-grained hemipelagic limestones in the geological record originated from aragonite ooze which was altered to low-Mg calcite during early diagenesis. She suggests that hemipelagic limestone/marl successions are the result of varying precursor mineralogy of the micrite (see Hüneke and Henrich, 2011, this volume; Henrich and Hüneke, 2011, this volume). Other authors interpret micrite as a biogenic carbonate with an open-marine planktic source. In some of the micrites, well-preserved calcispheres have been identified (e.g. Kälin and Bernoulli, 1984). These observations support the hypothesis that most of pelagic micrite deposited at distances beyond 100 km from carbonate platforms is of planktic origin, and it is possible that some of the micrite may have been formed by picoplanktic cyanobacteria (e.g. Robbins and Blackwelder, 1992).

A record of benthic life in deep-water environments is stored in the bioturbation of pelagic (Figs. 11.1C and 11.3C; see also Uchman and Wetzel, 2011, this volume). Researchers investigated benthic life in pelagic environments and they recognised that pelagic sediments deposited

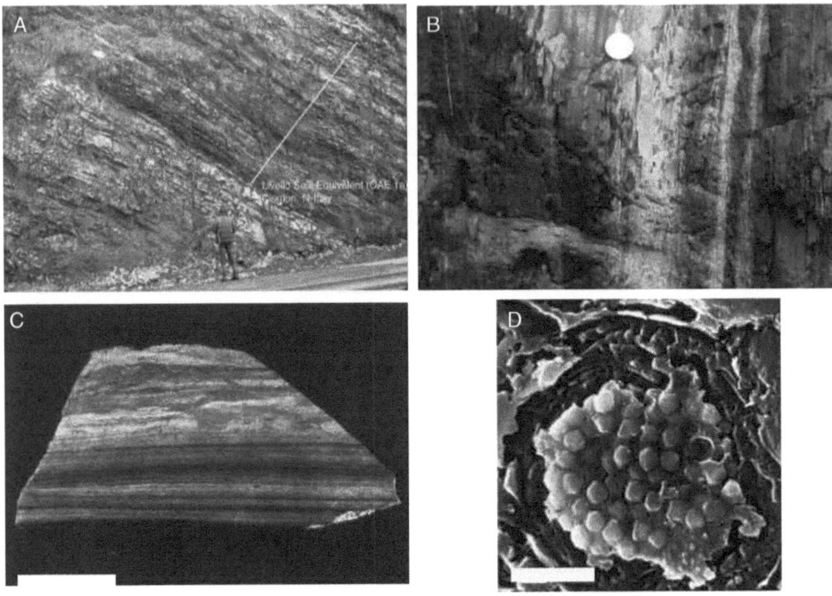

Figure 11.3 Early Cretaceous black shales. (A) OAE1a or Livello Selli Equivalent in contact with older Maiolica Formation, Cismon (N. Italy). (B) Black shales and intercalated pelagic limestone, Barremian, Maiolica Formation (S. Alps, N. Italy). (C) Transition from a laminated black marlstone to a bioturbated pelagic limestone, Maiolica Formation, Barremian (N. Italy): SEM photograph of framoidal pyrite filling a coccolith. White bar: 3 μm (Maiolica Formation, S. Alps, N. Italy). (A multi-colour version of this figure is on the included CD-ROM.)

under oxidising conditions with mean sedimentation rates of some mm per 1000 years are intensely bioturbated (e.g. Berger and Killingley, 1982). Burrowing organisms mix sediment down to depths of 10–20 cm. Bioturbation diminishes the stratigraphic resolution of pelagic archives; this poses a problem for high-resolution investigations ($<10^4$ years) in palaeoceanography (e.g. Berger and Killingley, 1982).

Even if pelagic sediments are considered as very good archives for palaeoceanography, discontinuous sedimentation resulting in sedimentary gaps can result in incomplete climate records. Sedimentary gaps can be caused by winnowing of pelagic sediments by currents (e.g. Nisbet and Price, 1974) or by erosion along submarine slopes (Figs. 11.1D and 11.2D; e.g. Bersezio and Fornaciari, 1987; Weissert, 1981a,b). Strong shelf or bottom currents can sort or redistribute pelagic and deep-sea sediments (Fig. 11.2). Hardgrounds or mineralized condensed beds can develop on current-swept shelves or on basin floors (e.g. Föllmi, 1996; James et al., 2001). Sediments redistributed by bottom currents or contour currents will form contourites (e.g. Stow et al., 1998; see also Faugères and Mulder, 2011, this volume). Contourite accumulations form drift deposits, which are characterised by sedimentation rates in deep-sea and pelagic environments of up to several tens of centimetres per 1000 years. These drift deposits serve as excellent palaeoceanographic archives since they provide climate records with a high-resolution ($< 10^4$ years) if stratigraphic control is assured (e.g. Mienert et al., 1994).

Bottom-current activity or sediment gravity flows (Fig. 11.2D, see also Mulder, 2011, this volume) can cause discontinuous sedimentation in both pelagic and hemipelagic settings. Resulting gaps in sedimentation are often difficult or impossible to identify with sedimentological methods. These gaps may be identified with a combined biostratigraphic and chemostratigraphic approach (Weissert et al., 2008). Wissler et al. (2003) recognised remarkable differences between two carbon-isotope records measured in a hemipelagic and a pelagic succession of Barremian age. They interpreted a missing positive excursion in the carbon-isotope record of the hemipelagic succession as evidence for a sedimentary gap. Their interpretation is supported by available palaeontological information.

The findings of DSDP and ODP also provided new insight into the diagenesis of pelagic sediments (see also Hesse and Schacht, 2011, this volume). While the diagenesis of shallow-water carbonates was studied already in the 1960s (e.g. Friedman, 1964), the diagenesis of pelagic and deep-sea deposits remained poorly understood before the start of the DSDP. In the first decade of pelagic sedimentology initiated by DSDP, a number of studies focused on the diagenesis of these sediments. The diagenetic pathway from a pelagic carbonate ooze to a limestone (Matter et al., 1975; Wise and Hsü, 1971) was traced in DSDP cores and the stepwise transition from a siliceous ooze to opal-CT and to chert (Fig. 11.1B–D) was reconstructed in deep-sea successions and in chert and porcelanite sediments on land (e.g. Calvert, 1974; Cecca et al., 1994; Pisciotto, 1981).

Sedimentologists first used a process-oriented approach when deciphering pelagic sediments. In a next step, pelagic sedimentary facies were traced through geological time. New developments in stratigraphy were essential for a reconstruction of ocean history at a higher time resolution and for comparison of pelagic sedimentary archives at a global scale. New developments in analytical techniques facilitated the measurement of magnetic properties of pelagic sediments, and a Mesozoic magnetostratigraphy was established in biostratigraphically dated pelagic sediments in the Alps and Apennines (e.g. Channell et al., 1979). Chemostratigraphy and cyclostratigraphy further improved time resolution in Phanerozoic sedimentary archives (Fig. 11.1A; Herbert, 1992; Kuhnt et al., 2005; Strasser et al., 2006; Weissert et al., 2008). Carbon-isotope stratigraphy, combined with biostratigraphy, developed into a very robust stratigraphic correlation tool (e.g. Channell et al., 1993). Pelagic sedimentary records could be correlated at high-resolution with successions from neritic environments or even with continental records (e.g. Gröcke et al., 1999).

5. The Alpine Tethys Succession—From Sedimentology to Palaeoceanography

Some of the best-studied Mesozoic pelagic sediments outcropping on land are the Middle Jurassic to Cretaceous pelagic successions of the Alpine Tethys Ocean (Figs. 11.1–11.3; Bernoulli and Jenkyns, 1974). The succession starts with the Middle Jurassic Rosso Ammonitico Inferiore, a nodular limestone marking the onset of pelagic conditions in a deepening Mesozoic Alpine Tethys Ocean (Fig. 11.1A; Bernoulli, 1964). It is overlain by the Radiolarite Formation (Fig. 11.1B and C), with up to several tens of metres of green and red carbonate-free radiolarites overlain by radiolarian limestones, known as Rosso ad Aptici (Bernoulli, 1964). White micritic limestones forming the Maiolica Formation were deposited in the deep parts of the Alpine Tethys from Tithonian to early Aptian times (Fig. 11.2; Weissert, 1979). Marlstones and siliceous shales of alternating black, green and red colour, described as Scaglia Variegata in the Southern Alps (N. Italy) or Scisti a Fucoidi in the Apennines were formed in the deep mid-Cretaceous Tethys Ocean between the Early Aptian and the Late Albian (Fig. 11.3; Arthur and Premoli-Silva, 1982). These variegated shales are overlain by the white limestones of the Scaglia Bianca and the red limestones of the Scaglia Rossa both rich in planktic foraminifers. They were formed during the Late Cretaceous over wide parts of the deep Alpine Tethys Ocean. The described Tethys succession has been compared with the pelagic sediments drilled by DSDP and ODP in the central North Atlantic by Bernoulli (1972), who recognised striking similarities between Atlantic and Tethys pelagic facies from the Late Jurassic to the Late Cretaceous.

The Mesozoic pelagic sedimentary rocks observed and studied in the Alps and Apennines were first used by geologists as a source of information for Tethys tectonics and the evolution of the 'geosyncline'. Radiolarites served as palaeodepth indicators, and the calcite compensation depth (CCD) was considered to have been fixed at a certain water depth throughout ocean history. Garrison and Fischer (1969) concluded that carbonate-free radiolarian-rich sediments had accumulated in the deepest part of the evolving Tethys geosyncline and that pelagic limestones of the Early Cretaceous record a shallowing of the Tethys basin. Palaeodepth interpretations of the 1960s came into conflict with plate tectonics and with depth reconstructions of evolving oceans based on cooling trends of new ocean crust (see the Sclater curve: Sclater et al., 1971). Ophiolites were reinterpreted as remnants of old oceanic lithosphere in mountain ranges. The ophiolite suite consists of peridotite, gabbro, a sheeted dike complex, basalts and the overlying deep-sea sediments. In the Alps, no complete ophiolite suite is preserved, and widespread serpentinites are interpreted as exhumed continental mantle rocks, which mark a fossilised ocean-continent transition (e.g. Desmurs et al., 2001). Despite of this, sediments accumulated on Jurassic serpentinites or basalts were interpreted as deposits formed in a truly deep-water setting (e.g. Winterer and Bosellini, 1981). Only a few authors questioned depositional depth reconstructions. Folk and McBride (1978), for example, proposed that the Late Jurassic radiolarian cherts deposited on serpentinites were formed in tidal settings and not in the deep sea. New investigations of pelagic sedimentary successions that had accumulated on submarine highs along the Jurassic southern continental margin of the alpine Tethys stimulated new discussions on the depositional environment of these deposits (see Winterer, 1998). Santantonio (1994) classified these submarine highs as pelagic carbonate platforms (PCP's). The submarine highs separating basinal settings along the Tethys margin experienced a rapid deepening controlled by tectonics during the Midde and Late Jurassic. Findings of pennular corals on drowned pelagic carbonates platforms suggest that palaeodepth estimates based on plate stratigraphy may be exaggerated (Gill et al., 2004). These corals colonised deeper parts of the marine photic zone. Such corals found in the Umbria-Marche and Sabina Apennines on top of Late Jurassic submarine highs and at basin margins suggest that some of the Tethyan pelagic sediments were deposited at shallower depths than expected based on plate stratigraphy and considerations of palaeo-CCD depths.

Alpine deep-sea sediments tell the story of a newly evolving seafloor and they contain information on 'palaeoceanography'. Hsü (1976) introduced the term 'palaeoceanography' into Mesozoic geology, 8 years after James Kennett (1968) used the term for the first time in scientific literature. Hsü (1976) combined modern oceanographic concepts with new plate tectonics and he used the Tethyan pelagic trilogy Rosso Ammonitico—Radiolarian chert/limestone—Maiolica limestone as an archive of evolving oceanography. He demonstrated that these deposits were accumulating on a continuously deepening distal continental margin and that radiolarites were formed when the Jurassic CCD was exceptionally shallow in the opening Tethys Ocean. The colour of the radiolarites serves as a source of information on bottom-water chemistry with green radiolarites formed under low-oxygen conditions and red radiolarites representing well-oxygenated deep-water masses. Radiolarians as rock-forming constituents serve as a proxy for nutrient-rich surface waters. Hsü (1976) and later Muttoni et al. (2005) reconstructed the oceanographic setting of radiolarite formation using palaeomagnetic information. The time of radiolarite formation coincided with a time of the opening Tethys seaway. Both Hsü (1976) and Muttoni et al. (2005) concluded that radiolarites are typically deposits formed under equatorial upwelling conditions. Weissert (1979) investigated the change from Radiolarite Formation to the nannofossil limestones of the Maiolica

Formation. He proposed that this change was related to a complete reorganisation of Tethyan oceanography. Upwelling conditions favouring radiolarian fauna were replaced by a gyre-controlled circulation pattern according to his interpretation. Low-productivity conditions favoured the expansion of calcareous nannoplankton, which became dominant as a rock-forming constituent. Decreased pelagic rain resulted in a drop of the CCD in the Tethys and the Atlantic Ocean.

The transition from siliceous sedimentation to carbonate sedimentation started in the middle Oxfordian. Louis-Schmid et al. (2007) and Rais et al. (2007) identified the *transversarium* ammonite zone as a turning point in Mesozoic palaeoceanography. The decrease in equatorial upwelling resulting in the end of the carbonate-free radiolarian ooze deposition that coincided with a weakening of upwelling along the northern margin of the Alpine Tethys. There, the Callovian–Oxfordian 'Terres Noires' sediments, enriched in organic matter, were replaced by the mid- to late-Oxfordian hemipelagic marlstones and limestones of the 'Argovian' (Fig. 11.4; Louis-Schmid et al., 2007). At the same time, the shelf-current intensity dropped along the northern Tethys shelf, and condensed sediments—formed under strong contour-current activity—were replaced by hemipelagic marlstones of late Oxfordian–Kimmeridigian age (Rais et al., 2007). The Tethys-wide changes in sedimentation coincided with a major reorganisation of the climate and with the establishment of a stable low-latitude ocean circulation pattern with weak upwelling and a stable thermocline. This stable 'Maiolica' ocean mode persisted from the Tithonian to the Valanginian. Several major perturbations of the carbon cycle between the Valanginian and Aptian triggered by volcanism shifted the Maiolica mode repeatedly into a black-shale ocean mode. Louis-Schmid et al. (2007) suggest that the deepening of the Hispanic corridor resulted in the formation of a robust circum-equatorial current system with weak upwelling activity favouring the expansion of calcareous nannoplankton into pelagic and even deep-sea environments. The Maiolica ocean mode resembles the permanent El Niño-like conditions during the Pliocene warm epoch in the Pacific (Wara et al., 2005).

6. Stable-Isotope Geochemistry—A New Tool in Palaeoceanography

In the early 1950s, when Cesare Emiliani started his oxygen-isotope studies in marine sediments, palaeoclimatologists were in search for a solution regarding a long-lasting controversy on causes of observed palaeoclimate change. Evidence for at least four ice ages had been found in continental records but no data existed which could confirm Milankovich's hypothesis of orbitally driven cyclic climate change in the Quaternary.

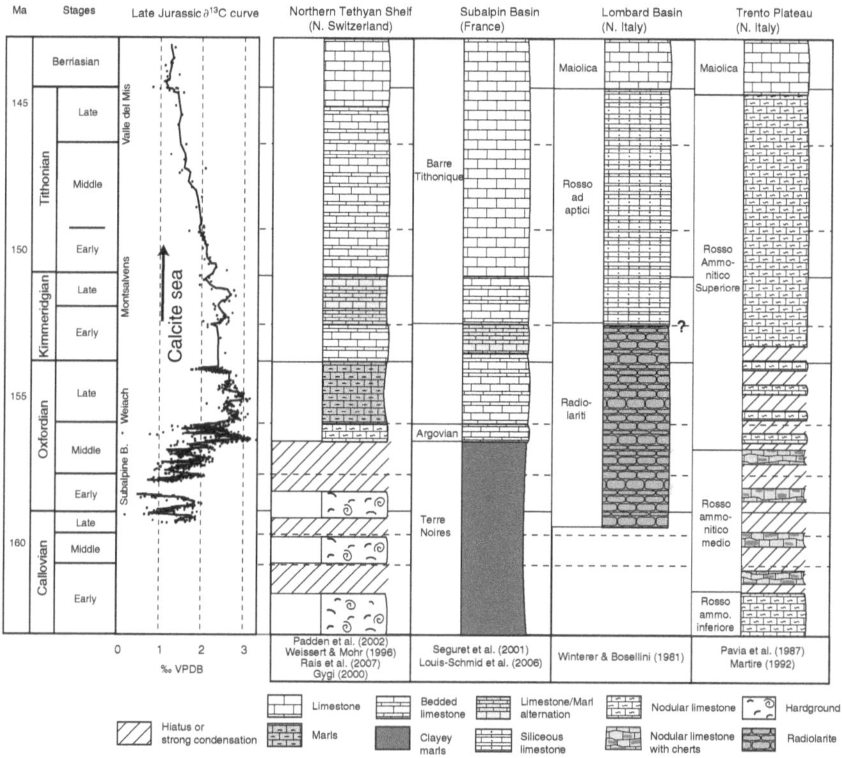

Figure 11.4 Correlation of Late Jurassic carbon-isotope stratigraphy with biostratigraphically dated lithostratigraphy across a Tethys North–South transect (modified after Rais et al., 2007). VPDB: Vienna-PDB standard.

Marine records and the new method of oxygen-isotope geochemistry opened a way out of this blocked climate controversy. Emiliani (1954, 1955) discovered how the oxygen-isotope composition in benthic and in planktic foraminifers fluctuated through time, and he concluded that only an orbitally driven climate change could explain the observed cyclic change in the marine oxygen-isotope record. In the years following Emiliani's cardinal work, oxygen-isotope geochemistry developed into a very powerful tool in palaeoceanography, and a large number of new oxygen-isotope curves were generated during the 1960s. Shackleton and Opdyke (1973) successfully combined oxygen-isotope geochemistry with palaeomagnetics. These new data, based on a new and sound stratigraphy, confirmed that orbitally driven changes in insolation controlled the Quaternary cold/warm cycles.

Despite the fundamental work by Craig (1953) and the detailed investigations of carbon-isotope fractionation processes in water/carbon systems (e.g. Deuser et al., 1968; Duplessy, 1972), carbon isotopes were rarely used as palaeoenvironmental proxies before the early 1970s. Climate research focused on the reconstruction of temperature records and only few palaeoceanographers started to recognise the potential of C-isotope geochemistry in their field (e.g. Berger et al., 1978; Tappan, 1968).

Due to a large kinetic fractionation during photosynthesis, organic matter of plants is strongly depleted in the heavy ^{13}C isotope, while biogenic calcite precipitates close to equilibrium with seawater. Calcite can therefore be used as a proxy for the past carbon-isotope composition of dissolved inorganic carbon (DIC) in seawater. Today, $\delta^{13}C$ of DIC in seawater ranges between a maximum of about +2‰ in surface waters and a minimum of around −0.5‰ at intermediate depths (Kroopnick, 1985). ^{13}C enrichment of surface water is explained by carbon-isotope fractionation during photosynthesis with marine organic matter strongly depleted in ^{13}C (approx. −25‰ $\delta^{13}C$). Isotopically light organic carbon is oxidised in intermediate and deep water, and CO_2 enriched in ^{12}C is added to this water. Therefore, the $\Delta\delta^{13}C_{surface/deep\ water}$ stored in planktic and benthic $\delta^{13}C$-values records biogenic productivity in surface water and the recycling rate of organic carbon within the oceanic water column (e.g. Berger et al., 1978; Shackleton, 1967). Of course, deep-water C-isotope signatures are not only controlled by the oxidation of organic carbon and the vertical recycling rate but also by deep-water circulation, as shown already in early palaeoceanographic studies (e.g. Boyle and Keigwin, 1982). Today 'old' deep water in the North Pacific is not only depleted in oxygen and enriched in nutrients, but it is also enriched in ^{12}C.

On long time scales (of 10^4–10^7 years), the C-isotope composition of the oceanic carbon pool is affected by changes in partitioning of carbon between the oxidised and the reduced marine carbon sink (Holland, 1984). The oceanic carbon reservoir is at isotopic steady state if the input from weathering and volcanism is balanced by output through sediment burial as organic and carbonate-mineral carbon. The carbon-isotopic composition of marine calcite serves as a proxy of the isotopic composition of the oceans through time. Measured changes in the isotopic composition of marine calcite record changes in the carbon-isotope composition of the oceans. These changes can best be interpreted as the result of a changing organic carbon flux from the ocean into the sedimentary carbon sink. Increased burial of organic carbon leads to a positive excursion in the organic and inorganic carbon-isotope record. Major changes in organic-carbon burial rates in the geological past were triggered by perturbations in the global carbon cycle causing changes in oceanography and/or by an altered carbon flux from continents to oceans, as shown in several isotope mass-balance models (e.g. Kump and Arthur, 1999; Louis-Schmid et al., 2007; Wissler, 2001).

7. Black Shales and the Carbon Cycle

The decades of early palaeoceanography ended in the late 1980s when conditions were mature for a new major step in historical geology: the step from oceanography to palaeoclimatology and to integrated earth-systems analysis. We will use black-shale research as a case study and follow this change from palaeoceanography to modern climate geology.

The DSDP was initiated with the aim to test the new seafloor-spreading hypothesis. Soon, the successful project also provided a wealth of new information for ocean history. With the new data from all the major oceans reaching back into the Mesozoic, palaeoceanography did not remain limited to Cenozoic environments. The Mesozoic with its peculiar climate became a promising target for the new field of palaeoceanography (Hsü, 1976). Most exciting were findings of Cretaceous black shales in all the major oceans and the recognition of globally occurring 'Oceanic Anoxic Events' (OAE) by Schlanger and Jenkyns (1976). Based on their DSDP and Tethys Ocean data, they defined two 'OAE', the 'Aptian–Albian OAE' and the latest Cenomanian OAE. During both OAEs, the oceanic conditions favoured the episodic deposition of sediments enriched in organic carbon (e.g. Arthur and Premoli-Silva, 1982; Jenkyns, 1980). The 'black shales' are dark coloured and they can be classified as marine hemipelagic or pelagic mudstones, with or without carbonate, sometimes siliceous, with organic carbon >1–2% of marine or terrestrial origin (Fig. 11.3; Arthur and Sageman, 1994). Many of the black shales are laminated, but some shales are bioturbated, indicating that bottom-water conditions were not always anoxic during the deposition of black shales (e.g. Menegatti et al., 1998; Fig. 11.3C). The Aptian–Albian OAE covered millions of years, whereas the latest Cenomanian OAE was of shorter duration (less than 1 million years: Kuhnt et al., 2005; Tsikos et al., 2004). The Aptian–Albian OAE is today subdivided into OAE 1a, b, c and d. OAE 1a is the most pronounced of these events (e.g. Bréhéret, 1985; Erbacher et al., 2001). Additional OAE's were identified later (Toarcian, Valanginian, Hauterivian, Coniacian–Santonian; see, for instance, Lini et al., 1992; Arthur and Sageman, 1994; Baudin et al., 1999; Erba et al., 2004; Hesselbo et al., 2007; Reboulet et al., 2003).

The sedimentology of the OAEs serves as a source of information of depositional conditions during OAEs (Wignall, 1994). Geochemical fingerprints in black shales and associated deep-sea sediments provide a wealth of information on past ocean conditions, on carbon cycling, on past pCO_2 levels and on palaeotemperature trends.

One of the best-studied OAEs in pelagic environments is OAE 1a in the Early Aptian. Coccioni et al. (1987) defined OAE1a as the 'Livello Selli' in the Apennines. There, the OAE 1a succession is 1–3 m thick, and it consists of a succession of alternating black-green and grey mudstones with intercalated radiolarian beds. Coccioni et al. (1987) subdivided the Livello Selli into a lower part dominated by green mudstones and an upper black-shale dominated part. Menegatti et al. (1998) studied the Livello Selli equivalent (LSE) at the Cismon locality (Northern Italy) in detail. There, the LSE is about 5 m thick and it consists of black, bioturbated marlstones or marly limestones alternating with black laminated siliceous marlstones or green-grey marly limestones. The burrows were identified as of type *Chondrites, Planolites* and *Zoophycos* (Menegatti et al., 1998). The succession contains more than 20% $CaCO_3$, and its organic-carbon content is highly variable. Four distinct black-shale horizons are strongly enriched in organic carbon with concentrations of up to 8%. Menegatti et al. (1998) interpreted burrows as proxies for bottom-water oxygenation and, based on bioturbation distribution patterns, he concluded that anoxic conditions only persisted at the Cismon locality during four short time intervals of a few tens of thousands of years.

Mesozoic black shales not only serve as proxies for peculiar conditions in palaeoceanography, but they also record perturbations of the global carbon cycle possibly triggered by changes in atmospheric CO_2 concentrations (see the review by Emeis and Weissert, 2009). These perturbations of the global carbon cycle are recorded as 'excursions' in the Mesozoic carbon-isotope record.

The use of C-isotope records as a proxy for pre-Cenozoic palaeoceanography is based on the assumption that diagenesis of the studied pelagic carbonate sediments did not alter an original marine δ^{13}C-signal. This assumption has been validated by numerous studies, which document how C-isotope stratigraphies can be reproduced in different sedimentary and diagenetic settings and in a variety of organic-carbon isotope carriers (e.g. Channell et al., 1993; Gröcke et al., 1999). A source of error may result from the original composition of fine micritic carbonate. Carbonate mud derived from carbonate platforms may preserve an isotope signature that significantly differs from an average open-marine value (e.g. Swart and Eberli, 2005).

The first Mesozoic carbon-isotope stratigraphies established were largely based on the analysis of bulk carbonate from pelagic limestones (e.g. Bartolini et al., 1996; Jenkyns et al., 1994; Scholle and Arthur, 1980; Weissert et al., 1985). Composite carbon-isotope curves established for the Mesozoic show repeated high-amplitude excursions to positive δ^{13}C values of up to +5‰ (e.g. Bartolini et al., 1996; Jenkyns et al., 2002; Voigt et al., 2007; Weissert and Erba, 2004) and several short-lived 'negative spikes' comparable to the one at the PETM (e.g. Dickens et al., 1995). If the Mesozoic carbon-isotope record is compared with the Palaeozoic carbon-isotope curve, a remarkable difference in the amplitude of the two curves can be recognised. The Palaeozoic carbon-isotope anomalies have amplitudes of up to 6‰ or more, whereas the Mesozoic anomalies are marked by lower amplitude changes of around 2–5‰. This change in the carbon-isotope pattern reflects an increasingly stable global carbon cycle. A shift from a Palaeozoic–Early Mesozoic 'neritic' ocean dominated by biogenic shallow-water carbonate precipitation to a Late Mesozoic 'Cretan' Ocean (Zeebe and Westbroek, 2003) with widespread biogenic deep-water carbonate sedimentation can explain this trend. The Cretan Ocean could best maintain a steady state between weathering and carbonate-mineral burial through a change of the size of the deep-sea sink. Increased/decreased burial of calcite at a global scale triggered by changes in weathering can be accomplished by deepening/shallowing of the CCD.

Many of the major positive $\delta^{13}C$ excursions coincided with OAEs, which are defined as episodes of widespread or even global black-shale deposition (Fig. 11.5; e.g. Arthur et al., 1985; Scholle and Arthur, 1980; Weissert and Bréheret, 1991). This lead most authors to the conclusion that positive, high amplitude $\delta^{13}C$ excursions correspond to higher C_{org}/C_{carb} burial ratios triggered by elevated C_{org} burial rates (e.g. Hochuli et al., 1999). Accelerated burial of organic carbon corresponds with an intensification of the 'biological carbon pump'. Widespread organic-carbon-rich sediments or 'black shales' serve as proxies for these peculiar conditions in Mesozoic palaeoceanography. Oceans were shifted into a 'black-shale mode' at times of global perturbations of the carbon cycle as recorded in shifts towards positive carbon-isotope excursions with amplitudes of up to several permil and durations of 10^5–10^6 years. Accelerated volcanic CO_2 degassing is regarded as the cause of the major carbon cycle perturbations (Arthur et al., 1985; Larson and Erba, 1999). The positive carbon-isotope excursions are, therefore, interpreted as the 'response signal' of the biosphere to a perturbation of the carbon cycle (Wortmann and Weissert, 2000).

We cannot only recognise positive carbon-isotope anomalies in the Mesozoic and Cenozoic carbon-isotope curves. Prominent 'negative spikes' of short duration (up to a few hundred thousands of years) sometimes preceded positive carbon-isotope excursions. These negative spikes could reflect a sudden addition of light volcanogenic carbon to the oceans and atmosphere, as suggested by Menegatti et al. (1998). Wissler (2001) simulated the impact of sudden volcanic degassing on the carbon-isotope record and he concluded that a negative carbon-isotope anomaly of up to 0.5‰ can be explained with extreme volcanic CO_2 degassing. If negative carbon-isotope anomalies are larger in their amplitude than 0.5‰, a volcanic source of light carbon becomes increasingly unlikely. Negative carbon-isotope spikes with an amplitude >0.5‰ have been identified in Toarcian, Oxfordian and Aptian successions (Hesselbo et al., 2000; Jahren et al., 2001; Padden et al., 2001). They have been identified on a global scale in both

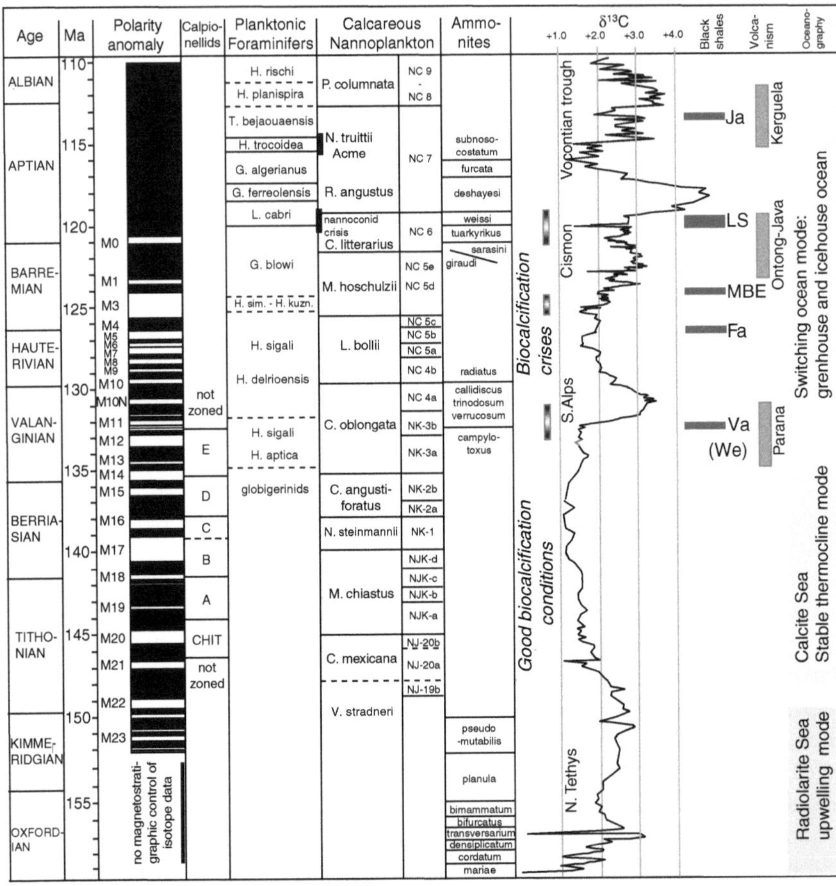

Figure 11.5 Carbon-isotope stratigraphy calibrated with magneto- and biostratigraphy (modified after Weissert and Erba, 2004). Oceanography with 'Oceanic Anoxic Events'. Va, Valanginian Event (or: We Weissert Event; Erba et al., 2004); FA, Faraoni Level (Cecca et al., 1994); MBE, Mid Barremian Event (Coccioni et al., 2004); LS, Livello Selli (Coccioni et al., 1987); Ja, Niveau Jacob (Bréhéret, 1985). (A multi-colour version of this figure is on the included CD-ROM.)

marine and terrestrial successions. The negative spikes detected in Mesozoic rocks are compared with the negative carbon-isotope spike marking the PETM. Most authors explain rapid negative shifts in the Cenozoic and Mesozoic C-isotope records as the result of a sudden addition of isotopically light methane to the marine carbon reservoir. This interpretation is based on the discovery of methane hydrates (clathrates) as an important carbon reservoir within the global carbon cycle (Kvenvolden, 1988). The trigger of sudden destabilisation of clathrates would have been a rapid climate warming related to volcanic degassing (e.g. Weissert and Erba, 2004) or, alternatively, volcanic activity could have mobilised thermogenic methane stored in organic-rich sediments (Svensen et al., 2004). Major volcanic episodes with an impact on the global carbon cycle and on the Cretaceous climate include those during which the formation of the Parana Flood Basalt (Valanginian), the Ontong-Java Large Igneous Province (OAE 1a), the Kerguelen Igneous Plateau (OAE 1b), and the Caribbean–Columbian Igneous Plateau (OAE 2) took place. Turgeon and Creaser (2008) used the osmium-isotope composition of sediments across the Bonarelli Level as a proxy for volcanic activity, whereas Kuroda et al. (2007) measured the lead-isotope composition of the silicate sediment fraction in the Bonarelli interval as an indicator of volcanic activity. Based on these data, these authors concluded that a major volcanic pulse triggered OAE 2. Methane/CO_2 pulses recorded in negative carbon-isotope anomalies had an impact on the ocean chemistry and on pelagic sedimentation. The negative carbon-isotope spike at the Paleocene/Eocene boundary coincides with a rapid shallowing of the CCD. The CCD shoaled within less than 10,000 years by about 2000 m. It recovered gradually over 100,000 years (Zachos et al., 2005). This change of the CCD reflects the sudden acidification of deep water during the PETM, which was caused by the addition of CO_2 possibly derived from methane oxidation in the oceans. A shallowing of the CCD has also been found for the base of OAE 1a, the Livello Selli of the Early Aptian (Coccioni et al., 1987). The shallowing of the Aptian CCD may serve as evidence for a rapid addition of carbon dioxide to Aptian oceans. This change in CCD coincides with a major calcification crisis recorded in calcareous nannoplankton (Erba and Tremolada, 2004) and in neritic carbonates (Wissler et al., 2003).

Palaeo-pCO_2 reconstructions for the mid-Cretaceous indicate indeed that pCO_2 levels were up to three times higher than modern values (e.g. Haworth et al., 2005; Robinson et al., 2002). Few geochemical high-resolution studies indicate that 'black-shale times' in the mid-Cretaceous coincided with greenhouse conditions. Reconstructions of past pCO_{2atm} using the C-isotope fractionation methods indicate that the concentration of atmospheric CO_2 fluctuated within a range of 500–1000 ppm; these values are smaller than numbers expected on the basis of geochemical mass-balance models (e.g. Heimhofer et al., 2004).

Increased organic-carbon burial resulted in a drawdown of atmospheric CO_2, leading to global cooling. Only few temperature records across OAEs exist so far. Oxygen-isotope compositions of pelagic sedimentary rocks were considered as unreliable temperature proxies because of diagenetic overprint. Stoll and Schrag (2000), however, compared Late Cretaceous bulk oxygen-isotope records from two Tethyan pelagic sections and they argue that the observed fluctuations are not an artefact of diagenesis but that they reflect changes in Cretaceous sea-water temperatures. Weissert and Erba (2004) drew a composite bulk oxygen-isotope curve for the Early

Cretaceous, and based on these data they proposed that the time just before OAE 1a was exceptionally warm. Ando et al. (2008) presented oxygen-isotope based temperature reconstructions for the OAE 1a in the Pacific Ocean. They confirmed that peak temperatures coincide with the negative C-isotope spike at the base of OAE 1a, while the positive carbon-isotope excursion coincides with lower temperatures (Ando et al., 2008). Palaeotemperature studies based on well-preserved foraminifers indicating that Albian sea-surface temperatures were higher than today are still missing for the Early Aptian (e.g. Norris and Wilson, 1998; Wilson et al., 2002).

Another promising carrier of palaeotemperature information is the oxygen-isotope composition of fish teeth. Puceat et al. (2003) succeeded in tracing surface water temperatures through the Early Cretaceous with the fish-teeth method, and they identified remarkable temperature fluctuations indicating episodes of cooling alternating with greenhouse pulses.

The novel Tex86 palaeotemperature method is based on the temperature-dependent composition of membrane lipids of marine Crenarchaeota ('TEX_{86}' is the tetraether index of 86 carbon atoms in Crenarchaeota). Organic matter from Cretaceous pelagic sediments from the Atlantic Ocean was analysed and the resulting TEX_{86} proxy indicates that tropical sea-surface temperatures in the proto-North Atlantic were 32–36 °C during the late Cenomanian to early Turonian, when OAE 2 occurred (Schouten et al., 2007). These results agree with palaeotemperature estimates based on ^{18}O palaeothermometry of well-preserved foraminifers. Forster et al. (2007) used the TEX_{86} method for their reconstruction of temperatures across OAE 2, and they recognised a similar trend as had been reported from OAE 1a. The temperature was highest at the very base of OAE 2, and increased organic-carbon burial during OAE 2 led to a marked cooling of about 4 °C.

8. Summary and Outlook

Depositional models for Mesozoic pelagic sediments outcropping nowadays in mountain ranges were developed in the second part of the twentieth century, the 'golden age' of pelagic-sediment investigation. Pelagic sediments today are used as proxies for palaeoceanography and palaeoclimatology. Geochemical fingerprints in pelagic sediments are deciphered and used as monitors of past ocean chemistry, of marine productivity or of past atmospheric CO_2 concentrations. Scientific controversies focus on the reliability of geochemical proxies in pre-Cenozoic sediments, but available data confirm that isotope geochemistry remains to be a very powerful tool in palaeoceanography.

Future research will combine new geochemical techniques with sedimentology and high-resolution (<20 ka) stratigraphy. Pelagic sediments will serve as carriers of a variety of 'new' isotope proxies including osmium, lead and neodymium isotopes. Marine and terrestrial biomarkers and their isotopic signature will provide essential information on past water chemistry, palaeotemperatures and pCO_2 (e.g. Koopmans et al., 1996; Schouten et al., 2007). Organic geochemistry will be combined with refined palynological investigations. Trace-element geochemistry will give new insight into physical and chemical palaeoceanography.

A major task for earth scientists will be the further improvement of the stratigraphy and the stratigraphic correlations between deep-marine, coastal and continental environments. Biostratigraphy and lithostratigraphy will provide the general framework. Cyclostratigraphy in pelagic sediments will be improved using more elaborated time-series analysis and an astrochronology will be established far back into the Mesozoic. Astrochronology, if combined with improved radiometric ages, will serve as the most accurate tool for measurement of time in earth history (d'Argenio et al., 2004). Chemostratigraphy and magnetostratigraphy will serve as strong tools for stratigraphic correlation. An accurate time frame established in pelagic sediments and a combination of marine and terrestrial archives will provide a more complete picture of the earth system's history.

ACKNOWLEDGEMENTS

I thank Heiko Hüneke for inviting me to write this review. The constructive comments by Elisabetta Erba and Thierry Mulder helped to improve the chapter. Continuous support by the Swiss Science Foundation and ETH is acknowledged.

REFERENCES

Ando, A., Kaiho, K., Kawahata, H., Kakegawa, T., 2008. Timing and magnitude of early Aptian extreme warming: unraveling primary $\delta^{18}O$ variation in indurated pelagic carbonates at Deep Sea Drilling Project Site 463, central Pacific Ocean. Palaeogeogr. Palaeoclimatol. Palaeoecol. 260, 463–476.

Arthur, M.A., Premoli-Silva, I., 1982. Development of widespread organic carbon-rich strata in the Mediterranean Tethys. In: Schlanger, S.O., Cita, M.B. (Eds.), Nature and Origin of Cretaceous Carbon-Rich Facies. Academic Press, London, pp. 7–54.

Arthur, M.A., Sageman, B.B., 1994. Marine black shales—depositional mechanisms and environments of ancient deposits. Annu. Rev. Earth Planet. Sci. Lett. 22, 499–551.

Arthur, M.A., Dean, W.E., Schlanger, S.O., 1985. Variations in the global carbon cycle during the Cretaceous related to climate, volcanism, and changes in atmospheric CO_2. In: Sundquist, E.T.S., Broecker, W.S. (Eds.), The Carbon Cycle and Atmospheric CO2: Natural Variations Archean to the Present, Geophys. Monogr., 504–529.

Bartolini, A., Baumgartner, P.O., Hunziker, J., 1996. Middle and Late Jurassic carbon stable isotope stratigraphy and radiolarite sedimentation of the Umbria-Marche Basin (Central Italy). Eclogae Geol. Helv. 89, 811–844.

Baudin, F., Bulot, L.G., Cecca, F., Coccioni, R., Gardin, S., Renard, M., 1999. An equivalent of the "Faraoni Level" in the South-East of France, possible evidence for a late Hauterivian anoxic event in the Mediterranean Tethys. Bull. Soc. Geol. Fr. 170, 487–498.
Berger, W.H., 1974. Deep-sea sedimentation. In: Burke, C.A., Drake, C.C. (Eds.), The Geology of Continental Margins. Springer, Berlin, pp. 213–241.
Berger, W.H., Killingley, J.S., 1982. Box cores from the Equatorial Pacific—C-14 sedimentation rates and benthic mixing. Mar. Geol. 45, 93–125.
Berger, W.H., Winterer, E.L., 1974. Plate stratigraphy and the fluctuating carbonate line. In: Hsu, K.J., Jenkyns, H.C. (Eds.), Pelagic Sediments: On Land and Under the Sea, Int. Ass. Sedimentologists, Spec. Publ., 1, 11–48.
Berger, W.H., Diester-Haas, L., Killingley, J.S., 1978. Upwelling off Northwest Africa: the Holocene decrease as seen in carbon isotopes and sedimentological indicators. Oceanol. Acta 1, 1–7.
Bernoulli, D., 1964. Zur Geologie des Monte Generoso (Lombardische Alpen). Mat. Carta Geol. 118, 134 Svizzera, Bern.
Bernoulli, D., 1972. North Atlantic and Mediterranean Mesozoic facies: a comparison. In: Hollister, C.D., Ewing, J.L. et al., Init. Reports of the Deep Sea Drilling Project, 11. U.S. Govt. Printing Office, Washington, D.C, pp. 801–822.
Bernoulli, D., Jenkyns, H.C., 1974. Alpine, Mediterranean, and central Atlantic Mesozoic facies in relation to the early evolution of the Tethys. Soc. Econ. Paleontol. Minerol. Spec. Publ. 19, 129–160.
Bersezio, R., Fornaciari, M., 1987. Cretaceous sequences in the Lombardy Basin: stratigraphic outine between the lakes of Lecco and Iseo. Mem. Soc. Geol. Ital. 40, 187–197.
Boyle, E., Keigwin, L., 1982. Deep circulation in the North Atlantic over the last 200 000 years: geochemical evidence. Science 237, 171–175.
Bréhéret, J.G., 1985. Sedimentology and diagenesis of organic-matter from the Paquier Level, a lowermost Albian key-bed of the Vocontian Basin. C. R. Acad. Sci. II 301, 1151–1156.
Calvert, S.E., 1974. Deposition and diagenesis of silica in marine sediments. In: Hsu, K.J., Jenkyns, H.C. (Eds.), Pelagic Sediments: On Land and Under the Sea, Int. Ass. Sedimentologists, Spec. Publ., 1, 273–300.
Cecca, F., Marini, A., Pallini, G., Baudin, F., Begouen, V., 1994. A guide-level of the uppermost Hauterivian (Lower Cretaceous) in the pelagic succession of Umbria-Marche Apennines (Central Italy): the Faraoni Level. Riv. Ital. Paleontol. Stratigr. 99, 551–568.
Channell, J.E.T., Lowrie, W., Medizza, F., 1979. Middle and Early Cretaceous magnetic stratigraphy from the Cismon section, Northern Italy. Earth Planet. Sci. Lett. 42, 153–166.
Channell, J.E.T., Erba, E., Lini, A., 1993. Magnetostratigraphic calibration of the Late Valanginian carbon-isotope event in pelagic limestones from Northern Italy and Switzerland. Earth Planet. Sci. Lett. 118, 145–166.
Coccioni, R., Nesci, O., Tramontana, M., Wezel, C.F., Moretti, E., 1987. Descrizione di un livello guida "radiolaritico—bituminoso—ittiolitico" alla base delle Marne a Fucoidi nell' Appennino umbro—marchigiano. Boll. Soc. Geol. Ital. 106, 183–192.
Coccioni, R., Galeotti, S., Marsili, M., Sprovieri, M., 2004. The Mid-Barremian event (MBE): the Prelude to OAE 1a. Geophys. Res. Abstr. 6, 05130.
Craig, H., 1953. The geochemistry of the stable carbon isotopes. Geochim. Cosmochim. Acta 3, 53–92.
D'Argenio, B., Fischer, A.G., Premoli Silva, I., Weissert, H., Ferreri, V. (Eds.), 2004. Cyclostratigraphy: Approaches and Case Histories, In: SEPM, Special Publication, 81, 311.
Desmurs, L., Manatschal, G., Bernoulli, D., 2001. The Steinmann Trinity revisited: mantle exhumation and magmatism along an ocean-continent transition: the Platta nappe,

eastern Switzerland, vol. 187. Geological Society, London Special Publications, pp. 235–266.
Deuser, W.G., Degens, E.T., Guillard, R.L., 1968. Carbon isotope relationships between plankton and seawater. Geochim. Cosmochim. Acta 32, 657–660.
Dickens, G.R., ÓNeil, J.R., Rea, D.K., Owen, R.M., 1995. Dissociation of oceanic methane hydrate as a cause of the carbon-isotope excursion at the end of the Paleocene. Paleoceanography 10, 965–971.
Dietz, R.S., 1961. Continent and ocean basin evolution by spreading of the sea-floor. Nature 190, 854–857.
Dott, R.H., 1997. James Dwight Dana's old tectonics—global contraction under divine direction. Am. J. Sci. 297, 288–311.
Duplessy, J.C., 1972. La géochimie des isotopes stables du carbon dans le mer. Commissariat à l'énergie atomique, Paris Thesis. CEA-N-1565.
Eberli, G.P., Swart, P.K., Malone, M., 1997. Proceedings of ODP Initial Reports, vol. 101. Ocean Drilling Program, College Station, TX.
Emeis, K., Weissert, H., 2009. Tethyan-Mediterranean organic carbon-rich sediments from Mesozoic black shales to sapropels. Sedimentology 56, 247–266.
Emiliani, C., 1954. Temperatures of Pacific Bottom Waters and Polar Superficial Waters during the Tertiary. Science 119, 853–855.
Emiliani, C., 1955. Pleistocene temperatures. J. Geol. 63, 538–578.
Erba, E., Tremolada, F., 2004. Nannofossil carbonate fluxes during Early Cretaceous: phytoplankton response to nutrification episodes, atmospheric CO_2 and anoxia. Paleoceanography 19, PA1008. doi:10.1029/2003PA000884.
Erba, E., Bartolini, A., Larson, R.L., 2004. Valanginian Weissert oceanic anoxic event. Geology 32, 149–152.
Erbacher, J., Huber, B.T., Norris, R.D., Markey, M., 2001. Increased thermohaline stratification as a possible cause for an ocean anoxic event in the Cretaceous period. Nature 409, 325–327.
Ewing, M., Heezen, B.C., 1960. Continuity of the mid-oceanic ridge and rift valley in the Southwestern Indian Ocean confirmed. Science 131, 1677–1679.
Faugères, J.-C., Mulder, T., 2011. Contour currents and contourite drifts. In: Hüneke, H., Mulder, T. (Eds.), Deep-Sea Sediments. Developments in Sedimentology, Elsevier, Amsterdam, vol. 63, pp. 149–214.
Folk, R., McBride, E.F., 1978. Radiolarites and their relation to subjacent oceanic crust in Liguria, Italy. J. Sed. Res. 48, 1069–1102.
Föllmi, K.B., 1996. The phosphorus cycle, phosphogenesis and marine phosphate-rich deposits. Earth Sci. Rev. 40, 55–124.
Forster, A., Schouten, S., Moriya, K., Wilson, P.A., Sinninghe Damsté, J.S., 2007. Tropical warming and intermittent cooling during the Cenomanian/Turonian oceanic anoxic event 2: sea surface temperature records from the equatorial Atlantic. Paleoceanography 22, PA1219. doi:10.1029/2006PA001349.
Friedman, G., 1964. Early diagenesis and lithification in carbonate sediments. J. Sediment. Petrol. 34, 777–813.
Fuchs, T., 1877. Ueber die Entstehung der Aptychenkalke. Sitzungsber. K. Akad. Wiss. Wien 76, 329–334.
Fuchs, T., 1898. Einige Bemerkungen über das Project eines internationalen schwimmenden Laboratoriums4, Verh. K.K.Geol. Reichsanstalt, Separatabdruck, 8pp.
Garrison, R.E., Fischer, A.G., 1969. Deep-water limestones and radiolarites of the alpine Jurassic. Soc. Ecom. Paleontol. Minerol. Spec. Publ., 14, 20–55.
Gill, G.A., Santantonio, M., Lathuiliere, B., 2004. The depth of pelagic deposits in the Tethyan Jurassic and the use of corals: an example from the Apennines. Sed. Geol. 166, 311–334.

Gröcke, D.R., Hesselbo, S., Jenkyns, H.C., 1999. Carbon-isotope composition of Lower Cretaceous fossil wood: ocean-atmosphere chemistry and relation to sea-level change. Geology 27, 155–158.

Grunau, H.R., 1947. Geologie von Arosa (Graubünden) mit besonderer Berücksichtigung des Radiolarit-Problems. Zürich, Aschmann und Scheller, 108pp.

Haug, E., 1900. Les géosynclinaux et les aires continentals. Contribution à l'étude des régressions et des transgressions marines. Bull. Soc. Géol. France, Sér. 3 28, 617–711.

Haworth, M., Hesselbo, S.P., McElwain, J.C., Robinson, S.A., Brunt, J.W., 2005. Mid-Cretaceous pCO_2 based on stomata of the extinct conifer *Pseudofrenelopsis* (Cheirolepidiaceae). Geology 33, 749–752.

Heim, A., 1924. Ueber submarine Denudation und chemische Sedimente. Geol. Rundsch. 15, 1–47.

Heimhofer, U., Hochuli, P.A., Herrle, J.O., Andersen, N., Weissert, H., 2004. Absence of major vegetation and palaeoatmospheric pCO_2 changes associated with oceanic anoxic event 1a (Early Aptian, SE France). Earth Planet. Sci. Lett. 223, 303–318.

Henrich, R., Hüneke, H., 2011. Hemipelagic advection and periplatform sedimentation. In: Hüneke, H., Mulder, T. (Eds.), Deep-Sea Sediments. Developments in Sedimentology, Elsevier, Amsterdam, vol. 63, pp. 353–396.

Herbert, T., 1992. Paleomagnetic calibration of Milankovich cyclicity in Lower Cretaceous sediments. Earth Planet. Sci. Lett. 112, 15–28.

Hesselbo, S.P., Grocke, D.R., Jenkyns, H.C., Bjerrum, C.J., Farrimond, P., Bell, H.S.M., et al., 2000. Massive dissociation of gas hydrate during a Jurassic oceanic anoxic event. Nature 406, 392–395.

Hesselbo, S.P., Jenkyns, H.C., Duarte, L.V., Oliveira, L.C.V., 2007. Carbon-isotope record of the Early Jurassic (Toarcian) Oceanic Anoxic Event from fossil wood and marine carbonate (Lusitanian Basin, Portugal). Earth Planet. Sci. Lett. 253, 455–470.

Hesse, R., Schacht, U., 2011. Early diagenesis of deep-sea sediments. In: Hüneke, H., Mulder, T. (Eds.), Deep-Sea Sediments. Developments in Sedimentology, Elsevier, Amsterdam, vol. 63, pp. 557–714.

Hochuli, P.A., Menegatti, A.P., Weissert, H., Riva, A., Erba, E., Silva, I.P., 1999. Episodes of high productivity and cooling in the Early Aptian Alpine Tethys. Geology 27, 657–660.

Holland, H.T., 1984. The chemical evolution of the atmosphere and the oceans. Princeton Series in Geochemistry. Princeton University Press, Princeton, NJ, 582pp.

Hsu, K.J., 1976. Paleoceanography of the Mesozoic Alpine Tethys. Geol. Soc. Am. Bull. Spec. Paper. 170, 44p.

Hsü, K.J., 1992. Challenger at Sea: The Ship that Revolutionized Earth Science. Princeton University Press, Princeton, NJ, 454pp.

Hsu, K.J., Jenkyns, H.C. (Eds.), 1974. Pelagic Sediments: On Land and Under the Sea, In: Int. Ass. Sedimentologists, Spec. Publ., 1, 447.

Hüneke, H., Henrich, R., 2011. Pelagic sedimentation in modern and ancient oceans. In: Hüneke, H., Mulder, T. (Eds.), Deep-Sea Sediments. Developments in Sedimentology, Elsevier, Amsterdam, vol. 63, pp. 215–351.

Jahren, A.H., Arens, N.C., Sarmiento, G., Guerrero, J., Amundson, R., 2001. Terrestrial record of methane hydrate dissociation in the Early Cretaceous. Geology 29, 159–162.

James, N.P., Bone, Y., Collins, L.B., Kyser, T.K., 2001. Surficial sediments of the Great Australian Bight: facies dynamics and oceanography on a vast cool-water carbonate shelf. J. Sed. Res. 71, 549–567.

Jenkyns, H.C., 1974. Origin of red nodular limestone (Ammonitico Rosso, Knollenkalke) in the Mediterranean Jurassic: a diagenetic model. In: Hsu, K.J., Jenkyns, H.C. (Eds.), Pelagic Sediments: On Land and Under the Sea, Int. Ass. Sedimentologists, Spec. Publ., 1, 249–272.

Jenkyns, H.C., 1980. Cretaceous anoxic events—from continents to oceans. J. Geol. Soc. London 137, 171–188.

Jenkyns, H.C., Hsü, K.J., 1974. Introduction. In: Hsu, K.J., Jenkyns, H.C. (Eds.), Pelagic Sediments: On Land and Under the Sea, Int. Ass. Sedimentologists, Spec. Publ., 1, 1–10.

Jenkyns, H.C., Gale, A.S., Corfield, R.M., 1994. Carbon-isotope and oxygen-isotope stratigraphy of the English Chalk and Italian Scaglia and its palaeoclimatic significance. Geol. Mag. 131, 1–34.

Jenkyns, H.C., Jones, C.E., Gröcke, D.R., Hesselbo, S.P., Parkinson, D.N., 2002. Chemostratigraphy of the Jurassic System: applications, limitations and implications for palaeoceanography. J. Geol. Soc. 159, 351–378.

Kaelin, O., Bernoulli, D., 1984. *Schizosphaerella* Deflandre and Dangeard in Jurassic deeper-water carbonate sediments, Mazagan continental margin (Hole-547b) and Mesozoic Tethys. Init. Rep. Deep Sea Drilling Proj. 79, 411–435.

Kennett, J.P., 1968. *Globorotalia truncatulinoides* as a paleoceanographic index. Science 159, 1461–1463.

Koopmans, M.P., Koster, J.R., Van Kaam-Peters, H.M.E., Kenig, F., Schouten, S., Hartgers, W.A., et al., 1996. Diagenetic and catagenetic products of isorenieratene: molecular indicators for photic zone anoxia. Geochim. Cosmochim. Acta 60, 4467–4496.

Kroopnick, P.M., 1985. The distribution of ^{13}C and $\sum CO_2$ in the world ocean. Deep-Sea Res. 32, 57–84.

Kuhnt, W., Luderer, F., Nederbragt, S., Thurow, J., Wagner, T., 2005. Orbital-scale record of the late Cenomanian–Turonian oceanic anoxic event (OAE-2) in the Tarfaya Basin (Morocco). Int. J. Earth Sci. 94, 147–159.

Kump, L.R., Arthur, M.A., 1999. Interpreting carbon-isotope excursions: carbonates and organic matter. Chem. Geol. 161, 181–198.

Kuroda, J., Ogawa, N.O., Tanimizu, M., Coffin, M.F., Tokuyama, H., Kitazato, H., et al., 2007. Contemporaneous massive subaerial volcanism and Late Cretaceous Oceanic Anoxic Event 2. Earth Planet. Sci. Lett. 256, 211–223.

Kvenvolden, K.A., 1988. Methane hydrate—a major reservoir of carbon in the shallow geosphere. Chem. Geol. 71, 41–51.

Larson, R.L., Erba, E., 1999. Onset of the mid-Cretaceous greenhouse in the Barremian–Aptian: igneous events and the biological, sedimentary, and geochemical responses. Paleoceanography 14, 663–678.

Lini, A., Weisssert, H., Erba, E., 1992. The Valanginian carbon isotope event: a first episode of greenhouse climate conditions during the Cretaceous. Terra Nova 4, 374–384.

Louis-Schmid, B., Rais, P., Schaeffer, P., Bernasconi, S.M., Weissert, H., 2007. Plate tectonic trigger of changes in pCO_2 and climate in the Oxfordian (Late Jurassic): carbon isotope and modeling evidence. Earth Planet. Sci. Lett. 258, 44–60.

Matter, A., Douglas, R.G., Perch-Nielsen, K., 1975. Fossil preservation, geochemistry, and diagenesis of pelagic carbonates from Shatsky Rise, Northwest Pacific. Initial Deep Sea Drilling Project 32, U.S. Government Printing Office, Washington 891–921.

Menegatti, A.P., Weissert, H., Brown, R.S., Tyson, R.V., Farrimond, P., Strasser, A., et al., 1998. High-resolution $\delta^{13}C$ stratigraphy through the Early Aptian 'Livello Selli' of the Alpine Tethys. Paleoceanography 13, 530–545.

Mienert, J., Flood, R.D., Dullo, W.C., 1994. Research perspectives of sediment waves and drifts—monitors of global change in deep-water circulation. Paleoceanography 9, 893–895.

Mulder, T., 2011. Gravity processes on continental slope, rise and abyssal plains. In: Hüneke, H., Mulder, T. (Eds.), Deep-Sea Sediments. Developments in Sedimentology, Elsevier, Amsterdam, vol. 63, pp. 25–148.

Murray, J., Renard, A.F., 1891. Voyage of H.M.S. Challenger during the years 1873–76. Report on the Deep-Sea deposits based on the specimens collected during the Voyage of H.M.S. Challenger in the years 1873–1876. Eyre and Spottiswoods, London 525pp.

Muttoni, G., Erba, E., Kent, D.V., Bachtadse, V., 2005. Mesozoic Alpine facies deposition as a result of past latitudinal plate motion. Nature 434, 59–63.
Neumayr, M., 1883. Üeber klimatische Zonen während der Jura- und Kreidezeit. Denkschr. Akad. Wiss. Wien (math.-naturw. KL) 47, 277–310, Wien.
Nisbet, E.G., Price, I., 1974. Siliceous turbidites: bedded cherts as redeposited ocean-ridge derived sediments. In: Hsü, K.J., Jenkyns, H.C. (Eds.), Pelagic Sediments: On Land and under the Sea. Blackwell Publishing Ltd, Oxford, UK doi: 10.1002/9781444304855.ch16.
Norris, R.D., Wilson, P.A., 1998. Low-latitude sea-surface temperatures for the mid-Cretaceous and the evolution of planktic foraminifera. Geology 26, 823–826.
Padden, M., Weissert, H., de Rafelis, M., 2001. Evidence for Late Jurassic release of methane from gas hydrate. Geology 223–226.
Pisciotto, K.A., 1981. Review of secondary carbonates in the Monterey Formation, California. In: Garrison, R.E., Douglas, R.G. (Eds.), The Monterey Formation and Related Siliceous Rocks of California, Pac. Sect. Soc. Econ. Paleontol. Mineral. Spec. Publ., 15, 273–283.
Puceat, E., Lecuyer, C., Sheppard, S.M.F., Dromart, G., Reboulet, S., Grandjean, P., 2003. Thermal evolution of Cretaceous Tethyan marine waters inferred from oxygen isotope composition of fish tooth enamels. Paleoceanography 18, 1029. doi:10.1029/2002PA000823.
Rais, P., Louis-Schmid, B., Bernasconi, S.M., Weissert, H., 2007. Palaeoceanographic and palaeoclimatic reorganization around the Middle-Late Jurassic transition. Palaeogeogr. Palaeoclimatol. Palaeoecol. 251, 527–546.
Reboulet, S., Mattioli, E., Pittet, B., Baudin, F., Olivero, D., Proux, O., 2003. Ammonoid and nannoplankton abundance in Valanginian (Early Cretaceous) limestone-marl successions from the southeast France Basin: carbonate dilution or productivity? Palaeogeogr. Palaeoclimatol. Palaeoecol. 201, 113–139.
Reuning, L., Reijmer, J.J.G., Betzler, C., Swart, P., Bauch, T., 2005. The use of paleoceanographic proxies in carbonate periplatform settings-opportunities and pitfalls. Sed. Geol. 175, 131–152.
Robbins, L.L., Blackwelder, P.L., 1992. Biochemical and ultrastructural evidence for the origin of whitings—a biologically induced calcium-carbonate precipitation mechanism. Geology 20, 464–468.
Robinson, S.A., Andrews, J.E., Hesselbo, S.P., Radley, J.D., Denni, P.F., Harding, I.C., et al., 2002. Atmospheric pCO_2 and depositional environment from stable-isotope geochemistry of calcrete nodules (Barremian, Lower Cretaceous, Wealden Beds, England). J. Geol. Soc. 159, 215–224.
Santantonio, M., 1994. Pelagic carbonate platforms in the geologic record—their classification, and sedimentary and paleotectonic evolution. AAPG Bull. 78, 122–141.
Schlanger, S.O., Douglas, R.G., 1974. Pelagic ooze-chalk-limestone transition and its implication for marine stratigraphy. In: Hsu, K.J., Jenkyns, H.C. (Eds.), Pelagic Sediments: On Land and Under the Sea, Int. Ass. Sedimentologists, Spec. Publ., 1, 117–148.
Schlanger, S.O., Jenkyns, H.C., 1976. Cretaceous anoxic events: causes and consequences. Geol. Mijnbouw 55, 179–184.
Scholle, P., Arthur, M.A., 1980. Carbon isotopic fluctuations in pelagic limestones: potential stratigraphic and petroleum exploration tool. Am. Assoc. Pet. Geol. Bull. 64, 67–87.
Schouten, S., Forster, A., Panoto, F.E., Sinninghe Damste, J.S., 2007. Towards calibration of the TEX86 palaeothermometer for tropical sea surface temperatures in ancient greenhouse worlds. Org. Geochem. 38, 1537–1546.
Sclater, J.G., Anderson, R.N., Bell, M.L., 1971. Elevation of ridges and evolution of the central eastern Pacific. J. Geophys. Res. 76, 7888–7915.

Shackleton, N.J., 1967. Oxygen isotope analyses and Pleistocene temperatures re-assessed. Nature 215, 15–17.
Shackleton, N.J., Opdyke, N.D., 1973. Oxygen isotope and palaeomagnetic stratigraphy of Equatorial Pacific core V28-238: oxygen isotope temperatures and ice volumes on a 10^5 year and 10^6 year scale. Quatern. Res. 3, 39–55.
Steinmann, G., 1905. Die geologische Bedeutung der Tiefseeabsätze und des ophiolithischen Massengesteins. Ber Naturforsch Ges Freibg 16, 44–65.
Steinmann, G., 1925. Gibt es Tiefseeablagerungen von erdgeschichtlicher Bedeutung? Geol. Rundsch. 16, 435–468.
Stoll, H.M., Schrag, D.P., 2000. High-resolution stable isotope records from the upper Cretaceous rocks of Italy and Spain: glacial episodes in a greenhouse planet? Geol. Soc. Am. Bull. 112, 308–319.
Stow, D.A.V., Faugères, J.C., Viana, A., Gonthier, E., 1998. Fossil contourites: a critical review. Sed. Geol. 115, 3–31.
Strasser, A., Hilgen, F.J., Heckel, P.H., 2006. Cyclostratigraphy – concepts, definitions, and applications. http://www.ingentaconnect.com/content/schweiz/nis; jsessionid=9o2f40rsfiqhj alexandra. Newsletters on Stratigraphy 42, 75–114.
Suess, E., 1885. Das Antlitz der Erde. Prag 1885.
Svensen, H., Planke, S., Malthe-Sorenssen, A., Jamtveit, B., Myklebust, R., Eidem, T.R., et al., 2004. Release of methane from a volcanic basin as a mechanism for initial Eocene global warming. Nature 429, 542–545.
Swart, P.K., Eberli, G., 2005. The nature of the $\delta^{13}C$ of periplatform sediments: implications for stratigraphy and the global carbon cycle. Sed. Geol. 175, 115–129.
Tappan, H., 1968. Primary production, isotopes, extinctions and the atmosphere. Palaeogeogr. Palaeoclimatol. Palaeoecol. 4, 187–210.
Tercier, J., 1939. Depôts marins actuels et séries géologiques. Eclogae Geol. Helv. 34, 47–100.
Thomson, W., Murray, J., 1911. Report on the scientific results of the voyage of H.M.S. Challenger during the years 1873–76. Narrative 1, .
Tsikos, H., Jenkyns, H.C., Walsworth-Bell, B., Petrizzo, M.R., Forster, A., Kolonic, S., et al., 2004. Carbon-isotope stratigraphy recorded by the Cenomanian-Turonian Oceanic Anoxic Event: correlation and implications based on three key localities. J. Geol. Soc. 161, 711–719.
Turgeon, S.C., Creaser, R.A., 2008. Cretaceous oceanic anoxic event 2 triggered by a massive magmatic episode. Nature 454, 323–326.
Uchman, A., Wetzel, A., 2011. Deep-sea ichnology: The relationships between depositional environment and endobenthic organisms. In: Hüneke, H., Mulder, T. (Eds.), Deep-Sea Sediments. Developments in Sedimentology, Elsevier, Amsterdam, 63 vol. 63, pp. 517–556.
Urey, H.C., Lowenstam, H.A., Epstein, S., McKinney, C.R., 1951. Measurements of paleotemperatures and temperatures of the Upper Cretaceous of England, Denmark and Southeastern U.S. Geol. Soc. Am. Bull. 62, 399–416.
Vine, F.J., 1966. Spreading of the ocean floor: new evidence. Science 154, 1405–1415.
Voigt, S., Aurag, A., Leis, F., Kaplan, U., 2007. Late Cenomanian to Middle Turonian high-resolution carbon isotope stratigraphy: New data from the Münsterland Cretaceous Basin, Germany. Earth Planet. Sci. Lett. 253, 196–210.
Wara, M.W., Ravelo, A.C., Delaney, M.L., 2005. Permanent El Nino-like conditions during the Pliocene warm period. Science 309, 758–761.
Weedon, G.P., 1989. The detection and illustration of regular sedimentary cycles using Walsh power spectra and filtering, with examples from the Lias of Switzerland. J. Geol. Soc. London 146 (1), 133–144.

Weissert, H., 1979. Die Paläoozeanographie der südwestlichen Tethys in der Unterkreide226, Mitt. geol Inst. ETH-Z und Univ.-Z 174pp.
Weissert, H., 1981a. Depositional processes in an ancient pelagic environment: the Lower Cretaceous Maiolica of the Southern Alps. Eclogae Geol. Helv. 74, 339–352.
Weissert, H., 1981b. The environment of deposition of black shales in the Early Cretaceous: an ongoing controversy. In: Warme, J.E., Douglas, R.G., Winterer, E.L. (Eds.), The Deep Sea Drilling Project: A Decade of Progress, Soc. Econ. Paleont. Min. Spec. Publ., 32, 547–560.
Weissert, H., Bréheret, J.G., 1991. A carbonate carbon-isotope record from Aptian–Albian sediments of the Vocontian Trough (Se France). Bull. Soc. Geol. Fr. 162, 1133–1140.
Weissert, H., Erba, E., 2004. Volcanism, CO_2 and palaeoclimate: a Late Jurassic–Early Cretaceous carbon and oxygen isotope record. J. Geol. Soc. 161, 695–702.
Weissert, H., McKenzie, J.A., Channell, J.E.T., 1985. Natural variations in the carbon cycle during the Early Cretaceous. In: The Carbon Cycle and Atmospheric CO_2: Natural Variations Archean to the Present, Geophys. Monogr., 32, 531–545.
Weissert, H., Joachimski, M., Sarnthein, M., 2008. Chemostratigraphy. Newsl. Stratigr. 42, 145–179.
Wendt, J., 1974. Encrusting organisms in deep-sea manganese nodules. In: Hsu, K.J., Jenkyns, H.C. (Eds.), Pelagic Sediments: On Land and Under the Sea, Int. Ass. Sedimentologists, Spec. Publ., 1, 437–447.
Westphal, H., 2006. Limestone-marl alternations as environmental archives and the role of early diagenesis: a critical review. Int. J. Earth Sci. 95, 947–961.
Wignall, P.B., 1994. Black shales. Clarendon, Oxford 144pp.
Wilson, P.A., Norris, R.D., Cooper, M.J., 2002. Testing the Cretaceous greenhouse hypothesis using glassy foraminiferal calcite from the core of the Turonian tropics on Demerara Rise. Geology 30, 607–610.
Winterer, E.L., 1998. Paleobathymetry of Mediterranean Tethyan Jurassic pelagic sediments. Mem. Soc. Geol. Ital. 53, 97–131.
Winterer, E.L., Bosellini, A., 1981. Subsidence and Sedimentation on Jurassic Passive Continental Margin, Southern Alps, Italy. AAPG Bulletin 65, 394–421.
Wise, S.W., Hsü, K.J., 1971. Genesis and lithification of a deep-sea chalk. Eclogae Geol. Helv. 64, 273–278.
Wissler, L., 2001. Response of Early Cretaceous sedimentary systems to perturbations in global carbon cycling. insights from stratigraphy, sedimentology and geochemical modeling. Diss. ETH Zürich. doi:10.3929/ethz-a-004282998.
Wissler, L., Funk, H., Weissert, H., 2003. Response of Early Cretaceous carbonate platforms to changes in atmospheric carbon dioxide levels. Palaeogeogr. Palaeoclimatol. Palaeoecol. 200, 187–205.
Wortmann, U.G., Weissert, H., 2000. Tying platform drowning to perturbations of the global carbon cycle with a $\delta^{13}C_{org}$-curve from the Valanginian of DSDP Site 416. Terra Nova 12, 289–294.
Zachos, J.C., Röhl, U., Schellenberg, S.A., Sluijs, A., Hodell, D.A., Kelly, D.C., et al., 2005. Rapid acidification of the ocean during the Paleocene–Eocene thermal maximum. Science 308, 1611–1615.
Zeebe, R.E., Westbroek, P., 2003. A simple model for the CaCO3 saturation state of the ocean: the "Strangelove", the "Neritan", and the "Cretan" Ocean. Geochem. Geophys. Geosyst. 4, 1104. doi:10.1029/2003GC000538.

CHAPTER 12

Climate Records of Deep-Sea Sediments: Towards the Cenozoic Ice House

Torsten Bickert*,[1] *and* Rüdiger Henrich†

Contents

1. Introduction

Cenozoic climate changes provide a great opportunity to study climate processes under boundary conditions different from today. The transition from the warm climates of the Paleocene to the glacial world of the Pleistocene offers the possibility to study long-term climate changes on a million-year scale as well as short-term events occurring within millennia or even decades. It allows to analyze cyclic climate dynamics as well as unidirectional developments, and to understand the complex interaction of external forcings and internal climate dynamics. The global ocean herein plays a major role as a distributor of heat and moisture, thus linking climate

* Zentrum für Marine Umweltwissenschaften, Universität Bremen, Germany
† Fachbereich Geowissenschaften, Universität Bremen, Germany
[1]Corresponding author.
E-mail address: tbickert@marum.de

Developments in Sedimentology, Volume 63
ISSN 0070-4571, DOI: 10.1016/S0070-4571(11)63012-4

variability of low and high latitudes, and as a major reservoir for dissolved matter and particles. Energy and matter are continuously exchanged in physical and biogeochemical processes; the history of these processes is recorded in the marine sediments.

What marks the boundary conditions of the Cenozoic? First of all, the moving lithosphere plates set up a changing geometry of ocean basins and the redistribution of land masses relative to climatic zones. The continued divergence of the southern continents gave way for the climatic isolation of Antarctica, initiated the opening of the Atlantic, and defined the closing of some tropical gateways like the Central American Isthmus, the eastern Tethys pathway, and the Indonesian throughflow, thereby constraining the distribution of salt and heat from a zonal into a meridional orientation (Fig. 12.1). The Cenozoic slow-down of sea-floor spreading induced the deepening of the ocean basins and hence the long-term lowering of the global sea level (Müeller et al., 2008). Continent collisions caused the elevation of mountain ranges (most of the major orogens uplifted during the Neogene), thereby influencing the atmospheric circulation, the precipitation, and hence the weathering of continental crust and the distribution of plants. The complex interaction of reduced volcanism, increased continental weathering, and the dynamics of the biosphere controlled the long-term decrease of atmospheric pCO_2 from the high level of the Paleocene to almost pre-industrial values over much of the Neogene (Fig. 12.2). As a consequence, glaciers and continental ice caps were formed, thereby increasing the surface albedo and allowing further lowering of the sea level and exposure of the land surface to continental weathering.

The understanding of the ocean's role in Cenozoic climate change is closely connected to the advantages of the Deep-Sea Drilling and Ocean Drilling Programs, which started in 1968 and which are still continued as an

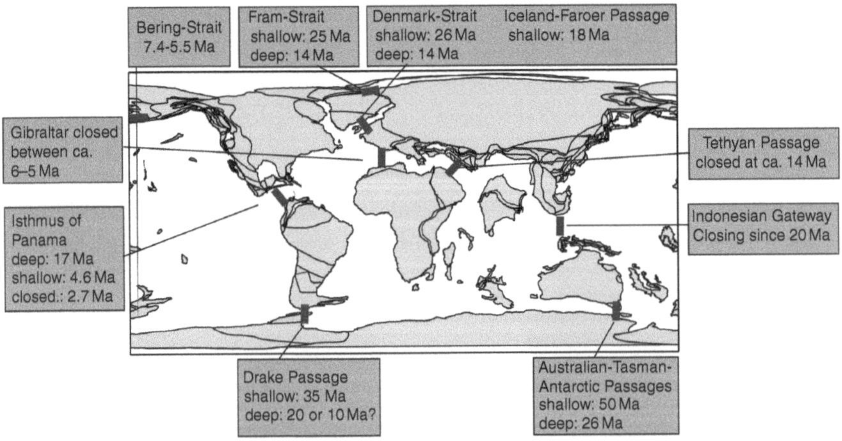

Figure 12.1 Eocene continent configuration with estimated opening and closing dates for ocean gateways (William Hay, 1998, unpublished figure).

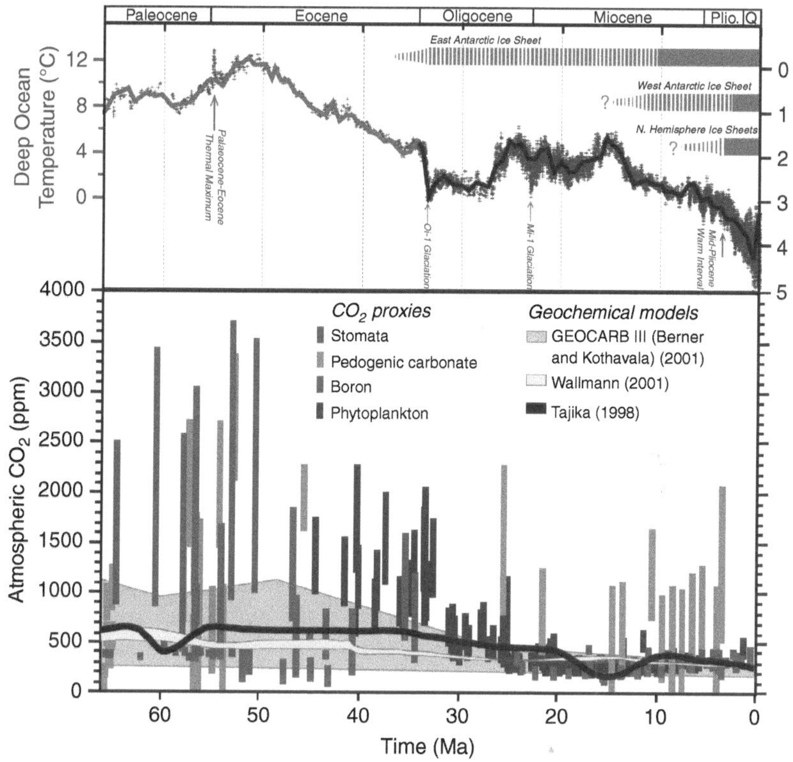

Figure 12.2 Global compilation (top figure) of the $\delta^{18}O$ isotope records of deep-sea benthic foraminifers from 40 Deep-Sea Drilling Program and Ocean Drilling Program sites (Zachos et al., 2001), compared with compiled reconstructions (bottom figure) of atmospheric CO_2 for the last 65 Ma. (modified from IPCC, 2007).

integrated effort of the international marine geoscience community. Substantial progress has been made since the 1990s due to high-resolution paleoclimatic studies of even the oldest sediments of the Cenozoic with a time resolution of only a few thousands of years or even better. In 1999, Nick Shackleton predicted during a meeting at the Royal Society of London that an astronomically calibrated timescale for the entire Cenozoic 'would be feasible over the next few years'. It was his revolutionizing work (Shackleton et al., 1990), co-authored by A. Berger and W. Peltier, which corrected the shortcoming of a too young K/Ar dating of the Brunhes–Matuyama geomagnetic reversal by means of tuning the cyclic $\delta^{18}O$ variations as a climate proxy to the variations of the changing orbital parameters of the Earth. Since then, numerous studies were added aiming at integrating the results of biostratigraphy, magnetic polarity, radiometric dating, and orbitally paced climate variations for establishing a continuous timescale for the Cenozoic (e.g. Gradstein et al., 2004). For the last 5 Myr, a stacked benthic oxygen-isotope record is now available (Lisiecki and Raymo, 2005), which extends and partly corrects the previous standard oxygen-isotope chronology for the late Pleistocene of the SPECMAP group (Imbrie et al., 1984).

A further advantage over earlier studies is the increasing implementation of earth-system modelling experiments to the understanding of the complex nature of interactions between external forcing and climate response as well as of internal oscillations of different climate systems. Key publications include models of long-term chemical processes (e.g. GEOCARB: Berner, 1991), coupled ocean/atmosphere models of intermediate complexity (e.g. Mikolajewicz and Crowley, 1997; Nisancioglu et al., 2003; Schneider and Schmittner, 2006) as well as simple box models to understand the role of ocean reservoirs and matter exchange (e.g. Merico, *et al.*, 2008).

In this chapter, selected episodes of the Cenozoic climate change will be discussed which are crucial for the understanding of various climate processes, that is, the Paleocene–Eocene thermal maximum as an example of methane hydrate outgassing (at 55 Ma), the long-term Eocene cooling followed by the first Antarctic glaciation (between 50 and 33.5 Ma), the Miocene warm climate and its relationship to the extraordinary Monterey carbon deposition (about 17–14 Ma), the final step into the icehouse world, including the Antarctic glaciation and the Arctic Ocean sea-ice covering (starting at 14 Ma), the Northern hemisphere glaciation (NHG) (starting at 3.5 Ma), and the development of the late Pleistocene 100 kyr glacial/interglacial cycles (starting at 1.0 Ma), which set the frame for abrupt warming and rapid melting events (Dansgaard/Oeschger cycles, Heinrich events). In addition, the evolution of the Neogene deep-water circulation will be discussed along with the observed large-scale changes in the carbon cycle and redistribution of nutrients and biogenic silica.

2. The Paleocene–Eocene Thermal Maximum (PETM): Large-Scale Carbon Release and Its Consequences for the Oceans' Carbonate Budget and the Global Climate System

Although life had changed dramatically at the end of the Cretaceous (one of the big five extinction events of the Phanerozoic), the global climate continued in its warm mode which persisted over much of the preceding Mesozoic, characterized by about 5 °C higher global temperatures than today, smaller temperature gradients between high and low latitudes, the absence of almost any continental ice, and atmospheric pCO_2 levels of 1500 ppm or even higher (Fig. 12.2; see also Weissert, 2011, this volume).

Deep-ocean temperatures were as high as 8–10 °C, compared to 0–4 °C in the modern-ocean. The benthic $\delta^{18}O$ record of deep-sea sediments exhibits decreasing values between 65 and 50 Ma, indicating slowly rising deep-ocean temperatures over much of the Paleocene and the early Eocene.

On top of this warm climate, the end of the Paleocene epoch at 55 Ma was marked by an abrupt episode of further global warming coincident with a large perturbation to the global carbon cycle. Within less than 30,000 years (Röhl et al., 2007), temperatures increased about 4–5 °C in the tropics, up to 8 °C in high latitudes, and about 4–5 °C in the deep-ocean. The warming was accompanied by a transient shift in precipitation, enhanced continental weathering, and increased ocean stratification, which may have caused the observed mass extinction of benthic foraminifers (see Higgins and Schrag, 2006, for a summary and references). This abrupt warming, the so-called PETM, was also associated with a 2.5‰ decrease in the carbon-isotopic composition of the global inorganic-carbon pool, first measured in planktonic and benthic foraminifers from the Southern Ocean (Kennett and Stott, 1991), followed by a gradual return to pre-excursion values. Later studies found a similar signal in the Atlantic, Pacific, and Tethys oceans (Fig. 12.3). A higher-amplitude carbon-isotope excursion of 5–6‰ in paleosol carbonates and fossil teeth has been attributed to enhanced ^{13}C-discrimination during carbon fixation in plants (Bowen et al., 2004), and has allowed for correlation between marine and terrestrial sections. In both shallow- and deep-marine records, the isotopic excursion is associated with a large increase in $CaCO_3$ dissolution (Zachos et al., 2005). Together, these observations indicate a massive injection of isotopically light carbon to the oceans and atmosphere, coincident with the onset of the PETM.

The carbon-isotope excursion was the starting point for hypotheses on the PETM. A simple mass-balance calculation neglecting effects such as carbon speciation and $CaCO_3$ dissolution implies that a 2.5‰ drop in the $\delta^{13}C$ of the modern global inorganic-carbon pool requires 40,000 Gt of carbon from the mantle ($\delta^{13}C = -5‰$), 4500 Gt of carbon as organic carbon ($\delta^{13}C = -25‰$), or 1700 Gt of carbon as methane ($\delta^{13}C = -60‰$). Because the mantle and organic source requirements seemed unreasonably large, Dickens et al. (1995) proposed that the PETM carbon-isotope excursion was due to the release of 1100–2100 Gt of methane associated with the destabilization of methane hydrate trapped below the sea-floor. They speculated that the released methane would be rapidly oxidized to produce ^{12}C-enriched CO_2, resulting in an abrupt drop in the carbon-isotopic composition of the ocean/atmosphere system and a shoaling in the depth of the lysocline. Several processes have been suggested to explain such a catastrophic release of methane hydrate. Dickens et al. (1995) estimated a 4–5 °C bottom-water warming sufficient to destabilize methane hydrates equivalent to 1000–2000 Gt C at water depths between 900 and 1400 m. However, this scenario requires an explanation for the initial warming

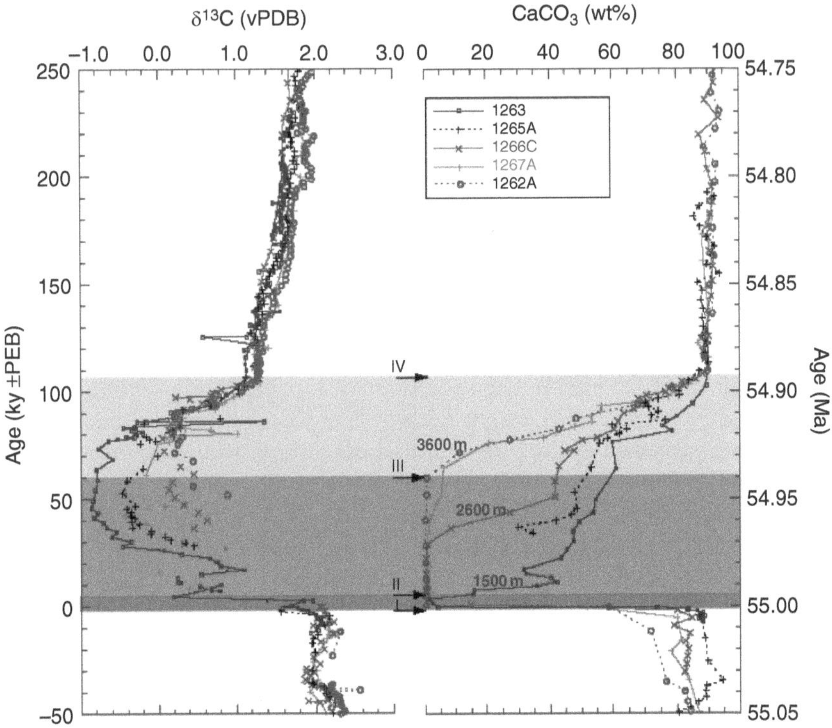

Figure 12.3 PETM carbon-isotope excursions and carbonate decreases in several cores from the Walvis Ridge, South Atlantic (Zachos et al., 2005). Reprinted with permission from AAAS.

independent of the methane. Further explanations included a sudden switch in the locus of deep-water formation, massive slope failures, and sediment slumping associated with seismic activity or erosion (see Higgins and Schrag, 2006). However, in addition to many obstacles against the proposed trigger mechanisms, the magnitude of the carbonate compensation depth (CCD) change at the onset of the PETM seems to require a much larger injection of carbon than consistent with the estimated addition of 1700 Gt C. Furthermore, the prolonged extreme warmth of the PETM precludes CH_4 as the greenhouse gas and requires much more CO_2 than would be supplied by the oxidation of the estimated amount of released CH_4.

Higgins and Schrag (2006) therefore tested other mechanisms for the abrupt warming at the end of the Paleocene. If an estimated oxidation of at least 5000 Gt C of organic carbon were the most likely explanation for the observed geochemical and climatic changes during the PETM, other potential mechanisms are needed to explain the larger amount of released carbon. The production of thermogenic CH_4 and CO_2 during contact metamorphism associated with the intrusion of a huge masses of magma belonging to a large igneous province into organic-rich sediments could have been capable of supplying such large amounts of carbon, but is inconsistent with the lack of extensive carbon loss in metamorphosed sediments, as well as with the abrupt onset and termination of carbon release during the PETM. A global burning of Paleocene peatlands would provide a large terrestrial carbon source, but massive carbon release by fire seems unlikely, as it would require that all peatlands burned at once and then for only 10–30 kyr. In addition, this hypothesis requires an order of magnitude increase in the amount of carbon stored in peat.

The isolation of a large epicontinental seaway by tectonic uplift associated with volcanism or continental collision, followed by desiccation and bacterial respiration of the aerated organic matter is another potential mechanism for the rapid release of large amounts of CO_2. In addition to the oxidation of the underlying marine sediments, the desiccation of a major epicontinental seaway would remove a large source of moisture for the continental interior, resulting in the desiccation and bacterial oxidation of adjacent terrestrial wetlands.

Whatever the result of the ongoing debate will be, it must consider that, apart from the carbon-isotope excursion at the onset of the PETM, several more anomalies are recognized within the Early Cenozoic, including the ELMO event at 53 Ma, the X-event at 51 Ma, the MECO event at 41 Ma, and some more (U. Röhl, personal communication). Although these events are all smaller in amplitude than that of the PETM, they might give a clue for the most plausible mechanism for the PETM, since they all share similarities with respect to carbon release and climate perturbation.

3. Eocene Cooling: Factors Causing the Antarctic Glaciation

The transition from the warm Eocene greenhouse climate some 50 Myr ago to the present glaciated state is one of the most prominent changes in Earth's climatic evolution. It is widely accepted that large ice sheets first appeared on Antarctica about 33.5 Myr ago, coincident with decreasing atmospheric carbon-dioxide concentrations and a deepening of the CCD in the world's oceans (Pagani et al., 1999, 2005; Tripati et al., 2005). The glaciation in the Northern Hemisphere is believed to have begun much later, between 12 and 6 Myr ago. However, there is a substantial controversy on the history of climate and atmospheric CO_2 concentrations during the greenhouse/icehouse transition. Long-term high-latitude cooling began in the Early Eocene about 50 Ma ago, as indicated by an increase in global benthic-foraminifer oxygen isotopes (Fig. 12.2), and a decrease in benthic-foraminiferal Mg/Ca ratios (Lear et al., 2000, 2008). Although some studies have concluded that tropical sea-surface temperatures also decreased gradually during the Eocene, most climate proxy records support warm and stable tropical sea-surface temperatures throughout the Eocene, with evidence for small cooling steps at 48, 45, and 42 Ma ago from planktonic foraminiferal Mg/Ca data (Tripati et al., 2003).

The earliest Oligocene (about 33.5 Ma) is widely accepted as the interval associated with the onset of continental glaciations. High-resolution reconstructions of seawater $\delta^{18}O$ show a +1‰ shift that is interpreted as recording a sudden and massive expansion of the ice volume, along with the occurrence of ice-rafted debris (IRD) in the Southern Ocean and a change in clay mineralogy consistent with increased glacial erosion on Antarctica (see the recent review by Shevenell and Kennett, 2007). Presently, the highest-resolution sedimentary record spanning the Eocene/Oligocene transition comes from equatorial Pacific Site 1218 (Coxall et al., 2005). The benthic-foraminifer $\delta^{18}O$ record from this site reveals that the so-called Oi-1 glaciation occurred in two distinct steps of each 40 kyr long, separated by a period of 200 kyr at a node of low eccentricity and obliquity. The increase in $\delta^{18}O$ is coincident with a 1-km deepening of the Pacific CCD inferred from $CaCO_3$ mass accumulation rates.

Glacial expansion has been linked to a decline in partial pressure of CO_2 on the basis of climate-modelling (DeConto et al., 2007). However, proxy constraints on atmospheric CO_2 levels before 33 Ma ago yield conflicting results. Plant stomata and carbon-isotope-based reconstructions support stable greenhouse conditions during the Eocene, whereas boron-isotope-based surface-water pH reconstructions for the early Eocene indicate highly variable CO_2 concentrations with values ranging from several hundreds to a few thousand parts per million by volume (Fig. 12.2). The question therefore arises whether atmospheric CO_2 plays a key role in the glaciation or is just responding to the drastic climatic change. Several lines of evidence suggest that changes in global carbon cycling were associated with Oi-1. Site 1218 in the Equatorial Pacific shows that the percentage of $CaCO_3$ at the site increases abruptly at 34 Ma in two steps of 40 kyr each, similar to that of the benthic-foraminifer $\delta^{18}O$ record. On the other hand, the Site 1218 carbonate records reveal a 10 kyr lag relative to $\delta^{18}O$ in the 40-kyr band, which suggests that changes in the global carbon cycle related to Earth's obliquity could both have forced the Oi-1 event (Fig. 12.4). The lag of the CCD records with respect to the $\delta^{18}O$ record indicates that the CCD increase was likely a result of Antarctic cryosphere expansion and related to a shift in global $CaCO_3$ sedimentation from the continental shelves to the deep-ocean (Coxall et al., 2005).

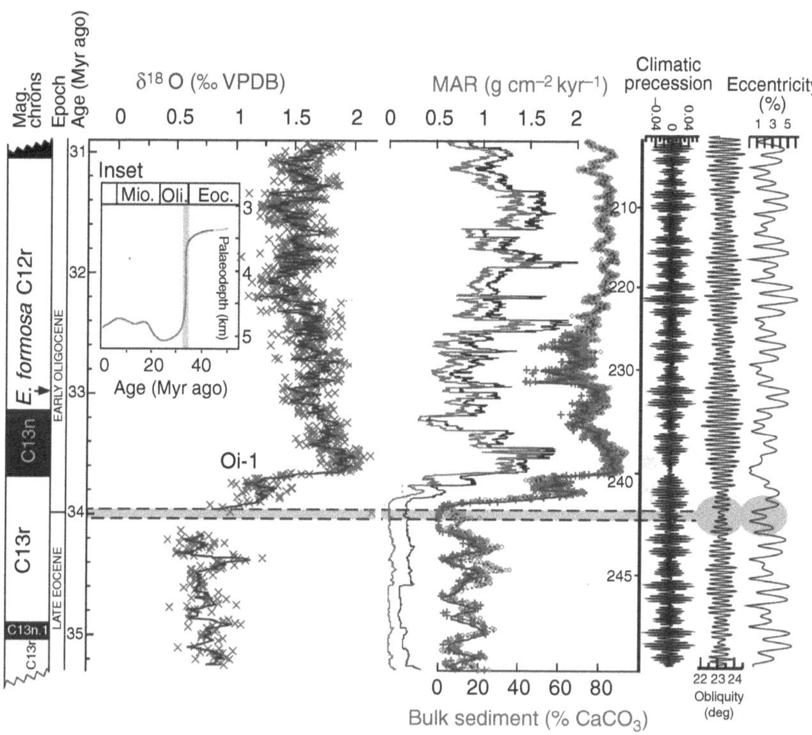

Figure 12.4 Paleoceanographic records from ODP Site 1218 showing that the $\delta^{13}C$ increase occurred in two steps of 40 ka each, synchronously with the stepwise onset of major permanent Cenozoic Antarctic ice-sheets ($\delta^{18}O$ increase in benthic-foraminiferal calcite) and during an eccentricity minimum and low-amplitude obliquity change (grey shading) favouring cool summers. $CaCO_3$ increases with a certain lag (green line). (modified from Coxall et al., 2005).

Other explanations for the rapid onset of the Antarctic glaciation invoke the tectonic opening of the ocean gateways between Antarctica and Australia (Tasmanian Passage), and Antarctica and South America (Drake Passage), leading to the organization of the Antarctic Circumpolar Current and the thermal isolation of Antarctica (Exon et al., 2000) (see Section 3). This notion is supported by model simulations of the oceans' general circulation, showing that the opening of Drake Passage and the organization of an Antarctic Circumpolar Current reduces southward oceanic heat transport and diminishes Southern Ocean sea-surface temperatures by approximately 3 °C (Nong et al., 2000). However, although most tectonic reconstructions place the opening of the Tasmanian Passage close to the Eocene/Oligocene boundary, the Drake Passage may not have provided a significant deep-water passage until several million years later (Lawver and Gahagan, 1998).

Additionally, these model simulations lack realistic atmospheric components, so that the effects on Antarctic climate are unresolved. Furthermore, it is noteworthy that during the Late Oligocene, decreasing benthic $\delta^{18}O$ values indicate an extensive loss of continental ice, and the recurrence of a warm climate, which persisted throughout the Early Miocene until the mid-Miocene climate optimum some 15 Ma ago.

4. The Middle Miocene Climate Transition

A further significant cooling of the Earth's climate system started in the Middle Miocene, as evidenced by an abrupt 1‰ increase in global benthic-foraminifer $\delta^{18}O$ at 14 Ma (Fig. 12.2; Zachos et al., 2001). As at the Eocene/Oligocene transition, this $\delta^{18}O$ increase occurred within two obliquity cycles, at a time of low eccentricity as well as obliquity amplitudes (Holbourn et al., 2005). This first step, the so-called Mi-3a event (Miller et al., 1991), was followed by further maxima in $\delta^{18}O$, occurring at 13.2, 11.7, and 10.4 (Mi-events 3b, 4, and 5, Fig. 12.5). Evidence for both ice growth and global cooling is found throughout the geological record of the Middle Miocene. Southern Ocean IRD is more abundant after 14 Ma; large fluctuations in global sea level are inferred from seismic sections, paleobotanical and faunal change occurred, and the East Antarctic Ice Sheet expanded across the Antarctic continental margin (see Shevenell and Kennett, 2007, for a summary and references). Surprisingly, the sea level rise as reconstructed from sequence stratigraphy (Fig. 12.5) was most significant at 10.4 Ma (Mi-5), and only minor at the first event at 14 Ma and at the other $\delta^{18}O$ maxima. The magnitudes of Middle Miocene Antarctic ice growth and temperature change have been estimated using paired measurements of Mg/Ca as a paleotemperature proxy and temperature-corrected $\delta^{18}O$ for ice volume estimates (e.g. Billups and Schrag, 2002; Shevenell et al., 2008). Results from these studies suggest that 70% of the global 1‰ benthic $\delta^{18}O$ increase at 14 Ma relates to an increase in Antarctic ice volume, and global deep waters are inferred to have cooled 1.8–2.5 °C between 14.2 and 13.8 Ma, consistent with similar estimates subtracting the sea level effect on $\delta^{18}O$ using the results of sequence stratigraphy. Time-series analysis of benthic $\delta^{18}O$ records from the western Pacific suggests that the Antarctic ice sheets entered an interval of eccentricity-modulated glacial advance and retreat at 14 Ma, whereas it was dominated by obliquity response in the interval before (Holbourn et al., 2005).

Glacial episodes increased in intensity between 15 and 13.8 Ma, revealing a central role for internal climate feedbacks (e.g. ice-albedo feedbacks) in this major Cenozoic climate transition (Shevenell et al., 2008). Subantarctic sea-surface temperatures have been shown to cool down since 14.2 Ma,

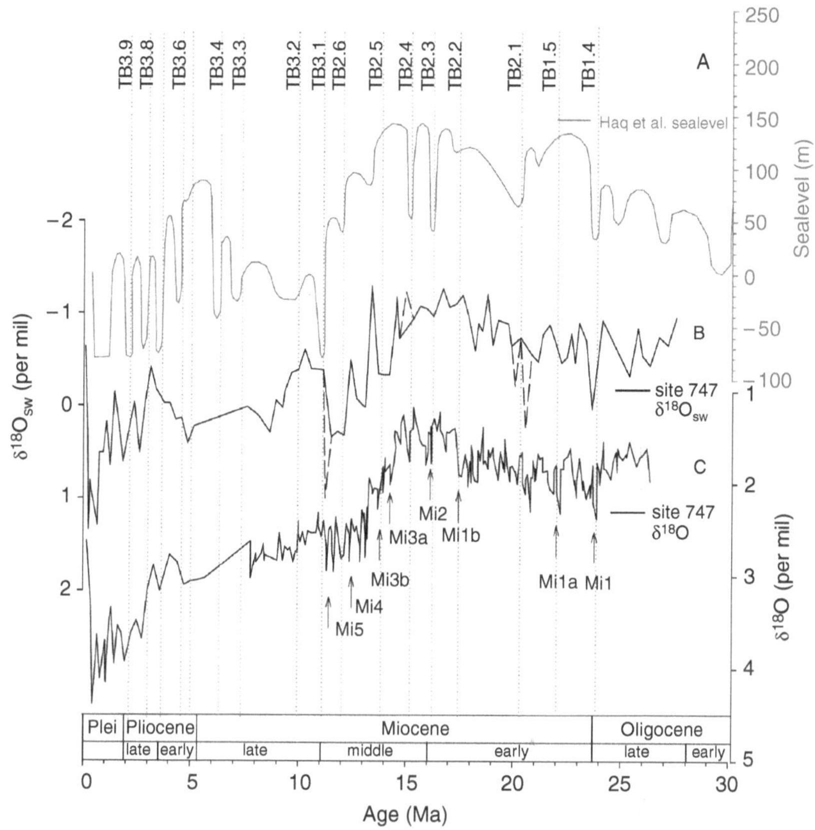

Figure 12.5 Comparison of (A) the Haq et al. (1987) sea level curve, (B) the $\delta^{18}O_{sw}$ derived from paired measurement of Mg/Ca ratios and benthic-foraminiferal $\delta^{18}O$ values at Site 747, and (C) the Site 747 benthic $\delta^{18}O$ record (modified from Billups and Schrag, 2002). Vertical dotted lines reflect the third-order sequence boundaries of Haq et al. (1987). Sequence boundaries coincide with times of increasing $\delta^{18}O$ values (Mi-events; Miller et al., 1991).

some 200 ka before the increase in benthic $\delta^{18}O$ and continental ice build-up (Kuhnert et al., 2009; Shevenell et al., 2004), heralding that surface processes might have triggered the rapid glaciation of Antarctica. However, reconstructions of pCO_2 do not show significant minima before the Mid-Miocene climate transition, but rather exhibit fairly constant low levels for the Oligocene and the entire Neogene (Pagani et al., 1999, 2005; Fig. 12.2).

Especially these low pCO_2 concentrations during the Neogene rule out previous ideas like the Monterey hypothesis of Vincent and Berger (1985), who suggested that a pCO_2 draw-down by sequestration of large amounts of carbon in organic-rich deposits surrounding the Pacific Rim may have triggered the mid-Miocene Antarctic glaciation. Furthermore, the 2.5-Myr offset between the changes in global carbon cycling (initial increase in $\delta^{13}C$ at 16.5 Ma) and the $\delta^{18}O$ step at 14 Ma seems to be unrealistic. However, modelling studies suggest that the low atmospheric pCO_2 concentrations may have preset the sensitivity of the Antarctic system to low-latitude-derived heat and moisture advection. Holbourn et al. (2005) showed from spectral analysis that, between 14.7 and 13.8 Ma, the ice volume was strongly modulated by obliquity-paced atmospheric moisture transfer. A change in eccentricity cadence (from 400 to 100 kyr) led initially to ice-sheet decay and instability from 14 to 13.9 Ma, and the ice sheet expanded rapidly from 13.9 Ma on during an extended interval of low seasonal contrasts at low eccentricity. The observed intensification of the Antarctic Circumpolar Current might have associated with the postulated changes in atmospheric circulation and acted as an amplifier for the internal climate feedback (Kuhnert et al., 2009). The $\delta^{13}C$ increase right after the Mi-3a event suggests a change in the carbon storage in the deep-sea, similar to that accompanying the Oi-1 glaciation (Holbourn et al., 2005). However, beside low-resolution CCD reconstructions, which indicate a further deepening of the carbonate compensation for most ocean basins, there is no specific study on the carbonate system available yet to confirm this hypothesis. It is remarkable that, like in the Late Oligocene, decreasing benthic $\delta^{18}O$ values and a rising sea level as reconstructed from sequence stratigraphy indicate also for the Late Miocene a significant loss of continental ice, and the recurrence of a warm climate, which persisted throughout the Early Pliocene: the so-called Pliocene warm phase some 5–3 Ma ago (see section 7).

5. Neogene Evolution of Deep-Water Circulation and Chemistry

During the Middle Miocene, the continents were already close to their modern position. However, lithosphere plate movements led to changes in the configuration of interoceanic passages, which significantly influenced the ocean circulation. The Central American Seaway (CAS) narrowed from 15 Ma on until its final closure around 2.7 Ma (Bartoli et al., 2005; Coates et al., 2003; Duque-Caro, 1990; Steph et al., 2006). Decisive for the interoceanic deep-water exchange was the emergence of the volcanic arc south of Central America and its collision with northern Columbia between 12.8 and 7.1 Ma (Coates et al., 2004), which hindered a deep-water exchange between the Pacific and the Atlantic. The connection from the Tethys to the Indian Ocean, which already narrowed between 18 and 16 Ma but reopened afterwards north of the Arabian Plate, finally closed around 14 Ma (Meulenkamp and Sissingh, 2003; Roegl, 1999). The Fram Strait widened since 17.5 Ma to a passage deeper than 2000 m at 13.7 Ma (Jakobsson et al., 2007; Jokat et al., 2007). The Greenland–Scotland Ridge deepened between 18 and 15.5 Ma and then again from 12.5 Ma on (Wright and Miller, 1996), perhaps leading to the initiation of a major deep-water outflow into the North Atlantic starting at 11.5 Ma (Wei and Peleo-Alampay, 1997). Erosional features south of the Denmark Strait like hardgrounds and glaucony rip-up clasts have been cited to indicate intensified bottom-water currents related to such an overflow. However, high sediment accumulation rates on the Eirik and Garda Drifts (see Section 3—Fig. 3.5) south of Greenland suggest a major overflow of bottom-water through the Denmark Strait not prior to 8 or 7 Ma (Wold, 1994). On the other hand, drift sediments from the deep North-east Atlantic indicate a major dense water flow over the Iceland-Scotland-Ridge already at about 17 Ma (Björn and Garda Drifts; see Section 3) and a compensating northward flow at 14 Ma (Hatton and Snorr Drifts) (Wold, 1994).

The major patterns of Miocene deep-water circulation have first been drawn in the synthesis by Woodruff and Savin (1989), using a global set of benthic carbon and oxygen-isotope records as well as foraminifer faunal counts from a set of deep-sea drilling cores. A major finding of this compilation was that Early Miocene intermediate- and deep-water masses were aging in a northward direction, with a common source somewhere around Antarctica. There was no evidence for North Atlantic Deep Water (NADW) formation prior to about 14.5 Ma. Furthermore, isotope and faunal data suggested the influx of warm and saline water from the Tethys into the northern Indian Ocean. From 14.5 Ma on, increasing $\delta^{13}C$ values in the North Atlantic indicate a weak flux of NADW, which intensified during the Late Miocene, initiating a modern-ocean-like circulation. However, Late Miocene NADW formation was still less than today (Fig. 12.6). The shift of silica deposition from the North Atlantic to the Indian and Pacific Oceans and later to the Southern Ocean (Keller and Barron, 1983) goes along with the Middle Miocene reorganisation of the global overturning.

Since the compilation by Woodruff and Savin (1989), many benthic isotope records have been added which allow for studying the Neogene climate response at a much higher resolution and hence for a better timing of events (e.g. Zachos et al., 2001, and references therein; Andersson and Jansen, 2003; Billups, 2002; Bickert et al., 2004; Holbourn et al., 2005; Poore et al., 2006; Shevenell and Kennett, 2004; Shevenell et al., 2004, 2008; Westerhold et al., 2005). The highest benthic $\delta^{18}O$ values in the Atlantic sector of the Southern Ocean hint to a common deep-water source north of the Weddell Sea during the Middle Miocene (Paulsen, 2005). However, it lasted until the Late Miocene for a geochemical basin-to-basin

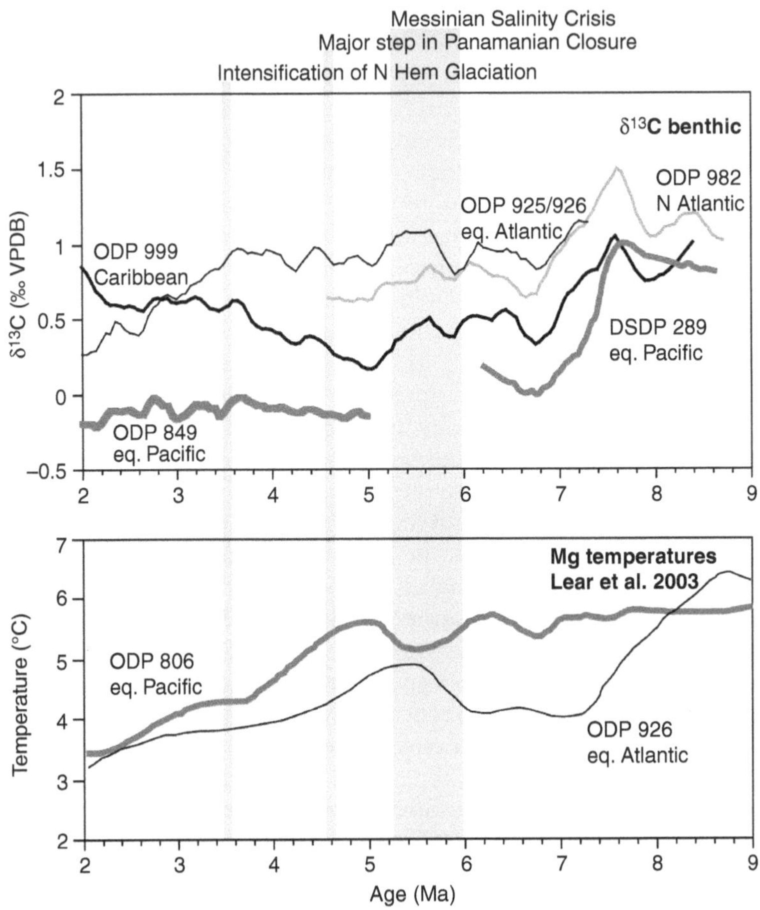

Figure 12.6 Late Miocene to Pliocene smoothed benthic carbon-isotope records of various sites representing the intermediate North Atlantic (Site 982: Hodell et al., 2001), the deep equatorial Atlantic (Sites 925 and 926: Bickert et al., 1997b; Billups et al., 1997; Tiedemann and Franz, 1997; Shackleton and Hall, 1997), the equatorial Pacific (Site 289: Wright et al., 1991; and Site 849: Mix et al., 1995), and the Caribbean (Site 999: Bickert et al., 2004). These trends are compared to the deep-water Mg temperatures of the deep Atlantic (Site 926) and Pacific (Site 806) published by Lear et al. (2003). See Bickert et al. (2004) for figure and references.

fractionation to develop, as indicated by diverging benthic $\delta^{13}C$ records of the deep Atlantic as well as the Southern Ocean (Bickert et al., 2004; Billups, 2002; Poore et al., 2006). This development of the interoceanic gradient between 9.5 and 7.5 Ma indicates that the Late Miocene intensification of the NADW production started much earlier than the onset of the final closure of the CAS, which is supposed to have begun around 4.6 Ma (Haug and Tiedemann, 1998).

Opposite to the results from benthic $\delta^{13}C$ data, Frank et al. (2002) concluded from Nd- and Pb-isotope time series that a continuous and strong export of NADW persisted throughout the interval from 14 to 3 Ma until the onset of NHG. However, the signatures of many sources for neodymium in the North Atlantic are indistinguishable from those of the Southern Ocean; only Labrador Sea water exhibits clearly negative εNd values (e.g. von Blanckenburg and Nägler, 2001). In a more recent study by Thomas and Via (2007), εNd data from a depth transect on Walvis Ridge suggest a change in deep-water stratification which began between about 10.6 and 7.3 Ma. A potential cause for the decrease in the deep South Atlantic εNd values is seen in the onset of deep-water convection in the Labrador Sea. Alternatively, an increase in the delivery of non-radiogenic Nd via glacial weathering could have produced the characteristic low Labrador Sea Nd-isotopic signature. Both the onset of deep convection in the Labrador Sea and the increase in glacial delivery of more non-radiogenic Nd to the region of downwelling might be explained by a first significant cooling of South Greenland during the Late Miocene. However, as stated above, a permanent ice cover on Greenland is not recognized before 7.3 Ma (St. John and Krissek, 2002).

Modelling experiments on the Neogene ocean circulation by Maier-Reimer et al. (1990), Mikolajewicz et al. (1993), Mikolajewicz and Crowley (1997), and—more recently—by Nisancioglu et al. (2003), Prange and Schulz (2004), and Von der Heydt and Dijkstra (2006) aimed to explore the physical response of ocean circulation to gateway changes, mostly related this response to the narrowing of the CAS. While former studies confirm the observation of a weak Atlantic overturning during the Early Miocene of a deep CAS sill (Maier-Reimer et al., 1990; Mikolajewicz and Crowley, 1997; Mikolajewicz et al., 1993), later studies indicate that small but significant NADW formation occurred even during times of an open and deep CAS. However, this proto-NADW was mostly exported through the CAS into the deep Pacific, where it traversed the basin from East to West in a relatively narrow zonal jet, then turned southwards along the western boundary, before it joined the Antarctic Circumpolar Current (Nisancioglu et al., 2003). Heinze and Crowley (1997), and later Heinze and Dittert (2005) studied the sedimentary response to certain ocean gateway changes with respect to carbonate dissolution and opal deposition. Modelling experiments, consistent with sedimentary data, showed a significant shallower calcite lysocline in the North Atlantic and a deeper saturation level in the Pacific for an open and deep CAS.

6. Middle to Late Miocene Carbonate Deposition

Another phenomenon of the Neogene, and maybe associated with the development of the ocean circulation, was the Middle to Late Miocene interval of carbonate crashes. Drastic reductions of calcium-carbonate contents, lower carbonate MARs, and poorer preservation of calcareous microfossils characterize pelagic sediment successions at the Middle/Late Miocene transition (Fig. 12.7). These phenomena have been first identified in several Ocean Drilling Program (ODP) Leg 138 sites of the eastern equatorial Pacific (Lyle et al., 1995). Similar occurrences of carbonate

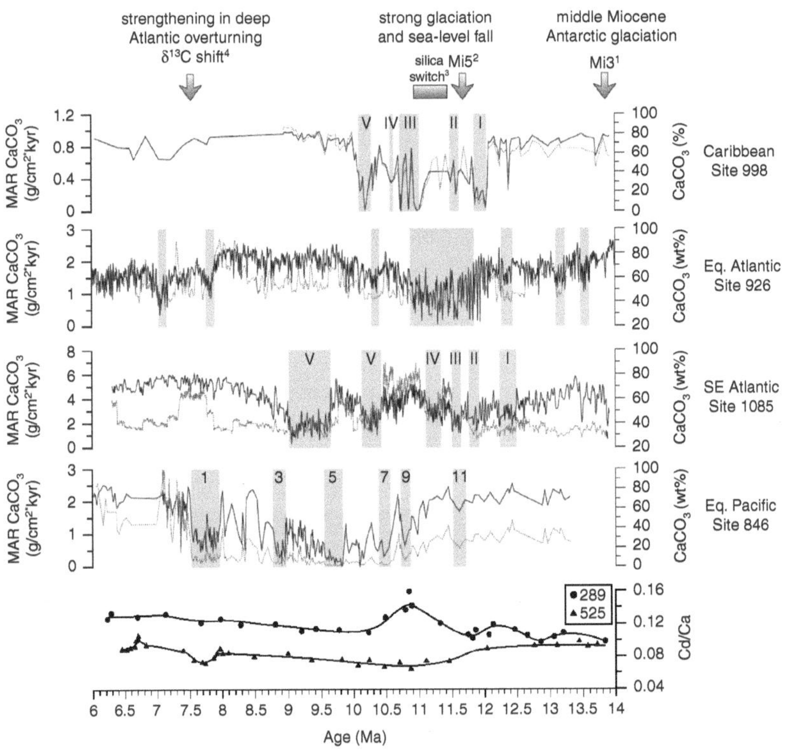

Figure 12.7 Variations of carbonate content (black line) and MAR $CaCO_3$ (grey dotted line) values between 14.0 and 6.0 Ma in the Caribbean Sea (Site 998: Roth et al., 2000), in the Equatorial Atlantic (Site 926: King et al., 1997), in the eastern South Atlantic (Site 1085: this study) and the East Equatorial Pacific (Site 846: Lyle et al., 1995) with carbonate-crash periods assigned by the above-mentioned authors. The lower panel shows Cd/Ca data of the Atlantic DSDP Site 525 and the Pacific DSDP Site 289 from Delaney (1990) with a moving average. (modified from Westerhold, 2003).

reductions have been recorded in other parts of the world oceans, for example, in the western equatorial Atlantic (King et al., 1997) and in the Caribbean Sea (Roth et al., 2000). Westerhold (2003) reported several events of unusually low carbonate contents from late Neogene successions at ODP site 1085 off Namibia. Based on high-resolution X-ray fluorescence (XRF) core logs, they showed that the carbonate reductions occurred within only a few thousands of years, followed by a long interval (of some hundreds of thousands of years) of low carbonate contents and a consequently slow rise of carbonate contents to previous values. During these events, the carbonate contents show cyclic variations related to the cyclicity of orbital parameters of insolation, where 41-kyr cycles dominate the interval of low carbonate contents and 100-kyr cycles dominate the intervals before and after the carbonate crash. This observation reveals a changing response of carbonate sedimentation to orbital forcing, with a stronger influence of sea level changes during times of low carbonate contents. Such an interpretation is corroborated by higher iron contents during periods of slow carbonate accumulation.

The comparable nature of carbonate reductions in the Pacific, Atlantic, and Caribbean suggests a common cause associated with changing ocean circulation. Lyle et al. (1995) attributed the carbonate crashes observed in the eastern equatorial Pacific successions to the emergence of the Isthmus of Panama. Although the isthmus was still below sea level for most of the time of the crash, they calculated that a restriction of 2 Sv of carbonate-rich deepwater from the Atlantic to the Pacific would account for the observed loss of carbonate accumulation west of the isthmus. These observations are consistent with sensitivity experiments carried out by Heinze and Crowley (1997) using a coupled ocean circulation, carbon-cycle, and sediment-chemistry model. Model-calculated differences from the case for an open Central American Isthmus to the present ocean revealed a weaker North Atlantic thermohaline cell and consequently a shallower lysocline in the North Atlantic and a deeper lysocline in the Pacific. Roth et al. (2000) explained the occurrence of the Caribbean carbonate crashes by a global reorganization of the thermohaline circulation at the Middle/Late Miocene transition. The re-establishment of the NADW production at this time caused an influx of corrosive intermediate water entering the Caribbean basins and ultimately resulted in a strong dissolution of calcareous sediments.

However, none of these models explains the different timing of the carbonate-reduction events in different regions. It is evident from Fig. 12.7 that the most distinctive reductions at Caribbean site 998 occurred between 12.3 and 10.7 Ma, while in the South Atlantic Site 1085 the strongest events occurred between 10.3 and 9.5 Ma. In the eastern Pacific, the main carbonate reductions showed up between 10.5 and 9.0 Ma, but even at 7.5 Ma a late event occurred. Furthermore, the congruent variability of the shallow-water Site 1085 (1750 m water depth) and the deep Atlantic Site

926 (3598 m water depth) carbonate records contradicts the assumed gradients between intermediate- and deep-water masses. So the question arises, whether the similar phenomenon of carbonate reduction might have various causes at different locations.

In a recent study, Kastanja and Henrich (2007) showed for Miocene South Atlantic sediments on the Walvis Ridge that short-term dissolution events were closely related to variations in NADW circulation in the deep circulation loop of the South Atlantic. Applying a new preservation index based on the abundance of the coarse calcareous silt fraction (10–63 μm) and a conventional foraminifer preservation index, Kastanja and Henrich (2007) registered overall good to moderate preservation in the Miocene sections, evidencing a persistent NADW supply to this southern location. However, significant decreases of preservation at 17.3, 16.3, 13.9, 13.2, 11.6, and 10.4 Ma were found to coincide with Miocene glacial events (Mi-events: Fig. 12.8), suggesting an increase of Southern Component Water influence during these intervals, which occurred as a response to the intensification of the Antarctic ice-sheet development. The Southern Component Water acts as the major deep-water source during the global Miocene cooling. At 10.4 Ma, a change to overall better preservation points to a weakening of this water mass that might have occurred as a response to the initiation or strengthening of NADW.

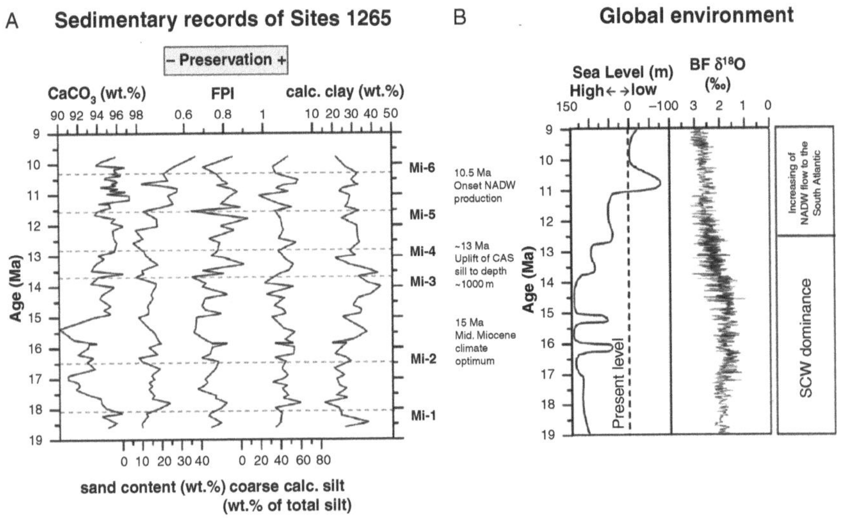

Figure 12.8 (A) Temporal variation of carbonate content, sand content, FPI (foraminifer preservation index), coarse calcareous silt fraction, and calcareous clay at Site 1265. Dashed lines are Mi-glaciation event (Miller et al., 1991). (B) The eustatic sea level curve (Hardenbol et al., 1998) and oxygen-isotope record (Zachos et al., 2001). (modified from Kastanja and Henrich, 2007).

At two other sites on the continental margin off Namibia, Kastanja et al. (2006) found that the drastic reductions in carbonate contents and carbonate accumulation rates were not related to dissolution but caused by drastic reductions of coccolith production rates. They showed that a first major drop in $CaCO_3$ concentration between 10.4 and 10.1 Ma was related mainly to changes in calcareous nannoplankton production, whereas they thought another drop between 9.6 and 9.0 Ma to have been triggered by a combination of production changes of calcareous nannoplankton and dilution, the latter presumably occurring in response to high supply to this region from the shelf during global lowering of the sea level (Fig. 12.9).

In conclusion, the Middle to Late Miocene carbonate-crash events mark a phase of major perturbations in the marine carbonate system, which were obviously associated with several steps in the reorganisation of global deep- and intermediate-water circulation, affecting the various parts of the global

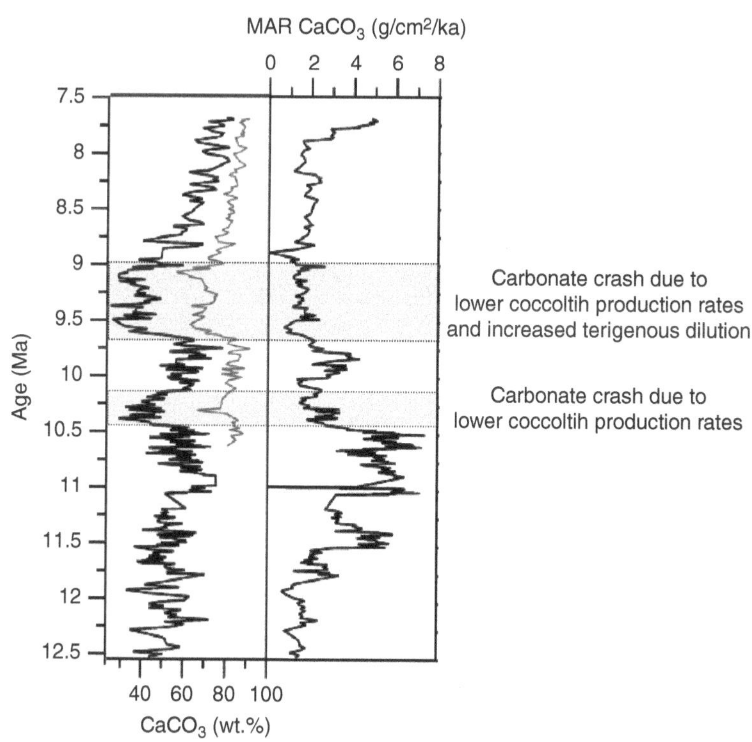

Figure 12.9 Record of carbonate contents ($CaCO_3$ wt%) and mass accumulation rates of carbonate (MAR $CaCO_3$) in sediments from Sites 1085 (black line) and 1087 (grey line) off Namibia. Carbonate-crash events are indicated by the light grey, shaded intervals; an interpretation of the assumed causes for each crash event is added. (modified from Kastanja et al., 2006).

ocean basins differently in time and space. Apparently, these reorganisations and perturbations were induced by significant changes in the plate tectonic. In particular, the Panama gateway may have played a major role. In addition, increased uplift of high mountain ranges on the continents and the associated increase in chemical weathering as well as the initiation, respectively strengthening, of coastal upwelling may have strongly altered the chemistry of intermediate- and potentially also surface-water masses. Interestingly, this may have affected the carbonate production rates of the main calcareous plankton groups. This in turn may have induced adaptations and evolutionary trends.

7. The Onset of the Northern Hemisphere Glaciation and Pleistocene Ice Ages

A widespread view is that ice build-up on the Northern Hemisphere was initiated during the latest Pliocene, very much delayed compared to the much earlier onset of the Oligocene ice growth on the Antarctic continent. However, there is clear evidence from the Arctic realm for a much earlier onset of the NHG from various climatic records on land and in the sea. Figure 12.10 displays a compilation of the IRD record and oxygen-isotope data of the past 13 Ma from the Norwegian-Greenland Sea for key locations on the Vøring Plateau (ODP sites 642 and 644) at the Mid-Norwegian margin and in the Iceland Sea (ODP 907). At the Norwegian margin, IRD grains were observed first at 12.6 Ma. In addition, there is evidence for early IRD occurrences from the Fram Strait at 14 Ma (ODP Site 909: Wolf-Welling et al., 1996). At the Vøring Plateau, the first IRD is followed by a long interval with only sporadic occurrences of a few IRD grains. An increase to the order of 100 grains per gram sediment is registered during the Messinian, with pulses at 6.9 Ma at Site 642 and more continuously from about 6.3 Ma at both sites. However, IRD-free intervals are observed in between, for example, at 5.3–5.0 and 4.4–4.2 Ma. At the western margin of the Norwegian-Greenland Sea, the IRD record dates back to 7.5 Ma in the Baffin Bay (Site 645: Korstgard and Nielsen, 1989) and diamictites and dropstones off south-east Greenland can be found from 7 Ma on (Larsen et al., 1994). However, the most significant increase in IRD, up to the order of several thousands of grains per gram sediment, is indicated at 2.9 Ma at Site 907 in the Iceland Sea, and at 2.75 Ma on the Vøring Plateau. Comparison with the benthic oxygen-isotope records of the same sites displays a clear trend of a gradual cooling of deep waters, with major events on the general cooling superimposed at approximately 11 and 6.4 Ma, almost simultaneously with increases in IRD.

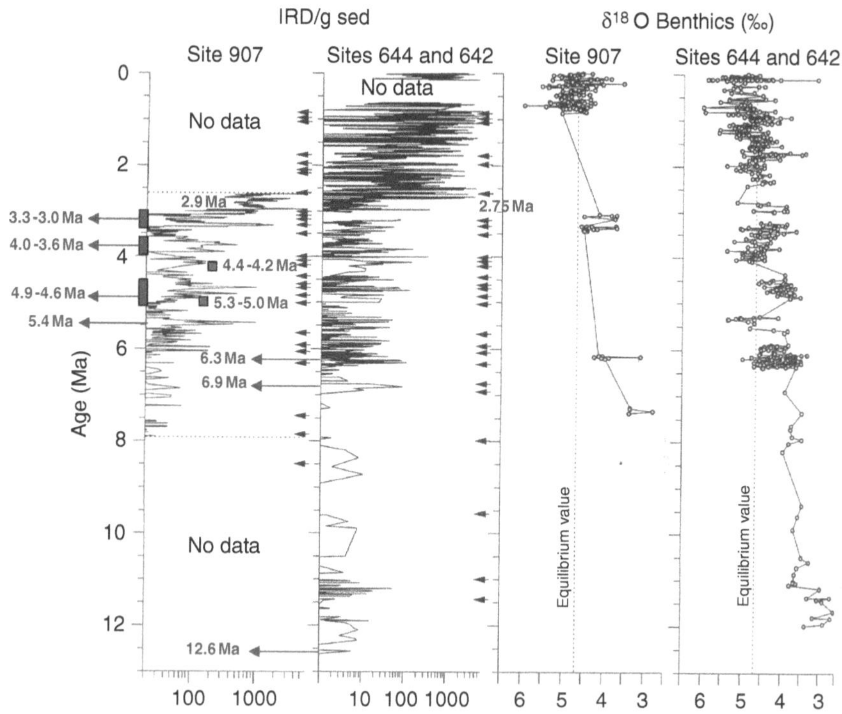

Figure 12.10 IRD and benthic oxygen-isotope records from key locations in the Nordic Seas. (modified from Fronval and Jansen, 1996).

Terrestrial climate records from the continents surrounding the Nordic seas provide additional information on the timing of the NHG. Early evidence comes from glaciomarine deposits of the Yakataga Formation in Alaska. However, the ages of these deposits vary between 16 Ma and less than 10 Ma (Marincovich, 1990). In addition, climate records from middle North America and Beringia (Wolfe, 1994) indicate a mean annual temperature of 6 °C between 14 and 12 Ma. Pollen data from eastern Iceland indicate a change from a warm temperate forest to a boreal forest, which reads as a cooling of approximately 10 °C around 9.6 Ma (Mudie and Helgason, 1983).

With respect to pelagic sedimentation, Bohrmann et al. (1990) could show from sediments of the Baffin Bay that, after a major hiatus from 12.6 to 10 Ma which correlates to the onset of the Iceland-Scotland overflow (see figures 3 and 4 in Bohrmann et al. (1990)), alternating phases of pelagic-carbonate and biogenic-silica sedimentation occurred during the Late Miocene and Early Pliocene. The carbonate-rich intervals represent an increased influx of warm Atlantic water associated with an overall increased temperature gradient within the Norwegian-Greenland Sea, evidenced by colder currents in the western regions. On the other hand, elevated opal accumulation rates from 9.3 to 8.7 Ma and from 5.4 to 4.8 Ma may be correlated with a reduced water mass exchange due to a lowering of the sea level around the critical sill depth at the Greenland–Scotland Ridge (Wright and Miller, 1996). South of the Greenland–Scotland Ridge, the record of ODP Site 646 in the Labrador Sea displays a significant increase in the bulk sediment accumulation rate, which Bohrmann et al. (1990) attributed to the onset of the Denmark Strait overflow. At 4 Ma, a pronounced increase in opal accumulation indicates the first large extension of sea-ice at the Greenland margin.

A major increase in the intensity of the NHG took place around 2.75 Ma. However, looking at the records very precisely, this sudden increase of IRD supply to the oceans appears to occur slightly time transgressive. On the Iceland Plateau, the first high-amplitude IRD peaks are found at 2.9 Ma (Site 907, Fig. 12.10), whereas they are registered at 2.75 Ma on the Vøring Plateau (Site 644). Up to 1.2 Ma ago, only low to moderate IRD input is recorded, with the exception of high peaks at 2.5 Ma on the Vøring Plateau and at 2.35 and 2.0 Ma on the Iceland Plateau. In addition, there is a pronounced increase in frequency and amplitude of IRD pulses at both sites since 1.55 Ma. Another significant shift towards higher amplitudes of terrigenous detritus is recorded at about 600 ka.

These stepwise increases of IRD supply are assumed to reflect consistent expansions of the Scandinavian ice sheet (Henrich and Baumann, 1994). This is supported by the occurrence pattern of a peculiar IRD component, for example, reworked nannofossils derived from Cretaceous chalk. Since chalk deposition in the Cretaceous was restricted to areas south of the NGS, for example, in the Baltic and North Sea region, its occurrence as ice-rafted nannofossils in glacial sediments in the NGS points to icebergs derived from these southern source regions. This feature can thus be used as an indicator of huge ice sheets expanding into these southern locations. In this respect, it is interesting that the first occurrence of reworked nannofossils (Fig. 12.11) coincides with the pronounced amplitude increase of the IRD records at about 1 Ma. In addition, significantly higher nannofossil grain percentages are observed during the glaciations between isotope stages 22 and 16, as well as within stages 12, 10, and 6. This observation might be taken as an indicator of extraordinary large ice sheets during these glacials.

Additional information about the dimension and flow dynamics of the northern ice sheets can be gained by investigating the seismic architecture of the outer shelves and slopes. The flow from the different Nordic ice sheets is focused into ice streams following the major troughs on the shelves and building up huge trough mouth fans, which are internally structured by seaward prograding thick glaciomarine sediment wedges. An obvious

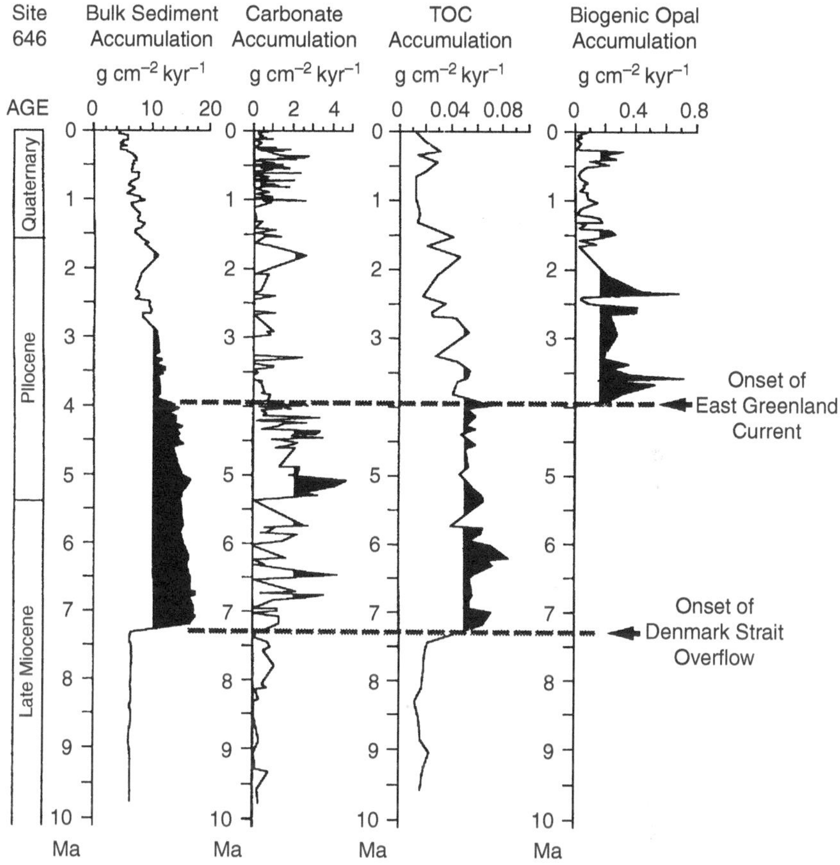

Figure 12.11 Accumulation rates of bulk sediment, carbonate, total organic carbon, and biogenic opal from the Eirk sediment drift area at OPD Site 646. Note (1) drastic changes of all parameters at 7.4 Ma assumed to be correlatable with the onset of the Denmark Strait overflow, and (2) onset of biogenic opal deposition around 4 Ma, which is interpreted to indicate the onset of the ice-covered East Greenland Current. (modified from Bohrmann et al., 1990).

feature comparing the western and eastern margin of the NGS is the much higher abundance of trough mouth fans at the eastern side. Whereas the relatively small-sized Scoresby Sund trough mouth fan is the only one in the West, three huge fans, viz. from South to North the North Sea trough mouth, the Bear Island trough mouth fans, and the Storfjorden trough mouth fans, as well as three additional small-sized trough mouth fans offshore Svalbard are developed. This asymmetric distribution of trough mouth fans clearly displays a much higher dynamic behaviour of the Scandinavian-Barents Sea ice sheet than of the Greenland ice sheet, which was probably triggered by increased seasonal contrasts in temperature regimes by intruding Atlantic waters (Vorren et al., 1998). Detailed seismic and sedimentological studies of the trough mouth fans revealed IRD in the Barents Sea since about 2.6 Ma, and off Svalbard glacial sedimentation since 2.3 Ma (Solheim et al., 1998). Svalbards glaciers reached the shelf inducing increased debris-flow activity on the slope at 1.6 Ma. At 1.2 Ma, a significant increase in debris-flow activity and mass wasting is observed all over the Svalbard/Barents Sea margin. This marks the transition from small, thin and more stable ice sheets before 1.2 Ma to thick, highly labile ice sheets during the glacials of the past 1.2 Ma. This pattern is also reflected by variations in the sediment yield of the Barents Sea and Svalbard trough mouth fans. The sediment yield in tonnes per km^2 per year increases drastically in all these fans with time. The highest yields regard the interval from 1.0 to 0.2 Ma, where the deep-sea IRD records also reveal the highest amplitudes.

A final perspective in the phase of significant strengthening of the NHG during the past 2.8 Ma is the interaction between ice-sheet dynamics and deep-water circulation, in particular NADW production. Henrich et al. (2002), applying carbonate accumulation and preservation proxies, revealed a close connection between NADW production, Atlantic water intrusion and ice-sheet dynamics in the NGS (Fig. 12.12). In the interval from 2.8 to 2.0 Ma, stable conditions in the NGS and Labrador Sea, characterised by small ice caps over Scandinavia and Greenland, caused severe dissolution,

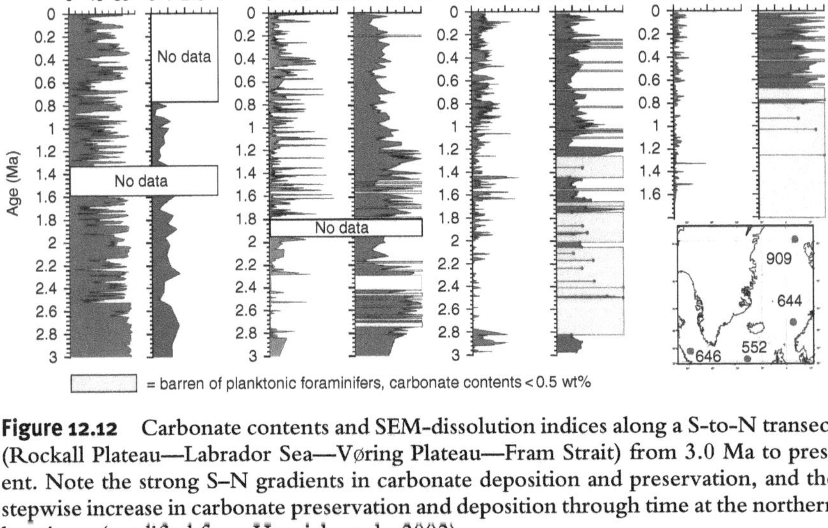

Figure 12.12 Carbonate contents and SEM-dissolution indices along a S-to-N transect (Rockall Plateau—Labrador Sea—Vøring Plateau—Fram Strait) from 3.0 Ma to present. Note the strong S–N gradients in carbonate deposition and preservation, and the stepwise increase in carbonate preservation and deposition through time at the northern locations. (modified from Henrich et al., 2002).

viz. the 'sea-ice dissolution mode' (Henrich et al., 2002) and prevented NADW formation in the NGS. The interval from 1.9 to 1.4 Ma displays, by elevated carbonate contents and improved preservation patterns (Fig. 12.12), the first intrusions of a Proto-Atlantic current into the Labrador Sea and the south-eastern NGS. The increased import of warm, saline surface waters triggered NADW formation, viz. the 'Atlantic water entrainment mode' (Henrich et al., 2002). During the past 1.1 Ma, huge ice sheets responded in a positive feedback loop to the climatic variations by orbital forcing. A generally good carbonate preservation was the most common pattern in interglacials and glacials implying a persistent production of various qualities of NADW in the NGS. During interglacials, dense Lower NADW was formed, whereas less dense Upper NADW was produced during glacials, enforcing the intermediate water circulation loop in the glacial Atlantic (Gröger et al., 2003). The only modifications are short-term dissolution spikes during glacials and terminations, viz. the 'ice sheet collapse and melt-water dissolution spike mode' (Henrich et al., 2002). In addition, long-term bad preservation in glacials occurred during the mid-Brunhes interval, evidencing the influence of the mid-Brunhes dissolution in the NGS. It may be speculated that this long dissolution phase occurred in response to the extensive, severe glaciomarine conditions in the NGS as a consequence of the largest ice sheets developed during the past 2.8 Ma.

There has been much speculation and controversy with regard to the onset of NHG. In a recent review, Molnar (2008) challenged the generally accepted view that the closure of the CAS played a decisive role in the onset of the NHG in the Pliocene. The contradicting results of palaeontological and geochemical investigations, the problem of exact timing of the different processes involved, as well as the contradictory results of various ocean circulation and climate-modelling experiments led Molnar (2008) to question whether the closure of the CAS did play any role. On one side, there are the advocates of such a cause and effect relationship (Cronin, 1988; Haug and Tiedemann, 1998; references are given by Molnar, 2008), whereas a second group questioned such a role (Dowsett et al., 1992; Mudelsee and Raymo, 2005); others hypothesized the opposite, for example, that the closure of the Panama Isthmus delayed the Ice Age (Berger and Wefer, 1996) or only caused gradual changes in ocean circulation (Poore et al., 2006). Molnar (2008) also questioned the hypothesis that a land bridge was established not earlier than 2.7 Ma, when the 'Great American Exchange' of vertebrates took place. This exchange not only implies the existence of the land bridge but requires also favourable climatic conditions for this exchange, viz. an aridity increase in central America enabling savannah-dwelling vertebrates to pass through this region (Marshall et al., 1982). Sarnthein et al. (2009) present a new line of evidence suggesting that the poleward heat and moisture transport may have triggered the onset of the NHG due to enhanced precipitation over NW Eurasia and a consequent decrease in sea-surface salinity of the Arctic Ocean, which in turn increased its sea-ice cover and hence its albedo. Furthermore, the doubled throughflow from the Bering Strait is supposed to have led to a cooling of the East Greenland Current, thereby promoting the formation of a Greenland ice sheet. Whatever the correct answer may be, this controversy clearly depicts that we are far away from understanding the dynamics and processes of the Earth's climate change.

REFERENCES

Andersson, C., Jansen, E., 2003. A Miocene (8–12 Ma) intermediate water benthic stable isotope record from the northeastern Atlantic. ODP Site 982, Paleoceanography 18 (1).

Bartoli, G., Sarnthein, M., Weinelt, M., Erlenkeuser, H., Garbe-Schönberg, D., Lea, W., 2005. Final closure of Panama and the onset of northern hemisphere glaciation. Earth Planet. Sci. Lett. 237, 33–44.

Berger, W.H., Wefer, G., 1996. Expeditions in the past: paleoceanographic studies in the South Atlantic. In: Wefer, G., Berger, W.H., Siedler, G., Webb, D.J. (Eds.), The South Atlantic: Present and Past Circulation. Springer Verlag, Berlin, pp. 363–410.

Berner, R.A., 1991. A model for atmospheric CO_2 over Phanerozoic time. Am. J. Sci. 291, 339–376.

Bickert, T., Haug, G.H., Tiedemann, R., 2004. Late Neogene benthic stable isotope record of Ocean Drilling Program Site 999: implications for Caribbean paleoceanography, organic carbon burial, and the Messinian salinity crisis. Paleoceanography 19, 1023.

Billups, K., Ravelo, A.C., Zachos, J.C., 1997. Early Pliocene deep-water circulation: Stable isotope evidence for enhanced Northern Component Deep Water. Proc. Ocean Drill. Program Sci. Results 154, 319–330.

Billups, K., 2002. Late Miocene through early Pliocene deep water circulation and climate change viewed from the sub-Antarctic South Atlantic. Palaeogeogr. Palaeoclimatol. Palaeoecol. 185, 287–307.

Billups, K., Schrag, D.P., 2002. Paleotemperatures and ice volume of the past 27 Myr revisited with paired Mg/Ca and $^{18}O/^{16}O$ measurements on benthic foraminifera. Paleoceanography 17-1, doi:10.1029/2000PA000567.

Bohrmann, G., Henrich, R., Thiede, J., 1990. Miocene to Quaternary paleoceanography in the northern North Atlantic: variability in carbonate and biogenic opal accumulation. In: Bleil, U., Thiede, J. (Eds.), Geological History of the Polar Oceans: Arctic versus Antarctic. Nato ASI Series C. Kluwer Academic Publisher, Dordrecht, pp. 647–675.

Bowen, G.J., Beerling, D.J., Koch, P.L., Zachos, J.C., Quattlebaum, T., 2004. A humid climate state during the Paleocene/Eocene thermal maximum. Nature 432, 495–499.

Coates, A.G., Aubry, M.-P., Berggren, W.A., Collins, L.S., Kunk, M., 2003. Early Neogene history of the Central American arc from Bocas del Toro, western Panama. Geol. Soc. Am. Bull. 115, 271–287.

Coates, A.G., Collins, L.S., Aubry, M.-P., Berggren, W.A., 2004. The geology of the Dar- ien, Panama, and the late Miocene-Pliocene col- lision of the Panama arc with northwestern South America, Geol. Soc. Am. Bull. 116, 1327–1344.

Coxall, H.K., Wilson, P.A., Pälike, H., Lear, C., Backman, J., 2005. Rapid stepwise onset of Antarctic glaciation and deeper calcite compensation in the Pacific Ocean. Nature 433, 53–57.

Cronin, T.M., 1988. Evolution of marine cli- mates of the U.S. Atlantic during the past 4 million years, Philos. Trans. R. Soc. London, Ser. B 318, 661–678.

DeConto, R., Pollard, D., Harwood, D., 2007. Sea ice feedback and Cenozoic evolution of Antarctic climate and ice sheets. Paleoceanography 22, PA3214. doi:10.1029/2006PA001350.

Delaney, M.L., 1990. Miocene Benthic Foraminiferal Cd/Ca Records: South Atlantic and Western Equatorial Pacific. Paleoceanography 5 (5), 743–760.

Dickens, G.R., O'Neil, J.R., Rea, D.K., Owen, R.M., 1995. Dissociation of oceanic methane hydrate as a cause of the carbon isotope excursion at the end of the Paleocene. Paleoceanography 10, 965–971.

Dowsett, H.J.T., Cronin, M., Poore, Z., Thompson, R.C., Whateley, R.C., Wood, A.M., 1992. Micropaleontological evidence for increased meridional heat transport in the North Atlantic Ocean during the Pliocene. Science 258, 1133–1135.

Duque-Caro, H., 1990. Neogene stratigraphy, paleoceanography and paleobiogeography in northwest South America and the evolution of the Panama Seaway. Palaeogeogr. Palaeoclimatol. Palaeoecol. 77, 203–234.

Exon, N., Kennett, J., Malone, M., the Leg 189 Shipboard Scientific Party, 2000. The opening of the Tasmanian gateway drove global Cenozoic paleoclimatic and paleoceanographic changes: results of Leg 189. JOIDES J. 26, 11–17.

Frank, M., Whiteley, N., Kasten, S., Hein, J.R., O'Nions, R.K., 2002. North Atlantic Deep Water export to the Southern Ocean over the past 14 Myr: evidence from Nd and Pb isotopes in ferromanganese crusts. Paleoceanography 17, PA1022. doi:10.1029/2000PA000606.

Fronval, T., Jansen, E., 1996. Late Neogene paleoclimates and paleoceanography in the Iceland-Norwegian Sea: evidence from the Iceland and Vøring Plateaus. Proc. Ocean Drill. Prog. Sci. Results 151, 455–468.

Gradstein, F., Ogg, J., Smith, A., 2004. A Geologic Time Scale 2004. Cambridge University Press, xix + 589pp.

Gröger, M., Bickert, T., Henrich, R., 2003. Variability of silt grain size and planktic foraminifer preservation in the Plio/Pleistocene sediments from the western equatorial Atlantic and Caribbean. Mar. Geol. 201, 307–320.

Haq, B.U., Hardenbol, J., Vail, P.R., 1987. Chronology of fluctuating sea levels since the Triassic. Science 235, 1156–1167.

Hardenbol, J., Thierry, J., Farley, M.B., Jacquin, T., De Graciansky, P.C., Vail, P.R., 1998. Mesozoic and Cenozoic sequence chronostratigraphic framework of European basins. In: De Graciansky, P.C., Hardenbol, J., Thierry, J., Vail, P.R. (Eds.), Mesozoic and Cenozoic Sequence Stratigraphy of European Basins. SEPM Special Publication, vol. 60, 3–13.

Haug, G.H., Tiedemann, R., 1998. Effect of the formation of the Isthmus of Panama on Atlantic Ocean thermohaline circulation. Nature 393, 673–676.

Heinze, C., Crowley, T.J., 1997. Sedimentary response to ocean gateway circulation changes. Paleoceanography 6, 742–754.

Heinze, C., Dittert, N., 2005. Impact of paleocirculations on the silicon redistribution in the world ocean. Mar. Geol. 214, 201–213.

Henrich, R., Baumann, K.H., 1994. Evolution of the Norwegian Current and the Scandinavian Ice Sheets during the past 2.6 My: evidence from ODP Leg 104 biogenic carbonate and terrigenous records. Palaeogeogr. Palaeoclimatol. Palaeoecol. 108, 75–94.

Henrich, R., Baumann, K.H., Huber, R., Meggers, H., 2002. Carbonate preservation records of the past 3 Myr in the Norwegian-Greenland Sea and the northern North Atlantic: implications for the history of NADW production. Mar. Geol. 184, 17–41.

Higgins, J.A., Schrag, D.P., 2006. Beyond methane: towards a theory for the Paleocene–Eocene Thermal Maximum. Earth Planet. Sci. Lett. 245, 523–537.

Hodell, D.A., Curtis, J.H., Sierro, F.J., Raymo, M.E., 2001. Correlation of late Miocene to early Pliocene sequences between the Mediterranean and North Atlantic. Paleoceanography 16, 164–178.

Holbourn, A., Kuhnt, W., Schulz, M., Erlenkeuser, H., 2005. Impacts of orbital forcing and atmospheric carbon dioxide on Miocene ice-sheet expansion. Nature 438, 483–487.

Imbrie, J., Hays, J.D., Martinson, D.G., McIntyre, A., Mix, A.C., Morley, J.J., et al., 1984. The orbital theory of Pleistocene climate: support from a revised chronology of the marine $\delta^{18}O$ record. In: Berger, A., Imbrie, J., Hays, J., Kukla, G., Saltzman, B. (Eds.), Milankovitch and Climate, Part ID. Reidel, Dordrecht, pp. 269–305.

IPCC, 2007. Climate Change 2007: The Physical Science Basis. Contribution of Working Group I to the Fourth Assessment Report of the Intergovernmental Panel on Climate Change. Cambridge University Press, Cambridge, UK; New York, NY, USA, 996pp.

Jakobsson, M., Backman, J., Rudels, B., Nycander, J., Frank, M., Mayer, L.A., et al., 2007. Opening of the Fram Strait in the early Miocene initiates ventilated circulation in the Arctic Ocean. Nature 447, 986–990.

Jokat, W., Leinweber, V., Ehlers, B.M., Schenke, H.W., 2007. The timing and geometry of the Fram Strait opening. Geophysical Research Abstracts 9.

Kastanja, M.-M., Henrich, R., 2007. Grain size variations in pelagic carbonate oozes from the Walvis Ridge- SE Atlantic (ODP Site 1265): a Miocene record of carbonate sedimentation and preservation. Mar. Geol. 237, 97–108.

Kastanja, M.-M., Diekman, B., Henrich, R., 2006. Controls on carbonate versus terrigenous deposition in the incipient Benguela upwelling system during the Middle to Late Miocene (ODP Sites 1085 and 1087). Palaeogeogr. Palaeoclimatol. Palaeoecol. 241, 515–530.

Keller, G., Barron, J.A., 1983. Paleoceanographic implications of Miocene deep-sea hiatuses. Geol. Soc. Am. Bull. 94, 590–613.

Kennett, J.P., Stott, L.D., 1991. Abrupt deep-sea warming, paleoceanographic changes and benthic extinctions at the end of the Paleocene. Nature 353, 225–228.

King, T.A., Ellis, W.G., Murray, D.W., Shackleton, N.J., Harris, S., 1997. Miocene evolution of carbonate sedimentation at the Ceara Rise: a multivariate data/proxy approach. Proc. Ocean Drill. Prog. Sci. Results 154, 349–365.

Korstgard, J.A., Nielsen, O.B., 1989. Provenance of dropstones in Baffin Bay and Labrador Sea. In: Srivastava, S.P., Arthur, M., Clement, B. et al., (Eds.), Proceedings Ocean Drilling Program, vol. 105, Scientific Results, College Station TX, pp. 65–69.

Kuhnert, H., Bickert, T., Paulsen, H., 2009. Southern Ocean frontal system changes precede Antarctic ice sheet growth during the middle Miocene. Earth Planet. Sci. Lett. (in press).

Larsen, H.C., Saunders, A.D., Clift, P.D., Beget, J., Wie, W., Spezzaferri, S., ODP Leg 152 Scientific Party, 1994. Seven million years of glaciation in Greenland. Science 264, 952–955.

Lawver, L.A., Gahagan, L.M., 1998. Opening of Drake Passage and its impact on Cenozoic ocean circulation. In: Crowley, T.J., Burke, K.C. (Eds.), Tectonic Boundary Conditions for Climate Reconstructions. Oxford University Press, New York, pp. 212–223.

Lear, C.H., Elderfield, H., Wilson, P.A., 2000. Cenozoic deep-sea temperatures and global ice volumes from Mg/Ca in benthic foraminiferal calcite. Science 287, 269–272.

Lear, C.H., Rosenthal, Y., Wright, J.D., 2003. The closing of a seaway: Ocean water masses and global climate change. Earth Planet. Sci. Lett. 210, 425–437.

Lear, C.H., Bailey, T.R., Pearson, P.N., Coxall, H.K., Rosenthal, Y., 2008. Cooling and ice growth across the Eocene–Oligocene transition. Geology 36, 251–254.

Lisiecki, L.E., Raymo, M.E., 2005. A Pliocene–Pleistocene stack of 57 globally distributed benthic $\delta^{18}O$ records. Paleoceanography 20, doi:10.1029/2004PA001071.

Lyle, M., Dadey, K.A., Farrell, J.W., 1995. The late Miocene (11–8 Ma) eastern Pacific carbonate crash: evidence for reorganisation of deep water circulation by the closure of the Panama gateway. Proc. Ocean Drill. Prog. Sci. Results 138, 821–838.
Maier-Reimer, E., Mikolajewicz, U., Crowley, T.J., 1990. Ocean general circulation model sensitivity experiment with an open central American isthmus. Paleoceanography 5, 349–366.
Marincovich, L., 1990. Molluscan evidence for early to middle Miocene marine glaciation in Southern Alaska. Geol. Soc. Am. Bull. 102, 1591–1599.
Marshall, L.G., Webb, S.D., Sepkoski, Jr. J., Raup, D.M., 1982. Mammalian evolution and the great American interchange. Science 215, 1351–1357.
Merico, A., Tyrrell, T., Wilson, P.A., 2008. Eocene/Oligocene ocean de-acidification linked to Antarctic glaciation by sea-level fall Nature, 425. doi:10.1038/nature06853.
Meulenkamp, J.E., Sissingh, W., 2003. Tertiary palaeogeography and tectonostratigraphic evolution of the Northern and Southern Peri-Tethys platforms and the intermediate domains of the African Eurasian convergent plate boundary zone. Palaeogeogr. Palaeoclimatol. Palaeoecol. 196, 209–228.
Mikolajewicz, U., Crowley, T.J., 1997. Response of a coupled ocean/energy balance model to restricted flow through the central American isthmus. Paleoceanography 12, 429–441.
Mikolajewicz, U., Maier-Reimer, E., Crowley, T.J., Kim, K.Y., 1993. Effect of Drake and Panamanian gateways on the circulation of an ocean model. Paleoceanography 8, 409–426.
Miller, K.G., Wright, J.D., Fairbanks, R.G., 1991. Unlocking the icehouse: Oligocene–Miocene oxygen isotope, eustacy, and margin erosion. J. Geophys. Res. 96, 6829–6848.
Mix, A.C., Pisias, N.G., Rugh, W., Wilson, J., Morey, A., Hagelberg, T.K., 1995. Benthic foraminifer stable isotope record from Site 849 (0–5 Ma): Local and global changes. Proc. Ocean Drill. Prog. Sci. Results 138, 371–412.
Molnar, P., 2008. Closing of the Central American Seaway and the Ice Age: a critical review. Paleoceanography 23, PA2201. doi:10.1029/2007PA001574.
Mudelsee, M., Raymo, M.E., 2005. Slow dynamics of the northern hemispher glaciation. Paleoceanography 20, PA4022. doi:10.1029/2005GL022844.
Mudie, P.J., Helgason, J., 1983. Palynological evidence for Miocene climate cooling in eastern Iceland about 9.8 Myr ago. Nature 303, 689–692.
Müller, R.D., Sdrolias, M., Gaina, C., Steinberger, B., Heine, C., 2008. Long-Term Sea-Level Fluctuations Driven by Basin Dynamics. Science 319, 1375. doi: 10.1126/science.1151540.
Nisancioglu, K., Raymo, M.E., Stone, P.H., 2003. Reorganization of Miocene deep water circulation in response to the shoaling of the Central American Seaway. Paleoceanography 18, 1006. doi:10.1029/2002PA000767.
Nong, G.T., Najjar, R.G., Seidov, D., Peterson, W., 2000. Simulation of ocean temperature change due to the opening of Drake Passage. Geophys. Res. Lett. 27, 2689–2692.
Pagani, M., Arthur, M.A., Freeman, K.H., 1999. Miocene evolution of atmospheric carbon dioxide. Paleoceanography 14, 273–292.
Pagani, M., Zachos, J., Freeman, K.H., Bohaty, S., Tipple, B., 2005. Marked change in atmospheric carbon dioxide concentrations during the Oligocene. Science 309, 600–603.
Paulsen, H., 2005. Miocene changes in the vertical structure of the near-surface water column: Influence on paleopruductivity. Ph.D.-thesis, Universität Bremen, p. 110.
Poore, H.R., Samworth, R., White, N.J., Jones, S.M., McCave, I.N., 2006. Neogene overflow of Northern Component water at the Greenland–Scotland Ridge. Geochem. Geophys. Geosyst. 7, Q06010. doi:10.1029/2005 GC 001085.

Prange, M., Schulz, M., 2004. A coastal upwelling seesaw in the Atlantic Ocean as a result of the closure of the Central American Seaway. Geophys. Res. Lett. 31, L17207. doi:10.1029/2004GL020073.

Roegl, F., 1999. Mediterranean and Paratethys. Facts and hypotheses of an Oligocene to Miocene Paleogeography (short overview). Geol. Carpath. 59, 339–349.

Röhl, U., Westerhold, T., Bralower, T.J., Zachos, J.C., 2007. On the duration of the Paleocene–Eocene thermal maximum (PETM). Geochem. Geophys. Geosyst. 8, Q12002. doi:10.1029/2007GC001784.

Roth, J.M., Droxler, A., Kameo, K., 2000. The Caribbean carbonate crash at the middle to late Miocene transition: linkage to the establishment of the modern global ocean conveyor. Proc. Ocean Drill. Prog. Sci. Results 165, 249–273.

Sarnthein, M., Bartoli, G., Prange, M., Schmittner, A., Schneider, B., Weinelt, M., Andersen, N., Garbe-Schönberg, D., 2009. Mid-Pliocene shifts in ocean overturning circulation and the onset of Quaternary-style climates. Clim. Past 5, 269–283. doi:10.5194/cp-5-269-2009.

Schneider, B., Schmittner, A., 2006. Simulating the impact of the Panamanian seaway closur eon ocean circulation, marine productivity and nutrient cycling. Earth Planet. Sci. Lett. 246, 367–380. doi:10.1016/j.epsl.2006.04.028.

Shackleton, N.J., Berger, A., Peltier, W.R., 1990. An alternative astronomical calibration of the lower Pleistocene timescale based on ODP Site 677. Trans. R. Soc. Edinb. Earth. Sci. 81, 251–261.

Shackleton, N.J., Hall, M.A., 1997. The late Miocene stable isotope record, Site 926. Proc. Ocean Drill. Program Sci. Results 154, 367–374.

Shevenell, A.E., Kennett, J.P., 2004. Paleoceanographic change during the middle Miocene climate revolution: an Antarctic stable isotope perspective. In: Exon, N., Kennett, J.P., Malone, M. (Eds.), The Cenozoic Southern Ocean: Tectonics, Sedimentation, and Climate Change between Australia and Antarctica. Geophys. Mon. Ser. 154, AGU, Washington, DC, 235–252.

Shevenell, A.E., Kennett, J.P., 2007. Cenozoic Antarctic cryosphere evolution: tales from deep-sea sedimentary records. Deep-Sea Res. II 54, 2308–2324.

Shevenell, A.E., Kennett, J.P., Lea, D.W., 2004. Middle Miocene Southern Ocean cooling and Antarctic cryosphere expansion. Science 305, 1766–1770.

Shevenell, A.E., Kennett, J.P., Lea, D.W., 2008. Middle Miocene ice sheet dynamics, deep-sea temperatures, and carbon cycling: a Southern Ocean perspective. Geochem. Geophys. Geosyst. 9, Q02006. doi:10.1029/2007GC001736.

Solheim, A., Faleide, J.I., Andersen, E.S., Elverhøi, A., Forsberg, C.F., Vanneste, K., et al., 1998. Late Cenozoic seismic stratigraphy and glacial geological development of the East Greenland and Svalbard-Barents Sea continental margins. Quat. Sci. Rev. 17, 155–184.

St. John, K., Krissek, L.A., 2002. The late Miocene to Pleistocene ice-rafting history of southeast Greenland. Boreas 31, 28–35.

Steph, S., Tiedemann, R., Prange, M., Groeneveld, J., Nürnberg, D., Reuning, L., et al., 2006. Changes in Caribbean surface hydrography during the Pliocene shoaling of the Central American Seaway. Paleoceanography 21, PA4221.

Thomas, D.J., Via, R.K., 2007. Neogene evolution of Atlantic thermohaline circulation: perspective from Walvis Ridge, southeastern Atlantic Ocean. Paleoceanography 22, PA2212.

Tiedemann, R., Franz, S.O., 1997. Deep water circulation, chemistry, and terrigenous sediment supply in the equatorial Atlantic during the Pliocene, 3.3–2.6 Ma and 5–4.5 Ma. Proc. Ocean Drill. Program Sci. Results, 154, 299–318..

Tripati, A.K., Delaney, M.L., Zachos, J.C., Anderson, L.D., Kelly, D.C., Elderfield, H., 2003. Tropical sea-surface temperature reconstruction for the early Paleogene using Mg/Ca ratios of planktonic foraminifera. Paleoceanography 18, doi:10.1029/2003PA000937.

Tripati, A., Backman, J., Elderfield, H., Ferretti, P., 2005. Eocene bipolar glaciation associated with global carbon cycle changes. Nature 436, 341–346.
Vincent, E., Berger, W.H., 1985. Carbon dioxide and polar cooling in the Miocene: the Monterey Hypothesis. In: Sundquist, E.T., Broecker, W.S. (Eds.), The Carbon Cycle and Atmospheric CO_2: Natural Variations Archean to Present. AGU, Washington, DC, pp. 455–468.
Von Blanckenburg, F., Nägler, T.F., 2001. Weathering versus circulation-controlled changes in radiogenic isotope tracer composition of the Labrador Sea and Northern Atlantic Deep Water. Paleoceanography 16, 424–434.
von der Heydt, A., Dijkstra, H.A., 2006. Effect of ocean gateways on the global ocean circulation in the late Oligocene and early Miocene. Paleoceanography 21, PA1011. doi:10.1029/2005PA001149.
Vorren, T., Laberg, J.S., Blaume, F., Dowdeswell, J.A., Kenyon, N.H., Mienert, J., Rumohr, J., Werner, F., 1998. The Norwegian-Greenland Sea continental margins: morphology and Late Quaternary sedimentary processes and environments. Quaternary. Science. Reviews. 17, 155–184.
Wei, W., Peleo-Alampay, A., 1997. Onset of North Atlantic deep Water 11.5 million years ago triggered by climate cooloing. In: Wang, P., Berggren, W.A. (Eds.), Proc. 30th International Geol. Congress, 57–64.
Weissert, H., 2011. Mesozoic pelagic sediments – archives for ocean and climate history during green-house conditions. In: Hüneke, H., Mulder, T. (Eds.), Deep-Sea Sediments. Developments in Sedimentology, Elsevier, Amsterdam, vol. 63, pp. 765–792.
Westerhold, T., 2003. The Middle Miocene Carbonate Crash: Relationship to Neogene Changes in Ocean Circulation and Global Climate. Ph.D.-thesis, Universität Bremen, p. 137.
Westerhold, T., Bickert, T., 2005. Middle to Late Miocene oxygen isotope stratigraphy of ODP Site 1085 (SE Atlantic): new constraints on Miocene climate variability and sea level fluctuations. Palaeogeog., Palaeoclimatol., Palaeoecol. 217, 205–222.
Wright, J.D., Miller, K.G., Fairbanks, R.G., 1991. Evolution of modern deepwater circulation: Evidence from the late Miocene Southern Ocean. Paleoceanography 6, 275–290.
Wold, C.N., 1994. Cenozoic sediment accumulation on drifts in the northern North Atlantic. Paleoceanography 9, 917–941.
Wolfe, J.A., 1994. An analysis of Neogene climates in Beringia. Palaeogeogr. Palaeoclimatol. Palaeoecol. 108, 207–216.
Wolf-Welling, T., Cremer, M., O'Connell, S., Winkler, A., Thiede, J., 1996. Cenozoic Arctic Gateway paleoclimate variability: indications from changes in coarse-fraction composition (ODP Leg 151). In: Thiede, J., Myhre, A.M., Firth, et al., (Eds.), Proceedings Ocean Drilling Program, vol. 151, Scientific Results, College Station TX, pp. 515–567.
Woodruff, F., Savin, S.M., 1989. Miocene deep water oceanography. Paleoceanography 4, 87–140.
Wright, J.D., Miller, K.G., 1996. Control of North Atlantic Deep Water circulation by the Greenland–Scotland Ridge. Paleoceanography 11, 157–169.
Zachos, J.C., Pagani, M., Sloan, L., Thomas, E., Billups, K., 2001. Trends, Rythms, and Aberrations in global Climate 65 Ma to Present, Science 292, 686–693.
Zachos, J.C., Röhl, U., Schellenberg, S.A., Sluijs, A., Hodell, D.A., Kelly, D.C., et al., 2005. Rapid acidification of the ocean during the Paleocene–Eocene Thermal Maximum. Science 308, 1611–1615.

Index

E

F

H

M

P

Q

R

S

U

V

W

MIX
Papier | Fördert gute Waldnutzung
FSC® C083411

Kolibri 360 GmbH
Ferdinand-Jühlke-Straße 7
99095 Erfurt, Deutschland
produktsicherheit@kolibri360.de